University Physics Volume 2

Senior Contributing Authors
Samuel J. Ling, Truman State University
Jeff Sanny, Loyola Marymount University
Bill Moebs, PhD

Rice University
6100 Main Street MS-375
Houston, Texas 77005

Changes to original document:
Non-numbered pages 1, 2, 3 and 4 were removed. The previous title page, this page, and the last two blank pages have been added. Color images and text were converted to grayscale.

Table of Contents

PREFACE

Welcome to *University Physics*, an OpenStax resource. This textbook was written to increase student access to high-quality learning materials, maintaining highest standards of academic rigor at little to no cost.

About OpenStax

OpenStax is a nonprofit based at Rice University, and it's our mission to improve student access to education. Our first openly licensed college textbook was published in 2012 and our library has since scaled to over 20 books used by hundreds of thousands of students across the globe. Our adaptive learning technology, designed to improve learning outcomes through personalized educational paths, is currently being piloted for K–12 and college. The OpenStax mission is made possible through the generous support of philanthropic foundations. Through these partnerships and with the help of additional low-cost resources from our OpenStax partners, OpenStax is breaking down the most common barriers to learning and empowering students and instructors to succeed.

About OpenStax Resources

Customization

University Physics is licensed under a Creative Commons Attribution 4.0 International (CC BY) license, which means that you can distribute, remix, and build upon the content, as long as you provide attribution to OpenStax and its content contributors.

Because our books are openly licensed, you are free to use the entire book or pick and choose the sections that are most relevant to the needs of your course. Feel free to remix the content by assigning your students certain chapters and sections in your syllabus in the order that you prefer. You can even provide a direct link in your syllabus to the sections in the web view of your book.

Faculty also have the option of creating a customized version of their OpenStax book through the aerSelect platform. The custom version can be made available to students in low-cost print or digital form through their campus bookstore. Visit your book page on openstax.org for a link to your book on aerSelect.

Errata

All OpenStax textbooks undergo a rigorous review process. However, like any professional-grade textbook, errors sometimes occur. Since our books are web based, we can make updates periodically when deemed pedagogically necessary. If you have a correction to suggest, submit it through the link on your book page on openstax.org. Subject matter experts review all errata suggestions. OpenStax is committed to remaining transparent about all updates, so you will also find a list of past errata changes on your book page on openstax.org.

Format

You can access this textbook for free in web view or PDF through openstax.org, and for a low cost in print.

About *University Physics*

University Physics is designed for the two- or three-semester calculus-based physics course. The text has been developed to meet the scope and sequence of most university physics courses and provides a foundation for a career in mathematics, science, or engineering. The book provides an important opportunity for students to learn the core concepts of physics and understand how those concepts apply to their lives and to the world around them.

Due to the comprehensive nature of the material, we are offering the book in three volumes for flexibility and efficiency.

Coverage and Scope

Our *University Physics* textbook adheres to the scope and sequence of most two- and three-semester physics courses nationwide. We have worked to make physics interesting and accessible to students while maintaining the mathematical rigor inherent in the subject. With this objective in mind, the content of this textbook has been developed and arranged to provide a logical progression from fundamental to more advanced concepts, building upon what students have already learned and emphasizing connections between topics and between theory and applications. The goal of each section is to enable students not just to recognize concepts, but to work with them in ways that will be useful in later courses and future careers. The organization and pedagogical features were developed and vetted with feedback from science educators dedicated to the project.

VOLUME I

Unit 1: Mechanics

Chapter 1: Units and Measurement

Chapter 2: Vectors

Chapter 3: Motion Along a Straight Line

Chapter 4: Motion in Two and Three Dimensions

Chapter 5: Newton's Laws of Motion

Chapter 6: Applications of Newton's Laws

Chapter 7: Work and Kinetic Energy

Chapter 8: Potential Energy and Conservation of Energy

Chapter 9: Linear Momentum and Collisions

Chapter 10: Fixed-Axis Rotation

Chapter 11: Angular Momentum

Chapter 12: Static Equilibrium and Elasticity

Chapter 13: Gravitation

Chapter 14: Fluid Mechanics

Unit 2: Waves and Acoustics

Chapter 15: Oscillations

Chapter 16: Waves

Chapter 17: Sound

VOLUME II

Unit 1: Thermodynamics

Chapter 1: Temperature and Heat

Chapter 2: The Kinetic Theory of Gases

Chapter 3: The First Law of Thermodynamics

Chapter 4: The Second Law of Thermodynamics

Unit 2: Electricity and Magnetism

Chapter 5: Electric Charges and Fields

Chapter 6: Gauss's Law

Chapter 7: Electric Potential

Chapter 8: Capacitance

Chapter 9: Current and Resistance

Chapter 10: Direct-Current Circuits

Chapter 11: Magnetic Forces and Fields

Chapter 12: Sources of Magnetic Fields

Chapter 13: Electromagnetic Induction

Chapter 14: Inductance

Chapter 15: Alternating-Current Circuits

Chapter 16: Electromagnetic Waves

VOLUME III

Unit 1: Optics

Chapter 1: The Nature of Light

Chapter 2: Geometric Optics and Image Formation

Pedagogical Foundation

Throughout *University Physics* you will find derivations of concepts that present classical ideas and techniques, as well as modern applications and methods. Most chapters start with observations or experiments that place the material in a context of physical experience. Presentations and explanations rely on years of classroom experience on the part of long-time physics professors, striving for a balance of clarity and rigor that has proven successful with their students. Throughout the text, links enable students to review earlier material and then return to the present discussion, reinforcing connections between topics. Key historical figures and experiments are discussed in the main text (rather than in boxes or sidebars), maintaining a focus on the development of physical intuition. Key ideas, definitions, and equations are highlighted in the text and listed in summary form at the end of each chapter. Examples and chapter-opening images often include contemporary applications from daily life or modern science and engineering that students can relate to, from smart phones to the internet to GPS devices.

Assessments That Reinforce Key Concepts

In-chapter **Examples** generally follow a three-part format of Strategy, Solution, and Significance to emphasize how to approach a problem, how to work with the equations, and how to check and generalize the result. Examples are often followed by **Check Your Understanding** questions and answers to help reinforce for students the important ideas of the examples. **Problem-Solving Strategies** in each chapter break down methods of approaching various types of problems into steps students can follow for guidance. The book also includes exercises at the end of each chapter so students can practice what they've learned.

Conceptual questions do not require calculation but test student learning of the key concepts.

Problems categorized by section test student problem-solving skills and the ability to apply ideas to practical situations.

Additional Problems apply knowledge across the chapter, forcing students to identify what concepts and equations are appropriate for solving given problems. Randomly located throughout the problems are **Unreasonable Results** exercises that ask students to evaluate the answer to a problem and explain why it is not reasonable and what assumptions made might not be correct.

Challenge Problems extend text ideas to interesting but difficult situations.

Answers for selected exercises are available in an **Answer Key** at the end of the book.

Additional Resources

Student and Instructor Resources

We've compiled additional resources for both students and instructors, including Getting Started Guides, PowerPoint slides, and answer and solution guides for instructors and students. Instructor resources require a verified instructor account, which can be requested on your openstax.org log-in. Take advantage of these resources to supplement your OpenStax book.

Partner Resources

OpenStax partners are our allies in the mission to make high-quality learning materials affordable and accessible to students and instructors everywhere. Their tools integrate seamlessly with our OpenStax titles at a low cost. To access the partner resources for your text, visit your book page on openstax.org.

About the Authors

Senior Contributing Authors

Samuel J. Ling, Truman State University

Dr. Samuel Ling has taught introductory and advanced physics for over 25 years at Truman State University, where he is currently Professor of Physics and the Department Chair. Dr. Ling has two PhDs from Boston University, one in Chemistry and the other in Physics, and he was a Research Fellow at the Indian Institute of Science, Bangalore, before joining Truman. Dr. Ling is also an author of *A First Course in Vibrations and Waves*, published by Oxford University Press. Dr. Ling has considerable experience with research in Physics Education and has published research on collaborative learning methods in physics teaching. He was awarded a Truman Fellow and a Jepson fellow in recognition of his innovative teaching methods. Dr. Ling's research publications have spanned Cosmology, Solid State Physics, and Nonlinear Optics.

Jeff Sanny, Loyola Marymount University

Dr. Jeff Sanny earned a BS in Physics from Harvey Mudd College in 1974 and a PhD in Solid State Physics from the University of California–Los Angeles in 1980. He joined the faculty at Loyola Marymount University in the fall of 1980. During his tenure, he has served as department Chair as well as Associate Dean. Dr. Sanny enjoys teaching introductory physics in particular. He is also passionate about providing students with research experience and has directed an active undergraduate student research group in space physics for many years.

Bill Moebs, PhD

Dr. William Moebs earned a BS and PhD (1959 and 1965) from the University of Michigan. He then joined their staff as a Research Associate for one year, where he continued his doctoral research in particle physics. In 1966, he accepted an appointment to the Physics Department of Indiana Purdue Fort Wayne (IPFW), where he served as Department Chair from 1971 to 1979. In 1979, he moved to Loyola Marymount University (LMU), where he served as Chair of the Physics Department from 1979 to 1986. He retired from LMU in 2000. He has published research in particle physics, chemical kinetics, cell division, atomic physics, and physics teaching.

Contributing Authors

David Anderson, Albion College

Daniel Bowman, Ferrum College

Dedra Demaree, Georgetown University

Gerald Friedman, Santa Fe Community College

Lev Gasparov, University of North Florida

Edw. S. Ginsberg, University of Massachusetts

Alice Kolakowska, University of Memphis

Lee LaRue, Paris Junior College

Mark Lattery, University of Wisconsin

Richard Ludlow, Daniel Webster College

Patrick Motl, Indiana University–Kokomo

Tao Pang, University of Nevada–Las Vegas

Kenneth Podolak, Plattsburgh State University

Takashi Sato, Kwantlen Polytechnic University

David Smith, University of the Virgin Islands

Joseph Trout, Richard Stockton College

Kevin Wheelock, Bellevue College

Reviewers

Salameh Ahmad, Rochester Institute of Technology–Dubai

John Aiken, University of Colorado–Boulder

Anand Batra, Howard University

Raymond Benge, Terrant County College

Gavin Buxton, Robert Morris University

Erik Christensen, South Florida State College

Clifton Clark, Fort Hays State University
Nelson Coates, California Maritime Academy
Herve Collin, Kapi'olani Community College
Carl Covatto, Arizona State University
Alexander Cozzani, Imperial Valley College
Danielle Dalafave, The College of New Jersey
Nicholas Darnton, Georgia Institute of Technology
Robert Edmonds, Tarrant County College
William Falls, Erie Community College
Stanley Forrester, Broward College
Umesh Garg, University of Notre Dame
Maurizio Giannotti, Barry University
Bryan Gibbs, Dallas County Community College
Mark Giroux, East Tennessee State University
Matthew Griffiths, University of New Haven
Alfonso Hinojosa, University of Texas–Arlington
Steuard Jensen, Alma College
David Kagan, University of Massachusetts
Jill Leggett, Florida State College–Jacksonville
Sergei Katsev, University of Minnesota–Duluth
Alfredo Louro, University of Calgary
James Maclaren, Tulane University
Ponn Maheswaranathan, Winthrop University
Seth Major, Hamilton College
Oleg Maksimov, Excelsior College
Aristides Marcano, Delaware State University
Marles McCurdy, Tarrant County College
James McDonald, University of Hartford
Ralph McGrew, SUNY–Broome Community College
Paul Miller, West Virginia University
Tamar More, University of Portland
Farzaneh Najmabadi, University of Phoenix
Richard Olenick, The University of Dallas
Christopher Porter, Ohio State University
Liza Pujji, Manakau Institute of Technology
Baishali Ray, Young Harris University
Andrew Robinson, Carleton University
Aruvana Roy, Young Harris University
Abhijit Sarkar, The Catholic University of America
Gajendra Tulsian, Daytona State College
Adria Updike, Roger Williams University
Clark Vangilder, Central Arizona University

Steven Wolf, Texas State University

Alexander Wurm, Western New England University

Lei Zhang, Winston Salem State University

Ulrich Zurcher, Cleveland State University

1 | TEMPERATURE AND HEAT

Figure 1.1 These snowshoers on Mount Hood in Oregon are enjoying the heat flow and light caused by high temperature. All three mechanisms of heat transfer are relevant to this picture. The heat flowing out of the fire also turns the solid snow to liquid water and vapor. (credit: "Mt. Hood Territory"/Flickr)

Chapter Outline

Introduction

Heat and temperature are important concepts for each of us, every day. How we dress in the morning depends on whether the day is hot or cold, and most of what we do requires energy that ultimately comes from the Sun. The study of heat and temperature is part of an area of physics known as thermodynamics. The laws of thermodynamics govern the flow of energy throughout the universe. They are studied in all areas of science and engineering, from chemistry to biology to environmental science.

In this chapter, we explore heat and temperature. It is not always easy to distinguish these terms. Heat is the flow of energy from one object to another. This flow of energy is caused by a difference in temperature. The transfer of heat can change temperature, as can work, another kind of energy transfer that is central to thermodynamics. We return to these basic ideas several times throughout the next four chapters, and you will see that they affect everything from the behavior of atoms and molecules to cooking to our weather on Earth to the life cycles of stars.

1.1 | Temperature and Thermal Equilibrium

Learning Objectives

By the end of this section, you will be able to:

- Define temperature and describe it qualitatively
- Explain thermal equilibrium
- Explain the zeroth law of thermodynamics

Heat is familiar to all of us. We can feel heat entering our bodies from the summer Sun or from hot coffee or tea after a winter stroll. We can also feel heat leaving our bodies as we feel the chill of night or the cooling effect of sweat after exercise.

What is heat? How do we define it and how is it related to temperature? What are the effects of heat and how does it flow from place to place? We will find that, in spite of the richness of the phenomena, a small set of underlying physical principles unites these subjects and ties them to other fields. We start by examining temperature and how to define and measure it.

Temperature

The concept of temperature has evolved from the common concepts of hot and cold. The scientific definition of temperature explains more than our senses of hot and cold. As you may have already learned, many physical quantities are defined solely in terms of how they are observed or measured, that is, they are defined *operationally*. **Temperature** is operationally defined as the quantity of what we measure with a thermometer. As we will see in detail in a later chapter on the kinetic theory of gases, temperature is proportional to the average kinetic energy of translation, a fact that provides a more physical definition. Differences in temperature maintain the transfer of heat, or *heat transfer*, throughout the universe. **Heat transfer** is the movement of energy from one place or material to another as a result of a difference in temperature. (You will learn more about heat transfer later in this chapter.)

Thermal Equilibrium

An important concept related to temperature is **thermal equilibrium**. Two objects are in thermal equilibrium if they are in close contact that allows either to gain energy from the other, but nevertheless, no net energy is transferred between them. Even when not in contact, they are in thermal equilibrium if, when they are placed in contact, no net energy is transferred between them. If two objects remain in contact for a long time, they typically come to equilibrium. In other words, two objects in thermal equilibrium do not exchange energy.

Experimentally, if object *A* is in equilibrium with object *B*, and object *B* is in equilibrium with object *C*, then (as you may have already guessed) object *A* is in equilibrium with object *C*. That statement of transitivity is called the **zeroth law of thermodynamics**. (The number "zeroth" was suggested by British physicist Ralph Fowler in the 1930s. The first, second, and third laws of thermodynamics were already named and numbered then. The zeroth law had seldom been stated, but it needs to be discussed before the others, so Fowler gave it a smaller number.) Consider the case where *A* is a thermometer. The zeroth law tells us that if *A* reads a certain temperature when in equilibrium with *B*, and it is then placed in contact with *C*, it will not exchange energy with *C*; therefore, its temperature reading will remain the same (Figure 1.2). In other words, *if two objects are in thermal equilibrium, they have the same temperature.*

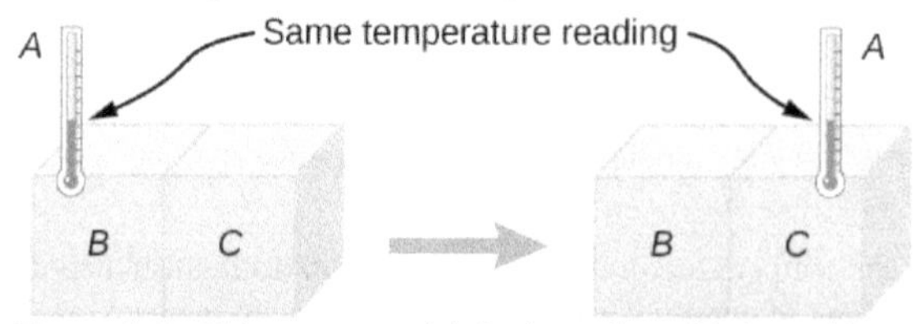

Figure 1.2 If thermometer *A* is in thermal equilibrium with object *B*, and *B* is in thermal equilibrium with *C*, then *A* is in thermal equilibrium with *C*. Therefore, the reading on *A* stays the same when *A* is moved over to make contact with *C*.

A thermometer measures its own temperature. It is through the concepts of thermal equilibrium and the zeroth law of thermodynamics that we can say that a thermometer measures the temperature of *something else*, and to make sense of the statement that two objects are at the same temperature.

In the rest of this chapter, we will often refer to "systems" instead of "objects." As in the chapter on linear momentum and collisions, a system consists of one or more objects—but in thermodynamics, we require a system to be macroscopic, that is, to consist of a huge number (such as 10^{23}) of molecules. Then we can say that a system is in thermal equilibrium with itself if all parts of it are at the same temperature. (We will return to the definition of a thermodynamic system in the chapter on the first law of thermodynamics.)

1.2 | Thermometers and Temperature Scales

Learning Objectives

By the end of this section, you will be able to:

- Describe several different types of thermometers
- Convert temperatures between the Celsius, Fahrenheit, and Kelvin scales

Any physical property that depends consistently and reproducibly on temperature can be used as the basis of a thermometer. For example, volume increases with temperature for most substances. This property is the basis for the common alcohol thermometer and the original mercury thermometers. Other properties used to measure temperature include electrical resistance, color, and the emission of infrared radiation (Figure 1.3).

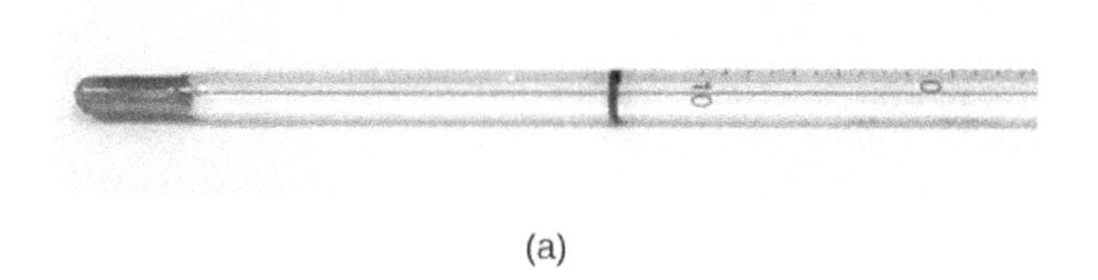

(a)

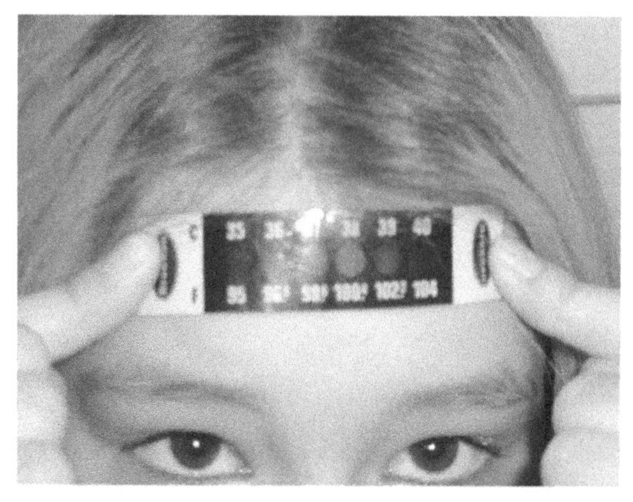

(b)

(c)

Figure 1.3 Because many physical properties depend on temperature, the variety of thermometers is remarkable. (a) In this common type of thermometer, the alcohol, containing a red dye, expands more rapidly than the glass encasing it. When the thermometer's temperature increases, the liquid from the bulb is forced into the narrow tube, producing a large change in the length of the column for a small change in temperature. (b) Each of the six squares on this plastic (liquid crystal) thermometer contains a film of a different heat-sensitive liquid crystal material. Below $95\ °F$, all six squares are black. When the plastic thermometer is exposed to a temperature of $95\ °F$, the first liquid crystal square changes color. When the temperature reaches above $96.8\ °F$, the second liquid crystal square also changes color, and so forth. (c) A firefighter uses a pyrometer to check the temperature of an aircraft carrier's ventilation system. The pyrometer measures infrared radiation (whose emission varies with temperature) from the vent and quickly produces a temperature readout. Infrared thermometers are also frequently used to measure body temperature by gently placing them in the ear canal. Such thermometers are more accurate than the alcohol thermometers placed under the tongue or in the armpit. (credit b: modification of work by Tess Watson; credit c: modification of work by Lamel J. Hinton)

Thermometers measure temperature according to well-defined scales of measurement. The three most common temperature scales are Fahrenheit, Celsius, and Kelvin. Temperature scales are created by identifying two reproducible temperatures. The freezing and boiling temperatures of water at standard atmospheric pressure are commonly used.

On the **Celsius scale**, the freezing point of water is $0\ °C$ and the boiling point is $100\ °C.$ The unit of temperature on this scale is the **degree Celsius** $(°C)$. The **Fahrenheit scale** (still the most frequently used for common purposes in the United States) has the freezing point of water at $32\ °F$ and the boiling point at $212\ °F.$ Its unit is the **degree Fahrenheit** $(°F)$. You can see that 100 Celsius degrees span the same range as 180 Fahrenheit degrees. Thus, a temperature difference of one degree on the Celsius scale is 1.8 times as large as a difference of one degree on the Fahrenheit scale, or $\Delta T_{\text{F}} = \frac{9}{5}\Delta T_{\text{C}}.$

The definition of temperature in terms of molecular motion suggests that there should be a lowest possible temperature, where the average kinetic energy of molecules is zero (or the minimum allowed by quantum mechanics). Experiments confirm the existence of such a temperature, called **absolute zero**. An **absolute temperature scale** is one whose zero point is absolute zero. Such scales are convenient in science because several physical quantities, such as the volume of an ideal gas, are directly related to absolute temperature.

The **Kelvin scale** is the absolute temperature scale that is commonly used in science. The SI temperature unit is the *kelvin*, which is abbreviated K (not accompanied by a degree sign). Thus 0 K is absolute zero. The freezing and boiling points of water are 273.15 K and 373.15 K, respectively. Therefore, temperature differences are the same in units of kelvins and degrees Celsius, or $\Delta T_C = \Delta T_K$.

The relationships between the three common temperature scales are shown in Figure 1.4. Temperatures on these scales can be converted using the equations in Table 1.1.

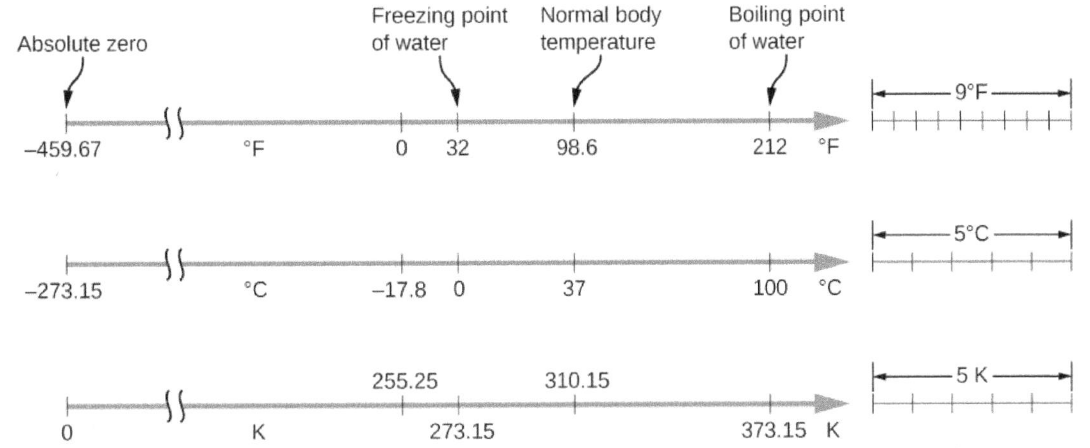

Figure 1.4 Relationships between the Fahrenheit, Celsius, and Kelvin temperature scales are shown. The relative sizes of the scales are also shown.

To convert from...	Use this equation...
Celsius to Fahrenheit	$T_F = \frac{9}{5}T_C + 32$
Fahrenheit to Celsius	$T_C = \frac{5}{9}(T_F - 32)$
Celsius to Kelvin	$T_K = T_C + 273.15$
Kelvin to Celsius	$T_C = T_K - 273.15$
Fahrenheit to Kelvin	$T_K = \frac{5}{9}(T_F - 32) + 273.15$
Kelvin to Fahrenheit	$T_F = \frac{9}{5}(T_K - 273.15) + 32$

Table 1.1 Temperature Conversions

To convert between Fahrenheit and Kelvin, convert to Celsius as an intermediate step.

Example 1.1

Converting between Temperature Scales: Room Temperature

"Room temperature" is generally defined in physics to be $25\ °C$. (a) What is room temperature in $°F$? (b) What is it in K?

Strategy

To answer these questions, all we need to do is choose the correct conversion equations and substitute the known values.

Solution

To convert from $°C$ to $°F$, use the equation

$$T_F = \frac{9}{5}T_C + 32.$$

Substitute the known value into the equation and solve:

$$T_F = \frac{9}{5}(25\ °C) + 32 = 77\ °F.$$

Similarly, we find that $T_K = T_C + 273.15 = 298\ K$.

The Kelvin scale is part of the SI system of units, so its actual definition is more complicated than the one given above. First, it is not defined in terms of the freezing and boiling points of water, but in terms of the **triple point**. The triple point is the unique combination of temperature and pressure at which ice, liquid water, and water vapor can coexist stably. As will be discussed in the section on phase changes, the coexistence is achieved by lowering the pressure and consequently the boiling point to reach the freezing point. The triple-point temperature is defined as 273.16 K. This definition has the advantage that although the freezing temperature and boiling temperature of water depend on pressure, there is only one triple-point temperature.

Second, even with two points on the scale defined, different thermometers give somewhat different results for other temperatures. Therefore, a standard thermometer is required. Metrologists (experts in the science of measurement) have chosen the *constant-volume gas thermometer* for this purpose. A vessel of constant volume filled with gas is subjected to temperature changes, and the measured temperature is proportional to the change in pressure. Using "TP" to represent the triple point,

$$T = \frac{p}{p_{TP}}T_{TP}.$$

The results depend somewhat on the choice of gas, but the less dense the gas in the bulb, the better the results for different gases agree. If the results are extrapolated to zero density, the results agree quite well, with zero pressure corresponding to a temperature of absolute zero.

Constant-volume gas thermometers are big and come to equilibrium slowly, so they are used mostly as standards to calibrate other thermometers.

Visit this site (https://openstaxcollege.org/l/21consvolgasth) to learn more about the constant-volume gas thermometer.

1.3 | Thermal Expansion

Learning Objectives

By the end of this section, you will be able to:

- Answer qualitative questions about the effects of thermal expansion
- Solve problems involving thermal expansion, including those involving thermal stress

The expansion of alcohol in a thermometer is one of many commonly encountered examples of **thermal expansion**, which is the change in size or volume of a given system as its temperature changes. The most visible example is the expansion of hot air. When air is heated, it expands and becomes less dense than the surrounding air, which then exerts an (upward) force on the hot air and makes steam and smoke rise, hot air balloons float, and so forth. The same behavior happens in all liquids and gases, driving natural heat transfer upward in homes, oceans, and weather systems, as we will discuss in an upcoming section. Solids also undergo thermal expansion. Railroad tracks and bridges, for example, have expansion joints to allow them to freely expand and contract with temperature changes, as shown in Figure 1.5.

(a) (b)

Figure 1.5 (a) Thermal expansion joints like these in the (b) Auckland Harbour Bridge in New Zealand allow bridges to change length without buckling. (credit: "ŠJů"/Wikimedia Commons)

What is the underlying cause of thermal expansion? As previously mentioned, an increase in temperature means an increase in the kinetic energy of individual atoms. In a solid, unlike in a gas, the molecules are held in place by forces from neighboring molecules; as we saw in Oscillations (http://cnx.org/content/m58360/latest/) , the forces can be modeled as in harmonic springs described by the Lennard-Jones potential. Energy in Simple Harmonic Motion (http://cnx.org/content/m58362/latest/#CNX_UPhysics_15_02_LennaJones) shows that such potentials are asymmetrical in that the potential energy increases more steeply when the molecules get closer to each other than when they get farther away. Thus, at a given kinetic energy, the distance moved is greater when neighbors move away from each other than when they move toward each other. The result is that increased kinetic energy (increased temperature) increases the average distance between molecules—the substance expands.

For most substances under ordinary conditions, it is an excellent approximation that there is no preferred direction (that is, the solid is "isotropic"), and an increase in temperature increases the solid's size by a certain fraction in each dimension. Therefore, if the solid is free to expand or contract, its proportions stay the same; only its overall size changes.

Linear Thermal Expansion

According to experiments, the dependence of thermal expansion on temperature, substance, and original length is summarized in the equation

$$\frac{dL}{dT} = \alpha L \tag{1.1}$$

where L is the original length, $\frac{dL}{dT}$ is the change in length with respect to temperature, and α is the **coefficient of linear expansion**, a material property that varies slightly with temperature. As α is nearly constant and also very small, for practical purposes, we use the linear approximation:

$$\Delta L = \alpha L \Delta T. \tag{1.2}$$

Table 1.2 lists representative values of the coefficient of linear expansion. As noted earlier, ΔT is the same whether it is expressed in units of degrees Celsius or kelvins; thus, α may have units of $1/°\text{C}$ or $1/\text{K}$ with the same value in either case. Approximating α as a constant is quite accurate for small changes in temperature and sufficient for most practical purposes, even for large changes in temperature. We examine this approximation more closely in the next example.

Material	Coefficient of Linear Expansion $\alpha(1/°C)$	Coefficient of Volume Expansion $\beta(1/°C)$
Solids		
Aluminum	25×10^{-6}	75×10^{-6}
Brass	19×10^{-6}	56×10^{-6}
Copper	17×10^{-6}	51×10^{-6}
Gold	14×10^{-6}	42×10^{-6}
Iron or steel	12×10^{-6}	35×10^{-6}
Invar (nickel-iron alloy)	0.9×10^{-6}	2.7×10^{-6}
Lead	29×10^{-6}	87×10^{-6}
Silver	18×10^{-6}	54×10^{-6}
Glass (ordinary)	9×10^{-6}	27×10^{-6}
Glass (Pyrex®)	3×10^{-6}	9×10^{-6}
Quartz	0.4×10^{-6}	1×10^{-6}
Concrete, brick	$\sim 12 \times 10^{-6}$	$\sim 36 \times 10^{-6}$
Marble (average)	2.5×10^{-6}	7.5×10^{-6}
Liquids		
Ether		1650×10^{-6}
Ethyl alcohol		1100×10^{-6}
Gasoline		950×10^{-6}
Glycerin		500×10^{-6}
Mercury		180×10^{-6}
Water		210×10^{-6}
Gases		
Air and most other gases at atmospheric pressure		3400×10^{-6}

Table 1.2 Thermal Expansion Coefficients

Thermal expansion is exploited in the bimetallic strip (Figure 1.6). This device can be used as a thermometer if the curving strip is attached to a pointer on a scale. It can also be used to automatically close or open a switch at a certain temperature, as in older or analog thermostats.

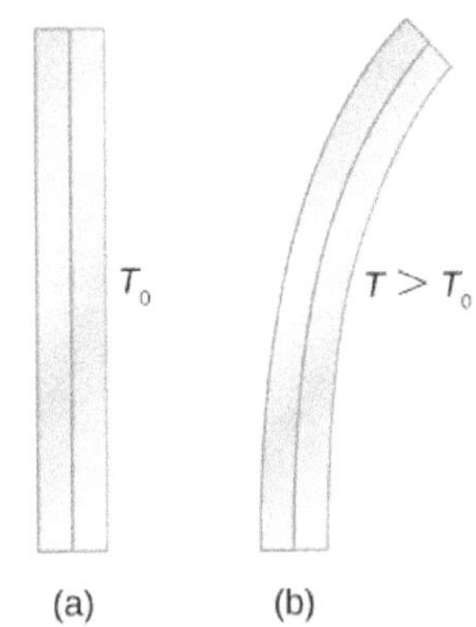

Figure 1.6 The curvature of a bimetallic strip depends on temperature. (a) The strip is straight at the starting temperature, where its two components have the same length. (b) At a higher temperature, this strip bends to the right, because the metal on the left has expanded more than the metal on the right. At a lower temperature, the strip would bend to the left.

Example 1.2

Calculating Linear Thermal Expansion

The main span of San Francisco's Golden Gate Bridge is 1275 m long at its coldest. The bridge is exposed to temperatures ranging from $-15\,°\text{C}$ to $40\,°\text{C}$. What is its change in length between these temperatures? Assume that the bridge is made entirely of steel.

Strategy

Use the equation for linear thermal expansion $\Delta L = \alpha L \Delta T$ to calculate the change in length, ΔL. Use the coefficient of linear expansion α for steel from Table 1.2, and note that the change in temperature ΔT is $55\,°\text{C}$.

Solution

Substitute all of the known values into the equation to solve for ΔL:

$$\Delta L = \alpha L \Delta T = \left(\frac{12 \times 10^{-6}}{°\text{C}}\right)(1275\text{ m})(55\,°\text{C}) = 0.84\text{ m}.$$

Significance

Although not large compared with the length of the bridge, this change in length is observable. It is generally spread over many expansion joints so that the expansion at each joint is small.

Thermal Expansion in Two and Three Dimensions

Unconstrained objects expand in all dimensions, as illustrated in Figure 1.7. That is, their areas and volumes, as well as their lengths, increase with temperature. Because the proportions stay the same, holes and container volumes also get larger with temperature. If you cut a hole in a metal plate, the remaining material will expand exactly as it would if the piece you removed were still in place. The piece would get bigger, so the hole must get bigger too.

Thermal Expansion in Two Dimensions

For small temperature changes, the change in area ΔA is given by

$$\Delta A = 2\alpha A \Delta T \tag{1.3}$$

where ΔA is the change in area A, ΔT is the change in temperature, and α is the coefficient of linear expansion, which varies slightly with temperature.

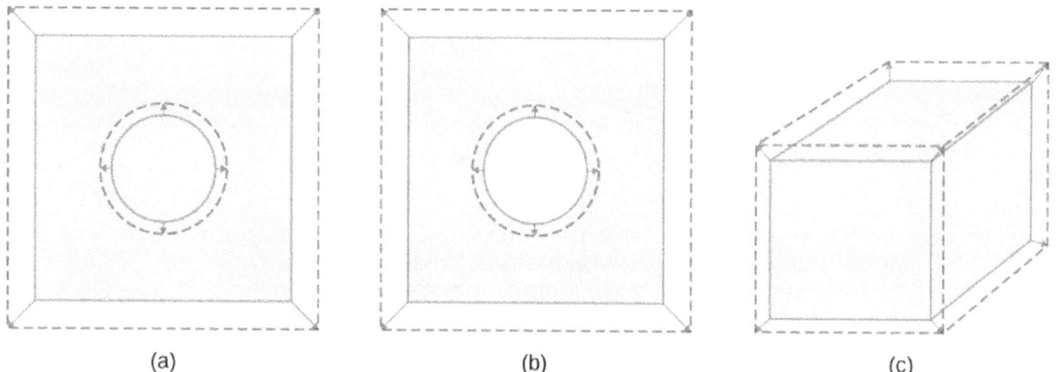

Figure 1.7 In general, objects expand in all directions as temperature increases. In these drawings, the original boundaries of the objects are shown with solid lines, and the expanded boundaries with dashed lines. (a) Area increases because both length and width increase. The area of a circular plug also increases. (b) If the plug is removed, the hole it leaves becomes larger with increasing temperature, just as if the expanding plug were still in place. (c) Volume also increases, because all three dimensions increase.

Thermal Expansion in Three Dimensions

The relationship between volume and temperature $\frac{dV}{dT}$ is given by $\frac{dV}{dT} = \beta V \Delta T$, where β is the **coefficient of volume expansion**. As you can show in Exercise 1.60, $\beta = 3\alpha$. This equation is usually written as

$$\Delta V = \beta V \Delta T. \tag{1.4}$$

Note that the values of β in Table 1.2 are equal to 3α except for rounding.

Volume expansion is defined for liquids, but linear and area expansion are not, as a liquid's changes in linear dimensions and area depend on the shape of its container. Thus, Table 1.2 shows liquids' values of β but not α.

In general, objects expand with increasing temperature. Water is the most important exception to this rule. Water does expand with increasing temperature (its density *decreases*) at temperatures greater than $4\ °\text{C}$ (40 °F). However, it is densest at $+4\ °\text{C}$ and expands with *decreasing* temperature between $+4\ °\text{C}$ and $0\ °\text{C}$ (40 °F to 32 °F), as shown in Figure 1.8. A striking effect of this phenomenon is the freezing of water in a pond. When water near the surface cools down to $4\ °\text{C}$, it is denser than the remaining water and thus sinks to the bottom. This "turnover" leaves a layer of warmer water near the surface, which is then cooled. However, if the temperature in the surface layer drops below $4\ °\text{C}$, that water is less dense than the water below, and thus stays near the top. As a result, the pond surface can freeze over. The layer of ice insulates the liquid water below it from low air temperatures. Fish and other aquatic life can survive in $4\ °\text{C}$ water beneath ice, due to this unusual characteristic of water.

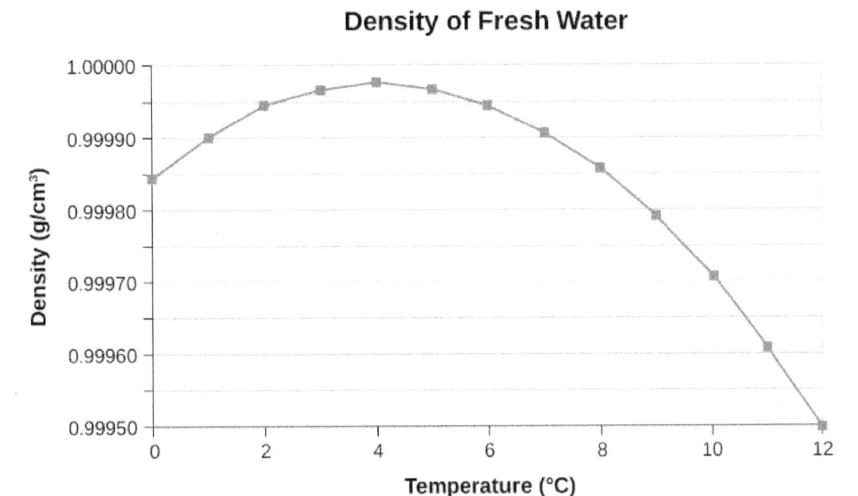

Figure 1.8 This curve shows the density of water as a function of temperature. Note that the thermal expansion at low temperatures is very small. The maximum density at $4\ °C$ is only 0.0075% greater than the density at $2\ °C$, and 0.012% greater than that at $0\ °C$. The decrease of density below $4\ °C$ occurs because the liquid water approachs the solid crystal form of ice, which contains more empty space than the liquid.

Example 1.3

Calculating Thermal Expansion

Suppose your 60.0-L (15.9 -gal -gal) steel gasoline tank is full of gas that is cool because it has just been pumped from an underground reservoir. Now, both the tank and the gasoline have a temperature of $15.0\ °C$. How much gasoline has spilled by the time they warm to $35.0\ °C$?

Strategy

The tank and gasoline increase in volume, but the gasoline increases more, so the amount spilled is the difference in their volume changes. We can use the equation for volume expansion to calculate the change in volume of the gasoline and of the tank. (The gasoline tank can be treated as solid steel.)

Solution

1. Use the equation for volume expansion to calculate the increase in volume of the steel tank:
$$\Delta V_s = \beta_s V_s \Delta T.$$
2. The increase in volume of the gasoline is given by this equation:
$$\Delta V_{gas} = \beta_{gas} V_{gas} \Delta T.$$
3. Find the difference in volume to determine the amount spilled as
$$V_{spill} = \Delta V_{gas} - \Delta V_s.$$

Alternatively, we can combine these three equations into a single equation. (Note that the original volumes are equal.)

$$\begin{aligned} V_{spill} &= (\beta_{ga\ s} - \beta_s) V \Delta T \\ &= \left[(950 - 35) \times 10^{-6} /°C\right](60.0\ L)(20.0\ °C) \\ &= 1.10\ L. \end{aligned}$$

Significance

This amount is significant, particularly for a 60.0-L tank. The effect is so striking because the gasoline and steel expand quickly. The rate of change in thermal properties is discussed later in this chapter.

If you try to cap the tank tightly to prevent overflow, you will find that it leaks anyway, either around the cap or by bursting the tank. Tightly constricting the expanding gas is equivalent to compressing it, and both liquids and solids resist compression with extremely large forces. To avoid rupturing rigid containers, these containers have air gaps, which allow them to expand and contract without stressing them.

1.1 Check Your Understanding Does a given reading on a gasoline gauge indicate more gasoline in cold weather or in hot weather, or does the temperature not matter?

Thermal Stress

If you change the temperature of an object while preventing it from expanding or contracting, the object is subjected to stress that is compressive if the object would expand in the absence of constraint and tensile if it would contract. This stress resulting from temperature changes is known as **thermal stress**. It can be quite large and can cause damage.

To avoid this stress, engineers may design components so they can expand and contract freely. For instance, in highways, gaps are deliberately left between blocks to prevent thermal stress from developing. When no gaps can be left, engineers must consider thermal stress in their designs. Thus, the reinforcing rods in concrete are made of steel because steel's coefficient of linear expansion is nearly equal to that of concrete.

To calculate the thermal stress in a rod whose ends are both fixed rigidly, we can think of the stress as developing in two steps. First, let the ends be free to expand (or contract) and find the expansion (or contraction). Second, find the stress necessary to compress (or extend) the rod to its original length by the methods you studied in Static Equilibrium and Elasticity (http://cnx.org/content/m58339/latest/) on static equilibrium and elasticity. In other words, the ΔL of the thermal expansion equals the ΔL of the elastic distortion (except that the signs are opposite).

Example 1.4

Calculating Thermal Stress

Concrete blocks are laid out next to each other on a highway without any space between them, so they cannot expand. The construction crew did the work on a winter day when the temperature was $5\,°\text{C}$. Find the stress in the blocks on a hot summer day when the temperature is $38\,°\text{C}$. The compressive Young's modulus of concrete is $Y = 20 \times 10^9\ \text{N/m}^2$.

Strategy

According to the chapter on static equilibrium and elasticity, the stress *F*/*A* is given by

$$\frac{F}{A} = Y\frac{\Delta L}{L_0},$$

where *Y* is the Young's modulus of the material—concrete, in this case. In thermal expansion, $\Delta L = \alpha L_0 \Delta T$. We combine these two equations by noting that the two ΔL's are equal, as stated above. Because we are not given L_0 or *A*, we can obtain a numerical answer only if they both cancel out.

Solution

We substitute the thermal-expansion equation into the elasticity equation to get

$$\frac{F}{A} = Y\frac{\alpha L_0 \Delta T}{L_0} = Y\alpha\Delta T,$$

and as we hoped, L_0 has canceled and *A* appears only in *F*/*A*, the notation for the quantity we are calculating.

Now we need only insert the numbers:

$$\frac{F}{A} = (20 \times 10^9 \text{ N/m}^2)(12 \times 10^{-6}/°\text{C})(38\text{ °C} - 5\text{ °C}) = 7.9 \times 10^6 \text{ N/m}^2.$$

Significance

The ultimate compressive strength of concrete is $20 \times 10^6 \text{ N/m}^2$, so the blocks are unlikely to break. However, the ultimate shear strength of concrete is only $2 \times 10^6 \text{ N/m}^2$, so some might chip off.

1.2 Check Your Understanding Two objects *A* and *B* have the same dimensions and are constrained identically. *A* is made of a material with a higher thermal expansion coefficient than *B*. If the objects are heated identically, will *A* feel a greater stress than *B*?

1.4 | Heat Transfer, Specific Heat, and Calorimetry

Learning Objectives

By the end of this section, you will be able to:

- Explain phenomena involving heat as a form of energy transfer
- Solve problems involving heat transfer

We have seen in previous chapters that energy is one of the fundamental concepts of physics. **Heat** is a type of energy transfer that is caused by a temperature difference, and it can change the temperature of an object. As we learned earlier in this chapter, heat transfer is the movement of energy from one place or material to another as a result of a difference in temperature. Heat transfer is fundamental to such everyday activities as home heating and cooking, as well as many industrial processes. It also forms a basis for the topics in the remainder of this chapter.

We also introduce the concept of internal energy, which can be increased or decreased by heat transfer. We discuss another way to change the internal energy of a system, namely doing work on it. Thus, we are beginning the study of the relationship of heat and work, which is the basis of engines and refrigerators and the central topic (and origin of the name) of thermodynamics.

Internal Energy and Heat

A thermal system has *internal energy* (also called thermal energy), which is the sum of the mechanical energies of its molecules. A system's internal energy is proportional to its temperature. As we saw earlier in this chapter, if two objects at different temperatures are brought into contact with each other, energy is transferred from the hotter to the colder object until the bodies reach thermal equilibrium (that is, they are at the same temperature). No work is done by either object because no force acts through a distance (as we discussed in Work and Kinetic Energy (http://cnx.org/content/m58307/latest/)). These observations reveal that heat is energy transferred spontaneously due to a temperature difference. Figure 1.9 shows an example of heat transfer.

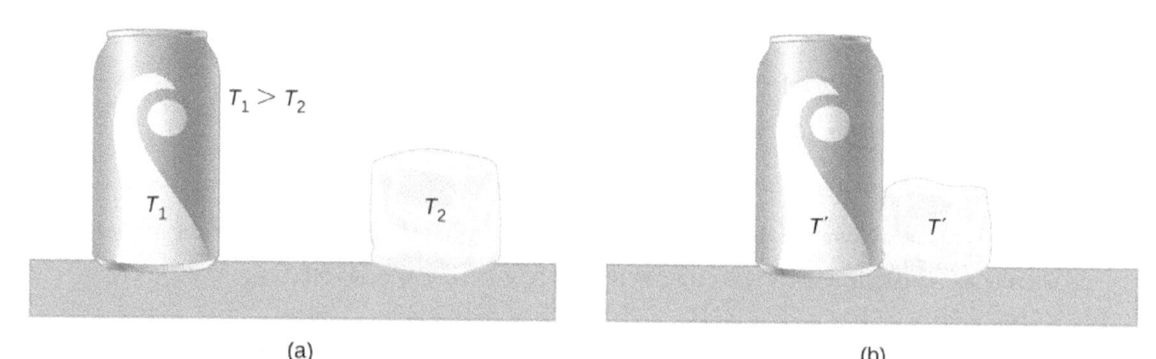

Figure 1.9 (a) Here, the soft drink has a higher temperature than the ice, so they are not in thermal equilibrium. (b) When the soft drink and ice are allowed to interact, heat is transferred from the drink to the ice due to the difference in temperatures until they reach the same temperature, T' , achieving equilibrium. In fact, since the soft drink and ice are both in contact with the surrounding air and the bench, the ultimate equilibrium temperature will be the same as that of the surroundings.

The meaning of "heat" in physics is different from its ordinary meaning. For example, in conversation, we may say "the heat was unbearable," but in physics, we would say that the temperature was high. Heat is a form of energy flow, whereas temperature is not. Incidentally, humans are sensitive to *heat flow* rather than to temperature.

Since heat is a form of energy, its SI unit is the joule (J). Another common unit of energy often used for heat is the **calorie** (cal), defined as the energy needed to change the temperature of 1.00 g of water by $1.00\,°\text{C}$ —specifically, between $14.5\,°\text{C}$ and $15.5\,°\text{C}$, since there is a slight temperature dependence. Also commonly used is the **kilocalorie** (kcal), which is the energy needed to change the temperature of 1.00 kg of water by $1.00\,°\text{C}$. Since mass is most often specified in kilograms, the kilocalorie is convenient. Confusingly, food calories (sometimes called "big calories," abbreviated Cal) are actually kilocalories, a fact not easily determined from package labeling.

Mechanical Equivalent of Heat

It is also possible to change the temperature of a substance by doing work, which transfers energy into or out of a system. This realization helped establish that heat is a form of energy. James Prescott Joule (1818–1889) performed many experiments to establish the **mechanical equivalent of heat**—*the work needed to produce the same effects as heat transfer*. In the units used for these two quantities, the value for this equivalence is

$$1.000\,\text{kcal} = 4186\,\text{J}.$$

We consider this equation to represent the conversion between two units of energy. (Other numbers that you may see refer to calories defined for temperature ranges other than $14.5\,°\text{C}$ to $15.5\,°\text{C}$.)

Figure 1.10 shows one of Joule's most famous experimental setups for demonstrating that work and heat can produce the same effects and measuring the mechanical equivalent of heat. It helped establish the principle of conservation of energy. Gravitational potential energy (*U*) was converted into kinetic energy (*K*), and then randomized by viscosity and turbulence into increased average kinetic energy of atoms and molecules in the system, producing a temperature increase. Joule's contributions to thermodynamics were so significant that the SI unit of energy was named after him.

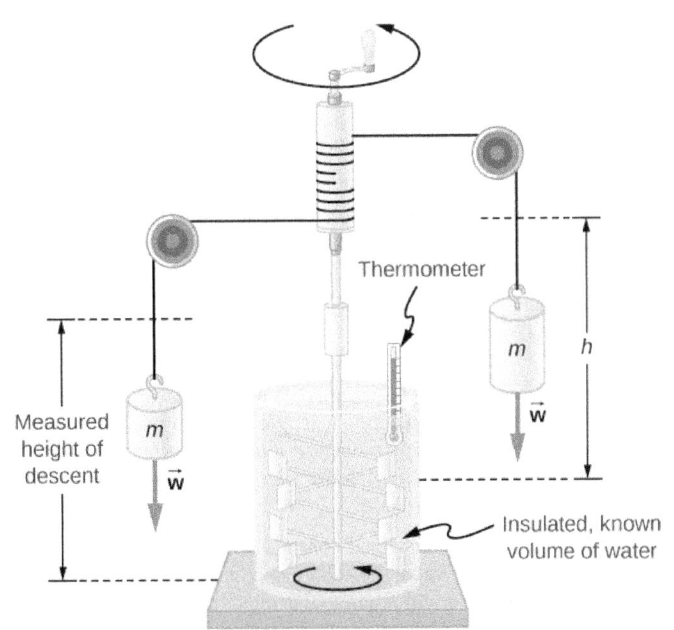

Figure 1.10 Joule's experiment established the equivalence of heat and work. As the masses descended, they caused the paddles to do work, $W = mgh$, on the water. The result was a temperature increase, ΔT, measured by the thermometer. Joule found that ΔT was proportional to W and thus determined the mechanical equivalent of heat.

Increasing internal energy by heat transfer gives the same result as increasing it by doing work. Therefore, although a system has a well-defined internal energy, we cannot say that it has a certain "heat content" or "work content." A well-defined quantity that depends only on the current state of the system, rather than on the history of that system, is known as a *state variable*. Temperature and internal energy are state variables. To sum up this paragraph, *heat and work are not state variables*.

Incidentally, increasing the internal energy of a system does not necessarily increase its temperature. As we'll see in the next section, the temperature does not change when a substance changes from one phase to another. An example is the melting of ice, which can be accomplished by adding heat or by doing frictional work, as when an ice cube is rubbed against a rough surface.

Temperature Change and Heat Capacity

We have noted that heat transfer often causes temperature change. Experiments show that with no phase change and no work done on or by the system, the transferred heat is typically directly proportional to the change in temperature and to the mass of the system, to a good approximation. (Below we show how to handle situations where the approximation is not valid.) The constant of proportionality depends on the substance and its phase, which may be gas, liquid, or solid. We omit discussion of the fourth phase, plasma, because although it is the most common phase in the universe, it is rare and short-lived on Earth.

We can understand the experimental facts by noting that the transferred heat is the change in the internal energy, which is the total energy of the molecules. Under typical conditions, the total kinetic energy of the molecules K_{total} is a constant fraction of the internal energy (for reasons and with exceptions that we'll see in the next chapter). The average kinetic energy of a molecule K_{ave} is proportional to the absolute temperature. Therefore, the change in internal energy of a system is typically proportional to the change in temperature and to the number of molecules, N. Mathematically, $\Delta U \propto \Delta K_{\text{total}} = NK_{\text{ave}} \propto N\Delta T$ The dependence on the substance results in large part from the different masses of atoms and molecules. We are considering its heat capacity in terms of its mass, but as we will see in the next chapter, in some cases, heat capacities *per molecule* are similar for different substances. The dependence on substance and phase also results from differences in the potential energy associated with interactions between atoms and molecules.

Heat Transfer and Temperature Change

A practical approximation for the relationship between heat transfer and temperature change is:

$$Q = mc\Delta T, \tag{1.5}$$

where Q is the symbol for heat transfer ("quantity of heat"), m is the mass of the substance, and ΔT is the change in temperature. The symbol c stands for the **specific heat** (also called "*specific heat capacity*") and depends on the material and phase. The specific heat is numerically equal to the amount of heat necessary to change the temperature of 1.00 kg of mass by $1.00\ °\text{C}$. The SI unit for specific heat is $\text{J/(kg} \times \text{K)}$ or $\text{J/(kg} \times °\text{C)}$. (Recall that the temperature change ΔT is the same in units of kelvin and degrees Celsius.)

Values of specific heat must generally be measured, because there is no simple way to calculate them precisely. Table 1.3 lists representative values of specific heat for various substances. We see from this table that the specific heat of water is five times that of glass and 10 times that of iron, which means that it takes five times as much heat to raise the temperature of water a given amount as for glass, and 10 times as much as for iron. In fact, water has one of the largest specific heats of any material, which is important for sustaining life on Earth.

The specific heats of gases depend on what is maintained constant during the heating—typically either the volume or the pressure. In the table, the first specific heat value for each gas is measured at constant volume, and the second (in parentheses) is measured at constant pressure. We will return to this topic in the chapter on the kinetic theory of gases.

Substances	**Specific Heat (*c*)**	
Solids	$\text{J/kg} \cdot °\text{C}$	$\text{kcal/kg} \cdot °\text{C}$[2]
Aluminum	900	0.215
Asbestos	800	0.19
Concrete, granite (average)	840	0.20
Copper	387	0.0924
Glass	840	0.20
Gold	129	0.0308
Human body (average at $37\ °\text{C}$)	3500	0.83
Ice (average, $-50\ °\text{C}$ to $0\ °\text{C}$)	2090	0.50
Iron, steel	452	0.108
Lead	128	0.0305
Silver	235	0.0562
Wood	1700	0.40
Liquids		
Benzene	1740	0.415
Ethanol	2450	0.586
Glycerin	2410	0.576

Table 1.3 Specific Heats of Various Substances[1] [1]The values for solids and liquids are at constant volume and $25\ °\text{C}$, except as noted. [2]These values are identical in units of $\text{cal/g} \cdot °\text{C}$. [3]Specific heats at constant volume and at $20.0\ °\text{C}$ except as noted, and at 1.00 atm pressure. Values in parentheses are specific heats at a constant pressure of 1.00 atm.

Substances	Specific Heat (*c*)	
Mercury	139	0.0333
Water (15.0 °C)	4186	1.000
Gases[3]		
Air (dry)	721 (1015)	0.172 (0.242)
Ammonia	1670 (2190)	0.399 (0.523)
Carbon dioxide	638 (833)	0.152 (0.199)
Nitrogen	739 (1040)	0.177 (0.248)
Oxygen	651 (913)	0.156 (0.218)
Steam (100 °C)	1520 (2020)	0.363 (0.482)

Table 1.3 Specific Heats of Various Substances[1] [1]The values for solids and liquids are at constant volume and 25 °C, except as noted. [2]These values are identical in units of cal/g · °C. [3]Specific heats at constant volume and at 20.0 °C except as noted, and at 1.00 atm pressure. Values in parentheses are specific heats at a constant pressure of 1.00 atm.

In general, specific heat also depends on temperature. Thus, a precise definition of c for a substance must be given in terms of an infinitesimal change in temperature. To do this, we note that $c = \frac{1}{m}\frac{\Delta Q}{\Delta T}$ and replace Δ with d:

$$c = \frac{1}{m}\frac{dQ}{dT}.$$

Except for gases, the temperature and volume dependence of the specific heat of most substances is weak at normal temperatures. Therefore, we will generally take specific heats to be constant at the values given in the table.

Example 1.5

Calculating the Required Heat

A 0.500-kg aluminum pan on a stove and 0.250 L of water in it are heated from 20.0 °C to 80.0 °C. (a) How much heat is required? What percentage of the heat is used to raise the temperature of (b) the pan and (c) the water?

Strategy

We can assume that the pan and the water are always at the same temperature. When you put the pan on the stove, the temperature of the water and that of the pan are increased by the same amount. We use the equation for the heat transfer for the given temperature change and mass of water and aluminum. The specific heat values for water and aluminum are given in Table 1.3.

Solution

1. Calculate the temperature difference:

$$\Delta T = T_f - T_i = 60.0\ °\text{C}.$$

2. Calculate the mass of water. Because the density of water is $1000\ \text{kg/m}^3$, 1 L of water has a mass of 1 kg, and the mass of 0.250 L of water is $m_w = 0.250\ \text{kg}$.

3. Calculate the heat transferred to the water. Use the specific heat of water in Table 1.3:

$$Q_w = m_w c_w \Delta T = (0.250\ \text{kg})(4186\ \text{J/kg °C})(60.0\ °\text{C}) = 62.8\ \text{kJ}.$$

4. Calculate the heat transferred to the aluminum. Use the specific heat for aluminum in Table 1.3:

$$Q_{\text{Al}} = m_{\text{Al}} c_{\text{Al}} \Delta T = (0.500 \text{ kg})(900 \text{ J/kg °C})(60.0 \text{ °C}) = 27.0 \text{ kJ}.$$

5. Find the total transferred heat:

$$Q_{\text{Total}} = Q_{\text{W}} + Q_{\text{Al}} = 89.8 \text{ kJ}.$$

Significance

In this example, the heat transferred to the container is a significant fraction of the total transferred heat. Although the mass of the pan is twice that of the water, the specific heat of water is over four times that of aluminum. Therefore, it takes a bit more than twice as much heat to achieve the given temperature change for the water as for the aluminum pan.

Example 1.6 illustrates a temperature rise caused by doing work. (The result is the same as if the same amount of energy had been added with a blowtorch instead of mechanically.)

Example 1.6

Calculating the Temperature Increase from the Work Done on a Substance

Truck brakes used to control speed on a downhill run do work, converting gravitational potential energy into increased internal energy (higher temperature) of the brake material (Figure 1.11). This conversion prevents the gravitational potential energy from being converted into kinetic energy of the truck. Since the mass of the truck is much greater than that of the brake material absorbing the energy, the temperature increase may occur too fast for sufficient heat to transfer from the brakes to the environment; in other words, the brakes may overheat.

Figure 1.11 The smoking brakes on a braking truck are visible evidence of the mechanical equivalent of heat.

Calculate the temperature increase of 10 kg of brake material with an average specific heat of $800 \text{ J/kg} \cdot \text{°C}$ if the material retains 10% of the energy from a 10,000-kg truck descending 75.0 m (in vertical displacement) at a constant speed.

Strategy

We calculate the gravitational potential energy (*Mgh*) that the entire truck loses in its descent, equate it to the increase in the brakes' internal energy, and then find the temperature increase produced in the brake material alone.

Solution

First we calculate the change in gravitational potential energy as the truck goes downhill:

$$Mgh = (10{,}000 \text{ kg})(9.80 \text{ m/s}^2)(75.0 \text{ m}) = 7.35 \times 10^6 \text{ J}.$$

Because the kinetic energy of the truck does not change, conservation of energy tells us the lost potential energy is dissipated, and we assume that 10% of it is transferred to internal energy of the brakes, so take $Q = Mgh/10$.

Then we calculate the temperature change from the heat transferred, using

$$\Delta T = \frac{Q}{mc},$$

where m is the mass of the brake material. Insert the given values to find

$$\Delta T = \frac{7.35 \times 10^5 \text{ J}}{(10 \text{ kg})(800 \text{ J/kg °C})} = 92 \text{ °C}.$$

Significance

If the truck had been traveling for some time, then just before the descent, the brake temperature would probably be higher than the ambient temperature. The temperature increase in the descent would likely raise the temperature of the brake material very high, so this technique is not practical. Instead, the truck would use the technique of engine braking. A different idea underlies the recent technology of hybrid and electric cars, where mechanical energy (kinetic and gravitational potential energy) is converted by the brakes into electrical energy in the battery, a process called regenerative braking.

In a common kind of problem, objects at different temperatures are placed in contact with each other but isolated from everything else, and they are allowed to come into equilibrium. A container that prevents heat transfer in or out is called a **calorimeter**, and the use of a calorimeter to make measurements (typically of heat or specific heat capacity) is called **calorimetry**.

We will use the term "calorimetry problem" to refer to any problem in which the objects concerned are thermally isolated from their surroundings. An important idea in solving calorimetry problems is that during a heat transfer between objects isolated from their surroundings, the heat gained by the colder object must equal the heat lost by the hotter object, due to conservation of energy:

$$Q_{\text{cold}} + Q_{\text{hot}} = 0. \tag{1.6}$$

We express this idea by writing that the sum of the heats equals zero because the heat gained is usually considered positive; the heat lost, negative.

Example 1.7

Calculating the Final Temperature in Calorimetry

Suppose you pour 0.250 kg of 20.0-°C water (about a cup) into a 0.500-kg aluminum pan off the stove with a temperature of 150 °C. Assume no heat transfer takes place to anything else: The pan is placed on an insulated pad, and heat transfer to the air is neglected in the short time needed to reach equilibrium. Thus, this is a calorimetry problem, even though no isolating container is specified. Also assume that a negligible amount of water boils off. What is the temperature when the water and pan reach thermal equilibrium?

Strategy

Originally, the pan and water are not in thermal equilibrium: The pan is at a higher temperature than the water. Heat transfer restores thermal equilibrium once the water and pan are in contact; it stops once thermal equilibrium between the pan and the water is achieved. The heat lost by the pan is equal to the heat gained by the water—that is the basic principle of calorimetry.

Solution

1. Use the equation for heat transfer $Q = mc\Delta T$ to express the heat lost by the aluminum pan in terms of the mass of the pan, the specific heat of aluminum, the initial temperature of the pan, and the final temperature:

$$Q_{\text{hot}} = m_{\text{Al}} c_{\text{Al}} (T_{\text{f}} - 150\ °\text{C}).$$

2. Express the heat gained by the water in terms of the mass of the water, the specific heat of water, the initial temperature of the water, and the final temperature:

$$Q_{\text{cold}} = m_{\text{w}} c_{\text{w}} (T_{\text{f}} - 20.0\ °\text{C}).$$

3. Note that $Q_{\text{hot}} < 0$ and $Q_{\text{cold}} > 0$ and that as stated above, they must sum to zero:

$$\begin{aligned} Q_{\text{cold}} + Q_{\text{hot}} &= 0 \\ Q_{\text{cold}} &= -Q_{\text{hot}} \\ m_{\text{w}} c_{\text{w}} (T_{\text{f}} - 20.0\ °\text{C}) &= -m_{\text{Al}} c_{\text{Al}} (T_{\text{f}} - 150\ °\text{C}). \end{aligned}$$

4. This a linear equation for the unknown final temperature, T_{f}. Solving for T_{f},

$$T_{\text{f}} = \frac{m_{\text{Al}} c_{\text{Al}} (150\ °\text{C}) + m_{\text{w}} c_{\text{w}} (20.0\ °\text{C})}{m_{\text{Al}} c_{\text{Al}} + m_{\text{w}} c_{\text{w}}},$$

and insert the numerical values:

$$T_{\text{f}} = \frac{(0.500\ \text{kg})(900\ \text{J/kg}\ °\text{C})(150\ °\text{C}) + (0.250\ \text{kg})(4186\ \text{J/kg}\ °\text{C})(20.0\ °\text{C})}{(0.500\ \text{kg})(900\ \text{J/kg}\ °\text{C}) + (0.250\ \text{kg})(4186\ \text{J/kg}\ °\text{C})} = 59.1\ °\text{C}.$$

Significance

Why is the final temperature so much closer to $20.0\ °\text{C}$ than to $150\ °\text{C}$? The reason is that water has a greater specific heat than most common substances and thus undergoes a smaller temperature change for a given heat transfer. A large body of water, such as a lake, requires a large amount of heat to increase its temperature appreciably. This explains why the temperature of a lake stays relatively constant during the day even when the temperature change of the air is large. However, the water temperature does change over longer times (e.g., summer to winter).

1.3 Check Your Understanding If 25 kJ is necessary to raise the temperature of a rock from $25\ °\text{C}$ to $30\ °\text{C}$, how much heat is necessary to heat the rock from $45\ °\text{C}$ to $50\ °\text{C}$?

Example 1.8

Temperature-Dependent Heat Capacity

At low temperatures, the specific heats of solids are typically proportional to T^3. The first understanding of this behavior was due to the Dutch physicist Peter Debye, who in 1912, treated atomic oscillations with the quantum theory that Max Planck had recently used for radiation. For instance, a good approximation for the specific heat of salt, NaCl, is $c = 3.33 \times 10^4 \frac{\text{J}}{\text{kg} \cdot \text{k}} \left(\frac{T}{321\ \text{K}}\right)^3$. The constant 321 K is called the *Debye temperature* of NaCl, Θ_{D}, and the formula works well when $T < 0.04\Theta_{\text{D}}$. Using this formula, how much heat is required to raise the temperature of 24.0 g of NaCl from 5 K to 15 K?

Solution

Because the heat capacity depends on the temperature, we need to use the equation

$$c = \frac{1}{m} \frac{dQ}{dT}.$$

We solve this equation for Q by integrating both sides: $Q = m \int_{T_1}^{T_2} c dT$.

Then we substitute the given values in and evaluate the integral:

$$Q = (0.024\text{ kg})\int_{T_1}^{T_2} 333 \times 10^4 \frac{\text{J}}{\text{kg}\cdot\text{K}}\left(\frac{T}{321\text{ K}}\right)^3 dT = \left(6.04 \times 10^{-4}\frac{\text{J}}{\text{K}^4}\right)T^4\Big|_{5\text{ K}}^{15\text{ K}} = 30.2\text{ J}.$$

Significance

If we had used the equation $Q = mc\Delta T$ and the room-temperature specific heat of salt, $880\text{ J/kg}\cdot\text{K}$, we would have gotten a very different value.

1.5 | Phase Changes

Learning Objectives

By the end of this section, you will be able to:

- Describe phase transitions and equilibrium between phases
- Solve problems involving latent heat
- Solve calorimetry problems involving phase changes

Phase transitions play an important theoretical and practical role in the study of heat flow. In melting (or "fusion"), a solid turns into a liquid; the opposite process is freezing. In evaporation, a liquid turns into a gas; the opposite process is condensation.

A substance melts or freezes at a temperature called its melting point, and boils (evaporates rapidly) or condenses at its boiling point. These temperatures depend on pressure. High pressure favors the denser form, so typically, high pressure raises the melting point and boiling point, and low pressure lowers them. For example, the boiling point of water is 100 °C at 1.00 atm. At higher pressure, the boiling point is higher, and at lower pressure, it is lower. The main exception is the melting and freezing of water, discussed in the next section.

Phase Diagrams

The phase of a given substance depends on the pressure and temperature. Thus, plots of pressure versus temperature showing the phase in each region provide considerable insight into thermal properties of substances. Such a pT graph is called a **phase diagram**.

Figure 1.12 shows the phase diagram for water. Using the graph, if you know the pressure and temperature, you can determine the phase of water. The solid curves—boundaries between phases—indicate phase transitions, that is, temperatures and pressures at which the phases coexist. For example, the boiling point of water is 100 °C at 1.00 atm. As the pressure increases, the boiling temperature rises gradually to 374 °C at a pressure of 218 atm. A pressure cooker (or even a covered pot) cooks food faster than an open pot, because the water can exist as a liquid at temperatures greater than 100 °C without all boiling away. (As we'll see in the next section, liquid water conducts heat better than steam or hot air.) The boiling point curve ends at a certain point called the **critical point**—that is, a **critical temperature**, above which the liquid and gas phases cannot be distinguished; the substance is called a *supercritical fluid*. At sufficiently high pressure above the critical point, the gas has the density of a liquid but does not condense. Carbon dioxide, for example, is supercritical at all temperatures above 31.0 °C. **Critical pressure** is the pressure of the critical point.

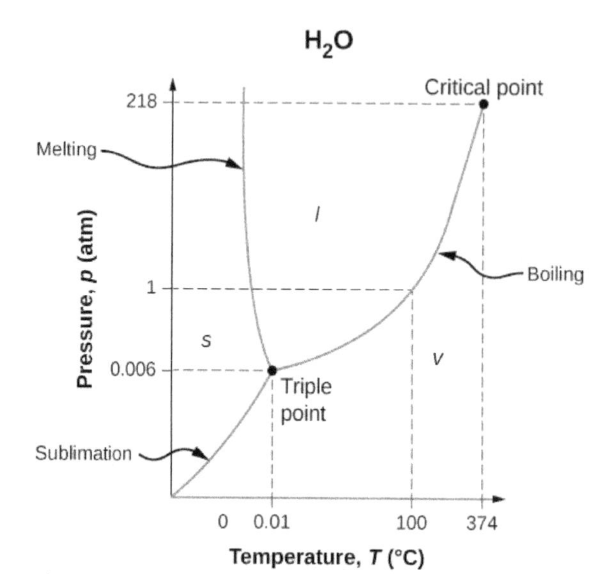

Figure 1.12 The phase diagram (pT graph) for water shows solid (s), liquid (l), and vapor (v) phases. At temperatures and pressure above those of the critical point, there is no distinction between liquid and vapor. Note that the axes are nonlinear and the graph is not to scale. This graph is simplified—it omits several exotic phases of ice at higher pressures. The phase diagram of water is unusual because the melting-point curve has a negative slope, showing that you can melt ice by *increasing* the pressure.

Similarly, the curve between the solid and liquid regions in Figure 1.12 gives the melting temperature at various pressures. For example, the melting point is $0\,°C$ at 1.00 atm, as expected. Water has the unusual property that ice is less dense than liquid water at the melting point, so at a fixed temperature, you can change the phase from solid (ice) to liquid (water) by increasing the pressure. That is, the melting temperature of ice falls with increased pressure, as the phase diagram shows. For example, when a car is driven over snow, the increased pressure from the tires melts the snowflakes; afterwards, the water refreezes and forms an ice layer.

As you learned in the earlier section on thermometers and temperature scales, the triple point is the combination of temperature and pressure at which ice, liquid water, and water vapor can coexist stably—that is, all three phases exist in equilibrium. For water, the triple point occurs at $273.16\,K\,(0.01\,°C)$ and 611.2 Pa; that is a more accurate calibration temperature than the melting point of water at 1.00 atm, or $273.15\,K\,(0.0\,°C)$.

View this video (https://openstaxcollege.org/l/21triplepoint) to see a substance at its triple point.

At pressures below that of the triple point, there is no liquid phase; the substance can exist as either gas or solid. For water, there is no liquid phase at pressures below 0.00600 atm. The phase change from solid to gas is called **sublimation**. You may have noticed that snow can disappear into thin air without a trace of liquid water, or that ice cubes can disappear in a freezer. Both are examples of sublimation. The reverse also happens: Frost can form on very cold windows without going through the liquid stage. Figure 1.13 shows the result, as well as showing a familiar example of sublimation. Carbon dioxide has no liquid phase at atmospheric pressure. Solid CO_2 is known as dry ice because instead of melting, it sublimes. Its sublimation temperature at atmospheric pressure is $-78\,°C$. Certain air fresheners use the sublimation of a solid to spread a perfume around a room. Some solids, such as osmium tetroxide, are so toxic that they must be kept in sealed containers to prevent human exposure to their sublimation-produced vapors.

(a) (b)

Figure 1.13 Direct transitions between solid and vapor are common, sometimes useful, and even beautiful. (a) Dry ice sublimes directly to carbon dioxide gas. The visible "smoke" consists of water droplets that condensed in the air cooled by the dry ice. (b) Frost forms patterns on a very cold window, an example of a solid formed directly from a vapor. (credit a: modification of work by Windell Oskay; credit b: modification of work by Liz West)

Equilibrium

At the melting temperature, the solid and liquid phases are in equilibrium. If heat is added, some of the solid will melt, and if heat is removed, some of the liquid will freeze. The situation is somewhat more complex for liquid-gas equilibrium. Generally, liquid and gas are in equilibrium at any temperature. We call the gas phase a **vapor** when it exists at a temperature below the boiling temperature, as it does for water at $20.0\,°\text{C}$. Liquid in a closed container at a fixed temperature evaporates until the pressure of the gas reaches a certain value, called the **vapor pressure**, which depends on the gas and the temperature. At this equilibrium, if heat is added, some of the liquid will evaporate, and if heat is removed, some of the gas will condense; molecules either join the liquid or form suspended droplets. If there is not enough liquid for the gas to reach the vapor pressure in the container, all the liquid eventually evaporates.

If the vapor pressure of the liquid is greater than the *total* ambient pressure, including that of any air (or other gas), the liquid evaporates rapidly; in other words, it boils. Thus, the boiling point of a liquid at a given pressure is the temperature at which its vapor pressure equals the ambient pressure. Liquid and gas phases are in equilibrium at the boiling temperature (Figure 1.14). If a substance is in a closed container at the boiling point, then the liquid is boiling and the gas is condensing at the same rate without net change in their amounts.

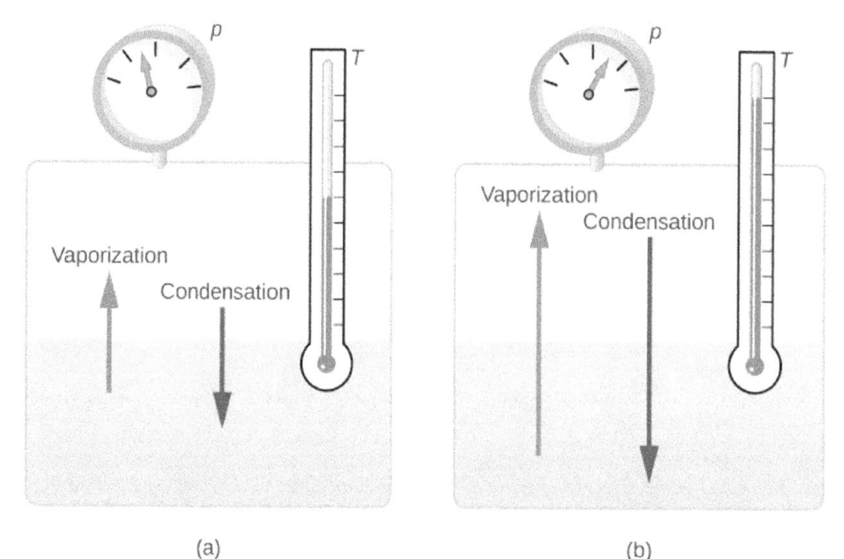

Figure 1.14 Equilibrium between liquid and gas at two different boiling points inside a closed container. (a) The rates of boiling and condensation are equal at this combination of temperature and pressure, so the liquid and gas phases are in equilibrium. (b) At a higher temperature, the boiling rate is faster, that is, the rate at which molecules leave the liquid and enter the gas is faster. This increases the number of molecules in the gas, which increases the gas pressure, which in turn increases the rate at which gas molecules condense and enter the liquid. The pressure stops increasing when it reaches the point where the boiling rate and the condensation rate are equal. The gas and liquid are in equilibrium again at this higher temperature and pressure.

For water, $100\ °C$ is the boiling point at 1.00 atm, so water and steam should exist in equilibrium under these conditions. Why does an open pot of water at $100\ °C$ boil completely away? The gas surrounding an open pot is not pure water: it is mixed with air. If pure water and steam are in a closed container at $100\ °C$ and 1.00 atm, they will coexist—but with air over the pot, there are fewer water molecules to condense, and water boils away. Another way to see this is that at the boiling point, the vapor pressure equals the ambient pressure. However, part of the ambient pressure is due to air, so the pressure of the steam is less than the vapor pressure at that temperature, and evaporation continues. Incidentally, the equilibrium vapor pressure of solids is not zero, a fact that accounts for sublimation.

1.4 Check Your Understanding Explain why a cup of water (or soda) with ice cubes stays at $0\ °C$, even on a hot summer day.

Phase Change and Latent Heat

So far, we have discussed heat transfers that cause temperature change. However, in a phase transition, heat transfer does not cause any temperature change.

For an example of phase changes, consider the addition of heat to a sample of ice at $-20\ °C$ (Figure 1.15) and atmospheric pressure. The temperature of the ice rises linearly, absorbing heat at a constant rate of $2090\ \text{J/kg} \cdot °C$ until it reaches $0\ °C$. Once at this temperature, the ice begins to melt and continues until it has all melted, absorbing 333 kJ/kg of heat. The temperature remains constant at $0\ °C$ during this phase change. Once all the ice has melted, the temperature of the liquid water rises, absorbing heat at a new constant rate of $4186\ \text{J/kg} \cdot °C$. At $100\ °C$, the water begins to boil. The temperature again remains constant during this phase change while the water absorbs 2256 kJ/kg of heat and turns into steam. When all the liquid has become steam, the temperature rises again, absorbing heat at a rate of $2020\ \text{J/kg} \cdot °C$. If we started with steam and cooled it to make it condense into liquid water and freeze into ice, the process would exactly reverse, with the temperature again constant during each phase transition.

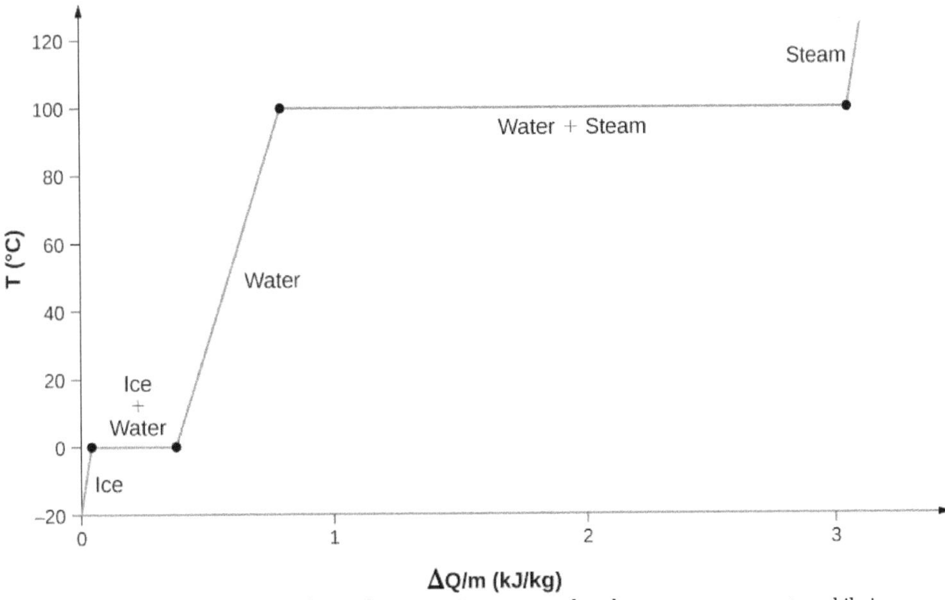

Figure 1.15 Temperature versus heat. The system is constructed so that no vapor evaporates while ice warms to become liquid water, and so that, when vaporization occurs, the vapor remains in the system. The long stretches of constant temperatures at $0\,°\text{C}$ and $100\,°\text{C}$ reflect the large amounts of heat needed to cause melting and vaporization, respectively.

Where does the heat added during melting or boiling go, considering that the temperature does not change until the transition is complete? Energy is required to melt a solid, because the attractive forces between the molecules in the solid must be broken apart, so that in the liquid, the molecules can move around at comparable kinetic energies; thus, there is no rise in temperature. Energy is needed to vaporize a liquid for similar reasons. Conversely, work is done by attractive forces when molecules are brought together during freezing and condensation. That energy must be transferred out of the system, usually in the form of heat, to allow the molecules to stay together (Figure 1.18). Thus, condensation occurs in association with cold objects—the glass in Figure 1.16, for example.

Figure 1.16 Condensation forms on this glass of iced tea because the temperature of the nearby air is reduced. The air cannot hold as much water as it did at room temperature, so water condenses. Energy is released when the water condenses, speeding the melting of the ice in the glass. (credit: Jenny Downing)

The energy released when a liquid freezes is used by orange growers when the temperature approaches $0\,°\text{C}$. Growers spray water on the trees so that the water freezes and heat is released to the growing oranges. This prevents the temperature inside the orange from dropping below freezing, which would damage the fruit (Figure 1.17).

Figure 1.17 The ice on these trees released large amounts of energy when it froze, helping to prevent the temperature of the trees from dropping below $0\,°\text{C}$. Water is intentionally sprayed on orchards to help prevent hard frosts. (credit: Hermann Hammer)

The energy involved in a phase change depends on the number of bonds or force pairs and their strength. The number of bonds is proportional to the number of molecules and thus to the mass of the sample. The energy per unit mass required to change a substance from the solid phase to the liquid phase, or released when the substance changes from liquid to solid, is known as the **heat of fusion**. The energy per unit mass required to change a substance from the liquid phase to the vapor phase is known as the **heat of vaporization**. The strength of the forces depends on the type of molecules. The heat Q absorbed or released in a phase change in a sample of mass m is given by

$$Q = mL_\text{f}\,(\text{melting/freezing}) \quad (1.7)$$

$$Q = mL_\text{v}\,(\text{vaporization/condensation}) \quad (1.8)$$

where the latent heat of fusion L_f and latent heat of vaporization L_v are material constants that are determined experimentally. (Latent heats are also called **latent heat coefficients** and heats of transformation.) These constants are "latent," or hidden, because in phase changes, energy enters or leaves a system without causing a temperature change in the system, so in effect, the energy is hidden.

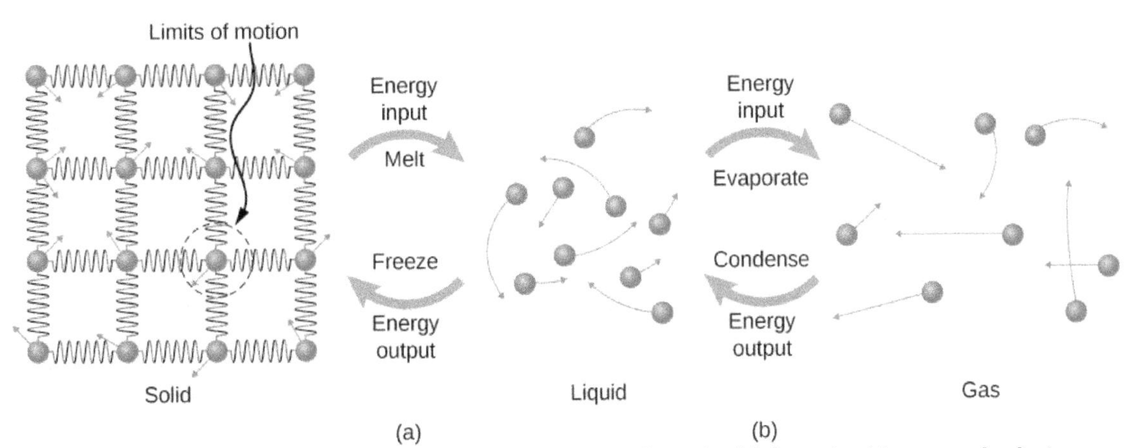

Figure 1.18 (a) Energy is required to partially overcome the attractive forces (modeled as springs) between molecules in a solid to form a liquid. That same energy must be removed from the liquid for freezing to take place. (b) Molecules become separated by large distances when going from liquid to vapor, requiring significant energy to completely overcome molecular attraction. The same energy must be removed from the vapor for condensation to take place.

Table 1.4 lists representative values of L_f and L_v in kJ/kg, together with melting and boiling points. Note that in general, $L_v > L_f$. The table shows that the amounts of energy involved in phase changes can easily be comparable to or greater than those involved in temperature changes, as Figure 1.15 and the accompanying discussion also showed.

		L_f			L_v	
Substance	**Melting Point** (°C)	**kJ/kg**	**kcal/kg**	**Boiling Point** (°C)	**kJ/kg**	**kcal/kg**
Helium[2]	−272.2 (0.95 K)	5.23	1.25	−268.9(4.2 K)	20.9	4.99
Hydrogen	−259.3(13.9 K)	58.6	14.0	−252.9(20.2 K)	452	108
Nitrogen	−210.0(63.2 K)	25.5	6.09	−195.8(77.4 K)	201	48.0
Oxygen	−218.8(54.4 K)	13.8	3.30	−183.0(90.2 K)	213	50.9
Ethanol	−114	104	24.9	78.3	854	204
Ammonia	−75	332	79.3	−33.4	1370	327
Mercury	−38.9	11.8	2.82	357	272	65.0
Water	0.00	334	79.8	100.0	2256[3]	539[4]
Sulfur	119	38.1	9.10	444.6	326	77.9
Lead	327	24.5	5.85	1750	871	208
Antimony	631	165	39.4	1440	561	134
Aluminum	660	380	90	2450	11400	2720
Silver	961	88.3	21.1	2193	2336	558

Table 1.4 Heats of Fusion and Vaporization[1] [1]Values quoted at the normal melting and boiling temperatures at standard atmospheric pressure (1 atm). [2]Helium has no solid phase at atmospheric pressure. The melting point given is at a pressure of 2.5 MPa. [3]At 37.0 °C (body temperature), the heat of vaporization L_v for water is 2430 kJ/kg or 580 kcal/kg. [4]At 37.0 °C (body temperature), the heat of vaporization, L_v for water is 2430 kJ/kg or 580 kcal/kg.

		L_f			L_v	
Gold	1063	64.5	15.4	2660	1578	377
Copper	1083	134	32.0	2595	5069	1211
Uranium	1133	84	20	3900	1900	454
Tungsten	3410	184	44	5900	4810	1150

Table 1.4 Heats of Fusion and Vaporization[1] [1]Values quoted at the normal melting and boiling temperatures at standard atmospheric pressure (1 atm). [2]Helium has no solid phase at atmospheric pressure. The melting point given is at a pressure of 2.5 MPa. [3]At $37.0\,°\text{C}$ (body temperature), the heat of vaporization L_v for water is 2430 kJ/kg or 580 kcal/kg. [4]At $37.0\,°\text{C}$ (body temperature), the heat of vaporization, L_v for water is 2430 kJ/kg or 580 kcal/kg.

Phase changes can have a strong stabilizing effect on temperatures that are not near the melting and boiling points, since evaporation and condensation occur even at temperatures below the boiling point. For example, air temperatures in humid climates rarely go above approximately $38.0\,°\text{C}$ because most heat transfer goes into evaporating water into the air. Similarly, temperatures in humid weather rarely fall below the dew point—the temperature where condensation occurs given the concentration of water vapor in the air—because so much heat is released when water vapor condenses.

More energy is required to evaporate water below the boiling point than at the boiling point, because the kinetic energy of water molecules at temperatures below $100\,°\text{C}$ is less than that at $100\,°\text{C}$, so less energy is available from random thermal motions. For example, at body temperature, evaporation of sweat from the skin requires a heat input of 2428 kJ/kg, which is about 10% higher than the latent heat of vaporization at $100\,°\text{C}$. This heat comes from the skin, and this evaporative cooling effect of sweating helps reduce the body temperature in hot weather. However, high humidity inhibits evaporation, so that body temperature might rise, while unevaporated sweat might be left on your brow.

Example 1.9

Calculating Final Temperature from Phase Change

Three ice cubes are used to chill a soda at $20\,°\text{C}$ with mass $m_{\text{soda}} = 0.25 \text{ kg}$. The ice is at $0\,°\text{C}$ and each ice cube has a mass of 6.0 g. Assume that the soda is kept in a foam container so that heat loss can be ignored and that the soda has the same specific heat as water. Find the final temperature when all ice has melted.

Strategy

The ice cubes are at the melting temperature of $0\,°\text{C}$. Heat is transferred from the soda to the ice for melting. Melting yields water at $0\,°\text{C}$, so more heat is transferred from the soda to this water until the water plus soda system reaches thermal equilibrium.

The heat transferred to the ice is

$$Q_{\text{ice}} = m_{\text{ice}} L_f + m_{\text{ice}} c_W (T_f - 0\,°\text{C}).$$

The heat given off by the soda is

$$Q_{\text{soda}} = m_{\text{soda}} c_W (T_f - 20\,°\text{C}).$$

Since no heat is lost, $Q_{\text{ice}} = -Q_{\text{soda}}$, as in Example 1.7, so that

$$m_{\text{ice}} L_f + m_{\text{ice}} c_W (T_f - 0\,°\text{C}) = -m_{\text{soda}} c_W (T_f - 20\,°\text{C}).$$

Solve for the unknown quantity T_f:

$$T_f = \frac{m_{\text{soda}} c_W (20\,°\text{C}) - m_{\text{ice}} L_f}{(m_{\text{soda}} + m_{\text{ice}}) c_W}.$$

Solution

First we identify the known quantities. The mass of ice is $m_{\text{ice}} = 3 \times 6.0 \text{ g} = 0.018 \text{ kg}$ and the mass of soda is $m_{\text{soda}} = 0.25 \text{ kg}$. Then we calculate the final temperature:

$$T_{\text{f}} = \frac{20{,}930 \text{ J} - 6012 \text{ J}}{1122 \text{ J/°C}} = 13 \text{ °C}.$$

Significance

This example illustrates the large energies involved during a phase change. The mass of ice is about 7% of the mass of the soda but leads to a noticeable change in the temperature of the soda. Although we assumed that the ice was at the freezing temperature, this is unrealistic for ice straight out of a freezer: The typical temperature is -6 °C. However, this correction makes no significant change from the result we found. Can you explain why?

Like solid-liquid and and liquid-vapor transitions, direct solid-vapor transitions or sublimations involve heat. The energy transferred is given by the equation $Q = mL_{\text{s}}$, where L_{s} is the **heat of sublimation**, analogous to L_{f} and L_{v}. The heat of sublimation at a given temperature is equal to the heat of fusion plus the heat of vaporization at that temperature.

We can now calculate any number of effects related to temperature and phase change. In each case, it is necessary to identify which temperature and phase changes are taking place. Keep in mind that heat transfer and work can cause both temperature and phase changes.

Problem-Solving Strategy: The Effects of Heat Transfer

1. Examine the situation to determine that there is a change in the temperature or phase. Is there heat transfer into or out of the system? When it is not obvious whether a phase change occurs or not, you may wish to first solve the problem as if there were no phase changes, and examine the temperature change obtained. If it is sufficient to take you past a boiling or melting point, you should then go back and do the problem in steps—temperature change, phase change, subsequent temperature change, and so on.
2. Identify and list all objects that change temperature or phase.
3. Identify exactly what needs to be determined in the problem (identify the unknowns). A written list is useful.
4. Make a list of what is given or what can be inferred from the problem as stated (identify the knowns). If there is a temperature change, the transferred heat depends on the specific heat of the substance (Heat Transfer, Specific Heat, and Calorimetry), and if there is a phase change, the transferred heat depends on the latent heat of the substance (Table 1.4).
5. Solve the appropriate equation for the quantity to be determined (the unknown).
6. Substitute the knowns along with their units into the appropriate equation and obtain numerical solutions complete with units. You may need to do this in steps if there is more than one state to the process, such as a temperature change followed by a phase change. However, in a calorimetry problem, each step corresponds to a term in the single equation $Q_{\text{hot}} + Q_{\text{cold}} = 0$.
7. Check the answer to see if it is reasonable. Does it make sense? As an example, be certain that any temperature change does not also cause a phase change that you have not taken into account.

1.5 Check Your Understanding Why does snow often remain even when daytime temperatures are higher than the freezing temperature?

1.6 | Mechanisms of Heat Transfer

Learning Objectives

By the end of this section, you will be able to:

- Explain some phenomena that involve conductive, convective, and radiative heat transfer
- Solve problems on the relationships between heat transfer, time, and rate of heat transfer
- Solve problems using the formulas for conduction and radiation

Just as interesting as the effects of heat transfer on a system are the methods by which it occurs. Whenever there is a temperature difference, heat transfer occurs. It may occur rapidly, as through a cooking pan, or slowly, as through the walls of a picnic ice chest. So many processes involve heat transfer that it is hard to imagine a situation where no heat transfer occurs. Yet every heat transfer takes place by only three methods:

1. **Conduction** is heat transfer through stationary matter by physical contact. (The matter is stationary on a macroscopic scale—we know that thermal motion of the atoms and molecules occurs at any temperature above absolute zero.) Heat transferred from the burner of a stove through the bottom of a pan to food in the pan is transferred by conduction.
2. **Convection** is the heat transfer by the macroscopic movement of a fluid. This type of transfer takes place in a forced-air furnace and in weather systems, for example.
3. Heat transfer by **radiation** occurs when microwaves, infrared radiation, visible light, or another form of electromagnetic radiation is emitted or absorbed. An obvious example is the warming of Earth by the Sun. A less obvious example is thermal radiation from the human body.

In the illustration at the beginning of this chapter, the fire warms the snowshoers' faces largely by radiation. Convection carries some heat to them, but most of the air flow from the fire is upward (creating the familiar shape of flames), carrying heat to the food being cooked and into the sky. The snowshoers wear clothes designed with low conductivity to prevent heat flow out of their bodies.

In this section, we examine these methods in some detail. Each method has unique and interesting characteristics, but all three have two things in common: They transfer heat solely because of a temperature difference, and the greater the temperature difference, the faster the heat transfer (Figure 1.19).

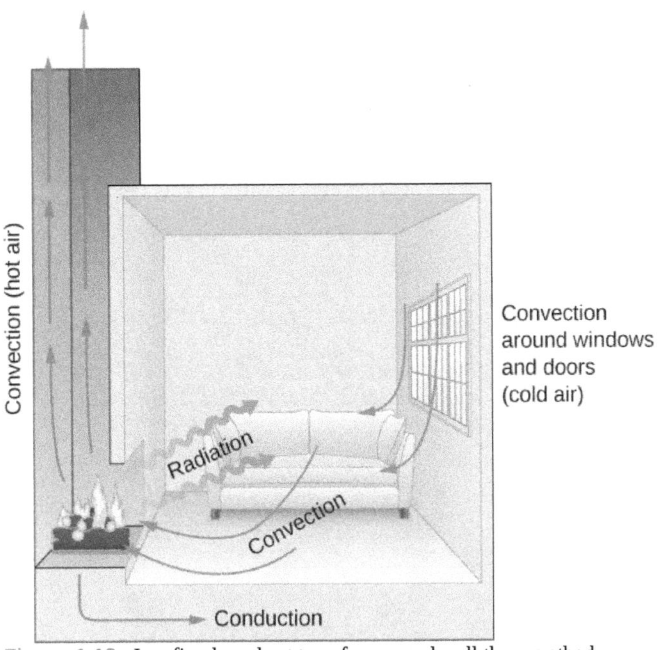

Figure 1.19 In a fireplace, heat transfer occurs by all three methods: conduction, convection, and radiation. Radiation is responsible for most of the heat transferred into the room. Heat transfer also occurs through conduction into the room, but much slower. Heat transfer by convection also occurs through cold air entering the room around windows and hot air leaving the room by rising up the chimney.

1.6 Check Your Understanding Name an example from daily life (different from the text) for each mechanism of heat transfer.

Conduction

As you walk barefoot across the living room carpet in a cold house and then step onto the kitchen tile floor, your feet feel colder on the tile. This result is intriguing, since the carpet and tile floor are both at the same temperature. The different sensation is explained by the different rates of heat transfer: The heat loss is faster for skin in contact with the tiles than with the carpet, so the sensation of cold is more intense.

Some materials conduct thermal energy faster than others. Figure 1.20 shows a material that conducts heat slowly—it is a good thermal insulator, or poor heat conductor—used to reduce heat flow into and out of a house.

Figure 1.20 Insulation is used to limit the conduction of heat from the inside to the outside (in winter) and from the outside to the inside (in summer). (credit: Giles Douglas)

A molecular picture of heat conduction will help justify the equation that describes it. Figure 1.21 shows molecules in two bodies at different temperatures, T_h and T_c, for "hot" and "cold." The average kinetic energy of a molecule in the hot body is higher than in the colder body. If two molecules collide, energy transfers from the high-energy to the low-energy molecule. In a metal, the picture would also include free valence electrons colliding with each other and with atoms, likewise transferring energy. The cumulative effect of all collisions is a net flux of heat from the hotter body to the colder body. Thus, the rate of heat transfer increases with increasing temperature difference $\Delta T = T_h - T_c$. If the temperatures are the same, the net heat transfer rate is zero. Because the number of collisions increases with increasing area, heat conduction is proportional to the cross-sectional area—a second factor in the equation.

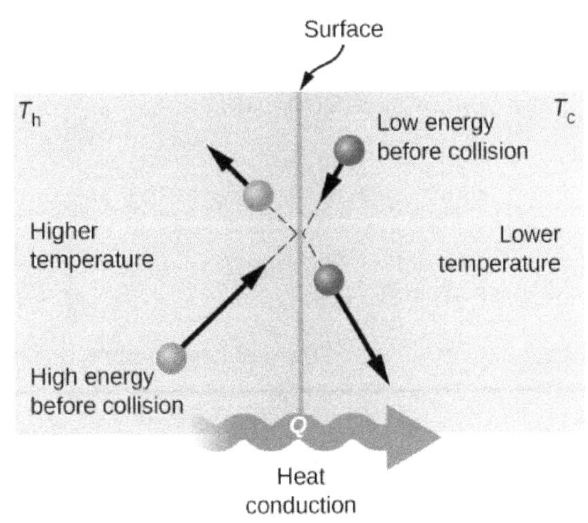

Figure 1.21 Molecules in two bodies at different temperatures have different average kinetic energies. Collisions occurring at the contact surface tend to transfer energy from high-temperature regions to low-temperature regions. In this illustration, a molecule in the lower-temperature region (right side) has low energy before collision, but its energy increases after colliding with a high-energy molecule at the contact surface. In contrast, a molecule in the higher-temperature region (left side) has high energy before collision, but its energy decreases after colliding with a low-energy molecule at the contact surface.

A third quantity that affects the conduction rate is the thickness of the material through which heat transfers. Figure 1.22 shows a slab of material with a higher temperature on the left than on the right. Heat transfers from the left to the right by a series of molecular collisions. The greater the distance between hot and cold, the more time the material takes to transfer the same amount of heat.

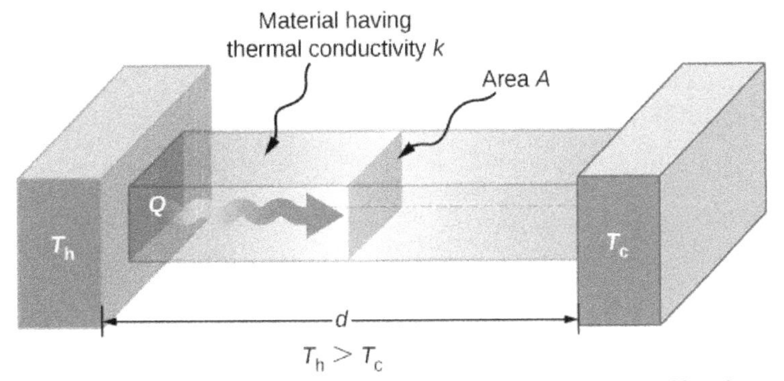

Figure 1.22 Heat conduction occurs through any material, represented here by a rectangular bar, whether window glass or walrus blubber.

All four of these quantities appear in a simple equation deduced from and confirmed by experiments. The **rate of conductive heat transfer** through a slab of material, such as the one in Figure 1.22, is given by

$$P = \frac{dQ}{dT} = \frac{kA(T_h - T_c)}{d} \tag{1.9}$$

where P is the power or rate of heat transfer in watts or in kilocalories per second, A and d are its surface area and thickness, as shown in Figure 1.22, $T_h - T_c$ is the temperature difference across the slab, and k is the **thermal conductivity** of the material. Table 1.5 gives representative values of thermal conductivity.

More generally, we can write

$$P = -kA\frac{dT}{dx},$$

where x is the coordinate in the direction of heat flow. Since in Figure 1.22, the power and area are constant, dT/dx is constant, and the temperature decreases linearly from T_h to T_c.

Substance	Thermal Conductivity k (W/m·°C)
Diamond	2000
Silver	420
Copper	390
Gold	318
Aluminum	220
Steel iron	80
Steel (stainless)	14
Ice	2.2
Glass (average)	0.84
Concrete brick	0.84
Water	0.6
Fatty tissue (without blood)	0.2
Asbestos	0.16
Plasterboard	0.16
Wood	0.08–0.16
Snow (dry)	0.10
Cork	0.042
Glass wool	0.042
Wool	0.04
Down feathers	0.025
Air	0.023
Polystyrene foam	0.010

Table 1.5 Thermal Conductivities of Common Substances Values are given for temperatures near $0\,°C$.

Example 1.10

Calculating Heat Transfer through Conduction

A polystyrene foam icebox has a total area of $0.950\,m^2$ and walls with an average thickness of 2.50 cm. The box contains ice, water, and canned beverages at $0\,°C$. The inside of the box is kept cold by melting ice. How much ice melts in one day if the icebox is kept in the trunk of a car at $35.0\,°C$?

Strategy

This question involves both heat for a phase change (melting of ice) and the transfer of heat by conduction. To find the amount of ice melted, we must find the net heat transferred. This value can be obtained by calculating the rate of heat transfer by conduction and multiplying by time.

Solution

First we identify the knowns.

$k = 0.010\ \text{W/m} \cdot {}^\circ\text{C}$ for polystyrene foam; $A = 0.950\ \text{m}^2$; $d = 2.50\ \text{cm} = 0.0250\ \text{m};$; $T_\text{c} = 0\ {}^\circ\text{C};$ $T_\text{h} = 35.0\ {}^\circ\text{C}$; $t = 1\ \text{day} = 24\ \text{hours - } 84{,}400\ \text{s}.$

Then we identify the unknowns. We need to solve for the mass of the ice, *m*. We also need to solve for the net heat transferred to melt the ice, *Q*. The rate of heat transfer by conduction is given by

$$P = \frac{dQ}{dT} = \frac{kA(T_\text{h} - T_\text{c})}{d}.$$

The heat used to melt the ice is $Q = mL_\text{f}$ .We insert the known values:

$$P = \frac{(0.010\ \text{W/m} \cdot {}^\circ\text{C})(0.950\ \text{m}^2)(35.0\ {}^\circ\text{C} - 0\ {}^\circ\text{C})}{0.0250\ \text{m}} = 13.3\ \text{W}.$$

Multiplying the rate of heat transfer by the time ($1\ \text{day} = 86{,}400\ \text{s}$), we obtain

$$Q = Pt = (13.3\ \text{W})(86.400\ \text{s}) = 1.15 \times 10^6\ \text{J}.$$

We set this equal to the heat transferred to melt the ice, $Q = mL_\text{f},$ and solve for the mass *m*:

$$m = \frac{Q}{L_\text{f}} = \frac{1.15 \times 10^6\ \text{J}}{334 \times 10^3\ \text{J/kg}} = 3.44\ \text{kg}.$$

Significance

The result of 3.44 kg, or about 7.6 lb, seems about right, based on experience. You might expect to use about a 4 kg (7–10 lb) bag of ice per day. A little extra ice is required if you add any warm food or beverages.

Table 1.5 shows that polystyrene foam is a very poor conductor and thus a good insulator. Other good insulators include fiberglass, wool, and goosedown feathers. Like polystyrene foam, these all contain many small pockets of air, taking advantage of air's poor thermal conductivity.

In developing insulation, the smaller the conductivity *k* and the larger the thickness *d*, the better. Thus, the ratio *d/k*, called the *R factor*, is large for a good insulator. The rate of conductive heat transfer is inversely proportional to *R*. *R* factors are most commonly quoted for household insulation, refrigerators, and the like. Unfortunately, in the United States, *R* is still in non-metric units of $\text{ft}^2 \cdot {}^\circ\text{F} \cdot \text{h/Btu}$, although the unit usually goes unstated [1 British thermal unit (Btu) is the amount of energy needed to change the temperature of 1.0 lb of water by $1.0\ {}^\circ\text{F}$, which is 1055.1 J]. A couple of representative values are an *R* factor of 11 for 3.5-inch-thick fiberglass batts (pieces) of insulation and an *R* factor of 19 for 6.5-inch-thick fiberglass batts (Figure 1.23). In the US, walls are usually insulated with 3.5-inch batts, whereas ceilings are usually insulated with 6.5-inch batts. In cold climates, thicker batts may be used.

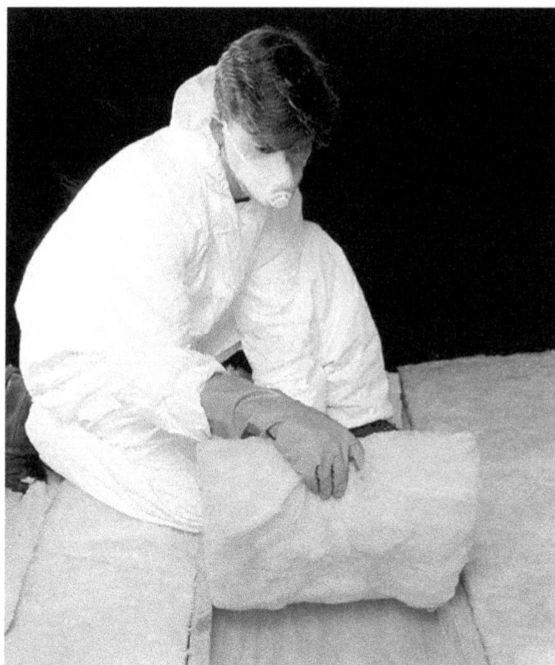

Figure 1.23 The fiberglass batt is used for insulation of walls and ceilings to prevent heat transfer between the inside of the building and the outside environment. (credit: Tracey Nicholls)

Note that in Table 1.5, most of the best thermal conductors—silver, copper, gold, and aluminum—are also the best electrical conductors, because they contain many free electrons that can transport thermal energy. (Diamond, an electrical insulator, conducts heat by atomic vibrations.) Cooking utensils are typically made from good conductors, but the handles of those used on the stove are made from good insulators (bad conductors).

Example 1.11

Two Conductors End to End

A steel rod and an aluminum rod, each of diameter 1.00 cm and length 25.0 cm, are welded end to end. One end of the steel rod is placed in a large tank of boiling water at $100\,°C$, while the far end of the aluminum rod is placed in a large tank of water at $20\,°C$. The rods are insulated so that no heat escapes from their surfaces. What is the temperature at the joint, and what is the rate of heat conduction through this composite rod?

Strategy

The heat that enters the steel rod from the boiling water has no place to go but through the steel rod, then through the aluminum rod, to the cold water. Therefore, we can equate the rate of conduction through the steel to the rate of conduction through the aluminum.

We repeat the calculation with a second method, in which we use the thermal resistance R of the rod, since it simply adds when two rods are joined end to end. (We will use a similar method in the chapter on direct-current circuits.)

Solution

1. Identify the knowns and convert them to SI units.
   The length of each rod is $L_{\text{Al}} = L_{\text{steel}} = 0.25\,\text{m}$, the cross-sectional area of each rod is $A_{\text{Al}} = A_{\text{steel}} = 7.85 \times 10^{-5}\,\text{m}^2$, the thermal conductivity of aluminum is $k_{\text{Al}} = 220\,\text{W/m} \cdot °\text{C}$, the thermal conductivity of steel is $k_{\text{steel}} = 80\,\text{W/m} \cdot °\text{C}$, the temperature at the hot end is $T = 100\,°\text{C}$, and the temperature at the cold end is $T = 20\,°\text{C}$.

2. Calculate the heat-conduction rate through the steel rod and the heat-conduction rate through the aluminum rod in terms of the unknown temperature T at the joint:

$$\begin{aligned} P_{\text{steel}} &= \frac{k_{\text{steel}} A_{\text{steel}} \Delta T_{\text{steel}}}{L_{\text{steel}}} \\ &= \frac{(80\ \text{W/m}\cdot{}^{\circ}\text{C})\left(7.85\times 10^{-5}\ \text{m}^2\right)(100\ {}^{\circ}\text{C} - T)}{0.25\ \text{m}} \\ &= (0.0251\ \text{W/}{}^{\circ}\text{C})(100\ {}^{\circ}\text{C} - T); \end{aligned}$$

$$\begin{aligned} P_{\text{Al}} &= \frac{k_{\text{Al}} A_{\text{Al}} \Delta T_{\text{Al}}}{L_{\text{Al}}} \\ &= \frac{(220\ \text{W/m}\cdot{}^{\circ}\text{C})\left(7.85\times 10^{-5}\ \text{m}^2\right)(T - 20\ {}^{\circ}\text{C})}{0.25\ \text{m}} \\ &= (0.0691\ \text{W/}{}^{\circ}\text{C})(T - 20\ {}^{\circ}\text{C}). \end{aligned}$$

3. Set the two rates equal and solve for the unknown temperature:

$$\begin{aligned} (0.0691\ \text{W/}{}^{\circ}\text{C})(T - 20\ {}^{\circ}\text{C}) &= (0.0251\ \text{W/}{}^{\circ}\text{C})(100\ {}^{\circ}\text{C} - T) \\ T &= 41.3\ {}^{\circ}\text{C}. \end{aligned}$$

4. Calculate either rate:

$$P_{\text{steel}} = (0.0251\ \text{W/}{}^{\circ}\text{C})(100\ {}^{\circ}\text{C} - 41.3\ {}^{\circ}\text{C}) = 1.47\ \text{W}.$$

5. If desired, check your answer by calculating the other rate.

Solution

1. Recall that $R = L/k$. Now $P = A\Delta T/R$, or $\Delta T = PR/A$.

2. We know that $\Delta T_{\text{steel}} + \Delta T_{\text{Al}} = 100\ {}^{\circ}\text{C} - 20\ {}^{\circ}\text{C} = 80\ {}^{\circ}\text{C}$. We also know that $P_{\text{steel}} = P_{\text{Al}}$, and we denote that rate of heat flow by P. Combine the equations:

$$\frac{PR_{\text{steel}}}{A} + \frac{PR_{\text{Al}}}{A} = 80\ {}^{\circ}\text{C}.$$

Thus, we can simply add R factors. Now, $P = \dfrac{80\ {}^{\circ}\text{C}}{A(R_{\text{steel}} + R_{\text{Al}})}$.

3. Find the R_s from the known quantities:

$$R_{\text{steel}} = 3.13\times 10^{-3}\ \text{m}^2\cdot{}^{\circ}\text{C/W}$$

and

$$R_{\text{Al}} = 1.14\times 10^{-3}\ \text{m}^2\cdot{}^{\circ}\text{C/W}.$$

4. Substitute these values in to find $P = 1.47\ \text{W}$ as before.

5. Determine ΔT for the aluminum rod (or for the steel rod) and use it to find T at the joint.

$$\Delta T_{\text{Al}} = \frac{PR_{\text{Al}}}{A} = \frac{(1.47\ \text{W})\left(1.14\times 10^{-3}\ \text{m}^2\cdot{}^{\circ}\text{C/W}\right)}{7.85\times 10^{-5}\ \text{m}^2} = 21.3\ {}^{\circ}\text{C},$$

so $T = 20\ {}^{\circ}\text{C} + 21.3\ {}^{\circ}\text{C} = 41.3\ {}^{\circ}\text{C}$, as in Solution 1.

6. If desired, check by determining ΔT for the other rod.

Significance

In practice, adding R values is common, as in calculating the R value of an insulated wall. In the analogous situation in electronics, the resistance corresponds to AR in this problem and is additive even when the areas are unequal, as is common in electronics. Our equation for heat conduction can be used only when the areas are equal; otherwise, we would have a problem in three-dimensional heat flow, which is beyond our scope.

1.7 Check Your Understanding How does the rate of heat transfer by conduction change when all spatial dimensions are doubled?

Conduction is caused by the random motion of atoms and molecules. As such, it is an ineffective mechanism for heat transport over macroscopic distances and short times. For example, the temperature on Earth would be unbearably cold during the night and extremely hot during the day if heat transport in the atmosphere were only through conduction. Also, car engines would overheat unless there was a more efficient way to remove excess heat from the pistons. The next module discusses the important heat-transfer mechanism in such situations.

Convection

In convection, thermal energy is carried by the large-scale flow of matter. It can be divided into two types. In *forced convection*, the flow is driven by fans, pumps, and the like. A simple example is a fan that blows air past you in hot surroundings and cools you by replacing the air heated by your body with cooler air. A more complicated example is the cooling system of a typical car, in which a pump moves coolant through the radiator and engine to cool the engine and a fan blows air to cool the radiator.

In *free* or *natural convection*, the flow is driven by buoyant forces: hot fluid rises and cold fluid sinks because density decreases as temperature increases. The house in Figure 1.24 is kept warm by natural convection, as is the pot of water on the stove in Figure 1.25. Ocean currents and large-scale atmospheric circulation, which result from the buoyancy of warm air and water, transfer hot air from the tropics toward the poles and cold air from the poles toward the tropics. (Earth's rotation interacts with those flows, causing the observed eastward flow of air in the temperate zones.)

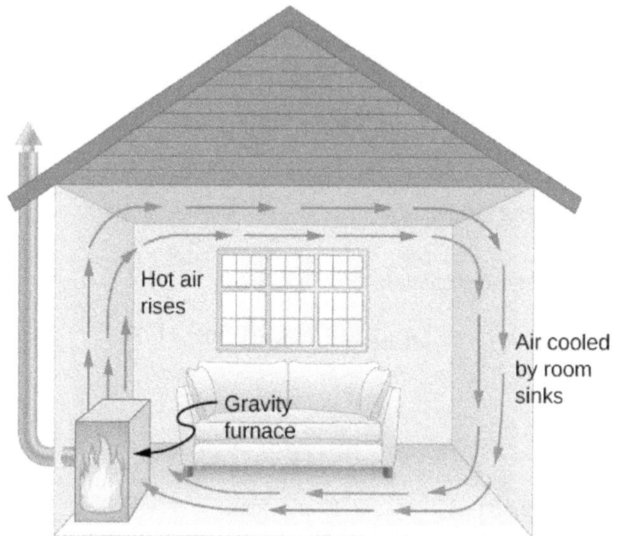

Figure 1.24 Air heated by a so-called gravity furnace expands and rises, forming a convective loop that transfers energy to other parts of the room. As the air is cooled at the ceiling and outside walls, it contracts, eventually becoming denser than room air and sinking to the floor. A properly designed heating system using natural convection, like this one, can heat a home quite efficiently.

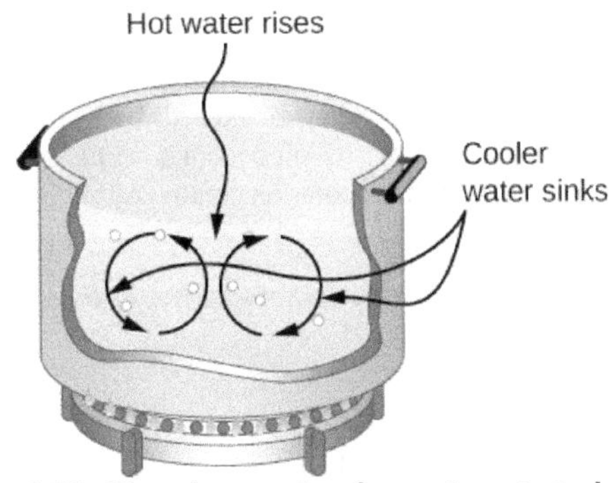

Figure 1.25 Natural convection plays an important role in heat transfer inside this pot of water. Once conducted to the inside, heat transfer to other parts of the pot is mostly by convection. The hotter water expands, decreases in density, and rises to transfer heat to other regions of the water, while colder water sinks to the bottom. This process keeps repeating.

Natural convection like that of Figure 1.24 and Figure 1.25, but acting on rock in Earth's mantle, drives plate tectonics (https://openstaxcollege.org/l/21platetecton) that are the motions that have shaped Earth's surface.

Convection is usually more complicated than conduction. Beyond noting that the convection rate is often approximately proportional to the temperature difference, we will not do any quantitative work comparable to the formula for conduction. However, we can describe convection qualitatively and relate convection rates to heat and time. However, air is a poor conductor. Therefore, convection dominates heat transfer by air, and the amount of available space for airflow determines whether air transfers heat rapidly or slowly. There is little heat transfer in a space filled with air with a small amount of other material that prevents flow. The space between the inside and outside walls of a typical American house, for example, is about 9 cm (3.5 in.)—large enough for convection to work effectively. The addition of wall insulation prevents airflow, so heat loss (or gain) is decreased. On the other hand, the gap between the two panes of a double-paned window is about 1 cm, which largely prevents convection and takes advantage of air's low conductivity reduce heat loss. Fur, cloth, and fiberglass also take advantage of the low conductivity of air by trapping it in spaces too small to support convection (Figure 1.26).

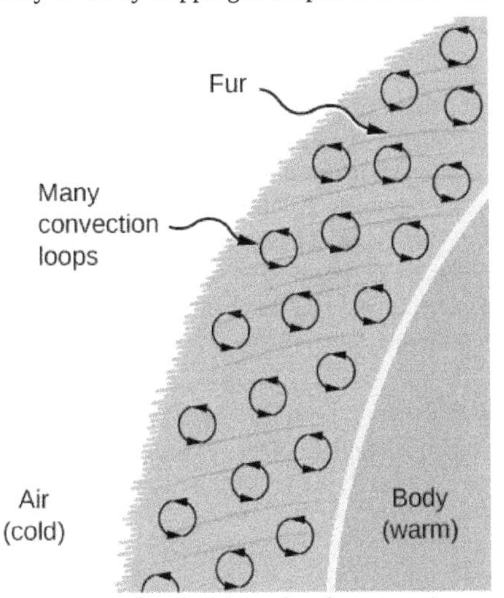

Figure 1.26 Fur is filled with air, breaking it up into many small pockets. Convection is very slow here, because the loops are so small. The low conductivity of air makes fur a very good lightweight insulator.

Some interesting phenomena happen when convection is accompanied by a phase change. The combination allows us to cool off by sweating even if the temperature of the surrounding air exceeds body temperature. Heat from the skin is required for sweat to evaporate from the skin, but without air flow, the air becomes saturated and evaporation stops. Air flow caused by convection replaces the saturated air by dry air and evaporation continues.

Example 1.12

Calculating the Flow of Mass during Convection

The average person produces heat at the rate of about 120 W when at rest. At what rate must water evaporate from the body to get rid of all this energy? (For simplicity, we assume this evaporation occurs when a person is sitting in the shade and surrounding temperatures are the same as skin temperature, eliminating heat transfer by other methods.)

Strategy

Energy is needed for this phase change ($Q = mL_v$). Thus, the energy loss per unit time is

$$\frac{Q}{t} = \frac{mL_V}{t} = 120\text{ W} = 120\text{ J/s}.$$

We divide both sides of the equation by L_v to find that the mass evaporated per unit time is

$$\frac{m}{t} = \frac{120\text{ J/s}}{L_V}.$$

Solution

Insert the value of the latent heat from Table 1.4, $L_v = 2430\text{ kJ/kg} = 2430\text{ J/g}$. This yields

$$\frac{m}{t} = \frac{120\text{ J/s}}{2430\text{ J/g}} = 0.0494\text{ g/s} = 2.96\text{ g/min}.$$

Significance

Evaporating about 3 g/min seems reasonable. This would be about 180 g (about 7 oz.) per hour. If the air is very dry, the sweat may evaporate without even being noticed. A significant amount of evaporation also takes place in the lungs and breathing passages.

Another important example of the combination of phase change and convection occurs when water evaporates from the oceans. Heat is removed from the ocean when water evaporates. If the water vapor condenses in liquid droplets as clouds form, possibly far from the ocean, heat is released in the atmosphere. Thus, there is an overall transfer of heat from the ocean to the atmosphere. This process is the driving power behind thunderheads, those great cumulus clouds that rise as much as 20.0 km into the stratosphere (Figure 1.27). Water vapor carried in by convection condenses, releasing tremendous amounts of energy. This energy causes the air to expand and rise to colder altitudes. More condensation occurs in these regions, which in turn drives the cloud even higher. This mechanism is an example of positive feedback, since the process reinforces and accelerates itself. It sometimes produces violent storms, with lightning and hail. The same mechanism drives hurricanes.

This time-lapse video (https://openstaxcollege.org/l/21convthuncurr) shows convection currents in a thunderstorm, including “rolling” motion similar to that of boiling water.

Figure 1.27 Cumulus clouds are caused by water vapor that rises because of convection. The rise of clouds is driven by a positive feedback mechanism. (credit: "Amada44"/Wikimedia Commons)

1.8 Check Your Understanding Explain why using a fan in the summer feels refreshing.

Radiation

You can feel the heat transfer from the Sun. The space between Earth and the Sun is largely empty, so the Sun warms us without any possibility of heat transfer by convection or conduction. Similarly, you can sometimes tell that the oven is hot without touching its door or looking inside—it may just warm you as you walk by. In these examples, heat is transferred by radiation (Figure 1.28). That is, the hot body emits electromagnetic waves that are absorbed by the skin. No medium is required for electromagnetic waves to propagate. Different names are used for electromagnetic waves of different wavelengths: radio waves, microwaves, infrared radiation, visible light, ultraviolet radiation, X-rays, and gamma rays.

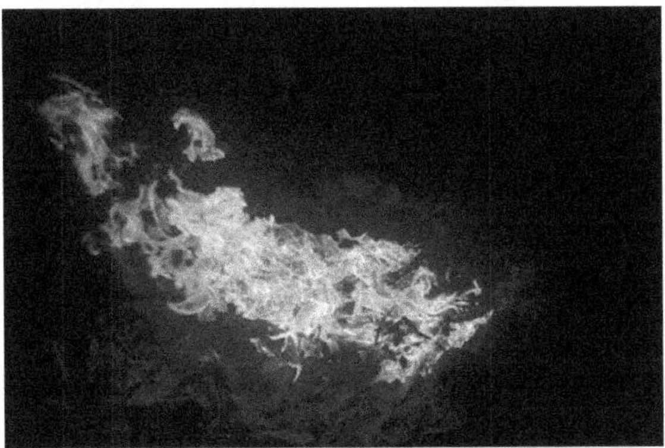

Figure 1.28 Most of the heat transfer from this fire to the observers occurs through infrared radiation. The visible light, although dramatic, transfers relatively little thermal energy. Convection transfers energy away from the observers as hot air rises, while conduction is negligibly slow here. Skin is very sensitive to infrared radiation, so you can sense the presence of a fire without looking at it directly. (credit: Daniel O'Neil)

The energy of electromagnetic radiation varies over a wide range, depending on the wavelength: A shorter wavelength (or higher frequency) corresponds to a higher energy. Because more heat is radiated at higher temperatures, higher temperatures produce more intensity at every wavelength but especially at shorter wavelengths. In visible light, wavelength determines color—red has the longest wavelength and violet the shortest—so a temperature change is accompanied by a color change. For example, an electric heating element on a stove glows from red to orange, while the higher-temperature steel in a blast furnace glows from yellow to white. Infrared radiation is the predominant form radiated by objects cooler than the electric element and the steel. The radiated energy as a function of wavelength depends on its intensity, which is represented in Figure 1.29 by the height of the distribution. (Electromagnetic Waves explains more about the electromagnetic spectrum, and Photons and Matter Waves (http://cnx.org/content/m58757/latest/) discusses why the decrease in wavelength corresponds to an increase in energy.)

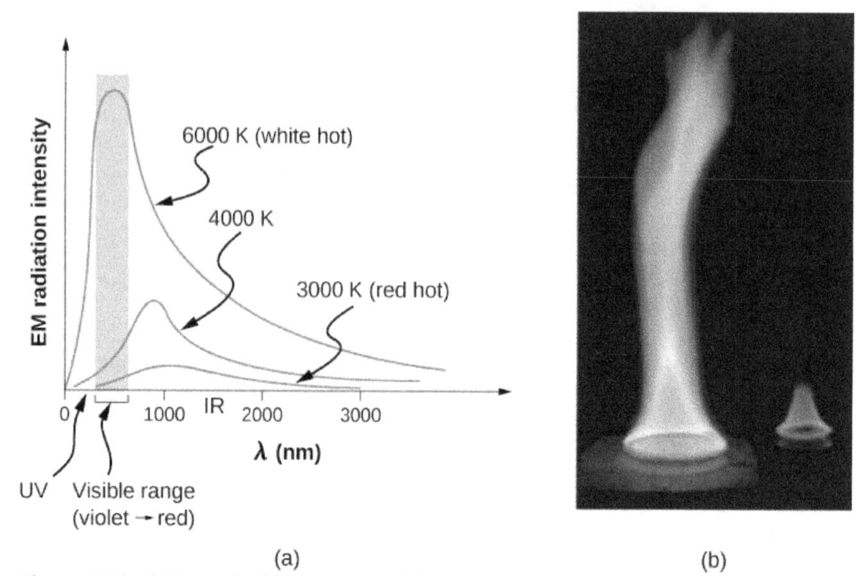

Figure 1.29 (a) A graph of the spectrum of electromagnetic waves emitted from an ideal radiator at three different temperatures. The intensity or rate of radiation emission increases dramatically with temperature, and the spectrum shifts down in wavelength toward the visible and ultraviolet parts of the spectrum. The shaded portion denotes the visible part of the spectrum. It is apparent that the shift toward the ultraviolet with temperature makes the visible appearance shift from red to white to blue as temperature increases. (b) Note the variations in color corresponding to variations in flame temperature.

The rate of heat transfer by radiation also depends on the object's color. Black is the most effective, and white is the least effective. On a clear summer day, black asphalt in a parking lot is hotter than adjacent gray sidewalk, because black absorbs better than gray (Figure 1.30). The reverse is also true—black radiates better than gray. Thus, on a clear summer night, the asphalt is colder than the gray sidewalk, because black radiates the energy more rapidly than gray. A perfectly black object would be an *ideal radiator* and an *ideal absorber*, as it would capture all the radiation that falls on it. In contrast, a perfectly white object or a perfect mirror would reflect all radiation, and a perfectly transparent object would transmit it all (Figure 1.31). Such objects would not emit any radiation. Mathematically, the color is represented by the **emissivity** *e*. A "blackbody" radiator would have an $e = 1$, whereas a perfect reflector or transmitter would have $e = 0$. For real examples, tungsten light bulb filaments have an *e* of about 0.5, and carbon black (a material used in printer toner) has an emissivity of about 0.95.

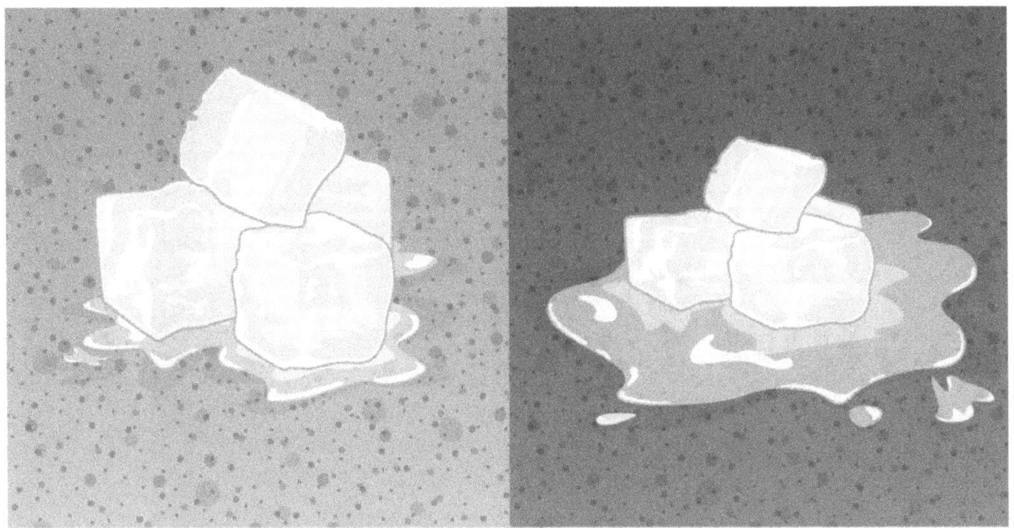

Figure 1.30 The darker pavement is hotter than the lighter pavement (much more of the ice on the right has melted), although both have been in the sunlight for the same time. The thermal conductivities of the pavements are the same.

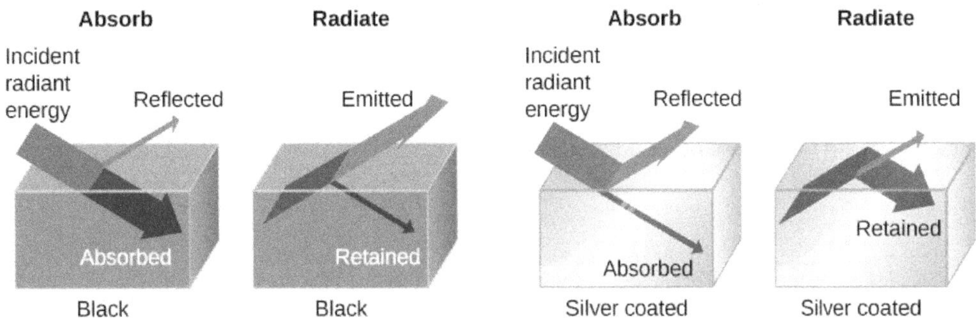

Figure 1.31 A black object is a good absorber and a good radiator, whereas a white, clear, or silver object is a poor absorber and a poor radiator.

To see that, consider a silver object and a black object that can exchange heat by radiation and are in thermal equilibrium. We know from experience that they will stay in equilibrium (the result of a principle that will be discussed at length in Second Law of Thermodynamics). For the black object's temperature to stay constant, it must emit as much radiation as it absorbs, so it must be as good at radiating as absorbing. Similar considerations show that the silver object must radiate as little as it absorbs. Thus, one property, emissivity, controls both radiation and absorption.

Finally, the radiated heat is proportional to the object's surface area, since every part of the surface radiates. If you knock apart the coals of a fire, the radiation increases noticeably due to an increase in radiating surface area.

The rate of heat transfer by emitted radiation is described by the **Stefan-Boltzmann law of radiation:**

$$P = \sigma A e T^4,$$

where $\sigma = 5.67 \times 10^{-8}\ \text{J/s} \cdot \text{m}^2 \cdot \text{K}^4$ is the Stefan-Boltzmann constant, a combination of fundamental constants of nature; A is the surface area of the object; and T is its temperature in kelvins.

The proportionality to the *fourth power* of the absolute temperature is a remarkably strong temperature dependence. It allows the detection of even small temperature variations. Images called *thermographs* can be used medically to detect regions of abnormally high temperature in the body, perhaps indicative of disease. Similar techniques can be used to detect heat leaks in homes (Figure 1.32), optimize performance of blast furnaces, improve comfort levels in work environments, and even remotely map Earth's temperature profile.

Figure 1.32 A thermograph of part of a building shows temperature variations, indicating where heat transfer to the outside is most severe. Windows are a major region of heat transfer to the outside of homes. (credit: US Army)

The Stefan-Boltzmann equation needs only slight refinement to deal with a simple case of an object's absorption of radiation from its surroundings. Assuming that an object with a temperature T_1 is surrounded by an environment with uniform temperature T_2, the **net rate of heat transfer by radiation** is

$$P_{\text{net}} = \sigma e A\left(T_2^{\ 4} - T_1^{\ 4}\right), \tag{1.10}$$

where e is the emissivity of the object alone. In other words, it does not matter whether the surroundings are white, gray, or black: The balance of radiation into and out of the object depends on how well it emits and absorbs radiation. When $T_2 > T_1,$ the quantity P_{net} is positive, that is, the net heat transfer is from hot to cold.

Before doing an example, we have a complication to discuss: different emissivities at different wavelengths. If the fraction of incident radiation an object reflects is the same at all visible wavelengths, the object is gray; if the fraction depends on the wavelength, the object has some other color. For instance, a red or reddish object reflects red light more strongly than other visible wavelengths. Because it absorbs less red, it radiates less red when hot. Differential reflection and absorption of wavelengths outside the visible range have no effect on what we see, but they may have physically important effects. Skin is a very good absorber and emitter of infrared radiation, having an emissivity of 0.97 in the infrared spectrum. Thus, in spite of the obvious variations in skin color, we are all nearly black in the infrared. This high infrared emissivity is why we can so easily feel radiation on our skin. It is also the basis for the effectiveness of night-vision scopes used by law enforcement and the military to detect human beings.

Example 1.13

Calculating the Net Heat Transfer of a Person

What is the rate of heat transfer by radiation of an unclothed person standing in a dark room whose ambient temperature is $22.0\ °\text{C}$? The person has a normal skin temperature of $33.0\ °\text{C}$ and a surface area of $1.50\ \text{m}^2$. The emissivity of skin is 0.97 in the infrared, the part of the spectrum where the radiation takes place.

Strategy

We can solve this by using the equation for the rate of radiative heat transfer.

Solution

Insert the temperature values $T_2 = 295\text{ K}$ and $T_1 = 306\text{ K}$, so that

$$\begin{aligned}\frac{Q}{t} &= \sigma eA\left(T_2{}^4 - T_1{}^4\right) \\ &= \left(5.67 \times 10^{-8}\text{ J/s}\cdot\text{m}^2\cdot\text{K}^4\right)(0.97)\left(1.50\text{ m}^2\right)\left[(295\text{ K})^4 - (306\text{ K})^4\right] \\ &= -99\text{ J/s} = -99\text{ W}.\end{aligned}$$

Significance

This value is a significant rate of heat transfer to the environment (note the minus sign), considering that a person at rest may produce energy at the rate of 125 W and that conduction and convection are also transferring energy to the environment. Indeed, we would probably expect this person to feel cold. Clothing significantly reduces heat transfer to the environment by all mechanisms, because clothing slows down both conduction and convection, and has a lower emissivity (especially if it is light-colored) than skin.

The average temperature of Earth is the subject of much current discussion. Earth is in radiative contact with both the Sun and dark space, so we cannot use the equation for an environment at a uniform temperature. Earth receives almost all its energy from radiation of the Sun and reflects some of it back into outer space. Conversely, dark space is very cold, about 3 K, so that Earth radiates energy into the dark sky. The rate of heat transfer from soil and grasses can be so rapid that frost may occur on clear summer evenings, even in warm latitudes.

The average temperature of Earth is determined by its energy balance. To a first approximation, it is the temperature at which Earth radiates heat to space as fast as it receives energy from the Sun.

An important parameter in calculating the temperature of Earth is its emissivity (e). On average, it is about 0.65, but calculation of this value is complicated by the great day-to-day variation in the highly reflective cloud coverage. Because clouds have lower emissivity than either oceans or land masses, they reflect some of the radiation back to the surface, greatly reducing heat transfer into dark space, just as they greatly reduce heat transfer into the atmosphere during the day. There is negative feedback (in which a change produces an effect that opposes that change) between clouds and heat transfer; higher temperatures evaporate more water to form more clouds, which reflect more radiation back into space, reducing the temperature.

The often-mentioned **greenhouse effect** is directly related to the variation of Earth's emissivity with wavelength (Figure 1.33). The greenhouse effect is a natural phenomenon responsible for providing temperatures suitable for life on Earth and for making Venus unsuitable for human life. Most of the infrared radiation emitted from Earth is absorbed by carbon dioxide (CO_2) and water (H_2O) in the atmosphere and then re-radiated into outer space or back to Earth. Re-radiation back to Earth maintains its surface temperature about $40\,°\text{C}$ higher than it would be if there were no atmosphere. (The glass walls and roof of a greenhouse increase the temperature inside by blocking convective heat losses, not radiative losses.)

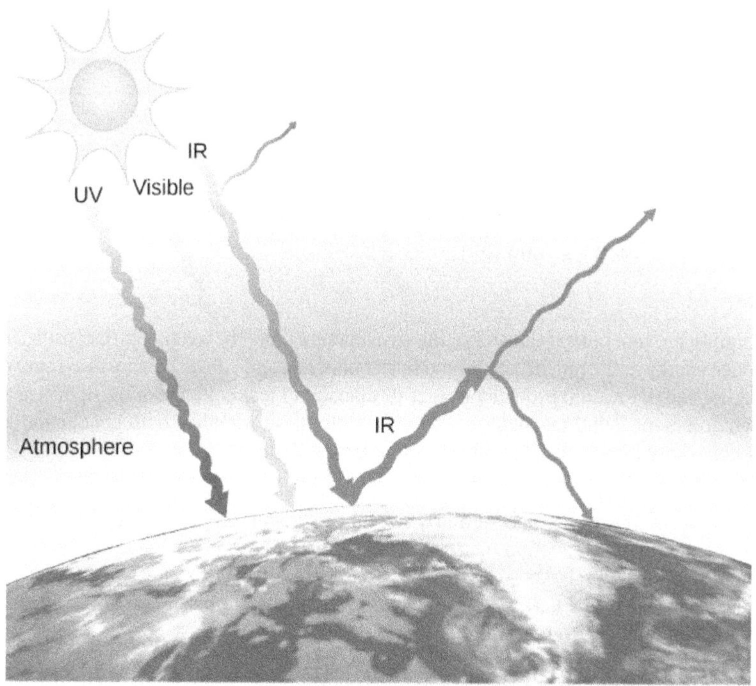

Figure 1.33 The greenhouse effect is the name given to the increase of Earth's temperature due to absorption of radiation in the atmosphere. The atmosphere is transparent to incoming visible radiation and most of the Sun's infrared. The Earth absorbs that energy and re-emits it. Since Earth's temperature is much lower than the Sun's, it re-emits the energy at much longer wavelengths, in the infrared. The atmosphere absorbs much of that infrared radiation and radiates about half of the energy back down, keeping Earth warmer than it would otherwise be. The amount of trapping depends on concentrations of trace gases such as carbon dioxide, and an increase in the concentration of these gases increases Earth's surface temperature.

The greenhouse effect is central to the discussion of global warming due to emission of carbon dioxide and methane (and other greenhouse gases) into Earth's atmosphere from industry, transportation, and farming. Changes in global climate could lead to more intense storms, precipitation changes (affecting agriculture), reduction in rain forest biodiversity, and rising sea levels.

You can explore a simulation of the greenhouse effect (https://openstaxcollege.org/l/21simgreeneff) that takes the point of view that the atmosphere scatters (redirects) infrared radiation rather than absorbing it and reradiating it. You may want to run the simulation first with no greenhouse gases in the atmosphere and then look at how adding greenhouse gases affects the infrared radiation from the Earth and the Earth's temperature.

Problem-Solving Strategy: Effects of Heat Transfer

1. Examine the situation to determine what type of heat transfer is involved.
2. Identify the type(s) of heat transfer—conduction, convection, or radiation.
3. Identify exactly what needs to be determined in the problem (identify the unknowns). A written list is useful.
4. Make a list of what is given or what can be inferred from the problem as stated (identify the knowns).
5. Solve the appropriate equation for the quantity to be determined (the unknown).

6. For conduction, use the equation $P = \frac{kA\Delta T}{d}$. Table 1.5 lists thermal conductivities. For convection, determine the amount of matter moved and the equation $Q = mc\Delta T$, along with $Q = mL_f$ or $Q = mL_V$ if a substance changes phase. For radiation, the equation $P_{net} = \sigma eA(T_2^4 - T_1^4)$ gives the net heat transfer rate.
7. Substitute the knowns along with their units into the appropriate equation and obtain numerical solutions complete with units.
8. Check the answer to see if it is reasonable. Does it make sense?

1.9 Check Your Understanding How much greater is the rate of heat radiation when a body is at the temperature $40\ °C$ than when it is at the temperature $20\ °C$?

CHAPTER 1 REVIEW

KEY TERMS

absolute temperature scale scale, such as Kelvin, with a zero point that is absolute zero

absolute zero temperature at which the average kinetic energy of molecules is zero

calorie (cal) energy needed to change the temperature of 1.00 g of water by $1.00\,°\text{C}$

calorimeter container that prevents heat transfer in or out

calorimetry study of heat transfer inside a container impervious to heat

Celsius scale temperature scale in which the freezing point of water is $0\,°\text{C}$ and the boiling point of water is $100\,°\text{C}$

coefficient of linear expansion (α) material property that gives the change in length, per unit length, per $1\text{-}°\text{C}$ change in temperature; a constant used in the calculation of linear expansion; the coefficient of linear expansion depends to some degree on the temperature of the material

coefficient of volume expansion (β) similar to α but gives the change in volume, per unit volume, per $1\text{-}°\text{C}$ change in temperature

conduction heat transfer through stationary matter by physical contact

convection heat transfer by the macroscopic movement of fluid

critical point for a given substance, the combination of temperature and pressure above which the liquid and gas phases are indistinguishable

critical pressure pressure at the critical point

critical temperature temperature at the critical point

degree Celsius ($°\text{C}$) unit on the Celsius temperature scale

degree Fahrenheit ($°\text{F}$) unit on the Fahrenheit temperature scale

emissivity measure of how well an object radiates

Fahrenheit scale temperature scale in which the freezing point of water is $32\,°\text{F}$ and the boiling point of water is $212\,°\text{F}$

greenhouse effect warming of the earth that is due to gases such as carbon dioxide and methane that absorb infrared radiation from Earth's surface and reradiate it in all directions, thus sending some of it back toward Earth

heat energy transferred solely due to a temperature difference

heat of fusion energy per unit mass required to change a substance from the solid phase to the liquid phase, or released when the substance changes from liquid to solid

heat of sublimation energy per unit mass required to change a substance from the solid phase to the vapor phase

heat of vaporization energy per unit mass required to change a substance from the liquid phase to the vapor phase

heat transfer movement of energy from one place or material to another as a result of a difference in temperature

Kelvin scale (K) temperature scale in which 0 K is the lowest possible temperature, representing absolute zero

kilocalorie (kcal) energy needed to change the temperature of 1.00 kg of water between $14.5\,°\text{C}$ and $15.5\,°\text{C}$

latent heat coefficient general term for the heats of fusion, vaporization, and sublimation

mechanical equivalent of heat work needed to produce the same effects as heat transfer

net rate of heat transfer by radiation $P_{\text{net}} = \sigma e A\left(T_2^{\ 4} - T_1^{\ 4}\right)$

phase diagram graph of pressure vs. temperature of a particular substance, showing at which pressures and temperatures the phases of the substance occur

radiation energy transferred by electromagnetic waves directly as a result of a temperature difference

rate of conductive heat transfer rate of heat transfer from one material to another

specific heat amount of heat necessary to change the temperature of 1.00 kg of a substance by $1.00\ °C$; also called "specific heat capacity"

Stefan-Boltzmann law of radiation $P = \sigma AeT^4$, where $\sigma = 5.67 \times 10^{-8}\ J/s \cdot m^2 \cdot K^4$ is the Stefan-Boltzmann constant, A is the surface area of the object, T is the absolute temperature, and e is the emissivity

sublimation phase change from solid to gas

temperature quantity measured by a thermometer, which reflects the mechanical energy of molecules in a system

thermal conductivity property of a material describing its ability to conduct heat

thermal equilibrium condition in which heat no longer flows between two objects that are in contact; the two objects have the same temperature

thermal expansion change in size or volume of an object with change in temperature

thermal stress stress caused by thermal expansion or contraction

triple point pressure and temperature at which a substance exists in equilibrium as a solid, liquid, and gas

vapor gas at a temperature below the boiling temperature

vapor pressure pressure at which a gas coexists with its solid or liquid phase

zeroth law of thermodynamics law that states that if two objects are in thermal equilibrium, and a third object is in thermal equilibrium with one of those objects, it is also in thermal equilibrium with the other object

KEY EQUATIONS

Linear thermal expansion	$\Delta L = \alpha L \Delta T$
Thermal expansion in two dimensions	$\Delta A = 2\alpha A \Delta T$
Thermal expansion in three dimensions	$\Delta V = \beta V \Delta T$
Heat transfer	$Q = mc\Delta T$
Transfer of heat in a calorimeter	$Q_{\text{cold}} + Q_{\text{hot}} = 0$
Heat due to phase change (melting and freezing)	$Q = mL_{\text{f}}$
Heat due to phase change (evaporation and condensation)	$Q = mL_{\text{v}}$
Rate of conductive heat transfer	$P = \frac{kA(T_h - T_c)}{d}$
Net rate of heat transfer by radiation	$P_{\text{net}} = \sigma eA(T_2{}^4 - T_1{}^4)$

SUMMARY

1.1 Temperature and Thermal Equilibrium

- Temperature is operationally defined as the quantity measured by a thermometer. It is proportional to the average kinetic energy of atoms and molecules in a system.
- Thermal equilibrium occurs when two bodies are in contact with each other and can freely exchange energy. Systems are in thermal equilibrium when they have the same temperature.
- The zeroth law of thermodynamics states that when two systems, *A* and *B*, are in thermal equilibrium with each other, and *B* is in thermal equilibrium with a third system *C*, then *A* is also in thermal equilibrium with *C*.

1.2 Thermometers and Temperature Scales

- Three types of thermometers are alcohol, liquid crystal, and infrared radiation (pyrometer).
- The three main temperature scales are Celsius, Fahrenheit, and Kelvin. Temperatures can be converted from one scale to another using temperature conversion equations.
- The three phases of water (ice, liquid water, and water vapor) can coexist at a single pressure and temperature known as the triple point.

1.3 Thermal Expansion

- Thermal expansion is the increase of the size (length, area, or volume) of a body due to a change in temperature, usually a rise. Thermal contraction is the decrease in size due to a change in temperature, usually a fall in temperature.
- Thermal stress is created when thermal expansion or contraction is constrained.

1.4 Heat Transfer, Specific Heat, and Calorimetry

- Heat and work are the two distinct methods of energy transfer.
- Heat transfer to an object when its temperature changes is often approximated well by $Q = mc\Delta T,$ where *m* is the object's mass and *c* is the specific heat of the substance.

1.5 Phase Changes

- Most substances have three distinct phases (under ordinary conditions on Earth), and they depend on temperature and pressure.
- Two phases coexist (i.e., they are in thermal equilibrium) at a set of pressures and temperatures.
- Phase changes occur at fixed temperatures for a given substance at a given pressure, and these temperatures are called boiling, freezing (or melting), and sublimation points.

1.6 Mechanisms of Heat Transfer

- Heat is transferred by three different methods: conduction, convection, and radiation.
- Heat conduction is the transfer of heat between two objects in direct contact with each other.
- The rate of heat transfer *P* (energy per unit time) is proportional to the temperature difference $T_h - T_c$ and the contact area *A* and inversely proportional to the distance *d* between the objects.
- Convection is heat transfer by the macroscopic movement of mass. Convection can be natural or forced, and generally transfers thermal energy faster than conduction. Convection that occurs along with a phase change can transfer energy from cold regions to warm ones.
- Radiation is heat transfer through the emission or absorption of electromagnetic waves.
- The rate of radiative heat transfer is proportional to the emissivity *e*. For a perfect blackbody, $e = 1$, whereas a perfectly white, clear, or reflective body has $e = 0$, with real objects having values of *e* between 1 and 0.
- The rate of heat transfer depends on the surface area and the fourth power of the absolute temperature:

  $$P = \sigma e A T^4,$$

  where $\sigma = 5.67 \times 10^{-8}\ \text{J/s} \cdot \text{m}^2 \cdot \text{K}^4$ is the Stefan-Boltzmann constant and *e* is the emissivity of the body. The net rate of heat transfer from an object by radiation is

  $$\frac{Q_{\text{net}}}{t} = \sigma e A\left(T_2{}^4 - T_1{}^4\right),$$

where T_1 is the temperature of the object surrounded by an environment with uniform temperature T_2 and e is the emissivity of the object.

CONCEPTUAL QUESTIONS

1.1 Temperature and Thermal Equilibrium

1. What does it mean to say that two systems are in thermal equilibrium?

2. Give an example in which *A* has some kind of non-thermal equilibrium relationship with *B*, and *B* has the same relationship with *C*, but *A* does not have that relationship with *C*.

1.2 Thermometers and Temperature Scales

3. If a thermometer is allowed to come to equilibrium with the air, and a glass of water is not in equilibrium with the air, what will happen to the thermometer reading when it is placed in the water?

4. Give an example of a physical property that varies with temperature and describe how it is used to measure temperature.

1.3 Thermal Expansion

5. Pouring cold water into hot glass or ceramic cookware can easily break it. What causes the breaking? Explain why Pyrex®, a glass with a small coefficient of linear expansion, is less susceptible.

6. One method of getting a tight fit, say of a metal peg in a hole in a metal block, is to manufacture the peg slightly larger than the hole. The peg is then inserted when at a different temperature than the block. Should the block be hotter or colder than the peg during insertion? Explain your answer.

7. Does it really help to run hot water over a tight metal lid on a glass jar before trying to open it? Explain your answer.

8. When a cold alcohol thermometer is placed in a hot liquid, the column of alcohol goes *down* slightly before going up. Explain why.

9. Calculate the length of a 1-meter rod of a material with thermal expansion coefficient α when the temperature is raised from 300 K to 600 K. Taking your answer as the new initial length, find the length after the rod is cooled back down to 300 K. Is your answer 1 meter? Should it be? How can you account for the result you got?

10. Noting the large stresses that can be caused by thermal expansion, an amateur weapon inventor decides to use it to make a new kind of gun. He plans to jam a bullet against an aluminum rod inside a closed invar tube. When he heats the tube, the rod will expand more than the tube and a very strong force will build up. Then, by a method yet to be determined, he will open the tube in a split second and let the force of the rod launch the bullet at very high speed. What is he overlooking?

1.4 Heat Transfer, Specific Heat, and Calorimetry

11. How is heat transfer related to temperature?

12. Describe a situation in which heat transfer occurs.

13. When heat transfers into a system, is the energy stored as heat? Explain briefly.

14. The brakes in a car increase in temperature by ΔT when bringing the car to rest from a speed *v*. How much greater would ΔT be if the car initially had twice the speed? You may assume the car stops fast enough that no heat transfers out of the brakes.

1.5 Phase Changes

15. A pressure cooker contains water and steam in equilibrium at a pressure greater than atmospheric pressure. How does this greater pressure increase cooking speed?

16. As shown below, which is the phase diagram for carbon dioxide, what is the vapor pressure of solid carbon dioxide (dry ice) at $-78.5\ °C$? (Note that the axes in the figure are nonlinear and the graph is not to scale.)

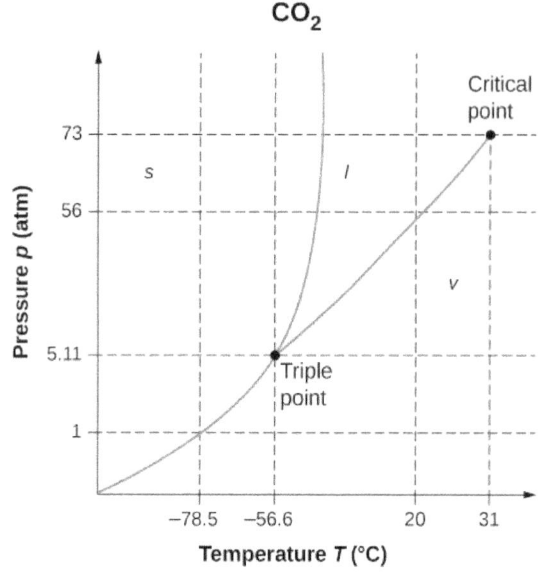

17. Can carbon dioxide be liquefied at room temperature ($20\ °C$)? If so, how? If not, why not? (See the phase diagram in the preceding problem.)

18. What is the distinction between gas and vapor?

19. Heat transfer can cause temperature and phase changes. What else can cause these changes?

20. How does the latent heat of fusion of water help slow the decrease of air temperatures, perhaps preventing temperatures from falling significantly below $0\ °C$, in the vicinity of large bodies of water?

21. What is the temperature of ice right after it is formed by freezing water?

22. If you place $0\ °C$ ice into $0\ °C$ water in an insulated container, what will the net result be? Will there be less ice and more liquid water, or more ice and less liquid water, or will the amounts stay the same?

23. What effect does condensation on a glass of ice water have on the rate at which the ice melts? Will the condensation speed up the melting process or slow it down?

24. In Miami, Florida, which has a very humid climate and numerous bodies of water nearby, it is unusual for temperatures to rise above about $38\ °C$ ($100\ °F$). In the desert climate of Phoenix, Arizona, however, temperatures rise above that almost every day in July and August. Explain how the evaporation of water helps limit high temperatures in humid climates.

25. In winter, it is often warmer in San Francisco than in Sacramento, 150 km inland. In summer, it is nearly always hotter in Sacramento. Explain how the bodies of water surrounding San Francisco moderate its extreme temperatures.

26. Freeze-dried foods have been dehydrated in a vacuum. During the process, the food freezes and must be heated to facilitate dehydration. Explain both how the vacuum speeds up dehydration and why the food freezes as a result.

27. In a physics classroom demonstration, an instructor inflates a balloon by mouth and then cools it in liquid nitrogen. When cold, the shrunken balloon has a small amount of light blue liquid in it, as well as some snow-like crystals. As it warms up, the liquid boils, and part of the crystals sublime, with some crystals lingering for a while and then producing a liquid. Identify the blue liquid and the two solids in the cold balloon. Justify your identifications using data from Table 1.4.

1.6 Mechanisms of Heat Transfer

28. What are the main methods of heat transfer from the hot core of Earth to its surface? From Earth's surface to outer space?

29. When our bodies get too warm, they respond by sweating and increasing blood circulation to the surface to transfer thermal energy away from the core. What effect will those processes have on a person in a $40.0\text{-}°C$ hot tub?

30. Shown below is a cut-away drawing of a thermos bottle (also known as a Dewar flask), which is a device designed specifically to slow down all forms of heat transfer. Explain the functions of the various parts, such as the vacuum, the silvering of the walls, the thin-walled long glass neck, the rubber support, the air layer, and the stopper.

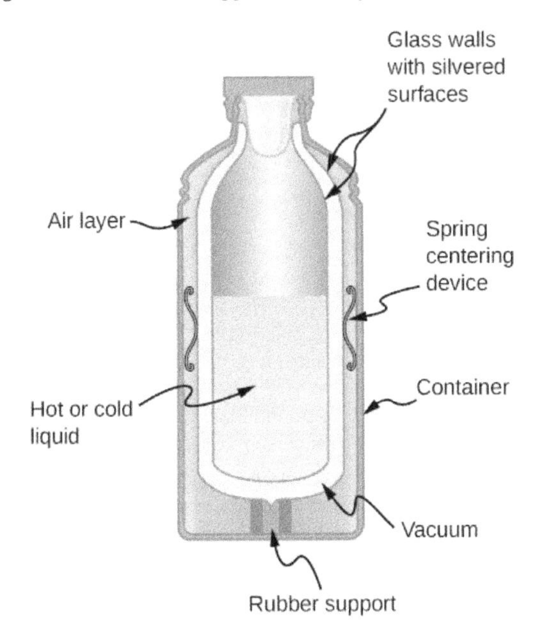

31. Some electric stoves have a flat ceramic surface with heating elements hidden beneath. A pot placed over a heating element will be heated, while the surface only a few centimeters away is safe to touch. Why is ceramic, with a conductivity less than that of a metal but greater than that of a good insulator, an ideal choice for the stove top?

32. Loose-fitting white clothing covering most of the body, shown below, is ideal for desert dwellers, both in the hot Sun and during cold evenings. Explain how such clothing is advantageous during both day and night.

33. One way to make a fireplace more energy-efficient is to have room air circulate around the outside of the fire box and back into the room. Detail the methods of heat transfer involved.

34. On cold, clear nights horses will sleep under the cover of large trees. How does this help them keep warm?

35. When watching a circus during the day in a large, dark-colored tent, you sense significant heat transfer from the tent. Explain why this occurs.

36. Satellites designed to observe the radiation from cold (3 K) dark space have sensors that are shaded from the Sun, Earth, and the Moon and are cooled to very low temperatures. Why must the sensors be at low temperature?

37. Why are thermometers that are used in weather stations shielded from the sunshine? What does a thermometer measure if it is shielded from the sunshine? What does it measure if it is not?

38. Putting a lid on a boiling pot greatly reduces the heat transfer necessary to keep it boiling. Explain why.

39. Your house will be empty for a while in cold weather, and you want to save energy and money. Should you turn the thermostat down to the lowest level that will protect the house from damage such as freezing pipes, or leave it at the normal temperature? (If you don't like coming back to a cold house, imagine that a timer controls the heating system so the house will be warm when you get back.) Explain your answer.

40. You pour coffee into an unlidded cup, intending to drink it 5 minutes later. You can add cream when you pour the cup or right before you drink it. (The cream is at the same temperature either way. Assume that the cream and coffee come into thermal equilibrium with each other very quickly.) Which way will give you hotter coffee? What feature of this question is different from the previous one?

41. Broiling is a method of cooking by radiation, which produces somewhat different results from cooking by conduction or convection. A gas flame or electric heating element produces a very high temperature close to the food and *above* it. Why is radiation the dominant heat-transfer method in this situation?

42. On a cold winter morning, why does the metal of a bike feel colder than the wood of a porch?

PROBLEMS

1.2 Thermometers and Temperature Scales

43. While traveling outside the United States, you feel sick. A companion gets you a thermometer, which says your temperature is 39. What scale is that on? What is your Fahrenheit temperature? Should you seek medical help?

44. What are the following temperatures on the Kelvin scale?

(a) $68.0\ °F$, an indoor temperature sometimes recommended for energy conservation in winter

(b) $134\ °F$, one of the highest atmospheric temperatures ever recorded on Earth (Death Valley, California, 1913)

(c) $9890\ °F$, the temperature of the surface of the Sun

45. (a) Suppose a cold front blows into your locale and drops the temperature by 40.0 Fahrenheit degrees. How many degrees Celsius does the temperature decrease when it decreases by $40.0\ °F$? (b) Show that any change in temperature in Fahrenheit degrees is nine-fifths the change in Celsius degrees

46. An Associated Press article on climate change said, "Some of the ice shelf's disappearance was probably during times when the planet was 36 degrees Fahrenheit (2 degrees Celsius) to 37 degrees Fahrenheit (3 degrees Celsius) warmer than it is today." What mistake did the reporter make?

47. (a) At what temperature do the Fahrenheit and Celsius scales have the same numerical value? (b) At what temperature do the Fahrenheit and Kelvin scales have the same numerical value?

48. A person taking a reading of the temperature in a freezer in Celsius makes two mistakes: first omitting the negative sign and then thinking the temperature is Fahrenheit. That is, the person reads $-x\ °C$ as $x\ °F$. Oddly enough, the result is the correct Fahrenheit temperature. What is the original Celsius reading? Round your answer to three significant figures.

1.3 Thermal Expansion

49. The height of the Washington Monument is measured to be 170.00 m on a day when the temperature is $35.0\ °C$. What will its height be on a day when the temperature falls to $-10.0\ °C$? Although the monument is made of limestone, assume that its coefficient of thermal expansion is the same as that of marble. Give your answer to five significant figures.

50. How much taller does the Eiffel Tower become at the end of a day when the temperature has increased by $15\ °C$? Its original height is 321 m and you can assume it is made of steel.

51. What is the change in length of a 3.00-cm-long column of mercury if its temperature changes from $37.0\ °C$ to $40.0\ °C$, assuming the mercury is constrained to a cylinder but unconstrained in length? Your answer will show why thermometers contain bulbs at the bottom instead of simple columns of liquid.

52. How large an expansion gap should be left between steel railroad rails if they may reach a maximum temperature $35.0\ °C$ greater than when they were laid? Their original length is 10.0 m.

53. You are looking to buy a small piece of land in Hong Kong. The price is “only” $60,000 per square meter. The land title says the dimensions are $20\text{ m} \times 30\text{ m}$. By how much would the total price change if you measured the parcel with a steel tape measure on a day when the temperature was 20 °C above the temperature that the tape measure was designed for? The dimensions of the land do not change.

54. Global warming will produce rising sea levels partly due to melting ice caps and partly due to the expansion of water as average ocean temperatures rise. To get some idea of the size of this effect, calculate the change in length of a column of water 1.00 km high for a temperature increase of 1.00 °C. Assume the column is not free to expand sideways. As a model of the ocean, that is a reasonable approximation, as only parts of the ocean very close to the surface can expand sideways onto land, and only to a limited degree. As another approximation, neglect the fact that ocean warming is not uniform with depth.

55. (a) Suppose a meter stick made of steel and one made of aluminum are the same length at 0 °C. What is their difference in length at 22.0 °C? (b) Repeat the calculation for two 30.0-m-long surveyor’s tapes.

56. (a) If a 500-mL glass beaker is filled to the brim with ethyl alcohol at a temperature of 5.00 °C, how much will overflow when the alcohol’s temperature reaches the room temperature of 22.0 °C? (b) How much less water would overflow under the same conditions?

57. Most cars have a coolant reservoir to catch radiator fluid that may overflow when the engine is hot. A radiator is made of copper and is filled to its 16.0-L capacity when at 10.0 °C. What volume of radiator fluid will overflow when the radiator and fluid reach a temperature of 95.0 °C, given that the fluid’s volume coefficient of expansion is $\beta = 400 \times 10^{-6}/\text{°C}$? (Your answer will be a conservative estimate, as most car radiators have operating temperatures greater than 95.0 °C).

58. A physicist makes a cup of instant coffee and notices that, as the coffee cools, its level drops 3.00 mm in the glass cup. Show that this decrease cannot be due to thermal contraction by calculating the decrease in level if the 350 cm^3 of coffee is in a 7.00-cm-diameter cup and decreases in temperature from 95.0 °C to 45.0 °C. (Most of the drop in level is actually due to escaping bubbles of air.)

59. The density of water at 0 °C is very nearly 1000 kg/m^3 (it is actually 999.84 kg/m^3), whereas the density of ice at 0 °C is 917 kg/m^3. Calculate the pressure necessary to keep ice from expanding when it freezes, neglecting the effect such a large pressure would have on the freezing temperature. (This problem gives you only an indication of how large the forces associated with freezing water might be.)

60. Show that $\beta = 3\alpha$, by calculating the infinitesimal change in volume dV of a cube with sides of length L when the temperature changes by dT.

1.4 Heat Transfer, Specific Heat, and Calorimetry

61. On a hot day, the temperature of an 80,000-L swimming pool increases by 1.50 °C. What is the net heat transfer during this heating? Ignore any complications, such as loss of water by evaporation.

62. To sterilize a 50.0-g glass baby bottle, we must raise its temperature from 22.0 °C to 95.0 °C. How much heat transfer is required?

63. The same heat transfer into identical masses of different substances produces different temperature changes. Calculate the final temperature when 1.00 kcal of heat transfers into 1.00 kg of the following, originally at 20.0 °C: (a) water; (b) concrete; (c) steel; and (d) mercury.

64. Rubbing your hands together warms them by converting work into thermal energy. If a woman rubs her hands back and forth for a total of 20 rubs, at a distance of 7.50 cm per rub, and with an average frictional force of 40.0 N, what is the temperature increase? The mass of tissues warmed is only 0.100 kg, mostly in the palms and fingers.

65. A 0.250-kg block of a pure material is heated from 20.0 °C to 65.0 °C by the addition of 4.35 kJ of energy. Calculate its specific heat and identify the substance of which it is most likely composed.

66. Suppose identical amounts of heat transfer into different masses of copper and water, causing identical changes in temperature. What is the ratio of the mass of copper to water?

67. (a) The number of kilocalories in food is determined by calorimetry techniques in which the food is burned and the amount of heat transfer is measured. How many kilocalories per gram are there in a 5.00-g peanut if the energy from burning it is transferred to 0.500 kg of water held in a 0.100-kg aluminum cup, causing a $54.9\text{-}°C$ temperature increase? Assume the process takes place in an ideal calorimeter, in other words a perfectly insulated container. (b) Compare your answer to the following labeling information found on a package of dry roasted peanuts: a serving of 33 g contains 200 calories. Comment on whether the values are consistent.

68. Following vigorous exercise, the body temperature of an 80.0 kg person is $40.0\,°C$. At what rate in watts must the person transfer thermal energy to reduce the body temperature to $37.0\,°C$ in 30.0 min, assuming the body continues to produce energy at the rate of 150 W? $(1 \text{ watt} = 1 \text{ joule/second or } 1 \text{ W} = 1 \text{ J/s})$

69. In a study of healthy young men[1], doing 20 push-ups in 1 minute burned an amount of energy per kg that for a 70.0-kg man corresponds to 8.06 calories (kcal). How much would a 70.0-kg man's temperature rise if he did not lose any heat during that time?

70. A 1.28-kg sample of water at $10.0\,°C$ is in a calorimeter. You drop a piece of steel with a mass of 0.385 kg at $215\,°C$ into it. After the sizzling subsides, what is the final equilibrium temperature? (Make the reasonable assumptions that any steam produced condenses into liquid water during the process of equilibration and that the evaporation and condensation don't affect the outcome, as we'll see in the next section.)

71. Repeat the preceding problem, assuming the water is in a glass beaker with a mass of 0.200 kg, which in turn is in a calorimeter. The beaker is initially at the same temperature as the water. Before doing the problem, should the answer be higher or lower than the preceding answer? Comparing the mass and specific heat of the beaker to those of the water, do you think the beaker will make much difference?

1.5 Phase Changes

72. How much heat transfer (in kilocalories) is required to thaw a 0.450-kg package of frozen vegetables originally at $0\,°C$ if their heat of fusion is the same as that of water?

73. A bag containing $0\,°C$ ice is much more effective in absorbing energy than one containing the same amount of $0\,°C$ water. (a) How much heat transfer is necessary to raise the temperature of 0.800 kg of water from $0\,°C$ to $30.0\,°C$? (b) How much heat transfer is required to first melt 0.800 kg of $0\,°C$ ice and then raise its temperature? (c) Explain how your answer supports the contention that the ice is more effective.

74. (a) How much heat transfer is required to raise the temperature of a 0.750-kg aluminum pot containing 2.50 kg of water from $30.0\,°C$ to the boiling point and then boil away 0.750 kg of water? (b) How long does this take if the rate of heat transfer is 500 W?

75. Condensation on a glass of ice water causes the ice to melt faster than it would otherwise. If 8.00 g of vapor condense on a glass containing both water and 200 g of ice, how many grams of the ice will melt as a result? Assume no other heat transfer occurs. Use L_v for water at $37\,°C$ as a better approximation than L_v for water at $100\,°C$.)

76. On a trip, you notice that a 3.50-kg bag of ice lasts an average of one day in your cooler. What is the average power in watts entering the ice if it starts at $0\,°C$ and completely melts to $0\,°C$ water in exactly one day?

77. On a certain dry sunny day, a swimming pool's temperature would rise by $1.50\,°C$ if not for evaporation. What fraction of the water must evaporate to carry away precisely enough energy to keep the temperature constant?

78. (a) How much heat transfer is necessary to raise the temperature of a 0.200-kg piece of ice from $-20.0\,°C$ to $130.0\,°C$, including the energy needed for phase changes? (b) How much time is required for each stage, assuming a constant 20.0 kJ/s rate of heat transfer? (c) Make a graph of temperature versus time for this process.

79. In 1986, an enormous iceberg broke away from the Ross Ice Shelf in Antarctica. It was an approximately rectangular prism 160 km long, 40.0 km wide, and 250 m thick. (a) What is the mass of this iceberg, given that the density of ice is 917 kg/m^3? (b) How much heat transfer (in joules) is needed to melt it? (c) How many years would it take sunlight alone to melt ice this thick, if the ice absorbs an average of 100 W/m^2, 12.00 h per day?

1. JW Vezina, "An examination of the differences between two methods of estimating energy expenditure in resistance training activities," *Journal of Strength and Conditioning Research*, April 28, 2014, http://www.ncbi.nlm.nih.gov/pubmed/24402448

80. How many grams of coffee must evaporate from 350 g of coffee in a 100-g glass cup to cool the coffee and the cup from $95.0\ ^\circ\text{C}$ to $45.0\ ^\circ\text{C}$? Assume the coffee has the same thermal properties as water and that the average heat of vaporization is 2340 kJ/kg (560 kcal/g). Neglect heat losses through processes other than evaporation, as well as the change in mass of the coffee as it cools. Do the latter two assumptions cause your answer to be higher or lower than the true answer?

81. (a) It is difficult to extinguish a fire on a crude oil tanker, because each liter of crude oil releases 2.80×10^7 J of energy when burned. To illustrate this difficulty, calculate the number of liters of water that must be expended to absorb the energy released by burning 1.00 L of crude oil, if the water's temperature rises from $20.0\ ^\circ\text{C}$ to $100\ ^\circ\text{C}$, it boils, and the resulting steam's temperature rises to $300\ ^\circ\text{C}$ at constant pressure. (b) Discuss additional complications caused by the fact that crude oil is less dense than water.

82. The energy released from condensation in thunderstorms can be very large. Calculate the energy released into the atmosphere for a small storm of radius 1 km, assuming that 1.0 cm of rain is precipitated uniformly over this area.

83. To help prevent frost damage, 4.00 kg of water at $0\ ^\circ\text{C}$ is sprayed onto a fruit tree. (a) How much heat transfer occurs as the water freezes? (b) How much would the temperature of the 200-kg tree decrease if this amount of heat transferred from the tree? Take the specific heat to be $3.35\ \text{kJ/kg} \cdot {}^\circ\text{C}$, and assume that no phase change occurs in the tree.

84. A 0.250-kg aluminum bowl holding $0.800\ \text{kg}$ of soup at $25.0\ ^\circ\text{C}$ is placed in a freezer. What is the final temperature if 388 kJ of energy is transferred from the bowl and soup, assuming the soup's thermal properties are the same as that of water?

85. A 0.0500-kg ice cube at $-30.0\ ^\circ\text{C}$ is placed in 0.400 kg of $35.0\text{-}^\circ\text{C}$ water in a very well-insulated container. What is the final temperature?

86. If you pour 0.0100 kg of $20.0\ ^\circ\text{C}$ water onto a 1.20-kg block of ice (which is initially at $-15.0\ ^\circ\text{C}$), what is the final temperature? You may assume that the water cools so rapidly that effects of the surroundings are negligible.

87. Indigenous people sometimes cook in watertight baskets by placing hot rocks into water to bring it to a boil. What mass of $500\text{-}^\circ\text{C}$ granite must be placed in 4.00 kg of $15.0\text{-}^\circ\text{C}$ water to bring its temperature to $100\ ^\circ\text{C}$, if 0.0250 kg of water escapes as vapor from the initial sizzle? You may neglect the effects of the surroundings.

88. What would the final temperature of the pan and water be in Example 1.7 if 0.260 kg of water were placed in the pan and 0.0100 kg of the water evaporated immediately, leaving the remainder to come to a common temperature with the pan?

1.6 Mechanisms of Heat Transfer

89. (a) Calculate the rate of heat conduction through house walls that are 13.0 cm thick and have an average thermal conductivity twice that of glass wool. Assume there are no windows or doors. The walls' surface area is $120\ \text{m}^2$ and their inside surface is at $18.0\ ^\circ\text{C}$, while their outside surface is at $5.00\ ^\circ\text{C}$. (b) How many 1-kW room heaters would be needed to balance the heat transfer due to conduction?

90. The rate of heat conduction out of a window on a winter day is rapid enough to chill the air next to it. To see just how rapidly the windows transfer heat by conduction, calculate the rate of conduction in watts through a 3.00-m^2 window that is 0.634 cm thick (1/4 in.) if the temperatures of the inner and outer surfaces are $5.00\ ^\circ\text{C}$ and $-10.0\ ^\circ\text{C}$, respectively. (This rapid rate will not be maintained—the inner surface will cool, even to the point of frost formation.)

91. Calculate the rate of heat conduction out of the human body, assuming that the core internal temperature is $37.0\ ^\circ\text{C}$, the skin temperature is $34.0\ ^\circ\text{C}$, the thickness of the fatty tissues between the core and the skin averages 1.00 cm, and the surface area is $1.40\ \text{m}^2$.

92. Suppose you stand with one foot on ceramic flooring and one foot on a wool carpet, making contact over an area of $80.0\ \text{cm}^2$ with each foot. Both the ceramic and the carpet are 2.00 cm thick and are $10.0\ ^\circ\text{C}$ on their bottom sides. At what rate must heat transfer occur from each foot to keep the top of the ceramic and carpet at $33.0\ ^\circ\text{C}$?

93. A man consumes 3000 kcal of food in one day, converting most of it to thermal energy to maintain body temperature. If he loses half this energy by evaporating water (through breathing and sweating), how many kilograms of water evaporate?

94. A firewalker runs across a bed of hot coals without sustaining burns. Calculate the heat transferred by conduction into the sole of one foot of a firewalker given that the bottom of the foot is a 3.00-mm-thick callus with a conductivity at the low end of the range for wood and its density is 300 kg/m^3. The area of contact is 25.0 cm^2, the temperature of the coals is 700 °C, and the time in contact is 1.00 s. Ignore the evaporative cooling of sweat.

95. (a) What is the rate of heat conduction through the 3.00-cm-thick fur of a large animal having a 1.40-m^2 surface area? Assume that the animal's skin temperature is 32.0 °C, that the air temperature is -5.00 °C, and that fur has the same thermal conductivity as air. (b) What food intake will the animal need in one day to replace this heat transfer?

96. A walrus transfers energy by conduction through its blubber at the rate of 150 W when immersed in -1.00 °C water. The walrus's internal core temperature is 37.0 °C, and it has a surface area of 2.00 m^2. What is the average thickness of its blubber, which has the conductivity of fatty tissues without blood?

97. Compare the rate of heat conduction through a 13.0-cm-thick wall that has an area of 10.0 m^2 and a thermal conductivity twice that of glass wool with the rate of heat conduction through a 0.750-cm-thick window that has an area of 2.00 m^2, assuming the same temperature difference across each.

98. Suppose a person is covered head to foot by wool clothing with average thickness of 2.00 cm and is transferring energy by conduction through the clothing at the rate of 50.0 W. What is the temperature difference across the clothing, given the surface area is 1.40 m^2?

99. Some stove tops are smooth ceramic for easy cleaning. If the ceramic is 0.600 cm thick and heat conduction occurs through the same area and at the same rate as computed in Example 1.11, what is the temperature difference across it? Ceramic has the same thermal conductivity as glass and brick.

100. One easy way to reduce heating (and cooling) costs is to add extra insulation in the attic of a house. Suppose a single-story cubical house already had 15 cm of fiberglass insulation in the attic and in all the exterior surfaces. If you added an extra 8.0 cm of fiberglass to the attic, by what percentage would the heating cost of the house drop? Take the house to have dimensions 10 m by 15 m by 3.0 m. Ignore air infiltration and heat loss through windows and doors, and assume that the interior is uniformly at one temperature and the exterior is uniformly at another.

101. Many decisions are made on the basis of the payback period: the time it will take through savings to equal the capital cost of an investment. Acceptable payback times depend upon the business or philosophy one has. (For some industries, a payback period is as small as 2 years.) Suppose you wish to install the extra insulation in the preceding problem. If energy cost $1.00 per million joules and the insulation was $4.00 per square meter, then calculate the simple payback time. Take the average ΔT for the 120-day heating season to be 15.0 °C.

ADDITIONAL PROBLEMS

102. In 1701, the Danish astronomer Ole Rømer proposed a temperature scale with two fixed points, freezing water at 7.5 degrees, and boiling water at 60.0 degrees. What is the boiling point of oxygen, 90.2 K, on the Rømer scale?

103. What is the percent error of thinking the melting point of tungsten is 3695 °C instead of the correct value of 3695 K?

104. An engineer wants to design a structure in which the difference in length between a steel beam and an aluminum beam remains at 0.500 m regardless of temperature, for ordinary temperatures. What must the lengths of the beams be?

105. How much stress is created in a steel beam if its temperature changes from -15 °C to 40 °C but it cannot expand? For steel, the Young's modulus $Y = 210 \times 10^9 \text{ N/m}^2$ from m58342 (http://cnx.org/content/m58342/latest/#fs-id1163713086230) . (Ignore the change in area resulting from the expansion.)

106. A brass rod $\left(Y = 90 \times 10^9 \text{ N/m}^2\right)$, with a diameter of 0.800 cm and a length of 1.20 m when the temperature is 25 °C, is fixed at both ends. At what temperature is the force in it at 36,000 N?

107. A mercury thermometer still in use for meteorology has a bulb with a volume of 0.780 cm^3 and a tube for the mercury to expand into of inside diameter 0.130 mm. (a) Neglecting the thermal expansion of the glass, what is the spacing between marks 1 °C apart? (b) If the thermometer is made of ordinary glass (not a good idea), what is the spacing?

108. Even when shut down after a period of normal use, a large commercial nuclear reactor transfers thermal energy at the rate of 150 MW by the radioactive decay of fission products. This heat transfer causes a rapid increase in temperature if the cooling system fails $(1 \text{ watt} = 1 \text{ joule/second or } 1 \text{ W} = 1 \text{ J/s and } 1 \text{ MW} = 1 \text{ megawatt})$. (a) Calculate the rate of temperature increase in degrees Celsius per second (°C/s) if the mass of the reactor core is $1.60 \times 10^5 \text{ kg}$ and it has an average specific heat of $0.3349 \text{ kJ/kg} \cdot \text{°C}$. (b) How long would it take to obtain a temperature increase of 2000 °C, which could cause some metals holding the radioactive materials to melt? (The initial rate of temperature increase would be greater than that calculated here because the heat transfer is concentrated in a smaller mass. Later, however, the temperature increase would slow down because the 500,000-kg steel containment vessel would also begin to heat up.)

109. You leave a pastry in the refrigerator on a plate and ask your roommate to take it out before you get home so you can eat it at room temperature, the way you like it. Instead, your roommate plays video games for hours. When you return, you notice that the pastry is still cold, but the game console has become hot. Annoyed, and knowing that the pastry will not be good if it is microwaved, you warm up the pastry by unplugging the console and putting it in a clean trash bag (which acts as a perfect calorimeter) with the pastry on the plate. After a while, you find that the equilibrium temperature is a nice, warm 38.3 °C. You know that the game console has a mass of 2.1 kg. Approximate it as having a uniform initial temperature of 45 °C. The pastry has a mass of 0.16 kg and a specific heat of $3.0 \text{ k J/(kg} \cdot \text{°C)}$, and is at a uniform initial temperature of 4.0 °C. The plate is at the same temperature and has a mass of 0.24 kg and a specific heat of $0.90 \text{ J/(kg} \cdot \text{°C)}$. What is the specific heat of the console?

110. Two solid spheres, *A* and *B*, made of the same material, are at temperatures of 0 °C and 100 °C, respectively. The spheres are placed in thermal contact in an ideal calorimeter, and they reach an equilibrium temperature of 20 °C. Which is the bigger sphere? What is the ratio of their diameters?

111. In some countries, liquid nitrogen is used on dairy trucks instead of mechanical refrigerators. A 3.00-hour delivery trip requires 200 L of liquid nitrogen, which has a density of 808 kg/m^3. (a) Calculate the heat transfer necessary to evaporate this amount of liquid nitrogen and raise its temperature to 3.00 °C. (Use c_P and assume it is constant over the temperature range.) This value is the amount of cooling the liquid nitrogen supplies. (b) What is this heat transfer rate in kilowatt-hours? (c) Compare the amount of cooling obtained from melting an identical mass of 0-°C ice with that from evaporating the liquid nitrogen.

112. Some gun fanciers make their own bullets, which involves melting lead and casting it into lead slugs. How much heat transfer is needed to raise the temperature and melt 0.500 kg of lead, starting from 25.0 °C?

113. A 0.800-kg iron cylinder at a temperature of $1.00 \times 10^3 \text{ °C}$ is dropped into an insulated chest of 1.00 kg of ice at its melting point. What is the final temperature, and how much ice has melted?

114. Repeat the preceding problem with 2.00 kg of ice instead of 1.00 kg.

115. Repeat the preceding problem with 0.500 kg of ice, assuming that the ice is initially in a copper container of mass 1.50 kg in equilibrium with the ice.

116. A 30.0-g ice cube at its melting point is dropped into an aluminum calorimeter of mass 100.0 g in equilibrium at 24.0 °C with 300.0 g of an unknown liquid. The final temperature is 4.0 °C. What is the heat capacity of the liquid?

117. (a) Calculate the rate of heat conduction through a double-paned window that has a 1.50-m^2 area and is made of two panes of 0.800-cm-thick glass separated by a 1.00-cm air gap. The inside surface temperature is 15.0 °C, while that on the outside is -10.0 °C. (*Hint:* There are identical temperature drops across the two glass panes. First find these and then the temperature drop across the air gap. This problem ignores the increased heat transfer in the air gap due to convection.) (b) Calculate the rate of heat conduction through a 1.60-cm-thick window of the same area and with the same temperatures. Compare your answer with that for part (a).

118. (a) An exterior wall of a house is 3 m tall and 10 m wide. It consists of a layer of drywall with an *R* factor of 0.56, a layer 3.5 inches thick filled with fiberglass batts, and a layer of insulated siding with an *R* factor of 2.6. The wall is built so well that there are no leaks of air through it. When the inside of the wall is at $22\ °C$ and the outside is at $-2\ °C$, what is the rate of heat flow through the wall? (b) More realistically, the 3.5-inch space also contains 2-by-4 studs—wooden boards 1.5 inches by 3.5 inches oriented so that 3.5-inch dimension extends from the drywall to the siding. They are "on 16-inch centers," that is, the centers of the studs are 16 inches apart. What is the heat current in this situation? Don't worry about one stud more or less.

119. For the human body, what is the rate of heat transfer by conduction through the body's tissue with the following conditions: the tissue thickness is 3.00 cm, the difference in temperature is $2.00\ °C$, and the skin area is $1.50\ m^2$. How does this compare with the average heat transfer rate to the body resulting from an energy intake of about 2400 kcal per day? (No exercise is included.)

120. You have a Dewar flask (a laboratory vacuum flask) that has an open top and straight sides, as shown below. You fill it with water and put it into the freezer. It is effectively a perfect insulator, blocking all heat transfer, except on the top. After a time, ice forms on the surface of the water. The liquid water and the bottom surface of the ice, in contact with the liquid water, are at $0\ °C$. The top surface of the ice is at the same temperature as the air in the freezer, $-18\ °C$. Set the rate of heat flow through the ice equal to the rate of loss of heat of fusion as the water freezes. When the ice layer is 0.700 cm thick, find the rate in m/s at which the ice is thickening.

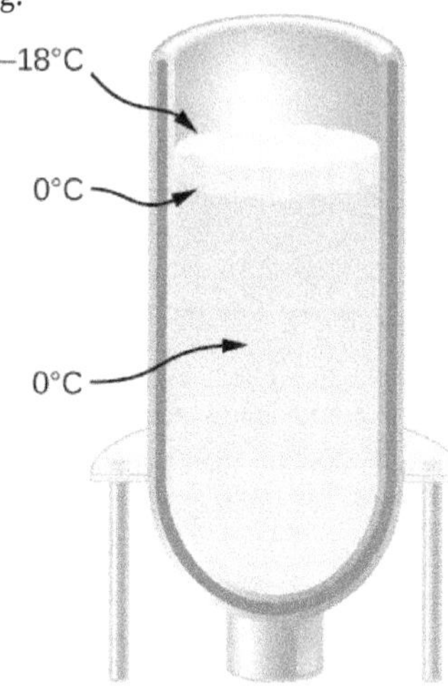

121. An infrared heater for a sauna has a surface area of $0.050\ m^2$ and an emissivity of 0.84. What temperature must it run at if the required power is 360 W? Neglect the temperature of the environment.

122. (a) Determine the power of radiation from the Sun by noting that the intensity of the radiation at the distance of Earth is $1370\ W/m^2$. *Hint:* That intensity will be found everywhere on a spherical surface with radius equal to that of Earth's orbit. (b) Assuming that the Sun's temperature is 5780 K and that its emissivity is 1, find its radius.

CHALLENGE PROBLEMS

123. A pendulum is made of a rod of length *L* and negligible mass, but capable of thermal expansion, and a weight of negligible size. (a) Show that when the temperature increases by *dT*, the period of the pendulum increases by a fraction $\alpha L dT/2$. (b) A clock controlled by a brass pendulum keeps time correctly at $10\ °C$. If the room temperature is $30\ °C$, does the clock run faster or slower? What is its error in seconds per day?

124. At temperatures of a few hundred kelvins the specific heat capacity of copper approximately follows the empirical formula $c = \alpha + \beta T + \delta T^{-2}$, where $\alpha = 349\ J/kg \cdot K$, $\beta = 0.107\ J/kg \cdot K^2$, and $\delta = 4.58 \times 10^5\ J \cdot kg \cdot K$. How much heat is needed to raise the temperature of a 2.00-kg piece of copper from $20\ °C$ to $250\ °C$?

125. In a calorimeter of negligible heat capacity, 200 g of steam at $150\,°C$ and 100 g of ice at $-40\,°C$ are mixed. The pressure is maintained at 1 atm. What is the final temperature, and how much steam, ice, and water are present?

126. An astronaut performing an extra-vehicular activity (space walk) shaded from the Sun is wearing a spacesuit that can be approximated as perfectly white $(e = 0)$ except for a $5\text{ cm} \times 8\text{ cm}$ patch in the form of the astronaut's national flag. The patch has emissivity 0.300. The spacesuit under the patch is 0.500 cm thick, with a thermal conductivity $k = 0.0600\text{ W/m °C}$, and its inner surface is at a temperature of $20.0\,°C$. What is the temperature of the patch, and what is the rate of heat loss through it? Assume the patch is so thin that its outer surface is at the same temperature as the outer surface of the spacesuit under it. Also assume the temperature of outer space is 0 K. You will get an equation that is very hard to solve in closed form, so you can solve it numerically with a graphing calculator, with software, or even by trial and error with a calculator.

127. The goal in this problem is to find the growth of an ice layer as a function of time. Call the thickness of the ice layer *L*. (a) Derive an equation for *dL*/*dt* in terms of *L*, the temperature *T* above the ice, and the properties of ice (which you can leave in symbolic form instead of substituting the numbers). (b) Solve this differential equation assuming that at $t = 0$, you have $L = 0$. If you have studied differential equations, you will know a technique for solving equations of this type: manipulate the equation to get *dL*/*dt* multiplied by a (very simple) function of *L* on one side, and integrate both sides with respect to time. Alternatively, you may be able to use your knowledge of the derivatives of various functions to guess the solution, which has a simple dependence on *t*. (c) Will the water eventually freeze to the bottom of the flask?

128. As the very first rudiment of climatology, estimate the temperature of Earth. Assume it is a perfect sphere and its temperature is uniform. Ignore the greenhouse effect. Thermal radiation from the Sun has an intensity (the "solar constant" *S*) of about 1370 W/m^2 at the radius of Earth's orbit. (a) Assuming the Sun's rays are parallel, what area must *S* be multiplied by to get the total radiation intercepted by Earth? It will be easiest to answer in terms of Earth's radius, *R*. (b) Assume that Earth reflects about 30% of the solar energy it intercepts. In other words, Earth has an albedo with a value of $A = 0.3$. In terms of *S*, *A*, and *R*, what is the rate at which Earth absorbs energy from the Sun? (c) Find the temperature at which Earth radiates energy at the same rate. Assume that at the infrared wavelengths where it radiates, the emissivity *e* is 1. Does your result show that the greenhouse effect is important? (d) How does your answer depend on the the area of Earth?

129. Let's stop ignoring the greenhouse effect and incorporate it into the previous problem in a very rough way. Assume the atmosphere is a single layer, a spherical shell around Earth, with an emissivity $e = 0.77$ (chosen simply to give the right answer) at infrared wavelengths emitted by Earth and by the atmosphere. However, the atmosphere is transparent to the Sun's radiation (that is, assume the radiation is at visible wavelengths with no infrared), so the Sun's radiation reaches the surface. The greenhouse effect comes from the difference between the atmosphere's transmission of visible light and its rather strong absorption of infrared. Note that the atmosphere's radius is not significantly different from Earth's, but since the atmosphere is a layer above Earth, it emits radiation both upward and downward, so it has twice Earth's area. There are three radiative energy transfers in this problem: solar radiation absorbed by Earth's surface; infrared radiation from the surface, which is absorbed by the atmosphere according to its emissivity; and infrared radiation from the atmosphere, half of which is absorbed by Earth and half of which goes out into space. Apply the method of the previous problem to get an equation for Earth's surface and one for the atmosphere, and solve them for the two unknown temperatures, surface and atmosphere.

a. In terms of Earth's radius, the constant σ, and the unknown temperature T_s of the surface, what is the power of the infrared radiation from the surface?
b. What is the power of Earth's radiation absorbed by the atmosphere?
c. In terms of the unknown temperature T_e of the atmosphere, what is the power radiated from the atmosphere?
d. Write an equation that says the power of the radiation the atmosphere absorbs from Earth equals the power of the radiation it emits.
e. Half of the power radiated by the atmosphere hits Earth. Write an equation that says that the power Earth absorbs from the atmosphere and the Sun equals the power that it emits.
f. Solve your two equations for the unknown temperature of Earth.
For steps that make this model less crude, see for example the lectures (https://openstaxcollege.org/l/21paulgormlec) by Paul O'Gorman.

2 | THE KINETIC THEORY OF GASES

Figure 2.1 A volcanic eruption releases tons of gas and dust into the atmosphere. Most of the gas is water vapor, but several other gases are common, including greenhouse gases such as carbon dioxide and acidic pollutants such as sulfur dioxide. However, the emission of volcanic gas is not all bad: Many geologists believe that in the earliest stages of Earth's formation, volcanic emissions formed the early atmosphere. (credit: modification of work by "Boaworm"/Wikimedia Commons)

Chapter Outline

Introduction

Gases are literally all around us—the air that we breathe is a mixture of gases. Other gases include those that make breads and cakes soft, those that make drinks fizzy, and those that burn to heat many homes. Engines and refrigerators depend on the behaviors of gases, as we will see in later chapters.

As we discussed in the preceding chapter, the study of heat and temperature is part of an area of physics known as thermodynamics, in which we require a system to be *macroscopic*, that is, to consist of a huge number (such as 10^{23}) of molecules. We begin by considering some macroscopic properties of gases: volume, pressure, and temperature. The simple model of a hypothetical "ideal gas" describes these properties of a gas very accurately under many conditions. We move from the ideal gas model to a more widely applicable approximation, called the Van der Waals model.

To understand gases even better, we must also look at them on the *microscopic* scale of molecules. In gases, the molecules interact weakly, so the microscopic behavior of gases is relatively simple, and they serve as a good introduction to systems of many molecules. The molecular model of gases is called the kinetic theory of gases and is one of the classic examples of a molecular model that explains everyday behavior.

2.1 | Molecular Model of an Ideal Gas

Learning Objectives

By the end of this section, you will be able to:

- Apply the ideal gas law to situations involving the pressure, volume, temperature, and the number of molecules of a gas
- Use the unit of moles in relation to numbers of molecules, and molecular and macroscopic masses
- Explain the ideal gas law in terms of moles rather than numbers of molecules
- Apply the van der Waals gas law to situations where the ideal gas law is inadequate

In this section, we explore the thermal behavior of gases. Our word "gas" comes from the Flemish word meaning "chaos," first used for vapors by the seventeenth-century chemist J. B. van Helmont. The term was more appropriate than he knew, because gases consist of molecules moving and colliding with each other at random. This randomness makes the connection between the microscopic and macroscopic domains simpler for gases than for liquids or solids.

How do gases differ from solids and liquids? Under ordinary conditions, such as those of the air around us, the difference is that the molecules of gases are much farther apart than those of solids and liquids. Because the typical distances between molecules are large compared to the size of a molecule, as illustrated in Figure 2.2, the forces between them are considered negligible, except when they come into contact with each other during collisions. Also, at temperatures well above the boiling temperature, the motion of molecules is fast, and the gases expand rapidly to occupy all of the accessible volume. In contrast, in liquids and solids, molecules are closer together, and the behavior of molecules in liquids and solids is highly constrained by the molecules' interactions with one another. The macroscopic properties of such substances depend strongly on the forces between the molecules, and since many molecules are interacting, the resulting "many-body problems" can be extremely complicated (see Condensed Matter Physics (http://cnx.org/content/m58591/latest/)).

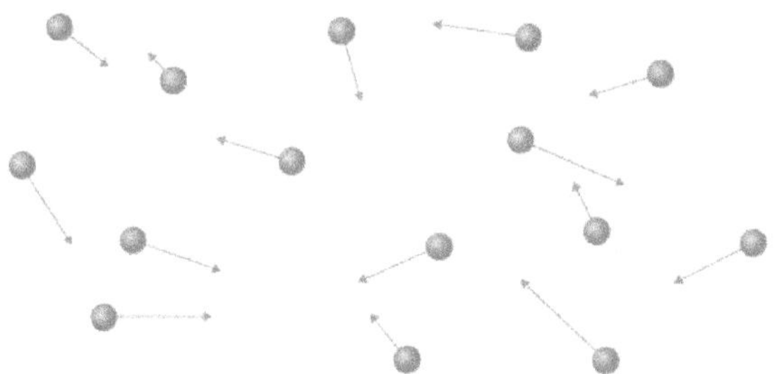

Figure 2.2 Atoms and molecules in a gas are typically widely separated. Because the forces between them are quite weak at these distances, the properties of a gas depend more on the number of atoms per unit volume and on temperature than on the type of atom.

The Gas Laws

In the previous chapter, we saw one consequence of the large intermolecular spacing in gases: Gases are easily compressed. Table 1.2 shows that gases have larger coefficients of volume expansion than either solids or liquids. These large coefficients mean that gases expand and contract very rapidly with temperature changes. We also saw (in the section on thermal expansion) that most gases expand at the same rate or have the same coefficient of volume expansion, β. This raises a question: Why do all gases act in nearly the same way, when all the various liquids and solids have widely varying expansion rates?

To study how the pressure, temperature, and volume of a gas relate to one another, consider what happens when you pump air into a deflated car tire. The tire's volume first increases in direct proportion to the amount of air injected, without much increase in the tire pressure. Once the tire has expanded to nearly its full size, the tire's walls limit its volume expansion. If we continue to pump air into the tire, the pressure increases. When the car is driven and the tires flex, their temperature increases, and therefore the pressure increases even further (Figure 2.3).

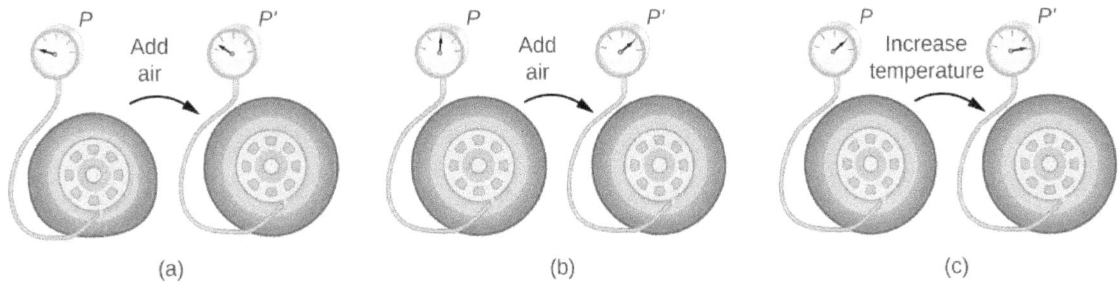

Figure 2.3 (a) When air is pumped into a deflated tire, its volume first increases without much increase in pressure. (b) When the tire is filled to a certain point, the tire walls resist further expansion, and the pressure increases with more air. (c) Once the tire is inflated, its pressure increases with temperature.

Figure 2.4 shows data from the experiments of Robert Boyle (1627–1691), illustrating what is now called Boyle's law: At constant temperature and number of molecules, the absolute pressure of a gas and its volume are inversely proportional. (Recall from Fluid Mechanics (http://cnx.org/content/m58624/latest/) that the absolute pressure is the true pressure and the gauge pressure is the absolute pressure minus the ambient pressure, typically atmospheric pressure.) The graph in Figure 2.4 displays this relationship as an inverse proportionality of volume to pressure.

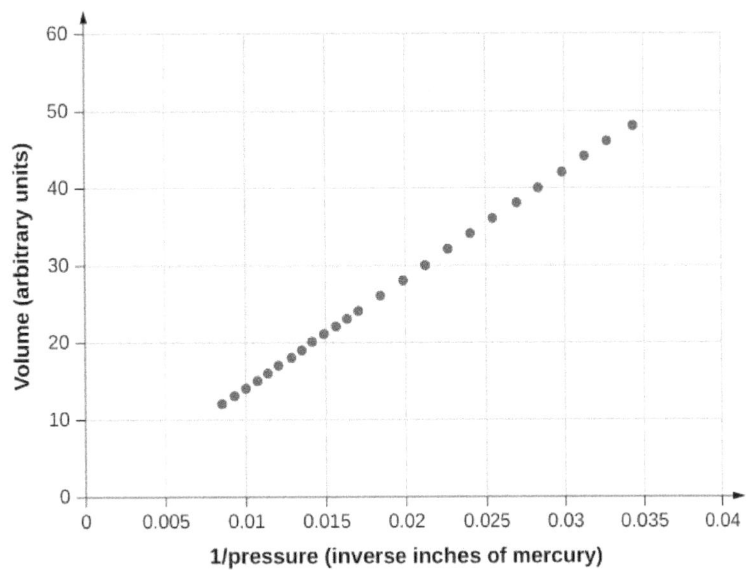

Figure 2.4 Robert Boyle and his assistant found that volume and pressure are inversely proportional. Here their data are plotted as V versus $1/p$; the linearity of the graph shows the inverse proportionality. The number shown as the volume is actually the height in inches of air in a cylindrical glass tube. The actual volume was that height multiplied by the cross-sectional area of the tube, which Boyle did not publish. The data are from Boyle's book *A Defence of the Doctrine Touching the Spring and Weight of the Air*..., p. 60.[1]

Figure 2.5 shows experimental data illustrating what is called Charles's law, after Jacques Charles (1746–1823). Charles's law states that at constant pressure and number of molecules, the volume of a gas is proportional to its absolute temperature.

1. http://bvpb.mcu.es/en/consulta/registro.cmd?id=406806

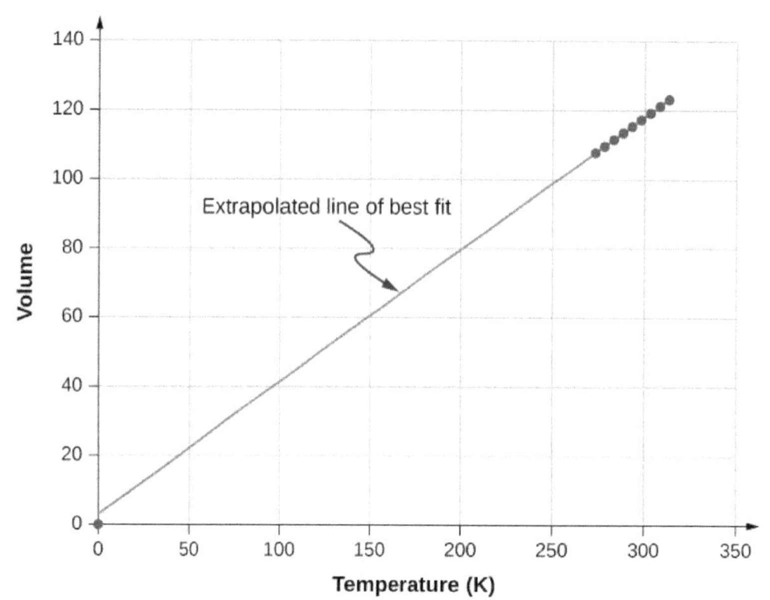

Figure 2.5 Experimental data showing that at constant pressure, volume is approximately proportional to temperature. The best-fit line passes approximately through the origin.[2]

Similar is Amonton's or Gay-Lussac's law, which states that at constant volume and number of molecules, the pressure is proportional to the temperature. That law is the basis of the constant-volume gas thermometer, discussed in the previous chapter. (The histories of these laws and the appropriate credit for them are more complicated than can be discussed here.)

It is known experimentally that for gases at low density (such that their molecules occupy a negligible fraction of the total volume) and at temperatures well above the boiling point, these proportionalities hold to a good approximation. Not surprisingly, with the other quantities held constant, either pressure or volume is proportional to the number of molecules. More surprisingly, when the proportionalities are combined into a single equation, the constant of proportionality is independent of the composition of the gas. The resulting equation for all gases applies in the limit of low density and high temperature; it's the same for oxygen as for helium or uranium hexafluoride. A gas at that limit is called an **ideal gas**; it obeys the **ideal gas law**, which is also called the equation of state of an ideal gas.

Ideal Gas Law

The ideal gas law states that

$$pV = Nk_B T, \tag{2.1}$$

where p is the absolute pressure of a gas, V is the volume it occupies, N is the number of molecules in the gas, and T is its absolute temperature.

The constant k_B is called the **Boltzmann constant** in honor of the Austrian physicist Ludwig Boltzmann (1844–1906) and has the value

$$k_B = 1.38 \times 10^{-23} \text{ J/K}.$$

The ideal gas law describes the behavior of any real gas when its density is low enough or its temperature high enough that it is far from liquefaction. This encompasses many practical situations. In the next section, we'll see why it's independent of the type of gas.

In many situations, the ideal gas law is applied to a sample of gas with a constant number of molecules; for instance, the gas may be in a sealed container. If N is constant, then solving for N shows that pV/T is constant. We can write that fact in a convenient form:

2. http://chemed.chem.purdue.edu/genchem/history/charles.html

$$\frac{p_1 V_1}{T_1} = \frac{p_2 V_2}{T_2}, \tag{2.2}$$

where the subscripts 1 and 2 refer to any two states of the gas at different times. Again, the temperature must be expressed in kelvin and the pressure must be absolute pressure, which is the sum of gauge pressure and atmospheric pressure.

Example 2.1

Calculating Pressure Changes Due to Temperature Changes

Suppose your bicycle tire is fully inflated, with an absolute pressure of 7.00×10^5 Pa (a gauge pressure of just under $90.0\ \text{lb/in.}^2$) at a temperature of $18.0\ °\text{C}$. What is the pressure after its temperature has risen to $35.0\ °\text{C}$ on a hot day? Assume there are no appreciable leaks or changes in volume.

Strategy

The pressure in the tire is changing only because of changes in temperature. We know the initial pressure $p_0 = 7.00 \times 10^5\ \text{Pa},$ the initial temperature $T_0 = 18.0\ °\text{C},$ and the final temperature $T_f = 35.0\ °\text{C}.$ We must find the final pressure p_f. Since the number of molecules is constant, we can use the equation

$$\frac{p_f V_f}{T_f} = \frac{p_0 V_0}{T_0}.$$

Since the volume is constant, V_f and V_0 are the same and they divide out. Therefore,

$$\frac{p_f}{T_f} = \frac{p_0}{T_0}.$$

We can then rearrange this to solve for p_f :

$$p_f = p_0 \frac{T_f}{T_0},$$

where the temperature must be in kelvin.

Solution

1. Convert temperatures from degrees Celsius to kelvin

$$T_0 = (18.0 + 273)\text{K} = 291\ \text{K},$$

$$T_f = (35.0 + 273)\text{K} = 308\ \text{K}.$$

2. Substitute the known values into the equation,

$$p_f = p_0 \frac{T_f}{T_0} = 7.00 \times 10^5\ \text{Pa}\left(\frac{308\ \text{K}}{291\ \text{K}}\right) = 7.41 \times 10^5\ \text{Pa}.$$

Significance

The final temperature is about 6% greater than the original temperature, so the final pressure is about 6% greater as well. Note that *absolute pressure* (see Fluid Mechanics (http://cnx.org/content/m58624/latest/)) and *absolute temperature* (see Temperature and Heat) must be used in the ideal gas law.

Example 2.2

Calculating the Number of Molecules in a Cubic Meter of Gas

How many molecules are in a typical object, such as gas in a tire or water in a glass? This calculation can give us an idea of how large N typically is. Let's calculate the number of molecules in the air that a typical healthy young adult inhales in one breath, with a volume of 500 mL, at *standard temperature and pressure* (STP), which is defined as $0\,°\text{C}$ and atmospheric pressure. (Our young adult is apparently outside in winter.)

Strategy

Because pressure, volume, and temperature are all specified, we can use the ideal gas law, $pV = Nk_\text{B}T,$ to find $N.$

Solution

1. Identify the knowns.

$$T = 0\,°\text{C} = 273\,\text{K},\ \ p = 1.01 \times 10^5\ \text{Pa},\ \ V = 500\,\text{mL} = 5 \times 10^{-4}\ \text{m}^3,\ \ k_\text{B} = 1.38 \times 10^{-23}\ \text{J/K}$$

2. Substitute the known values into the equation and solve for N.

$$N = \frac{pV}{k_\text{B}T} = \frac{(1.01 \times 10^5\ \text{Pa})\,(5 \times 10^{-4}\ \text{m}^3)}{(1.38 \times 10^{-23}\ \text{J/K})\,(273\ \text{K})} = 1.34 \times 10^{22}\ \text{molecules}$$

Significance

N is huge, even in small volumes. For example, $1\ \text{cm}^3$ of a gas at STP contains 2.68×10^{19} molecules. Once again, note that our result for N is the same for all types of gases, including mixtures.

As we observed in the chapter on fluid mechanics, pascals are N/m^2, so $\text{Pa} \cdot \text{m}^3 = \text{N} \cdot \text{m} = \text{J}.$ Thus, our result for N is dimensionless, a pure number that could be obtained by counting (in principle) rather than measuring. As it is the number of molecules, we put "molecules" after the number, keeping in mind that it is an aid to communication rather than a unit.

Moles and Avogadro's Number

It is often convenient to measure the amount of substance with a unit on a more human scale than molecules. The SI unit for this purpose was developed by the Italian scientist Amedeo Avogadro (1776–1856). (He worked from the hypothesis that equal volumes of gas at equal pressure and temperature contain equal numbers of molecules, independent of the type of gas. As mentioned above, this hypothesis has been confirmed when the ideal gas approximation applies.) A **mole** (abbreviated mol) is defined as the amount of any substance that contains as many molecules as there are atoms in exactly 12 grams (0.012 kg) of carbon-12. (Technically, we should say "formula units," not "molecules," but this distinction is irrelevant for our purposes.) The number of molecules in one mole is called **Avogadro's number** $(N_\text{A}),$ and the value of Avogadro's number is now known to be

$$N_\text{A} = 6.02 \times 10^{23}\ \text{mol}^{-1}.$$

We can now write $N = N_A n$, where n represents the number of moles of a substance.

Avogadro's number relates the mass of an amount of substance in grams to the number of protons and neutrons in an atom or molecule (12 for a carbon-12 atom), which roughly determine its mass. It's natural to define a unit of mass such that the mass of an atom is approximately equal to its number of neutrons and protons. The unit of that kind accepted for use with the SI is the *unified atomic mass unit* (u), also called the *dalton*. Specifically, a carbon-12 atom has a mass of exactly 12 u, so that its molar mass M in grams per mole is numerically equal to the mass of one carbon-12 atom in u. That equality holds for any substance. In other words, N_A is not only the conversion from numbers of molecules to moles, but it is also the conversion from u to grams: $6.02 \times 10^{23}\ \text{u} = 1\ \text{g}.$ See Figure 2.6.

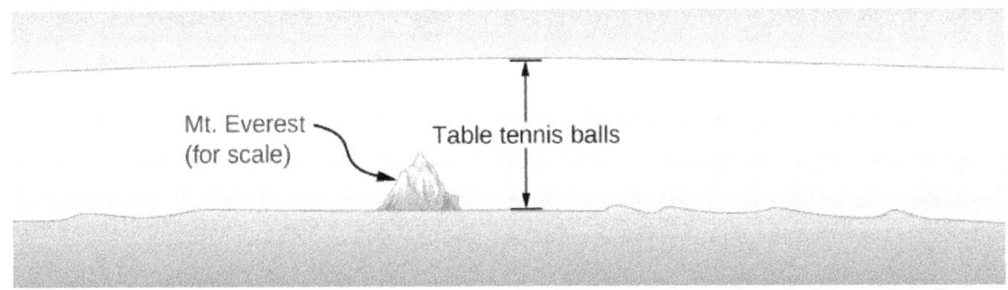

Figure 2.6 How big is a mole? On a macroscopic level, Avogadro's number of table tennis balls would cover Earth to a depth of about 40 km.

Now letting m_s stand for the mass of a sample of a substance, we have $m_s = nM$. Letting m stand for the mass of a molecule, we have $M = N_A m$.

2.1 Check Your Understanding The recommended daily amount of vitamin B_3 or niacin, $C_6NH_5O_2$, for women who are not pregnant or nursing, is 14 mg. Find the number of molecules of niacin in that amount.

2.2 Check Your Understanding The density of air in a classroom ($p = 1.00\ \text{atm}$ and $T = 20\ °\text{C}$) is $1.28\ \text{kg/m}^3$. At what pressure is the density $0.600\ \text{kg/m}^3$ if the temperature is kept constant?

The Ideal Gas Law Restated using Moles

A very common expression of the ideal gas law uses the number of moles in a sample, *n*, rather than the number of molecules, *N*. We start from the ideal gas law,

$$pV = Nk_B T,$$

and multiply and divide the right-hand side of the equation by Avogadro's number N_A. This gives us

$$pV = \frac{N}{N_A} N_A k_B T.$$

Note that $n = N/N_A$ is the number of moles. We define the **universal gas constant** as $R = N_A k_B$, and obtain the ideal gas law in terms of moles.

Ideal Gas Law (in terms of moles)

In terms of number of moles *n*, the ideal gas law is written as

$$pV = nRT. \tag{2.3}$$

In SI units,

$$R = N_A k_B = (6.02 \times 10^{23}\ \text{mol}^{-1})\left(1.38 \times 10^{-23}\ \frac{\text{J}}{\text{K}}\right) = 8.31 \frac{\text{J}}{\text{mol} \cdot \text{K}}.$$

In other units,

$$R = 1.99 \frac{\text{cal}}{\text{mol} \cdot \text{K}} = 0.0821 \frac{\text{L} \cdot \text{atm}}{\text{mol} \cdot \text{K}}.$$

You can use whichever value of *R* is most convenient for a particular problem.

Example 2.3

Density of Air at STP and in a Hot Air Balloon

Calculate the density of dry air (a) under standard conditions and (b) in a hot air balloon at a temperature of $120\,°\text{C}$. Dry air is approximately $78\%\ \text{N}_2$, $21\%\ \text{O}_2$, and $1\%\ \text{Ar}$.

Strategy and Solution

a. We are asked to find the density, or mass per cubic meter. We can begin by finding the molar mass. If we have a hundred molecules, of which 78 are nitrogen, 21 are oxygen, and 1 is argon, the average molecular mass is $\frac{78\,m_{\text{N}_2} + 21\,m_{\text{O}_2} + m_{\text{Ar}}}{100}$, or the mass of each constituent multiplied by its percentage. The same applies to the molar mass, which therefore is

$$M = 0.78\,M_{\text{N}_2} + 0.21\,M_{\text{O}_2} + 0.01\,M_{\text{Ar}} = 29.0\ \text{g/mol}.$$

Now we can find the number of moles per cubic meter. We use the ideal gas law in terms of moles, $pV = nRT$, with $p = 1.00\ \text{atm}$, $T = 273\ \text{K}$, $V = 1\ \text{m}^3$, and $R = 8.31\ \text{J/mol}\cdot\text{K}$. The most convenient choice for R in this case is $R = 8.31\ \text{J/mol}\cdot\text{K}$ because the known quantities are in SI units:

$$n = \frac{pV}{RT} = \frac{(1.01\times10^5\ \text{Pa})\,(1\ \text{m}^3)}{(8.31\ \text{J/mol}\cdot\text{K})\,(273\ \text{K})} = 44.5\ \text{mol}.$$

Then, the mass m_s of that air is

$$m_s = nM = (44.5\ \text{mol})(29.0\ \text{g/mol}) = 1290\ \text{g} = 1.29\ \text{kg}.$$

Finally the density of air at STP is

$$\rho = \frac{m_s}{V} = \frac{1.29\ \text{kg}}{1\ \text{m}^3} = 1.29\ \text{kg/m}^3.$$

b. The air pressure inside the balloon is still 1 atm because the bottom of the balloon is open to the atmosphere. The calculation is the same except that we use a temperature of $120\,°\text{C}$, which is 393 K. We can repeat the calculation in (a), or simply observe that the density is proportional to the number of moles, which is inversely proportional to the temperature. Then using the subscripts 1 for air at STP and 2 for the hot air, we have

$$\rho_2 = \frac{T_1}{T_2}\rho_1 = \frac{273\ \text{K}}{393\ \text{K}}(1.29\ \text{kg/m}^3) = 0.896\ \text{kg/m}^3.$$

Significance

Using the methods of Archimedes' Principle and Buoyancy (http://cnx.org/content/m58356/latest/), we can find the net force on $2200\ \text{m}^3$ of air at $120\,°\text{C}$ is $F_b - F_g = \rho_{\text{atmosphere}}Vg - \rho_{\text{hot air}}Vg = 8.49\times10^3\ \text{N}$, or enough to lift about 867 kg. The mass density and molar density of air at STP, found above, are often useful numbers. From the molar density, we can easily determine another useful number, the volume of a mole of any ideal gas at STP, which is 22.4 L.

2.3 Check Your Understanding Liquids and solids have densities on the order of 1000 times greater than gases. Explain how this implies that the distances between molecules in gases are on the order of 10 times greater than the size of their molecules.

The ideal gas law is closely related to energy: The units on both sides of the equation are joules. The right-hand side of the ideal gas law equation is $Nk_\text{B}T$. This term is roughly the total translational kinetic energy (which, when discussing gases, refers to the energy of translation of a molecule, not that of vibration of its atoms or rotation) of N molecules at an absolute temperature T, as we will see formally in the next section. The left-hand side of the ideal gas law equation is pV. As mentioned in the example on the number of molecules in an ideal gas, pressure multiplied by volume has units of energy. The energy of a gas can be changed when the gas does work as it increases in volume, something we explored in the preceding chapter, and the amount of work is related to the pressure. This is the process that occurs in gasoline or steam engines and turbines, as we'll see in the next chapter.

Problem-Solving Strategy: The Ideal Gas Law

Step 1. Examine the situation to determine that an ideal gas is involved. Most gases are nearly ideal unless they are close to the boiling point or at pressures far above atmospheric pressure.

Step 2. Make a list of what quantities are given or can be inferred from the problem as stated (identify the known quantities).

Step 3. Identify exactly what needs to be determined in the problem (identify the unknown quantities). A written list is useful.

Step 4. Determine whether the number of molecules or the number of moles is known or asked for to decide whether to use the ideal gas law as $pV = Nk_B T$, where N is the number of molecules, or $pV = nRT$, where n is the number of moles.

Step 5. Convert known values into proper SI units (K for temperature, Pa for pressure, m^3 for volume, molecules for N, and moles for n). If the units of the knowns are consistent with one of the non-SI values of R, you can leave them in those units. Be sure to use absolute temperature and absolute pressure.

Step 6. Solve the ideal gas law for the quantity to be determined (the unknown quantity). You may need to take a ratio of final states to initial states to eliminate the unknown quantities that are kept fixed.

Step 7. Substitute the known quantities, along with their units, into the appropriate equation and obtain numerical solutions complete with units.

Step 8. Check the answer to see if it is reasonable: Does it make sense?

The Van der Waals Equation of State

We have repeatedly noted that the ideal gas law is an approximation. How can it be improved upon? The **van der Waals equation of state** (named after the Dutch physicist Johannes van der Waals, 1837−1923) improves it by taking into account two factors. First, the attractive forces between molecules, which are stronger at higher density and reduce the pressure, are taken into account by adding to the pressure a term equal to the square of the molar density multiplied by a positive coefficient a. Second, the volume of the molecules is represented by a positive constant b, which can be thought of as the volume of a mole of molecules. This is subtracted from the total volume to give the remaining volume that the molecules can move in. The constants a and b are determined experimentally for each gas. The resulting equation is

$$\left[p + a\left(\frac{n}{V}\right)^2\right](V - nb) = nRT. \tag{2.4}$$

In the limit of low density (small n), the a and b terms are negligible, and we have the ideal gas law, as we should for low density. On the other hand, if $V - nb$ is small, meaning that the molecules are very close together, the pressure must be higher to give the same nRT, as we would expect in the situation of a highly compressed gas. However, the increase in pressure is less than that argument would suggest, because at high density the $(n/V)^2$ term is significant. Since it's positive, it causes a lower pressure to give the same nRT.

The van der Waals equation of state works well for most gases under a wide variety of conditions. As we'll see in the next module, it even predicts the gas-liquid transition.

pV Diagrams

We can examine aspects of the behavior of a substance by plotting a ***pV* diagram**, which is a graph of pressure versus volume. When the substance behaves like an ideal gas, the ideal gas law $pV = nRT$ describes the relationship between its pressure and volume. On a *pV* diagram, it's common to plot an *isotherm*, which is a curve showing *p* as a function of *V* with the number of molecules and the temperature fixed. Then, for an ideal gas, $pV = \text{constant}.$ For example, the volume of the gas decreases as the pressure increases. The resulting graph is a hyperbola.

However, if we assume the van der Waals equation of state, the isotherms become more interesting, as shown in Figure 2.7. At high temperatures, the curves are approximately hyperbolas, representing approximately ideal behavior at various fixed temperatures. At lower temperatures, the curves look less and less like hyperbolas—that is, the gas is not behaving ideally. There is a **critical temperature** T_c at which the curve has a point with zero slope. Below that temperature, the curves do not decrease monotonically; instead, they each have a "hump," meaning that for a certain range of volume, increasing the volume increases the pressure.

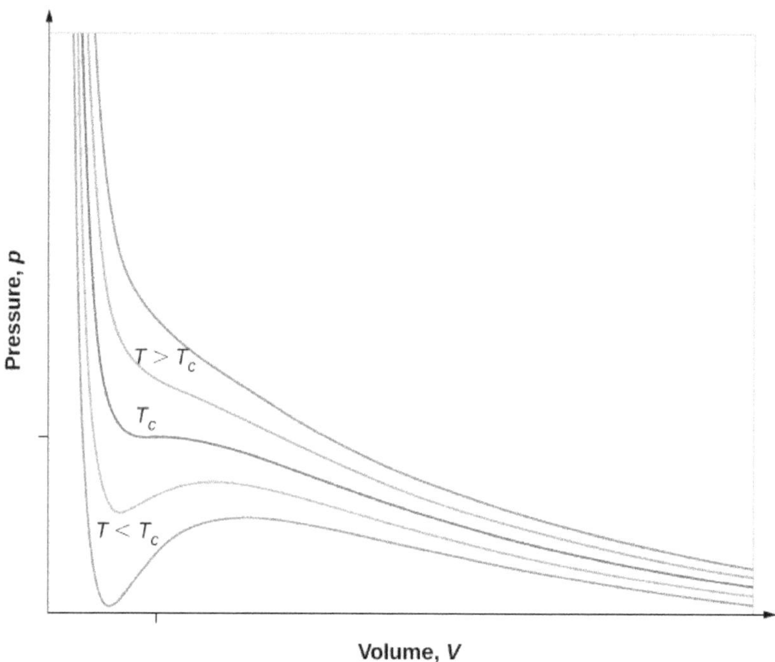

Figure 2.7 *pV* diagram for a Van der Waals gas at various temperatures. The red curves are calculated at temperatures above the critical temperature and the blue curves at temperatures below it. The blue curves have an oscillation in which volume (*V*) increases with increasing temperature (*T*), an impossible situation, so they must be corrected as in Figure 2.8. (credit: "Eman"/Wikimedia Commons)

Such behavior would be completely unphysical. Instead, the curves are understood as describing a liquid-gas phase transition. The oscillating part of the curve is replaced by a horizontal line, showing that as the volume increases at constant temperature, the pressure stays constant. That behavior corresponds to boiling and condensation; when a substance is at its boiling temperature for a particular pressure, it can increase in volume as some of the liquid turns to gas, or decrease as some of the gas turns to liquid, without any change in temperature or pressure.

Figure 2.8 shows similar isotherms that are more realistic than those based on the van der Waals equation. The steep parts of the curves to the left of the transition region show the liquid phase, which is almost incompressible—a slight decrease in volume requires a large increase in pressure. The flat parts show the liquid-gas transition; the blue regions that they define represent combinations of pressure and volume where liquid and gas can coexist.

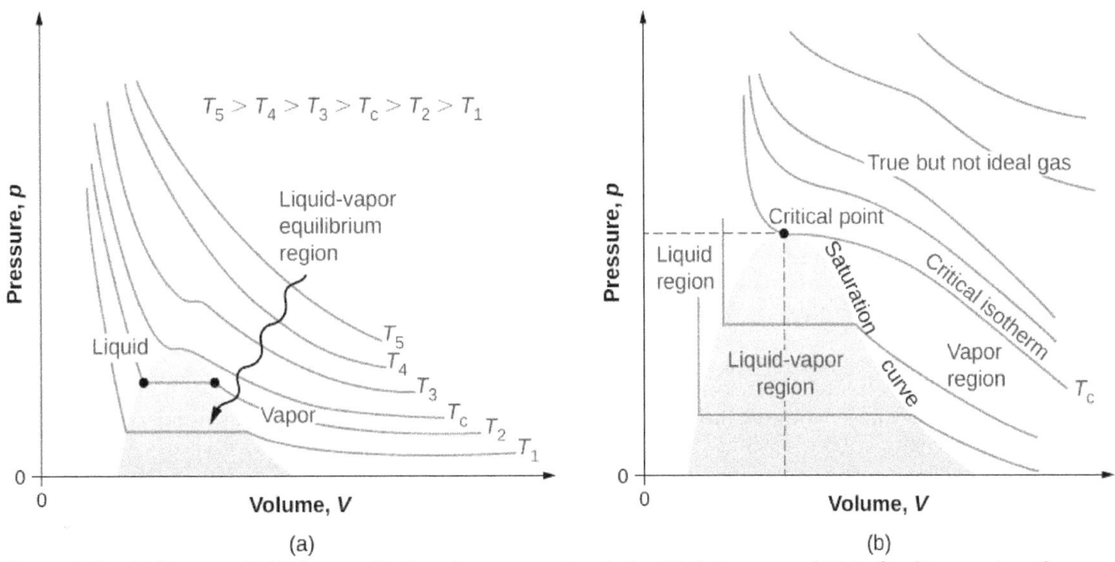

Figure 2.8 *pV* diagrams. (a) Each curve (isotherm) represents the relationship between *p* and *V* at a fixed temperature; the upper curves are at higher temperatures. The lower curves are not hyperbolas because the gas is no longer an ideal gas. (b) An expanded portion of the *pV* diagram for low temperatures, where the phase can change from a gas to a liquid. The term "vapor" refers to the gas phase when it exists at a temperature below the boiling temperature.

The isotherms above T_c do not go through the liquid-gas transition. Therefore, liquid cannot exist above that temperature, which is the critical temperature (described in the chapter on temperature and heat). At sufficiently low pressure above that temperature, the gas has the density of a liquid but will not condense; the gas is said to be **supercritical**. At higher pressure, it is solid. Carbon dioxide, for example, has no liquid phase at a temperature above $31.0\,°C$. The critical pressure is the maximum pressure at which the liquid can exist. The point on the *pV* diagram at the critical pressure and temperature is the critical point (which you learned about in the chapter on temperature and heat). Table 2.1 lists representative critical temperatures and pressures.

Substance	Critical temperature		Critical pressure	
	K	°C	Pa	atm
Water	647.4	374.3	22.12×10^6	219.0
Sulfur dioxide	430.7	157.6	7.88×10^6	78.0
Ammonia	405.5	132.4	11.28×10^6	111.7
Carbon dioxide	304.2	31.1	7.39×10^6	73.2
Oxygen	154.8	–118.4	5.08×10^6	50.3
Nitrogen	126.2	–146.9	3.39×10^6	33.6
Hydrogen	33.3	–239.9	1.30×10^6	12.9
Helium	5.3	–267.9	0.229×10^6	2.27

Table 2.1 Critical Temperatures and Pressures for Various Substances

2.2 | Pressure, Temperature, and RMS Speed

Learning Objectives

By the end of this section, you will be able to:

- Explain the relations between microscopic and macroscopic quantities in a gas
- Solve problems involving mixtures of gases
- Solve problems involving the distance and time between a gas molecule's collisions

We have examined pressure and temperature based on their macroscopic definitions. Pressure is the force divided by the area on which the force is exerted, and temperature is measured with a thermometer. We can gain a better understanding of pressure and temperature from the **kinetic theory of gases**, the theory that relates the macroscopic properties of gases to the motion of the molecules they consist of. First, we make two assumptions about molecules in an ideal gas.

1. There is a very large number N of molecules, all identical and each having mass m.
2. The molecules obey Newton's laws and are in continuous motion, which is random and isotropic, that is, the same in all directions.

To derive the ideal gas law and the connection between microscopic quantities such as the energy of a typical molecule and macroscopic quantities such as temperature, we analyze a sample of an ideal gas in a rigid container, about which we make two further assumptions:

3. The molecules are much smaller than the average distance between them, so their total volume is much less than that of their container (which has volume V). In other words, we take the Van der Waals constant b, the volume of a mole of gas molecules, to be negligible compared to the volume of a mole of gas in the container.
4. The molecules make perfectly elastic collisions with the walls of the container and with each other. Other forces on them, including gravity and the attractions represented by the Van der Waals constant a, are negligible (as is necessary for the assumption of isotropy).

The collisions between molecules do not appear in the derivation of the ideal gas law. They do not disturb the derivation either, since collisions between molecules moving with random velocities give new random velocities. Furthermore, if the velocities of gas molecules in a container are initially not random and isotropic, molecular collisions are what make them random and isotropic.

We make still further assumptions that simplify the calculations but do not affect the result. First, we let the container be a rectangular box. Second, we begin by considering *monatomic* gases, those whose molecules consist of single atoms, such as helium. Then, we can assume that the atoms have no energy except their translational kinetic energy; for instance, they have neither rotational nor vibrational energy. (Later, we discuss the validity of this assumption for real monatomic gases and dispense with it to consider diatomic and polyatomic gases.)

Figure 2.9 shows a collision of a gas molecule with the wall of a container, so that it exerts a force on the wall (by Newton's third law). These collisions are the source of pressure in a gas. As the number of molecules increases, the number of collisions, and thus the pressure, increases. Similarly, if the average velocity of the molecules is higher, the gas pressure is higher.

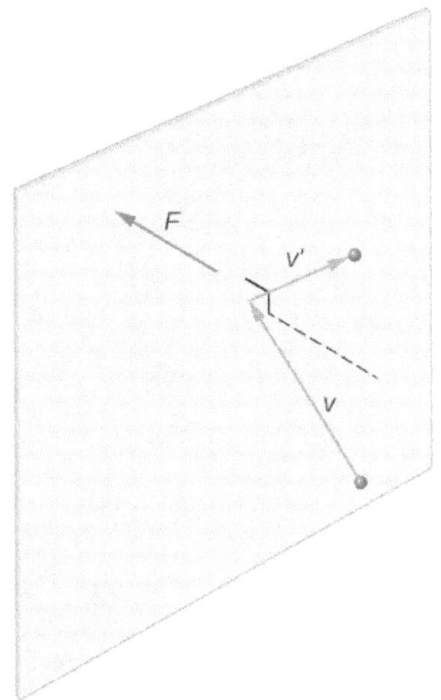

Figure 2.9 When a molecule collides with a rigid wall, the component of its momentum perpendicular to the wall is reversed. A force is thus exerted on the wall, creating pressure.

In a sample of gas in a container, the randomness of the molecular motion causes the number of collisions of molecules with any part of the wall in a given time to fluctuate. However, because a huge number of molecules collide with the wall in a short time, the number of collisions on the scales of time and space we measure fluctuates by only a tiny, usually unobservable fraction from the average. We can compare this situation to that of a casino, where the outcomes of the bets are random and the casino's takings fluctuate by the minute and the hour. However, over long times such as a year, the casino's takings are very close to the averages expected from the odds. A tank of gas has enormously more molecules than a casino has bettors in a year, and the molecules make enormously more collisions in a second than a casino has bets.

A calculation of the average force exerted by molecules on the walls of the box leads us to the ideal gas law and to the connection between temperature and molecular kinetic energy. (In fact, we will take two averages: one over time to get the average force exerted by one molecule with a given velocity, and then another average over molecules with different velocities.) This approach was developed by Daniel Bernoulli (1700–1782), who is best known in physics for his work on fluid flow (hydrodynamics). Remarkably, Bernoulli did this work before Dalton established the view of matter as consisting of atoms.

Figure 2.10 shows a container full of gas and an expanded view of an elastic collision of a gas molecule with a wall of the container, broken down into components. We have assumed that a molecule is small compared with the separation of molecules in the gas, and that its interaction with other molecules can be ignored. Under these conditions, the ideal gas law is experimentally valid. Because we have also assumed the wall is rigid and the particles are points, the collision is elastic (by conservation of energy—there's nowhere for a particle's kinetic energy to go). Therefore, the molecule's kinetic energy remains constant, and hence, its speed and the magnitude of its momentum remain constant as well. This assumption is not always valid, but the results in the rest of this module are also obtained in models that let the molecules exchange energy and momentum with the wall.

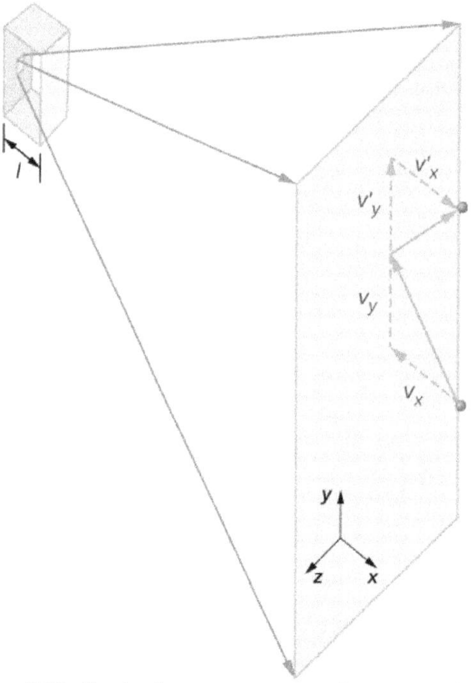

Figure 2.10 Gas in a box exerts an outward pressure on its walls. A molecule colliding with a rigid wall has its velocity and momentum in the *x*-direction reversed. This direction is perpendicular to the wall. The components of its velocity momentum in the *y*- and *z*-directions are not changed, which means there is no force parallel to the wall.

If the molecule's velocity changes in the *x*-direction, its momentum changes from $-mv_x$ to $+mv_x$. Thus, its change in momentum is $\Delta mv = +mv_x - (-mv_x) = 2mv_x$. According to the impulse-momentum theorem given in the chapter on linear momentum and collisions, the force exerted on the *i*th molecule, where *i* labels the molecules from 1 to *N*, is given by

$$F_i = \frac{\Delta p_i}{\Delta t} = \frac{2mv_{ix}}{\Delta t}.$$

(In this equation alone, *p* represents momentum, not pressure.) There is no force between the wall and the molecule except while the molecule is touching the wall. During the short time of the collision, the force between the molecule and wall is relatively large, but that is not the force we are looking for. We are looking for the average force, so we take Δt to be the average time between collisions of the given molecule with this wall, which is the time in which we expect to find one collision. Let *l* represent the length of the box in the *x*-direction. Then Δt is the time the molecule would take to go across the box and back, a distance 2*l*, at a speed of v_x. Thus $\Delta t = 2l/v_x$, and the expression for the force becomes

$$F_i = \frac{2mv_{ix}}{2l/v_{ix}} = \frac{mv_{ix}^2}{l}.$$

This force is due to *one* molecule. To find the total force on the wall, *F*, we need to add the contributions of all *N* molecules:

$$F = \sum_{i=1}^{N} F_i = \sum_{i=1}^{N} \frac{mv_{ix}^2}{l} = \frac{m}{l}\sum_{i=1}^{N} v_{ix}^2.$$

We now use the definition of the average, which we denote with a bar, to find the force:

$$F = N\frac{m}{l}\left(\frac{1}{N}\sum_{i=1}^{N} v_{ix}^2\right) = N\frac{m\overline{v_x^2}}{l}.$$

We want the force in terms of the speed v, rather than the x-component of the velocity. Note that the total velocity squared is the sum of the squares of its components, so that

$$\overline{v^2} = \overline{v_x^2} + \overline{v_y^2} + \overline{v_z^2}.$$

With the assumption of isotropy, the three averages on the right side are equal, so

$$\overline{v^2} = 3\overline{v_{ix}^2}.$$

Substituting this into the expression for F gives

$$F = N\frac{m\overline{v^2}}{3l}.$$

The pressure is F/A, so we obtain

$$p = \frac{F}{A} = N\frac{m\overline{v^2}}{3Al} = \frac{Nm\overline{v^2}}{3V},$$

where we used $V = Al$ for the volume. This gives the important result

$$pV = \frac{1}{3}Nm\overline{v^2}. \tag{2.5}$$

Combining this equation with $pV = Nk_B T$ gives

$$\frac{1}{3}Nm\overline{v^2} = Nk_B T.$$

We can get the average kinetic energy of a molecule, $\frac{1}{2}m\overline{v^2}$, from the left-hand side of the equation by dividing out N and multiplying by 3/2.

Average Kinetic Energy per Molecule

The average kinetic energy of a molecule is directly proportional to its absolute temperature:

$$\overline{K} = \frac{1}{2}m\overline{v^2} = \frac{3}{2}k_B T. \tag{2.6}$$

The equation $\overline{K} = \frac{3}{2}k_B T$ is the average kinetic energy per molecule. Note in particular that nothing in this equation depends on the molecular mass (or any other property) of the gas, the pressure, or anything but the temperature. If samples of helium and xenon gas, with very different molecular masses, are at the same temperature, the molecules have the same average kinetic energy.

The **internal energy** of a thermodynamic system is the sum of the mechanical energies of all of the molecules in it. We can now give an equation for the internal energy of a monatomic ideal gas. In such a gas, the molecules' only energy is their translational kinetic energy. Therefore, denoting the internal energy by E_{int}, we simply have $E_{int} = N\overline{K}$, or

$$E_{int} = \frac{3}{2}Nk_B T. \tag{2.7}$$

Often we would like to use this equation in terms of moles:

$$E_{int} = \frac{3}{2}nRT.$$

We can solve $\bar{K} = \frac{1}{2}m\bar{v^2} = \frac{3}{2}k_B T$ for a typical speed of a molecule in an ideal gas in terms of temperature to determine what is known as the *root-mean-square (rms) speed* of a molecule.

RMS Speed of a Molecule

The **root-mean-square (rms) speed** of a molecule, or the square root of the average of the square of the speed $\bar{v^2}$, is

$$v_{rms} = \sqrt{\bar{v^2}} = \sqrt{\frac{3k_B T}{m}}. \quad (2.8)$$

The rms speed is not the average or the most likely speed of molecules, as we will see in Distribution of Molecular Speeds, but it provides an easily calculated estimate of the molecules' speed that is related to their kinetic energy. Again we can write this equation in terms of the gas constant R and the molar mass M in kg/mol:

$$v_{rms} = \sqrt{\frac{3RT}{M}}. \quad (2.9)$$

We digress for a moment to answer a question that may have occurred to you: When we apply the model to atoms instead of theoretical point particles, does rotational kinetic energy change our results? To answer this question, we have to appeal to quantum mechanics. In quantum mechanics, rotational kinetic energy cannot take on just any value; it's limited to a discrete set of values, and the smallest value is inversely proportional to the rotational inertia. The rotational inertia of an atom is tiny because almost all of its mass is in the nucleus, which typically has a radius less than 10^{-14} m. Thus the minimum rotational energy of an atom is much more than $\frac{1}{2}k_B T$ for any attainable temperature, and the energy available is not enough to make an atom rotate. We will return to this point when discussing diatomic and polyatomic gases in the next section.

Example 2.4

Calculating Kinetic Energy and Speed of a Gas Molecule

(a) What is the average kinetic energy of a gas molecule at $20.0\,°C$ (room temperature)? (b) Find the rms speed of a nitrogen molecule (N_2) at this temperature.

Strategy

(a) The known in the equation for the average kinetic energy is the temperature:

$$\bar{K} = \frac{1}{2}m\bar{v^2} = \frac{3}{2}k_B T.$$

Before substituting values into this equation, we must convert the given temperature into kelvin: $T = (20.0 + 273)\text{ K} = 293\text{ K}$. We can find the rms speed of a nitrogen molecule by using the equation

$$v_{rms} = \sqrt{\bar{v^2}} = \sqrt{\frac{3k_B T}{m}},$$

but we must first find the mass of a nitrogen molecule. Obtaining the molar mass of nitrogen N_2 from the periodic table, we find

$$m = \frac{M}{N_A} = \frac{2\,(14.0067) \times 10^{-3}\text{ kg/mol})}{6.02 \times 10^{23}\text{ mol}^{-1}} = 4.65 \times 10^{-26}\text{ kg}.$$

Solution

a. The temperature alone is sufficient for us to find the average translational kinetic energy. Substituting the temperature into the translational kinetic energy equation gives

$$\bar{K} = \frac{3}{2}k_B T = \frac{3}{2}(1.38 \times 10^{-23}\ \text{J/K})(293\ \text{K}) = 6.07 \times 10^{-21}\ \text{J}.$$

b. Substituting this mass and the value for k_B into the equation for v_{rms} yields

$$v_{rms} = \sqrt{\frac{3k_B T}{m}} = \sqrt{\frac{3(1.38 \times 10^{-23}\ \text{J/K})(293\ \text{K})}{4.65 \times 10^{-26}\ \text{kg}}} = 511\ \text{m/s}.$$

Significance

Note that the average kinetic energy of the molecule is independent of the type of molecule. The average translational kinetic energy depends only on absolute temperature. The kinetic energy is very small compared to macroscopic energies, so that we do not feel when an air molecule is hitting our skin. On the other hand, it is much greater than the typical difference in gravitational potential energy when a molecule moves from, say, the top to the bottom of a room, so our neglect of gravitation is justified in typical real-world situations. The rms speed of the nitrogen molecule is surprisingly large. These large molecular velocities do not yield macroscopic movement of air, since the molecules move in all directions with equal likelihood. The *mean free path* (the distance a molecule moves on average between collisions, discussed a bit later in this section) of molecules in air is very small, so the molecules move rapidly but do not get very far in a second. The high value for rms speed is reflected in the speed of sound, which is about 340 m/s at room temperature. The higher the rms speed of air molecules, the faster sound vibrations can be transferred through the air. The speed of sound increases with temperature and is greater in gases with small molecular masses, such as helium (see Figure 2.11).

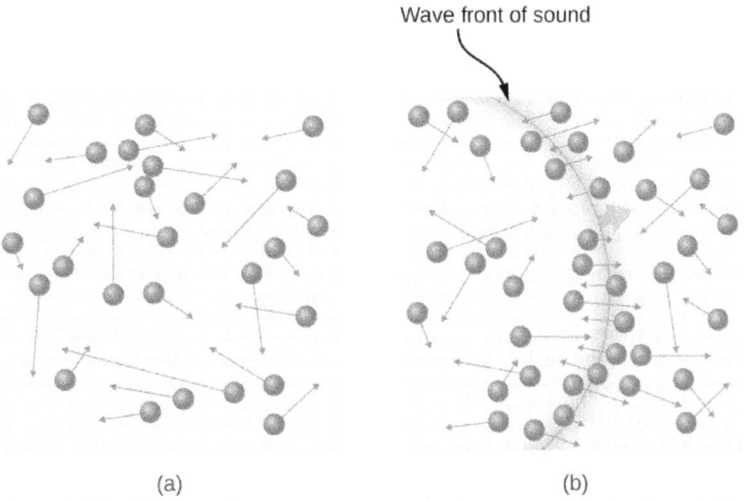

Figure 2.11 (a) In an ordinary gas, so many molecules move so fast that they collide billions of times every second. (b) Individual molecules do not move very far in a small amount of time, but disturbances like sound waves are transmitted at speeds related to the molecular speeds.

Example 2.5

Calculating Temperature: Escape Velocity of Helium Atoms

To escape Earth's gravity, an object near the top of the atmosphere (at an altitude of 100 km) must travel away from Earth at 11.1 km/s. This speed is called the *escape velocity*. At what temperature would helium atoms have an rms speed equal to the escape velocity?

Strategy

Identify the knowns and unknowns and determine which equations to use to solve the problem.

Solution

1. Identify the knowns: v is the escape velocity, 11.1 km/s.

2. Identify the unknowns: We need to solve for temperature, T. We also need to solve for the mass m of the helium atom.
3. Determine which equations are needed.
    - To get the mass m of the helium atom, we can use information from the periodic table:
    $$m = \frac{M}{N_A}.$$
    - To solve for temperature T, we can rearrange
    $$\frac{1}{2}m\overline{v^2} = \frac{3}{2}k_B T$$
    to yield
    $$T = \frac{m\overline{v^2}}{3k_B}.$$
4. Substitute the known values into the equations and solve for the unknowns,
$$m = \frac{M}{N_A} = \frac{4.0026 \times 10^{-3}\ \text{kg/mol}}{6.02 \times 10^{23}\ \text{mol}} = 6.65 \times 10^{-27}\ \text{kg}$$
and
$$T = \frac{(6.65 \times 10^{-27}\ \text{kg})\left(11.1 \times 10^3\ \text{m/s}\right)^2}{3\,(1.38 \times 10^{-23}\ \text{J/K})} = 1.98 \times 10^4\ \text{K}.$$

Significance

This temperature is much higher than atmospheric temperature, which is approximately 250 K ($-25\ °\text{C}$ or $-10\ °\text{F}$) at high elevation. Very few helium atoms are left in the atmosphere, but many were present when the atmosphere was formed, and more are always being created by radioactive decay (see the chapter on nuclear physics). The reason for the loss of helium atoms is that a small number of helium atoms have speeds higher than Earth's escape velocity even at normal temperatures. The speed of a helium atom changes from one collision to the next, so that at any instant, there is a small but nonzero chance that the atom's speed is greater than the escape velocity. The chance is high enough that over the lifetime of Earth, almost all the helium atoms that have been in the atmosphere have reached escape velocity at high altitudes and escaped from Earth's gravitational pull. Heavier molecules, such as oxygen, nitrogen, and water, have smaller rms speeds, and so it is much less likely that any of them will have speeds greater than the escape velocity. In fact, the likelihood is so small that billions of years are required to lose significant amounts of heavier molecules from the atmosphere. Figure 2.12 shows the effect of a lack of an atmosphere on the Moon. Because the gravitational pull of the Moon is much weaker, it has lost almost its entire atmosphere. The atmospheres of Earth and other bodies are compared in this chapter's exercises.

Figure 2.12 This photograph of Apollo 17 Commander Eugene Cernan driving the lunar rover on the Moon in 1972 looks as though it was taken at night with a large spotlight. In fact, the light is coming from the Sun. Because the acceleration due to gravity on the Moon is so low (about 1/6 that of Earth), the Moon's escape velocity is much smaller. As a result, gas molecules escape very easily from the Moon, leaving it with virtually no atmosphere. Even during the daytime, the sky is black because there is no gas to scatter sunlight. (credit: Harrison H. Schmitt/NASA)

2.4 Check Your Understanding If you consider a very small object, such as a grain of pollen, in a gas, then the number of molecules striking its surface would also be relatively small. Would you expect the grain of pollen to experience any fluctuations in pressure due to statistical fluctuations in the number of gas molecules striking it in a given amount of time?

Vapor Pressure, Partial Pressure, and Dalton's Law

The pressure a gas would create if it occupied the total volume available is called the gas's **partial pressure**. If two or more gases are mixed, they will come to thermal equilibrium as a result of collisions between molecules; the process is analogous to heat conduction as described in the chapter on temperature and heat. As we have seen from kinetic theory, when the gases have the same temperature, their molecules have the same average kinetic energy. Thus, each gas obeys the ideal gas law separately and exerts the same pressure on the walls of a container that it would if it were alone. Therefore, in a mixture of gases, *the total pressure is the sum of partial pressures of the component gases*, assuming ideal gas behavior and no chemical reactions between the components. This law is known as **Dalton's law of partial pressures**, after the English scientist John Dalton (1766–1844) who proposed it. Dalton's law is consistent with the fact that pressures add according to Pascal's principle.

In a mixture of ideal gases in thermal equilibrium, the number of molecules of each gas is proportional to its partial pressure. This result follows from applying the ideal gas law to each in the form $p/n = RT/V$. Because the right-hand side is the same for any gas at a given temperature in a container of a given volume, the left-hand side is the same as well.

- Partial pressure is the pressure a gas would create if it existed alone.
- Dalton's law states that the total pressure is the sum of the partial pressures of all of the gases present.
- For any two gases (labeled 1 and 2) in equilibrium in a container, $\frac{p_1}{n_1} = \frac{p_2}{n_2}$.

An important application of partial pressure is that, in chemistry, it functions as the concentration of a gas in determining the rate of a reaction. Here, we mention only that the partial pressure of oxygen in a person's lungs is crucial to life and health. Breathing air that has a partial pressure of oxygen below 0.16 atm can impair coordination and judgment, particularly in people not acclimated to a high elevation. Lower partial pressures of O_2 have more serious effects; partial pressures below 0.06 atm can be quickly fatal, and permanent damage is likely even if the person is rescued. However, the sensation of needing to breathe, as when holding one's breath, is caused much more by high concentrations of carbon dioxide in the blood than by low concentrations of oxygen. Thus, if a small room or closet is filled with air having a low concentration of oxygen, perhaps because a leaking cylinder of some compressed gas is stored there, a person will not feel any "choking" sensation and may go into convulsions or lose consciousness without noticing anything wrong. Safety engineers give considerable attention to this danger.

Another important application of partial pressure is **vapor pressure**, which is the partial pressure of a vapor at which it is in equilibrium with the liquid (or solid, in the case of sublimation) phase of the same substance. At any temperature, the partial pressure of the water in the air cannot exceed the vapor pressure of the water at that temperature, because whenever the partial pressure reaches the vapor pressure, water condenses out of the air. Dew is an example of this condensation. The temperature at which condensation occurs for a sample of air is called the *dew point*. It is easily measured by slowly cooling a metal ball; the dew point is the temperature at which condensation first appears on the ball.

The vapor pressures of water at some temperatures of interest for meteorology are given in Table 2.2.

T (°C)	Vapor Pressure (Pa)
0	610.5
3	757.9
5	872.3
8	1073
10	1228
13	1497
15	1705
18	2063
20	2338
23	2809
25	3167
30	4243
35	5623
40	7376

Table 2.2 Vapor Pressure of Water at Various Temperatures

The *relative humidity* (R.H.) at a temperature *T* is defined by

$$\text{R.H.} = \frac{\text{Partial pressure of water vapor at } T}{\text{Vapor pressure of water at } T} \times 100\%.$$

A relative humidity of 100% means that the partial pressure of water is equal to the vapor pressure; in other words, the air is saturated with water.

Example 2.6

Calculating Relative Humidity

What is the relative humidity when the air temperature is $25\,°\text{C}$ and the dew point is $15\,°\text{C}$?

Strategy

We simply look up the vapor pressure at the given temperature and that at the dew point and find the ratio.

Solution

$$\text{R.H.} = \frac{\text{Partial pressure of water vapor at } 15\ ^\circ\text{C}}{\text{Partial pressure of water vapor at } 25\ ^\circ\text{C}} \times 100\% = \frac{1705\ \text{Pa}}{3167\ \text{Pa}} \times 100\% = 53.8\%.$$

Significance

R.H. is important to our comfort. The value of 53.8% is within the range of 40% to 60% recommended for comfort indoors.

As noted in the chapter on temperature and heat, the temperature seldom falls below the dew point, because when it reaches the dew point or frost point, water condenses and releases a relatively large amount of latent heat of vaporization.

Mean Free Path and Mean Free Time

We now consider collisions explicitly. The usual first step (which is all we'll take) is to calculate the **mean free path**, λ, the average distance a molecule travels between collisions with other molecules, and the *mean free time* τ, the average time between the collisions of a molecule. If we assume all the molecules are spheres with a radius *r*, then a molecule will collide with another if their centers are within a distance 2*r* of each other. For a given particle, we say that the area of a circle with that radius, $4\pi r^2$, is the "cross-section" for collisions. As the particle moves, it traces a cylinder with that cross-sectional area. The mean free path is the length λ such that the expected number of other molecules in a cylinder of length λ and cross-section $4\pi r^2$ is 1. If we temporarily ignore the motion of the molecules other than the one we're looking at, the expected number is the number density of molecules, *N*/*V*, times the volume, and the volume is $4\pi r^2 \lambda$, so we have $(N/V)4\pi r^2 \lambda = 1$, or

$$\lambda = \frac{V}{4\pi r^2 N}.$$

Taking the motion of all the molecules into account makes the calculation much harder, but the only change is a factor of $\sqrt{2}$. The result is

$$\lambda = \frac{V}{4\sqrt{2}\pi r^2 N}. \tag{2.10}$$

In an ideal gas, we can substitute $V/N = k_B T/p$ to obtain

$$\lambda = \frac{k_B T}{4\sqrt{2}\pi r^2 p}. \tag{2.11}$$

The **mean free time** τ is simply the mean free path divided by a typical speed, and the usual choice is the rms speed. Then

$$\tau = \frac{k_B T}{4\sqrt{2}\pi r^2 p v_{\text{rms}}}. \tag{2.12}$$

Example 2.7

Calculating Mean Free Time

Find the mean free time for argon atoms $(M = 39.9\ \text{g/mol})$ at a temperature of $0\ °\text{C}$ and a pressure of 1.00 atm. Take the radius of an argon atom to be 1.70×10^{-10} m.

Solution

1. Identify the knowns and convert into SI units. We know the molar mass is 0.0399 kg/mol, the temperature is 273 K, the pressure is 1.01×10^5 Pa, and the radius is 1.70×10^{-10} m.

2. Find the rms speed: $v_{\text{rms}} = \sqrt{\frac{3RT}{M}} = 413\,\frac{\text{m}}{\text{s}}$.

3. Substitute into the equation for the mean free time:

$$\tau = \frac{k_B T}{4\sqrt{2}\pi r^2 p v_{\text{rms}}} = \frac{(1.38 \times 10^{-23}\ \text{J/K})(273\ \text{K})}{4\sqrt{2}\pi(1.70 \times 10^{-10}\ \text{m})^2(1.01 \times 10^5\ \text{Pa})(413\ \text{m/s})} = 1.76 \times 10^{-10}\ \text{s}.$$

Significance

We can hardly compare this result with our intuition about gas molecules, but it gives us a picture of molecules colliding with extremely high frequency.

2.5 Check Your Understanding Which has a longer mean free path, liquid water or water vapor in the air?

2.3 | Heat Capacity and Equipartition of Energy

Learning Objectives

By the end of this section, you will be able to:

- Solve problems involving heat transfer to and from ideal monatomic gases whose volumes are held constant
- Solve similar problems for non-monatomic ideal gases based on the number of degrees of freedom of a molecule
- Estimate the heat capacities of metals using a model based on degrees of freedom

In the chapter on temperature and heat, we defined the specific heat capacity with the equation $Q = mc\Delta T$, or $c = (1/m)Q/\Delta T$. However, the properties of an ideal gas depend directly on the number of moles in a sample, so here we define specific heat capacity in terms of the number of moles, not the mass. Furthermore, when talking about solids and liquids, we ignored any changes in volume and pressure with changes in temperature—a good approximation for solids and liquids, but for gases, we have to make some condition on volume or pressure changes. Here, we focus on the heat capacity with the volume held constant. We can calculate it for an ideal gas.

Heat Capacity of an Ideal Monatomic Gas at Constant Volume

We define the *molar heat capacity at constant volume* C_V as

$$C_V = \frac{1}{n}\frac{Q}{\Delta T}, \text{ with } V \text{ held constant.}$$

This is often expressed in the form

$$Q = nC_V \Delta T. \quad (2.13)$$

If the volume does not change, there is no overall displacement, so no work is done, and the only change in internal energy is due to the heat flow $\Delta E_{int} = Q$. (This statement is discussed further in the next chapter.) We use the equation $E_{int} = 3nRT/2$ to write $\Delta E_{int} = 3nR\Delta T/2$ and substitute ΔE for Q to find $Q = 3nR\Delta T/2$, which gives the following simple result for an ideal monatomic gas:

$$C_V = \frac{3}{2}R.$$

It is independent of temperature, which justifies our use of finite differences instead of a derivative. This formula agrees well with experimental results.

In the next chapter we discuss the molar specific heat at constant pressure C_p, which is always greater than C_V.

Example 2.8

Calculating Temperature

A sample of 0.125 kg of xenon is contained in a rigid metal cylinder, big enough that the xenon can be modeled as an ideal gas, at a temperature of $20.0\ °C$. The cylinder is moved outside on a hot summer day. As the xenon comes into equilibrium by reaching the temperature of its surroundings, 180 J of heat are conducted to it through the cylinder walls. What is the equilibrium temperature? Ignore the expansion of the metal cylinder.

Solution

1. Identify the knowns: We know the initial temperature T_1 is $20.0\ °C$, the heat Q is 180 J, and the mass m of the xenon is 0.125 kg.
2. Identify the unknown. We need the final temperature, so we'll need ΔT.
3. Determine which equations are needed. Because xenon gas is monatomic, we can use $Q = 3nR\Delta T/2$. Then we need the number of moles, $n = m/M$.
4. Substitute the known values into the equations and solve for the unknowns.
   The molar mass of xenon is 131.3 g, so we obtain

$$n = \frac{125\ \text{g}}{131.3\ \text{g/mol}} = 0.952\ \text{mol},$$

$$\Delta T = \frac{2Q}{3nR} = \frac{2(180\ \text{J})}{3(0.952\ \text{mol})(8.31\ \text{J/mol} \cdot °\text{C})} = 15.2\ °\text{C}.$$

Therefore, the final temperature is $35.2\ °C$. The problem could equally well be solved in kelvin; as a kelvin is the same size as a degree Celsius of temperature change, you would get $\Delta T = 15.2$ K.

Significance

The heating of an ideal or almost ideal gas at constant volume is important in car engines and many other practical systems.

2.6 Check Your Understanding Suppose 2 moles of helium gas at 200 K are mixed with 2 moles of krypton gas at 400 K in a calorimeter. What is the final temperature?

We would like to generalize our results to ideal gases with more than one atom per molecule. In such systems, the molecules can have other forms of energy beside translational kinetic energy, such as rotational kinetic energy and vibrational kinetic and potential energies. We will see that a simple rule lets us determine the average energies present in these forms and solve problems in much the same way as we have for monatomic gases.

Degrees of Freedom

In the previous section, we found that $\frac{1}{2}m\overline{v^2} = \frac{3}{2}k_B T$ and $\overline{v^2} = 3\overline{v_x^2}$, from which it follows that $\frac{1}{2}m\overline{v_x^2} = \frac{1}{2}k_B T$. The same equation holds for $\overline{v_y^2}$ and for $\overline{v_z^2}$. Thus, we can look at our energy of $\frac{3}{2}k_B T$ as the sum of contributions of $\frac{1}{2}k_B T$ from each of the three dimensions of translational motion. Shifting to the gas as a whole, we see that the 3 in the formula $C_V = \frac{3}{2}R$ also reflects those three dimensions. We define a **degree of freedom** as an independent possible motion of a molecule, such as each of the three dimensions of translation. Then, letting *d* represent the number of degrees of freedom, the molar heat capacity at constant volume of a monatomic ideal gas is $C_V = \frac{d}{2}R,$ where $d = 3$.

The branch of physics called *statistical mechanics* tells us, and experiment confirms, that C_V of any ideal gas is given by this equation, regardless of the number of degrees of freedom. This fact follows from a more general result, the **equipartition theorem**, which holds in classical (non-quantum) thermodynamics for systems in thermal equilibrium under technical conditions that are beyond our scope. Here, we mention only that in a system, the energy is shared among the degrees of freedom by collisions.

Equipartition Theorem

The energy of a thermodynamic system in equilibrium is partitioned equally among its degrees of freedom. Accordingly, the molar heat capacity of an ideal gas is proportional to its number of degrees of freedom, *d*:

$$C_V = \frac{d}{2}R. \tag{2.14}$$

This result is due to the Scottish physicist James Clerk Maxwell (1831−1871), whose name will appear several more times in this book.

For example, consider a diatomic ideal gas (a good model for nitrogen, N_2, and oxygen, O_2). Such a gas has more degrees of freedom than a monatomic gas. In addition to the three degrees of freedom for translation, it has two degrees of freedom for rotation perpendicular to its axis. Furthermore, the molecule can vibrate along its axis. This motion is often modeled by imagining a spring connecting the two atoms, and we know from simple harmonic motion that such motion has both kinetic and potential energy. Each of these forms of energy corresponds to a degree of freedom, giving two more.

We might expect that for a diatomic gas, we should use 7 as the number of degrees of freedom; classically, if the molecules of a gas had only translational kinetic energy, collisions between molecules would soon make them rotate and vibrate. However, as explained in the previous module, quantum mechanics controls which degrees of freedom are active. The result is shown in Figure 2.13. Both rotational and vibrational energies are limited to discrete values. For temperatures below about 60 K, the energies of hydrogen molecules are too low for a collision to bring the rotational state or vibrational state of a molecule from the lowest energy to the second lowest, so the only form of energy is translational kinetic energy, and $d = 3$ or $C_V = 3R/2$ as in a monatomic gas. Above that temperature, the two rotational degrees of freedom begin to contribute, that is, some molecules are excited to the rotational state with the second-lowest energy. (This temperature is much lower than that where rotations of monatomic gases contribute, because diatomic molecules have much higher rotational inertias and hence much lower rotational energies.) From about room temperature (a bit less than 300 K) to about 600 K, the rotational degrees of freedom are fully active, but the vibrational ones are not, and $d = 5$. Then, finally, above about 3000 K, the vibrational degrees of freedom are fully active, and $d = 7$ as the classical theory predicted.

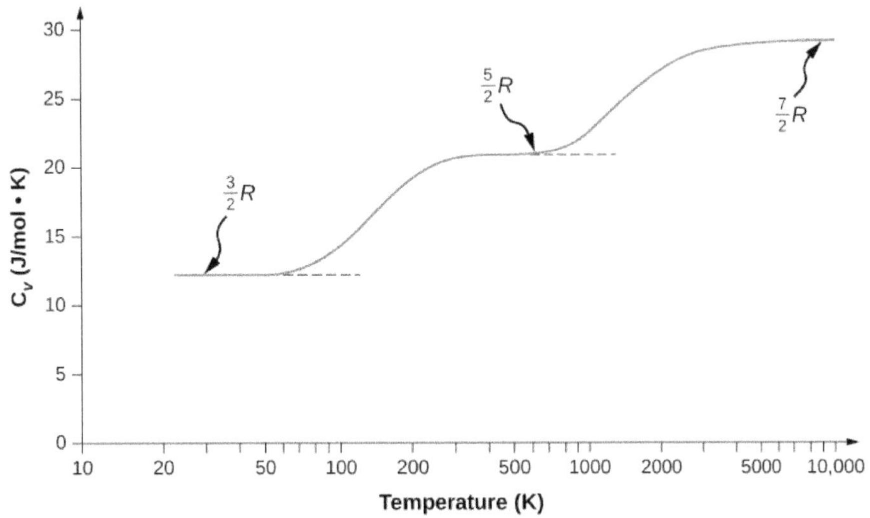

Figure 2.13 The molar heat capacity of hydrogen as a function of temperature (on a logarithmic scale). The three "steps" or "plateaus" show different numbers of degrees of freedom that the typical energies of molecules must achieve to activate. Translational kinetic energy corresponds to three degrees of freedom, rotational to another two, and vibrational to yet another two.

Polyatomic molecules typically have one additional rotational degree of freedom at room temperature, since they have comparable moments of inertia around any axis. Thus, at room temperature, they have $d = 6,$ and at high temperature, $d = 8.$ We usually assume that gases have the theoretical room-temperature values of d.

As shown in Table 2.3, the results agree well with experiments for many monatomic and diatomic gases, but the agreement for triatomic gases is only fair. The differences arise from interactions that we have ignored between and within molecules.

Gas	C_V/R at $25\,°\text{C}$ and 1 atm
Ar	1.50
He	1.50
Ne	1.50
CO	2.50
H_2	2.47
N_2	2.50
O_2	2.53
F_2	2.8
CO_2	3.48
H_2S	3.13
N_2O	3.66

Table 2.3 C_V/R for Various Monatomic, Diatomic, and Triatomic Gases

What about internal energy for diatomic and polyatomic gases? For such gases, C_V is a function of temperature (Figure 2.13), so we do not have the kind of simple result we have for monatomic ideal gases.

Molar Heat Capacity of Solid Elements

The idea of equipartition leads to an estimate of the molar heat capacity of solid elements at ordinary temperatures. We can model the atoms of a solid as attached to neighboring atoms by springs (Figure 2.14).

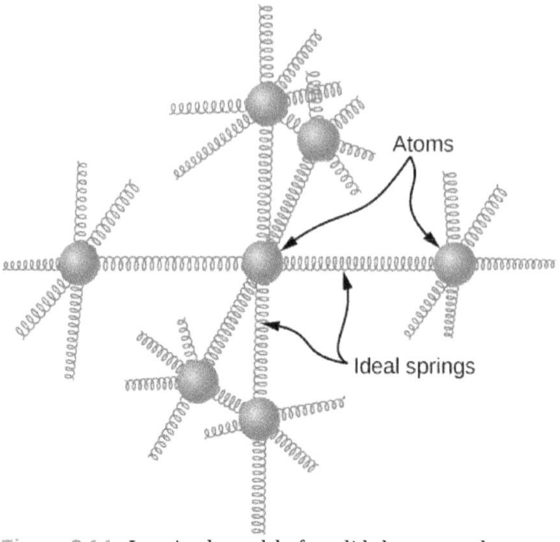

Figure 2.14 In a simple model of a solid element, each atom is attached to others by six springs, two for each possible motion: x, y, and z. Each of the three motions corresponds to two degrees of freedom, one for kinetic energy and one for potential energy. Thus $d = 6$.

Analogously to the discussion of vibration in the previous module, each atom has six degrees of freedom: one kinetic and one potential for each of the x-, y-, and z-directions. Accordingly, the molar specific heat of a metal should be $3R$. This result, known as the Law of Dulong and Petit, works fairly well experimentally at room temperature. (For every element, it fails at low temperatures for quantum-mechanical reasons. Since quantum effects are particularly important for low-mass particles, the Law of Dulong and Petit already fails at room temperature for some light elements, such as beryllium and carbon. It also fails for some heavier elements for various reasons beyond what we can cover.)

Problem-Solving Strategy: Heat Capacity and Equipartition

The strategy for solving these problems is the same as the one in Phase Changes for the effects of heat transfer. The only new feature is that you should determine whether the case just presented—ideal gases at constant volume—applies to the problem. (For solid elements, looking up the specific heat capacity is generally better than estimating it from the Law of Dulong and Petit.) In the case of an ideal gas, determine the number d of degrees of freedom from the number of atoms in the gas molecule and use it to calculate C_V (or use C_V to solve for d).

Example 2.9

Calculating Temperature: Calorimetry with an Ideal Gas

A 300-g piece of solid gallium (a metal used in semiconductor devices) at its melting point of only $30.0\,°\text{C}$ is in contact with 12.0 moles of air (assumed diatomic) at $95.0\,°\text{C}$ in an insulated container. When the air reaches equilibrium with the gallium, 202 g of the gallium have melted. Based on those data, what is the heat of fusion of gallium? Assume the volume of the air does not change and there are no other heat transfers.

Strategy

We'll use the equation $Q_{hot} + Q_{cold} = 0$. As some of the gallium doesn't melt, we know the final temperature is still the melting point. Then the only Q_{hot} is the heat lost as the air cools, $Q_{hot} = n_{air} C_V \Delta T$, where $C_V = 5R/2$. The only Q_{cold} is the latent heat of fusion of the gallium, $Q_{cold} = m_{Ga} L_f$. It is positive because heat flows into the gallium.

Solution

1. Set up the equation:

$$n_{air} C_V \Delta T + m_{Ga} L_f = 0.$$

2. Substitute the known values and solve:

$$(12.0 \text{ mol})\left(\frac{5}{2}\right)\left(8.31\frac{\text{J}}{\text{mol} \cdot {}^\circ\text{C}}\right)(30.0\,{}^\circ\text{C} - 95.0\,{}^\circ\text{C}) + (0.202 \text{ kg})L_f = 0.$$

We solve to find that the heat of fusion of gallium is 80.2 kJ/kg.

2.4 | Distribution of Molecular Speeds

Learning Objectives

By the end of this section, you will be able to:

- Describe the distribution of molecular speeds in an ideal gas
- Find the average and most probable molecular speeds in an ideal gas

Particles in an ideal gas all travel at relatively high speeds, but they do not travel at the same speed. The rms speed is one kind of average, but many particles move faster and many move slower. The actual distribution of speeds has several interesting implications for other areas of physics, as we will see in later chapters.

The Maxwell-Boltzmann Distribution

The motion of molecules in a gas is random in magnitude and direction for individual molecules, but a gas of many molecules has a predictable distribution of molecular speeds. This predictable distribution of molecular speeds is known as the **Maxwell-Boltzmann distribution**, after its originators, who calculated it based on kinetic theory, and it has since been confirmed experimentally (Figure 2.15).

To understand this figure, we must define a distribution function of molecular speeds, since with a finite number of molecules, the probability that a molecule will have exactly a given speed is 0.

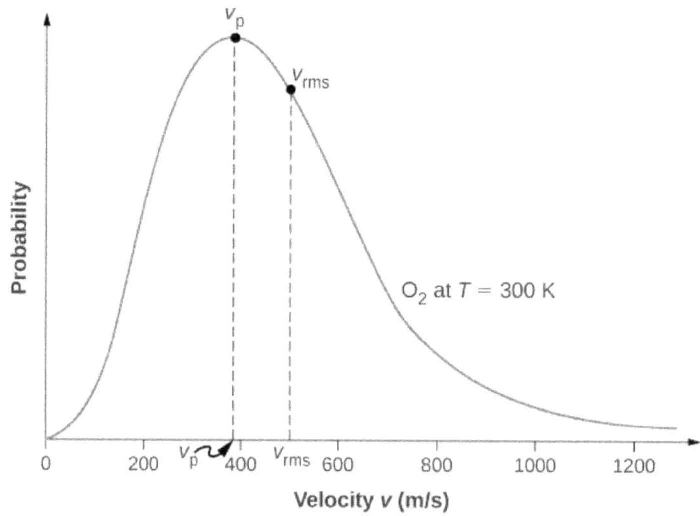

Figure 2.15 The Maxwell-Boltzmann distribution of molecular speeds in an ideal gas. The most likely speed v_p is less than the rms speed v_{rms}. Although very high speeds are possible, only a tiny fraction of the molecules have speeds that are an order of magnitude greater than v_{rms}.

We define the distribution function $f(v)$ by saying that the expected number $N(v_1, v_2)$ of particles with speeds between v_1 and v_2 is given by

$$N(v_1, v_2) = N\int_{v_1}^{v_2} f(v)dv.$$

[Since N is dimensionless, the unit of $f(v)$ is seconds per meter.] We can write this equation conveniently in differential form:

$$dN = Nf(v)dv.$$

In this form, we can understand the equation as saying that the number of molecules with speeds between v and $v + dv$ is the total number of molecules in the sample times $f(v)$ times dv. That is, the probability that a molecule's speed is between v and $v + dv$ is $f(v)dv$.

We can now quote Maxwell's result, although the proof is beyond our scope.

Maxwell-Boltzmann Distribution of Speeds

The distribution function for speeds of particles in an ideal gas at temperature T is

$$f(v) = \frac{4}{\sqrt{\pi}}\left(\frac{m}{2k_B T}\right)^{3/2} v^2 e^{-mv^2/2k_B T}. \tag{2.15}$$

The factors before the v^2 are a normalization constant; they make sure that $N(0, \infty) = N$ by making sure that $\int_0^{\infty} f(v)dv = 1$. Let's focus on the dependence on v. The factor of v^2 means that $f(0) = 0$ and for small v, the curve looks like a parabola. The factor of $e^{-m_0 v^2/2k_B T}$ means that $\lim_{v\to\infty} f(v) = 0$ and the graph has an exponential tail, which indicates that a few molecules may move at several times the rms speed. The interaction of these factors gives the function the single-peaked shape shown in the figure.

Example 2.10

Calculating the Ratio of Numbers of Molecules Near Given Speeds

In a sample of nitrogen (N_2, with a molar mass of 28.0 g/mol) at a temperature of $273\ °C$, find the ratio of the number of molecules with a speed very close to 300 m/s to the number with a speed very close to 100 m/s.

Strategy

Since we're looking at a small range, we can approximate the number of molecules near 100 m/s as $dN_{100} = f(100\text{ m/s})dv$. Then the ratio we want is

$$\frac{dN_{300}}{dN_{100}} = \frac{f(300\text{ m/s})dv}{f(100\text{ m/s})dv} = \frac{f(300\text{ m/s})}{f(100\text{ m/s})}.$$

All we have to do is take the ratio of the two f values.

Solution

1. Identify the knowns and convert to SI units if necessary.

$$T = 300\text{ K},\ k_B = 1.38 \times 10^{-23}\text{ J/K}$$

$$M = 0.0280\text{ kg/mol so } m = 4.65 \times 10^{-26}\text{ kg}$$

2. Substitute the values and solve.

$$\begin{aligned}\frac{f(300\text{ m/s})}{f(100\text{ m/s})} &= \frac{\frac{4}{\sqrt{\pi}}\left(\frac{m}{2k_BT}\right)^{3/2}(300\text{ m/s})^2\exp[-m(300\text{ m/s})^2/2k_BT]}{\frac{4}{\sqrt{\pi}}\left(\frac{m}{2k_BT}\right)^{3/2}(100\text{ m/s})^2\exp[-m(100\text{ m/s})^2/2k_BT]}\\ &= \frac{(300\text{ m/s})^2\exp[-(4.65\times10^{-26}\text{ kg})(300\text{ m/s})^2/2(1.38\times10^{-23}\text{ J/K})(300\text{ K})]}{(100\text{ m/s})^2\exp[-(4.65\times10^{-26}\text{ kg})(100\text{ m/s})^2/2(1.38\times10^{-23}\text{ J/K})(300\text{ K})]}\\ &= 3^2\exp\left[-\frac{(4.65\times10^{-26}\text{ kg})[(300\text{ m/s})^2-(100\text{ ms})^2]}{2(1.38\times10^{-23}\text{ J/K})(300\text{ K})}\right]\\ &= 5.74\end{aligned}$$

Figure 2.16 shows that the curve is shifted to higher speeds at higher temperatures, with a broader range of speeds.

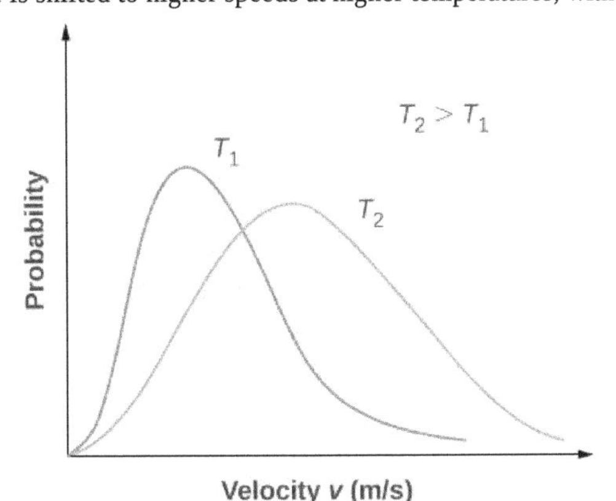

Figure 2.16 The Maxwell-Boltzmann distribution is shifted to higher speeds and broadened at higher temperatures.

With only a relatively small number of molecules, the distribution of speeds fluctuates around the Maxwell-Boltzmann distribution. However, you can view this simulation (https://openstaxcollege.org/l/21maxboltzdisim) to see the essential features that more massive molecules move slower and have a narrower distribution. Use the set-up "2 Gases, Random Speeds". Note the display at the bottom comparing histograms of the speed distributions with the theoretical curves.

We can use a probability distribution to calculate average values by multiplying the distribution function by the quantity to be averaged and integrating the product over all possible speeds. (This is analogous to calculating averages of discrete distributions, where you multiply each value by the number of times it occurs, add the results, and divide by the number of values. The integral is analogous to the first two steps, and the normalization is analogous to dividing by the number of values.) Thus the average velocity is

$$\bar{v} = \int_0^{\infty} v f(v) dv = \sqrt{\frac{8}{\pi}\frac{k_B T}{m}} = \sqrt{\frac{8}{\pi}\frac{RT}{M}}. \tag{2.16}$$

Similarly,

$$v_{\text{rms}} = \sqrt{\overline{v^2}} = \sqrt{\int_0^{\infty} v^2 f(v) dv} = \sqrt{\frac{3k_B T}{m}} = \sqrt{\frac{3RT}{M}}$$

as in Pressure, Temperature, and RMS Speed. The **most probable speed**, also called the **peak speed** v_p, is the speed at the peak of the velocity distribution. (In statistics it would be called the mode.) It is less than the rms speed v_{rms}. The most probable speed can be calculated by the more familiar method of setting the derivative of the distribution function, with respect to *v*, equal to 0. The result is

$$v_p = \sqrt{\frac{2k_B T}{m}} = \sqrt{\frac{2RT}{M}}, \tag{2.17}$$

which is less than v_{rms}. In fact, the rms speed is greater than both the most probable speed and the average speed.

The peak speed provides a sometimes more convenient way to write the Maxwell-Boltzmann distribution function:

$$f(v) = \frac{4v^2}{\sqrt{\pi} v_p^3} e^{-v^2/v_p^2} \tag{2.18}$$

In the factor $e^{-mv^2/2k_B T}$, it is easy to recognize the translational kinetic energy. Thus, that expression is equal to $e^{-K/k_B T}$. The distribution *f*(*v*) can be transformed into a kinetic energy distribution by requiring that $f(K)dK = f(v)dv$. Boltzmann showed that the resulting formula is much more generally applicable if we replace the kinetic energy of translation with the total mechanical energy *E*. Boltzmann's result is

$$f(E) = \frac{2}{\sqrt{\pi}}(k_B T)^{-3/2}\sqrt{E}e^{-E/k_B T} = \frac{2}{\sqrt{\pi}(k_B T)^{3/2}}\frac{\sqrt{E}}{e^{E/k_B T}}.$$

The first part of this equation, with the negative exponential, is the usual way to write it. We give the second part only to remark that $e^{E/k_B T}$ in the denominator is ubiquitous in quantum as well as classical statistical mechanics.

Problem-Solving Strategy: Speed Distribution

Step 1. Examine the situation to determine that it relates to the distribution of molecular speeds.

Step 2. Make a list of what quantities are given or can be inferred from the problem as stated (identify the known quantities).

Step 3. Identify exactly what needs to be determined in the problem (identify the unknown quantities). A written list is useful.

Step 4. Convert known values into proper SI units (K for temperature, Pa for pressure, m^3 for volume, molecules for *N*, and moles for *n*). In many cases, though, using *R* and the molar mass will be more convenient than using k_B and the molecular mass.

Step 5. Determine whether you need the distribution function for velocity or the one for energy, and whether you are using a formula for one of the characteristic speeds (average, most probably, or rms), finding a ratio of values of the distribution function, or approximating an integral.

Step 6. Solve the appropriate equation for the ideal gas law for the quantity to be determined (the unknown quantity). Note that if you are taking a ratio of values of the distribution function, the normalization factors divide out. Or if approximating an integral, use the method asked for in the problem.

Step 7. Substitute the known quantities, along with their units, into the appropriate equation and obtain numerical solutions complete with units.

We can now gain a qualitative understanding of a puzzle about the composition of Earth's atmosphere. Hydrogen is by far the most common element in the universe, and helium is by far the second-most common. Moreover, helium is constantly produced on Earth by radioactive decay. Why are those elements so rare in our atmosphere? The answer is that gas molecules that reach speeds above Earth's escape velocity, about 11 km/s, can escape from the atmosphere into space. Because of the lower mass of hydrogen and helium molecules, they move at higher speeds than other gas molecules, such as nitrogen and oxygen. Only a few exceed escape velocity, but far fewer heavier molecules do. Thus, over the billions of years that Earth has existed, far more hydrogen and helium molecules have escaped from the atmosphere than other molecules, and hardly any of either is now present.

We can also now take another look at evaporative cooling, which we discussed in the chapter on temperature and heat. Liquids, like gases, have a distribution of molecular energies. The highest-energy molecules are those that can escape from the intermolecular attractions of the liquid. Thus, when some liquid evaporates, the molecules left behind have a lower average energy, and the liquid has a lower temperature.

CHAPTER 2 REVIEW

KEY TERMS

Avogadro's number N_A, the number of molecules in one mole of a substance; $N_A = 6.02 \times 10^{23}$ particles/mole

Boltzmann constant k_B, a physical constant that relates energy to temperature and appears in the ideal gas law; $k_B = 1.38 \times 10^{-23}$ J/K

critical temperature T_c at which the isotherm has a point with zero slope

Dalton's law of partial pressures physical law that states that the total pressure of a gas is the sum of partial pressures of the component gases

degree of freedom independent kind of motion possessing energy, such as the kinetic energy of motion in one of the three orthogonal spatial directions

equipartition theorem theorem that the energy of a classical thermodynamic system is shared equally among its degrees of freedom

ideal gas gas at the limit of low density and high temperature

ideal gas law physical law that relates the pressure and volume of a gas, far from liquefaction, to the number of gas molecules or number of moles of gas and the temperature of the gas

internal energy sum of the mechanical energies of all of the molecules in it

kinetic theory of gases theory that derives the macroscopic properties of gases from the motion of the molecules they consist of

Maxwell-Boltzmann distribution function that can be integrated to give the probability of finding ideal gas molecules with speeds in the range between the limits of integration

mean free path average distance between collisions of a particle

mean free time average time between collisions of a particle

mole quantity of a substance whose mass (in grams) is equal to its molecular mass

most probable speed speed near which the speeds of most molecules are found, the peak of the speed distribution function

partial pressure pressure a gas would create if it occupied the total volume of space available

peak speed same as "most probable speed"

***pV* diagram** graph of pressure vs. volume

root-mean-square (rms) speed square root of the average of the square (of a quantity)

supercritical condition of a fluid being at such a high temperature and pressure that the liquid phase cannot exist

universal gas constant R, the constant that appears in the ideal gas law expressed in terms of moles, given by $R = N_A k_B$

van der Waals equation of state equation, typically approximate, which relates the pressure and volume of a gas to the number of gas molecules or number of moles of gas and the temperature of the gas

vapor pressure partial pressure of a vapor at which it is in equilibrium with the liquid (or solid, in the case of sublimation) phase of the same substance

KEY EQUATIONS

Ideal gas law in terms of molecules	$pV = Nk_B T$

Ideal gas law ratios if the amount of gas is constant	$\frac{p_1 V_1}{T_1} = \frac{p_2 V_2}{T_2}$
Ideal gas law in terms of moles	$pV = nRT$
Van der Waals equation	$\left[p + a\left(\frac{n}{V}\right)^2\right](V - nb) = nRT$
Pressure, volume, and molecular speed	$pV = \frac{1}{3}Nm\overline{v^2}$
Root-mean-square speed	$v_{\text{rms}} = \sqrt{\frac{3RT}{M}} = \sqrt{\frac{3k_B T}{m}}$
Mean free path	$\lambda = \frac{V}{4\sqrt{2}\pi r^2 N} = \frac{k_B T}{4\sqrt{2}\pi r^2 p}$
Mean free time	$\tau = \frac{k_B T}{4\sqrt{2}\pi r^2 p v_{\text{rms}}}$
The following two equations apply only to a monatomic ideal gas:	
Average kinetic energy of a molecule	$\overline{K} = \frac{3}{2}k_B T$
Internal energy	$E_{\text{int}} = \frac{3}{2}Nk_B T.$
Heat in terms of molar heat capacity at constant volume	$Q = nC_V \Delta T$
Molar heat capacity at constant volume for an ideal gas with *d* degrees of freedom	$C_V = \frac{d}{2}R$
Maxwell–Boltzmann speed distribution	$f(v) = \frac{4}{\sqrt{\pi}}\left(\frac{m}{2k_B T}\right)^{3/2} v^2 e^{-mv^2/2k_B T}$
Average velocity of a molecule	$\bar{v} = \sqrt{\frac{8}{\pi}\frac{k_B T}{m}} = \sqrt{\frac{8}{\pi}\frac{RT}{M}}$
Peak velocity of a molecule	$v_p = \sqrt{\frac{2k_B T}{m}} = \sqrt{\frac{2RT}{M}}$

SUMMARY

2.1 Molecular Model of an Ideal Gas

- The ideal gas law relates the pressure and volume of a gas to the number of gas molecules and the temperature of the gas.
- A mole of any substance has a number of molecules equal to the number of atoms in a 12-g sample of carbon-12. The number of molecules in a mole is called Avogadro's number N_A,

$$N_A = 6.02 \times 10^{23}\ \text{mol}^{-1}.$$

- A mole of any substance has a mass in grams numerically equal to its molecular mass in unified mass units, which can be determined from the periodic table of elements. The ideal gas law can also be written and solved in terms of the number of moles of gas:

$$pV = nRT,$$

where n is the number of moles and R is the universal gas constant,

$$R = 8.31 \text{ J/mol} \cdot \text{K}.$$

- The ideal gas law is generally valid at temperatures well above the boiling temperature.
- The van der Waals equation of state for gases is valid closer to the boiling point than the ideal gas law.
- Above the critical temperature and pressure for a given substance, the liquid phase does not exist, and the sample is "supercritical."

2.2 Pressure, Temperature, and RMS Speed

- Kinetic theory is the atomic description of gases as well as liquids and solids. It models the properties of matter in terms of continuous random motion of molecules.
- The ideal gas law can be expressed in terms of the mass of the gas's molecules and $\overline{v^2}$, the average of the molecular speed squared, instead of the temperature.
- The temperature of gases is proportional to the average translational kinetic energy of molecules. Hence, the typical speed of gas molecules v_{rms} is proportional to the square root of the temperature and inversely proportional to the square root of the molecular mass.
- In a mixture of gases, each gas exerts a pressure equal to the total pressure times the fraction of the mixture that the gas makes up.
- The mean free path (the average distance between collisions) and the mean free time of gas molecules are proportional to the temperature and inversely proportional to the molar density and the molecules' cross-sectional area.

2.3 Heat Capacity and Equipartition of Energy

- Every degree of freedom of an ideal gas contributes $\frac{1}{2}k_{\text{B}}T$ per atom or molecule to its changes in internal energy.
- Every degree of freedom contributes $\frac{1}{2}R$ to its molar heat capacity at constant volume C_V.
- Degrees of freedom do not contribute if the temperature is too low to excite the minimum energy of the degree of freedom as given by quantum mechanics. Therefore, at ordinary temperatures, $d = 3$ for monatomic gases, $d = 5$ for diatomic gases, and $d \approx 6$ for polyatomic gases.

2.4 Distribution of Molecular Speeds

- The motion of individual molecules in a gas is random in magnitude and direction. However, a gas of many molecules has a predictable distribution of molecular speeds, known as the Maxwell-Boltzmann distribution.
- The average and most probable velocities of molecules having the Maxwell-Boltzmann speed distribution, as well as the rms velocity, can be calculated from the temperature and molecular mass.

CONCEPTUAL QUESTIONS

2.1 Molecular Model of an Ideal Gas

1. Two H_2 molecules can react with one O_2 molecule to produce two H_2O molecules. How many moles of hydrogen molecules are needed to react with one mole of oxygen molecules?

2. Under what circumstances would you expect a gas to behave significantly differently than predicted by the ideal gas law?

3. A constant-volume gas thermometer contains a fixed amount of gas. What property of the gas is measured to indicate its temperature?

4. Inflate a balloon at room temperature. Leave the inflated balloon in the refrigerator overnight. What happens to the balloon, and why?

5. In the last chapter, free convection was explained as the result of buoyant forces on hot fluids. Explain the upward motion of air in flames based on the ideal gas law.

2.2 Pressure, Temperature, and RMS Speed

6. How is momentum related to the pressure exerted by a gas? Explain on the molecular level, considering the behavior of molecules.

7. If one kind of molecule has double the radius of another and eight times the mass, how do their mean free paths under the same conditions compare? How do their mean free times compare?

8. What is the average *velocity* of the air molecules in the room where you are right now?

9. Why do the atmospheres of Jupiter, Saturn, Uranus, and Neptune, which are much more massive and farther from the Sun than Earth is, contain large amounts of hydrogen and helium?

10. Statistical mechanics says that in a gas maintained at a constant temperature through thermal contact with a bigger system (a "reservoir") at that temperature, the fluctuations in internal energy are typically a fraction $1/\sqrt{N}$ of the internal energy. As a fraction of the total internal energy of a mole of gas, how big are the fluctuations in the internal energy? Are we justified in ignoring them?

11. Which is more dangerous, a closet where tanks of nitrogen are stored, or one where tanks of carbon dioxide are stored?

2.3 Heat Capacity and Equipartition of Energy

12. Experimentally it appears that many polyatomic molecules' vibrational degrees of freedom can contribute to some extent to their energy at room temperature. Would you expect that fact to increase or decrease their heat capacity from the value *R*? Explain.

13. One might think that the internal energy of diatomic gases is given by $E_{\text{int}} = 5RT/2$. Do diatomic gases near room temperature have more or less internal energy than that? *Hint:* Their internal energy includes the total energy added in raising the temperature from the boiling point (very low) to room temperature.

14. You mix 5 moles of H_2 at 300 K with 5 moles of He at 360 K in a perfectly insulated calorimeter. Is the final temperature higher or lower than 330 K?

2.4 Distribution of Molecular Speeds

15. One cylinder contains helium gas and another contains krypton gas at the same temperature. Mark each of these statements true, false, or impossible to determine from the given information. (a) The rms speeds of atoms in the two gases are the same. (b) The average kinetic energies of atoms in the two gases are the same. (c) The internal energies of 1 mole of gas in each cylinder are the same. (d) The pressures in the two cylinders are the same.

16. Repeat the previous question if one gas is still helium but the other is changed to fluorine, F_2.

17. An ideal gas is at a temperature of 300 K. To double the average speed of its molecules, what does the temperature need to be changed to?

PROBLEMS

2.1 Molecular Model of an Ideal Gas

18. The gauge pressure in your car tires is $2.50 \times 10^5 \text{ N/m}^2$ at a temperature of 35.0 °C when you drive it onto a ship in Los Angeles to be sent to Alaska. What is their gauge pressure on a night in Alaska when their temperature has dropped to −40.0 °C ? Assume the tires have not gained or lost any air.

19. Suppose a gas-filled incandescent light bulb is manufactured so that the gas inside the bulb is at atmospheric pressure when the bulb has a temperature of 20.0 °C. (a) Find the gauge pressure inside such a bulb when it is hot, assuming its average temperature is 60.0 °C (an approximation) and neglecting any change in volume due to thermal expansion or gas leaks. (b) The actual final pressure for the light bulb will be less than calculated in part (a) because the glass bulb will expand. Is this effect significant?

20. People buying food in sealed bags at high elevations often notice that the bags are puffed up because the air inside has expanded. A bag of pretzels was packed at a pressure of 1.00 atm and a temperature of 22.0 °C. When opened at a summer picnic in Santa Fe, New Mexico, at a temperature of 32.0 °C, the volume of the air in the bag is 1.38 times its original volume. What is the pressure of the air?

21. How many moles are there in (a) 0.0500 g of N_2 gas ($M = 28.0$ g/mol)? (b) 10.0 g of CO_2 gas ($M = 44.0$ g/mol)? (c) How many molecules are present in each case?

22. A cubic container of volume 2.00 L holds 0.500 mol of nitrogen gas at a temperature of 25.0 °C. What is the net force due to the nitrogen on one wall of the container? Compare that force to the sample's weight.

23. Calculate the number of moles in the 2.00-L volume of air in the lungs of the average person. Note that the air is at 37.0 °C (body temperature) and that the total volume in the lungs is several times the amount inhaled in a typical breath as given in Example 2.2.

24. An airplane passenger has $100\,\text{cm}^3$ of air in his stomach just before the plane takes off from a sea-level airport. What volume will the air have at cruising altitude if cabin pressure drops to $7.50 \times 10^4\ \text{N/m}^2$?

25. A company advertises that it delivers helium at a gauge pressure of 1.72×10^7 Pa in a cylinder of volume 43.8 L. How many balloons can be inflated to a volume of 4.00 L with that amount of helium? Assume the pressure inside the balloons is 1.01×10^5 Pa and the temperature in the cylinder and the balloons is 25.0 °C.

26. According to http://hyperphysics.phy-astr.gsu.edu/hbase/solar/venusenv.html, the atmosphere of Venus is approximately 96.5% CO_2 and 3.5% N_2 by volume. On the surface, where the temperature is about 750 K and the pressure is about 90 atm, what is the density of the atmosphere?

27. An expensive vacuum system can achieve a pressure as low as $1.00 \times 10^{-7}\ \text{N/m}^2$ at 20.0 °C. How many molecules are there in a cubic centimeter at this pressure and temperature?

28. The number density *N*/*V* of gas molecules at a certain location in the space above our planet is about $1.00 \times 10^{11}\ \text{m}^{-3}$, and the pressure is $2.75 \times 10^{-10}\ \text{N/m}^2$ in this space. What is the temperature there?

29. A bicycle tire contains 2.00 L of gas at an absolute pressure of $7.00 \times 10^5\ \text{N/m}^2$ and a temperature of 18.0 °C. What will its pressure be if you let out an amount of air that has a volume of $100\,\text{cm}^3$ at atmospheric pressure? Assume tire temperature and volume remain constant.

30. In a common demonstration, a bottle is heated and stoppered with a hard-boiled egg that's a little bigger than the bottle's neck. When the bottle is cooled, the pressure difference between inside and outside forces the egg into the bottle. Suppose the bottle has a volume of 0.500 L and the temperature inside it is raised to 80.0 °C while the pressure remains constant at 1.00 atm because the bottle is open. (a) How many moles of air are inside? (b) Now the egg is put in place, sealing the bottle. What is the gauge pressure inside after the air cools back to the ambient temperature of 25 °C but before the egg is forced into the bottle?

31. A high-pressure gas cylinder contains 50.0 L of toxic gas at a pressure of $1.40 \times 10^7\ \text{N/m}^2$ and a temperature of 25.0 °C. The cylinder is cooled to dry ice temperature (−78.5 °C) to reduce the leak rate and pressure so that it can be safely repaired. (a) What is the final pressure in the tank, assuming a negligible amount of gas leaks while being cooled and that there is no phase change? (b) What is the final pressure if one-tenth of the gas escapes? (c) To what temperature must the tank be cooled to reduce the pressure to 1.00 atm (assuming the gas does not change phase and that there is no leakage during cooling)? (d) Does cooling the tank as in part (c) appear to be a practical solution?

32. Find the number of moles in 2.00 L of gas at 35.0 °C and under $7.41 \times 10^7\ \text{N/m}^2$ of pressure.

33. Calculate the depth to which Avogadro's number of table tennis balls would cover Earth. Each ball has a diameter of 3.75 cm. Assume the space between balls adds an extra 25.0% to their volume and assume they are not crushed by their own weight.

34. (a) What is the gauge pressure in a $25.0\ °C$ car tire containing 3.60 mol of gas in a 30.0-L volume? (b) What will its gauge pressure be if you add 1.00 L of gas originally at atmospheric pressure and $25.0\ °C$? Assume the temperature remains at $25.0\ °C$ and the volume remains constant.

2.2 Pressure, Temperature, and RMS Speed

In the problems in this section, assume all gases are ideal.

35. A person hits a tennis ball with a mass of 0.058 kg against a wall. The average component of the ball's velocity perpendicular to the wall is 11 m/s, and the ball hits the wall every 2.1 s on average, rebounding with the opposite perpendicular velocity component. (a) What is the average force exerted on the wall? (b) If the part of the wall the person hits has an area of $3.0\ \text{m}^2$, what is the average pressure on that area?

36. A person is in a closed room (a racquetball court) with $V = 453\ \text{m}^3$ hitting a ball $(m = 42.0\ \text{g})$ around at random without any pauses. The average kinetic energy of the ball is 2.30 J. (a) What is the average value of v_x^2? Does it matter which direction you take to be x? (b) Applying the methods of this chapter, find the average pressure on the walls? (c) Aside from the presence of only one "molecule" in this problem, what is the main assumption in Pressure, Temperature, and RMS Speed that does not apply here?

37. Five bicyclists are riding at the following speeds: 5.4 m/s, 5.7 m/s, 5.8 m/s, 6.0 m/s, and 6.5 m/s. (a) What is their average speed? (b) What is their rms speed?

38. Some incandescent light bulbs are filled with argon gas. What is v_{rms} for argon atoms near the filament, assuming their temperature is 2500 K?

39. Typical molecular speeds (v_{rms}) are large, even at low temperatures. What is v_{rms} for helium atoms at 5.00 K, less than one degree above helium's liquefaction temperature?

40. What is the average kinetic energy in joules of hydrogen atoms on the $5500\ °C$ surface of the Sun? (b) What is the average kinetic energy of helium atoms in a region of the solar corona where the temperature is $6.00 \times 10^5\ \text{K}$?

41. What is the ratio of the average translational kinetic energy of a nitrogen molecule at a temperature of 300 K to the gravitational potential energy of a nitrogen-molecule−Earth system at the ceiling of a 3-m-tall room with respect to the same system with the molecule at the floor?

42. What is the total translational kinetic energy of the air molecules in a room of volume $23\ \text{m}^3$ if the pressure is $9.5 \times 10^4\ \text{Pa}$ (the room is at fairly high elevation) and the temperature is $21\ °C$? Is any item of data unnecessary for the solution?

43. The product of the pressure and volume of a sample of hydrogen gas at $0.00\ °C$ is 80.0 J. (a) How many moles of hydrogen are present? (b) What is the average translational kinetic energy of the hydrogen molecules? (c) What is the value of the product of pressure and volume at $200\ °C$?

44. What is the gauge pressure inside a tank of 4.86×10^4 mol of compressed nitrogen with a volume of $6.56\ \text{m}^3$ if the rms speed is 514 m/s?

45. If the rms speed of oxygen molecules inside a refrigerator of volume $22.0\ \text{ft.}^3$ is 465 m/s, what is the partial pressure of the oxygen? There are 5.71 moles of oxygen in the refrigerator, and the molar mass of oxygen is 32.0 g/mol.

46. The escape velocity of any object from Earth is 11.1 km/s. At what temperature would oxygen molecules (molar mass is equal to 32.0 g/mol) have root-mean-square velocity v_{rms} equal to Earth's escape velocity of 11.1 km/s?

47. The escape velocity from the Moon is much smaller than that from the Earth, only 2.38 km/s. At what temperature would hydrogen molecules (molar mass is equal to 2.016 g/mol) have a root-mean-square velocity v_{rms} equal to the Moon's escape velocity?

48. Nuclear fusion, the energy source of the Sun, hydrogen bombs, and fusion reactors, occurs much more readily when the average kinetic energy of the atoms is high—that is, at high temperatures. Suppose you want the atoms in your fusion experiment to have average kinetic energies of $6.40 \times 10^{-14}\ \text{J}$. What temperature is needed?

49. Suppose that the typical speed (v_{rms}) of carbon dioxide molecules (molar mass is 44.0 g/mol) in a flame is found to be 1350 m/s. What temperature does this indicate?

50. (a) Hydrogen molecules (molar mass is equal to 2.016 g/mol) have v_{rms} equal to 193 m/s. What is the temperature? (b) Much of the gas near the Sun is atomic hydrogen (H rather than H_2). Its temperature would have to be 1.5×10^7 K for the rms speed v_{rms} to equal the escape velocity from the Sun. What is that velocity?

51. There are two important isotopes of uranium, ^{235}U and ^{238}U; these isotopes are nearly identical chemically but have different atomic masses. Only ^{235}U is very useful in nuclear reactors. Separating the isotopes is called uranium enrichment (and is often in the news as of this writing, because of concerns that some countries are enriching uranium with the goal of making nuclear weapons.) One of the techniques for enrichment, gas diffusion, is based on the different molecular speeds of uranium hexafluoride gas, UF_6. (a) The molar masses of ^{235}U and $^{238}UF_6$ are 349.0 g/mol and 352.0 g/mol, respectively. What is the ratio of their typical speeds v_{rms}? (b) At what temperature would their typical speeds differ by 1.00 m/s? (c) Do your answers in this problem imply that this technique may be difficult?

52. The partial pressure of carbon dioxide in the lungs is about 470 Pa when the total pressure in the lungs is 1.0 atm. What percentage of the air molecules in the lungs is carbon dioxide? Compare your result to the percentage of carbon dioxide in the atmosphere, about 0.033%.

53. Dry air consists of approximately 78% nitrogen, 21% oxygen, and 1% argon by mole, with trace amounts of other gases. A tank of compressed dry air has a volume of 1.76 cubic feet at a gauge pressure of 2200 pounds per square inch and a temperature of 293 K. How much oxygen does it contain in moles?

54. (a) Using data from the previous problem, find the mass of nitrogen, oxygen, and argon in 1 mol of dry air. The molar mass of N_2 is 28.0 g/mol, that of O_2 is 32.0 g/mol, and that of argon is 39.9 g/mol. (b) Dry air is mixed with pentane (C_5H_{12}, molar mass 72.2 g/mol), an important constituent of gasoline, in an air-fuel ratio of 15:1 by mass (roughly typical for car engines). Find the partial pressure of pentane in this mixture at an overall pressure of 1.00 atm.

55. (a) Given that air is 21% oxygen, find the minimum atmospheric pressure that gives a relatively safe partial pressure of oxygen of 0.16 atm. (b) What is the minimum pressure that gives a partial pressure of oxygen above the quickly fatal level of 0.06 atm? (c) The air pressure at the summit of Mount Everest (8848 m) is 0.334 atm. Why have a few people climbed it without oxygen, while some who have tried, even though they had trained at high elevation, had to turn back?

56. (a) If the partial pressure of water vapor is 8.05 torr, what is the dew point? (760 torr = 1 atm = 101, 325 Pa) (b) On a warm day when the air temperature is 35 °C and the dew point is 25 °C, what are the partial pressure of the water in the air and the relative humidity?

2.3 Heat Capacity and Equipartition of Energy

57. To give a helium atom nonzero angular momentum requires about 21.2 eV of energy (that is, 21.2 eV is the difference between the energies of the lowest-energy or ground state and the lowest-energy state with angular momentum). The electron-volt or eV is defined as 1.60×10^{-19} J. Find the temperature T where this amount of energy equals $k_B T/2$. Does this explain why we can ignore the rotational energy of helium for most purposes? (The results for other monatomic gases, and for diatomic gases rotating around the axis connecting the two atoms, have comparable orders of magnitude.)

58. (a) How much heat must be added to raise the temperature of 1.5 mol of air from 25.0 °C to 33.0 °C at constant volume? Assume air is completely diatomic. (b) Repeat the problem for the same number of moles of xenon, Xe.

59. A sealed, rigid container of 0.560 mol of an unknown ideal gas at a temperature of 30.0 °C is cooled to −40.0 °C. In the process, 980 J of heat are removed from the gas. Is the gas monatomic, diatomic, or polyatomic?

60. A sample of neon gas (Ne, molar mass $M = 20.2$ g/mol) at a temperature of 13.0 °C is put into a steel container of mass 47.2 g that's at a temperature of −40.0 °C. The final temperature is −28.0 °C. (No heat is exchanged with the surroundings, and you can neglect any change in the volume of the container.) What is the mass of the sample of neon?

61. A steel container of mass 135 g contains 24.0 g of ammonia, NH_3, which has a molar mass of 17.0 g/mol. The container and gas are in equilibrium at 12.0 °C. How much heat has to be removed to reach a temperature of −20.0 °C ? Ignore the change in volume of the steel.

62. A sealed room has a volume of $24\ m^3$. It's filled with air, which may be assumed to be diatomic, at a temperature of 24 °C and a pressure of 9.83×10^4 Pa. A 1.00-kg block of ice at its melting point is placed in the room. Assume the walls of the room transfer no heat. What is the equilibrium temperature?

63. Heliox, a mixture of helium and oxygen, is sometimes given to hospital patients who have trouble breathing, because the low mass of helium makes it easier to breathe than air. Suppose helium at 25 °C is mixed with oxygen at 35 °C to make a mixture that is 70% helium by mole. What is the final temperature? Ignore any heat flow to or from the surroundings, and assume the final volume is the sum of the initial volumes.

64. Professional divers sometimes use heliox, consisting of 79% helium and 21% oxygen by mole. Suppose a perfectly rigid scuba tank with a volume of 11 L contains heliox at an absolute pressure of 2.1×10^7 Pa at a temperature of 31 °C. (a) How many moles of helium and how many moles of oxygen are in the tank? (b) The diver goes down to a point where the sea temperature is 27 °C while using a negligible amount of the mixture. As the gas in the tank reaches this new temperature, how much heat is removed from it?

65. In car racing, one advantage of mixing liquid nitrous oxide (N_2O) with air is that the boiling of the "nitrous" absorbs latent heat of vaporization and thus cools the air and ultimately the fuel-air mixture, allowing more fuel-air mixture to go into each cylinder. As a very rough look at this process, suppose 1.0 mol of nitrous oxide gas at its boiling point, −88 °C, is mixed with 4.0 mol of air (assumed diatomic) at 30 °C. What is the final temperature of the mixture? Use the measured heat capacity of N_2O at 25 °C, which is 30.4 J/mol °C. (The primary advantage of nitrous oxide is that it consists of 1/3 oxygen, which is more than air contains, so it supplies more oxygen to burn the fuel. Another advantage is that its decomposition into nitrogen and oxygen releases energy in the cylinder.)

2.4 Distribution of Molecular Speeds

66. In a sample of hydrogen sulfide ($M = 34.1$ g/mol) at a temperature of 3.00×10^2 K, estimate the ratio of the number of molecules that have speeds very close to v_{rms} to the number that have speeds very close to $2v_{\text{rms}}$.

67. Using the approximation $\int_{v_1}^{v_1+\Delta v} f(v)dv \approx f(v_1)\Delta v$ for small Δv, estimate the fraction of nitrogen molecules at a temperature of 3.00×10^2 K that have speeds between 290 m/s and 291 m/s.

68. Using the method of the preceding problem, estimate the fraction of nitric oxide (NO) molecules at a temperature of 250 K that have energies between 3.45×10^{-21} J and 3.50×10^{-21} J.

69. By counting squares in the following figure, estimate the fraction of argon atoms at $T = 300$ K that have speeds between 600 m/s and 800 m/s. The curve is correctly normalized. The value of a square is its length as measured on the *x*-axis times its height as measured on the *y*-axis, with the units given on those axes.

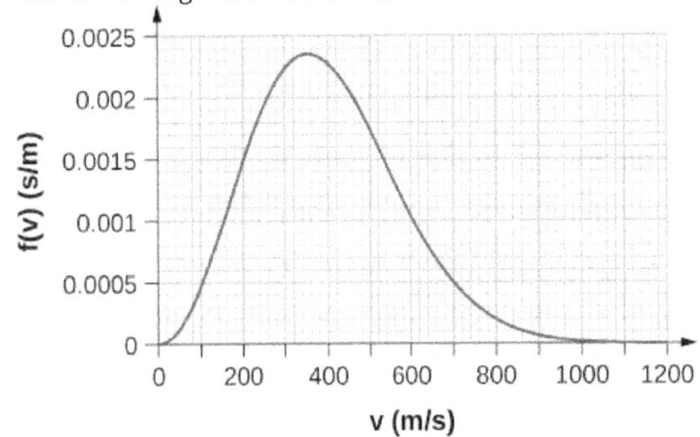

70. Using a numerical integration method such as Simpson's rule, find the fraction of molecules in a sample of oxygen gas at a temperature of 250 K that have speeds between 100 m/s and 150 m/s. The molar mass of oxygen (O_2) is 32.0 g/mol. A precision to two significant digits is enough.

71. Find (a) the most probable speed, (b) the average speed, and (c) the rms speed for nitrogen molecules at 295 K.

72. Repeat the preceding problem for nitrogen molecules at 2950 K.

73. At what temperature is the average speed of carbon dioxide molecules $(M = 44.0\ \text{g/mol})$ 510 m/s?

74. The most probable speed for molecules of a gas at 296 K is 263 m/s. What is the molar mass of the gas? (You might like to figure out what the gas is likely to be.)

75. a) At what temperature do oxygen molecules have the same average speed as helium atoms $(M = 4.00\ \text{g/mol})$ have at 300 K? b) What is the answer to the same question about most probable speeds? c) What is the answer to the same question about rms speeds?

ADDITIONAL PROBLEMS

76. In the deep space between galaxies, the density of molecules (which are mostly single atoms) can be as low as $10^6\ \text{atoms/m}^3$, and the temperature is a frigid 2.7 K. What is the pressure? (b) What volume (in m^3) is occupied by 1 mol of gas? (c) If this volume is a cube, what is the length of its sides in kilometers?

77. (a) Find the density in SI units of air at a pressure of 1.00 atm and a temperature of $20\ °\text{C}$, assuming that air is $78\%\ \text{N}_2$, $21\%\ \text{O}_2$, and $1\%\ \text{Ar}$, (b) Find the density of the atmosphere on Venus, assuming that it's $96\%\ \text{CO}_2$ and $4\%\ \text{N}_2$, with a temperature of 737 K and a pressure of 92.0 atm.

78. The air inside a hot-air balloon has a temperature of 370 K and a pressure of 101.3 kPa, the same as that of the air outside. Using the composition of air as $78\%\ \text{N}_2$, $21\%\text{O}_2$, and $1\%\ \text{Ar}$, find the density of the air inside the balloon.

79. When an air bubble rises from the bottom to the top of a freshwater lake, its volume increases by 80%. If the temperatures at the bottom and the top of the lake are 4.0 and 10 $°\text{C}$, respectively, how deep is the lake?

80. (a) Use the ideal gas equation to estimate the temperature at which 1.00 kg of steam (molar mass $M = 18.0\ \text{g/mol}$) at a pressure of $1.50 \times 10^6\ \text{Pa}$ occupies a volume of $0.220\ \text{m}^3$. (b) The van der Waals constants for water are $a = 0.5537\ \text{Pa} \cdot \text{m}^6/\text{mol}^2$ and $b = 3.049 \times 10^{-5}\ \text{m}^3/\text{mol}$. Use the Van der Waals equation of state to estimate the temperature under the same conditions. (c) The actual temperature is 779 K. Which estimate is better?

81. One process for decaffeinating coffee uses carbon dioxide $(M = 44.0\ \text{g/mol})$ at a molar density of about $14{,}600\ \text{mol/m}^3$ and a temperature of about $60\ °\text{C}$. (a) Is CO_2 a solid, liquid, gas, or supercritical fluid under those conditions? (b) The van der Waals constants for carbon dioxide are $a = 0.3658\ \text{Pa} \cdot \text{m}^6/\text{mol}^2$ and $b = 4.286 \times 10^{-5}\ \text{m}^3/\text{mol}$. Using the van der Waals equation, estimate the pressure of CO_2 at that temperature and density.

82. On a winter day when the air temperature is $0\ °\text{C}$, the relative humidity is 50%. Outside air comes inside and is heated to a room temperature of $20\ °\text{C}$. What is the relative humidity of the air inside the room. (Does this problem show why inside air is so dry in winter?)

83. On a warm day when the air temperature is $30\ °\text{C}$, a metal can is slowly cooled by adding bits of ice to liquid water in it. Condensation first appears when the can reaches $15\ °\text{C}$. What is the relative humidity of the air?

84. (a) People often think of humid air as "heavy." Compare the densities of air with 0% relative humidity and 100% relative humidity when both are at 1 atm and $30\ °\text{C}$. Assume that the dry air is an ideal gas composed of molecules with a molar mass of 29.0 g/mol and the moist air is the same gas mixed with water vapor. (b) As discussed in the chapter on the applications of Newton's laws, the air resistance felt by projectiles such as baseballs and golf balls is approximately $F_\text{D} = C\rho A v^2/2$, where ρ is the mass density of the air, *A* is the cross-sectional area of the projectile, and *C* is the projectile's drag coefficient. For a fixed air pressure, describe qualitatively how the range of a projectile changes with the relative humidity. (c) When a thunderstorm is coming, usually the humidity is high and the air pressure is low. Do those conditions give an advantage or disadvantage to home-run hitters?

85. The mean free path for helium at a certain temperature and pressure is 2.10×10^{-7} m. The radius of a helium atom can be taken as $1.10 \times 10^{-11}\ \text{m}$. What is the measure of the density of helium under those conditions (a) in molecules per cubic meter and (b) in moles per cubic meter?

86. The mean free path for methane at a temperature of 269 K and a pressure of 1.11×10^5 Pa is 4.81×10^{-8} m. Find the effective radius *r* of the methane molecule.

87. In the chapter on fluid mechanics, Bernoulli's equation for the flow of incompressible fluids was explained in terms of changes affecting a small volume *dV* of fluid. Such volumes are a fundamental idea in the study of the flow of compressible fluids such as gases as well. For the equations of hydrodynamics to apply, the mean free path must be much less than the linear size of such a volume, $a \approx dV^{1/3}$. For air in the stratosphere at a temperature of 220 K and a pressure of 5.8 kPa, how big should *a* be for it to be 100 times the mean free path? Take the effective radius of air molecules to be 1.88×10^{-11} m, which is roughly correct for N_2.

88. Find the total number of collisions between molecules in 1.00 s in 1.00 L of nitrogen gas at standard temperature and pressure ($0\,°C$, 1.00 atm). Use 1.88×10^{-10} m as the effective radius of a nitrogen molecule. (The number of collisions per second is the reciprocal of the collision time.) Keep in mind that each collision involves two molecules, so if one molecule collides once in a certain period of time, the collision of the molecule it hit cannot be counted.

89. (a) Estimate the specific heat capacity of sodium from the Law of Dulong and Petit. The molar mass of sodium is 23.0 g/mol. (b) What is the percent error of your estimate from the known value, $1230\ \text{J/kg} \cdot °\text{C}$?

90. A sealed, perfectly insulated container contains 0.630 mol of air at $20.0\,°C$ and an iron stirring bar of mass 40.0 g. The stirring bar is magnetically driven to a kinetic energy of 50.0 J and allowed to slow down by air resistance. What is the equilibrium temperature?

91. Find the ratio $f(v_p)/f(v_{rms})$ for hydrogen gas $(M = 2.02\ \text{g/mol})$ at a temperature of 77.0 K.

92. **Unreasonable results.** (a) Find the temperature of 0.360 kg of water, modeled as an ideal gas, at a pressure of 1.01×10^5 Pa if it has a volume of $0.615\ \text{m}^3$. (b) What is unreasonable about this answer? How could you get a better answer?

93. **Unreasonable results.** (a) Find the average speed of hydrogen sulfide, H_2S, molecules at a temperature of 250 K. Its molar mass is 31.4 g/mol (b) The result isn't *very* unreasonable, but why is it less reliable than those for, say, neon or nitrogen?

CHALLENGE PROBLEMS

94. An airtight dispenser for drinking water is $25\ \text{cm} \times 10\ \text{cm}$ in horizontal dimensions and 20 cm tall. It has a tap of negligible volume that opens at the level of the bottom of the dispenser. Initially, it contains water to a level 3.0 cm from the top and air at the ambient pressure, 1.00 atm, from there to the top. When the tap is opened, water will flow out until the gauge pressure at the bottom of the dispenser, and thus at the opening of the tap, is 0. What volume of water flows out? Assume the temperature is constant, the dispenser is perfectly rigid, and the water has a constant density of $1000\ \text{kg/m}^3$.

95. Eight bumper cars, each with a mass of 322 kg, are running in a room 21.0 m long and 13.0 m wide. They have no drivers, so they just bounce around on their own. The rms speed of the cars is 2.50 m/s. Repeating the arguments of Pressure, Temperature, and RMS Speed, find the average force per unit length (analogous to pressure) that the cars exert on the walls.

96. Verify that $v_p = \sqrt{\frac{2k_B T}{m}}$.

97. Verify the normalization equation $\int_0^\infty f(v)dv = 1$. In doing the integral, first make the substitution $u = \sqrt{\frac{m}{2k_B T}}v = \frac{v}{v_p}$. This "scaling" transformation gives you all features of the answer except for the integral, which is a dimensionless numerical factor. You'll need the formula

$$\int_0^\infty x^2 e^{-x^2} dx = \frac{\sqrt{\pi}}{4}$$

to find the numerical factor and verify the normalization.

98. Verify that $\bar{v} = \sqrt{\frac{8}{\pi}\frac{k_B T}{m}}$. Make the same scaling transformation as in the preceding problem.

99. Verify that $v_{rms} = \sqrt{\bar{v^2}} = \sqrt{\frac{3k_B T}{m}}$.

3 | THE FIRST LAW OF THERMODYNAMICS

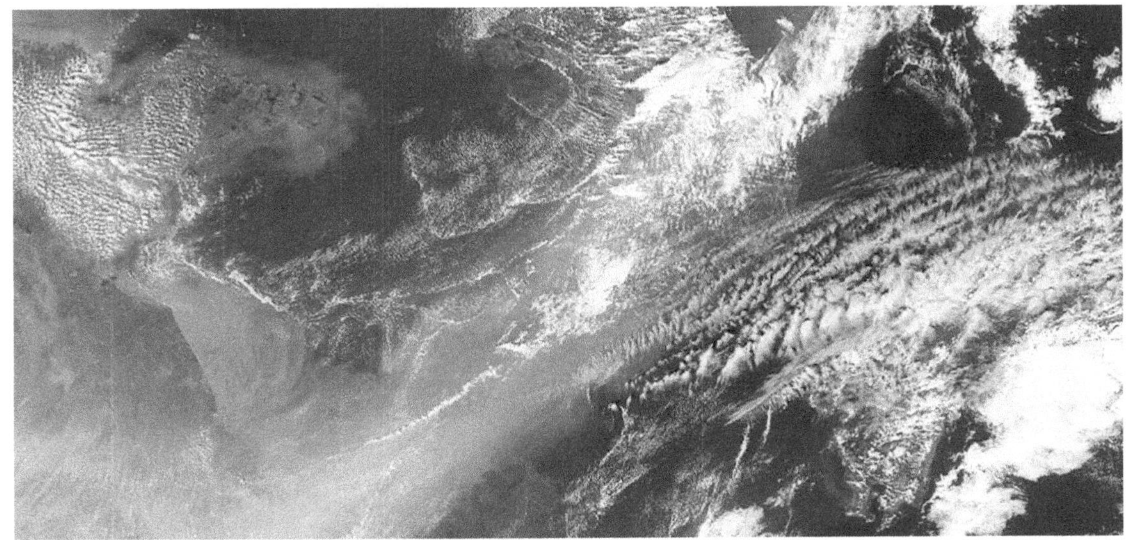

Figure 3.1 A weak cold front of air pushes all the smog in northeastern China into a giant smog blanket over the Yellow Sea, as captured by NASA's Terra satellite in 2012. To understand changes in weather and climate, such as the event shown here, you need a thorough knowledge of thermodynamics. (credit: modification of work by NASA)

Chapter Outline

Introduction

Heat is energy in transit, and it can be used to do work. It can also be converted into any other form of energy. A car engine, for example, burns gasoline. Heat is produced when the burned fuel is chemically transformed into mostly CO_2 and H_2O, which are gases at the combustion temperature. These gases exert a force on a piston through a displacement, doing work and converting the piston's kinetic energy into a variety of other forms—into the car's kinetic energy; into electrical energy to run the spark plugs, radio, and lights; and back into stored energy in the car's battery.

Energy is conserved in all processes, including those associated with thermodynamic systems. The roles of heat transfer and internal energy change vary from process to process and affect how work is done by the system in that process. We will see that the first law of thermodynamics puts a limit on the amount of work that can be delivered by the system when the amount of internal energy change or heat transfer is constrained. Understanding the laws that govern thermodynamic processes and the relationship between the system and its surroundings is therefore paramount in gaining scientific knowledge of energy and energy consumption.

3.1 | Thermodynamic Systems

Learning Objectives

By the end of this section, you will be able to:

- Define a thermodynamic system, its boundary, and its surroundings
- Explain the roles of all the components involved in thermodynamics
- Define thermal equilibrium and thermodynamic temperature
- Link an equation of state to a system

A **thermodynamic system** includes anything whose thermodynamic properties are of interest. It is embedded in its **surroundings** or **environment**; it can exchange heat with, and do work on, its environment through a **boundary**, which is the imagined wall that separates the system and the environment (Figure 3.2). In reality, the immediate surroundings of the system are interacting with it directly and therefore have a much stronger influence on its behavior and properties. For example, if we are studying a car engine, the burning gasoline inside the cylinder of the engine is the thermodynamic system; the piston, exhaust system, radiator, and air outside form the surroundings of the system. The boundary then consists of the inner surfaces of the cylinder and piston.

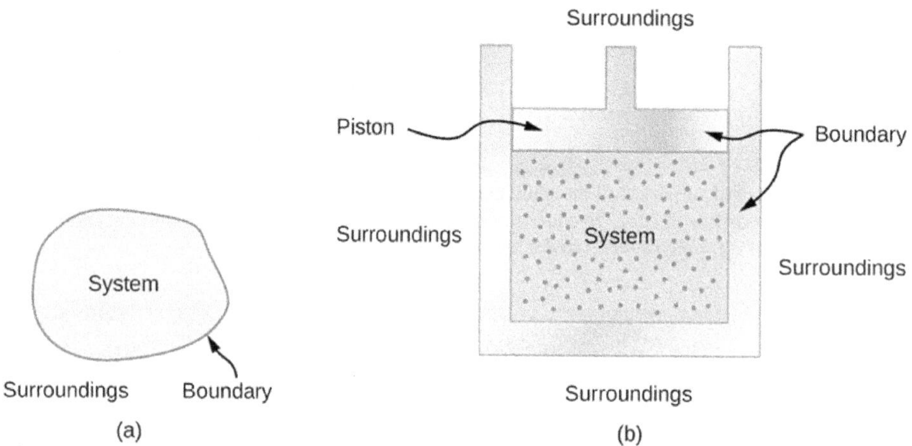

Figure 3.2 (a) A system, which can include any relevant process or value, is self-contained in an area. The surroundings may also have relevant information; however, the surroundings are important to study only if the situation is an open system. (b) The burning gasoline in the cylinder of a car engine is an example of a thermodynamic system.

Normally, a system must have some interactions with its surroundings. A system is called an isolated or **closed system** if it is completely separated from its environment—for example, a gas that is surrounded by immovable and thermally insulating walls. In reality, a closed system does not exist unless the entire universe is treated as the system, or it is used as a model for an actual system that has minimal interactions with its environment. Most systems are known as an **open system**, which can exchange energy and/or matter with its surroundings (Figure 3.3).

(a) (b)

Figure 3.3 (a) This boiling tea kettle is an open thermodynamic system. It transfers heat and matter (steam) to its surroundings. (b) A pressure cooker is a good approximation to a closed system. A little steam escapes through the top valve to prevent explosion. (credit a: modification of work by Gina Hamilton)

When we examine a thermodynamic system, we ignore the difference in behavior from place to place inside the system for a given moment. In other words, we concentrate on the macroscopic properties of the system, which are the averages of the microscopic properties of all the molecules or entities in the system. Any thermodynamic system is therefore treated as a continuum that has the same behavior everywhere inside. We assume the system is in **equilibrium**. You could have, for example, a temperature gradient across the system. However, when we discuss a thermodynamic system in this chapter, we study those that have uniform properties throughout the system.

Before we can carry out any study on a thermodynamic system, we need a fundamental characterization of the system. When we studied a mechanical system, we focused on the forces and torques on the system, and their balances dictated the mechanical equilibrium of the system. In a similar way, we should examine the heat transfer between a thermodynamic system and its environment or between the different parts of the system, and its balance should dictate the thermal equilibrium of the system. Intuitively, such a balance is reached if the temperature becomes the same for different objects or parts of the system in thermal contact, and the net heat transfer over time becomes zero.

Thus, when we say two objects (a thermodynamic system and its environment, for example) are in thermal equilibrium, we mean that they are at the same temperature, as we discussed in Temperature and Heat. Let us consider three objects at temperatures $T_1, T_2,$ and $T_3,$ respectively. How do we know whether they are in thermal equilibrium? The governing principle here is the zeroth law of thermodynamics, as described in Temperature and Heat on temperature and heat:

If object 1 is in thermal equilibrium with objects 2 and 3, respectively, then objects 2 and 3 must also be in thermal equilibrium.

Mathematically, we can simply write the zeroth law of thermodynamics as

$$\text{If } T_1 = T_2 \text{ and } T_1 = T_3, \text{ then } T_2 = T_3. \tag{3.1}$$

This is the most fundamental way of defining temperature: Two objects must be at the same temperature thermodynamically if the net heat transfer between them is zero when they are put in thermal contact and have reached a thermal equilibrium.

The zeroth law of thermodynamics is equally applicable to the different parts of a closed system and requires that the temperature everywhere inside the system be the same if the system has reached a thermal equilibrium. To simplify our discussion, we assume the system is uniform with only one type of material—for example, water in a tank. The measurable properties of the system at least include its volume, pressure, and temperature. The range of specific relevant variables depends upon the system. For example, for a stretched rubber band, the relevant variables would be length, tension, and temperature. The relationship between these three basic properties of the system is called the **equation of state** of the system and is written symbolically *for a closed system* as

$$f(p,\,V,\,T) = 0, \tag{3.2}$$

where V, p, and T are the volume, pressure, and temperature of the system at a given condition.

In principle, this equation of state exists for any thermodynamic system but is not always readily available. The forms of $f(p,\,V,\,T) = 0$ for many materials have been determined either experimentally or theoretically. In the preceding chapter, we saw an example of an equation of state for an ideal gas, $f(p,\,V,\,T) = pV - nRT = 0.$

We have so far introduced several physical properties that are relevant to the thermodynamics of a thermodynamic system, such as its volume, pressure, and temperature. We can separate these quantities into two generic categories. The quantity associated with an amount of matter is an **extensive variable**, such as the volume and the number of moles. The other properties of a system are **intensive variables**, such as the pressure and temperature. An extensive variable doubles its value if the amount of matter in the system doubles, provided all the intensive variables remain the same. For example, the volume or total energy of the system doubles if we double the amount of matter in the system while holding the temperature and pressure of the system unchanged.

3.2 | Work, Heat, and Internal Energy

Learning Objectives

By the end of this section, you will be able to:

- Describe the work done by a system, heat transfer between objects, and internal energy change of a system
- Calculate the work, heat transfer, and internal energy change in a simple process

We discussed the concepts of work and energy earlier in mechanics. Examples and related issues of heat transfer between different objects have also been discussed in the preceding chapters. Here, we want to expand these concepts to a thermodynamic system and its environment. Specifically, we elaborated on the concepts of heat and heat transfer in the previous two chapters. Here, we want to understand how work is done by or to a thermodynamic system; how heat is transferred between a system and its environment; and how the total energy of the system changes under the influence of the work done and heat transfer.

Work Done by a System

A force created from any source can do work by moving an object through a displacement. Then how does a thermodynamic system do work? Figure 3.4 shows a gas confined to a cylinder that has a movable piston at one end. If the gas expands against the piston, it exerts a force through a distance and does work on the piston. If the piston compresses the gas as it is moved inward, work is also done—in this case, on the gas. The work associated with such volume changes can be determined as follows: Let the gas pressure on the piston face be p. Then the force on the piston due to the gas is pA, where A is the area of the face. When the piston is pushed outward an infinitesimal distance dx, the magnitude of the work done by the gas is

$$dW = F\,dx = pA\,dx.$$

Since the change in volume of the gas is $dV = A\,dx,$ this becomes

$$dW = pdV. \tag{3.3}$$

For a finite change in volume from V_1 to $V_2,$ we can integrate this equation from V_1 to V_2 to find the net work:

$$W = \int_{V_1}^{V_2} pdV. \tag{3.4}$$

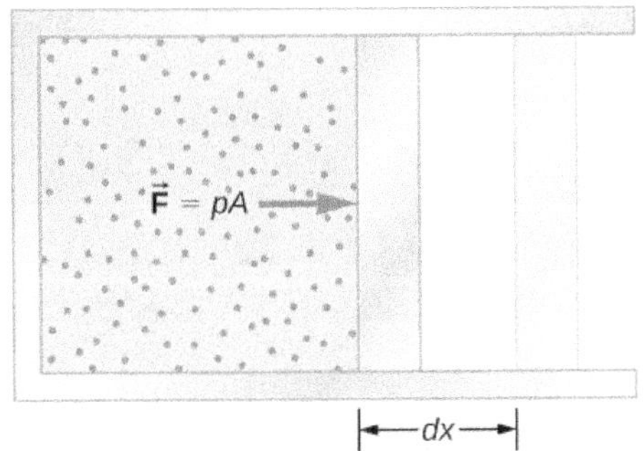

Figure 3.4 The work done by a confined gas in moving a piston a distance dx is given by $dW = Fdx = pdV$.

This integral is only meaningful for a **quasi-static process**, which means a process that takes place in infinitesimally small steps, keeping the system at thermal equilibrium. (We examine this idea in more detail later in this chapter.) Only then does a well-defined mathematical relationship (the equation of state) exist between the pressure and volume. This relationship can be plotted on a pV diagram of pressure versus volume, where the curve is the change of state. We can approximate such a process as one that occurs slowly, through a series of equilibrium states. The integral is interpreted graphically as the area under the pV curve (the shaded area of Figure 3.5). Work done by the gas is positive for expansion and negative for compression.

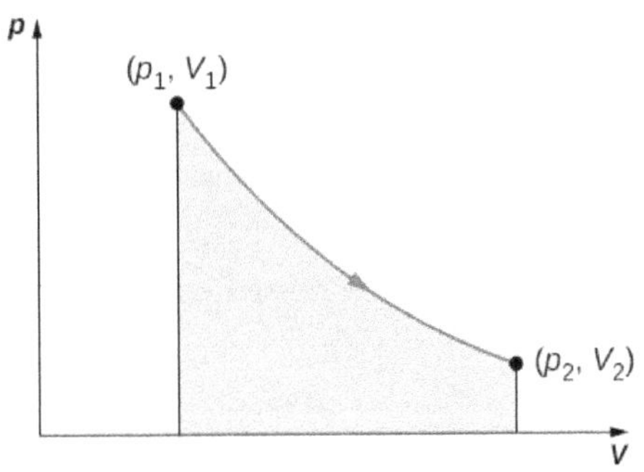

Figure 3.5 When a gas expands slowly from V_1 to V_2, the work done by the system is represented by the shaded area under the pV curve.

Consider the two processes involving an ideal gas that are represented by paths AC and ABC in Figure 3.6. The first process is an isothermal expansion, with the volume of the gas changing its volume from V_1 to V_2. This isothermal process is represented by the curve between points A and C. The gas is kept at a constant temperature T by keeping it in thermal equilibrium with a heat reservoir at that temperature. From Equation 3.4 and the ideal gas law,

$$W = \int_{V_1}^{V_2} pdV = \int_{V_1}^{V_2} \left(\frac{nRT}{V}\right) dV.$$

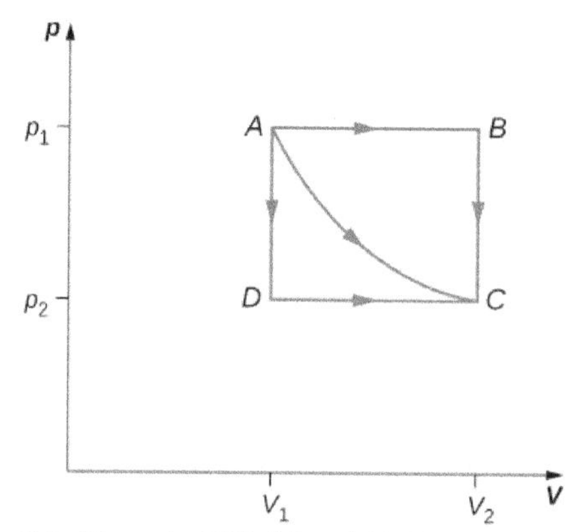

Figure 3.6 The paths *ABC*, *AC*, and *ADC* represent three different quasi-static transitions between the equilibrium states *A* and *C*.

The expansion is isothermal, so *T* remains constant over the entire process. Since *n* and *R* are also constant, the only variable in the integrand is *V*, so the work done by an ideal gas in an isothermal process is

$$W = nRT\int_{V_1}^{V_2}\frac{dV}{V} = nRT\ln\frac{V_2}{V_1}.$$

Notice that if $V_2 > V_1$ (expansion), *W* is positive, as expected.

The straight lines from *A* to *B* and then from *B* to *C* represent a different process. Here, a gas at a pressure p_1 first expands isobarically (constant pressure) and quasi-statically from V_1 to V_2, after which it cools quasi-statically at the constant volume V_2 until its pressure drops to p_2. From *A* to *B*, the pressure is constant at *p*, so the work over this part of the path is

$$W = \int_{V_1}^{V_2} pdV = p_1\int_{V_1}^{V_2} dV = p_1(V_2 - V_1).$$

From *B* to *C*, there is no change in volume and therefore no work is done. The net work over the path *ABC* is then

$$W = p_1(V_2 - V_1) + 0 = p_1(V_2 - V_1).$$

A comparison of the expressions for the work done by the gas in the two processes of Figure 3.6 shows that they are quite different. This illustrates a very important property of thermodynamic work: It is *path dependent*. We cannot determine the work done by a system as it goes from one equilibrium state to another unless we know its thermodynamic path. Different values of the work are associated with different paths.

Example 3.1

Isothermal Expansion of a van der Waals Gas

Studies of a van der Waals gas require an adjustment to the ideal gas law that takes into consideration that gas molecules have a definite volume (see The Kinetic Theory of Gases). One mole of a van der Waals gas has an equation of state

$$\left(p + \frac{a}{V^2}\right)(V - b) = RT,$$

where *a* and *b* are two parameters for a specific gas. Suppose the gas expands isothermally and quasi-statically from volume V_1 to volume V_2. How much work is done by the gas during the expansion?

Strategy

Because the equation of state is given, we can use Equation 3.4 to express the pressure in terms of V and T. Furthermore, temperature T is a constant under the isothermal condition, so V becomes the only changing variable under the integral.

Solution

To evaluate this integral, we must express p as a function of V. From the given equation of state, the gas pressure is

$$p = \frac{RT}{V-b} - \frac{a}{V^2}.$$

Because T is constant under the isothermal condition, the work done by 1 mol of a van der Waals gas in expanding from a volume V_1 to a volume V_2 is thus

$$\begin{aligned} W &= \int_{V_1}^{V_2} \left(\frac{RT}{V-b} - \frac{a}{V^2}\right) = \left| RT\ln(V-b) + \frac{a}{V} \right|_{V_1}^{V_2} \\ &= RT\ln\left(\frac{V_2-b}{V_1-b}\right) + a\left(\frac{1}{V_2} - \frac{1}{V_1}\right). \end{aligned}$$

Significance

By taking into account the volume of molecules, the expression for work is much more complex. If, however, we set $a = 0$ and $b = 0,$ we see that the expression for work matches exactly the work done by an isothermal process for one mole of an ideal gas.

3.1 Check Your Understanding How much work is done by the gas, as given in Figure 3.6, when it expands quasi-statically along the path *ADC*?

Internal Energy

The **internal energy** E_{int} of a thermodynamic system is, by definition, the sum of the mechanical energies of all the molecules or entities in the system. If the kinetic and potential energies of molecule *i* are K_i and $U_i,$ respectively, then the internal energy of the system is the average of the total mechanical energy of all the entities:

$$E_{\text{int}} = \sum_i (\bar{K}_i + \bar{U}_i), \tag{3.5}$$

where the summation is over all the molecules of the system, and the bars over K and U indicate average values. The kinetic energy K_i of an individual molecule includes contributions due to its rotation and vibration, as well as its translational energy $m_i v_i^2/2,$ where v_i is the molecule's speed measured relative to the center of mass of the system. The potential energy U_i is associated only with the interactions between molecule *i* and the other molecules of the system. In fact, neither the system's location nor its motion is of any consequence as far as the internal energy is concerned. The internal energy of the system is not affected by moving it from the basement to the roof of a 100-story building or by placing it on a moving train.

In an ideal monatomic gas, each molecule is a single atom. Consequently, there is no rotational or vibrational kinetic energy and $K_i = m_i v_i^2/2$. Furthermore, there are no interatomic interactions (collisions notwithstanding), so $U_i = \text{constant}$, which we set to zero. The internal energy is therefore due to translational kinetic energy only and

$$E_{\text{int}} = \sum_i \bar{K}_i = \sum_i \frac{1}{2} m_i \overline{v_i^2}.$$

From the discussion in the preceding chapter, we know that the average kinetic energy of a molecule in an ideal monatomic gas is

$$\frac{1}{2} m_i \overline{v_i^2} = \frac{3}{2} k_B T,$$

where T is the Kelvin temperature of the gas. Consequently, the average mechanical energy per molecule of an ideal monatomic gas is also $3k_B T/2$, that is,

$$\overline{K_i + U_i} = \bar{K}_i = \frac{3}{2} k_B T.$$

The internal energy is just the number of molecules multiplied by the average mechanical energy per molecule. Thus for n moles of an ideal monatomic gas,

$$E_{\text{int}} = nN_A \left(\frac{3}{2} k_B T\right) = \frac{3}{2} nRT. \tag{3.6}$$

Notice that the internal energy of a given quantity of an ideal monatomic gas depends on just the temperature and is completely independent of the pressure and volume of the gas. For other systems, the internal energy cannot be expressed so simply. However, an increase in internal energy can often be associated with an increase in temperature.

We know from the zeroth law of thermodynamics that when two systems are placed in thermal contact, they eventually reach thermal equilibrium, at which point they are at the same temperature. As an example, suppose we mix two monatomic ideal gases. Now, the energy per molecule of an ideal monatomic gas is proportional to its temperature. Thus, when the two gases are mixed, the molecules of the hotter gas must lose energy and the molecules of the colder gas must gain energy. This continues until thermal equilibrium is reached, at which point, the temperature, and therefore the average translational kinetic energy per molecule, is the same for both gases. The approach to equilibrium for real systems is somewhat more complicated than for an ideal monatomic gas. Nevertheless, we can still say that energy is exchanged between the systems until their temperatures are the same.

3.3 | First Law of Thermodynamics

Learning Objectives

By the end of this section, you will be able to:

- State the first law of thermodynamics and explain how it is applied
- Explain how heat transfer, work done, and internal energy change are related in any thermodynamic process

Now that we have seen how to calculate internal energy, heat, and work done for a thermodynamic system undergoing change during some process, we can see how these quantities interact to affect the amount of change that can occur. This interaction is given by the first law of thermodynamics. British scientist and novelist C. P. Snow (1905–1980) is credited with a joke about the four laws of thermodynamics. His humorous statement of the first law of thermodynamics is stated "you can't win," or in other words, you cannot get more energy out of a system than you put into it. We will see in this chapter how internal energy, heat, and work all play a role in the first law of thermodynamics.

Suppose Q represents the heat exchanged between a system and the environment, and W is the work done by or on the system. The first law states that the change in internal energy of that system is given by $Q - W$. Since added heat increases the internal energy of a system, Q is positive when it is added to the system and negative when it is removed from the system.

When a gas expands, it does work and its internal energy decreases. Thus, W is positive when work is done by the system and negative when work is done on the system. This sign convention is summarized in Table 3.1. The **first law of thermodynamics** is stated as follows:

First Law of Thermodynamics

Associated with every equilibrium state of a system is its internal energy E_{int}. The change in E_{int} for any transition between two equilibrium states is

$$\Delta E_{int} = Q - W \quad (3.7)$$

where Q and W represent, respectively, the heat exchanged by the system and the work done by or on the system.

Thermodynamic Sign Conventions for Heat and Work

Process	Convention
Heat added to system	$Q > 0$
Heat removed from system	$Q < 0$
Work done by system	$W > 0$
Work done on system	$W < 0$

Table 3.1

The first law is a statement of energy conservation. It tells us that a system can exchange energy with its surroundings by the transmission of heat and by the performance of work. The net energy exchanged is then equal to the change in the total mechanical energy of the molecules of the system (i.e., the system's internal energy). Thus, if a system is isolated, its internal energy must remain constant.

Although Q and W both depend on the thermodynamic path taken between two equilibrium states, their difference $Q - W$ does not. Figure 3.7 shows the pV diagram of a system that is making the transition from A to B repeatedly along different thermodynamic paths. Along path 1, the system absorbs heat Q_1 and does work W_1; along path 2, it absorbs heat Q_2 and does work W_2, and so on. The values of Q_i and W_i may vary from path to path, but we have

$$Q_1 - W_1 = Q_2 - W_2 = \cdots = Q_i - W_i = \cdots,$$

or

$$\Delta E_{int1} = \Delta E_{int2} = \cdots = \Delta E_{inti} = \cdots.$$

That is, the change in the internal energy of the system between A and B is path independent. In the chapter on potential energy and the conservation of energy, we encountered another path-independent quantity: the change in potential energy between two arbitrary points in space. This change represents the negative of the work done by a conservative force between the two points. The potential energy is a function of spatial coordinates, whereas the internal energy is a function of thermodynamic variables. For example, we might write $E_{int}(T, p)$ for the internal energy. Functions such as internal energy and potential energy are known as *state functions* because their values depend solely on the state of the system.

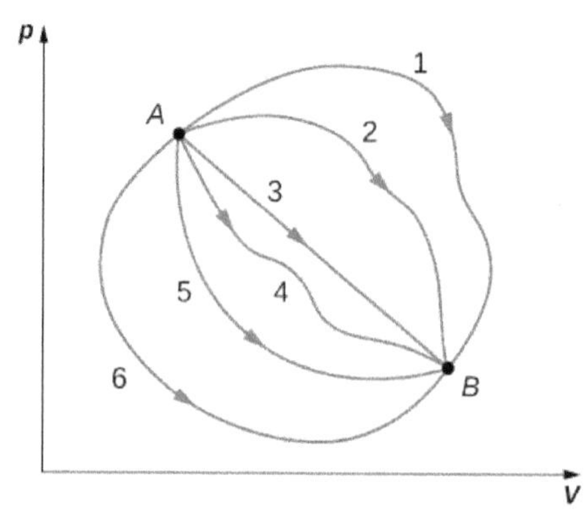

Figure 3.7 Different thermodynamic paths taken by a system in going from state A to state B. For all transitions, the change in the internal energy of the system $\Delta E_{\text{int}} = Q - W$ is the same.

Often the first law is used in its differential form, which is

$$dE_{\text{int}} = dQ - dW. \tag{3.8}$$

Here dE_{int} is an infinitesimal change in internal energy when an infinitesimal amount of heat dQ is exchanged with the system and an infinitesimal amount of work dW is done by (positive in sign) or on (negative in sign) the system.

Example 3.2

Changes of State and the First Law

During a thermodynamic process, a system moves from state A to state B, it is supplied with 400 J of heat and does 100 J of work. (a) For this transition, what is the system's change in internal energy? (b) If the system then moves from state B back to state A, what is its change in internal energy? (c) If in moving from A to B along a different path, $W'_{AB} = 400\,\text{J}$ of work is done on the system, how much heat does it absorb?

Strategy

The first law of thermodynamics relates the internal energy change, work done by the system, and the heat transferred to the system in a simple equation. The internal energy is a function of state and is therefore fixed at any given point regardless of how the system reaches the state.

Solution

a. From the first law, the change in the system's internal energy is

$$\Delta E_{\text{int}AB} = Q_{AB} - W_{AB} = 400\,\text{J} - 100\,\text{J} = 300\,\text{J}.$$

b. Consider a closed path that passes through the states A and B. Internal energy is a state function, so ΔE_{int} is zero for a closed path. Thus

$$\Delta E_{\text{int}} = \Delta E_{\text{int}AB} + \Delta E_{\text{int}BA} = 0,$$

and

$$\Delta E_{\text{int}AB} = -\Delta E_{\text{int}BA}.$$

This yields

$$\Delta E_{\text{int}BA} = -300\,\text{J}.$$

c. The change in internal energy is the same for any path, so

$$\begin{aligned}\Delta E_{\text{int}AB} &= \Delta E'_{\text{int}AB} = Q'_{AB} - W'_{AB};\\ 300\,\text{J} &= Q'_{AB} - (-400\,\text{J}),\end{aligned}$$

and the heat exchanged is

$$Q'_{AB} = -100\,\text{J}.$$

The negative sign indicates that the system loses heat in this transition.

Significance

When a closed cycle is considered for the first law of thermodynamics, the change in internal energy around the whole path is equal to zero. If friction were to play a role in this example, less work would result from this heat added. Example 3.3 takes into consideration what happens if friction plays a role.

Notice that in Example 3.2, we did not assume that the transitions were quasi-static. This is because the first law is not subject to such a restriction. It describes transitions between equilibrium states but is not concerned with the intermediate states. The system does not have to pass through only equilibrium states. For example, if a gas in a steel container at a well-defined temperature and pressure is made to explode by means of a spark, some of the gas may condense, different gas molecules may combine to form new compounds, and there may be all sorts of turbulence in the container—but eventually, the system will settle down to a new equilibrium state. This system is clearly not in equilibrium during its transition; however, its behavior is still governed by the first law because the process starts and ends with the system in equilibrium states.

Example 3.3

Polishing a Fitting

A machinist polishes a 0.50-kg copper fitting with a piece of emery cloth for 2.0 min. He moves the cloth across the fitting at a constant speed of 1.0 m/s by applying a force of 20 N, tangent to the surface of the fitting. (a) What is the total work done on the fitting by the machinist? (b) What is the increase in the internal energy of the fitting? Assume that the change in the internal energy of the cloth is negligible and that no heat is exchanged between the fitting and its environment. (c) What is the increase in the temperature of the fitting?

Strategy

The machinist's force over a distance that can be calculated from the speed and time given is the work done on the system. The work, in turn, increases the internal energy of the system. This energy can be interpreted as the heat that raises the temperature of the system via its heat capacity. Be careful with the sign of each quantity.

Solution

a. The power created by a force on an object or the rate at which the machinist does frictional work on the fitting is $\vec{\mathbf{F}} \cdot \vec{\mathbf{v}} = -Fv$. Thus, in an elapsed time Δt (2.0 min), the work done on the fitting is

$$\begin{aligned}W &= -Fv\Delta t = -(20\,\text{N})(0.1\,\text{m/s})(1.2 \times 10^2\,\text{s})\\ &= -2.4 \times 10^3\,\text{J}.\end{aligned}$$

b. By assumption, no heat is exchanged between the fitting and its environment, so the first law gives for the change in the internal energy of the fitting:

$$\Delta E_{\text{int}} = -W = 2.4 \times 10^3\,\text{J}.$$

c. Since ΔE_{int} is path independent, the effect of the 2.4×10^3 J of work is the same as if it were supplied at atmospheric pressure by a transfer of heat. Thus,

$$2.4 \times 10^3\,\text{J} = mc\Delta T = (0.50\,\text{kg})(3.9 \times 10^2\,\text{J/kg}\cdot{}^\circ\text{C})\Delta T,$$

and the increase in the temperature of the fitting is

$$\Delta T = 12\ °\text{C},$$

where we have used the value for the specific heat of copper, $c = 3.9 \times 10^2\ \text{J/kg} \cdot °\text{C}$.

Significance

If heat were released, the change in internal energy would be less and cause less of a temperature change than what was calculated in the problem.

3.2 Check Your Understanding The quantities below represent four different transitions between the same initial and final state. Fill in the blanks.

Q (J)	*W* (J)	ΔE_{int}(J)
–80	–120	
90		
	40	
	–40	

Table 3.2

Example 3.4

An Ideal Gas Making Transitions between Two States

Consider the quasi-static expansions of an ideal gas between the equilibrium states *A* and *C* of Figure 3.6. If 515 J of heat are added to the gas as it traverses the path *ABC*, how much heat is required for the transition along *ADC*? Assume that $p_1 = 2.10 \times 10^5\ \text{N/m}^2, p_2 = 1.05 \times 10^5\ \text{N/m}^2, V_1 = 2.25 \times 10^{-3}\ \text{m}^3$, and $V_2 = 4.50 \times 10^{-3}\ \text{m}^3$.

Strategy

The difference in work done between process *ABC* and process *ADC* is the area enclosed by *ABCD*. Because the change of the internal energy (a function of state) is the same for both processes, the difference in work is thus the same as the difference in heat transferred to the system.

Solution

For path *ABC*, the heat added is $Q_{ABC} = 515\ \text{J}$ and the work done by the gas is the area under the path on the *pV* diagram, which is

$$W_{ABC} = p_1(V_2 - V_1) = 473\ \text{J}.$$

Along *ADC*, the work done by the gas is again the area under the path:

$$W_{ADC} = p_2(V_2 - V_1) = 236\ \text{J}.$$

Then using the strategy we just described, we have

$$Q_{ADC} - Q_{ABC} = W_{ADC} - W_{ABC},$$

which leads to

$$Q_{ADC} = Q_{ABC} + W_{ADC} - W_{ABC} = (515 + 236 - 473)\ \text{J} = 278\ \text{J}.$$

Significance

The work calculations in this problem are made simple since no work is done along *AD* and *BC* and along *AB* and *DC*; the pressure is constant over the volume change, so the work done is simply $p\Delta V$. An isothermal line could also have been used, as we have derived the work for an isothermal process as $W = nRT\ln\frac{V_2}{V_1}$.

Example 3.5

Isothermal Expansion of an Ideal Gas

Heat is added to 1 mol of an ideal monatomic gas confined to a cylinder with a movable piston at one end. The gas expands quasi-statically at a constant temperature of 300 K until its volume increases from *V* to 3*V*. (a) What is the change in internal energy of the gas? (b) How much work does the gas do? (c) How much heat is added to the gas?

Strategy

(a) Because the system is an ideal gas, the internal energy only changes when the temperature changes. (b) The heat added to the system is therefore purely used to do work that has been calculated in Work, Heat, and Internal Energy. (c) Lastly, the first law of thermodynamics can be used to calculate the heat added to the gas.

Solution

a. We saw in the preceding section that the internal energy of an ideal monatomic gas is a function only of temperature. Since $\Delta T = 0$, for this process, $\Delta E_{\text{int}} = 0$.

b. The quasi-static isothermal expansion of an ideal gas was considered in the preceding section and was found to be

$$\begin{aligned} W &= nRT\ln\frac{V_2}{V_1} = nRT\ln\frac{3V}{V} \\ &= (1.00\text{ mol})(8.314\text{ J/K}\cdot\text{mol})(300\text{ K})(\ln 3) = 2.74 \times 10^3\text{ J}. \end{aligned}$$

c. With the results of parts (a) and (b), we can use the first law to determine the heat added:

$$\Delta E_{\text{int}} = Q - W = 0,$$

which leads to

$$Q = W = 2.74 \times 10^3\text{ J}.$$

Significance

An isothermal process has no change in the internal energy. Based on that, the first law of thermodynamics reduces to $Q = W$.

3.3 Check Your Understanding Why was it necessary to state that the process of Example 3.5 is quasi-static?

Example 3.6

Vaporizing Water

When 1.00 g of water at 100 °C changes from the liquid to the gas phase at atmospheric pressure, its change in volume is $1.67 \times 10^{-3}\text{ m}^3$. (a) How much heat must be added to vaporize the water? (b) How much work is done by the water against the atmosphere in its expansion? (c) What is the change in the internal energy of the water?

Strategy

We can first figure out how much heat is needed from the latent heat of vaporization of the water. From the volume change, we can calculate the work done from $W = p\Delta V$ because the pressure is constant. Then, the first law of thermodynamics provides us with the change in the internal energy.

Solution

a. With L_v representing the latent heat of vaporization, the heat required to vaporize the water is

$$Q = mL_v = (1.00\ \text{g})(2.26 \times 10^3\ \text{J/g}) = 2.26 \times 10^3\ \text{J}.$$

b. Since the pressure on the system is constant at $1.00\ \text{atm} = 1.01 \times 10^5\ \text{N/m}^2$, the work done by the water as it is vaporized is

$$W = p\Delta V = (1.01 \times 10^5\ \text{N/m}^2)(1.67 \times 10^{-3}\ \text{m}^3) = 169\ \text{J}.$$

c. From the first law, the thermal energy of the water during its vaporization changes by

$$\Delta E_{\text{int}} = Q - W = 2.26 \times 10^3\ \text{J} - 169\ \text{J} = 2.09 \times 10^3\ \text{J}.$$

Significance

We note that in part (c), we see a change in internal energy, yet there is no change in temperature. Ideal gases that are not undergoing phase changes have the internal energy proportional to temperature. Internal energy in general is the sum of all energy in the system.

3.4 Check Your Understanding When 1.00 g of ammonia boils at atmospheric pressure and $-33.0\ °\text{C}$, its volume changes from 1.47 to $1130\ \text{cm}^3$. Its heat of vaporization at this pressure is $1.37 \times 10^6\ \text{J/kg}$. What is the change in the internal energy of the ammonia when it vaporizes?

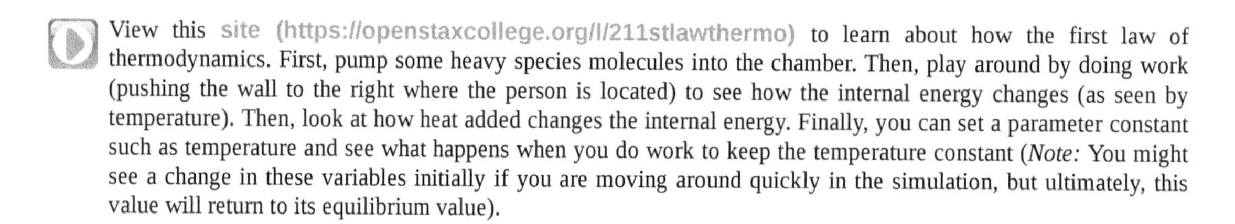

View this site (https://openstaxcollege.org/l/211stlawthermo) to learn about how the first law of thermodynamics. First, pump some heavy species molecules into the chamber. Then, play around by doing work (pushing the wall to the right where the person is located) to see how the internal energy changes (as seen by temperature). Then, look at how heat added changes the internal energy. Finally, you can set a parameter constant such as temperature and see what happens when you do work to keep the temperature constant (*Note:* You might see a change in these variables initially if you are moving around quickly in the simulation, but ultimately, this value will return to its equilibrium value).

3.4 | Thermodynamic Processes

Learning Objectives

By the end of this section, you will be able to:

- Define a thermodynamic process
- Distinguish between quasi-static and non-quasi-static processes
- Calculate physical quantities, such as the heat transferred, work done, and internal energy change for isothermal, adiabatic, and cyclical thermodynamic processes

In solving mechanics problems, we isolate the body under consideration, analyze the external forces acting on it, and then use Newton's laws to predict its behavior. In thermodynamics, we take a similar approach. We start by identifying the part of the universe we wish to study; it is also known as our system. (We defined a system at the beginning of this chapter as anything whose properties are of interest to us; it can be a single atom or the entire Earth.) Once our system is selected, we determine how the environment, or surroundings, interact with the system. Finally, with the interaction understood, we study the thermal behavior of the system with the help of the laws of thermodynamics.

The thermal behavior of a system is described in terms of *thermodynamic variables*. For an ideal gas, these variables are pressure, volume, temperature, and the number of molecules or moles of the gas. Different types of systems are generally characterized by different sets of variables. For example, the thermodynamic variables for a stretched rubber band are tension, length, temperature, and mass.

The state of a system can change as a result of its interaction with the environment. The change in a system can be fast or slow and large or small. The manner in which a state of a system can change from an initial state to a final state is called a **thermodynamic process**. For analytical purposes in thermodynamics, it is helpful to divide up processes as either *quasi-static* or *non-quasi-static*, as we now explain.

Quasi-static and Non-quasi-static Processes

A quasi-static process refers to an idealized or imagined process where the change in state is made infinitesimally slowly so that at each instant, the system can be assumed to be at a thermodynamic equilibrium with itself and with the environment. For instance, imagine heating 1 kg of water from a temperature $20\ °C$ to $21\ °C$ at a constant pressure of 1 atmosphere. To heat the water very slowly, we may imagine placing the container with water in a large bath that can be slowly heated such that the temperature of the bath can rise infinitesimally slowly from $20\ °C$ to $21\ °C$. If we put 1 kg of water at $20\ °C$ directly into a bath at $21\ °C$, the temperature of the water will rise rapidly to $21\ °C$ in a non-quasi-static way.

Quasi-static processes are done slowly enough that the system remains at thermodynamic equilibrium at each instant, despite the fact that the system changes over time. The thermodynamic equilibrium of the system is necessary for the system to have well-defined values of macroscopic properties such as the temperature and the pressure of the system at each instant of the process. Therefore, quasi-static processes can be shown as well-defined paths in state space of the system.

Since quasi-static processes cannot be completely realized for any finite change of the system, all processes in nature are non-quasi-static. Examples of quasi-static and non-quasi-static processes are shown in Figure 3.8. Despite the fact that all finite changes must occur essentially non-quasi-statically at some stage of the change, we can imagine performing infinitely many quasi-static process corresponding to every quasi-static process. Since quasi-static processes can be analyzed analytically, we mostly study quasi-static processes in this book. We have already seen that in a quasi-static process the work by a gas is given by pdV.

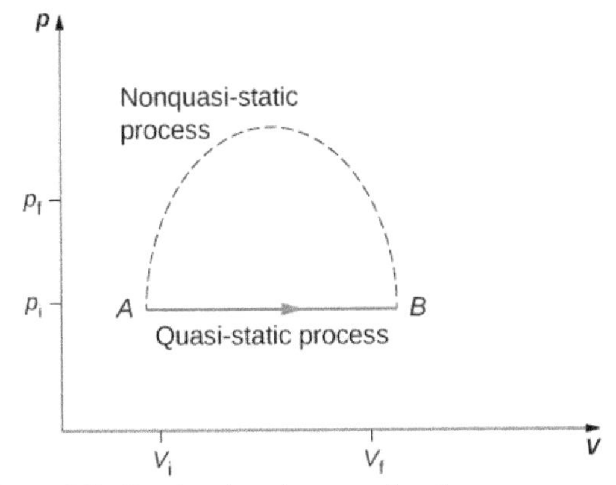

Figure 3.8 Quasi-static and non-quasi-static processes between states *A* and *B* of a gas. In a quasi-static process, the path of the process between *A* and *B* can be drawn in a state diagram since all the states that the system goes through are known. In a non-quasi-static process, the states between *A* and *B* are not known, and hence no path can be drawn. It may follow the dashed line as shown in the figure or take a very different path.

Isothermal Processes

An **isothermal process** is a change in the state of the system at a constant temperature. This process is accomplished by keeping the system in thermal equilibrium with a large heat bath during the process. Recall that a heat bath is an idealized "infinitely" large system whose temperature does not change. In practice, the temperature of a finite bath is controlled by either adding or removing a finite amount of energy as the case may be.

As an illustration of an isothermal process, consider a cylinder of gas with a movable piston immersed in a large water tank whose temperature is maintained constant. Since the piston is freely movable, the pressure inside P_{in} is balanced by the pressure outside P_{out} by some weights on the piston, as in Figure 3.9.

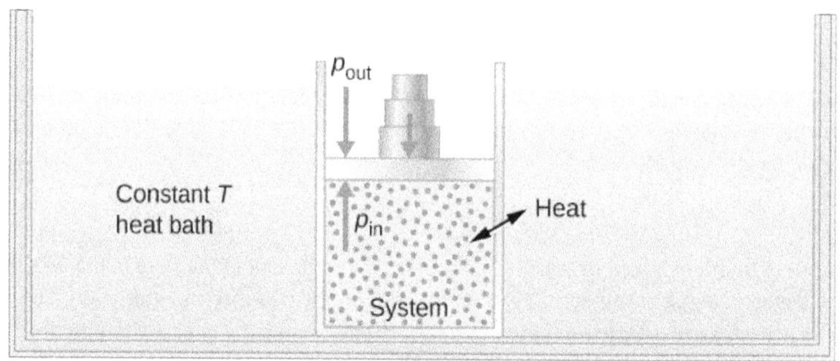

Figure 3.9 Expanding a system at a constant temperature. Removing weights on the piston leads to an imbalance of forces on the piston, which causes the piston to move up. As the piston moves up, the temperature is lowered momentarily, which causes heat to flow from the heat bath to the system. The energy to move the piston eventually comes from the heat bath.

As weights on the piston are removed, an imbalance of forces on the piston develops. The net nonzero force on the piston would cause the piston to accelerate, resulting in an increase in volume. The expansion of the gas cools the gas to a lower temperature, which makes it possible for the heat to enter from the heat bath into the system until the temperature of the gas is reset to the temperature of the heat bath. If weights are removed in infinitesimal steps, the pressure in the system decreases infinitesimally slowly. This way, an isothermal process can be conducted quasi-statically. An isothermal line on a (p, V) diagram is represented by a curved line from starting point A to finishing point B, as seen in Figure 3.10. For an ideal gas, an isothermal process is hyperbolic, since for an ideal gas at constant temperature, $p \propto \frac{1}{V}$.

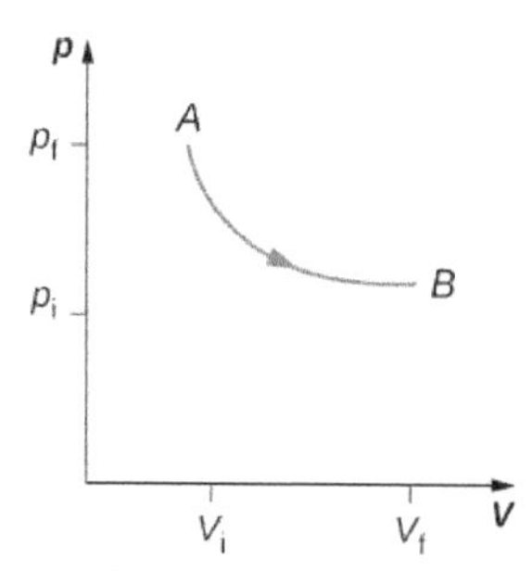

Figure 3.10 An isothermal expansion from a state labeled A to another state labeled B on a pV diagram. The curve represents the relation between pressure and volume in an ideal gas at constant temperature.

An isothermal process studied in this chapter is quasi-statically performed, since to be isothermal throughout the change of volume, you must be able to state the temperature of the system at each step, which is possible only if the system is in thermal equilibrium continuously. The system must go out of equilibrium for the state to change, but for quasi-static processes, we imagine that the process is conducted in infinitesimal steps such that these departures from equilibrium can be made as brief and as small as we like.

Other quasi-static processes of interest for gases are isobaric and isochoric processes. An **isobaric process** is a process where the pressure of the system does not change, whereas an **isochoric process** is a process where the volume of the system does not change.

Adiabatic Processes

In an **adiabatic process**, the system is insulated from its environment so that although the state of the system changes, no heat is allowed to enter or leave the system, as seen in Figure 3.11. An adiabatic process can be conducted either quasi-statically or non-quasi-statically. When a system expands adiabatically, it must do work against the outside world, and therefore its energy goes down, which is reflected in the lowering of the temperature of the system. An adiabatic expansion leads to a lowering of temperature, and an adiabatic compression leads to an increase of temperature. We discuss adiabatic expansion again in Adiabatic Processes for an ideal Gas.

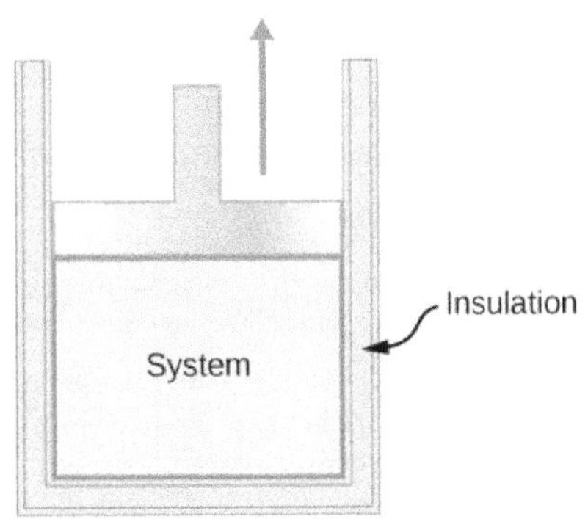

Figure 3.11 An insulated piston with a hot, compressed gas is released. The piston moves up, the volume expands, and the pressure and temperature decrease. The internal energy goes into work. If the expansion occurs within a time frame in which negligible heat can enter the system, then the process is called adiabatic. Ideally, during an adiabatic process no heat enters or exits the system.

Cyclic Processes

We say that a system goes through a **cyclic process** if the state of the system at the end is same as the state at the beginning. Therefore, state properties such as temperature, pressure, volume, and internal energy of the system do not change over a complete cycle:

$$\Delta E_{\text{int}} = 0.$$

When the first law of thermodynamics is applied to a cyclic process, we obtain a simple relation between heat into the system and the work done by the system over the cycle:

$$Q = W \text{ (cyclic process)}.$$

Thermodynamic processes are also distinguished by whether or not they are reversible. A **reversible process** is one that can be made to retrace its path by differential changes in the environment. Such a process must therefore also be quasi-static. Note, however, that a quasi-static process is not necessarily reversible, since there may be dissipative forces involved. For example, if friction occurred between the piston and the walls of the cylinder containing the gas, the energy lost to friction would prevent us from reproducing the original states of the system.

We considered several thermodynamic processes:

1. An isothermal process, during which the system's temperature remains constant
2. An adiabatic process, during which no heat is transferred to or from the system
3. An isobaric process, during which the system's pressure does not change
4. An isochoric process, during which the system's volume does not change

Many other processes also occur that do not fit into any of these four categories.

View this site (https://openstaxcollege.org/l/21idegaspvdiag) to set up your own process in a pV diagram. See if you can calculate the values predicted by the simulation for heat, work, and change in internal energy.

3.5 | Heat Capacities of an Ideal Gas

Learning Objectives

By the end of this section, you will be able to:

- Define heat capacity of an ideal gas for a specific process
- Calculate the specific heat of an ideal gas for either an isobaric or isochoric process
- Explain the difference between the heat capacities of an ideal gas and a real gas
- Estimate the change in specific heat of a gas over temperature ranges

We learned about specific heat and molar heat capacity in Temperature and Heat; however, we have not considered a process in which heat is added. We do that in this section. First, we examine a process where the system has a constant volume, then contrast it with a system at constant pressure and show how their specific heats are related.

Let's start with looking at Figure 3.12, which shows two vessels *A* and *B*, each containing 1 mol of the same type of ideal gas at a temperature *T* and a volume *V*. The only difference between the two vessels is that the piston at the top of *A* is fixed, whereas the one at the top of *B* is free to move against a constant external pressure *p*. We now consider what happens when the temperature of the gas in each vessel is slowly increased to $T + dT$ with the addition of heat.

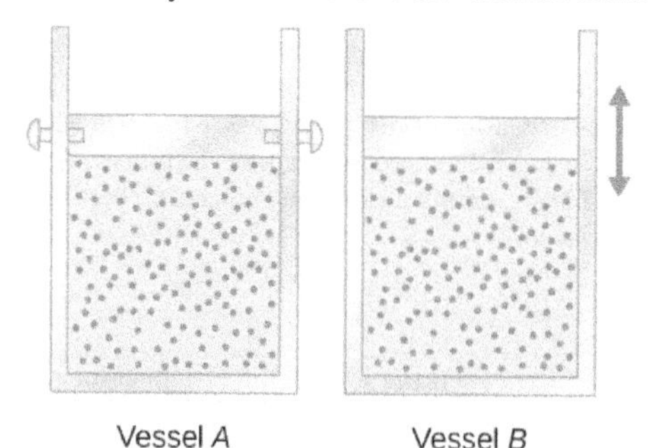

Figure 3.12 Two vessels are identical except that the piston at the top of *A* is fixed, whereas that atop *B* is free to move against a constant external pressure *p*.

Since the piston of vessel *A* is fixed, the volume of the enclosed gas does not change. Consequently, the gas does no work, and we have from the first law

$$dE_{\text{int}} = dQ - dW = dQ.$$

We represent the fact that the heat is exchanged at constant volume by writing

$$dQ = C_V dT,$$

where C_V is the **molar heat capacity at constant volume** of the gas. In addition, since $dE_{\text{int}} = dQ$ for this particular process,

$$dE_{\text{int}} = C_V dT. \quad (3.9)$$

We obtained this equation assuming the volume of the gas was fixed. However, internal energy is a state function that depends on only the temperature of an ideal gas. Therefore, $dE_{\text{int}} = C_V dT$ gives the change in internal energy of an ideal gas for any process involving a temperature change *dT*.

When the gas in vessel *B* is heated, it expands against the movable piston and does work $dW = pdV$. In this case, the heat is added at constant pressure, and we write

$$dQ = C_p dT,$$

where C_p is the **molar heat capacity at constant pressure** of the gas. Furthermore, since the ideal gas expands against a constant pressure,

$$d(pV) = d(RT)$$

becomes

$$pdV = RdT.$$

Finally, inserting the expressions for dQ and pdV into the first law, we obtain

$$dE_{\text{int}} = dQ - pdV = (C_p - R)dT.$$

We have found dE_{int} for both an isochoric and an isobaric process. Because the internal energy of an ideal gas depends only on the temperature, dE_{int} must be the same for both processes. Thus,

$$C_V dT = (C_p - R)dT,$$

and

$$C_p = C_V + R. \tag{3.10}$$

The derivation of Equation 3.10 was based only on the ideal gas law. Consequently, this relationship is approximately valid for all dilute gases, whether monatomic like He, diatomic like O_2, or polyatomic like CO_2 or NH_3.

In the preceding chapter, we found the molar heat capacity of an ideal gas under constant volume to be

$$C_V = \frac{d}{2}R,$$

where d is the number of degrees of freedom of a molecule in the system. Table 3.3 shows the molar heat capacities of some dilute ideal gases at room temperature. The heat capacities of real gases are somewhat higher than those predicted by the expressions of C_V and C_p given in Equation 3.10. This indicates that vibrational motion in polyatomic molecules is significant, even at room temperature. Nevertheless, the difference in the molar heat capacities, $C_p - C_V$, is very close to R, even for the polyatomic gases.

Molar Heat Capacities of Dilute Ideal Gases at Room Temperature

Type of Molecule	Gas	C_p (J/mol K)	C_V (J/mol K)	$C_p - C_V$ (J/mol K)
Monatomic	Ideal	$\frac{5}{2}R = 20.79$	$\frac{3}{2}R = 12.47$	$R = 8.31$
Diatomic	Ideal	$\frac{7}{2}R = 29.10$	$\frac{5}{2}R = 20.79$	$R = 8.31$
Polyatomic	Ideal	$4R = 33.26$	$3R = 24.94$	$R = 8.31$

Table 3.3

3.6 | Adiabatic Processes for an Ideal Gas

Learning Objectives

By the end of this section, you will be able to:

- Define adiabatic expansion of an ideal gas
- Demonstrate the qualitative difference between adiabatic and isothermal expansions

When an ideal gas is compressed adiabatically $(Q = 0)$, work is done on it and its temperature increases; in an adiabatic expansion, the gas does work and its temperature drops. Adiabatic compressions actually occur in the cylinders of a car, where the compressions of the gas-air mixture take place so quickly that there is no time for the mixture to exchange heat with its environment. Nevertheless, because work is done on the mixture during the compression, its temperature does rise significantly. In fact, the temperature increases can be so large that the mixture can explode without the addition of a spark. Such explosions, since they are not timed, make a car run poorly—it usually "knocks." Because ignition temperature rises with the octane of gasoline, one way to overcome this problem is to use a higher-octane gasoline.

Another interesting adiabatic process is the free expansion of a gas. Figure 3.13 shows a gas confined by a membrane to one side of a two-compartment, thermally insulated container. When the membrane is punctured, gas rushes into the empty side of the container, thereby expanding freely. Because the gas expands "against a vacuum" $(p = 0)$, it does no work, and because the vessel is thermally insulated, the expansion is adiabatic. With $Q = 0$ and $W = 0$ in the first law, $\Delta E_{\text{int}} = 0$, so $E_{\text{int}\,i} = E_{\text{int}\,f}$ for the free expansion.

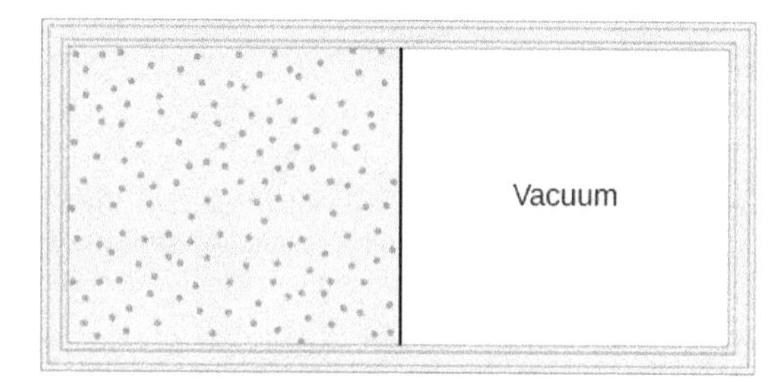

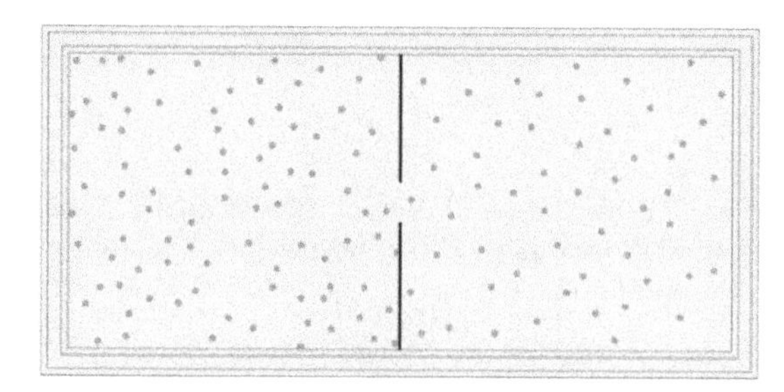

Initial equilibrium state Final equilibrium state

Figure 3.13 The gas in the left chamber expands freely into the right chamber when the membrane is punctured.

If the gas is ideal, the internal energy depends only on the temperature. Therefore, when an ideal gas expands freely, its temperature does not change.

A quasi-static, adiabatic expansion of an ideal gas is represented in Figure 3.14, which shows an insulated cylinder that contains 1 mol of an ideal gas. The gas is made to expand quasi-statically by removing one grain of sand at a time from the top of the piston. When the gas expands by dV, the change in its temperature is dT. The work done by the gas in the expansion is $dW = pdV$; $dQ = 0$ because the cylinder is insulated; and the change in the internal energy of the gas is, from Equation 3.9, $dE_{\text{int}} = C_V dT$. Therefore, from the first law,

$$C_V dT = 0 - pdV = -pdV,$$

so

$$dT = -\frac{pdV}{C_V}.$$

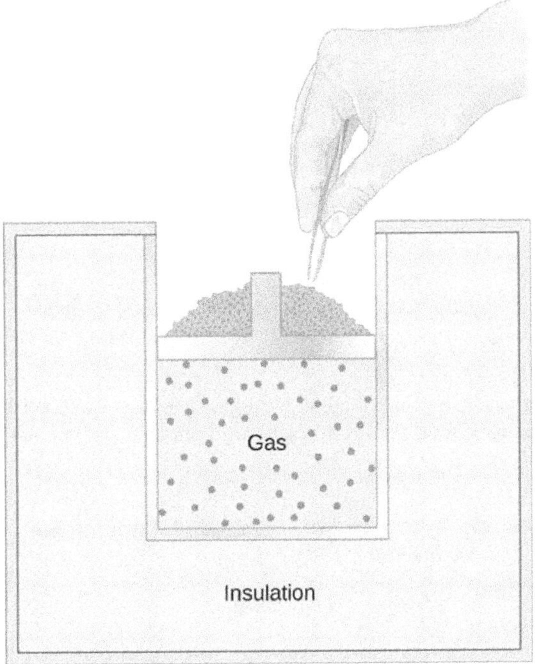

Figure 3.14 When sand is removed from the piston one grain at a time, the gas expands adiabatically and quasi-statically in the insulated vessel.

Also, for 1 mol of an ideal gas,

$$d(pV) = d(RT),$$

so

$$pdV + Vdp = RdT$$

and

$$dT = \frac{pdV + Vdp}{R}.$$

We now have two equations for dT. Upon equating them, we find that

$$C_V Vdp + (C_V + R)pdV = 0.$$

Now, we divide this equation by pV and use $C_p = C_V + R$. We are then left with

$$C_V \frac{dp}{p} + C_p \frac{dV}{V} = 0,$$

which becomes

$$\frac{dp}{p} + \gamma \frac{dV}{V} = 0,$$

where we define γ as the ratio of the molar heat capacities:

$$\gamma = \frac{C_p}{C_V}. \quad (3.11)$$

Thus,

$$\int \frac{dp}{p} + \gamma \int \frac{dV}{V} = 0$$

and

$$\ln p + \gamma \ln V = \text{constant.}$$

Finally, using $\ln(A^x) = x\ln A$ and $\ln AB = \ln A + \ln B$, we can write this in the form

$$pV^{\gamma} = \text{constant.} \tag{3.12}$$

This equation is the condition that must be obeyed by an ideal gas in a quasi-static adiabatic process. For example, if an ideal gas makes a quasi-static adiabatic transition from a state with pressure and volume p_1 and V_1 to a state with p_2 and V_2, then it must be true that $p_1 V_1^{\gamma} = p_2 V_2^{\gamma}$.

The adiabatic condition of Equation 3.12 can be written in terms of other pairs of thermodynamic variables by combining it with the ideal gas law. In doing this, we find that

$$p^{1-\gamma} T^{\gamma} = \text{constant} \tag{3.13}$$

and

$$TV^{\gamma - 1} = \text{constant.} \tag{3.14}$$

A reversible adiabatic expansion of an ideal gas is represented on the pV diagram of Figure 3.15. The slope of the curve at any point is

$$\frac{dp}{dV} = \frac{d}{dV}\left(\frac{\text{constant}}{V^{\gamma}}\right) = -\gamma \frac{p}{V}.$$

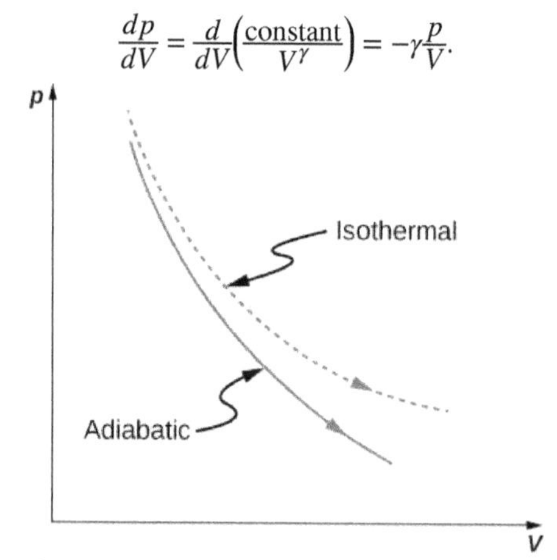

Figure 3.15 Quasi-static adiabatic and isothermal expansions of an ideal gas.

The dashed curve shown on this pV diagram represents an isothermal expansion where T (and therefore pV) is constant. The slope of this curve is useful when we consider the second law of thermodynamics in the next chapter. This slope is

$$\frac{dp}{dV} = \frac{d}{dV}\frac{nRT}{V} = -\frac{p}{V}.$$

Because $\gamma > 1$, the isothermal curve is not as steep as that for the adiabatic expansion.

Example 3.7

Compression of an Ideal Gas in an Automobile Engine

Gasoline vapor is injected into the cylinder of an automobile engine when the piston is in its expanded position. The temperature, pressure, and volume of the resulting gas-air mixture are $20\,°\text{C}$, $1.00 \times 10^5\ \text{N/m}^2$, and $240\ \text{cm}^3$, respectively. The mixture is then compressed adiabatically to a volume of $40\ \text{cm}^3$. Note that in the actual operation of an automobile engine, the compression is not quasi-static, although we are making that assumption here. (a) What are the pressure and temperature of the mixture after the compression? (b) How much work is done by the mixture during the compression?

Strategy

Because we are modeling the process as a quasi-static adiabatic compression of an ideal gas, we have $pV^\gamma = \text{constant}$ and $pV = nRT$. The work needed can then be evaluated with $W = \int_{V_1}^{V_2} p dV$.

Solution

a. For an adiabatic compression we have

$$p_2 = p_1\left(\frac{V_1}{V_2}\right)^\gamma,$$

so after the compression, the pressure of the mixture is

$$p_2 = (1.00 \times 10^5\ \text{N/m}^2)\left(\frac{240 \times 10^{-6}\ \text{m}^3}{40 \times 10^{-6}\ \text{m}^3}\right)^{1.40} = 1.23 \times 10^6\ \text{N/m}^2.$$

From the ideal gas law, the temperature of the mixture after the compression is

$$\begin{aligned} T_2 &= \left(\frac{p_2 V_2}{p_1 V_1}\right) T_1 \\ &= \frac{(1.23 \times 10^6\ \text{N/m}^2)(40 \times 10^{-6}\ \text{m}^3)}{(1.00 \times 10^5\ \text{N/m}^2)(240 \times 10^{-6}\ \text{m}^3)} \cdot 293\ \text{K} \\ &= 600\ \text{K} = 328\ °\text{C}. \end{aligned}$$

b. The work done by the mixture during the compression is

$$W = \int_{V_1}^{V_2} p dV.$$

With the adiabatic condition of Equation 3.12, we may write p as K/V^γ, where $K = p_1 V_1^\gamma = p_2 V_2^\gamma$. The work is therefore

$$\begin{aligned} W &= \int_{V_1}^{V_2} \frac{K}{V^\gamma} dV \\ &= \frac{K}{1-\gamma}\left(\frac{1}{V_2{}^{\gamma-1}} - \frac{1}{V_1{}^{\gamma-1}}\right) \\ &= \frac{1}{1-\gamma}\left(\frac{p_2 V_2^\gamma}{V_2{}^{\gamma-1}} - \frac{p_1 V_1^\gamma}{V_1{}^{\gamma-1}}\right) \\ &= \frac{1}{1-\gamma}(p_2 V_2 - p_1 V_1) \\ &= \frac{1}{1-1.40}[(1.23 \times 10^6 \text{ N/m}^2)(40 \times 10^{-6} \text{ m}^3) \\ &\quad -(1.00 \times 10^5 \text{ N/m}^2)(240 \times 10^{-6} \text{ m}^3)] \\ &= -63 \text{ J}. \end{aligned}$$

Significance

The negative sign on the work done indicates that the piston does work on the gas-air mixture. The engine would not work if the gas-air mixture did work on the piston.

CHAPTER 3 REVIEW

KEY TERMS

adiabatic process process during which no heat is transferred to or from the system

boundary imagined walls that separate the system and its surroundings

closed system system that is mechanically and thermally isolated from its environment

cyclic process process in which the state of the system at the end is same as the state at the beginning

environment outside of the system being studied

equation of state describes properties of matter under given physical conditions

equilibrium thermal balance established between two objects or parts within a system

extensive variable variable that is proportional to the amount of matter in the system

first law of thermodynamics the change in internal energy for any transition between two equilibrium states is $\Delta E_{\text{int}} = Q - W$

intensive variable variable that is independent of the amount of matter in the system

internal energy average of the total mechanical energy of all the molecules or entities in the system

isobaric process process during which the system's pressure does not change

isochoric process process during which the system's volume does not change

isothermal process process during which the system's temperature remains constant

molar heat capacity at constant pressure quantifies the ratio of the amount of heat added removed to the temperature while measuring at constant pressure

molar heat capacity at constant volume quantifies the ratio of the amount of heat added removed to the temperature while measuring at constant volume

open system system that can exchange energy and/or matter with its surroundings

quasi-static process evolution of a system that goes so slowly that the system involved is always in thermodynamic equilibrium

reversible process process that can be reverted to restore both the system and its environment back to their original states together

surroundings environment that interacts with an open system

thermodynamic process manner in which a state of a system can change from initial state to final state

thermodynamic system object and focus of thermodynamic study

KEY EQUATIONS

Equation of state for a closed system	$f(p, V, T) = 0$
Net work for a finite change in volume	$W = \int_{V_1}^{V_2} p dV$
Internal energy of a system (average total energy)	$E_{\text{int}} = \sum_i (\bar{K}_i + \bar{U}_i),$
Internal energy of a monatomic ideal gas	$E_{\text{int}} = nN_A\left(\frac{3}{2}k_B T\right) = \frac{3}{2}nRT$
First law of thermodynamics	$\Delta E_{\text{int}} = Q - W$

Molar heat capacity at constant pressure	$C_p = C_V + R$
Ratio of molar heat capacities	$\gamma = C_p / C_V$
Condition for an ideal gas in a quasi-static adiabatic process	$pV^{\gamma} = \text{constant}$

SUMMARY

3.1 Thermodynamic Systems

- A thermodynamic system, its boundary, and its surroundings must be defined with all the roles of the components fully explained before we can analyze a situation.
- Thermal equilibrium is reached with two objects if a third object is in thermal equilibrium with the other two separately.
- A general equation of state for a closed system has the form $f(p, V, T) = 0,$ with an ideal gas as an illustrative example.

3.2 Work, Heat, and Internal Energy

- Positive (negative) work is done by a thermodynamic system when it expands (contracts) under an external pressure.
- Heat is the energy transferred between two objects (or two parts of a system) because of a temperature difference.
- Internal energy of a thermodynamic system is its total mechanical energy.

3.3 First Law of Thermodynamics

- The internal energy of a thermodynamic system is a function of state and thus is unique for every equilibrium state of the system.
- The increase in the internal energy of the thermodynamic system is given by the heat added to the system less the work done by the system in any thermodynamics process.

3.4 Thermodynamic Processes

- The thermal behavior of a system is described in terms of thermodynamic variables. For an ideal gas, these variables are pressure, volume, temperature, and number of molecules or moles of the gas.
- For systems in thermodynamic equilibrium, the thermodynamic variables are related by an equation of state.
- A heat reservoir is so large that when it exchanges heat with other systems, its temperature does not change.
- A quasi-static process takes place so slowly that the system involved is always in thermodynamic equilibrium.
- A reversible process is one that can be made to retrace its path and both the temperature and pressure are uniform throughout the system.
- There are several types of thermodynamic processes, including (a) isothermal, where the system's temperature is constant; (b) adiabatic, where no heat is exchanged by the system; (c) isobaric, where the system's pressure is constant; and (d) isochoric, where the system's volume is constant.
- As a consequence of the first law of thermodymanics, here is a summary of the thermodymaic processes: (a) isothermal: $\Delta E_{\text{int}} = 0, Q = W;$ (b) adiabatic: $Q = 0, \Delta E_{\text{int}} = -W;$ (c) isobaric: $\Delta E_{\text{int}} = Q - W;$ and (d) isochoric: $W = 0, \Delta E_{\text{int}} = Q.$

3.5 Heat Capacities of an Ideal Gas

- For an ideal gas, the molar capacity at constant pressure C_p is given by $C_p = C_V + R = dR/2 + R$, where d is the number of degrees of freedom of each molecule/entity in the system.

- A real gas has a specific heat close to but a little bit higher than that of the corresponding ideal gas with $C_p \simeq C_V + R$.

3.6 Adiabatic Processes for an Ideal Gas

- A quasi-static adiabatic expansion of an ideal gas produces a steeper pV curve than that of the corresponding isotherm.
- A realistic expansion can be adiabatic but rarely quasi-static.

CONCEPTUAL QUESTIONS

3.1 Thermodynamic Systems

1. Consider these scenarios and state whether work is done by the system on the environment (SE) or by the environment on the system (ES): (a) opening a carbonated beverage; (b) filling a flat tire; (c) a sealed empty gas can expands on a hot day, bowing out the walls.

3.2 Work, Heat, and Internal Energy

2. Is it possible to determine whether a change in internal energy is caused by heat transferred, by work performed, or by a combination of the two?

3. When a liquid is vaporized, its change in internal energy is not equal to the heat added. Why?

4. Why does a bicycle pump feel warm as you inflate your tire?

5. Is it possible for the temperature of a system to remain constant when heat flows into or out of it? If so, give examples.

3.3 First Law of Thermodynamics

6. What does the first law of thermodynamics tell us about the energy of the universe?

7. Does adding heat to a system always increase its internal energy?

8. A great deal of effort, time, and money has been spent in the quest for a so-called perpetual-motion machine, which is defined as a hypothetical machine that operates or produces useful work indefinitely and/or a hypothetical machine that produces more work or energy than it consumes. Explain, in terms of the first law of thermodynamics, why or why not such a machine is likely to be constructed.

3.4 Thermodynamic Processes

9. When a gas expands isothermally, it does work. What is the source of energy needed to do this work?

10. If the pressure and volume of a system are given, is the temperature always uniquely determined?

11. It is unlikely that a process can be isothermal unless it is a very slow process. Explain why. Is the same true for isobaric and isochoric processes? Explain your answer.

3.5 Heat Capacities of an Ideal Gas

12. How can an object transfer heat if the object does not possess a discrete quantity of heat?

13. Most materials expand when heated. One notable exception is water between $0\,°\text{C}$ and $4\,°\text{C}$, which actually decreases in volume with the increase in temperature. Which is greater for water in this temperature region, C_p or C_V ?

14. Why are there two specific heats for gases C_p and C_V, yet only one given for solid?

3.6 Adiabatic Processes for an Ideal Gas

15. Is it possible for γ to be smaller than unity?

16. Would you expect γ to be larger for a gas or a solid? Explain.

17. There is no change in the internal energy of an ideal gas undergoing an isothermal process since the internal energy depends only on the temperature. Is it therefore correct to say that an isothermal process is the same as an adiabatic process for an ideal gas? Explain your answer.

18. Does a gas do any work when it expands adiabatically? If so, what is the source of the energy needed to do this work?

PROBLEMS

3.1 Thermodynamic Systems

19. A gas follows $pV = bp + c_T$ on an isothermal curve, where p is the pressure, V is the volume, b is a constant, and c is a function of temperature. Show that a temperature scale under an isochoric process can be established with this gas and is identical to that of an ideal gas.

20. A mole of gas has isobaric expansion coefficient $dV/dT = R/p$ and isochoric pressure-temperature coefficient $dp/dT = p/T$. Find the equation of state of the gas.

21. Find the equation of state of a solid that has an isobaric expansion coefficient $dV/dT = 2cT - bp$ and an isothermal pressure-volume coefficient $dV/dp = -bT$.

3.2 Work, Heat, and Internal Energy

22. A gas at a pressure of 2.00 atm undergoes a quasi-static isobaric expansion from 3.00 to 5.00 L. How much work is done by the gas?

23. It takes 500 J of work to compress quasi-statically 0.50 mol of an ideal gas to one-fifth its original volume. Calculate the temperature of the gas, assuming it remains constant during the compression.

24. It is found that, when a dilute gas expands quasi-statically from 0.50 to 4.0 L, it does 250 J of work. Assuming that the gas temperature remains constant at 300 K, how many moles of gas are present?

25. In a quasi-static isobaric expansion, 500 J of work are done by the gas. If the gas pressure is 0.80 atm, what is the fractional increase in the volume of the gas, assuming it was originally at 20.0 L?

26. When a gas undergoes a quasi-static isobaric change in volume from 10.0 to 2.0 L, 15 J of work from an external source are required. What is the pressure of the gas?

27. An ideal gas expands quasi-statically and isothermally from a state with pressure p and volume V to a state with volume 4V. Show that the work done by the gas in the expansion is pV(ln 4).

28. As shown below, calculate the work done by the gas in the quasi-static processes represented by the paths (a) AB; (b) ADB; (c) ACB; and (d) ADCB.

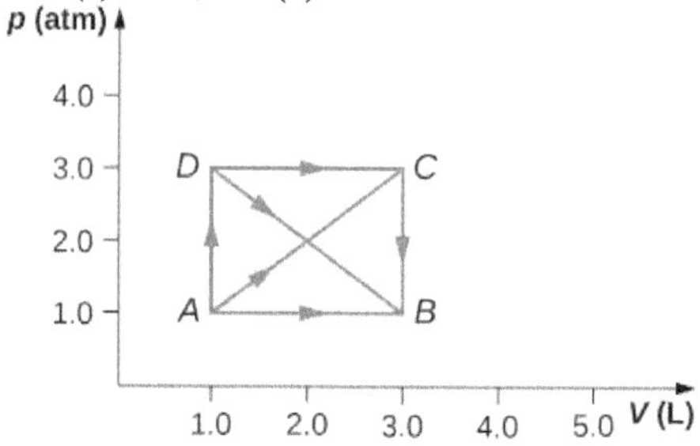

29. (a) Calculate the work done by the gas along the closed path shown below. The curved section between R and S is semicircular. (b) If the process is carried out in the opposite direction, what is the work done by the gas?

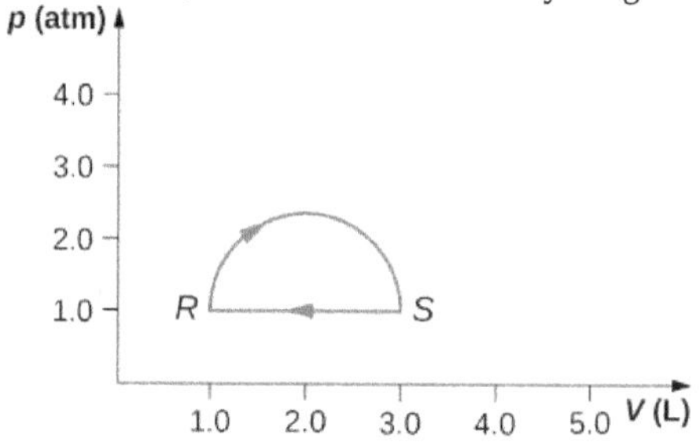

30. An ideal gas expands quasi-statically to three times its original volume. Which process requires more work from the gas, an isothermal process or an isobaric one? Determine the ratio of the work done in these processes.

31. A dilute gas at a pressure of 2.0 atm and a volume of 4.0 L is taken through the following quasi-static steps: (a) an isobaric expansion to a volume of 10.0 L, (b) an isochoric change to a pressure of 0.50 atm, (c) an isobaric compression to a volume of 4.0 L, and (d) an isochoric change to a pressure of 2.0 atm. Show these steps on a pV diagram and determine from your graph the net work done by the gas.

32. What is the average mechanical energy of the atoms of an ideal monatomic gas at 300 K?

33. What is the internal energy of 6.00 mol of an ideal monatomic gas at 200 °C ?

34. Calculate the internal energy of 15 mg of helium at a temperature of 0 °C.

35. Two monatomic ideal gases A and B are at the same temperature. If 1.0 g of gas A has the same internal energy as 0.10 g of gas B, what are (a) the ratio of the number of moles of each gas and (b) the ration of the atomic masses of the two gases?

36. The van der Waals coefficients for oxygen are $a = 0.138\,\text{J} \cdot \text{m}^3/\text{mol}^2$ and $b = 3.18 \times 10^{-5}\ \text{m}^3/\text{mol}$. Use these values to draw a van der Waals isotherm of oxygen at 100 K. On the same graph, draw isotherms of one mole of an ideal gas.

37. Find the work done in the quasi-static processes shown below. The states are given as (p, V) values for the points in the pV plane: 1 (3 atm, 4 L), 2 (3 atm, 6 L), 3 (5 atm, 4 L), 4 (2 atm, 6 L), 5 (4 atm, 2 L), 6 (5 atm, 5 L), and 7 (2 atm, 5 L).

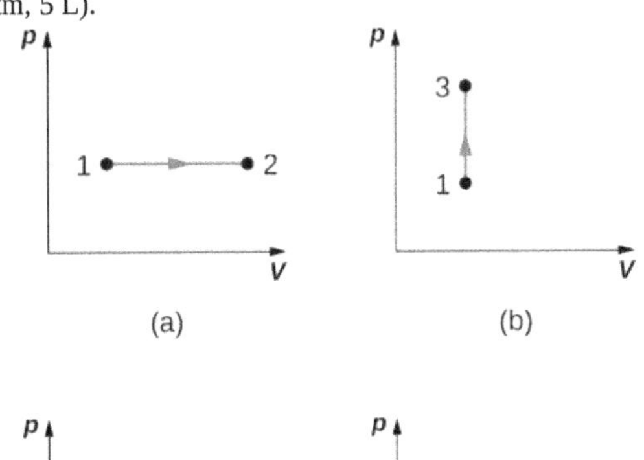

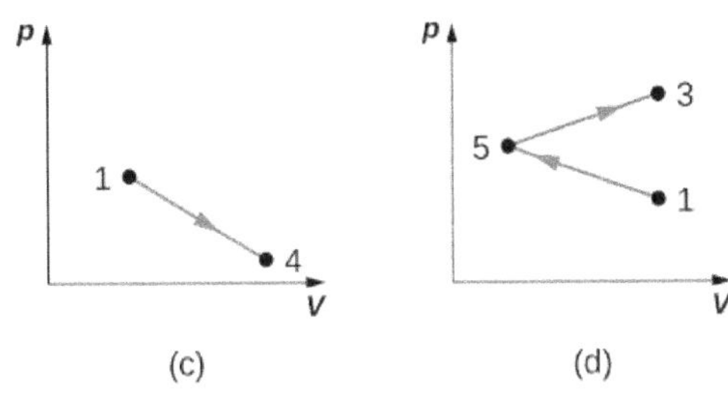

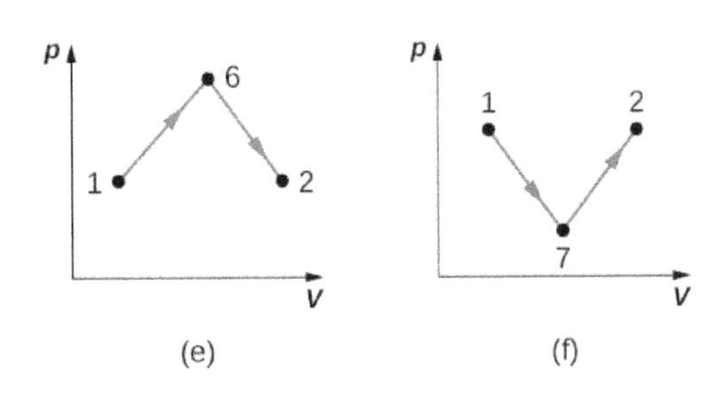

3.3 First Law of Thermodynamics

38. When a dilute gas expands quasi-statically from 0.50 to 4.0 L, it does 250 J of work. Assuming that the gas temperature remains constant at 300 K, (a) what is the change in the internal energy of the gas? (b) How much heat is absorbed by the gas in this process?

39. In a quasi-static isobaric expansion, 500 J of work are done by the gas. The gas pressure is 0.80 atm and it was originally at 20.0 L. If the internal energy of the gas increased by 80 J in the expansion, how much heat does the gas absorb?

40. An ideal gas expands quasi-statically and isothermally from a state with pressure p and volume V to a state with volume 4V. How much heat is added to the expanding gas?

41. As shown below, if the heat absorbed by the gas along AB is 400 J, determine the quantities of heat absorbed along (a) ADB; (b) ACB; and (c) ADCB.

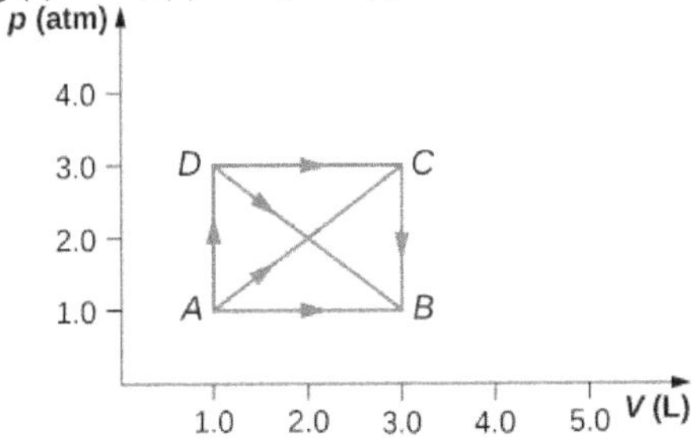

42. During the isobaric expansion from A to B represented below, 130 J of heat are removed from the gas. What is the change in its internal energy?

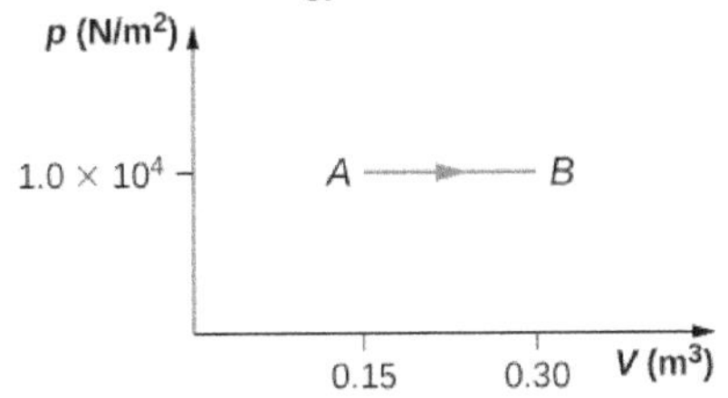

43. (a) What is the change in internal energy for the process represented by the closed path shown below? (b) How much heat is exchanged? (c) If the path is traversed in the opposite direction, how much heat is exchanged?

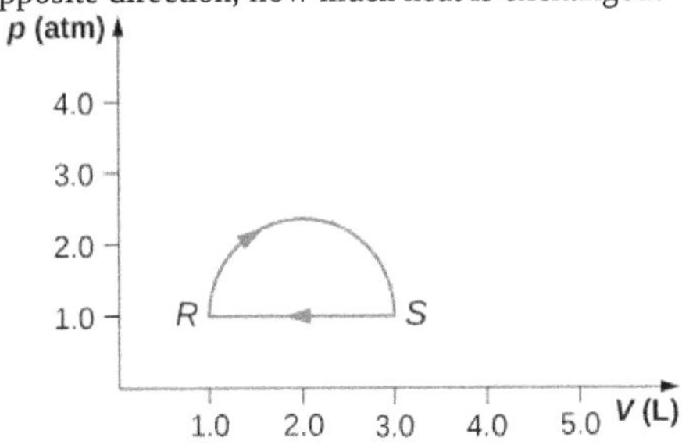

44. When a gas expands along path AC shown below, it does 400 J of work and absorbs either 200 or 400 J of heat. (a) Suppose you are told that along path ABC, the gas absorbs either 200 or 400 J of heat. Which of these values is correct? (b) Give the correct answer from part (a), how much work is done by the gas along ABC? (c) Along CD, the internal energy of the gas decreases by 50 J. How much heat is exchanged by the gas along this path?

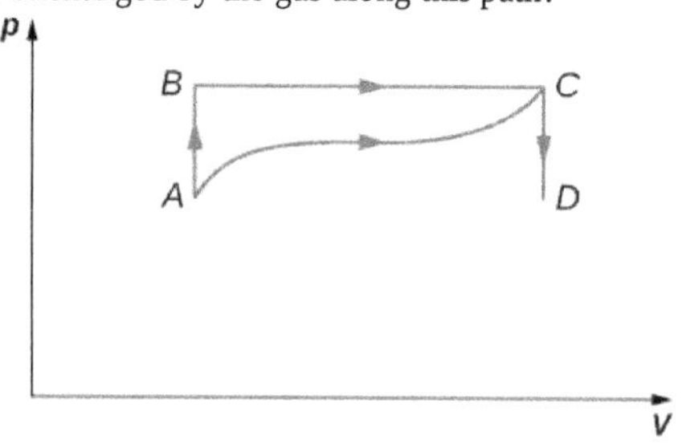

45. When a gas expands along *AB* (see below), it does 500 J of work and absorbs 250 J of heat. When the gas expands along *AC*, it does 700 J of work and absorbs 300 J of heat. (a) How much heat does the gas exchange along *BC*? (b) When the gas makes the transmission from *C* to *A* along *CDA*, 800 J of work are done on it from *C* to *D*. How much heat does it exchange along *CDA*?

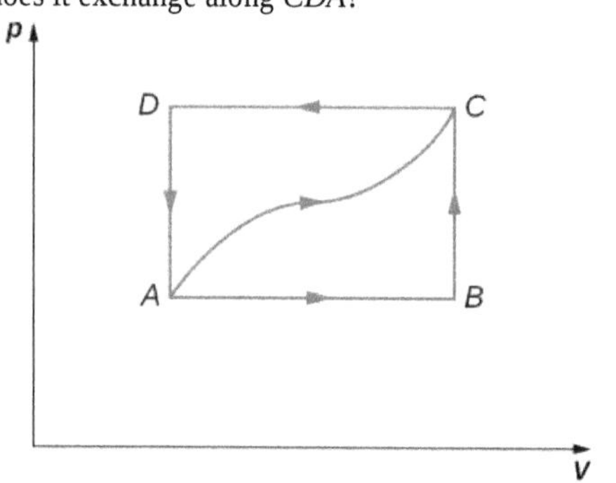

46. A dilute gas is stored in the left chamber of a container whose walls are perfectly insulating (see below), and the right chamber is evacuated. When the partition is removed, the gas expands and fills the entire container. Calculate the work done by the gas. Does the internal energy of the gas change in this process?

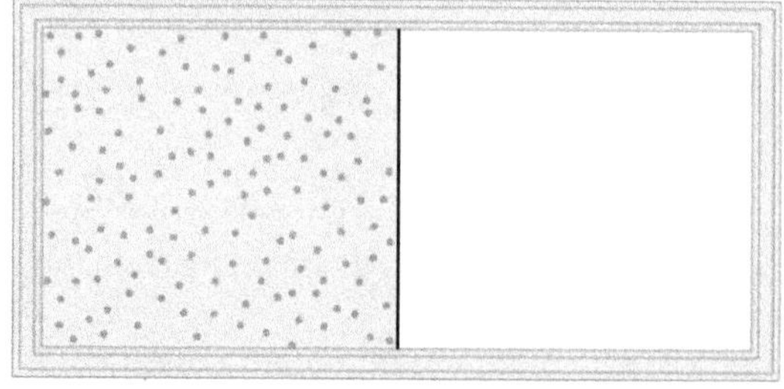

47. Ideal gases A and B are stored in the left and right chambers of an insulated container, as shown below. The partition is removed and the gases mix. Is any work done in this process? If the temperatures of A and B are initially equal, what happens to their common temperature after they are mixed?

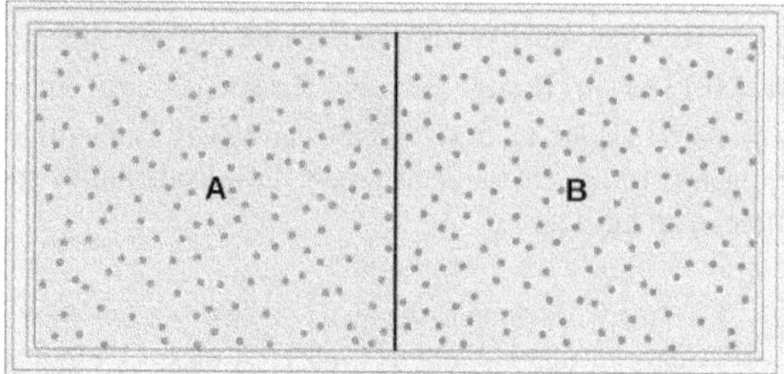

48. An ideal monatomic gas at a pressure of $2.0 \times 10^5\ \text{N/m}^2$ and a temperature of 300 K undergoes a quasi-static isobaric expansion from 2.0×10^3 to $4.0 \times 10^3\ \text{cm}^3$. (a) What is the work done by the gas? (b) What is the temperature of the gas after the expansion? (c) How many moles of gas are there? (d) What is the change in internal energy of the gas? (e) How much heat is added to the gas?

49. Consider the process for steam in a cylinder shown below. Suppose the change in the internal energy in this process is 30 kJ. Find the heat entering the system.

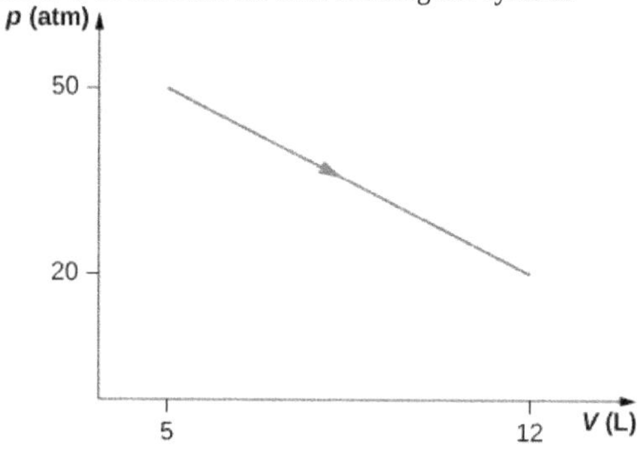

50. The state of 30 moles of steam in a cylinder is changed in a cyclic manner from a-b-c-a, where the pressure and volume of the states are: a (30 atm, 20 L), b (50 atm, 20 L), and c (50 atm, 45 L). Assume each change takes place along the line connecting the initial and final states in the pV plane. (a) Display the cycle in the pV plane. (b) Find the net work done by the steam in one cycle. (c) Find the net amount of heat flow in the steam over the course of one cycle.

51. A monatomic ideal gas undergoes a quasi-static process that is described by the function $p(V) = p_1 + 3(V - V_1)$, where the starting state is (p_1, V_1) and the final state (p_2, V_2). Assume the system consists of n moles of the gas in a container that can exchange heat with the environment and whose volume can change freely. (a) Evaluate the work done by the gas during the change in the state. (b) Find the change in internal energy of the gas. (c) Find the heat input to the gas during the change. (d) What are initial and final temperatures?

52. A metallic container of fixed volume of $2.5 \times 10^{-3}\ \text{m}^3$ immersed in a large tank of temperature $27\ °\text{C}$ contains two compartments separated by a freely movable wall. Initially, the wall is kept in place by a stopper so that there are 0.02 mol of the nitrogen gas on one side and 0.03 mol of the oxygen gas on the other side, each occupying half the volume. When the stopper is removed, the wall moves and comes to a final position. The movement of the wall is controlled so that the wall moves in infinitesimal quasi-static steps. (a) Find the final volumes of the two sides assuming the ideal gas behavior for the two gases. (b) How much work does each gas do on the other? (c) What is the change in the internal energy of each gas? (d) Find the amount of heat that enters or leaves each gas.

53. A gas in a cylindrical closed container is adiabatically and quasi-statically expanded from a state *A* (3 MPa, 2 L) to a state *B* with volume of 6 L along the path $1.8\,pV = \text{constant}$. (a) Plot the path in the *pV* plane. (b) Find the amount of work done by the gas and the change in the internal energy of the gas during the process.

3.4 Thermodynamic Processes

54. Two moles of a monatomic ideal gas at (5 MPa, 5 L) is expanded isothermally until the volume is doubled (step 1). Then it is cooled isochorically until the pressure is 1 MPa (step 2). The temperature drops in this process. The gas is now compressed isothermally until its volume is back to 5 L, but its pressure is now 2 MPa (step 3). Finally, the gas is heated isochorically to return to the initial state (step 4). (a) Draw the four processes in the pV plane. (b) Find the total work done by the gas.

55. Consider a transformation from point *A* to *B* in a two-step process. First, the pressure is lowered from 3 MPa at point *A* to a pressure of 1 MPa, while keeping the volume at 2 L by cooling the system. The state reached is labeled *C*. Then the system is heated at a constant pressure to reach a volume of 6 L in the state *B*. (a) Find the amount of work done on the *ACB* path. (b) Find the amount of heat exchanged by the system when it goes from *A* to *B* on the *ACB* path. (c) Compare the change in the internal energy when the *AB* process occurs adiabatically with the *AB* change through the two-step process on the *ACB* path.

56. Consider a cylinder with a movable piston containing n moles of an ideal gas. The entire apparatus is immersed in a constant temperature bath of temperature T kelvin. The piston is then pushed slowly so that the pressure of the gas changes quasi-statically from p_1 to p_2 at constant temperature T. Find the work done by the gas in terms of n, R, T, p_1, and p_2.

57. An ideal gas expands isothermally along AB and does 700 J of work (see below). (a) How much heat does the gas exchange along AB? (b) The gas then expands adiabatically along BC and does 400 J of work. When the gas returns to A along CA, it exhausts 100 J of heat to its surroundings. How much work is done on the gas along this path?

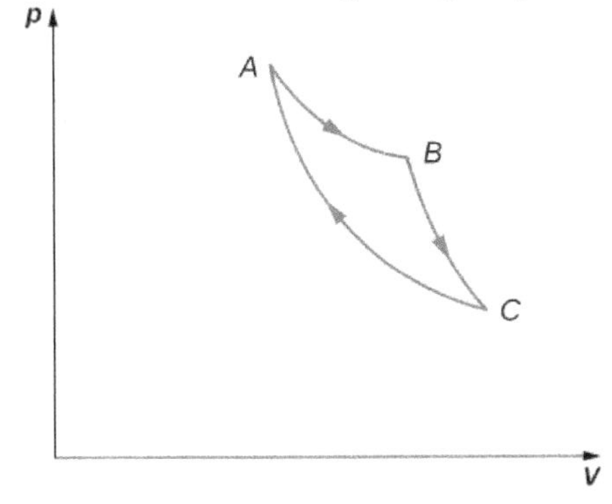

58. Consider the processes shown below. In the processes AB and BC, 3600 J and 2400 J of heat are added to the system, respectively. (a) Find the work done in each of the processes AB, BC, AD, and DC. (b) Find the internal energy change in processes AB and BC. (c) Find the internal energy difference between states C and A. (d) Find the total heat added in the ADC process. (e) From the information give, can you find the heat added in process AD? Why or why not?

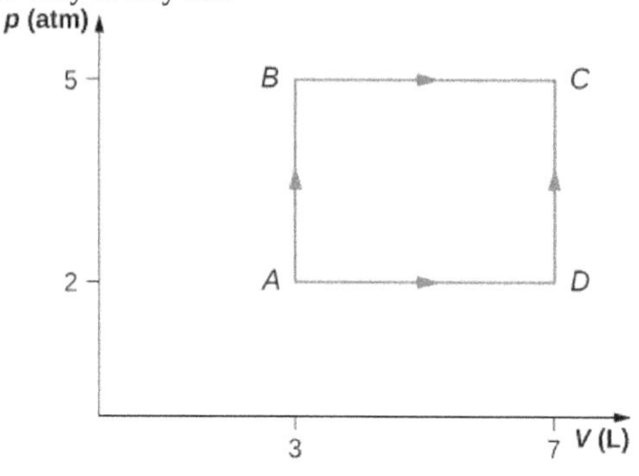

59. Two moles of helium gas are placed in a cylindrical container with a piston. The gas is at room temperature 25 °C and under a pressure of 3.0×10^5 Pa. When the pressure from the outside is decreased while keeping the temperature the same as the room temperature, the volume of the gas doubles. (a) Find the work the external agent does on the gas in the process. (b) Find the heat exchanged by the gas and indicate whether the gas takes in or gives up heat. Assume ideal gas behavior.

60. An amount of n moles of a monatomic ideal gas in a conducting container with a movable piston is placed in a large thermal heat bath at temperature T_1 and the gas is allowed to come to equilibrium. After the equilibrium is reached, the pressure on the piston is lowered so that the gas expands at constant temperature. The process is continued quasi-statically until the final pressure is 4/3 of the initial pressure p_1. (a) Find the change in the internal energy of the gas. (b) Find the work done by the gas. (c) Find the heat exchanged by the gas, and indicate, whether the gas takes in or gives up heat.

3.5 Heat Capacities of an Ideal Gas

61. The temperature of an ideal monatomic gas rises by 8.0 K. What is the change in the internal energy of 1 mol of the gas at constant volume?

62. For a temperature increase of 10 °C at constant volume, what is the heat absorbed by (a) 3.0 mol of a dilute monatomic gas; (b) 0.50 mol of a dilute diatomic gas; and (c) 15 mol of a dilute polyatomic gas?

63. If the gases of the preceding problem are initially at 300 K, what are their internal energies after they absorb the heat?

64. Consider 0.40 mol of dilute carbon dioxide at a pressure of 0.50 atm and a volume of 50 L. What is the internal energy of the gas?

65. When 400 J of heat are slowly added to 10 mol of an ideal monatomic gas, its temperature rises by 10 °C. What is the work done on the gas?

66. One mole of a dilute diatomic gas occupying a volume of 10.00 L expands against a constant pressure of 2.000 atm when it is slowly heated. If the temperature of the gas rises by 10.00 K and 400.0 J of heat are added in the process, what is its final volume?

3.6 Adiabatic Processes for an Ideal Gas

67. A monatomic ideal gas undergoes a quasi-static adiabatic expansion in which its volume is doubled. How is the pressure of the gas changed?

68. An ideal gas has a pressure of 0.50 atm and a volume of 10 L. It is compressed adiabatically and quasi-statically until its pressure is 3.0 atm and its volume is 2.8 L. Is the gas monatomic, diatomic, or polyatomic?

69. Pressure and volume measurements of a dilute gas undergoing a quasi-static adiabatic expansion are shown below. Plot ln p vs. V and determine γ for this gas from your graph.

P (atm)	V (L)
20.0	1.0
17.0	1.1
14.0	1.3
11.0	1.5
8.0	2.0
5.0	2.6
2.0	5.2
1.0	8.4

70. An ideal monatomic gas at 300 K expands adiabatically and reversibly to twice its volume. What is its final temperature?

71. An ideal diatomic gas at 80 K is slowly compressed adiabatically and reversibly to twice its volume. What is its final temperature?

72. An ideal diatomic gas at 80 K is slowly compressed adiabatically to one-third its original volume. What is its final temperature?

73. Compare the charge in internal energy of an ideal gas for a quasi-static adiabatic expansion with that for a quasi-static isothermal expansion. What happens to the temperature of an ideal gas in an adiabatic expansion?

74. The temperature of n moles of an ideal gas changes from T_1 to T_2 in a quasi-static adiabatic transition. Show that the work done by the gas is given by

$$W = \frac{nR}{\gamma - 1}(T_1 - T_2).$$

75. A dilute gas expands quasi-statically to three times its initial volume. Is the final gas pressure greater for an isothermal or an adiabatic expansion? Does your answer depend on whether the gas is monatomic, diatomic, or polyatomic?

76. (a) An ideal gas expands adiabatically from a volume of $2.0 \times 10^{-3}\ \text{m}^3$ to $2.5 \times 10^{-3}\ \text{m}^3$. If the initial pressure and temperature were 5.0×10^5 Pa and 300 K, respectively, what are the final pressure and temperature of the gas? Use $\gamma = 5/3$ for the gas. (b) In an isothermal process, an ideal gas expands from a volume of $2.0 \times 10^{-3}\ \text{m}^3$ to $2.5 \times 10^{-3}\ \text{m}^3$. If the initial pressure and temperature were 5.0×10^5 Pa and 300 K, respectively, what are the final pressure and temperature of the gas?

77. On an adiabatic process of an ideal gas pressure, volume and temperature change such that pV^{γ} is constant with $\gamma = 5/3$ for monatomic gas such as helium and $\gamma = 7/5$ for diatomic gas such as hydrogen at room temperature. Use numerical values to plot two isotherms of 1 mol of helium gas using ideal gas law and two adiabatic processes mediating between them. Use $T_1 = 500\ \text{K}$, $V_1 = 1\ \text{L}$, and $T_2 = 300\ \text{K}$ for your plot.

78. Two moles of a monatomic ideal gas such as helium is compressed adiabatically and reversibly from a state (3 atm, 5 L) to a state with pressure 4 atm. (a) Find the volume and temperature of the final state. (b) Find the temperature of the initial state of the gas. (c) Find the work done by the gas in the process. (d) Find the change in internal energy of the gas in the process.

ADDITIONAL PROBLEMS

79. Consider the process shown below. During steps *AB* and *BC*, 3600 J and 2400 J of heat, respectively, are added to the system. (a) Find the work done in each of the processes *AB*, *BC*, *AD*, and *DC*. (b) Find the internal energy change in processes *AB* and *BC*. (c) Find the internal energy difference between states *C* and *A*. (d) Find the total heat added in the *ADC* process. (e) From the information given, can you find the heat added in process *AD*? Why or why not?

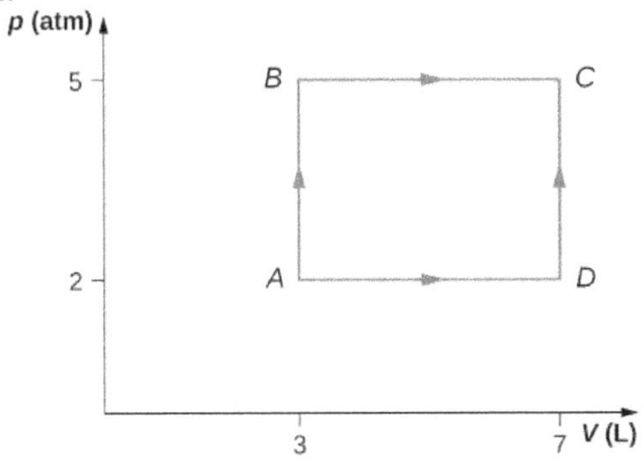

80. A car tire contains $0.0380\ \text{m}^3$ of air at a pressure of 2.20×10^5 Pa (about 32 psi). How much more internal energy does this gas have than the same volume has at zero gauge pressure (which is equivalent to normal atmospheric pressure)?

81. A helium-filled toy balloon has a gauge pressure of 0.200 atm and a volume of 10.0 L. How much greater is the internal energy of the helium in the balloon than it would be at zero gauge pressure?

82. Steam to drive an old-fashioned steam locomotive is supplied at a constant gauge pressure of $1.75 \times 10^6\ \text{N/m}^2$ (about 250 psi) to a piston with a 0.200-m radius. (a) By calculating $p\Delta V$, find the work done by the steam when the piston moves 0.800 m. Note that this is the net work output, since gauge pressure is used. (b) Now find the amount of work by calculating the force exerted times the distance traveled. Is the answer the same as in part (a)?

83. A hand-driven tire pump has a piston with a 2.50-cm diameter and a maximum stroke of 30.0 cm. (a) How much work do you do in one stroke if the average gauge pressure is $2.4 \times 10^5\ \text{N/m}^2$ (about 35 psi)? (b) What average force do you exert on the piston, neglecting friction and gravitational force?

84. Calculate the net work output of a heat engine following path *ABCDA* as shown below.

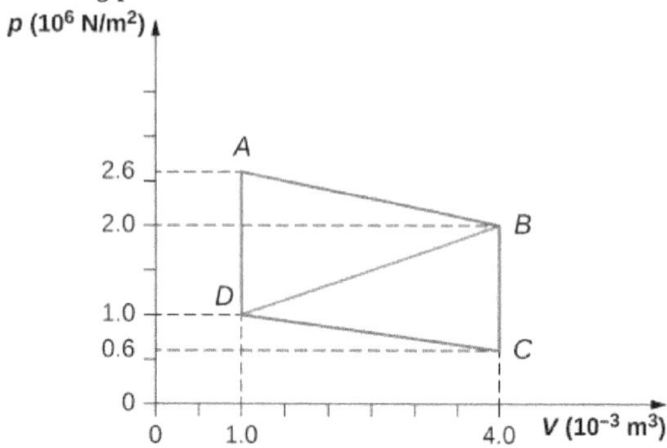

85. What is the net work output of a heat engine that follows path *ABDA* in the preceding problem with a straight line from *B* to *D*? Why is the work output less than for path *ABCDA*?

86. Five moles of a monatomic ideal gas in a cylinder at $27\ °\text{C}$ is expanded isothermally from a volume of 5 L to 10 L. (a) What is the change in internal energy? (b) How much work was done on the gas in the process? (c) How much heat was transferred to the gas?

87. Four moles of a monatomic ideal gas in a cylinder at $27\ °\text{C}$ is expanded at constant pressure equal to 1 atm until its volume doubles. (a) What is the change in internal energy? (b) How much work was done by the gas in the process? (c) How much heat was transferred to the gas?

88. Helium gas is cooled from $20\ °\text{C}$ to $10\ °\text{C}$ by expanding from 40 atm to 1 atm. If there is 1.4 mol of helium, (a) What is the final volume of helium? (b) What is the change in internal energy?

89. In an adiabatic process, oxygen gas in a container is compressed along a path that can be described by the following pressure in atm as a function of volume V, with $V_0 = 1L$: $p = (3.0\ \text{atm})(V/V_0)^{-1.2}$. The initial and final volumes during the process were 2 L and 1.5 L, respectively. Find the amount of work done on the gas.

90. A cylinder containing three moles of a monatomic ideal gas is heated at a constant pressure of 2 atm. The temperature of the gas changes from 300 K to 350 K as a result of the expansion. Find work done (a) on the gas; and (b) by the gas.

91. A cylinder containing three moles of nitrogen gas is heated at a constant pressure of 2 atm. The temperature of the gas changes from 300 K to 350 K as a result of the expansion. Find work done (a) on the gas, and (b) by the gas by using van der Waals equation of state instead of ideal gas law.

92. Two moles of a monatomic ideal gas such as oxygen is compressed adiabatically and reversibly from a state (3 atm, 5 L) to a state with a pressure of 4 atm. (a) Find the volume and temperature of the final state. (b) Find the temperature of the initial state. (c) Find work done by the gas in the process. (d) Find the change in internal energy in the process. Assume $C_V = 5R$ and $C_p = C_V + R$ for the diatomic ideal gas in the conditions given.

93. An insulated vessel contains 1.5 moles of argon at 2 atm. The gas initially occupies a volume of 5 L. As a result of the adiabatic expansion the pressure of the gas is reduced to 1 atm. (a) Find the volume and temperature of the final state. (b) Find the temperature of the gas in the initial state. (c) Find the work done by the gas in the process. (d) Find the change in the internal energy of the gas in the process.

CHALLENGE PROBLEMS

94. One mole of an ideal monatomic gas occupies a volume of $1.0 \times 10^{-2}\ \text{m}^3$ at a pressure of $2.0 \times 10^5\ \text{N/m}^2$. (a) What is the temperature of the gas? (b) The gas undergoes a quasi-static adiabatic compression until its volume is decreased to $5.0 \times 10^{-3}\ \text{m}^3$. What is the new gas temperature? (c) How much work is done on the gas during the compression? (d) What is the change in the internal energy of the gas?

95. One mole of an ideal gas is initially in a chamber of volume $1.0 \times 10^{-2}\ \text{m}^3$ and at a temperature of $27\ °\text{C}$. (a) How much heat is absorbed by the gas when it slowly expands isothermally to twice its initial volume? (b) Suppose the gas is slowly transformed to the same final state by first decreasing the pressure at constant volume and then expanding it isobarically. What is the heat transferred for this case? (c) Calculate the heat transferred when the gas is transformed quasi-statically to the same final state by expanding it isobarically, then decreasing its pressure at constant volume.

96. A bullet of mass 10 g is traveling horizontally at 200 m/s when it strikes and embeds in a pendulum bob of mass 2.0 kg. (a) How much mechanical energy is dissipated in the collision? (b) Assuming that C_v for the bob plus bullet is 3R, calculate the temperature increase of the system due to the collision. Take the molecular mass of the system to be 200 g/mol.

97. The insulated cylinder shown below is closed at both ends and contains an insulating piston that is free to move on frictionless bearings. The piston divides the chamber into two compartments containing gases A and B. Originally, each compartment has a volume of $5.0 \times 10^{-2}\ m^3$ and contains a monatomic ideal gas at a temperature of $0\,°C$ and a pressure of 1.0 atm. (a) How many moles of gas are in each compartment? (b) Heat Q is slowly added to A so that it expands and B is compressed until the pressure of both gases is 3.0 atm. Use the fact that the compression of B is adiabatic to determine the final volume of both gases. (c) What are their final temperatures? (d) What is the value of Q?

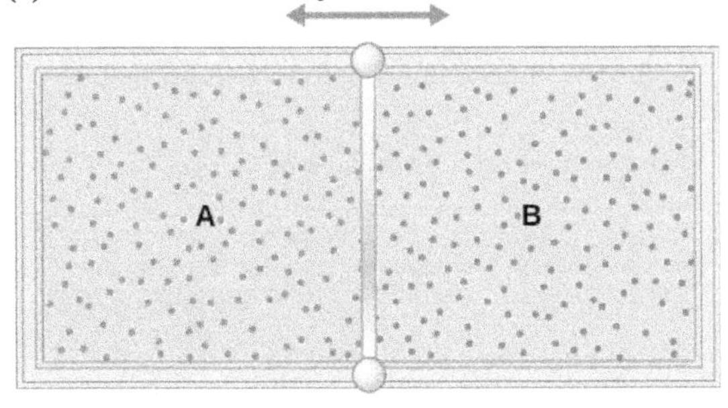

98. In a diesel engine, the fuel is ignited without a spark plug. Instead, air in a cylinder is compressed adiabatically to a temperature above the ignition temperature of the fuel; at the point of maximum compression, the fuel is injected into the cylinder. Suppose that air at $20\,°C$ is taken into the cylinder at a volume V_1 and then compressed adiabatically and quasi-statically to a temperature of $600\,°C$ and a volume V_2. If $\gamma = 1.4$, what is the ratio V_1/V_2? (Note: In an operating diesel engine, the compression is not quasi-static.)

4 | THE SECOND LAW OF THERMODYNAMICS

Figure 4.1 A xenon ion engine from the Jet Propulsion Laboratory shows the faint blue glow of charged atoms emitted from the engine. The ion propulsion engine is the first nonchemical propulsion to be used as the primary means of propelling a spacecraft.

Chapter Outline

Introduction

According to the first law of thermodynamics, the only processes that can occur are those that conserve energy. But this cannot be the only restriction imposed by nature, because many seemingly possible thermodynamic processes that would conserve energy do not occur. For example, when two bodies are in thermal contact, heat never flows from the colder body to the warmer one, even though this is not forbidden by the first law. So some other thermodynamic principles must be controlling the behavior of physical systems.

One such principle is the *second law of thermodynamics*, which limits the use of energy within a source. Energy cannot arbitrarily pass from one object to another, just as we cannot transfer heat from a cold object to a hot one without doing any work. We cannot unmix cream from coffee without a chemical process that changes the physical characteristics of the system or its environment. We cannot use internal energy stored in the air to propel a car, or use the energy of the ocean to run a ship, without disturbing something around that object.

In the chapter covering the first law of thermodynamics, we started our discussion with a joke by C. P. Snow stating that the first law means "you can't win." He paraphrased the second law as "you can't break even, except on a very cold day." Unless you are at zero kelvin, you cannot convert 100% of thermal energy into work. We start by discussing spontaneous processes and explain why some processes require work to occur even if energy would have been conserved.

4.1 | Reversible and Irreversible Processes

Learning Objectives

By the end of this section, you will be able to:

- Define reversible and irreversible processes
- State the second law of thermodynamics via an irreversible process

Consider an ideal gas that is held in half of a thermally insulated container by a wall in the middle of the container. The other half of the container is under vacuum with no molecules inside. Now, if we remove the wall in the middle quickly, the gas expands and fills up the entire container immediately, as shown in Figure 4.2.

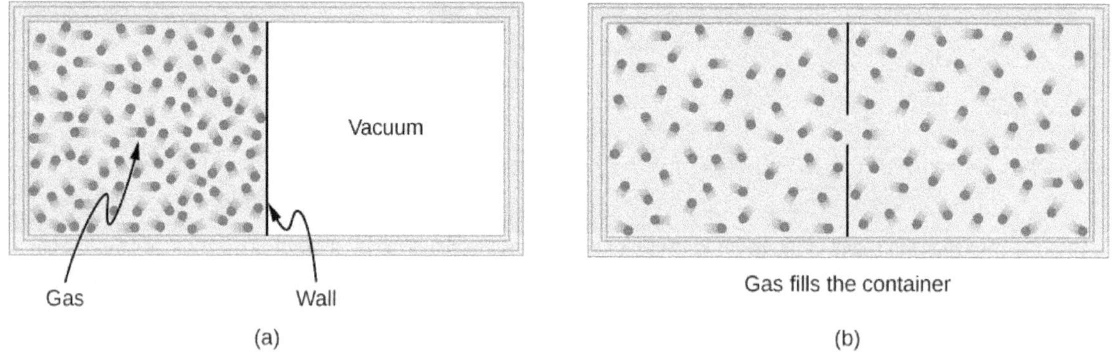

Figure 4.2 A gas expanding from half of a container to the entire container (a) before and (b) after the wall in the middle is removed.

Because half of the container is under vacuum before the gas expands there, we do not expect any work to be done by the system—that is, $W = 0$—because no force from the vacuum is exerted on the gas during the expansion. If the container is thermally insulated from the rest of the environment, we do not expect any heat transfer to the system either, so $Q = 0$. Then the first law of thermodynamics leads to the change of the internal energy of the system,

$$\Delta E_{\text{int}} = Q - W = 0.$$

For an ideal gas, if the internal energy doesn't change, then the temperature stays the same. Thus, the equation of state of the ideal gas gives us the final pressure of the gas, $p = nRT/V = p_0/2$, where p_0 is the pressure of the gas before the expansion. The volume is doubled and the pressure is halved, but nothing else seems to have changed during the expansion.

All of this discussion is based on what we have learned so far and makes sense. Here is what puzzles us: Can all the molecules go backward to the original half of the container in some future time? Our intuition tells us that this is going to be very unlikely, even though nothing we have learned so far prevents such an event from happening, regardless of how small the probability is. What we are really asking is whether the expansion into the vacuum half of the container is *reversible*.

A **reversible process** is a process in which the system and environment can be restored to exactly the same initial states that they were in before the process occurred, if we go backward along the path of the process. The necessary condition for a reversible process is therefore the quasi-static requirement. Note that it is quite easy to restore a system to its original state; the hard part is to have its environment restored to its original state at the same time. For example, in the example of an ideal gas expanding into vacuum to twice its original volume, we can easily push it back with a piston and restore its temperature and pressure by removing some heat from the gas. The problem is that we cannot do it without changing something in its surroundings, such as dumping some heat there.

A reversible process is truly an ideal process that rarely happens. We can make certain processes close to reversible and therefore use the consequences of the corresponding reversible processes as a starting point or reference. In reality, almost all processes are irreversible, and some properties of the environment are altered when the properties of the system are restored. The expansion of an ideal gas, as we have just outlined, is irreversible because the process is not even quasi-static, that is, not in an equilibrium state at any moment of the expansion.

From the microscopic point of view, a particle described by Newton's second law can go backward if we flip the direction of time. But this is not the case, in practical terms, in a macroscopic system with more than 10^{23} particles or molecules, where numerous collisions between these molecules tend to erase any trace of memory of the initial trajectory of each of the particles. For example, we can actually estimate the chance for all the particles in the expanded gas to go back to the original half of the container, but the current age of the universe is still not long enough for it to happen even once.

An **irreversible process** is what we encounter in reality almost all the time. The system and its environment cannot be restored to their original states at the same time. Because this is what happens in nature, it is also called a natural process. The sign of an irreversible process comes from the finite gradient between the states occurring in the actual process. For example, when heat flows from one object to another, there is a finite temperature difference (gradient) between the two objects. More importantly, at any given moment of the process, the system most likely is not at equilibrium or in a well-defined state. This phenomenon is called **irreversibility**.

Let us see another example of irreversibility in thermal processes. Consider two objects in thermal contact: one at temperature T_1 and the other at temperature $T_2 > T_1$, as shown in Figure 4.3.

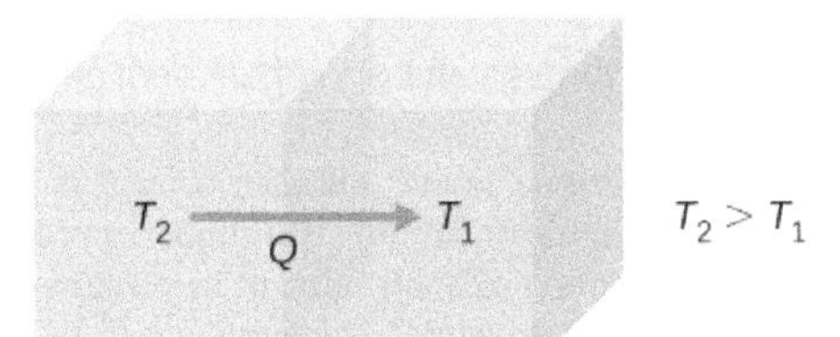

Figure 4.3 Spontaneous heat flow from an object at higher temperature T_2 to another at lower temperature T_1.

We know from common personal experience that heat flows from a hotter object to a colder one. For example, when we hold a few pieces of ice in our hands, we feel cold because heat has left our hands into the ice. The opposite is true when we hold one end of a metal rod while keeping the other end over a fire. Based on all of the experiments that have been done on spontaneous heat transfer, the following statement summarizes the governing principle:

Second Law of Thermodynamics (Clausius statement)

Heat never flows spontaneously from a colder object to a hotter object.

This statement turns out to be one of several different ways of stating the second law of thermodynamics. The form of this statement is credited to German physicist Rudolf Clausius (1822−1888) and is referred to as the **Clausius statement of the second law of thermodynamics**. The word "spontaneously" here means no other effort has been made by a third party, or one that is neither the hotter nor colder object. We will introduce some other major statements of the second law and show that they imply each other. In fact, all the different statements of the second law of thermodynamics can be shown to be equivalent, and all lead to the irreversibility of spontaneous heat flow between macroscopic objects of a very large number of molecules or particles.

Both isothermal and adiabatic processes sketched on a *pV* graph (discussed in The First Law of Thermodynamics) are reversible in principle because the system is always at an equilibrium state at any point of the processes and can go forward or backward along the given curves. Other idealized processes can be represented by *pV* curves; Table 4.1 summarizes the most common reversible processes.

Process	Constant Quantity and Resulting Fact
Isobaric	Constant pressure $W = p\Delta V$
Isochoric	Constant volume $W = 0$
Isothermal	Constant temperature $\Delta T = 0$
Adiabatic	No heat transfer $Q = 0$

Table 4.1 Summary of Simple Thermodynamic Processes

4.2 | Heat Engines

Learning Objectives

By the end of this section, you will be able to:

- Describe the function and components of a heat engine
- Explain the efficiency of an engine
- Calculate the efficiency of an engine for a given cycle of an ideal gas

A **heat engine** is a device used to extract heat from a source and then convert it into mechanical work that is used for all sorts of applications. For example, a steam engine on an old-style train can produce the work needed for driving the train. Several questions emerge from the construction and application of heat engines. For example, what is the maximum percentage of the heat extracted that can be used to do work? This turns out to be a question that can only be answered through the second law of thermodynamics.

The second law of thermodynamics can be formally stated in several ways. One statement presented so far is about the direction of spontaneous heat flow, known as the Clausius statement. A couple of other statements are based on heat engines. *Whenever we consider heat engines and associated devices such as refrigerators and heat pumps, we do not use the normal sign convention for heat and work*. For convenience, we assume that the symbols Q_h, Q_c, and W represent only the amounts of heat transferred and work delivered, regardless what the givers or receivers are. Whether heat is entering or leaving a system and work is done to or by a system are indicated by proper signs in front of the symbols and by the directions of arrows in diagrams.

It turns out that we need more than one heat source/sink to construct a heat engine. We will come back to this point later in the chapter, when we compare different statements of the second law of thermodynamics. For the moment, we assume that a heat engine is constructed between a heat source (high-temperature reservoir or hot reservoir) and a heat sink (low-temperature reservoir or cold reservoir), represented schematically in Figure 4.4. The engine absorbs heat Q_h from a heat source (**hot reservoir**) of Kelvin temperature T_h, uses some of that energy to produce useful work W, and then discards the remaining energy as heat Q_c into a heat sink (**cold reservoir**) of Kelvin temperature T_c. Power plants and internal combustion engines are examples of heat engines. Power plants use steam produced at high temperature to drive electric generators, while exhausting heat to the atmosphere or a nearby body of water in the role of the heat sink. In an internal combustion engine, a hot gas-air mixture is used to push a piston, and heat is exhausted to the nearby atmosphere in a similar manner.

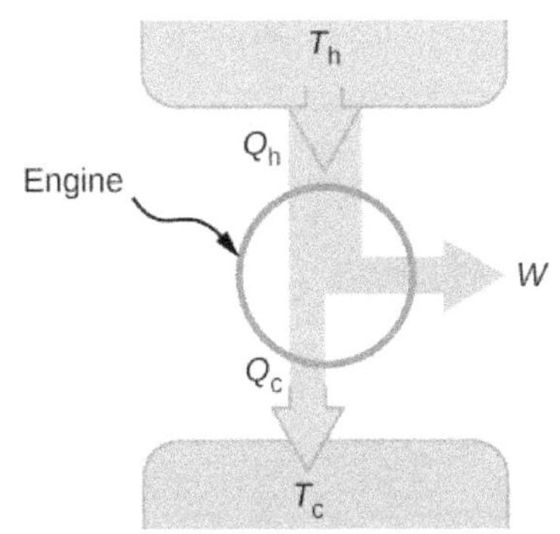

Figure 4.4 A schematic representation of a heat engine. Energy flows from the hot reservoir to the cold reservoir while doing work.

Actual heat engines have many different designs. Examples include internal combustion engines, such as those used in most cars today, and external combustion engines, such as the steam engines used in old steam-engine trains. Figure 4.5 shows a photo of a nuclear power plant in operation. The atmosphere around the reactors acts as the cold reservoir, and the heat generated from the nuclear reaction provides the heat from the hot reservoir.

Figure 4.5 The heat exhausted from a nuclear power plant goes to the cooling towers, where it is released into the atmosphere.

Heat engines operate by carrying a *working substance* through a cycle. In a steam power plant, the working substance is water, which starts as a liquid, becomes vaporized, is then used to drive a turbine, and is finally condensed back into the liquid state. As is the case for all working substances in cyclic processes, once the water returns to its initial state, it repeats the same sequence.

For now, we assume that the cycles of heat engines are reversible, so there is no energy loss to friction or other irreversible effects. Suppose that the engine of Figure 4.4 goes through one complete cycle and that $Q_h, Q_c,$ and W represent the heats exchanged and the work done for that cycle. Since the initial and final states of the system are the same, $\Delta E_{int} = 0$ for the cycle. We therefore have from the first law of thermodynamics,

$$W = Q - \Delta E_{int} = (Q_h - Q_c) - 0,$$

so that

$$W = Q_h - Q_c. \quad (4.1)$$

The most important measure of a heat engine is its **efficiency (*e*)**, which is simply "what we get out" divided by "what we put in" during each cycle, as defined by $e = W_{out}/Q_{in}$.

With a heat engine working between two heat reservoirs, we get out W and put in Q_h, so the efficiency of the engine is

$$e = \frac{W}{Q_h} = 1 - \frac{Q_c}{Q_h}. \tag{4.2}$$

Here, we used Equation 4.1, $W = Q_h - Q_c$, in the final step of this expression for the efficiency.

Example 4.1

A Lawn Mower

A lawn mower is rated to have an efficiency of 25.0% and an average power of 3.00 kW. What are (a) the average work and (b) the minimum heat discharge into the air by the lawn mower in one minute of use?

Strategy

From the average power—that is, the rate of work production—we can figure out the work done in the given elapsed time. Then, from the efficiency given, we can figure out the minimum heat discharge $Q_c = Q_h(1 - e)$ with $Q_h = Q_c + W$.

Solution

a. The average work delivered by the lawn mower is

$$W = P\Delta t = 3.00 \times 10^3 \times 60 \times 1.00\,\text{J} = 180\,\text{kJ}.$$

b. The minimum heat discharged into the air is given by

$$Q_c = Q_h(1 - e) = (Q_c + W)(1 - e),$$

which leads to

$$Q_c = W(1/e - 1) = 180 \times (1/0.25 - 1)\,\text{kJ} = 540\,\text{kJ}.$$

Significance

As the efficiency rises, the minimum heat discharged falls. This helps our environment and atmosphere by not having as much waste heat expelled.

4.3 | Refrigerators and Heat Pumps

Learning Objectives

By the end of this section, you will be able to:

- Describe a refrigerator and a heat pump and list their differences
- Calculate the performance coefficients of simple refrigerators and heat pumps

The cycles we used to describe the engine in the preceding section are all reversible, so each sequence of steps can just as easily be performed in the opposite direction. In this case, the engine is known as a refrigerator or a heat pump, depending on what is the focus: the heat removed from the cold reservoir or the heat dumped to the hot reservoir. Either a refrigerator or a heat pump is an engine running in reverse. For a **refrigerator**, the focus is on removing heat from a specific area. For a **heat pump**, the focus is on dumping heat to a specific area.

We first consider a refrigerator (Figure 4.6). The purpose of this engine is to remove heat from the cold reservoir, which is the space inside the refrigerator for an actual household refrigerator or the space inside a building for an air-conditioning unit.

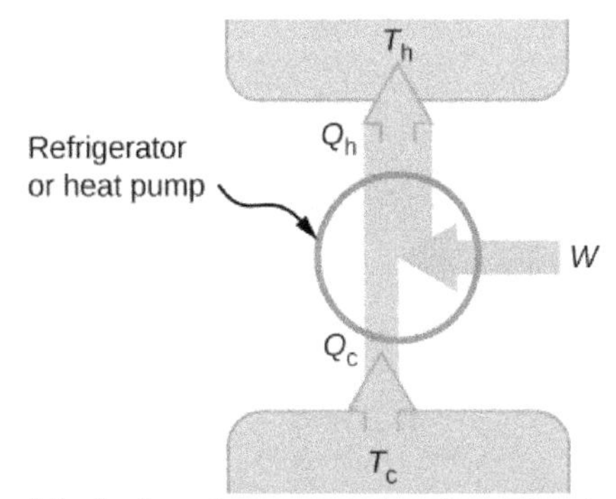

Figure 4.6 A schematic representation of a refrigerator (or a heat pump). The arrow next to work (*W*) indicates work being put into the system.

A refrigerator (or heat pump) absorbs heat Q_c from the cold reservoir at Kelvin temperature T_c and discards heat Q_h to the hot reservoir at Kelvin temperature T_h, while work *W* is done on the engine's working substance, as shown by the arrow pointing toward the system in the figure. A household refrigerator removes heat from the food within it while exhausting heat to the surrounding air. The required work, for which we pay in our electricity bill, is performed by the motor that moves a coolant through the coils. A schematic sketch of a household refrigerator is given in Figure 4.7.

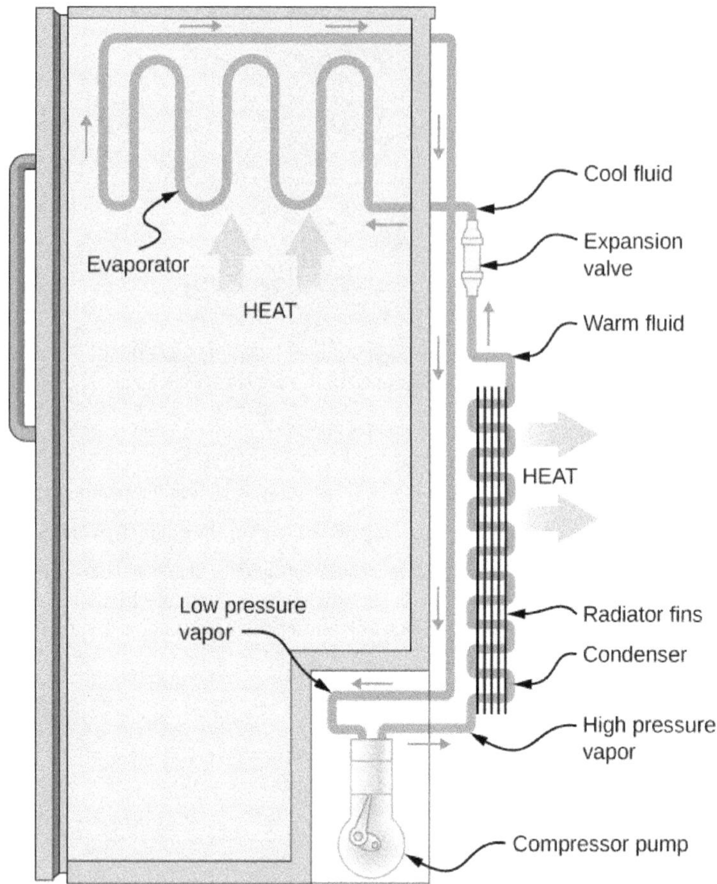

Figure 4.7 A schematic diagram of a household refrigerator. A coolant with a boiling temperature below the freezing point of water is sent through the cycle (clockwise in this diagram). The coolant extracts heat from the refrigerator at the evaporator, causing coolant to vaporize. It is then compressed and sent through the condenser, where it exhausts heat to the outside.

The effectiveness or **coefficient of performance** K_R of a refrigerator is measured by the heat removed from the cold reservoir divided by the work done by the working substance cycle by cycle:

$$K_R = \frac{Q_c}{W} = \frac{Q_c}{Q_h - Q_c}. \tag{4.3}$$

Note that we have used the condition of energy conservation, $W = Q_h - Q_c$, in the final step of this expression.

The effectiveness or coefficient of performance K_P of a heat pump is measured by the heat dumped to the hot reservoir divided by the work done to the engine on the working substance cycle by cycle:

$$K_P = \frac{Q_h}{W} = \frac{Q_h}{Q_h - Q_c}. \tag{4.4}$$

Once again, we use the energy conservation condition $W = Q_h - Q_c$ to obtain the final step of this expression.

4.4 | Statements of the Second Law of Thermodynamics

Learning Objectives

By the end of this section, you will be able to:

- Contrast the second law of thermodynamics statements according to Kelvin and Clausius formulations
- Interpret the second of thermodynamics via irreversibility

Earlier in this chapter, we introduced the Clausius statement of the second law of thermodynamics, which is based on the irreversibility of spontaneous heat flow. As we remarked then, the second law of thermodynamics can be stated in several different ways, and all of them can be shown to imply the others. In terms of heat engines, the second law of thermodynamics may be stated as follows:

Second Law of Thermodynamics (Kelvin statement)

It is impossible to convert the heat from a single source into work without any other effect.

This is known as the **Kelvin statement of the second law of thermodynamics**. This statement describes an unattainable " **perfect engine**," as represented schematically in Figure 4.8(a). Note that "without any other effect" is a very strong restriction. For example, an engine can absorb heat and turn it all into work, *but not if it completes a cycle*. Without completing a cycle, the substance in the engine is not in its original state and therefore an "other effect" has occurred. Another example is a chamber of gas that can absorb heat from a heat reservoir and do work isothermally against a piston as it expands. However, if the gas were returned to its initial state (that is, made to complete a cycle), it would have to be compressed and heat would have to be extracted from it.

The Kelvin statement is a manifestation of a well-known engineering problem. Despite advancing technology, we are not able to build a heat engine that is 100% efficient. The first law does not exclude the possibility of constructing a perfect engine, but the second law forbids it.

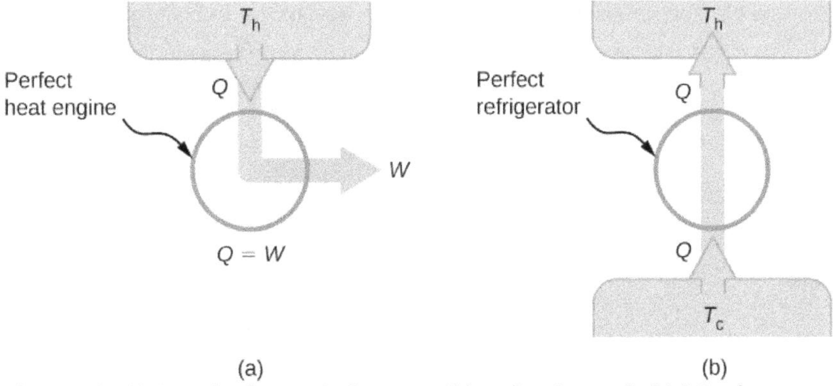

Figure 4.8 (a) A "perfect heat engine" converts all input heat into work. (b) A "perfect refrigerator" transports heat from a cold reservoir to a hot reservoir without work input. Neither of these devices is achievable in reality.

We can show that the Kelvin statement is equivalent to the Clausius statement if we view the two objects in the Clausius statement as a cold reservoir and a hot reservoir. Thus, the Clausius statement becomes: *It is impossible to construct a refrigerator that transfers heat from a cold reservoir to a hot reservoir without aid from an external source*. The Clausius statement is related to the everyday observation that heat never flows spontaneously from a cold object to a hot object. *Heat transfer in the direction of increasing temperature always requires some energy input*. A " **perfect refrigerator**," shown in Figure 4.8(b), which works without such external aid, is impossible to construct.

To prove the equivalence of the Kelvin and Clausius statements, we show that if one statement is false, it necessarily follows that the other statement is also false. Let us first assume that the Clausius statement is false, so that the perfect refrigerator of Figure 4.8(b) does exist. The refrigerator removes heat Q from a cold reservoir at a temperature T_c and transfers all of it to a hot reservoir at a temperature T_h. Now consider a real heat engine working in the same temperature range. It extracts heat $Q + \Delta Q$ from the hot reservoir, does work W, and discards heat Q to the cold reservoir. From the first law, these quantities are related by $W = (Q + \Delta Q) - Q = \Delta Q$.

Suppose these two devices are combined as shown in Figure 4.9. The net heat removed from the hot reservoir is ΔQ, no net heat transfer occurs to or from the cold reservoir, and work W is done on some external body. Since $W = \Delta Q$, the combination of a perfect refrigerator and a real heat engine is itself a perfect heat engine, thereby contradicting the Kelvin statement. Thus, if the Clausius statement is false, the Kelvin statement must also be false.

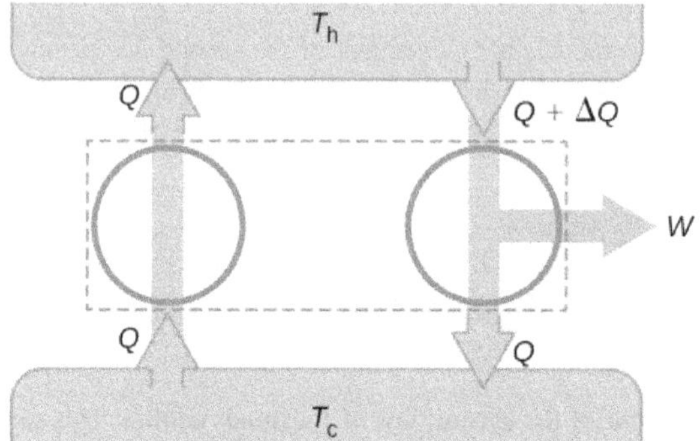

Figure 4.9 Combining a perfect refrigerator and a real heat engine yields a perfect heat engine because $W = \Delta Q$.

Using the second law of thermodynamics, we now prove two important properties of heat engines operating between two heat reservoirs. The first property is that *any reversible engine operating between two reservoirs has a greater efficiency than any irreversible engine operating between the same two reservoirs.*

The second property to be demonstrated is that *all reversible engines operating between the same two reservoirs have the same efficiency*. To show this, we start with the two engines D and E of Figure 4.10(a), which are operating between two common heat reservoirs at temperatures T_h and T_c. First, we assume that D is a reversible engine and that E is a hypothetical irreversible engine that has a higher efficiency than D. If both engines perform the same amount of work W per cycle, it follows from Equation 4.2 that $Q_h > Q'_h$. It then follows from the first law that $Q_c > Q'_c$.

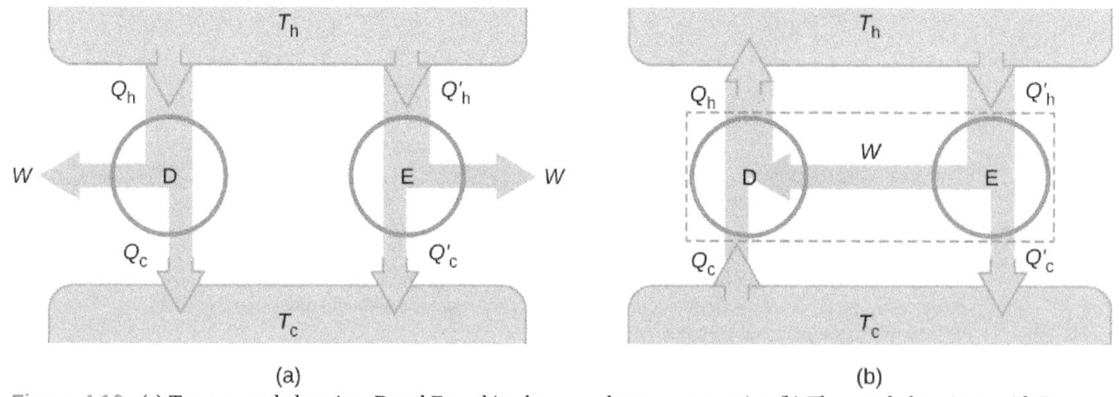

Figure 4.10 (a) Two uncoupled engines D and E working between the same reservoirs. (b) The coupled engines, with D working in reverse.

Suppose the cycle of D is reversed so that it operates as a refrigerator, and the two engines are coupled such that the work output of E is used to drive D, as shown in Figure 4.10(b). Since $Q_h > Q'_h$ and $Q_c > Q'_c$, the net result of each cycle is equivalent to a spontaneous transfer of heat from the cold reservoir to the hot reservoir, a process the second law does not allow. The original assumption must therefore be wrong, and it is impossible to construct an irreversible engine such that E is more efficient than the reversible engine D.

Now it is quite easy to demonstrate that the efficiencies of all reversible engines operating between the same reservoirs are equal. Suppose that D and E are both reversible engines. If they are coupled as shown in Figure 4.10(b), the efficiency of E cannot be greater than the efficiency of D, or the second law would be violated. If both engines are then reversed, the same reasoning implies that the efficiency of D cannot be greater than the efficiency of E. Combining these results leads to the conclusion that all reversible engines working between the same two reservoirs have the same efficiency.

4.1 Check Your Understanding What is the efficiency of a perfect heat engine? What is the coefficient of performance of a perfect refrigerator?

4.2 Check Your Understanding Show that $Q_{\text{h}} - Q'_{\text{h}} = Q_{\text{c}} - Q'_{\text{c}}$ for the hypothetical engine of Figure 4.10(b).

4.5 | The Carnot Cycle

Learning Objectives

- Describe the Carnot cycle with the roles of all four processes involved
- Outline the Carnot principle and its implications
- Demonstrate the equivalence of the Carnot principle and the second law of thermodynamics

In the early 1820s, Sadi Carnot (1786−1832), a French engineer, became interested in improving the efficiencies of practical heat engines. In 1824, his studies led him to propose a hypothetical working cycle with the highest possible efficiency between the same two reservoirs, known now as the **Carnot cycle**. An engine operating in this cycle is called a **Carnot engine**. The Carnot cycle is of special importance for a variety of reasons. At a practical level, this cycle represents a reversible model for the steam power plant and the refrigerator or heat pump. Yet, it is also very important theoretically, for it plays a major role in the development of another important statement of the second law of thermodynamics. Finally, because only two reservoirs are involved in its operation, it can be used along with the second law of thermodynamics to define an absolute temperature scale that is truly independent of any substance used for temperature measurement.

With an ideal gas as the working substance, the steps of the Carnot cycle, as represented by Figure 4.11, are as follows.

1. *Isothermal expansion.* The gas is placed in thermal contact with a heat reservoir at a temperature T_{h}. The gas absorbs heat Q_{h} from the heat reservoir and is allowed to expand isothermally, doing work W_1. Because the internal energy E_{int} of an ideal gas is a function of the temperature only, the change of the internal energy is zero, that is, $\Delta E_{\text{int}} = 0$ during this isothermal expansion. With the first law of thermodynamics, $\Delta E_{\text{int}} = Q - W$, we find that the heat absorbed by the gas is

$$Q_{\text{h}} = W_1 = nRT_{\text{h}} \ln\frac{V_N}{V_M}.$$

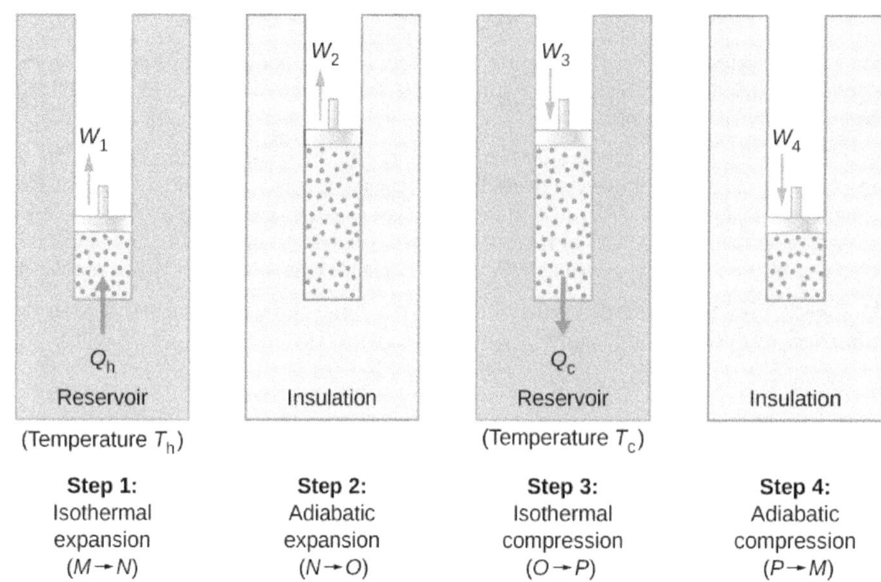

Figure 4.11 The four processes of the Carnot cycle. The working substance is assumed to be an ideal gas whose thermodynamic path *MNOP* is represented in Figure 4.12.

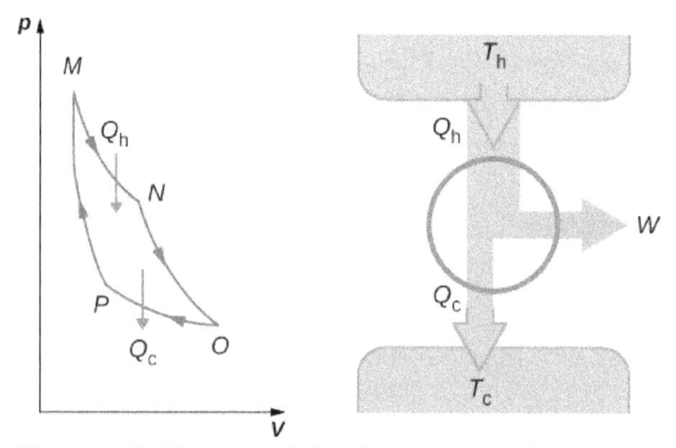

Figure 4.12 The total work done by the gas in the Carnot cycle is shown and given by the area enclosed by the loop *MNOPM*.

2. *Adiabatic expansion*. The gas is thermally isolated and allowed to expand further, doing work W_2. Because this expansion is adiabatic, the temperature of the gas falls—in this case, from T_h to T_c. From $pV^\gamma =$ constant and the equation of state for an ideal gas, $pV = nRT$, we have

$$TV^{\gamma - 1} = \text{constant},$$

so that

$$T_h V_N{}^{\gamma - 1} = T_c V_O{}^{\gamma - 1}.$$

3. *Isothermal compression*. The gas is placed in thermal contact with a cold reservoir at temperature T_c and compressed isothermally. During this process, work W_3 is done on the gas and it gives up heat Q_c to the cold reservoir. The reasoning used in step 1 now yields

$$Q_c = nRT_c \ln\frac{V_O}{V_P},$$

where Q_c is the heat dumped to the cold reservoir by the gas.

4. *Adiabatic compression*. The gas is thermally isolated and returned to its initial state by compression. In this process, work W_4 is done on the gas. Because the compression is adiabatic, the temperature of the gas rises—from T_c to T_h in this particular case. The reasoning of step 2 now gives

$$T_c V_P{}^{\gamma-1} = T_h V_M{}^{\gamma-1}.$$

The total work done by the gas in the Carnot cycle is given by

$$W = W_1 + W_2 - W_3 - W_4.$$

This work is equal to the area enclosed by the loop shown in the *pV* diagram of Figure 4.12. Because the initial and final states of the system are the same, the change of the internal energy of the gas in the cycle must be zero, that is, $\Delta E_{int} = 0$. The first law of thermodynamics then gives

$$W = Q - \Delta E_{int} = (Q_h - Q_c) - 0,$$

and

$$W = Q_h - Q_c.$$

To find the efficiency of this engine, we first divide Q_c by Q_h:

$$\frac{Q_c}{Q_h} = \frac{T_c}{T_h}\frac{\ln V_O/V_P}{\ln V_N/V_M}.$$

When the adiabatic constant from step 2 is divided by that of step 4, we find

$$\frac{V_O}{V_P} = \frac{V_N}{V_M}.$$

Substituting this into the equation for Q_c/Q_h, we obtain

$$\frac{Q_c}{Q_h} = \frac{T_c}{T_h}.$$

Finally, with Equation 4.2, we find that the efficiency of this ideal gas Carnot engine is given by

$$e = 1 - \frac{T_c}{T_h}. \quad (4.5)$$

An engine does not necessarily have to follow a Carnot engine cycle. All engines, however, have the same *net* effect, namely the absorption of heat from a hot reservoir, the production of work, and the discarding of heat to a cold reservoir. This leads us to ask: Do all reversible cycles operating between the same two reservoirs have the same efficiency? The answer to this question comes from the second law of thermodynamics discussed earlier: *All reversible engine cycles produce exactly the same efficiency*. Also, as you might expect, all real engines operating between two reservoirs are less efficient than reversible engines operating between the same two reservoirs. This too is a consequence of the second law of thermodynamics shown earlier.

The cycle of an ideal gas Carnot refrigerator is represented by the *pV* diagram of Figure 4.13. It is a Carnot engine operating in reverse. The refrigerator extracts heat Q_c from a cold-temperature reservoir at T_c when the ideal gas expands isothermally. The gas is then compressed adiabatically until its temperature reaches T_h, after which an isothermal compression of the gas results in heat Q_h being discarded to a high-temperature reservoir at T_h. Finally, the cycle is completed by an adiabatic expansion of the gas, causing its temperature to drop to T_c.

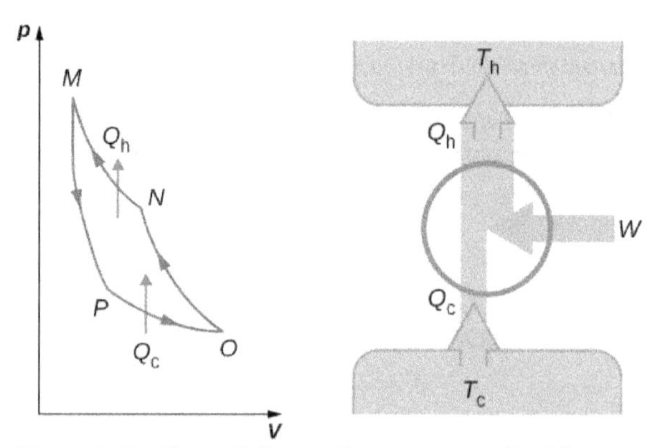

Figure 4.13 The work done on the gas in one cycle of the Carnot refrigerator is shown and given by the area enclosed by the loop *MPONM*.

The work done on the ideal gas is equal to the area enclosed by the path of the *pV* diagram. From the first law, this work is given by

$$W = Q_h - Q_c.$$

An analysis just like the analysis done for the Carnot engine gives

$$\frac{Q_c}{T_c} = \frac{Q_h}{T_h}.$$

When combined with Equation 4.3, this yields

$$K_R = \frac{T_c}{T_h - T_c} \tag{4.6}$$

for the coefficient of performance of the ideal-gas Carnot refrigerator. Similarly, we can work out the coefficient of performance for a Carnot heat pump as

$$K_P = \frac{Q_h}{Q_h - Q_c} = \frac{T_h}{T_h - T_c}. \tag{4.7}$$

We have just found equations representing the efficiency of a Carnot engine and the coefficient of performance of a Carnot refrigerator or a Carnot heat pump, assuming an ideal gas for the working substance in both devices. However, these equations are more general than their derivations imply. We will soon show that they are both valid no matter what the working substance is.

Carnot summarized his study of the Carnot engine and Carnot cycle into what is now known as **Carnot's principle**:

Carnot's Principle

No engine working between two reservoirs at constant temperatures can have a greater efficiency than a reversible engine.

This principle can be viewed as another statement of the second law of thermodynamics and can be shown to be equivalent to the Kelvin statement and the Clausius statement.

Example 4.2

The Carnot Engine

A Carnot engine has an efficiency of 0.60 and the temperature of its cold reservoir is 300 K. (a) What is the temperature of the hot reservoir? (b) If the engine does 300 J of work per cycle, how much heat is removed from the high-temperature reservoir per cycle? (c) How much heat is exhausted to the low-temperature reservoir per cycle?

Strategy

From the temperature dependence of the thermal efficiency of the Carnot engine, we can find the temperature of the hot reservoir. Then, from the definition of the efficiency, we can find the heat removed when the work done by the engine is given. Finally, energy conservation will lead to how much heat must be dumped to the cold reservoir.

Solution

a. From $e = 1 - T_c / T_h$ we have

$$0.60 = 1 - \frac{300\text{ K}}{T_h},$$

so that the temperature of the hot reservoir is

$$T_h = \frac{300\text{ K}}{1 - 0.60} = 750\text{ K}.$$

b. By definition, the efficiency of the engine is $e = W/Q$, so that the heat removed from the high-temperature reservoir per cycle is

$$Q_h = \frac{W}{e} = \frac{300\text{ J}}{0.60} = 500\text{ J}.$$

c. From the first law, the heat exhausted to the low-temperature reservoir per cycle by the engine is

$$Q_c = Q_h - W = 500\text{ J} - 300\text{ J} = 200\text{ J}.$$

Significance

A Carnot engine has the maximum possible efficiency of converting heat into work between two reservoirs, but this does not necessarily mean it is 100% efficient. As the difference in temperatures of the hot and cold reservoir increases, the efficiency of a Carnot engine increases.

Example 4.3

A Carnot Heat Pump

Imagine a Carnot heat pump operates between an outside temperature of $0\,°\text{C}$ and an inside temperature of $20.0\,°\text{C}$. What is the work needed if the heat delivered to the inside of the house is 30.0 kJ?

Strategy

Because the heat pump is assumed to be a Carnot pump, its performance coefficient is given by $K_P = Q_h / W = T_h / (T_h - T_c)$. Thus, we can find the work *W* from the heat delivered Q_h.

Solution

The work needed is obtained from

$$W = Q_h / K_P = Q_h(T_h - T_c)/T_h = 30\text{ kJ} \times (293\text{ K} - 273\text{ K})/293\text{ K} = 2\text{ kJ}.$$

Significance

We note that this work depends not only on the heat delivered to the house but also on the temperatures outside and inside. The dependence on the temperature outside makes them impractical to use in areas where the temperature is much colder outside than room temperature.

In terms of energy costs, the heat pump is a very economical means for heating buildings (Figure 4.14). Contrast this method with turning electrical energy directly into heat with resistive heating elements. In this case, one unit of electrical energy furnishes at most only one unit of heat. Unfortunately, heat pumps have problems that do limit their usefulness. They are quite expensive to purchase compared to resistive heating elements, and, as the performance coefficient for a Carnot heat pump shows, they become less effective as the outside temperature decreases. In fact, below about $-10\ °C$, the heat they furnish is less than the energy used to operate them.

Figure 4.14 A photograph of a heat pump (large box) located outside a house. This heat pump is located in a warm climate area, like the southern United States, since it would be far too inefficient located in the northern half of the United States. (credit: modification of work by Peter Stevens)

4.3 Check Your Understanding A Carnot engine operates between reservoirs at $400\ °C$ and $30\ °C$. (a) What is the efficiency of the engine? (b) If the engine does 5.0 J of work per cycle, how much heat per cycle does it absorb from the high-temperature reservoir? (c) How much heat per cycle does it exhaust to the cold-temperature reservoir? (d) What temperatures at the cold reservoir would give the minimum and maximum efficiency?

4.4 Check Your Understanding A Carnot refrigerator operates between two heat reservoirs whose temperatures are $0\ °C$ and $25\ °C$. (a) What is the coefficient of performance of the refrigerator? (b) If 200 J of work are done on the working substance per cycle, how much heat per cycle is extracted from the cold reservoir? (c) How much heat per cycle is discarded to the hot reservoir?

4.6 | Entropy

Learning Objectives

By the end of this section you will be able to:

- Describe the meaning of entropy
- Calculate the change of entropy for some simple processes

The second law of thermodynamics is best expressed in terms of a *change* in the thermodynamic variable known as **entropy**, which is represented by the symbol S. Entropy, like internal energy, is a state function. This means that when a system makes a transition from one state into another, the change in entropy ΔS is independent of path and depends only on the thermodynamic variables of the two states.

We first consider ΔS for a system undergoing a reversible process at a constant temperature. In this case, the change in entropy of the system is given by

$$\Delta S = \frac{Q}{T}, \tag{4.8}$$

where Q is the heat exchanged by the system kept at a temperature T (in kelvin). If the system absorbs heat—that is, with $Q > 0$—the entropy of the system increases. As an example, suppose a gas is kept at a constant temperature of 300 K while it absorbs 10 J of heat in a reversible process. Then from Equation 4.8, the entropy change of the gas is

$$\Delta S = \frac{10 \text{ J}}{300 \text{ K}} = 0.033 \text{ J/K}.$$

Similarly, if the gas loses 5.0 J of heat; that is, $Q = -5.0 \text{ J}$, at temperature $T = 200 \text{ K}$, we have the entropy change of the system given by

$$\Delta S = \frac{-5.0 \text{ J}}{200 \text{ K}} = -0.025 \text{ J/K}.$$

Example 4.4

Entropy Change of Melting Ice

Heat is slowly added to a 50-g chunk of ice at 0 °C until it completely melts into water at the same temperature. What is the entropy change of the ice?

Strategy

Because the process is slow, we can approximate it as a reversible process. The temperature is a constant, and we can therefore use Equation 4.8 in the calculation.

Solution

The ice is melted by the addition of heat:

$$Q = mL_f = 50 \text{ g} \times 335 \text{ J/g} = 16.8 \text{ kJ}.$$

In this reversible process, the temperature of the ice-water mixture is fixed at 0 °C or 273 K. Now from $\Delta S = Q/T$, the entropy change of the ice is

$$\Delta S = \frac{16.8 \text{ kJ}}{273 \text{ K}} = 61.5 \text{ J/K}$$

when it melts to water at 0 °C.

Significance

During a phase change, the temperature is constant, allowing us to use Equation 4.8 to solve this problem. The same equation could also be used if we changed from a liquid to a gas phase, since the temperature does not change during that process either.

The change in entropy of a system for an arbitrary, reversible transition for which the temperature is not necessarily constant is defined by modifying $\Delta S = Q/T$. Imagine a system making a transition from state A to B in small, discrete steps. The temperatures associated with these states are T_A and T_B, respectively. During each step of the transition, the system exchanges heat ΔQ_i reversibly at a temperature T_i. This can be accomplished experimentally by placing the system in thermal contact with a large number of heat reservoirs of varying temperatures T_i, as illustrated in Figure 4.15. The change in entropy for each step is $\Delta S_i = Q_i/T_i$. The net change in entropy of the system for the transition is

$$\Delta S = S_B - S_A = \sum_i \Delta S_i = \sum_i \frac{\Delta Q_i}{T_i}. \tag{4.9}$$

We now take the limit as $\Delta Q_i \to 0$, and the number of steps approaches infinity. Then, replacing the summation by an integral, we obtain

$$\Delta S = S_B - S_A = \int_A^B \frac{dQ}{T}, \tag{4.10}$$

where the integral is taken between the initial state A and the final state B. This equation is valid only if the transition from A to B is reversible.

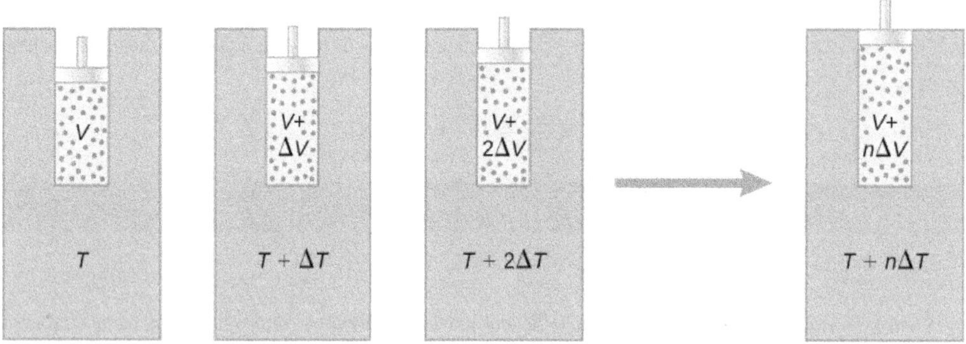

Figure 4.15 The gas expands at constant pressure as its temperature is increased in small steps through the use of a series of heat reservoirs.

As an example, let us determine the net entropy change of a reversible engine while it undergoes a single Carnot cycle. In the adiabatic steps 2 and 4 of the cycle shown in Figure 4.11, no heat exchange takes place, so $\Delta S_2 = \Delta S_4 = \int dQ/T = 0$. In step 1, the engine absorbs heat Q_h at a temperature T_h, so its entropy change is $\Delta S_1 = Q_\text{h}/T_\text{h}$. Similarly, in step 3, $\Delta S_3 = -Q_\text{c}/T_\text{c}$. The net entropy change of the engine in one cycle of operation is then

$$\Delta S_E = \Delta S_1 + \Delta S_2 + \Delta S_3 + \Delta S_4 = \frac{Q_\text{h}}{T_\text{h}} - \frac{Q_\text{c}}{T_\text{c}}.$$

However, we know that for a Carnot engine,

$$\frac{Q_\text{h}}{T_\text{h}} = \frac{Q_\text{c}}{T_\text{c}},$$

so

$$\Delta S_E = 0.$$

There is no net change in the entropy of the Carnot engine over a complete cycle. Although this result was obtained for a particular case, its validity can be shown to be far more general: There is no net change in the entropy of a system undergoing any complete reversible cyclic process. Mathematically, we write this statement as

$$\oint dS = \oint \frac{dQ}{T} = 0 \tag{4.11}$$

where $\oint$ represents the integral over a *closed reversible path*.

We can use Equation 4.11 to show that the entropy change of a system undergoing a reversible process between two given states is path independent. An arbitrary, closed path for a reversible cycle that passes through the states *A* and *B* is shown in Figure 4.16. From Equation 4.11, $\oint dS = 0$ for this closed path. We may split this integral into two segments, one along I, which leads from *A* to *B*, the other along II, which leads from *B* to *A*. Then

$$\left[\int_A^B dS\right]_{\text{I}} + \left[\int_B^A dS\right]_{\text{II}} = 0.$$

Since the process is reversible,

$$\left[\int_A^B dS\right]_{\text{I}} = \left[\int_A^B dS\right]_{\text{II}}.$$

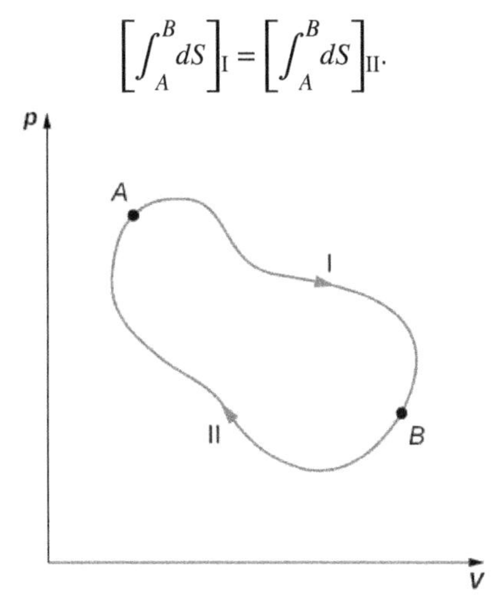

Figure 4.16 The closed loop passing through states *A* and *B* represents a reversible cycle.

Hence, the entropy change in going from *A* to *B* is the same for paths I and II. Since paths I and II are arbitrary, reversible paths, the entropy change in a transition between two equilibrium states is the same for all the reversible processes joining these states. Entropy, like internal energy, is therefore a state function.

What happens if the process is irreversible? When the process is irreversible, we expect the entropy of a closed system, or the system and its environment (the universe), to increase. Therefore we can rewrite this expression as

$$\Delta S \geq 0, \tag{4.12}$$

where *S* is the total entropy of the closed system or the entire universe, and the equal sign is for a reversible process. The fact is the **entropy statement of the second law of thermodynamics**:

Second Law of Thermodynamics (Entropy statement)

The entropy of a closed system and the entire universe never decreases.

We can show that this statement is consistent with the Kelvin statement, the Clausius statement, and the Carnot principle.

Example 4.5

Entropy Change of a System during an Isobaric Process

Determine the entropy change of an object of mass *m* and specific heat *c* that is cooled rapidly (and irreversibly) at constant pressure from T_h to T_c.

Strategy

The process is clearly stated as an irreversible process; therefore, we cannot simply calculate the entropy change from the actual process. However, because entropy of a system is a function of state, we can imagine a reversible process that starts from the same initial state and ends at the given final state. Then, the entropy change of the system is given by Equation 4.10, $\Delta S = \int_A^B dQ/T$.

Solution

To replace this rapid cooling with a process that proceeds reversibly, we imagine that the hot object is put into thermal contact with successively cooler heat reservoirs whose temperatures range from T_h to T_c. Throughout the substitute transition, the object loses infinitesimal amounts of heat *dQ*, so we have

$$\Delta S = \int_{T_h}^{T_c} \frac{dQ}{T}.$$

From the definition of heat capacity, an infinitesimal exchange *dQ* for the object is related to its temperature change *dT* by

$$dQ = mc\,dT.$$

Substituting this *dQ* into the expression for ΔS, we obtain the entropy change of the object as it is cooled at constant pressure from T_h to T_c :

$$\Delta S = \int_{T_h}^{T_c} \frac{mc\,dT}{T} = mc \ln\frac{T_c}{T_h}.$$

Note that $\Delta S < 0$ here because $T_c < T_h$. In other words, the object has lost some entropy. But if we count whatever is used to remove the heat from the object, we would still end up with $\Delta S_{universe} > 0$ because the process is irreversible.

Significance

If the temperature changes during the heat flow, you must keep it inside the integral to solve for the change in entropy. If, however, the temperature is constant, you can simply calculate the entropy change as the heat flow divided by the temperature.

Example 4.6

Stirling Engine

The steps of a reversible Stirling engine are as follows. For this problem, we will use 0.0010 mol of a monatomic gas that starts at a temperature of $133\,°C$ and a volume of $0.10\,m^3$, which will be called point *A*. Then it goes through the following steps:

1. Step *AB*: isothermal expansion at $133\,°C$ from $0.10\,m^3$ to $0.20\,m^3$
2. Step *BC*: isochoric cooling to $33\,°C$
3. Step *CD*: isothermal compression at $33\,°C$ from $0.20\,m^3$ to $0.10\,m^3$

4. Step *DA*: isochoric heating back to $133\ °\text{C}$ and $0.10\ \text{m}^3$

(a) Draw the *pV* diagram for the Stirling engine with proper labels.

(b) Fill in the following table.

Step	*W* (J)	*Q* (J)	ΔS (J/K)
Step *AB*			
Step *BC*			
Step *CD*			
Step *DA*			
Complete cycle			

(c) How does the efficiency of the Stirling engine compare to the Carnot engine working within the same two heat reservoirs?

Strategy

Using the ideal gas law, calculate the pressure at each point so that they can be labeled on the *pV* diagram. Isothermal work is calculated using $W = nRT\ln\left(\frac{V_2}{V_1}\right)$, and an isochoric process has no work done. The heat flow is calculated from the first law of thermodynamics, $Q = \Delta E_{\text{int}} - W$ where $\Delta E_{\text{int}} = \frac{3}{2}nR\Delta T$ for monatomic gasses. Isothermal steps have a change in entropy of *Q*/*T*, whereas isochoric steps have $\Delta S = \frac{3}{2}nR\ln\left(\frac{T_2}{T_1}\right)$. The efficiency of a heat engine is calculated by using $e_{\text{Stir}} = W/Q_{\text{h}}$.

Solution

a. The graph is shown below.

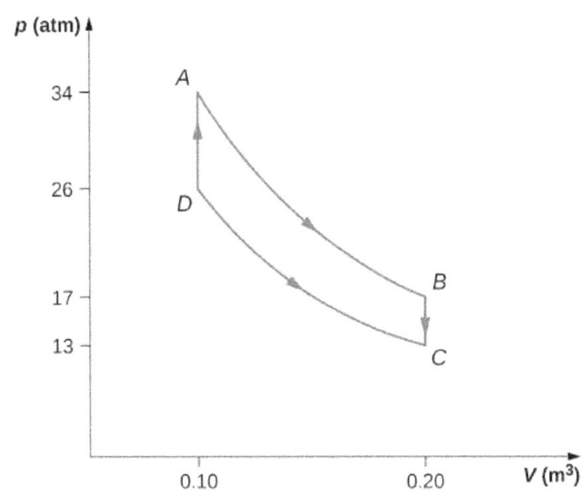

b. The completed table is shown below.

Step	*W* (J)	*Q* (J)	ΔS (J/K)
Step *AB* Isotherm	2.3	2.3	0.0057
Step *BC* Isochoric	0	–1.2	0.0035
Step *CD* Isotherm	–1.8	–1.8	–0.0059

Step	*W* (J)	*Q* (J)	ΔS (J/K)
Step *DA* Isochoric	0	1.2	–0.0035
Complete cycle	0.5	0.5	~ 0

c. The efficiency of the Stirling heat engine is

$$e_{\text{Stir}} = W/Q_{\text{h}} = (Q_{AB} + Q_{CD})/(Q_{AB} + Q_{DA}) = 0.5/4.5 = 0.11.$$

If this were a Carnot engine operating between the same heat reservoirs, its efficiency would be

$$e_{\text{Car}} = 1 - \left(\frac{T_{\text{c}}}{T_{\text{h}}}\right) = 0.25.$$

Therefore, the Carnot engine would have a greater efficiency than the Stirling engine.

Significance

In the early days of steam engines, accidents would occur due to the high pressure of the steam in the boiler. Robert Stirling developed an engine in 1816 that did not use steam and therefore was safer. The Stirling engine was commonly used in the nineteenth century, but developments in steam and internal combustion engines have made it difficult to broaden the use of the Stirling engine.

The Stirling engine uses compressed air as the working substance, which passes back and forth between two chambers with a porous plug, called the regenerator, which is made of material that does not conduct heat as well. In two of the steps, pistons in the two chambers move in phase.

4.7 | Entropy on a Microscopic Scale

Learning Objectives

By the end of this section you will be able to:

- Interpret the meaning of entropy at a microscopic scale
- Calculate a change in entropy for an irreversible process of a system and contrast with the change in entropy of the universe
- Explain the third law of thermodynamics

We have seen how entropy is related to heat exchange at a particular temperature. In this section, we consider entropy from a statistical viewpoint. Although the details of the argument are beyond the scope of this textbook, it turns out that entropy can be related to how disordered or randomized a system is—the more it is disordered, the higher is its entropy. For example, a new deck of cards is very ordered, as the cards are arranged numerically by suit. In shuffling this new deck, we randomize the arrangement of the cards and therefore increase its entropy (Figure 4.17). Thus, by picking one card off the top of the deck, there would be no indication of what the next selected card will be.

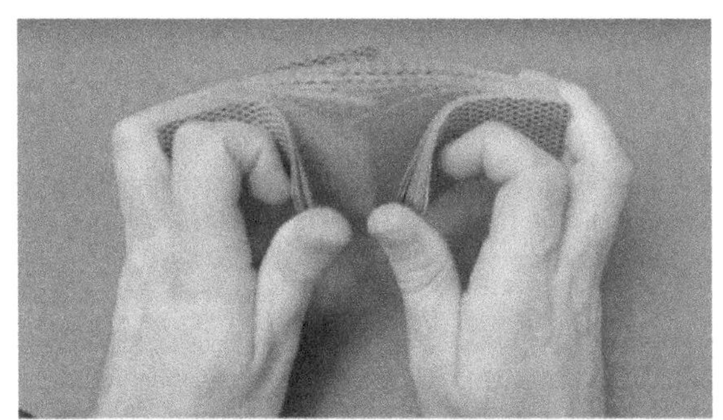

Figure 4.17 The entropy of a new deck of cards goes up after the dealer shuffles them. (credit: "Rommel SK"/YouTube)

The second law of thermodynamics requires that the entropy of the universe increase in any irreversible process. Thus, in terms of order, the second law may be stated as follows:

In any irreversible process, the universe becomes more disordered. For example, the irreversible free expansion of an ideal gas, shown in Figure 4.2, results in a larger volume for the gas molecules to occupy. A larger volume means more possible arrangements for the same number of atoms, so disorder is also increased. As a result, the entropy of the gas has gone up. The gas in this case is a closed system, and the process is irreversible. Changes in phase also illustrate the connection between entropy and **disorder**.

Example 4.7

Entropy Change of the Universe

Suppose we place 50 g of ice at $0\,°\text{C}$ in contact with a heat reservoir at $20\,°\text{C}$. Heat spontaneously flows from the reservoir to the ice, which melts and eventually reaches a temperature of $20\,°\text{C}$. Find the change in entropy of (a) the ice and (b) the universe.

Strategy

Because the entropy of a system is a function of its state, we can imagine two reversible processes for the ice: (1) ice is melted at $0\,°\text{C}(T_A)$; and (2) melted ice (water) is warmed up from $0\,°\text{C}$ to $20\,°\text{C}(T_B)$ under constant pressure. Then, we add the change in entropy of the reservoir when we calculate the change in entropy of the universe.

Solution

a. From Equation 4.10, the increase in entropy of the ice is

$$\begin{aligned}\Delta S_{\text{ice}} &= \Delta S_1 + \Delta S_2 \\ &= \frac{mL_f}{T_A} + mc\int_A^B \frac{dT}{T} \\ &= \left(\frac{50 \times 335}{273} + 50 \times 4.19 \times \ln\frac{293}{273}\right) \text{J/K} \\ &= 76.3\ \text{J/K}.\end{aligned}$$

b. During this transition, the reservoir gives the ice an amount of heat equal to

$$\begin{aligned}Q &= mL_f + mc(T_B - T_A) \\ &= 50 \times (335 + 4.19 \times 20)\ \text{J} \\ &= 2.10 \times 10^4\ \text{J}.\end{aligned}$$

This leads to a change (decrease) in entropy of the reservoir:

$$\Delta S_{\text{reservoir}} = \frac{-Q}{T_B} = -71.7\ \text{J/K}.$$

The increase in entropy of the universe is therefore

$$\Delta S_{\text{universe}} = 76.3 \text{ J/K} - 71.7 \text{ J/K} = 4.6 \text{ J/K} > 0.$$

Significance

The entropy of the universe therefore is greater than zero since the ice gains more entropy than the reservoir loses. If we considered only the phase change of the ice into water and not the temperature increase, the entropy change of the ice and reservoir would be the same, resulting in the universe gaining no entropy.

This process also results in a more disordered universe. The ice changes from a solid with molecules located at specific sites to a liquid whose molecules are much freer to move. The molecular arrangement has therefore become more randomized. Although the change in average kinetic energy of the molecules of the heat reservoir is negligible, there is nevertheless a significant decrease in the entropy of the reservoir because it has many more molecules than the melted ice cube. However, the reservoir's decrease in entropy is still not as large as the increase in entropy of the ice. The increased disorder of the ice more than compensates for the increased order of the reservoir, and the entropy of the universe increases by 4.6 J/K.

You might suspect that the growth of different forms of life might be a net ordering process and therefore a violation of the second law. After all, a single cell gathers molecules and eventually becomes a highly structured organism, such as a human being. However, this ordering process is more than compensated for by the disordering of the rest of the universe. The net result is an increase in entropy and an increase in the disorder of the universe.

4.5 Check Your Understanding In Example 4.7, the spontaneous flow of heat from a hot object to a cold object results in a net increase in entropy of the universe. Discuss how this result can be related to an increase in disorder of the system.

The second law of thermodynamics makes clear that the entropy of the universe never decreases during any thermodynamic process. For any other thermodynamic system, when the process is reversible, the change of the entropy is given by $\Delta S = Q/T$. But what happens if the temperature goes to zero, $T \to 0$? It turns out this is not a question that can be answered by the second law.

A fundamental issue still remains: Is it possible to cool a system all the way down to zero kelvin? We understand that the system must be at its lowest energy state because lowering temperature reduces the kinetic energy of the constituents in the system. What happens to the entropy of a system at the absolute zero temperature? It turns out the absolute zero temperature is not reachable—at least, not though a finite number of cooling steps. This is a statement of the **third law of thermodynamics**, whose proof requires quantum mechanics that we do not present here. In actual experiments, physicists have continuously pushed that limit downward, with the lowest temperature achieved at about 1×10^{-10} K in a low-temperature lab at the Helsinki University of Technology in 2008.

Like the second law of thermodynamics, the third law of thermodynamics can be stated in different ways. One of the common statements of the third law of thermodynamics is: *The absolute zero temperature cannot be reached through any finite number of cooling steps.*

In other words, the temperature of any given physical system must be finite, that is, $T > 0$. This produces a very interesting question in physics: Do we know how a system would behave if it were at the absolute zero temperature?

The reason a system is unable to reach 0 K is fundamental and requires quantum mechanics to fully understand its origin. But we can certainly ask what happens to the entropy of a system when we try to cool it down to 0 K. Because the amount of heat that can be removed from the system becomes vanishingly small, we expect that the change in entropy of the system along an isotherm approaches zero, that is,

$$\lim_{T \to 0} (\Delta S)_T = 0. \tag{4.13}$$

This can be viewed as another statement of the third law, with all the isotherms becoming **isentropic**, or into a reversible ideal adiabat. We can put this expression in words: *A system becomes perfectly ordered when its temperature approaches absolute zero and its entropy approaches its absolute minimum.*

The third law of thermodynamics puts another limit on what can be done when we look for energy resources. If there could be a reservoir at the absolute zero temperature, we could have engines with efficiency of 100%, which would, of course, violate the second law of thermodynamics.

Example 4.8

Entropy Change of an Ideal Gas in Free Expansion

An ideal gas occupies a partitioned volume V_1 inside a box whose walls are thermally insulating, as shown in Figure 4.18(a). When the partition is removed, the gas expands and fills the entire volume V_2 of the box, as shown in part (b). What is the entropy change of the universe (the system plus its environment)?

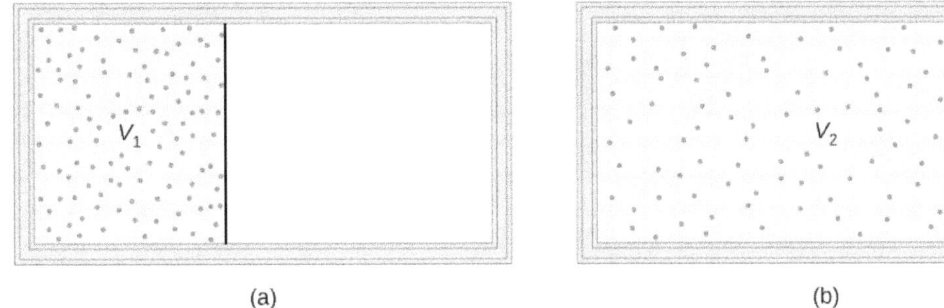

Figure 4.18 The adiabatic free expansion of an ideal gas from volume V_1 to volume V_2.

Strategy

The adiabatic free expansion of an ideal gas is an irreversible process. There is no change in the internal energy (and hence temperature) of the gas in such an expansion because no work or heat transfer has happened. Thus, a convenient reversible path connecting the same two equilibrium states is a slow, isothermal expansion from V_1 to V_2. In this process, the gas could be expanding against a piston while in thermal contact with a heat reservoir, as in step 1 of the Carnot cycle.

Solution

Since the temperature is constant, the entropy change is given by $\Delta S = Q/T$, where

$$Q = W = \int_{V_1}^{V_2} p dV$$

because $\Delta E_{\text{int}} = 0$. Now, with the help of the ideal gas law, we have

$$Q = nRT \int_{V_1}^{V_2} \frac{dV}{V} = nRT \ln \frac{V_2}{V_1},$$

so the change in entropy of the gas is

$$\Delta S = \frac{Q}{T} = nR \ln \frac{V_2}{V_1}.$$

Because $V_2 > V_1$, ΔS is positive, and the entropy of the gas has gone up during the free expansion.

Significance

What about the environment? The walls of the container are thermally insulating, so no heat exchange takes place between the gas and its surroundings. The entropy of the environment is therefore constant during the expansion.

The net entropy change of the universe is then simply the entropy change of the gas. Since this is positive, the entropy of the universe increases in the free expansion of the gas.

Example 4.9

Entropy Change during Heat Transfer

Heat flows from a steel object of mass 4.00 kg whose temperature is 400 K to an identical object at 300 K. Assuming that the objects are thermally isolated from the environment, what is the net entropy change of the universe after thermal equilibrium has been reached?

Strategy

Since the objects are identical, their common temperature at equilibrium is 350 K. To calculate the entropy changes associated with their transitions, we substitute the irreversible process of the heat transfer by two isobaric, reversible processes, one for each of the two objects. The entropy change for each object is then given by $\Delta S = mc\ln(T_B/T_A)$.

Solution

Using $c = 450\,\text{J/kg}\cdot\text{K}$, the specific heat of steel, we have for the hotter object

$$\begin{aligned}\Delta S_\text{h} &= \int_{T_1}^{T_2}\frac{mc\,dT}{T} = mc\ln\frac{T_2}{T_1}\\ &= (4.00\,\text{kg})(450\,\text{J/kg}\cdot\text{K})\ln\frac{350\,\text{K}}{400\,\text{K}} = -240\,\text{J/K}.\end{aligned}$$

Similarly, the entropy change of the cooler object is

$$\Delta S_\text{c} = (4.00\,\text{kg})(450\,\text{J/kg}\cdot\text{K})\ln\frac{350\,\text{K}}{300\,\text{K}} = 277\,\text{J/K}.$$

The net entropy change of the two objects during the heat transfer is then

$$\Delta S_\text{h} + \Delta S_\text{c} = 37\,\text{J/K}.$$

Significance

The objects are thermally isolated from the environment, so its entropy must remain constant. Thus, the entropy of the universe also increases by 37 J/K.

4.6 Check Your Understanding A quantity of heat Q is absorbed from a reservoir at a temperature T_h by a cooler reservoir at a temperature T_c. What is the entropy change of the hot reservoir, the cold reservoir, and the universe?

4.7 Check Your Understanding A 50-g copper piece at a temperature of $20\,°\text{C}$ is placed into a large insulated vat of water at $100\,°\text{C}$. (a) What is the entropy change of the copper piece when it reaches thermal equilibrium with the water? (b) What is the entropy change of the water? (c) What is the entropy change of the universe?

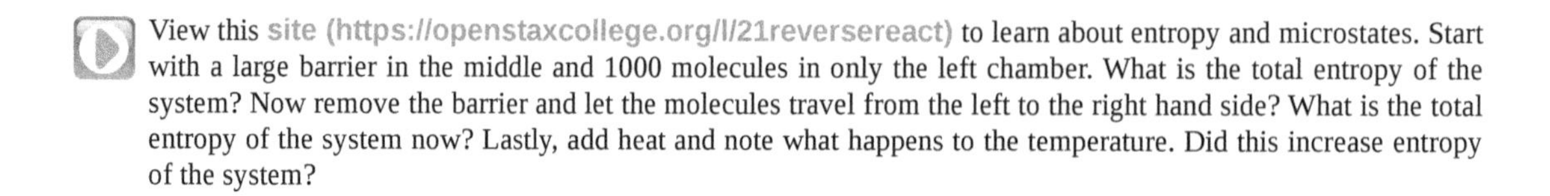

View this site (https://openstaxcollege.org/l/21reversereact) to learn about entropy and microstates. Start with a large barrier in the middle and 1000 molecules in only the left chamber. What is the total entropy of the system? Now remove the barrier and let the molecules travel from the left to the right hand side? What is the total entropy of the system now? Lastly, add heat and note what happens to the temperature. Did this increase entropy of the system?

CHAPTER 4 REVIEW

KEY TERMS

Carnot cycle cycle that consists of two isotherms at the temperatures of two reservoirs and two adiabatic processes connecting the isotherms

Carnot engine Carnot heat engine, refrigerator, or heat pump that operates on a Carnot cycle

Carnot principle principle governing the efficiency or performance of a heat device operating on a Carnot cycle: any reversible heat device working between two reservoirs must have the same efficiency or performance coefficient, greater than that of an irreversible heat device operating between the same two reservoirs

Clausius statement of the second law of thermodynamics heat never flows spontaneously from a colder object to a hotter object

coefficient of performance measure of effectiveness of a refrigerator or heat pump

cold reservoir sink of heat used by a heat engine

disorder measure of order in a system; the greater the disorder is, the higher the entropy

efficiency (*e*) output work from the engine over the input heat to the engine from the hot reservoir

entropy state function of the system that changes when heat is transferred between the system and the environment

entropy statement of the second law of thermodynamics entropy of a closed system or the entire universe never decreases

heat engine device that converts heat into work

heat pump device that delivers heat to a hot reservoir

hot reservoir source of heat used by a heat engine

irreversibility phenomenon associated with a natural process

irreversible process process in which neither the system nor its environment can be restored to their original states at the same time

isentropic reversible adiabatic process where the process is frictionless and no heat is transferred

Kelvin statement of the second law of thermodynamics it is impossible to convert the heat from a single source into work without any other effect

perfect engine engine that can convert heat into work with 100% efficiency

perfect refrigerator (heat pump) refrigerator (heat pump) that can remove (dump) heat without any input of work

refrigerator device that removes heat from a cold reservoir

reversible process process in which both the system and the external environment theoretically can be returned to their original states

third law of thermodynamics absolute zero temperature cannot be reached through any finite number of cooling steps

KEY EQUATIONS

Result of energy conservation	$W = Q_h - Q_c$
Efficiency of a heat engine	$e = \frac{W}{Q_h} = 1 - \frac{Q_c}{Q_h}$
Coefficient of performance of a refrigerator	$K_R = \frac{Q_c}{W} = \frac{Q_c}{Q_h - Q_c}$

Coefficient of performance of a heat pump

$$K_{\text{P}} = \frac{Q_{\text{h}}}{W} = \frac{Q_{\text{h}}}{Q_{\text{h}} - Q_{\text{c}}}$$

Resulting efficiency of a Carnot cycle

$$e = 1 - \frac{T_{\text{c}}}{T_{\text{h}}}$$

Performance coefficient of a reversible refrigerator

$$K_{\text{R}} = \frac{T_{\text{c}}}{T_{\text{h}} - T_{\text{c}}}$$

Performance coefficient of a reversible heat pump

$$K_{\text{P}} = \frac{T_{\text{h}}}{T_{\text{h}} - T_{\text{c}}}$$

Entropy of a system undergoing a reversible process at a constant temperature

$$\Delta S = \frac{Q}{T}$$

Change of entropy of a system under a reversible process

$$\Delta S = S_B - S_A = \int_A^B dQ/T$$

Entropy of a system undergoing any complete reversible cyclic process

$$\oint dS = \oint \frac{dQ}{T} = 0$$

Change of entropy of a closed system under an irreversible process

$$\Delta S \geq 0$$

Change in entropy of the system along an isotherm

$$\lim_{T \to 0} (\Delta S)_T = 0$$

SUMMARY

4.1 Reversible and Irreversible Processes

- A reversible process is one in which both the system and its environment can return to exactly the states they were in by following the reverse path.
- An irreversible process is one in which the system and its environment cannot return together to exactly the states that they were in.
- The irreversibility of any natural process results from the second law of thermodynamics.

4.2 Heat Engines

- The work done by a heat engine is the difference between the heat absorbed from the hot reservoir and the heat discharged to the cold reservoir, that is, $W = Q_{\text{h}} - Q_{\text{c}}$.
- The ratio of the work done by the engine and the heat absorbed from the hot reservoir provides the efficiency of the engine, that is, $e = W/Q_{\text{h}} = 1 - Q_{\text{c}}/Q_{\text{h}}$.

4.3 Refrigerators and Heat Pumps

- A refrigerator or a heat pump is a heat engine run in reverse.
- The focus of a refrigerator is on removing heat from the cold reservoir with a coefficient of performance K_{R}.
- The focus of a heat pump is on dumping heat to the hot reservoir with a coefficient of performance K_{P}.

4.4 Statements of the Second Law of Thermodynamics

- The Kelvin statement of the second law of thermodynamics: It is impossible to convert the heat from a single source into work without any other effect.
- The Kelvin statement and Clausius statement of the second law of thermodynamics are equivalent.

4.5 The Carnot Cycle

- The Carnot cycle is the most efficient engine for a reversible cycle designed between two reservoirs.
- The Carnot principle is another way of stating the second law of thermodynamics.

4.6 Entropy

- The change in entropy for a reversible process at constant temperature is equal to the heat divided by the temperature. The entropy change of a system under a reversible process is given by $\Delta S = \int_A^B dQ/T$.
- A system's change in entropy between two states is independent of the reversible thermodynamic path taken by the system when it makes a transition between the states.

4.7 Entropy on a Microscopic Scale

- Entropy can be related to how disordered a system is—the more it is disordered, the higher is its entropy. In any irreversible process, the universe becomes more disordered.
- According to the third law of thermodynamics, absolute zero temperature is unreachable.

CONCEPTUAL QUESTIONS

4.1 Reversible and Irreversible Processes

1. State an example of a process that occurs in nature that is as close to reversible as it can be.

4.2 Heat Engines

2. Explain in practical terms why efficiency is defined as W/Q_h.

4.3 Refrigerators and Heat Pumps

3. If the refrigerator door is left open, what happens to the temperature of the kitchen?

4. Is it possible for the efficiency of a reversible engine to be greater than 1.0? Is it possible for the coefficient of performance of a reversible refrigerator to be less than 1.0?

4.4 Statements of the Second Law of Thermodynamics

5. In the text, we showed that if the Clausius statement is false, the Kelvin statement must also be false. Now show the reverse, such that if the Kelvin statement is false, it follows that the Clausius statement is false.

6. Why don't we operate ocean liners by extracting heat from the ocean or operate airplanes by extracting heat from the atmosphere?

7. Discuss the practical advantages and disadvantages of heat pumps and electric heating.

8. The energy output of a heat pump is greater than the energy used to operate the pump. Why doesn't this statement violate the first law of thermodynamics?

9. Speculate as to why nuclear power plants are less efficient than fossil-fuel plants based on temperature arguments.

10. An ideal gas goes from state (p_i, V_i) to state (p_f, V_f) when it is allowed to expand freely. Is it possible to represent the actual process on a pV diagram? Explain.

4.5 The Carnot Cycle

11. To increase the efficiency of a Carnot engine, should the temperature of the hot reservoir be raised or lowered? What about the cold reservoir?

12. How could you design a Carnot engine with 100% efficiency?

13. What type of processes occur in a Carnot cycle?

4.6 Entropy

14. Does the entropy increase for a Carnot engine for each cycle?

15. Is it possible for a system to have an entropy change if it neither absorbs nor emits heat during a reversible transition? What happens if the process is irreversible?

4.7 Entropy on a Microscopic Scale

16. Are the entropy changes of the *systems* in the following processes positive or negative? (a) *water vapor* that condenses on a cold surface; (b) gas in a container that leaks into the surrounding atmosphere; (c) an *ice cube* that melts in a glass of lukewarm water; (d) the *lukewarm water* of part (c); (e) a *real heat engine* performing a cycle; (f) *food* cooled in a refrigerator.

17. Discuss the entropy changes in the systems of Question 21.10 in terms of disorder.

PROBLEMS

4.1 Reversible and Irreversible Processes

18. A tank contains 111.0 g chlorine gas (Cl_2), which is at temperature $82.0\,°C$ and absolute pressure 5.70×10^5 Pa. The temperature of the air outside the tank is $20.0\,°C$. The molar mass of Cl_2 is 70.9 g/mol. (a) What is the volume of the tank? (b) What is the internal energy of the gas? (c) What is the work done by the gas if the temperature and pressure inside the tank drop to $31.0\,°C$ and 3.80×10^5 Pa, respectively, due to a leak?

19. A mole of ideal monatomic gas at $0\,°C$ and 1.00 atm is warmed up to expand isobarically to triple its volume. How much heat is transferred during the process?

20. A mole of an ideal gas at pressure 4.00 atm and temperature 298 K expands isothermally to double its volume. What is the work done by the gas?

21. After a free expansion to quadruple its volume, a mole of ideal diatomic gas is compressed back to its original volume isobarically and then cooled down to its original temperature. What is the minimum heat removed from the gas in the final step to restoring its state?

4.2 Heat Engines

22. An engine is found to have an efficiency of 0.40. If it does 200 J of work per cycle, what are the corresponding quantities of heat absorbed and rejected?

23. In performing 100.0 J of work, an engine rejects 50.0 J of heat. What is the efficiency of the engine?

24. An engine with an efficiency of 0.30 absorbs 500 J of heat per cycle. (a) How much work does it perform per cycle? (b) How much heat does it reject per cycle?

25. It is found that an engine rejects 100.0 J while absorbing 125.0 J each cycle of operation. (a) What is the efficiency of the engine? (b) How much work does it perform per cycle?

26. The temperature of the cold reservoir of the engine is 300 K. It has an efficiency of 0.30 and absorbs 500 J of heat per cycle. (a) How much work does it perform per cycle? (b) How much heat does it reject per cycle?

27. The Kelvin temperature of the hot reservoir of an engine is twice that of the cold reservoir, and work done by the engine per cycle is 50 J. Calculate (a) the efficiency of the engine, (b) the heat absorbed per cycle, and (c) the heat rejected per cycle.

28. A coal power plant consumes 100,000 kg of coal per hour and produces 500 MW of power. If the heat of combustion of coal is 30 MJ/kg, what is the efficiency of the power plant?

4.3 Refrigerators and Heat Pumps

29. A refrigerator has a coefficient of performance of 3.0. (a) If it requires 200 J of work per cycle, how much heat per cycle does it remove the cold reservoir? (b) How much heat per cycle is discarded to the hot reservoir?

30. During one cycle, a refrigerator removes 500 J from a cold reservoir and rejects 800 J to its hot reservoir. (a) What is its coefficient of performance? (b) How much work per cycle does it require to operate?

31. If a refrigerator discards 80 J of heat per cycle and its coefficient of performance is 6.0, what are (a) the quantity off heat it removes per cycle from a cold reservoir and (b) the amount of work per cycle required for its operation?

32. A refrigerator has a coefficient of performance of 3.0. (a) If it requires 200 J of work per cycle, how much heat per cycle does it remove the cold reservoir? (b) How much heat per cycle is discarded to the hot reservoir?

4.5 The Carnot Cycle

33. The temperature of the cold and hot reservoirs between which a Carnot refrigerator operates are $-73\ °C$ and $270\ °C$, respectively. Which is its coefficient of performance?

34. Suppose a Carnot refrigerator operates between T_c and T_h. Calculate the amount of work required to extract 1.0 J of heat from the cold reservoir if (a) $T_c = 7\ °C$, $T_h = 27\ °C$; (b) $T_c = -73\ °C$, $T_h = 27\ °C$; (c) $T_c = -173\ °C$, $T_h = 27\ °C$; and (d) $T_c = -273\ °C$, $T_h = 27\ °C$.

35. A Carnot engine operates between reservoirs at 600 and 300 K. If the engine absorbs 100 J per cycle at the hot reservoir, what is its work output per cycle?

36. A 500-W motor operates a Carnot refrigerator between $-5\ °C$ and $30\ °C$. (a) What is the amount of heat per second extracted from the inside of the refrigerator? (b) How much heat is exhausted to the outside air per second?

37. Sketch a Carnot cycle on a temperature-volume diagram.

38. A Carnot heat pump operates between $0\ °C$ and $20\ °C$. How much heat is exhausted into the interior of a house for every 1.0 J of work done by the pump?

39. An engine operating between heat reservoirs at $20\ °C$ and $200\ °C$ extracts 1000 J per cycle from the hot reservoir. (a) What is the maximum possible work that engine can do per cycle? (b) For this maximum work, how much heat is exhausted to the cold reservoir per cycle?

40. Suppose a Carnot engine can be operated between two reservoirs as either a heat engine or a refrigerator. How is the coefficient of performance of the refrigerator related to the efficiency of the heat engine?

41. A Carnot engine is used to measure the temperature of a heat reservoir. The engine operates between the heat reservoir and a reservoir consisting of water at its triple point. (a) If 400 J per cycle are removed from the heat reservoir while 200 J per cycle are deposited in the triple-point reservoir, what is the temperature of the heat reservoir? (b) If 400 J per cycle are removed from the triple-point reservoir while 200 J per cycle are deposited in the heat reservoir, what is the temperature of the heat reservoir?

42. What is the minimum work required of a refrigerator if it is to extract 50 J per cycle from the inside of a freezer at $-10\ °C$ and exhaust heat to the air at $25\ °C$?

4.6 Entropy

43. Two hundred joules of heat are removed from a heat reservoir at a temperature of 200 K. What is the entropy change of the reservoir?

44. In an isothermal reversible expansion at $27\ °C$, an ideal gas does 20 J of work. What is the entropy change of the gas?

45. An ideal gas at 300 K is compressed isothermally to one-fifth its original volume. Determine the entropy change per mole of the gas.

46. What is the entropy change of 10 g of steam at $100\ °C$ when it condenses to water at the same temperature?

47. A metal rod is used to conduct heat between two reservoirs at temperatures T_h and T_c, respectively. When an amount of heat Q flows through the rod from the hot to the cold reservoir, what is the net entropy change of the rod, the hot reservoir, the cold reservoir, and the universe?

48. For the Carnot cycle of Figure 4.12, what is the entropy change of the hot reservoir, the cold reservoir, and the universe?

49. A 5.0-kg piece of lead at a temperature of $600\ °C$ is placed in a lake whose temperature is $15\ °C$. Determine the entropy change of (a) the lead piece, (b) the lake, and (c) the universe.

50. One mole of an ideal gas doubles its volume in a reversible isothermal expansion. (a) What is the change in entropy of the gas? (b) If 1500 J of heat are added in this process, what is the temperature of the gas?

51. One mole of an ideal monatomic gas is confined to a rigid container. When heat is added reversibly to the gas, its temperature changes from T_1 to T_2. (a) How much heat is added? (b) What is the change in entropy of the gas?

52. (a) A 5.0-kg rock at a temperature of $20\ °C$ is dropped into a shallow lake also at $20\ °C$ from a height of 1.0×10^3 m. What is the resulting change in entropy of the universe? (b) If the temperature of the rock is $100\ °C$ when it is dropped, what is the change of entropy of the universe? Assume that air friction is negligible (not a good assumption) and that $c = 860\ \text{J/kg} \cdot \text{K}$ is the specific heat of the rock.

4.7 Entropy on a Microscopic Scale

53. A copper rod of cross-sectional area $5.0\ \text{cm}^2$ and length 5.0 m conducts heat from a heat reservoir at 373 K to one at 273 K. What is the time rate of change of the universe's entropy for this process?

54. Fifty grams of water at $20\ °C$ is heated until it becomes vapor at $100\ °C$. Calculate the change in entropy of the water in this process.

55. Fifty grams of water at $0\ °C$ are changed into vapor at $100\ °C$. What is the change in entropy of the water in this process?

56. In an isochoric process, heat is added to 10 mol of monoatomic ideal gas whose temperature increases from 273 to 373 K. What is the entropy change of the gas?

57. Two hundred grams of water at $0\ °C$ is brought into contact with a heat reservoir at $80\ °C$. After thermal equilibrium is reached, what is the temperature of the water? Of the reservoir? How much heat has been transferred in the process? What is the entropy change of the water? Of the reservoir? What is the entropy change of the universe?

58. Suppose that the temperature of the water in the previous problem is raised by first bringing it to thermal equilibrium with a reservoir at a temperature of $40\ °C$ and then with a reservoir at $80\ °C$. Calculate the entropy changes of (a) each reservoir, (b) of the water, and (c) of the universe.

59. Two hundred grams of water at $0\ °C$ is brought into contact into thermal equilibrium successively with reservoirs at $20\ °C$, $40\ °C$, $60\ °C$, and $80\ °C$. (a) What is the entropy change of the water? (b) Of the reservoir? (c) What is the entropy change of the universe?

60. (a) Ten grams of H_2O starts as ice at $0\ °C$. The ice absorbs heat from the air (just above $0\ °C$) until all of it melts. Calculate the entropy change of the H_2O, of the air, and of the universe. (b) Suppose that the air in part (a) is at $20\ °C$ rather than $0\ °C$ and that the ice absorbs heat until it becomes water at $20\ °C$. Calculate the entropy change of the H_2O, of the air, and of the universe. (c) Is either of these processes reversible?

61. The Carnot cycle is represented by the temperature-entropy diagram shown below. (a) How much heat is absorbed per cycle at the high-temperature reservoir? (b) How much heat is exhausted per cycle at the low-temperature reservoir? (c) How much work is done per cycle by the engine? (d) What is the efficiency of the engine?

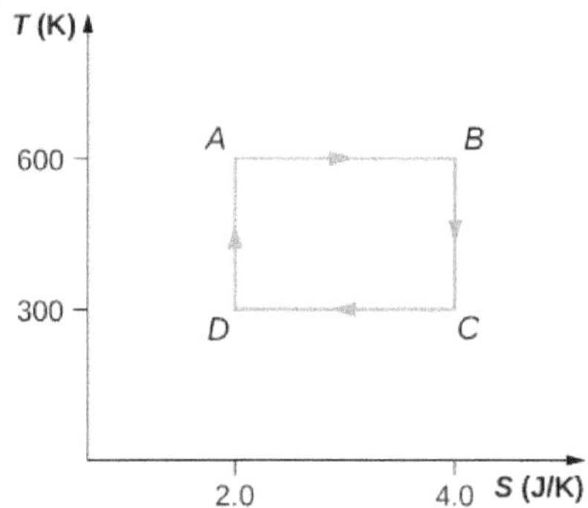

62. A Carnot engine operating between heat reservoirs at 500 and 300 K absorbs 1500 J per cycle at the high-temperature reservoir. (a) Represent the engine's cycle on a temperature-entropy diagram. (b) How much work per cycle is done by the engine?

63. A monoatomic ideal gas (n moles) goes through a cyclic process shown below. Find the change in entropy of the gas in each step and the total entropy change over the entire cycle.

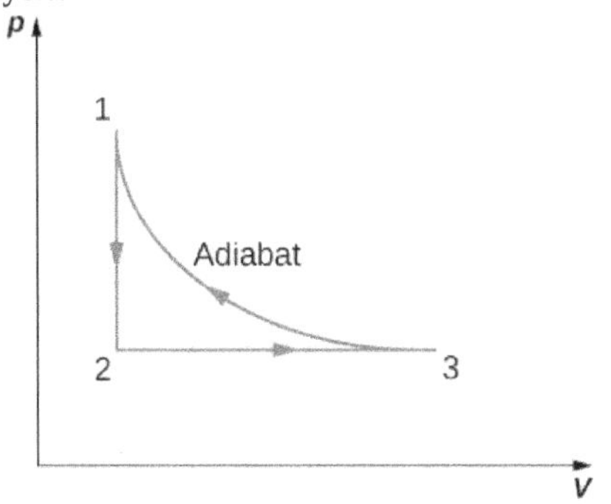

64. A Carnot engine has an efficiency of 0.60. When the temperature of its cold reservoir changes, the efficiency drops to 0.55. If initially $T_c = 27\ °C$, determine (a) the constant value of T_h and (b) the final value of T_c.

65. A Carnot engine performs 100 J of work while rejecting 200 J of heat each cycle. After the temperature of the hot reservoir only is adjusted, it is found that the engine now does 130 J of work while discarding the same quantity of heat. (a) What are the initial and final efficiencies of the engine? (b) What is the fractional change in the temperature of the hot reservoir?

66. A Carnot refrigerator exhausts heat to the air, which is at a temperature of $25\ °C$. How much power is used by the refrigerator if it freezes 1.5 g of water per second? Assume the water is at $0\ °C$.

ADDITIONAL PROBLEMS

67. A 300-W heat pump operates between the ground, whose temperature is $0\ °C$, and the interior of a house at $22\ °C$. What is the maximum amount of heat per hour that the heat pump can supply to the house?

68. An engineer must design a refrigerator that does 300 J of work per cycle to extract 2100 J of heat per cycle from a freezer whose temperature is $-10\ °C$. What is the maximum air temperature for which this condition can be met? Is this a reasonable condition to impose on the design?

69. A Carnot engine employs 1.5 mol of nitrogen gas as a working substance, which is considered as an ideal diatomic gas with $\gamma = 7.5$ at the working temperatures of the engine. The Carnot cycle goes in the cycle *ABCDA* with *AB* being an isothermal expansion. The volume at points *A* and *C* of the cycle are $5.0 \times 10^{-3}\ \text{m}^3$ and 0.15 L, respectively. The engine operates between two thermal baths of temperature 500 K and 300 K. (a) Find the values of volume at *B* and *D*. (b) How much heat is absorbed by the gas in the *AB* isothermal expansion? (c) How much work is done by the gas in the *AB* isothermal expansion? (d) How much heat is given up by the gas in the *CD* isothermal expansion? (e) How much work is done by the gas in the *CD* isothermal compression? (f) How much work is done by the gas in the *BC* adiabatic expansion? (g) How much work is done by the gas in the *DA* adiabatic compression? (h) Find the value of efficiency of the engine based on the net work and heat input. Compare this value to the efficiency of a Carnot engine based on the temperatures of the two baths.

70. A 5.0-kg wood block starts with an initial speed of 8.0 m/s and slides across the floor until friction stops it. Estimate the resulting change in entropy of the universe. Assume that everything stays at a room temperature of $20\ °C$.

71. A system consisting of 20.0 mol of a monoatomic ideal gas is cooled at constant pressure from a volume of 50.0 L to 10.0 L. The initial temperature was 300 K. What is the change in entropy of the gas?

72. A glass beaker of mass 400 g contains 500 g of water at $27\ °C$. The beaker is heated reversibly so that the temperature of the beaker and water rise gradually to $57\ °C$. Find the change in entropy of the beaker and water together.

73. A Carnot engine operates between $550\ °C$ and $20\ °C$ baths and produces 300 kJ of energy in each cycle. Find the change in entropy of the (a) hot bath and (b) cold bath, in each Carnot cycle?

74. An ideal gas at temperature T is stored in the left half of an insulating container of volume V using a partition of negligible volume (see below). What is the entropy change per mole of the gas in each of the following cases? (a) The partition is suddenly removed and the gas quickly fills the entire container. (b) A tiny hole is punctured in the partition and after a long period, the gas reaches an equilibrium state such that there is no net flow through the hole. (c) The partition is moved very slowly and adiabatically all the way to the right wall so that the gas finally fills the entire container.

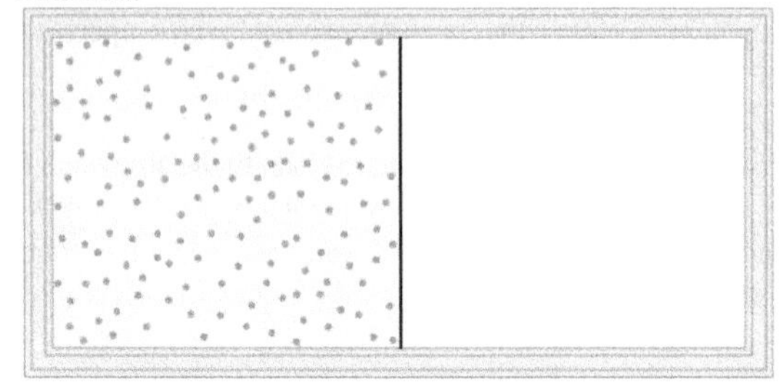

75. A 0.50-kg piece of aluminum at 250 °C is dropped into 1.0 kg of water at 20 °C. After equilibrium is reached, what is the net entropy change of the system?

76. Suppose 20 g of ice at 0 °C is added to 300 g of water at 60 °C. What is the total change in entropy of the mixture after it reaches thermal equilibrium?

77. A heat engine operates between two temperatures such that the working substance of the engine absorbs 5000 J of heat from the high-temperature bath and rejects 3000 J to the low-temperature bath. The rest of the energy is converted into mechanical energy of the turbine. Find (a) the amount of work produced by the engine and (b) the efficiency of the engine.

78. A thermal engine produces 4 MJ of electrical energy while operating between two thermal baths of different temperatures. The working substance of the engine rejects 5 MJ of heat to the cold temperature bath. What is the efficiency of the engine?

79. A coal power plant consumes 100,000 kg of coal per hour and produces 500 MW of power. If the heat of combustion of coal is 30 MJ/kg, what is the efficiency of the power plant?

80. A Carnot engine operates in a Carnot cycle between a heat source at 550 °C and a heat sink at 20 °C. Find the efficiency of the Carnot engine.

81. A Carnot engine working between two heat baths of temperatures 600 K and 273 K completes each cycle in 5 sec. In each cycle, the engine absorbs 10 kJ of heat. Find the power of the engine.

82. A Carnot cycle working between 100 °C and 30 °C is used to drive a refrigerator between −10 °C and 30 °C. How much energy must the Carnot engine produce per second so that the refrigerator is able to discard 10 J of energy per second?

CHALLENGE PROBLEMS

83. (a) An infinitesimal amount of heat is added reversibly to a system. By combining the first and second laws, show that $dU = TdS - dW$. (b) When heat is added to an ideal gas, its temperature and volume change from T_1 and V_1 to T_2 and V_2. Show that the entropy change of n moles of the gas is given by

$$\Delta S = nC_v \ln\frac{T_2}{T_1} + nR \ln\frac{V_2}{V_1}.$$

84. Using the result of the preceding problem, show that for an ideal gas undergoing an adiabatic process, $TV^{\gamma - 1}$ is constant.

85. With the help of the two preceding problems, show that ΔS between states 1 and 2 of n moles an ideal gas is given by

$$\Delta S = nC_p \ln\frac{T_2}{T_1} - nR \ln\frac{p_2}{p_1}.$$

86. A cylinder contains 500 g of helium at 120 atm and 20 °C. The valve is leaky, and all the gas slowly escapes isothermally into the atmosphere. Use the results of the preceding problem to determine the resulting change in entropy of the universe.

87. A diatomic ideal gas is brought from an initial equilibrium state at $p_1 = 0.50$ atm and $T_1 = 300$ K to a final stage with $p_2 = 0.20$ atm and $T_1 = 500$ K. Use the results of the previous problem to determine the entropy change per mole of the gas.

88. The gasoline internal combustion engine operates in a cycle consisting of six parts. Four of these parts involve, among other things, friction, heat exchange through finite temperature differences, and accelerations of the piston; it is irreversible. Nevertheless, it is represented by the ideal reversible *Otto cycle*, which is illustrated below. The working substance of the cycle is assumed to be air. The six steps of the Otto cycle are as follows:

i. Isobaric intake stroke (*OA*). A mixture of gasoline and air is drawn into the combustion chamber at atmospheric pressure p_0 as the piston expands, increasing the volume of the cylinder from zero to V_A.

ii. Adiabatic compression stroke (*AB*). The temperature of the mixture rises as the piston compresses it adiabatically from a volume V_A to V_B.

iii. Ignition at constant volume (*BC*). The mixture is ignited by a spark. The combustion happens so fast that there is essentially no motion of the piston. During this process, the added heat Q_1 causes the pressure to increase from p_B to p_C at the constant volume $V_B(=V_C)$.

iv. Adiabatic expansion (*CD*). The heated mixture of gasoline and air expands against the piston, increasing the volume from V_C to V_D. This is called the *power stroke*, as it is the part of the cycle that delivers most of the power to the crankshaft.

v. Constant-volume exhaust (*DA*). When the exhaust valve opens, some of the combustion products escape. There is almost no movement of the piston during this part of the cycle, so the volume remains constant at $V_A(=V_D)$. Most of the available energy is lost here, as represented by the heat exhaust Q_2.

vi. Isobaric compression (*AO*). The exhaust valve remains open, and the compression from V_A to zero drives out the remaining combustion products.

(a) Using (*i*) $e = W/Q_1$; (*ii*) $W = Q_1 - Q_2$; and (*iii*) $Q_1 = nC_v(T_C - T_B)$, $Q_2 = nC_v(T_D - T_A)$, show that

$$e = 1 - \frac{T_D - T_A}{T_C - T_B}.$$

(b) Use the fact that steps (ii) and (iv) are adiabatic to show that

$$e = 1 - \frac{1}{r^{\gamma - 1}},$$

where $r = V_A/V_B$. The quantity r is called the *compression ratio* of the engine.

(c) In practice, r is kept less than around 7. For larger values, the gasoline-air mixture is compressed to temperatures so high that it explodes before the finely timed spark is delivered. This *preignition* causes engine knock and loss of power. Show that for $r = 6$ and $\gamma = 1.4$ (the value for air), $e = 0.51$, or an efficiency of 51%. Because of the many irreversible processes, an actual internal combustion engine has an efficiency much less than this ideal value. A typical efficiency for a tuned engine is about 25% to 30%.

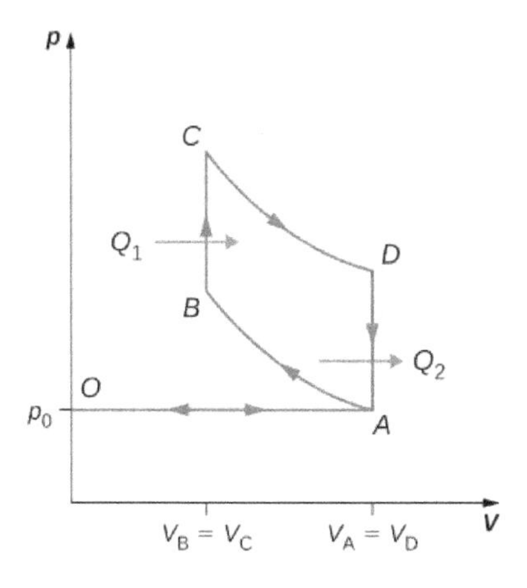

89. An ideal *diesel* cycle is shown below. This cycle consists of five strokes. In this case, only air is drawn into the chamber during the intake stroke *OA*. The air is then compressed adiabatically from state *A* to state *B*, raising its temperature high enough so that when fuel is added during the power stroke *BC*, it ignites. After ignition ends at *C*, there is a further adiabatic power stroke *CD*. Finally, there is an exhaust at constant volume as the pressure drops from p_D to p_A, followed by a further exhaust when the piston compresses the chamber volume to zero.

(a) Use $W = Q_1 - Q_2$, $Q_1 = nC_p(T_C - T_B)$, and $Q_2 = nC_v(T_D - T_A)$ to show that $e = \frac{W}{Q_1} = 1 - \frac{T_D - T_A}{\gamma(T_C - T_B)}$.

(b) Use the fact that $A \to B$ and $C \to D$ are adiabatic to show that

$$e = 1 - \frac{1}{\gamma}\frac{\left(\frac{V_C}{V_D}\right)^\gamma - \left(\frac{V_B}{V_A}\right)^\gamma}{\left(\frac{V_C}{V_D}\right) - \left(\frac{V_B}{V_A}\right)}.$$

(c) Since there is no preignition (remember, the chamber does not contain any fuel during the compression), the compression ratio can be larger than that for a gasoline engine. Typically, $V_A/V_B = 15$ and $V_D/V_C = 5$. For these values and $\gamma = 1.4$, show that $\varepsilon = 0.56$, or an efficiency of 56%. Diesel engines actually operate at an efficiency of about 30% to 35% compared with 25% to 30% for gasoline engines.

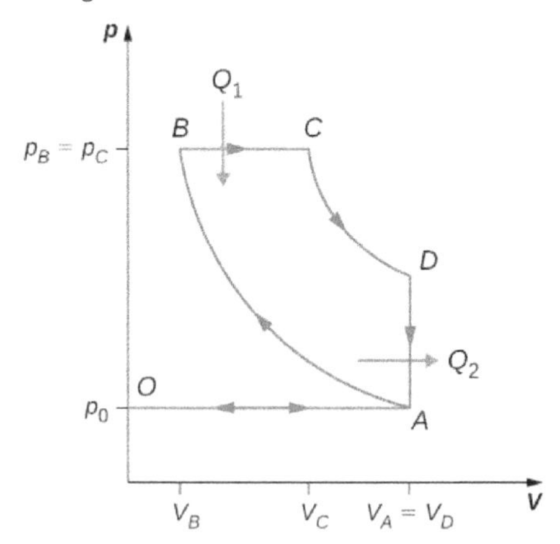

90. Consider an ideal gas Joule cycle, also called the Brayton cycle, shown below. Find the formula for efficiency of the engine using this cycle in terms of P_1, P_2, and γ.

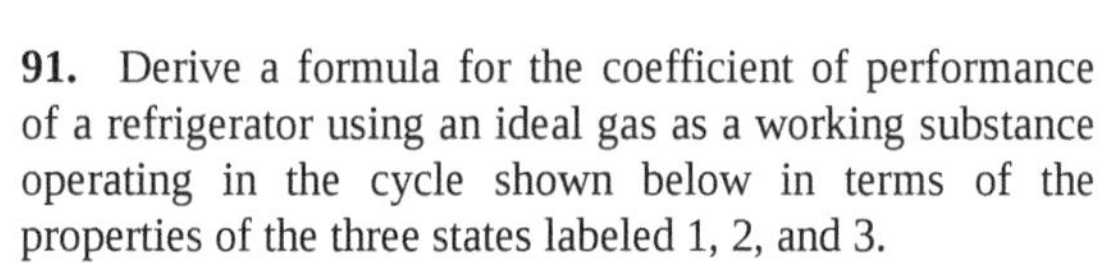

91. Derive a formula for the coefficient of performance of a refrigerator using an ideal gas as a working substance operating in the cycle shown below in terms of the properties of the three states labeled 1, 2, and 3.

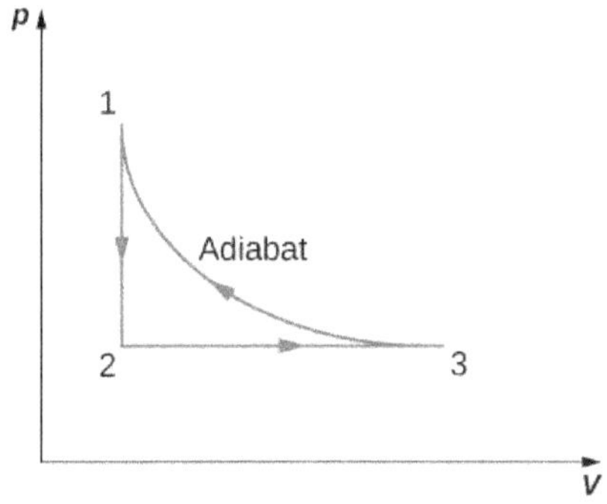

92. Two moles of nitrogen gas, with $\gamma = 7/5$ for ideal diatomic gases, occupies a volume of $10^{-2}\,\text{m}^3$ in an insulated cylinder at temperature 300 K. The gas is adiabatically and reversibly compressed to a volume of 5 L. The piston of the cylinder is locked in its place, and the insulation around the cylinder is removed. The heat-conducting cylinder is then placed in a 300-K bath. Heat from the compressed gas leaves the gas, and the temperature of the gas becomes 300 K again. The gas is then slowly expanded at the fixed temperature 300 K until the volume of the gas becomes $10^{-2}\,\text{m}^3$, thus making a complete cycle for the gas. For the entire cycle, calculate (a) the work done by the gas, (b) the heat into or out of the gas, (c) the change in the internal energy of the gas, and (d) the change in entropy of the gas.

93. A Carnot refrigerator, working between $0\,°\text{C}$ and $30\,°\text{C}$ is used to cool a bucket of water containing $10^{-2}\,\text{m}^3$ of water at $30\,°\text{C}$ to $5\,°\text{C}$ in 2 hours. Find the total amount of work needed.

5 | ELECTRIC CHARGES AND FIELDS

Figure 5.1 Electric charges exist all around us. They can cause objects to be repelled from each other or to be attracted to each other. (credit: modification of work by Sean McGrath)

Chapter Outline

Introduction

Back when we were studying Newton's laws, we identified several physical phenomena as forces. We did so based on the effect they had on a physical object: Specifically, they caused the object to accelerate. Later, when we studied impulse and momentum, we expanded this idea to identify a force as any physical phenomenon that changed the momentum of an object. In either case, the result is the same: We recognize a force by the effect that it has on an object.

In Gravitation (http://cnx.org/content/m58344/latest/) , we examined the force of gravity, which acts on all objects with mass. In this chapter, we begin the study of the electric force, which acts on all objects with a property called charge. The electric force is much stronger than gravity (in most systems where both appear), but it can be a force of attraction or a force of repulsion, which leads to very different effects on objects. The electric force helps keep atoms together, so it is of fundamental importance in matter. But it also governs most everyday interactions we deal with, from chemical interactions to biological processes.

5.1 | Electric Charge

Learning Objectives

By the end of this section, you will be able to:

- Describe the concept of electric charge
- Explain qualitatively the force electric charge creates

You are certainly familiar with electronic devices that you activate with the click of a switch, from computers to cell phones to television. And you have certainly seen electricity in a flash of lightning during a heavy thunderstorm. But you have also most likely experienced electrical effects in other ways, maybe without realizing that an electric force was involved. Let's take a look at some of these activities and see what we can learn from them about electric charges and forces.

Discoveries

You have probably experienced the phenomenon of **static electricity**: When you first take clothes out of a dryer, many (not all) of them tend to stick together; for some fabrics, they can be very difficult to separate. Another example occurs if you take a woolen sweater off quickly—you can feel (and hear) the static electricity pulling on your clothes, and perhaps even your hair. If you comb your hair on a dry day and then put the comb close to a thin stream of water coming out of a faucet, you will find that the water stream bends toward (is attracted to) the comb (Figure 5.2).

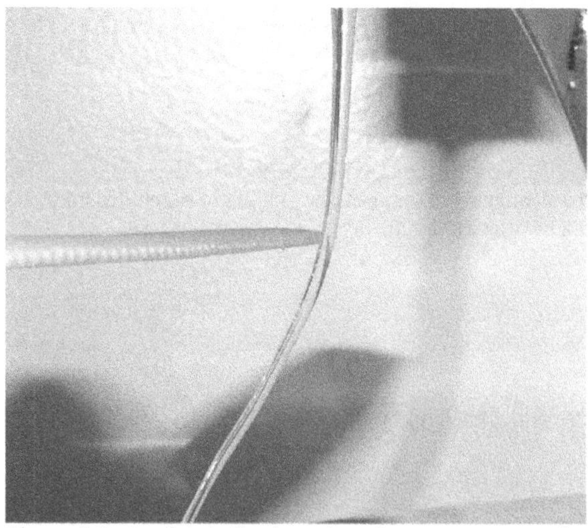

Figure 5.2 An electrically charged comb attracts a stream of water from a distance. Note that the water is not touching the comb. (credit: Jane Whitney)

Suppose you bring the comb close to some small strips of paper; the strips of paper are attracted to the comb and even cling to it (Figure 5.3). In the kitchen, quickly pull a length of plastic cling wrap off the roll; it will tend to cling to most any nonmetallic material (such as plastic, glass, or food). If you rub a balloon on a wall for a few seconds, it will stick to the wall. Probably the most annoying effect of static electricity is getting shocked by a doorknob (or a friend) after shuffling your feet on some types of carpeting.

Figure 5.3 After being used to comb hair, this comb attracts small strips of paper from a distance, without physical contact. Investigation of this behavior helped lead to the concept of the electric force.

Many of these phenomena have been known for centuries. The ancient Greek philosopher Thales of Miletus (624–546 BCE) recorded that when amber (a hard, translucent, fossilized resin from extinct trees) was vigorously rubbed with a piece of fur, a force was created that caused the fur and the amber to be attracted to each other (Figure 5.4). Additionally, he found that the rubbed amber would not only attract the fur, and the fur attract the amber, but they both could affect other (nonmetallic) objects, even if not in contact with those objects (Figure 5.5).

Figure 5.4 Borneo amber is mined in Sabah, Malaysia, from shale-sandstone-mudstone veins. When a piece of amber is rubbed with a piece of fur, the amber gains more electrons, giving it a net negative charge. At the same time, the fur, having lost electrons, becomes positively charged. (credit: “Sebakoamber”/Wikimedia Commons)

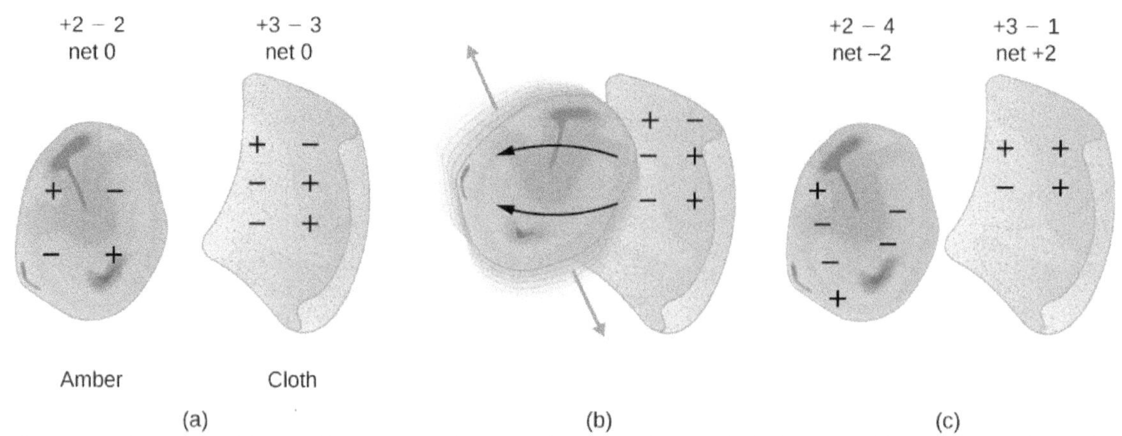

Figure 5.5 When materials are rubbed together, charges can be separated, particularly if one material has a greater affinity for electrons than another. (a) Both the amber and cloth are originally neutral, with equal positive and negative charges. Only a tiny fraction of the charges are involved, and only a few of them are shown here. (b) When rubbed together, some negative charge is transferred to the amber, leaving the cloth with a net positive charge. (c) When separated, the amber and cloth now have net charges, but the absolute value of the net positive and negative charges will be equal.

The English physicist William Gilbert (1544–1603) also studied this attractive force, using various substances. He worked with amber, and, in addition, he experimented with rock crystal and various precious and semi-precious gemstones. He also experimented with several metals. He found that the metals never exhibited this force, whereas the minerals did. Moreover, although an electrified amber rod would attract a piece of fur, it would repel another electrified amber rod; similarly, two electrified pieces of fur would repel each other.

This suggested there were two types of an electric property; this property eventually came to be called **electric charge**. The difference between the two types of electric charge is in the directions of the electric forces that each type of charge causes: These forces are repulsive when the same type of charge exists on two interacting objects and attractive when the charges are of opposite types. The SI unit of electric charge is the **coulomb** (C), after the French physicist Charles Augustine de Coulomb (1736–1806).

The most peculiar aspect of this new force is that it does not require physical contact between the two objects in order to cause an acceleration. This is an example of a so-called "long-range" force. (Or, as Albert Einstein later phrased it, "action at a distance.") With the exception of gravity, all other forces we have discussed so far act only when the two interacting objects actually touch.

The American physicist and statesman Benjamin Franklin found that he could concentrate charge in a " Leyden jar," which was essentially a glass jar with two sheets of metal foil, one inside and one outside, with the glass between them (Figure 5.6). This created a large electric force between the two foil sheets.

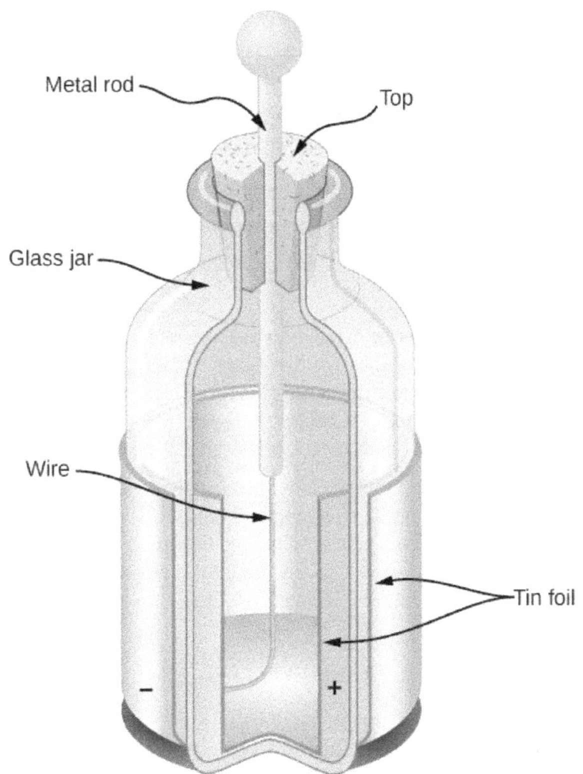

Figure 5.6 A Leyden jar (an early version of what is now called a capacitor) allowed experimenters to store large amounts of electric charge. Benjamin Franklin used such a jar to demonstrate that lightning behaved exactly like the electricity he got from the equipment in his laboratory.

Franklin pointed out that the observed behavior could be explained by supposing that one of the two types of charge remained motionless, while the other type of charge flowed from one piece of foil to the other. He further suggested that an excess of what he called this "electrical fluid" be called "positive electricity" and the deficiency of it be called "negative electricity." His suggestion, with some minor modifications, is the model we use today. (With the experiments that he was able to do, this was a pure guess; he had no way of actually determining the sign of the moving charge. Unfortunately, he guessed wrong; we now know that the charges that flow are the ones Franklin labeled negative, and the positive charges remain largely motionless. Fortunately, as we'll see, it makes no practical or theoretical difference which choice we make, as long as we stay consistent with our choice.)

Let's list the specific observations that we have of this **electric force**:

- The force acts without physical contact between the two objects.
- The force can be either attractive or repulsive: If two interacting objects carry the same sign of charge, the force is repulsive; if the charges are of opposite sign, the force is attractive. These interactions are referred to as **electrostatic repulsion** and **electrostatic attraction**, respectively.
- Not all objects are affected by this force.
- The magnitude of the force decreases (rapidly) with increasing separation distance between the objects.

To be more precise, we find experimentally that the magnitude of the force decreases as the square of the distance between the two interacting objects increases. Thus, for example, when the distance between two interacting objects is doubled, the force between them decreases to one fourth what it was in the original system. We can also observe that the surroundings of the charged objects affect the magnitude of the force. However, we will explore this issue in a later chapter.

Properties of Electric Charge

In addition to the existence of two types of charge, several other properties of charge have been discovered.

- **Charge is quantized.** This means that electric charge comes in discrete amounts, and there is a smallest possible amount of charge that an object can have. In the SI system, this smallest amount is $e \equiv 1.602 \times 10^{-19}\text{ C}$. No free particle can have less charge than this, and, therefore, the charge on any object—the charge on all objects—must be an integer multiple of this amount. All macroscopic, charged objects have charge because electrons have either been added or taken away from them, resulting in a net charge.
- **The magnitude of the charge is independent of the type.** Phrased another way, the smallest possible positive charge (to four significant figures) is $+1.602 \times 10^{-19}\text{ C}$, and the smallest possible negative charge is $-1.602 \times 10^{-19}\text{ C}$; these values are exactly equal. This is simply how the laws of physics in our universe turned out.
- **Charge is conserved.** Charge can neither be created nor destroyed; it can only be transferred from place to place, from one object to another. Frequently, we speak of two charges "canceling"; this is verbal shorthand. It means that if two objects that have equal and opposite charges are physically close to each other, then the (oppositely directed) forces they apply on some other charged object cancel, for a net force of zero. It is important that you understand that the charges on the objects by no means disappear, however. The net charge of the universe is constant.
- **Charge is conserved in closed systems.** In principle, if a negative charge disappeared from your lab bench and reappeared on the Moon, conservation of charge would still hold. However, this never happens. If the total charge you have in your local system on your lab bench is changing, there will be a measurable flow of charge into or out of the system. Again, charges can and do move around, and their effects can and do cancel, but the net charge in your local environment (if closed) is conserved. The last two items are both referred to as the **law of conservation of charge**.

The Source of Charges: The Structure of the Atom

Once it became clear that all matter was composed of particles that came to be called atoms, it also quickly became clear that the constituents of the atom included both positively charged particles and negatively charged particles. The next question was, what are the physical properties of those electrically charged particles?

The negatively charged particle was the first one to be discovered. In 1897, the English physicist J. J. Thomson was studying what was then known as *cathode rays*. Some years before, the English physicist William Crookes had shown that these "rays" were negatively charged, but his experiments were unable to tell any more than that. (The fact that they carried a negative electric charge was strong evidence that these were not rays at all, but particles.) Thomson prepared a pure beam of these particles and sent them through crossed electric and magnetic fields, and adjusted the various field strengths until the net deflection of the beam was zero. With this experiment, he was able to determine the charge-to-mass ratio of the particle. This ratio showed that the mass of the particle was much smaller than that of any other previously known particle—1837 times smaller, in fact. Eventually, this particle came to be called the **electron**.

Since the atom as a whole is electrically neutral, the next question was to determine how the positive and negative charges are distributed within the atom. Thomson himself imagined that his electrons were embedded within a sort of positively charged paste, smeared out throughout the volume of the atom. However, in 1908, the New Zealand physicist Ernest Rutherford showed that the positive charges of the atom existed within a tiny core—called a nucleus—that took up only a very tiny fraction of the overall volume of the atom, but held over 99% of the mass. (See Linear Momentum and Collisions (http://cnx.org/content/m58317/latest/) .) In addition, he showed that the negatively charged electrons perpetually orbited about this nucleus, forming a sort of electrically charged cloud that surrounds the nucleus (Figure 5.7). Rutherford concluded that the nucleus was constructed of small, massive particles that he named **protons**.

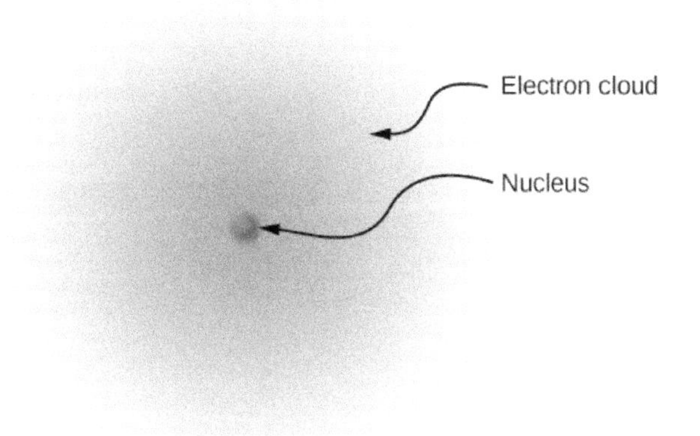

Figure 5.7 This simplified model of a hydrogen atom shows a positively charged nucleus (consisting, in the case of hydrogen, of a single proton), surrounded by an electron "cloud." The charge of the electron cloud is equal (and opposite in sign) to the charge of the nucleus, but the electron does not have a definite location in space; hence, its representation here is as a cloud. Normal macroscopic amounts of matter contain immense numbers of atoms and molecules, and, hence, even greater numbers of individual negative and positive charges.

Since it was known that different atoms have different masses, and that ordinarily atoms are electrically neutral, it was natural to suppose that different atoms have different numbers of protons in their nucleus, with an equal number of negatively charged electrons orbiting about the positively charged nucleus, thus making the atoms overall electrically neutral. However, it was soon discovered that although the lightest atom, hydrogen, did indeed have a single proton as its nucleus, the next heaviest atom—helium—has twice the number of protons (two), but *four* times the mass of hydrogen.

This mystery was resolved in 1932 by the English physicist James Chadwick, with the discovery of the **neutron**. The neutron is, essentially, an electrically neutral twin of the proton, with no electric charge, but (nearly) identical mass to the proton. The helium nucleus therefore has two neutrons along with its two protons. (Later experiments were to show that although the neutron is electrically neutral overall, it does have an internal charge *structure*. Furthermore, although the masses of the neutron and the proton are *nearly* equal, they aren't exactly equal: The neutron's mass is very slightly larger than the mass of the proton. That slight mass excess turned out to be of great importance. That, however, is a story that will have to wait until our study of modern physics in Nuclear Physics (http://cnx.org/content/m58606/latest/) .)

Thus, in 1932, the picture of the atom was of a small, massive nucleus constructed of a combination of protons and neutrons, surrounded by a collection of electrons whose combined motion formed a sort of negatively charged "cloud" around the nucleus (Figure 5.8). In an electrically neutral atom, the total negative charge of the collection of electrons is equal to the total positive charge in the nucleus. The very low-mass electrons can be more or less easily removed or added to an atom, changing the net charge on the atom (though without changing its type). An atom that has had the charge altered in this way is called an **ion**. Positive ions have had electrons removed, whereas negative ions have had excess electrons added. We also use this term to describe molecules that are not electrically neutral.

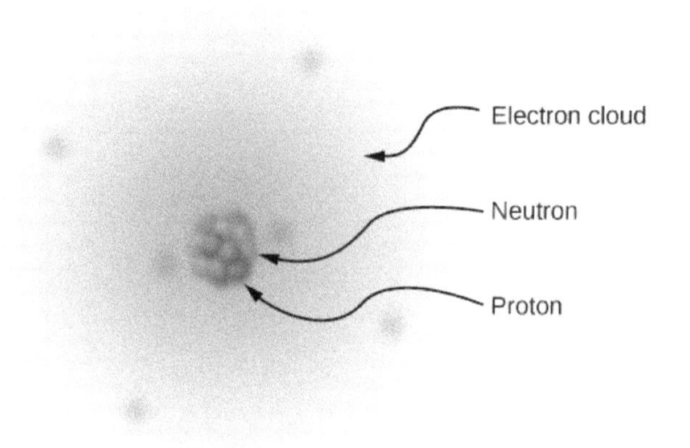

Figure 5.8 The nucleus of a carbon atom is composed of six protons and six neutrons. As in hydrogen, the surrounding six electrons do not have definite locations and so can be considered to be a sort of cloud surrounding the nucleus.

The story of the atom does not stop there, however. In the latter part of the twentieth century, many more subatomic particles were discovered in the nucleus of the atom: pions, neutrinos, and quarks, among others. With the exception of the photon, none of these particles are directly relevant to the study of electromagnetism, so we defer further discussion of them until the chapter on particle physics (Particle Physics and Cosmology (http://cnx.org/content/m58767/latest/)).

A Note on Terminology

As noted previously, electric charge is a property that an object can have. This is similar to how an object can have a property that we call mass, a property that we call density, a property that we call temperature, and so on. Technically, we should always say something like, "Suppose we have a particle that carries a charge of $3\ \mu\text{C}$. " However, it is very common to say instead, "Suppose we have a $3\text{-}\mu\text{C}$ charge." Similarly, we often say something like, "Six charges are located at the vertices of a regular hexagon." A charge is not a particle; rather, it is a *property* of a particle. Nevertheless, this terminology is extremely common (and is frequently used in this book, as it is everywhere else). So, keep in the back of your mind what we really mean when we refer to a "charge."

5.2 | Conductors, Insulators, and Charging by Induction

Learning Objectives

By the end of this section, you will be able to:

- Explain what a conductor is
- Explain what an insulator is
- List the differences and similarities between conductors and insulators
- Describe the process of charging by induction

In the preceding section, we said that scientists were able to create electric charge only on nonmetallic materials and never on metals. To understand why this is the case, you have to understand more about the nature and structure of atoms. In this section, we discuss how and why electric charges do—or do not—move through materials (Figure 5.9). A more complete description is given in a later chapter.

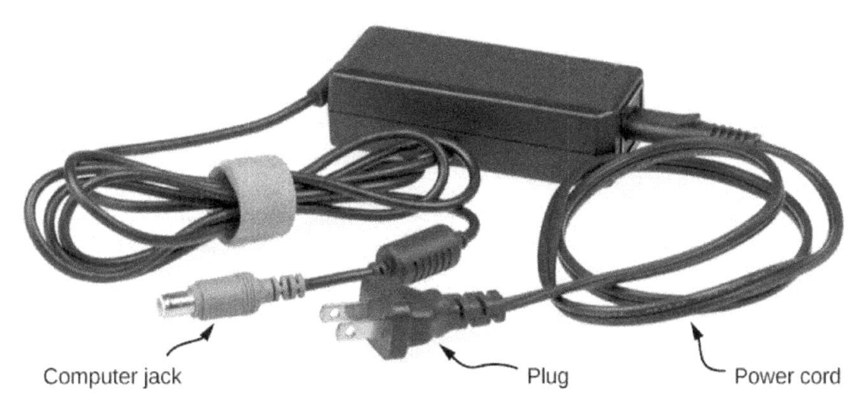

Figure 5.9 This power adapter uses metal wires and connectors to conduct electricity from the wall socket to a laptop computer. The conducting wires allow electrons to move freely through the cables, which are shielded by rubber and plastic. These materials act as insulators that don't allow electric charge to escape outward. (credit: modification of work by "Evan-Amos"/Wikimedia Commons)

Conductors and Insulators

As discussed in the previous section, electrons surround the tiny nucleus in the form of a (comparatively) vast cloud of negative charge. However, this cloud does have a definite structure to it. Let's consider an atom of the most commonly used conductor, copper.

For reasons that will become clear in Atomic Structure (http://cnx.org/content/m58583/latest/) , there is an outermost electron that is only loosely bound to the atom's nucleus. It can be easily dislodged; it then moves to a neighboring atom. In a large mass of copper atoms (such as a copper wire or a sheet of copper), these vast numbers of outermost electrons (one per atom) wander from atom to atom, and are the electrons that do the moving when electricity flows. These wandering, or "free," electrons are called **conduction electrons**, and copper is therefore an excellent **conductor** (of electric charge). All conducting elements have a similar arrangement of their electrons, with one or two conduction electrons. This includes most metals.

Insulators, in contrast, are made from materials that lack conduction electrons; charge flows only with great difficulty, if at all. Even if excess charge is added to an insulating material, it cannot move, remaining indefinitely in place. This is why insulating materials exhibit the electrical attraction and repulsion forces described earlier, whereas conductors do not; any excess charge placed on a conductor would instantly flow away (due to mutual repulsion from existing charges), leaving no excess charge around to create forces. Charge cannot flow along or through an **insulator**, so its electric forces remain for long periods of time. (Charge will dissipate from an insulator, given enough time.) As it happens, amber, fur, and most semi-precious gems are insulators, as are materials like wood, glass, and plastic.

Charging by Induction

Let's examine in more detail what happens in a conductor when an electrically charged object is brought close to it. As mentioned, the conduction electrons in the conductor are able to move with nearly complete freedom. As a result, when a charged insulator (such as a positively charged glass rod) is brought close to the conductor, the (total) charge on the insulator exerts an electric force on the conduction electrons. Since the rod is positively charged, the conduction electrons (which themselves are negatively charged) are attracted, flowing toward the insulator to the near side of the conductor (Figure 5.10).

Now, the conductor is still overall electrically neutral; the conduction electrons have changed position, but they are still in the conducting material. However, the conductor now has a charge *distribution*; the near end (the portion of the conductor closest to the insulator) now has more negative charge than positive charge, and the reverse is true of the end farthest from the insulator. The relocation of negative charges to the near side of the conductor results in an overall positive charge in the part of the conductor farthest from the insulator. We have thus created an electric charge distribution where one did not exist before. This process is referred to as *inducing polarization*—in this case, polarizing the conductor. The resulting separation of positive and negative charge is called **polarization**, and a material, or even a molecule, that exhibits polarization is said to be polarized. A similar situation occurs with a negatively charged insulator, but the resulting polarization is in the opposite direction.

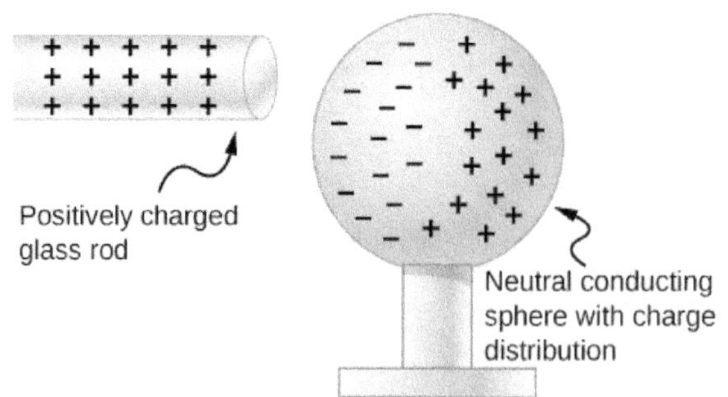

Figure 5.10 Induced polarization. A positively charged glass rod is brought near the left side of the conducting sphere, attracting negative charge and leaving the other side of the sphere positively charged. Although the sphere is overall still electrically neutral, it now has a charge distribution, so it can exert an electric force on other nearby charges. Furthermore, the distribution is such that it will be attracted to the glass rod.

The result is the formation of what is called an electric **dipole**, from a Latin phrase meaning "two ends." The presence of electric charges on the insulator—and the electric forces they apply to the conduction electrons—creates, or "induces," the dipole in the conductor.

Neutral objects can be attracted to any charged object. The pieces of straw attracted to polished amber are neutral, for example. If you run a plastic comb through your hair, the charged comb can pick up neutral pieces of paper. Figure 5.11 shows how the polarization of atoms and molecules in neutral objects results in their attraction to a charged object.

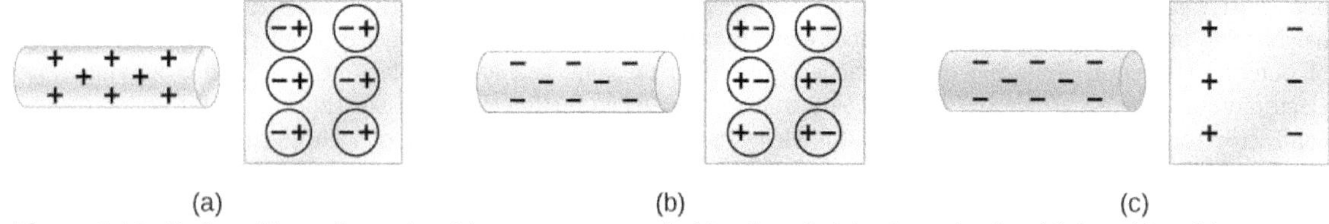

Figure 5.11 Both positive and negative objects attract a neutral object by polarizing its molecules. (a) A positive object brought near a neutral insulator polarizes its molecules. There is a slight shift in the distribution of the electrons orbiting the molecule, with unlike charges being brought nearer and like charges moved away. Since the electrostatic force decreases with distance, there is a net attraction. (b) A negative object produces the opposite polarization, but again attracts the neutral object. (c) The same effect occurs for a conductor; since the unlike charges are closer, there is a net attraction.

When a charged rod is brought near a neutral substance, an insulator in this case, the distribution of charge in atoms and molecules is shifted slightly. Opposite charge is attracted nearer the external charged rod, while like charge is repelled. Since the electrostatic force decreases with distance, the repulsion of like charges is weaker than the attraction of unlike charges, and so there is a net attraction. Thus, a positively charged glass rod attracts neutral pieces of paper, as will a negatively charged rubber rod. Some molecules, like water, are polar molecules. Polar molecules have a natural or inherent separation of charge, although they are neutral overall. Polar molecules are particularly affected by other charged objects and show greater polarization effects than molecules with naturally uniform charge distributions.

When the two ends of a dipole can be separated, this method of **charging by induction** may be used to create charged objects without transferring charge. In Figure 5.12, we see two neutral metal spheres in contact with one another but insulated from the rest of the world. A positively charged rod is brought near one of them, attracting negative charge to that side, leaving the other sphere positively charged.

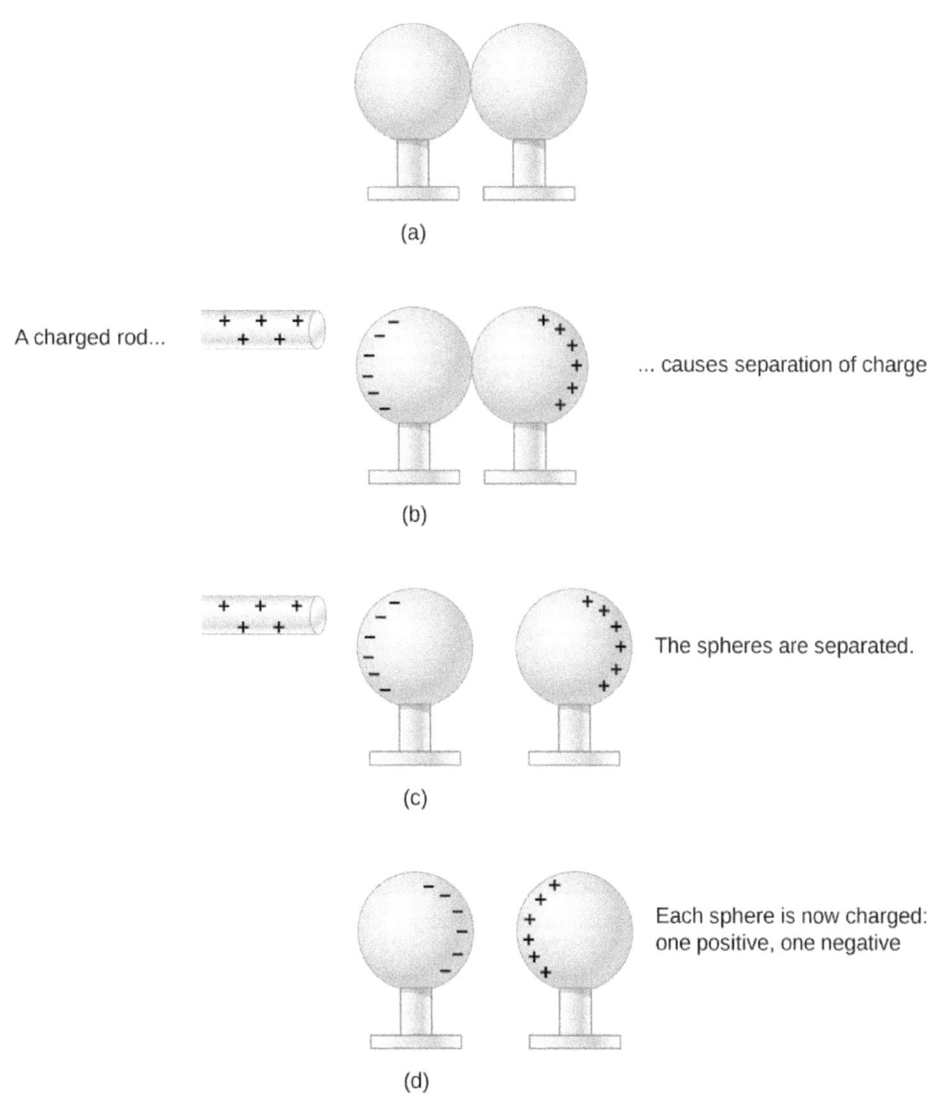

Figure 5.12 Charging by induction. (a) Two uncharged or neutral metal spheres are in contact with each other but insulated from the rest of the world. (b) A positively charged glass rod is brought near the sphere on the left, attracting negative charge and leaving the other sphere positively charged. (c) The spheres are separated before the rod is removed, thus separating negative and positive charges. (d) The spheres retain net charges after the inducing rod is removed—without ever having been touched by a charged object.

Another method of charging by induction is shown in Figure 5.13. The neutral metal sphere is polarized when a charged rod is brought near it. The sphere is then grounded, meaning that a conducting wire is run from the sphere to the ground. Since Earth is large and most of the ground is a good conductor, it can supply or accept excess charge easily. In this case, electrons are attracted to the sphere through a wire called the ground wire, because it supplies a conducting path to the ground. The ground connection is broken before the charged rod is removed, leaving the sphere with an excess charge opposite to that of the rod. Again, an opposite charge is achieved when charging by induction, and the charged rod loses none of its excess charge.

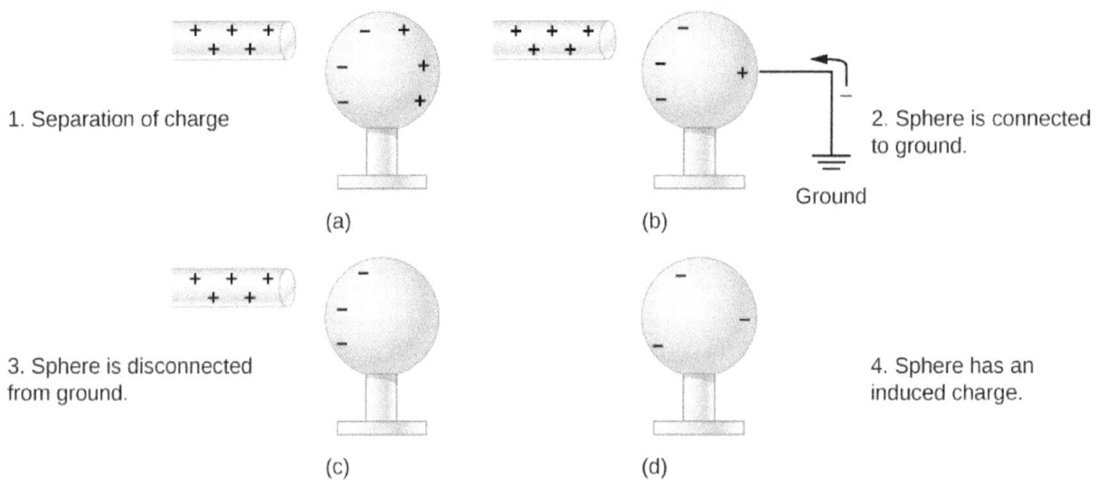

Figure 5.13 Charging by induction using a ground connection. (a) A positively charged rod is brought near a neutral metal sphere, polarizing it. (b) The sphere is grounded, allowing electrons to be attracted from Earth's ample supply. (c) The ground connection is broken. (d) The positive rod is removed, leaving the sphere with an induced negative charge.

5.3 | Coulomb's Law

Learning Objectives

By the end of this section, you will be able to:

- Describe the electric force, both qualitatively and quantitatively
- Calculate the force that charges exert on each other
- Determine the direction of the electric force for different source charges
- Correctly describe and apply the superposition principle for multiple source charges

Experiments with electric charges have shown that if two objects each have electric charge, then they exert an electric force on each other. The magnitude of the force is linearly proportional to the net charge on each object and inversely proportional to the square of the distance between them. (Interestingly, the force does not depend on the mass of the objects.) The direction of the force vector is along the imaginary line joining the two objects and is dictated by the signs of the charges involved.

Let

- $q_1, q_2 =$ the net electric charges of the two objects;
- $\vec{\mathbf{r}}_{12} =$ the vector displacement from q_1 to q_2.

The electric force $\vec{\mathbf{F}}$ on one of the charges is proportional to the magnitude of its own charge and the magnitude of the other charge, and is inversely proportional to the square of the distance between them:

$$F \propto \frac{q_1 q_2}{r_{12}^2}.$$

This proportionality becomes an equality with the introduction of a proportionality constant. For reasons that will become clear in a later chapter, the proportionality constant that we use is actually a collection of constants. (We discuss this constant shortly.)

Coulomb's Law

The electric force (or **Coulomb force**) between two electrically charged particles is equal to

$$\vec{\mathbf{F}}_{12}(r) = \frac{1}{4\pi\varepsilon_0} \frac{|q_1 q_2|}{r_{12}^2} \hat{\mathbf{r}}_{12} \tag{5.1}$$

We use absolute value signs around the product $q_1 q_2$ because one of the charges may be negative, but the magnitude of the force is always positive. The unit vector $\hat{\mathbf{r}}$ points directly from the charge q_1 toward q_2. If q_1 and q_2 have the same sign, the force vector on q_2 points away from q_1; if they have opposite signs, the force on q_2 points toward q_1 (Figure 5.14).

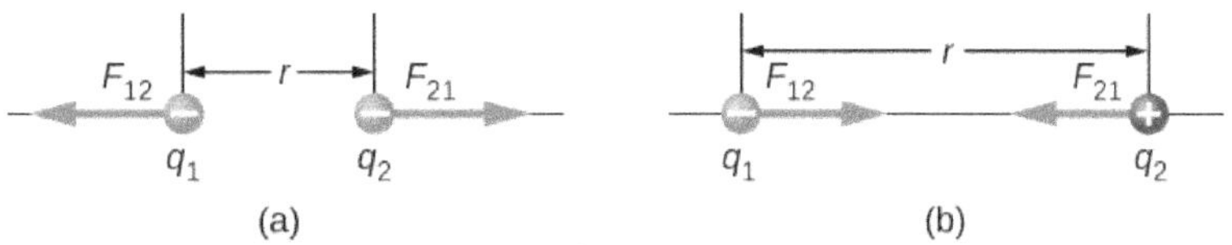

Figure 5.14 The electrostatic force $\vec{\mathbf{F}}$ between point charges q_1 and q_2 separated by a distance *r* is given by Coulomb's law. Note that Newton's third law (every force exerted creates an equal and opposite force) applies as usual—the force on q_1 is equal in magnitude and opposite in direction to the force it exerts on q_2. (a) Like charges; (b) unlike charges.

It is important to note that the electric force is not constant; it is a function of the separation distance between the two charges. If either the test charge or the source charge (or both) move, then $\vec{\mathbf{r}}$ changes, and therefore so does the force. An immediate consequence of this is that direct application of Newton's laws with this force can be mathematically difficult, depending on the specific problem at hand. It can (usually) be done, but we almost always look for easier methods of calculating whatever physical quantity we are interested in. (Conservation of energy is the most common choice.)

Finally, the new constant ε_0 in Coulomb's law is called the *permittivity of free space*, or (better) the **permittivity of vacuum**. It has a very important physical meaning that we will discuss in a later chapter; for now, it is simply an empirical proportionality constant. Its numerical value (to three significant figures) turns out to be

$$\varepsilon_0 = 8.85 \times 10^{-12} \frac{\text{C}^2}{\text{N} \cdot \text{m}^2}.$$

These units are required to give the force in Coulomb's law the correct units of newtons. Note that in Coulomb's law, the permittivity of vacuum is only part of the proportionality constant. For convenience, we often define a Coulomb's constant:

$$k_e = \frac{1}{4\pi\varepsilon_0} = 8.99 \times 10^9 \frac{\text{N} \cdot \text{m}^2}{\text{C}^2}.$$

Example 5.1

The Force on the Electron in Hydrogen

A hydrogen atom consists of a single proton and a single electron. The proton has a charge of $+e$ and the electron has $-e$. In the "ground state" of the atom, the electron orbits the proton at most probable distance of 5.29×10^{-11} m (Figure 5.15). Calculate the electric force on the electron due to the proton.

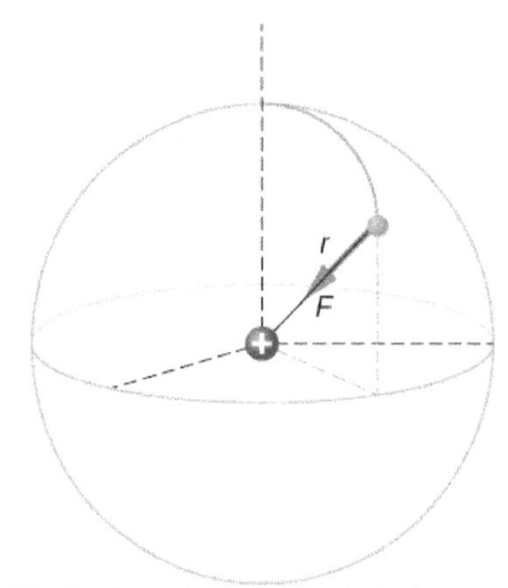

Figure 5.15 A schematic depiction of a hydrogen atom, showing the force on the electron. This depiction is only to enable us to calculate the force; the hydrogen atom does not really look like this. Recall Figure 5.7.

Strategy

For the purposes of this example, we are treating the electron and proton as two point particles, each with an electric charge, and we are told the distance between them; we are asked to calculate the force on the electron. We thus use Coulomb's law.

Solution

Our two charges and the distance between them are,

$$\begin{aligned} q_1 &= +e = +1.602 \times 10^{-19}\ \text{C} \\ q_2 &= -e = -1.602 \times 10^{-19}\ \text{C} \\ r &= 5.29 \times 10^{-11}\ \text{m}. \end{aligned}$$

The magnitude of the force on the electron is

$$F = \frac{1}{4\pi\epsilon_0}\frac{|e|^2}{r^2} = \frac{1}{4\pi\left(8.85 \times 10^{-12}\ \frac{\text{C}^2}{\text{N}\cdot\text{m}^2}\right)}\frac{\left(1.602 \times 10^{-19}\ \text{C}\right)^2}{\left(5.29 \times 10^{-11}\ \text{m}\right)^2} = 8.25 \times 10^{-8}\ \text{N}.$$

As for the direction, since the charges on the two particles are opposite, the force is attractive; the force on the electron points radially directly toward the proton, everywhere in the electron's orbit. The force is thus expressed as

$$\vec{\mathbf{F}} = \left(8.25 \times 10^{-8}\ \text{N}\right)\hat{\mathbf{r}}.$$

Significance

This is a three-dimensional system, so the electron (and therefore the force on it) can be anywhere in an imaginary spherical shell around the proton. In this "classical" model of the hydrogen atom, the electrostatic force on the electron points in the inward centripetal direction, thus maintaining the electron's orbit. But note that the quantum mechanical model of hydrogen (discussed in Quantum Mechanics (http://cnx.org/content/m58573/latest/)) is utterly different.

5.1 Check Your Understanding What would be different if the electron also had a positive charge?

Multiple Source Charges

The analysis that we have done for two particles can be extended to an arbitrary number of particles; we simply repeat the analysis, two charges at a time. Specifically, we ask the question: Given N charges (which we refer to as source charge), what is the net electric force that they exert on some other point charge (which we call the test charge)? Note that we use these terms because we can think of the test charge being used to test the strength of the force provided by the source charges.

Like all forces that we have seen up to now, the net electric force on our test charge is simply the vector sum of each individual electric force exerted on it by each of the individual test charges. Thus, we can calculate the net force on the test charge Q by calculating the force on it from each source charge, taken one at a time, and then adding all those forces together (as vectors). This ability to simply add up individual forces in this way is referred to as the **principle of superposition**, and is one of the more important features of the electric force. In mathematical form, this becomes

$$\vec{\mathbf{F}}(r) = \frac{1}{4\pi\varepsilon_0} Q \sum_{i=1}^{N} \frac{q_i}{r_i^2} \hat{\mathbf{r}}_i. \quad (5.2)$$

In this expression, Q represents the charge of the particle that is experiencing the electric force $\vec{\mathbf{F}}$, and is located at $\vec{\mathbf{r}}$ from the origin; the q_i's are the N source charges, and the vectors $\vec{\mathbf{r}}_i = r_i \hat{\mathbf{r}}_i$ are the displacements from the position of the ith charge to the position of Q. Each of the N unit vectors points directly from its associated source charge toward the test charge. All of this is depicted in Figure 5.16. Please note that there is no physical difference between Q and q_i; the difference in labels is merely to allow clear discussion, with Q being the charge we are determining the force on.

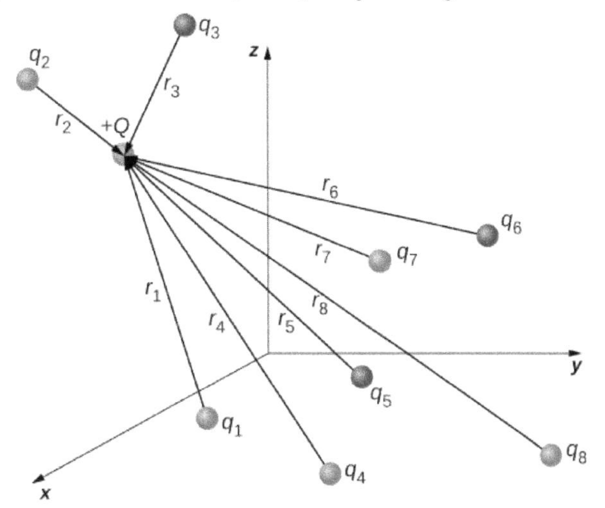

Figure 5.16 The eight source charges each apply a force on the single test charge Q. Each force can be calculated independently of the other seven forces. This is the essence of the superposition principle.

(Note that the force vector $\vec{\mathbf{F}}_i$ does not necessarily point in the same direction as the unit vector $\hat{\mathbf{r}}_i$; it may point in the opposite direction, $-\hat{\mathbf{r}}_i$. The signs of the source charge and test charge determine the direction of the force on the test charge.)

There is a complication, however. Just as the source charges each exert a force on the test charge, so too (by Newton's third law) does the test charge exert an equal and opposite force on each of the source charges. As a consequence, each source charge would change position. However, by Equation 5.2, the force on the test charge is a function of position; thus, as the positions of the source charges change, the net force on the test charge necessarily changes, which changes the force, which again changes the positions. Thus, the entire mathematical analysis quickly becomes intractable. Later, we will learn techniques for handling this situation, but for now, we make the simplifying assumption that the source charges are fixed in place somehow, so that their positions are constant in time. (The test charge is allowed to move.) With this restriction in place, the analysis of charges is known as **electrostatics**, where "statics" refers to the constant (that is, static) positions of the source charges and the force is referred to as an **electrostatic force**.

Example 5.2

The Net Force from Two Source Charges

Three different, small charged objects are placed as shown in Figure 5.17. The charges q_1 and q_3 are fixed in place; q_2 is free to move. Given $q_1 = 2e$, $q_2 = -3e$, and $q_3 = -5e$, and that $d = 2.0 \times 10^{-7}$ m, what is the net force on the middle charge q_2?

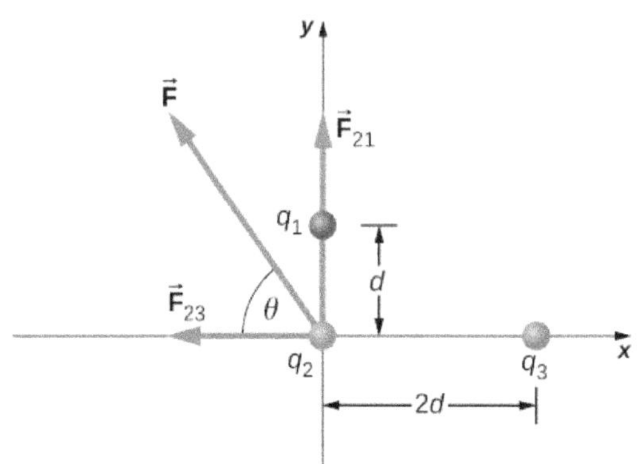

Figure 5.17 Source charges q_1 and q_3 each apply a force on q_2.

Strategy

We use Coulomb's law again. The way the question is phrased indicates that q_2 is our test charge, so that q_1 and q_3 are source charges. The principle of superposition says that the force on q_2 from each of the other charges is unaffected by the presence of the other charge. Therefore, we write down the force on q_2 from each and add them together as vectors.

Solution

We have two source charges $(q_1$ and $q_3)$, a test charge (q_2), distances $(r_{21}$ and $r_{23})$, and we are asked to find a force. This calls for Coulomb's law and superposition of forces. There are two forces:

$$\vec{\mathbf{F}} = \vec{\mathbf{F}}_{21} + \vec{\mathbf{F}}_{23} = \frac{1}{4\pi\varepsilon_0}\left[\frac{q_2 q_1}{r_{21}^2}\hat{\mathbf{j}} + \left(-\frac{q_2 q_3}{r_{23}^2}\hat{\mathbf{i}}\right)\right].$$

We can't add these forces directly because they don't point in the same direction: $\vec{\mathbf{F}}_{12}$ points only in the $-x$-direction, while $\vec{\mathbf{F}}_{13}$ points only in the $+y$-direction. The net force is obtained from applying the Pythagorean theorem to its x- and y-components:

$$F = \sqrt{F_x^2 + F_y^2}$$

where

$$F_x = -F_{23} = -\frac{1}{4\pi\varepsilon_0}\frac{q_2 q_3}{r_{23}^2}$$
$$= -\left(8.99 \times 10^9 \frac{\text{N}\cdot\text{m}^2}{\text{C}^2}\right)\frac{(4.806 \times 10^{-19}\ \text{C})(8.01 \times 10^{-19}\ \text{C})}{(4.00 \times 10^{-7}\ \text{m})^2}$$
$$= -2.16 \times 10^{-14}\ \text{N}$$

and

$$F_y = F_{21} = \frac{1}{4\pi\varepsilon_0}\frac{q_2 q_1}{r_{21}^2}$$
$$= \left(8.99 \times 10^9 \frac{\text{N}\cdot\text{m}^2}{\text{C}^2}\right)\frac{(4.806 \times 10^{-19}\ \text{C})(3.204 \times 10^{-19}\ \text{C})}{(2.00 \times 10^{-7}\ \text{m})^2}$$
$$= 3.46 \times 10^{-14}\ \text{N}.$$

We find that

$$F = \sqrt{F_x^2 + F_y^2} = 4.08 \times 10^{-14}\ \text{N}$$

at an angle of

$$\phi = \tan^{-1}\left(\frac{F_y}{F_x}\right) = \tan^{-1}\left(\frac{3.46 \times 10^{-14}\ \text{N}}{-2.16 \times 10^{-14}\ \text{N}}\right) = -58°,$$

that is, $58°$ above the $-x$-axis, as shown in the diagram.

Significance

Notice that when we substituted the numerical values of the charges, we did not include the negative sign of either q_2 or q_3. Recall that negative signs on vector quantities indicate a reversal of direction of the vector in question. But for electric forces, the direction of the force is determined by the types (signs) of both interacting charges; we determine the force directions by considering whether the signs of the two charges are the same or are opposite. If you also include negative signs from negative charges when you substitute numbers, you run the risk of mathematically reversing the direction of the force you are calculating. Thus, the safest thing to do is to calculate just the magnitude of the force, using the absolute values of the charges, and determine the directions physically.

It's also worth noting that the only new concept in this example is how to calculate the electric forces; everything else (getting the net force from its components, breaking the forces into their components, finding the direction of the net force) is the same as force problems you have done earlier.

5.2 Check Your Understanding What would be different if q_1 were negative?

5.4 | Electric Field

Learning Objectives

By the end of this section, you will be able to:

- Explain the purpose of the electric field concept
- Describe the properties of the electric field
- Calculate the field of a collection of source charges of either sign

As we showed in the preceding section, the net electric force on a test charge is the vector sum of all the electric forces acting on it, from all of the various source charges, located at their various positions. But what if we use a different test charge, one with a different magnitude, or sign, or both? Or suppose we have a dozen different test charges we wish to try at the same location? We would have to calculate the sum of the forces from scratch. Fortunately, it is possible to define a quantity, called the **electric field**, which is independent of the test charge. It only depends on the configuration of the source charges, and once found, allows us to calculate the force on any test charge.

Defining a Field

Suppose we have N source charges $q_1, q_2, q_3, \ldots, q_N$ located at positions $\vec{\mathbf{r}}_1, \vec{\mathbf{r}}_2, \vec{\mathbf{r}}_3, \ldots, \vec{\mathbf{r}}_N$, applying N electrostatic forces on a test charge Q. The net force on Q is (see Equation 5.2)

$$\begin{aligned}\vec{\mathbf{F}} &= \vec{\mathbf{F}}_1 + \vec{\mathbf{F}}_2 + \vec{\mathbf{F}}_3 + \cdots + \vec{\mathbf{F}}_N \\ &= \frac{1}{4\pi\varepsilon_0}\left(\frac{Qq_1}{r_1^2}\hat{\mathbf{r}}_1 + \frac{Qq_2}{r_2^2}\hat{\mathbf{r}}_2 + \frac{Qq_3}{r_3^2}\hat{\mathbf{r}}_3 + \cdots + \frac{Qq_N}{r_1^2}\hat{\mathbf{r}}_N\right) \\ &= Q\left[\frac{1}{4\pi\varepsilon_0}\left(\frac{q_1}{r_1^2}\hat{\mathbf{r}}_1 + \frac{q_2}{r_2^2}\hat{\mathbf{r}}_2 + \frac{q_3}{r_3^2}\hat{\mathbf{r}}_3 + \cdots + \frac{q_N}{r_1^2}\hat{\mathbf{r}}_N\right)\right].\end{aligned}$$

We can rewrite this as

$$\vec{\mathbf{F}} = Q\vec{\mathbf{E}} \tag{5.3}$$

where

$$\vec{\mathbf{E}} \equiv \frac{1}{4\pi\varepsilon_0}\left(\frac{q_1}{r_1^2}\hat{\mathbf{r}}_1 + \frac{q_2}{r_2^2}\hat{\mathbf{r}}_2 + \frac{q_3}{r_3^2}\hat{\mathbf{r}}_3 + \cdots + \frac{q_N}{r_1^2}\hat{\mathbf{r}}_N\right)$$

or, more compactly,

$$\vec{\mathbf{E}}(P) \equiv \frac{1}{4\pi\varepsilon_0}\sum_{i=1}^{N}\frac{q_i}{r_i^2}\hat{\mathbf{r}}_i. \tag{5.4}$$

This expression is called the electric field at position $P = P(x, y, z)$ of the N source charges. Here, P is the location of the point in space where you are calculating the field and is relative to the positions $\vec{\mathbf{r}}_i$ of the source charges (Figure 5.18). Note that we have to impose a coordinate system to solve actual problems.

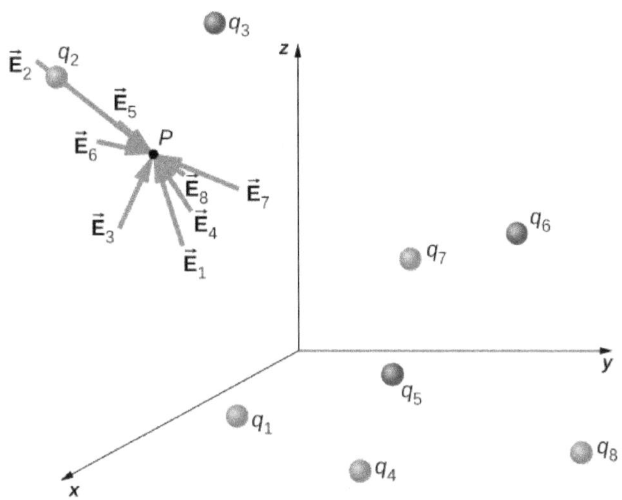

Figure 5.18 Each of these eight source charges creates its own electric field at every point in space; shown here are the field vectors at an arbitrary point P. Like the electric force, the net electric field obeys the superposition principle.

Notice that the calculation of the electric field makes no reference to the test charge. Thus, the physically useful approach is to calculate the electric field and then use it to calculate the force on some test charge later, if needed. Different test charges experience different forces Equation 5.3, but it is the same electric field Equation 5.4. That being said, recall that there is no fundamental difference between a test charge and a source charge; these are merely convenient labels for the system of interest. Any charge produces an electric field; however, just as Earth's orbit is not affected by Earth's own gravity, a charge is not subject to a force due to the electric field it generates. Charges are only subject to forces from the electric fields of other charges.

In this respect, the electric field $\vec{\mathbf{E}}$ of a point charge is similar to the gravitational field $\vec{\mathbf{g}}$ of Earth; once we have calculated the gravitational field at some point in space, we can use it any time we want to calculate the resulting force on any mass we choose to place at that point. In fact, this is exactly what we do when we say the gravitational field of Earth (near Earth's surface) has a value of 9.81 m/s^2, and then we calculate the resulting force (i.e., weight) on different masses. Also, the general expression for calculating $\vec{\mathbf{g}}$ at arbitrary distances from the center of Earth (i.e., not just near Earth's surface) is very similar to the expression for $\vec{\mathbf{E}}$: $\vec{\mathbf{g}} = G\frac{M}{r^2}\hat{\mathbf{r}}$, where G is a proportionality constant, playing the same role for $\vec{\mathbf{g}}$ as $\frac{1}{4\pi\varepsilon_0}$ does for $\vec{\mathbf{E}}$. The value of $\vec{\mathbf{g}}$ is calculated once and is then used in an endless number of problems.

To push the analogy further, notice the units of the electric field: From $F = QE$, the units of E are newtons per coulomb, N/C, that is, the electric field applies a force on each unit charge. Now notice the units of g: From $w = mg$, the units of g are newtons per kilogram, N/kg, that is, the gravitational field applies a force on each unit mass. We could say that the gravitational field of Earth, near Earth's surface, has a value of 9.81 N/kg.

The Meaning of "Field"

Recall from your studies of gravity that the word "field" in this context has a precise meaning. A field, in physics, is a physical quantity whose value depends on (is a function of) position, relative to the source of the field. In the case of the electric field, Equation 5.4 shows that the value of $\vec{\mathbf{E}}$ (both the magnitude and the direction) depends on where in space the point P is located, measured from the locations $\vec{\mathbf{r}}_i$ of the source charges q_i.

In addition, since the electric field is a vector quantity, the electric field is referred to as a *vector field*. (The gravitational field is also a vector field.) In contrast, a field that has only a magnitude at every point is a *scalar field*. The temperature in a room is an example of a scalar field. It is a field because the temperature, in general, is different at different locations in the room, and it is a scalar field because temperature is a scalar quantity.

Also, as you did with the gravitational field of an object with mass, you should picture the electric field of a charge-bearing object (the source charge) as a continuous, immaterial substance that surrounds the source charge, filling all of space—in principle, to $\pm\infty$ in all directions. The field exists at every physical point in space. To put it another way, the electric charge on an object alters the space around the charged object in such a way that all other electrically charged objects in space experience an electric force as a result of being in that field. The electric field, then, is the mechanism by which the electric properties of the source charge are transmitted to and through the rest of the universe. (Again, the range of the electric force is infinite.)

We will see in subsequent chapters that the speed at which electrical phenomena travel is the same as the speed of light. There is a deep connection between the electric field and light.

Superposition

Yet another experimental fact about the field is that it obeys the superposition principle. In this context, that means that we can (in principle) calculate the total electric field of many source charges by calculating the electric field of only q_1 at position *P*, then calculate the field of q_2 at *P*, while—and this is the crucial idea—ignoring the field of, and indeed even the existence of, q_1. We can repeat this process, calculating the field of each individual source charge, independently of the existence of any of the other charges. The total electric field, then, is the vector sum of all these fields. That, in essence, is what Equation 5.4 says.

In the next section, we describe how to determine the shape of an electric field of a source charge distribution and how to sketch it.

The Direction of the Field

Equation 5.4 enables us to determine the magnitude of the electric field, but we need the direction also. We use the convention that the direction of any electric field vector is the same as the direction of the electric force vector that the field would apply to a positive test charge placed in that field. Such a charge would be repelled by positive source charges (the force on it would point away from the positive source charge) but attracted to negative charges (the force points toward the negative source).

Direction of the Electric Field

By convention, all electric fields $\vec{\mathbf{E}}$ point away from positive source charges and point toward negative source charges.

Add charges to the Electric Field of Dreams (https://openstaxcollege.org/l/21elefiedream) and see how they react to the electric field. Turn on a background electric field and adjust the direction and magnitude.

Example 5.3

The *E*-field of an Atom

In an ionized helium atom, the most probable distance between the nucleus and the electron is $r = 26.5 \times 10^{-12}\ \text{m}$. What is the electric field due to the nucleus at the location of the electron?

Strategy

Note that although the electron is mentioned, it is not used in any calculation. The problem asks for an electric field, not a force; hence, there is only one charge involved, and the problem specifically asks for the field due to the nucleus. Thus, the electron is a red herring; only its distance matters. Also, since the distance between the two protons in the nucleus is much, much smaller than the distance of the electron from the nucleus, we can treat the two protons as a single charge +2*e* (Figure 5.19).

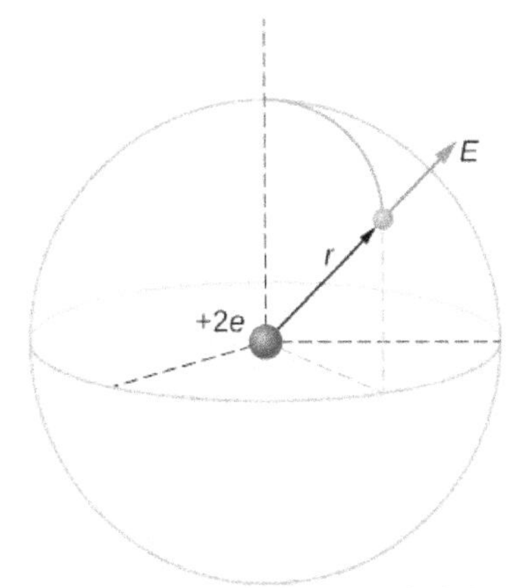

Figure 5.19 A schematic representation of a helium atom. Again, helium physically looks nothing like this, but this sort of diagram is helpful for calculating the electric field of the nucleus.

Solution

The electric field is calculated by

$$\vec{\mathbf{E}} = \frac{1}{4\pi\varepsilon_0}\sum_{i=1}^{N}\frac{q_i}{r_i^2}\hat{\mathbf{r}}_i.$$

Since there is only one source charge (the nucleus), this expression simplifies to

$$\vec{\mathbf{E}} = \frac{1}{4\pi\varepsilon_0}\frac{q}{r^2}\hat{r}.$$

Here $q = 2e = 2\left(1.6 \times 10^{-19}\ \text{C}\right)$ (since there are two protons) and r is given; substituting gives

$$\vec{\mathbf{E}} = \frac{1}{4\pi\left(8.85 \times 10^{-12}\frac{\text{C}^2}{\text{N}\cdot\text{m}^2}\right)}\frac{2\left(1.6 \times 10^{-19}\ \text{C}\right)}{\left(26.5 \times 10^{-12}\ \text{m}\right)^2}\hat{\mathbf{r}} = 4.1 \times 10^{12}\frac{\text{N}}{\text{C}}\hat{\mathbf{r}}.$$

The direction of $\vec{\mathbf{E}}$ is radially away from the nucleus in all directions. Why? Because a positive test charge placed in this field would accelerate radially away from the nucleus (since it is also positively charged), and again, the convention is that the direction of the electric field vector is defined in terms of the direction of the force it would apply to positive test charges.

Example 5.4

The *E*-Field above Two Equal Charges

(a) Find the electric field (magnitude and direction) a distance z above the midpoint between two equal charges $+q$ that are a distance d apart (Figure 5.20). Check that your result is consistent with what you'd expect when $z \gg d$.

(b) The same as part (a), only this time make the right-hand charge $-q$ instead of $+q$.

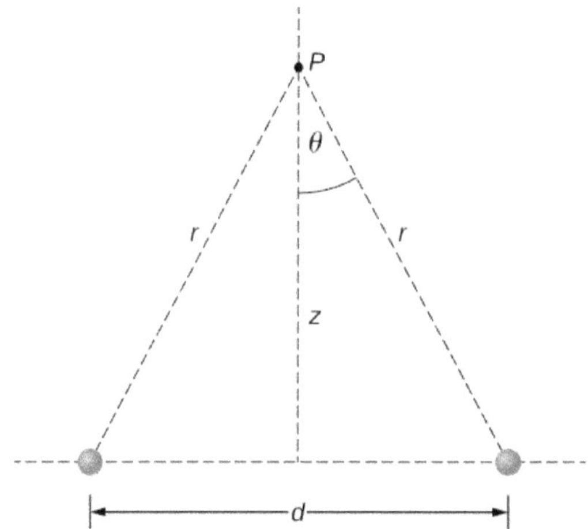

Figure 5.20 Finding the field of two identical source charges at the point *P*. Due to the symmetry, the net field at *P* is entirely vertical. (Notice that this is *not* true away from the midline between the charges.)

Strategy

We add the two fields as vectors, per Equation 5.4. Notice that the system (and therefore the field) is symmetrical about the vertical axis; as a result, the horizontal components of the field vectors cancel. This simplifies the math. Also, we take care to express our final answer in terms of only quantities that are given in the original statement of the problem: *q*, *z*, *d*, and constants $(\pi,\ \varepsilon_0)$.

Solution

a. By symmetry, the horizontal (*x*)-components of $\vec{\mathbf{E}}$ cancel (Figure 5.21); $E_x = \frac{1}{4\pi\varepsilon_0}\frac{q}{r^2}\sin\theta - \frac{1}{4\pi\varepsilon_0}\frac{q}{r^2}\sin\theta = 0$.

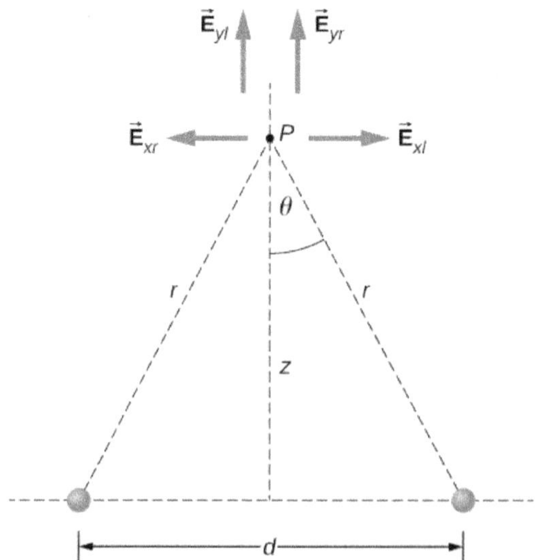

Figure 5.21 Note that the horizontal components of the electric fields from the two charges cancel each other out, while the vertical components add together.

The vertical (z)-component is given by

$$E_z = \frac{1}{4\pi\varepsilon_0}\frac{q}{r^2}\cos\theta + \frac{1}{4\pi\varepsilon_0}\frac{q}{r^2}\cos\theta = \frac{1}{4\pi\varepsilon_0}\frac{2q}{r^2}\cos\theta.$$

Since none of the other components survive, this is the entire electric field, and it points in the $\hat{\mathbf{k}}$ direction. Notice that this calculation uses the principle of **superposition**; we calculate the fields of the two charges independently and then add them together.

What we want to do now is replace the quantities in this expression that we don't know (such as r), or can't easily measure (such as $\cos\theta$) with quantities that we do know, or can measure. In this case, by geometry,

$$r^2 = z^2 + \left(\frac{d}{2}\right)^2$$

and

$$\cos\theta = \frac{z}{r} = \frac{z}{\left[z^2 + \left(\frac{d}{2}\right)^2\right]^{1/2}}.$$

Thus, substituting,

$$\vec{\mathbf{E}}(z) = \frac{1}{4\pi\varepsilon_0}\frac{2q}{\left[z^2 + \left(\frac{d}{2}\right)^2\right]^2}\frac{z}{\left[z^2 + \left(\frac{d}{2}\right)^2\right]^{1/2}}\hat{\mathbf{k}}.$$

Simplifying, the desired answer is

$$\vec{\mathbf{E}}(z) = \frac{1}{4\pi\varepsilon_0}\frac{2qz}{\left[z^2 + \left(\frac{d}{2}\right)^2\right]^{3/2}}\hat{\mathbf{k}}. \tag{5.5}$$

b. If the source charges are equal and opposite, the vertical components cancel because $E_z = \frac{1}{4\pi\varepsilon_0}\frac{q}{r^2}\cos\theta - \frac{1}{4\pi\varepsilon_0}\frac{q}{r^2}\cos\theta = 0$

and we get, for the horizontal component of $\vec{\mathbf{E}}$,

$$\begin{aligned}\vec{\mathbf{E}}(z) &= \frac{1}{4\pi\varepsilon_0}\frac{q}{r^2}\sin\theta\,\hat{\mathbf{i}} - \frac{1}{4\pi\varepsilon_0}\frac{-q}{r^2}\sin\theta\,\hat{\mathbf{i}} \\ &= \frac{1}{4\pi\varepsilon_0}\frac{2q}{r^2}\sin\theta\,\hat{\mathbf{i}} \\ &= \frac{1}{4\pi\varepsilon_0}\frac{2q}{\left[z^2 + \left(\frac{d}{2}\right)^2\right]^2}\frac{\left(\frac{d}{2}\right)}{\left[z^2 + \left(\frac{d}{2}\right)^2\right]^{1/2}}\hat{\mathbf{i}}.\end{aligned}$$

This becomes

$$\vec{\mathbf{E}}(z) = \frac{1}{4\pi\varepsilon_0}\frac{qd}{\left[z^2 + \left(\frac{d}{2}\right)^2\right]^{3/2}}\hat{\mathbf{i}}. \tag{5.6}$$

Significance

It is a very common and very useful technique in physics to check whether your answer is reasonable by evaluating it at extreme cases. In this example, we should evaluate the field expressions for the cases $d = 0$, $z \gg d$, and $z \to \infty$, and confirm that the resulting expressions match our physical expectations. Let's do so:

Let's start with Equation 5.5, the field of two identical charges. From far away (i.e., $z \gg d$), the two source charges should "merge" and we should then "see" the field of just one charge, of size $2q$. So, let $z \gg d$; then we can neglect d^2 in Equation 5.5 to obtain

$$\begin{aligned}\lim_{d \to 0} \vec{\mathbf{E}} &= \frac{1}{4\pi\varepsilon_0}\frac{2qz}{\left[z^2\right]^{3/2}}\hat{\mathbf{k}} \\ &= \frac{1}{4\pi\varepsilon_0}\frac{2qz}{z^3}\hat{\mathbf{k}} \\ &= \frac{1}{4\pi\varepsilon_0}\frac{(2q)}{z^2}\hat{\mathbf{k}},\end{aligned}$$

which is the correct expression for a field at a distance z away from a charge $2q$.

Next, we consider the field of equal and opposite charges, Equation 5.6. It can be shown (via a Taylor expansion) that for $d \ll z \ll \infty$, this becomes

$$\vec{\mathbf{E}}(z) = \frac{1}{4\pi\varepsilon_0}\frac{qd}{z^3}\hat{\mathbf{i}}, \tag{5.7}$$

which is the field of a dipole, a system that we will study in more detail later. (Note that the units of $\vec{\mathbf{E}}$ are still correct in this expression, since the units of d in the numerator cancel the unit of the "extra" z in the denominator.) If z is *very* large $(z \to \infty)$, then $E \to 0$, as it should; the two charges "merge" and so cancel out.

5.3 Check Your Understanding What is the electric field due to a single point particle?

Try this simulation of electric field hockey (https://openstaxcollege.org/l/21elefielhocke) to get the charge in the goal by placing other charges on the field.

5.5 | Calculating Electric Fields of Charge Distributions

Learning Objectives

By the end of this section, you will be able to:

- Explain what a continuous source charge distribution is and how it is related to the concept of quantization of charge
- Describe line charges, surface charges, and volume charges
- Calculate the field of a continuous source charge distribution of either sign

The charge distributions we have seen so far have been discrete: made up of individual point particles. This is in contrast with a **continuous charge distribution**, which has at least one nonzero dimension. If a charge distribution is continuous rather than discrete, we can generalize the definition of the electric field. We simply divide the charge into infinitesimal pieces and treat each piece as a point charge.

Note that because charge is quantized, there is no such thing as a "truly" continuous charge distribution. However, in most practical cases, the total charge creating the field involves such a huge number of discrete charges that we can safely ignore the discrete nature of the charge and consider it to be continuous. This is exactly the kind of approximation we make when we deal with a bucket of water as a continuous fluid, rather than a collection of H_2O molecules.

Our first step is to define a charge density for a charge distribution along a line, across a surface, or within a volume, as shown in Figure 5.22.

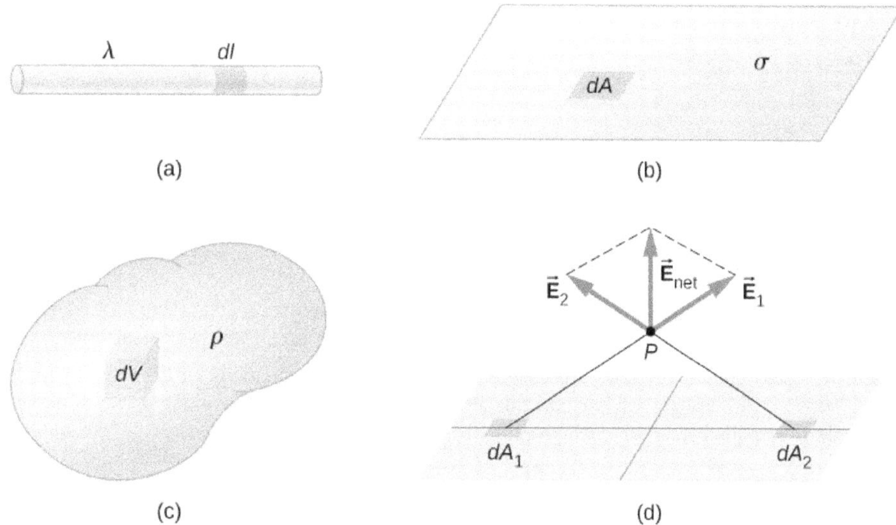

Figure 5.22 The configuration of charge differential elements for a (a) line charge, (b) sheet of charge, and (c) a volume of charge. Also note that (d) some of the components of the total electric field cancel out, with the remainder resulting in a net electric field.

Definitions of charge density:

- $\lambda \equiv$ charge per unit length (**linear charge density**); units are coulombs per meter (C/m)
- $\sigma \equiv$ charge per unit area (**surface charge density**); units are coulombs per square meter (C/m^2)
- $\rho \equiv$ charge per unit volume (**volume charge density**); units are coulombs per cubic meter (C/m^3)

Then, for a line charge, a surface charge, and a volume charge, the summation in Equation 5.4 becomes an integral and q_i is replaced by $dq = \lambda dl$, σdA, or ρdV, respectively:

$$\text{Point charge:} \quad \vec{\mathbf{E}}(P) = \frac{1}{4\pi\varepsilon_0}\sum_{i=1}^{N}\left(\frac{q_i}{r^2}\right)\hat{\mathbf{r}} \tag{5.8}$$

$$\text{Line charge:} \quad \vec{\mathbf{E}}(P) = \frac{1}{4\pi\varepsilon_0}\int_{\text{line}}\left(\frac{\lambda dl}{r^2}\right)\hat{\mathbf{r}} \tag{5.9}$$

$$\text{Surface charge:} \quad \vec{\mathbf{E}}(P) = \frac{1}{4\pi\varepsilon_0}\int_{\text{surface}}\left(\frac{\sigma dA}{r^2}\right)\hat{\mathbf{r}} \tag{5.10}$$

$$\text{Volume charge:} \quad \vec{\mathbf{E}}(P) = \frac{1}{4\pi\varepsilon_0}\int_{\text{volume}}\left(\frac{\rho dV}{r^2}\right)\hat{\mathbf{r}} \tag{5.11}$$

The integrals are generalizations of the expression for the field of a point charge. They implicitly include and assume the principle of superposition. The "trick" to using them is almost always in coming up with correct expressions for *dl, dA*, or *dV*, as the case may be, expressed in terms of *r*, and also expressing the charge density function appropriately. It may be constant; it might be dependent on location.

Note carefully the meaning of r in these equations: It is the distance from the charge element $(q_i, \lambda dl, \sigma dA, \rho dV)$ to the location of interest, $P(x, y, z)$ (the point in space where you want to determine the field). However, don't confuse this with the meaning of $\hat{\mathbf{r}}$; we are using it and the vector notation $\vec{\mathbf{E}}$ to write three integrals at once. That is, Equation 5.9 is actually

$$E_x(P) = \frac{1}{4\pi\varepsilon_0}\int_{\text{line}}\left(\frac{\lambda dl}{r^2}\right)_x, \quad E_y(P) = \frac{1}{4\pi\varepsilon_0}\int_{\text{line}}\left(\frac{\lambda dl}{r^2}\right)_y, \quad E_z(P) = \frac{1}{4\pi\varepsilon_0}\int_{\text{line}}\left(\frac{\lambda dl}{r^2}\right)_z.$$

Example 5.5

Electric Field of a Line Segment

Find the electric field a distance z above the midpoint of a straight line segment of length L that carries a uniform line charge density λ.

Strategy

Since this is a continuous charge distribution, we conceptually break the wire segment into differential pieces of length dl, each of which carries a differential amount of charge $dq = \lambda dl$. Then, we calculate the differential field created by two symmetrically placed pieces of the wire, using the symmetry of the setup to simplify the calculation (Figure 5.23). Finally, we integrate this differential field expression over the length of the wire (half of it, actually, as we explain below) to obtain the complete electric field expression.

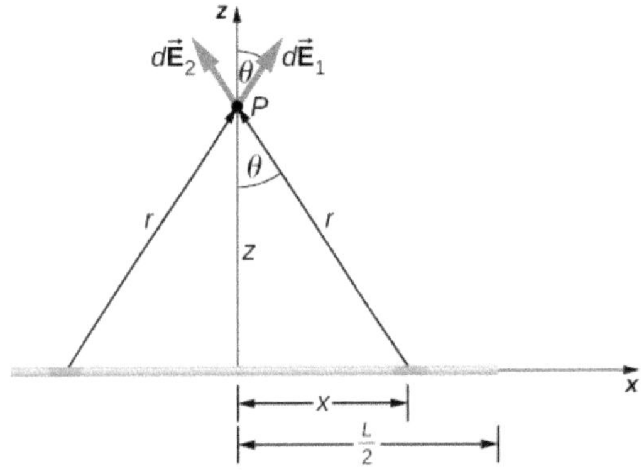

Figure 5.23 A uniformly charged segment of wire. The electric field at point P can be found by applying the superposition principle to symmetrically placed charge elements and integrating.

Solution

Before we jump into it, what do we expect the field to "look like" from far away? Since it is a finite line segment, from far away, it should look like a point charge. We will check the expression we get to see if it meets this expectation.

The electric field for a line charge is given by the general expression

$$\vec{\mathbf{E}}(P) = \frac{1}{4\pi\varepsilon_0}\int_{\text{line}}\frac{\lambda dl}{r^2}\hat{\mathbf{r}}.$$

The symmetry of the situation (our choice of the two identical differential pieces of charge) implies the horizontal (x)-components of the field cancel, so that the net field points in the z-direction. Let's check this formally.

The total field $\vec{\mathbf{E}}(P)$ is the vector sum of the fields from each of the two charge elements (call them $\vec{\mathbf{E}}_1$ and $\vec{\mathbf{E}}_2$, for now):

$$\vec{\mathbf{E}}(P) = \vec{\mathbf{E}}_1 + \vec{\mathbf{E}}_2 = E_{1x}\hat{\mathbf{i}} + E_{1z}\hat{\mathbf{k}} + E_{2x}\left(-\hat{\mathbf{i}}\right) + E_{2z}\hat{\mathbf{k}}.$$

Because the two charge elements are identical and are the same distance away from the point *P* where we want to calculate the field, $E_{1x} = E_{2x}$, so those components cancel. This leaves

$$\vec{\mathbf{E}}(P) = E_{1z}\hat{\mathbf{k}} + E_{2z}\hat{\mathbf{k}} = E_1 \cos\theta\hat{\mathbf{k}} + E_2 \cos\theta\hat{\mathbf{k}}.$$

These components are also equal, so we have

$$\begin{aligned}\vec{\mathbf{E}}(P) &= \frac{1}{4\pi\varepsilon_0}\int\frac{\lambda dl}{r^2}\cos\theta\hat{\mathbf{k}} + \frac{1}{4\pi\varepsilon_0}\int\frac{\lambda dl}{r^2}\cos\theta\hat{\mathbf{k}} \\ &= \frac{1}{4\pi\varepsilon_0}\int_0^{L/2}\frac{2\lambda dx}{r^2}\cos\theta\hat{\mathbf{k}}\end{aligned}$$

where our differential line element *dl* is *dx*, in this example, since we are integrating along a line of charge that lies on the *x*-axis. (The limits of integration are 0 to $\frac{L}{2}$, not $-\frac{L}{2}$ to $+\frac{L}{2}$, because we have constructed the net field from two differential pieces of charge *dq*. If we integrated along the entire length, we would pick up an erroneous factor of 2.)

In principle, this is complete. However, to actually calculate this integral, we need to eliminate all the variables that are not given. In this case, both *r* and θ change as we integrate outward to the end of the line charge, so those are the variables to get rid of. We can do that the same way we did for the two point charges: by noticing that

$$r = \left(z^2 + x^2\right)^{1/2}$$

and

$$\cos\theta = \frac{z}{r} = \frac{z}{\left(z^2 + x^2\right)^{1/2}}.$$

Substituting, we obtain

$$\begin{aligned}\vec{\mathbf{E}}(P) &= \frac{1}{4\pi\varepsilon_0}\int_0^{L/2}\frac{2\lambda dx}{\left(z^2 + x^2\right)}\frac{z}{\left(z^2 + x^2\right)^{1/2}}\hat{\mathbf{k}} \\ &= \frac{1}{4\pi\varepsilon_0}\int_0^{L/2}\frac{2\lambda z}{\left(z^2 + x^2\right)^{3/2}}dx\hat{\mathbf{k}} \\ &= \frac{2\lambda z}{4\pi\varepsilon_0}\left[\frac{x}{z^2\sqrt{z^2 + x^2}}\right]_0^{L/2}\hat{\mathbf{k}}\end{aligned}$$

which simplifies to

$$\vec{\mathbf{E}}(z) = \frac{1}{4\pi\varepsilon_0}\frac{\lambda L}{z\sqrt{z^2 + \frac{L^2}{4}}}\hat{\mathbf{k}}. \tag{5.12}$$

Significance

Notice, once again, the use of symmetry to simplify the problem. This is a very common strategy for calculating electric fields. The fields of nonsymmetrical charge distributions have to be handled with multiple integrals and may need to be calculated numerically by a computer.

5.4 Check Your Understanding How would the strategy used above change to calculate the electric field at a point a distance *z* above one end of the finite line segment?

Example 5.6

Electric Field of an Infinite Line of Charge

Find the electric field a distance *z* above the midpoint of an infinite line of charge that carries a uniform line charge density λ.

Strategy

This is exactly like the preceding example, except the limits of integration will be $-\infty$ to $+\infty$.

Solution

Again, the horizontal components cancel out, so we wind up with

$$\vec{\mathbf{E}}(P) = \frac{1}{4\pi\varepsilon_0}\int_{-\infty}^{\infty}\frac{\lambda dx}{r^2}\cos\theta\,\hat{\mathbf{k}}$$

where our differential line element *dl* is *dx*, in this example, since we are integrating along a line of charge that lies on the *x*-axis. Again,

$$\cos\theta = \frac{z}{r} = \frac{z}{\left(z^2+x^2\right)^{1/2}}.$$

Substituting, we obtain

$$\begin{aligned}\vec{\mathbf{E}}(P) &= \frac{1}{4\pi\varepsilon_0}\int_{-\infty}^{\infty}\frac{\lambda dx}{\left(z^2+x^2\right)}\frac{z}{\left(z^2+x^2\right)^{1/2}}\hat{\mathbf{k}} \\ &= \frac{1}{4\pi\varepsilon_0}\int_{-\infty}^{\infty}\frac{\lambda z}{\left(z^2+x^2\right)^{3/2}}dx\,\hat{\mathbf{k}} \\ &= \frac{\lambda z}{4\pi\varepsilon_0}\left[\frac{x}{z^2\sqrt{z^2+x^2}}\right]_{-\infty}^{\infty}\hat{\mathbf{k}},\end{aligned}$$

which simplifies to

$$\vec{\mathbf{E}}(z) = \frac{1}{4\pi\varepsilon_0}\frac{2\lambda}{z}\hat{\mathbf{k}}.$$

Significance

Our strategy for working with continuous charge distributions also gives useful results for charges with infinite dimension.

In the case of a finite line of charge, note that for $z \gg L$, z^2 dominates the L in the denominator, so that Equation 5.12 simplifies to

$$\vec{\mathbf{E}} \approx \frac{1}{4\pi\varepsilon_0}\frac{\lambda L}{z^2}\hat{\mathbf{k}}.$$

If you recall that $\lambda L = q$, the total charge on the wire, we have retrieved the expression for the field of a point charge, as expected.

In the limit $L \to \infty$, on the other hand, we get the field of an **infinite straight wire**, which is a straight wire whose length is much, much greater than either of its other dimensions, and also much, much greater than the distance at which the field is to be calculated:

$$\vec{\mathbf{E}}(z) = \frac{1}{4\pi\varepsilon_0}\frac{2\lambda}{z}\hat{\mathbf{k}}. \quad (5.13)$$

An interesting artifact of this infinite limit is that we have lost the usual $1/r^2$ dependence that we are used to. This will become even more intriguing in the case of an infinite plane.

Example 5.7

Electric Field due to a Ring of Charge

A ring has a uniform charge density λ, with units of coulomb per unit meter of arc. Find the electric potential at a point on the axis passing through the center of the ring.

Strategy

We use the same procedure as for the charged wire. The difference here is that the charge is distributed on a circle. We divide the circle into infinitesimal elements shaped as arcs on the circle and use polar coordinates shown in Figure 5.24.

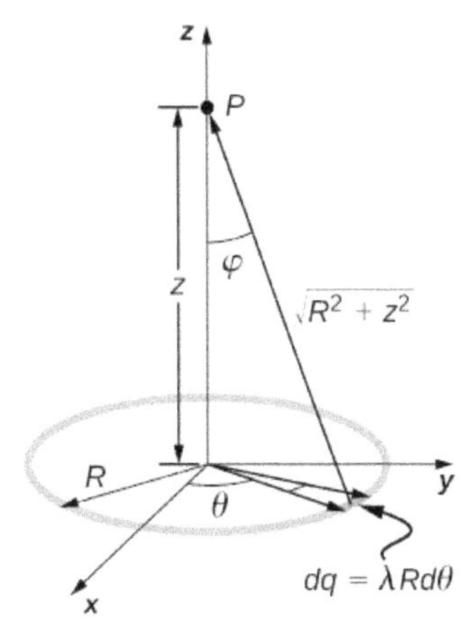

Figure 5.24 The system and variable for calculating the electric field due to a ring of charge.

Solution

The electric field for a line charge is given by the general expression

$$\vec{\mathbf{E}}(P) = \frac{1}{4\pi\varepsilon_0}\int_{\text{line}}\frac{\lambda dl}{r^2}\hat{\mathbf{r}}.$$

A general element of the arc between θ and $\theta + d\theta$ is of length $Rd\theta$ and therefore contains a charge equal to $\lambda Rd\theta$. The element is at a distance of $r = \sqrt{z^2 + R^2}$ from P, the angle is $\cos\phi = \frac{z}{\sqrt{z^2 + R^2}}$, and therefore the electric field is

$$\begin{aligned}\vec{\mathbf{E}}(P) &= \frac{1}{4\pi\varepsilon_0}\int_{\text{line}} \frac{\lambda dl}{r^2}\hat{\mathbf{r}} = \frac{1}{4\pi\varepsilon_0}\int_0^{2\pi} \frac{\lambda Rd\theta}{z^2 + R^2}\frac{z}{\sqrt{z^2 + R^2}}\hat{\mathbf{z}} \\ &= \frac{1}{4\pi\varepsilon_0}\frac{\lambda Rz}{\left(z^2 + R^2\right)^{3/2}}\hat{\mathbf{z}}\int_0^{2\pi} d\theta = \frac{1}{4\pi\varepsilon_0}\frac{2\pi\lambda Rz}{\left(z^2 + R^2\right)^{3/2}}\hat{\mathbf{z}} \\ &= \frac{1}{4\pi\varepsilon_0}\frac{q_{\text{tot}} z}{\left(z^2 + R^2\right)^{3/2}}\hat{\mathbf{z}}.\end{aligned}$$

Significance

As usual, symmetry simplified this problem, in this particular case resulting in a trivial integral. Also, when we take the limit of $z >> R$, we find that

$$\vec{\mathbf{E}} \approx \frac{1}{4\pi\varepsilon_0}\frac{q_{\text{tot}}}{z^2}\hat{\mathbf{z}},$$

as we expect.

Example 5.8

The Field of a Disk

Find the electric field of a circular thin disk of radius R and uniform charge density at a distance z above the center of the disk (Figure 5.25)

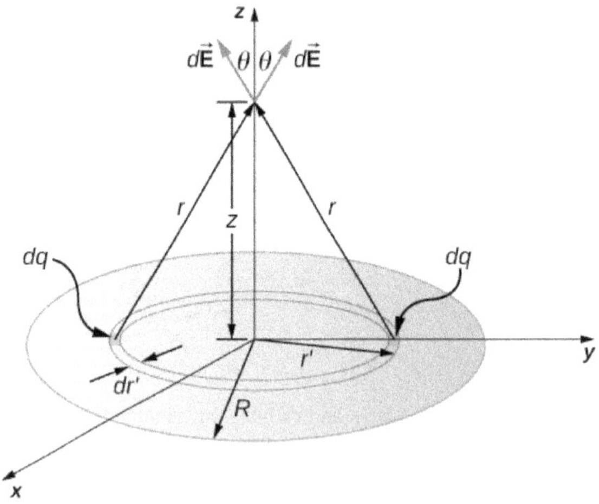

Figure 5.25 A uniformly charged disk. As in the line charge example, the field above the center of this disk can be calculated by taking advantage of the symmetry of the charge distribution.

Strategy

The electric field for a surface charge is given by

$$\vec{\mathbf{E}}(P) = \frac{1}{4\pi\varepsilon_0}\int_{\text{surface}} \frac{\sigma dA}{r^2}\hat{\mathbf{r}}.$$

To solve surface charge problems, we break the surface into symmetrical differential "stripes" that match the shape of the surface; here, we'll use rings, as shown in the figure. Again, by symmetry, the horizontal components cancel and the field is entirely in the vertical $(\hat{\mathbf{k}})$ direction. The vertical component of the electric field is extracted by multiplying by $\cos\theta$, so

$$\vec{\mathbf{E}}(P) = \frac{1}{4\pi\varepsilon_0}\int_{\text{surface}} \frac{\sigma dA}{r^2}\cos\theta\,\hat{\mathbf{k}}.$$

As before, we need to rewrite the unknown factors in the integrand in terms of the given quantities. In this case,

$$\begin{aligned} dA &= 2\pi r' dr' \\ r^2 &= r'^2 + z^2 \\ \cos\theta &= \frac{z}{\left(r'^2 + z^2\right)^{1/2}}. \end{aligned}$$

(Please take note of the two different "*r*'s" here; *r* is the distance from the differential ring of charge to the point *P* where we wish to determine the field, whereas r' is the distance from the center of the disk to the differential ring of charge.) Also, we already performed the polar angle integral in writing down *dA*.

Solution

Substituting all this in, we get

$$\begin{aligned} \vec{\mathbf{E}}(P) &= \vec{\mathbf{E}}(z) = \frac{1}{4\pi\varepsilon_0}\int_0^R \frac{\sigma(2\pi r' dr')z}{\left(r'^2 + z^2\right)^{3/2}}\hat{\mathbf{k}} \\ &= \frac{1}{4\pi\varepsilon_0}(2\pi\sigma z)\left(\frac{1}{z} - \frac{1}{\sqrt{R^2 + z^2}}\right)\hat{\mathbf{k}} \end{aligned}$$

or, more simply,

$$\vec{\mathbf{E}}(z) = \frac{1}{4\pi\varepsilon_0}\left(2\pi\sigma - \frac{2\pi\sigma z}{\sqrt{R^2 + z^2}}\right)\hat{\mathbf{k}}. \tag{5.14}$$

Significance

Again, it can be shown (via a Taylor expansion) that when $z \gg R$, this reduces to

$$\vec{\mathbf{E}}(z) \approx \frac{1}{4\pi\varepsilon_0}\frac{\sigma\pi R^2}{z^2}\hat{\mathbf{k}},$$

which is the expression for a point charge $Q = \sigma\pi R^2$.

5.5 Check Your Understanding How would the above limit change with a uniformly charged rectangle instead of a disk?

As $R \to \infty$, Equation 5.14 reduces to the field of an **infinite plane**, which is a flat sheet whose area is much, much greater than its thickness, and also much, much greater than the distance at which the field is to be calculated:

$$\vec{\mathbf{E}} = \frac{\sigma}{2\varepsilon_0}\hat{\mathbf{k}}. \tag{5.15}$$

Note that this field is constant. This surprising result is, again, an artifact of our limit, although one that we will make use of repeatedly in the future. To understand why this happens, imagine being placed above an infinite plane of constant charge. Does the plane look any different if you vary your altitude? No—you still see the plane going off to infinity, no matter how far you are from it. It is important to note that Equation 5.15 is because we are above the plane. If we were below, the field would point in the $-\hat{\mathbf{k}}$ direction.

Example 5.9

The Field of Two Infinite Planes

Find the electric field everywhere resulting from two infinite planes with equal but opposite charge densities (Figure 5.26).

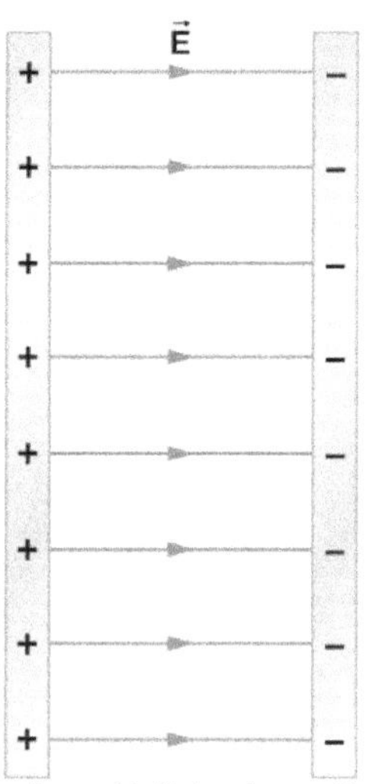

Figure 5.26 Two charged infinite planes. Note the direction of the electric field.

Strategy

We already know the electric field resulting from a single infinite plane, so we may use the principle of superposition to find the field from two.

Solution

The electric field points away from the positively charged plane and toward the negatively charged plane. Since the σ are equal and opposite, this means that in the region outside of the two planes, the electric fields cancel each other out to zero.

However, in the region between the planes, the electric fields add, and we get

$$\vec{\mathbf{E}} = \frac{\sigma}{\varepsilon_0}\hat{\mathbf{i}}$$

for the electric field. The $\hat{\mathbf{i}}$ is because in the figure, the field is pointing in the +x-direction.

Significance

Systems that may be approximated as two infinite planes of this sort provide a useful means of creating uniform electric fields.

5.6 Check Your Understanding What would the electric field look like in a system with two parallel positively charged planes with equal charge densities?

5.6 | Electric Field Lines

Learning Objectives

By the end of this section, you will be able to:

- Explain the purpose of an electric field diagram
- Describe the relationship between a vector diagram and a field line diagram
- Explain the rules for creating a field diagram and why these rules make physical sense
- Sketch the field of an arbitrary source charge

Now that we have some experience calculating electric fields, let's try to gain some insight into the geometry of electric fields. As mentioned earlier, our model is that the charge on an object (the source charge) alters space in the region around it in such a way that when another charged object (the test charge) is placed in that region of space, that test charge experiences an electric force. The concept of electric **field lines**, and of electric field line diagrams, enables us to visualize the way in which the space is altered, allowing us to visualize the field. The purpose of this section is to enable you to create sketches of this geometry, so we will list the specific steps and rules involved in creating an accurate and useful sketch of an electric field.

It is important to remember that electric fields are three-dimensional. Although in this book we include some pseudo-three-dimensional images, several of the diagrams that you'll see (both here, and in subsequent chapters) will be two-dimensional projections, or cross-sections. Always keep in mind that in fact, you're looking at a three-dimensional phenomenon.

Our starting point is the physical fact that the electric field of the source charge causes a test charge in that field to experience a force. By definition, electric field vectors point in the same direction as the electric force that a (hypothetical) positive test charge would experience, if placed in the field (Figure 5.27)

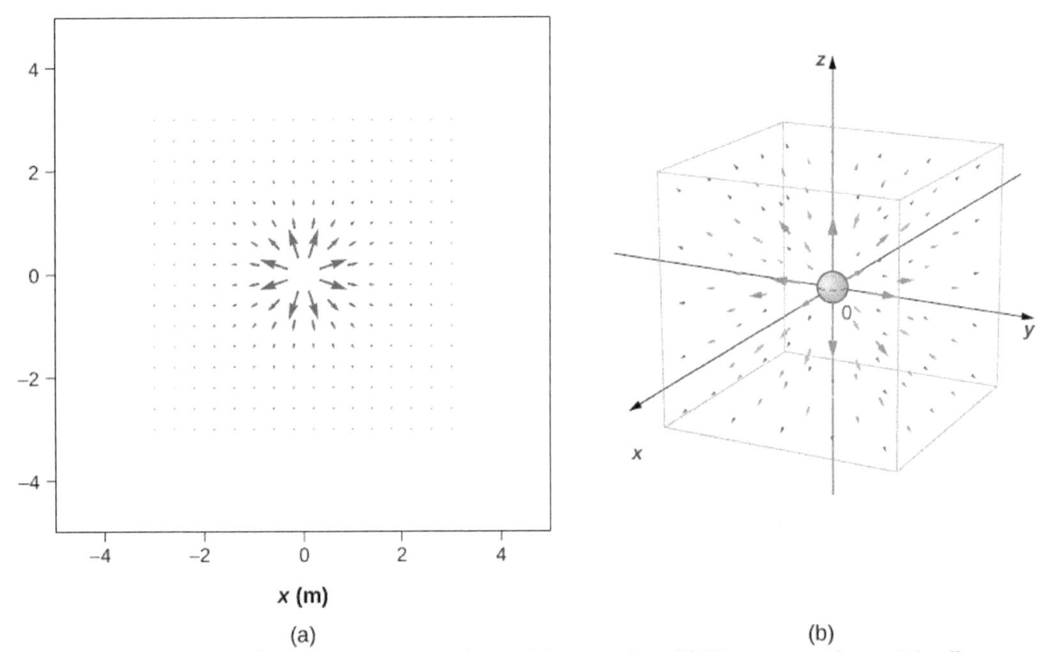

Figure 5.27 The electric field of a positive point charge. A large number of field vectors are shown. Like all vector arrows, the length of each vector is proportional to the magnitude of the field at each point. (a) Field in two dimensions; (b) field in three dimensions.

We've plotted many field vectors in the figure, which are distributed uniformly around the source charge. Since the electric field is a vector, the arrows that we draw correspond at every point in space to both the magnitude and the direction of the field at that point. As always, the length of the arrow that we draw corresponds to the magnitude of the field vector at that point. For a point source charge, the length decreases by the square of the distance from the source charge. In addition, the direction of the field vector is radially away from the source charge, because the direction of the electric field is defined by the direction of the force that a positive test charge would experience in that field. (Again, keep in mind that the actual field is three-dimensional; there are also field lines pointing out of and into the page.)

This diagram is correct, but it becomes less useful as the source charge distribution becomes more complicated. For example, consider the vector field diagram of a dipole (Figure 5.28).

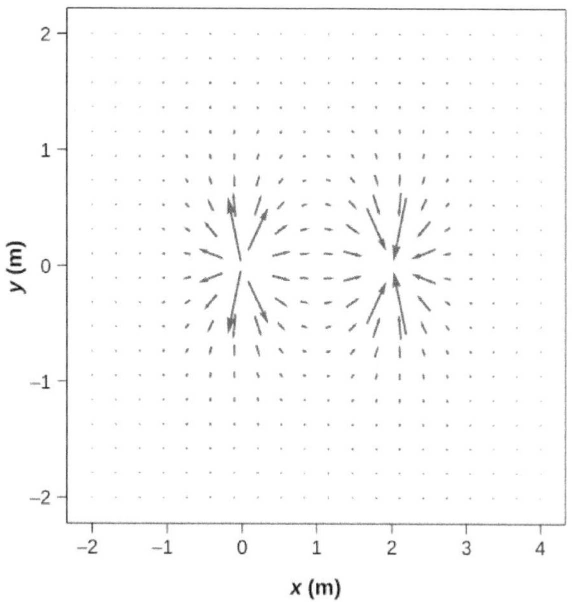

Figure 5.28 The vector field of a dipole. Even with just two identical charges, the vector field diagram becomes difficult to understand.

There is a more useful way to present the same information. Rather than drawing a large number of increasingly smaller vector arrows, we instead connect all of them together, forming continuous lines and curves, as shown in Figure 5.29.

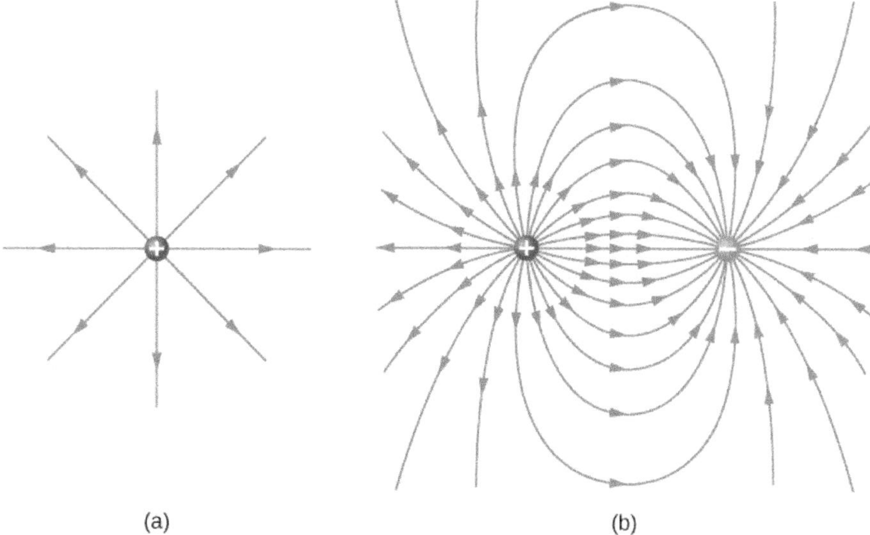

Figure 5.29 (a) The electric field line diagram of a positive point charge. (b) The field line diagram of a dipole. In both diagrams, the magnitude of the field is indicated by the field line density. The field *vectors* (not shown here) are everywhere tangent to the field lines.

Although it may not be obvious at first glance, these field diagrams convey the same information about the electric field as do the vector diagrams. First, the direction of the field at every point is simply the direction of the field vector at that same point. In other words, at any point in space, the field vector at each point is tangent to the field line at that same point. The arrowhead placed on a field line indicates its direction.

As for the magnitude of the field, that is indicated by the **field line density**—that is, the number of field lines per unit area passing through a small cross-sectional area perpendicular to the electric field. This field line density is drawn to be proportional to the magnitude of the field at that cross-section. As a result, if the field lines are close together (that is, the field line density is greater), this indicates that the magnitude of the field is large at that point. If the field lines are far apart at the cross-section, this indicates the magnitude of the field is small. Figure 5.30 shows the idea.

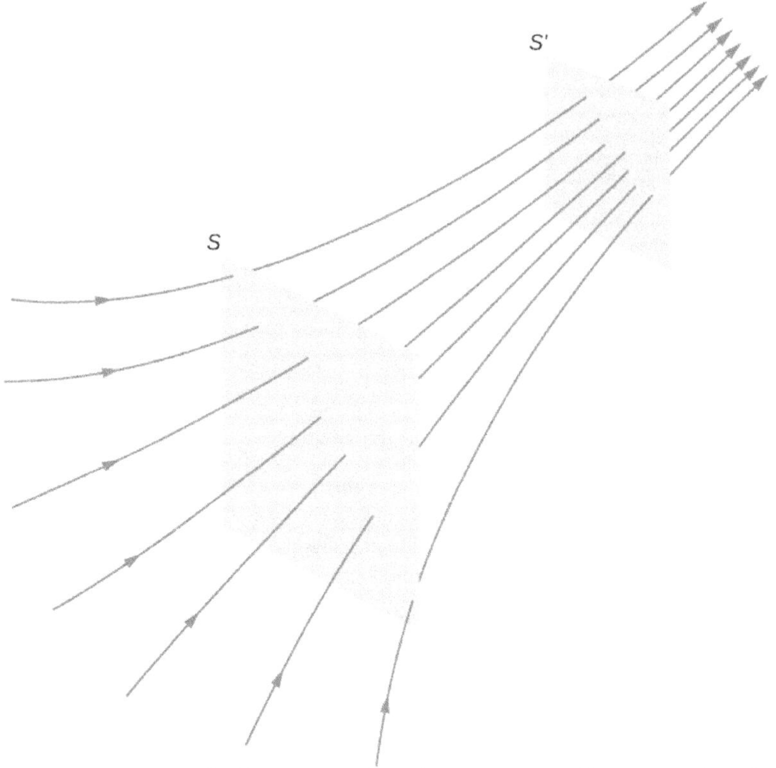

Figure 5.30 Electric field lines passing through imaginary areas. Since the number of lines passing through each area is the same, but the areas themselves are different, the field line density is different. This indicates different magnitudes of the electric field at these points.

In Figure 5.30, the same number of field lines passes through both surfaces (S and S'), but the surface S is larger than surface S'. Therefore, the density of field lines (number of lines per unit area) is larger at the location of S', indicating that the electric field is stronger at the location of S' than at S. The rules for creating an electric field diagram are as follows.

Problem-Solving Strategy: Drawing Electric Field Lines

1. Electric field lines either originate on positive charges or come in from infinity, and either terminate on negative charges or extend out to infinity.
2. The number of field lines originating or terminating at a charge is proportional to the magnitude of that charge. A charge of $2q$ will have twice as many lines as a charge of q.
3. At every point in space, the field vector at that point is tangent to the field line at that same point.
4. The field line density at any point in space is proportional to (and therefore is representative of) the magnitude of the field at that point in space.
5. Field lines can never cross. Since a field line represents the direction of the field at a given point, if two field lines crossed at some point, that would imply that the electric field was pointing in two different directions at a single point. This in turn would suggest that the (net) force on a test charge placed at that point would point in two different directions. Since this is obviously impossible, it follows that field lines must never cross.

Always keep in mind that field lines serve only as a convenient way to visualize the electric field; they are not physical entities. Although the direction and relative intensity of the electric field can be deduced from a set of field lines, the lines can also be misleading. For example, the field lines drawn to represent the electric field in a region must, by necessity, be discrete. However, the actual electric field in that region exists at every point in space.

Field lines for three groups of discrete charges are shown in Figure 5.31. Since the charges in parts (a) and (b) have the same magnitude, the same number of field lines are shown starting from or terminating on each charge. In (c), however, we draw three times as many field lines leaving the $+3q$ charge as entering the $-q$. The field lines that do not terminate at $-q$ emanate outward from the charge configuration, to infinity.

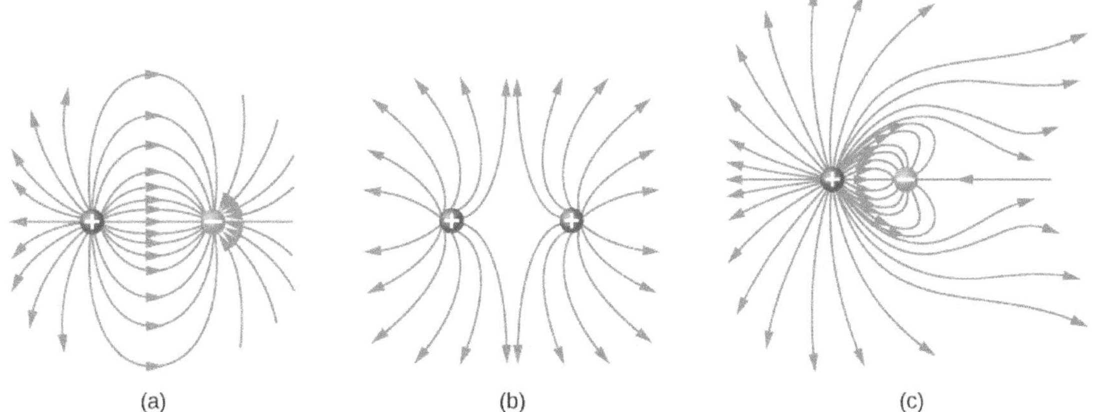

Figure 5.31 Three typical electric field diagrams. (a) A dipole. (b) Two identical charges. (c) Two charges with opposite signs and different magnitudes. Can you tell from the diagram which charge has the larger magnitude?

The ability to construct an accurate electric field diagram is an important, useful skill; it makes it much easier to estimate, predict, and therefore calculate the electric field of a source charge. The best way to develop this skill is with software that allows you to place source charges and then will draw the net field upon request. We strongly urge you to search the Internet for a program. Once you've found one you like, run several simulations to get the essential ideas of field diagram construction. Then practice drawing field diagrams, and checking your predictions with the computer-drawn diagrams.

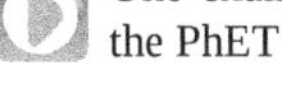

One example of a field-line drawing program (https://openstaxcollege.org/l/21fieldlindrapr) is from the PhET "Charges and Fields" simulation.

5.7 | Electric Dipoles

Learning Objectives

By the end of this section, you will be able to:

- Describe a permanent dipole
- Describe an induced dipole
- Define and calculate an electric dipole moment
- Explain the physical meaning of the dipole moment

Earlier we discussed, and calculated, the electric field of a dipole: two equal and opposite charges that are "close" to each other. (In this context, "close" means that the distance *d* between the two charges is much, much less than the distance of the field point *P*, the location where you are calculating the field.) Let's now consider what happens to a dipole when it is placed in an external field $\vec{\mathbf{E}}$. We assume that the dipole is a **permanent dipole**; it exists without the field, and does not break apart in the external field.

Rotation of a Dipole due to an Electric Field

For now, we deal with only the simplest case: The external field is uniform in space. Suppose we have the situation depicted in Figure 5.32, where we denote the distance between the charges as the vector $\vec{\mathbf{d}}$, pointing from the negative charge to the positive charge. The forces on the two charges are equal and opposite, so there is no net force on the dipole. However, there is a torque:

$$\begin{aligned}\vec{\tau} &= \left(\frac{\vec{\mathbf{d}}}{2} \times \vec{\mathbf{F}}_{+}\right) + \left(-\frac{\vec{\mathbf{d}}}{2} \times \vec{\mathbf{F}}_{-}\right) \\ &= \left[\left(\frac{\vec{\mathbf{d}}}{2}\right) \times \left(+q\vec{\mathbf{E}}\right) + \left(-\frac{\vec{\mathbf{d}}}{2}\right) \times \left(-q\vec{\mathbf{E}}\right)\right] \\ &= q\vec{\mathbf{d}} \times \vec{\mathbf{E}}.\end{aligned}$$

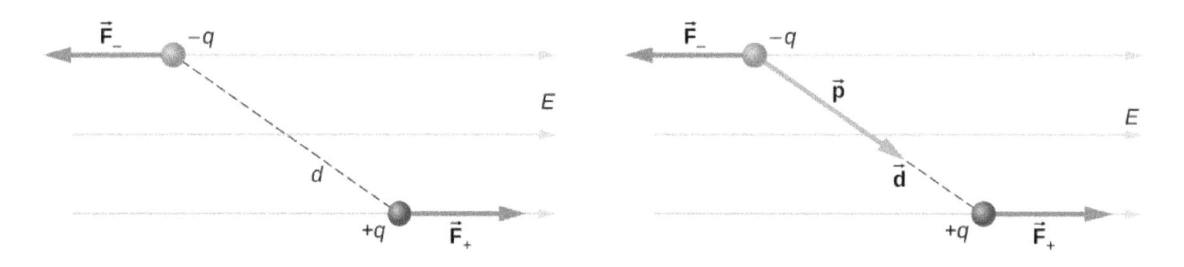

Figure 5.32 A dipole in an external electric field. (a) The net force on the dipole is zero, but the net torque is not. As a result, the dipole rotates, becoming aligned with the external field. (b) The dipole moment is a convenient way to characterize this effect. The $\vec{\mathbf{d}}$ points in the same direction as $\vec{\mathbf{p}}$.

The quantity $q\vec{\mathbf{d}}$ (the magnitude of each charge multiplied by the vector distance between them) is a property of the dipole; its value, as you can see, determines the torque that the dipole experiences in the external field. It is useful, therefore, to define this product as the so-called **dipole moment** of the dipole:

$$\vec{\mathbf{p}} \equiv q\vec{\mathbf{d}}. \tag{5.16}$$

We can therefore write

$$\vec{\tau} = \vec{\mathbf{p}} \times \vec{\mathbf{E}}. \tag{5.17}$$

Recall that a torque changes the angular velocity of an object, the dipole, in this case. In this situation, the effect is to rotate the dipole (that is, align the direction of $\vec{\mathbf{p}}$) so that it is parallel to the direction of the external field.

Induced Dipoles

Neutral atoms are, by definition, electrically neutral; they have equal amounts of positive and negative charge. Furthermore, since they are spherically symmetrical, they do not have a "built-in" dipole moment the way most asymmetrical molecules do. They obtain one, however, when placed in an external electric field, because the external field causes oppositely directed forces on the positive nucleus of the atom versus the negative electrons that surround the nucleus. The result is a new charge distribution of the atom, and therefore, an **induced dipole** moment (Figure 5.33).

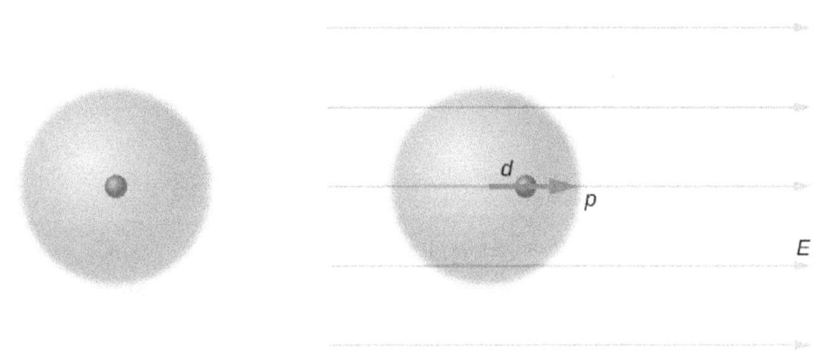

Figure 5.33 A dipole is induced in a neutral atom by an external electric field. The induced dipole moment is aligned with the external field.

An important fact here is that, just as for a rotated polar molecule, the result is that the dipole moment ends up aligned parallel to the external electric field. Generally, the magnitude of an induced dipole is much smaller than that of an inherent dipole. For both kinds of dipoles, notice that once the alignment of the dipole (rotated or induced) is complete, the net effect is to decrease the total electric field $\vec{\mathbf{E}}_{\text{total}} = \vec{\mathbf{E}}_{\text{external}} + \vec{\mathbf{E}}_{\text{dipole}}$ in the regions outside the dipole charges (Figure 5.34). By "outside" we mean further from the charges than they are from each other. This effect is crucial for capacitors, as you will see in Capacitance.

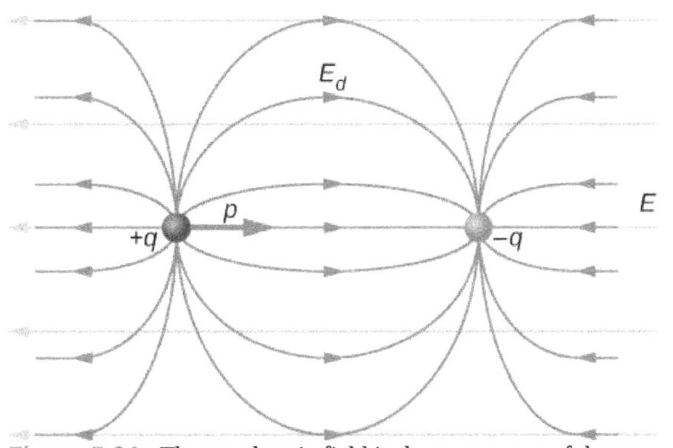

Figure 5.34 The net electric field is the vector sum of the field of the dipole plus the external field.

Recall that we found the electric field of a dipole in Equation 5.7. If we rewrite it in terms of the dipole moment we get:

$$\vec{\mathbf{E}}(z) = \frac{1}{4\pi\varepsilon_0}\frac{\vec{\mathbf{p}}}{z^3}.$$

The form of this field is shown in Figure 5.34. Notice that along the plane perpendicular to the axis of the dipole and midway between the charges, the direction of the electric field is opposite that of the dipole and gets weaker the further from the axis one goes. Similarly, on the axis of the dipole (but outside it), the field points in the same direction as the dipole, again getting weaker the further one gets from the charges.

CHAPTER 5 REVIEW

KEY TERMS

charging by induction process by which an electrically charged object brought near a neutral object creates a charge separation in that object

conduction electron electron that is free to move away from its atomic orbit

conductor material that allows electrons to move separately from their atomic orbits; object with properties that allow charges to move about freely within it

continuous charge distribution total source charge composed of so large a number of elementary charges that it must be treated as continuous, rather than discrete

coulomb SI unit of electric charge

Coulomb force another term for the electrostatic force

Coulomb's law mathematical equation calculating the electrostatic force vector between two charged particles

dipole two equal and opposite charges that are fixed close to each other

dipole moment property of a dipole; it characterizes the combination of distance between the opposite charges, and the magnitude of the charges

electric charge physical property of an object that causes it to be attracted toward or repelled from another charged object; each charged object generates and is influenced by a force called an electric force

electric field physical phenomenon created by a charge; it "transmits" a force between a two charges

electric force noncontact force observed between electrically charged objects

electron particle surrounding the nucleus of an atom and carrying the smallest unit of negative charge

electrostatic attraction phenomenon of two objects with opposite charges attracting each other

electrostatic force amount and direction of attraction or repulsion between two charged bodies; the assumption is that the source charges remain motionless

electrostatic repulsion phenomenon of two objects with like charges repelling each other

electrostatics study of charged objects which are not in motion

field line smooth, usually curved line that indicates the direction of the electric field

field line density number of field lines per square meter passing through an imaginary area; its purpose is to indicate the field strength at different points in space

induced dipole typically an atom, or a spherically symmetric molecule; a dipole created due to opposite forces displacing the positive and negative charges

infinite plane flat sheet in which the dimensions making up the area are much, much greater than its thickness, and also much, much greater than the distance at which the field is to be calculated; its field is constant

infinite straight wire straight wire whose length is much, much greater than either of its other dimensions, and also much, much greater than the distance at which the field is to be calculated

insulator material that holds electrons securely within their atomic orbits

ion atom or molecule with more or fewer electrons than protons

law of conservation of charge net electric charge of a closed system is constant

linear charge density amount of charge in an element of a charge distribution that is essentially one-dimensional (the width and height are much, much smaller than its length); its units are C/m

neutron neutral particle in the nucleus of an atom, with (nearly) the same mass as a proton

permanent dipole typically a molecule; a dipole created by the arrangement of the charged particles from which the dipole is created

permittivity of vacuum also called the permittivity of free space, and constant describing the strength of the electric force in a vacuum

polarization slight shifting of positive and negative charges to opposite sides of an object

principle of superposition useful fact that we can simply add up all of the forces due to charges acting on an object

proton particle in the nucleus of an atom and carrying a positive charge equal in magnitude to the amount of negative charge carried by an electron

static electricity buildup of electric charge on the surface of an object; the arrangement of the charge remains constant ("static")

superposition concept that states that the net electric field of multiple source charges is the vector sum of the field of each source charge calculated individually

surface charge density amount of charge in an element of a two-dimensional charge distribution (the thickness is small); its units are C/m^2

volume charge density amount of charge in an element of a three-dimensional charge distribution; its units are C/m^3

KEY EQUATIONS

Coulomb's law	$\vec{\mathbf{F}}_{12}(r) = \frac{1}{4\pi\varepsilon_0}\frac{q_1 q_2}{r_{12}^2}\hat{\mathbf{r}}_{12}$
Superposition of electric forces	$\vec{\mathbf{F}}(r) = \frac{1}{4\pi\varepsilon_0}Q\sum_{i=1}^{N}\frac{q_i}{r_i^2}\hat{\mathbf{r}}_i$
Electric force due to an electric field	$\vec{\mathbf{F}} = Q\vec{\mathbf{E}}$
Electric field at point P	$\vec{\mathbf{E}}(P) \equiv \frac{1}{4\pi\varepsilon_0}\sum_{i=1}^{N}\frac{q_i}{r_i^2}\hat{\mathbf{r}}_i$
Field of an infinite wire	$\vec{\mathbf{E}}(z) = \frac{1}{4\pi\varepsilon_0}\frac{2\lambda}{z}\hat{\mathbf{k}}$
Field of an infinite plane	$\vec{\mathbf{E}} = \frac{\sigma}{2\varepsilon_0}\hat{\mathbf{k}}$
Dipole moment	$\vec{\mathbf{p}} \equiv q\vec{\mathbf{d}}$
Torque on dipole in external E-field	$\vec{\tau} = \vec{\mathbf{p}} \times \vec{\mathbf{E}}$

SUMMARY

5.1 Electric Charge

- There are only two types of charge, which we call positive and negative. Like charges repel, unlike charges attract, and the force between charges decreases with the square of the distance.
- The vast majority of positive charge in nature is carried by protons, whereas the vast majority of negative charge is carried by electrons. The electric charge of one electron is equal in magnitude and opposite in sign to the charge of one proton.
- An ion is an atom or molecule that has nonzero total charge due to having unequal numbers of electrons and protons.
- The SI unit for charge is the coulomb (C), with protons and electrons having charges of opposite sign but equal magnitude; the magnitude of this basic charge is $e \equiv 1.602 \times 10^{-19}$ C

- Both positive and negative charges exist in neutral objects and can be separated by bringing the two objects into physical contact; rubbing the objects together can remove electrons from the bonds in one object and place them on the other object, increasing the charge separation.
- For macroscopic objects, negatively charged means an excess of electrons and positively charged means a depletion of electrons.
- The law of conservation of charge states that the net charge of a closed system is constant.

5.2 Conductors, Insulators, and Charging by Induction

- A conductor is a substance that allows charge to flow freely through its atomic structure.
- An insulator holds charge fixed in place.
- Polarization is the separation of positive and negative charges in a neutral object. Polarized objects have their positive and negative charges concentrated in different areas, giving them a charge distribution.

5.3 Coulomb's Law

- Coulomb's law gives the magnitude of the force between point charges. It is

$$\vec{\mathbf{F}}_{12}(r) = \frac{1}{4\pi\varepsilon_0}\frac{q_1 q_2}{r_{12}^2}\hat{\mathbf{r}}_{12}$$

where q_2 and q_2 are two point charges separated by a distance r. This Coulomb force is extremely basic, since most charges are due to point-like particles. It is responsible for all electrostatic effects and underlies most macroscopic forces.

5.4 Electric Field

- The electric field is an alteration of space caused by the presence of an electric charge. The electric field mediates the electric force between a source charge and a test charge.
- The electric field, like the electric force, obeys the superposition principle
- The field is a vector; by definition, it points away from positive charges and toward negative charges.

5.5 Calculating Electric Fields of Charge Distributions

- A very large number of charges can be treated as a continuous charge distribution, where the calculation of the field requires integration. Common cases are:
    - one-dimensional (like a wire); uses a line charge density λ
    - two-dimensional (metal plate); uses surface charge density σ
    - three-dimensional (metal sphere); uses volume charge density ρ
- The "source charge" is a differential amount of charge dq. Calculating dq depends on the type of source charge distribution:

$$dq = \lambda dl; \quad dq = \sigma dA; \quad dq = \rho dV.$$

- Symmetry of the charge distribution is usually key.
- Important special cases are the field of an "infinite" wire and the field of an "infinite" plane.

5.6 Electric Field Lines

- Electric field diagrams assist in visualizing the field of a source charge.
- The magnitude of the field is proportional to the field line density.
- Field vectors are everywhere tangent to field lines.

5.7 Electric Dipoles

- If a permanent dipole is placed in an external electric field, it results in a torque that aligns it with the external field.
- If a nonpolar atom (or molecule) is placed in an external field, it gains an induced dipole that is aligned with the external field.
- The net field is the vector sum of the external field plus the field of the dipole (physical or induced).
- The strength of the polarization is described by the dipole moment of the dipole, $\vec{\mathbf{p}} = q\vec{\mathbf{d}}$.

CONCEPTUAL QUESTIONS

5.1 Electric Charge

1. There are very large numbers of charged particles in most objects. Why, then, don't most objects exhibit static electricity?

2. Why do most objects tend to contain nearly equal numbers of positive and negative charges?

3. A positively charged rod attracts a small piece of cork. (a) Can we conclude that the cork is negatively charged? (b) The rod repels another small piece of cork. Can we conclude that this piece is positively charged?

4. Two bodies attract each other electrically. Do they both have to be charged? Answer the same question if the bodies repel one another.

5. How would you determine whether the charge on a particular rod is positive or negative?

5.2 Conductors, Insulators, and Charging by Induction

6. An eccentric inventor attempts to levitate a cork ball by wrapping it with foil and placing a large negative charge on the ball and then putting a large positive charge on the ceiling of his workshop. Instead, while attempting to place a large negative charge on the ball, the foil flies off. Explain.

7. When a glass rod is rubbed with silk, it becomes positive and the silk becomes negative—yet both attract dust. Does the dust have a third type of charge that is attracted to both positive and negative? Explain.

8. Why does a car always attract dust right after it is polished? (Note that car wax and car tires are insulators.)

9. Does the uncharged conductor shown below experience a net electric force?

10. While walking on a rug, a person frequently becomes charged because of the rubbing between his shoes and the rug. This charge then causes a spark and a slight shock when the person gets close to a metal object. Why are these shocks so much more common on a dry day?

11. Compare charging by conduction to charging by induction.

12. Small pieces of tissue are attracted to a charged comb. Soon after sticking to the comb, the pieces of tissue are repelled from it. Explain.

13. Trucks that carry gasoline often have chains dangling from their undercarriages and brushing the ground. Why?

14. Why do electrostatic experiments work so poorly in humid weather?

15. Why do some clothes cling together after being removed from the clothes dryer? Does this happen if they're still damp?

16. Can induction be used to produce charge on an insulator?

17. Suppose someone tells you that rubbing quartz with cotton cloth produces a third kind of charge on the quartz. Describe what you might do to test this claim.

18. A handheld copper rod does not acquire a charge when you rub it with a cloth. Explain why.

19. Suppose you place a charge q near a large metal plate. (a) If q is attracted to the plate, is the plate necessarily charged? (b) If q is repelled by the plate, is the plate necessarily charged?

5.3 Coulomb's Law

20. Would defining the charge on an electron to be positive have any effect on Coulomb's law?

21. An atomic nucleus contains positively charged protons and uncharged neutrons. Since nuclei do stay together, what must we conclude about the forces between these nuclear particles?

22. Is the force between two fixed charges influenced by the presence of other charges?

5.4 Electric Field

23. When measuring an electric field, could we use a negative rather than a positive test charge?

24. During fair weather, the electric field due to the net charge on Earth points downward. Is Earth charged positively or negatively?

25. If the electric field at a point on the line between two charges is zero, what do you know about the charges?

26. Two charges lie along the x-axis. Is it true that the net electric field always vanishes at some point (other than infinity) along the x-axis?

5.5 Calculating Electric Fields of Charge Distributions

27. Give a plausible argument as to why the electric field outside an infinite charged sheet is constant.

28. Compare the electric fields of an infinite sheet of charge, an infinite, charged conducting plate, and infinite, oppositely charged parallel plates.

29. Describe the electric fields of an infinite charged plate and of two infinite, charged parallel plates in terms of the electric field of an infinite sheet of charge.

30. A negative charge is placed at the center of a ring of uniform positive charge. What is the motion (if any) of the charge? What if the charge were placed at a point on the axis of the ring other than the center?

5.6 Electric Field Lines

31. If a point charge is released from rest in a uniform electric field, will it follow a field line? Will it do so if the electric field is not uniform?

32. Under what conditions, if any, will the trajectory of a charged particle not follow a field line?

33. How would you experimentally distinguish an electric field from a gravitational field?

34. A representation of an electric field shows 10 field lines perpendicular to a square plate. How many field lines should pass perpendicularly through the plate to depict a field with twice the magnitude?

35. What is the ratio of the number of electric field lines leaving a charge $10q$ and a charge q?

5.7 Electric Dipoles

36. What are the stable orientation(s) for a dipole in an external electric field? What happens if the dipole is slightly perturbed from these orientations?

PROBLEMS

5.1 Electric Charge

37. Common static electricity involves charges ranging from nanocoulombs to microcoulombs. (a) How many electrons are needed to form a charge of -2.00 nC? (b) How many electrons must be removed from a neutral object to leave a net charge of $0.500\ \mu\text{C}$?

38. If 1.80×10^{20} electrons move through a pocket calculator during a full day's operation, how many coulombs of charge moved through it?

39. To start a car engine, the car battery moves 3.75×10^{21} electrons through the starter motor. How many coulombs of charge were moved?

40. A certain lightning bolt moves 40.0 C of charge. How many fundamental units of charge is this?

41. A 2.5-g copper penny is given a charge of -2.0×10^{-9} C. (a) How many excess electrons are on the penny? (b) By what percent do the excess electrons change the mass of the penny?

42. A 2.5-g copper penny is given a charge of 4.0×10^{-9} C. (a) How many electrons are removed from the penny? (b) If no more than one electron is removed from an atom, what percent of the atoms are ionized by this charging process?

5.2 Conductors, Insulators, and Charging by Induction

43. Suppose a speck of dust in an electrostatic precipitator has 1.0000×10^{12} protons in it and has a net charge of -5.00 nC (a very large charge for a small speck). How many electrons does it have?

44. An amoeba has 1.00×10^{16} protons and a net charge of 0.300 pC. (a) How many fewer electrons are there than protons? (b) If you paired them up, what fraction of the protons would have no electrons?

45. A 50.0-g ball of copper has a net charge of $2.00\ \mu$C. What fraction of the copper's electrons has been removed? (Each copper atom has 29 protons, and copper has an atomic mass of 63.5.)

46. What net charge would you place on a 100-g piece of sulfur if you put an extra electron on 1 in 10^{12} of its atoms? (Sulfur has an atomic mass of 32.1 u.)

47. How many coulombs of positive charge are there in 4.00 kg of plutonium, given its atomic mass is 244 and that each plutonium atom has 94 protons?

5.3 Coulomb's Law

48. Two point particles with charges $+3\ \mu$C and $+5\ \mu$C are held in place by 3-N forces on each charge in appropriate directions. (a) Draw a free-body diagram for each particle. (b) Find the distance between the charges.

49. Two charges $+3\ \mu$C and $+12\ \mu$C are fixed 1 m apart, with the second one to the right. Find the magnitude and direction of the net force on a −2-nC charge when placed at the following locations: (a) halfway between the two (b) half a meter to the left of the $+3\ \mu$C charge (c) half a meter above the $+12\ \mu$C charge in a direction perpendicular to the line joining the two fixed charges

50. In a salt crystal, the distance between adjacent sodium and chloride ions is 2.82×10^{-10} m. What is the force of attraction between the two singly charged ions?

51. Protons in an atomic nucleus are typically 10^{-15} m apart. What is the electric force of repulsion between nuclear protons?

52. Suppose Earth and the Moon each carried a net negative charge $-Q$. Approximate both bodies as point masses and point charges.

(a) What value of Q is required to balance the gravitational attraction between Earth and the Moon?

(b) Does the distance between Earth and the Moon affect your answer? Explain.

(c) How many electrons would be needed to produce this charge?

53. Point charges $q_1 = 50\ \mu$C and $q_2 = -25\ \mu$C are placed 1.0 m apart. What is the force on a third charge $q_3 = 20\ \mu$C placed midway between q_1 and q_2?

54. Where must q_3 of the preceding problem be placed so that the net force on it is zero?

55. Two small balls, each of mass 5.0 g, are attached to silk threads 50 cm long, which are in turn tied to the same point on the ceiling, as shown below. When the balls are given the same charge Q, the threads hang at $5.0°$ to the vertical, as shown below. What is the magnitude of Q? What are the signs of the two charges?

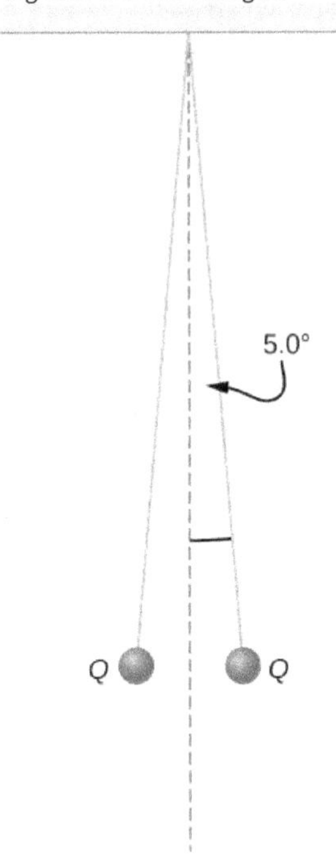

56. Point charges $Q_1 = 2.0\,\mu\text{C}$ and $Q_2 = 4.0\,\mu\text{C}$ are located at $\vec{\mathbf{r}}_1 = (4.0\,\hat{\mathbf{i}} - 2.0\,\hat{\mathbf{j}} + 5.0\,\hat{\mathbf{k}})\text{m}$ and $\vec{\mathbf{r}}_2 = (8.0\,\hat{\mathbf{i}} + 5.0\,\hat{\mathbf{j}} - 9.0\,\hat{\mathbf{k}})\text{m}$. What is the force of Q_2 on Q_1?

57. The net excess charge on two small spheres (small enough to be treated as point charges) is Q. Show that the force of repulsion between the spheres is greatest when each sphere has an excess charge $Q/2$. Assume that the distance between the spheres is so large compared with their radii that the spheres can be treated as point charges.

58. Two small, identical conducting spheres repel each other with a force of 0.050 N when they are 0.25 m apart. After a conducting wire is connected between the spheres and then removed, they repel each other with a force of 0.060 N. What is the original charge on each sphere?

59. A charge $q = 2.0\,\mu\text{C}$ is placed at the point P shown below. What is the force on q?

2.0 m   1.0 m
1.0 μC   −3.0 μC P

60. What is the net electric force on the charge located at the lower right-hand corner of the triangle shown here?

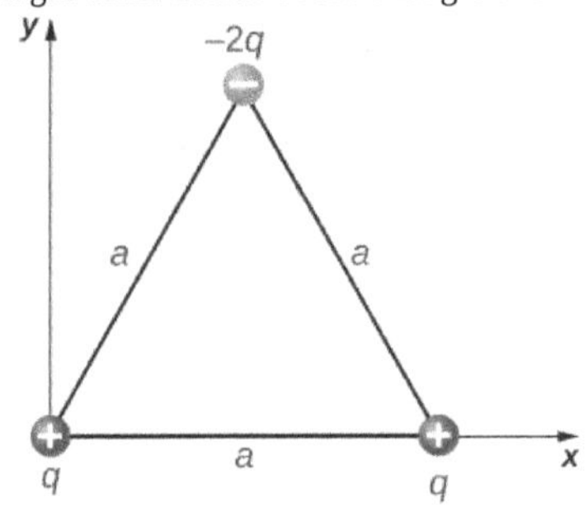

61. Two fixed particles, each of charge 5.0×10^{-6} C, are 24 cm apart. What force do they exert on a third particle of charge -2.5×10^{-6} C that is 13 cm from each of them?

62. The charges $q_1 = 2.0 \times 10^{-7}$ C, $q_2 = -4.0 \times 10^{-7}$ C, and $q_3 = -1.0 \times 10^{-7}$ C are placed at the corners of the triangle shown below. What is the force on q_1?

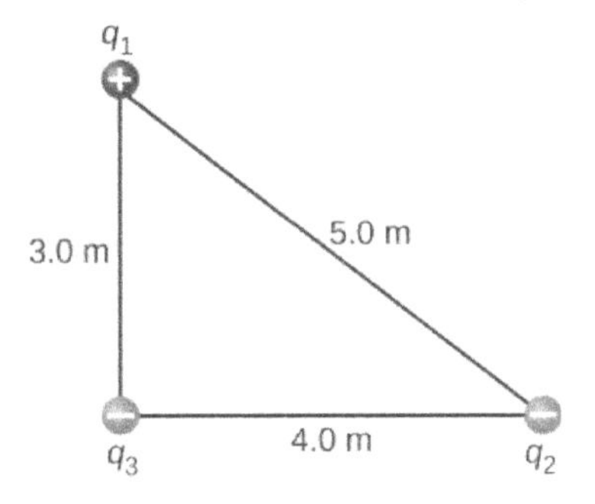

63. What is the force on the charge q at the lower-right-hand corner of the square shown here?

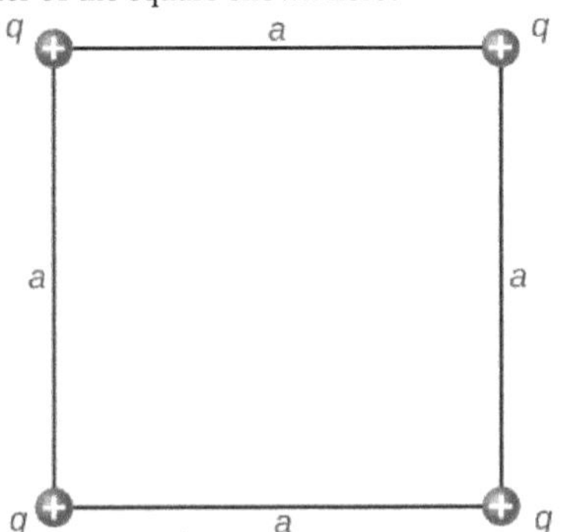

64. Point charges $q_1 = 10\,\mu C$ and $q_2 = -30\,\mu C$ are fixed at $r_1 = \left(3.0\,\hat{\mathbf{i}} - 4.0\,\hat{\mathbf{j}}\right)\text{m}$ and $r_2 = \left(9.0\,\hat{\mathbf{i}} + 6.0\,\hat{\mathbf{j}}\right)\text{m}$. What is the force of q_2 on q_1?

5.4 Electric Field

65. A particle of charge 2.0×10^{-8} C experiences an upward force of magnitude 4.0×10^{-6} N when it is placed in a particular point in an electric field. (a) What is the electric field at that point? (b) If a charge $q = -1.0 \times 10^{-8}$ C is placed there, what is the force on it?

66. On a typical clear day, the atmospheric electric field points downward and has a magnitude of approximately 100 N/C. Compare the gravitational and electric forces on a small dust particle of mass 2.0×10^{-15} g that carries a single electron charge. What is the acceleration (both magnitude and direction) of the dust particle?

67. Consider an electron that is 10^{-10} m from an alpha particle $(q = 3.2 \times 10^{-19}\text{ C})$. (a) What is the electric field due to the alpha particle at the location of the electron? (b) What is the electric field due to the electron at the location of the alpha particle? (c) What is the electric force on the alpha particle? On the electron?

68. Each the balls shown below carries a charge q and has a mass m. The length of each thread is l, and at equilibrium, the balls are separated by an angle 2θ. How does θ vary with q and l? Show that θ satisfies

$$\sin(\theta)^2 \tan(\theta) = \frac{q^2}{16\pi\varepsilon_0 g l^2 m}.$$

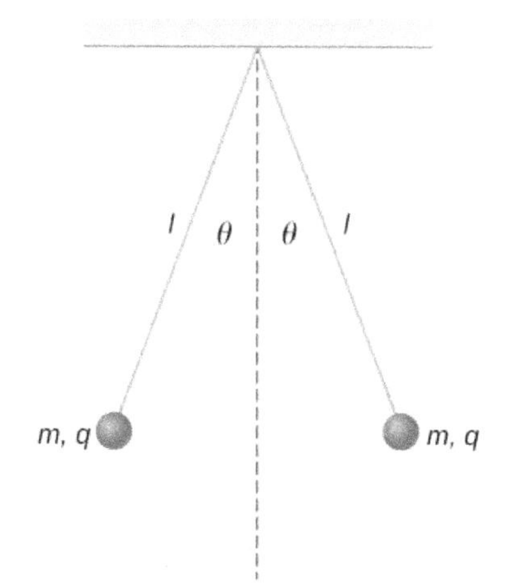

69. What is the electric field at a point where the force on a -2.0×10^{-6} –C charge is $\left(4.0\,\hat{\mathbf{i}} - 6.0\,\hat{\mathbf{j}}\right) \times 10^{-6}$ N?

70. A proton is suspended in the air by an electric field at the surface of Earth. What is the strength of this electric field?

71. The electric field in a particular thundercloud is 2.0×10^5 N/C. What is the acceleration of an electron in this field?

72. A small piece of cork whose mass is 2.0 g is given a charge of 5.0×10^{-7} C. What electric field is needed to place the cork in equilibrium under the combined electric and gravitational forces?

73. If the electric field is 100 N/C at a distance of 50 cm from a point charge q, what is the value of q?

74. What is the electric field of a proton at the first Bohr orbit for hydrogen $(r = 5.29 \times 10^{-11}\text{ m})$? What is the force on the electron in that orbit?

75. (a) What is the electric field of an oxygen nucleus at a point that is 10^{-10} m from the nucleus? (b) What is the force this electric field exerts on a second oxygen nucleus placed at that point?

76. Two point charges, $q_1 = 2.0 \times 10^{-7}$ C and $q_2 = -6.0 \times 10^{-8}$ C, are held 25.0 cm apart. (a) What is the electric field at a point 5.0 cm from the negative charge and along the line between the two charges? (b)What is the force on an electron placed at that point?

77. Point charges $q_1 = 50\,\mu C$ and $q_2 = -25\,\mu C$ are placed 1.0 m apart. (a) What is the electric field at a point midway between them? (b) What is the force on a charge $q_3 = 20\,\mu C$ situated there?

78. Can you arrange the two point charges $q_1 = -2.0 \times 10^{-6}$ C and $q_2 = 4.0 \times 10^{-6}$ C along the x-axis so that $E = 0$ at the origin?

79. Point charges $q_1 = q_2 = 4.0 \times 10^{-6}$ C are fixed on the x-axis at $x = -3.0$ m and $x = 3.0$ m. What charge q must be placed at the origin so that the electric field vanishes at $x = 0,\ y = 3.0$ m?

5.5 Calculating Electric Fields of Charge Distributions

80. A thin conducting plate 1.0 m on the side is given a charge of -2.0×10^{-6} C. An electron is placed 1.0 cm above the center of the plate. What is the acceleration of the electron?

81. Calculate the magnitude and direction of the electric field 2.0 m from a long wire that is charged uniformly at $\lambda = 4.0 \times 10^{-6}$ C/m.

82. Two thin conducting plates, each 25.0 cm on a side, are situated parallel to one another and 5.0 mm apart. If 10^{-11} electrons are moved from one plate to the other, what is the electric field between the plates?

83. The charge per unit length on the thin rod shown below is λ. What is the electric field at the point *P*? (*Hint*: Solve this problem by first considering the electric field $d\vec{\mathbf{E}}$ at *P* due to a small segment *dx* of the rod, which contains charge $dq = \lambda dx$. Then find the net field by integrating $d\vec{\mathbf{E}}$ over the length of the rod.)

84. The charge per unit length on the thin semicircular wire shown below is λ. What is the electric field at the point *P*?

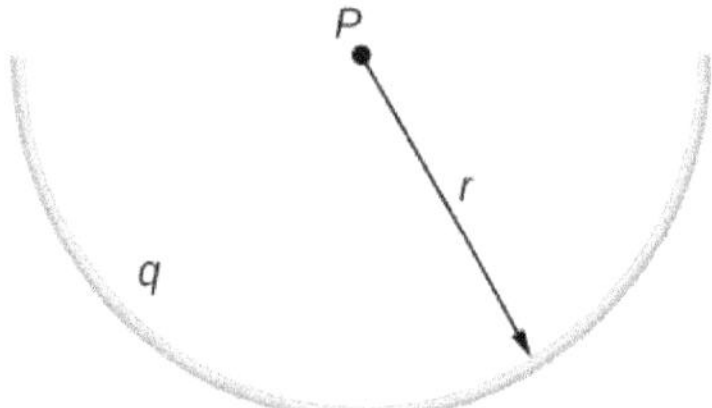

85. Two thin parallel conducting plates are placed 2.0 cm apart. Each plate is 2.0 cm on a side; one plate carries a net charge of $8.0\,\mu$C, and the other plate carries a net charge of $-8.0\,\mu$C. What is the charge density on the inside surface of each plate? What is the electric field between the plates?

86. A thin conducing plate 2.0 m on a side is given a total charge of $-10.0\,\mu$C. (a) What is the electric field 1.0 cm above the plate? (b) What is the force on an electron at this point? (c) Repeat these calculations for a point 2.0 cm above the plate. (d) When the electron moves from 1.0 to 2,0 cm above the plate, how much work is done on it by the electric field?

87. A total charge *q* is distributed uniformly along a thin, straight rod of length *L* (see below). What is the electric field at P_1? At P_2?

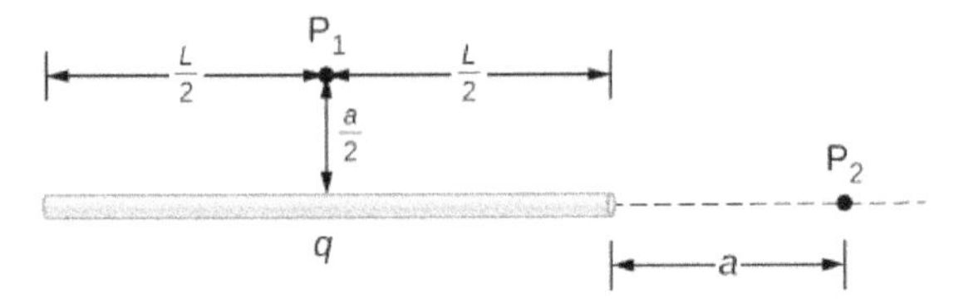

88. Charge is distributed along the entire *x*-axis with uniform density λ. How much work does the electric field of this charge distribution do on an electron that moves along the *y*-axis from $y = a$ to $y = b$?

89. Charge is distributed along the entire *x*-axis with uniform density λ_x and along the entire *y*-axis with uniform density λ_y. Calculate the resulting electric field at (a) $\vec{\mathbf{r}} = a\hat{\mathbf{i}} + b\hat{\mathbf{j}}$ and (b) $\vec{\mathbf{r}} = c\hat{\mathbf{k}}$.

90. A rod bent into the arc of a circle subtends an angle 2θ at the center *P* of the circle (see below). If the rod is charged uniformly with a total charge *Q*, what is the electric field at *P*?

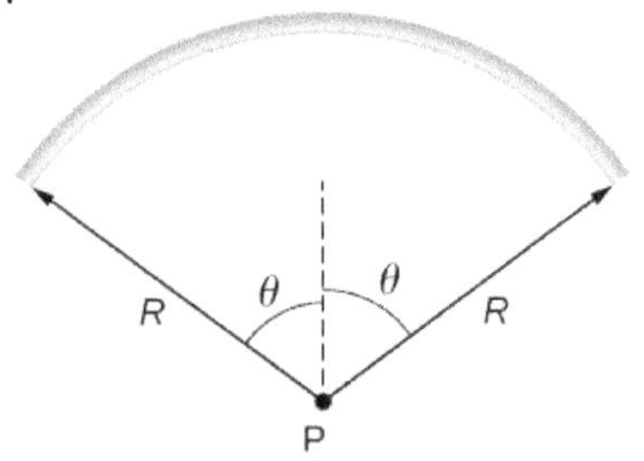

91. A proton moves in the electric field $\vec{\mathbf{E}} = 200\,\hat{\mathbf{i}}$ N/C. (a) What are the force on and the acceleration of the proton? (b) Do the same calculation for an electron moving in this field.

92. An electron and a proton, each starting from rest, are accelerated by the same uniform electric field of 200 N/C. Determine the distance and time for each particle to acquire a kinetic energy of 3.2×10^{-16} J.

93. A spherical water droplet of radius 25 μm carries an excess 250 electrons. What vertical electric field is needed to balance the gravitational force on the droplet at the surface of the earth?

94. A proton enters the uniform electric field produced by the two charged plates shown below. The magnitude of the electric field is 4.0×10^5 N/C, and the speed of the proton when it enters is 1.5×10^7 m/s. What distance d has the proton been deflected downward when it leaves the plates?

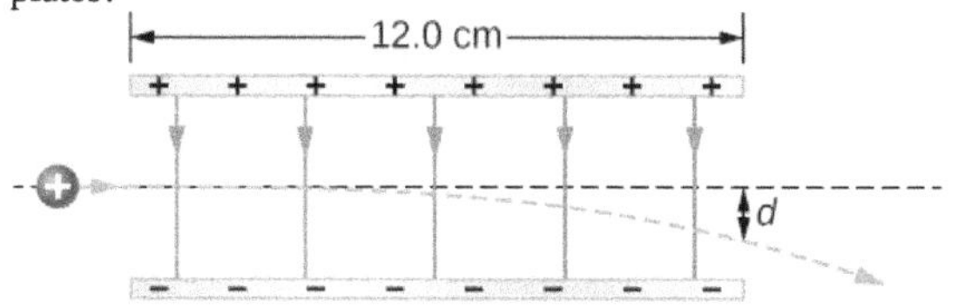

95. Shown below is a small sphere of mass 0.25 g that carries a charge of 9.0×10^{-10} C. The sphere is attached to one end of a very thin silk string 5.0 cm long. The other end of the string is attached to a large vertical conducting plate that has a charge density of 30×10^{-6} C/m^2. What is the angle that the string makes with the vertical?

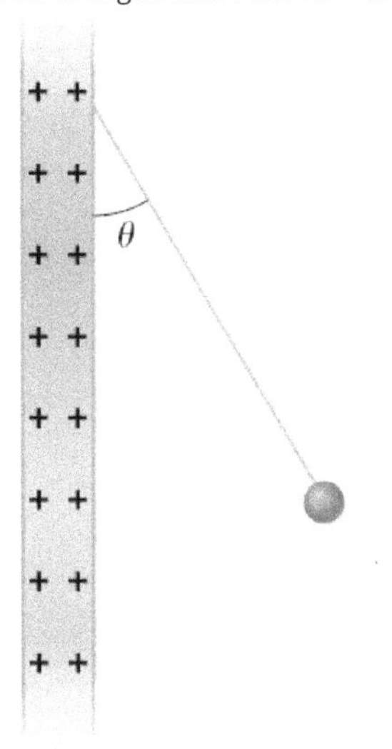

96. Two infinite rods, each carrying a uniform charge density λ, are parallel to one another and perpendicular to the plane of the page. (See below.) What is the electrical field at P_1? At P_2?

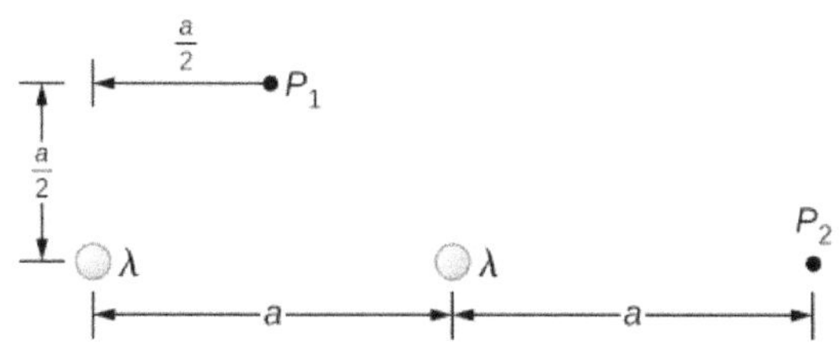

97. Positive charge is distributed with a uniform density λ along the positive x-axis from r to ∞, along the positive y-axis from r to ∞, and along a 90° arc of a circle of radius r, as shown below. What is the electric field at O?

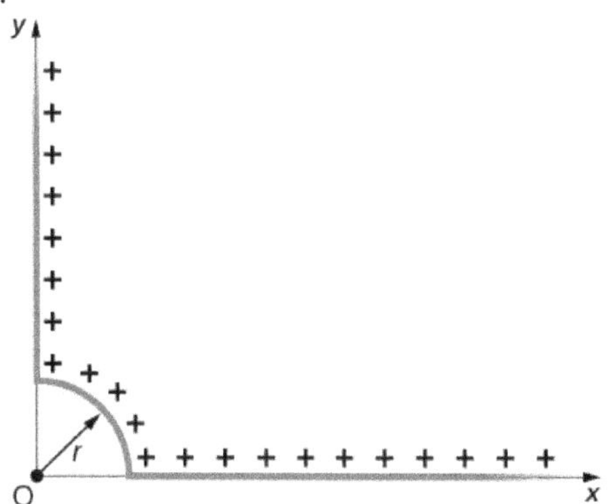

98. From a distance of 10 cm, a proton is projected with a speed of $v = 4.0 \times 10^6$ m/s directly at a large, positively charged plate whose charge density is $\sigma = 2.0 \times 10^{-5}$ C/m^2. (See below.) (a) Does the proton reach the plate? (b) If not, how far from the plate does it turn around?

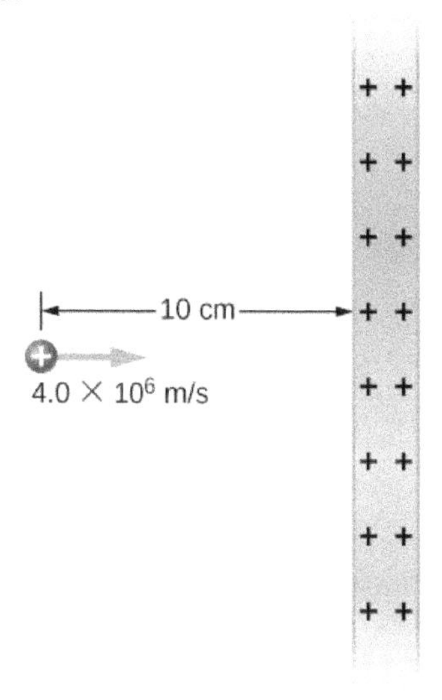

99. A particle of mass m and charge $-q$ moves along a straight line away from a fixed particle of charge Q. When the distance between the two particles is r_0, $-q$ is moving with a speed v_0. (a) Use the work-energy theorem to calculate the maximum separation of the charges. (b) What do you have to assume about v_0 to make this calculation? (c) What is the minimum value of v_0 such that $-q$ escapes from Q?

5.6 Electric Field Lines

100. Which of the following electric field lines are incorrect for point charges? Explain why.

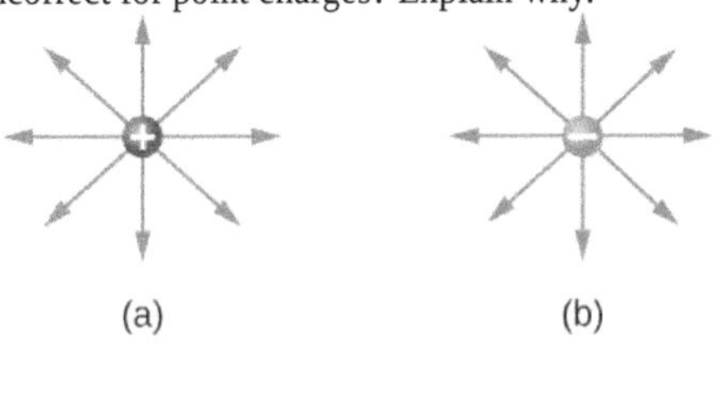

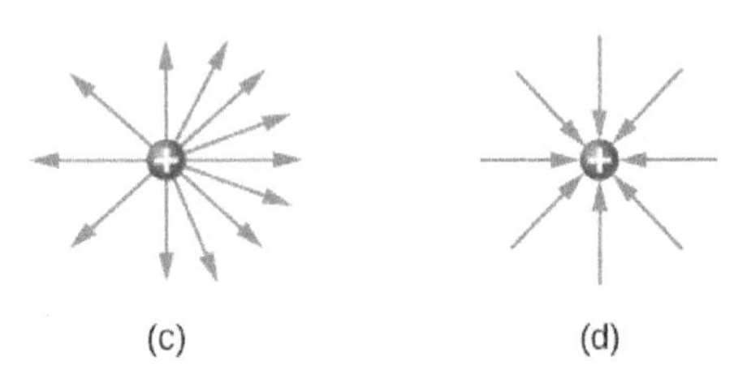

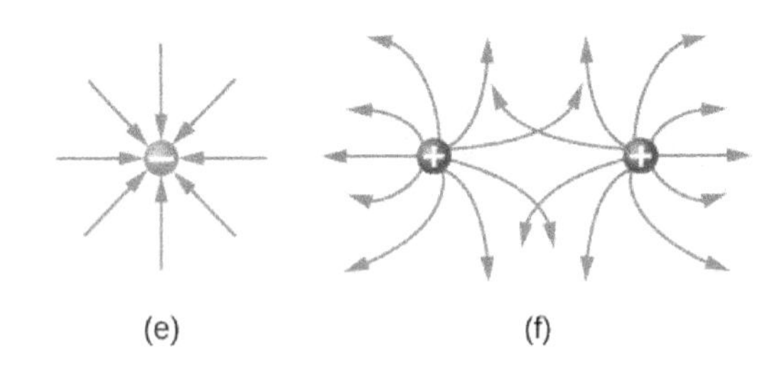

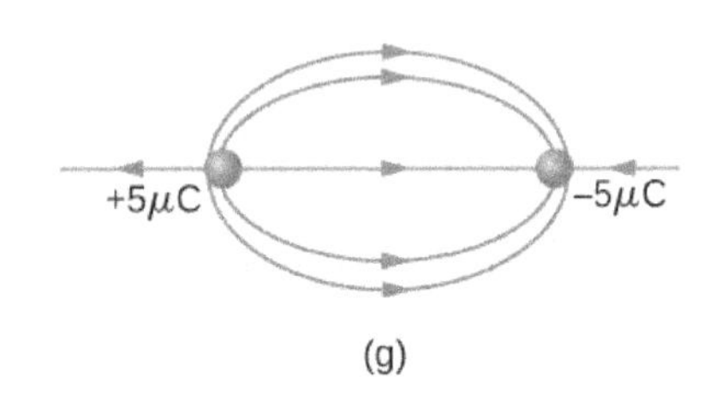

101. In this exercise, you will practice drawing electric field lines. Make sure you represent both the magnitude and direction of the electric field adequately. Note that the number of lines into or out of charges is proportional to the charges.

(a) Draw the electric field lines map for two charges $+20\,\mu\text{C}$ and $-20\,\mu\text{C}$ situated 5 cm from each other.

(b) Draw the electric field lines map for two charges $+20\,\mu\text{C}$ and $+20\,\mu\text{C}$ situated 5 cm from each other.

(c) Draw the electric field lines map for two charges $+20\,\mu\text{C}$ and $-30\,\mu\text{C}$ situated 5 cm from each other.

102. Draw the electric field for a system of three particles of charges $+1\,\mu\text{C}$, $+2\,\mu\text{C}$, and $-3\,\mu\text{C}$ fixed at the corners of an equilateral triangle of side 2 cm.

103. Two charges of equal magnitude but opposite sign make up an electric dipole. A quadrupole consists of two electric dipoles are placed anti-parallel at two edges of a square as shown.

+10 nC −10 nC

−10 nC +10 nC

Draw the electric field of the charge distribution.

104. Suppose the electric field of an isolated point charge decreased with distance as $1/r^{2+\delta}$ rather than as $1/r^2$. Show that it is then impossible to draw continous field lines so that their number per unit area is proportional to E.

5.7 Electric Dipoles

105. Consider the equal and opposite charges shown below. (a) Show that at all points on the x-axis for which $|x| \gg a$, $E \approx Qa/2\pi\varepsilon_0 x^3$. (b) Show that at all points on the y-axis for which $|y| \gg a$, $E \approx Qa/\pi\varepsilon_0 y^3$.

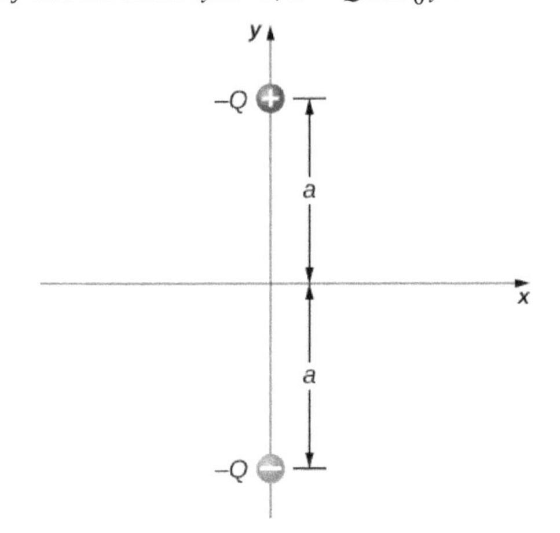

106. (a) What is the dipole moment of the configuration shown above? If $Q = 4.0\,\mu\text{C}$, (b) what is the torque on this dipole with an electric field of $4.0 \times 10^5\ \text{N/C}\ \hat{\mathbf{i}}$? (c) What is the torque on this dipole with an electric field of $-4.0 \times 10^5\ \text{N/C}\ \hat{\mathbf{i}}$? (d) What is the torque on this dipole with an electric field of $\pm 4.0 \times 10^5\ \text{N/C}\ \hat{\mathbf{j}}$?

107. A water molecule consists of two hydrogen atoms bonded with one oxygen atom. The bond angle between the two hydrogen atoms is $104°$ (see below). Calculate the net dipole moment of a water molecule that is placed in a uniform, horizontal electric field of magnitude 2.3×10^{-8} N/C. (You are missing some information for solving this problem; you will need to determine what information you need, and look it up.)

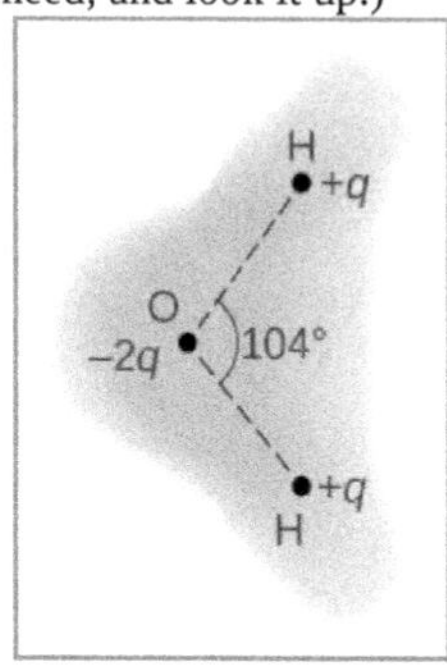

ADDITIONAL PROBLEMS

108. Point charges $q_1 = 2.0\,\mu\text{C}$ and $q_1 = 4.0\,\mu\text{C}$ are located at $r_1 = \left(4.0\,\hat{\mathbf{i}} - 2.0\,\hat{\mathbf{j}} + 2.0\,\hat{\mathbf{k}}\right)\text{m}$ and $r_2 = \left(8.0\,\hat{\mathbf{i}} + 5.0\,\hat{\mathbf{j}} - 9.0\,\hat{\mathbf{k}}\right)\text{m}$. What is the force of q_2 on q_1?

109. What is the force on the 5.0-μC charges shown below?

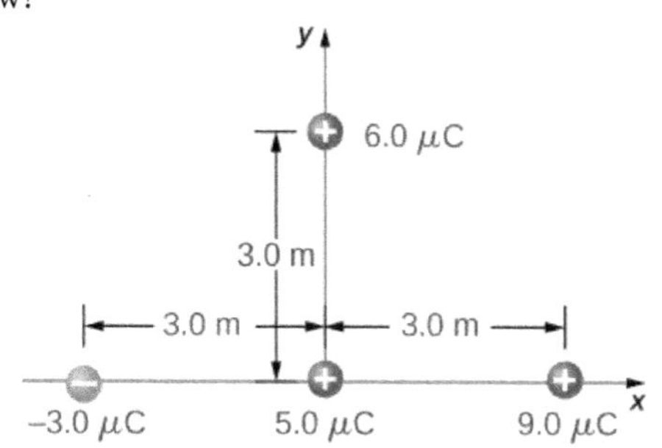

110. What is the force on the 2.0-μC charge placed at the center of the square shown below?

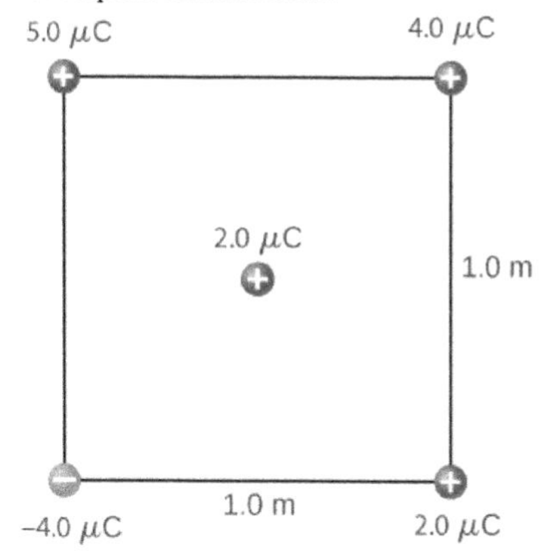

111. Four charged particles are positioned at the corners of a parallelogram as shown below. If $q = 5.0\,\mu\text{C}$ and $Q = 8.0\,\mu\text{C}$, what is the net force on q?

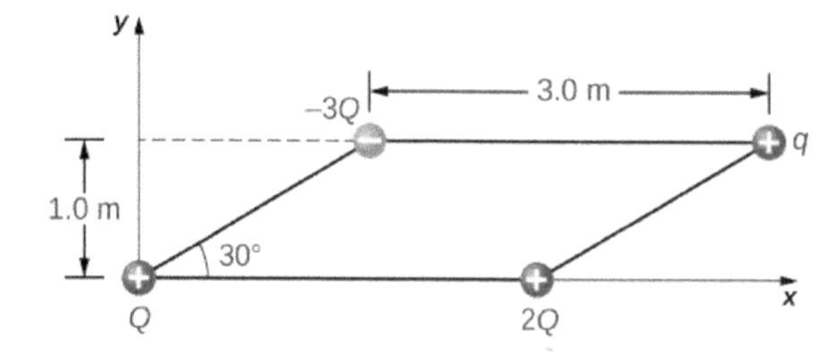

112. A charge Q is fixed at the origin and a second charge q moves along the x-axis, as shown below. How much work is done on q by the electric force when q moves from x_1 to x_2?

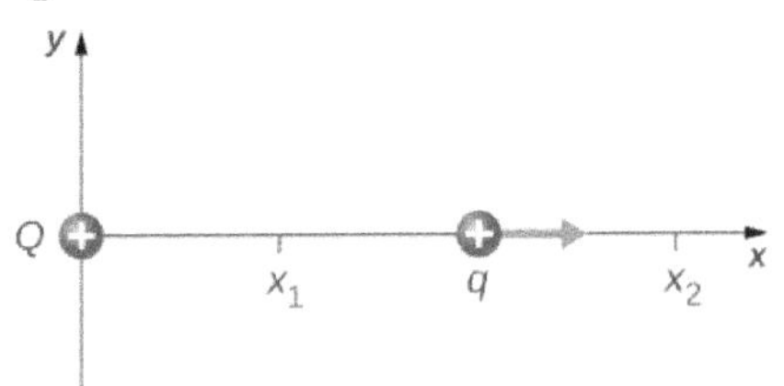

113. A charge $q = -2.0\,\mu\text{C}$ is released from rest when it is 2.0 m from a fixed charge $Q = 6.0\,\mu\text{C}$. What is the kinetic energy of q when it is 1.0 m from Q?

114. What is the electric field at the midpoint M of the hypotenuse of the triangle shown below?

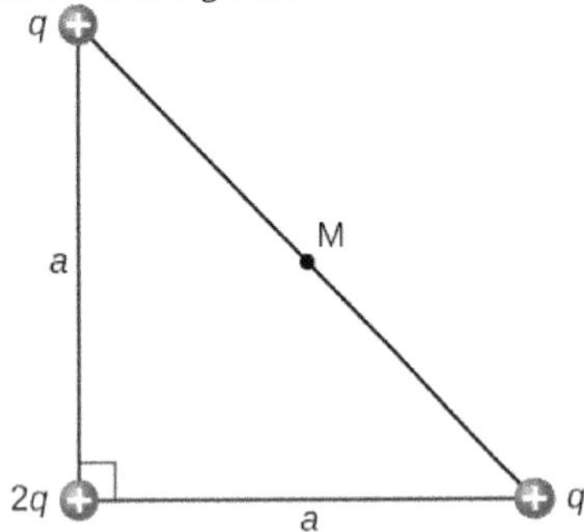

115. Find the electric field at P for the charge configurations shown below.

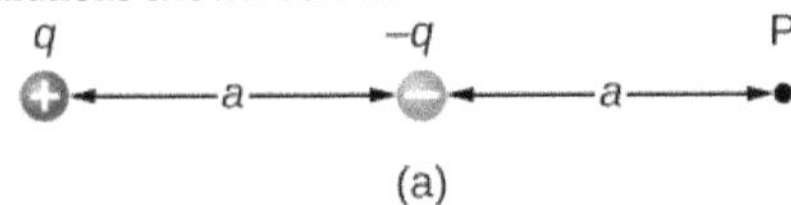

(a)

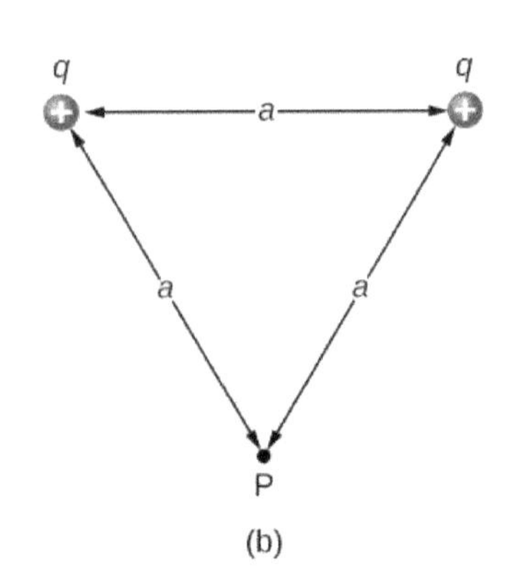

(b)

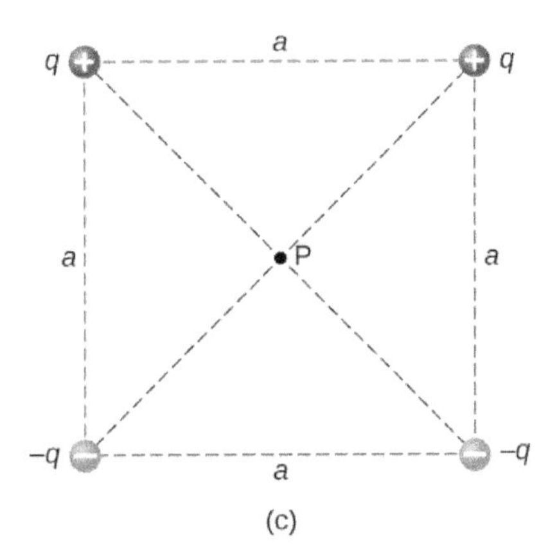

(c)

116. (a) What is the electric field at the lower-right-hand corner of the square shown below? (b) What is the force on a charge q placed at that point?

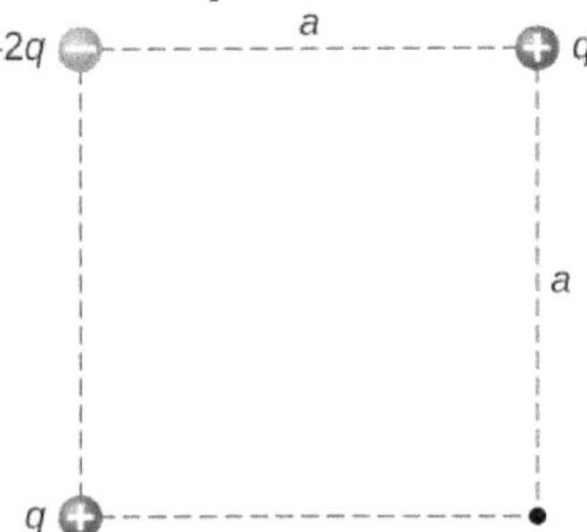

117. Point charges are placed at the four corners of a rectangle as shown below: $q_1 = 2.0 \times 10^{-6}$ C, $q_2 = -2.0 \times 10^{-6}$ C, $q_3 = 4.0 \times 10^{-6}$ C, and $q_4 = 1.0 \times 10^{-6}$ C. What is the electric field at *P*?

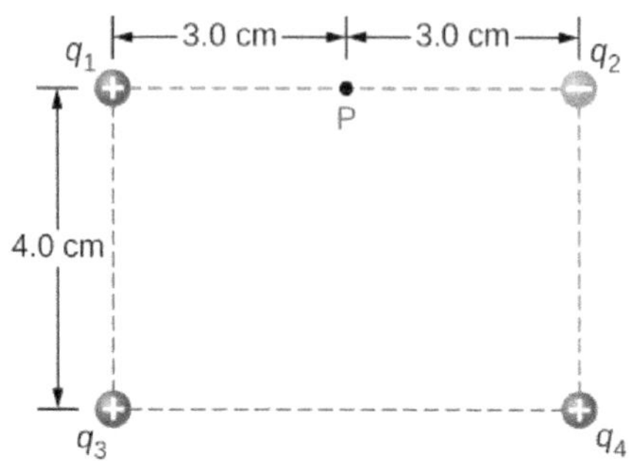

118. Three charges are positioned at the corners of a parallelogram as shown below. (a) If $Q = 8.0\ \mu\text{C}$, what is the electric field at the unoccupied corner? (b) What is the force on a 5.0-μC charge placed at this corner?

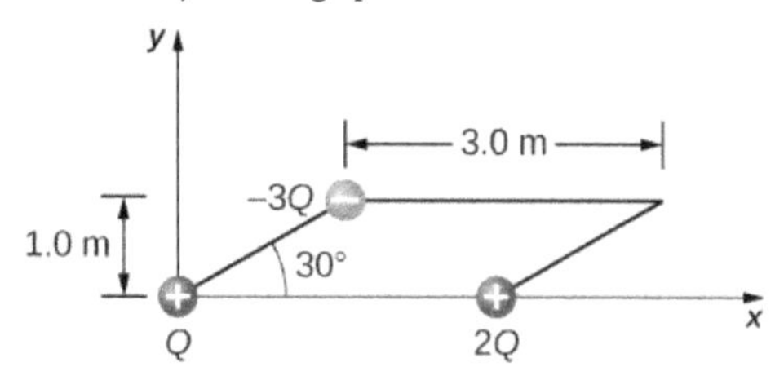

119. A positive charge *q* is released from rest at the origin of a rectangular coordinate system and moves under the influence of the electric field $\vec{\mathbf{E}} = E_0(1 + x/a)\,\hat{\mathbf{i}}$. What is the kinetic energy of *q* when it passes through $x = 3a$?

120. A particle of charge $-q$ and mass *m* is placed at the center of a uniformaly charged ring of total charge *Q* and radius *R*. The particle is displaced a small distance along the axis perpendicular to the plane of the ring and released. Assuming that the particle is constrained to move along the axis, show that the particle oscillates in simple harmonic motion with a frequency $f = \frac{1}{2\pi}\sqrt{\frac{qQ}{4\pi\varepsilon_0 mR^3}}$.

121. Charge is distributed uniformly along the entire *y*-axis with a density λ_y and along the positive *x*-axis from $x = a$ to $x = b$ with a density λ_x. What is the force between the two distributions?

122. The circular arc shown below carries a charge per unit length $\lambda = \lambda_0 \cos\theta$, where θ is measured from the *x*-axis. What is the electric field at the origin?

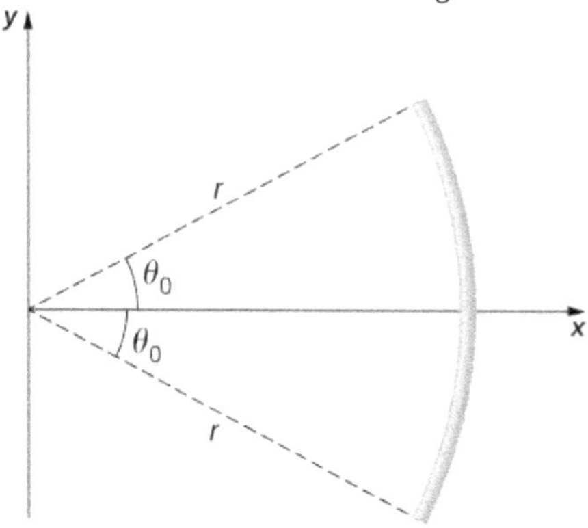

123. Calculate the electric field due to a uniformly charged rod of length *L*, aligned with the *x*-axis with one end at the origin; at a point *P* on the *z*-axis.

124. The charge per unit length on the thin rod shown below is λ. What is the electric force on the point charge *q*? Solve this problem by first considering the electric force $d\vec{\mathbf{F}}$ on *q* due to a small segment *dx* of the rod, which contains charge λdx. Then, find the net force by integrating $d\vec{\mathbf{F}}$ over the length of the rod.

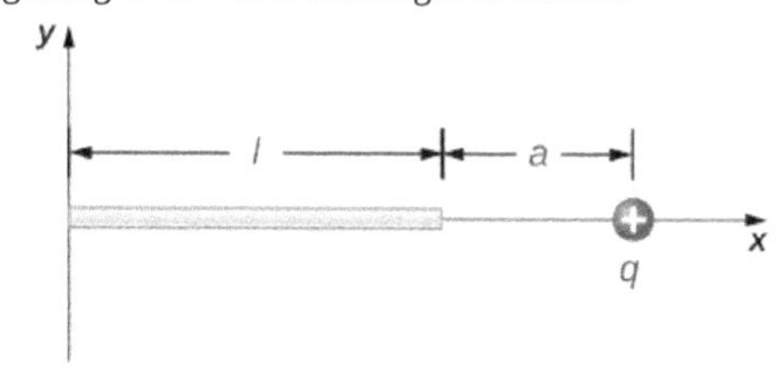

125. The charge per unit length on the thin rod shown here is λ. What is the electric force on the point charge *q*? (See the preceding problem.)

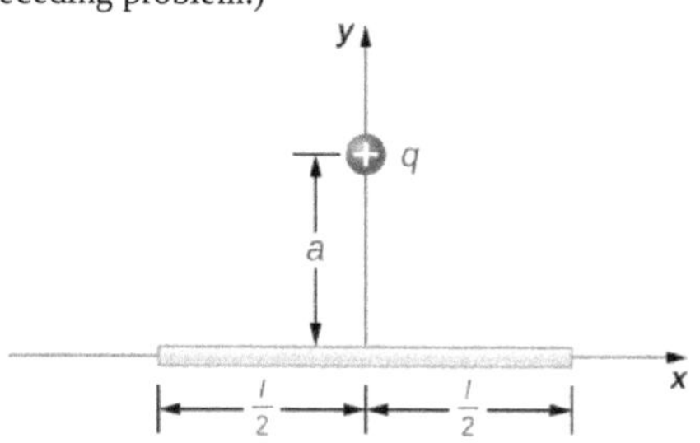

126. The charge per unit length on the thin semicircular wire shown below is λ. What is the electric force on the point charge q? (See the preceding problems.)

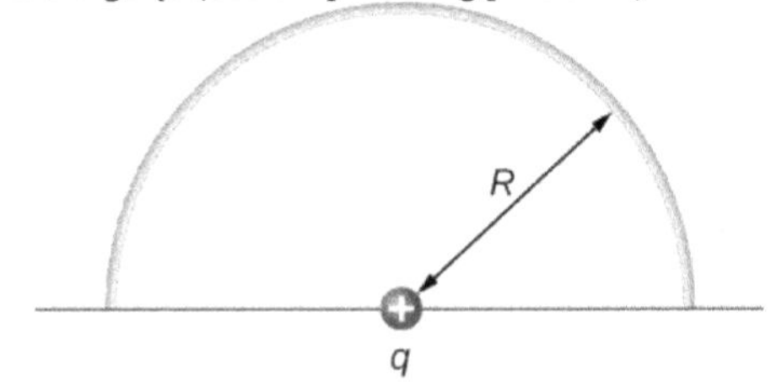

6 | GAUSS'S LAW

Figure 6.1 This chapter introduces the concept of flux, which relates a physical quantity and the area through which it is flowing. Although we introduce this concept with the electric field, the concept may be used for many other quantities, such as fluid flow. (credit: modification of work by "Alessandro"/Flickr)

Chapter Outline

Introduction

Flux is a general and broadly applicable concept in physics. However, in this chapter, we concentrate on the flux of the electric field. This allows us to introduce Gauss's law, which is particularly useful for finding the electric fields of charge distributions exhibiting spatial symmetry. The main topics discussed here are

1. **Electric flux.** We define electric flux for both open and closed surfaces.
2. **Gauss's law.** We derive Gauss's law for an arbitrary charge distribution and examine the role of electric flux in Gauss's law.
3. **Calculating electric fields with Gauss's law.** The main focus of this chapter is to explain how to use Gauss's law to find the electric fields of spatially symmetrical charge distributions. We discuss the importance of choosing a Gaussian surface and provide examples involving the applications of Gauss's law.
4. **Electric fields in conductors.** Gauss's law provides useful insight into the absence of electric fields in conducting materials.

So far, we have found that the electrostatic field begins and ends at point charges and that the field of a point charge varies inversely with the square of the distance from that charge. These characteristics of the electrostatic field lead to an important mathematical relationship known as Gauss's law. This law is named in honor of the extraordinary German mathematician and scientist Karl Friedrich Gauss (Figure 6.2). Gauss's law gives us an elegantly simple way of finding the electric field, and, as you will see, it can be much easier to use than the integration method described in the previous chapter. However, there is a catch—Gauss's law has a limitation in that, while always true, it can be readily applied only for charge distributions with certain symmetries.

Figure 6.2 Karl Friedrich Gauss (1777–1855) was a legendary mathematician of the nineteenth century. Although his major contributions were to the field of mathematics, he also did important work in physics and astronomy.

6.1 | Electric Flux

Learning Objectives

By the end of this section, you will be able to:

- Define the concept of flux
- Describe electric flux
- Calculate electric flux for a given situation

The concept of **flux** describes how much of something goes through a given area. More formally, it is the dot product of a vector field (in this chapter, the electric field) with an area. You may conceptualize the flux of an electric field as a measure of the number of electric field lines passing through an area (Figure 6.3). The larger the area, the more field lines go through it and, hence, the greater the flux; similarly, the stronger the electric field is (represented by a greater density of lines), the greater the flux. On the other hand, if the area rotated so that the plane is aligned with the field lines, none will pass through and there will be no flux.

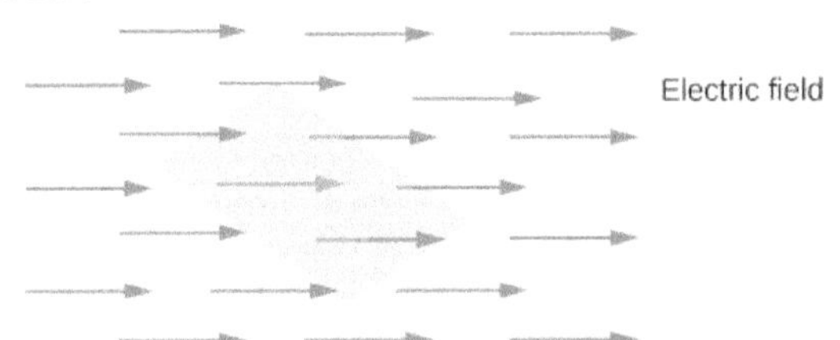

Figure 6.3 The flux of an electric field through the shaded area captures information about the "number" of electric field lines passing through the area. The numerical value of the electric flux depends on the magnitudes of the electric field and the area, as well as the relative orientation of the area with respect to the direction of the electric field.

A macroscopic analogy that might help you imagine this is to put a hula hoop in a flowing river. As you change the angle of the hoop relative to the direction of the current, more or less of the flow will go through the hoop. Similarly, the amount of flow through the hoop depends on the strength of the current and the size of the hoop. Again, flux is a general concept; we can also use it to describe the amount of sunlight hitting a solar panel or the amount of energy a telescope receives from a distant star, for example.

To quantify this idea, Figure 6.4(a) shows a planar surface S_1 of area A_1 that is perpendicular to the uniform electric field $\vec{\mathbf{E}} = E\hat{\mathbf{y}}$. If N field lines pass through S_1, then we know from the definition of electric field lines (Electric Charges and Fields) that $N/A_1 \propto E$, or $N \propto EA_1$.

The quantity EA_1 is the **electric flux** through S_1. We represent the electric flux through an open surface like S_1 by the symbol Φ. Electric flux is a scalar quantity and has an SI unit of newton-meters squared per coulomb ($\text{N} \cdot \text{m}^2/\text{C}$). Notice that $N \propto EA_1$ may also be written as $N \propto \Phi$, demonstrating that *electric flux is a measure of the number of field lines crossing a surface*.

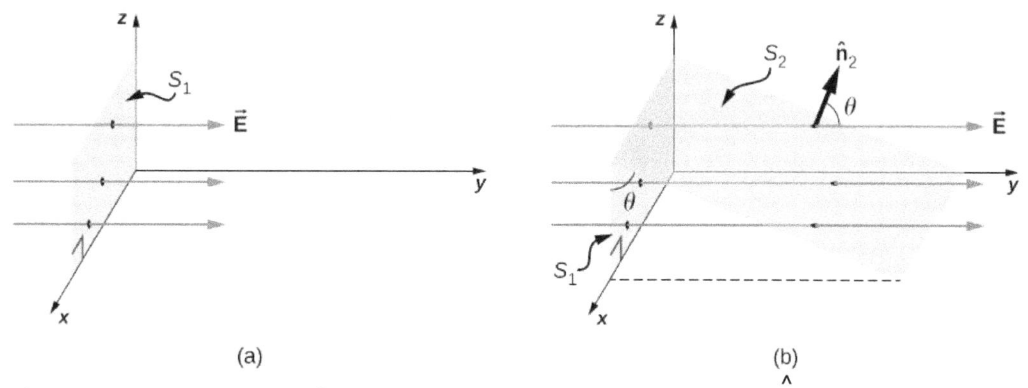

Figure 6.4 (a) A planar surface S_1 of area A_1 is perpendicular to the electric field $E\hat{\mathbf{j}}$. N field lines cross surface S_1. (b) A surface S_2 of area A_2 whose projection onto the *xz*-plane is S_1. The same number of field lines cross each surface.

Now consider a planar surface that is not perpendicular to the field. How would we represent the electric flux? Figure 6.4(b) shows a surface S_2 of area A_2 that is inclined at an angle θ to the *xz*-plane and whose projection in that plane is S_1 (area A_1). The areas are related by $A_2 \cos\theta = A_1$. Because the same number of field lines crosses both S_1 and S_2, the fluxes through both surfaces must be the same. The flux through S_2 is therefore $\Phi = EA_1 = EA_2 \cos\theta$. Designating $\hat{\mathbf{n}}_2$ as a unit vector normal to S_2 (see Figure 6.4(b)), we obtain

$$\Phi = \vec{\mathbf{E}} \cdot \hat{\mathbf{n}}_2 A_2.$$

Check out this video (https://openstaxcollege.org/l/21fluxsizeangl) to observe what happens to the flux as the area changes in size and angle, or the electric field changes in strength.

Area Vector

For discussing the flux of a vector field, it is helpful to introduce an area vector $\vec{\mathbf{A}}$. This allows us to write the last equation in a more compact form. What should the magnitude of the area vector be? What should the direction of the area vector be? What are the implications of how you answer the previous question?

The **area vector** of a flat surface of area A has the following magnitude and direction:

- Magnitude is equal to area (A)

- Direction is along the normal to the surface ($\hat{\mathbf{n}}$); that is, perpendicular to the surface.

Since the normal to a flat surface can point in either direction from the surface, the direction of the area vector of an open surface needs to be chosen, as shown in Figure 6.5.

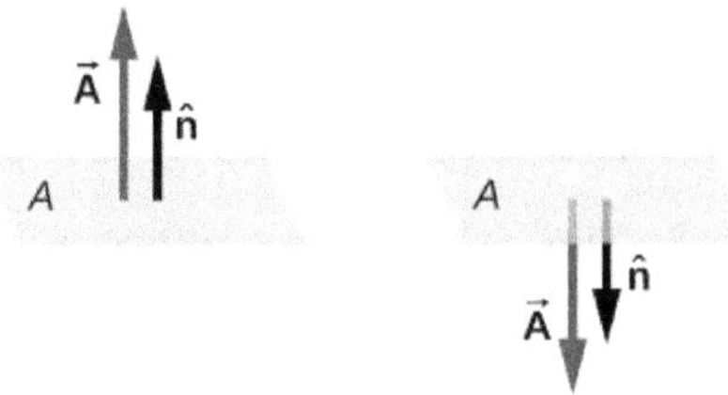

Figure 6.5 The direction of the area vector of an open surface needs to be chosen; it could be either of the two cases displayed here. The area vector of a part of a closed surface is defined to point from the inside of the closed space to the outside. This rule gives a unique direction.

Since $\hat{\mathbf{n}}$ is a unit normal to a surface, it has two possible directions at every point on that surface (Figure 6.6(a)). For an open surface, we can use either direction, as long as we are consistent over the entire surface. Part (c) of the figure shows several cases.

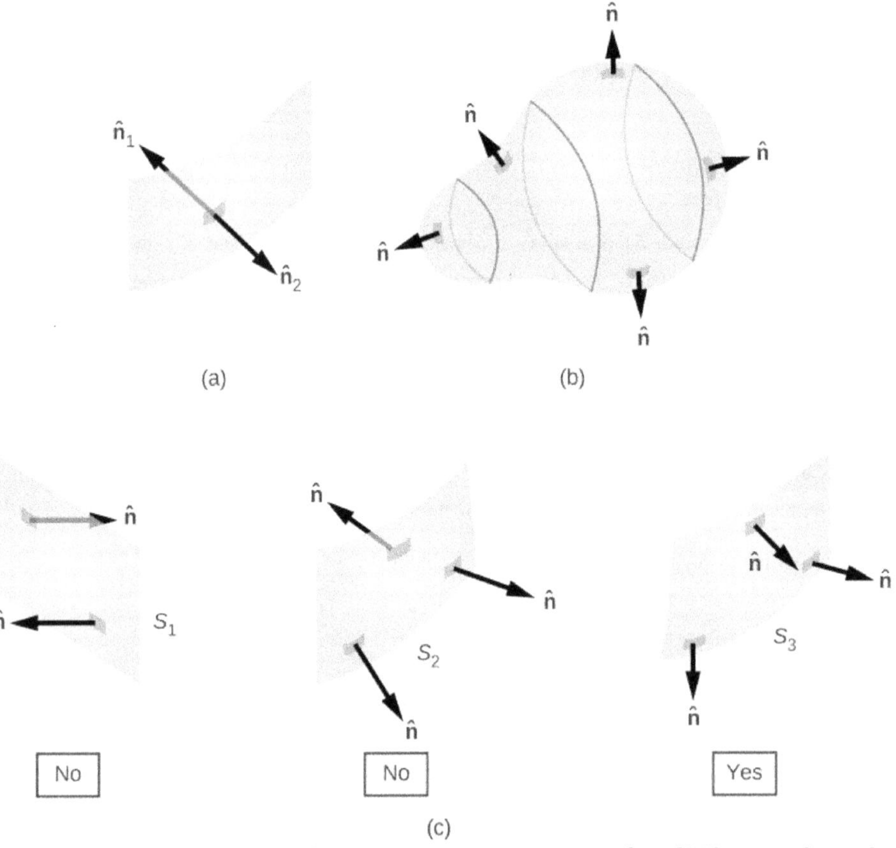

Figure 6.6 (a) Two potential normal vectors arise at every point on a surface. (b) The outward normal is used to calculate the flux through a closed surface. (c) Only S_3 has been given a consistent set of normal vectors that allows us to define the flux through the surface.

However, if a surface is closed, then the surface encloses a volume. In that case, the direction of the normal vector at any point on the surface points from the inside to the outside. On a *closed surface* such as that of Figure 6.6(b), $\hat{\mathbf{n}}$ is chosen to be the *outward normal* at every point, to be consistent with the sign convention for electric charge.

Electric Flux

Now that we have defined the area vector of a surface, we can define the electric flux of a uniform electric field through a flat area as the scalar product of the electric field and the area vector, as defined in Products of Vectors (http://cnx.org/content/m58280/latest/) :

$$\Phi = \vec{\mathbf{E}} \cdot \vec{\mathbf{A}} \text{ (uniform } \vec{\mathbf{E}}\text{, flat su face).} \tag{6.1}$$

Figure 6.7 shows the electric field of an oppositely charged, parallel-plate system and an imaginary box between the plates. The electric field between the plates is uniform and points from the positive plate toward the negative plate. A calculation of the flux of this field through various faces of the box shows that the net flux through the box is zero. Why does the flux cancel out here?

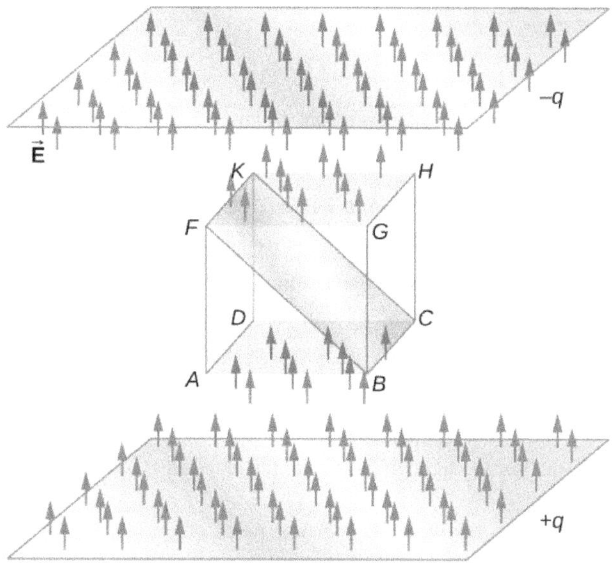

Figure 6.7 Electric flux through a cube, placed between two charged plates. Electric flux through the bottom face (*ABCD*) is negative, because $\vec{\mathbf{E}}$ is in the opposite direction to the normal to the surface. The electric flux through the top face (*FGHK*) is positive, because the electric field and the normal are in the same direction. The electric flux through the other faces is zero, since the electric field is perpendicular to the normal vectors of those faces. The net electric flux through the cube is the sum of fluxes through the six faces. Here, the net flux through the cube is equal to zero. The magnitude of the flux through rectangle *BCKF* is equal to the magnitudes of the flux through both the top and bottom faces.

The reason is that the sources of the electric field are outside the box. Therefore, if any electric field line enters the volume of the box, it must also exit somewhere on the surface because there is no charge inside for the lines to land on. Therefore, quite generally, electric flux through a closed surface is zero if there are no sources of electric field, whether positive or negative charges, inside the enclosed volume. In general, when field lines leave (or "flow out of") a closed surface, Φ is positive; when they enter (or "flow into") the surface, Φ is negative.

Any smooth, non-flat surface can be replaced by a collection of tiny, approximately flat surfaces, as shown in Figure 6.8. If we divide a surface S into small patches, then we notice that, as the patches become smaller, they can be approximated by flat surfaces. This is similar to the way we treat the surface of Earth as locally flat, even though we know that globally, it is approximately spherical.

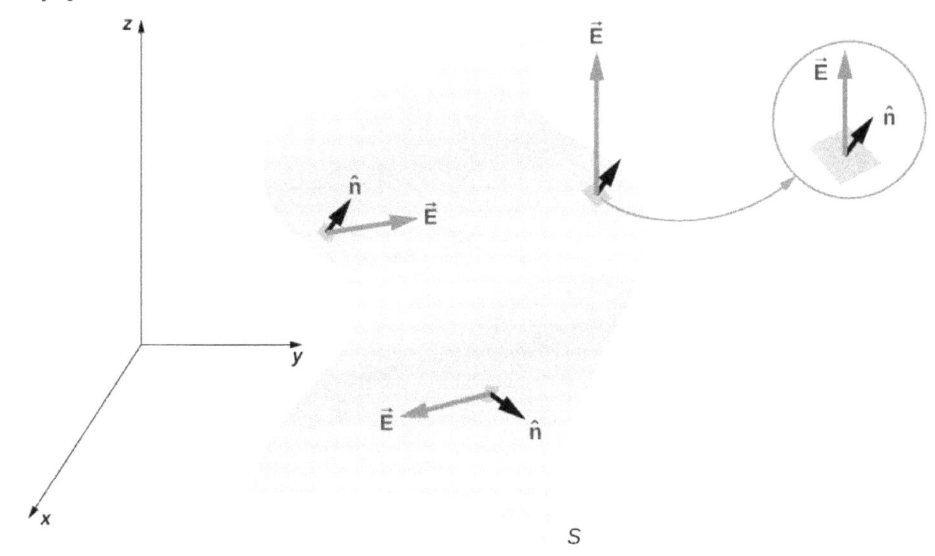

Figure 6.8 A surface is divided into patches to find the flux.

To keep track of the patches, we can number them from 1 through N. Now, we define the area vector for each patch as the area of the patch pointed in the direction of the normal. Let us denote the area vector for the ith patch by $\delta \vec{\mathbf{A}}_i$. (We have used the symbol δ to remind us that the area is of an arbitrarily small patch.) With sufficiently small patches, we may approximate the electric field over any given patch as uniform. Let us denote the average electric field at the location of the ith patch by $\vec{\mathbf{E}}_i$.

$$\vec{\mathbf{E}}_i = \text{average electric field ver the } i\text{th patch.}$$

Therefore, we can write the electric flux Φ_i through the area of the ith patch as

$$\Phi_i = \vec{\mathbf{E}}_i \cdot \delta \vec{\mathbf{A}}_i \ (i\text{th patch}).$$

The flux through each of the individual patches can be constructed in this manner and then added to give us an estimate of the net flux through the entire surface S, which we denote simply as Φ.

$$\Phi = \sum_{i=1}^{N} \Phi_i = \sum_{i=1}^{N} \vec{\mathbf{E}}_i \cdot \delta \vec{\mathbf{A}}_i \ (N \text{ patch estimate}).$$

This estimate of the flux gets better as we decrease the size of the patches. However, when you use smaller patches, you need more of them to cover the same surface. In the limit of infinitesimally small patches, they may be considered to have area dA and unit normal $\hat{\mathbf{n}}$. Since the elements are infinitesimal, they may be assumed to be planar, and $\vec{\mathbf{E}}_i$ may be taken as constant over any element. Then the flux $d\Phi$ through an area dA is given by $d\Phi = \vec{\mathbf{E}} \cdot \hat{\mathbf{n}} \, dA$. It is positive when the angle between $\vec{\mathbf{E}}_i$ and $\hat{\mathbf{n}}$ is less than $90°$ and negative when the angle is greater than $90°$. The net flux is the sum of the infinitesimal flux elements over the entire surface. With infinitesimally small patches, you need infinitely many patches, and the limit of the sum becomes a surface integral. With $\int_S$ representing the integral over S,

$$\Phi = \int_S \vec{\mathbf{E}} \cdot \hat{\mathbf{n}}\, dA = \int_S \vec{\mathbf{E}} \cdot d\vec{\mathbf{A}} \text{ (open surface).} \tag{6.2}$$

In practical terms, surface integrals are computed by taking the antiderivatives of both dimensions defining the area, with the edges of the surface in question being the bounds of the integral.

To distinguish between the flux through an open surface like that of Figure 6.4 and the flux through a closed surface (one that completely bounds some volume), we represent flux through a closed surface by

$$\Phi = \oint_S \vec{\mathbf{E}} \cdot \hat{\mathbf{n}}\, dA = \oint_S \vec{\mathbf{E}} \cdot d\vec{\mathbf{A}} \text{ (closed surface)} \tag{6.3}$$

where the circle through the integral symbol simply means that the surface is closed, and we are integrating over the entire thing. If you only integrate over a portion of a closed surface, that means you are treating a subset of it as an open surface.

Example 6.1

Flux of a Uniform Electric Field

A constant electric field of magnitude E_0 points in the direction of the positive *z*-axis (Figure 6.9). What is the electric flux through a rectangle with sides *a* and *b* in the (a) *xy*-plane and in the (b) *xz*-plane?

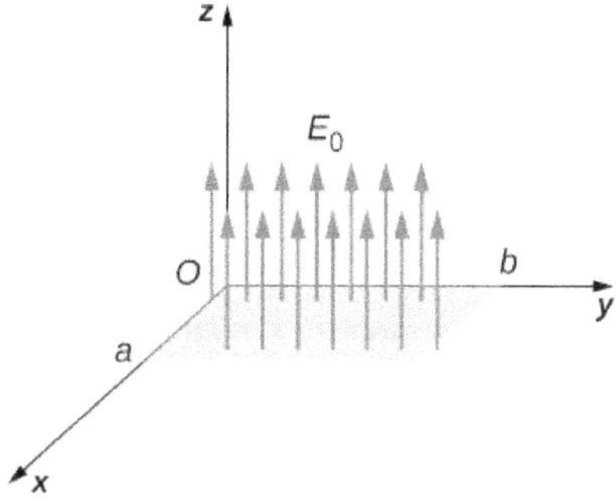

Figure 6.9 Calculating the flux of E_0 through a rectangular surface.

Strategy

Apply the definition of flux: $\Phi = \vec{\mathbf{E}} \cdot \vec{\mathbf{A}}$ (uniform $\vec{\mathbf{E}}$), where the definition of dot product is crucial.

Solution

a. In this case, $\Phi = \vec{\mathbf{E}}_0 \cdot \vec{\mathbf{A}} = E_0 A = E_0 ab.$

b. Here, the direction of the area vector is either along the positive *y*-axis or toward the negative *y*-axis. Therefore, the scalar product of the electric field with the area vector is zero, giving zero flux.

Significance

The relative directions of the electric field and area can cause the flux through the area to be zero.

Example 6.2

Flux of a Uniform Electric Field through a Closed Surface

A constant electric field of magnitude E_0 points in the direction of the positive *z*-axis (Figure 6.10). What is the net electric flux through a cube?

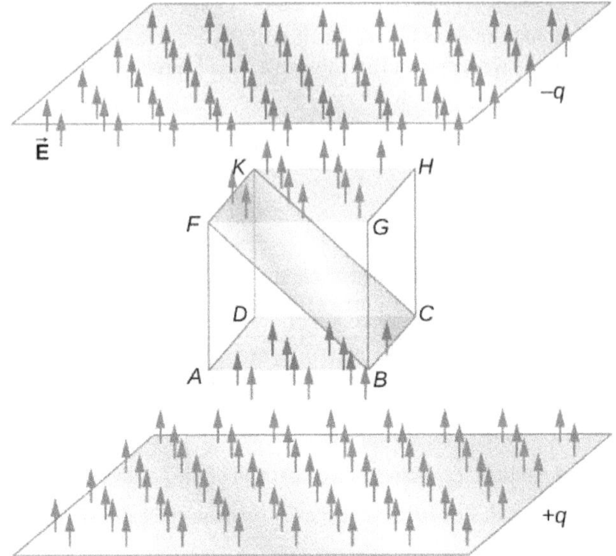

Figure 6.10 Calculating the flux of E_0 through a closed cubic surface.

Strategy

Apply the definition of flux: $\Phi = \vec{\mathbf{E}} \cdot \vec{\mathbf{A}}$ (uniform $\vec{\mathbf{E}}$), noting that a closed surface eliminates the ambiguity in the direction of the area vector.

Solution

Through the top face of the cube, $\Phi = \vec{\mathbf{E}}_0 \cdot \vec{\mathbf{A}} = E_0 A.$

Through the bottom face of the cube, $\Phi = \vec{\mathbf{E}}_0 \cdot \vec{\mathbf{A}} = -E_0 A,$ because the area vector here points downward.

Along the other four sides, the direction of the area vector is perpendicular to the direction of the electric field. Therefore, the scalar product of the electric field with the area vector is zero, giving zero flux.

The net flux is $\Phi_{\text{net}} = E_0 A - E_0 A + 0 + 0 + 0 + 0 = 0$.

Significance

The net flux of a uniform electric field through a closed surface is zero.

Example 6.3

Electric Flux through a Plane, Integral Method

A uniform electric field $\vec{\mathbf{E}}$ of magnitude 10 N/C is directed parallel to the *yz*-plane at $30°$ above the *xy*-plane, as shown in Figure 6.11. What is the electric flux through the plane surface of area $6.0\,\text{m}^2$ located in the *xz*-plane? Assume that $\hat{\mathbf{n}}$ points in the positive *y*-direction.

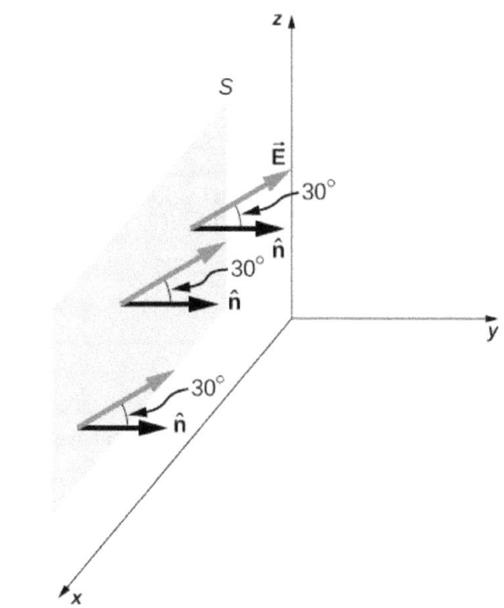

Figure 6.11 The electric field produces a net electric flux through the surface *S*.

Strategy

Apply $\Phi = \int_S \vec{\mathbf{E}} \cdot \hat{\mathbf{n}}\, dA$, where the direction and magnitude of the electric field are constant.

Solution

The angle between the uniform electric field $\vec{\mathbf{E}}$ and the unit normal $\hat{\mathbf{n}}$ to the planar surface is $30°$. Since both the direction and magnitude are constant, *E* comes outside the integral. All that is left is a surface integral over *dA*, which is *A*. Therefore, using the open-surface equation, we find that the electric flux through the surface is

$$\begin{aligned} \Phi &= \int_S \vec{\mathbf{E}} \cdot \hat{\mathbf{n}}\, dA = EA\cos\theta \\ &= (10\,\text{N/C})(6.0\,\text{m}^2)(\cos 30°) = 52\,\text{N}\cdot\text{m}^2/\text{C}. \end{aligned}$$

Significance

Again, the relative directions of the field and the area matter, and the general equation with the integral will simplify to the simple dot product of area and electric field.

6.1 Check Your Understanding What angle should there be between the electric field and the surface shown in Figure 6.11 in the previous example so that no electric flux passes through the surface?

Example 6.4

Inhomogeneous Electric Field

What is the total flux of the electric field $\vec{\mathbf{E}} = cy^2\hat{\mathbf{k}}$ through the rectangular surface shown in Figure 6.12?

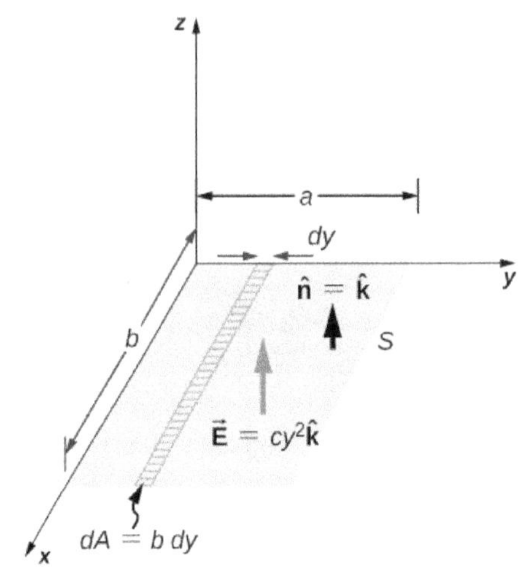

Figure 6.12 Since the electric field is not constant over the surface, an integration is necessary to determine the flux.

Strategy

Apply $\Phi = \int_S \vec{\mathbf{E}} \cdot \hat{\mathbf{n}}\,dA$. We assume that the unit normal $\hat{\mathbf{n}}$ to the given surface points in the positive *z*-direction, so $\hat{\mathbf{n}} = \hat{\mathbf{k}}$. Since the electric field is not uniform over the surface, it is necessary to divide the surface into infinitesimal strips along which $\vec{\mathbf{E}}$ is essentially constant. As shown in Figure 6.12, these strips are parallel to the *x*-axis, and each strip has an area $dA = b\,dy$.

Solution

From the open surface integral, we find that the net flux through the rectangular surface is

$$\begin{aligned}\Phi &= \int_S \vec{\mathbf{E}} \cdot \hat{\mathbf{n}}\,dA = \int_0^a (cy^2\hat{\mathbf{k}}) \cdot \hat{\mathbf{k}}(b\,dy) \\ &= cb\int_0^a y^2\,dy = \frac{1}{3}a^3bc.\end{aligned}$$

Significance

For a non-constant electric field, the integral method is required.

6.2 Check Your Understanding If the electric field in Example 6.4 is $\vec{\mathbf{E}} = mx\hat{\mathbf{k}}$, what is the flux through the rectangular area?

6.2 | Explaining Gauss's Law

Learning Objectives

By the end of this section, you will be able to:

- State Gauss's law
- Explain the conditions under which Gauss's law may be used
- Apply Gauss's law in appropriate systems

We can now determine the electric flux through an arbitrary closed surface due to an arbitrary charge distribution. We found that if a closed surface does not have any charge inside where an electric field line can terminate, then any electric field line entering the surface at one point must necessarily exit at some other point of the surface. Therefore, if a closed surface does not have any charges inside the enclosed volume, then the electric flux through the surface is zero. Now, what happens to the electric flux if there are some charges inside the enclosed volume? Gauss's law gives a quantitative answer to this question.

To get a feel for what to expect, let's calculate the electric flux through a spherical surface around a positive point charge q, since we already know the electric field in such a situation. Recall that when we place the point charge at the origin of a coordinate system, the electric field at a point P that is at a distance r from the charge at the origin is given by

$$\vec{\mathbf{E}}_P = \frac{1}{4\pi\varepsilon_0}\frac{1}{r^2}\hat{\mathbf{r}},$$

where $\hat{\mathbf{r}}$ is the radial vector from the charge at the origin to the point P. We can use this electric field to find the flux through the spherical surface of radius r, as shown in Figure 6.13.

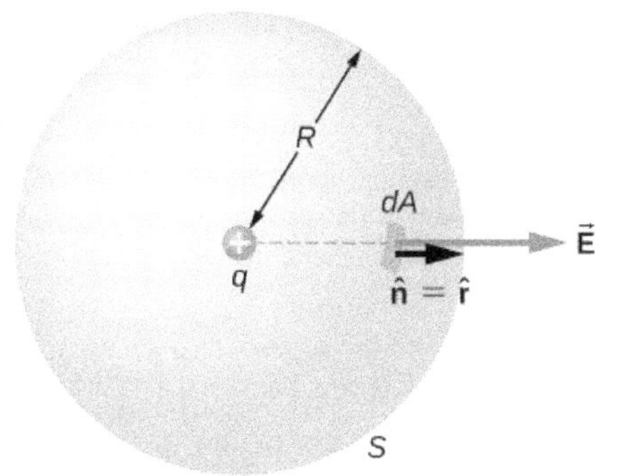

Figure 6.13 A closed spherical surface surrounding a point charge q.

Then we apply $\Phi = \int_S \vec{\mathbf{E}} \cdot \hat{\mathbf{n}}\, dA$ to this system and substitute known values. On the sphere, $\hat{\mathbf{n}} = \hat{\mathbf{r}}$ and $r = R$, so for an infinitesimal area dA,

$$d\Phi = \vec{\mathbf{E}} \cdot \hat{\mathbf{n}}\, dA = \frac{1}{4\pi\varepsilon_0}\frac{q}{R^2}\hat{\mathbf{r}} \cdot \hat{\mathbf{r}}\, dA = \frac{1}{4\pi\varepsilon_0}\frac{q}{R^2}dA.$$

We now find the net flux by integrating this flux over the surface of the sphere:

$$\Phi = \frac{1}{4\pi\varepsilon_0}\frac{q}{R^2}\oint_S dA = \frac{1}{4\pi\varepsilon_0}\frac{q}{R^2}(4\pi R^2) = \frac{q}{\varepsilon_0}.$$

where the total surface area of the spherical surface is $4\pi R^2$. This gives the flux through the closed spherical surface at radius r as

$$\Phi = \frac{q}{\varepsilon_0}. \tag{6.4}$$

A remarkable fact about this equation is that the flux is independent of the size of the spherical surface. This can be directly attributed to the fact that the electric field of a point charge decreases as $1/r^2$ with distance, which just cancels the r^2 rate of increase of the surface area.

Electric Field Lines Picture

An alternative way to see why the flux through a closed spherical surface is independent of the radius of the surface is to look at the electric field lines. Note that every field line from q that pierces the surface at radius R_1 also pierces the surface at R_2 (Figure 6.14).

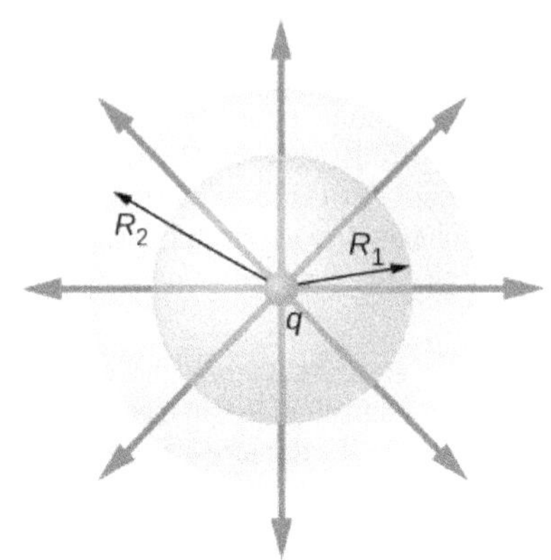

Figure 6.14 Flux through spherical surfaces of radii R_1 and R_2 enclosing a charge q are equal, independent of the size of the surface, since all E-field lines that pierce one surface from the inside to outside direction also pierce the other surface in the same direction.

Therefore, the net number of electric field lines passing through the two surfaces from the inside to outside direction is equal. This net number of electric field lines, which is obtained by subtracting the number of lines in the direction from outside to inside from the number of lines in the direction from inside to outside gives a visual measure of the electric flux through the surfaces.

You can see that if no charges are included within a closed surface, then the electric flux through it must be zero. A typical field line enters the surface at dA_1 and leaves at dA_2. Every line that enters the surface must also leave that surface. Hence the net "flow" of the field lines into or out of the surface is zero (Figure 6.15(a)). The same thing happens if charges of equal and opposite sign are included inside the closed surface, so that the total charge included is zero (part (b)). A surface that includes the same amount of charge has the same number of field lines crossing it, regardless of the shape or size of the surface, as long as the surface encloses the same amount of charge (part (c)).

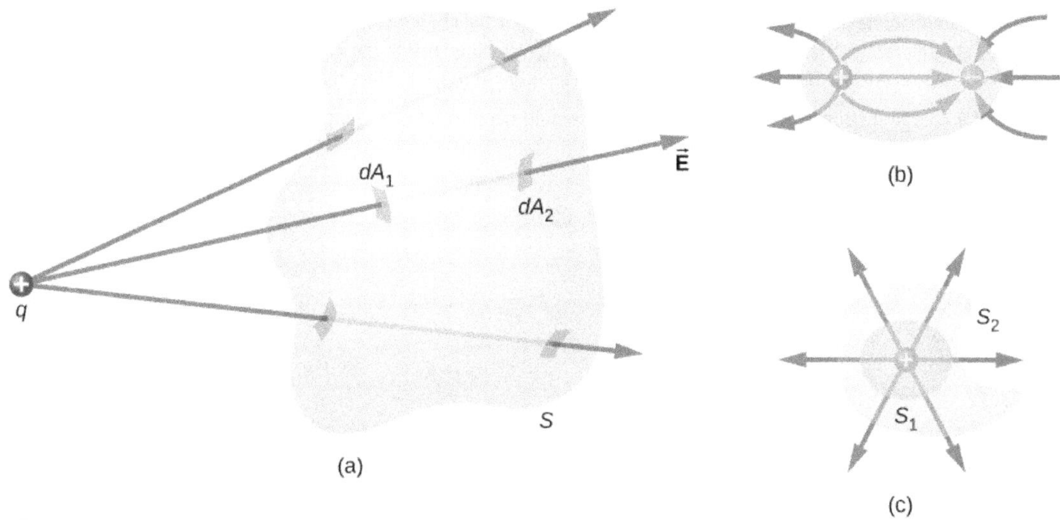

Figure 6.15 Understanding the flux in terms of field lines. (a) The electric flux through a closed surface due to a charge outside that surface is zero. (b) Charges are enclosed, but because the net charge included is zero, the net flux through the closed surface is also zero. (c) The shape and size of the surfaces that enclose a charge does not matter because all surfaces enclosing the same charge have the same flux.

Statement of Gauss's Law

Gauss's law generalizes this result to the case of any number of charges and any location of the charges in the space inside the closed surface. According to Gauss's law, the flux of the electric field $\vec{\mathbf{E}}$ through any closed surface, also called a **Gaussian surface**, is equal to the net charge enclosed (q_{enc}) divided by the permittivity of free space (ε_0):

$$\Phi_{\text{Closed Surface}} = \frac{q_{\text{enc}}}{\varepsilon_0}.$$

This equation holds for *charges of either sign*, because we define the area vector of a closed surface to point outward. If the enclosed charge is negative (see Figure 6.16(b)), then the flux through either S or S' is negative.

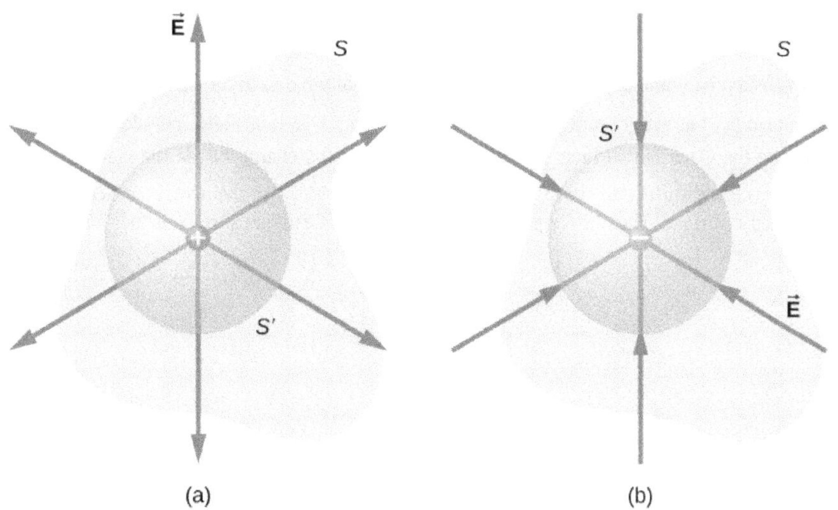

Figure 6.16 The electric flux through any closed surface surrounding a point charge q is given by Gauss's law. (a) Enclosed charge is positive. (b) Enclosed charge is negative.

The Gaussian surface does not need to correspond to a real, physical object; indeed, it rarely will. It is a mathematical construct that may be of any shape, provided that it is closed. However, since our goal is to integrate the flux over it, we tend to choose shapes that are highly symmetrical.

If the charges are discrete point charges, then we just add them. If the charge is described by a continuous distribution, then we need to integrate appropriately to find the total charge that resides inside the enclosed volume. For example, the flux through the Gaussian surface S of Figure 6.17 is $\Phi = (q_1 + q_2 + q_5)/\varepsilon_0$. Note that q_{enc} is simply the sum of the point charges. If the charge distribution were continuous, we would need to integrate appropriately to compute the total charge within the Gaussian surface.

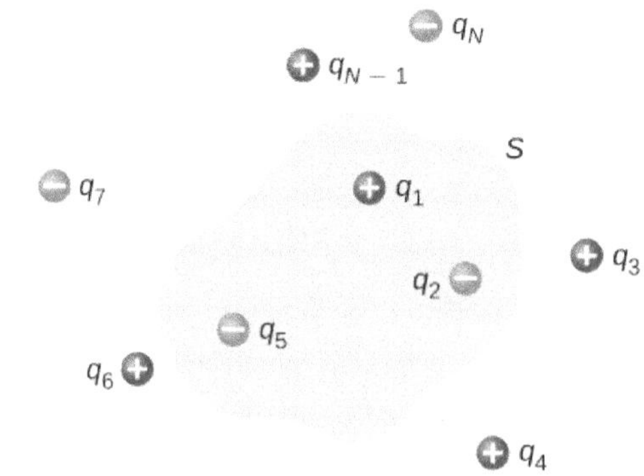

Figure 6.17 The flux through the Gaussian surface shown, due to the charge distribution, is $\Phi = (q_1 + q_2 + q_5)/\varepsilon_0$.

Recall that the principle of superposition holds for the electric field. Therefore, the total electric field at any point, including those on the chosen Gaussian surface, is the sum of all the electric fields present at this point. This allows us to write Gauss's law in terms of the total electric field.

Gauss's Law

The flux Φ of the electric field $\vec{\mathbf{E}}$ through any closed surface S (a Gaussian surface) is equal to the net charge enclosed (q_{enc}) divided by the permittivity of free space (ε_0) :

$$\Phi = \oint_S \vec{\mathbf{E}} \cdot \hat{\mathbf{n}}\, dA = \frac{q_{\text{enc}}}{\varepsilon_0}. \tag{6.5}$$

To use Gauss's law effectively, you must have a clear understanding of what each term in the equation represents. The field $\vec{\mathbf{E}}$ is the *total electric field* at every point on the Gaussian surface. This total field includes contributions from charges both inside and outside the Gaussian surface. However, q_{enc} is just the charge *inside* the Gaussian surface. Finally, the Gaussian surface is any closed surface in space. That surface can coincide with the actual surface of a conductor, or it can be an imaginary geometric surface. The only requirement imposed on a Gaussian surface is that it be closed (Figure 6.18).

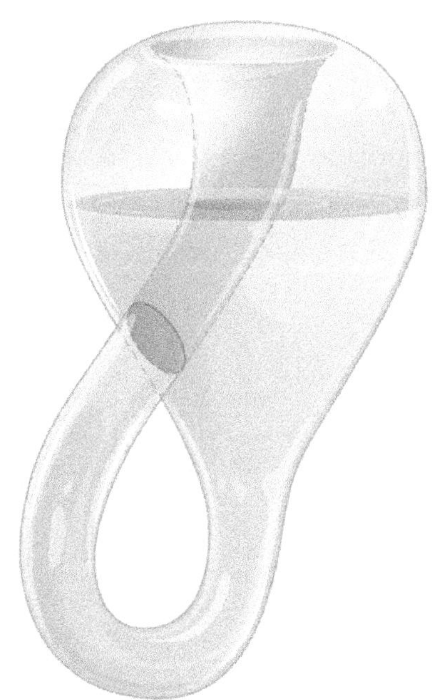

Figure 6.18 A Klein bottle partially filled with a liquid. Could the Klein bottle be used as a Gaussian surface?

Example 6.5

Electric Flux through Gaussian Surfaces

Calculate the electric flux through each Gaussian surface shown in Figure 6.19.

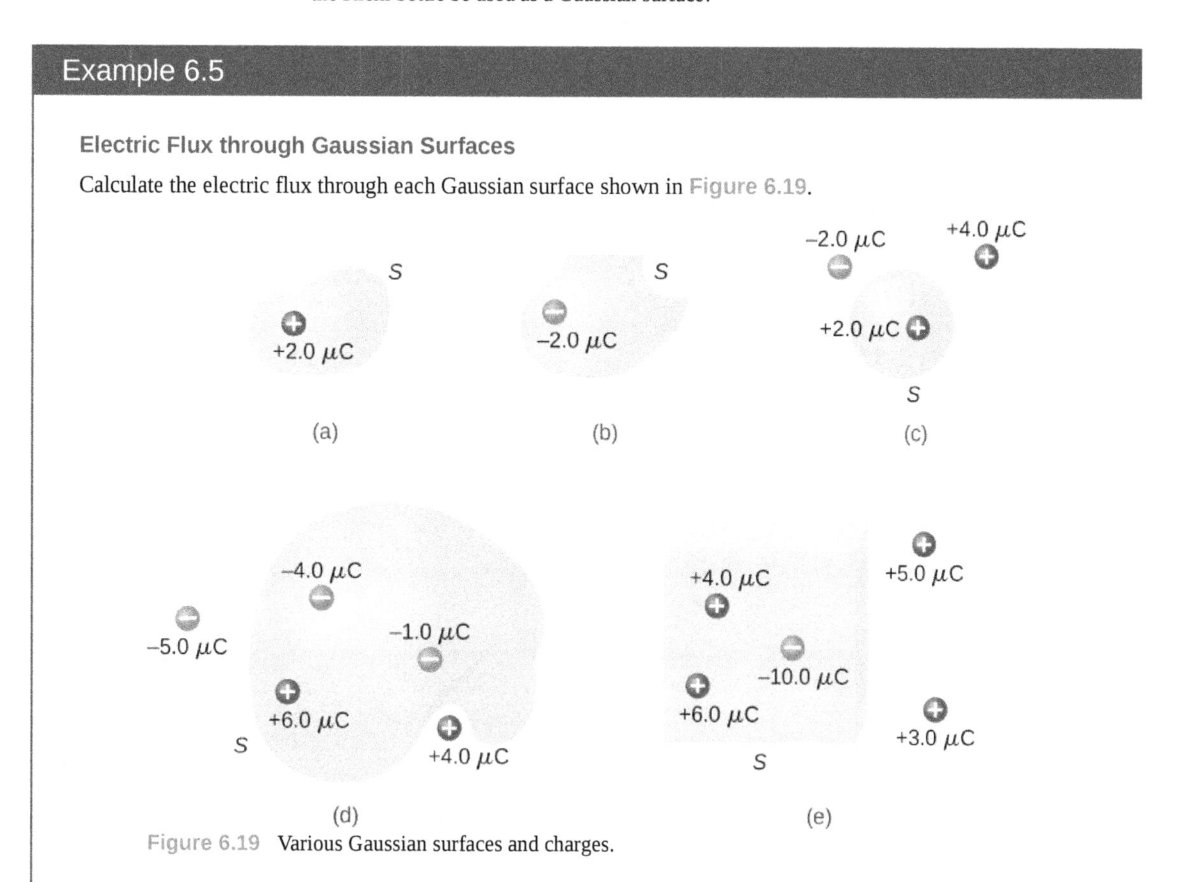

Figure 6.19 Various Gaussian surfaces and charges.

Strategy

From Gauss's law, the flux through each surface is given by q_{enc}/ε_0, where q_{enc} is the charge enclosed by that surface.

Solution

For the surfaces and charges shown, we find

a. $\Phi = \frac{2.0\,\mu C}{\varepsilon_0} = 2.3 \times 10^5\ N \cdot m^2/C.$

b. $\Phi = \frac{-2.0\,\mu C}{\varepsilon_0} = -2.3 \times 10^5\ N \cdot m^2/C.$

c. $\Phi = \frac{2.0\,\mu C}{\varepsilon_0} = 2.3 \times 10^5\ N \cdot m^2/C.$

d. $\Phi = \frac{-4.0\,\mu C + 6.0\,\mu C - 1.0\,\mu C}{\varepsilon_0} = 1.1 \times 10^5\ N \cdot m^2/C.$

e. $\Phi = \frac{4.0\,\mu C + 6.0\,\mu C - 10.0\,\mu C}{\varepsilon_0} = 0.$

Significance

In the special case of a closed surface, the flux calculations become a sum of charges. In the next section, this will allow us to work with more complex systems.

6.3 Check Your Understanding Calculate the electric flux through the closed cubical surface for each charge distribution shown in Figure 6.20.

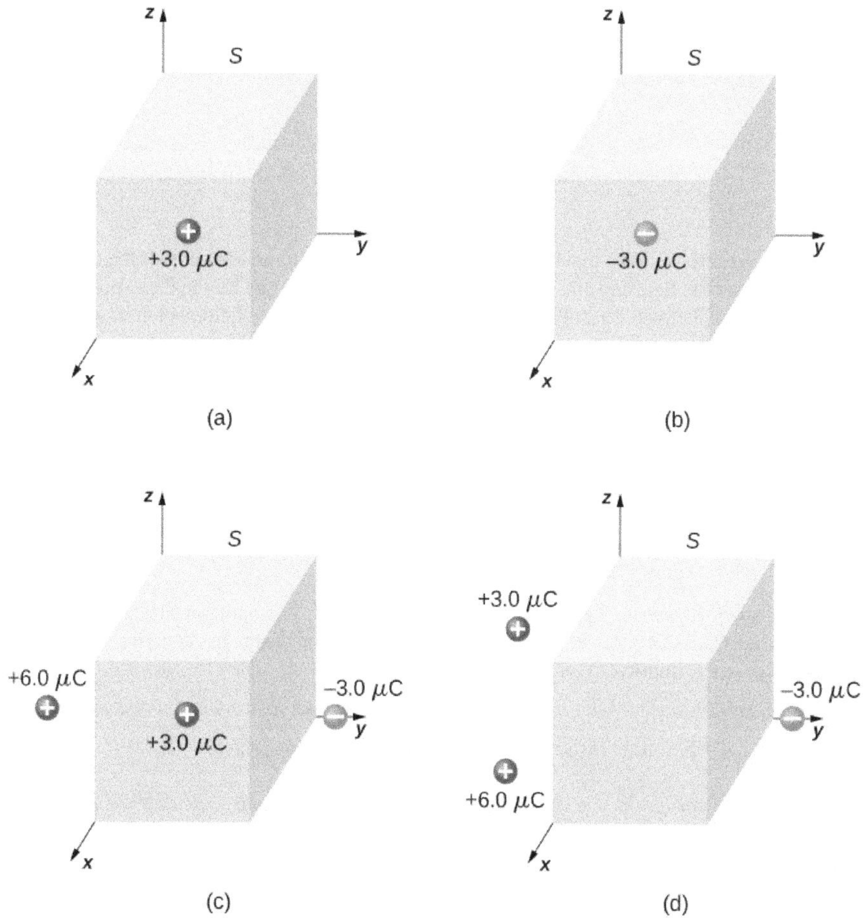

Figure 6.20 A cubical Gaussian surface with various charge distributions.

Use this simulation (https://openstaxcollege.org/l/21gaussimulat) to adjust the magnitude of the charge and the radius of the Gaussian surface around it. See how this affects the total flux and the magnitude of the electric field at the Gaussian surface.

6.3 | Applying Gauss's Law

Learning Objectives

By the end of this section, you will be able to:

- Explain what spherical, cylindrical, and planar symmetry are
- Recognize whether or not a given system possesses one of these symmetries
- Apply Gauss's law to determine the electric field of a system with one of these symmetries

Gauss's law is very helpful in determining expressions for the electric field, even though the law is not directly about the electric field; it is about the electric flux. It turns out that in situations that have certain symmetries (spherical, cylindrical, or planar) in the charge distribution, we can deduce the electric field based on knowledge of the electric flux. In these systems, we can find a Gaussian surface S over which the electric field has constant magnitude. Furthermore, if $\vec{\mathbf{E}}$ is parallel to $\hat{\mathbf{n}}$ everywhere on the surface, then $\vec{\mathbf{E}} \cdot \hat{\mathbf{n}} = E$. (If $\vec{\mathbf{E}}$ and $\hat{\mathbf{n}}$ are antiparallel everywhere on the surface, then $\vec{\mathbf{E}} \cdot \hat{\mathbf{n}} = -E$.) Gauss's law then simplifies to

$$\Phi = \oint_S \vec{\mathbf{E}} \cdot \hat{\mathbf{n}}\, dA = E \oint_S dA = EA = \frac{q_{\text{enc}}}{\varepsilon_0}, \tag{6.6}$$

where A is the area of the surface. Note that these symmetries lead to the transformation of the flux integral into a product of the magnitude of the electric field and an appropriate area. When you use this flux in the expression for Gauss's law, you obtain an algebraic equation that you can solve for the magnitude of the electric field, which looks like

$$E \sim \frac{q_{\text{enc}}}{\varepsilon_0 \text{ area}}.$$

The direction of the electric field at the field point P is obtained from the symmetry of the charge distribution and the type of charge in the distribution. Therefore, Gauss's law can be used to determine $\vec{\mathbf{E}}$. Here is a summary of the steps we will follow:

Problem-Solving Strategy: Gauss's Law

1. *Identify the spatial symmetry of the charge distribution.* This is an important first step that allows us to choose the appropriate Gaussian surface. As examples, an isolated point charge has spherical symmetry, and an infinite line of charge has cylindrical symmetry.
2. *Choose a Gaussian surface with the same symmetry as the charge distribution and identify its consequences.* With this choice, $\vec{\mathbf{E}} \cdot \hat{\mathbf{n}}$ is easily determined over the Gaussian surface.
3. *Evaluate the integral* $\oint_S \vec{\mathbf{E}} \cdot \hat{\mathbf{n}}\, dA$ *over the Gaussian surface, that is, calculate the flux through the surface.* The symmetry of the Gaussian surface allows us to factor $\vec{\mathbf{E}} \cdot \hat{\mathbf{n}}$ outside the integral.
4. *Determine the amount of charge enclosed by the Gaussian surface.* This is an evaluation of the right-hand side of the equation representing Gauss's law. It is often necessary to perform an integration to obtain the net enclosed charge.
5. *Evaluate the electric field of the charge distribution.* The field may now be found using the results of steps 3 and 4.

Basically, there are only three types of symmetry that allow Gauss's law to be used to deduce the electric field. They are

- A charge distribution with spherical symmetry
- A charge distribution with cylindrical symmetry
- A charge distribution with planar symmetry

To exploit the symmetry, we perform the calculations in appropriate coordinate systems and use the right kind of Gaussian surface for that symmetry, applying the remaining four steps.

Charge Distribution with Spherical Symmetry

A charge distribution has **spherical symmetry** if the density of charge depends only on the distance from a point in space and not on the direction. In other words, if you rotate the system, it doesn't look different. For instance, if a sphere of radius R is uniformly charged with charge density ρ_0 then the distribution has spherical symmetry (Figure 6.21(a)). On the other hand, if a sphere of radius R is charged so that the top half of the sphere has uniform charge density ρ_1 and the bottom half has a uniform charge density $\rho_2 \neq \rho_1$, then the sphere does not have spherical symmetry because the charge density depends on the direction (Figure 6.21(b)). Thus, it is not the shape of the object but rather the shape of the charge distribution that determines whether or not a system has spherical symmetry.

Figure 6.21(c) shows a sphere with four different shells, each with its own uniform charge density. Although this is a situation where charge density in the full sphere is not uniform, the charge density function depends only on the distance from the center and not on the direction. Therefore, this charge distribution does have spherical symmetry.

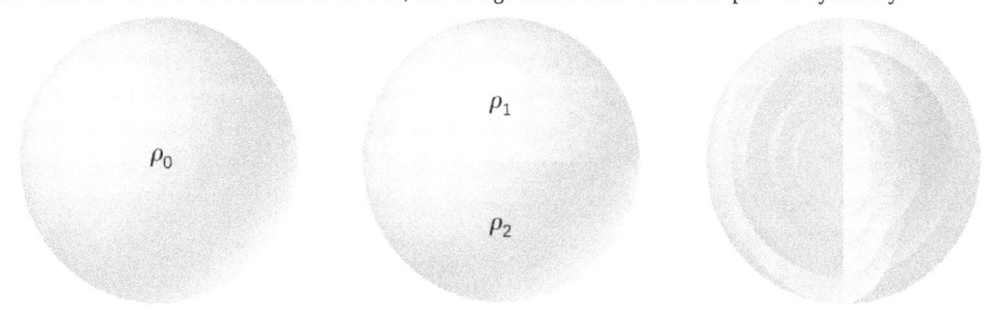

Figure 6.21 Illustrations of spherically symmetrical and nonsymmetrical systems. Different shadings indicate different charge densities. Charges on spherically shaped objects do not necessarily mean the charges are distributed with spherical symmetry. The spherical symmetry occurs only when the charge density does not depend on the direction. In (a), charges are distributed uniformly in a sphere. In (b), the upper half of the sphere has a different charge density from the lower half; therefore, (b) does not have spherical symmetry. In (c), the charges are in spherical shells of different charge densities, which means that charge density is only a function of the radial distance from the center; therefore, the system has spherical symmetry.

One good way to determine whether or not your problem has spherical symmetry is to look at the charge density function in spherical coordinates, $\rho(r,\ \theta,\ \phi)$. If the charge density is only a function of *r*, that is $\rho = \rho(r)$, then you have spherical symmetry. If the density depends on θ or ϕ, you could change it by rotation; hence, you would not have spherical symmetry.

Consequences of symmetry

In all spherically symmetrical cases, the electric field at any point must be radially directed, because the charge and, hence, the field must be invariant under rotation. Therefore, using spherical coordinates with their origins at the center of the spherical charge distribution, we can write down the expected form of the electric field at a point *P* located at a distance *r* from the center:

$$\text{Spherical symmetry: } \vec{\mathbf{E}}_P = E_P(r)\,\hat{\mathbf{r}}, \tag{6.7}$$

where $\hat{\mathbf{r}}$ is the unit vector pointed in the direction from the origin to the field point *P*. The radial component E_P of the electric field can be positive or negative. When $E_P > 0,$ the electric field at *P* points away from the origin, and when $E_P < 0,$ the electric field at *P* points toward the origin.

Gaussian surface and flux calculations

We can now use this form of the electric field to obtain the flux of the electric field through the Gaussian surface. For spherical symmetry, the Gaussian surface is a closed spherical surface that has the same center as the center of the charge distribution. Thus, the direction of the area vector of an area element on the Gaussian surface at any point is parallel to the direction of the electric field at that point, since they are both radially directed outward (Figure 6.22).

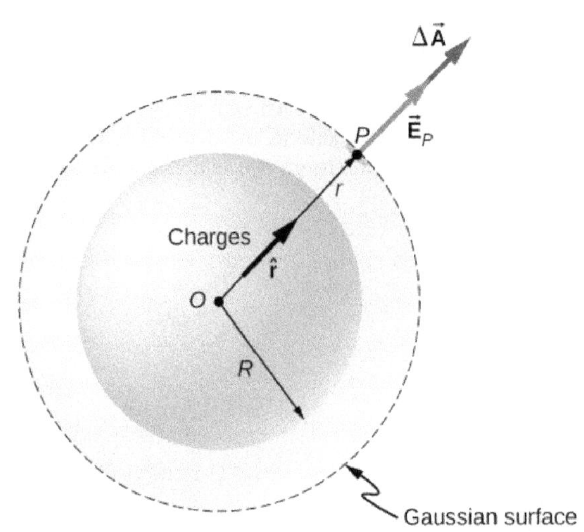

Figure 6.22 The electric field at any point of the spherical Gaussian surface for a spherically symmetrical charge distribution is parallel to the area element vector at that point, giving flux as the product of the magnitude of electric field and the value of the area. Note that the radius R of the charge distribution and the radius r of the Gaussian surface are different quantities.

The magnitude of the electric field $\vec{\mathbf{E}}$ must be the same everywhere on a spherical Gaussian surface concentric with the distribution. For a spherical surface of radius r,

$$\Phi = \oint_S \vec{\mathbf{E}}_P \cdot \hat{\mathbf{n}}\, dA = E_P \oint_S dA = E_P\, 4\pi r^2.$$

Using Gauss's law

According to Gauss's law, the flux through a closed surface is equal to the total charge enclosed within the closed surface divided by the permittivity of vacuum ε_0. Let q_{enc} be the total charge enclosed inside the distance r from the origin, which is the space inside the Gaussian spherical surface of radius r. This gives the following relation for Gauss's law:

$$4\pi r^2 E = \frac{q_{\text{enc}}}{\varepsilon_0}.$$

Hence, the electric field at point P that is a distance r from the center of a spherically symmetrical charge distribution has the following magnitude and direction:

$$\text{Magnitude: } E(r) = \frac{1}{4\pi\varepsilon_0}\frac{q_{\text{enc}}}{r^2} \tag{6.8}$$

Direction: radial from O to P or from P to O.

The direction of the field at point P depends on whether the charge in the sphere is positive or negative. For a net positive charge enclosed within the Gaussian surface, the direction is from O to P, and for a net negative charge, the direction is from P to O. This is all we need for a point charge, and you will notice that the result above is identical to that for a point charge. However, Gauss's law becomes truly useful in cases where the charge occupies a finite volume.

Computing enclosed charge

The more interesting case is when a spherical charge distribution occupies a volume, and asking what the electric field inside the charge distribution is thus becomes relevant. In this case, the charge enclosed depends on the distance r of the field point relative to the radius of the charge distribution R, such as that shown in Figure 6.23.

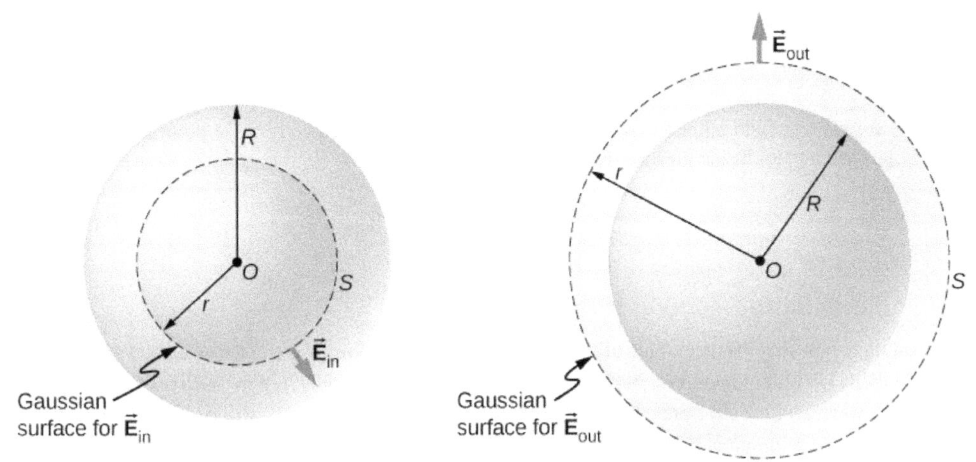

Figure 6.23 A spherically symmetrical charge distribution and the Gaussian surface used for finding the field (a) inside and (b) outside the distribution.

If point *P* is located outside the charge distribution—that is, if $r \geq R$—then the Gaussian surface containing *P* encloses all charges in the sphere. In this case, q_{enc} equals the total charge in the sphere. On the other hand, if point *P* is within the spherical charge distribution, that is, if $r < R$, then the Gaussian surface encloses a smaller sphere than the sphere of charge distribution. In this case, q_{enc} is less than the total charge present in the sphere. Referring to Figure 6.23, we can write q_{enc} as

$$q_{enc} = \begin{cases} q_{tot}(\text{total charge}) \text{ if } r \geq R \\ q_{\text{within } r < R}(\text{only charge within } r < R) \text{ if } r < R \end{cases}.$$

The field at a point outside the charge distribution is also called $\vec{\mathbf{E}}_{out}$, and the field at a point inside the charge distribution is called $\vec{\mathbf{E}}_{in}$. Focusing on the two types of field points, either inside or outside the charge distribution, we can now write the magnitude of the electric field as

$$P \text{ outside sphere } E_{out} = \frac{1}{4\pi\varepsilon_0} \frac{q_{tot}}{r^2} \tag{6.9}$$

$$P \text{ inside sphere } E_{in} = \frac{1}{4\pi\varepsilon_0} \frac{q_{\text{within } r < R}}{r^2}. \tag{6.10}$$

Note that the electric field outside a spherically symmetrical charge distribution is identical to that of a point charge at the center that has a charge equal to the total charge of the spherical charge distribution. This is remarkable since the charges are not located at the center only. We now work out specific examples of spherical charge distributions, starting with the case of a uniformly charged sphere.

Example 6.6

Uniformly Charged Sphere

A sphere of radius *R*, such as that shown in Figure 6.23, has a uniform volume charge density ρ_0. Find the electric field at a point outside the sphere and at a point inside the sphere.

Strategy

Apply the Gauss's law problem-solving strategy, where we have already worked out the flux calculation.

Solution

The charge enclosed by the Gaussian surface is given by

$$q_{\text{enc}} = \int \rho_0 dV = \int_0^r \rho_0 4\pi r'^2 dr' = \rho_0 \left(\frac{4}{3}\pi r^3\right).$$

The answer for electric field amplitude can then be written down immediately for a point outside the sphere, labeled E_{out}, and a point inside the sphere, labeled E_{in}.

$$\begin{aligned} E_{\text{out}} &= \frac{1}{4\pi\varepsilon_0}\frac{q_{\text{tot}}}{r^2},\ q_{\text{tot}} = \frac{4}{3}\pi R^3\ \rho_0, \\ E_{\text{in}} &= \frac{q_{\text{enc}}}{4\pi\varepsilon_0 r^2} = \frac{\rho_0 r}{3\varepsilon_0},\ \text{since}\ q_{\text{enc}} = \frac{4}{3}\pi r^3\ \rho_0. \end{aligned}$$

It is interesting to note that the magnitude of the electric field increases inside the material as you go out, since the amount of charge enclosed by the Gaussian surface increases with the volume. Specifically, the charge enclosed grows $\propto r^3$, whereas the field from each infinitesimal element of charge drops off $\propto 1/r^2$ with the net result that the electric field within the distribution increases in strength linearly with the radius. The magnitude of the electric field outside the sphere decreases as you go away from the charges, because the included charge remains the same but the distance increases. Figure 6.24 displays the variation of the magnitude of the electric field with distance from the center of a uniformly charged sphere.

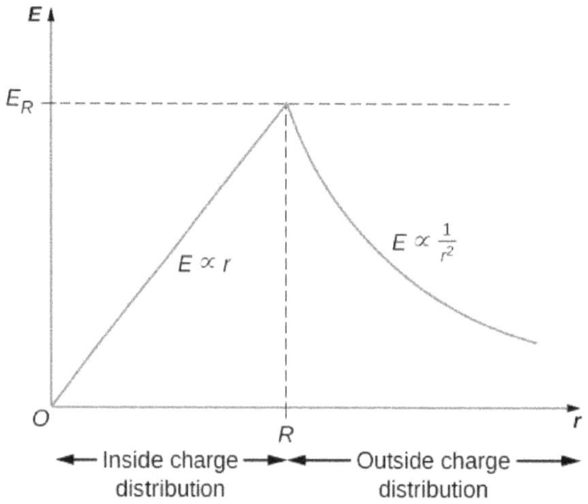

Figure 6.24 Electric field of a uniformly charged, non-conducting sphere increases inside the sphere to a maximum at the surface and then decreases as $1/r^2$. Here, $E_R = \frac{\rho_0 R}{3\varepsilon_0}$. The electric field is due to a spherical charge distribution of uniform charge density and total charge Q as a function of distance from the center of the distribution.

The direction of the electric field at any point *P* is radially outward from the origin if ρ_0 is positive, and inward (i.e., toward the center) if ρ_0 is negative. The electric field at some representative space points are displayed in Figure 6.25 whose radial coordinates *r* are $r = R/2$, $r = R$, and $r = 2R$.

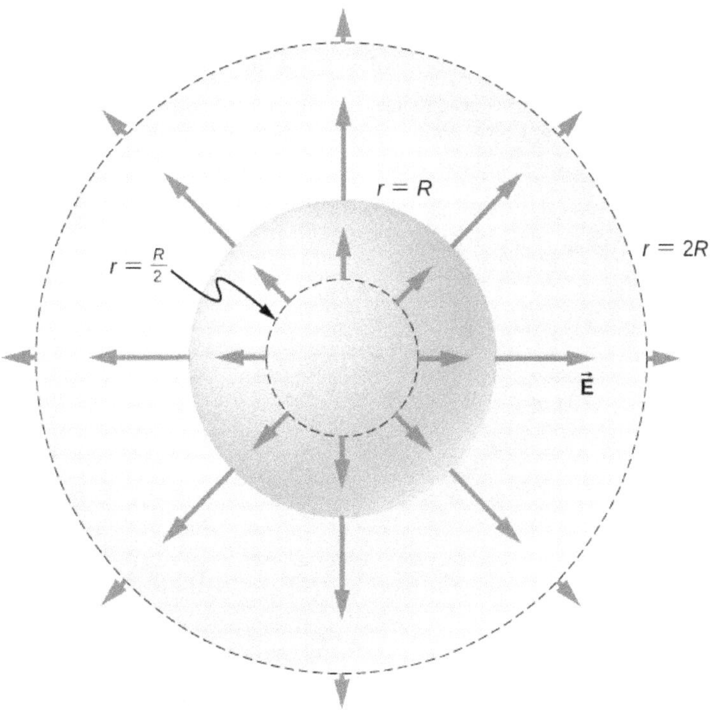

Figure 6.25 Electric field vectors inside and outside a uniformly charged sphere.

Significance

Notice that E_{out} has the same form as the equation of the electric field of an isolated point charge. In determining the electric field of a uniform spherical charge distribution, we can therefore assume that all of the charge inside the appropriate spherical Gaussian surface is located at the center of the distribution.

Example 6.7

Non-Uniformly Charged Sphere

A non-conducting sphere of radius R has a non-uniform charge density that varies with the distance from its center as given by

$$\rho(r) = ar^n \ (r \leq R; n \geq 0),$$

where a is a constant. We require $n \geq 0$ so that the charge density is not undefined at $r = 0$. Find the electric field at a point outside the sphere and at a point inside the sphere.

Strategy

Apply the Gauss's law strategy given above, where we work out the enclosed charge integrals separately for cases inside and outside the sphere.

Solution

Since the given charge density function has only a radial dependence and no dependence on direction, we have a spherically symmetrical situation. Therefore, the magnitude of the electric field at any point is given above and the direction is radial. We just need to find the enclosed charge q_{enc}, which depends on the location of the field point.

A note about symbols: We use r' for locating charges in the charge distribution and r for locating the field point(s) at the Gaussian surface(s). The letter R is used for the radius of the charge distribution.

As charge density is not constant here, we need to integrate the charge density function over the volume enclosed by the Gaussian surface. Therefore, we set up the problem for charges in one spherical shell, say between r' and $r' + dr'$, as shown in Figure 6.26. The volume of charges in the shell of infinitesimal width is equal to the product of the area of surface $4\pi r'^2$ and the thickness dr'. Multiplying the volume with the density at this location, which is ar'^n, gives the charge in the shell:

$$dq = ar'^n \, 4\pi r'^2 dr'.$$

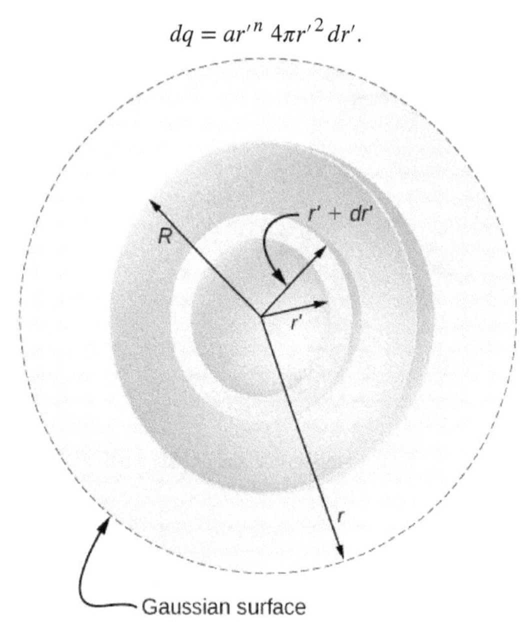

Figure 6.26 Spherical symmetry with non-uniform charge distribution. In this type of problem, we need four radii: R is the radius of the charge distribution, r is the radius of the Gaussian surface, r' is the inner radius of the spherical shell, and $r' + dr'$ is the outer radius of the spherical shell. The spherical shell is used to calculate the charge enclosed within the Gaussian surface. The range for r' is from 0 to r for the field at a point inside the charge distribution and from 0 to R for the field at a point outside the charge distribution. If $r > R$, then the Gaussian surface encloses more volume than the charge distribution, but the additional volume does not contribute to q_{enc}.

(a) **Field at a point outside the charge distribution.** In this case, the Gaussian surface, which contains the field point P, has a radius r that is greater than the radius R of the charge distribution, $r > R$. Therefore, all charges of the charge distribution are enclosed within the Gaussian surface. Note that the space between $r' = R$ and $r' = r$ is empty of charges and therefore does not contribute to the integral over the volume enclosed by the Gaussian surface:

$$q_{\text{enc}} = \int dq = \int_0^R ar'^n \, 4\pi r'^2 dr' = \frac{4\pi a}{n+3} R^{n+3}.$$

This is used in the general result for $\vec{\mathbf{E}}_{\text{out}}$ above to obtain the electric field at a point outside the charge distribution as

$$\vec{\mathbf{E}}_{\text{out}} = \left[\frac{aR^{n+3}}{\varepsilon_0(n+3)}\right]\frac{1}{r^2}\hat{\mathbf{r}},$$

where $\hat{\mathbf{r}}$ is a unit vector in the direction from the origin to the field point at the Gaussian surface.

(b) **Field at a point inside the charge distribution.** The Gaussian surface is now buried inside the charge distribution, with $r < R$. Therefore, only those charges in the distribution that are within a distance *r* of the center of the spherical charge distribution count in r_{enc}:

$$q_{\text{enc}} = \int_0^r ar'^n 4\pi r'^2\, dr' = \frac{4\pi a}{n+3}r^{n+3}.$$

Now, using the general result above for $\vec{\mathbf{E}}_{\text{in}},$ we find the electric field at a point that is a distance *r* from the center and lies within the charge distribution as

$$\vec{\mathbf{E}}_{\text{in}} = \left[\frac{a}{\varepsilon_0(n+3)}\right]r^{n+1}\hat{\mathbf{r}},$$

where the direction information is included by using the unit radial vector.

6.4 Check Your Understanding Check that the electric fields for the sphere reduce to the correct values for a point charge.

Charge Distribution with Cylindrical Symmetry

A charge distribution has **cylindrical symmetry** if the charge density depends only upon the distance *r* from the axis of a cylinder and must not vary along the axis or with direction about the axis. In other words, if your system varies if you rotate it around the axis, or shift it along the axis, you do not have cylindrical symmetry.

Figure 6.27 shows four situations in which charges are distributed in a cylinder. A uniform charge density ρ_0. in an infinite straight wire has a cylindrical symmetry, and so does an infinitely long cylinder with constant charge density ρ_0. An infinitely long cylinder that has different charge densities along its length, such as a charge density ρ_1 for $z > 0$ and $\rho_2 \neq \rho_1$ for $z < 0$, does not have a usable cylindrical symmetry for this course. Neither does a cylinder in which charge density varies with the direction, such as a charge density ρ_1 for $0 \leq \theta < \pi$ and $\rho_2 \neq \rho_1$ for $\pi \leq \theta < 2\pi$. A system with concentric cylindrical shells, each with uniform charge densities, albeit different in different shells, as in Figure 6.27(d), does have cylindrical symmetry if they are infinitely long. The infinite length requirement is due to the charge density changing along the axis of a finite cylinder. In real systems, we don't have infinite cylinders; however, if the cylindrical object is considerably longer than the radius from it that we are interested in, then the approximation of an infinite cylinder becomes useful.

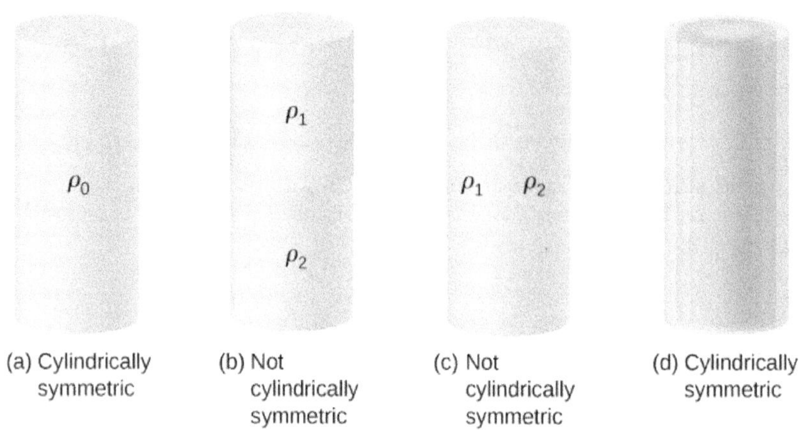

Figure 6.27 To determine whether a given charge distribution has cylindrical symmetry, look at the cross-section of an "infinitely long" cylinder. If the charge density does not depend on the polar angle of the cross-section or along the axis, then you have cylindrical symmetry. (a) Charge density is constant in the cylinder; (b) upper half of the cylinder has a different charge density from the lower half; (c) left half of the cylinder has a different charge density from the right half; (d) charges are constant in different cylindrical rings, but the density does not depend on the polar angle. Cases (a) and (d) have cylindrical symmetry, whereas (b) and (c) do not.

Consequences of symmetry

In all cylindrically symmetrical cases, the electric field $\vec{\mathbf{E}}_P$ at any point P must also display cylindrical symmetry.

$$\text{Cylindrical symmetry:}\quad \vec{\mathbf{E}}_P = E_P(r)\,\hat{\mathbf{r}},$$

where r is the distance from the axis and $\hat{\mathbf{r}}$ is a unit vector directed perpendicularly away from the axis (Figure 6.28).

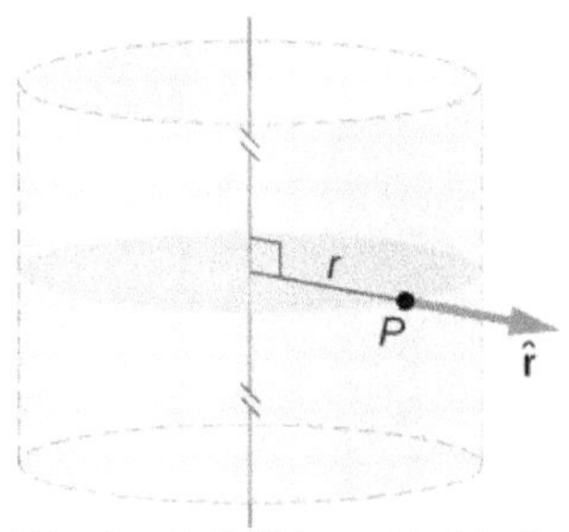

Figure 6.28 The electric field in a cylindrically symmetrical situation depends only on the distance from the axis. The direction of the electric field is pointed away from the axis for positive charges and toward the axis for negative charges.

Gaussian surface and flux calculation

To make use of the direction and functional dependence of the electric field, we choose a closed Gaussian surface in the shape of a cylinder with the same axis as the axis of the charge distribution. The flux through this surface of radius s and height L is easy to compute if we divide our task into two parts: (a) a flux through the flat ends and (b) a flux through the curved surface (Figure 6.29).

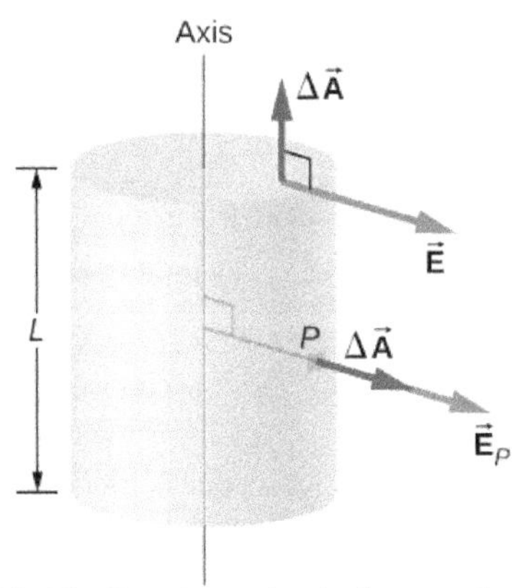

Figure 6.29 The Gaussian surface in the case of cylindrical symmetry. The electric field at a patch is either parallel or perpendicular to the normal to the patch of the Gaussian surface.

The electric field is perpendicular to the cylindrical side and parallel to the planar end caps of the surface. The flux through the cylindrical part is

$$\int_S \vec{\mathbf{E}} \cdot \hat{\mathbf{n}}\, dA = E\int_S dA = E(2\pi rL),$$

whereas the flux through the end caps is zero because $\vec{\mathbf{E}} \cdot \hat{\mathbf{n}} = 0$ there. Thus, the flux is

$$\int_S \vec{\mathbf{E}} \cdot \hat{\mathbf{n}}\, dA = E(2\pi rL) + 0 + 0 = 2\pi rLE.$$

Using Gauss's law

According to Gauss's law, the flux must equal the amount of charge within the volume enclosed by this surface, divided by the permittivity of free space. When you do the calculation for a cylinder of length *L*, you find that q_{enc} of Gauss's law is directly proportional to *L*. Let us write it as charge per unit length (λ_{enc}) times length *L*:

$$q_{\text{enc}} = \lambda_{\text{enc}} L.$$

Hence, Gauss's law for any cylindrically symmetrical charge distribution yields the following magnitude of the electric field a distance *s* away from the axis:

$$\text{Magnitude: } E(r) = \frac{\lambda_{\text{enc}}}{2\pi\varepsilon_0}\frac{1}{r}.$$

The charge per unit length λ_{enc} depends on whether the field point is inside or outside the cylinder of charge distribution, just as we have seen for the spherical distribution.

Computing enclosed charge

Let *R* be the radius of the cylinder within which charges are distributed in a cylindrically symmetrical way. Let the field point *P* be at a distance *s* from the axis. (The side of the Gaussian surface includes the field point *P*.) When $r > R$ (that is, when *P* is outside the charge distribution), the Gaussian surface includes all the charge in the cylinder of radius *R* and length *L*. When $r < R$ (*P* is located inside the charge distribution), then only the charge within a cylinder of radius *s* and length *L* is enclosed by the Gaussian surface:

$$\lambda_{\text{enc}} L = \begin{cases} (\text{total charge}) \text{ if } r \geq R \\ (\text{only charge within } r < R) \text{ if } r < R \end{cases}.$$

Example 6.8

Uniformly Charged Cylindrical Shell

A very long non-conducting cylindrical shell of radius R has a uniform surface charge density σ_0. Find the electric field (a) at a point outside the shell and (b) at a point inside the shell.

Strategy

Apply the Gauss's law strategy given earlier, where we treat the cases inside and outside the shell separately.

Solution

a. **Electric field at a point outside the shell.** For a point outside the cylindrical shell, the Gaussian surface is the surface of a cylinder of radius $r > R$ and length L, as shown in Figure 6.30. The charge enclosed by the Gaussian cylinder is equal to the charge on the cylindrical shell of length L. Therefore, λ_{enc} is given by

$$\lambda_{\text{enc}} = \frac{\sigma_0 2\pi RL}{L} = 2\pi R\sigma_0.$$

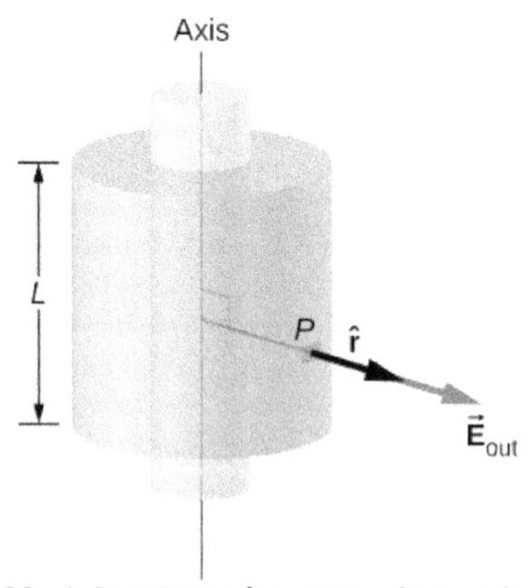

Figure 6.30 A Gaussian surface surrounding a cylindrical shell.

Hence, the electric field at a point P outside the shell at a distance r away from the axis is

$$\vec{\mathbf{E}} = \frac{2\pi R\sigma_0}{2\pi\varepsilon_o}\frac{1}{r}\hat{\mathbf{r}} = \frac{R\sigma_0}{\varepsilon_o}\frac{1}{r}\hat{\mathbf{r}}\ (r > R)$$

where $\hat{\mathbf{r}}$ is a unit vector, perpendicular to the axis and pointing away from it, as shown in the figure. The electric field at P points in the direction of $\hat{\mathbf{r}}$ given in Figure 6.30 if $\sigma_0 > 0$ and in the opposite direction to $\hat{\mathbf{r}}$ if $\sigma_0 < 0$.

b. **Electric field at a point inside the shell.** For a point inside the cylindrical shell, the Gaussian surface is a cylinder whose radius r is less than R (Figure 6.31). This means no charges are included inside the Gaussian surface:

$$\lambda_{\text{enc}} = 0.$$

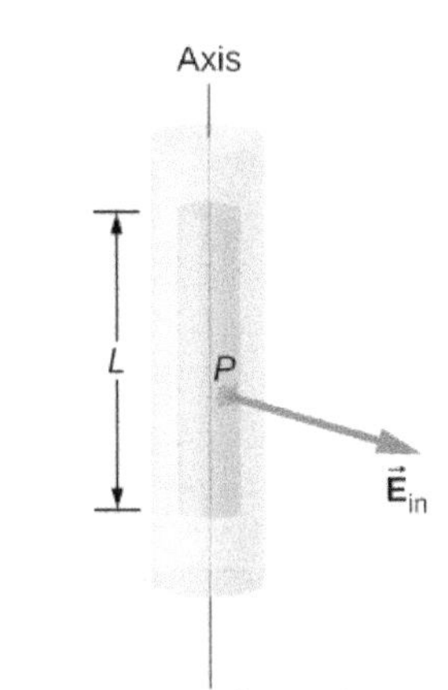

Figure 6.31 A Gaussian surface within a cylindrical shell.

This gives the following equation for the magnitude of the electric field E_{in} at a point whose r is less than R of the shell of charges.

$$E_{\text{in}}\, 2\pi rL = 0\,(r < R),$$

This gives us

$$E_{\text{in}} = 0\,(r < R).$$

Significance

Notice that the result inside the shell is exactly what we should expect: No enclosed charge means zero electric field. Outside the shell, the result becomes identical to a wire with uniform charge $R\sigma_0$.

6.5 Check Your Understanding A thin straight wire has a uniform linear charge density λ_0. Find the electric field at a distance d from the wire, where d is much less than the length of the wire.

Charge Distribution with Planar Symmetry

A **planar symmetry** of charge density is obtained when charges are uniformly spread over a large flat surface. In planar symmetry, all points in a plane parallel to the plane of charge are identical with respect to the charges.

Consequences of symmetry

We take the plane of the charge distribution to be the *xy*-plane and we find the electric field at a space point *P* with coordinates (*x*, *y*, *z*). Since the charge density is the same at all (*x*, *y*)-coordinates in the $z = 0$ plane, by symmetry, the electric field at *P* cannot depend on the *x*- or *y*-coordinates of point *P*, as shown in Figure 6.32. Therefore, the electric field at *P* can only depend on the distance from the plane and has a direction either toward the plane or away from the plane. That is, the electric field at *P* has only a nonzero *z*-component.

Uniform charges in *xy* plane: $\vec{\mathbf{E}} = E(z)\,\hat{\mathbf{z}}$

where z is the distance from the plane and $\hat{\mathbf{z}}$ is the unit vector normal to the plane. Note that in this system, $E(z) = E(-z)$, although of course they point in opposite directions.

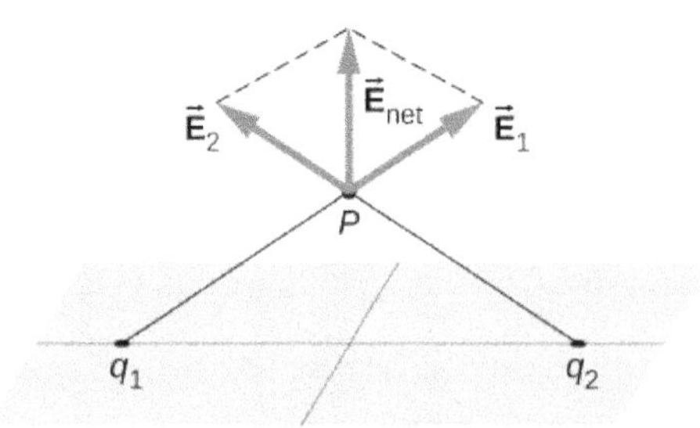

Figure 6.32 The components of the electric field parallel to a plane of charges cancel out the two charges located symmetrically from the field point P. Therefore, the field at any point is pointed vertically from the plane of charges. For any point P and charge q_1, we can always find a q_2 with this effect.

Gaussian surface and flux calculation

In the present case, a convenient Gaussian surface is a box, since the expected electric field points in one direction only. To keep the Gaussian box symmetrical about the plane of charges, we take it to straddle the plane of the charges, such that one face containing the field point P is taken parallel to the plane of the charges. In Figure 6.33, sides I and II of the Gaussian surface (the box) that are parallel to the infinite plane have been shaded. They are the only surfaces that give rise to nonzero flux because the electric field and the area vectors of the other faces are perpendicular to each other.

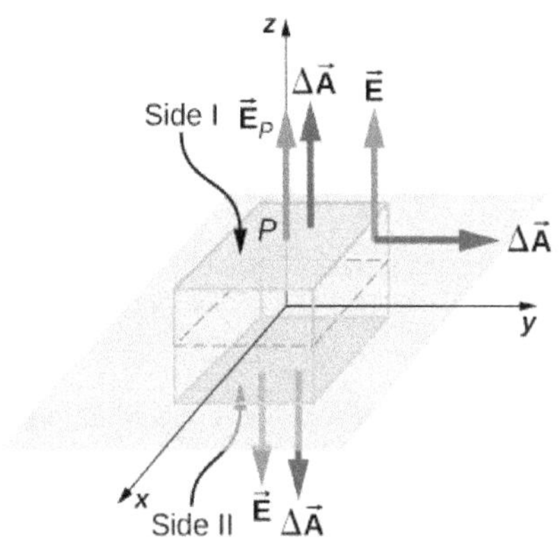

Figure 6.33 A thin charged sheet and the Gaussian box for finding the electric field at the field point P. The normal to each face of the box is from inside the box to outside. On two faces of the box, the electric fields are parallel to the area vectors, and on the other four faces, the electric fields are perpendicular to the area vectors.

Let A be the area of the shaded surface on each side of the plane and E_P be the magnitude of the electric field at point P. Since sides I and II are at the same distance from the plane, the electric field has the same magnitude at points in these planes, although the directions of the electric field at these points in the two planes are opposite to each other.

Magnitude at I or II: $E(z) = E_P$.

If the charge on the plane is positive, then the direction of the electric field and the area vectors are as shown in Figure 6.33. Therefore, we find for the flux of electric field through the box

$$\Phi = \oint_S \vec{\mathbf{E}}_P \cdot \hat{\mathbf{n}}\, dA = E_P A + E_P A + 0 + 0 + 0 + 0 = 2E_P A \tag{6.11}$$

where the zeros are for the flux through the other sides of the box. Note that if the charge on the plane is negative, the directions of electric field and area vectors for planes I and II are opposite to each other, and we get a negative sign for the flux. According to Gauss's law, the flux must equal $q_{\text{enc}}/\varepsilon_0$. From Figure 6.33, we see that the charges inside the volume enclosed by the Gaussian box reside on an area A of the xy-plane. Hence,

$$q_{\text{enc}} = \sigma_0 A. \tag{6.12}$$

Using the equations for the flux and enclosed charge in Gauss's law, we can immediately determine the electric field at a point at height z from a uniformly charged plane in the xy-plane:

$$\vec{\mathbf{E}}_P = \frac{\sigma_0}{2\varepsilon_0}\hat{\mathbf{n}}.$$

The direction of the field depends on the sign of the charge on the plane and the side of the plane where the field point P is located. Note that above the plane, $\hat{\mathbf{n}} = +\hat{\mathbf{z}}$, while below the plane, $\hat{\mathbf{n}} = -\hat{\mathbf{z}}$.

You may be surprised to note that the electric field does not actually depend on the distance from the plane; this is an effect of the assumption that the plane is infinite. In practical terms, the result given above is still a useful approximation for finite planes near the center.

6.4 | Conductors in Electrostatic Equilibrium

Learning Objectives

By the end of this section, you will be able to:

- Describe the electric field within a conductor at equilibrium
- Describe the electric field immediately outside the surface of a charged conductor at equilibrium
- Explain why if the field is not as described in the first two objectives, the conductor is not at equilibrium

So far, we have generally been working with charges occupying a volume within an insulator. We now study what happens when free charges are placed on a conductor. Generally, in the presence of a (generally external) electric field, the free charge in a conductor redistributes and very quickly reaches electrostatic equilibrium. The resulting charge distribution and its electric field have many interesting properties, which we can investigate with the help of Gauss's law and the concept of electric potential.

The Electric Field inside a Conductor Vanishes

If an electric field is present inside a conductor, it exerts forces on the **free electrons** (also called conduction electrons), which are electrons in the material that are not bound to an atom. These free electrons then accelerate. However, moving charges by definition means nonstatic conditions, contrary to our assumption. Therefore, when electrostatic equilibrium is reached, the charge is distributed in such a way that the electric field inside the conductor vanishes.

If you place a piece of a metal near a positive charge, the free electrons in the metal are attracted to the external positive charge and migrate freely toward that region. The region the electrons move to then has an excess of electrons over the protons in the atoms and the region from where the electrons have migrated has more protons than electrons. Consequently, the metal develops a negative region near the charge and a positive region at the far end (Figure 6.34). As we saw in the preceding chapter, this separation of equal magnitude and opposite type of electric charge is called polarization. If you remove the external charge, the electrons migrate back and neutralize the positive region.

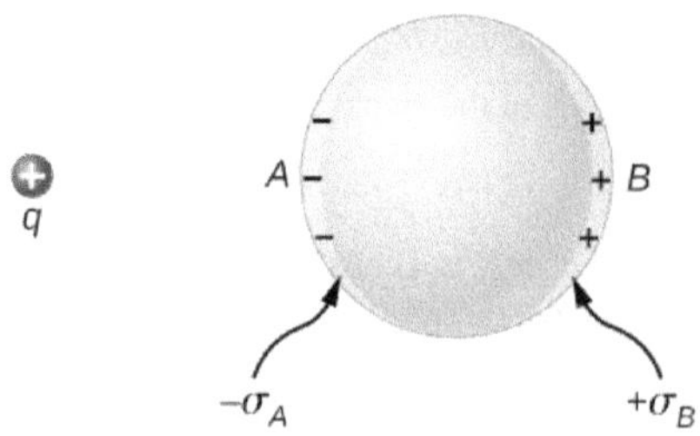

Figure 6.34 Polarization of a metallic sphere by an external point charge $+q$. The near side of the metal has an opposite surface charge compared to the far side of the metal. The sphere is said to be polarized. When you remove the external charge, the polarization of the metal also disappears.

The polarization of the metal happens only in the presence of external charges. You can think of this in terms of electric fields. The external charge creates an external electric field. When the metal is placed in the region of this electric field, the electrons and protons of the metal experience electric forces due to this external electric field, but only the conduction electrons are free to move in the metal over macroscopic distances. The movement of the conduction electrons leads to the polarization, which creates an induced electric field in addition to the external electric field (Figure 6.35). The net electric field is a vector sum of the fields of $+q$ and the surface charge densities $-\sigma_A$ and $+\sigma_B$. This means that the net field inside the conductor is different from the field outside the conductor.

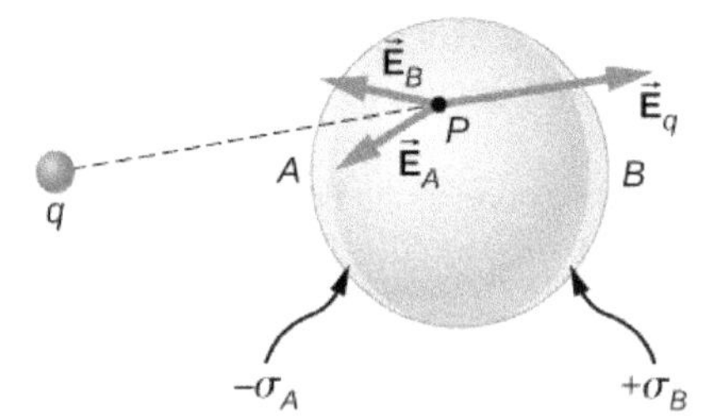

Figure 6.35 In the presence of an external charge q, the charges in a metal redistribute. The electric field at any point has three contributions, from $+q$ and the induced charges $-\sigma_A$ and $+\sigma_B$. Note that the surface charge distribution will not be uniform in this case.

The redistribution of charges is such that the sum of the three contributions at any point P inside the conductor is

$$\vec{\mathbf{E}}_P = \vec{\mathbf{E}}_q + \vec{\mathbf{E}}_B + \vec{\mathbf{E}}_A = \vec{\mathbf{0}}.$$

Now, thanks to Gauss's law, we know that there is no net charge enclosed by a Gaussian surface that is solely within the volume of the conductor at equilibrium. That is, $q_{\text{enc}} = 0$ and hence

$$\vec{\mathbf{E}}_{\text{net}} = \vec{\mathbf{0}} \text{ (at points inside a conductor).} \tag{6.13}$$

Charge on a Conductor

An interesting property of a conductor in static equilibrium is that extra charges on the conductor end up on the outer surface of the conductor, regardless of where they originate. Figure 6.36 illustrates a system in which we bring an external positive charge inside the cavity of a metal and then touch it to the inside surface. Initially, the inside surface of the cavity is negatively charged and the outside surface of the conductor is positively charged. When we touch the inside surface of the cavity, the induced charge is neutralized, leaving the outside surface and the whole metal charged with a net positive charge.

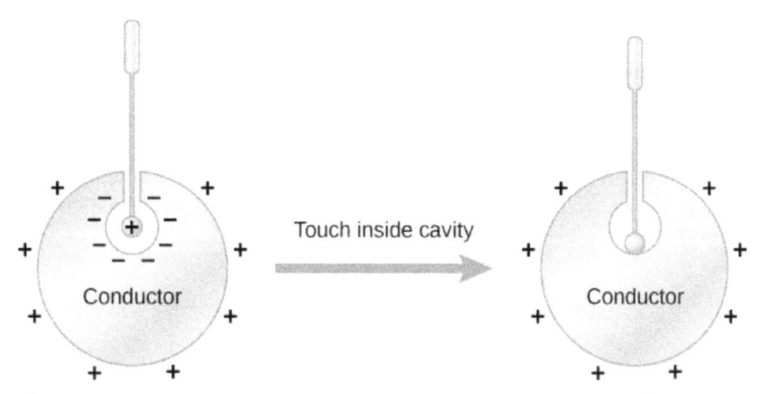

Figure 6.36 Electric charges on a conductor migrate to the outside surface no matter where you put them initially.

To see why this happens, note that the Gaussian surface in Figure 6.37 (the dashed line) follows the contour of the actual surface of the conductor and is located an infinitesimal distance *within* it. Since $E = 0$ everywhere inside a conductor,

$$\oint_S \vec{\mathbf{E}} \cdot \hat{\mathbf{n}} dA = 0.$$

Thus, from Gauss' law, there is no net charge inside the Gaussian surface. But the Gaussian surface lies just below the actual surface of the conductor; consequently, there is no net charge inside the conductor. Any excess charge must lie on its surface.

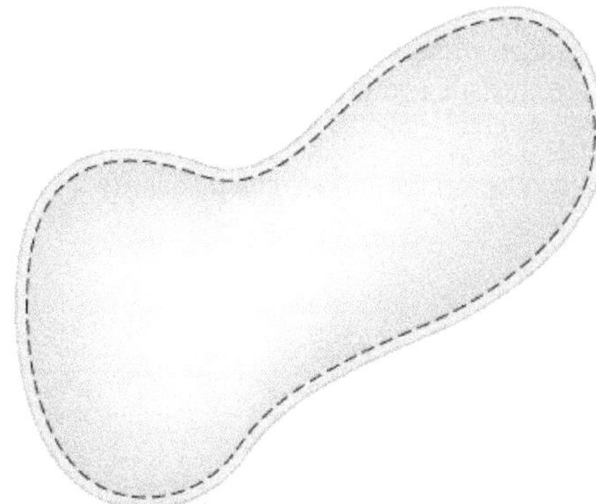

Figure 6.37 The dashed line represents a Gaussian surface that is just beneath the actual surface of the conductor.

This particular property of conductors is the basis for an extremely accurate method developed by Plimpton and Lawton in 1936 to verify Gauss's law and, correspondingly, Coulomb's law. A sketch of their apparatus is shown in Figure 6.38. Two spherical shells are connected to one another through an electrometer E, a device that can detect a very slight amount of charge flowing from one shell to the other. When switch S is thrown to the left, charge is placed on the outer shell by the battery B. Will charge flow through the electrometer to the inner shell?

No. Doing so would mean a violation of Gauss's law. Plimpton and Lawton did not detect any flow and, knowing the sensitivity of their electrometer, concluded that if the radial dependence in Coulomb's law were $1/r^{2+\delta}$, δ would be less than 2×10^{-9} [1]. More recent measurements place δ at less than 3×10^{-16} [2], a number so small that the validity of Coulomb's law seems indisputable.

---

1. S. Plimpton and W. Lawton. 1936. "A Very Accurate Test of Coulomb's Law of Force between Charges." *Physical Review* 50, No. 11: 1066, doi:10.1103/PhysRev.50.1066
2. E. Williams, J. Faller, and H. Hill. 1971. "New Experimental Test of Coulomb's Law: A Laboratory Upper Limit on the Photon Rest Mass." *Physical Review Letters* 26 , No. 12: 721, doi:10.1103/PhysRevLett.26.721

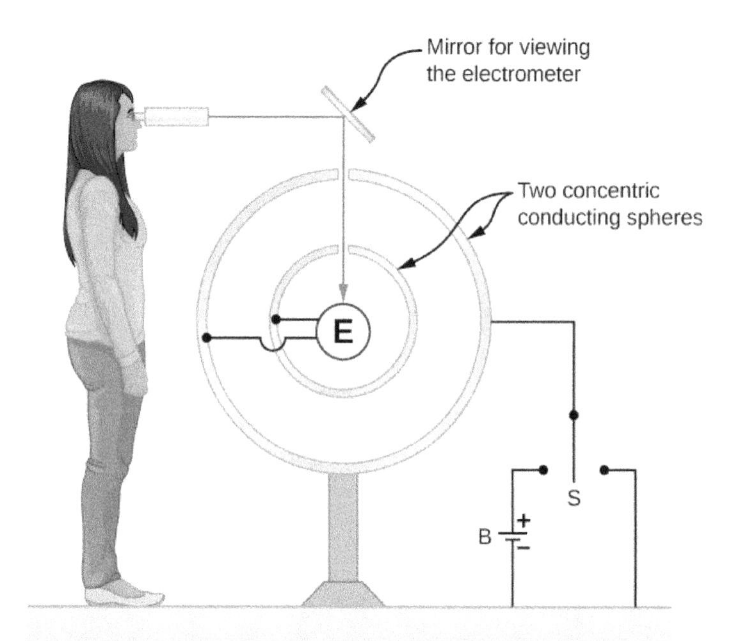

Figure 6.38 A representation of the apparatus used by Plimpton and Lawton. Any transfer of charge between the spheres is detected by the electrometer E.

The Electric Field at the Surface of a Conductor

If the electric field had a component parallel to the surface of a conductor, free charges on the surface would move, a situation contrary to the assumption of electrostatic equilibrium. Therefore, the electric field is always perpendicular to the surface of a conductor.

At any point just above the surface of a conductor, the surface charge density δ and the magnitude of the electric field E are related by

$$E = \frac{\sigma}{\varepsilon_0}. \quad (6.14)$$

To see this, consider an infinitesimally small Gaussian cylinder that surrounds a point on the surface of the conductor, as in Figure 6.39. The cylinder has one end face inside and one end face outside the surface. The height and cross-sectional area of the cylinder are δ and ΔA, respectively. The cylinder's sides are perpendicular to the surface of the conductor, and its end faces are parallel to the surface. Because the cylinder is infinitesimally small, the charge density σ is essentially constant over the surface enclosed, so the total charge inside the Gaussian cylinder is $\sigma \Delta A$. Now E is perpendicular to the surface of the conductor outside the conductor and vanishes within it, because otherwise, the charges would accelerate, and we would not be in equilibrium. Electric flux therefore crosses only the outer end face of the Gaussian surface and may be written as $E\Delta A$, since the cylinder is assumed to be small enough that E is approximately constant over that area. From Gauss' law,

$$E\Delta A = \frac{\sigma \Delta A}{\varepsilon_0}.$$

Thus,

$$E = \frac{\sigma}{\varepsilon_0}.$$

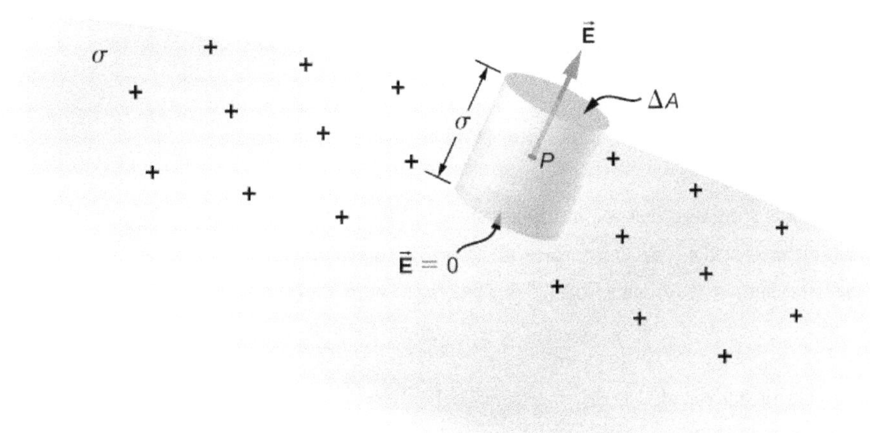

Figure 6.39 An infinitesimally small cylindrical Gaussian surface surrounds point P, which is on the surface of the conductor. The field $\vec{\mathbf{E}}$ is perpendicular to the surface of the conductor outside the conductor and vanishes within it.

Example 6.9

Electric Field of a Conducting Plate

The infinite conducting plate in Figure 6.40 has a uniform surface charge density σ. Use Gauss' law to find the electric field outside the plate. Compare this result with that previously calculated directly.

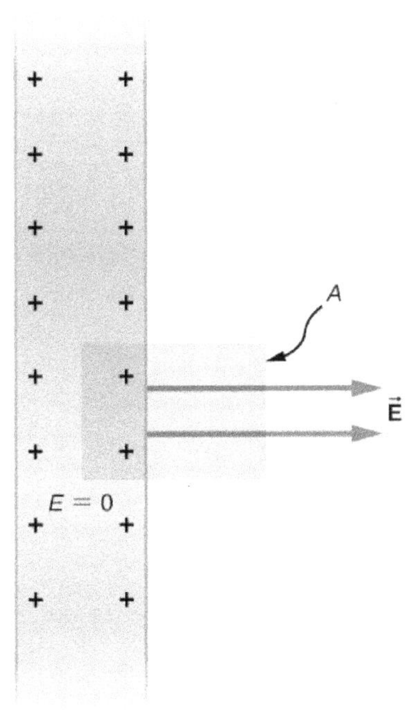

Figure 6.40 A side view of an infinite conducting plate and Gaussian cylinder with cross-sectional area A.

Strategy

For this case, we use a cylindrical Gaussian surface, a side view of which is shown.

Solution

The flux calculation is similar to that for an infinite sheet of charge from the previous chapter with one major exception: The left face of the Gaussian surface is inside the conductor where $\vec{\mathbf{E}} = \vec{\mathbf{0}}$, so the total flux through the Gaussian surface is *EA* rather than 2*EA*. Then from Gauss' law,

$$EA = \frac{\sigma A}{\varepsilon_0}$$

and the electric field outside the plate is

$$E = \frac{\sigma}{\varepsilon_0}.$$

Significance

This result is in agreement with the result from the previous section, and consistent with the rule stated above.

Example 6.10

Electric Field between Oppositely Charged Parallel Plates

Two large conducting plates carry equal and opposite charges, with a surface charge density σ of magnitude $6.81 \times 10^{-7}\ \text{C/m}^2$, as shown in Figure 6.41. The separation between the plates is $l = 6.50\ \text{mm}$. What is the electric field between the plates?

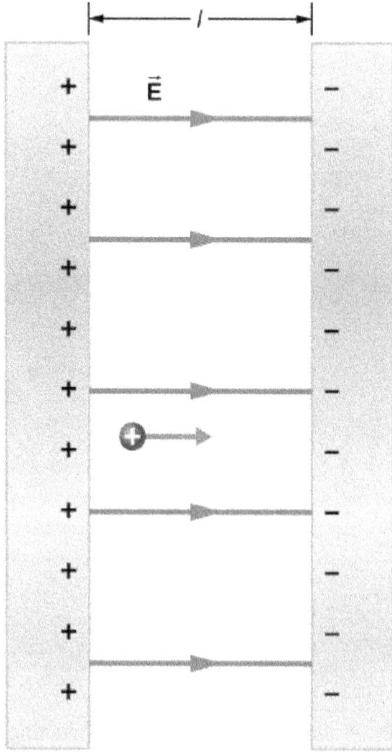

Figure 6.41 The electric field between oppositely charged parallel plates. A test charge is released at the positive plate.

Strategy

Note that the electric field at the surface of one plate only depends on the charge on that plate. Thus, apply $E = \sigma/\varepsilon_0$ with the given values.

Solution

The electric field is directed from the positive to the negative plate, as shown in the figure, and its magnitude is given by

$$E = \frac{\sigma}{\varepsilon_0} = \frac{6.81 \times 10^{-7}\ \text{C/m}^2}{8.85 \times 10^{-12}\ \text{C}^2/\text{N m}^2} = 7.69 \times 10^4\ \text{N/C}.$$

Significance

This formula is applicable to more than just a plate. Furthermore, two-plate systems will be important later.

Example 6.11

A Conducting Sphere

The isolated conducting sphere (Figure 6.42) has a radius R and an excess charge q. What is the electric field both inside and outside the sphere?

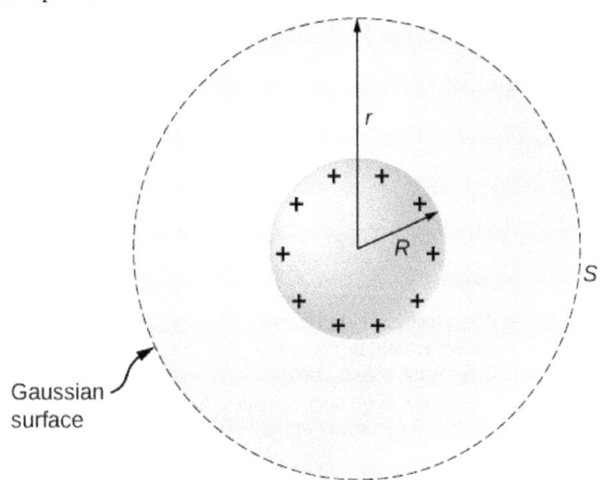

Figure 6.42 An isolated conducting sphere.

Strategy

The sphere is isolated, so its surface change distribution and the electric field of that distribution are spherically symmetrical. We can therefore represent the field as $\vec{\mathbf{E}} = E(r)\hat{\mathbf{r}}$. To calculate $E(r)$, we apply Gauss's law over a closed spherical surface S of radius r that is concentric with the conducting sphere.

Solution

Since r is constant and $\hat{\mathbf{n}} = \hat{\mathbf{r}}$ on the sphere,

$$\oint_S \vec{\mathbf{E}} \cdot \hat{\mathbf{n}}\, dA = E(r) \oint_S dA = E(r)\, 4\pi r^2.$$

For $r < R$, S is within the conductor, so $q_{\text{enc}} = 0$, and Gauss's law gives

$$E(r) = 0,$$

as expected inside a conductor. If $r > R$, S encloses the conductor so $q_{enc} = q$. From Gauss's law,

$$E(r)\,4\pi r^2 = \frac{q}{\varepsilon_0}.$$

The electric field of the sphere may therefore be written as

$$\begin{aligned} \vec{\mathbf{E}} &= \vec{\mathbf{0}} && (r < R), \\ \vec{\mathbf{E}} &= \frac{1}{4\pi\varepsilon_0}\frac{q}{r^2}\hat{\mathbf{r}} && (r \geq R). \end{aligned}$$

Significance

Notice that in the region $r \geq R$, the electric field due to a charge q placed on an isolated conducting sphere of radius R is identical to the electric field of a point charge q located at the center of the sphere. The difference between the charged metal and a point charge occurs only at the space points inside the conductor. For a point charge placed at the center of the sphere, the electric field is not zero at points of space occupied by the sphere, but a conductor with the same amount of charge has a zero electric field at those points (Figure 6.43). However, there is no distinction at the outside points in space where $r > R$, and we can replace the isolated charged spherical conductor by a point charge at its center with impunity.

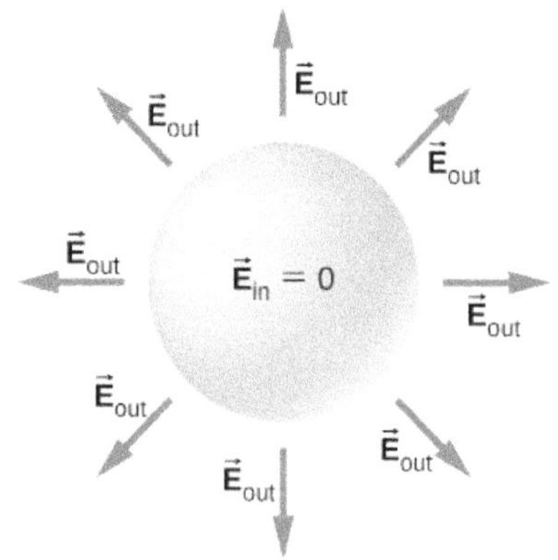

Figure 6.43 Electric field of a positively charged metal sphere. The electric field inside is zero, and the electric field outside is same as the electric field of a point charge at the center, although the charge on the metal sphere is at the surface.

6.6 Check Your Understanding How will the system above change if there are charged objects external to the sphere?

For a conductor with a cavity, if we put a charge $+q$ inside the cavity, then the charge separation takes place in the conductor, with $-q$ amount of charge on the inside surface and a $+q$ amount of charge at the outside surface (Figure 6.44(a)). For the same conductor with a charge $+q$ outside it, there is no excess charge on the inside surface; both the positive and negative induced charges reside on the outside surface (Figure 6.44(b)).

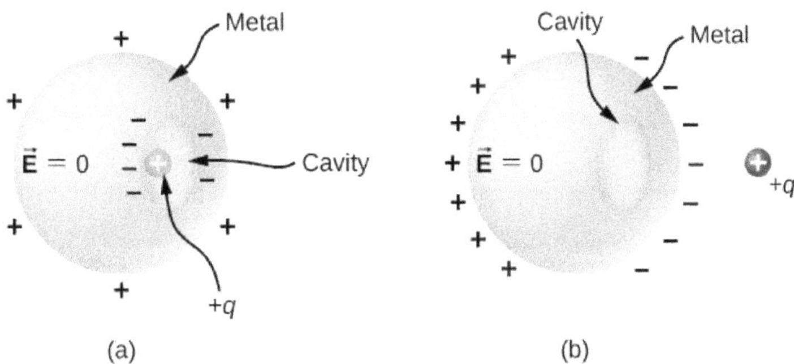

Figure 6.44 (a) A charge inside a cavity in a metal. The distribution of charges at the outer surface does not depend on how the charges are distributed at the inner surface, since the *E*-field inside the body of the metal is zero. That magnitude of the charge on the outer surface does depend on the magnitude of the charge inside, however. (b) A charge outside a conductor containing an inner cavity. The cavity remains free of charge. The polarization of charges on the conductor happens at the surface.

If a conductor has two cavities, one of them having a charge $+q_a$ inside it and the other a charge $-q_b$, the polarization of the conductor results in $-q_a$ on the inside surface of the cavity *a*, $+q_b$ on the inside surface of the cavity *b*, and $q_a - q_b$ on the outside surface (Figure 6.45). The charges on the surfaces may not be uniformly spread out; their spread depends upon the geometry. The only rule obeyed is that when the equilibrium has been reached, the charge distribution in a conductor is such that the electric field by the charge distribution in the conductor cancels the electric field of the external charges at all space points inside the body of the conductor.

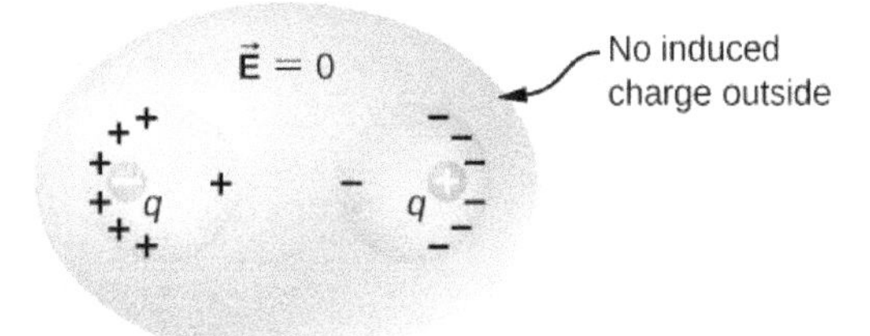

Figure 6.45 The charges induced by two equal and opposite charges in two separate cavities of a conductor. If the net charge on the cavity is nonzero, the external surface becomes charged to the amount of the net charge.

CHAPTER 6 REVIEW

KEY TERMS

area vector vector with magnitude equal to the area of a surface and direction perpendicular to the surface

cylindrical symmetry system only varies with distance from the axis, not direction

electric flux dot product of the electric field and the area through which it is passing

flux quantity of something passing through a given area

free electrons also called conduction electrons, these are the electrons in a conductor that are not bound to any particular atom, and hence are free to move around

Gaussian surface any enclosed (usually imaginary) surface

planar symmetry system only varies with distance from a plane

spherical symmetry system only varies with the distance from the origin, not in direction

KEY EQUATIONS

Definition of electric flux, for uniform electric field	$\Phi = \vec{\mathbf{E}} \cdot \vec{\mathbf{A}} \rightarrow EA\cos\theta$
Electric flux through an open surface	$\Phi = \int_S \vec{\mathbf{E}} \cdot \hat{\mathbf{n}}\, dA = \int_S \vec{\mathbf{E}} \cdot d\vec{\mathbf{A}}$
Electric flux through a closed surface	$\Phi = \oint_S \vec{\mathbf{E}} \cdot \hat{\mathbf{n}}\, dA = \oint_S \vec{\mathbf{E}} \cdot d\vec{\mathbf{A}}$
Gauss's law	$\Phi = \oint_S \vec{\mathbf{E}} \cdot \hat{\mathbf{n}}\, dA = \frac{q_{\text{enc}}}{\varepsilon_0}$
Gauss's Law for systems with symmetry	$\Phi = \oint_S \vec{\mathbf{E}} \cdot \hat{\mathbf{n}}\, dA = E\oint_S dA = EA = \frac{q_{\text{enc}}}{\varepsilon_0}$
The magnitude of the electric field just outside the surface of a conductor	$E = \frac{\sigma}{\varepsilon_0}$

SUMMARY

6.1 Electric Flux

- The electric flux through a surface is proportional to the number of field lines crossing that surface. Note that this means the magnitude is proportional to the portion of the field perpendicular to the area.
- The electric flux is obtained by evaluating the surface integral

$$\Phi = \oint_S \vec{\mathbf{E}} \cdot \hat{\mathbf{n}}\, dA = \oint_S \vec{\mathbf{E}} \cdot d\vec{\mathbf{A}},$$

where the notation used here is for a closed surface *S*.

6.2 Explaining Gauss's Law

- Gauss's law relates the electric flux through a closed surface to the net charge within that surface,

$$\Phi = \oint_S \vec{\mathbf{E}} \cdot \hat{\mathbf{n}}\, dA = \frac{q_{\text{enc}}}{\varepsilon_0},$$

where q_{enc} is the total charge inside the Gaussian surface *S*.

- All surfaces that include the same amount of charge have the same number of field lines crossing it, regardless of the shape or size of the surface, as long as the surfaces enclose the same amount of charge.

6.3 Applying Gauss's Law

- For a charge distribution with certain spatial symmetries (spherical, cylindrical, and planar), we can find a Gaussian surface over which $\vec{\mathbf{E}} \cdot \hat{\mathbf{n}} = E$, where E is constant over the surface. The electric field is then determined with Gauss's law.
- For spherical symmetry, the Gaussian surface is also a sphere, and Gauss's law simplifies to $4\pi r^2 E = \frac{q_{\text{enc}}}{\varepsilon_0}$.
- For cylindrical symmetry, we use a cylindrical Gaussian surface, and find that Gauss's law simplifies to $2\pi rLE = \frac{q_{\text{enc}}}{\varepsilon_0}$.
- For planar symmetry, a convenient Gaussian surface is a box penetrating the plane, with two faces parallel to the plane and the remainder perpendicular, resulting in Gauss's law being $2AE = \frac{q_{\text{enc}}}{\varepsilon_0}$.

6.4 Conductors in Electrostatic Equilibrium

- The electric field inside a conductor vanishes.
- Any excess charge placed on a conductor resides entirely on the surface of the conductor.
- The electric field is perpendicular to the surface of a conductor everywhere on that surface.
- The magnitude of the electric field just above the surface of a conductor is given by $E = \frac{\sigma}{\varepsilon_0}$.

CONCEPTUAL QUESTIONS

6.1 Electric Flux

1. Discuss how would orient a planar surface of area A in a uniform electric field of magnitude E_0 to obtain (a) the maximum flux and (b) the minimum flux through the area.

2. What are the maximum and minimum values of the flux in the preceding question?

3. The net electric flux crossing a closed surface is always zero. True or false?

4. The net electric flux crossing an open surface is never zero. True or false?

6.2 Explaining Gauss's Law

5. Two concentric spherical surfaces enclose a point charge q. The radius of the outer sphere is twice that of the inner one. Compare the electric fluxes crossing the two surfaces.

6. Compare the electric flux through the surface of a cube of side length a that has a charge q at its center to the flux through a spherical surface of radius a with a charge q at its center.

7. (a) If the electric flux through a closed surface is zero, is the electric field necessarily zero at all points on the surface? (b) What is the net charge inside the surface?

8. Discuss how Gauss's law would be affected if the electric field of a point charge did not vary as $1/r^2$.

9. Discuss the similarities and differences between the gravitational field of a point mass m and the electric field of a point charge q.

10. Discuss whether Gauss's law can be applied to other forces, and if so, which ones.

11. Is the term $\vec{\mathbf{E}}$ in Gauss's law the electric field produced by just the charge inside the Gaussian surface?

12. Reformulate Gauss's law by choosing the unit normal of the Gaussian surface to be the one directed inward.

6.3 Applying Gauss's Law

13. Would Gauss's law be helpful for determining the electric field of two equal but opposite charges a fixed distance apart?

14. Discuss the role that symmetry plays in the application of Gauss's law. Give examples of continuous charge distributions in which Gauss's law is useful and not useful in determining the electric field.

15. Discuss the restrictions on the Gaussian surface used to discuss planar symmetry. For example, is its length important? Does the cross-section have to be square? Must the end faces be on opposite sides of the sheet?

6.4 Conductors in Electrostatic Equilibrium

16. Is the electric field inside a metal always zero?

17. Under electrostatic conditions, the excess charge on a conductor resides on its surface. Does this mean that all the conduction electrons in a conductor are on the surface?

18. A charge q is placed in the cavity of a conductor as shown below. Will a charge outside the conductor experience an electric field due to the presence of q?

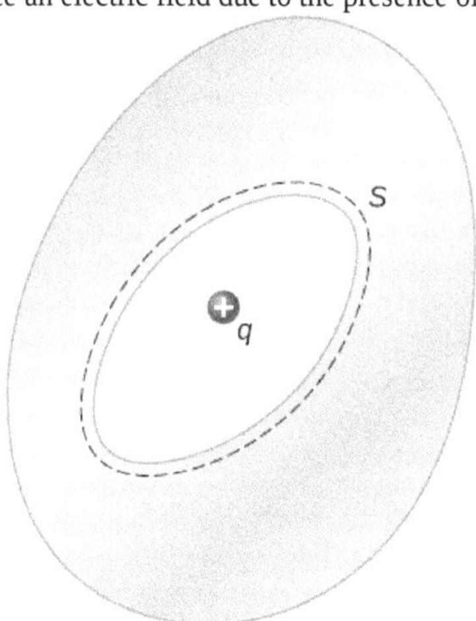

19. The conductor in the preceding figure has an excess charge of $-5.0\,\mu\text{C}$. If a $2.0\text{-}\mu\text{C}$ point charge is placed in the cavity, what is the net charge on the surface of the cavity and on the outer surface of the conductor?

PROBLEMS

6.1 Electric Flux

20. A uniform electric field of magnitude 1.1×10^4 N/C is perpendicular to a square sheet with sides 2.0 m long. What is the electric flux through the sheet?

21. Calculate the flux through the sheet of the previous problem if the plane of the sheet is at an angle of $60°$ to the field. Find the flux for both directions of the unit normal to the sheet.

22. Find the electric flux through a rectangular area $3\text{ cm} \times 2\text{ cm}$ between two parallel plates where there is a constant electric field of 30 N/C for the following orientations of the area: (a) parallel to the plates, (b) perpendicular to the plates, and (c) the normal to the area making a $30°$ angle with the direction of the electric field. Note that this angle can also be given as $180° + 30°$.

23. The electric flux through a square-shaped area of side 5 cm near a large charged sheet is found to be $3 \times 10^{-5}\ \text{N} \cdot \text{m}^2/\text{C}$ when the area is parallel to the plate. Find the charge density on the sheet.

24. Two large rectangular aluminum plates of area 150 cm^2 face each other with a separation of 3 mm between them. The plates are charged with equal amount of opposite charges, $\pm 20\,\mu\text{C}$. The charges on the plates face each other. Find the flux through a circle of radius 3 cm between the plates when the normal to the circle makes an angle of $5°$ with a line perpendicular to the plates. Note that this angle can also be given as $180° + 5°$.

25. A square surface of area 2 cm^2 is in a space of uniform electric field of magnitude 10^3 N/C. The amount of flux through it depends on how the square is oriented relative to the direction of the electric field. Find the electric flux through the square, when the normal to it makes the following angles with electric field: (a) $30°$, (b) $90°$, and (c) $0°$. Note that these angles can also be given as $180° + \theta$.

26. A vector field is pointed along the z-axis, $\vec{\mathbf{v}} = \dfrac{\alpha}{x^2 + y^2}\hat{\mathbf{z}}$. (a) Find the flux of the vector field through a rectangle in the xy-plane between $a < x < b$ and $c < y < d$. (b) Do the same through a rectangle in the yz-plane between $a < z < b$ and $c < y < d$. (Leave your answer as an integral.)

27. Consider the uniform electric field $\vec{\mathbf{E}} = (4.0\,\hat{\mathbf{j}} + 3.0\,\hat{\mathbf{k}}) \times 10^3$ N/C. What is its electric flux through a circular area of radius 2.0 m that lies in the *xy*-plane?

28. Repeat the previous problem, given that the circular area is (a) in the *yz*-plane and (b) 45° above the *xy*-plane.

29. An infinite charged wire with charge per unit length λ lies along the central axis of a cylindrical surface of radius r and length l. What is the flux through the surface due to the electric field of the charged wire?

6.2 Explaining Gauss's Law

30. Determine the electric flux through each surface whose cross-section is shown below.

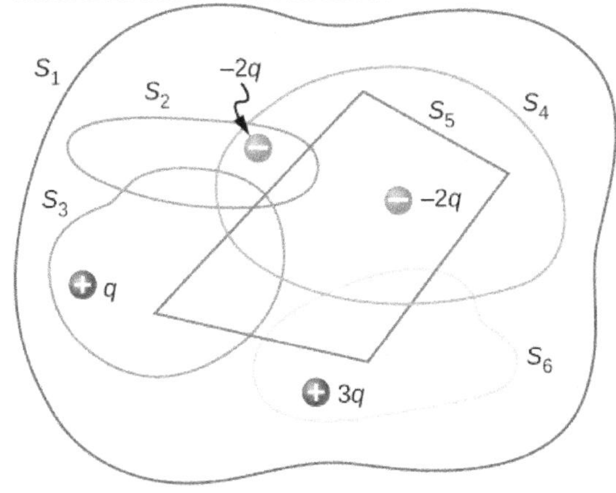

31. Find the electric flux through the closed surface whose cross-sections are shown below.

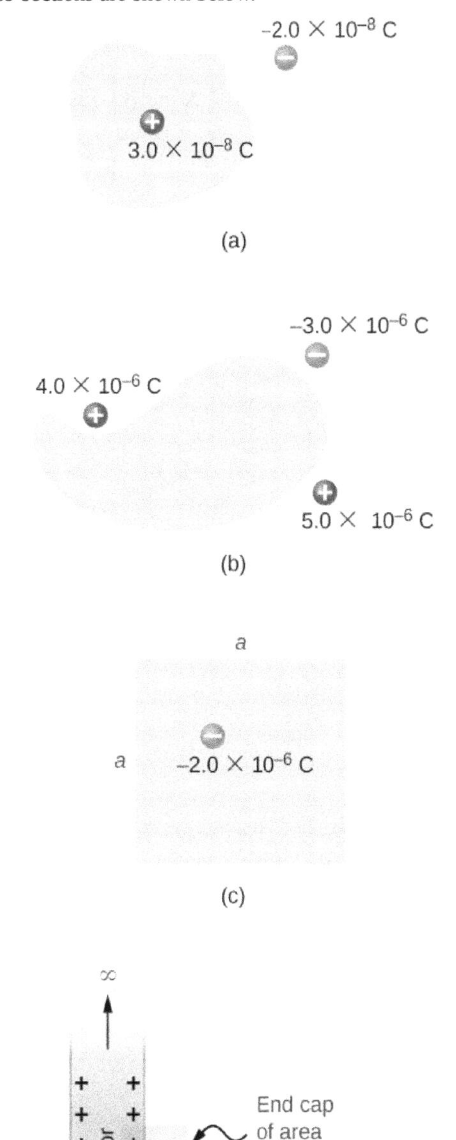

32. A point charge q is located at the center of a cube whose sides are of length a. If there are no other charges in this system, what is the electric flux through one face of the cube?

33. A point charge of $10\ \mu\text{C}$ is at an unspecified location inside a cube of side 2 cm. Find the net electric flux though the surfaces of the cube.

34. A net flux of $1.0 \times 10^4\ \text{N} \cdot \text{m}^2/\text{C}$ passes inward through the surface of a sphere of radius 5 cm. (a) How much charge is inside the sphere? (b) How precisely can we determine the location of the charge from this information?

35. A charge q is placed at one of the corners of a cube of side a, as shown below. Find the magnitude of the electric flux through the shaded face due to q. Assume $q > 0$.

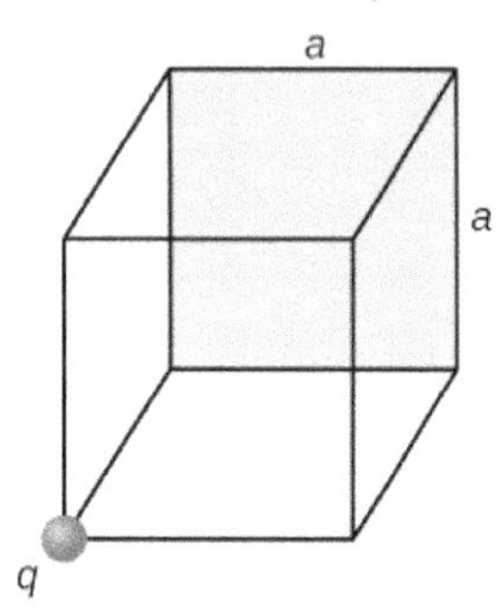

36. The electric flux through a cubical box 8.0 cm on a side is $1.2 \times 10^3\ \text{N} \cdot \text{m}^2/\text{C}$. What is the total charge enclosed by the box?

37. The electric flux through a spherical surface is $4.0 \times 10^4\ \text{N} \cdot \text{m}^2/\text{C}$. What is the net charge enclosed by the surface?

38. A cube whose sides are of length d is placed in a uniform electric field of magnitude $E = 4.0 \times 10^3\ \text{N/C}$ so that the field is perpendicular to two opposite faces of the cube. What is the net flux through the cube?

39. Repeat the previous problem, assuming that the electric field is directed along a body diagonal of the cube.

40. A total charge $5.0 \times 10^{-6}\ \text{C}$ is distributed uniformly throughout a cubical volume whose edges are 8.0 cm long. (a) What is the charge density in the cube? (b) What is the electric flux through a cube with 12.0-cm edges that is concentric with the charge distribution? (c) Do the same calculation for cubes whose edges are 10.0 cm long and 5.0 cm long. (d) What is the electric flux through a spherical surface of radius 3.0 cm that is also concentric with the charge distribution?

6.3 Applying Gauss's Law

41. Recall that in the example of a uniform charged sphere, $\rho_0 = Q/(\frac{4}{3}\pi R^3)$. Rewrite the answers in terms of the total charge Q on the sphere.

42. Suppose that the charge density of the spherical charge distribution shown in Figure 6.23 is $\rho(r) = \rho_0 r/R$ for $r \leq R$ and zero for $r > R$. Obtain expressions for the electric field both inside and outside the distribution.

43. A very long, thin wire has a uniform linear charge density of $50\ \mu\text{C/m}$. What is the electric field at a distance 2.0 cm from the wire?

44. A charge of $-30\ \mu\text{C}$ is distributed uniformly throughout a spherical volume of radius 10.0 cm. Determine the electric field due to this charge at a distance of (a) 2.0 cm, (b) 5.0 cm, and (c) 20.0 cm from the center of the sphere.

45. Repeat your calculations for the preceding problem, given that the charge is distributed uniformly over the surface of a spherical conductor of radius 10.0 cm.

46. A total charge Q is distributed uniformly throughout a spherical shell of inner and outer radii r_1 and r_2, respectively. Show that the electric field due to the charge is

$$\vec{\mathbf{E}} = \vec{\mathbf{0}} \qquad (r \leq r_1);$$

$$\vec{\mathbf{E}} = \frac{Q}{4\pi\varepsilon_0 r^2}\left(\frac{r^3 - r_1{}^3}{r_2{}^3 - r_1{}^3}\right)\hat{\mathbf{r}} \qquad (r_1 \leq r \leq r_2);$$

$$\vec{\mathbf{E}} = \frac{Q}{4\pi\varepsilon_0 r^2}\hat{\mathbf{r}} \qquad (r \geq r_2).$$

47. When a charge is placed on a metal sphere, it ends up in equilibrium at the outer surface. Use this information to determine the electric field of $+3.0\ \mu\text{C}$ charge put on a 5.0-cm aluminum spherical ball at the following two points in space: (a) a point 1.0 cm from the center of the ball (an inside point) and (b) a point 10 cm from the center of the ball (an outside point).

48. A large sheet of charge has a uniform charge density of $10\ \mu\text{C/m}^2$. What is the electric field due to this charge at a point just above the surface of the sheet?

49. Determine if approximate cylindrical symmetry holds for the following situations. State why or why not. (a) A 300-cm long copper rod of radius 1 cm is charged with +500 nC of charge and we seek electric field at a point 5 cm from the center of the rod. (b) A 10-cm long copper rod of radius 1 cm is charged with +500 nC of charge and we seek electric field at a point 5 cm from the center of the rod. (c) A 150-cm wooden rod is glued to a 150-cm plastic rod to make a 300-cm long rod, which is then painted with a charged paint so that one obtains a uniform charge density. The radius of each rod is 1 cm, and we seek an electric field at a point that is 4 cm from the center of the rod. (d) Same rod as (c), but we seek electric field at a point that is 500 cm from the center of the rod.

50. A long silver rod of radius 3 cm has a charge of $-5\ \mu\text{C/cm}$ on its surface. (a) Find the electric field at a point 5 cm from the center of the rod (an outside point). (b) Find the electric field at a point 2 cm from the center of the rod (an inside point).

51. The electric field at 2 cm from the center of long copper rod of radius 1 cm has a magnitude 3 N/C and directed outward from the axis of the rod. (a) How much charge per unit length exists on the copper rod? (b) What would be the electric flux through a cube of side 5 cm situated such that the rod passes through opposite sides of the cube perpendicularly?

52. A long copper cylindrical shell of inner radius 2 cm and outer radius 3 cm surrounds concentrically a charged long aluminum rod of radius 1 cm with a charge density of 4 pC/m. All charges on the aluminum rod reside at its surface. The inner surface of the copper shell has exactly opposite charge to that of the aluminum rod while the outer surface of the copper shell has the same charge as the aluminum rod. Find the magnitude and direction of the electric field at points that are at the following distances from the center of the aluminum rod: (a) 0.5 cm, (b) 1.5 cm, (c) 2.5 cm, (d) 3.5 cm, and (e) 7 cm.

53. Charge is distributed uniformly with a density ρ throughout an infinitely long cylindrical volume of radius R. Show that the field of this charge distribution is directed radially with respect to the cylinder and that

$$E = \frac{\rho r}{2\varepsilon_0} \quad (r \le R);$$
$$E = \frac{\rho R^2}{2\varepsilon_0 r} \quad (r \ge R).$$

54. Charge is distributed throughout a very long cylindrical volume of radius R such that the charge density increases with the distance r from the central axis of the cylinder according to $\rho = \alpha r$, where α is a constant. Show that the field of this charge distribution is directed radially with respect to the cylinder and that

$$E = \frac{\alpha r^2}{3\varepsilon_0} \quad (r \le R);$$
$$E = \frac{\alpha R^3}{3\varepsilon_0 r} \quad (r \ge R).$$

55. The electric field 10.0 cm from the surface of a copper ball of radius 5.0 cm is directed toward the ball's center and has magnitude 4.0×10^2 N/C. How much charge is on the surface of the ball?

56. Charge is distributed throughout a spherical shell of inner radius r_1 and outer radius r_2 with a volume density given by $\rho = \rho_0 r_1/r$, where ρ_0 is a constant. Determine the electric field due to this charge as a function of r, the distance from the center of the shell.

57. Charge is distributed throughout a spherical volume of radius R with a density $\rho = \alpha r^2$, where α is a constant. Determine the electric field due to the charge at points both inside and outside the sphere.

58. Consider a uranium nucleus to be sphere of radius $R = 7.4 \times 10^{-15}$ m with a charge of $92e$ distributed uniformly throughout its volume. (a) What is the electric force exerted on an electron when it is 3.0×10^{-15} m from the center of the nucleus? (b) What is the acceleration of the electron at this point?

59. The volume charge density of a spherical charge distribution is given by $\rho(r) = \rho_0 e^{-\alpha r}$, where ρ_0 and α are constants. What is the electric field produced by this charge distribution?

6.4 Conductors in Electrostatic Equilibrium

60. An uncharged conductor with an internal cavity is shown in the following figure. Use the closed surface S along with Gauss' law to show that when a charge q is placed in the cavity a total charge $-q$ is induced on the inner surface of the conductor. What is the charge on the outer surface of the conductor?

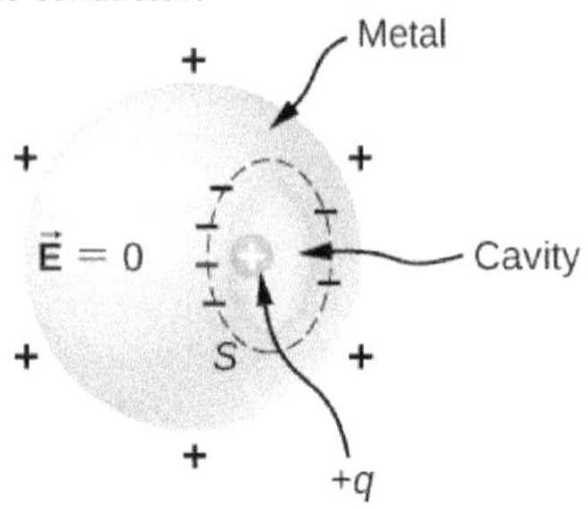

Figure 6.46 A charge inside a cavity of a metal. Charges at the outer surface do not depend on how the charges are distributed at the inner surface since E field inside the body of the metal is zero.

61. An uncharged spherical conductor S of radius R has two spherical cavities A and B of radii a and b, respectively as shown below. Two point charges $+q_a$ and $+q_b$ are placed at the center of the two cavities by using non-conducting supports. In addition, a point charge $+q_0$ is placed outside at a distance r from the center of the sphere. (a) Draw approximate charge distributions in the metal although metal sphere has no net charge. (b) Draw electric field lines. Draw enough lines to represent all distinctly different places.

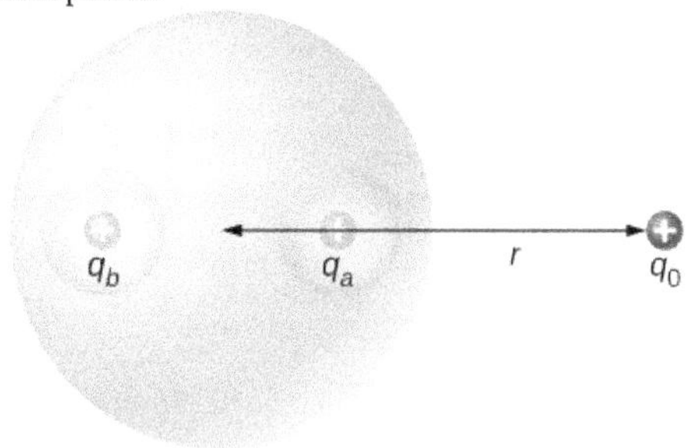

62. A positive point charge is placed at the angle bisector of two uncharged plane conductors that make an angle of 45°. See below. Draw the electric field lines.

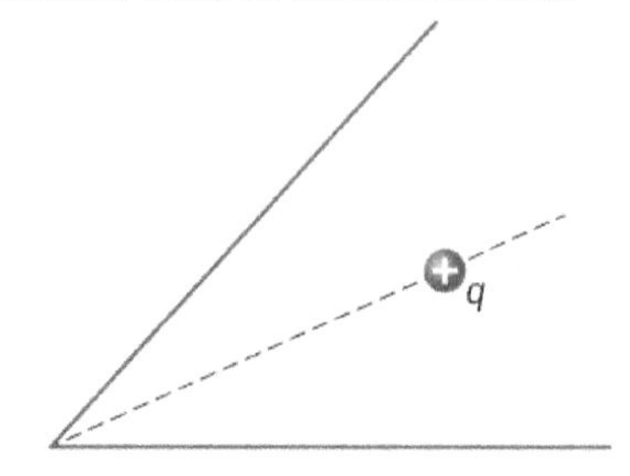

63. A long cylinder of copper of radius 3 cm is charged so that it has a uniform charge per unit length on its surface of 3 C/m. (a) Find the electric field inside and outside the cylinder. (b) Draw electric field lines in a plane perpendicular to the rod.

64. An aluminum spherical ball of radius 4 cm is charged with $5\,\mu\text{C}$ of charge. A copper spherical shell of inner radius 6 cm and outer radius 8 cm surrounds it. A total charge of $-8\,\mu\text{C}$ is put on the copper shell. (a) Find the electric field at all points in space, including points inside the aluminum and copper shell when copper shell and aluminum sphere are concentric. (b) Find the electric field at all points in space, including points inside the aluminum and copper shell when the centers of copper shell and aluminum sphere are 1 cm apart.

65. A long cylinder of aluminum of radius R meters is charged so that it has a uniform charge per unit length on its surface of λ. (a) Find the electric field inside and outside the cylinder. (b) Plot electric field as a function of distance from the center of the rod.

66. At the surface of any conductor in electrostatic equilibrium, $E = \sigma/\varepsilon_0$. Show that this equation is consistent with the fact that $E = kq/r^2$ at the surface of a spherical conductor.

67. Two parallel plates 10 cm on a side are given equal and opposite charges of magnitude 5.0×10^{-9} C. The plates are 1.5 mm apart. What is the electric field at the center of the region between the plates?

68. Two parallel conducting plates, each of cross-sectional area $400\,\text{cm}^2$, are 2.0 cm apart and uncharged. If 1.0×10^{12} electrons are transferred from one plate to the other, what are (a) the charge density on each plate? (b) The electric field between the plates?

69. The surface charge density on a long straight metallic pipe is σ. What is the electric field outside and inside the pipe? Assume the pipe has a diameter of $2a$.

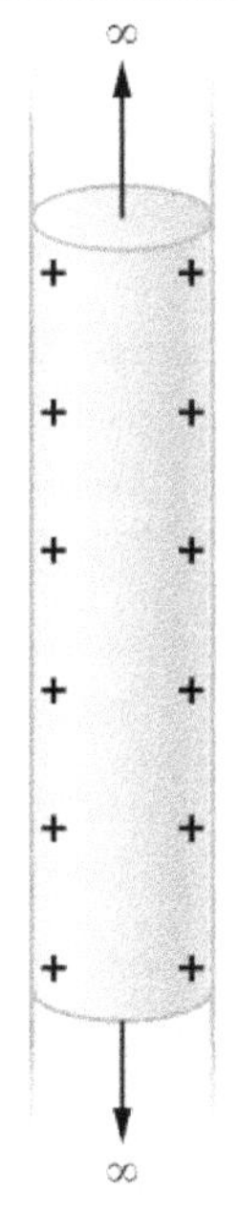

70. A point charge $q = -5.0 \times 10^{-12}$ C is placed at the center of a spherical conducting shell of inner radius 3.5 cm and outer radius 4.0 cm. The electric field just above the surface of the conductor is directed radially outward and has magnitude 8.0 N/C. (a) What is the charge density on the inner surface of the shell? (b) What is the charge density on the outer surface of the shell? (c) What is the net charge on the conductor?

71. A solid cylindrical conductor of radius a is surrounded by a concentric cylindrical shell of inner radius b. The solid cylinder and the shell carry charges $+Q$ and $-Q$, respectively. Assuming that the length L of both conductors is much greater than a or b, determine the electric field as a function of r, the distance from the common central axis of the cylinders, for (a) $r < a$; (b) $a < r < b$; and (c) $r > b$.

ADDITIONAL PROBLEMS

72. A vector field $\vec{\mathbf{E}}$ (not necessarily an electric field; note units) is given by $\vec{\mathbf{E}} = 3x^2\hat{\mathbf{k}}$. Calculate $\int_S \vec{\mathbf{E}} \cdot \hat{\mathbf{n}}\, da$, where S is the area shown below. Assume that $\hat{\mathbf{n}} = \hat{\mathbf{k}}$.

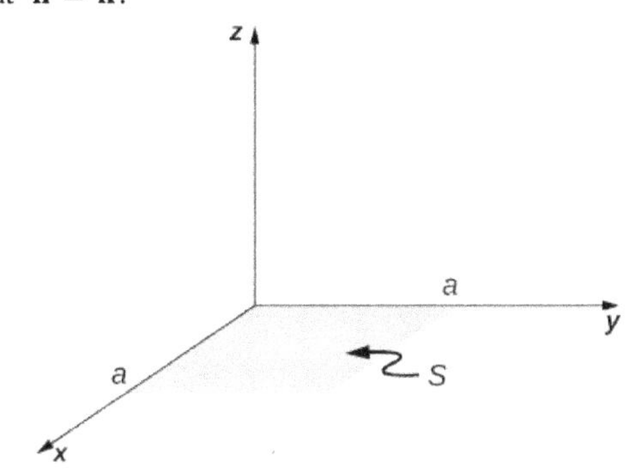

73. Repeat the preceding problem, with $\vec{\mathbf{E}} = 2x\hat{\mathbf{i}} + 3x^2\hat{\mathbf{k}}$.

74. A circular area S is concentric with the origin, has radius a, and lies in the yz-plane. Calculate $\int_S \vec{\mathbf{E}} \cdot \hat{\mathbf{n}}\, dA$ for $\vec{\mathbf{E}} = 3z^2\hat{\mathbf{i}}$.

75. (a) Calculate the electric flux through the open hemispherical surface due to the electric field $\vec{\mathbf{E}} = E_0\hat{\mathbf{k}}$ (see below). (b) If the hemisphere is rotated by 90° around the x-axis, what is the flux through it?

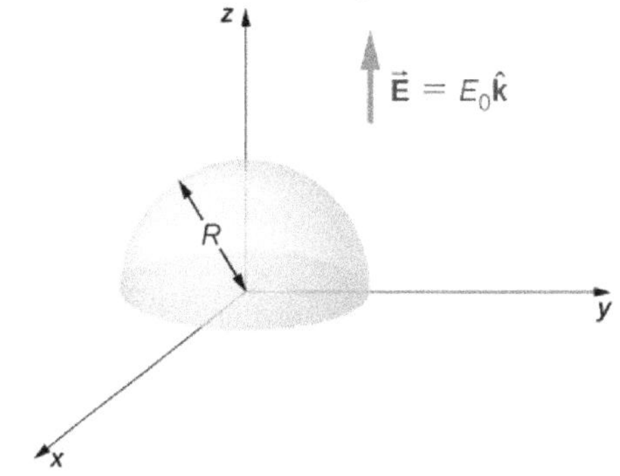

76. Suppose that the electric field of an isolated point charge were proportional to $1/r^{2+\sigma}$ rather than $1/r^2$. Determine the flux that passes through the surface of a sphere of radius R centered at the charge. Would Gauss's law remain valid?

77. The electric field in a region is given by $\vec{\mathbf{E}} = a/(b + cx)\,\hat{\mathbf{i}}$, where $a = 200\,\text{N} \cdot \text{m/C}$, $b = 2.0\,\text{m}$, and $c = 2.0$. What is the net charge enclosed by the shaded volume shown below?

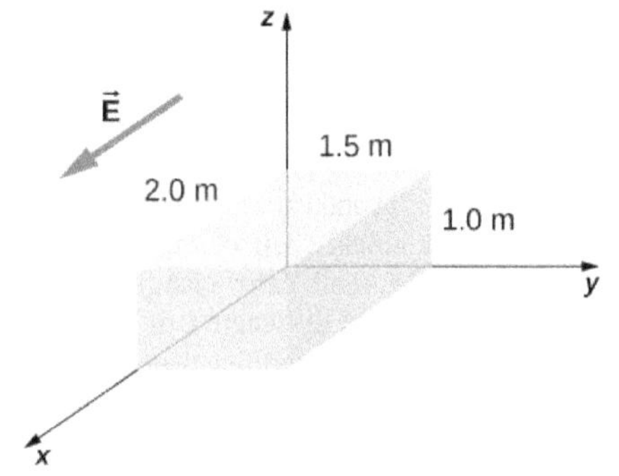

78. Two equal and opposite charges of magnitude Q are located on the x-axis at the points $+a$ and $-a$, as shown below. What is the net flux due to these charges through a square surface of side $2a$ that lies in the yz-plane and is centered at the origin? (*Hint:* Determine the flux due to each charge separately, then use the principle of superposition. You may be able to make a symmetry argument.)

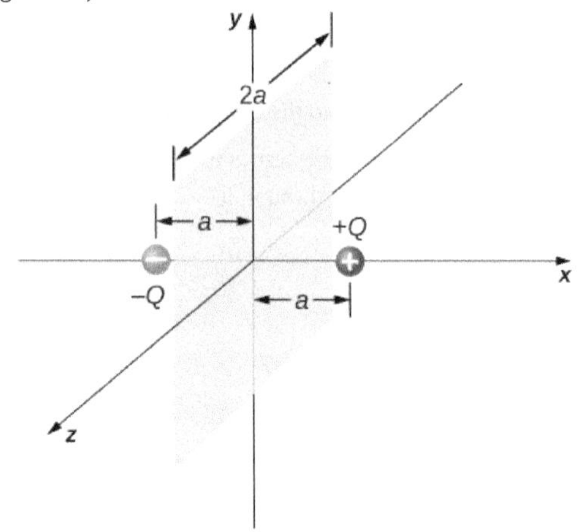

79. A fellow student calculated the flux through the square for the system in the preceding problem and got 0. What went wrong?

80. A $10\,\text{cm} \times 10\,\text{cm}$ piece of aluminum foil of 0.1 mm thickness has a charge of $20\,\mu\text{C}$ that spreads on both wide side surfaces evenly. You may ignore the charges on the thin sides of the edges. (a) Find the charge density. (b) Find the electric field 1 cm from the center, assuming approximate planar symmetry.

81. Two $10\,\text{cm} \times 10\,\text{cm}$ pieces of aluminum foil of thickness 0.1 mm face each other with a separation of 5 mm. One of the foils has a charge of $+30\,\mu\text{C}$ and the other has $-30\,\mu\text{C}$. (a) Find the charge density at all surfaces, i.e., on those facing each other and those facing away. (b) Find the electric field between the plates near the center assuming planar symmetry.

82. Two large copper plates facing each other have charge densities $\pm 4.0\,\text{C/m}^2$ on the surface facing the other plate, and zero in between the plates. Find the electric flux through a $3\,\text{cm} \times 4\,\text{cm}$ rectangular area between the plates, as shown below, for the following orientations of the area. (a) If the area is parallel to the plates, and (b) if the area is tilted $\theta = 30°$ from the parallel direction. Note, this angle can also be $\theta = 180° + 30°$.

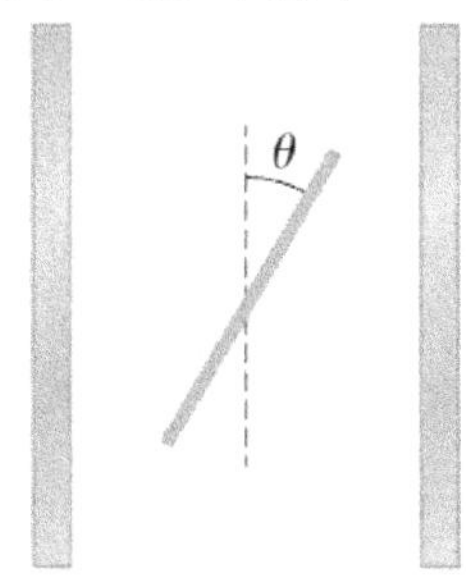

83. The infinite slab between the planes defined by $z = -a/2$ and $z = a/2$ contains a uniform volume charge density ρ (see below). What is the electric field produced by this charge distribution, both inside and outside the distribution?

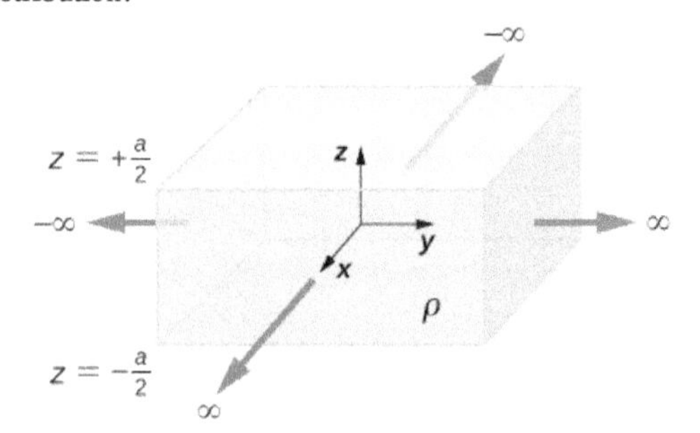

84. A total charge Q is distributed uniformly throughout a spherical volume that is centered at O_1 and has a radius R. Without disturbing the charge remaining, charge is removed from the spherical volume that is centered at O_2 (see below). Show that the electric field everywhere in the empty region is given by

$$\vec{\mathbf{E}} = \frac{Q\vec{\mathbf{r}}}{4\pi\varepsilon_0 R^3},$$

where $\vec{\mathbf{r}}$ is the displacement vector directed from O_1 to O_2.

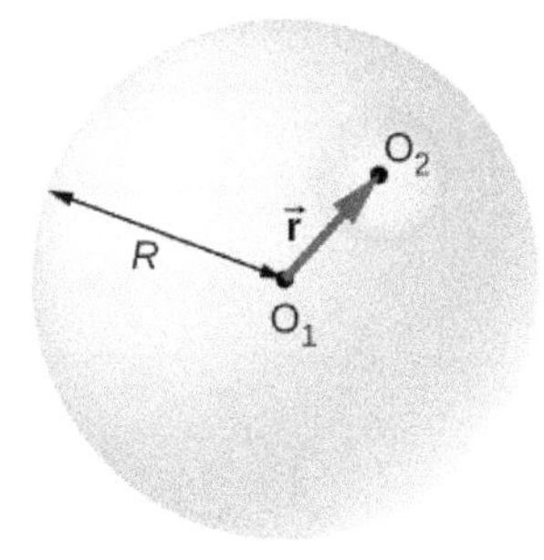

85. A non-conducting spherical shell of inner radius a_1 and outer radius b_1 is uniformly charged with charged density ρ_1 inside another non-conducting spherical shell of inner radius a_2 and outer radius b_2 that is also uniformly charged with charge density ρ_2. See below. Find the electric field at space point P at a distance r from the common center such that (a) $r > b_2$, (b) $a_2 < r < b_2$, (c) $b_1 < r < a_2$, (d) $a_1 < r < b_1$, and (e) $r < a_1$.

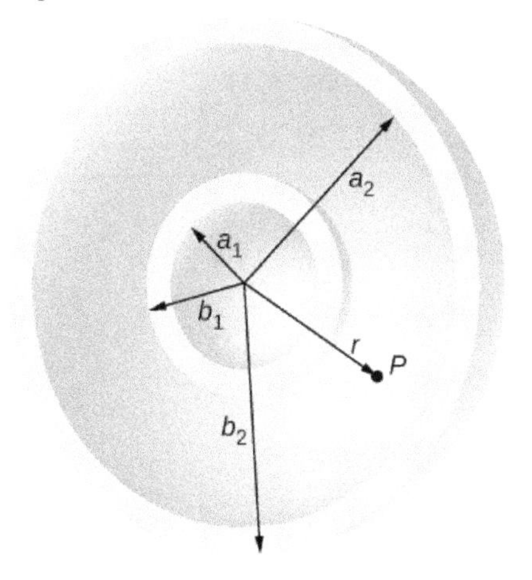

86. Two non-conducting spheres of radii R_1 and R_2 are uniformly charged with charge densities ρ_1 and ρ_2, respectively. They are separated at center-to-center distance a (see below). Find the electric field at point P located at a distance r from the center of sphere 1 and is in the direction θ from the line joining the two spheres assuming their charge densities are not affected by the presence of the other sphere. (*Hint:* Work one sphere at a time and use the superposition principle.)

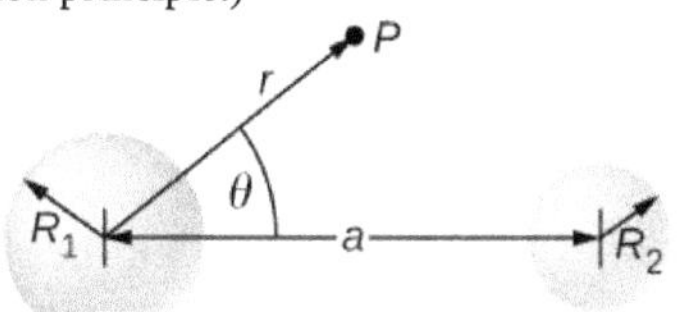

87. A disk of radius R is cut in a non-conducting large plate that is uniformly charged with charge density σ (coulomb per square meter). See below. Find the electric field at a height h above the center of the disk. $(h >> R, h << l \text{ or } w)$. (*Hint:* Fill the hole with $\pm\sigma$.)

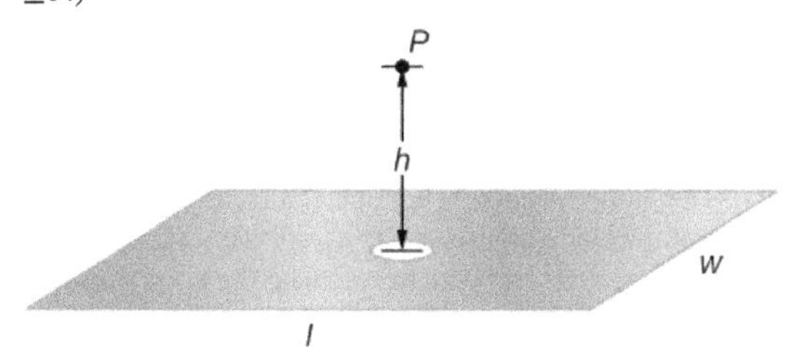

88. Concentric conducting spherical shells carry charges Q and $-Q$, respectively (see below). The inner shell has negligible thickness. Determine the electric field for (a) $r < a$; (b) $a < r < b$; (c) $b < r < c$; and (d) $r > c$.

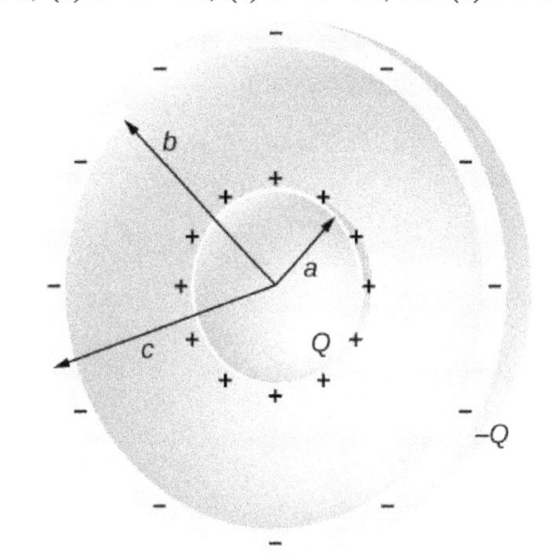

89. Shown below are two concentric conducting spherical shells of radii R_1 and R_2, each of finite thickness much less than either radius. The inner and outer shell carry net charges q_1 and q_2, respectively, where both q_1 and q_2 are positive. What is the electric field for (a) $r < R_1$; (b) $R_1 < r < R_2$; and (c) $r > R_2$? (d) What is the net charge on the inner surface of the inner shell, the outer surface of the inner shell, the inner surface of the outer shell, and the outer surface of the outer shell?

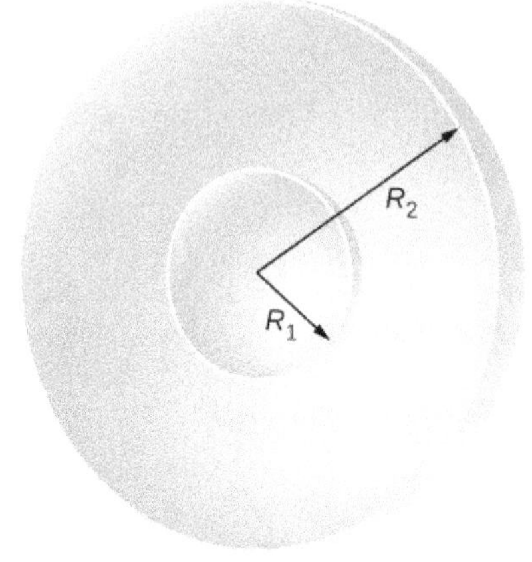

90. A point charge of $q = 5.0 \times 10^{-8}$ C is placed at the center of an uncharged spherical conducting shell of inner radius 6.0 cm and outer radius 9.0 cm. Find the electric field at (a) $r = 4.0\,\text{cm}$, (b) $r = 8.0\,\text{cm}$, and (c) $r = 12.0\,\text{cm}$. (d) What are the charges induced on the inner and outer surfaces of the shell?

CHALLENGE PROBLEMS

91. The Hubble Space Telescope can measure the energy flux from distant objects such as supernovae and stars. Scientists then use this data to calculate the energy emitted by that object. Choose an interstellar object which scientists have observed the flux at the Hubble with (for example, Vega[3]), find the distance to that object and the size of Hubble's primary mirror, and calculate the total energy flux. (*Hint:* The Hubble intercepts only a small part of the total flux.)

92. Re-derive Gauss's law for the gravitational field, with $\vec{g}$ directed positively outward.

93. An infinite plate sheet of charge of surface charge density σ is shown below. What is the electric field at a distance x from the sheet? Compare the result of this calculation with that of worked out in the text.

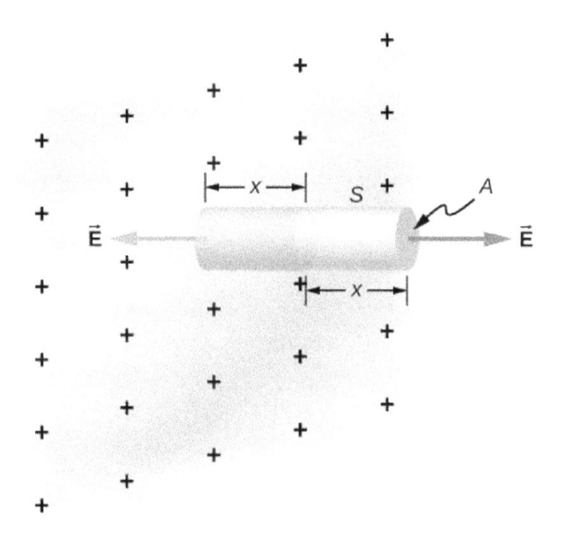

3. http://adsabs.harvard.edu/abs/2004AJ....127.3508B

94. A spherical rubber balloon carries a total charge Q distributed uniformly over its surface. At $t = 0$, the radius of the balloon is R. The balloon is then slowly inflated until its radius reaches $2R$ at the time t_0. Determine the electric field due to this charge as a function of time (a) at the surface of the balloon, (b) at the surface of radius R, and (c) at the surface of radius $2R$. Ignore any effect on the electric field due to the material of the balloon and assume that the radius increases uniformly with time.

95. Find the electric field of a large conducting plate containing a net charge q. Let A be area of one side of the plate and h the thickness of the plate (see below). The charge on the metal plate will distribute mostly on the two planar sides and very little on the edges if the plate is thin.

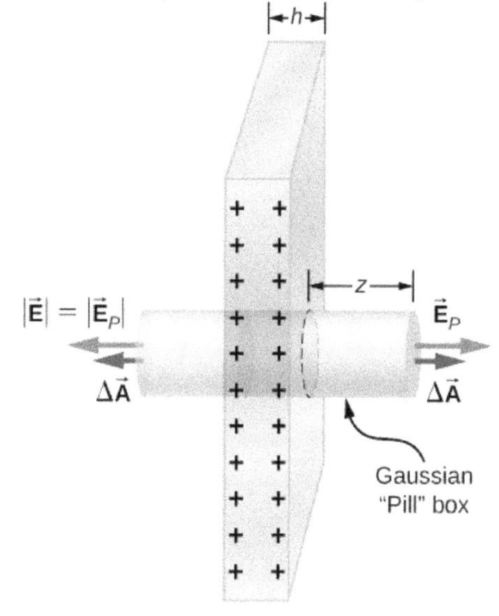

7 | ELECTRIC POTENTIAL

Figure 7.1 The energy released in a lightning strike is an excellent illustration of the vast quantities of energy that may be stored and released by an electric potential difference. In this chapter, we calculate just how much energy can be released in a lightning strike and how this varies with the height of the clouds from the ground. (credit: Anthony Quintano)

Chapter Outline

Introduction

In Electric Charges and Fields, we just scratched the surface (or at least rubbed it) of electrical phenomena. Two terms commonly used to describe electricity are its energy and *voltage*, which we show in this chapter is directly related to the potential energy in a system.

We know, for example, that great amounts of electrical energy can be stored in batteries, are transmitted cross-country via currents through power lines, and may jump from clouds to explode the sap of trees. In a similar manner, at the molecular level, ions cross cell membranes and transfer information.

We also know about voltages associated with electricity. Batteries are typically a few volts, the outlets in your home frequently produce 120 volts, and power lines can be as high as hundreds of thousands of volts. But energy and voltage are not the same thing. A motorcycle battery, for example, is small and would not be very successful in replacing a much larger car battery, yet each has the same voltage. In this chapter, we examine the relationship between voltage and electrical energy, and begin to explore some of the many applications of electricity.

7.1 | Electric Potential Energy

Learning Objectives

By the end of this section, you will be able to:

- Define the work done by an electric force
- Define electric potential energy
- Apply work and potential energy in systems with electric charges

When a free positive charge q is accelerated by an electric field, it is given kinetic energy (Figure 7.2). The process is analogous to an object being accelerated by a gravitational field, as if the charge were going down an electrical hill where its electric potential energy is converted into kinetic energy, although of course the sources of the forces are very different. Let us explore the work done on a charge q by the electric field in this process, so that we may develop a definition of electric potential energy.

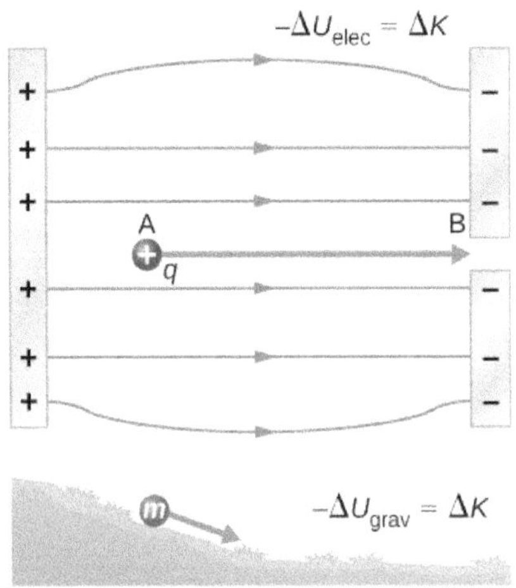

Figure 7.2 A charge accelerated by an electric field is analogous to a mass going down a hill. In both cases, potential energy decreases as kinetic energy increases, $-\Delta U = \Delta K$. Work is done by a force, but since this force is conservative, we can write $W = -\Delta U$.

The electrostatic or Coulomb force is conservative, which means that the work done on q is independent of the path taken, as we will demonstrate later. This is exactly analogous to the gravitational force. When a force is conservative, it is possible to define a potential energy associated with the force. It is usually easier to work with the potential energy (because it depends only on position) than to calculate the work directly.

To show this explicitly, consider an electric charge $+q$ fixed at the origin and move another charge $+Q$ toward q in such a manner that, at each instant, the applied force $\vec{\mathbf{F}}$ exactly balances the electric force $\vec{\mathbf{F}}_e$ on Q (Figure 7.3). The work done by the applied force $\vec{\mathbf{F}}$ on the charge Q changes the potential energy of Q. We call this potential energy the **electrical potential energy** of Q.

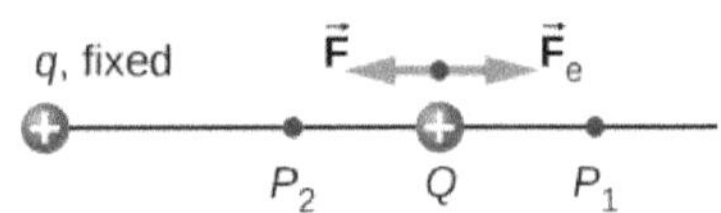

Figure 7.3 Displacement of “test” charge Q in the presence of fixed “source” charge q.

The work W_{12} done by the applied force $\vec{\mathbf{F}}$ when the particle moves from P_1 to P_2 may be calculated by

$$W_{12} = \int_{P_1}^{P_2} \vec{\mathbf{F}} \cdot d\vec{\mathbf{l}}.$$

Since the applied force $\vec{\mathbf{F}}$ balances the electric force $\vec{\mathbf{F}}_e$ on Q, the two forces have equal magnitude and opposite directions. Therefore, the applied force is

$$\vec{\mathbf{F}} = -\vec{\mathbf{F}}_e = -\frac{kqQ}{r^2}\hat{\mathbf{r}},$$

where we have defined positive to be pointing away from the origin and r is the distance from the origin. The directions of both the displacement and the applied force in the system in Figure 7.3 are parallel, and thus the work done on the system is positive.

We use the letter U to denote electric potential energy, which has units of joules (J). When a conservative force does negative work, the system gains potential energy. When a conservative force does positive work, the system loses potential energy, $\Delta U = -W$. In the system in Figure 7.3, the Coulomb force acts in the opposite direction to the displacement; therefore, the work is negative. However, we have increased the potential energy in the two-charge system.

Example 7.1

Kinetic Energy of a Charged Particle

A $+3.0\text{-nC}$ charge Q is initially at rest a distance of 10 cm (r_1) from a $+5.0\text{-nC}$ charge q fixed at the origin (Figure 7.4). Naturally, the Coulomb force accelerates Q away from q, eventually reaching 15 cm (r_2).

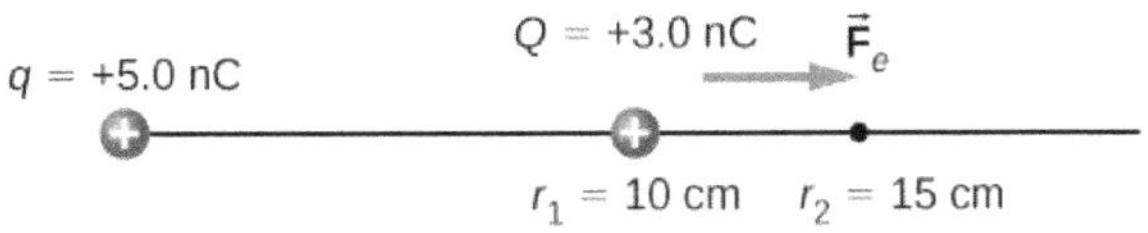

Figure 7.4 The charge Q is repelled by q, thus having work done on it and gaining kinetic energy.

a. What is the work done by the electric field between r_1 and r_2?

b. How much kinetic energy does Q have at r_2?

Strategy

Calculate the work with the usual definition. Since Q started from rest, this is the same as the kinetic energy.

Solution

Integrating force over distance, we obtain

$$\begin{aligned} W_{12} &= \int_{r_1}^{r_2} \vec{\mathbf{F}} \cdot d\vec{\mathbf{r}} = \int_{r_1}^{r_2} \frac{kqQ}{r^2} dr = \left[-\frac{kqQ}{r}\right]_{r_1}^{r_2} = kqQ\left[\frac{-1}{r_2} + \frac{1}{r_1}\right] \\ &= \left(8.99 \times 10^9 \text{ Nm}^2/\text{C}^2\right)\left(5.0 \times 10^{-9} \text{ C}\right)\left(3.0 \times 10^{-9} \text{ C}\right)\left[\frac{-1}{0.15 \text{ m}} + \frac{1}{0.10 \text{ m}}\right] \\ &= 4.5 \times 10^{-7} \text{ J}. \end{aligned}$$

This is also the value of the kinetic energy at r_2.

Significance

Charge Q was initially at rest; the electric field of q did work on Q, so now Q has kinetic energy equal to the work done by the electric field.

7.1 Check Your Understanding If Q has a mass of $4.00\,\mu\text{g}$, what is the speed of Q at r_2?

In this example, the work W done to accelerate a positive charge from rest is positive and results from a loss in U, or a negative ΔU. A value for U can be found at any point by taking one point as a reference and calculating the work needed to move a charge to the other point.

Electric Potential Energy

Work W done to accelerate a positive charge from rest is positive and results from a loss in U, or a negative ΔU. Mathematically,

$$W = -\Delta U. \tag{7.1}$$

Gravitational potential energy and electric potential energy are quite analogous. Potential energy accounts for work done by a conservative force and gives added insight regarding energy and energy transformation without the necessity of dealing with the force directly. It is much more common, for example, to use the concept of electric potential energy than to deal with the Coulomb force directly in real-world applications.

In polar coordinates with q at the origin and Q located at r, the displacement element vector is $d\vec{\mathbf{l}} = \hat{\mathbf{r}}\,dr$ and thus the work becomes

$$W_{12} = -kqQ\int_{r_1}^{r_2} \frac{1}{r^2}\hat{\mathrm{r}} \cdot \hat{\mathrm{r}}\,dr = kqQ\frac{1}{r_2} - kqQ\frac{1}{r_1}.$$

Notice that this result only depends on the endpoints and is otherwise independent of the path taken. To explore this further, compare path P_1 to P_2 with path $P_1P_3P_4P_2$ in Figure 7.5.

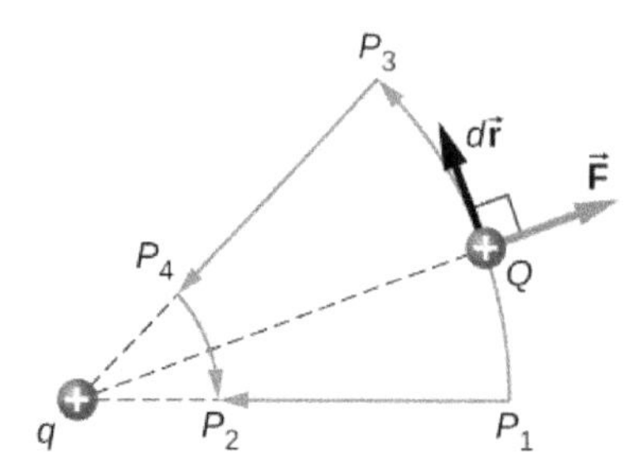

Figure 7.5 Two paths for displacement P_1 to P_2. The work on segments P_1P_3 and P_4P_2 are zero due to the electrical force being perpendicular to the displacement along these paths. Therefore, work on paths P_1P_2 and $P_1P_3P_4P_2$ are equal.

The segments P_1P_3 and P_4P_2 are arcs of circles centered at q. Since the force on Q points either toward or away from q, no work is done by a force balancing the electric force, because it is perpendicular to the displacement along these arcs. Therefore, the only work done is along segment P_3P_4, which is identical to P_1P_2.

One implication of this work calculation is that if we were to go around the path $P_1P_3P_4P_2P_1$, the net work would be zero (Figure 7.6). Recall that this is how we determine whether a force is conservative or not. Hence, because the electric force is related to the electric field by $\vec{\mathbf{F}} = q\vec{\mathbf{E}}$, the electric field is itself conservative. That is,

$$\oint \vec{\mathbf{E}} \cdot d\vec{\mathbf{l}} = 0.$$

Note that Q is a constant.

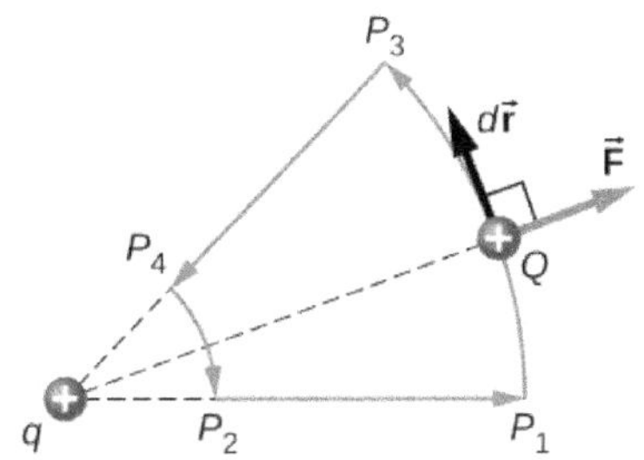

Figure 7.6 A closed path in an electric field. The net work around this path is zero.

Another implication is that we may define an electric potential energy. Recall that the work done by a conservative force is also expressed as the difference in the potential energy corresponding to that force. Therefore, the work W_{ref} to bring a charge from a reference point to a point of interest may be written as

$$W_{\text{ref}} = \int_{r_{\text{ref}}}^{r} \vec{\mathbf{F}} \cdot d\vec{\mathbf{l}}$$

and, by Equation 7.1, the difference in potential energy $(U_2 - U_1)$ of the test charge Q between the two points is

$$\Delta U = -\int_{r_{\text{ref}}}^{r} \vec{\mathbf{F}} \cdot d\vec{\mathbf{l}}.$$

Therefore, we can write a general expression for the potential energy of two point charges (in spherical coordinates):

$$\Delta U = -\int_{r_{\text{ref}}}^{r} \frac{kqQ}{r^2}dr = -\left[-\frac{kqQ}{r}\right]_{r_{\text{ref}}}^{r} = kqQ\left[\frac{1}{r} - \frac{1}{r_{\text{ref}}}\right].$$

We may take the second term to be an arbitrary constant reference level, which serves as the zero reference:

$$U(r) = k\frac{qQ}{r} - U_{\text{ref}}.$$

A convenient choice of reference that relies on our common sense is that when the two charges are infinitely far apart, there is no interaction between them. (Recall the discussion of reference potential energy in Potential Energy and Conservation of Energy (http://cnx.org/content/m58311/latest/) .) Taking the potential energy of this state to be zero removes the term U_{ref} from the equation (just like when we say the ground is zero potential energy in a gravitational potential energy problem), and the potential energy of Q when it is separated from q by a distance r assumes the form

$$U(r) = k\frac{qQ}{r} \text{ (zero reference at } r = \infty). \tag{7.2}$$

This formula is symmetrical with respect to q and Q, so it is best described as the potential energy of the two-charge system.

Example 7.2

Potential Energy of a Charged Particle

A $+3.0\text{-nC}$ charge Q is initially at rest a distance of 10 cm (r_1) from a $+5.0\text{-nC}$ charge q fixed at the origin (Figure 7.7). Naturally, the Coulomb force accelerates Q away from q, eventually reaching 15 cm (r_2).

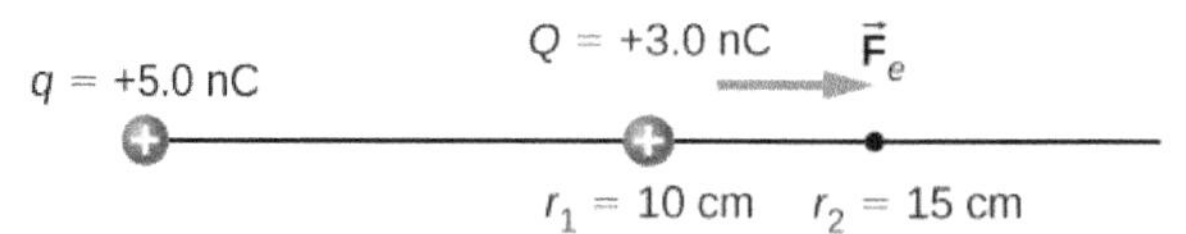

Figure 7.7 The charge Q is repelled by q, thus having work done on it and losing potential energy.

What is the change in the potential energy of the two-charge system from r_1 to r_2?

Strategy

Calculate the potential energy with the definition given above: $\Delta U_{12} = -\int_{r_1}^{r_2} \vec{\mathbf{F}} \cdot d\vec{\mathbf{r}}$. Since Q started from rest, this is the same as the kinetic energy.

Solution

We have

$$\Delta U_{12} = -\int_{r_1}^{r_2} \vec{\mathbf{F}} \cdot d\vec{\mathbf{r}} = -\int_{r_1}^{r_2} \frac{kqQ}{r^2} dr = -\left[-\frac{kqQ}{r}\right]_{r_1}^{r_2} = kqQ\left[\frac{1}{r_2} - \frac{1}{r_1}\right]$$

$$= \left(8.99 \times 10^9 \text{ Nm}^2/\text{C}^2\right)\left(5.0 \times 10^{-9} \text{ C}\right)\left(3.0 \times 10^{-9} \text{ C}\right)\left[\frac{1}{0.15 \text{ m}} - \frac{1}{0.10 \text{ m}}\right]$$

$$= -4.5 \times 10^{-7} \text{ J}.$$

Significance

The change in the potential energy is negative, as expected, and equal in magnitude to the change in kinetic energy in this system. Recall from Example 7.1 that the change in kinetic energy was positive.

7.2 Check Your Understanding What is the potential energy of Q relative to the zero reference at infinity at r_2 in the above example?

Due to Coulomb's law, the forces due to multiple charges on a test charge Q superimpose; they may be calculated individually and then added. This implies that the work integrals and hence the resulting potential energies exhibit the same behavior. To demonstrate this, we consider an example of assembling a system of four charges.

Example 7.3

Assembling Four Positive Charges

Find the amount of work an external agent must do in assembling four charges $+2.0\,\mu\text{C}$, $+3.0\,\mu\text{C}$, $+4.0\,\mu\text{C}$, and $+5.0\,\mu\text{C}$ at the vertices of a square of side 1.0 cm, starting each charge from infinity (Figure 7.8).

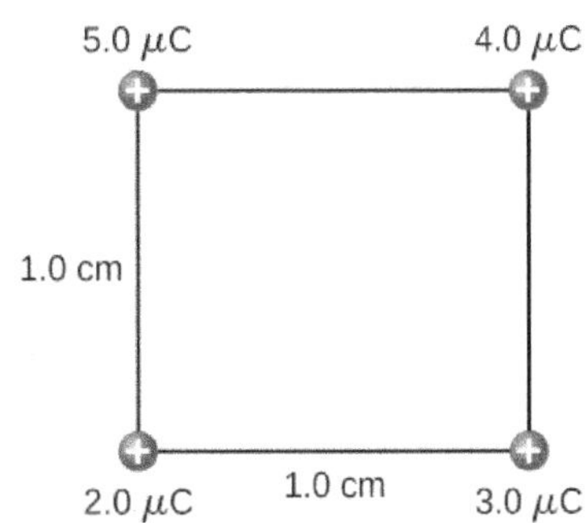

Figure 7.8 How much work is needed to assemble this charge configuration?

Strategy

We bring in the charges one at a time, giving them starting locations at infinity and calculating the work to bring them in from infinity to their final location. We do this in order of increasing charge.

Solution

Step 1. First bring the $+2.0\text{-}\mu\text{C}$ charge to the origin. Since there are no other charges at a finite distance from this charge yet, no work is done in bringing it from infinity,

$$W_1 = 0.$$

Step 2. While keeping the $+2.0\text{-}\mu\text{C}$ charge fixed at the origin, bring the $+3.0\text{-}\mu\text{C}$ charge to $(x, y, z) = (1.0\text{ cm}, 0, 0)$ (Figure 7.9). Now, the applied force must do work against the force exerted by the $+2.0\text{-}\mu\text{C}$ charge fixed at the origin. The work done equals the change in the potential energy of the $+3.0\text{-}\mu\text{C}$ charge:

$$W_2 = k\frac{q_1 q_2}{r_{12}} = \left(9.0\times 10^9 \frac{\text{N}\cdot\text{m}^2}{\text{C}^2}\right)\frac{\left(2.0\times 10^{-6}\text{ C}\right)\left(3.0\times 10^{-6}\text{ C}\right)}{1.0\times 10^{-2}\text{ m}} = 5.4\text{ J}.$$

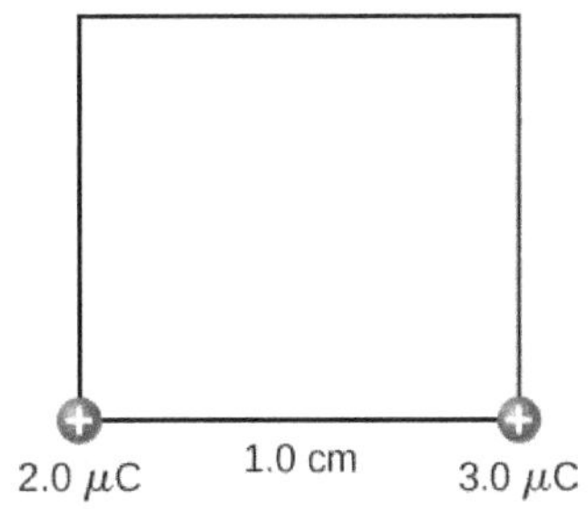

Figure 7.9 Step 2. Work W_2 to bring the $+3.0\text{-}\mu\text{C}$ charge from infinity.

Step 3. While keeping the charges of $+2.0\,\mu\text{C}$ and $+3.0\,\mu\text{C}$ fixed in their places, bring in the $+4.0\text{-}\mu\text{C}$ charge to $(x, y, z) = (1.0\text{ cm}, 1.0\text{ cm}, 0)$ (Figure 7.10). The work done in this step is

$$\begin{aligned} W_3 &= k\frac{q_1 q_3}{r_{13}} + k\frac{q_2 q_3}{r_{23}} \\ &= \left(9.0\times 10^9 \frac{\text{N}\cdot\text{m}^2}{\text{C}^2}\right)\left[\frac{\left(2.0\times 10^{-6}\text{ C}\right)\left(4.0\times 10^{-6}\text{ C}\right)}{\sqrt{2}\times 10^{-2}\text{ m}} + \frac{\left(3.0\times 10^{-6}\text{ C}\right)\left(4.0\times 10^{-6}\text{ C}\right)}{1.0\times 10^{-2}\text{ m}}\right] = 15.9\text{ J}. \end{aligned}$$

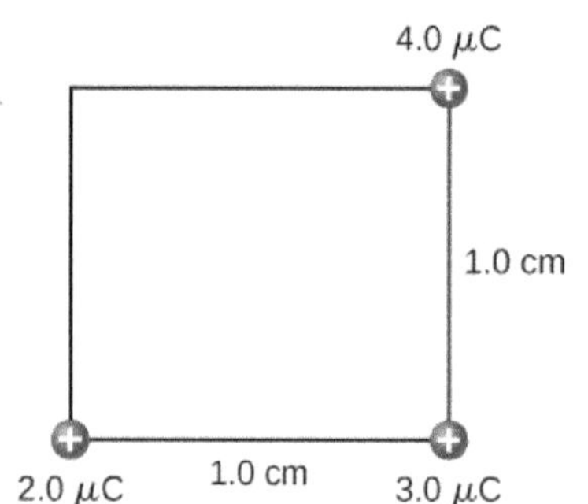

Figure 7.10 Step 3. The work W_3 to bring the $+4.0\text{-}\mu\text{C}$ charge from infinity.

Step 4. Finally, while keeping the first three charges in their places, bring the $+5.0\text{-}\mu\text{C}$ charge to $(x, y, z) = (0,\ 1.0\,\text{cm},\ 0)$ (Figure 7.11). The work done here is

$$\begin{aligned} W_4 &= kq_4\left[\frac{q_1}{r_{14}} + \frac{q_2}{r_{24}} + \frac{q_3}{r_{34}}\right], \\ &= \left(9.0 \times 10^9 \frac{\text{N} \cdot \text{m}^2}{\text{C}^2}\right)\left(5.0 \times 10^{-6}\ \text{C}\right)\left[\frac{\left(2.0 \times 10^{-6}\ \text{C}\right)}{1.0 \times 10^{-2}\ \text{m}} + \frac{\left(3.0 \times 10^{-6}\ \text{C}\right)}{\sqrt{2} \times 10^{-2}\ \text{m}} + \frac{\left(4.0 \times 10^{-6}\ \text{C}\right)}{1.0 \times 10^{-2}\ \text{m}}\right] = 36.5\ \text{J}. \end{aligned}$$

5.0 μC
4.0 μC
1.0 cm
2.0 μC
1.0 cm
3.0 μC

Figure 7.11 Step 4. The work W_4 to bring the $+5.0\text{-}\mu\text{C}$ charge from infinity.

Hence, the total work done by the applied force in assembling the four charges is equal to the sum of the work in bringing each charge from infinity to its final position:

$$W_\text{T} = W_1 + W_2 + W_3 + W_4 = 0 + 5.4\ \text{J} + 15.9\ \text{J} + 36.5\ \text{J} = 57.8\ \text{J}.$$

Significance

The work on each charge depends only on its pairwise interactions with the other charges. No more complicated interactions need to be considered; the work on the third charge only depends on its interaction with the first and second charges, the interaction between the first and second charge does not affect the third.

7.3 Check Your Understanding Is the electrical potential energy of two point charges positive or negative if the charges are of the same sign? Opposite signs? How does this relate to the work necessary to bring the charges into proximity from infinity?

Note that the electrical potential energy is positive if the two charges are of the same type, either positive or negative, and negative if the two charges are of opposite types. This makes sense if you think of the change in the potential energy ΔU as you bring the two charges closer or move them farther apart. Depending on the relative types of charges, you may have to work on the system or the system would do work on you, that is, your work is either positive or negative. If you have to do positive work on the system (actually push the charges closer), then the energy of the system should increase. If you bring two positive charges or two negative charges closer, you have to do positive work on the system, which raises their potential energy. Since potential energy is proportional to $1/r$, the potential energy goes up when r goes down between two positive or two negative charges.

On the other hand, if you bring a positive and a negative charge nearer, you have to do negative work on the system (the charges are pulling you), which means that you take energy away from the system. This reduces the potential energy. Since potential energy is negative in the case of a positive and a negative charge pair, the increase in $1/r$ makes the potential energy more negative, which is the same as a reduction in potential energy.

The result from Example 7.1 may be extended to systems with any arbitrary number of charges. In this case, it is most convenient to write the formula as

$$W_{12 \ldots N} = \frac{k}{2}\sum_{i}^{N}\sum_{j}^{N}\frac{q_i q_j}{r_{ij}} \text{ for } i \neq j. \tag{7.3}$$

The factor of 1/2 accounts for adding each pair of charges twice.

7.2 | Electric Potential and Potential Difference

Learning Objectives

By the end of this section, you will be able to:

- Define electric potential, voltage, and potential difference
- Define the electron-volt
- Calculate electric potential and potential difference from potential energy and electric field
- Describe systems in which the electron-volt is a useful unit
- Apply conservation of energy to electric systems

Recall that earlier we defined electric field to be a quantity independent of the test charge in a given system, which would nonetheless allow us to calculate the force that would result on an arbitrary test charge. (The default assumption in the absence of other information is that the test charge is positive.) We briefly defined a field for gravity, but gravity is always attractive, whereas the electric force can be either attractive or repulsive. Therefore, although potential energy is perfectly adequate in a gravitational system, it is convenient to define a quantity that allows us to calculate the work on a charge independent of the magnitude of the charge. Calculating the work directly may be difficult, since $W = \vec{\mathbf{F}} \cdot \vec{\mathbf{d}}$ and the direction and magnitude of $\vec{\mathbf{F}}$ can be complex for multiple charges, for odd-shaped objects, and along arbitrary paths. But we do know that because $\vec{\mathbf{F}} = q\vec{\mathbf{E}}$, the work, and hence ΔU, is proportional to the test charge q. To have a physical quantity that is independent of test charge, we define **electric potential** V (or simply potential, since electric is understood) to be the potential energy per unit charge:

Electric Potential

The electric potential energy per unit charge is

$$V = \frac{U}{q}. \tag{7.4}$$

Since U is proportional to q, the dependence on q cancels. Thus, V does not depend on q. The change in potential energy ΔU is crucial, so we are concerned with the difference in potential or potential difference ΔV between two points, where

$$\Delta V = V_B - V_A = \frac{\Delta U}{q}.$$

Electric Potential Difference

The **electric potential difference** between points *A* and *B*, $V_B - V_A$, is defined to be the change in potential energy of a charge *q* moved from *A* to *B*, divided by the charge. Units of potential difference are joules per coulomb, given the name volt (V) after Alessandro Volta.

$$1\ \text{V} = 1\ \text{J/C}$$

The familiar term **voltage** is the common name for electric potential difference. Keep in mind that whenever a voltage is quoted, it is understood to be the potential difference between two points. For example, every battery has two terminals, and its voltage is the potential difference between them. More fundamentally, the point you choose to be zero volts is arbitrary. This is analogous to the fact that gravitational potential energy has an arbitrary zero, such as sea level or perhaps a lecture hall floor. It is worthwhile to emphasize the distinction between potential difference and electrical potential energy.

Potential Difference and Electrical Potential Energy

The relationship between potential difference (or voltage) and electrical potential energy is given by

$$\Delta V = \frac{\Delta U}{q} \text{ or } \Delta U = q\Delta V. \tag{7.5}$$

Voltage is not the same as energy. Voltage is the energy per unit charge. Thus, a motorcycle battery and a car battery can both have the same voltage (more precisely, the same potential difference between battery terminals), yet one stores much more energy than the other because $\Delta U = q\Delta V$. The car battery can move more charge than the motorcycle battery, although both are 12-V batteries.

Example 7.4

Calculating Energy

You have a 12.0-V motorcycle battery that can move 5000 C of charge, and a 12.0-V car battery that can move 60,000 C of charge. How much energy does each deliver? (Assume that the numerical value of each charge is accurate to three significant figures.)

Strategy

To say we have a 12.0-V battery means that its terminals have a 12.0-V potential difference. When such a battery moves charge, it puts the charge through a potential difference of 12.0 V, and the charge is given a change in potential energy equal to $\Delta U = q\Delta V$. To find the energy output, we multiply the charge moved by the potential difference.

Solution

For the motorcycle battery, $q = 5000\ \text{C}$ and $\Delta V = 12.0\ \text{V}$. The total energy delivered by the motorcycle battery is

$$\Delta U_{\text{cycle}} = (5000\ \text{C})(12.0\ \text{V}) = (5000\ \text{C})(12.0\ \text{J/C}) = 6.00 \times 10^4\ \text{J}.$$

Similarly, for the car battery, $q = 60{,}000\ \text{C}$ and

$$\Delta U_{\text{car}} = (60{,}000\ \text{C})(12.0\ \text{V}) = 7.20 \times 10^5\ \text{J}.$$

Significance

Voltage and energy are related, but they are not the same thing. The voltages of the batteries are identical, but the energy supplied by each is quite different. A car battery has a much larger engine to start than a motorcycle. Note also that as a battery is discharged, some of its energy is used internally and its terminal voltage drops, such as when headlights dim because of a depleted car battery. The energy supplied by the battery is still calculated as in this example, but not all of the energy is available for external use.

7.4 Check Your Understanding How much energy does a 1.5-V AAA battery have that can move 100 C?

Note that the energies calculated in the previous example are absolute values. The change in potential energy for the battery is negative, since it loses energy. These batteries, like many electrical systems, actually move negative charge—electrons in particular. The batteries repel electrons from their negative terminals (*A*) through whatever circuitry is involved and attract them to their positive terminals (*B*), as shown in Figure 7.12. The change in potential is $\Delta V = V_B - V_A = +12\text{ V}$ and the charge q is negative, so that $\Delta U = q\Delta V$ is negative, meaning the potential energy of the battery has decreased when q has moved from *A* to *B*.

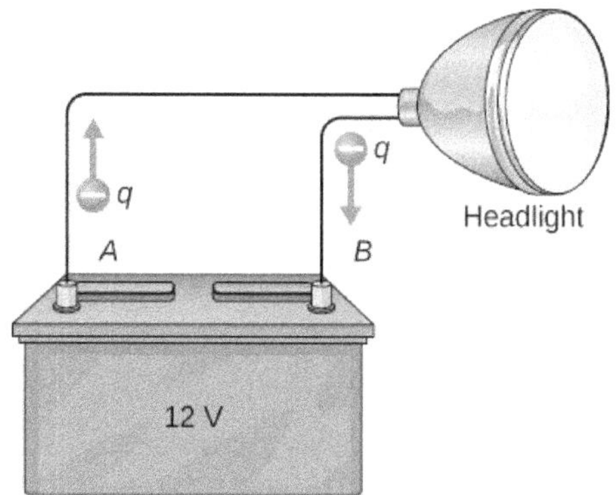

Figure 7.12 A battery moves negative charge from its negative terminal through a headlight to its positive terminal. Appropriate combinations of chemicals in the battery separate charges so that the negative terminal has an excess of negative charge, which is repelled by it and attracted to the excess positive charge on the other terminal. In terms of potential, the positive terminal is at a higher voltage than the negative terminal. Inside the battery, both positive and negative charges move.

Example 7.5

How Many Electrons Move through a Headlight Each Second?

When a 12.0-V car battery powers a single 30.0-W headlight, how many electrons pass through it each second?

Strategy

To find the number of electrons, we must first find the charge that moves in 1.00 s. The charge moved is related to voltage and energy through the equations $\Delta U = q\Delta V$. A 30.0-W lamp uses 30.0 joules per second. Since the battery loses energy, we have $\Delta U = -30\text{ J}$ and, since the electrons are going from the negative terminal to the positive, we see that $\Delta V = +12.0\text{ V}$.

Solution

To find the charge q moved, we solve the equation $\Delta U = q\Delta V$:

$$q = \frac{\Delta U}{\Delta V}.$$

Entering the values for ΔU and ΔV, we get

$$q = \frac{-30.0\,\text{J}}{+12.0\,\text{V}} = \frac{-30.0\,\text{J}}{+12.0\,\text{J/C}} = -2.50\,\text{C}.$$

The number of electrons n_e is the total charge divided by the charge per electron. That is,

$$n_e = \frac{-2.50\,\text{C}}{-1.60 \times 10^{-19}\,\text{C/e}^-} = 1.56 \times 10^{19}\ \text{electrons}.$$

Significance

This is a very large number. It is no wonder that we do not ordinarily observe individual electrons with so many being present in ordinary systems. In fact, electricity had been in use for many decades before it was determined that the moving charges in many circumstances were negative. Positive charge moving in the opposite direction of negative charge often produces identical effects; this makes it difficult to determine which is moving or whether both are moving.

7.5 Check Your Understanding How many electrons would go through a 24.0-W lamp?

The Electron-Volt

The energy per electron is very small in macroscopic situations like that in the previous example—a tiny fraction of a joule. But on a submicroscopic scale, such energy per particle (electron, proton, or ion) can be of great importance. For example, even a tiny fraction of a joule can be great enough for these particles to destroy organic molecules and harm living tissue. The particle may do its damage by direct collision, or it may create harmful X-rays, which can also inflict damage. It is useful to have an energy unit related to submicroscopic effects.

Figure 7.13 shows a situation related to the definition of such an energy unit. An electron is accelerated between two charged metal plates, as it might be in an old-model television tube or oscilloscope. The electron gains kinetic energy that is later converted into another form—light in the television tube, for example. (Note that in terms of energy, "downhill" for the electron is "uphill" for a positive charge.) Since energy is related to voltage by $\Delta U = q\Delta V$, we can think of the joule as a coulomb-volt.

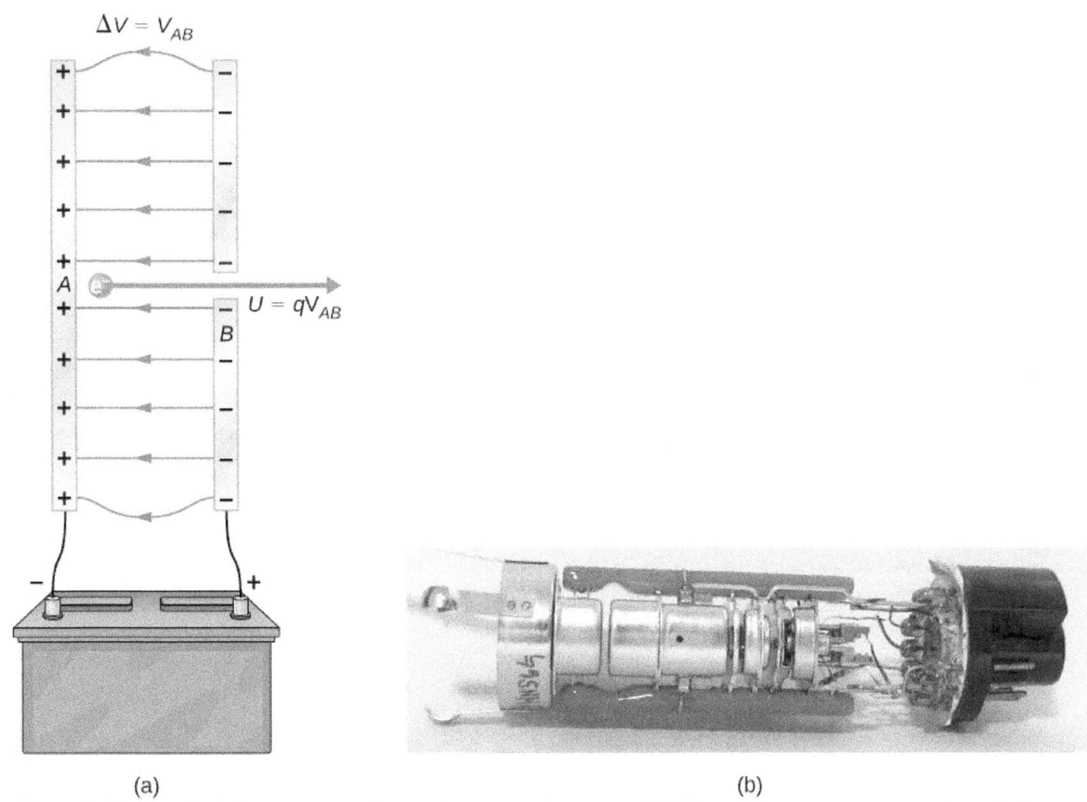

Figure 7.13 A typical electron gun accelerates electrons using a potential difference between two separated metal plates. By conservation of energy, the kinetic energy has to equal the change in potential energy, so $KE = qV$. The energy of the electron in electron-volts is numerically the same as the voltage between the plates. For example, a 5000-V potential difference produces 5000-eV electrons. The conceptual construct, namely two parallel plates with a hole in one, is shown in (a), while a real electron gun is shown in (b).

Electron-Volt

On the submicroscopic scale, it is more convenient to define an energy unit called the **electron-volt** (eV), which is the energy given to a fundamental charge accelerated through a potential difference of 1 V. In equation form,

$$1\text{ eV} = (1.60 \times 10^{-19}\text{ C})(1\text{ V}) = (1.60 \times 10^{-19}\text{ C})(1\text{ J/C}) = 1.60 \times 10^{-19}\text{ J}.$$

An electron accelerated through a potential difference of 1 V is given an energy of 1 eV. It follows that an electron accelerated through 50 V gains 50 eV. A potential difference of 100,000 V (100 kV) gives an electron an energy of 100,000 eV (100 keV), and so on. Similarly, an ion with a double positive charge accelerated through 100 V gains 200 eV of energy. These simple relationships between accelerating voltage and particle charges make the electron-volt a simple and convenient energy unit in such circumstances.

The electron-volt is commonly employed in submicroscopic processes—chemical valence energies and molecular and nuclear binding energies are among the quantities often expressed in electron-volts. For example, about 5 eV of energy is required to break up certain organic molecules. If a proton is accelerated from rest through a potential difference of 30 kV, it acquires an energy of 30 keV (30,000 eV) and can break up as many as 6000 of these molecules $(30{,}000\text{ eV} \div 5\text{ eV per molecule} = 6000\text{ molecules})$. Nuclear decay energies are on the order of 1 MeV (1,000,000 eV) per event and can thus produce significant biological damage.

Conservation of Energy

The total energy of a system is conserved if there is no net addition (or subtraction) due to work or heat transfer. For conservative forces, such as the electrostatic force, conservation of energy states that mechanical energy is a constant.

Mechanical energy is the sum of the kinetic energy and potential energy of a system; that is, $K + U = \text{constant}$. A loss of U for a charged particle becomes an increase in its K. Conservation of energy is stated in equation form as

$$K + U = \text{constant}$$

or

$$K_\text{i} + U_\text{i} = K_\text{f} + U_\text{f}$$

where i and f stand for initial and final conditions. As we have found many times before, considering energy can give us insights and facilitate problem solving.

Example 7.6

Electrical Potential Energy Converted into Kinetic Energy

Calculate the final speed of a free electron accelerated from rest through a potential difference of 100 V. (Assume that this numerical value is accurate to three significant figures.)

Strategy

We have a system with only conservative forces. Assuming the electron is accelerated in a vacuum, and neglecting the gravitational force (we will check on this assumption later), all of the electrical potential energy is converted into kinetic energy. We can identify the initial and final forms of energy to be $K_\text{i} = 0,\ K_\text{f} = \frac{1}{2}mv^2,\ U_\text{i} = qV,\ U_\text{f} = 0.$

Solution

Conservation of energy states that

$$K_\text{i} + U_\text{i} = K_\text{f} + U_\text{f}.$$

Entering the forms identified above, we obtain

$$qV = \frac{mv^2}{2}.$$

We solve this for v:

$$v = \sqrt{\frac{2qV}{m}}.$$

Entering values for q, V, and m gives

$$v = \sqrt{\frac{2(-1.60 \times 10^{-19}\ \text{C})(-100\ \text{J/C})}{9.11 \times 10^{-31}\ \text{kg}}} = 5.93 \times 10^6\ \text{m/s}.$$

Significance

Note that both the charge and the initial voltage are negative, as in Figure 7.13. From the discussion of electric charge and electric field, we know that electrostatic forces on small particles are generally very large compared with the gravitational force. The large final speed confirms that the gravitational force is indeed negligible here. The large speed also indicates how easy it is to accelerate electrons with small voltages because of their very small mass. Voltages much higher than the 100 V in this problem are typically used in electron guns. These higher voltages produce electron speeds so great that effects from special relativity must be taken into account and hence are reserved for a later chapter (Relativity (http://cnx.org/content/m58555/latest/)). That is why we consider a low voltage (accurately) in this example.

7.6 Check Your Understanding How would this example change with a positron? A positron is identical to an electron except the charge is positive.

Voltage and Electric Field

So far, we have explored the relationship between voltage and energy. Now we want to explore the relationship between voltage and electric field. We will start with the general case for a non-uniform $\vec{\mathbf{E}}$ field. Recall that our general formula for the potential energy of a test charge q at point P relative to reference point R is

$$U_P = -\int_R^P \vec{\mathbf{F}} \cdot d\vec{\mathbf{l}}.$$

When we substitute in the definition of electric field $(\vec{\mathbf{E}} = \vec{\mathbf{F}}/q),$ this becomes

$$U_P = -q\int_R^P \vec{\mathbf{E}} \cdot d\vec{\mathbf{l}}.$$

Applying our definition of potential $(V = U/q)$ to this potential energy, we find that, in general,

$$V_P = -\int_R^P \vec{\mathbf{E}} \cdot d\vec{\mathbf{l}}. \quad (7.6)$$

From our previous discussion of the potential energy of a charge in an electric field, the result is independent of the path chosen, and hence we can pick the integral path that is most convenient.

Consider the special case of a positive point charge q at the origin. To calculate the potential caused by q at a distance r from the origin relative to a reference of 0 at infinity (recall that we did the same for potential energy), let $P = r$ and $R = \infty$, with $d\vec{\mathbf{l}} = d\vec{\mathbf{r}} = \hat{\mathbf{r}}dr$ and use $\vec{\mathbf{E}} = \frac{kq}{r^2}\hat{\mathbf{r}}$. When we evaluate the integral

$$V_P = -\int_R^P \vec{\mathbf{E}} \cdot d\vec{\mathbf{l}}$$

for this system, we have

$$V_r = -\int_\infty^r \frac{kq}{r^2}\hat{\mathbf{r}} \cdot \hat{\mathbf{r}}dr,$$

which simplifies to

$$V_r = -\int_\infty^r \frac{kq}{r^2}dr = \frac{kq}{r} - \frac{kq}{\infty} = \frac{kq}{r}.$$

This result,

$$V_r = \frac{kq}{r}$$

is the standard form of the potential of a point charge. This will be explored further in the next section.

To examine another interesting special case, suppose a uniform electric field $\vec{\mathbf{E}}$ is produced by placing a potential difference (or voltage) ΔV across two parallel metal plates, labeled A and B (Figure 7.14). Examining this situation will tell us what voltage is needed to produce a certain electric field strength. It will also reveal a more fundamental relationship between electric potential and electric field.

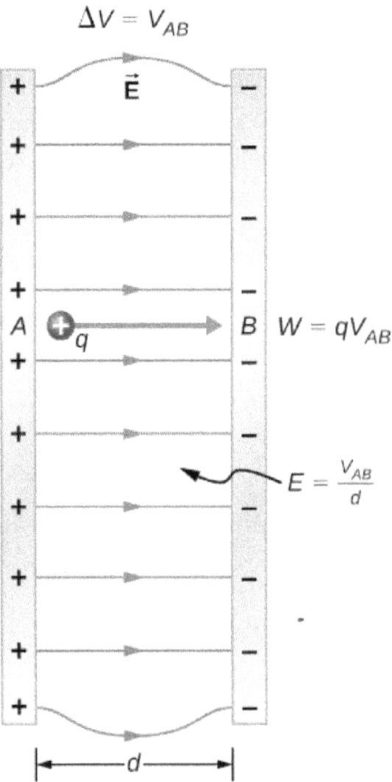

Figure 7.14 The relationship between V and E for parallel conducting plates is $E = V / d$. (Note that $\Delta V = V_{AB}$ in magnitude. For a charge that is moved from plate A at higher potential to plate B at lower potential, a minus sign needs to be included as follows: $-\Delta V = V_A - V_B = V_{AB}$.)

From a physicist's point of view, either ΔV or $\vec{\mathbf{E}}$ can be used to describe any interaction between charges. However, ΔV is a scalar quantity and has no direction, whereas $\vec{\mathbf{E}}$ is a vector quantity, having both magnitude and direction. (Note that the magnitude of the electric field, a scalar quantity, is represented by E.) The relationship between ΔV and $\vec{\mathbf{E}}$ is revealed by calculating the work done by the electric force in moving a charge from point A to point B. But, as noted earlier, arbitrary charge distributions require calculus. We therefore look at a uniform electric field as an interesting special case.

The work done by the electric field in Figure 7.14 to move a positive charge q from A, the positive plate, higher potential, to B, the negative plate, lower potential, is

$$W = -\Delta U = -q\Delta V.$$

The potential difference between points A and B is

$$-\Delta V = -(V_B - V_A) = V_A - V_B = V_{AB}.$$

Entering this into the expression for work yields

$$W = qV_{AB}.$$

Work is $W = \vec{\mathbf{F}} \cdot \vec{\mathbf{d}} = Fd\cos\theta$; here $\cos\theta = 1$, since the path is parallel to the field. Thus, $W = Fd$. Since $F = qE$, we see that $W = qEd$.

Substituting this expression for work into the previous equation gives

$$qEd = qV_{AB}.$$

The charge cancels, so we obtain for the voltage between points A and B

$$\left.\begin{array}{l} V_{AB} = Ed \\ E = \dfrac{V_{AB}}{d} \end{array}\right\} \text{(uniform } E\text{-field on y)}$$

where d is the distance from A to B, or the distance between the plates in Figure 7.14. Note that this equation implies that the units for electric field are volts per meter. We already know the units for electric field are newtons per coulomb; thus, the following relation among units is valid:

$$1\ \text{N/C} = 1\ \text{V/m}.$$

Furthermore, we may extend this to the integral form. Substituting Equation 7.5 into our definition for the potential difference between points A and B, we obtain

$$V_{AB} = V_B - V_A = -\int_R^B \vec{\mathbf{E}} \cdot d\vec{\mathbf{l}} + \int_R^A \vec{\mathbf{E}} \cdot d\vec{\mathbf{l}}$$

which simplifies to

$$V_B - V_A = -\int_A^B \vec{\mathbf{E}} \cdot d\vec{\mathbf{l}}.$$

As a demonstration, from this we may calculate the potential difference between two points (A and B) equidistant from a point charge q at the origin, as shown in Figure 7.15.

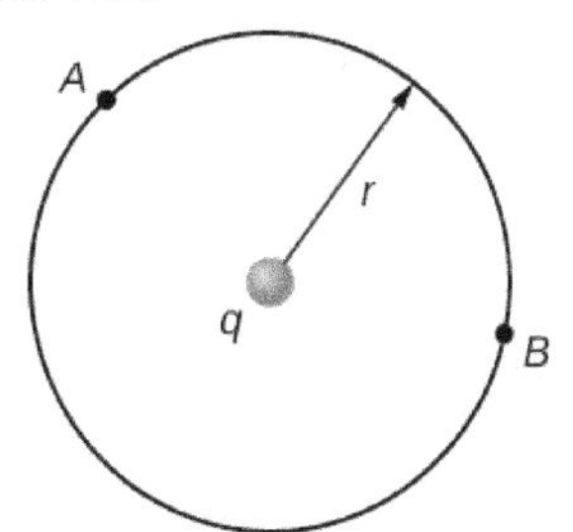

Figure 7.15 The arc for calculating the potential difference between two points that are equidistant from a point charge at the origin.

To do this, we integrate around an arc of the circle of constant radius r between A and B, which means we let $d\vec{\mathbf{l}} = r\hat{\varphi}d\varphi$, while using $\vec{\mathbf{E}} = \frac{kq}{r^2}\hat{\mathbf{r}}$. Thus,

$$\Delta V_{AB} = V_B - V_A = -\int_A^B \vec{\mathbf{E}} \cdot d\vec{\mathbf{l}} \tag{7.7}$$

for this system becomes

$$V_B - V_A = -\int_A^B \frac{kq}{r^2}\hat{\mathbf{r}} \cdot r\hat{\boldsymbol{\varphi}}d\varphi.$$

However, $\hat{\mathbf{r}} \cdot \hat{\boldsymbol{\varphi}} = 0$ and therefore

$$V_B - V_A = 0.$$

This result, that there is no difference in potential along a constant radius from a point charge, will come in handy when we map potentials.

Example 7.7

What Is the Highest Voltage Possible between Two Plates?

Dry air can support a maximum electric field strength of about 3.0×10^6 V/m. Above that value, the field creates enough ionization in the air to make the air a conductor. This allows a discharge or spark that reduces the field. What, then, is the maximum voltage between two parallel conducting plates separated by 2.5 cm of dry air?

Strategy

We are given the maximum electric field E between the plates and the distance d between them. We can use the equation $V_{AB} = Ed$ to calculate the maximum voltage.

Solution

The potential difference or voltage between the plates is

$$V_{AB} = Ed.$$

Entering the given values for E and d gives

$$V_{AB} = (3.0 \times 10^6 \text{ V/m})(0.025 \text{ m}) = 7.5 \times 10^4 \text{ V}$$

or

$$V_{AB} = 75 \text{ kV}.$$

(The answer is quoted to only two digits, since the maximum field strength is approximate.)

Significance

One of the implications of this result is that it takes about 75 kV to make a spark jump across a 2.5-cm (1-in.) gap, or 150 kV for a 5-cm spark. This limits the voltages that can exist between conductors, perhaps on a power transmission line. A smaller voltage can cause a spark if there are spines on the surface, since sharp points have larger field strengths than smooth surfaces. Humid air breaks down at a lower field strength, meaning that a smaller voltage will make a spark jump through humid air. The largest voltages can be built up with static electricity on dry days (Figure 7.16).

Figure 7.16 A spark chamber is used to trace the paths of high-energy particles. Ionization created by the particles as they pass through the gas between the plates allows a spark to jump. The sparks are perpendicular to the plates, following electric field lines between them. The potential difference between adjacent plates is not high enough to cause sparks without the ionization produced by particles from accelerator experiments (or cosmic rays). This form of detector is now archaic and no longer in use except for demonstration purposes. (credit b: modification of work by Jack Collins)

Example 7.8

Field and Force inside an Electron Gun

An electron gun (Figure 7.13) has parallel plates separated by 4.00 cm and gives electrons 25.0 keV of energy. (a) What is the electric field strength between the plates? (b) What force would this field exert on a piece of plastic with a $0.500\text{-}\mu\text{C}$ charge that gets between the plates?

Strategy

Since the voltage and plate separation are given, the electric field strength can be calculated directly from the expression $E = \frac{V_{AB}}{d}$. Once we know the electric field strength, we can find the force on a charge by using $\vec{\mathbf{F}} = q\vec{\mathbf{E}}$. Since the electric field is in only one direction, we can write this equation in terms of the magnitudes, $F = qE$.

Solution

a. The expression for the magnitude of the electric field between two uniform metal plates is

$$E = \frac{V_{AB}}{d}.$$

Since the electron is a single charge and is given 25.0 keV of energy, the potential difference must be 25.0 kV. Entering this value for V_{AB} and the plate separation of 0.0400 m, we obtain

$$E = \frac{25.0\,\text{kV}}{0.0400\,\text{m}} = 6.25 \times 10^5\ \text{V/m}.$$

b. The magnitude of the force on a charge in an electric field is obtained from the equation

$$F = qE.$$

Substituting known values gives

$$F = (0.500 \times 10^{-6}\ \text{C})(6.25 \times 10^5\ \text{V/m}) = 0.313\ \text{N}.$$

Significance

Note that the units are newtons, since $1\ \text{V/m} = 1\ \text{N/C}$. Because the electric field is uniform between the plates, the force on the charge is the same no matter where the charge is located between the plates.

Example 7.9

Calculating Potential of a Point Charge

Given a point charge $q = +2.0\,\text{nC}$ at the origin, calculate the potential difference between point P_1 a distance $a = 4.0\,\text{cm}$ from q, and P_2 a distance $b = 12.0\,\text{cm}$ from q, where the two points have an angle of $\varphi = 24°$ between them (Figure 7.17).

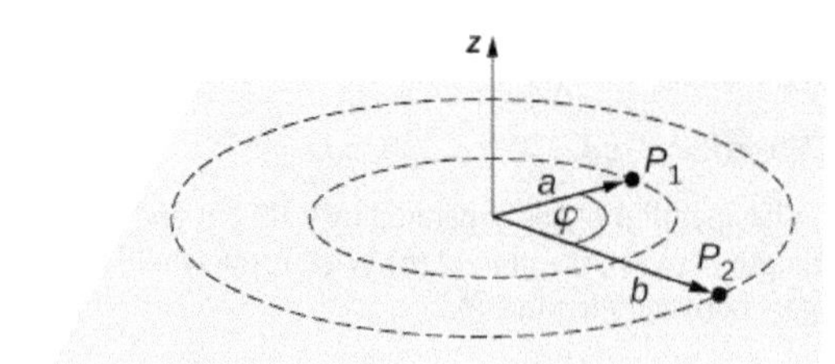

Figure 7.17 Find the difference in potential between P_1 and P_2.

Strategy

Do this in two steps. The first step is to use $V_B - V_A = -\int_A^B \vec{\mathbf{E}} \cdot d\vec{\mathbf{l}}$ and let $A = a = 4.0\,\text{cm}$ and $B = b = 12.0\,\text{cm}$, with $d\vec{\mathbf{l}} = d\vec{\mathbf{r}} = \hat{\mathbf{r}}dr$ and $\vec{\mathbf{E}} = \frac{kq}{r^2}\hat{\mathbf{r}}$. Then perform the integral. The second step is to integrate $V_B - V_A = -\int_A^B \vec{\mathbf{E}} \cdot d\vec{\mathbf{l}}$ around an arc of constant radius *r*, which means we let $d\vec{\mathbf{l}} = r\hat{\varphi}d\varphi$ with limits $0 \le \varphi \le 24°$, still using $\vec{\mathbf{E}} = \frac{kq}{r^2}\hat{\mathbf{r}}$. Then add the two results together.

Solution

For the first part, $V_B - V_A = -\int_A^B \vec{\mathbf{E}} \cdot d\vec{\mathbf{l}}$ for this system becomes $V_b - V_a = -\int_a^b \frac{kq}{r^2}\hat{\mathbf{r}} \cdot \hat{\mathbf{r}}dr$ which computes to

$$\begin{aligned}\Delta V &= -\int_a^b \frac{kq}{r^2}dr = kq\left[\frac{1}{a} - \frac{1}{b}\right] \\ &= \left(8.99 \times 10^9\ \text{Nm}^2/\text{C}^2\right)\left(2.0 \times 10^{-9}\ \text{C}\right)\left[\frac{1}{0.040\ \text{m}} - \frac{1}{0.12\ \text{m}}\right] = 300\ \text{V}.\end{aligned}$$

For the second step, $V_B - V_A = -\int_A^B \vec{\mathbf{E}} \cdot d\vec{\mathbf{l}}$ becomes $\Delta V = -\int_0^{24°} \frac{kq}{r^2}\hat{\mathbf{r}} \cdot r\hat{\boldsymbol{\varphi}}d\varphi$, but $\hat{\mathbf{r}} \cdot \hat{\boldsymbol{\varphi}} = 0$ and therefore $\Delta V = 0$. Adding the two parts together, we get 300 V.

Significance

We have demonstrated the use of the integral form of the potential difference to obtain a numerical result. Notice that, in this particular system, we could have also used the formula for the potential due to a point charge at the two points and simply taken the difference.

7.7 Check Your Understanding From the examples, how does the energy of a lightning strike vary with the height of the clouds from the ground? Consider the cloud-ground system to be two parallel plates.

Before presenting problems involving electrostatics, we suggest a problem-solving strategy to follow for this topic.

Problem-Solving Strategy: Electrostatics

1. Examine the situation to determine if static electricity is involved; this may concern separated stationary charges, the forces among them, and the electric fields they create.
2. Identify the system of interest. This includes noting the number, locations, and types of charges involved.
3. Identify exactly what needs to be determined in the problem (identify the unknowns). A written list is useful. Determine whether the Coulomb force is to be considered directly—if so, it may be useful to draw a free-body diagram, using electric field lines.
4. Make a list of what is given or can be inferred from the problem as stated (identify the knowns). It is important to distinguish the Coulomb force *F* from the electric field *E*, for example.
5. Solve the appropriate equation for the quantity to be determined (the unknown) or draw the field lines as requested.
6. Examine the answer to see if it is reasonable: Does it make sense? Are units correct and the numbers involved reasonable?

7.3 | Calculations of Electric Potential

Learning Objectives

By the end of this section, you will be able to:

- Calculate the potential due to a point charge
- Calculate the potential of a system of multiple point charges
- Describe an electric dipole
- Define dipole moment
- Calculate the potential of a continuous charge distribution

Point charges, such as electrons, are among the fundamental building blocks of matter. Furthermore, spherical charge distributions (such as charge on a metal sphere) create external electric fields exactly like a point charge. The electric potential due to a point charge is, thus, a case we need to consider.

We can use calculus to find the work needed to move a test charge *q* from a large distance away to a distance of *r* from a point charge *q*. Noting the connection between work and potential $W = -q\Delta V,$ as in the last section, we can obtain the following result.

Electric Potential *V* of a Point Charge

The electric potential *V* of a point charge is given by

$$V = \frac{kq}{r} \text{(point charge)} \tag{7.8}$$

where *k* is a constant equal to $9.0 \times 10^9 \text{ N} \cdot \text{m}^2/\text{C}^2.$

The potential at infinity is chosen to be zero. Thus, *V* for a point charge decreases with distance, whereas $\vec{\mathbf{E}}$ for a point charge decreases with distance squared:

$$E = \frac{F}{q_t} = \frac{kq}{r^2}.$$

Recall that the electric potential *V* is a scalar and has no direction, whereas the electric field $\vec{\mathbf{E}}$ is a vector. To find the voltage due to a combination of point charges, you add the individual voltages as numbers. To find the total electric field, you must add the individual fields as vectors, taking magnitude and direction into account. This is consistent with the fact that V is closely associated with energy, a scalar, whereas $\vec{\mathbf{E}}$ is closely associated with force, a vector.

Example 7.10

What Voltage Is Produced by a Small Charge on a Metal Sphere?

Charges in static electricity are typically in the nanocoulomb (nC) to microcoulomb (μC) range. What is the voltage 5.00 cm away from the center of a 1-cm-diameter solid metal sphere that has a –3.00-nC static charge?

Strategy

As we discussed in Electric Charges and Fields, charge on a metal sphere spreads out uniformly and produces a field like that of a point charge located at its center. Thus, we can find the voltage using the equation $V = \frac{kq}{r}$.

Solution

Entering known values into the expression for the potential of a point charge, we obtain

$$V = k\frac{q}{r} = \left(8.99 \times 10^9 \text{ N} \cdot \text{m}^2/\text{C}^2\right)\left(\frac{-3.00 \times 10^{-9} \text{ C}}{5.00 \times 10^{-2} \text{ m}}\right) = -539 \text{ V}.$$

Significance

The negative value for voltage means a positive charge would be attracted from a larger distance, since the potential is lower (more negative) than at larger distances. Conversely, a negative charge would be repelled, as expected.

Example 7.11

What Is the Excess Charge on a Van de Graaff Generator?

A demonstration Van de Graaff generator has a 25.0-cm-diameter metal sphere that produces a voltage of 100 kV near its surface (Figure 7.18). What excess charge resides on the sphere? (Assume that each numerical value here is shown with three significant figures.)

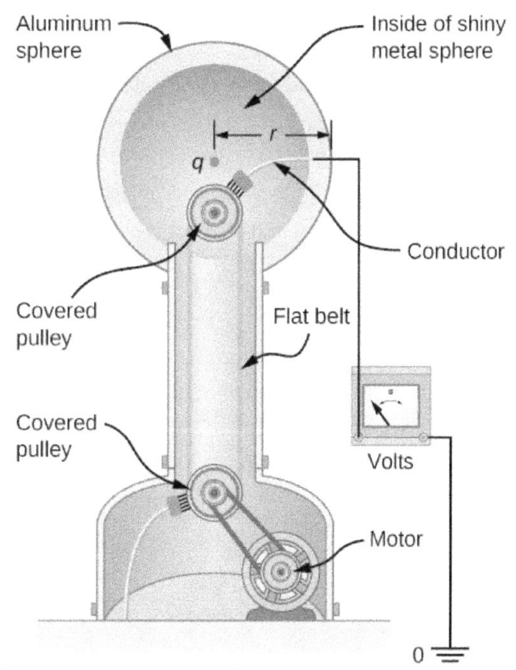

Figure 7.18 The voltage of this demonstration Van de Graaff generator is measured between the charged sphere and ground. Earth's potential is taken to be zero as a reference. The potential of the charged conducting sphere is the same as that of an equal point charge at its center.

Strategy

The potential on the surface is the same as that of a point charge at the center of the sphere, 12.5 cm away. (The radius of the sphere is 12.5 cm.) We can thus determine the excess charge using the equation

$$V = \frac{kq}{r}.$$

Solution

Solving for q and entering known values gives

$$q = \frac{rV}{k} = \frac{(0.125\ \text{m})(100 \times 10^3\ \text{V})}{8.99 \times 10^9\ \text{N} \cdot \text{m}^2/\text{C}^2} = 1.39 \times 10^{-6}\ \text{C} = 1.39\ \mu\text{C}.$$

Significance

This is a relatively small charge, but it produces a rather large voltage. We have another indication here that it is difficult to store isolated charges.

7.8 Check Your Understanding What is the potential inside the metal sphere in Example 7.10?

The voltages in both of these examples could be measured with a meter that compares the measured potential with ground potential. Ground potential is often taken to be zero (instead of taking the potential at infinity to be zero). It is the potential difference between two points that is of importance, and very often there is a tacit assumption that some reference point, such as Earth or a very distant point, is at zero potential. As noted earlier, this is analogous to taking sea level as $h = 0$ when considering gravitational potential energy $U_g = mgh$.

Systems of Multiple Point Charges

Just as the electric field obeys a superposition principle, so does the electric potential. Consider a system consisting of N charges $q_1, q_2, \ldots, q_N$. What is the net electric potential V at a space point P from these charges? Each of these charges is a source charge that produces its own electric potential at point P, independent of whatever other changes may be doing. Let $V_1, V_2, \ldots, V_N$ be the electric potentials at P produced by the charges $q_1, q_2, \ldots, q_N$, respectively. Then, the net electric potential V_P at that point is equal to the sum of these individual electric potentials. You can easily show this by calculating the potential energy of a test charge when you bring the test charge from the reference point at infinity to point P:

$$V_P = V_1 + V_2 + \cdots + V_N = \sum_1^N V_i.$$

Note that electric potential follows the same principle of superposition as electric field and electric potential energy. To show this more explicitly, note that a test charge q_i at the point P in space has distances of $r_1, r_2, \ldots, r_N$ from the N charges fixed in space above, as shown in Figure 7.19. Using our formula for the potential of a point charge for each of these (assumed to be point) charges, we find that

$$V_P = \sum_1^N k\frac{q_i}{r_i} = k\sum_1^N \frac{q_i}{r_i}. \tag{7.9}$$

Therefore, the electric potential energy of the test charge is

$$U_P = q_t V_P = q_t k \sum_1^N \frac{q_i}{r_i},$$

which is the same as the work to bring the test charge into the system, as found in the first section of the chapter.

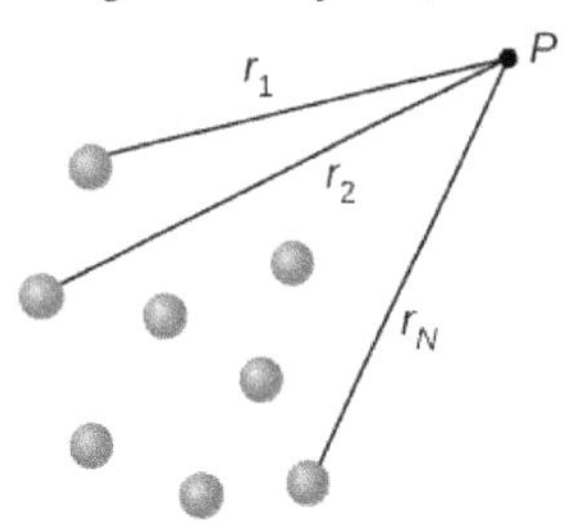

Figure 7.19 Notation for direct distances from charges to a space point P.

The Electric Dipole

An **electric dipole** is a system of two equal but opposite charges a fixed distance apart. This system is used to model many real-world systems, including atomic and molecular interactions. One of these systems is the water molecule, under certain circumstances. These circumstances are met inside a microwave oven, where electric fields with alternating directions make the water molecules change orientation. This vibration is the same as heat at the molecular level.

Example 7.12

Electric Potential of a Dipole

Consider the dipole in Figure 7.20 with the charge magnitude of $q = 3.0\,\text{nC}$ and separation distance $d = 4.0\,\text{cm}$. What is the potential at the following locations in space? (a) (0, 0, 1.0 cm); (b) (0, 0, –5.0 cm); (c) (3.0 cm, 0, 2.0 cm).

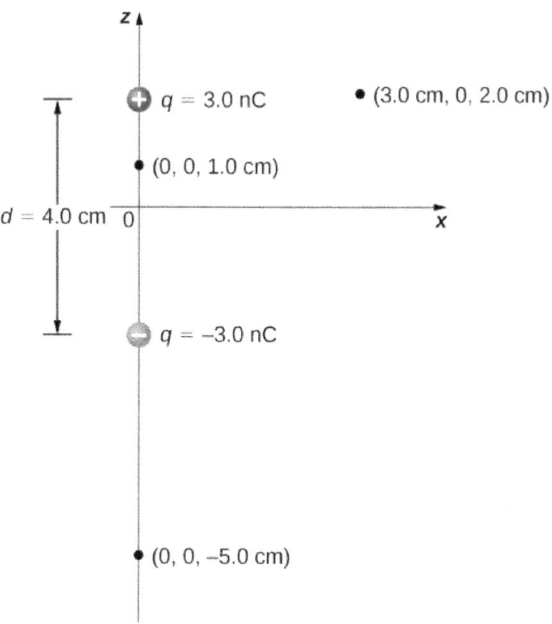

Figure 7.20 A general diagram of an electric dipole, and the notation for the distances from the individual charges to a point P in space.

Strategy

Apply $V_P = k\sum_1^N \frac{q_i}{r_i}$ to each of these three points.

Solution

a. $V_P = k\sum_1^N \frac{q_i}{r_i} = (9.0\times 10^9\ \text{N}\cdot\text{m}^2/\text{C}^2)\left(\frac{3.0\ \text{nC}}{0.010\ \text{m}} - \frac{3.0\ \text{nC}}{0.030\ \text{m}}\right) = 1.8\times 10^3\ \text{V}$

b. $V_P = k\sum_1^N \frac{q_i}{r_i} = (9.0\times 10^9\ \text{N}\cdot\text{m}^2/\text{C}^2)\left(\frac{3.0\ \text{nC}}{0.070\ \text{m}} - \frac{3.0\ \text{nC}}{0.030\ \text{m}}\right) = -5.1\times 10^2\ \text{V}$

c. $V_P = k\sum_1^N \frac{q_i}{r_i} = (9.0\times 10^9\ \text{N}\cdot\text{m}^2/\text{C}^2)\left(\frac{3.0\ \text{nC}}{0.030\ \text{m}} - \frac{3.0\ \text{nC}}{0.050\ \text{m}}\right) = 3.6\times 10^2\ \text{V}$

Significance

Note that evaluating potential is significantly simpler than electric field, due to potential being a scalar instead of a vector.

7.9 Check Your Understanding What is the potential on the x-axis? The z-axis?

Now let us consider the special case when the distance of the point P from the dipole is much greater than the distance between the charges in the dipole, $r \gg d$; for example, when we are interested in the electric potential due to a polarized molecule such as a water molecule. This is not so far (infinity) that we can simply treat the potential as zero, but the distance is great enough that we can simplify our calculations relative to the previous example.

We start by noting that in Figure 7.21 the potential is given by

$$V_P = V_+ + V_- = k\left(\frac{q}{r_+} - \frac{q}{r_-}\right)$$

where

$$r_\pm = \sqrt{x^2 + \left(z \mp \frac{d}{2}\right)^2}.$$

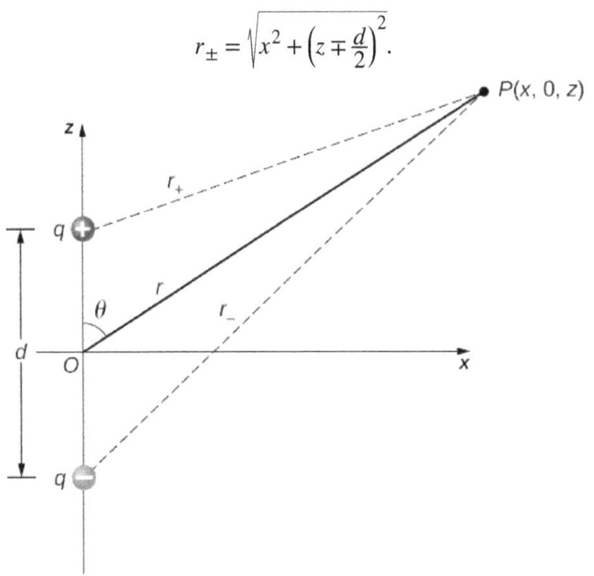

Figure 7.21 A general diagram of an electric dipole, and the notation for the distances from the individual charges to a point P in space.

This is still the exact formula. To take advantage of the fact that $r \gg d,$ we rewrite the radii in terms of polar coordinates, with $x = r\sin\theta$ and $z = r\cos\theta$. This gives us

$$r_\pm = \sqrt{r^2 \sin^2\theta + \left(r\cos\theta \mp \frac{d}{2}\right)^2}.$$

We can simplify this expression by pulling r out of the root,

$$r_\pm = r\sqrt{\sin^2\theta + \left(\cos\theta \mp \frac{d}{2r}\right)^2}$$

and then multiplying out the parentheses

$$r_\pm = r\sqrt{\sin^2\theta + \cos^2\theta \mp \cos\theta\frac{d}{r} + \left(\frac{d}{2r}\right)^2} = r\sqrt{1 \mp \cos\theta\frac{d}{r} + \left(\frac{d}{2r}\right)^2}.$$

The last term in the root is small enough to be negligible (remember $r \gg d,$ and hence $(d/r)^2$ is extremely small, effectively zero to the level we will probably be measuring), leaving us with

$$r_\pm = r\sqrt{1 \mp \cos\theta\frac{d}{r}}.$$

Using the binomial approximation (a standard result from the mathematics of series, when α is small)

$$\frac{1}{\sqrt{1 \mp \alpha}} \approx 1 \pm \frac{\alpha}{2}$$

and substituting this into our formula for V_P , we get

$$V_P = k\left[\frac{q}{r}\left(1 + \frac{d\cos\theta}{2r}\right) - \frac{q}{r}\left(1 - \frac{d\cos\theta}{2r}\right)\right] = k\frac{qd\cos\theta}{r^2}.$$

This may be written more conveniently if we define a new quantity, the **electric dipole moment**,

$$\vec{\mathbf{p}} = q\vec{\mathbf{d}}, \tag{7.10}$$

where these vectors point from the negative to the positive charge. Note that this has magnitude *qd*. This quantity allows us to write the potential at point *P* due to a dipole at the origin as

$$V_P = k\frac{\vec{\mathbf{p}} \cdot \hat{\mathbf{r}}}{r^2}. \tag{7.11}$$

A diagram of the application of this formula is shown in Figure 7.22.

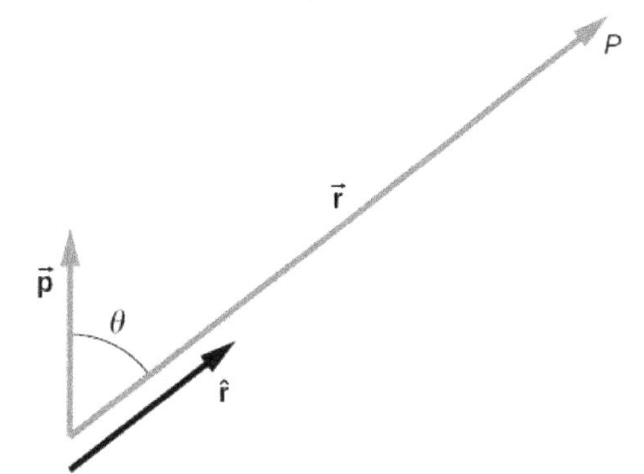

Figure 7.22 The geometry for the application of the potential of a dipole.

There are also higher-order moments, for quadrupoles, octupoles, and so on. You will see these in future classes.

Potential of Continuous Charge Distributions

We have been working with point charges a great deal, but what about continuous charge distributions? Recall from Equation 7.9 that

$$V_P = k\sum\frac{q_i}{r_i}.$$

We may treat a continuous charge distribution as a collection of infinitesimally separated individual points. This yields the integral

$$V_P = k\int\frac{dq}{r} \tag{7.12}$$

for the potential at a point *P*. Note that *r* is the distance from each individual point in the charge distribution to the point *P*. As we saw in Electric Charges and Fields, the infinitesimal charges are given by

$$dq = \begin{cases} \lambda\, dl & \text{(one dimension)} \\ \sigma\, dA & \text{(two dimensions)} \\ \rho\, dV & \text{(three dimensions)} \end{cases}$$

where λ is linear charge density, σ is the charge per unit area, and ρ is the charge per unit volume.

Example 7.13

Potential of a Line of Charge

Find the electric potential of a uniformly charged, nonconducting wire with linear density λ (coulomb/meter) and length L at a point that lies on a line that divides the wire into two equal parts.

Strategy

To set up the problem, we choose Cartesian coordinates in such a way as to exploit the symmetry in the problem as much as possible. We place the origin at the center of the wire and orient the *y*-axis along the wire so that the ends of the wire are at $y = \pm L/2$. The field point *P* is in the *xy*-plane and since the choice of axes is up to us, we choose the *x*-axis to pass through the field point *P*, as shown in Figure 7.23.

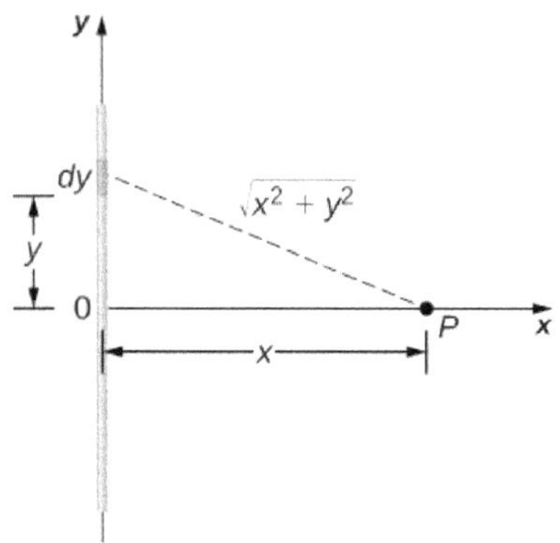

Figure 7.23 We want to calculate the electric potential due to a line of charge.

Solution

Consider a small element of the charge distribution between y and $y + dy$. The charge in this cell is $dq = \lambda\, dy$ and the distance from the cell to the field point *P* is $\sqrt{x^2 + y^2}$. Therefore, the potential becomes

$$\begin{aligned} V_P &= k\int \frac{dq}{r} = k\int_{-L/2}^{L/2} \frac{\lambda dy}{\sqrt{x^2+y^2}} = k\lambda\left[\ln\left(y + \sqrt{y^2 + x^2}\right)\right]_{-L/2}^{L/2} \\ &= k\lambda\left[\ln\left(\left(\frac{L}{2}\right) + \sqrt{\left(\frac{L}{2}\right)^2 + x^2}\right) - \ln\left(\left(-\frac{L}{2}\right) + \sqrt{\left(-\frac{L}{2}\right)^2 + x^2}\right)\right] \\ &= k\lambda \ln\left[\frac{L + \sqrt{L^2 + 4x^2}}{-L + \sqrt{L^2 + 4x^2}}\right]. \end{aligned}$$

Significance

Note that this was simpler than the equivalent problem for electric field, due to the use of scalar quantities. Recall that we expect the zero level of the potential to be at infinity, when we have a finite charge. To examine this, we take the limit of the above potential as *x* approaches infinity; in this case, the terms inside the natural log approach one, and hence the potential approaches zero in this limit. Note that we could have done this problem equivalently in cylindrical coordinates; the only effect would be to substitute *r* for *x* and *z* for *y*.

Example 7.14

Potential Due to a Ring of Charge

A ring has a uniform charge density λ, with units of coulomb per unit meter of arc. Find the electric potential at a point on the axis passing through the center of the ring.

Strategy

We use the same procedure as for the charged wire. The difference here is that the charge is distributed on a circle. We divide the circle into infinitesimal elements shaped as arcs on the circle and use cylindrical coordinates shown in Figure 7.24.

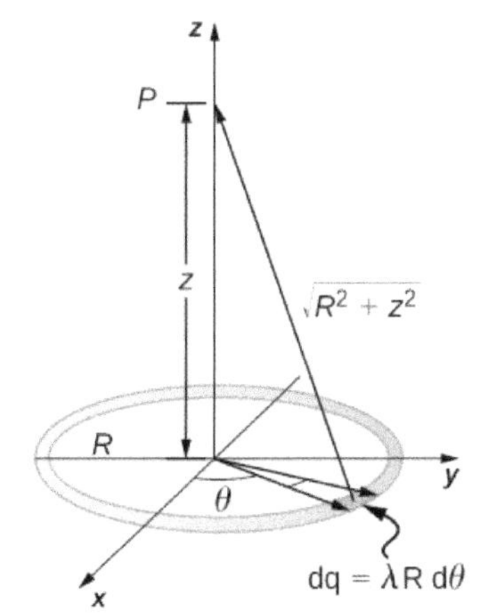

Figure 7.24 We want to calculate the electric potential due to a ring of charge.

Solution

A general element of the arc between θ and $\theta + d\theta$ is of length $Rd\theta$ and therefore contains a charge equal to $\lambda Rd\theta$. The element is at a distance of $\sqrt{z^2 + R^2}$ from P, and therefore the potential is

$$V_P = k\int \frac{dq}{r} = k\int_0^{2\pi} \frac{\lambda Rd\theta}{\sqrt{z^2 + R^2}} = \frac{k\lambda R}{\sqrt{z^2 + R^2}}\int_0^{2\pi} d\theta = \frac{2\pi k\lambda R}{\sqrt{z^2 + R^2}} = k\frac{q_{\text{tot}}}{\sqrt{z^2 + R^2}}.$$

Significance

This result is expected because every element of the ring is at the same distance from point P. The net potential at P is that of the total charge placed at the common distance, $\sqrt{z^2 + R^2}$.

Example 7.15

Potential Due to a Uniform Disk of Charge

A disk of radius R has a uniform charge density σ, with units of coulomb meter squared. Find the electric potential at any point on the axis passing through the center of the disk.

Strategy

We divide the disk into ring-shaped cells, and make use of the result for a ring worked out in the previous example, then integrate over r in addition to θ. This is shown in Figure 7.25.

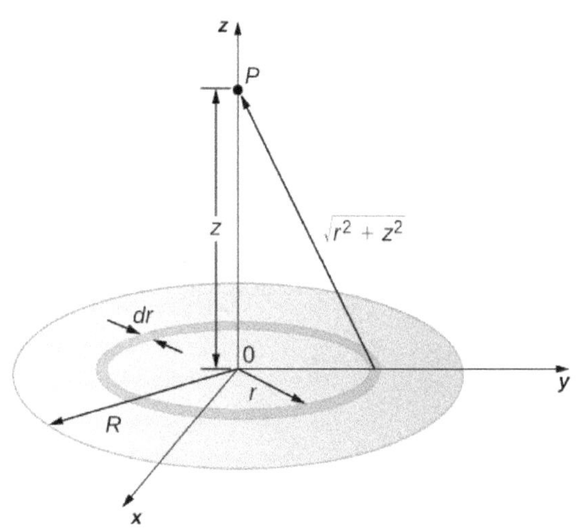

Figure 7.25 We want to calculate the electric potential due to a disk of charge.

Solution

An infinitesimal width cell between cylindrical coordinates r and $r + dr$ shown in Figure 7.25 will be a ring of charges whose electric potential dV_P at the field point has the following expression

$$dV_P = k\frac{dq}{\sqrt{z^2 + r^2}}$$

where

$$dq = \sigma 2\pi r dr.$$

The superposition of potential of all the infinitesimal rings that make up the disk gives the net potential at point P. This is accomplished by integrating from $r = 0$ to $r = R$:

$$\begin{aligned} V_P &= \int dV_P = k2\pi\sigma\int_0^R \frac{r\,dr}{\sqrt{z^2 + r^2}}, \\ &= k2\pi\sigma\left(\sqrt{z^2 + R^2} - \sqrt{z^2}\right). \end{aligned}$$

Significance

The basic procedure for a disk is to first integrate around θ and then over r. This has been demonstrated for uniform (constant) charge density. Often, the charge density will vary with r, and then the last integral will give different results.

Example 7.16

Potential Due to an Infinite Charged Wire

Find the electric potential due to an infinitely long uniformly charged wire.

Strategy

Since we have already worked out the potential of a finite wire of length L in Example 7.7, we might wonder if taking $L \to \infty$ in our previous result will work:

$$V_P = \lim_{L \to \infty} k\lambda \ln\left(\frac{L + \sqrt{L^2 + 4x^2}}{-L + \sqrt{L^2 + 4x^2}}\right).$$

However, this limit does not exist because the argument of the logarithm becomes [2/0] as $L \to \infty$, so this way of finding V of an infinite wire does not work. The reason for this problem may be traced to the fact that the charges are not localized in some space but continue to infinity in the direction of the wire. Hence, our (unspoken) assumption that zero potential must be an infinite distance from the wire is no longer valid.

To avoid this difficulty in calculating limits, let us use the definition of potential by integrating over the electric field from the previous section, and the value of the electric field from this charge configuration from the previous chapter.

Solution

We use the integral

$$V_P = -\int_R^P \vec{\mathbf{E}} \cdot d\vec{\mathbf{l}}$$

where R is a finite distance from the line of charge, as shown in Figure 7.26.

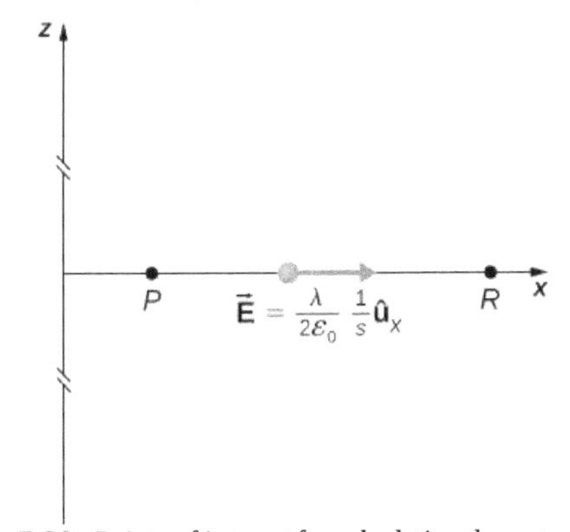

Figure 7.26 Points of interest for calculating the potential of an infinite line of charge.

With this setup, we use $\vec{\mathbf{E}}_P = 2k\lambda\frac{1}{s}\hat{\mathbf{s}}$ and $d\vec{\mathbf{l}} = d\vec{\mathbf{s}}$ to obtain

$$V_P - V_R = -\int_R^P 2k\lambda\frac{1}{s}ds = -2k\lambda\ln\frac{s_P}{s_R}.$$

Now, if we define the reference potential $V_R = 0$ at $s_R = 1$ m, this simplifies to

$$V_P = -2k\lambda \ln s_P.$$

Note that this form of the potential is quite usable; it is 0 at 1 m and is undefined at infinity, which is why we could not use the latter as a reference.

Significance

Although calculating potential directly can be quite convenient, we just found a system for which this strategy does not work well. In such cases, going back to the definition of potential in terms of the electric field may offer a way forward.

7.10 Check Your Understanding What is the potential on the axis of a nonuniform ring of charge, where the charge density is $\lambda(\theta) = \lambda\cos\theta$?

7.4 | Determining Field from Potential

Learning Objectives

By the end of this section, you will be able to:

- Explain how to calculate the electric field in a system from the given potential
- Calculate the electric field in a given direction from a given potential
- Calculate the electric field throughout space from a given potential

Recall that we were able, in certain systems, to calculate the potential by integrating over the electric field. As you may already suspect, this means that we may calculate the electric field by taking derivatives of the potential, although going from a scalar to a vector quantity introduces some interesting wrinkles. We frequently need $\vec{\mathbf{E}}$ to calculate the force in a system; since it is often simpler to calculate the potential directly, there are systems in which it is useful to calculate *V* and then derive $\vec{\mathbf{E}}$ from it.

In general, regardless of whether the electric field is uniform, it points in the direction of decreasing potential, because the force on a positive charge is in the direction of $\vec{\mathbf{E}}$ and also in the direction of lower potential *V*. Furthermore, the magnitude of $\vec{\mathbf{E}}$ equals the rate of decrease of *V* with distance. The faster *V* decreases over distance, the greater the electric field. This gives us the following result.

Relationship between Voltage and Uniform Electric Field

In equation form, the relationship between voltage and uniform electric field is

$$E = -\frac{\Delta V}{\Delta s}$$

where Δs is the distance over which the change in potential ΔV takes place. The minus sign tells us that *E* points in the direction of decreasing potential. The electric field is said to be the gradient (as in grade or slope) of the electric potential.

For continually changing potentials, ΔV and Δs become infinitesimals, and we need differential calculus to determine the electric field. As shown in Figure 7.27, if we treat the distance Δs as very small so that the electric field is essentially constant over it, we find that

$$E_s = -\frac{dV}{ds}.$$

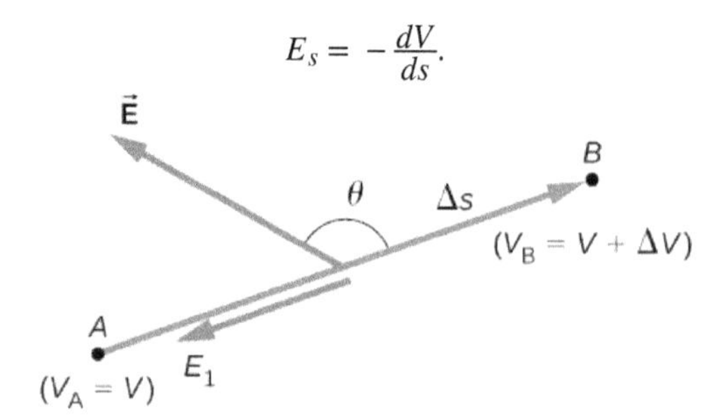

Figure 7.27 The electric field component along the displacement Δs is given by $E = -\frac{\Delta V}{\Delta s}$. Note that *A* and *B* are assumed to be so close together that the field is constant along Δs.

Therefore, the electric field components in the Cartesian directions are given by

$$E_x = -\frac{\partial V}{\partial x},\ E_y = -\frac{\partial V}{\partial y},\ E_z = -\frac{\partial V}{\partial z}. \quad (7.13)$$

This allows us to define the "grad" or "del" vector operator, which allows us to compute the gradient in one step. In Cartesian coordinates, it takes the form

$$\vec{\nabla} = \hat{\mathbf{i}}\frac{\partial}{\partial x} + \hat{\mathbf{j}}\frac{\partial}{\partial y} + \hat{\mathbf{k}}\frac{\partial}{\partial z}. \quad (7.14)$$

With this notation, we can calculate the electric field from the potential with

$$\vec{\mathbf{E}} = -\vec{\nabla} V, \quad (7.15)$$

a process we call calculating the gradient of the potential.

If we have a system with either cylindrical or spherical symmetry, we only need to use the del operator in the appropriate coordinates:

$$\text{Cylindrical: } \vec{\nabla} = \hat{\mathbf{r}}\frac{\partial}{\partial r} + \hat{\boldsymbol{\varphi}}\frac{1}{r}\frac{\partial}{\partial \varphi} + \hat{\mathbf{z}}\frac{\partial}{\partial z} \quad (7.16)$$

$$\text{Spherical: } \vec{\nabla} = \hat{\mathbf{r}}\frac{\partial}{\partial r} + \hat{\boldsymbol{\theta}}\frac{1}{r}\frac{\partial}{\partial \theta} + \hat{\boldsymbol{\varphi}}\frac{1}{r\sin\theta}\frac{\partial}{\partial \varphi} \quad (7.17)$$

Example 7.17

Electric Field of a Point Charge

Calculate the electric field of a point charge from the potential.

Strategy

The potential is known to be $V = k\frac{q}{r}$, which has a spherical symmetry. Therefore, we use the spherical del operator in the formula $\vec{\mathbf{E}} = -\vec{\nabla} V$.

Solution

Performing this calculation gives us

$$\vec{\mathbf{E}} = -\left(\hat{\mathbf{r}}\frac{\partial}{\partial r} + \hat{\boldsymbol{\theta}}\frac{1}{r}\frac{\partial}{\partial \theta} + \hat{\boldsymbol{\varphi}}\frac{1}{r\sin\theta}\frac{\partial}{\partial \varphi}\right)k\frac{q}{r} = -kq\left(\hat{\mathbf{r}}\frac{\partial}{\partial r}\frac{1}{r} + \hat{\boldsymbol{\theta}}\frac{1}{r}\frac{\partial}{\partial \theta}\frac{1}{r} + \hat{\boldsymbol{\varphi}}\frac{1}{r\sin\theta}\frac{\partial}{\partial \varphi}\frac{1}{r}\right).$$

This equation simplifies to

$$\vec{\mathbf{E}} = -kq\left(\hat{\mathbf{r}}\frac{-1}{r^2} + \hat{\boldsymbol{\theta}}0 + \hat{\boldsymbol{\varphi}}0\right) = k\frac{q}{r^2}\hat{\mathbf{r}}$$

as expected.

Significance

We not only obtained the equation for the electric field of a point particle that we've seen before, we also have a demonstration that $\vec{\mathbf{E}}$ points in the direction of decreasing potential, as shown in Figure 7.28.

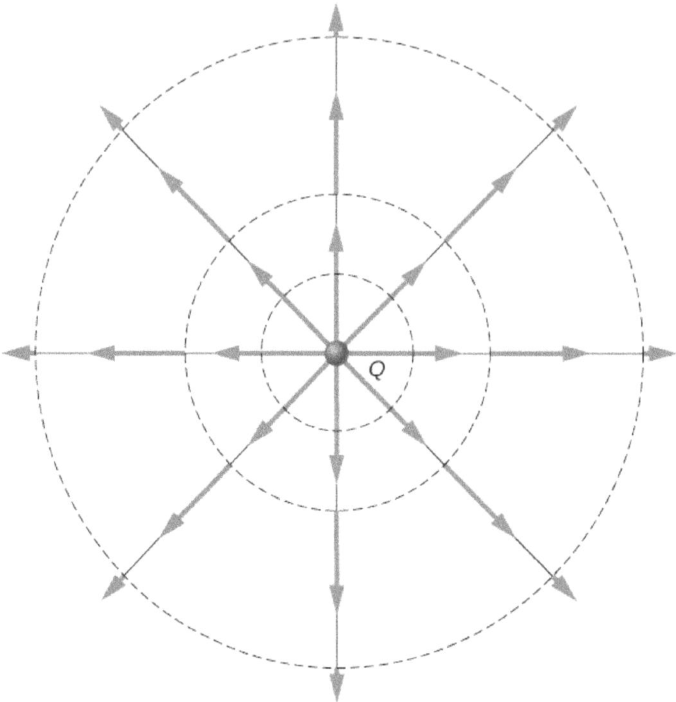

Figure 7.28 Electric field vectors inside and outside a uniformly charged sphere.

Example 7.18

Electric Field of a Ring of Charge

Use the potential found in Example 7.8 to calculate the electric field along the axis of a ring of charge (Figure 7.29).

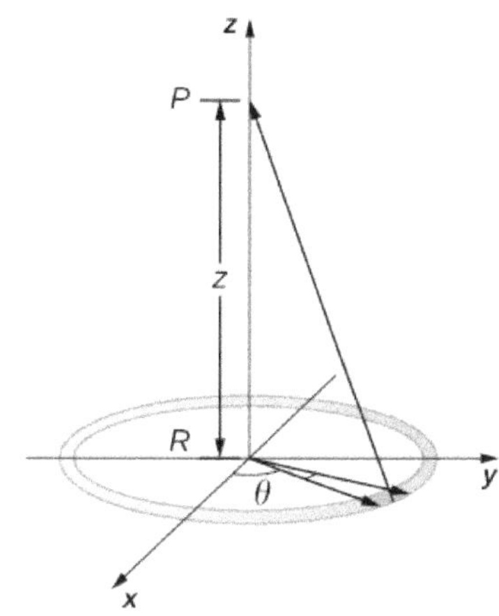

Figure 7.29 We want to calculate the electric field from the electric potential due to a ring charge.

Strategy

In this case, we are only interested in one dimension, the *z*-axis. Therefore, we use $E_z = -\frac{\partial V}{\partial z}$ with the potential $V = k\frac{q_{\text{tot}}}{\sqrt{z^2 + R^2}}$ found previously.

Solution

Taking the derivative of the potential yields

$$E_z = -\frac{\partial}{\partial z}\frac{kq_{\text{tot}}}{\sqrt{z^2 + R^2}} = k\frac{q_{\text{tot}}\,z}{\left(z^2 + R^2\right)^{3/2}}.$$

Significance

Again, this matches the equation for the electric field found previously. It also demonstrates a system in which using the full del operator is not necessary.

7.11 Check Your Understanding Which coordinate system would you use to calculate the electric field of a dipole?

7.5 | Equipotential Surfaces and Conductors

Learning Objectives

By the end of this section, you will be able to:

- Define equipotential surfaces and equipotential lines
- Explain the relationship between equipotential lines and electric field lines
- Map equipotential lines for one or two point charges
- Describe the potential of a conductor
- Compare and contrast equipotential lines and elevation lines on topographic maps

We can represent electric potentials (voltages) pictorially, just as we drew pictures to illustrate electric fields. This is not surprising, since the two concepts are related. Consider Figure 7.30, which shows an isolated positive point charge and its electric field lines, which radiate out from a positive charge and terminate on negative charges. We use blue arrows to represent the magnitude and direction of the electric field, and we use green lines to represent places where the electric potential is constant. These are called **equipotential surfaces** in three dimensions, or **equipotential lines** in two dimensions. The term *equipotential* is also used as a noun, referring to an equipotential line or surface. The potential for a point charge is the same anywhere on an imaginary sphere of radius r surrounding the charge. This is true because the potential for a point charge is given by $V = kq/r$ and thus has the same value at any point that is a given distance r from the charge. An equipotential sphere is a circle in the two-dimensional view of Figure 7.30. Because the electric field lines point radially away from the charge, they are perpendicular to the equipotential lines.

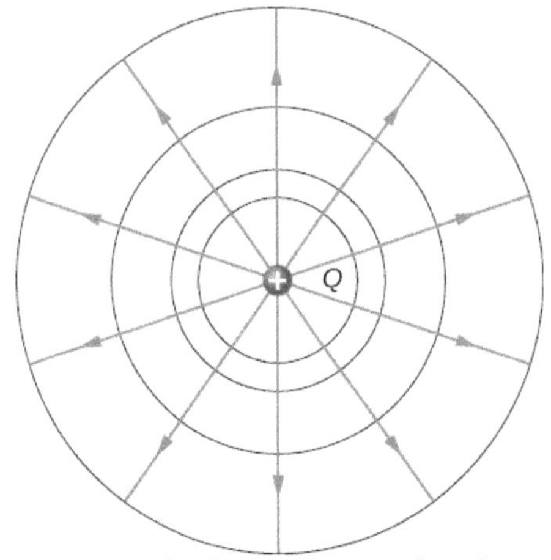

Figure 7.30 An isolated point charge Q with its electric field lines in blue and equipotential lines in green. The potential is the same along each equipotential line, meaning that no work is required to move a charge anywhere along one of those lines. Work is needed to move a charge from one equipotential line to another. Equipotential lines are perpendicular to electric field lines in every case. For a three-dimensional version, explore the first media link.

It is important to note that *equipotential lines are always perpendicular to electric field lines*. No work is required to move a charge along an equipotential, since $\Delta V = 0$. Thus, the work is

$$W = -\Delta U = -q\Delta V = 0.$$

Work is zero if the direction of the force is perpendicular to the displacement. Force is in the same direction as E, so motion along an equipotential must be perpendicular to E. More precisely, work is related to the electric field by

$$W = \vec{\mathbf{F}} \cdot \vec{\mathbf{d}} = q\vec{\mathbf{E}} \cdot \vec{\mathbf{d}} = qEd\cos\theta = 0.$$

Note that in this equation, E and F symbolize the magnitudes of the electric field and force, respectively. Neither q nor E is zero; d is also not zero. So $\cos\theta$ must be 0, meaning θ must be $90°$. In other words, motion along an equipotential is perpendicular to E.

One of the rules for static electric fields and conductors is that the electric field must be perpendicular to the surface of any conductor. This implies that a *conductor is an equipotential surface in static situations*. There can be no voltage difference across the surface of a conductor, or charges will flow. One of the uses of this fact is that a conductor can be fixed at what we consider zero volts by connecting it to the earth with a good conductor—a process called **grounding**. Grounding can be a useful safety tool. For example, grounding the metal case of an electrical appliance ensures that it is at zero volts relative to Earth.

Because a conductor is an equipotential, it can replace any equipotential surface. For example, in Figure 7.30, a charged spherical conductor can replace the point charge, and the electric field and potential surfaces outside of it will be unchanged, confirming the contention that a spherical charge distribution is equivalent to a point charge at its center.

Figure 7.31 shows the electric field and equipotential lines for two equal and opposite charges. Given the electric field lines, the equipotential lines can be drawn simply by making them perpendicular to the electric field lines. Conversely, given the equipotential lines, as in Figure 7.32(a), the electric field lines can be drawn by making them perpendicular to the equipotentials, as in Figure 7.32(b).

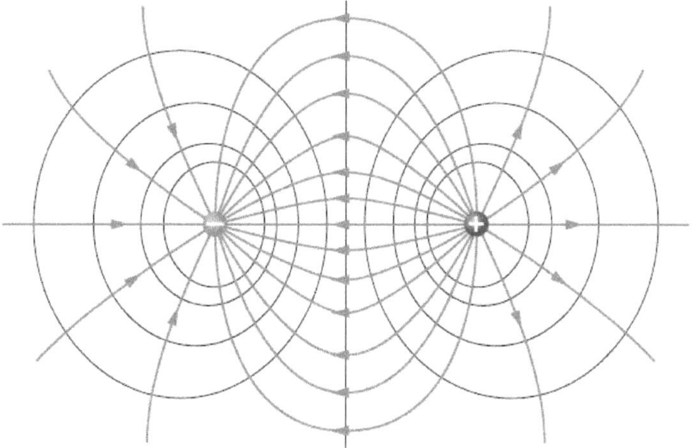

Figure 7.31 The electric field lines and equipotential lines for two equal but opposite charges. The equipotential lines can be drawn by making them perpendicular to the electric field lines, if those are known. Note that the potential is greatest (most positive) near the positive charge and least (most negative) near the negative charge. For a three-dimensional version, explore the first media link.

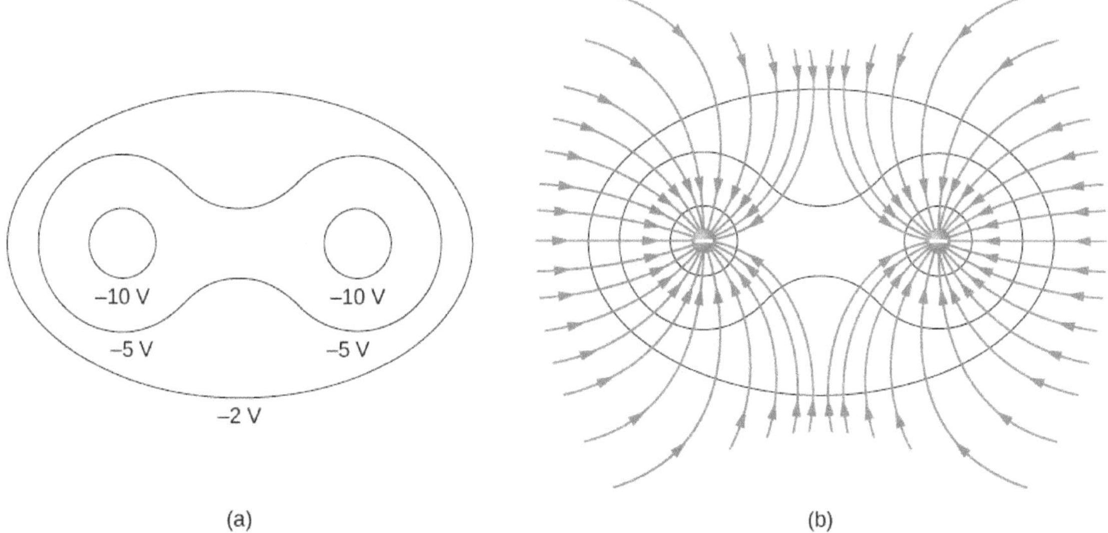

Figure 7.32 (a) These equipotential lines might be measured with a voltmeter in a laboratory experiment. (b) The corresponding electric field lines are found by drawing them perpendicular to the equipotentials. Note that these fields are consistent with two equal negative charges. For a three-dimensional version, play with the first media link.

To improve your intuition, we show a three-dimensional variant of the potential in a system with two opposing charges. Figure 7.33 displays a three-dimensional map of electric potential, where lines on the map are for equipotential surfaces. The hill is at the positive charge, and the trough is at the negative charge. The potential is zero far away from the charges. Note that the cut off at a particular potential implies that the charges are on conducting spheres with a finite radius.

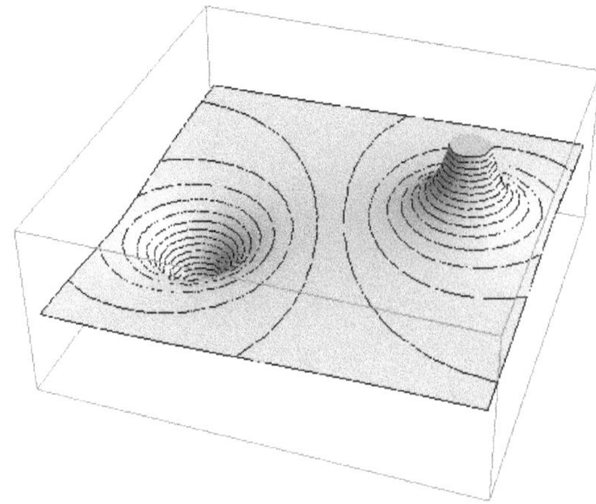

Figure 7.33 Electric potential map of two opposite charges of equal magnitude on conducting spheres. The potential is negative near the negative charge and positive near the positive charge.

A two-dimensional map of the cross-sectional plane that contains both charges is shown in Figure 7.34. The line that is equidistant from the two opposite charges corresponds to zero potential, since at the points on the line, the positive potential from the positive charge cancels the negative potential from the negative charge. Equipotential lines in the cross-sectional plane are closed loops, which are not necessarily circles, since at each point, the net potential is the sum of the potentials from each charge.

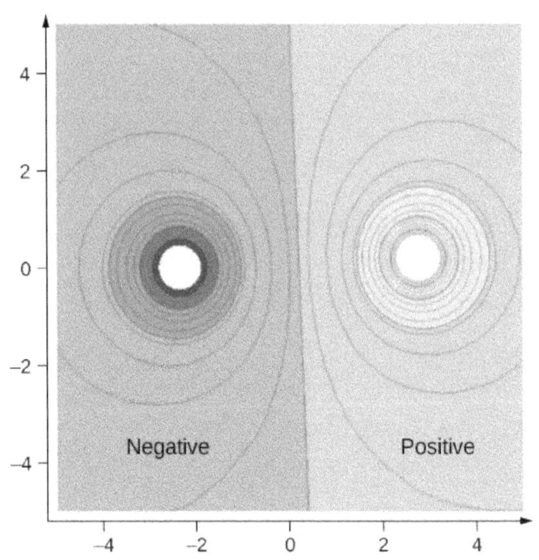

Figure 7.34 A cross-section of the electric potential map of two opposite charges of equal magnitude. The potential is negative near the negative charge and positive near the positive charge.

View this simulation (https://openstaxcollege.org/l/21equipsurelec) to observe and modify the equipotential surfaces and electric fields for many standard charge configurations. There's a lot to explore.

One of the most important cases is that of the familiar parallel conducting plates shown in Figure 7.35. Between the plates, the equipotentials are evenly spaced and parallel. The same field could be maintained by placing conducting plates at the equipotential lines at the potentials shown.

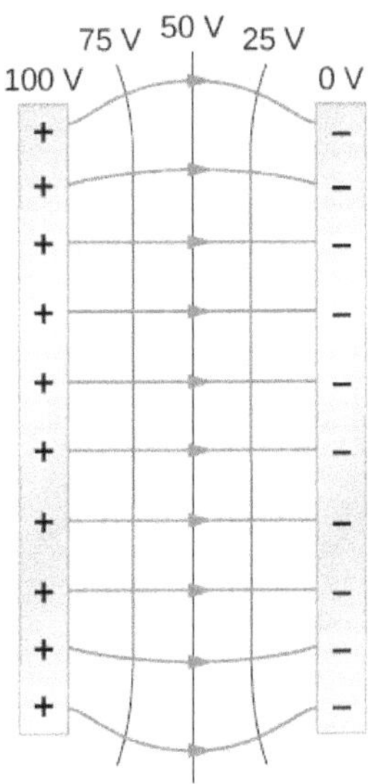

Figure 7.35 The electric field and equipotential lines between two metal plates. Note that the electric field is perpendicular to the equipotentials and hence normal to the plates at their surface as well as in the center of the region between them.

Consider the parallel plates in Figure 7.2. These have equipotential lines that are parallel to the plates in the space between and evenly spaced. An example of this (with sample values) is given in Figure 7.35. We could draw a similar set of equipotential isolines for gravity on the hill shown in Figure 7.2. If the hill has any extent at the same slope, the isolines along that extent would be parallel to each other. Furthermore, in regions of constant slope, the isolines would be evenly spaced. An example of real topographic lines is shown in Figure 7.36.

Figure 7.36 A topographical map along a ridge has roughly parallel elevation lines, similar to the equipotential lines in Figure 7.35. (a) A topographical map of Devil's Tower, Wyoming. Lines that are close together indicate very steep terrain. (b) A perspective photo of Devil's Tower shows just how steep its sides are. Notice the top of the tower has the same shape as the center of the topographical map.

Example 7.19

Calculating Equipotential Lines

You have seen the equipotential lines of a point charge in Figure 7.30. How do we calculate them? For example, if we have a $+10\text{-nC}$ charge at the origin, what are the equipotential surfaces at which the potential is (a) 100 V, (b) 50 V, (c) 20 V, and (d) 10 V?

Strategy

Set the equation for the potential of a point charge equal to a constant and solve for the remaining variable(s). Then calculate values as needed.

Solution

In $V = k\frac{q}{r}$, let V be a constant. The only remaining variable is r; hence, $r = k\frac{q}{V} = \text{constant}$. Thus, the equipotential surfaces are spheres about the origin. Their locations are:

a. $r = k\frac{q}{V} = \left(8.99 \times 10^9 \text{ Nm}^2/\text{C}^2\right)\frac{\left(10 \times 10^{-9} \text{ C}\right)}{100 \text{ V}} = 0.90 \text{ m};$

b. $r = k\frac{q}{V} = \left(8.99 \times 10^9 \text{ Nm}^2/\text{C}^2\right)\frac{\left(10 \times 10^{-9} \text{ C}\right)}{50 \text{ V}} = 1.8 \text{ m};$

c. $r = k\frac{q}{V} = \left(8.99 \times 10^9 \text{ Nm}^2/\text{C}^2\right)\frac{\left(10 \times 10^{-9} \text{ C}\right)}{20 \text{ V}} = 4.5 \text{ m};$

d. $r = k\frac{q}{V} = \left(8.99 \times 10^9 \text{ Nm}^2/\text{C}^2\right)\frac{\left(10 \times 10^{-9} \text{ C}\right)}{10 \text{ V}} = 9.0 \text{ m}.$

Significance

This means that equipotential surfaces around a point charge are spheres of constant radius, as shown earlier, with well-defined locations.

Example 7.20

Potential Difference between Oppositely Charged Parallel Plates

Two large conducting plates carry equal and opposite charges, with a surface charge density σ of magnitude $6.81 \times 10^{-7} \text{ C/m}^2$, as shown in Figure 7.37. The separation between the plates is $l = 6.50 \text{ mm}$. (a) What is the electric field between the plates? (b) What is the potential difference between the plates? (c) What is the distance between equipotential planes which differ by 100 V?

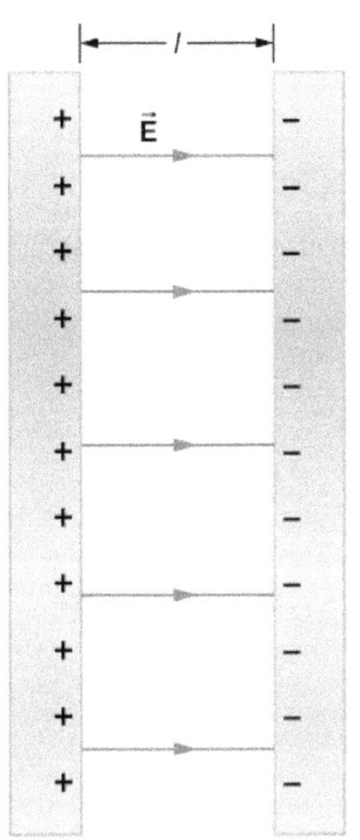

Figure 7.37 The electric field between oppositely charged parallel plates. A portion is released at the positive plate.

Strategy

(a) Since the plates are described as "large" and the distance between them is not, we will approximate each of them as an infinite plane, and apply the result from Gauss's law in the previous chapter.

(b) Use $\Delta V_{AB} = -\int_A^B \vec{\mathbf{E}} \cdot d\vec{\mathbf{l}}$.

(c) Since the electric field is constant, find the ratio of 100 V to the total potential difference; then calculate this fraction of the distance.

Solution

a. The electric field is directed from the positive to the negative plate as shown in the figure, and its magnitude is given by

$$E = \frac{\sigma}{\varepsilon_0} = \frac{6.81 \times 10^{-7} \text{ C/m}^2}{8.85 \times 10^{-12} \text{ C}^2/\text{N} \cdot \text{m}^2} = 7.69 \times 10^4 \text{ V/m}.$$

b. To find the potential difference ΔV between the plates, we use a path from the negative to the positive plate that is directed against the field. The displacement vector $d\vec{\mathbf{l}}$ and the electric field $\vec{\mathbf{E}}$ are antiparallel so $\vec{\mathbf{E}} \cdot d\vec{\mathbf{l}} = -E\,dl.$ The potential difference between the positive plate and the negative plate is then

$$\Delta V = -\int E \cdot dl = E\int dl = El = (7.69 \times 10^4 \text{ V/m})(6.50 \times 10^{-3} \text{ m}) = 500 \text{ V}.$$

c. The total potential difference is 500 V, so 1/5 of the distance between the plates will be the distance between 100-V potential differences. The distance between the plates is 6.5 mm, so there will be 1.3 mm between 100-V potential differences.

Significance

You have now seen a numerical calculation of the locations of equipotentials between two charged parallel plates.

7.12 Check Your Understanding What are the equipotential surfaces for an infinite line charge?

Distribution of Charges on Conductors

In Example 7.19 with a point charge, we found that the equipotential surfaces were in the form of spheres, with the point charge at the center. Given that a conducting sphere in electrostatic equilibrium is a spherical equipotential surface, we should expect that we could replace one of the surfaces in Example 7.19 with a conducting sphere and have an identical solution outside the sphere. Inside will be rather different, however.

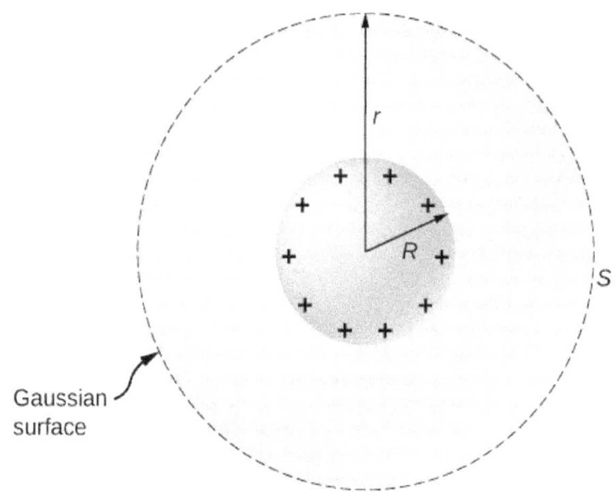

Figure 7.38 An isolated conducting sphere.

To investigate this, consider the isolated conducting sphere of Figure 7.38 that has a radius R and an excess charge q. To find the electric field both inside and outside the sphere, note that the sphere is isolated, so its surface change distribution and the electric field of that distribution are spherically symmetric. We can therefore represent the field as $\vec{\mathbf{E}} = E(r)\hat{\mathbf{r}}$. To calculate $E(r)$, we apply Gauss's law over a closed spherical surface S of radius r that is concentric with the conducting sphere. Since r is constant and $\hat{\mathbf{n}} = \hat{\mathbf{r}}$ on the sphere,

$$\oint_S \vec{\mathbf{E}} \cdot \hat{\mathbf{n}}\, da = E(r)\oint da = E(r)\, 4\pi r^2.$$

For $r < R$, S is within the conductor, so recall from our previous study of Gauss's law that $q_{\text{enc}} = 0$ and Gauss's law gives $E(r) = 0$, as expected inside a conductor at equilibrium. If $r > R$, S encloses the conductor so $q_{\text{enc}} = q$. From Gauss's law,

$$E(r)\, 4\pi r^2 = \frac{q}{\varepsilon_0}.$$

The electric field of the sphere may therefore be written as

$$\begin{aligned} E &= 0 && (r < R), \\ E &= \frac{1}{4\pi\varepsilon_0}\frac{q}{r^2}\hat{\mathbf{r}} && (r \geq R). \end{aligned}$$

As expected, in the region $r \geq R$, the electric field due to a charge q placed on an isolated conducting sphere of radius R is identical to the electric field of a point charge q located at the center of the sphere.

To find the electric potential inside and outside the sphere, note that for $r \geq R$, the potential must be the same as that of an isolated point charge q located at $r = 0$,

$$V(r) = \frac{1}{4\pi r\varepsilon_0}\frac{q}{r}\,(r \geq R)$$

simply due to the similarity of the electric field.

For $r < R$, $E = 0$, so $V(r)$ is constant in this region. Since $V(R) = q/4\pi\varepsilon_0 R$,

$$V(r) = \frac{1}{4\pi r\varepsilon_0}\frac{q}{R}(r < R).$$

We will use this result to show that

$$\sigma_1 R_1 = \sigma_2 R_2,$$

for two conducting spheres of radii R_1 and R_2, with surface charge densities σ_1 and σ_2 respectively, that are connected by a thin wire, as shown in Figure 7.39. The spheres are sufficiently separated so that each can be treated as if it were isolated (aside from the wire). Note that the connection by the wire means that this entire system must be an equipotential.

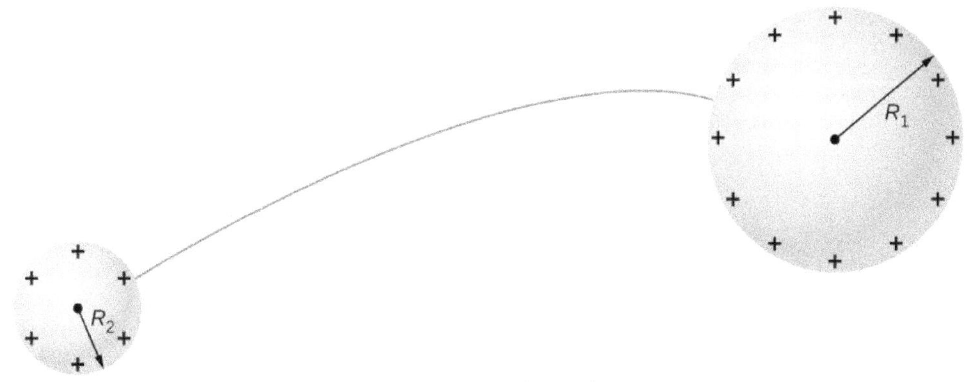

Figure 7.39 Two conducting spheres are connected by a thin conducting wire.

We have just seen that the electrical potential at the surface of an isolated, charged conducting sphere of radius R is

$$V = \frac{1}{4\pi r\varepsilon_0}\frac{q}{R}.$$

Now, the spheres are connected by a conductor and are therefore at the same potential; hence

$$\frac{1}{4\pi r\varepsilon_0}\frac{q_1}{R_1} = \frac{1}{4\pi r\varepsilon_0}\frac{q_2}{R_2},$$

and

$$\frac{q_1}{R_1} = \frac{q_2}{R_2}.$$

The net charge on a conducting sphere and its surface charge density are related by $q = \sigma(4\pi R^2)$. Substituting this equation into the previous one, we find

$$\sigma_1 R_1 = \sigma_2 R_2.$$

Obviously, two spheres connected by a thin wire do not constitute a typical conductor with a variable radius of curvature. Nevertheless, this result does at least provide a qualitative idea of how charge density varies over the surface of a conductor. The equation indicates that where the radius of curvature is large (points B and D in Figure 7.40), σ and E are small.

Similarly, the charges tend to be denser where the curvature of the surface is greater, as demonstrated by the charge distribution on oddly shaped metal (Figure 7.40). The surface charge density is higher at locations with a small radius of curvature than at locations with a large radius of curvature.

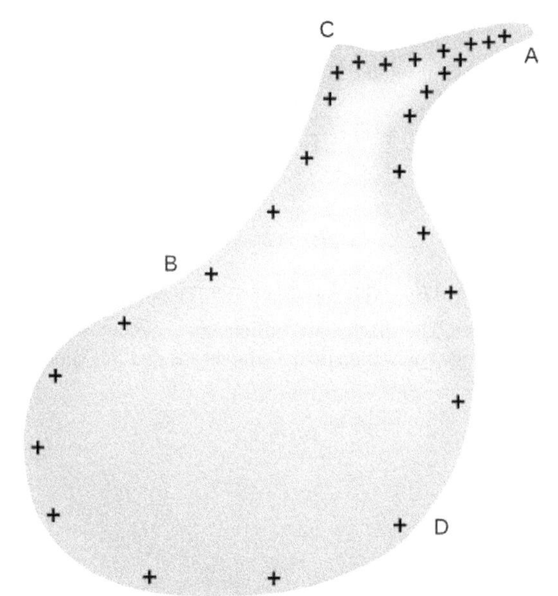

Figure 7.40 The surface charge density and the electric field of a conductor are greater at regions with smaller radii of curvature.

A practical application of this phenomenon is the lightning rod, which is simply a grounded metal rod with a sharp end pointing upward. As positive charge accumulates in the ground due to a negatively charged cloud overhead, the electric field around the sharp point gets very large. When the field reaches a value of approximately 3.0×10^6 N/C (the *dielectric strength* of the air), the free ions in the air are accelerated to such high energies that their collisions with air molecules actually ionize the molecules. The resulting free electrons in the air then flow through the rod to Earth, thereby neutralizing some of the positive charge. This keeps the electric field between the cloud and the ground from getting large enough to produce a lightning bolt in the region around the rod.

An important application of electric fields and equipotential lines involves the heart. The heart relies on electrical signals to maintain its rhythm. The movement of electrical signals causes the chambers of the heart to contract and relax. When a person has a heart attack, the movement of these electrical signals may be disturbed. An artificial pacemaker and a defibrillator can be used to initiate the rhythm of electrical signals. The equipotential lines around the heart, the thoracic region, and the axis of the heart are useful ways of monitoring the structure and functions of the heart. An electrocardiogram (ECG) measures the small electric signals being generated during the activity of the heart.

Play around with this simulation (https://openstaxcollege.org/l/21pointcharsim) to move point charges around on the playing field and then view the electric field, voltages, equipotential lines, and more.

7.6 | Applications of Electrostatics

Learning Objectives

By the end of this section, you will be able to:

- Describe some of the many practical applications of electrostatics, including several printing technologies
- Relate these applications to Newton's second law and the electric force

The study of electrostatics has proven useful in many areas. This module covers just a few of the many applications of electrostatics.

The Van de Graaff Generator

Van de Graaff generators (or Van de Graaffs) are not only spectacular devices used to demonstrate high voltage due to static electricity—they are also used for serious research. The first was built by Robert Van de Graaff in 1931 (based on original suggestions by Lord Kelvin) for use in nuclear physics research. Figure 7.41 shows a schematic of a large research version. Van de Graaffs use both smooth and pointed surfaces, and conductors and insulators to generate large static charges and, hence, large voltages.

A very large excess charge can be deposited on the sphere because it moves quickly to the outer surface. Practical limits arise because the large electric fields polarize and eventually ionize surrounding materials, creating free charges that neutralize excess charge or allow it to escape. Nevertheless, voltages of 15 million volts are well within practical limits.

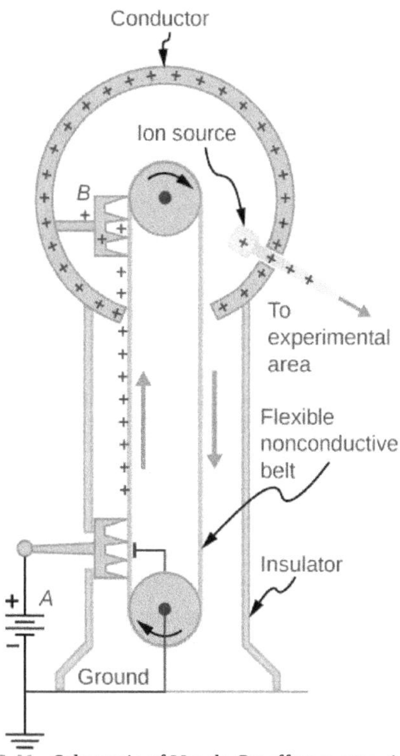

Figure 7.41 Schematic of Van de Graaff generator. A battery (*A*) supplies excess positive charge to a pointed conductor, the points of which spray the charge onto a moving insulating belt near the bottom. The pointed conductor (*B*) on top in the large sphere picks up the charge. (The induced electric field at the points is so large that it removes the charge from the belt.) This can be done because the charge does not remain inside the conducting sphere but moves to its outside surface. An ion source inside the sphere produces positive ions, which are accelerated away from the positive sphere to high velocities.

Xerography

Most copy machines use an electrostatic process called **xerography**—a word coined from the Greek words *xeros* for dry and *graphos* for writing. The heart of the process is shown in simplified form in Figure 7.42.

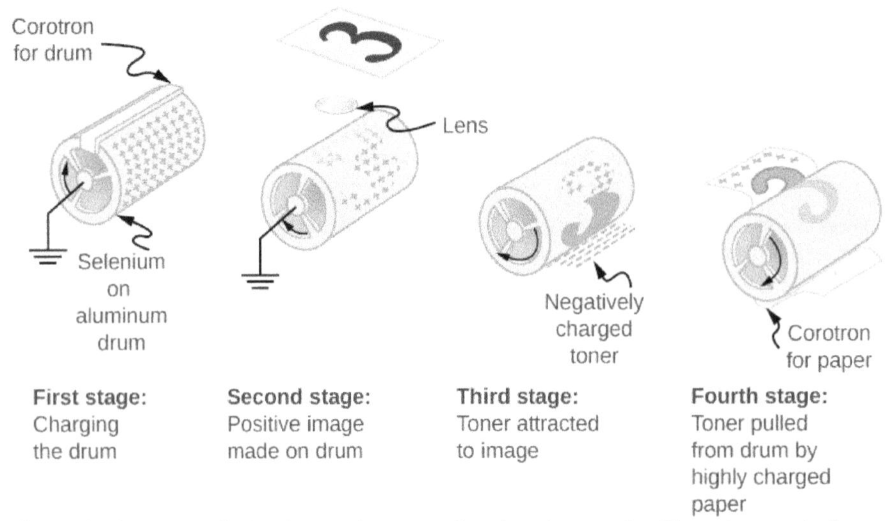

Figure 7.42 Xerography is a dry copying process based on electrostatics. The major steps in the process are the charging of the photoconducting drum, transfer of an image, creating a positive charge duplicate, attraction of toner to the charged parts of the drum, and transfer of toner to the paper. Not shown are heat treatment of the paper and cleansing of the drum for the next copy.

A selenium-coated aluminum drum is sprayed with positive charge from points on a device called a corotron. Selenium is a substance with an interesting property—it is a **photoconductor**. That is, selenium is an insulator when in the dark and a conductor when exposed to light.

In the first stage of the xerography process, the conducting aluminum drum is grounded so that a negative charge is induced under the thin layer of uniformly positively charged selenium. In the second stage, the surface of the drum is exposed to the image of whatever is to be copied. In locations where the image is light, the selenium becomes conducting, and the positive charge is neutralized. In dark areas, the positive charge remains, so the image has been transferred to the drum.

The third stage takes a dry black powder, called toner, and sprays it with a negative charge so that it is attracted to the positive regions of the drum. Next, a blank piece of paper is given a greater positive charge than on the drum so that it will pull the toner from the drum. Finally, the paper and electrostatically held toner are passed through heated pressure rollers, which melt and permanently adhere the toner to the fibers of the paper.

Laser Printers

Laser printers use the xerographic process to make high-quality images on paper, employing a laser to produce an image on the photoconducting drum as shown in Figure 7.43. In its most common application, the laser printer receives output from a computer, and it can achieve high-quality output because of the precision with which laser light can be controlled. Many laser printers do significant information processing, such as making sophisticated letters or fonts, and in the past may have contained a computer more powerful than the one giving them the raw data to be printed.

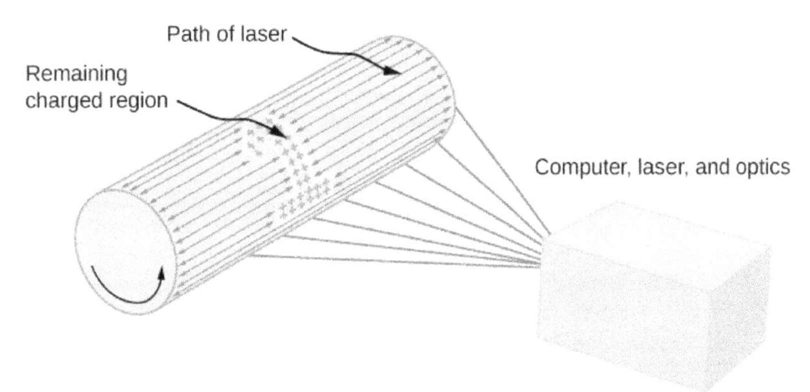

Figure 7.43 In a laser printer, a laser beam is scanned across a photoconducting drum, leaving a positively charged image. The other steps for charging the drum and transferring the image to paper are the same as in xerography. Laser light can be very precisely controlled, enabling laser printers to produce high-quality images.

Ink Jet Printers and Electrostatic Painting

The **ink jet printer**, commonly used to print computer-generated text and graphics, also employs electrostatics. A nozzle makes a fine spray of tiny ink droplets, which are then given an electrostatic charge (Figure 7.44).

Once charged, the droplets can be directed, using pairs of charged plates, with great precision to form letters and images on paper. Ink jet printers can produce color images by using a black jet and three other jets with primary colors, usually cyan, magenta, and yellow, much as a color television produces color. (This is more difficult with xerography, requiring multiple drums and toners.)

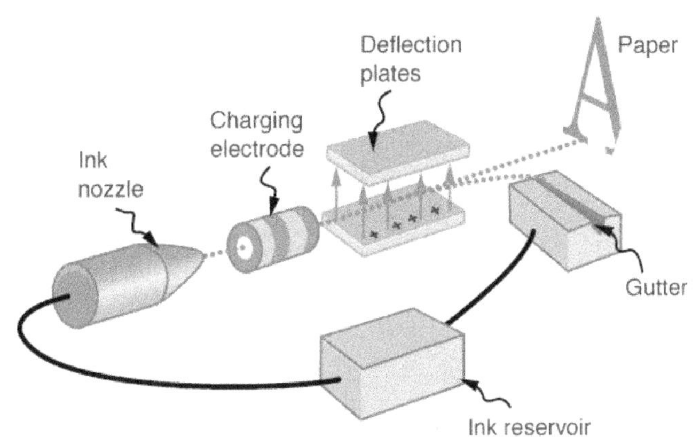

Figure 7.44 The nozzle of an ink-jet printer produces small ink droplets, which are sprayed with electrostatic charge. Various computer-driven devices are then used to direct the droplets to the correct positions on a page.

Electrostatic painting employs electrostatic charge to spray paint onto oddly shaped surfaces. Mutual repulsion of like charges causes the paint to fly away from its source. Surface tension forms drops, which are then attracted by unlike charges to the surface to be painted. Electrostatic painting can reach hard-to-get-to places, applying an even coat in a controlled manner. If the object is a conductor, the electric field is perpendicular to the surface, tending to bring the drops in perpendicularly. Corners and points on conductors will receive extra paint. Felt can similarly be applied.

Smoke Precipitators and Electrostatic Air Cleaning

Another important application of electrostatics is found in air cleaners, both large and small. The electrostatic part of the process places excess (usually positive) charge on smoke, dust, pollen, and other particles in the air and then passes the air through an oppositely charged grid that attracts and retains the charged particles (Figure 7.45)

Large **electrostatic precipitators** are used industrially to remove over 99% of the particles from stack gas emissions associated with the burning of coal and oil. Home precipitators, often in conjunction with the home heating and air conditioning system, are very effective in removing polluting particles, irritants, and allergens.

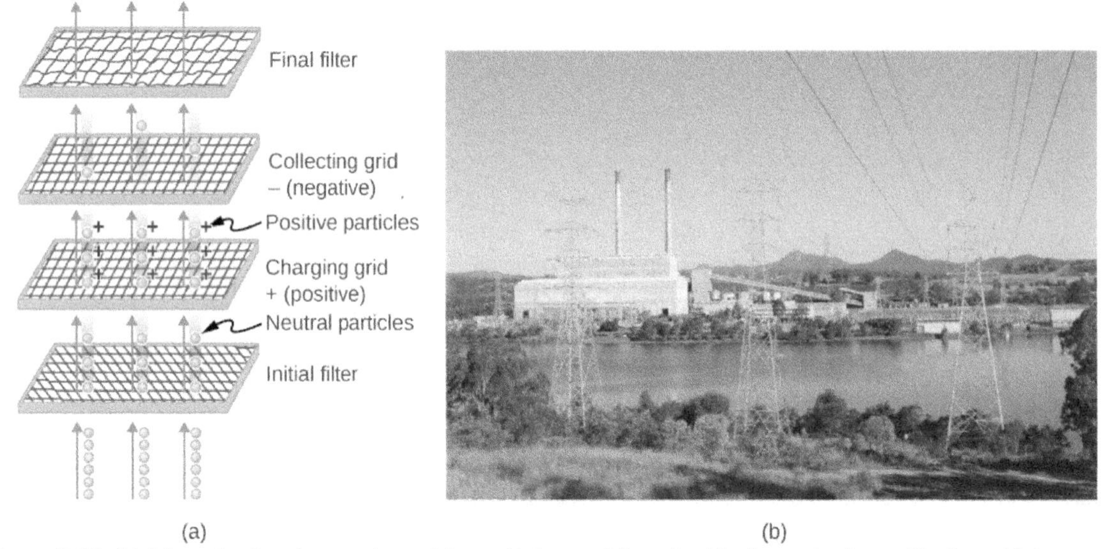

Figure 7.45 (a) Schematic of an electrostatic precipitator. Air is passed through grids of opposite charge. The first grid charges airborne particles, while the second attracts and collects them. (b) The dramatic effect of electrostatic precipitators is seen by the absence of smoke from this power plant. (credit b: modification of work by "Cmdalgleish"/Wikimedia Commons)

CHAPTER 7 REVIEW

KEY TERMS

electric dipole system of two equal but opposite charges a fixed distance apart

electric dipole moment quantity defined as $\vec{\mathbf{p}} = q\vec{\mathbf{d}}$ for all dipoles, where the vector points from the negative to positive charge

electric potential potential energy per unit charge

electric potential difference the change in potential energy of a charge q moved between two points, divided by the charge.

electric potential energy potential energy stored in a system of charged objects due to the charges

electron-volt energy given to a fundamental charge accelerated through a potential difference of one volt

electrostatic precipitators filters that apply charges to particles in the air, then attract those charges to a filter, removing them from the airstream

equipotential line two-dimensional representation of an equipotential surface

equipotential surface surface (usually in three dimensions) on which all points are at the same potential

grounding process of attaching a conductor to the earth to ensure that there is no potential difference between it and Earth

ink jet printer small ink droplets sprayed with an electric charge are controlled by electrostatic plates to create images on paper

photoconductor substance that is an insulator until it is exposed to light, when it becomes a conductor

Van de Graaff generator machine that produces a large amount of excess charge, used for experiments with high voltage

voltage change in potential energy of a charge moved from one point to another, divided by the charge; units of potential difference are joules per coulomb, known as volt

xerography dry copying process based on electrostatics

KEY EQUATIONS

Potential energy of a two-charge system	$U(r) = k\frac{qQ}{r}$
Work done to assemble a system of charges	$W_{12 \ldots N} = \frac{k}{2}\sum_{i}^{N}\sum_{j}^{N}\frac{q_i q_j}{r_{ij}}$ for $i \neq j$
Potential difference	$\Delta V = \frac{\Delta U}{q}$ or $\Delta U = q\Delta V$
Electric potential	$V = \frac{U}{q} = -\int_R^P \vec{\mathbf{E}} \cdot d\vec{\mathbf{l}}$
Potential difference between two points	$\Delta V_{AB} = V_B - V_A = -\int_A^B \vec{\mathbf{E}} \cdot d\vec{\mathbf{l}}$
Electric potential of a point charge	$V = \frac{kq}{r}$
Electric potential of a system of point charges	$V_P = k\sum_1^N \frac{q_i}{r_i}$

Electric dipole moment	$\vec{\mathbf{p}} = q\vec{\mathbf{d}}$
Electric potential due to a dipole	$V_P = k\frac{\vec{\mathbf{p}} \cdot \hat{\mathbf{r}}}{r^2}$
Electric potential of a continuous charge distribution	$V_P = k\int\frac{dq}{r}$
Electric field components	$E_x = -\frac{\partial V}{\partial x},\ E_y = -\frac{\partial V}{\partial y},\ E_z = -\frac{\partial V}{\partial z}$
Del operator in Cartesian coordinates	$\vec{\nabla} = \hat{\mathbf{i}}\frac{\partial}{\partial x} + \hat{\mathbf{j}}\frac{\partial}{\partial y} + \hat{\mathbf{k}}\frac{\partial}{\partial z}$
Electric field as gradient of potential	$\vec{\mathbf{E}} = -\vec{\nabla} V$
Del operator in cylindrical coordinates	$\vec{\nabla} = \hat{\mathbf{r}}\frac{\partial}{\partial r} + \hat{\boldsymbol{\varphi}}\frac{1}{r}\frac{\partial}{\partial \varphi} + \hat{\mathbf{z}}\frac{\partial}{\partial z}$
Del operator in spherical coordinates	$\vec{\nabla} = \hat{\mathbf{r}}\frac{\partial}{\partial r} + \hat{\boldsymbol{\theta}}\frac{1}{r}\frac{\partial}{\partial \theta} + \hat{\boldsymbol{\varphi}}\frac{1}{r\sin\theta}\frac{\partial}{\partial \varphi}$

SUMMARY

7.1 Electric Potential Energy

- The work done to move a charge from point *A* to *B* in an electric field is path independent, and the work around a closed path is zero. Therefore, the electric field and electric force are conservative.
- We can define an electric potential energy, which between point charges is $U(r) = k\frac{qQ}{r}$, with the zero reference taken to be at infinity.
- The superposition principle holds for electric potential energy; the potential energy of a system of multiple charges is the sum of the potential energies of the individual pairs.

7.2 Electric Potential and Potential Difference

- Electric potential is potential energy per unit charge.
- The potential difference between points *A* and *B*, $V_B - V_A$, that is, the change in potential of a charge *q* moved from *A* to *B*, is equal to the change in potential energy divided by the charge.
- Potential difference is commonly called voltage, represented by the symbol ΔV:
  $\Delta V = \frac{\Delta U}{q}$ or $\Delta U = q\Delta V$.
- An electron-volt is the energy given to a fundamental charge accelerated through a potential difference of 1 V. In equation form,
  $1\text{ eV} = (1.60 \times 10^{-19}\text{ C})(1\text{ V}) = (1.60 \times 10^{-19}\text{ C})(1\text{ J/C}) = 1.60 \times 10^{-19}\text{ J}$.

7.3 Calculations of Electric Potential

- Electric potential is a scalar whereas electric field is a vector.
- Addition of voltages as numbers gives the voltage due to a combination of point charges, allowing us to use the principle of superposition: $V_P = k\sum_1^N \frac{q_i}{r_i}$.

- An electric dipole consists of two equal and opposite charges a fixed distance apart, with a dipole moment $\vec{\mathbf{p}} = q\vec{\mathbf{d}}$.
- Continuous charge distributions may be calculated with $V_P = k\int \frac{dq}{r}$.

7.4 Determining Field from Potential

- Just as we may integrate over the electric field to calculate the potential, we may take the derivative of the potential to calculate the electric field.
- This may be done for individual components of the electric field, or we may calculate the entire electric field vector with the gradient operator.

7.5 Equipotential Surfaces and Conductors

- An equipotential surface is the collection of points in space that are all at the same potential. Equipotential lines are the two-dimensional representation of equipotential surfaces.
- Equipotential surfaces are always perpendicular to electric field lines.
- Conductors in static equilibrium are equipotential surfaces.
- Topographic maps may be thought of as showing gravitational equipotential lines.

7.6 Applications of Electrostatics

- Electrostatics is the study of electric fields in static equilibrium.
- In addition to research using equipment such as a Van de Graaff generator, many practical applications of electrostatics exist, including photocopiers, laser printers, ink jet printers, and electrostatic air filters.

CONCEPTUAL QUESTIONS

7.1 Electric Potential Energy

1. Would electric potential energy be meaningful if the electric field were not conservative?

2. Why do we need to be careful about work done *on* the system versus work done *by* the system in calculations?

3. Does the order in which we assemble a system of point charges affect the total work done?

7.2 Electric Potential and Potential Difference

4. Discuss how potential difference and electric field strength are related. Give an example.

5. What is the strength of the electric field in a region where the electric potential is constant?

6. If a proton is released from rest in an electric field, will it move in the direction of increasing or decreasing potential? Also answer this question for an electron and a neutron. Explain why.

7. Voltage is the common word for potential difference. Which term is more descriptive, voltage or potential difference?

8. If the voltage between two points is zero, can a test charge be moved between them with zero net work being done? Can this necessarily be done without exerting a force? Explain.

9. What is the relationship between voltage and energy? More precisely, what is the relationship between potential difference and electric potential energy?

10. Voltages are always measured between two points. Why?

11. How are units of volts and electron-volts related? How do they differ?

12. Can a particle move in a direction of increasing electric potential, yet have its electric potential energy decrease? Explain

7.3 Calculations of Electric Potential

13. Compare the electric dipole moments of charges $\pm Q$ separated by a distance *d* and charges $\pm Q/2$ separated by a distance *d*/2.

14. Would Gauss's law be helpful for determining the electric field of a dipole? Why?

15. In what region of space is the potential due to a uniformly charged sphere the same as that of a point charge? In what region does it differ from that of a point charge?

16. Can the potential of a nonuniformly charged sphere be the same as that of a point charge? Explain.

7.4 Determining Field from Potential

17. If the electric field is zero throughout a region, must the electric potential also be zero in that region?

18. Explain why knowledge of $\vec{\mathbf{E}}(x, y, z)$ is not sufficient to determine *V*(*x*,*y*,*z*). What about the other way around?

7.5 Equipotential Surfaces and Conductors

19. If two points are at the same potential, are there any electric field lines connecting them?

20. Suppose you have a map of equipotential surfaces spaced 1.0 V apart. What do the distances between the surfaces in a particular region tell you about the strength of the $\vec{\mathbf{E}}$ in that region?

21. Is the electric potential necessarily constant over the surface of a conductor?

22. Under electrostatic conditions, the excess charge on a conductor resides on its surface. Does this mean that all of the conduction electrons in a conductor are on the surface?

23. Can a positively charged conductor be at a negative potential? Explain.

24. Can equipotential surfaces intersect?

7.6 Applications of Electrostatics

25. Why are the metal support rods for satellite network dishes generally grounded?

26. (a) Why are fish reasonably safe in an electrical storm? (b) Why are swimmers nonetheless ordered to get out of the water in the same circumstance?

27. What are the similarities and differences between the processes in a photocopier and an electrostatic precipitator?

28. About what magnitude of potential is used to charge the drum of a photocopy machine? A web search for "xerography" may be of use.

PROBLEMS

7.1 Electric Potential Energy

29. Consider a charge $Q_1(+5.0\,\mu\text{C})$ fixed at a site with another charge Q_2 (charge $+3.0\,\mu\text{C}$, mass $6.0\,\mu\text{g}$) moving in the neighboring space. (a) Evaluate the potential energy of Q_2 when it is 4.0 cm from Q_1. (b) If Q_2 starts from rest from a point 4.0 cm from Q_1, what will be its speed when it is 8.0 cm from Q_1? (*Note:* Q_1 is held fixed in its place.)

30. Two charges $Q_1(+2.00\,\mu\text{C})$ and $Q_2(+2.00\,\mu\text{C})$ are placed symmetrically along the *x*-axis at $x = \pm 3.00\,\text{cm}$. Consider a charge Q_3 of charge $+4.00\,\mu\text{C}$ and mass 10.0 mg moving along the *y*-axis. If Q_3 starts from rest at $y = 2.00\,\text{cm}$, what is its speed when it reaches $y = 4.00\,\text{cm}$?

31. To form a hydrogen atom, a proton is fixed at a point and an electron is brought from far away to a distance of 0.529×10^{-10} m, the average distance between proton and electron in a hydrogen atom. How much work is done?

32. (a) What is the average power output of a heart defibrillator that dissipates 400 J of energy in 10.0 ms? (b) Considering the high-power output, why doesn't the defibrillator produce serious burns?

7.2 Electric Potential and Potential Difference

33. Find the ratio of speeds of an electron and a negative hydrogen ion (one having an extra electron) accelerated through the same voltage, assuming non-relativistic final speeds. Take the mass of the hydrogen ion to be 1.67×10^{-27} kg.

34. An evacuated tube uses an accelerating voltage of 40 kV to accelerate electrons to hit a copper plate and produce X-rays. Non-relativistically, what would be the maximum speed of these electrons?

35. Show that units of V/m and N/C for electric field strength are indeed equivalent.

36. What is the strength of the electric field between two parallel conducting plates separated by 1.00 cm and having a potential difference (voltage) between them of 1.50×10^4 V?

37. The electric field strength between two parallel conducting plates separated by 4.00 cm is 7.50×10^4 V. (a) What is the potential difference between the plates? (b) The plate with the lowest potential is taken to be zero volts. What is the potential 1.00 cm from that plate and 3.00 cm from the other?

38. The voltage across a membrane forming a cell wall is 80.0 mV and the membrane is 9.00 nm thick. What is the electric field strength? (The value is surprisingly large, but correct.) You may assume a uniform electric field.

39. Two parallel conducting plates are separated by 10.0 cm, and one of them is taken to be at zero volts. (a) What is the electric field strength between them, if the potential 8.00 cm from the zero volt plate (and 2.00 cm from the other) is 450 V? (b) What is the voltage between the plates?

40. Find the maximum potential difference between two parallel conducting plates separated by 0.500 cm of air, given the maximum sustainable electric field strength in air to be 3.0×10^6 V/m.

41. An electron is to be accelerated in a uniform electric field having a strength of 2.00×10^6 V/m. (a) What energy in keV is given to the electron if it is accelerated through 0.400 m? (b) Over what distance would it have to be accelerated to increase its energy by 50.0 GeV?

42. Use the definition of potential difference in terms of electric field to deduce the formula for potential difference between $r = r_a$ and $r = r_b$ for a point charge located at the origin. Here r is the spherical radial coordinate.

43. The electric field in a region is pointed away from the z-axis and the magnitude depends upon the distance s from the axis. The magnitude of the electric field is given as $E = \frac{\alpha}{s}$ where α is a constant. Find the potential difference between points P_1 and P_2, explicitly stating the path over which you conduct the integration for the line integral.

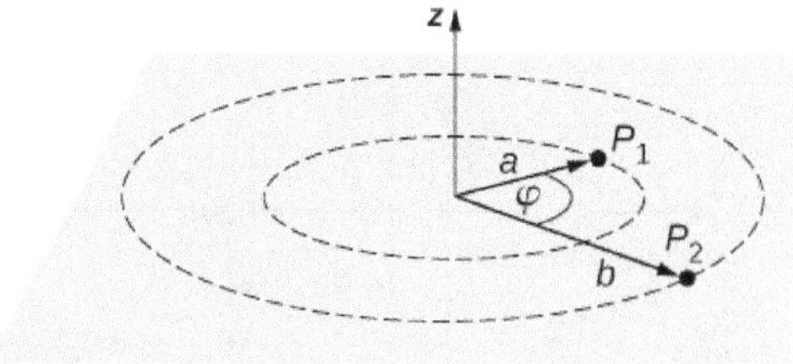

44. Singly charged gas ions are accelerated from rest through a voltage of 13.0 V. At what temperature will the average kinetic energy of gas molecules be the same as that given these ions?

7.3 Calculations of Electric Potential

45. A 0.500-cm-diameter plastic sphere, used in a static electricity demonstration, has a uniformly distributed 40.0-pC charge on its surface. What is the potential near its surface?

46. How far from a 1.00-μC point charge is the potential 100 V? At what distance is it 2.00×10^2 V?

47. If the potential due to a point charge is 5.00×10^2 V at a distance of 15.0 m, what are the sign and magnitude of the charge?

48. In nuclear fission, a nucleus splits roughly in half. (a) What is the potential 2.00×10^{-14} m from a fragment that has 46 protons in it? (b) What is the potential energy in MeV of a similarly charged fragment at this distance?

49. A research Van de Graaff generator has a 2.00-m-diameter metal sphere with a charge of 5.00 mC on it. (a) What is the potential near its surface? (b) At what distance from its center is the potential 1.00 MV? (c) An oxygen atom with three missing electrons is released near the Van de Graaff generator. What is its energy in MeV when the atom is at the distance found in part b?

50. An electrostatic paint sprayer has a 0.200-m-diameter metal sphere at a potential of 25.0 kV that repels paint droplets onto a grounded object.

(a) What charge is on the sphere? (b) What charge must a 0.100-mg drop of paint have to arrive at the object with a speed of 10.0 m/s?

51. (a) What is the potential between two points situated 10 cm and 20 cm from a $3.0\text{-}\mu\text{C}$ point charge? (b) To what location should the point at 20 cm be moved to increase this potential difference by a factor of two?

52. Find the potential at points P_1, P_2, P_3, and P_4 in the diagram due to the two given charges.

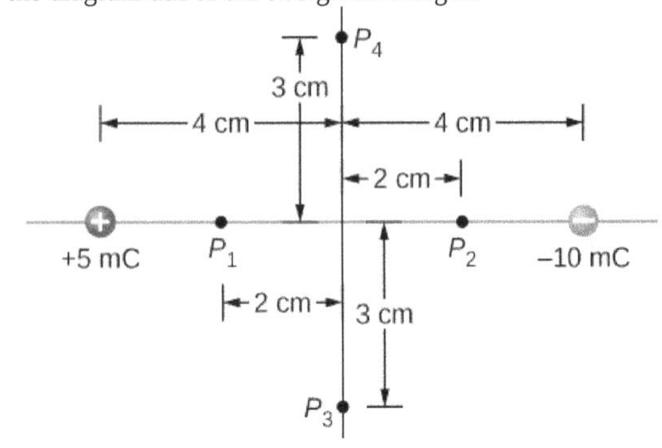

53. Two charges $-2.0\,\mu\text{C}$ and $+2.0\,\mu\text{C}$ are separated by 4.0 cm on the *z*-axis symmetrically about origin, with the positive one uppermost. Two space points of interest P_1 and P_2 are located 3.0 cm and 30 cm from origin at an angle $30°$ with respect to the *z*-axis. Evaluate electric potentials at P_1 and P_2 in two ways: (a) Using the exact formula for point charges, and (b) using the approximate dipole potential formula.

54. (a) Plot the potential of a uniformly charged 1-m rod with 1 C/m charge as a function of the perpendicular distance from the center. Draw your graph from $\text{s} = 0.1\text{ m}$ to $\text{s} = 1.0\text{ m}$. (b) On the same graph, plot the potential of a point charge with a 1-C charge at the origin. (c) Which potential is stronger near the rod? (d) What happens to the difference as the distance increases? Interpret your result.

7.4 Determining Field from Potential

55. Throughout a region, equipotential surfaces are given by $z = \text{constant}$. The surfaces are equally spaced with $V = 100\text{ V}$ for $z = 0.00\text{ m}, V = 200\text{ V}$ for $z = 0.50\text{ m}, V = 300\text{ V}$ for $z = 1.00\text{ m}$. What is the electric field in this region?

56. In a particular region, the electric potential is given by $V = -xy^2z + 4xy$. What is the electric field in this region?

57. Calculate the electric field of an infinite line charge, throughout space.

7.5 Equipotential Surfaces and Conductors

58. Two very large metal plates are placed 2.0 cm apart, with a potential difference of 12 V between them. Consider one plate to be at 12 V, and the other at 0 V. (a) Sketch the equipotential surfaces for 0, 4, 8, and 12 V. (b) Next sketch in some electric field lines, and confirm that they are perpendicular to the equipotential lines.

59. A very large sheet of insulating material has had an excess of electrons placed on it to a surface charge density of $-3.00\,\text{nC/m}^2$. (a) As the distance from the sheet increases, does the potential increase or decrease? Can you explain why without any calculations? Does the location of your reference point matter? (b) What is the shape of the equipotential surfaces? (c) What is the spacing between surfaces that differ by 1.00 V?

60. A metallic sphere of radius 2.0 cm is charged with $+5.0\text{-}\mu\text{C}$ charge, which spreads on the surface of the sphere uniformly. The metallic sphere stands on an insulated stand and is surrounded by a larger metallic spherical shell, of inner radius 5.0 cm and outer radius 6.0 cm. Now, a charge of $-5.0\text{-}\mu\text{C}$ is placed on the inside of the spherical shell, which spreads out uniformly on the inside surface of the shell. If potential is zero at infinity, what is the potential of (a) the spherical shell, (b) the sphere, (c) the space between the two, (d) inside the sphere, and (e) outside the shell?

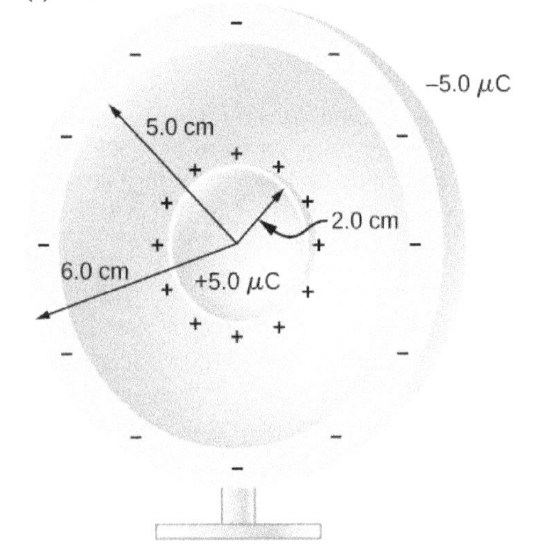

61. Two large charged plates of charge density $\pm 30\,\mu\text{C/m}^2$ face each other at a separation of 5.0 mm. (a) Find the electric potential everywhere. (b) An electron is released from rest at the negative plate; with what speed will it strike the positive plate?

62. A long cylinder of aluminum of radius R meters is charged so that it has a uniform charge per unit length on its surface of λ.

(a) Find the electric field inside and outside the cylinder. (b) Find the electric potential inside and outside the cylinder. (c) Plot electric field and electric potential as a function of distance from the center of the rod.

63. Two parallel plates 10 cm on a side are given equal and opposite charges of magnitude 5.0×10^{-9} C. The plates are 1.5 mm apart. What is the potential difference between the plates?

64. The surface charge density on a long straight metallic pipe is σ. What is the electric potential outside and inside the pipe? Assume the pipe has a diameter of $2a$.

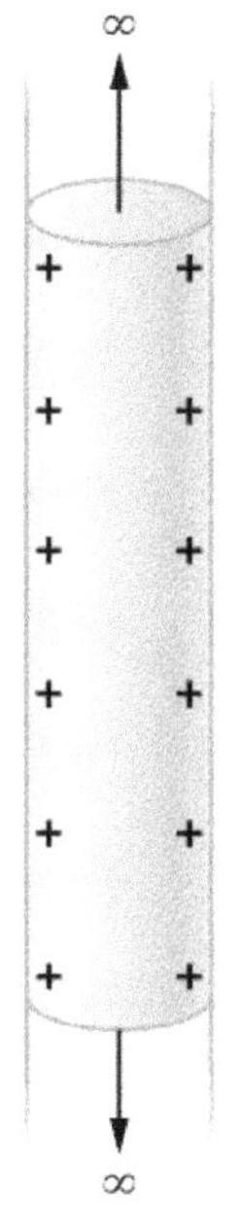

65. Concentric conducting spherical shells carry charges Q and $–Q$, respectively. The inner shell has negligible thickness. What is the potential difference between the shells?

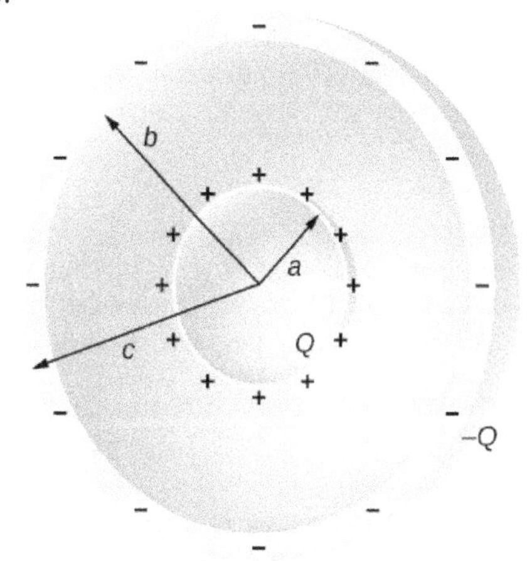

66. Shown below are two concentric spherical shells of negligible thicknesses and radii R_1 and R_2. The inner and outer shell carry net charges q_1 and q_2, respectively, where both q_1 and q_2 are positive. What is the electric potential in the regions (a) $r < R_1$, (b) $R_1 < r < R_2$, and (c) $r > R_2$?

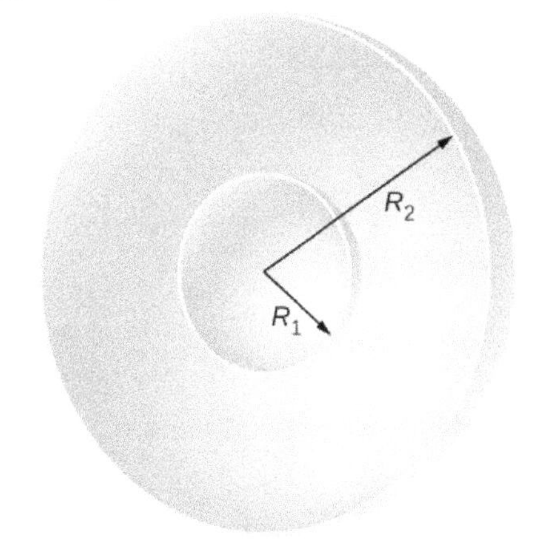

67. A solid cylindrical conductor of radius a is surrounded by a concentric cylindrical shell of inner radius b. The solid cylinder and the shell carry charges Q and $–Q$, respectively. Assuming that the length L of both conductors is much greater than a or b, what is the potential difference between the two conductors?

7.6 Applications of Electrostatics

68. (a) What is the electric field 5.00 m from the center of the terminal of a Van de Graaff with a 3.00-mC charge, noting that the field is equivalent to that of a point charge at the center of the terminal? (b) At this distance, what force does the field exert on a $2.00\text{-}\mu\text{C}$ charge on the Van de Graaff's belt?

69. (a) What is the direction and magnitude of an electric field that supports the weight of a free electron near the surface of Earth? (b) Discuss what the small value for this field implies regarding the relative strength of the gravitational and electrostatic forces.

70. A simple and common technique for accelerating electrons is shown in Figure 7.46, where there is a uniform electric field between two plates. Electrons are released, usually from a hot filament, near the negative plate, and there is a small hole in the positive plate that allows the electrons to continue moving. (a) Calculate the acceleration of the electron if the field strength is $2.50 \times 10^4\ \text{N/C}$. (b) Explain why the electron will not be pulled back to the positive plate once it moves through the hole.

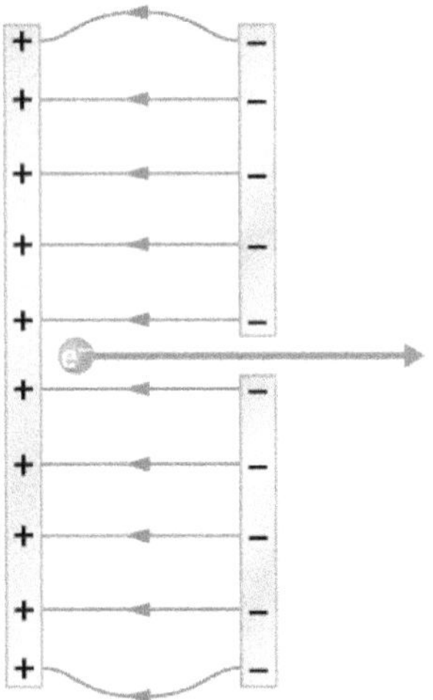

Figure 7.46 Parallel conducting plates with opposite charges on them create a relatively uniform electric field used to accelerate electrons to the right. Those that go through the hole can be used to make a TV or computer screen glow or to produce X- rays.

71. In a Geiger counter, a thin metallic wire at the center of a metallic tube is kept at a high voltage with respect to the metal tube. Ionizing radiation entering the tube knocks electrons off gas molecules or sides of the tube that then accelerate towards the center wire, knocking off even more electrons. This process eventually leads to an avalanche that is detectable as a current. A particular Geiger counter has a tube of radius R and the inner wire of radius a is at a potential of V_0 volts with respect to the outer metal tube. Consider a point P at a distance s from the center wire and far away from the ends. (a) Find a formula for the electric field at a point P inside using the infinite wire approximation. (b) Find a formula for the electric potential at a point P inside. (c) Use $V_0 = 900\ \text{V}$, $a = 3.00\ \text{mm}$, $R = 2.00\ \text{cm}$, and find the value of the electric field at a point 1.00 cm from the center.

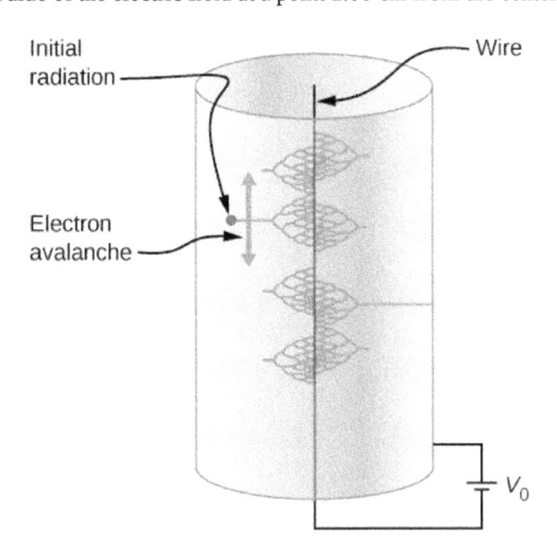

72. The practical limit to an electric field in air is about $3.00 \times 10^6\ \text{N/C}$. Above this strength, sparking takes place because air begins to ionize. (a) At this electric field strength, how far would a proton travel before hitting the speed of light (ignore relativistic effects)? (b) Is it practical to leave air in particle accelerators?

73. To form a helium atom, an alpha particle that contains two protons and two neutrons is fixed at one location, and two electrons are brought in from far away, one at a time. The first electron is placed at $0.600 \times 10^{-10}\ \text{m}$ from the alpha particle and held there while the second electron is brought to $0.600 \times 10^{-10}\ \text{m}$ from the alpha particle on the other side from the first electron. See the final configuration below. (a) How much work is done in each step? (b) What is the electrostatic energy of the alpha particle and two electrons in the final configuration?

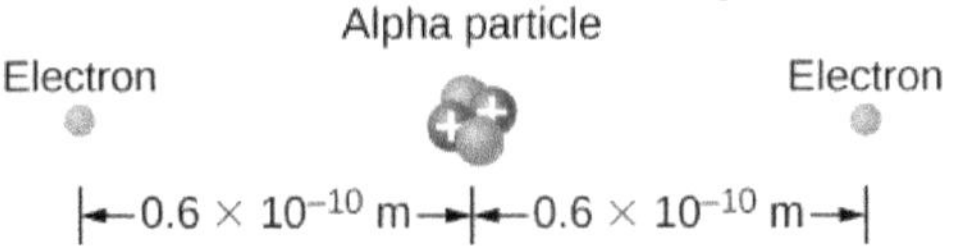

74. Find the electrostatic energy of eight equal charges ($+3\ \mu C$) each fixed at the corners of a cube of side 2 cm.

75. The probability of fusion occurring is greatly enhanced when appropriate nuclei are brought close together, but mutual Coulomb repulsion must be overcome. This can be done using the kinetic energy of high-temperature gas ions or by accelerating the nuclei toward one another. (a) Calculate the potential energy of two singly charged nuclei separated by 1.00×10^{-12} m. (b) At what temperature will atoms of a gas have an average kinetic energy equal to this needed electrical potential energy?

76. A bare helium nucleus has two positive charges and a mass of 6.64×10^{-27} kg. (a) Calculate its kinetic energy in joules at 2.00% of the speed of light. (b) What is this in electron-volts? (c) What voltage would be needed to obtain this energy?

77. An electron enters a region between two large parallel plates made of aluminum separated by a distance of 2.0 cm and kept at a potential difference of 200 V. The electron enters through a small hole in the negative plate and moves toward the positive plate. At the time the electron is near the negative plate, its speed is 4.0×10^5 m/s. Assume the electric field between the plates to be uniform, and find the speed of electron at (a) 0.10 cm, (b) 0.50 cm, (c) 1.0 cm, and (d) 1.5 cm from the negative plate, and (e) immediately before it hits the positive plate.

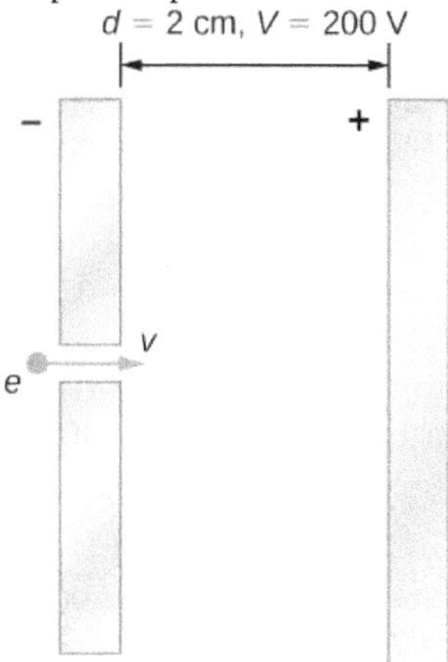

78. How far apart are two conducting plates that have an electric field strength of 4.50×10^3 V/m between them, if their potential difference is 15.0 kV?

79. (a) Will the electric field strength between two parallel conducting plates exceed the breakdown strength of dry air, which is 3.00×10^6 V/m, if the plates are separated by 2.00 mm and a potential difference of 5.0×10^3 V is applied? (b) How close together can the plates be with this applied voltage?

80. Membrane walls of living cells have surprisingly large electric fields across them due to separation of ions. What is the voltage across an 8.00-nm-thick membrane if the electric field strength across it is 5.50 MV/m? You may assume a uniform electric field.

81. A double charged ion is accelerated to an energy of 32.0 keV by the electric field between two parallel conducting plates separated by 2.00 cm. What is the electric field strength between the plates?

82. The temperature near the center of the Sun is thought to be 15 million degrees Celsius (1.5×10^7 °C) (or kelvin). Through what voltage must a singly charged ion be accelerated to have the same energy as the average kinetic energy of ions at this temperature?

83. A lightning bolt strikes a tree, moving 20.0 C of charge through a potential difference of 1.00×10^2 MV. (a) What energy was dissipated? (b) What mass of water could be raised from 15 °C to the boiling point and then boiled by this energy? (c) Discuss the damage that could be caused to the tree by the expansion of the boiling steam.

84. What is the potential 0.530×10^{-10} m from a proton (the average distance between the proton and electron in a hydrogen atom)?

85. (a) A sphere has a surface uniformly charged with 1.00 C. At what distance from its center is the potential 5.00 MV? (b) What does your answer imply about the practical aspect of isolating such a large charge?

86. What are the sign and magnitude of a point charge that produces a potential of –2.00 V at a distance of 1.00 mm?

87. In one of the classic nuclear physics experiments at the beginning of the twentieth century, an alpha particle was accelerated toward a gold nucleus, and its path was substantially deflected by the Coulomb interaction. If the energy of the doubly charged alpha nucleus was 5.00 MeV, how close to the gold nucleus (79 protons) could it come before being deflected?

ADDITIONAL PROBLEMS

88. A 12.0-V battery-operated bottle warmer heats 50.0 g of glass, 2.50×10^2 g of baby formula, and 2.00×10^2 g of aluminum from $20.0\,°\text{C}$ to $90.0\,°\text{C}$. (a) How much charge is moved by the battery? (b) How many electrons per second flow if it takes 5.00 min to warm the formula? (*Hint:* Assume that the specific heat of baby formula is about the same as the specific heat of water.)

89. A battery-operated car uses a 12.0-V system. Find the charge the batteries must be able to move in order to accelerate the 750 kg car from rest to 25.0 m/s, make it climb a 2.00×10^2-m high hill, and finally cause it to travel at a constant 25.0 m/s while climbing with 5.00×10^2-N force for an hour.

90. (a) Find the voltage near a 10.0 cm diameter metal sphere that has 8.00 C of excess positive charge on it. (b) What is unreasonable about this result? (c) Which assumptions are responsible?

91. A uniformly charged ring of radius 10 cm is placed on a nonconducting table. It is found that 3.0 cm above the center of the half-ring the potential is –3.0 V with respect to zero potential at infinity. How much charge is in the half-ring?

92. A glass ring of radius 5.0 cm is painted with a charged paint such that the charge density around the ring varies continuously given by the following function of the polar angle θ, $\lambda = \left(3.0 \times 10^{-6}\ \text{C/m}\right)\cos^2\theta$. Find the potential at a point 15 cm above the center.

93. A CD disk of radius ($R = 3.0\,\text{cm}$) is sprayed with a charged paint so that the charge varies continually with radial distance *r* from the center in the following manner: $\sigma = -(6.0\,\text{C/m})r/R$.

Find the potential at a point 4 cm above the center.

94. (a) What is the final speed of an electron accelerated from rest through a voltage of 25.0 MV by a negatively charged Van de Graff terminal? (b) What is unreasonable about this result? (c) Which assumptions are responsible?

95. A large metal plate is charged uniformly to a density of $\sigma = 2.0 \times 10^{-9}\ \text{C/m}^2$. How far apart are the equipotential surfaces that represent a potential difference of 25 V?

96. Your friend gets really excited by the idea of making a lightning rod or maybe just a sparking toy by connecting two spheres as shown in Figure 7.39, and making R_2 so small that the electric field is greater than the dielectric strength of air, just from the usual 150 V/m electric field near the surface of the Earth. If R_1 is 10 cm, how small does R_2 need to be, and does this seem practical? (*Hint:* recall the calculation for electric field at the surface of a conductor from Gauss's Law.)

97. (a) Find $x >> L$ limit of the potential of a finite uniformly charged rod and show that it coincides with that of a point charge formula. (b) Why would you expect this result?

98. A small spherical pith ball of radius 0.50 cm is painted with a silver paint and then $-10\,\mu\text{C}$ of charge is placed on it. The charged pith ball is put at the center of a gold spherical shell of inner radius 2.0 cm and outer radius 2.2 cm. (a) Find the electric potential of the gold shell with respect to zero potential at infinity. (b) How much charge should you put on the gold shell if you want to make its potential 100 V?

99. Two parallel conducting plates, each of cross-sectional area $400\,\text{cm}^2$, are 2.0 cm apart and uncharged. If 1.0×10^{12} electrons are transferred from one plate to the other, (a) what is the potential difference between the plates? (b) What is the potential difference between the positive plate and a point 1.25 cm from it that is between the plates?

100. A point charge of $q = 5.0 \times 10^{-8}$ C is placed at the center of an uncharged spherical conducting shell of inner radius 6.0 cm and outer radius 9.0 cm. Find the electric potential at (a) $r = 4.0\,\text{cm}$, (b) $r = 8.0\,\text{cm}$, (c) $r = 12.0\,\text{cm}$.

101. Earth has a net charge that produces an electric field of approximately 150 N/C downward at its surface. (a) What is the magnitude and sign of the excess charge, noting the electric field of a conducting sphere is equivalent to a point charge at its center? (b) What acceleration will the field produce on a free electron near Earth's surface? (c) What mass object with a single extra electron will have its weight supported by this field?

102. Point charges of $25.0\,\mu\text{C}$ and $45.0\,\mu\text{C}$ are placed 0.500 m apart.

(a) At what point along the line between them is the electric field zero?

(b) What is the electric field halfway between them?

103. What can you say about two charges q_1 and q_2, if the electric field one-fourth of the way from q_1 to q_2 is zero?

104. Calculate the angular velocity ω of an electron orbiting a proton in the hydrogen atom, given the radius of the orbit is 0.530×10^{-10} m. You may assume that the proton is stationary and the centripetal force is supplied by Coulomb attraction.

105. An electron has an initial velocity of 5.00×10^6 m/s in a uniform 2.00×10^5 -N/C electric field. The field accelerates the electron in the direction opposite to its initial velocity. (a) What is the direction of the electric field? (b) How far does the electron travel before coming to rest? (c) How long does it take the electron to come to rest? (d) What is the electron's velocity when it returns to its starting point?

CHALLENGE PROBLEMS

106. Three Na^+ and three Cl^- ions are placed alternately and equally spaced around a circle of radius 50 nm. Find the electrostatic energy stored.

107. Look up (presumably online, or by dismantling an old device and making measurements) the magnitude of the potential deflection plates (and the space between them) in an ink jet printer. Then look up the speed with which the ink comes out the nozzle. Can you calculate the typical mass of an ink drop?

108. Use the electric field of a finite sphere with constant volume charge density to calculate the electric potential, throughout space. Then check your results by calculating the electric field from the potential.

109. Calculate the electric field of a dipole throughout space from the potential.

8 | CAPACITANCE

Figure 8.1 The tree-like branch patterns in this clear Plexiglas® block are known as a Lichtenberg figure, named for the German physicist Georg Christof Lichtenberg (1742–1799), who was the first to study these patterns. The "branches" are created by the dielectric breakdown produced by a strong electric field. (credit: modification of work by Bert Hickman)

Chapter Outline

Introduction

Capacitors are important components of electrical circuits in many electronic devices, including pacemakers, cell phones, and computers. In this chapter, we study their properties, and, over the next few chapters, we examine their function in combination with other circuit elements. By themselves, capacitors are often used to store electrical energy and release it when needed; with other circuit components, capacitors often act as part of a filter that allows some electrical signals to pass while blocking others. You can see why capacitors are considered one of the fundamental components of electrical circuits.

8.1 | Capacitors and Capacitance

Learning Objectives

By the end of this section, you will be able to:

- Explain the concepts of a capacitor and its capacitance
- Describe how to evaluate the capacitance of a system of conductors

A **capacitor** is a device used to store electrical charge and electrical energy. It consists of at least two electrical conductors separated by a distance. (Note that such electrical conductors are sometimes referred to as "electrodes," but more correctly, they are “capacitor plates.”) The space between capacitors may simply be a vacuum, and, in that case, a capacitor is then known as a “vacuum capacitor.” However, the space is usually filled with an insulating material known as a **dielectric**. (You will learn more about dielectrics in the sections on dielectrics later in this chapter.) The amount of storage in a capacitor is determined by a property called *capacitance*, which you will learn more about a bit later in this section.

Capacitors have applications ranging from filtering static from radio reception to energy storage in heart defibrillators. Typically, commercial capacitors have two conducting parts close to one another but not touching, such as those in Figure 8.2. Most of the time, a dielectric is used between the two plates. When battery terminals are connected to an initially uncharged capacitor, the battery potential moves a small amount of charge of magnitude Q from the positive plate to the negative plate. The capacitor remains neutral overall, but with charges $+Q$ and $-Q$ residing on opposite plates.

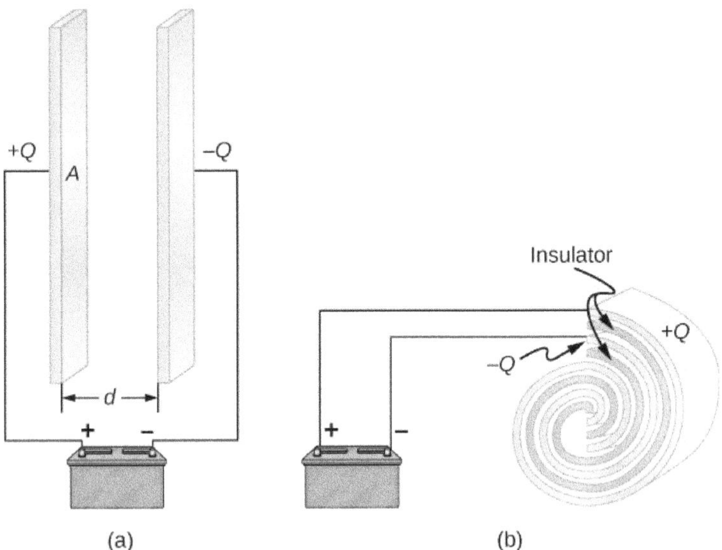

Figure 8.2 Both capacitors shown here were initially uncharged before being connected to a battery. They now have charges of $+Q$ and $-Q$ (respectively) on their plates. (a) A parallel-plate capacitor consists of two plates of opposite charge with area *A* separated by distance *d*. (b) A rolled capacitor has a dielectric material between its two conducting sheets (plates).

A system composed of two identical parallel-conducting plates separated by a distance is called a **parallel-plate capacitor** (Figure 8.3). The magnitude of the electrical field in the space between the parallel plates is $E = \sigma/\varepsilon_0$, where σ denotes the surface charge density on one plate (recall that σ is the charge *Q* per the surface area *A*). Thus, the magnitude of the field is directly proportional to *Q*.

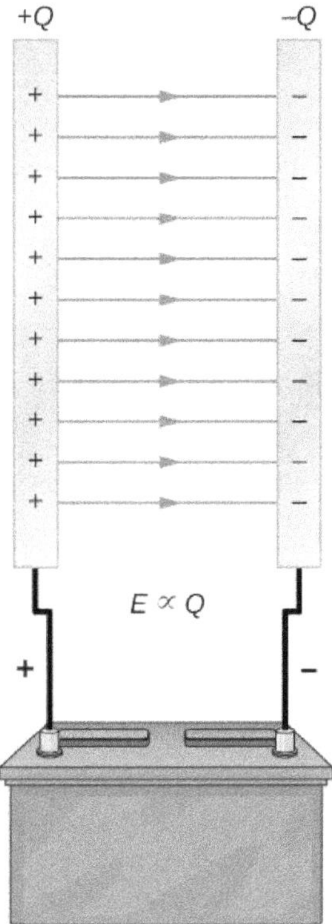

Figure 8.3 The charge separation in a capacitor shows that the charges remain on the surfaces of the capacitor plates. Electrical field lines in a parallel-plate capacitor begin with positive charges and end with negative charges. The magnitude of the electrical field in the space between the plates is in direct proportion to the amount of charge on the capacitor.

Capacitors with different physical characteristics (such as shape and size of their plates) store different amounts of charge for the same applied voltage V across their plates. The **capacitance** C of a capacitor is defined as the ratio of the maximum charge Q that can be stored in a capacitor to the applied voltage V across its plates. In other words, capacitance is the largest amount of charge per volt that can be stored on the device:

$$C = \frac{Q}{V}. \tag{8.1}$$

The SI unit of capacitance is the farad (F), named after Michael Faraday (1791–1867). Since capacitance is the charge per unit voltage, one farad is one coulomb per one volt, or

$$1\,\text{F} = \frac{1\,\text{C}}{1\,\text{V}}.$$

By definition, a 1.0-F capacitor is able to store 1.0 C of charge (a very large amount of charge) when the potential difference between its plates is only 1.0 V. One farad is therefore a very large capacitance. Typical capacitance values range from picofarads $(1\text{ pF} = 10^{-12}\text{ F})$ to millifarads $(1\text{ mF} = 10^{-3}\text{ F})$, which also includes microfarads $(1\,\mu\text{F} = 10^{-6}\text{ F})$. Capacitors can be produced in various shapes and sizes (Figure 8.4).

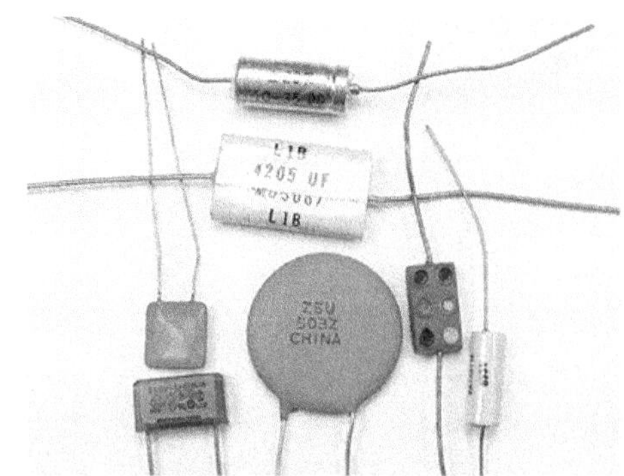

Figure 8.4 These are some typical capacitors used in electronic devices. A capacitor's size is not necessarily related to its capacitance value.

Calculation of Capacitance

We can calculate the capacitance of a pair of conductors with the standard approach that follows.

Problem-Solving Strategy: Calculating Capacitance

1. Assume that the capacitor has a charge Q.
2. Determine the electrical field $\vec{\mathbf{E}}$ between the conductors. If symmetry is present in the arrangement of conductors, you may be able to use Gauss's law for this calculation.
3. Find the potential difference between the conductors from

$$V_B - V_A = -\int_A^B \vec{\mathbf{E}} \cdot d\vec{\mathbf{l}}, \tag{8.2}$$

where the path of integration leads from one conductor to the other. The magnitude of the potential difference is then $V = |V_B - V_A|$.

4. With V known, obtain the capacitance directly from Equation 8.1.

To show how this procedure works, we now calculate the capacitances of parallel-plate, spherical, and cylindrical capacitors. In all cases, we assume vacuum capacitors (empty capacitors) with no dielectric substance in the space between conductors.

Parallel-Plate Capacitor

The parallel-plate capacitor (Figure 8.5) has two identical conducting plates, each having a surface area A, separated by a distance d. When a voltage V is applied to the capacitor, it stores a charge Q, as shown. We can see how its capacitance may depend on A and d by considering characteristics of the Coulomb force. We know that force between the charges increases with charge values and decreases with the distance between them. We should expect that the bigger the plates are, the more charge they can store. Thus, C should be greater for a larger value of A. Similarly, the closer the plates are together, the greater the attraction of the opposite charges on them. Therefore, C should be greater for a smaller d.

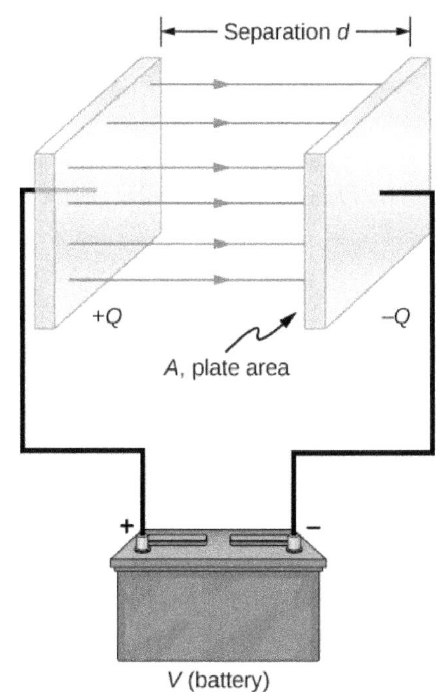

Figure 8.5 In a parallel-plate capacitor with plates separated by a distance d, each plate has the same surface area A.

We define the surface charge density σ on the plates as

$$\sigma = \frac{Q}{A}.$$

We know from previous chapters that when d is small, the electrical field between the plates is fairly uniform (ignoring edge effects) and that its magnitude is given by

$$E = \frac{\sigma}{\varepsilon_0},$$

where the constant ε_0 is the permittivity of free space, $\varepsilon_0 = 8.85 \times 10^{-12}$ F/m. The SI unit of F/m is equivalent to $\mathrm{C^2/N \cdot m^2}$. Since the electrical field $\vec{\mathbf{E}}$ between the plates is uniform, the potential difference between the plates is

$$V = Ed = \frac{\sigma d}{\varepsilon_0} = \frac{Qd}{\varepsilon_0 A}.$$

Therefore Equation 8.1 gives the capacitance of a parallel-plate capacitor as

$$C = \frac{Q}{V} = \frac{Q}{Qd/\varepsilon_0 A} = \varepsilon_0 \frac{A}{d}. \tag{8.3}$$

Notice from this equation that capacitance is a function *only of the geometry* and what material fills the space between the plates (in this case, vacuum) of this capacitor. In fact, this is true not only for a parallel-plate capacitor, but for all capacitors: The capacitance is independent of Q or V. If the charge changes, the potential changes correspondingly so that Q/V remains constant.

Example 8.1

Capacitance and Charge Stored in a Parallel-Plate Capacitor

(a) What is the capacitance of an empty parallel-plate capacitor with metal plates that each have an area of 1.00 m^2, separated by 1.00 mm? (b) How much charge is stored in this capacitor if a voltage of $3.00 \times 10^3 \text{ V}$ is applied to it?

Strategy

Finding the capacitance C is a straightforward application of Equation 8.3. Once we find C, we can find the charge stored by using Equation 8.1.

Solution

a. Entering the given values into Equation 8.3 yields

$$C = \varepsilon_0 \frac{A}{d} = \left(8.85 \times 10^{-12} \frac{\text{F}}{\text{m}}\right) \frac{1.00 \text{ m}^2}{1.00 \times 10^{-3} \text{ m}} = 8.85 \times 10^{-9} \text{ F} = 8.85 \text{ nF}.$$

This small capacitance value indicates how difficult it is to make a device with a large capacitance.

b. Inverting Equation 8.1 and entering the known values into this equation gives

$$Q = CV = (8.85 \times 10^{-9} \text{ F})(3.00 \times 10^3 \text{ V}) = 26.6 \,\mu\text{C}.$$

Significance

This charge is only slightly greater than those found in typical static electricity applications. Since air breaks down (becomes conductive) at an electrical field strength of about 3.0 MV/m, no more charge can be stored on this capacitor by increasing the voltage.

Example 8.2

A 1-F Parallel-Plate Capacitor

Suppose you wish to construct a parallel-plate capacitor with a capacitance of 1.0 F. What area must you use for each plate if the plates are separated by 1.0 mm?

Solution

Rearranging Equation 8.3, we obtain

$$A = \frac{Cd}{\varepsilon_0} = \frac{(1.0 \text{ F})(1.0 \times 10^{-3} \text{ m})}{8.85 \times 10^{-12} \text{ F/m}} = 1.1 \times 10^8 \text{ m}^2.$$

Each square plate would have to be 10 km across. It used to be a common prank to ask a student to go to the laboratory stockroom and request a 1-F parallel-plate capacitor, until stockroom attendants got tired of the joke.

8.1 Check Your Understanding The capacitance of a parallel-plate capacitor is 2.0 pF. If the area of each plate is 2.4 cm^2, what is the plate separation?

8.2 Check Your Understanding Verify that σ/V and ε_0/d have the same physical units.

Spherical Capacitor

A spherical capacitor is another set of conductors whose capacitance can be easily determined (Figure 8.6). It consists of two concentric conducting spherical shells of radii R_1 (inner shell) and R_2 (outer shell). The shells are given equal and opposite charges $+Q$ and $-Q$, respectively. From symmetry, the electrical field between the shells is directed radially outward. We can obtain the magnitude of the field by applying Gauss's law over a spherical Gaussian surface of radius r concentric with the shells. The enclosed charge is $+Q$; therefore we have

$$\oint_S \vec{\mathbf{E}} \cdot \hat{\mathbf{n}} dA = E(4\pi r^2) = \frac{Q}{\varepsilon_0}.$$

Thus, the electrical field between the conductors is

$$\vec{\mathbf{E}} = \frac{1}{4\pi\varepsilon_0}\frac{Q}{r^2}\hat{\mathbf{r}}.$$

We substitute this $\vec{\mathbf{E}}$ into Equation 8.2 and integrate along a radial path between the shells:

$$V = \int_{R_1}^{R_2} \vec{\mathbf{E}} \cdot d\vec{\mathbf{l}} = \int_{R_1}^{R_2} \left(\frac{1}{4\pi\varepsilon_0}\frac{Q}{r^2}\hat{\mathbf{r}}\right) \cdot (\hat{\mathbf{r}}\, dr) = \frac{Q}{4\pi\varepsilon_0}\int_{R_1}^{R_2} \frac{dr}{r^2} = \frac{Q}{4\pi\varepsilon_0}\left(\frac{1}{R_1} - \frac{1}{R_2}\right).$$

In this equation, the potential difference between the plates is $V = -(V_2 - V_1) = V_1 - V_2$. We substitute this result into Equation 8.1 to find the capacitance of a spherical capacitor:

$$C = \frac{Q}{V} = 4\pi\varepsilon_0\frac{R_1 R_2}{R_2 - R_1}. \quad (8.4)$$

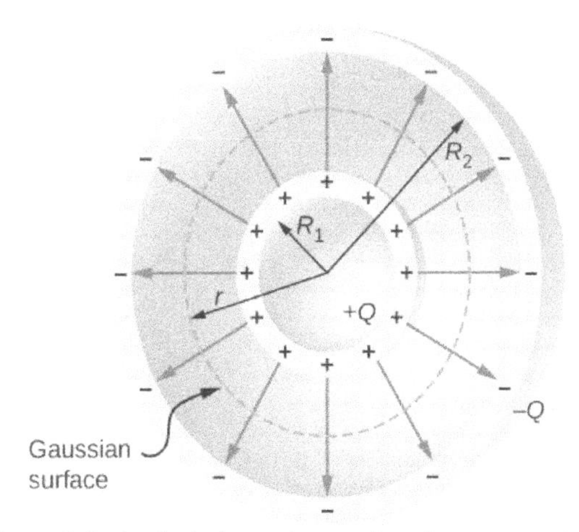

Figure 8.6 A spherical capacitor consists of two concentric conducting spheres. Note that the charges on a conductor reside on its surface.

Example 8.3

Capacitance of an Isolated Sphere

Calculate the capacitance of a single isolated conducting sphere of radius R_1 and compare it with Equation 8.4 in the limit as $R_2 \to \infty$.

Strategy

We assume that the charge on the sphere is Q, and so we follow the four steps outlined earlier. We also assume the other conductor to be a concentric hollow sphere of infinite radius.

Solution

On the outside of an isolated conducting sphere, the electrical field is given by Equation 8.2. The magnitude of the potential difference between the surface of an isolated sphere and infinity is

$$V = \int_{R_1}^{+\infty} \vec{\mathbf{E}} \cdot d\vec{\mathbf{l}} = \frac{Q}{4\pi\varepsilon_0}\int_{R_1}^{+\infty} \frac{1}{r^2}\hat{\mathbf{r}} \cdot (\hat{\mathbf{r}}\, dr) = \frac{Q}{4\pi\varepsilon_0}\int_{R_1}^{+\infty}\frac{dr}{r^2} = \frac{1}{4\pi\varepsilon_0}\frac{Q}{R_1}.$$

The capacitance of an isolated sphere is therefore

$$C = \frac{Q}{V} = Q\frac{4\pi\varepsilon_0 R_1}{Q} = 4\pi\varepsilon_0 R_1.$$

Significance

The same result can be obtained by taking the limit of Equation 8.4 as $R_2 \to \infty$. A single isolated sphere is therefore equivalent to a spherical capacitor whose outer shell has an infinitely large radius.

8.3 Check Your Understanding The radius of the outer sphere of a spherical capacitor is five times the radius of its inner shell. What are the dimensions of this capacitor if its capacitance is 5.00 pF?

Cylindrical Capacitor

A cylindrical capacitor consists of two concentric, conducting cylinders (Figure 8.7). The inner cylinder, of radius R_1, may either be a shell or be completely solid. The outer cylinder is a shell of inner radius R_2. We assume that the length of each cylinder is l and that the excess charges $+Q$ and $-Q$ reside on the inner and outer cylinders, respectively.

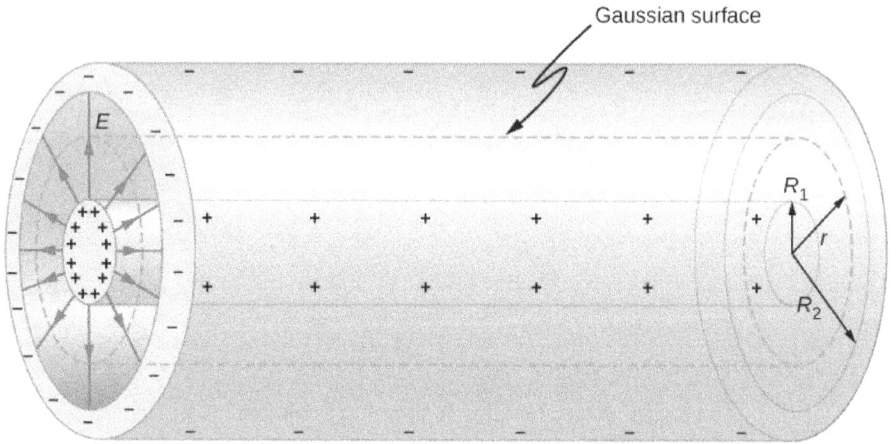

Figure 8.7 A cylindrical capacitor consists of two concentric, conducting cylinders. Here, the charge on the outer surface of the inner cylinder is positive (indicated by $+$) and the charge on the inner surface of the outer cylinder is negative (indicated by $-$).

With edge effects ignored, the electrical field between the conductors is directed radially outward from the common axis of the cylinders. Using the Gaussian surface shown in Figure 8.7, we have

$$\oint_S \vec{\mathbf{E}} \cdot \hat{\mathbf{n}}\, dA = E(2\pi r l) = \frac{Q}{\varepsilon_0}.$$

Therefore, the electrical field between the cylinders is

$$\vec{\mathbf{E}} = \frac{1}{2\pi\varepsilon_0} \frac{Q}{r\,l} \hat{\mathbf{r}}. \tag{8.5}$$

Here $\hat{\mathbf{r}}$ is the unit radial vector along the radius of the cylinder. We can substitute into Equation 8.2 and find the potential difference between the cylinders:

$$V = \int_{R_1}^{R_2} \vec{\mathbf{E}} \cdot d\vec{\mathbf{l}}_p = \frac{Q}{2\pi\varepsilon_0 l} \int_{R_1}^{R_2} \frac{1}{r} \hat{\mathbf{r}} \cdot (\hat{\mathbf{r}}\, dr) = \frac{Q}{2\pi\varepsilon_0 l} \int_{R_1}^{R_2} \frac{dr}{r} = \frac{Q}{2\pi\varepsilon_0 l} \ln r |_{R_1}^{R_2} = \frac{Q}{2\pi\varepsilon_0 l} \ln \frac{R_2}{R_1}.$$

Thus, the capacitance of a cylindrical capacitor is

$$C = \frac{Q}{V} = \frac{2\pi\varepsilon_0 l}{\ln(R_2/R_1)}. \tag{8.6}$$

As in other cases, this capacitance depends only on the geometry of the conductor arrangement. An important application of Equation 8.6 is the determination of the capacitance per unit length of a *coaxial cable*, which is commonly used to transmit time-varying electrical signals. A coaxial cable consists of two concentric, cylindrical conductors separated by an insulating material. (Here, we assume a vacuum between the conductors, but the physics is qualitatively almost the same when the space between the conductors is filled by a dielectric.) This configuration shields the electrical signal propagating down the inner conductor from stray electrical fields external to the cable. Current flows in opposite directions in the inner and the outer conductors, with the outer conductor usually grounded. Now, from Equation 8.6, the capacitance per unit length of the coaxial cable is given by

$$\frac{C}{l} = \frac{2\pi\varepsilon_0}{\ln(R_2/R_1)}.$$

In practical applications, it is important to select specific values of C/l. This can be accomplished with appropriate choices of radii of the conductors and of the insulating material between them.

8.4 Check Your Understanding When a cylindrical capacitor is given a charge of 0.500 nC, a potential difference of 20.0 V is measured between the cylinders. (a) What is the capacitance of this system? (b) If the cylinders are 1.0 m long, what is the ratio of their radii?

Several types of practical capacitors are shown in Figure 8.4. Common capacitors are often made of two small pieces of metal foil separated by two small pieces of insulation (see Figure 8.2(b)). The metal foil and insulation are encased in a protective coating, and two metal leads are used for connecting the foils to an external circuit. Some common insulating materials are mica, ceramic, paper, and Teflon™ non-stick coating.

Another popular type of capacitor is an electrolytic capacitor. It consists of an oxidized metal in a conducting paste. The main advantage of an electrolytic capacitor is its high capacitance relative to other common types of capacitors. For example, capacitance of one type of aluminum electrolytic capacitor can be as high as 1.0 F. However, you must be careful when using an electrolytic capacitor in a circuit, because it only functions correctly when the metal foil is at a higher potential than the conducting paste. When reverse polarization occurs, electrolytic action destroys the oxide film. This type of capacitor cannot be connected across an alternating current source, because half of the time, ac voltage would have the wrong polarity, as an alternating current reverses its polarity (see Alternating-Current Circuts on alternating-current circuits).

A variable air capacitor (Figure 8.8) has two sets of parallel plates. One set of plates is fixed (indicated as “stator”), and the other set of plates is attached to a shaft that can be rotated (indicated as “rotor”). By turning the shaft, the cross-sectional area in the overlap of the plates can be changed; therefore, the capacitance of this system can be tuned to a desired value. Capacitor tuning has applications in any type of radio transmission and in receiving radio signals from electronic devices. Any time you tune your car radio to your favorite station, think of capacitance.

Figure 8.8 In a variable air capacitor, capacitance can be tuned by changing the effective area of the plates. (credit: modification of work by Robbie Sproule)

The symbols shown in Figure 8.9 are circuit representations of various types of capacitors. We generally use the symbol shown in Figure 8.9(a). The symbol in Figure 8.9(c) represents a variable-capacitance capacitor. Notice the similarity of these symbols to the symmetry of a parallel-plate capacitor. An electrolytic capacitor is represented by the symbol in part Figure 8.9(b), where the curved plate indicates the negative terminal.

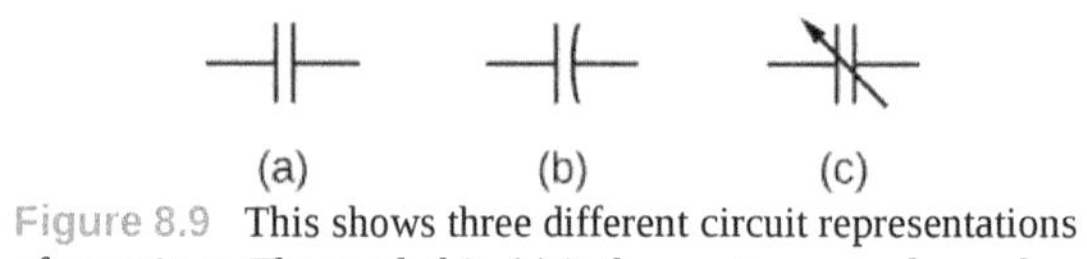

Figure 8.9 This shows three different circuit representations of capacitors. The symbol in (a) is the most commonly used one. The symbol in (b) represents an electrolytic capacitor. The symbol in (c) represents a variable-capacitance capacitor.

An interesting applied example of a capacitor model comes from cell biology and deals with the electrical potential in the plasma membrane of a living cell (Figure 8.10). Cell membranes separate cells from their surroundings but allow some selected ions to pass in or out of the cell. The potential difference across a membrane is about 70 mV. The cell membrane may be 7 to 10 nm thick. Treating the cell membrane as a nano-sized capacitor, the estimate of the smallest electrical field strength across its 'plates' yields the value $E = \frac{V}{d} = \frac{70 \times 10^{-3}\,\text{V}}{10 \times 10^{-9}\,\text{m}} = 7 \times 10^{6}\ \text{V/m} > 3\ \text{MV/m}$.

This magnitude of electrical field is great enough to create an electrical spark in the air.

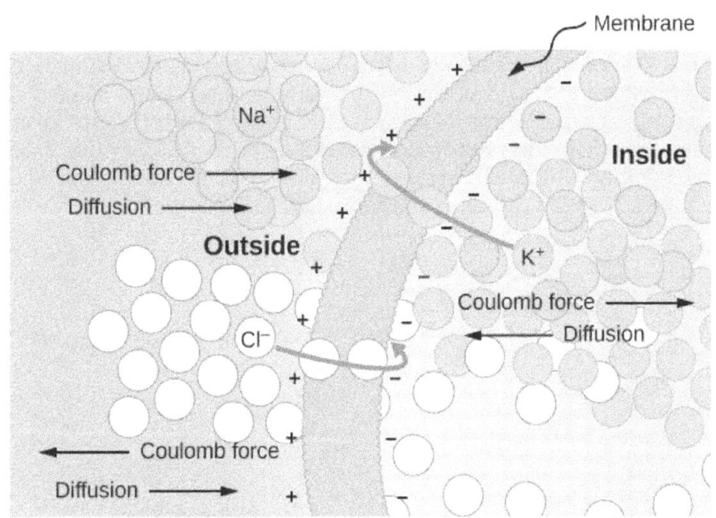

Figure 8.10 The semipermeable membrane of a biological cell has different concentrations of ions on its interior surface than on its exterior. Diffusion moves the K^+ (potassium) and Cl^- (chloride) ions in the directions shown, until the Coulomb force halts further transfer. In this way, the exterior of the membrane acquires a positive charge and its interior surface acquires a negative charge, creating a potential difference across the membrane. The membrane is normally impermeable to Na+ (sodium ions).

Visit the PhET Explorations: Capacitor Lab (https://openstaxcollege.org/l/21phetcapacitor) to explore how a capacitor works. Change the size of the plates and add a dielectric to see the effect on capacitance. Change the voltage and see charges built up on the plates. Observe the electrical field in the capacitor. Measure the voltage and the electrical field.

8.2 | Capacitors in Series and in Parallel

Learning Objectives

By the end of this section, you will be able to:

- Explain how to determine the equivalent capacitance of capacitors in series and in parallel combinations
- Compute the potential difference across the plates and the charge on the plates for a capacitor in a network and determine the net capacitance of a network of capacitors

Several capacitors can be connected together to be used in a variety of applications. Multiple connections of capacitors behave as a single equivalent capacitor. The total capacitance of this equivalent single capacitor depends both on the individual capacitors and how they are connected. Capacitors can be arranged in two simple and common types of connections, known as *series* and *parallel*, for which we can easily calculate the total capacitance. These two basic combinations, series and parallel, can also be used as part of more complex connections.

The Series Combination of Capacitors

Figure 8.11 illustrates a series combination of three capacitors, arranged in a row within the circuit. As for any capacitor, the capacitance of the combination is related to the charge and voltage by using Equation 8.1. When this series combination is connected to a battery with voltage *V*, each of the capacitors acquires an identical charge *Q*. To explain, first note that the charge on the plate connected to the positive terminal of the battery is $+Q$ and the charge on the plate connected to the negative terminal is $-Q$. Charges are then induced on the other plates so that the sum of the charges on all plates, and the sum of charges on any pair of capacitor plates, is zero. However, the potential drop $V_1 = Q/C_1$ on one capacitor may be different from the potential drop $V_2 = Q/C_2$ on another capacitor, because, generally, the capacitors may have different capacitances. The series combination of two or three capacitors resembles a single capacitor with a smaller capacitance. Generally, any number of capacitors connected in series is equivalent to one capacitor whose capacitance (called the *equivalent capacitance*) is smaller than the smallest of the capacitances in the series combination. Charge on this equivalent capacitor is the same as the charge on any capacitor in a series combination: That is, *all capacitors of a series combination have the same charge*. This occurs due to the conservation of charge in the circuit. When a charge Q in a series circuit is removed from a plate of the first capacitor (which we denote as $-Q$), it must be placed on a plate of the second capacitor (which we denote as $+Q$), and so on.

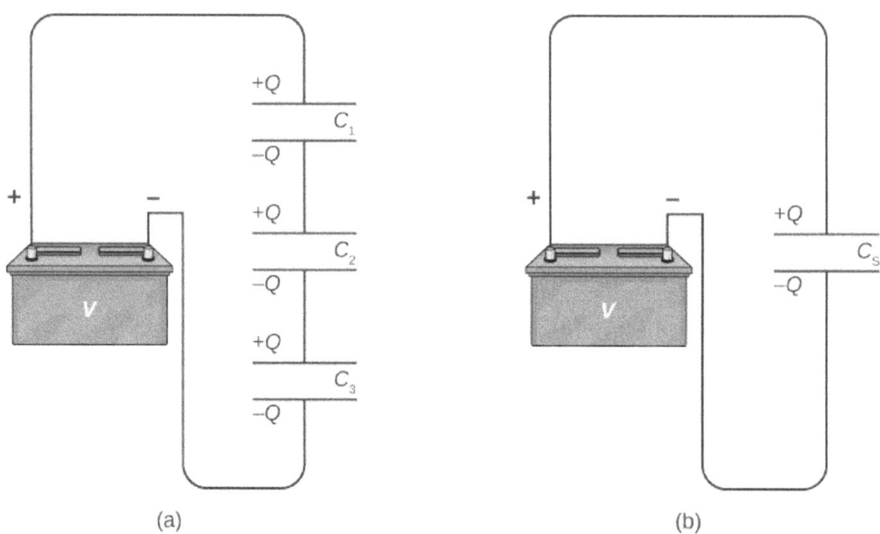

Figure 8.11 (a) Three capacitors are connected in series. The magnitude of the charge on each plate is Q. (b) The network of capacitors in (a) is equivalent to one capacitor that has a smaller capacitance than any of the individual capacitances in (a), and the charge on its plates is Q.

We can find an expression for the total (equivalent) capacitance by considering the voltages across the individual capacitors. The potentials across capacitors 1, 2, and 3 are, respectively, $V_1 = Q/C_1$, $V_2 = Q/C_2$, and $V_3 = Q/C_3$. These potentials must sum up to the voltage of the battery, giving the following potential balance:

$$V = V_1 + V_2 + V_3.$$

Potential V is measured across an equivalent capacitor that holds charge Q and has an equivalent capacitance C_S. Entering the expressions for V_1, V_2, and V_3, we get

$$\frac{Q}{C_S} = \frac{Q}{C_1} + \frac{Q}{C_2} + \frac{Q}{C_3}.$$

Canceling the charge Q, we obtain an expression containing the equivalent capacitance, C_S, of three capacitors connected in series:

$$\frac{1}{C_S} = \frac{1}{C_1} + \frac{1}{C_2} + \frac{1}{C_3}.$$

This expression can be generalized to any number of capacitors in a series network.

Series Combination

For capacitors connected in a **series combination**, the reciprocal of the equivalent capacitance is the sum of reciprocals of individual capacitances:

$$\frac{1}{C_S} = \frac{1}{C_1} + \frac{1}{C_2} + \frac{1}{C_3} + \cdots. \quad (8.7)$$

Example 8.4

Equivalent Capacitance of a Series Network

Find the total capacitance for three capacitors connected in series, given their individual capacitances are $1.000\ \mu F$, $5.000\ \mu F$, and $8.000\ \mu F$.

Strategy

Because there are only three capacitors in this network, we can find the equivalent capacitance by using Equation 8.7 with three terms.

Solution

We enter the given capacitances into Equation 8.7:

$$\begin{aligned} \frac{1}{C_S} &= \frac{1}{C_1} + \frac{1}{C_2} + \frac{1}{C_3} \\ &= \frac{1}{1.000\ \mu F} + \frac{1}{5.000\ \mu F} + \frac{1}{8.000\ \mu F} \\ \frac{1}{C_S} &= \frac{1.325}{\mu F}. \end{aligned}$$

Now we invert this result and obtain $C_S = \frac{\mu F}{1.325} = 0.755\ \mu F.$

Significance

Note that in a series network of capacitors, the equivalent capacitance is always less than the smallest individual capacitance in the network.

The Parallel Combination of Capacitors

A parallel combination of three capacitors, with one plate of each capacitor connected to one side of the circuit and the other plate connected to the other side, is illustrated in Figure 8.12(a). Since the capacitors are connected in parallel, *they all have the same voltage V across their plates*. However, each capacitor in the parallel network may store a different charge. To find the equivalent capacitance C_P of the parallel network, we note that the total charge Q stored by the network is the sum of all the individual charges:

$$Q = Q_1 + Q_2 + Q_3.$$

On the left-hand side of this equation, we use the relation $Q = C_P V$, which holds for the entire network. On the right-hand side of the equation, we use the relations $Q_1 = C_1 V, Q_2 = C_2 V,$ and $Q_3 = C_3 V$ for the three capacitors in the network. In this way we obtain

$$C_P V = C_1 V + C_2 V + C_3 V.$$

This equation, when simplified, is the expression for the equivalent capacitance of the parallel network of three capacitors:

$$C_P = C_1 + C_2 + C_3.$$

This expression is easily generalized to any number of capacitors connected in parallel in the network.

Parallel Combination

For capacitors connected in a **parallel combination**, the equivalent (net) capacitance is the sum of all individual capacitances in the network,

$$C_P = C_1 + C_2 + C_3 + \cdots. \tag{8.8}$$

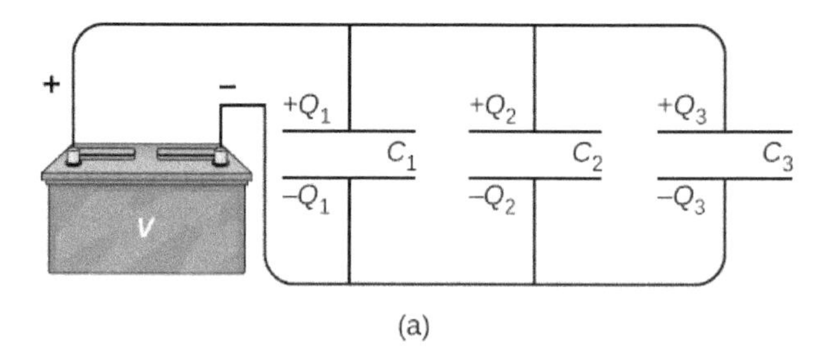

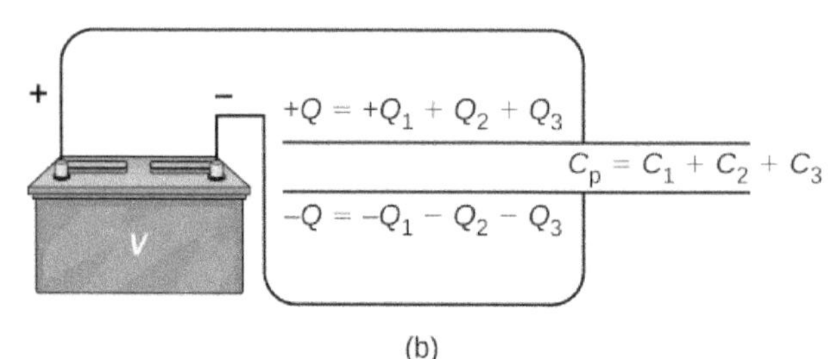

Figure 8.12 (a) Three capacitors are connected in parallel. Each capacitor is connected directly to the battery. (b) The charge on the equivalent capacitor is the sum of the charges on the individual capacitors.

Example 8.5

Equivalent Capacitance of a Parallel Network

Find the net capacitance for three capacitors connected in parallel, given their individual capacitances are $1.0\,\mu\text{F},\ 5.0\,\mu\text{F},$ and $8.0\,\mu\text{F}$.

Strategy

Because there are only three capacitors in this network, we can find the equivalent capacitance by using Equation 8.8 with three terms.

Solution

Entering the given capacitances into Equation 8.8 yields

$$\begin{aligned} C_\text{P} &= C_1 + C_2 + C_3 = 1.0\,\mu\text{F} + 5.0\,\mu\text{F} + 8.0\,\mu\text{F} \\ C_\text{P} &= 14.0\,\mu\text{F}. \end{aligned}$$

Significance

Note that in a parallel network of capacitors, the equivalent capacitance is always larger than any of the individual capacitances in the network.

Capacitor networks are usually some combination of series and parallel connections, as shown in Figure 8.13. To find the net capacitance of such combinations, we identify parts that contain only series or only parallel connections, and find their equivalent capacitances. We repeat this process until we can determine the equivalent capacitance of the entire network. The following example illustrates this process.

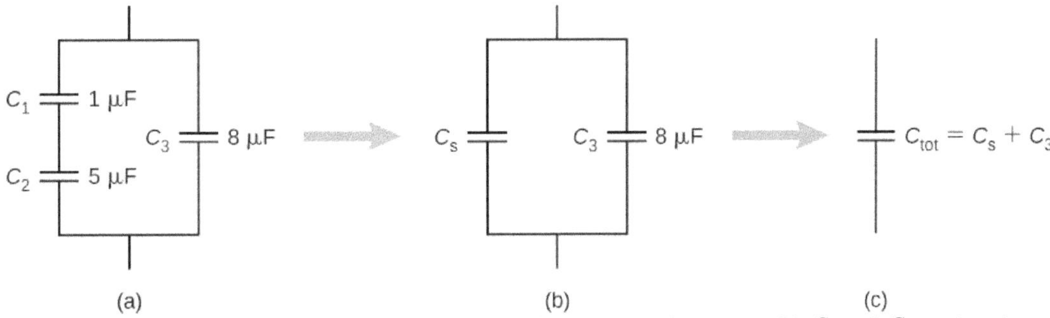

Figure 8.13 (a) This circuit contains both series and parallel connections of capacitors. (b) C_1 and C_2 are in series; their equivalent capacitance is C_S. (c) The equivalent capacitance C_S is connected in parallel with C_3. Thus, the equivalent capacitance of the entire network is the sum of C_S and C_3.

Example 8.6

Equivalent Capacitance of a Network

Find the total capacitance of the combination of capacitors shown in Figure 8.13. Assume the capacitances are known to three decimal places $(C_1 = 1.000\,\mu\text{F},\ C_2 = 5.000\,\mu\text{F},\ C_3 = 8.000\,\mu\text{F})$. Round your answer to three decimal places.

Strategy

We first identify which capacitors are in series and which are in parallel. Capacitors C_1 and C_2 are in series. Their combination, labeled C_S, is in parallel with C_3.

Solution

Since C_1 and C_2 are in series, their equivalent capacitance C_S is obtained with Equation 8.7:

$$\frac{1}{C_S} = \frac{1}{C_1} + \frac{1}{C_2} = \frac{1}{1.000\,\mu\text{F}} + \frac{1}{5.000\,\mu\text{F}} = \frac{1.200}{\mu\text{F}} \Rightarrow C_S = 0.833\,\mu\text{F}.$$

Capacitance C_S is connected in parallel with the third capacitance C_3, so we use Equation 8.8 to find the equivalent capacitance C of the entire network:

$$C = C_S + C_3 = 0.833\,\mu\text{F} + 8.000\,\mu\text{F} = 8.833\,\mu\text{F}.$$

Example 8.7

Network of Capacitors

Determine the net capacitance C of the capacitor combination shown in Figure 8.14 when the capacitances are $C_1 = 12.0\,\mu\text{F},\ C_2 = 2.0\,\mu\text{F},$ and $C_3 = 4.0\,\mu\text{F}$. When a 12.0-V potential difference is maintained across the combination, find the charge and the voltage across each capacitor.

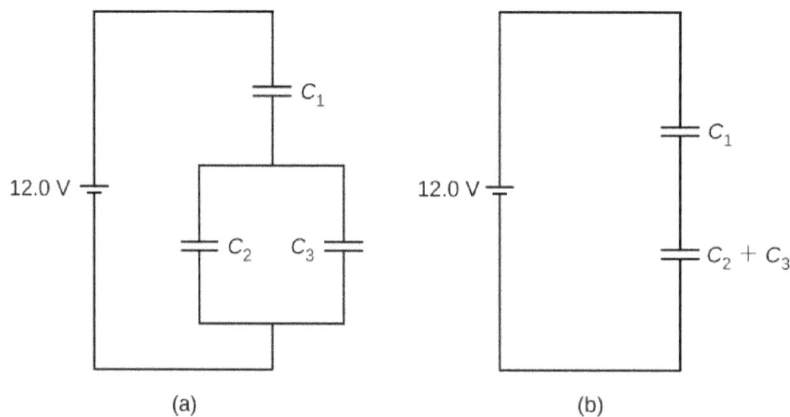

Figure 8.14 (a) A capacitor combination. (b) An equivalent two-capacitor combination.

Strategy

We first compute the net capacitance C_{23} of the parallel connection C_2 and C_3. Then C is the net capacitance of the series connection C_1 and C_{23}. We use the relation $C = Q/V$ to find the charges Q_1, Q_2, and Q_3, and the voltages V_1, V_2, and V_3, across capacitors 1, 2, and 3, respectively.

Solution

The equivalent capacitance for C_2 and C_3 is

$$C_{23} = C_2 + C_3 = 2.0\,\mu\text{F} + 4.0\,\mu\text{F} = 6.0\,\mu\text{F}.$$

The entire three-capacitor combination is equivalent to two capacitors in series,

$$\frac{1}{C} = \frac{1}{12.0\,\mu\text{F}} + \frac{1}{6.0\,\mu\text{F}} = \frac{1}{4.0\,\mu\text{F}} \Rightarrow C = 4.0\,\mu\text{F}.$$

Consider the equivalent two-capacitor combination in Figure 8.14(b). Since the capacitors are in series, they have the same charge, $Q_1 = Q_{23}$. Also, the capacitors share the 12.0-V potential difference, so

$$12.0\,\text{V} = V_1 + V_{23} = \frac{Q_1}{C_1} + \frac{Q_{23}}{C_{23}} = \frac{Q_1}{12.0\,\mu\text{F}} + \frac{Q_1}{6.0\,\mu\text{F}} \Rightarrow Q_1 = 48.0\,\mu\text{C}.$$

Now the potential difference across capacitor 1 is

$$V_1 = \frac{Q_1}{C_1} = \frac{48.0\,\mu\text{C}}{12.0\,\mu\text{F}} = 4.0\,\text{V}.$$

Because capacitors 2 and 3 are connected in parallel, they are at the same potential difference:

$$V_2 = V_3 = 12.0\,\text{V} - 4.0\,\text{V} = 8.0\,\text{V}.$$

Hence, the charges on these two capacitors are, respectively,

$$Q_2 = C_2 V_2 = (2.0\,\mu\text{F})(8.0\,\text{V}) = 16.0\,\mu\text{C},$$
$$Q_3 = C_3 V_3 = (4.0\,\mu\text{F})(8.0\,\text{V}) = 32.0\,\mu\text{C}.$$

Significance

As expected, the net charge on the parallel combination of C_2 and C_3 is $Q_{23} = Q_2 + Q_3 = 48.0\,\mu\text{C}$.

8.5 Check Your Understanding Determine the net capacitance C of each network of capacitors shown below. Assume that $C_1 = 1.0\,\text{pF}$, $C_2 = 2.0\,\text{pF}$, $C_3 = 4.0\,\text{pF}$, and $C_4 = 5.0\,\text{pF}$. Find the charge on each capacitor, assuming there is a potential difference of 12.0 V across each network.

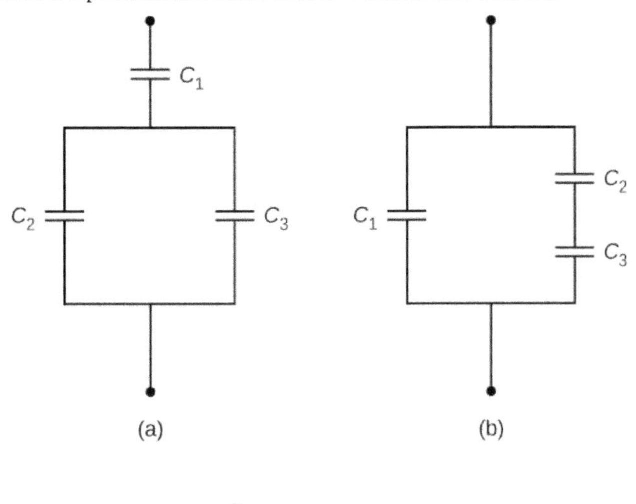

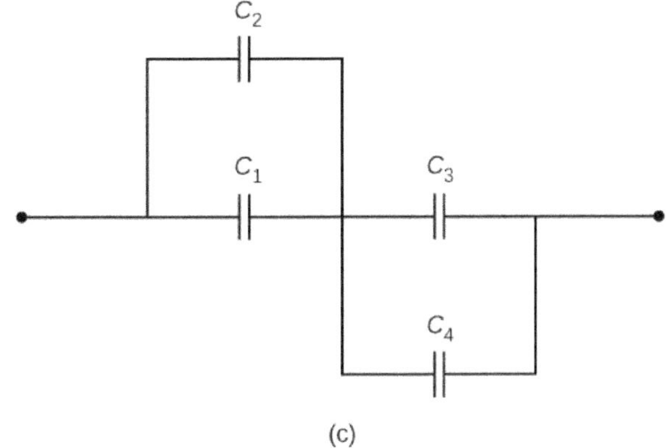

8.3 | Energy Stored in a Capacitor

Learning Objectives

By the end of this section, you will be able to:

- Explain how energy is stored in a capacitor
- Use energy relations to determine the energy stored in a capacitor network

Most of us have seen dramatizations of medical personnel using a defibrillator to pass an electrical current through a patient's heart to get it to beat normally. Often realistic in detail, the person applying the shock directs another person to "make it 400 joules this time." The energy delivered by the defibrillator is stored in a capacitor and can be adjusted to fit the situation. SI units of joules are often employed. Less dramatic is the use of capacitors in microelectronics to supply energy when batteries are charged (Figure 8.15). Capacitors are also used to supply energy for flash lamps on cameras.

Figure 8.15 The capacitors on the circuit board for an electronic device follow a labeling convention that identifies each one with a code that begins with the letter "C."

The energy U_C stored in a capacitor is electrostatic potential energy and is thus related to the charge Q and voltage V between the capacitor plates. A charged capacitor stores energy in the electrical field between its plates. As the capacitor is being charged, the electrical field builds up. When a charged capacitor is disconnected from a battery, its energy remains in the field in the space between its plates.

To gain insight into how this energy may be expressed (in terms of Q and V), consider a charged, empty, parallel-plate capacitor; that is, a capacitor without a dielectric but with a vacuum between its plates. The space between its plates has a volume Ad, and it is filled with a uniform electrostatic field E. The total energy U_C of the capacitor is contained within this space. The **energy density** u_E in this space is simply U_C divided by the volume Ad. If we know the energy density, the energy can be found as $U_C = u_E(Ad)$. We will learn in Electromagnetic Waves (after completing the study of Maxwell's equations) that the energy density u_E in a region of free space occupied by an electrical field E depends only on the magnitude of the field and is

$$u_E = \frac{1}{2}\varepsilon_0 E^2. \tag{8.9}$$

If we multiply the energy density by the volume between the plates, we obtain the amount of energy stored between the plates of a parallel-plate capacitor: $U_C = u_E(Ad) = \frac{1}{2}\varepsilon_0 E^2 Ad = \frac{1}{2}\varepsilon_0 \frac{V^2}{d^2} Ad = \frac{1}{2}V^2 \varepsilon_0 \frac{A}{d} = \frac{1}{2}V^2 C$.

In this derivation, we used the fact that the electrical field between the plates is uniform so that $E = V/d$ and $C = \varepsilon_0 A/d$. Because $C = Q/V$, we can express this result in other equivalent forms:

$$U_C = \frac{1}{2}V^2 C = \frac{1}{2}\frac{Q^2}{C} = \frac{1}{2}QV. \tag{8.10}$$

The expression in Equation 8.10 for the energy stored in a parallel-plate capacitor is generally valid for all types of capacitors. To see this, consider any uncharged capacitor (not necessarily a parallel-plate type). At some instant, we connect it across a battery, giving it a potential difference $V = q/C$ between its plates. Initially, the charge on the plates is $Q = 0$. As the capacitor is being charged, the charge gradually builds up on its plates, and after some time, it reaches the value Q. To move an infinitesimal charge dq from the negative plate to the positive plate (from a lower to a higher potential), the amount of work dW that must be done on dq is $dW = Vdq = \frac{q}{C}dq$.

This work becomes the energy stored in the electrical field of the capacitor. In order to charge the capacitor to a charge Q, the total work required is

$$W = \int_0^{W(Q)} dW = \int_0^Q \frac{q}{C} dq = \frac{1}{2}\frac{Q^2}{C}.$$

Since the geometry of the capacitor has not been specified, this equation holds for any type of capacitor. The total work W needed to charge a capacitor is the electrical potential energy U_C stored in it, or $U_C = W$. When the charge is expressed in coulombs, potential is expressed in volts, and the capacitance is expressed in farads, this relation gives the energy in joules.

Knowing that the energy stored in a capacitor is $U_C = Q^2/(2C)$, we can now find the energy density u_E stored in a vacuum between the plates of a charged parallel-plate capacitor. We just have to divide U_C by the volume Ad of space between its plates and take into account that for a parallel-plate capacitor, we have $E = \sigma/\varepsilon_0$ and $C = \varepsilon_0 A/d$. Therefore, we obtain

$$u_E = \frac{U_C}{Ad} = \frac{1}{2}\frac{Q^2}{C}\frac{1}{Ad} = \frac{1}{2}\frac{Q^2}{\varepsilon_0 A/d}\frac{1}{Ad} = \frac{1}{2}\frac{1}{\varepsilon_0}\left(\frac{Q}{A}\right)^2 = \frac{\sigma^2}{2\varepsilon_0} = \frac{(E\varepsilon_0)^2}{2\varepsilon_0} = \frac{\varepsilon_0}{2}E^2.$$

We see that this expression for the density of energy stored in a parallel-plate capacitor is in accordance with the general relation expressed in Equation 8.9. We could repeat this calculation for either a spherical capacitor or a cylindrical capacitor—or other capacitors—and in all cases, we would end up with the general relation given by Equation 8.9.

Example 8.8

Energy Stored in a Capacitor

Calculate the energy stored in the capacitor network in Figure 8.14(a) when the capacitors are fully charged and when the capacitances are $C_1 = 12.0\,\mu\text{F}$, $C_2 = 2.0\,\mu\text{F}$, and $C_3 = 4.0\,\mu\text{F}$, respectively.

Strategy

We use Equation 8.10 to find the energy U_1, U_2, and U_3 stored in capacitors 1, 2, and 3, respectively. The total energy is the sum of all these energies.

Solution

We identify $C_1 = 12.0\,\mu\text{F}$ and $V_1 = 4.0\,\text{V}$, $C_2 = 2.0\,\mu\text{F}$ and $V_2 = 8.0\,\text{V}$, $C_3 = 4.0\,\mu\text{F}$ and $V_3 = 8.0\,\text{V}$. The energies stored in these capacitors are

$$\begin{aligned} U_1 &= \tfrac{1}{2}C_1 V_1^2 = \tfrac{1}{2}(12.0\,\mu\text{F})(4.0\,\text{V})^2 = 96\,\mu\text{J}, \\ U_2 &= \tfrac{1}{2}C_2 V_2^2 = \tfrac{1}{2}(2.0\,\mu\text{F})(8.0\,\text{V})^2 = 64\,\mu\text{J}, \\ U_3 &= \tfrac{1}{2}C_3 V_3^2 = \tfrac{1}{2}(4.0\,\mu\text{F})(8.0\,\text{V})^2 = 130\,\mu\text{J}. \end{aligned}$$

The total energy stored in this network is

$$U_C = U_1 + U_2 + U_3 = 96\,\mu\text{J} + 64\,\mu\text{J} + 130\,\mu\text{J} = 0.29\,\text{mJ}.$$

Significance

We can verify this result by calculating the energy stored in the single 4.0-μF capacitor, which is found to be equivalent to the entire network. The voltage across the network is 12.0 V. The total energy obtained in this way agrees with our previously obtained result, $U_C = \frac{1}{2}CV^2 = \frac{1}{2}(4.0\,\mu\text{F})(12.0\,\text{V})^2 = 0.29\,\text{mJ}$.

8.6 Check Your Understanding The potential difference across a 5.0-pF capacitor is 0.40 V. (a) What is the energy stored in this capacitor? (b) The potential difference is now increased to 1.20 V. By what factor is the stored energy increased?

In a cardiac emergency, a portable electronic device known as an automated external defibrillator (AED) can be a lifesaver. A **defibrillator** (Figure 8.16) delivers a large charge in a short burst, or a shock, to a person's heart to correct abnormal heart rhythm (an arrhythmia). A heart attack can arise from the onset of fast, irregular beating of the heart—called cardiac or ventricular fibrillation. Applying a large shock of electrical energy can terminate the arrhythmia and allow the body's natural pacemaker to resume its normal rhythm. Today, it is common for ambulances to carry AEDs. AEDs are also found in many public places. These are designed to be used by lay persons. The device automatically diagnoses the patient's heart rhythm and then applies the shock with appropriate energy and waveform. CPR (cardiopulmonary resuscitation) is recommended in many cases before using a defibrillator.

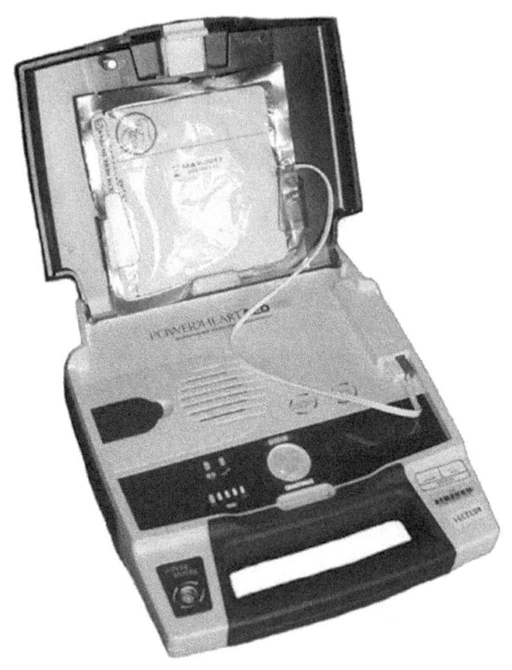

Figure 8.16 Automated external defibrillators are found in many public places. These portable units provide verbal instructions for use in the important first few minutes for a person suffering a cardiac attack.

Example 8.9

Capacitance of a Heart Defibrillator

A heart defibrillator delivers $4.00 \times 10^2\,\text{J}$ of energy by discharging a capacitor initially at $1.00 \times 10^4\ \text{V}$. What is its capacitance?

Strategy

We are given U_C and V, and we are asked to find the capacitance C. We solve Equation 8.10 for C and substitute.

Solution

Solving this expression for C and entering the given values yields $C = 2\frac{U_C}{V^2} = 2\frac{4.00 \times 10^2\ \text{J}}{(1.00 \times 10^4\ \text{V})^2} = 8.00\ \mu\text{F}$.

8.4 | Capacitor with a Dielectric

Learning Objectives

By the end of this section, you will be able to:

- Describe the effects a dielectric in a capacitor has on capacitance and other properties
- Calculate the capacitance of a capacitor containing a dielectric

As we discussed earlier, an insulating material placed between the plates of a capacitor is called a dielectric. Inserting a dielectric between the plates of a capacitor affects its capacitance. To see why, let's consider an experiment described in Figure 8.17. Initially, a capacitor with capacitance C_0 when there is air between its plates is charged by a battery to voltage V_0. When the capacitor is fully charged, the battery is disconnected. A charge Q_0 then resides on the plates, and the potential difference between the plates is measured to be V_0. Now, suppose we insert a dielectric that *totally* fills the gap between the plates. If we monitor the voltage, we find that the voltmeter reading has dropped to a *smaller* value *V*. We write this new voltage value as a fraction of the original voltage V_0, with a positive number κ, $\kappa > 1$:

$$V = \frac{1}{\kappa}V_0.$$

The constant κ in this equation is called the **dielectric constant** of the material between the plates, and its value is characteristic for the material. A detailed explanation for why the dielectric reduces the voltage is given in the next section. Different materials have different dielectric constants (a table of values for typical materials is provided in the next section). Once the battery becomes disconnected, there is no path for a charge to flow to the battery from the capacitor plates. Hence, the insertion of the dielectric has no effect on the charge on the plate, which remains at a value of Q_0. Therefore, we find that the capacitance of the capacitor with a dielectric is

$$C = \frac{Q_0}{V} = \frac{Q_0}{V_0/\kappa} = \kappa\frac{Q_0}{V_0} = \kappa C_0. \quad (8.11)$$

This equation tells us that the *capacitance* C_0 *of an empty (vacuum) capacitor can be increased by a factor of* κ *when we insert a dielectric material to completely fill the space between its plates*. Note that Equation 8.11 can also be used for an empty capacitor by setting $\kappa = 1$. In other words, we can say that the dielectric constant of the vacuum is 1, which is a reference value.

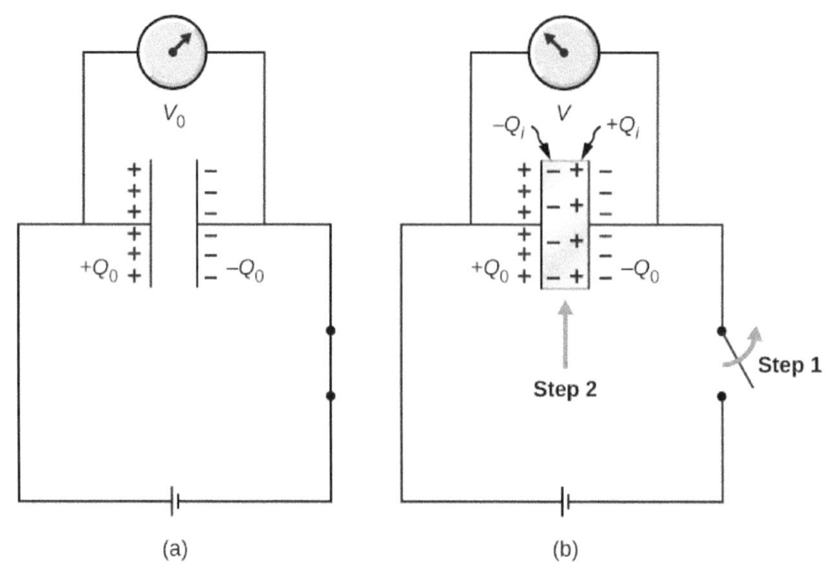

Figure 8.17 (a) When fully charged, a vacuum capacitor has a voltage V_0 and charge Q_0 (the charges remain on plate's inner surfaces; the schematic indicates the sign of charge on each plate). (b) In step 1, the battery is disconnected. Then, in step 2, a dielectric (that is electrically neutral) is inserted into the charged capacitor. When the voltage across the capacitor is now measured, it is found that the voltage value has decreased to $V = V_0/\kappa$. The schematic indicates the sign of the induced charge that is now present on the surfaces of the dielectric material between the plates.

The principle expressed by Equation 8.11 is widely used in the construction industry (Figure 8.18). Metal plates in an electronic stud finder act effectively as a capacitor. You place a stud finder with its flat side on the wall and move it continually in the horizontal direction. When the finder moves over a wooden stud, the capacitance of its plates changes, because wood has a different dielectric constant than a gypsum wall. This change triggers a signal in a circuit, and thus the stud is detected.

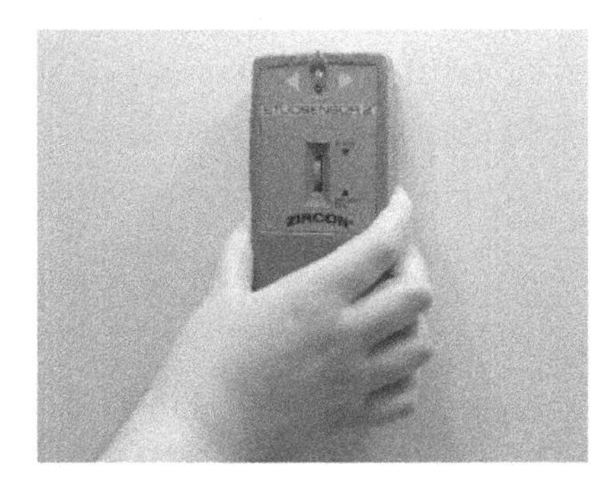

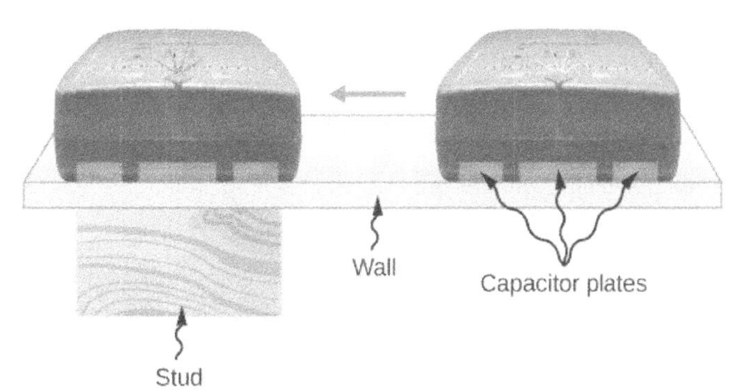

Figure 8.18 An electronic stud finder is used to detect wooden studs behind drywall.

The electrical energy stored by a capacitor is also affected by the presence of a dielectric. When the energy stored in an empty capacitor is U_0, the energy U stored in a capacitor with a dielectric is smaller by a factor of κ,

$$U = \frac{1}{2}\frac{Q^2}{C} = \frac{1}{2}\frac{Q_0^2}{\kappa C_0} = \frac{1}{\kappa}U_0. \tag{8.12}$$

As a dielectric material sample is brought near an empty charged capacitor, the sample reacts to the electrical field of the charges on the capacitor plates. Just as we learned in Electric Charges and Fields on electrostatics, there will be the induced charges on the surface of the sample; however, they are not free charges like in a conductor, because a perfect insulator does not have freely moving charges. These induced charges on the dielectric surface are of an opposite sign to the free charges on the plates of the capacitor, and so they are attracted by the free charges on the plates. Consequently, the dielectric is "pulled" into the gap, and the work to polarize the dielectric material between the plates is done at the expense of the stored electrical energy, which is reduced, in accordance with Equation 8.12.

Example 8.10

Inserting a Dielectric into an Isolated Capacitor

An empty 20.0-pF capacitor is charged to a potential difference of 40.0 V. The charging battery is then disconnected, and a piece of Teflon™ with a dielectric constant of 2.1 is inserted to completely fill the space between the capacitor plates (see Figure 8.17). What are the values of (a) the capacitance, (b) the charge of the plate, (c) the potential difference between the plates, and (d) the energy stored in the capacitor with and without dielectric?

Strategy

We identify the original capacitance $C_0 = 20.0\,\text{pF}$ and the original potential difference $V_0 = 40.0\,\text{V}$ between the plates. We combine Equation 8.11 with other relations involving capacitance and substitute.

Solution

a. The capacitance increases to

$$C = \kappa C_0 = 2.1(20.0\,\text{pF}) = 42.0\,\text{pF}.$$

b. Without dielectric, the charge on the plates is

$$Q_0 = C_0 V_0 = (20.0\,\text{pF})(40.0\,\text{V}) = 0.8\,\text{nC}.$$

Since the battery is disconnected before the dielectric is inserted, the plate charge is unaffected by the dielectric and remains at 0.8 nC.

c. With the dielectric, the potential difference becomes

$$V = \frac{1}{\kappa}V_0 = \frac{1}{2.1}40.0\,\text{V} = 19.0\,\text{V}.$$

d. The stored energy without the dielectric is

$$U_0 = \frac{1}{2}C_0 V_0^2 = \frac{1}{2}(20.0\,\text{pF})(40.0\,\text{V})^2 = 16.0\,\text{nJ}.$$

With the dielectric inserted, we use Equation 8.12 to find that the stored energy decreases to

$$U = \frac{1}{\kappa}U_0 = \frac{1}{2.1}16.0\,\text{nJ} = 7.6\,\text{nJ}.$$

Significance

Notice that the effect of a dielectric on the capacitance of a capacitor is a drastic increase of its capacitance. This effect is far more profound than a mere change in the geometry of a capacitor.

8.7 Check Your Understanding When a dielectric is inserted into an isolated and charged capacitor, the stored energy decreases to 33% of its original value. (a) What is the dielectric constant? (b) How does the capacitance change?

8.5 | Molecular Model of a Dielectric

Learning Objectives

By the end of this section, you will be able to:

- Explain the polarization of a dielectric in a uniform electrical field
- Describe the effect of a polarized dielectric on the electrical field between capacitor plates
- Explain dielectric breakdown

We can understand the effect of a dielectric on capacitance by looking at its behavior at the molecular level. As we have seen in earlier chapters, in general, all molecules can be classified as either *polar* or *nonpolar*. There is a net separation of positive and negative charges in an isolated polar molecule, whereas there is no charge separation in an isolated nonpolar molecule (Figure 8.19). In other words, polar molecules have permanent *electric-dipole moments* and nonpolar molecules do not. For example, a molecule of water is polar, and a molecule of oxygen is nonpolar. Nonpolar molecules can become polar in the presence of an external electrical field, which is called *induced polarization*.

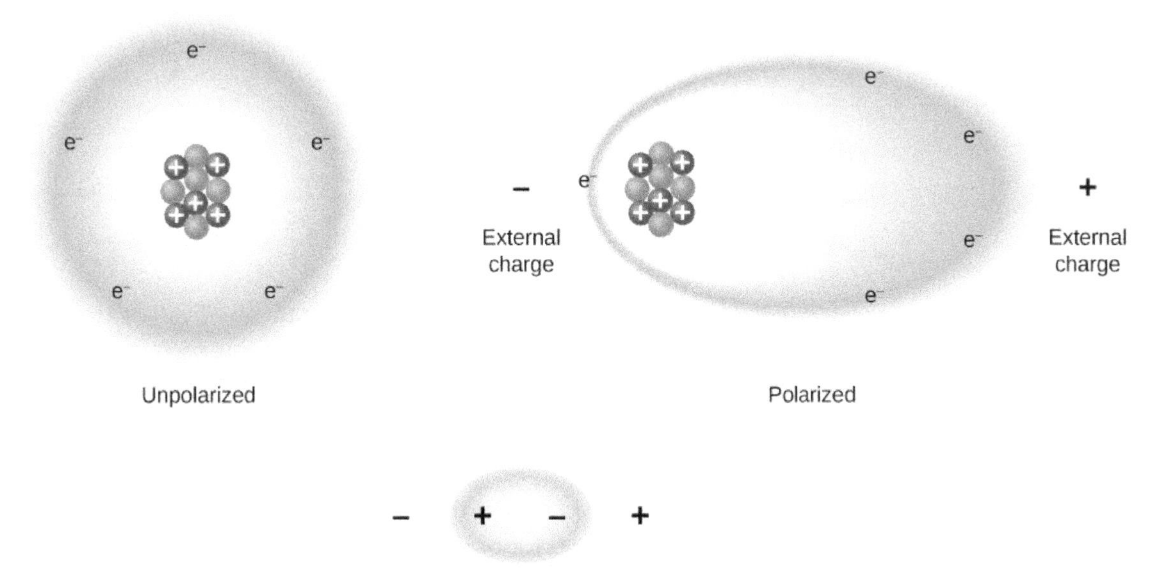

Figure 8.19 The concept of polarization: In an unpolarized atom or molecule, a negatively charged electron cloud is evenly distributed around positively charged centers, whereas a polarized atom or molecule has an excess of negative charge at one side so that the other side has an excess of positive charge. However, the entire system remains electrically neutral. The charge polarization may be caused by an external electrical field. Some molecules and atoms are permanently polarized (electric dipoles) even in the absence of an external electrical field (polar molecules and atoms).

Let's first consider a dielectric composed of polar molecules. In the absence of any external electrical field, the electric dipoles are oriented randomly, as illustrated in Figure 8.20(a). However, if the dielectric is placed in an external electrical field $\vec{\mathbf{E}}_0$, the polar molecules align with the external field, as shown in part (b) of the figure. Opposite charges on adjacent dipoles within the volume of dielectric neutralize each other, so there is no net charge within the dielectric (see the dashed circles in part (b)). However, this is not the case very close to the upper and lower surfaces that border the dielectric (the region enclosed by the dashed rectangles in part (b)), where the alignment does produce a net charge. Since the external electrical field merely aligns the dipoles, the dielectric as a whole is neutral, and the surface charges induced on its opposite faces are equal and opposite. These **induced surface charges** $+Q_i$ and $-Q_i$ produce an additional electrical field $\vec{\mathbf{E}}_{\text{i}}$ (an **induced electrical field**), which *opposes* the external field $\vec{\mathbf{E}}_0$, as illustrated in part (c).

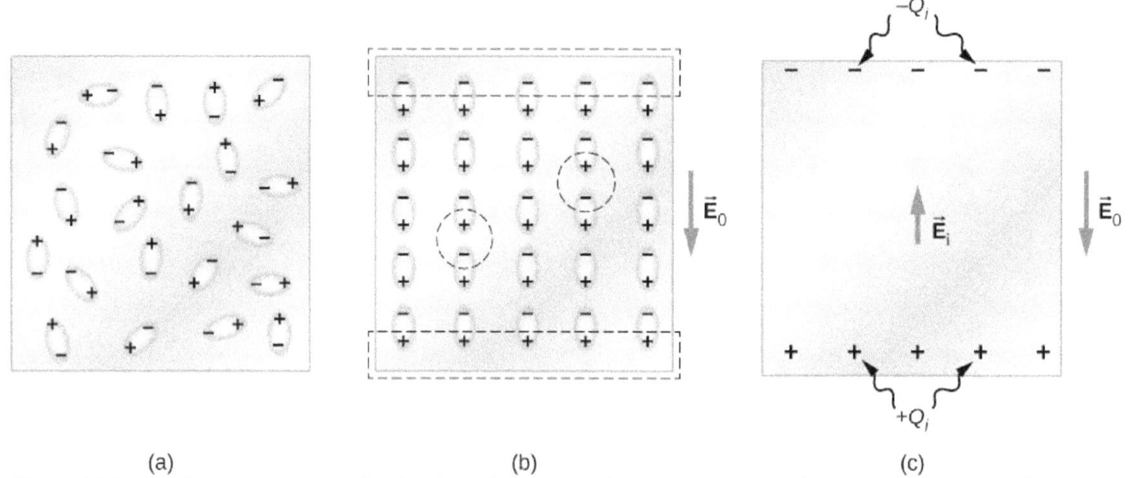

Figure 8.20 A dielectric with polar molecules: (a) In the absence of an external electrical field; (b) in the presence of an external electrical field $\vec{E}_0$. The dashed lines indicate the regions immediately adjacent to the capacitor plates. (c) The induced electrical field $\vec{E}_i$ inside the dielectric produced by the induced surface charge Q_i of the dielectric. Note that, in reality, the individual molecules are not perfectly aligned with an external field because of thermal fluctuations; however, the *average* alignment is along the field lines as shown.

The same effect is produced when the molecules of a dielectric are nonpolar. In this case, a nonpolar molecule acquires an **induced electric-dipole moment** because the external field $\vec{E}_0$ causes a separation between its positive and negative charges. The induced dipoles of the nonpolar molecules align with $\vec{E}_0$ in the same way as the permanent dipoles of the polar molecules are aligned (shown in part (b)). Hence, the electrical field within the dielectric is weakened regardless of whether its molecules are polar or nonpolar.

Therefore, when the region between the parallel plates of a charged capacitor, such as that shown in Figure 8.21(a), is filled with a dielectric, within the dielectric there is an electrical field $\vec{E}_0$ due to the *free charge* Q_0 on the capacitor plates and an electrical field $\vec{E}_i$ due to the induced charge Q_i on the surfaces of the dielectric. Their vector sum gives the net electrical field $\vec{E}$ within the dielectric between the capacitor plates (shown in part (b) of the figure):

$$\vec{E} = \vec{E}_0 + \vec{E}_i. \tag{8.13}$$

This net field can be considered to be the field produced by an *effective charge* $Q_0 - Q_i$ on the capacitor.

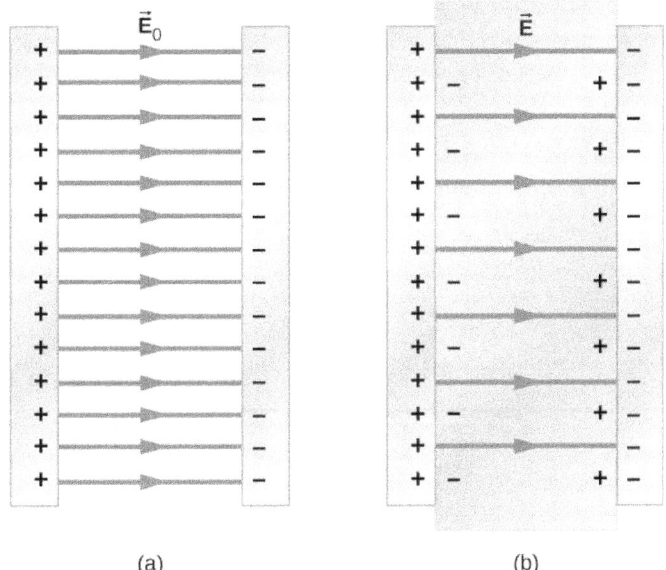

Figure 8.21 Electrical field: (a) In an empty capacitor, electrical field $\vec{\mathbf{E}}_0$. (b) In a dielectric-filled capacitor, electrical field $\vec{\mathbf{E}}$.

In most dielectrics, the net electrical field $\vec{\mathbf{E}}$ is proportional to the field $\vec{\mathbf{E}}_0$ produced by the free charge. In terms of these two electrical fields, the dielectric constant κ of the material is defined as

$$\kappa = \frac{E_0}{E}. \tag{8.14}$$

Since $\vec{\mathbf{E}}_0$ and $\vec{\mathbf{E}}_i$ point in opposite directions, the magnitude E is smaller than the magnitude E_0 and therefore $\kappa > 1$. Combining Equation 8.14 with Equation 8.13, and rearranging the terms, yields the following expression for the induced electrical field in a dielectric:

$$\vec{\mathbf{E}}_i = \left(\frac{1}{\kappa} - 1\right)\vec{\mathbf{E}}_0. \tag{8.15}$$

When the magnitude of an external electrical field becomes too large, the molecules of dielectric material start to become ionized. A molecule or an atom is ionized when one or more electrons are removed from it and become free electrons, no longer bound to the molecular or atomic structure. When this happens, the material can conduct, thereby allowing charge to move through the dielectric from one capacitor plate to the other. This phenomenon is called **dielectric breakdown**. (Figure 8.1 shows typical random-path patterns of electrical discharge during dielectric breakdown.) The critical value, E_c, of the electrical field at which the molecules of an insulator become ionized is called the **dielectric strength** of the material. The dielectric strength imposes a limit on the voltage that can be applied for a given plate separation in a capacitor. For example, the dielectric strength of air is $E_c = 3.0\,\text{MV/m}$, so for an air-filled capacitor with a plate separation of $d = 1.00\,\text{mm}$, the limit on the potential difference that can be safely applied across its plates without causing dielectric breakdown is $V = E_c d = (3.0 \times 10^6\ \text{V/m})(1.00 \times 10^{-3}\ \text{m}) = 3.0\,\text{kV}$.

However, this limit becomes 60.0 kV when the same capacitor is filled with Teflon™, whose dielectric strength is about $60.0\,\mathrm{MV/m}$. Because of this limit imposed by the dielectric strength, the amount of charge that an air-filled capacitor can store is only $Q_0 = \kappa_{\text{air}} C_0(3.0\,\mathrm{kV})$ and the charge stored on the same Teflon™-filled capacitor can be as much as

$$Q = \kappa_{\text{teflo}}\ C_0(60.0\,\mathrm{kV}) = \kappa_{\text{teflo}}\ \frac{Q_0}{\kappa_{\text{air}}(3.0\,\mathrm{kV})}(60.0\,\mathrm{kV}) = 20\frac{\kappa_{\text{teflo}}}{\kappa_{\text{air}}}Q_0 = 20\frac{2.1}{1.00059}Q_0 \cong 42\,Q_0,$$

which is about 42 times greater than a charge stored on an air-filled capacitor. Typical values of dielectric constants and dielectric strengths for various materials are given in Table 8.1. Notice that the dielectric constant κ is exactly 1.0 for a vacuum (the empty space serves as a reference condition) and very close to 1.0 for air under normal conditions (normal pressure at room temperature). These two values are so close that, in fact, the properties of an air-filled capacitor are essentially the same as those of an empty capacitor.

Material	Dielectric constant κ	Dielectric strength $E_c[\times 10^6\,\mathrm{V/m}]$
Vacuum	1	∞
Dry air (1 atm)	1.00059	3.0
Teflon™	2.1	60 to 173
Paraffin	2.3	11
Silicon oil	2.5	10 to 15
Polystyrene	2.56	19.7
Nylon	3.4	14
Paper	3.7	16
Fused quartz	3.78	8
Glass	4 to 6	9.8 to 13.8
Concrete	4.5	–
Bakelite	4.9	24
Diamond	5.5	2,000
Pyrex glass	5.6	14
Mica	6.0	118
Neoprene rubber	6.7	15.7 to 26.7
Water	80	–
Sulfuric acid	84 to 100	–
Titanium dioxide	86 to 173	–
Strontium titanate	310	8
Barium titanate	1,200 to 10,000	–
Calcium copper titanate	> 250,000	–

Table 8.1 Representative Values of Dielectric Constants and Dielectric Strengths of Various Materials at Room Temperature

Not all substances listed in the table are good insulators, despite their high dielectric constants. Water, for example, consists of polar molecules and has a large dielectric constant of about 80. In a water molecule, electrons are more likely found around the oxygen nucleus than around the hydrogen nuclei. This makes the oxygen end of the molecule slightly negative and leaves the hydrogens end slightly positive, which makes the molecule easy to align along an external electrical field, and thus water has a large dielectric constant. However, the polar nature of water molecules also makes water a good solvent for many substances, which produces undesirable effects, because any concentration of free ions in water conducts electricity.

Example 8.11

Electrical Field and Induced Surface Charge

Suppose that the distance between the plates of the capacitor in Example 8.10 is 2.0 mm and the area of each plate is $4.5 \times 10^{-3}\ \text{m}^2$. Determine: (a) the electrical field between the plates before and after the Teflon™ is inserted, and (b) the surface charge induced on the Teflon™ surfaces.

Strategy

In part (a), we know that the voltage across the empty capacitor is $V_0 = 40\ \text{V}$, so to find the electrical fields we use the relation $V = Ed$ and Equation 8.14. In part (b), knowing the magnitude of the electrical field, we use the expression for the magnitude of electrical field near a charged plate $E = \sigma/\varepsilon_0$, where σ is a uniform surface charge density caused by the surface charge. We use the value of free charge $Q_0 = 8.0 \times 10^{-10}\ \text{C}$ obtained in Example 8.10.

Solution

a. The electrical field E_0 between the plates of an empty capacitor is

$$E_0 = \frac{V_0}{d} = \frac{40\ \text{V}}{2.0 \times 10^{-3}\,\text{m}} = 2.0 \times 10^4\ \text{V/m}.$$

The electrical field E with the Teflon™ in place is

$$E = \frac{1}{\kappa}E_0 = \frac{1}{2.1}2.0 \times 10^4\ \text{V/m} = 9.5 \times 10^3\ \text{V/m}.$$

b. The effective charge on the capacitor is the difference between the free charge Q_0 and the induced charge Q_i. The electrical field in the Teflon™ is caused by this effective charge. Thus

$$E = \frac{1}{\varepsilon_0}\sigma = \frac{1}{\varepsilon_0}\frac{Q_0 - Q_\text{i}}{A}.$$

We invert this equation to obtain Q_i, which yields

$$\begin{aligned} Q_\text{i} &= Q_0 - \varepsilon_0 AE \\ &= 8.0 \times 10^{-10}\text{C} - \left(8.85 \times 10^{-12}\ \frac{\text{C}^2}{\text{N}\cdot\text{m}^2}\right)\left(4.5 \times 10^{-3}\ \text{m}^2\right)\left(9.5 \times 10^3\ \frac{\text{V}}{\text{m}}\right) \\ &= 4.2 \times 10^{-10}\text{C} = 0.42\ \text{nC}. \end{aligned}$$

Example 8.12

Inserting a Dielectric into a Capacitor Connected to a Battery

When a battery of voltage V_0 is connected across an empty capacitor of capacitance C_0, the charge on its plates is Q_0, and the electrical field between its plates is E_0. A dielectric of dielectric constant κ is inserted between the plates *while the battery remains in place*, as shown in Figure 8.22. (a) Find the capacitance C, the voltage V across the capacitor, and the electrical field E between the plates after the dielectric is inserted. (b) Obtain an expression for the free charge Q on the plates of the filled capacitor and the induced charge Q_i on the dielectric surface in terms of the original plate charge Q_0.

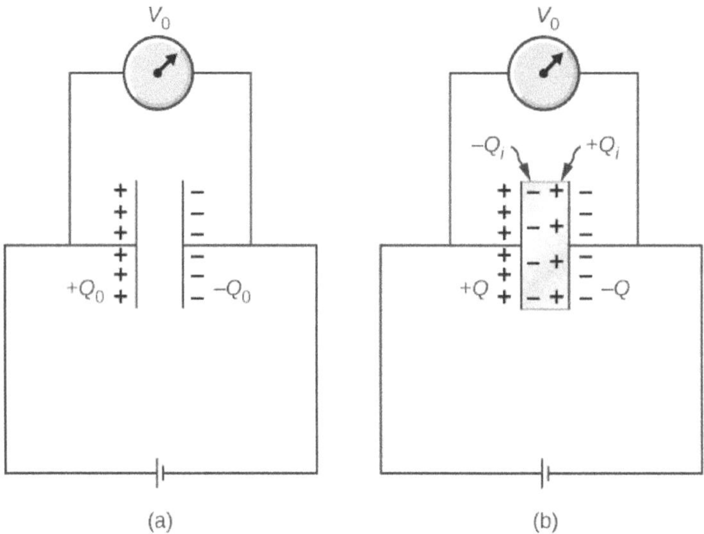

Figure 8.22 A dielectric is inserted into the charged capacitor while the capacitor remains connected to the battery.

Strategy

We identify the known values: V_0, C_0, E_0, κ, and Q_0. Our task is to express the unknown values in terms of these known values.

Solution

(a) The capacitance of the filled capacitor is $C = \kappa C_0$. Since the battery is always connected to the capacitor plates, the potential difference between them does not change; hence, $V = V_0$. Because of that, the electrical field in the filled capacitor is the same as the field in the empty capacitor, so we can obtain directly that

$$E = \frac{V}{d} = \frac{V_0}{d} = E_0.$$

(b) For the filled capacitor, the free charge on the plates is

$$Q = CV = (\kappa C_0)V_0 = \kappa(C_0 V_0) = \kappa Q_0.$$

The electrical field E in the filled capacitor is due to the effective charge $Q - Q_\text{i}$ (Figure 8.22(b)). Since $E = E_0$, we have

$$\frac{Q - Q_\text{i}}{\varepsilon_0 A} = \frac{Q_0}{\varepsilon_0 A}.$$

Solving this equation for Q_i, we obtain for the induced charge

$$Q_\text{i} = Q - Q_0 = \kappa Q_0 - Q_0 = (\kappa - 1)Q_0.$$

Significance

Notice that for materials with dielectric constants larger than 2 (see Table 8.1), the induced charge on the surface of dielectric is larger than the charge on the plates of a vacuum capacitor. The opposite is true for gasses like air whose dielectric constant is smaller than 2.

8.8 Check Your Understanding Continuing with Example 8.12, show that when the battery is connected across the plates the energy stored in dielectric-filled capacitor is $U = \kappa U_0$ (larger than the energy U_0 of an empty capacitor kept at the same voltage). Compare this result with the result $U = U_0/\kappa$ found previously for an isolated, charged capacitor.

8.9 Check Your Understanding Repeat the calculations of Example 8.10 for the case in which the battery remains connected while the dielectric is placed in the capacitor.

CHAPTER 8 REVIEW

KEY TERMS

capacitance amount of charge stored per unit volt

capacitor device that stores electrical charge and electrical energy

dielectric insulating material used to fill the space between two plates

dielectric breakdown phenomenon that occurs when an insulator becomes a conductor in a strong electrical field

dielectric constant factor by which capacitance increases when a dielectric is inserted between the plates of a capacitor

dielectric strength critical electrical field strength above which molecules in insulator begin to break down and the insulator starts to conduct

energy density energy stored in a capacitor divided by the volume between the plates

induced electric-dipole moment dipole moment that a nonpolar molecule may acquire when it is placed in an electrical field

induced electrical field electrical field in the dielectric due to the presence of induced charges

induced surface charges charges that occur on a dielectric surface due to its polarization

parallel combination components in a circuit arranged with one side of each component connected to one side of the circuit and the other sides of the components connected to the other side of the circuit

parallel-plate capacitor system of two identical parallel conducting plates separated by a distance

series combination components in a circuit arranged in a row one after the other in a circuit

KEY EQUATIONS

Capacitance	$C = \frac{Q}{V}$
Capacitance of a parallel-plate capacitor	$C = \varepsilon_0 \frac{A}{d}$
Capacitance of a vacuum spherical capacitor	$C = 4\pi\varepsilon_0 \frac{R_1 R_2}{R_2 - R_1}$
Capacitance of a vacuum cylindrical capacitor	$C = \frac{2\pi\varepsilon_0 l}{\ln(R_2/R_1)}$
Capacitance of a series combination	$\frac{1}{C_S} = \frac{1}{C_1} + \frac{1}{C_2} + \frac{1}{C_3} + \cdots$
Capacitance of a parallel combination	$C_P = C_1 + C_2 + C_3 + \cdots$
Energy density	$u_E = \frac{1}{2}\varepsilon_0 E^2$
Energy stored in a capacitor	$U_C = \frac{1}{2}V^2 C = \frac{1}{2}\frac{Q^2}{C} = \frac{1}{2}QV$
Capacitance of a capacitor with dielectric	$C = \kappa C_0$
Energy stored in an isolated capacitor with dielectric	$U = \frac{1}{\kappa}U_0$
Dielectric constant	$\kappa = \frac{E_0}{E}$

Induced electrical field in a dielectric $$\vec{\mathbf{E}}_{\text{i}} = \left(\frac{1}{\kappa} - 1\right)\vec{\mathbf{E}}_0$$

SUMMARY

8.1 Capacitors and Capacitance

- A capacitor is a device that stores an electrical charge and electrical energy. The amount of charge a vacuum capacitor can store depends on two major factors: the voltage applied and the capacitor's physical characteristics, such as its size and geometry.
- The capacitance of a capacitor is a parameter that tells us how much charge can be stored in the capacitor per unit potential difference between its plates. Capacitance of a system of conductors depends only on the geometry of their arrangement and physical properties of the insulating material that fills the space between the conductors. The unit of capacitance is the farad, where $1\text{ F} = 1\text{ C}/1\text{ V}$.

8.2 Capacitors in Series and in Parallel

- When several capacitors are connected in a series combination, the reciprocal of the equivalent capacitance is the sum of the reciprocals of the individual capacitances.
- When several capacitors are connected in a parallel combination, the equivalent capacitance is the sum of the individual capacitances.
- When a network of capacitors contains a combination of series and parallel connections, we identify the series and parallel networks, and compute their equivalent capacitances step by step until the entire network becomes reduced to one equivalent capacitance.

8.3 Energy Stored in a Capacitor

- Capacitors are used to supply energy to a variety of devices, including defibrillators, microelectronics such as calculators, and flash lamps.
- The energy stored in a capacitor is the work required to charge the capacitor, beginning with no charge on its plates. The energy is stored in the electrical field in the space between the capacitor plates. It depends on the amount of electrical charge on the plates and on the potential difference between the plates.
- The energy stored in a capacitor network is the sum of the energies stored on individual capacitors in the network. It can be computed as the energy stored in the equivalent capacitor of the network.

8.4 Capacitor with a Dielectric

- The capacitance of an empty capacitor is increased by a factor of κ when the space between its plates is completely filled by a dielectric with dielectric constant κ.
- Each dielectric material has its specific dielectric constant.
- The energy stored in an empty isolated capacitor is decreased by a factor of κ when the space between its plates is completely filled with a dielectric with dielectric constant κ.

8.5 Molecular Model of a Dielectric

- When a dielectric is inserted between the plates of a capacitor, equal and opposite surface charge is induced on the two faces of the dielectric. The induced surface charge produces an induced electrical field that opposes the field of the free charge on the capacitor plates.
- The dielectric constant of a material is the ratio of the electrical field in vacuum to the net electrical field in the material. A capacitor filled with dielectric has a larger capacitance than an empty capacitor.
- The dielectric strength of an insulator represents a critical value of electrical field at which the molecules in an insulating material start to become ionized. When this happens, the material can conduct and dielectric breakdown is observed.

CONCEPTUAL QUESTIONS

8.1 Capacitors and Capacitance

1. Does the capacitance of a device depend on the applied voltage? Does the capacitance of a device depend on the charge residing on it?

2. Would you place the plates of a parallel-plate capacitor closer together or farther apart to increase their capacitance?

3. The value of the capacitance is zero if the plates are not charged. True or false?

4. If the plates of a capacitor have different areas, will they acquire the same charge when the capacitor is connected across a battery?

5. Does the capacitance of a spherical capacitor depend on which sphere is charged positively or negatively?

8.2 Capacitors in Series and in Parallel

6. If you wish to store a large amount of charge in a capacitor bank, would you connect capacitors in series or in parallel? Explain.

7. What is the maximum capacitance you can get by connecting three $1.0\text{-}\mu\text{F}$ capacitors? What is the minimum capacitance?

8.3 Energy Stored in a Capacitor

8. If you wish to store a large amount of energy in a capacitor bank, would you connect capacitors in series or parallel? Explain.

8.4 Capacitor with a Dielectric

9. Discuss what would happen if a conducting slab rather than a dielectric were inserted into the gap between the capacitor plates.

10. Discuss how the energy stored in an empty but charged capacitor changes when a dielectric is inserted if (a) the capacitor is isolated so that its charge does not change; (b) the capacitor remains connected to a battery so that the potential difference between its plates does not change.

8.5 Molecular Model of a Dielectric

11. Distinguish between dielectric strength and dielectric constant.

12. Water is a good solvent because it has a high dielectric constant. Explain.

13. Water has a high dielectric constant. Explain why it is then not used as a dielectric material in capacitors.

14. Elaborate on why molecules in a dielectric material experience net forces on them in a non-uniform electrical field but not in a uniform field.

15. Explain why the dielectric constant of a substance containing permanent molecular electric dipoles decreases with increasing temperature.

16. Give a reason why a dielectric material increases capacitance compared with what it would be with air between the plates of a capacitor. How does a dielectric material also allow a greater voltage to be applied to a capacitor? (The dielectric thus increases *C* and permits a greater *V*.)

17. Elaborate on the way in which the polar character of water molecules helps to explain water's relatively large dielectric constant.

18. Sparks will occur between the plates of an air-filled capacitor at a lower voltage when the air is humid than when it is dry. Discuss why, considering the polar character of water molecules.

PROBLEMS

8.1 Capacitors and Capacitance

19. What charge is stored in a $180.0\text{-}\mu\text{F}$ capacitor when 120.0 V is applied to it?

20. Find the charge stored when 5.50 V is applied to an 8.00-pF capacitor.

21. Calculate the voltage applied to a $2.00\text{-}\mu\text{F}$ capacitor when it holds $3.10\,\mu\text{C}$ of charge.

22. What voltage must be applied to an 8.00-nF capacitor to store 0.160 mC of charge?

23. What capacitance is needed to store $3.00\,\mu\text{C}$ of charge at a voltage of 120 V?

24. What is the capacitance of a large Van de Graaff generator's terminal, given that it stores 8.00 mC of charge at a voltage of 12.0 MV?

25. The plates of an empty parallel-plate capacitor of capacitance 5.0 pF are 2.0 mm apart. What is the area of each plate?

26. A 60.0-pF vacuum capacitor has a plate area of $0.010\ \text{m}^2$. What is the separation between its plates?

27. A set of parallel plates has a capacitance of $5.0\mu\text{F}$. How much charge must be added to the plates to increase the potential difference between them by 100 V?

28. Consider Earth to be a spherical conductor of radius 6400 km and calculate its capacitance.

29. If the capacitance per unit length of a cylindrical capacitor is 20 pF/m, what is the ratio of the radii of the two cylinders?

30. An empty parallel-plate capacitor has a capacitance of $20\ \mu\text{F}$. How much charge must leak off its plates before the voltage across them is reduced by 100 V?

8.2 Capacitors in Series and in Parallel

31. A 4.00-pF is connected in series with an 8.00-pF capacitor and a 400-V potential difference is applied across the pair. (a) What is the charge on each capacitor? (b) What is the voltage across each capacitor?

32. Three capacitors, with capacitances of $C_1 = 2.0\ \mu\text{F}$, $C_2 = 3.0\ \mu\text{F}$, and $C_3 = 6.0\ \mu\text{F}$, respectively, are connected in parallel. A 500-V potential difference is applied across the combination. Determine the voltage across each capacitor and the charge on each capacitor.

33. Find the total capacitance of this combination of series and parallel capacitors shown below.

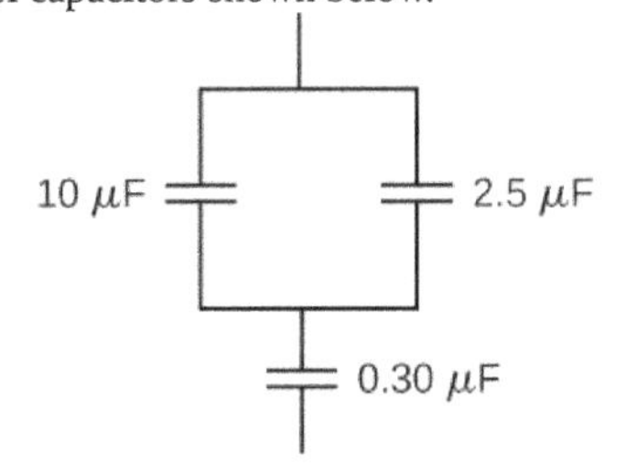

34. Suppose you need a capacitor bank with a total capacitance of 0.750 F but you have only 1.50-mF capacitors at your disposal. What is the smallest number of capacitors you could connect together to achieve your goal, and how would you connect them?

35. What total capacitances can you make by connecting a 5.00-μF and a 8.00-μF capacitor?

36. Find the equivalent capacitance of the combination of series and parallel capacitors shown below.

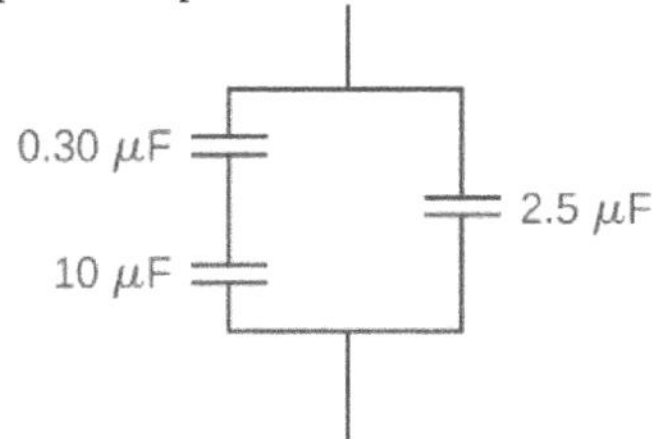

37. Find the net capacitance of the combination of series and parallel capacitors shown below.

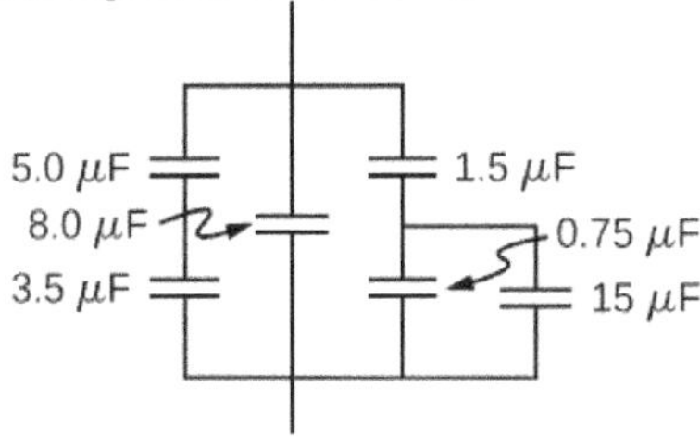

38. A 40-pF capacitor is charged to a potential difference of 500 V. Its terminals are then connected to those of an uncharged 10-pF capacitor. Calculate: (a) the original charge on the 40-pF capacitor; (b) the charge on each capacitor after the connection is made; and (c) the potential difference across the plates of each capacitor after the connection.

39. A 2.0-μF capacitor and a 4.0-μF capacitor are connected in series across a 1.0-kV potential. The charged capacitors are then disconnected from the source and connected to each other with terminals of like sign together. Find the charge on each capacitor and the voltage across each capacitor.

8.3 Energy Stored in a Capacitor

40. How much energy is stored in an $8.00\text{-}\mu\text{F}$ capacitor whose plates are at a potential difference of 6.00 V?

41. A capacitor has a charge of $2.5\ \mu\text{C}$ when connected to a 6.0-V battery. How much energy is stored in this capacitor?

42. How much energy is stored in the electrical field of a metal sphere of radius 2.0 m that is kept at a 10.0-V potential?

43. (a) What is the energy stored in the 10.0-μF capacitor of a heart defibrillator charged to 9.00×10^3 V ? (b) Find the amount of the stored charge.

44. In open-heart surgery, a much smaller amount of energy will defibrillate the heart. (a) What voltage is applied to the 8.00-μF capacitor of a heart defibrillator that stores 40.0 J of energy? (b) Find the amount of the stored charge.

45. A 165-μF capacitor is used in conjunction with a dc motor. How much energy is stored in it when 119 V is applied?

46. Suppose you have a 9.00-V battery, a 2.00-μF capacitor, and a 7.40-μF capacitor. (a) Find the charge and energy stored if the capacitors are connected to the battery in series. (b) Do the same for a parallel connection.

47. An anxious physicist worries that the two metal shelves of a wood frame bookcase might obtain a high voltage if charged by static electricity, perhaps produced by friction. (a) What is the capacitance of the empty shelves if they have area $1.00 \times 10^2\ \text{m}^2$ and are 0.200 m apart? (b) What is the voltage between them if opposite charges of magnitude 2.00 nC are placed on them? (c) To show that this voltage poses a small hazard, calculate the energy stored. (d) The actual shelves have an area 100 times smaller than these hypothetical shelves. Are his fears justified?

48. A parallel-plate capacitor is made of two square plates 25 cm on a side and 1.0 mm apart. The capacitor is connected to a 50.0-V battery. With the battery still connected, the plates are pulled apart to a separation of 2.00 mm. What are the energies stored in the capacitor before and after the plates are pulled farther apart? Why does the energy decrease even though work is done in separating the plates?

49. Suppose that the capacitance of a variable capacitor can be manually changed from 100 pF to 800 pF by turning a dial, connected to one set of plates by a shaft, from $0°$ to $180°$. With the dial set at $180°$ (corresponding to $C = 800\ \text{pF}$), the capacitor is connected to a 500-V source. After charging, the capacitor is disconnected from the source, and the dial is turned to $0°$. If friction is negligible, how much work is required to turn the dial from $180°$ to $0°$?

8.4 Capacitor with a Dielectric

50. Show that for a given dielectric material, the maximum energy a parallel-plate capacitor can store is directly proportional to the volume of dielectric.

51. An air-filled capacitor is made from two flat parallel plates 1.0 mm apart. The inside area of each plate is $8.0\ \text{cm}^2$. (a) What is the capacitance of this set of plates? (b) If the region between the plates is filled with a material whose dielectric constant is 6.0, what is the new capacitance?

52. A capacitor is made from two concentric spheres, one with radius 5.00 cm, the other with radius 8.00 cm. (a) What is the capacitance of this set of conductors? (b) If the region between the conductors is filled with a material whose dielectric constant is 6.00, what is the capacitance of the system?

53. A parallel-plate capacitor has charge of magnitude $9.00\ \mu\text{C}$ on each plate and capacitance $3.00\ \mu\text{C}$ when there is air between the plates. The plates are separated by 2.00 mm. With the charge on the plates kept constant, a dielectric with $\kappa = 5$ is inserted between the plates, completely filling the volume between the plates. (a) What is the potential difference between the plates of the capacitor, before and after the dielectric has been inserted? (b) What is the electrical field at the point midway between the plates before and after the dielectric is inserted?

54. Some cell walls in the human body have a layer of negative charge on the inside surface. Suppose that the surface charge densities are $\pm 0.50 \times 10^{-3}\ \text{C/m}^2$, the cell wall is $5.0 \times 10^{-9}\ \text{m}$ thick, and the cell wall material has a dielectric constant of $\kappa = 5.4$. (a) Find the magnitude of the electric field in the wall between two charge layers. (b) Find the potential difference between the inside and the outside of the cell. Which is at higher potential? (c) A typical cell in the human body has volume $10^{-16}\ \text{m}^3$. Estimate the total electrical field energy stored in the wall of a cell of this size when assuming that the cell is spherical. (*Hint*: Calculate the volume of the cell wall.)

55. A parallel-plate capacitor with only air between its plates is charged by connecting the capacitor to a battery. The capacitor is then disconnected from the battery, without any of the charge leaving the plates. (a) A voltmeter reads 45.0 V when placed across the capacitor. When a dielectric is inserted between the plates, completely filling the space, the voltmeter reads 11.5 V. What is the dielectric constant of the material? (b) What will the voltmeter read if the dielectric is now pulled away out so it fills only one-third of the space between the plates?

8.5 Molecular Model of a Dielectric

56. Two flat plates containing equal and opposite charges are separated by material 4.0 mm thick with a dielectric constant of 5.0. If the electrical field in the dielectric is 1.5 MV/m, what are (a) the charge density on the capacitor plates, and (b) the induced charge density on the surfaces of the dielectric?

57. For a Teflon™-filled, parallel-plate capacitor, the area of the plate is $50.0\ \text{cm}^2$ and the spacing between the plates is 0.50 mm. If the capacitor is connected to a 200-V battery, find (a) the free charge on the capacitor plates, (b) the electrical field in the dielectric, and (c) the induced charge on the dielectric surfaces.

58. Find the capacitance of a parallel-plate capacitor having plates with a surface area of $5.00\ m^2$ and separated by 0.100 mm of Teflon™.

59. (a) What is the capacitance of a parallel-plate capacitor with plates of area $1.50\ \text{m}^2$ that are separated by 0.0200 mm of neoprene rubber? (b) What charge does it hold when 9.00 V is applied to it?

60. Two parallel plates have equal and opposite charges. When the space between the plates is evacuated, the electrical field is $E = 3.20 \times 10^5\ \text{V/m}$. When the space is filled with dielectric, the electrical field is $E = 2.50 \times 10^5\ \text{V/m}$. (a) What is the surface charge density on each surface of the dielectric? (b) What is the dielectric constant?

61. The dielectric to be used in a parallel-plate capacitor has a dielectric constant of 3.60 and a dielectric strength of $1.60 \times 10^7\ \text{V/m}$. The capacitor has to have a capacitance of 1.25 nF and must be able to withstand a maximum potential difference 5.5 kV. What is the minimum area the plates of the capacitor may have?

62. When a 360-nF air capacitor is connected to a power supply, the energy stored in the capacitor is $18.5\ \mu\text{J}$. While the capacitor is connected to the power supply, a slab of dielectric is inserted that completely fills the space between the plates. This increases the stored energy by $23.2\ \mu\text{J}$. (a) What is the potential difference between the capacitor plates? (b) What is the dielectric constant of the slab?

63. A parallel-plate capacitor has square plates that are 8.00 cm on each side and 3.80 mm apart. The space between the plates is completely filled with two square slabs of dielectric, each 8.00 cm on a side and 1.90 mm thick. One slab is Pyrex glass and the other slab is polystyrene. If the potential difference between the plates is 86.0 V, find how much electrical energy can be stored in this capacitor.

ADDITIONAL PROBLEMS

64. A capacitor is made from two flat parallel plates placed 0.40 mm apart. When a charge of $0.020\ \mu\text{C}$ is placed on the plates the potential difference between them is 250 V. (a) What is the capacitance of the plates? (b) What is the area of each plate? (c) What is the charge on the plates when the potential difference between them is 500 V? (d) What maximum potential difference can be applied between the plates so that the magnitude of electrical fields between the plates does not exceed 3.0 MV/m?

65. An air-filled (empty) parallel-plate capacitor is made from two square plates that are 25 cm on each side and 1.0 mm apart. The capacitor is connected to a 50-V battery and fully charged. It is then disconnected from the battery and its plates are pulled apart to a separation of 2.00 mm. (a) What is the capacitance of this new capacitor? (b) What is the charge on each plate? (c) What is the electrical field between the plates?

66. Suppose that the capacitance of a variable capacitor can be manually changed from 100 to 800 pF by turning a dial connected to one set of plates by a shaft, from $0°$ to $180°$. With the dial set at $180°$ (corresponding to $C = 800\ \text{pF}$), the capacitor is connected to a 500-V source. After charging, the capacitor is disconnected from the source, and the dial is turned to $0°$. (a) What is the charge on the capacitor? (b) What is the voltage across the capacitor when the dial is set to $0°$?

67. Earth can be considered as a spherical capacitor with two plates, where the negative plate is the surface of Earth and the positive plate is the bottom of the ionosphere, which is located at an altitude of approximately 70 km. The potential difference between Earth's surface and the ionosphere is about 350,000 V. (a) Calculate the capacitance of this system. (b) Find the total charge on this capacitor. (c) Find the energy stored in this system.

68. A $4.00\text{-}\mu\text{F}$ capacitor and a $6.00\text{-}\mu\text{F}$ capacitor are connected in parallel across a 600-V supply line. (a) Find the charge on each capacitor and voltage across each. (b) The charged capacitors are disconnected from the line and from each other. They are then reconnected to each other with terminals of unlike sign together. Find the final charge on each capacitor and the voltage across each.

69. Three capacitors having capacitances of 8.40, 8.40, and 4.20 μF, respectively, are connected in series across a 36.0-V potential difference. (a) What is the charge on the $4.20\text{-}\mu\text{F}$ capacitor? (b) The capacitors are disconnected from the potential difference without allowing them to discharge. They are then reconnected in parallel with each other with the positively charged plates connected together. What is the voltage across each capacitor in the parallel combination?

70. A parallel-plate capacitor with capacitance $5.0\,\mu\text{F}$ is charged with a 12.0-V battery, after which the battery is disconnected. Determine the minimum work required to increase the separation between the plates by a factor of 3.

71. (a) How much energy is stored in the electrical fields in the capacitors (in total) shown below? (b) Is this energy equal to the work done by the 400-V source in charging the capacitors?

72. Three capacitors having capacitances 8.4, 8.4, and 4.2 μF are connected in series across a 36.0-V potential difference. (a) What is the total energy stored in all three capacitors? (b) The capacitors are disconnected from the potential difference without allowing them to discharge. They are then reconnected in parallel with each other with the positively charged plates connected together. What is the total energy now stored in the capacitors?

73. (a) An $8.00\text{-}\mu\text{F}$ capacitor is connected in parallel to another capacitor, producing a total capacitance of $5.00\,\mu\text{F}$. What is the capacitance of the second capacitor? (b) What is unreasonable about this result? (c) Which assumptions are unreasonable or inconsistent?

74. (a) On a particular day, it takes 9.60×10^3 J of electrical energy to start a truck's engine. Calculate the capacitance of a capacitor that could store that amount of energy at 12.0 V. (b) What is unreasonable about this result? (c) Which assumptions are responsible?

75. (a) A certain parallel-plate capacitor has plates of area $4.00\,\text{m}^2$, separated by 0.0100 mm of nylon, and stores 0.170 C of charge. What is the applied voltage? (b) What is unreasonable about this result? (c) Which assumptions are responsible or inconsistent?

76. A prankster applies 450 V to an $80.0\text{-}\mu\text{F}$ capacitor and then tosses it to an unsuspecting victim. The victim's finger is burned by the discharge of the capacitor through 0.200 g of flesh. Estimate, what is the temperature increase of the flesh? Is it reasonable to assume that no thermodynamic phase change happened?

CHALLENGE PROBLEMS

77. A spherical capacitor is formed from two concentric spherical conducting spheres separated by vacuum. The inner sphere has radius 12.5 cm and the outer sphere has radius 14.8 cm. A potential difference of 120 V is applied to the capacitor. (a) What is the capacitance of the capacitor? (b) What is the magnitude of the electrical field at $r = 12.6\,\text{cm}$, just outside the inner sphere? (c) What is the magnitude of the electrical field at $r = 14.7\,\text{cm}$, just inside the outer sphere? (d) For a parallel-plate capacitor the electrical field is uniform in the region between the plates, except near the edges of the plates. Is this also true for a spherical capacitor?

78. The network of capacitors shown below are all uncharged when a 300-V potential is applied between points A and B with the switch S open. (a) What is the potential difference $V_E - V_D$? (b) What is the potential at point E after the switch is closed? (c) How much charge flows through the switch after it is closed?

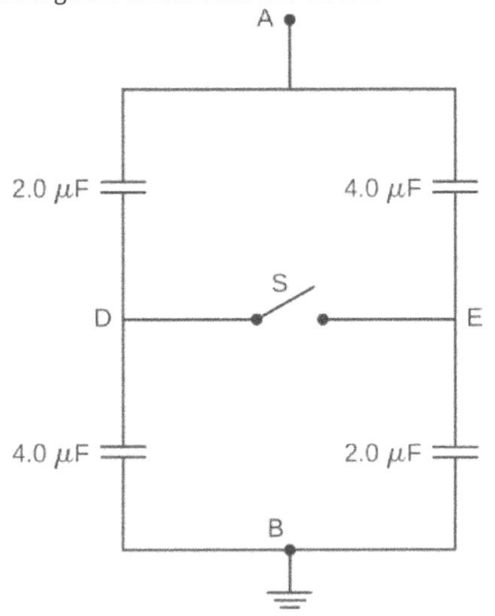

79. Electronic flash units for cameras contain a capacitor for storing the energy used to produce the flash. In one such unit the flash lasts for 1/675 fraction of a second with an average light power output of 270 kW. (a) If the conversion of electrical energy to light is 95% efficient (because the rest of the energy goes to thermal energy), how much energy must be stored in the capacitor for one flash? (b) The capacitor has a potential difference between its plates of 125 V when the stored energy equals the value stored in part (a). What is the capacitance?

80. A spherical capacitor is formed from two concentric spherical conducting shells separated by a vacuum. The inner sphere has radius 12.5 cm and the outer sphere has radius 14.8 cm. A potential difference of 120 V is applied to the capacitor. (a) What is the energy density at $r = 12.6\,\text{cm}$, just outside the inner sphere? (b) What is the energy density at $r = 14.7\,\text{cm}$, just inside the outer sphere? (c) For the parallel-plate capacitor the energy density is uniform in the region between the plates, except near the edges of the plates. Is this also true for the spherical capacitor?

81. A metal plate of thickness t is held in place between two capacitor plates by plastic pegs, as shown below. The effect of the pegs on the capacitance is negligible. The area of each capacitor plate and the area of the top and bottom surfaces of the inserted plate are all A. What is the capacitance of this system?

82. A parallel-plate capacitor is filled with two dielectrics, as shown below. When the plate area is A and separation between plates is d, show that the capacitance is given by

$$C = \varepsilon_0 \frac{A}{d} \frac{\kappa_1 + \kappa_2}{2}.$$

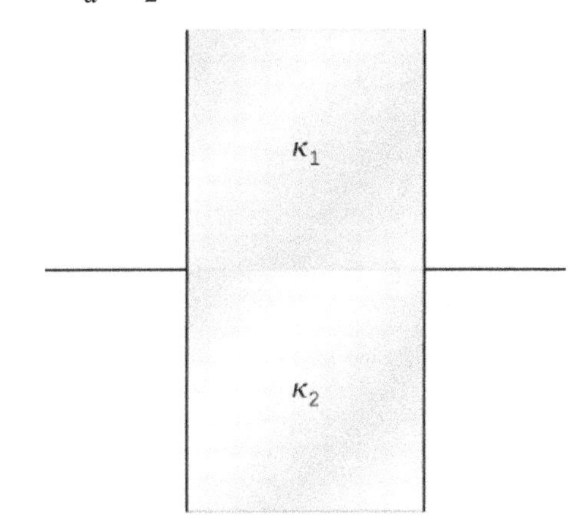

83. A parallel-plate capacitor is filled with two dielectrics, as shown below. Show that the capacitance is given by

$$C = 2\varepsilon_0 \frac{A}{d} \frac{\kappa_1 \kappa_2}{\kappa_1 + \kappa_2}.$$

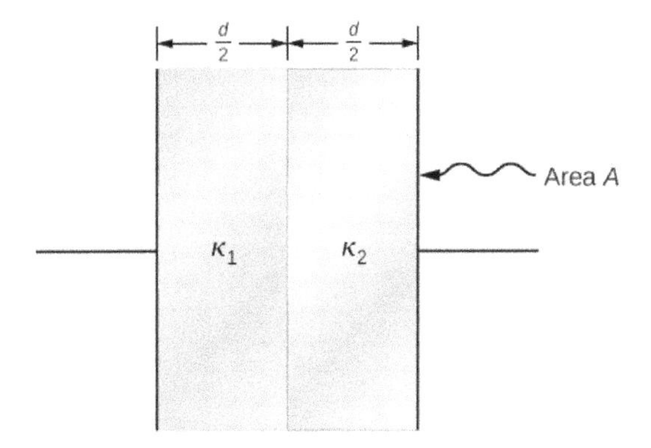

84. A capacitor has parallel plates of area $12\,\text{cm}^2$ separated by 2.0 mm. The space between the plates is filled with polystyrene. (a) Find the maximum permissible voltage across the capacitor to avoid dielectric breakdown. (b) When the voltage equals the value found in part (a), find the surface charge density on the surface of the dielectric.

9 | CURRENT AND RESISTANCE

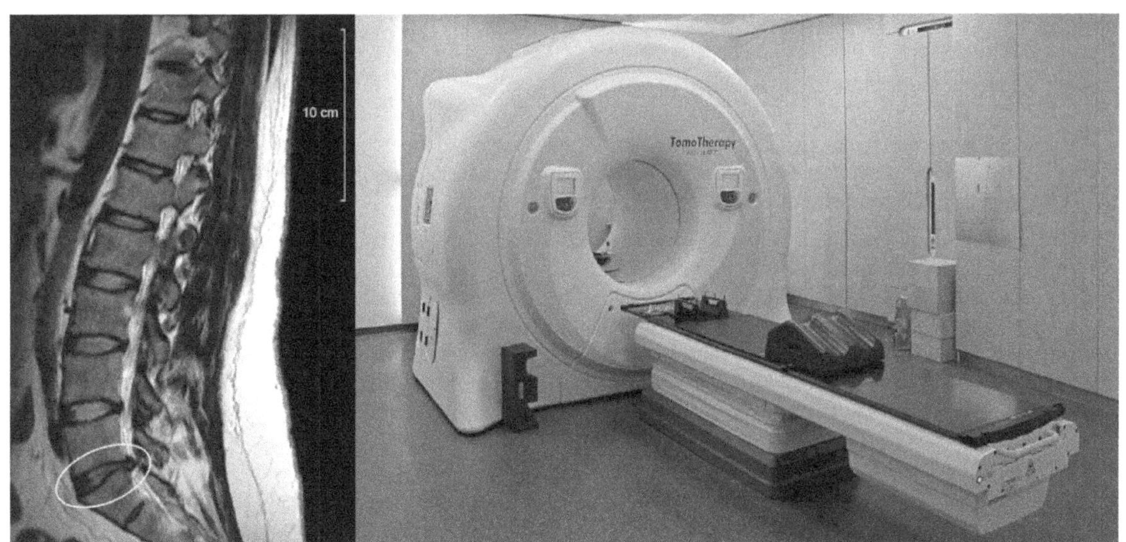

Figure 9.1 Magnetic resonance imaging (MRI) uses superconducting magnets and produces high-resolution images without the danger of radiation. The image on the left shows the spacing of vertebrae along a human spinal column, with the circle indicating where the vertebrae are too close due to a ruptured disc. On the right is a picture of the MRI instrument, which surrounds the patient on all sides. A large amount of electrical current is required to operate the electromagnets (credit right: modification of work by "digital cat"/Flickr).

Chapter Outline

Introduction

In this chapter, we study the electrical current through a material, where the electrical current is the rate of flow of charge. We also examine a characteristic of materials known as the resistance. Resistance is a measure of how much a material impedes the flow of charge, and it will be shown that the resistance depends on temperature. In general, a good conductor, such as copper, gold, or silver, has very low resistance. Some materials, called superconductors, have zero resistance at very low temperatures.

High currents are required for the operation of electromagnets. Superconductors can be used to make electromagnets that are 10 times stronger than the strongest conventional electromagnets. These superconducting magnets are used in the construction of magnetic resonance imaging (MRI) devices that can be used to make high-resolution images of the human body. The chapter-opening picture shows an MRI image of the vertebrae of a human subject and the MRI device itself. Superconducting magnets have many other uses. For example, superconducting magnets are used in the Large Hadron Collider (LHC) to curve the path of protons in the ring.

9.1 | Electrical Current

Learning Objectives

By the end of this section, you will be able to:

- Describe an electrical current
- Define the unit of electrical current
- Explain the direction of current flow

Up to now, we have considered primarily static charges. When charges did move, they were accelerated in response to an electrical field created by a voltage difference. The charges lost potential energy and gained kinetic energy as they traveled through a potential difference where the electrical field did work on the charge.

Although charges do not require a material to flow through, the majority of this chapter deals with understanding the movement of charges through a material. The rate at which the charges flow past a location—that is, the amount of charge per unit time—is known as the *electrical current*. When charges flow through a medium, the current depends on the voltage applied, the material through which the charges flow, and the state of the material. Of particular interest is the motion of charges in a conducting wire. In previous chapters, charges were accelerated due to the force provided by an electrical field, losing potential energy and gaining kinetic energy. In this chapter, we discuss the situation of the force provided by an electrical field in a conductor, where charges lose kinetic energy to the material reaching a constant velocity, known as the "*drift velocity*." This is analogous to an object falling through the atmosphere and losing kinetic energy to the air, reaching a constant terminal velocity.

If you have ever taken a course in first aid or safety, you may have heard that in the event of electric shock, it is the current, not the voltage, which is the important factor on the severity of the shock and the amount of damage to the human body. Current is measured in units called amperes; you may have noticed that circuit breakers in your home and fuses in your car are rated in amps (or amperes). But what is the ampere and what does it measure?

Defining Current and the Ampere

Electrical current is defined to be the rate at which charge flows. When there is a large current present, such as that used to run a refrigerator, a large amount of charge moves through the wire in a small amount of time. If the current is small, such as that used to operate a handheld calculator, a small amount of charge moves through the circuit over a long period of time.

Electrical Current

The average electrical current I is the rate at which charge flows,

$$I_{\text{ave}} = \frac{\Delta Q}{\Delta t}, \quad (9.1)$$

where ΔQ is the amount of charge passing through a given area in time Δt (Figure 9.2). The SI unit for current is the **ampere** (A), named for the French physicist André-Marie Ampère (1775–1836). Since $I = \frac{\Delta Q}{\Delta t}$, we see that an ampere is defined as one coulomb of charge passing through a given area per second:

$$1\text{A} \equiv 1\frac{\text{C}}{\text{s}}. \quad (9.2)$$

The instantaneous electrical current, or simply the **electrical current**, is the time derivative of the charge that flows and is found by taking the limit of the average electrical current as $\Delta t \to 0$:

$$I = \lim_{\Delta t \to 0} \frac{\Delta Q}{\Delta t} = \frac{dQ}{dt}. \quad (9.3)$$

Most electrical appliances are rated in amperes (or amps) required for proper operation, as are fuses and circuit breakers.

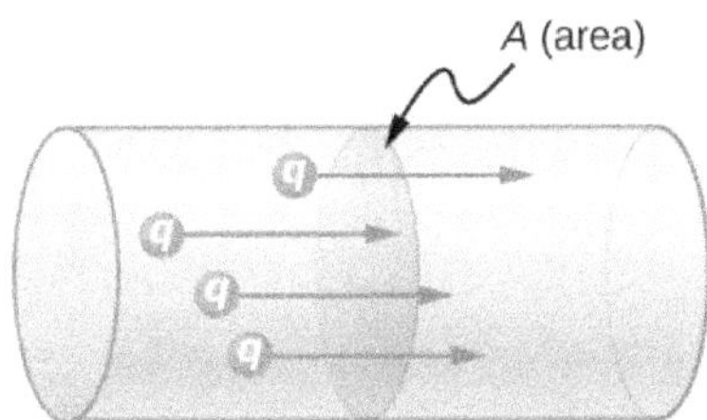

Figure 9.2 The rate of flow of charge is current. An ampere is the flow of one coulomb of charge through an area in one second. A current of one amp would result from 6.25×10^{18} electrons flowing through the area *A* each second.

Example 9.1

Calculating the Average Current

The main purpose of a battery in a car or truck is to run the electric starter motor, which starts the engine. The operation of starting the vehicle requires a large current to be supplied by the battery. Once the engine starts, a device called an alternator takes over supplying the electric power required for running the vehicle and for charging the battery.

(a) What is the average current involved when a truck battery sets in motion 720 C of charge in 4.00 s while starting an engine? (b) How long does it take 1.00 C of charge to flow from the battery?

Strategy

We can use the definition of the average current in the equation $I = \frac{\Delta Q}{\Delta t}$ to find the average current in part (a), since charge and time are given. For part (b), once we know the average current, we can its definition $I = \frac{\Delta Q}{\Delta t}$ to find the time required for 1.00 C of charge to flow from the battery.

Solution

a. Entering the given values for charge and time into the definition of current gives

$$I = \frac{\Delta Q}{\Delta t} = \frac{720\ \text{C}}{4.00\ \text{s}} = 180\ \text{C/s} = 180\ \text{A}.$$

b. Solving the relationship $I = \frac{\Delta Q}{\Delta t}$ for time Δt and entering the known values for charge and current gives

$$\Delta t = \frac{\Delta Q}{I} = \frac{1.00\ \text{C}}{180\ \text{C/s}} = 5.56 \times 10^{-3}\ \text{s} = 5.56\ \text{ms}.$$

Significance

a. This large value for current illustrates the fact that a large charge is moved in a small amount of time. The currents in these “starter motors” are fairly large to overcome the inertia of the engine. b. A high current requires a short time to supply a large amount of charge. This large current is needed to supply the large amount of energy needed to start the engine.

Example 9.2

Calculating Instantaneous Currents

Consider a charge moving through a cross-section of a wire where the charge is modeled as $Q(t) = Q_M\left(1 - e^{-t/\tau}\right)$. Here, Q_M is the charge after a long period of time, as time approaches infinity, with units of coulombs, and τ is a time constant with units of seconds (see Figure 9.3). What is the current through the wire?

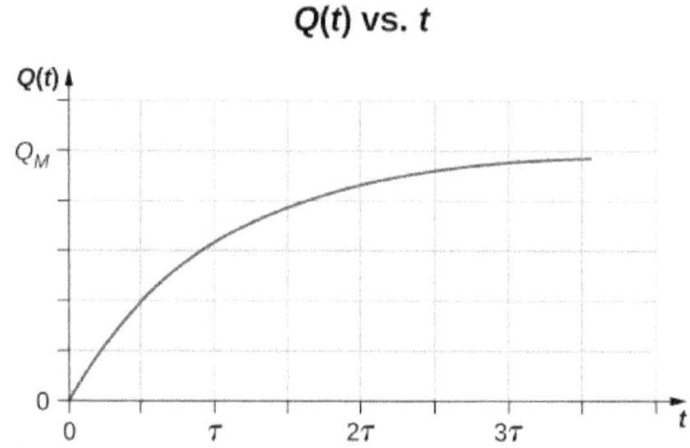

Figure 9.3 A graph of the charge moving through a cross-section of a wire over time.

Strategy

The current through the cross-section can be found from $I = \frac{dQ}{dt}$. Notice from the figure that the charge increases to Q_M and the derivative decreases, approaching zero, as time increases (Figure 9.4).

Solution

The derivative can be found using $\frac{d}{dx}e^u = e^u \frac{du}{dx}$.

$$I = \frac{dQ}{dt} = \frac{d}{dt}\left[Q_M\left(1 - e^{-t/\tau}\right)\right] = \frac{Q_M}{\tau}e^{-t/\tau}.$$

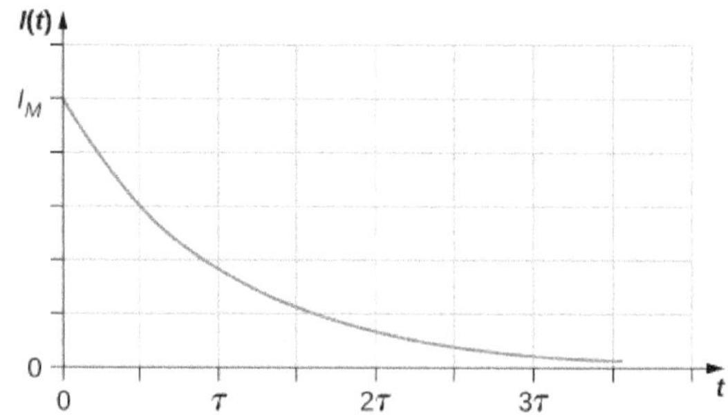

Figure 9.4 A graph of the current flowing through the wire over time.

Significance

The current through the wire in question decreases exponentially, as shown in Figure 9.4. In later chapters, it will be shown that a time-dependent current appears when a capacitor charges or discharges through a resistor. Recall that a capacitor is a device that stores charge. You will learn about the resistor in Model of Conduction in Metals.

9.1 Check Your Understanding Handheld calculators often use small solar cells to supply the energy required to complete the calculations needed to complete your next physics exam. The current needed to run your calculator can be as small as 0.30 mA. How long would it take for 1.00 C of charge to flow from the solar cells? Can solar cells be used, instead of batteries, to start traditional internal combustion engines presently used in most cars and trucks?

9.2 Check Your Understanding Circuit breakers in a home are rated in amperes, normally in a range from 10 amps to 30 amps, and are used to protect the residents from harm and their appliances from damage due to large currents. A single 15-amp circuit breaker may be used to protect several outlets in the living room, whereas a single 20-amp circuit breaker may be used to protect the refrigerator in the kitchen. What can you deduce from this about current used by the various appliances?

Current in a Circuit

In the previous paragraphs, we defined the current as the charge that flows through a cross-sectional area per unit time. In order for charge to flow through an appliance, such as the headlight shown in Figure 9.5, there must be a complete path (or **circuit**) from the positive terminal to the negative terminal. Consider a simple circuit of a car battery, a switch, a headlight lamp, and wires that provide a current path between the components. In order for the lamp to light, there must be a complete path for current flow. In other words, a charge must be able to leave the positive terminal of the battery, travel through the component, and back to the negative terminal of the battery. The switch is there to control the circuit. Part (a) of the figure shows the simple circuit of a car battery, a switch, a conducting path, and a headlight lamp. Also shown is the **schematic** of the circuit [part (b)]. A schematic is a graphical representation of a circuit and is very useful in visualizing the main features of a circuit. Schematics use standardized symbols to represent the components in a circuits and solid lines to represent the wires connecting the components. The battery is shown as a series of long and short lines, representing the historic voltaic pile. The lamp is shown as a circle with a loop inside, representing the filament of an incandescent bulb. The switch is shown as two points with a conducting bar to connect the two points and the wires connecting the components are shown as solid lines. The schematic in part (c) shows the direction of current flow when the switch is closed.

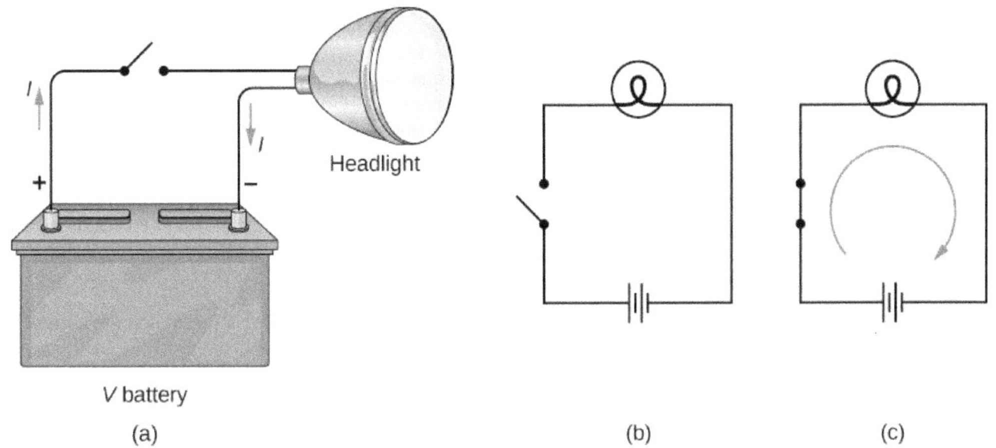

Figure 9.5 (a) A simple electric circuit of a headlight (lamp), a battery, and a switch. When the switch is closed, an uninterrupted path for current to flow through is supplied by conducting wires connecting a load to the terminals of a battery. (b) In this schematic, the battery is represented by parallel lines, which resemble plates in the original design of a battery. The longer lines indicate the positive terminal. The conducting wires are shown as solid lines. The switch is shown, in the open position, as two terminals with a line representing a conducting bar that can make contact between the two terminals. The lamp is represented by a circle encompassing a filament, as would be seen in an incandescent light bulb. (c) When the switch is closed, the circuit is complete and current flows from the positive terminal to the negative terminal of the battery.

When the switch is closed in Figure 9.5(c), there is a complete path for charges to flow, from the positive terminal of the battery, through the switch, then through the headlight and back to the negative terminal of the battery. Note that the direction of current flow is from positive to negative. The direction of **conventional current** is always represented in the direction that positive charge would flow, from the positive terminal to the negative terminal.

The conventional current flows from the positive terminal to the negative terminal, but depending on the actual situation, positive charges, negative charges, or both may move. In metal wires, for example, current is carried by electrons—that is, negative charges move. In ionic solutions, such as salt water, both positive and negative charges move. This is also true in nerve cells. A Van de Graaff generator, used for nuclear research, can produce a current of pure positive charges, such as protons. In the Tevatron Accelerator at Fermilab, before it was shut down in 2011, beams of protons and antiprotons traveling in opposite directions were collided. The protons are positive and therefore their current is in the same direction as they travel. The antiprotons are negativity charged and thus their current is in the opposite direction that the actual particles travel.

A closer look at the current flowing through a wire is shown in Figure 9.6. The figure illustrates the movement of charged particles that compose a current. The fact that conventional current is taken to be in the direction that positive charge would flow can be traced back to American scientist and statesman Benjamin Franklin in the 1700s. Having no knowledge of the particles that make up the atom (namely the proton, electron, and neutron), Franklin believed that electrical current flowed from a material that had more of an "electrical fluid" and to a material that had less of this "electrical fluid." He coined the term *positive* for the material that had more of this electrical fluid and *negative* for the material that lacked the electrical fluid. He surmised that current would flow from the material with more electrical fluid—the positive material—to the negative material, which has less electrical fluid. Franklin called this direction of current a positive current flow. This was pretty advanced thinking for a man who knew nothing about the atom.

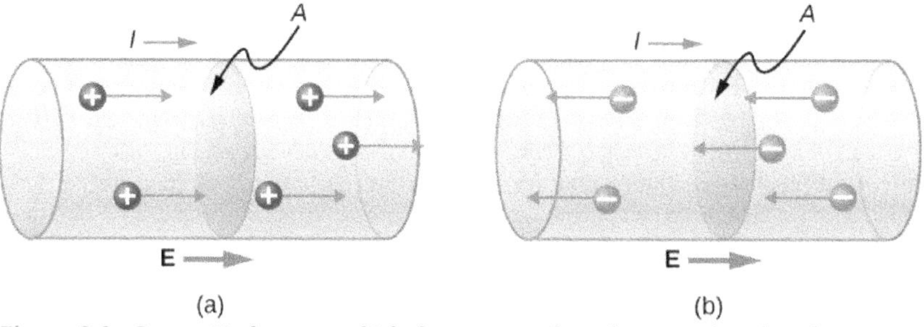

Figure 9.6 Current I is the rate at which charge moves through an area A, such as the cross-section of a wire. Conventional current is defined to move in the direction of the electrical field. (a) Positive charges move in the direction of the electrical field, which is the same direction as conventional current. (b) Negative charges move in the direction opposite to the electrical field. Conventional current is in the direction opposite to the movement of negative charge. The flow of electrons is sometimes referred to as electronic flow.

We now know that a material is positive if it has a greater number of protons than electrons, and it is negative if it has a greater number of electrons than protons. In a conducting metal, the current flow is due primarily to electrons flowing from the negative material to the positive material, but for historical reasons, we consider the positive current flow and the current is shown to flow from the positive terminal of the battery to the negative terminal.

It is important to realize that an electrical field is present in conductors and is responsible for producing the current (Figure 9.6). In previous chapters, we considered the static electrical case, where charges in a conductor quickly redistribute themselves on the surface of the conductor in order to cancel out the external electrical field and restore equilibrium. In the case of an electrical circuit, the charges are prevented from ever reaching equilibrium by an external source of electric potential, such as a battery. The energy needed to move the charge is supplied by the electric potential from the battery.

Although the electrical field is responsible for the motion of the charges in the conductor, the work done on the charges by the electrical field does not increase the kinetic energy of the charges. We will show that the electrical field is responsible for keeping the electric charges moving at a "drift velocity."

9.2 | Model of Conduction in Metals

Learning Objectives

By the end of this section, you will be able to:

- Define the drift velocity of charges moving through a metal
- Define the vector current density
- Describe the operation of an incandescent lamp

When electrons move through a conducting wire, they do not move at a constant velocity, that is, the electrons do not move in a straight line at a constant speed. Rather, they interact with and collide with atoms and other free electrons in the conductor. Thus, the electrons move in a zig-zag fashion and drift through the wire. We should also note that even though it is convenient to discuss the direction of current, current is a scalar quantity. When discussing the velocity of charges in a current, it is more appropriate to discuss the current density. We will come back to this idea at the end of this section.

Drift Velocity

Electrical signals move very rapidly. Telephone conversations carried by currents in wires cover large distances without noticeable delays. Lights come on as soon as a light switch is moved to the 'on' position. Most electrical signals carried by currents travel at speeds on the order of 10^8 m/s, a significant fraction of the speed of light. Interestingly, the individual charges that make up the current move much slower on average, typically drifting at speeds on the order of 10^{-4} m/s. How do we reconcile these two speeds, and what does it tell us about standard conductors?

The high speed of electrical signals results from the fact that the force between charges acts rapidly at a distance. Thus, when a free charge is forced into a wire, as in Figure 9.7, the incoming charge pushes other charges ahead of it due to the repulsive force between like charges. These moving charges push on charges farther down the line. The density of charge in a system cannot easily be increased, so the signal is passed on rapidly. The resulting electrical shock wave moves through the system at nearly the speed of light. To be precise, this fast-moving signal, or shock wave, is a rapidly propagating change in the electrical field.

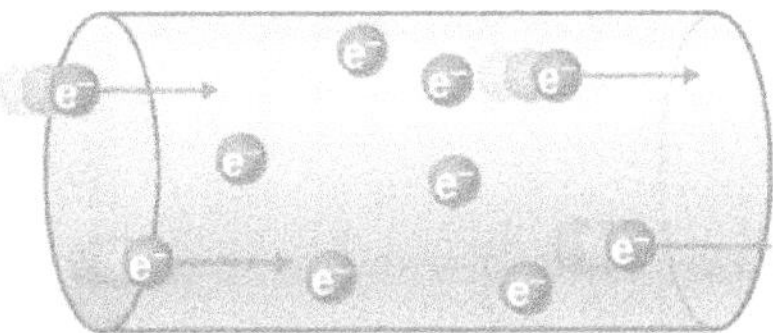

Figure 9.7 When charged particles are forced into this volume of a conductor, an equal number are quickly forced to leave. The repulsion between like charges makes it difficult to increase the number of charges in a volume. Thus, as one charge enters, another leaves almost immediately, carrying the signal rapidly forward.

Good conductors have large numbers of free charges. In metals, the free charges are free electrons. (In fact, good electrical conductors are often good heat conductors too, because large numbers of free electrons can transport thermal energy as well as carry electrical current.) Figure 9.8 shows how free electrons move through an ordinary conductor. The distance that an individual electron can move between collisions with atoms or other electrons is quite small. The electron paths thus appear nearly random, like the motion of atoms in a gas. But there is an electrical field in the conductor that causes the electrons to drift in the direction shown (opposite to the field, since they are negative). The **drift velocity** $\vec{v}_d$ is the average velocity of the free charges. Drift velocity is quite small, since there are so many free charges. If we have an estimate of the density of free electrons in a conductor, we can calculate the drift velocity for a given current. The larger the density, the lower the velocity required for a given current.

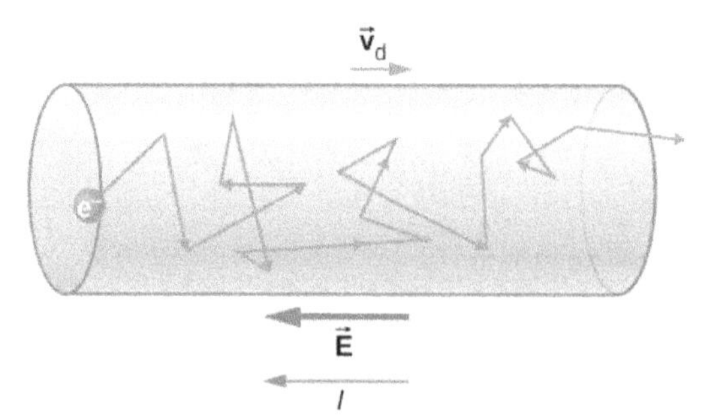

Figure 9.8 Free electrons moving in a conductor make many collisions with other electrons and other particles. A typical path of one electron is shown. The average velocity of the free charges is called the drift velocity $\vec{v}_d$ and for electrons, it is in the direction opposite to the electrical field. The collisions normally transfer energy to the conductor, requiring a constant supply of energy to maintain a steady current.

Free-electron collisions transfer energy to the atoms of the conductor. The electrical field does work in moving the electrons through a distance, but that work does not increase the kinetic energy (nor speed) of the electrons. The work is transferred to the conductor's atoms, often increasing temperature. Thus, a continuous power input is required to keep a current flowing. (An exception is superconductors, for reasons we shall explore in a later chapter. Superconductors can have a steady current without a continual supply of energy—a great energy savings.) For a conductor that is not a superconductor, the supply of energy can be useful, as in an incandescent light bulb filament (Figure 9.9). The supply of energy is necessary to increase the temperature of the tungsten filament, so that the filament glows.

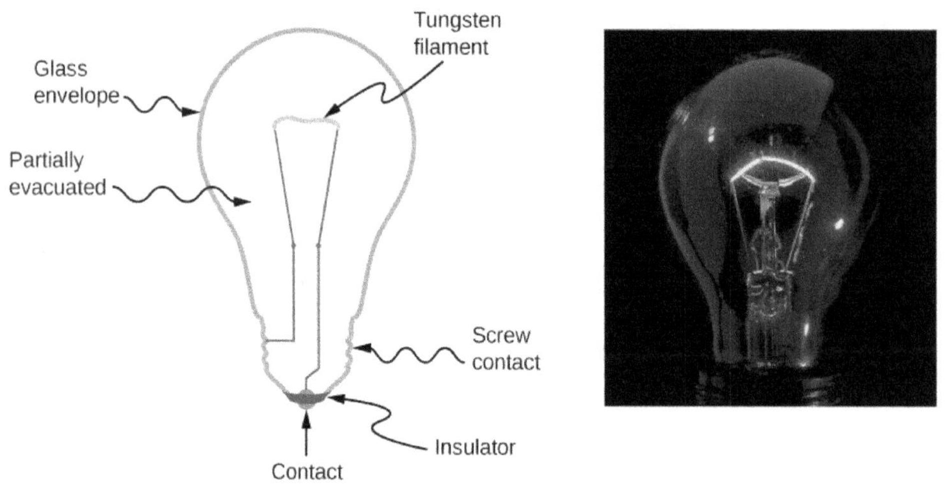

Figure 9.9 The incandescent lamp is a simple design. A tungsten filament is placed in a partially evacuated glass envelope. One end of the filament is attached to the screw base, which is made out of a conducting material. The second end of the filament is attached to a second contact in the base of the bulb. The two contacts are separated by an insulating material. Current flows through the filament, and the temperature of the filament becomes large enough to cause the filament to glow and produce light. However, these bulbs are not very energy efficient, as evident from the heat coming from the bulb. In the year 2012, the United States, along with many other countries, began to phase out incandescent lamps in favor of more energy-efficient lamps, such as light-emitting diode (LED) lamps and compact fluorescent lamps (CFL) (credit right: modification of work by Serge Saint).

We can obtain an expression for the relationship between current and drift velocity by considering the number of free charges in a segment of wire, as illustrated in Figure 9.10. The number of free charges per unit volume, or the number density of free charges, is given the symbol n where $n = \frac{\text{number of charges}}{\text{volume}}$. The value of n depends on the material. The shaded segment has a volume $Av_{\mathrm{d}}dt$, so that the number of free charges in the volume is $nAv_{\mathrm{d}}dt$. The charge dQ in this segment is thus $qnAv_{\mathrm{d}}dt$, where q is the amount of charge on each carrier. (The magnitude of the charge of electrons is $q = 1.60 \times 10^{-19}\ \mathrm{C}$.) Current is charge moved per unit time; thus, if all the original charges move out of this segment in time dt, the current is

$$I = \frac{dQ}{dt} = qnAv_{\mathrm{d}}.$$

Rearranging terms gives

$$v_{\mathrm{d}} = \frac{nqA}{I} \quad (9.4)$$

where v_{d} is the drift velocity, n is the free charge density, A is the cross-sectional area of the wire, and I is the current through the wire. The carriers of the current each have charge q and move with a drift velocity of magnitude v_{d}.

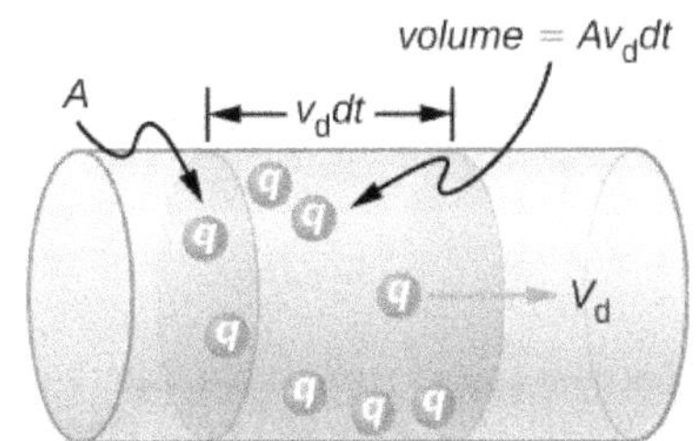

Figure 9.10 All the charges in the shaded volume of this wire move out in a time dt, having a drift velocity of magnitude v_{d}.

Note that simple drift velocity is not the entire story. The speed of an electron is sometimes much greater than its drift velocity. In addition, not all of the electrons in a conductor can move freely, and those that do move might move somewhat faster or slower than the drift velocity. So what do we mean by free electrons?

Atoms in a metallic conductor are packed in the form of a lattice structure. Some electrons are far enough away from the atomic nuclei that they do not experience the attraction of the nuclei as strongly as the inner electrons do. These are the free electrons. They are not bound to a single atom but can instead move freely among the atoms in a "sea" of electrons. When an electrical field is applied, these free electrons respond by accelerating. As they move, they collide with the atoms in the lattice and with other electrons, generating thermal energy, and the conductor gets warmer. In an insulator, the organization of the atoms and the structure do not allow for such free electrons.

As you know, electric power is usually supplied to equipment and appliances through round wires made of a conducting material (copper, aluminum, silver, or gold) that are stranded or solid. The diameter of the wire determines the current-carrying capacity—the larger the diameter, the greater the current-carrying capacity. Even though the current-carrying capacity is determined by the diameter, wire is not normally characterized by the diameter directly. Instead, wire is commonly sold in a unit known as "gauge." Wires are manufactured by passing the material through circular forms called "drawing dies." In order to make thinner wires, manufacturers draw the wires through multiple dies of successively thinner diameter. Historically, the gauge of the wire was related to the number of drawing processes required to manufacture the wire. For this reason, the larger the gauge, the smaller the diameter. In the United States, the American Wire Gauge (AWG) was developed to standardize the system. Household wiring commonly consists of 10-gauge (2.588-mm diameter) to 14-gauge (1.628-mm diameter) wire. A device used to measure the gauge of wire is shown in Figure 9.11.

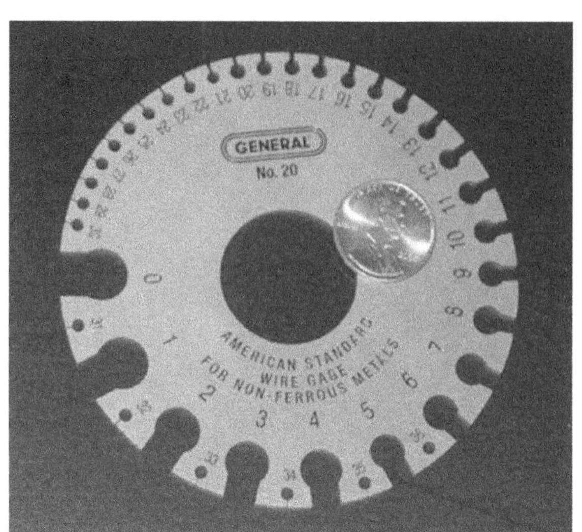

Figure 9.11 A device for measuring the gauge of electrical wire. As you can see, higher gauge numbers indicate thinner wires.

Example 9.3

Calculating Drift Velocity in a Common Wire

Calculate the drift velocity of electrons in a copper wire with a diameter of 2.053 mm (12-gauge) carrying a 20.0-A current, given that there is one free electron per copper atom. (Household wiring often contains 12-gauge copper wire, and the maximum current allowed in such wire is usually 20.0 A.) The density of copper is $8.80 \times 10^3\ \text{kg/m}^3$ and the atomic mass of copper is 63.54 g/mol.

Strategy

We can calculate the drift velocity using the equation $I = nqAv_{\text{d}}$. The current is $I = 20.00\ \text{A}$ and $q = 1.60 \times 10^{-19}\,\text{C}$ is the charge of an electron. We can calculate the area of a cross-section of the wire using the formula $A = \pi r^2$, where *r* is one-half the diameter. The given diameter is 2.053 mm, so *r* is 1.0265 mm. We are given the density of copper, $8.80 \times 10^3\ \text{kg/m}^3$, and the atomic mass of copper is $63.54\ \text{g/mol}$. We can use these two quantities along with Avogadro's number, $6.02 \times 10^{23}\ \text{atoms/mol}$, to determine *n*, the number of free electrons per cubic meter.

Solution

First, we calculate the density of free electrons in copper. There is one free electron per copper atom. Therefore, the number of free electrons is the same as the number of copper atoms per m^3. We can now find *n* as follows:

$$\begin{aligned} n &= \frac{1\,e^-}{\text{atom}} \times \frac{6.02 \times 10^{23}\,\text{atoms}}{\text{mol}} \times \frac{1\ \text{mol}}{63.54\ \text{g}} \times \frac{1000\ \text{g}}{\text{kg}} \times \frac{8.80 \times 10^3\,\text{kg}}{1\ \text{m}^3} \\ &= 8.34 \times 10^{28}\,e^-/\text{m}^3. \end{aligned}$$

The cross-sectional area of the wire is

$$A = \pi r^2 = \pi\left(\frac{2.05 \times 10^{-3}\,\text{m}}{2}\right)^2 = 3.30 \times 10^{-6}\,\text{m}^2.$$

Rearranging $I = nqAv_{\text{d}}$ to isolate drift velocity gives

$$v_{\text{d}} = \frac{I}{nqA} = \frac{20.00 \text{ A}}{(8.34 \times 10^{28}/\text{m}^3)(-1.60 \times 10^{-19}\text{C})(3.30 \times 10^{-6}\text{m}^2)} = -4.54 \times 10^{-4} \text{ m/s}.$$

Significance

The minus sign indicates that the negative charges are moving in the direction opposite to conventional current. The small value for drift velocity (on the order of 10^{-4} m/s) confirms that the signal moves on the order of 10^{12} times faster (about 10^8 m/s) than the charges that carry it.

9.3 Check Your Understanding In Example 9.4, the drift velocity was calculated for a 2.053-mm diameter (12-gauge) copper wire carrying a 20-amp current. Would the drift velocity change for a 1.628-mm diameter (14-gauge) wire carrying the same 20-amp current?

Current Density

Although it is often convenient to attach a negative or positive sign to indicate the overall direction of motion of the charges, current is a scalar quantity, $I = \frac{dQ}{dt}$. It is often necessary to discuss the details of the motion of the charge, instead of discussing the overall motion of the charges. In such cases, it is necessary to discuss the current density, $\vec{\mathbf{J}}$, a vector quantity. The **current density** is the flow of charge through an infinitesimal area, divided by the area. The current density must take into account the local magnitude and direction of the charge flow, which varies from point to point. The unit of current density is ampere per meter squared, and the direction is defined as the direction of net flow of positive charges through the area.

The relationship between the current and the current density can be seen in Figure 9.12. The differential current flow through the area $d\vec{\mathbf{A}}$ is found as

$$dI = \vec{\mathbf{J}} \cdot d\vec{\mathbf{A}} = JdA\cos\theta,$$

where θ is the angle between the area and the current density. The total current passing through area $d\vec{\mathbf{A}}$ can be found by integrating over the area,

$$I = \iint_{\text{area}} \vec{\mathbf{J}} \cdot d\vec{\mathbf{A}}. \tag{9.5}$$

Consider the magnitude of the current density, which is the current divided by the area:

$$J = \frac{I}{A} = \frac{n|q|Av_{\text{d}}}{A} = n|q|v_{\text{d}}.$$

Thus, the current density is $\vec{\mathbf{J}} = nq\vec{\mathbf{v}}_{\text{d}}$. If q is positive, $\vec{\mathbf{v}}_{\text{d}}$ is in the same direction as the electrical field $\vec{\mathbf{E}}$. If q is negative, $\vec{\mathbf{v}}_{\text{d}}$ is in the opposite direction of $\vec{\mathbf{E}}$. Either way, the direction of the current density $\vec{\mathbf{J}}$ is in the direction of the electrical field $\vec{\mathbf{E}}$.

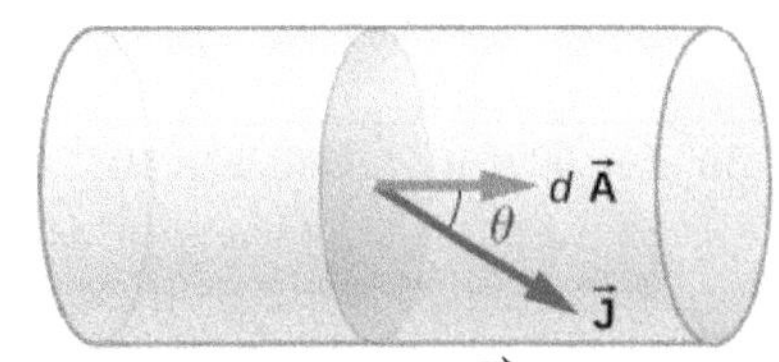

Figure 9.12 The current density $\vec{J}$ is defined as the current passing through an infinitesimal cross-sectional area divided by the area. The direction of the current density is the direction of the net flow of positive charges and the magnitude is equal to the current divided by the infinitesimal area.

Example 9.4

Calculating the Current Density in a Wire

The current supplied to a lamp with a 100-W light bulb is 0.87 amps. The lamp is wired using a copper wire with diameter 2.588 mm (10-gauge). Find the magnitude of the current density.

Strategy

The current density is the current moving through an infinitesimal cross-sectional area divided by the area. We can calculate the magnitude of the current density using $J = \frac{I}{A}$. The current is given as 0.87 A. The cross-sectional area can be calculated to be $A = 5.26\,\text{mm}^2$.

Solution

Calculate the current density using the given current $I = 0.87\text{ A}$ and the area, found to be $A = 5.26\,\text{mm}^2$.

$$J = \frac{I}{A} = \frac{0.87\text{ A}}{5.26 \times 10^{-6}\,\text{m}^2} = 1.65 \times 10^5 \frac{\text{A}}{\text{m}^2}.$$

Significance

The current density in a conducting wire depends on the current through the conducting wire and the cross-sectional area of the wire. For a given current, as the diameter of the wire increases, the charge density decreases.

9.4 Check Your Understanding The current density is proportional to the current and inversely proportional to the area. If the current density in a conducting wire increases, what would happen to the drift velocity of the charges in the wire?

What is the significance of the current density? The current density is proportional to the current, and the current is the number of charges that pass through a cross-sectional area per second. The charges move through the conductor, accelerated by the electric force provided by the electrical field. The electrical field is created when a voltage is applied across the conductor. In Ohm's Law, we will use this relationship between the current density and the electrical field to examine the relationship between the current through a conductor and the voltage applied.

9.3 | Resistivity and Resistance

Learning Objectives

By the end of this section, you will be able to:

- Differentiate between resistance and resistivity
- Define the term conductivity
- Describe the electrical component known as a resistor
- State the relationship between resistance of a resistor and its length, cross-sectional area, and resistivity
- State the relationship between resistivity and temperature

What drives current? We can think of various devices—such as batteries, generators, wall outlets, and so on—that are necessary to maintain a current. All such devices create a potential difference and are referred to as voltage sources. When a voltage source is connected to a conductor, it applies a potential difference V that creates an electrical field. The electrical field, in turn, exerts force on free charges, causing current. The amount of current depends not only on the magnitude of the voltage, but also on the characteristics of the material that the current is flowing through. The material can resist the flow of the charges, and the measure of how much a material resists the flow of charges is known as the *resistivity*. This resistivity is crudely analogous to the friction between two materials that resists motion.

Resistivity

When a voltage is applied to a conductor, an electrical field $\vec{\mathbf{E}}$ is created, and charges in the conductor feel a force due to the electrical field. The current density $\vec{\mathbf{J}}$ that results depends on the electrical field and the properties of the material. This dependence can be very complex. In some materials, including metals at a given temperature, the current density is approximately proportional to the electrical field. In these cases, the current density can be modeled as

$$\vec{\mathbf{J}} = \sigma \vec{\mathbf{E}},$$

where σ is the **electrical conductivity**. The electrical conductivity is analogous to thermal conductivity and is a measure of a material's ability to conduct or transmit electricity. Conductors have a higher electrical conductivity than insulators. Since the electrical conductivity is $\sigma = J/E$, the units are

$$\sigma = \frac{[J]}{[E]} = \frac{\text{A/m}^2}{\text{V/m}} = \frac{\text{A}}{\text{V} \cdot \text{m}}.$$

Here, we define a unit named the **ohm** with the Greek symbol uppercase omega, Ω. The unit is named after Georg Simon Ohm, whom we will discuss later in this chapter. The Ω is used to avoid confusion with the number 0. One ohm equals one volt per amp: $1\ \Omega = 1\ \text{V/A}$. The units of electrical conductivity are therefore $(\Omega \cdot \text{m})^{-1}$.

Conductivity is an intrinsic property of a material. Another intrinsic property of a material is the **resistivity**, or electrical resistivity. The resistivity of a material is a measure of how strongly a material opposes the flow of electrical current. The symbol for resistivity is the lowercase Greek letter rho, ρ, and resistivity is the reciprocal of electrical conductivity:

$$\rho = \frac{1}{\sigma}.$$

The unit of resistivity in SI units is the ohm-meter $(\Omega \cdot \text{m})$. We can define the resistivity in terms of the electrical field and the current density,

$$\rho = \frac{E}{J}. \tag{9.6}$$

The greater the resistivity, the larger the field needed to produce a given current density. The lower the resistivity, the larger the current density produced by a given electrical field. Good conductors have a high conductivity and low resistivity. Good insulators have a low conductivity and a high resistivity. Table 9.1 lists resistivity and conductivity values for various materials.

Material	Conductivity, σ $(\Omega \cdot \text{m})^{-1}$	Resistivity, ρ $(\Omega \cdot \text{m})$	Temperature Coefficient, α $(°\text{C})^{-1}$
Conductors			
Silver	6.29×10^7	1.59×10^{-8}	0.0038
Copper	5.95×10^7	1.68×10^{-8}	0.0039
Gold	4.10×10^7	2.44×10^{-8}	0.0034
Aluminum	3.77×10^7	2.65×10^{-8}	0.0039
Tungsten	1.79×10^7	5.60×10^{-8}	0.0045
Iron	1.03×10^7	9.71×10^{-8}	0.0065
Platinum	0.94×10^7	10.60×10^{-8}	0.0039
Steel	0.50×10^7	20.00×10^{-8}	
Lead	0.45×10^7	22.00×10^{-8}	
Manganin (Cu, Mn, Ni alloy)	0.21×10^7	48.20×10^{-8}	0.000002
Constantan (Cu, Ni alloy)	0.20×10^7	49.00×10^{-8}	0.00003
Mercury	0.10×10^7	98.00×10^{-8}	0.0009
Nichrome (Ni, Fe, Cr alloy)	0.10×10^7	100.00×10^{-8}	0.0004
Semiconductors[1]			
Carbon (pure)	2.86×10^{-6}	3.50×10^{-5}	−0.0005
Carbon	$(2.86 - 1.67) \times 10^{-6}$	$(3.5 - 60) \times 10^{-5}$	−0.0005
Germanium (pure)		600×10^{-3}	−0.048
Germanium		$(1 - 600) \times 10^{-3}$	−0.050
Silicon (pure)		2300	−0.075
Silicon		$0.1 - 2300$	−0.07
Insulators			
Amber	2.00×10^{-15}	5×10^{14}	
Glass	$10^{-9} - 10^{-14}$	$10^9 - 10^{14}$	
Lucite	$<10^{-13}$	$>10^{13}$	

Table 9.1 Resistivities and Conductivities of Various Materials at 20 °C [1] Values depend strongly on amounts and types of impurities.

Material	Conductivity, σ $(\Omega \cdot \text{m})^{-1}$	Resistivity, ρ $(\Omega \cdot \text{m})$	Temperature Coefficient, α $(°\text{C})^{-1}$
Mica	$10^{-11} - 10^{-15}$	$10^{11} - 10^{15}$	
Quartz (fused)	2.00×10^{-15}	75×10^{16}	
Rubber (hard)	$10^{-13} - 10^{-16}$	$10^{13} - 10^{16}$	
Sulfur	10^{-15}	10^{15}	
Teflon™	$<10^{-13}$	$>10^{13}$	
Wood	$10^{-8} - 10^{-11}$	$10^{8} - 10^{11}$	

Table 9.1 Resistivities and Conductivities of Various Materials at 20 °C [1] Values depend strongly on amounts and types of impurities.

The materials listed in the table are separated into categories of conductors, semiconductors, and insulators, based on broad groupings of resistivity. Conductors have the smallest resistivity, and insulators have the largest; semiconductors have intermediate resistivity. Conductors have varying but large, free charge densities, whereas most charges in insulators are bound to atoms and are not free to move. Semiconductors are intermediate, having far fewer free charges than conductors, but having properties that make the number of free charges depend strongly on the type and amount of impurities in the semiconductor. These unique properties of semiconductors are put to use in modern electronics, as we will explore in later chapters.

Example 9.5

Current Density, Resistance, and Electrical field for a Current-Carrying Wire

Calculate the current density, resistance, and electrical field of a 5-m length of copper wire with a diameter of 2.053 mm (12-gauge) carrying a current of $I = 10\ \text{mA}$.

Strategy

We can calculate the current density by first finding the cross-sectional area of the wire, which is $A = 3.31\ \text{mm}^2$, and the definition of current density $J = \frac{I}{A}$. The resistance can be found using the length of the wire $L = 5.00\ \text{m}$, the area, and the resistivity of copper $\rho = 1.68 \times 10^{-8}\ \Omega \cdot \text{m}$, where $R = \rho\frac{L}{A}$. The resistivity and current density can be used to find the electrical field.

Solution

First, we calculate the current density:

$$J = \frac{I}{A} = \frac{10 \times 10^{-3}\,\text{A}}{3.31 \times 10^{-6}\,\text{m}^2} = 3.02 \times 10^{3}\frac{\text{A}}{\text{m}^2}.$$

The resistance of the wire is

$$R = \rho\frac{L}{A} = \left(1.68 \times 10^{-8}\ \Omega \cdot \text{m}\right)\frac{5.00\ \text{m}}{3.31 \times 10^{-6}\,\text{m}^2} = 0.025\ \Omega.$$

Finally, we can find the electrical field:

$$E = \rho J = 1.68 \times 10^{-8}\ \Omega \cdot \text{m}\left(3.02 \times 10^{3}\frac{\text{A}}{\text{m}^2}\right) = 5.07 \times 10^{-5}\frac{\text{V}}{\text{m}}.$$

Significance

From these results, it is not surprising that copper is used for wires for carrying current because the resistance is quite small. Note that the current density and electrical field are independent of the length of the wire, but the voltage depends on the length.

9.5 Check Your Understanding Copper wires use routinely used for extension cords and house wiring for several reasons. Copper has the highest electrical conductivity rating, and therefore the lowest resistivity rating, of all nonprecious metals. Also important is the tensile strength, where the tensile strength is a measure of the force required to pull an object to the point where it breaks. The tensile strength of a material is the maximum amount of tensile stress it can take before breaking. Copper has a high tensile strength, $2 \times 10^8 \frac{\text{N}}{\text{m}^2}$. A third important characteristic is ductility. Ductility is a measure of a material's ability to be drawn into wires and a measure of the flexibility of the material, and copper has a high ductility. Summarizing, for a conductor to be a suitable candidate for making wire, there are at least three important characteristics: low resistivity, high tensile strength, and high ductility. What other materials are used for wiring and what are the advantages and disadvantages?

View this interactive simulation (https://openstaxcollege.org/l/21resistwire) to see what the effects of the cross-sectional area, the length, and the resistivity of a wire are on the resistance of a conductor. Adjust the variables using slide bars and see if the resistance becomes smaller or larger.

Temperature Dependence of Resistivity

Looking back at Table 9.1, you will see a column labeled "Temperature Coefficient." The resistivity of some materials has a strong temperature dependence. In some materials, such as copper, the resistivity increases with increasing temperature. In fact, in most conducting metals, the resistivity increases with increasing temperature. The increasing temperature causes increased vibrations of the atoms in the lattice structure of the metals, which impede the motion of the electrons. In other materials, such as carbon, the resistivity decreases with increasing temperature. In many materials, the dependence is approximately linear and can be modeled using a linear equation:

$$\rho \approx \rho_0[1 + \alpha(T - T_0)], \tag{9.7}$$

where ρ is the resistivity of the material at temperature T, α is the temperature coefficient of the material, and ρ_0 is the resistivity at T_0, usually taken as $T_0 = 20.00\ °\text{C}$.

Note also that the temperature coefficient α is negative for the semiconductors listed in Table 9.1, meaning that their resistivity decreases with increasing temperature. They become better conductors at higher temperature, because increased thermal agitation increases the number of free charges available to carry current. This property of decreasing ρ with temperature is also related to the type and amount of impurities present in the semiconductors.

Resistance

We now consider the resistance of a wire or component. The resistance is a measure of how difficult it is to pass current through a wire or component. Resistance depends on the resistivity. The resistivity is a characteristic of the material used to fabricate a wire or other electrical component, whereas the resistance is a characteristic of the wire or component.

To calculate the resistance, consider a section of conducting wire with cross-sectional area A, length L, and resistivity ρ. A battery is connected across the conductor, providing a potential difference ΔV across it (Figure 9.13). The potential difference produces an electrical field that is proportional to the current density, according to $\vec{\mathbf{E}} = \rho \vec{\mathbf{J}}$.

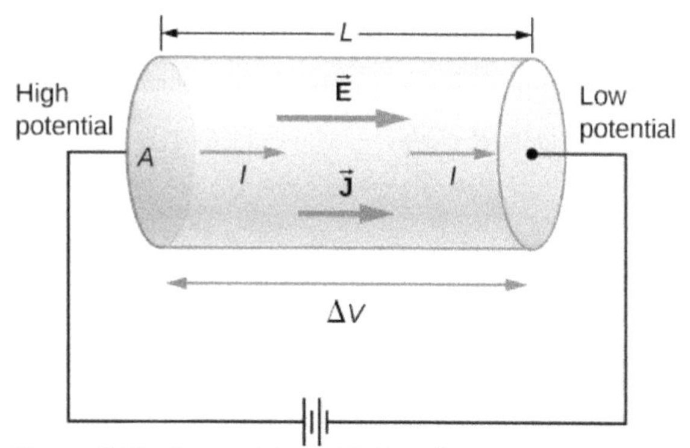

Figure 9.13 A potential provided by a battery is applied to a segment of a conductor with a cross-sectional area A and a length L.

The magnitude of the electrical field across the segment of the conductor is equal to the voltage divided by the length, $E = V/L$, and the magnitude of the current density is equal to the current divided by the cross-sectional area, $J = I/A$. Using this information and recalling that the electrical field is proportional to the resistivity and the current density, we can see that the voltage is proportional to the current:

$$E = \rho J$$
$$\frac{V}{L} = \rho \frac{I}{A}$$
$$V = \left(\rho \frac{L}{A}\right) I.$$

Resistance

The ratio of the voltage to the current is defined as the **resistance** R:

$$R \equiv \frac{V}{I}. \quad (9.8)$$

The resistance of a cylindrical segment of a conductor is equal to the resistivity of the material times the length divided by the area:

$$R \equiv \frac{V}{I} = \rho \frac{L}{A}. \quad (9.9)$$

The unit of resistance is the ohm, Ω. For a given voltage, the higher the resistance, the lower the current.

Resistors

A common component in electronic circuits is the resistor. The resistor can be used to reduce current flow or provide a voltage drop. Figure 9.14 shows the symbols used for a resistor in schematic diagrams of a circuit. Two commonly used standards for circuit diagrams are provided by the American National Standard Institute (ANSI, pronounced "AN-see") and the International Electrotechnical Commission (IEC). Both systems are commonly used. We use the ANSI standard in this text for its visual recognition, but we note that for larger, more complex circuits, the IEC standard may have a cleaner presentation, making it easier to read.

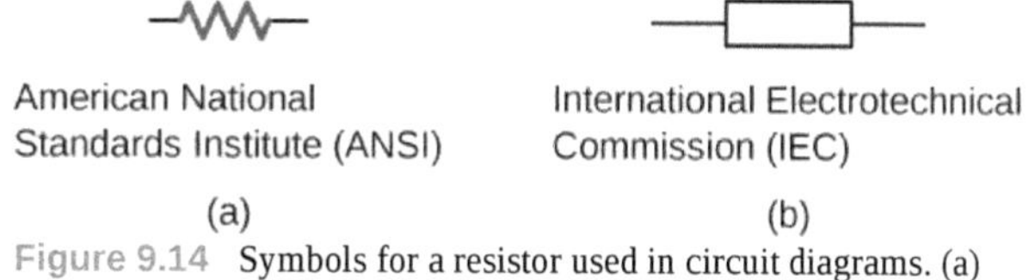

Figure 9.14 Symbols for a resistor used in circuit diagrams. (a) The ANSI symbol; (b) the IEC symbol.

Material and shape dependence of resistance

A resistor can be modeled as a cylinder with a cross-sectional area A and a length L, made of a material with a resistivity ρ (Figure 9.15). The resistance of the resistor is $R = \rho \frac{L}{A}$.

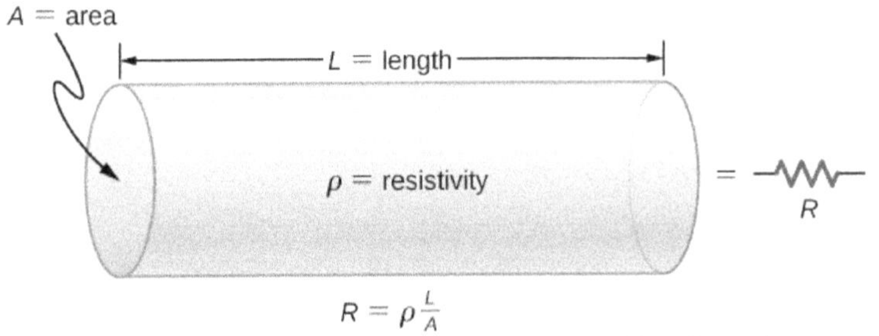

Figure 9.15 A model of a resistor as a uniform cylinder of length L and cross-sectional area A. Its resistance to the flow of current is analogous to the resistance posed by a pipe to fluid flow. The longer the cylinder, the greater its resistance. The larger its cross-sectional area A, the smaller its resistance.

The most common material used to make a resistor is carbon. A carbon track is wrapped around a ceramic core, and two copper leads are attached. A second type of resistor is the metal film resistor, which also has a ceramic core. The track is made from a metal oxide material, which has semiconductive properties similar to carbon. Again, copper leads are inserted into the ends of the resistor. The resistor is then painted and marked for identification. A resistor has four colored bands, as shown in Figure 9.16.

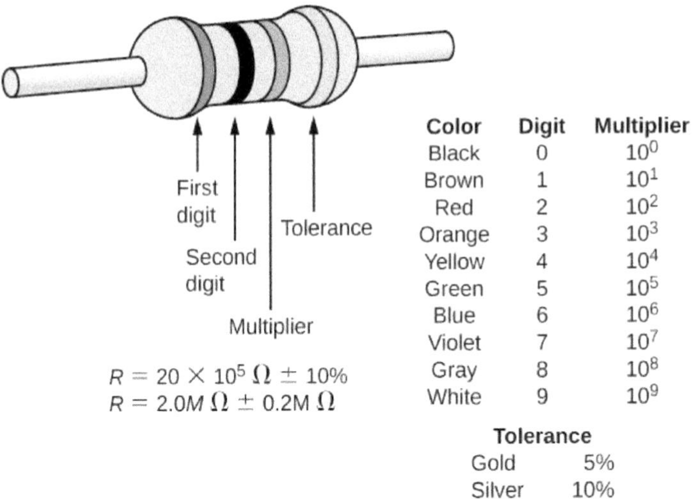

Color	Digit	Multiplier
Black	0	10^0
Brown	1	10^1
Red	2	10^2
Orange	3	10^3
Yellow	4	10^4
Green	5	10^5
Blue	6	10^6
Violet	7	10^7
Gray	8	10^8
White	9	10^9

Tolerance	
Gold	5%
Silver	10%

Figure 9.16 Many resistors resemble the figure shown above. The four bands are used to identify the resistor. The first two colored bands represent the first two digits of the resistance of the resistor. The third color is the multiplier. The fourth color represents the tolerance of the resistor. The resistor shown has a resistance of $20 \times 10^5\ \Omega \pm 10\%$.

Resistances range over many orders of magnitude. Some ceramic insulators, such as those used to support power lines, have resistances of $10^{12}\ \Omega$ or more. A dry person may have a hand-to-foot resistance of $10^5\ \Omega$, whereas the resistance of the human heart is about $10^3\ \Omega$. A meter-long piece of large-diameter copper wire may have a resistance of $10^{-5}\ \Omega$, and superconductors have no resistance at all at low temperatures. As we have seen, resistance is related to the shape of an object and the material of which it is composed.

The resistance of an object also depends on temperature, since R_0 is directly proportional to ρ. For a cylinder, we know $R = \rho \frac{L}{A}$, so if L and A do not change greatly with temperature, R has the same temperature dependence as ρ. (Examination of the coefficients of linear expansion shows them to be about two orders of magnitude less than typical temperature coefficients of resistivity, so the effect of temperature on L and A is about two orders of magnitude less than on ρ.) Thus,

$$R = R_0(1 + \alpha \Delta T) \tag{9.10}$$

is the temperature dependence of the resistance of an object, where R_0 is the original resistance (usually taken to be $20.00\ °\text{C}$) and R is the resistance after a temperature change ΔT. The color code gives the resistance of the resistor at a temperature of $T = 20.00\ °\text{C}$.

Numerous thermometers are based on the effect of temperature on resistance (Figure 9.17). One of the most common thermometers is based on the thermistor, a semiconductor crystal with a strong temperature dependence, the resistance of which is measured to obtain its temperature. The device is small, so that it quickly comes into thermal equilibrium with the part of a person it touches.

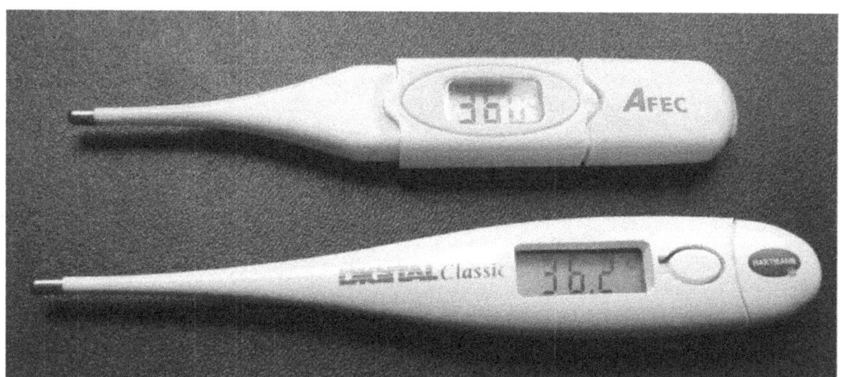

Figure 9.17 These familiar thermometers are based on the automated measurement of a thermistor's temperature-dependent resistance.

Example 9.6

Calculating Resistance

Although caution must be used in applying $\rho = \rho_0(1 + \alpha \Delta T)$ and $R = R_0(1 + \alpha \Delta T)$ for temperature changes greater than $100\ °\text{C}$, for tungsten, the equations work reasonably well for very large temperature changes. A tungsten filament at $20\ °\text{C}$ has a resistance of $0.350\ \Omega$. What would the resistance be if the temperature is increased to $2850\ °\text{C}$?

Strategy

This is a straightforward application of $R = R_0(1 + \alpha \Delta T)$, since the original resistance of the filament is given as $R_0 = 0.350\ \Omega$ and the temperature change is $\Delta T = 2830\ °\text{C}$.

Solution

The resistance of the hotter filament R is obtained by entering known values into the above equation:

$$R = R_0(1 + \alpha \Delta T) = (0.350\ \Omega)\left[1 + \left(\frac{4.5 \times 10^{-3}}{°\text{C}}\right)(2830\ °\text{C})\right] = 4.8\ \Omega.$$

Significance

Notice that the resistance changes by more than a factor of 10 as the filament warms to the high temperature and the current through the filament depends on the resistance of the filament and the voltage applied. If the filament is used in an incandescent light bulb, the initial current through the filament when the bulb is first energized will be higher than the current after the filament reaches the operating temperature.

9.6 Check Your Understanding A strain gauge is an electrical device to measure strain, as shown below. It consists of a flexible, insulating backing that supports a conduction foil pattern. The resistance of the foil changes as the backing is stretched. How does the strain gauge resistance change? Is the strain gauge affected by temperature changes?

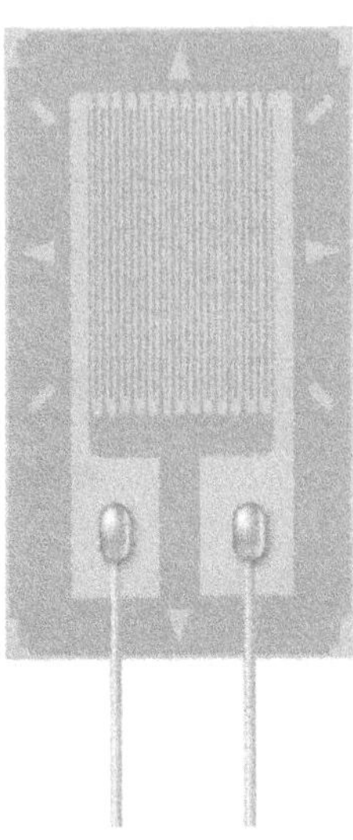

Example 9.7

The Resistance of Coaxial Cable

Long cables can sometimes act like antennas, picking up electronic noise, which are signals from other equipment and appliances. Coaxial cables are used for many applications that require this noise to be eliminated. For example, they can be found in the home in cable TV connections or other audiovisual connections. Coaxial cables consist of an inner conductor of radius r_i surrounded by a second, outer concentric conductor with radius r_o (Figure 9.18). The space between the two is normally filled with an insulator such as polyethylene plastic. A small amount of radial leakage current occurs between the two conductors. Determine the resistance of a coaxial cable of length L.

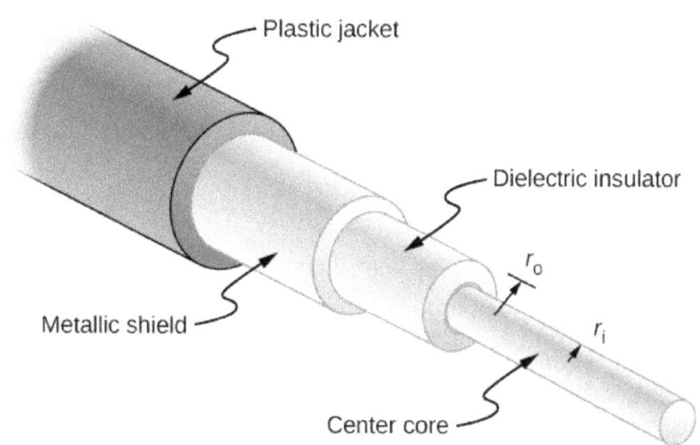

Figure 9.18 Coaxial cables consist of two concentric conductors separated by insulation. They are often used in cable TV or other audiovisual connections.

Strategy

We cannot use the equation $R = \rho \frac{L}{A}$ directly. Instead, we look at concentric cylindrical shells, with thickness *dr*, and integrate.

Solution

We first find an expression for *dR* and then integrate from r_i to r_o,

$$\begin{aligned} dR &= \frac{\rho}{A}dr = \frac{\rho}{2\pi rL}dr, \\ R &= \int_{r_i}^{r_o} dR = \int_{r_i}^{r_o} \frac{\rho}{2\pi rL}dr = \frac{\rho}{2\pi L}\int_{r_i}^{r_o} \frac{1}{r}dr = \frac{\rho}{2\pi L}\ln\frac{r_o}{r_i}. \end{aligned}$$

Significance

The resistance of a coaxial cable depends on its length, the inner and outer radii, and the resistivity of the material separating the two conductors. Since this resistance is not infinite, a small leakage current occurs between the two conductors. This leakage current leads to the attenuation (or weakening) of the signal being sent through the cable.

9.7 Check Your Understanding The resistance between the two conductors of a coaxial cable depends on the resistivity of the material separating the two conductors, the length of the cable and the inner and outer radius of the two conductor. If you are designing a coaxial cable, how does the resistance between the two conductors depend on these variables?

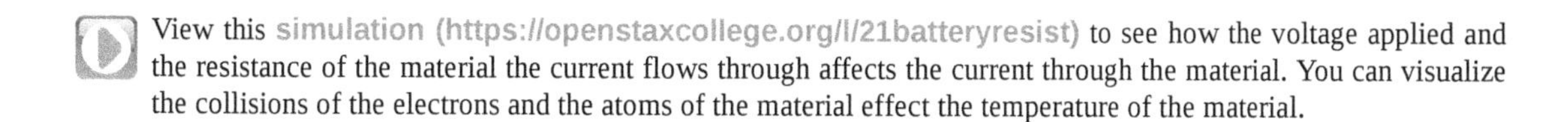

View this simulation (https://openstaxcollege.org/l/21batteryresist) to see how the voltage applied and the resistance of the material the current flows through affects the current through the material. You can visualize the collisions of the electrons and the atoms of the material effect the temperature of the material.

9.4 | Ohm's Law

Learning Objectives

By the end of this section, you will be able to:

- Describe Ohm's law
- Recognize when Ohm's law applies and when it does not

We have been discussing three electrical properties so far in this chapter: current, voltage, and resistance. It turns out that many materials exhibit a simple relationship among the values for these properties, known as Ohm's law. Many other materials do not show this relationship, so despite being called Ohm's law, it is not considered a law of nature, like Newton's laws or the laws of thermodynamics. But it is very useful for calculations involving materials that do obey Ohm's law.

Description of Ohm's Law

The current that flows through most substances is directly proportional to the voltage V applied to it. The German physicist Georg Simon Ohm (1787–1854) was the first to demonstrate experimentally that the current in a metal wire is *directly proportional to the voltage applied*:

$$I \propto V.$$

This important relationship is the basis for **Ohm's law**. It can be viewed as a cause-and-effect relationship, with voltage the cause and current the effect. This is an empirical law, which is to say that it is an experimentally observed phenomenon, like friction. Such a linear relationship doesn't always occur. Any material, component, or device that obeys Ohm's law, where the current through the device is proportional to the voltage applied, is known as an **ohmic** material or ohmic component. Any material or component that does not obey Ohm's law is known as a **nonohmic** material or nonohmic component.

Ohm's Experiment

In a paper published in 1827, Georg Ohm described an experiment in which he measured voltage across and current through various simple electrical circuits containing various lengths of wire. A similar experiment is shown in Figure 9.19. This experiment is used to observe the current through a resistor that results from an applied voltage. In this simple circuit, a resistor is connected in series with a battery. The voltage is measured with a voltmeter, which must be placed across the resistor (in parallel with the resistor). The current is measured with an ammeter, which must be in line with the resistor (in series with the resistor).

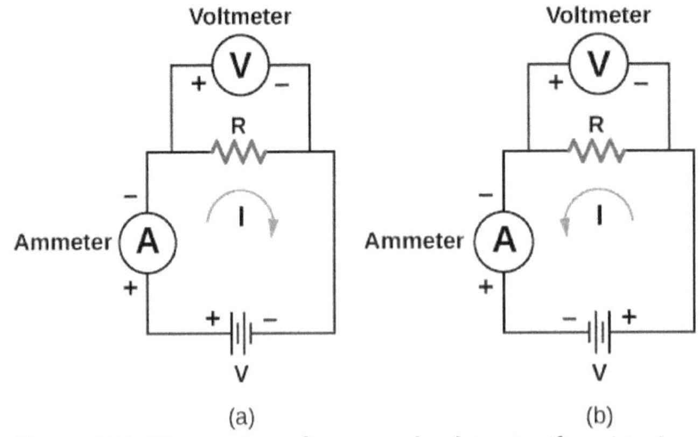

Figure 9.19 The experimental set-up used to determine if a resistor is an ohmic or nonohmic device. (a) When the battery is attached, the current flows in the clockwise direction and the voltmeter and ammeter have positive readings. (b) When the leads of the battery are switched, the current flows in the counterclockwise direction and the voltmeter and ammeter have negative readings.

In this updated version of Ohm's original experiment, several measurements of the current were made for several different voltages. When the battery was hooked up as in Figure 9.19(a), the current flowed in the clockwise direction and the readings of the voltmeter and ammeter were positive. Does the behavior of the current change if the current flowed in the opposite direction? To get the current to flow in the opposite direction, the leads of the battery can be switched. When the leads of the battery were switched, the readings of the voltmeter and ammeter readings were negative because the current flowed in the opposite direction, in this case, counterclockwise. Results of a similar experiment are shown in Figure 9.20.

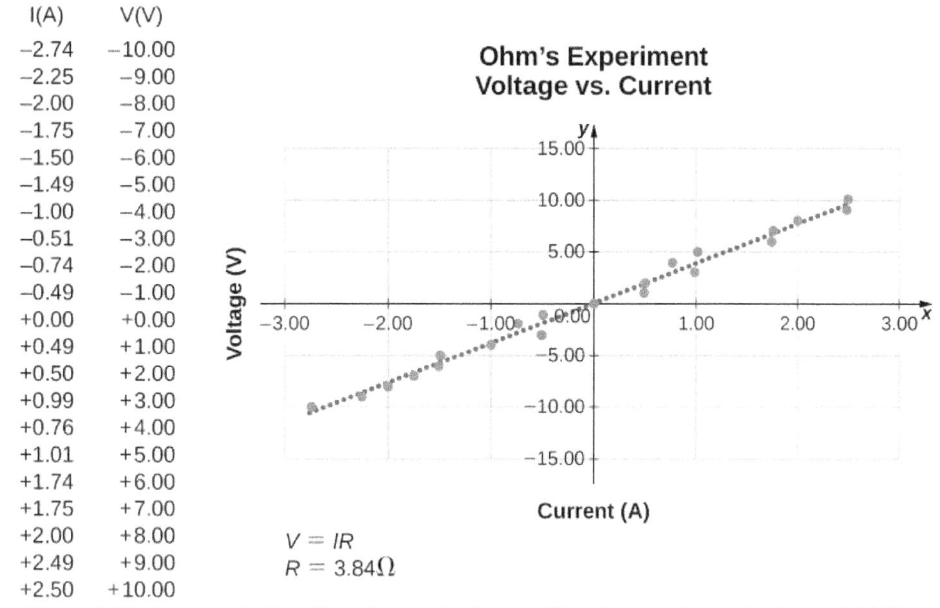

I(A)	V(V)
−2.74	−10.00
−2.25	−9.00
−2.00	−8.00
−1.75	−7.00
−1.50	−6.00
−1.49	−5.00
−1.00	−4.00
−0.51	−3.00
−0.74	−2.00
−0.49	−1.00
+0.00	+0.00
+0.49	+1.00
+0.50	+2.00
+0.99	+3.00
+0.76	+4.00
+1.01	+5.00
+1.74	+6.00
+1.75	+7.00
+2.00	+8.00
+2.49	+9.00
+2.50	+10.00

Figure 9.20 A resistor is placed in a circuit with a battery. The voltage applied varies from −10.00 V to +10.00 V, increased by 1.00-V increments. A plot shows values of the voltage versus the current typical of what a casual experimenter might find.

In this experiment, the voltage applied across the resistor varies from −10.00 to +10.00 V, by increments of 1.00 V. The current through the resistor and the voltage across the resistor are measured. A plot is made of the voltage versus the current, and the result is approximately linear. The slope of the line is the resistance, or the voltage divided by the current. This result is known as Ohm's law:

$$V = IR, \tag{9.11}$$

where V is the voltage measured in volts across the object in question, I is the current measured through the object in amps, and R is the resistance in units of ohms. As stated previously, any device that shows a linear relationship between the voltage and the current is known as an ohmic device. A resistor is therefore an ohmic device.

Example 9.8

Measuring Resistance

A carbon resistor at room temperature $(20\ °\text{C})$ is attached to a 9.00-V battery and the current measured through the resistor is 3.00 mA. (a) What is the resistance of the resistor measured in ohms? (b) If the temperature of the resistor is increased to $60\ °\text{C}$ by heating the resistor, what is the current through the resistor?

Strategy

(a) The resistance can be found using Ohm's law. Ohm's law states that $V = IR$, so the resistance can be found using $R = V/I$.

(b) First, the resistance is temperature dependent so the new resistance after the resistor has been heated can be found using $R = R_0(1 + \alpha\Delta T)$. The current can be found using Ohm's law in the form $I = V/R$.

Solution

a. Using Ohm's law and solving for the resistance yields the resistance at room temperature:

$$R = \frac{V}{I} = \frac{9.00 \text{ V}}{3.00 \times 10^{-3} \text{ A}} = 3.00 \times 10^3 \ \Omega = 3.00 \text{ k}\Omega.$$

b. The resistance at $60\ °\text{C}$ can be found using $R = R_0(1 + \alpha\Delta T)$ where the temperature coefficient for carbon is $\alpha = -0.0005$. $R = R_0(1 + \alpha\Delta T) = 3.00 \times 10^3(1 - 0.0005(60\ °\text{C} - 20\ °\text{C})) = 2.94 \text{ k}\Omega$. The current through the heated resistor is

$$I = \frac{V}{R} = \frac{9.00 \text{ V}}{2.94 \times 10^3 \ \Omega} = 3.06 \times 10^{-3} \text{ A} = 3.06 \text{ mA}.$$

Significance

A change in temperature of $40\ °\text{C}$ resulted in a 2.00% change in current. This may not seem like a very great change, but changing electrical characteristics can have a strong effect on the circuits. For this reason, many electronic appliances, such as computers, contain fans to remove the heat dissipated by components in the electric circuits.

9.8 Check Your Understanding The voltage supplied to your house varies as $V(t) = V_{max} \sin(2\pi ft)$. If a resistor is connected across this voltage, will Ohm's law $V = IR$ still be valid?

See how the equation form of Ohm's law (https://openstaxcollege.org/l/21ohmslaw) relates to a simple circuit. Adjust the voltage and resistance, and see the current change according to Ohm's law. The sizes of the symbols in the equation change to match the circuit diagram.

Nonohmic devices do not exhibit a linear relationship between the voltage and the current. One such device is the semiconducting circuit element known as a diode. A **diode** is a circuit device that allows current flow in only one direction. A diagram of a simple circuit consisting of a battery, a diode, and a resistor is shown in Figure 9.21. Although we do not cover the theory of the diode in this section, the diode can be tested to see if it is an ohmic or a nonohmic device.

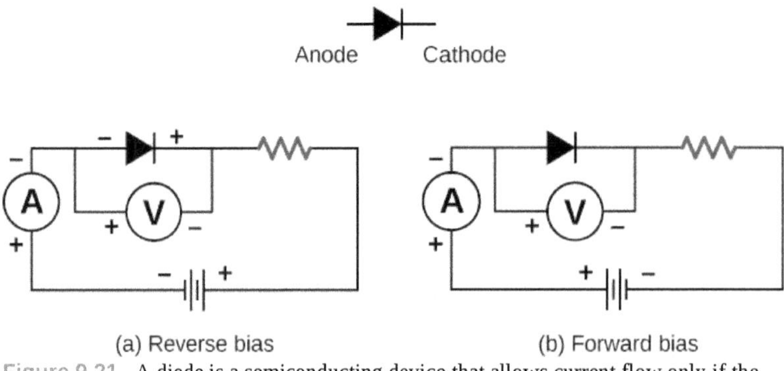

Figure 9.21 A diode is a semiconducting device that allows current flow only if the diode is forward biased, which means that the anode is positive and the cathode is negative.

A plot of current versus voltage is shown in Figure 9.22. Note that the behavior of the diode is shown as current versus voltage, whereas the resistor operation was shown as voltage versus current. A diode consists of an anode and a cathode. When the anode is at a negative potential and the cathode is at a positive potential, as shown in part (a), the diode is said to have reverse bias. With reverse bias, the diode has an extremely large resistance and there is very little current flow—essentially zero current—through the diode and the resistor. As the voltage applied to the circuit increases, the current remains essentially zero, until the voltage reaches the breakdown voltage and the diode conducts current, as shown in Figure 9.22. When the battery and the potential across the diode are reversed, making the anode positive and the cathode negative, the diode conducts and current flows through the diode if the voltage is greater than 0.7 V. The resistance of the diode is close to zero. (This is the reason for the resistor in the circuit; if it were not there, the current would become very large.) You can see from the graph in Figure 9.22 that the voltage and the current do not have a linear relationship. Thus, the diode is an example of a nonohmic device.

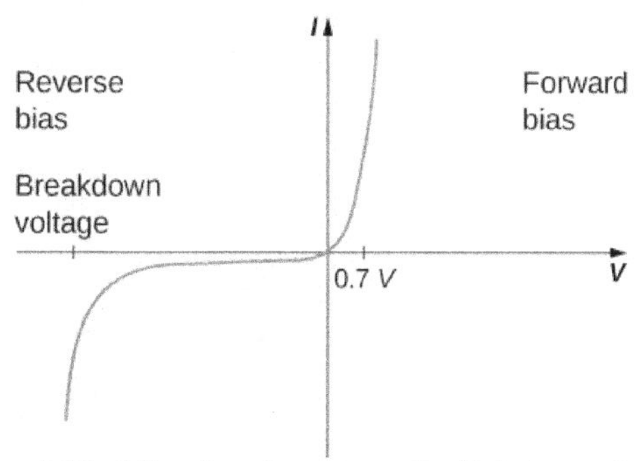

Figure 9.22 When the voltage across the diode is negative and small, there is very little current flow through the diode. As the voltage reaches the breakdown voltage, the diode conducts. When the voltage across the diode is positive and greater than 0.7 V (the actual voltage value depends on the diode), the diode conducts. As the voltage applied increases, the current through the diode increases, but the voltage across the diode remains approximately 0.7 V.

Ohm's law is commonly stated as $V = IR$, but originally it was stated as a microscopic view, in terms of the current density, the conductivity, and the electrical field. This microscopic view suggests the proportionality $V \propto I$ comes from the drift velocity of the free electrons in the metal that results from an applied electrical field. As stated earlier, the current density is proportional to the applied electrical field. The reformulation of Ohm's law is credited to Gustav Kirchhoff, whose name we will see again in the next chapter.

9.5 | Electrical Energy and Power

Learning Objectives

By the end of this section, you will be able to:

- Express electrical power in terms of the voltage and the current
- Describe the power dissipated by a resistor in an electric circuit
- Calculate the energy efficiency and cost effectiveness of appliances and equipment

In an electric circuit, electrical energy is continuously converted into other forms of energy. For example, when a current flows in a conductor, electrical energy is converted into thermal energy within the conductor. The electrical field, supplied by the voltage source, accelerates the free electrons, increasing their kinetic energy for a short time. This increased kinetic energy is converted into thermal energy through collisions with the ions of the lattice structure of the conductor. In Work and Kinetic Energy (http://cnx.org/content/m58307/latest/) , we defined power as the rate at which work is done by a force measured in watts. Power can also be defined as the rate at which energy is transferred. In this section, we discuss the time rate of energy transfer, or power, in an electric circuit.

Power in Electric Circuits

Power is associated by many people with electricity. Power transmission lines might come to mind. We also think of light bulbs in terms of their power ratings in watts. What is the expression for **electric power**?

Let us compare a 25-W bulb with a 60-W bulb (Figure 9.23(a)). The 60-W bulb glows brighter than the 25-W bulb. Although it is not shown, a 60-W light bulb is also warmer than the 25-W bulb. The heat and light is produced by from the conversion of electrical energy. The kinetic energy lost by the electrons in collisions is converted into the internal energy of the conductor and radiation. How are voltage, current, and resistance related to electric power?

Figure 9.23 (a) Pictured above are two incandescent bulbs: a 25-W bulb (left) and a 60-W bulb (right). The 60-W bulb provides a higher intensity light than the 25-W bulb. The electrical energy supplied to the light bulbs is converted into heat and light. (b) This compact fluorescent light (CFL) bulb puts out the same intensity of light as the 60-W bulb, but at 1/4 to 1/10 the input power. (credit a: modification of works by "Dickbauch"/Wikimedia Commons and Greg Westfall; credit b: modification of work by "dbgg1979"/Flickr)

To calculate electric power, consider a voltage difference existing across a material (Figure 9.24). The electric potential V_1 is higher than the electric potential at V_2, and the voltage difference is negative $V = V_2 - V_1$. As discussed in Electric Potential, an electrical field exists between the two potentials, which points from the higher potential to the lower potential. Recall that the electrical potential is defined as the potential energy per charge, $V = \Delta U/q$, and the charge ΔQ loses potential energy moving through the potential difference.

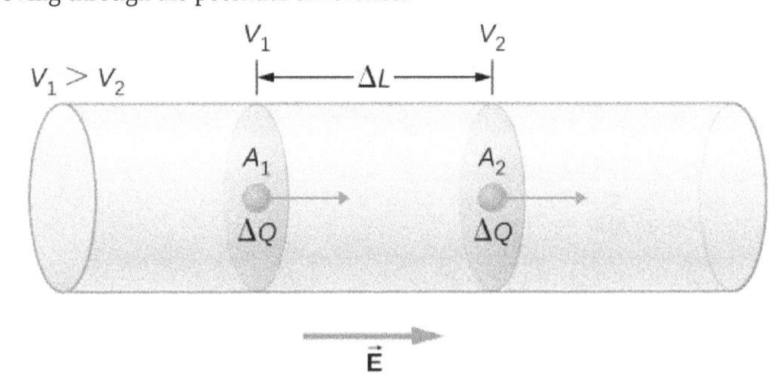

Figure 9.24 When there is a potential difference across a conductor, an electrical field is present that points in the direction from the higher potential to the lower potential.

If the charge is positive, the charge experiences a force due to the electrical field $\vec{F} = m\vec{a} = \Delta Q\vec{E}$. This force is necessary to keep the charge moving. This force does not act to accelerate the charge through the entire distance ΔL

because of the interactions of the charge with atoms and free electrons in the material. The speed, and therefore the kinetic energy, of the charge do not increase during the entire trip across ΔL, and charge passing through area A_2 has the same drift velocity v_d as the charge that passes through area A_1. However, work is done on the charge, by the electrical field, which changes the potential energy. Since the change in the electrical potential difference is negative, the electrical field is found to be

$$E = -\frac{(V_2 - V_1)}{\Delta L} = \frac{V}{\Delta L}.$$

The work done on the charge is equal to the electric force times the length at which the force is applied,

$$W = F\Delta L = (\Delta QE)\Delta L = \left(\Delta Q\frac{V}{\Delta L}\right)\Delta L = \Delta QV = \Delta U.$$

The charge moves at a drift velocity v_d so the work done on the charge results in a loss of potential energy, but the average kinetic energy remains constant. The lost electrical potential energy appears as thermal energy in the material. On a microscopic scale, the energy transfer is due to collisions between the charge and the molecules of the material, which leads to an increase in temperature in the material. The loss of potential energy results in an increase in the temperature of the material, which is dissipated as radiation. In a resistor, it is dissipated as heat, and in a light bulb, it is dissipated as heat and light.

The power dissipated by the material as heat and light is equal to the time rate of change of the work:

$$P = \frac{\Delta U}{\Delta t} = -\frac{\Delta QV}{\Delta t} = IV.$$

With a resistor, the voltage drop across the resistor is dissipated as heat. Ohm's law states that the voltage across the resistor is equal to the current times the resistance, $V = IR$. The power dissipated by the resistor is therefore

$$P = IV = I(IR) = I^2R \text{ or } P = IV = \left(\frac{V}{R}\right)V = \frac{V^2}{R}.$$

If a resistor is connected to a battery, the power dissipated as radiant energy by the wires and the resistor is equal to $P = IV = I^2R = \frac{V^2}{R}$. The power supplied from the battery is equal to current times the voltage, $P = IV$.

Electric Power

The electric power gained or lost by any device has the form

$$P = IV. \tag{9.12}$$

The power dissipated by a resistor has the form

$$P = I^2R = \frac{V^2}{R}. \tag{9.13}$$

Different insights can be gained from the three different expressions for electric power. For example, $P = V^2/R$ implies that the lower the resistance connected to a given voltage source, the greater the power delivered. Furthermore, since voltage is squared in $P = V^2/R$, the effect of applying a higher voltage is perhaps greater than expected. Thus, when the voltage is doubled to a 25-W bulb, its power nearly quadruples to about 100 W, burning it out. If the bulb's resistance remained constant, its power would be exactly 100 W, but at the higher temperature, its resistance is higher, too.

Example 9.9

Calculating Power in Electric Devices

A DC winch motor is rated at 20.00 A with a voltage of 115 V. When the motor is running at its maximum power, it can lift an object with a weight of 4900.00 N a distance of 10.00 m, in 30.00 s, at a constant speed. (a) What is the power consumed by the motor? (b) What is the power used in lifting the object? Ignore air resistance. (c) Assuming that the difference in the power consumed by the motor and the power used lifting the object are dissipated as heat by the resistance of the motor, estimate the resistance of the motor?

Strategy

(a) The power consumed by the motor can be found using $P = IV$. (b) The power used in lifting the object at a constant speed can be found using $P = Fv$, where the speed is the distance divided by the time. The upward force supplied by the motor is equal to the weight of the object because the acceleration is constant. (c) The resistance of the motor can be found using $P = I^2 R$.

Solution

a. The power consumed by the motor is equal to $P = IV$ and the current is given as 20.00 A and the voltage is 115.00 V:

$$P = IV = (20.00\ \text{A})115.00\ \text{V} = 2300.00\ \text{W}.$$

b. The power used lifting the object is equal to $P = Fv$ where the force is equal to the weight of the object (1960 N) and the magnitude of the velocity is $v = \frac{10.00\ \text{m}}{30.00\ \text{s}} = 0.33\frac{\text{m}}{\text{s}}$,

$$P = Fv = (4900\ \text{N})0.33\ \text{m/s} = 1633.33\ \text{W}.$$

c. The difference in the power equals $2300.00\ \text{W} - 1633.33\ \text{W} = 666.67\ \text{W}$ and the resistance can be found using $P = I^2 R$:

$$R = \frac{P}{I^2} = \frac{666.67\ \text{W}}{(20.00\ \text{A})^2} = 1.67\ \Omega.$$

Significance

The resistance of the motor is quite small. The resistance of the motor is due to many windings of copper wire. The power dissipated by the motor can be significant since the thermal power dissipated by the motor is proportional to the square of the current $\left(P = I^2 R\right)$.

9.9 Check Your Understanding Electric motors have a reasonably high efficiency. A 100-hp motor can have an efficiency of 90% and a 1-hp motor can have an efficiency of 80%. Why is it important to use high-performance motors?

A fuse (Figure 9.25) is a device that protects a circuit from currents that are too high. A fuse is basically a short piece of wire between two contacts. As we have seen, when a current is running through a conductor, the kinetic energy of the charge carriers is converted into thermal energy in the conductor. The piece of wire in the fuse is under tension and has a low melting point. The wire is designed to heat up and break at the rated current. The fuse is destroyed and must be replaced, but it protects the rest of the circuit. Fuses act quickly, but there is a small time delay while the wire heats up and breaks.

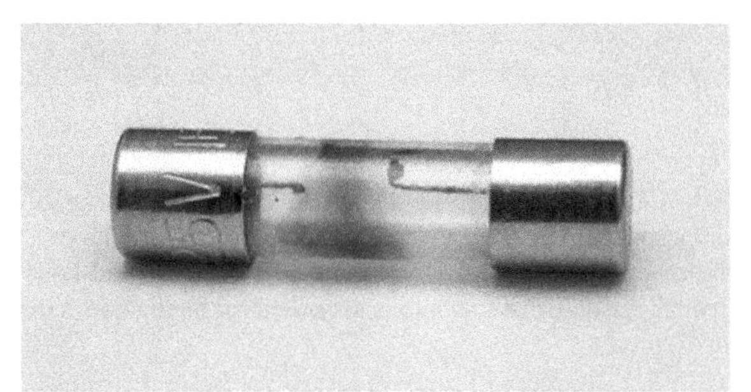

Figure 9.25 A fuse consists of a piece of wire between two contacts. When a current passes through the wire that is greater than the rated current, the wire melts, breaking the connection. Pictured is a "blown" fuse where the wire broke protecting a circuit (credit: modification of work by "Shardayyy"/Flickr).

Circuit breakers are also rated for a maximum current, and open to protect the circuit, but can be reset. Circuit breakers react much faster. The operation of circuit breakers is not within the scope of this chapter and will be discussed in later chapters. Another method of protecting equipment and people is the ground fault circuit interrupter (GFCI), which is common in bathrooms and kitchens. The GFCI outlets respond very quickly to changes in current. These outlets open when there is a change in magnetic field produced by current-carrying conductors, which is also beyond the scope of this chapter and is covered in a later chapter.

The Cost of Electricity

The more electric appliances you use and the longer they are left on, the higher your electric bill. This familiar fact is based on the relationship between energy and power. You pay for the energy used. Since $P = \frac{dE}{dt}$, we see that

$$E = \int P dt$$

is the energy used by a device using power P for a time interval t. If power is delivered at a constant rate, then then the energy can be found by $E = Pt$. For example, the more light bulbs burning, the greater P used; the longer they are on, the greater t is.

The energy unit on electric bills is the kilowatt-hour $(\text{kW} \cdot \text{h})$, consistent with the relationship $E = Pt$. It is easy to estimate the cost of operating electrical appliances if you have some idea of their power consumption rate in watts or kilowatts, the time they are on in hours, and the cost per kilowatt-hour for your electric utility. Kilowatt-hours, like all other specialized energy units such as food calories, can be converted into joules. You can prove to yourself that $1 \, \text{kW} \cdot \text{h} = 3.6 \times 10^6 \, \text{J}$.

The electrical energy (E) used can be reduced either by reducing the time of use or by reducing the power consumption of that appliance or fixture. This not only reduces the cost but also results in a reduced impact on the environment. Improvements to lighting are some of the fastest ways to reduce the electrical energy used in a home or business. About 20% of a home's use of energy goes to lighting, and the number for commercial establishments is closer to 40%. Fluorescent lights are about four times more efficient than incandescent lights—this is true for both the long tubes and the compact fluorescent lights (CFLs). (See Figure 9.23(b).) Thus, a 60-W incandescent bulb can be replaced by a 15-W CFL, which has the same brightness and color. CFLs have a bent tube inside a globe or a spiral-shaped tube, all connected to a standard screw-in base that fits standard incandescent light sockets. (Original problems with color, flicker, shape, and high initial investment for CFLs have been addressed in recent years.)

The heat transfer from these CFLs is less, and they last up to 10 times longer than incandescent bulbs. The significance of an investment in such bulbs is addressed in the next example. New white LED lights (which are clusters of small LED bulbs) are even more efficient (twice that of CFLs) and last five times longer than CFLs.

Example 9.10

Calculating the Cost Effectiveness of LED Bulb

The typical replacement for a 100-W incandescent bulb is a 20-W LED bulb. The 20-W LED bulb can provide the same amount of light output as the 100-W incandescent light bulb. What is the cost savings for using the LED bulb in place of the incandescent bulb for one year, assuming $0.10 per kilowatt-hour is the average energy rate charged by the power company? Assume that the bulb is turned on for three hours a day.

Strategy

(a) Calculate the energy used during the year for each bulb, using $E = Pt$.

(b) Multiply the energy by the cost.

Solution

a. Calculate the power for each bulb.

$$\begin{aligned} E_{\text{Incandescent}} &= Pt = 100\text{ W}\left(\frac{1\text{ kW}}{1000\text{ W}}\right)\left(\frac{3\text{ h}}{\text{day}}\right)(365\text{ days}) = 109.5\text{ kW}\cdot\text{h} \\ E_{\text{LED}} &= Pt = 20\text{ W}\left(\frac{1\text{ kW}}{1000\text{ W}}\right)\left(\frac{3\text{ h}}{\text{day}}\right)(365\text{ days}) = 21.90\text{ kW}\cdot\text{h} \end{aligned}.$$

b. Calculate the cost for each.

$$\begin{aligned} \text{cost}_{\text{Incandescent}} &= 109.5\text{ kW-h}\left(\frac{\$0.10}{\text{kW}\cdot\text{h}}\right) = \$10.95 \\ \text{cost}_{\text{LED}} &= 21.90\text{ kW-h}\left(\frac{\$0.10}{\text{kW}\cdot\text{h}}\right) = \$2.19 \end{aligned}.$$

Significance

A LED bulb uses 80% less energy than the incandescent bulb, saving $8.76 over the incandescent bulb for one year. The LED bulb can cost $20.00 and the 100-W incandescent bulb can cost $0.75, which should be calculated into the computation. A typical lifespan of an incandescent bulb is 1200 hours and is 50,000 hours for the LED bulb. The incandescent bulb would last 1.08 years at 3 hours a day and the LED bulb would last 45.66 years. The initial cost of the LED bulb is high, but the cost to the home owner will be $0.69 for the incandescent bulbs versus $0.44 for the LED bulbs per year. (Note that the LED bulbs are coming down in price.) The cost savings per year is approximately $8.50, and that is just for one bulb.

9.10 Check Your Understanding Is the efficiency of the various light bulbs the only consideration when comparing the various light bulbs?

Changing light bulbs from incandescent bulbs to CFL or LED bulbs is a simple way to reduce energy consumption in homes and commercial sites. CFL bulbs operate with a much different mechanism than do incandescent lights. The mechanism is complex and beyond the scope of this chapter, but here is a very general description of the mechanism. CFL bulbs contain argon and mercury vapor housed within a spiral-shaped tube. The CFL bulbs use a "ballast" that increases the voltage used by the CFL bulb. The ballast produce an electrical current, which passes through the gas mixture and excites the gas molecules. The excited gas molecules produce ultraviolet (UV) light, which in turn stimulates the fluorescent coating on the inside of the tube. This coating fluoresces in the visible spectrum, emitting visible light. Traditional fluorescent tubes and CFL bulbs had a short time delay of up to a few seconds while the mixture was being "warmed up" and the molecules reached an excited state. It should be noted that these bulbs do contain mercury, which is poisonous, but if the bulb is broken, the mercury is never released. Even if the bulb is broken, the mercury tends to remain in the fluorescent coating. The amount is also quite small and the advantage of the energy saving may outweigh the disadvantage of using mercury.

The CFL light bulbs are being replaced with LED light bulbs, where LED stands for "light-emitting diode." The diode was briefly discussed as a nonohmic device, made of semiconducting material, which essentially permits current flow in one direction. LEDs are a special type of diode made of semiconducting materials infused with impurities in combinations and concentrations that enable the extra energy from the movement of the electrons during electrical excitation to be converted into visible light. Semiconducting devices will be explained in greater detail in Condensed Matter Physics (http://cnx.org/content/m58591/latest/) .

Commercial LEDs are quickly becoming the standard for commercial and residential lighting, replacing incandescent and CFL bulbs. They are designed for the visible spectrum and are constructed from gallium doped with arsenic and phosphorous atoms. The color emitted from an LED depends on the materials used in the semiconductor and the current. In the early years of LED development, small LEDs found on circuit boards were red, green, and yellow, but LED light bulbs can now be programmed to produce millions of colors of light as well as many different hues of white light.

Comparison of Incandescent, CFL, and LED Light Bulbs

The energy savings can be significant when replacing an incandescent light bulb or a CFL light bulb with an LED light. Light bulbs are rated by the amount of power that the bulb consumes, and the amount of light output is measured in lumens. The lumen (lm) is the SI -derived unit of luminous flux and is a measure of the total quantity of visible light emitted by a source. A 60-W incandescent light bulb can be replaced with a 13- to 15-W CFL bulb or a 6- to 8-W LED bulb, all three of which have a light output of approximately 800 lm. A table of light output for some commonly used light bulbs appears in Table 9.2.

The life spans of the three types of bulbs are significantly different. An LED bulb has a life span of 50,000 hours, whereas the CFL has a lifespan of 8000 hours and the incandescent lasts a mere 1200 hours. The LED bulb is the most durable, easily withstanding rough treatment such as jarring and bumping. The incandescent light bulb has little tolerance to the same treatment since the filament and glass can easily break. The CFL bulb is also less durable than the LED bulb because of its glass construction. The amount of heat emitted is 3.4 btu/h for the 8-W LED bulb, 85 btu/h for the 60-W incandescent bulb, and 30 btu/h for the CFL bulb. As mentioned earlier, a major drawback of the CFL bulb is that it contains mercury, a neurotoxin, and must be disposed of as hazardous waste. From these data, it is easy to understand why the LED light bulb is quickly becoming the standard in lighting.

Light Output (lumens)	LED Light Bulbs (watts)	Incandescent Light Bulbs (watts)	CFL Light Bulbs (watts)
450	4–5	40	9–13
800	6–8	60	13–15
1100	9–13	75	18–25
1600	16–20	100	23–30
2600	25–28	150	30–55

Table 9.2 Light Output of LED, Incandescent, and CFL Light Bulbs

Summary of Relationships

In this chapter, we have discussed relationships between voltages, current, resistance, and power. Figure 9.26 shows a summary of the relationships between these measurable quantities for ohmic devices. (Recall that ohmic devices follow Ohm's law $V = IR$.) For example, if you need to calculate the power, use the pink section, which shows that $P = VI$, $P = \frac{V^2}{R}$, and $P = I^2 R$.

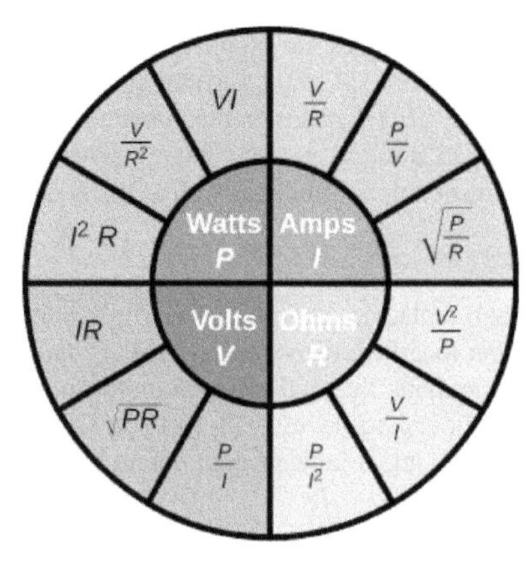

P = Power I = Current
V = Voltage R = Resistance

Figure 9.26 This circle shows a summary of the equations for the relationships between power, current, voltage, and resistance.

Which equation you use depends on what values you are given, or you measure. For example if you are given the current and the resistance, use $P = I^2 R$. Although all the possible combinations may seem overwhelming, don't forget that they all are combinations of just two equations, Ohm's law $(V = IR)$ and power $(P = IV)$.

9.6 | Superconductors

Learning Objectives

By the end of this section, you will be able to:

- Describe the phenomenon of superconductivity
- List applications of superconductivity

Touch the power supply of your laptop computer or some other device. It probably feels slightly warm. That heat is an unwanted byproduct of the process of converting household electric power into a current that can be used by your device. Although electric power is reasonably efficient, other losses are associated with it. As discussed in the section on power and energy, transmission of electric power produces $I^2 R$ line losses. These line losses exist whether the power is generated from conventional power plants (using coal, oil, or gas), nuclear plants, solar plants, hydroelectric plants, or wind farms. These losses can be reduced, but not eliminated, by transmitting using a higher voltage. It would be wonderful if these line losses could be eliminated, but that would require transmission lines that have zero resistance. In a world that has a global interest in not wasting energy, the reduction or elimination of this unwanted thermal energy would be a significant achievement. Is this possible?

The Resistance of Mercury

In 1911, Heike Kamerlingh Onnes of Leiden University, a Dutch physicist, was looking at the temperature dependence of the resistance of the element mercury. He cooled the sample of mercury and noticed the familiar behavior of a linear dependence of resistance on temperature; as the temperature decreased, the resistance decreased. Kamerlingh Onnes continued to cool the sample of mercury, using liquid helium. As the temperature approached $4.2\,\text{K}(-269.2\,°\text{C})$, the resistance abruptly went to zero (Figure 9.27). This temperature is known as the **critical temperature** T_c for mercury. The sample of mercury entered into a phase where the resistance was absolutely zero. This phenomenon is known as **superconductivity**. (*Note:* If you connect the leads of a three-digit ohmmeter across a conductor, the reading commonly shows up as $0.00\,\Omega$. The resistance of the conductor is not actually zero, it is less than $0.01\,\Omega$.) There are various methods to measure very small resistances, such as the four-point method, but an ohmmeter is not an acceptable method to use for testing resistance in superconductivity.

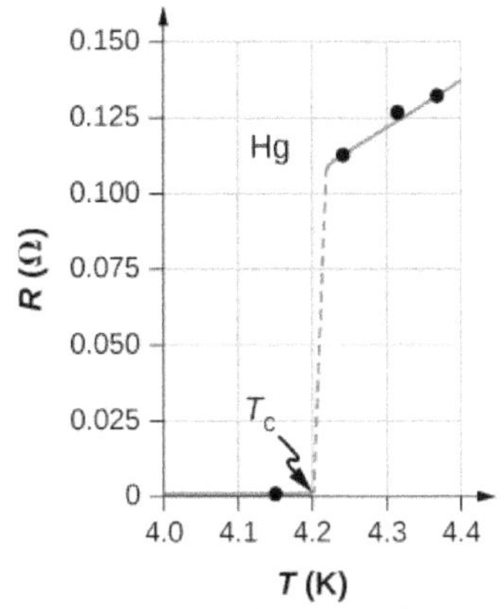

Figure 9.27 The resistance of a sample of mercury is zero at very low temperatures—it is a superconductor up to the temperature of about 4.2 K. Above that critical temperature, its resistance makes a sudden jump and then increases nearly linearly with temperature.

Other Superconducting Materials

As research continued, several other materials were found to enter a superconducting phase, when the temperature reached near absolute zero. In 1941, an alloy of niobium-nitride was found that could become superconducting at $T_c = 16\,\text{K}(-257\,°\text{C})$ and in 1953, vanadium-silicon was found to become superconductive at $T_c = 17.5\,\text{K}(-255.7\,°\text{C})$. The temperatures for the transition into superconductivity were slowly creeping higher. Strangely, many materials that make good conductors, such as copper, silver, and gold, do not exhibit superconductivity. Imagine the energy savings if transmission lines for electric power-generating stations could be made to be superconducting at temperatures near room temperature! A resistance of zero ohms means no I^2R losses and a great boost to reducing energy consumption. The problem is that $T_c = 17.5\text{ K}$ is still very cold and in the range of liquid helium temperatures. At this temperature, it is not cost effective to transmit electrical energy because of the cooling requirements.

A large jump was seen in 1986, when a team of researchers, headed by Dr. Ching Wu Chu of Houston University, fabricated a brittle, ceramic compound with a transition temperature of $T_c = 92\,\text{K}(-181\,°\text{C})$. The ceramic material, composed of yttrium barium copper oxide (YBCO), was an insulator at room temperature. Although this temperature still seems quite cold, it is near the boiling point of liquid nitrogen, a liquid commonly used in refrigeration. You may have noticed refrigerated trucks traveling down the highway labeled as "Liquid Nitrogen Cooled."

YBCO ceramic is a material that could be useful for transmitting electrical energy because the cost saving of reducing the I^2R losses are larger than the cost of cooling the superconducting cable, making it financially feasible. There were and are many engineering problems to overcome. For example, unlike traditional electrical cables, which are flexible and have a decent tensile strength, ceramics are brittle and would break rather than stretch under pressure. Processes that are rather simple with traditional cables, such as making connections, become difficult when working with ceramics. The problems are difficult and complex, and material scientists and engineers are coming up with innovative solutions.

An interesting consequence of the resistance going to zero is that once a current is established in a superconductor, it persists without an applied voltage source. Current loops in a superconductor have been set up and the current loops have been observed to persist for years without decaying.

Zero resistance is not the only interesting phenomenon that occurs as the materials reach their transition temperatures. A second effect is the exclusion of magnetic fields. This is known as the **Meissner effect** (Figure 9.28). A light, permanent magnet placed over a superconducting sample will levitate in a stable position above the superconductor. High-speed trains have been developed that levitate on strong superconducting magnets, eliminating the friction normally experienced between the train and the tracks. In Japan, the Yamanashi Maglev test line opened on April 3, 1997. In April 2015, the MLX01 test vehicle attained a speed of 374 mph (603 km/h).

Figure 9.28 A small, strong magnet levitates over a superconductor cooled to liquid nitrogen temperature. The magnet levitates because the superconductor excludes magnetic fields.

Table 9.3 shows a select list of elements, compounds, and high-temperature superconductors, along with the critical temperatures for which they become superconducting. Each section is sorted from the highest critical temperature to the lowest. Also listed is the critical magnetic field for some of the materials. This is the strength of the magnetic field that destroys superconductivity. Finally, the type of the superconductor is listed.

There are two types of superconductors. There are 30 pure metals that exhibit zero resistivity below their critical temperature and exhibit the Meissner effect, the property of excluding magnetic fields from the interior of the superconductor while the superconductor is at a temperature below the critical temperature. These metals are called Type I superconductors. The superconductivity exists only below their critical temperatures and below a critical magnetic field strength. Type I superconductors are well described by the BCS theory (described next). Type I superconductors have limited practical applications because the strength of the critical magnetic field needed to destroy the superconductivity is quite low.

Type II superconductors are found to have much higher critical magnetic fields and therefore can carry much higher current densities while remaining in the superconducting state. A collection of various ceramics containing barium-copper-oxide have much higher critical temperatures for the transition into a superconducting state. Superconducting materials that belong to this subcategory of the Type II superconductors are often categorized as high-temperature superconductors.

Introduction to BCS Theory

Type I superconductors, along with some Type II superconductors can be modeled using the BCS theory, proposed by John Bardeen, Leon Cooper, and Robert Schrieffer. Although the theory is beyond the scope of this chapter, a short summary of the theory is provided here. (More detail is provided in Condensed Matter Physics (http://cnx.org/content/m58591/latest/) .) The theory considers pairs of electrons and how they are coupled together through lattice-vibration interactions. Through the interactions with the crystalline lattice, electrons near the Fermi energy level feel a small attractive force and form pairs (Cooper pairs), and the coupling is known as a phonon interaction. Single electrons are fermions, which are particles that obey the Pauli exclusion principle. The Pauli exclusion principle in quantum mechanics states that two identical fermions (particles with half-integer spin) cannot occupy the same quantum state simultaneously. Each electron has four quantum numbers (n, l, m_l, m_s) . The principal quantum number (n) describes the energy of the electron, the orbital angular momentum quantum number (l) indicates the most probable distance from the nucleus, the magnetic quantum number (m_l) describes the energy levels in the subshell, and the electron spin quantum number (m_s) describes the orientation of the spin of the electron, either up or down. As the material enters a superconducting state, pairs of electrons act more like bosons, which can condense into the same energy level and need not obey the Pauli exclusion principle. The electron pairs have a slightly lower energy and leave an energy gap above them on the order of 0.001 eV. This energy gap inhibits collision interactions that lead to ordinary resistivity. When the material is below the critical temperature, the thermal energy is less than the band gap and the material exhibits zero resistivity.

Material	Symbol or Formula	Critical Temperature T_c (K)	Critical Magnetic Field H_c (T)	Type
Elements				
Lead	Pb	7.19	0.08	I
Lanthanum	La	(α) 4.90 – (β) 6.30		I
Tantalum	Ta	4.48	0.09	I
Mercury	Hg	(α) 4.15 – (β) 3.95	0.04	I
Tin	Sn	3.72	0.03	I
Indium	In	3.40	0.03	I
Thallium	Tl	2.39	0.03	I
Rhenium	Re	2.40	0.03	I
Thorium	Th	1.37	0.013	I
Protactinium	Pa	1.40		I
Aluminum	Al	1.20	0.01	I
Gallium	Ga	1.10	0.005	I
Zinc	Zn	0.86	0.014	I
Titanium	Ti	0.39	0.01	I
Uranium	U	(α) 0.68 – (β) 1.80		I
Cadmium	Cd	11.4	4.00	I
Compounds				
Niobium-germanium	Nb_3Ge	23.20	37.00	II
Niobium-tin	Nb_3Sn	18.30	30.00	II
Niobium-nitrite	NbN	16.00		II
Niobium-titanium	NbTi	10.00	15.00	II
High-Temperature Oxides				
	$HgBa_2CaCu_2O_8$	134.00		II
	$Tl_2Ba_2Ca_2Cu_3O_{10}$	125.00		II
	$YBa_2Cu_3O_7$	92.00	120.00	II

Table 9.3 Superconductor Critical Temperatures

Applications of Superconductors

Superconductors can be used to make superconducting magnets. These magnets are 10 times stronger than the strongest electromagnets. These magnets are currently in use in magnetic resonance imaging (MRI), which produces high-quality images of the body interior without dangerous radiation.

Another interesting application of superconductivity is the **SQUID** (superconducting quantum interference device). A SQUID is a very sensitive magnetometer used to measure extremely subtle magnetic fields. The operation of the SQUID is based on superconducting loops containing Josephson junctions. A **Josephson junction** is the result of a theoretical prediction made by B. D. Josephson in an article published in 1962. In the article, Josephson described how a supercurrent can flow between two pieces of superconductor separated by a thin layer of insulator. This phenomenon is now called the Josephson effect. The SQUID consists of a superconducting current loop containing two Josephson junctions, as shown in

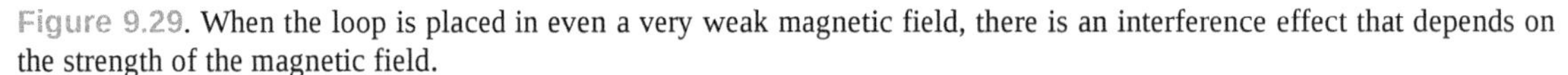

Figure 9.29. When the loop is placed in even a very weak magnetic field, there is an interference effect that depends on the strength of the magnetic field.

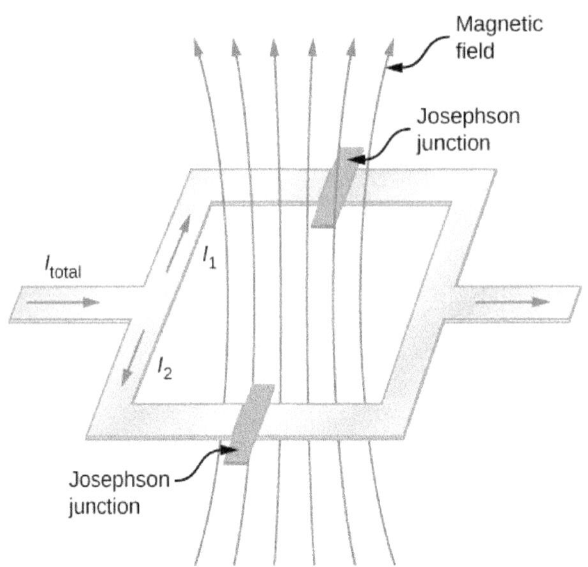

Figure 9.29 The SQUID (superconducting quantum interference device) uses a superconducting current loop and two Josephson junctions to detect magnetic fields as low as 10^{-14} T (Earth's magnet field is on the order of 0.3×10^{-5} T).

Superconductivity is a fascinating and useful phenomenon. At critical temperatures near the boiling point of liquid nitrogen, superconductivity has special applications in MRIs, particle accelerators, and high-speed trains. Will we reach a state where we can have materials enter the superconducting phase at near room temperatures? It seems a long way off, but if scientists in 1911 were asked if we would reach liquid-nitrogen temperatures with a ceramic, they might have thought it implausible.

CHAPTER 9 REVIEW

KEY TERMS

ampere (amp) SI unit for current; $1 \text{ A} = 1 \text{ C/s}$

circuit complete path that an electrical current travels along

conventional current current that flows through a circuit from the positive terminal of a battery through the circuit to the negative terminal of the battery

critical temperature temperature at which a material reaches superconductivity

current density flow of charge through a cross-sectional area divided by the area

diode nonohmic circuit device that allows current flow in only one direction

drift velocity velocity of a charge as it moves nearly randomly through a conductor, experiencing multiple collisions, averaged over a length of a conductor, whose magnitude is the length of conductor traveled divided by the time it takes for the charges to travel the length

electrical conductivity measure of a material's ability to conduct or transmit electricity

electrical current rate at which charge flows, $I = \frac{dQ}{dt}$

electrical power time rate of change of energy in an electric circuit

Josephson junction junction of two pieces of superconducting material separated by a thin layer of insulating material, which can carry a supercurrent

Meissner effect phenomenon that occurs in a superconducting material where all magnetic fields are expelled

nonohmic type of a material for which Ohm's law is not valid

ohm (Ω) unit of electrical resistance, $1 \, \Omega = 1 \text{ V/A}$

ohmic type of a material for which Ohm's law is valid, that is, the voltage drop across the device is equal to the current times the resistance

Ohm's law empirical relation stating that the current *I* is proportional to the potential difference *V*; it is often written as $V = IR$, where *R* is the resistance

resistance electric property that impedes current; for ohmic materials, it is the ratio of voltage to current, $R = V/I$

resistivity intrinsic property of a material, independent of its shape or size, directly proportional to the resistance, denoted by ρ

schematic graphical representation of a circuit using standardized symbols for components and solid lines for the wire connecting the components

SQUID (Superconducting Quantum Interference Device) device that is a very sensitive magnetometer, used to measure extremely subtle magnetic fields

superconductivity phenomenon that occurs in some materials where the resistance goes to exactly zero and all magnetic fields are expelled, which occurs dramatically at some low critical temperature (T_C)

KEY EQUATIONS

Average electrical current	$I_{\text{ave}} = \frac{\Delta Q}{\Delta t}$
Definition of an ampere	$1 \text{ A} = 1 \text{ C/s}$
Electrical current	$I = \frac{dQ}{dt}$

Drift velocity	$v_d = \frac{nqA}{I}$
Current density	$I = \iint_{\text{area}} \vec{\mathbf{J}} \cdot d\vec{\mathbf{A}}$
Resistivity	$\rho = \frac{E}{J}$
Common expression of Ohm's law	$V = IR$
Resistivity as a function of temperature	$\rho = \rho_0[1 + \alpha(T - T_0)]$
Definition of resistance	$R \equiv \frac{V}{I}$
Resistance of a cylinder of material	$R = \rho\frac{L}{A}$
Temperature dependence of resistance	$R = R_0(1 + \alpha\Delta T)$
Electric power	$P = IV$
Power dissipated by a resistor	$P = I^2 R = \frac{V^2}{R}$

SUMMARY

9.1 Electrical Current

- The average electrical current I_{ave} is the rate at which charge flows, given by $I_{\text{ave}} = \frac{\Delta Q}{\Delta t}$, where ΔQ is the amount of charge passing through an area in time Δt.
- The instantaneous electrical current, or simply the current *I*, is the rate at which charge flows. Taking the limit as the change in time approaches zero, we have $I = \frac{dQ}{dt}$, where $\frac{dQ}{dt}$ is the time derivative of the charge.
- The direction of conventional current is taken as the direction in which positive charge moves. In a simple direct-current (DC) circuit, this will be from the positive terminal of the battery to the negative terminal.
- The SI unit for current is the ampere, or simply the amp (A), where $1\text{ A} = 1\text{ C/s}$.
- Current consists of the flow of free charges, such as electrons, protons, and ions.

9.2 Model of Conduction in Metals

- The current through a conductor depends mainly on the motion of free electrons.
- When an electrical field is applied to a conductor, the free electrons in a conductor do not move through a conductor at a constant speed and direction; instead, the motion is almost random due to collisions with atoms and other free electrons.
- Even though the electrons move in a nearly random fashion, when an electrical field is applied to the conductor, the overall velocity of the electrons can be defined in terms of a drift velocity.
- The current density is a vector quantity defined as the current through an infinitesimal area divided by the area.
- The current can be found from the current density, $I = \iint_{\text{area}} \vec{\mathbf{J}} \cdot d\vec{\mathbf{A}}$.
- An incandescent light bulb is a filament of wire enclosed in a glass bulb that is partially evacuated. Current runs through the filament, where the electrical energy is converted to light and heat.

9.3 Resistivity and Resistance

- Resistance has units of ohms (Ω), related to volts and amperes by $1\ \Omega = 1\ \text{V/A}$.
- The resistance *R* of a cylinder of length *L* and cross-sectional area *A* is $R = \frac{\rho L}{A}$, where ρ is the resistivity of the material.
- Values of ρ in Table 9.1 show that materials fall into three groups—conductors, semiconductors, and insulators.
- Temperature affects resistivity; for relatively small temperature changes ΔT, resistivity is $\rho = \rho_0(1 + \alpha\Delta T)$, where ρ_0 is the original resistivity and α is the temperature coefficient of resistivity.
- The resistance *R* of an object also varies with temperature: $R = R_0(1 + \alpha\Delta T)$, where R_0 is the original resistance, and *R* is the resistance after the temperature change.

9.4 Ohm's Law

- Ohm's law is an empirical relationship for current, voltage, and resistance for some common types of circuit elements, including resistors. It does not apply to other devices, such as diodes.
- One statement of Ohm's law gives the relationship among current *I*, voltage *V*, and resistance *R* in a simple circuit as $V = IR$.
- Another statement of Ohm's law, on a microscopic level, is $J = \sigma E$.

9.5 Electrical Energy and Power

- Electric power is the rate at which electric energy is supplied to a circuit or consumed by a load.
- Power dissipated by a resistor depends on the square of the current through the resistor and is equal to $P = I^2 R = \frac{V^2}{R}$.
- The SI unit for electric power is the watt and the SI unit for electric energy is the joule. Another common unit for electric energy, used by power companies, is the kilowatt-hour (kW · h).
- The total energy used over a time interval can be found by $E = \int P dt$.

9.6 Superconductors

- Superconductivity is a phenomenon that occurs in some materials when cooled to very low critical temperatures, resulting in a resistance of exactly zero and the expulsion of all magnetic fields.
- Materials that are normally good conductors (such as copper, gold, and silver) do not experience superconductivity.
- Superconductivity was first observed in mercury by Heike Kamerlingh Onnes in 1911. In 1986, Dr. Ching Wu Chu of Houston University fabricated a brittle, ceramic compound with a critical temperature close to the temperature of liquid nitrogen.
- Superconductivity can be used in the manufacture of superconducting magnets for use in MRIs and high-speed, levitated trains.

CONCEPTUAL QUESTIONS

9.1 Electrical Current

1. Can a wire carry a current and still be neutral—that is, have a total charge of zero? Explain.

2. Car batteries are rated in ampere-hours $(\text{A} \cdot \text{h})$. To what physical quantity do ampere-hours correspond (voltage, current, charge, energy, power,...)?

3. When working with high-power electric circuits, it is advised that whenever possible, you work "one-handed" or "keep one hand in your pocket." Why is this a sensible suggestion?

9.2 Model of Conduction in Metals

4. Incandescent light bulbs are being replaced with more efficient LED and CFL light bulbs. Is there any obvious evidence that incandescent light bulbs might not be that energy efficient? Is energy converted into anything but visible light?

5. It was stated that the motion of an electron appears nearly random when an electrical field is applied to the conductor. What makes the motion nearly random and differentiates it from the random motion of molecules in a gas?

6. Electric circuits are sometimes explained using a conceptual model of water flowing through a pipe. In this conceptual model, the voltage source is represented as a pump that pumps water through pipes and the pipes connect components in the circuit. Is a conceptual model of water flowing through a pipe an adequate representation of the circuit? How are electrons and wires similar to water molecules and pipes? How are they different?

7. An incandescent light bulb is partially evacuated. Why do you suppose that is?

9.3 Resistivity and Resistance

8. The IR drop across a resistor means that there is a change in potential or voltage across the resistor. Is there any change in current as it passes through a resistor? Explain.

9. Do impurities in semiconducting materials listed in Table 9.1 supply free charges? (*Hint*: Examine the range of resistivity for each and determine whether the pure semiconductor has the higher or lower conductivity.)

10. Does the resistance of an object depend on the path current takes through it? Consider, for example, a rectangular bar—is its resistance the same along its length as across its width?

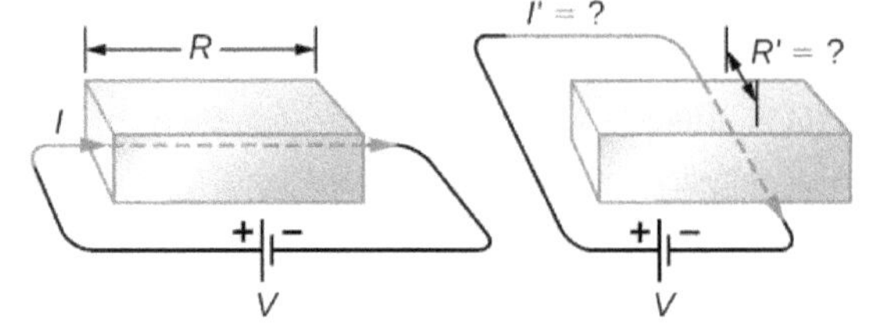

11. If aluminum and copper wires of the same length have the same resistance, which has the larger diameter? Why?

9.4 Ohm's Law

12. In Determining Field from Potential, resistance was defined as $R \equiv \frac{V}{I}$. In this section, we presented Ohm's law, which is commonly expressed as $V = IR$. The equations look exactly alike. What is the difference between Ohm's law and the definition of resistance?

13. Shown below are the results of an experiment where four devices were connected across a variable voltage source. The voltage is increased and the current is measured. Which device, if any, is an ohmic device?

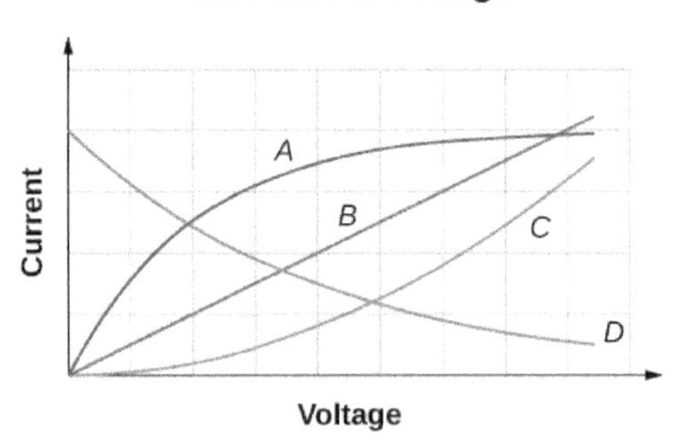

14. The current I is measured through a sample of an ohmic material as a voltage V is applied. (a) What is the current when the voltage is doubled to $2V$ (assume the change in temperature of the material is negligible)? (b) What is the voltage applied is the current measured is $0.2I$ (assume the change in temperature of the material is negligible)? What will happen to the current if the material if the voltage remains constant, but the temperature of the material increases significantly?

9.5 Electrical Energy and Power

15. Common household appliances are rated at 110 V, but power companies deliver voltage in the kilovolt range and then step the voltage down using transformers to 110 V to be used in homes. You will learn in later chapters that transformers consist of many turns of wire, which warm up as current flows through them, wasting some of the energy that is given off as heat. This sounds inefficient. Why do the power companies transport electric power using this method?

16. Your electric bill gives your consumption in units of kilowatt-hour (kW · h). Does this unit represent the amount of charge, current, voltage, power, or energy you buy?

17. Resistors are commonly rated at $\frac{1}{8}$W, $\frac{1}{4}$W, $\frac{1}{2}$W, 1 W and 2 W for use in electrical circuits. If a current of $I = 2.00\text{ A}$ is accidentally passed through a $R = 1.00\,\Omega$ resistor rated at 1 W, what would be the most probable outcome? Is there anything that can be done to prevent such an accident?

18. An immersion heater is a small appliance used to heat a cup of water for tea by passing current through a resistor. If the voltage applied to the appliance is doubled, will the time required to heat the water change? By how much? Is this a good idea?

9.6 Superconductors

19. What requirement for superconductivity makes current superconducting devices expensive to operate?

20. Name two applications for superconductivity listed in this section and explain how superconductivity is used in the application. Can you think of a use for superconductivity that is not listed?

PROBLEMS

9.1 Electrical Current

21. A Van de Graaff generator is one of the original particle accelerators and can be used to accelerate charged particles like protons or electrons. You may have seen it used to make human hair stand on end or produce large sparks. One application of the Van de Graaff generator is to create X-rays by bombarding a hard metal target with the beam. Consider a beam of protons at 1.00 keV and a current of 5.00 mA produced by the generator. (a) What is the speed of the protons? (b) How many protons are produced each second?

22. A cathode ray tube (CRT) is a device that produces a focused beam of electrons in a vacuum. The electrons strike a phosphor-coated glass screen at the end of the tube, which produces a bright spot of light. The position of the bright spot of light on the screen can be adjusted by deflecting the electrons with electrical fields, magnetic fields, or both. Although the CRT tube was once commonly found in televisions, computer displays, and oscilloscopes, newer appliances use a liquid crystal display (LCD) or plasma screen. You still may come across a CRT in your study of science. Consider a CRT with an electron beam average current of $25.00\mu\text{ A}$. How many electrons strike the screen every minute?

23. How many electrons flow through a point in a wire in 3.00 s if there is a constant current of $I = 4.00\text{ A}$?

24. A conductor carries a current that is decreasing exponentially with time. The current is modeled as $I = I_0 e^{-t/\tau}$, where $I_0 = 3.00\text{ A}$ is the current at time $t = 0.00\text{ s}$ and $\tau = 0.50\text{ s}$ is the time constant. How much charge flows through the conductor between $t = 0.00\text{ s}$ and $t = 3\tau$?

25. The quantity of charge through a conductor is modeled as $Q = 4.00\frac{\text{C}}{\text{s}^4}t^4 - 1.00\frac{\text{C}}{\text{s}}t + 6.00\text{ mC}$.

What is the current at time $t = 3.00\text{ s}$?

26. The current through a conductor is modeled as $I(t) = I_m \sin(2\pi[60\text{ Hz}]t)$. Write an equation for the charge as a function of time.

27. The charge on a capacitor in a circuit is modeled as $Q(t) = Q_{\text{max}} \cos(\omega t + \phi)$. What is the current through the circuit as a function of time?

9.2 Model of Conduction in Metals

28. An aluminum wire 1.628 mm in diameter (14-gauge) carries a current of 3.00 amps. (a) What is the absolute value of the charge density in the wire? (b) What is the drift velocity of the electrons? (c) What would be the drift velocity if the same gauge copper were used instead of aluminum? The density of copper is 8.96 g/cm^3 and the density of aluminum is 2.70 g/cm^3. The molar mass of aluminum is 26.98 g/mol and the molar mass of copper is 63.5 g/mol. Assume each atom of metal contributes one free electron.

29. The current of an electron beam has a measured current of $I = 50.00\,\mu\text{A}$ with a radius of 1.00 mm^2. What is the magnitude of the current density of the beam?

30. A high-energy proton accelerator produces a proton beam with a radius of $r = 0.90\text{ mm}$. The beam current is $I = 9.00\,\mu\text{A}$ and is constant. The charge density of the beam is $n = 6.00 \times 10^{11}$ protons per cubic meter. (a) What is the current density of the beam? (b) What is the drift velocity of the beam? (c) How much time does it take for 1.00×10^{10} protons to be emitted by the accelerator?

31. Consider a wire of a circular cross-section with a radius of $R = 3.00\text{ mm}$. The magnitude of the current density is modeled as $J = cr^2 = 5.00 \times 10^6 \frac{\text{A}}{\text{m}^4} r^2$. What is the current through the inner section of the wire from the center to $r = 0.5R$?

32. The current of an electron beam has a measured current of $I = 50.00\ \mu\text{A}$ with a radius of 1.00 mm^2. What is the magnitude of the current density of the beam?

33. The current supplied to an air conditioner unit is 4.00 amps. The air conditioner is wired using a 10-gauge (diameter 2.588 mm) wire. The charge density is $n = 8.48 \times 10^{28} \frac{\text{electrons}}{\text{m}^3}$. Find the magnitude of (a) current density and (b) the drift velocity.

9.3 Resistivity and Resistance

34. What current flows through the bulb of a 3.00-V flashlight when its hot resistance is $3.60\ \Omega$?

35. Calculate the effective resistance of a pocket calculator that has a 1.35-V battery and through which 0.200 mA flows.

36. How many volts are supplied to operate an indicator light on a DVD player that has a resistance of $140\ \Omega$, given that 25.0 mA passes through it?

37. What is the resistance of a 20.0-m-long piece of 12-gauge copper wire having a 2.053-mm diameter?

38. The diameter of 0-gauge copper wire is 8.252 mm. Find the resistance of a 1.00-km length of such wire used for power transmission.

39. If the 0.100-mm-diameter tungsten filament in a light bulb is to have a resistance of $0.200\ \Omega$ at $20.0\ °\text{C}$, how long should it be?

40. A lead rod has a length of 30.00 cm and a resistance of $5.00\ \mu\Omega$. What is the radius of the rod?

41. Find the ratio of the diameter of aluminum to copper wire, if they have the same resistance per unit length (as they might in household wiring).

42. What current flows through a 2.54-cm-diameter rod of pure silicon that is 20.0 cm long, when 1.00×10^3 V is applied to it? (Such a rod may be used to make nuclear-particle detectors, for example.)

43. (a) To what temperature must you raise a copper wire, originally at $20.0\ °\text{C}$, to double its resistance, neglecting any changes in dimensions? (b) Does this happen in household wiring under ordinary circumstances?

44. A resistor made of nichrome wire is used in an application where its resistance cannot change more than 1.00% from its value at $20.0\ °\text{C}$. Over what temperature range can it be used?

45. Of what material is a resistor made if its resistance is 40.0% greater at $100.0\ °\text{C}$ than at $20.0\ °\text{C}$?

46. An electronic device designed to operate at any temperature in the range from $-10.0\ °\text{C}$ to $55.0\ °\text{C}$ contains pure carbon resistors. By what factor does their resistance increase over this range?

47. (a) Of what material is a wire made, if it is 25.0 m long with a diameter of 0.100 mm and has a resistance of $77.7\ \Omega$ at $20.0\ °\text{C}$? (b) What is its resistance at $150.0\ °\text{C}$?

48. Assuming a constant temperature coefficient of resistivity, what is the maximum percent decrease in the resistance of a constantan wire starting at $20.0\ °\text{C}$?

49. A copper wire has a resistance of $0.500\ \Omega$ at $20.0\ °\text{C}$, and an iron wire has a resistance of $0.525\ \Omega$ at the same temperature. At what temperature are their resistances equal?

9.4 Ohm's Law

50. A $2.2\text{-k}\Omega$ resistor is connected across a D cell battery (1.5 V). What is the current through the resistor?

51. A resistor rated at $250\text{ k}\Omega$ is connected across two D cell batteries (each 1.50 V) in series, with a total voltage of 3.00 V. The manufacturer advertises that their resistors are within 5% of the rated value. What are the possible minimum current and maximum current through the resistor?

52. A resistor is connected in series with a power supply of 20.00 V. The current measure is 0.50 A. What is the resistance of the resistor?

53. A resistor is placed in a circuit with an adjustable voltage source. The voltage across and the current through the resistor and the measurements are shown below. Estimate the resistance of the resistor.

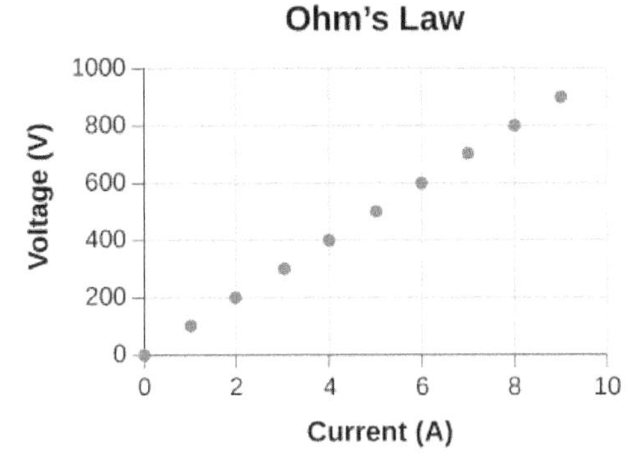

54. The following table show the measurements of a current through and the voltage across a sample of material. Plot the data, and assuming the object is an ohmic device, estimate the resistance.

I(A)	*V*(V)
0	3
2	23
4	39
6	58
8	77
10	100
12	119
14	142
16	162

9.5 Electrical Energy and Power

55. A 20.00-V battery is used to supply current to a 10-kΩ resistor. Assume the voltage drop across any wires used for connections is negligible. (a) What is the current through the resistor? (b) What is the power dissipated by the resistor? (c) What is the power input from the battery, assuming all the electrical power is dissipated by the resistor? (d) What happens to the energy dissipated by the resistor?

56. What is the maximum voltage that can be applied to a 10-kΩ resistor rated at $\frac{1}{4}$W ?

57. A heater is being designed that uses a coil of 14-gauge nichrome wire to generate 300 W using a voltage of $V = 110\,\text{V}$. How long should the engineer make the wire?

58. An alternative to CFL bulbs and incandescent bulbs are light-emitting diode (LED) bulbs. A 100-W incandescent bulb can be replaced by a 16-W LED bulb. Both produce 1600 lumens of light. Assuming the cost of electricity is $0.10 per kilowatt-hour, how much does it cost to run the bulb for one year if it runs for four hours a day?

59. The power dissipated by a resistor with a resistance of $R = 100\,\Omega$ is $P = 2.0\,\text{W}$. What are the current through and the voltage drop across the resistor?

60. Running late to catch a plane, a driver accidentally leaves the headlights on after parking the car in the airport parking lot. During takeoff, the driver realizes the mistake. Having just replaced the battery, the driver knows that the battery is a 12-V automobile battery, rated at 100 $\text{A} \cdot \text{h}$. The driver, knowing there is nothing that can be done, estimates how long the lights will shine, assuming there are two 12-V headlights, each rated at 40 W. What did the driver conclude?

61. A physics student has a single-occupancy dorm room. The student has a small refrigerator that runs with a current of 3.00 A and a voltage of 110 V, a lamp that contains a 100-W bulb, an overhead light with a 60-W bulb, and various other small devices adding up to 3.00 W. (a) Assuming the power plant that supplies 110 V electricity to the dorm is 10 km away and the two aluminum transmission cables use 0-gauge wire with a diameter of 8.252 mm, estimate the percentage of the total power supplied by the power company that is lost in the transmission. (b) What would be the result is the power company delivered the electric power at 110 kV?

62. A 0.50-W, 220-Ω resistor carries the maximum current possible without damaging the resistor. If the current were reduced to half the value, what would be the power consumed?

9.6 Superconductors

63. Consider a power plant is located 60 km away from a residential area uses 0-gauge $\left(A = 42.40\,\text{mm}^2\right)$ wire of copper to transmit power at a current of $I = 100.00\,\text{A}$. How much more power is dissipated in the copper wires than it would be in superconducting wires?

64. A wire is drawn through a die, stretching it to four times its original length. By what factor does its resistance increase?

65. Digital medical thermometers determine temperature by measuring the resistance of a semiconductor device called a thermistor (which has $\alpha = -0.06/°C$) when it is at the same temperature as the patient. What is a patient's temperature if the thermistor's resistance at that temperature is 82.0% of its value at 37 °C (normal body temperature)?

66. Electrical power generators are sometimes "load tested" by passing current through a large vat of water. A similar method can be used to test the heat output of a resistor. A $R = 30\,\Omega$ resistor is connected to a 9.0-V battery and the resistor leads are waterproofed and the resistor is placed in 1.0 kg of room temperature water ($T = 20\,°C$). Current runs through the resistor for 20 minutes. Assuming all the electrical energy dissipated by the resistor is converted to heat, what is the final temperature of the water?

67. A 12-guage gold wire has a length of 1 meter. (a) What would be the length of a silver 12-gauge wire with the same resistance? (b) What are their respective resistances at the temperature of boiling water?

68. What is the change in temperature required to decrease the resistance for a carbon resistor by 10%?

ADDITIONAL PROBLEMS

69. A coaxial cable consists of an inner conductor with radius $r_i = 0.25\,\text{cm}$ and an outer radius of $r_o = 0.5\,\text{cm}$ and has a length of 10 meters. Plastic, with a resistivity of $\rho = 2.00 \times 10^{13}\,\Omega \cdot \text{m}$, separates the two conductors. What is the resistance of the cable?

70. A 10.00-meter long wire cable that is made of copper has a resistance of 0.051 ohms. (a) What is the weight if the wire was made of copper? (b) What is the weight of a 10.00-meter-long wire of the same gauge made of aluminum? (c)What is the resistance of the aluminum wire? The density of copper is $8960\,\text{kg/m}^3$ and the density of aluminum is $2760\,\text{kg/m}^3$.

71. A nichrome rod that is 3.00 mm long with a cross-sectional area of $1.00\,\text{mm}^2$ is used for a digital thermometer. (a) What is the resistance at room temperature? (b) What is the resistance at body temperature?

72. The temperature in Philadelphia, PA can vary between $68.00\,°F$ and $100.00\,°F$ in one summer day. By what percentage will an aluminum wire's resistance change during the day?

73. When 100.0 V is applied across a 5-gauge (diameter 4.621 mm) wire that is 10 m long, the magnitude of the current density is $2.0 \times 10^8\,\text{A/m}^2$. What is the resistivity of the wire?

74. A wire with a resistance of $5.0\,\Omega$ is drawn out through a die so that its new length is twice times its original length. Find the resistance of the longer wire. You may assume that the resistivity and density of the material are unchanged.

75. What is the resistivity of a wire of 5-gauge wire ($A = 16.8 \times 10^{-6}\,\text{m}^2$), 5.00 m length, and $5.10\,\text{m}\Omega$ resistance?

76. Coils are often used in electrical and electronic circuits. Consider a coil which is formed by winding 1000 turns of insulated 20-gauge copper wire (area $0.52\,\text{mm}^2$) in a single layer on a cylindrical non-conducting core of radius 2.0 mm. What is the resistance of the coil? Neglect the thickness of the insulation.

77. Currents of approximately 0.06 A can be potentially fatal. Currents in that range can make the heart fibrillate (beat in an uncontrolled manner). The resistance of a dry human body can be approximately $100\,\text{k}\Omega$. (a) What voltage can cause 0.2 A through a dry human body? (b) When a human body is wet, the resistance can fall to $100\,\Omega$. What voltage can cause harm to a wet body?

78. A 20.00-ohm, 5.00-watt resistor is placed in series with a power supply. (a) What is the maximum voltage that can be applied to the resistor without harming the resistor? (b) What would be the current through the resistor?

79. A battery with an emf of 24.00 V delivers a constant current of 2.00 mA to an appliance. How much work does the battery do in three minutes?

80. A 12.00-V battery has an internal resistance of a tenth of an ohm. (a) What is the current if the battery terminals are momentarily shorted together? (b) What is the terminal voltage if the battery delivers 0.25 amps to a circuit?

CHALLENGE PROBLEMS

81. A 10-gauge copper wire has a cross-sectional area $A = 5.26\,\text{mm}^2$ and carries a current of $I = 5.00\,\text{A}$. The density of copper is $\rho = 89.50\,\text{g/cm}^3$. One mole of copper atoms $\left(6.02 \times 10^{23}\,\text{atoms}\right)$ has a mass of approximately 63.50 g. What is the magnitude of the drift velocity of the electrons, assuming that each copper atom contributes one free electron to the current?

82. The current through a 12-gauge wire is given as $I(t) = (5.00\,\text{A})\sin(2\pi 60\,\text{Hz}\,t)$. What is the current density at time 15.00 ms?

83. A particle accelerator produces a beam with a radius of 1.25 mm with a current of 2.00 mA. Each proton has a kinetic energy of 10.00 MeV. (a) What is the velocity of the protons? (b) What is the number (*n*) of protons per unit volume? (b) How many electrons pass a cross sectional area each second?

84. In this chapter, most examples and problems involved direct current (DC). DC circuits have the current flowing in one direction, from positive to negative. When the current was changing, it was changed linearly from $I = -I_{\text{max}}$ to $I = +I_{\text{max}}$ and the voltage changed linearly from $V = -V_{\text{max}}$ to $V = +V_{\text{max}}$, where $V_{\text{max}} = I_{\text{max}}R$. Suppose a voltage source is placed in series with a resistor of $R = 10\,\Omega$ that supplied a current that alternated as a sine wave, for example, $I(t) = (3.00\,\text{A})\sin\left(\frac{2\pi}{4.00\,\text{s}}t\right)$. (a) What would a graph of the voltage drop across the resistor *V*(*t*) versus time look like? (b) What would a plot of *V*(*t*) versus *I*(*t*) for one period look like? (*Hint*: If you are not sure, try plotting *V*(*t*) versus *I*(*t*) using a spreadsheet.)

85. A current of $I = 25A$ is drawn from a 100-V battery for 30 seconds. By how much is the chemical energy reduced?

86. Consider a square rod of material with sides of length $L = 3.00\,\text{cm}$ with a current density of $\vec{\mathbf{J}} = J_0 e^{\alpha x}\hat{k} = \left(0.35\frac{\text{A}}{\text{m}^2}\right)e^{\left(2.1 \times 10^{-3}\,\text{m}^{-1}\right)x}\hat{\mathbf{k}}$ as shown below. Find the current that passes through the face of the rod.

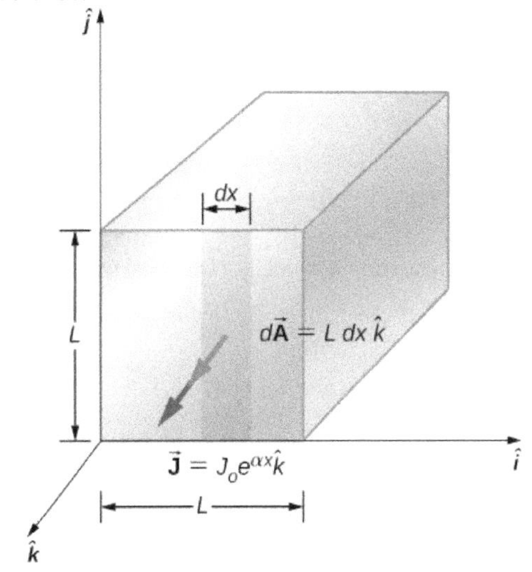

87. A resistor of an unknown resistance is placed in an insulated container filled with 0.75 kg of water. A voltage source is connected in series with the resistor and a current of 1.2 amps flows through the resistor for 10 minutes. During this time, the temperature of the water is measured and the temperature change during this time is $\Delta T = 10.00\,°\text{C}$. (a) What is the resistance of the resistor? (b) What is the voltage supplied by the power supply?

88. The charge that flows through a point in a wire as a function of time is modeled as $q(t) = q_0 e^{-t/T} = 10.0\,\text{C}e^{-t/5\,\text{s}}$. (a) What is the initial current through the wire at time $t = 0.00\,\text{s}$? (b) Find the current at time $t = \frac{1}{2}T$. (c) At what time *t* will the current be reduced by one-half $I = \frac{1}{2}I_0$?

89. Consider a resistor made from a hollow cylinder of carbon as shown below. The inner radius of the cylinder is $R_i = 0.20\text{ mm}$ and the outer radius is $R_0 = 0.30\text{ mm}$. The length of the resistor is $L = 0.90\text{ mm}$. The resistivity of the carbon is $\rho = 3.5 \times 10^{-5}\ \Omega \cdot \text{m}$. (a) Prove that the resistance perpendicular from the axis is $R = \frac{\rho}{2\pi L}\ln\left(\frac{R_0}{R_i}\right)$. (b) What is the resistance?

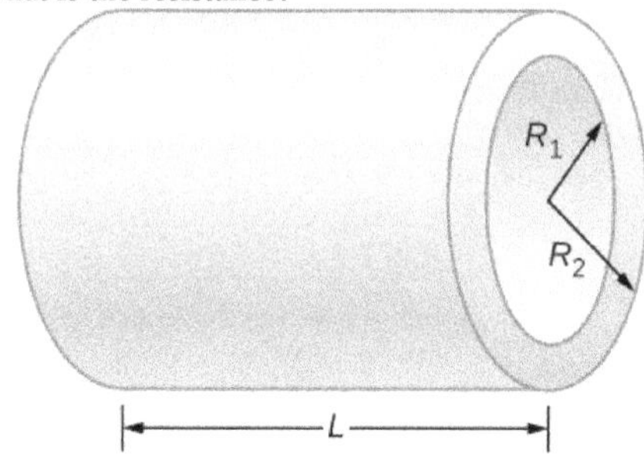

90. What is the current through a cylindrical wire of radius $R = 0.1\text{ mm}$ if the current density is $J = \frac{J_0}{R}r$, where $J_0 = 32000\frac{\text{A}}{\text{m}^2}$?

91. A student uses a 100.00-W, 115.00-V radiant heater to heat the student's dorm room, during the hours between sunset and sunrise, 6:00 p.m. to 7:00 a.m. (a) What current does the heater operate at? (b) How many electrons move through the heater? (c) What is the resistance of the heater? (d) How much heat was added to the dorm room?

92. A 12-V car battery is used to power a 20.00-W, 12.00-V lamp during the physics club camping trip/star party. The cable to the lamp is 2.00 meters long, 14-gauge copper wire with a charge density of $n = 9.50 \times 10^{28}\ \text{m}^{-3}$. (a) What is the current draw by the lamp? (b) How long would it take an electron to get from the battery to the lamp?

93. A physics student uses a 115.00-V immersion heater to heat 400.00 grams (almost two cups) of water for herbal tea. During the two minutes it takes the water to heat, the physics student becomes bored and decides to figure out the resistance of the heater. The student starts with the assumption that the water is initially at the temperature of the room $T_i = 25.00\ °\text{C}$ and reaches $T_f = 100.00\ °\text{C}$. The specific heat of the water is $c = 4180\frac{\text{J}}{\text{kg}}$. What is the resistance of the heater?

10 | DIRECT-CURRENT CIRCUITS

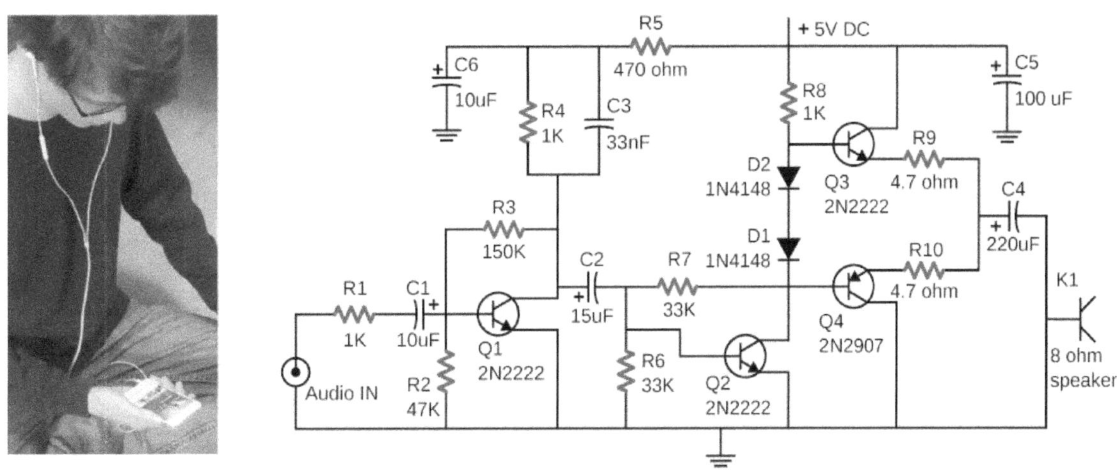

Figure 10.1 This circuit shown is used to amplify small signals and power the earbud speakers attached to a cellular phone. This circuit's components include resistors, capacitors, and diodes, all of which have been covered in previous chapters, as well as transistors, which are semi-conducting devices covered in Condensed Matter Physics (http://cnx.org/content/m58591/latest/) . Circuits using similar components are found in all types of equipment and appliances you encounter in everyday life, such as alarm clocks, televisions, computers, and refrigerators. (credit: Jane Whitney)

Chapter Outline

Introduction

In the preceding few chapters, we discussed electric components, including capacitors, resistors, and diodes. In this chapter, we use these electric components in circuits. A circuit is a collection of electrical components connected to accomplish a specific task. Figure 10.1 shows an amplifier circuit, which takes a small-amplitude signal and amplifies it to power the speakers in earbuds. Although the circuit looks complex, it actually consists of a set of series, parallel, and series-parallel circuits. The second section of this chapter covers the analysis of series and parallel circuits that consist of resistors. Later in this chapter, we introduce the basic equations and techniques to analyze any circuit, including those that are not reducible through simplifying parallel and series elements. But first, we need to understand how to power a circuit.

10.1 | Electromotive Force

Learning Objectives

By the end of the section, you will be able to:

- Describe the electromotive force (emf) and the internal resistance of a battery
- Explain the basic operation of a battery

If you forget to turn off your car lights, they slowly dim as the battery runs down. Why don't they suddenly blink off when the battery's energy is gone? Their gradual dimming implies that the battery output voltage decreases as the battery is depleted. The reason for the decrease in output voltage for depleted batteries is that all voltage sources have two fundamental parts—a source of electrical energy and an internal resistance. In this section, we examine the energy source and the internal resistance.

Introduction to Electromotive Force

Voltage has many sources, a few of which are shown in Figure 10.2. All such devices create a **potential difference** and can supply current if connected to a circuit. A special type of potential difference is known as **electromotive force (emf)**. The emf is not a force at all, but the term 'electromotive force' is used for historical reasons. It was coined by Alessandro Volta in the 1800s, when he invented the first battery, also known as the voltaic pile. Because the electromotive force is not a force, it is common to refer to these sources simply as sources of emf (pronounced as the letters "ee-em-eff"), instead of sources of electromotive force.

Figure 10.2 A variety of voltage sources. (a) The Brazos Wind Farm in Fluvanna, Texas; (b) the Krasnoyarsk Dam in Russia; (c) a solar farm; (d) a group of nickel metal hydride batteries. The voltage output of each device depends on its construction and load. The voltage output equals emf only if there is no load. (credit a: modification of work by "Leaflet"/Wikimedia Commons; credit b: modification of work by Alex Polezhaev; credit c: modification of work by US Department of Energy; credit d: modification of work by Tiaa Monto)

If the electromotive force is not a force at all, then what is the emf and what is a source of emf? To answer these questions, consider a simple circuit of a 12-V lamp attached to a 12-V battery, as shown in Figure 10.3. The battery can be modeled as a two-terminal device that keeps one terminal at a higher electric potential than the second terminal. The higher electric potential is sometimes called the positive terminal and is labeled with a plus sign. The lower-potential terminal is sometimes called the negative terminal and labeled with a minus sign. This is the source of the emf.

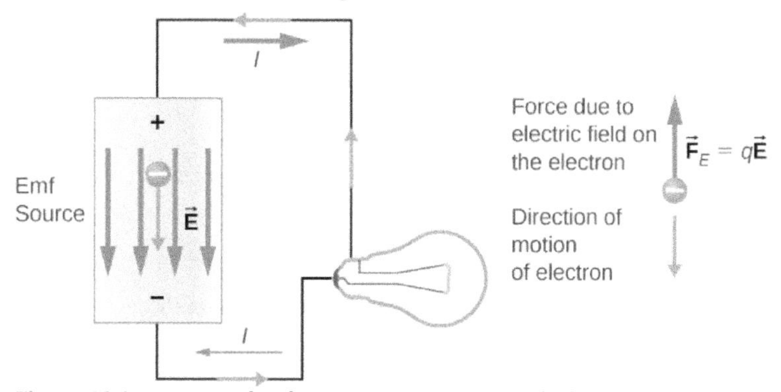

Figure 10.3 A source of emf maintains one terminal at a higher electric potential than the other terminal, acting as a source of current in a circuit.

When the emf source is not connected to the lamp, there is no net flow of charge within the emf source. Once the battery is connected to the lamp, charges flow from one terminal of the battery, through the lamp (causing the lamp to light), and back to the other terminal of the battery. If we consider positive (conventional) current flow, positive charges leave the positive terminal, travel through the lamp, and enter the negative terminal.

Positive current flow is useful for most of the circuit analysis in this chapter, but in metallic wires and resistors, electrons contribute the most to current, flowing in the opposite direction of positive current flow. Therefore, it is more realistic to consider the movement of electrons for the analysis of the circuit in Figure 10.3. The electrons leave the negative terminal, travel through the lamp, and return to the positive terminal. In order for the emf source to maintain the potential difference between the two terminals, negative charges (electrons) must be moved from the positive terminal to the negative terminal. The emf source acts as a charge pump, moving negative charges from the positive terminal to the negative terminal to maintain the potential difference. This increases the potential energy of the charges and, therefore, the electric potential of the charges.

The force on the negative charge from the electric field is in the opposite direction of the electric field, as shown in Figure 10.3. In order for the negative charges to be moved to the negative terminal, work must be done on the negative charges. This requires energy, which comes from chemical reactions in the battery. The potential is kept high on the positive terminal and low on the negative terminal to maintain the potential difference between the two terminals. The emf is equal to the work done on the charge per unit charge $\left(\varepsilon = \frac{dW}{dq}\right)$ when there is no current flowing. Since the unit for work is the joule and the unit for charge is the coulomb, the unit for emf is the volt $(1 \text{ V} = 1 \text{ J/C})$.

The **terminal voltage** V_{terminal} of a battery is voltage measured across the terminals of the battery when there is no load connected to the terminal. An ideal battery is an emf source that maintains a constant terminal voltage, independent of the current between the two terminals. An ideal battery has no internal resistance, and the terminal voltage is equal to the emf of the battery. In the next section, we will show that a real battery does have internal resistance and the terminal voltage is always less than the emf of the battery.

The Origin of Battery Potential

The combination of chemicals and the makeup of the terminals in a battery determine its emf. The lead acid battery used in cars and other vehicles is one of the most common combinations of chemicals. Figure 10.4 shows a single cell (one of six) of this battery. The cathode (positive) terminal of the cell is connected to a lead oxide plate, whereas the anode (negative) terminal is connected to a lead plate. Both plates are immersed in sulfuric acid, the electrolyte for the system.

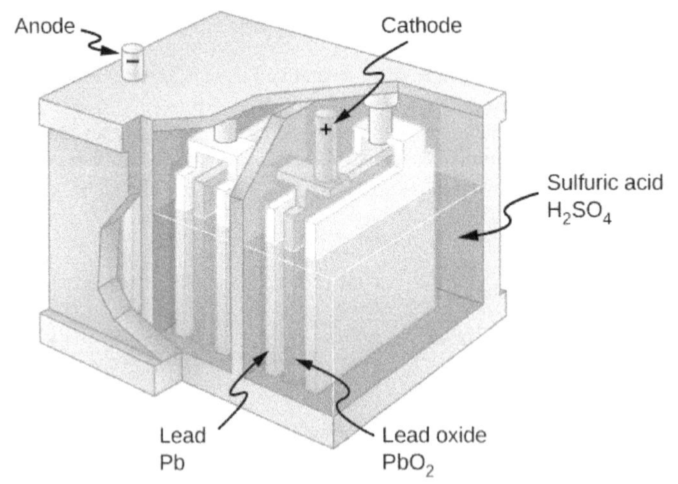

Figure 10.4 Chemical reactions in a lead-acid cell separate charge, sending negative charge to the anode, which is connected to the lead plates. The lead oxide plates are connected to the positive or cathode terminal of the cell. Sulfuric acid conducts the charge, as well as participates in the chemical reaction.

Knowing a little about how the chemicals in a lead-acid battery interact helps in understanding the potential created by the battery. Figure 10.5 shows the result of a single chemical reaction. Two electrons are placed on the anode, making it negative, provided that the cathode supplies two electrons. This leaves the cathode positively charged, because it has lost two electrons. In short, a separation of charge has been driven by a chemical reaction.

Note that the reaction does not take place unless there is a complete circuit to allow two electrons to be supplied to the cathode. Under many circumstances, these electrons come from the anode, flow through a resistance, and return to the cathode. Note also that since the chemical reactions involve substances with resistance, it is not possible to create the emf without an internal resistance.

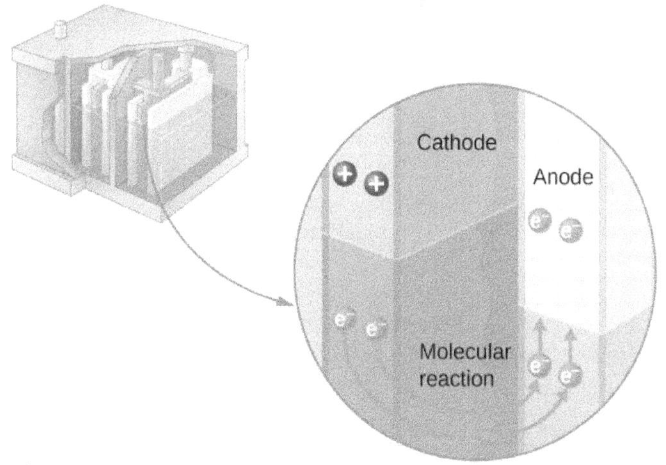

Figure 10.5 In a lead-acid battery, two electrons are forced onto the anode of a cell, and two electrons are removed from the cathode of the cell. The chemical reaction in a lead-acid battery places two electrons on the anode and removes two from the cathode. It requires a closed circuit to proceed, since the two electrons must be supplied to the cathode.

Internal Resistance and Terminal Voltage

The amount of resistance to the flow of current within the voltage source is called the **internal resistance**. The internal resistance r of a battery can behave in complex ways. It generally increases as a battery is depleted, due to the oxidation of the plates or the reduction of the acidity of the electrolyte. However, internal resistance may also depend on the magnitude and direction of the current through a voltage source, its temperature, and even its history. The internal resistance of rechargeable nickel-cadmium cells, for example, depends on how many times and how deeply they have been depleted. A simple model for a battery consists of an idealized emf source ε and an internal resistance r (Figure 10.6).

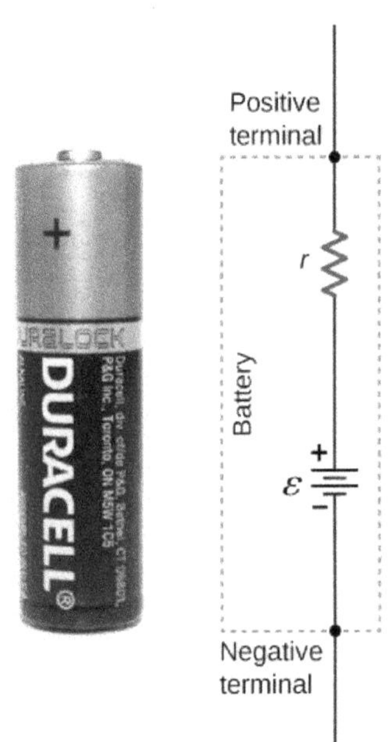

Figure 10.6 A battery can be modeled as an idealized emf (ε) with an internal resistance (r). The terminal voltage of the battery is $V_{\text{terminal}} = \varepsilon - Ir$.

Suppose an external resistor, known as the load resistance R, is connected to a voltage source such as a battery, as in Figure 10.7. The figure shows a model of a battery with an emf ε, an internal resistance r, and a load resistor R connected across its terminals. Using conventional current flow, positive charges leave the positive terminal of the battery, travel through the resistor, and return to the negative terminal of the battery. The terminal voltage of the battery depends on the emf, the internal resistance, and the current, and is equal to

$$V_{\text{terminal}} = \varepsilon - Ir. \tag{10.1}$$

For a given emf and internal resistance, the terminal voltage decreases as the current increases due to the potential drop Ir of the internal resistance.

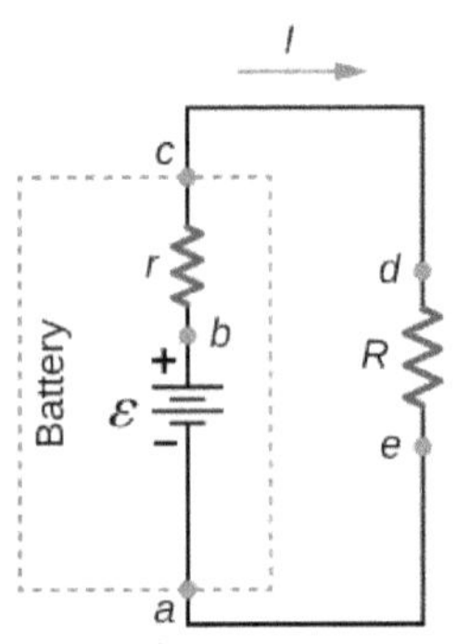

Figure 10.7 Schematic of a voltage source and its load resistor R. Since the internal resistance r is in series with the load, it can significantly affect the terminal voltage and the current delivered to the load.

A graph of the potential difference across each element the circuit is shown in Figure 10.8. A current I runs through the circuit, and the potential drop across the internal resistor is equal to Ir. The terminal voltage is equal to $\varepsilon - Ir$, which is equal to the **potential drop** across the load resistor $IR = \varepsilon - Ir$. As with potential energy, it is the change in voltage that is important. When the term "voltage" is used, we assume that it is actually the change in the potential, or ΔV. However, Δ is often omitted for convenience.

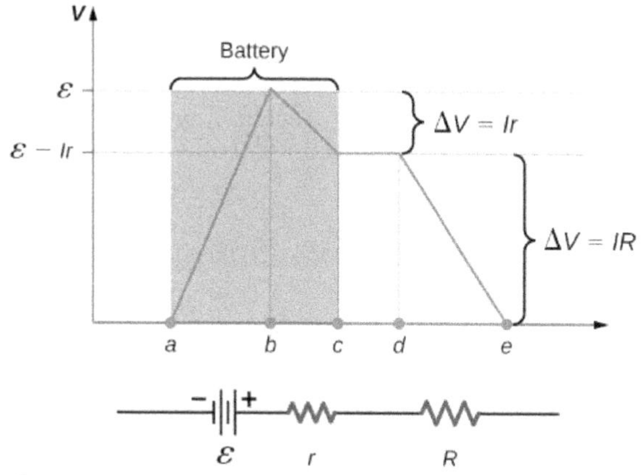

Figure 10.8 A graph of the voltage through the circuit of a battery and a load resistance. The electric potential increases the emf of the battery due to the chemical reactions doing work on the charges. There is a decrease in the electric potential in the battery due to the internal resistance. The potential decreases due to the internal resistance $(-Ir)$, making the terminal voltage of the battery equal to $(\varepsilon - Ir)$. The voltage then decreases by (IR). The current is equal to $I = \frac{\varepsilon}{r + R}$.

The current through the load resistor is $I = \frac{\varepsilon}{r + R}$. We see from this expression that the smaller the internal resistance r, the greater the current the voltage source supplies to its load R. As batteries are depleted, r increases. If r becomes a significant fraction of the load resistance, then the current is significantly reduced, as the following example illustrates.

Example 10.1

Analyzing a Circuit with a Battery and a Load

A given battery has a 12.00-V emf and an internal resistance of $0.100\,\Omega$. (a) Calculate its terminal voltage when connected to a $10.00\text{-}\Omega$ load. (b) What is the terminal voltage when connected to a $0.500\text{-}\Omega$ load? (c) What power does the $0.500\text{-}\Omega$ load dissipate? (d) If the internal resistance grows to $0.500\,\Omega$, find the current, terminal voltage, and power dissipated by a $0.500\text{-}\Omega$ load.

Strategy

The analysis above gave an expression for current when internal resistance is taken into account. Once the current is found, the terminal voltage can be calculated by using the equation $V_{\text{terminal}} = \varepsilon - Ir$. Once current is found, we can also find the power dissipated by the resistor.

Solution

a. Entering the given values for the emf, load resistance, and internal resistance into the expression above yields

$$I = \frac{\varepsilon}{R + r} = \frac{12.00\text{ V}}{10.10\,\Omega} = 1.188\text{ A}.$$

Enter the known values into the equation $V_{\text{terminal}} = \varepsilon - Ir$ to get the terminal voltage:

$$V_{\text{terminal}} = \varepsilon - Ir = 12.00\text{ V} - (1.188\text{ A})(0.100\ \Omega) = 11.90\text{ V}.$$

The terminal voltage here is only slightly lower than the emf, implying that the current drawn by this light load is not significant.

b. Similarly, with $R_{\text{load}} = 0.500\ \Omega$, the current is

$$I = \frac{\varepsilon}{R + r} = \frac{12.00\text{ V}}{0.600\ \Omega} = 20.00\text{ A}.$$

The terminal voltage is now

$$V_{\text{terminal}} = \varepsilon - Ir = 12.00\text{ V} - (20.00\text{ A})(0.100\ \Omega) = 10.00\text{ V}.$$

The terminal voltage exhibits a more significant reduction compared with emf, implying $0.500\ \Omega$ is a heavy load for this battery. A "heavy load" signifies a larger draw of current from the source but not a larger resistance.

c. The power dissipated by the $0.500\text{-}\Omega$ load can be found using the formula $P = I^2R$. Entering the known values gives

$$P = I^2R = (20.0\text{ A})^2(0.500\ \Omega) = 2.00 \times 10^2\text{ W}.$$

Note that this power can also be obtained using the expression $\frac{V^2}{R}$ or IV, where V is the terminal voltage (10.0 V in this case).

d. Here, the internal resistance has increased, perhaps due to the depletion of the battery, to the point where it is as great as the load resistance. As before, we first find the current by entering the known values into the expression, yielding

$$I = \frac{\varepsilon}{R + r} = \frac{12.00\text{ V}}{1.00\ \Omega} = 12.00\text{ A}.$$

Now the terminal voltage is

$$V_{\text{terminal}} = \varepsilon - Ir = 12.00\text{ V} - (12.00\text{ A})(0.500\ \Omega) = 6.00\text{ V},$$

and the power dissipated by the load is

$$P = I^2R = (12.00\text{ A})^2(0.500\ \Omega) = 72.00\text{ W}.$$

We see that the increased internal resistance has significantly decreased the terminal voltage, current, and power delivered to a load.

Significance

The internal resistance of a battery can increase for many reasons. For example, the internal resistance of a rechargeable battery increases as the number of times the battery is recharged increases. The increased internal resistance may have two effects on the battery. First, the terminal voltage will decrease. Second, the battery may overheat due to the increased power dissipated by the internal resistance.

10.1 Check Your Understanding If you place a wire directly across the two terminal of a battery, effectively shorting out the terminals, the battery will begin to get hot. Why do you suppose this happens?

Battery Testers

Battery testers, such as those in Figure 10.9, use small load resistors to intentionally draw current to determine whether the terminal potential drops below an acceptable level. Although it is difficult to measure the internal resistance of a battery, battery testers can provide a measurement of the internal resistance of the battery. If internal resistance is high, the battery is weak, as evidenced by its low terminal voltage.

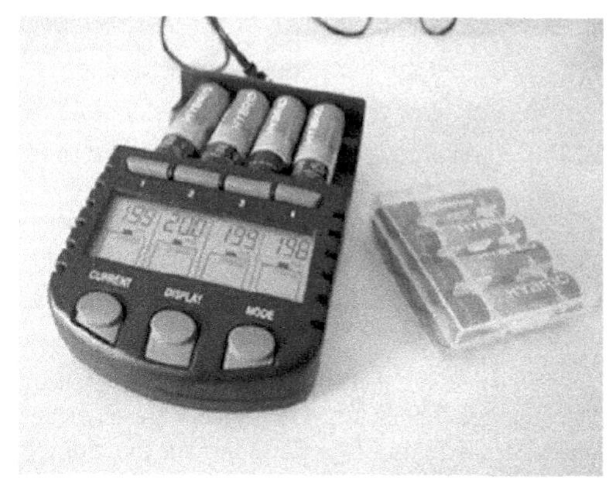

(a) (b)

Figure 10.9 Battery testers measure terminal voltage under a load to determine the condition of a battery. (a) A US Navy electronics technician uses a battery tester to test large batteries aboard the aircraft carrier USS *Nimitz*. The battery tester she uses has a small resistance that can dissipate large amounts of power. (b) The small device shown is used on small batteries and has a digital display to indicate the acceptability of the terminal voltage. (credit a: modification of work by Jason A. Johnston; credit b: modification of work by Keith Williamson)

Some batteries can be recharged by passing a current through them in the direction opposite to the current they supply to an appliance. This is done routinely in cars and in batteries for small electrical appliances and electronic devices (Figure 10.10). The voltage output of the battery charger must be greater than the emf of the battery to reverse the current through it. This causes the terminal voltage of the battery to be greater than the emf, since $V = \varepsilon - Ir$ and I is now negative.

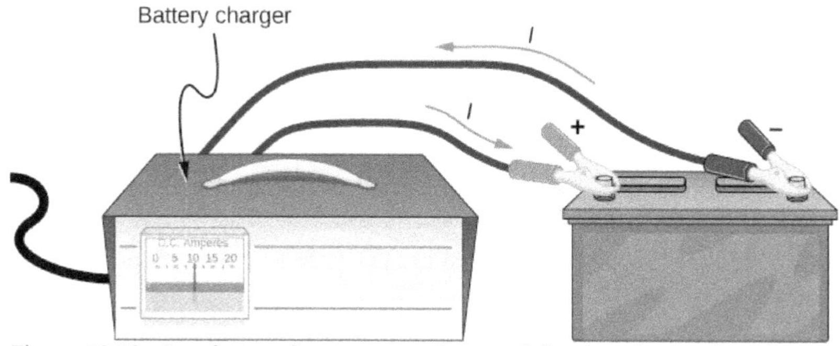

Figure 10.10 A car battery charger reverses the normal direction of current through a battery, reversing its chemical reaction and replenishing its chemical potential.

It is important to understand the consequences of the internal resistance of emf sources, such as batteries and solar cells, but often, the analysis of circuits is done with the terminal voltage of the battery, as we have done in the previous sections. The terminal voltage is referred to as simply as *V*, dropping the subscript "terminal." This is because the internal resistance of the battery is difficult to measure directly and can change over time.

10.2 | Resistors in Series and Parallel

Learning Objectives

By the end of the section, you will be able to:

- Define the term equivalent resistance
- Calculate the equivalent resistance of resistors connected in series
- Calculate the equivalent resistance of resistors connected in parallel

In Current and Resistance, we described the term 'resistance' and explained the basic design of a resistor. Basically, a resistor limits the flow of charge in a circuit and is an ohmic device where $V = IR$. Most circuits have more than one resistor. If several resistors are connected together and connected to a battery, the current supplied by the battery depends on the **equivalent resistance** of the circuit.

The equivalent resistance of a combination of resistors depends on both their individual values and how they are connected. The simplest combinations of resistors are series and parallel connections (Figure 10.11). In a series circuit, the output current of the first resistor flows into the input of the second resistor; therefore, the current is the same in each resistor. In a parallel circuit, all of the resistor leads on one side of the resistors are connected together and all the leads on the other side are connected together. In the case of a parallel configuration, each resistor has the same potential drop across it, and the currents through each resistor may be different, depending on the resistor. The sum of the individual currents equals the current that flows into the parallel connections.

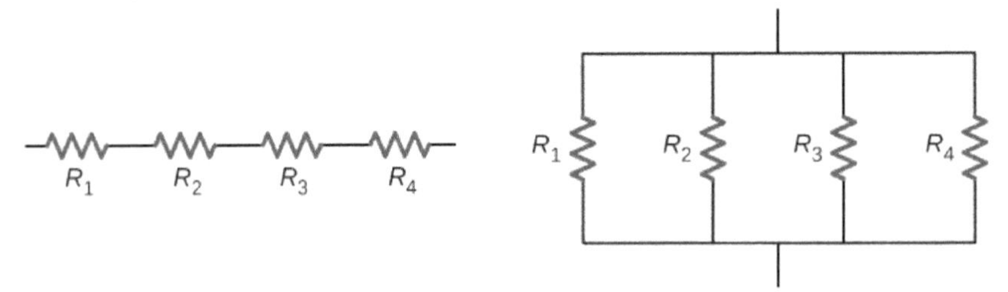

(a) Resistors connected in series (b) Resistors connected in parallel

Figure 10.11 (a) For a series connection of resistors, the current is the same in each resistor. (b) For a parallel connection of resistors, the voltage is the same across each resistor.

Resistors in Series

Resistors are said to be in series whenever the current flows through the resistors sequentially. Consider Figure 10.12, which shows three resistors in series with an applied voltage equal to V_{ab}. Since there is only one path for the charges to flow through, the current is the same through each resistor. The equivalent resistance of a set of resistors in a series connection is equal to the algebraic sum of the individual resistances.

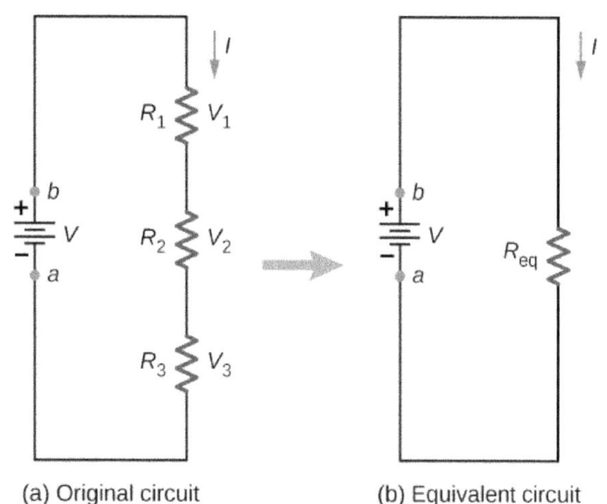

Figure 10.12 (a) Three resistors connected in series to a voltage source. (b) The original circuit is reduced to an equivalent resistance and a voltage source.

In Figure 10.12, the current coming from the voltage source flows through each resistor, so the current through each resistor is the same. The current through the circuit depends on the voltage supplied by the voltage source and the resistance of the resistors. For each resistor, a potential drop occurs that is equal to the loss of electric potential energy as a current travels through each resistor. According to Ohm's law, the potential drop V across a resistor when a current flows through it is calculated using the equation $V = IR$, where I is the current in amps (A) and R is the resistance in ohms (Ω). Since energy is conserved, and the voltage is equal to the potential energy per charge, the sum of the voltage applied to the circuit by the source and the potential drops across the individual resistors around a loop should be equal to zero:

$$\sum_{i=1}^{N} V_i = 0.$$

This equation is often referred to as Kirchhoff's loop law, which we will look at in more detail later in this chapter. For Figure 10.12, the sum of the potential drop of each resistor and the voltage supplied by the voltage source should equal zero:

$$\begin{aligned} V - V_1 - V_2 - V_3 &= 0, \\ V &= V_1 + V_2 + V_3, \\ &= IR_1 + IR_2 + IR_3, \\ I &= \frac{V}{R_1 + R_2 + R_3} = \frac{V}{R_{\text{eq}}}. \end{aligned}$$

Since the current through each component is the same, the equality can be simplified to an equivalent resistance, which is just the sum of the resistances of the individual resistors.

Any number of resistors can be connected in series. If N resistors are connected in series, the equivalent resistance is

$$R_{\text{eq}} = R_1 + R_2 + R_3 + \cdots + R_{N-1} + R_N = \sum_{i=1}^{N} R_i. \tag{10.2}$$

One result of components connected in a series circuit is that if something happens to one component, it affects all the other components. For example, if several lamps are connected in series and one bulb burns out, all the other lamps go dark.

Example 10.2

Equivalent Resistance, Current, and Power in a Series Circuit

A battery with a terminal voltage of 9 V is connected to a circuit consisting of four $20\text{-}\Omega$ and one $10\text{-}\Omega$ resistors all in series (Figure 10.13). Assume the battery has negligible internal resistance. (a) Calculate the equivalent resistance of the circuit. (b) Calculate the current through each resistor. (c) Calculate the potential drop across each resistor. (d) Determine the total power dissipated by the resistors and the power supplied by the battery.

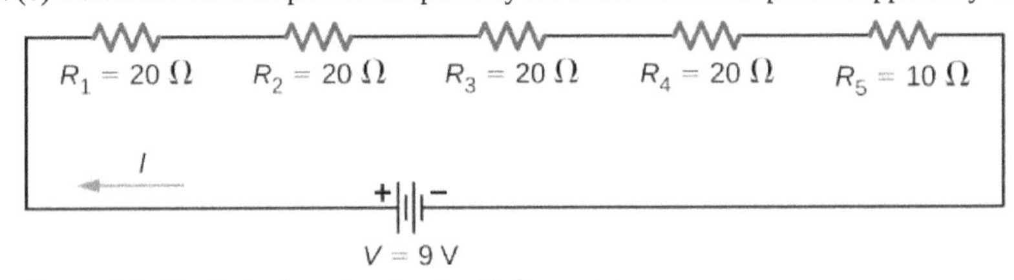

Figure 10.13 A simple series circuit with five resistors.

Strategy

In a series circuit, the equivalent resistance is the algebraic sum of the resistances. The current through the circuit can be found from Ohm's law and is equal to the voltage divided by the equivalent resistance. The potential drop across each resistor can be found using Ohm's law. The power dissipated by each resistor can be found using $P = I^2 R$, and the total power dissipated by the resistors is equal to the sum of the power dissipated by each resistor. The power supplied by the battery can be found using $P = I\varepsilon$.

Solution

a. The equivalent resistance is the algebraic sum of the resistances:

$$R_{\text{eq}} = R_1 + R_2 + R_3 + R_4 + R_5 = 20\,\Omega + 20\,\Omega + 20\,\Omega + 20\,\Omega + 10\,\Omega = 90\,\Omega.$$

b. The current through the circuit is the same for each resistor in a series circuit and is equal to the applied voltage divided by the equivalent resistance:

$$I = \frac{V}{R_{\text{eq}}} = \frac{9\text{ V}}{90\,\Omega} = 0.1\text{ A}.$$

c. The potential drop across each resistor can be found using Ohm's law:

$$\begin{aligned} V_1 = V_2 = V_3 = V_4 &= (0.1\text{ A})20\,\Omega = 2\text{ V}, \\ V_5 &= (0.1\text{ A})10\,\Omega = 1\text{ V}, \\ V_1 + V_2 + V_3 + V_4 + V_5 &= 9\text{ V}. \end{aligned}$$

Note that the sum of the potential drops across each resistor is equal to the voltage supplied by the battery.

d. The power dissipated by a resistor is equal to $P = I^2 R$, and the power supplied by the battery is equal to $P = I\varepsilon$:

$$\begin{aligned} P_1 = P_2 = P_3 = P_4 &= (0.1\text{ A})^2(20\,\Omega) = 0.2\text{ W}, \\ P_5 &= (0.1\text{ A})^2(10\,\Omega) = 0.1\text{ W}, \\ P_{\text{dissipated}} &= 0.2\text{ W} + 0.2\text{ W} + 0.2\text{ W} + 0.2\text{ W} + 0.1\text{ W} = 0.9\text{ W}, \\ P_{\text{source}} &= I\varepsilon = (0.1\text{ A})(9\text{ V}) = 0.9\text{ W}. \end{aligned}$$

Significance

There are several reasons why we would use multiple resistors instead of just one resistor with a resistance equal to the equivalent resistance of the circuit. Perhaps a resistor of the required size is not available, or we need to dissipate the heat generated, or we want to minimize the cost of resistors. Each resistor may cost a few cents to a few dollars, but when multiplied by thousands of units, the cost saving may be appreciable.

10.2 Check Your Understanding Some strings of miniature holiday lights are made to short out when a bulb burns out. The device that causes the short is called a shunt, which allows current to flow around the open circuit. A "short" is like putting a piece of wire across the component. The bulbs are usually grouped in series of nine bulbs. If too many bulbs burn out, the shunts eventually open. What causes this?

Let's briefly summarize the major features of resistors in series:

1. Series resistances add together to get the equivalent resistance:

$$R_{eq} = R_1 + R_2 + R_3 + \cdots + R_{N-1} + R_N = \sum_{i=1}^{N} R_i.$$

2. The same current flows through each resistor in series.
3. Individual resistors in series do not get the total source voltage, but divide it. The total potential drop across a series configuration of resistors is equal to the sum of the potential drops across each resistor.

Resistors in Parallel

Figure 10.14 shows resistors in parallel, wired to a voltage source. Resistors are in parallel when one end of all the resistors are connected by a continuous wire of negligible resistance and the other end of all the resistors are also connected to one another through a continuous wire of negligible resistance. The potential drop across each resistor is the same. Current through each resistor can be found using Ohm's law $I = V/R,$ where the voltage is constant across each resistor. For example, an automobile's headlights, radio, and other systems are wired in parallel, so that each subsystem utilizes the full voltage of the source and can operate completely independently. The same is true of the wiring in your house or any building.

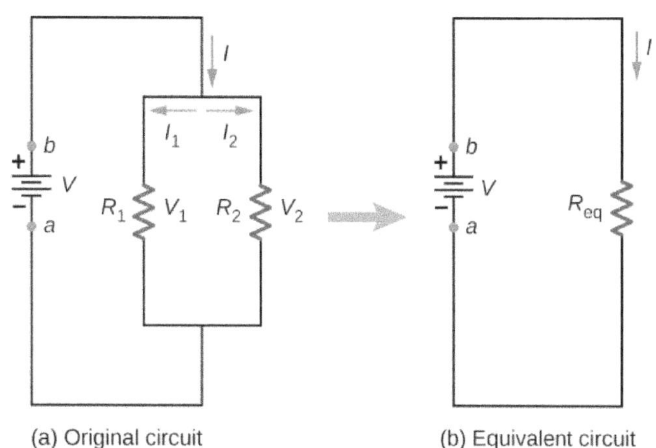

Figure 10.14 (a) Two resistors connected in parallel to a voltage source. (b) The original circuit is reduced to an equivalent resistance and a voltage source.

The current flowing from the voltage source in Figure 10.14 depends on the voltage supplied by the voltage source and the equivalent resistance of the circuit. In this case, the current flows from the voltage source and enters a junction, or node, where the circuit splits flowing through resistors R_1 and R_2. As the charges flow from the battery, some go through resistor R_1 and some flow through resistor R_2. The sum of the currents flowing into a junction must be equal to the sum of the currents flowing out of the junction:

$$\sum I_{in} = \sum I_{out}.$$

This equation is referred to as Kirchhoff's junction rule and will be discussed in detail in the next section. In Figure 10.14, the junction rule gives $I = I_1 + I_2$. There are two loops in this circuit, which leads to the equations $V = I_1 R_1$ and $I_1 R_1 = I_2 R_2$. Note the voltage across the resistors in parallel are the same $(V = V_1 = V_2)$ and the current is additive:

$$\begin{aligned} I &= I_1 + I_2 \\ &= \frac{V_1}{R_1} + \frac{V_2}{R_2} \\ &= \frac{V}{R_1} + \frac{V}{R_2} \\ &= V\left(\frac{1}{R_1} + \frac{1}{R_2}\right) = \frac{V}{R_{\text{eq}}} \\ R_{\text{eq}} &= \left(\frac{1}{R_1} + \frac{1}{R_2}\right)^{-1}. \end{aligned}$$

Generalizing to any number of N resistors, the equivalent resistance R_{eq} of a parallel connection is related to the individual resistances by

$$R_{\text{eq}} = \left(\frac{1}{R_1} + \frac{1}{R_2} + \frac{1}{R_3} + \cdots + \frac{1}{R_{N-1}} + \frac{1}{R_N}\right)^{-1} = \left(\sum_{i=1}^{N} \frac{1}{R_i}\right)^{-1}. \tag{10.3}$$

This relationship results in an equivalent resistance R_{eq} that is less than the smallest of the individual resistances. When resistors are connected in parallel, more current flows from the source than would flow for any of them individually, so the total resistance is lower.

Example 10.3

Analysis of a Parallel Circuit

Three resistors $R_1 = 1.00\,\Omega$, $R_2 = 2.00\,\Omega$, and $R_3 = 2.00\,\Omega$, are connected in parallel. The parallel connection is attached to a $V = 3.00\,\text{V}$ voltage source. (a) What is the equivalent resistance? (b) Find the current supplied by the source to the parallel circuit. (c) Calculate the currents in each resistor and show that these add together to equal the current output of the source. (d) Calculate the power dissipated by each resistor. (e) Find the power output of the source and show that it equals the total power dissipated by the resistors.

Strategy

(a) The total resistance for a parallel combination of resistors is found using $R_{\text{eq}} = \left(\sum_i \frac{1}{R_i}\right)^{-1}$.

(Note that in these calculations, each intermediate answer is shown with an extra digit.)

(b) The current supplied by the source can be found from Ohm's law, substituting R_{eq} for the total resistance $I = \frac{V}{R_{\text{eq}}}$.

(c) The individual currents are easily calculated from Ohm's law $\left(I_i = \frac{V_i}{R_i}\right)$, since each resistor gets the full voltage. The total current is the sum of the individual currents: $I = \sum_i I_i$.

(d) The power dissipated by each resistor can be found using any of the equations relating power to current, voltage, and resistance, since all three are known. Let us use $P_i = V^2/R_i$, since each resistor gets full voltage.

(e) The total power can also be calculated in several ways, use $P = IV$.

Solution

a. The total resistance for a parallel combination of resistors is found using Equation 10.3. Entering known values gives

$$R_{\text{eq}} = \left(\frac{1}{R_1} + \frac{1}{R_2} + \frac{1}{R_3}\right)^{-1} = \left(\frac{1}{1.00\,\Omega} + \frac{1}{2.00\,\Omega} + \frac{1}{2.00\,\Omega}\right)^{-1} = 0.50\,\Omega.$$

The total resistance with the correct number of significant digits is $R_{\text{eq}} = 0.50\,\Omega$. As predicted, R_{eq} is less than the smallest individual resistance.

b. The total current can be found from Ohm's law, substituting R_{eq} for the total resistance. This gives

$$I = \frac{V}{R_{\text{eq}}} = \frac{3.00\text{ V}}{0.50\,\Omega} = 6.00\text{ A}.$$

Current I for each device is much larger than for the same devices connected in series (see the previous example). A circuit with parallel connections has a smaller total resistance than the resistors connected in series.

c. The individual currents are easily calculated from Ohm's law, since each resistor gets the full voltage. Thus,

$$I_1 = \frac{V}{R_1} = \frac{3.00\text{ V}}{1.00\,\Omega} = 3.00\text{ A}.$$

Similarly,

$$I_2 = \frac{V}{R_2} = \frac{3.00\text{ V}}{2.00\,\Omega} = 1.50\text{ A}$$

and

$$I_3 = \frac{V}{R_3} = \frac{6.00\text{ V}}{2.00\,\Omega} = 1.50\text{ A}.$$

The total current is the sum of the individual currents:

$$I_1 + I_2 + I_3 = 6.00\text{ A}.$$

d. The power dissipated by each resistor can be found using any of the equations relating power to current, voltage, and resistance, since all three are known. Let us use $P = V^2/R$, since each resistor gets full voltage. Thus,

$$P_1 = \frac{V^2}{R_1} = \frac{(3.00\text{ V})^2}{1.00\,\Omega} = 9.00\text{ W}.$$

Similarly,

$$P_2 = \frac{V^2}{R_2} = \frac{(3.00\text{ V})^2}{2.00\,\Omega} = 4.50\text{ W}$$

and

$$P_3 = \frac{V^2}{R_3} = \frac{(3.00\text{ V})^2}{2.00\,\Omega} = 4.50\text{ W}.$$

e. The total power can also be calculated in several ways. Choosing $P = IV$ and entering the total current yields

$$P = IV = (6.00\text{ A})(3.00\text{ V}) = 18.00\text{ W}.$$

Significance

Total power dissipated by the resistors is also 18.00 W:

$$P_1 + P_2 + P_3 = 9.00\text{ W} + 4.50\text{ W} + 4.50\text{ W} = 18.00\text{ W}.$$

Notice that the total power dissipated by the resistors equals the power supplied by the source.

10.3 Check Your Understanding Consider the same potential difference $(V = 3.00\text{ V})$ applied to the same three resistors connected in series. Would the equivalent resistance of the series circuit be higher, lower, or equal to the three resistor in parallel? Would the current through the series circuit be higher, lower, or equal to the current provided by the same voltage applied to the parallel circuit? How would the power dissipated by the resistor in series compare to the power dissipated by the resistors in parallel?

10.4 Check Your Understanding How would you use a river and two waterfalls to model a parallel configuration of two resistors? How does this analogy break down?

Let us summarize the major features of resistors in parallel:

1. Equivalent resistance is found from

$$R_{\text{eq}} = \left(\frac{1}{R_1} + \frac{1}{R_2} + \frac{1}{R_3} + \cdots + \frac{1}{R_{N-1}} + \frac{1}{R_N}\right)^{-1} = \left(\sum_{i=1}^{N} \frac{1}{R_i}\right)^{-1},$$

   and is smaller than any individual resistance in the combination.
2. The potential drop across each resistor in parallel is the same.
3. Parallel resistors do not each get the total current; they divide it. The current entering a parallel combination of resistors is equal to the sum of the current through each resistor in parallel.

In this chapter, we introduced the equivalent resistance of resistors connect in series and resistors connected in parallel. You may recall that in Capacitance, we introduced the equivalent capacitance of capacitors connected in series and parallel. Circuits often contain both capacitors and resistors. Table 10.1 summarizes the equations used for the equivalent resistance and equivalent capacitance for series and parallel connections.

	Series combination	Parallel combination
Equivalent capacitance	$\frac{1}{C_{\text{eq}}} = \frac{1}{C_1} + \frac{1}{C_2} + \frac{1}{C_3} + \cdots$	$C_{\text{eq}} = C_1 + C_2 + C_3 + \cdots$
Equivalent resistance	$R_{\text{eq}} = R_1 + R_2 + R_3 + \cdots = \sum_{i=1}^{N} R_i$	$\frac{1}{R_{\text{eq}}} = \frac{1}{R_1} + \frac{1}{R_2} + \frac{1}{R_3} + \cdots$

Table 10.1 Summary for Equivalent Resistance and Capacitance in Series and Parallel Combinations

Combinations of Series and Parallel

More complex connections of resistors are often just combinations of series and parallel connections. Such combinations are common, especially when wire resistance is considered. In that case, wire resistance is in series with other resistances that are in parallel.

Combinations of series and parallel can be reduced to a single equivalent resistance using the technique illustrated in Figure 10.15. Various parts can be identified as either series or parallel connections, reduced to their equivalent resistances, and then further reduced until a single equivalent resistance is left. The process is more time consuming than difficult. Here, we note the equivalent resistance as R_{eq}.

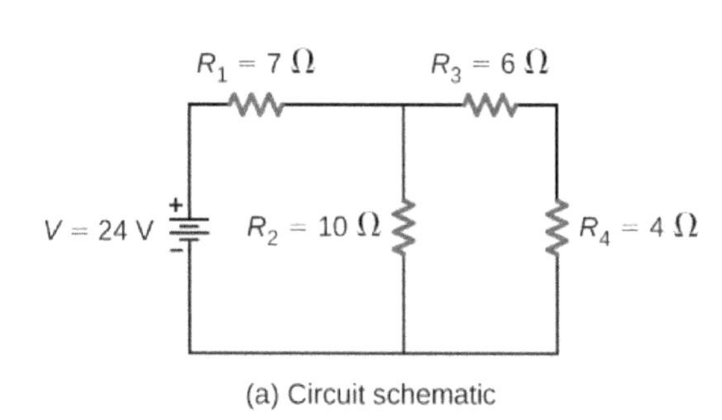

(a) Circuit schematic

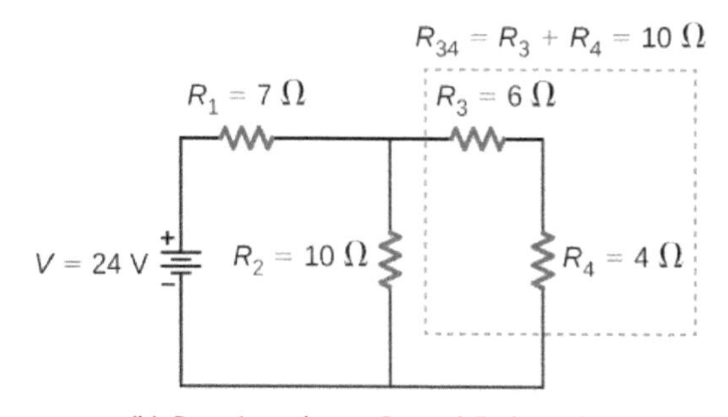

(b) Step 1: resistors R_3 and R_4 in series

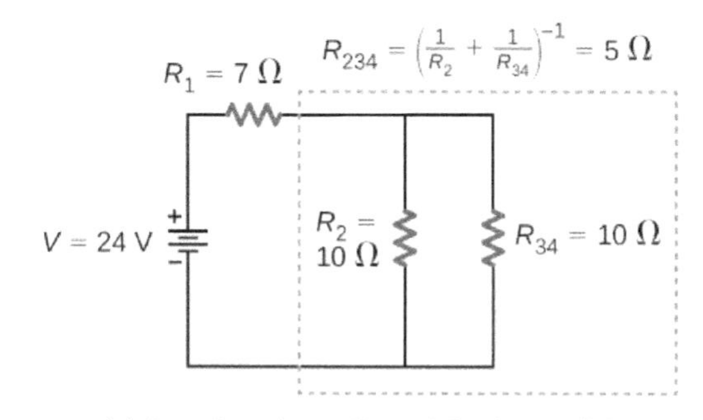

(c) Step 2: resistors R_2 and R_{34} in parallel

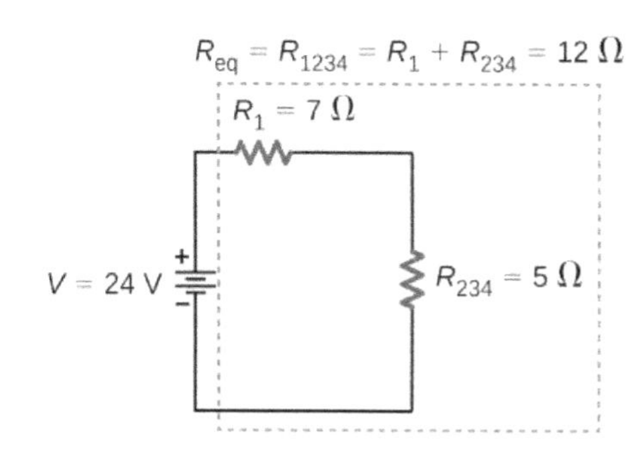

(d) Step 3: resistors R_1 and R_{234} in series

$V = 24\text{ V}$ $R_{eq} = 12\,\Omega$

(e) Simplified schematic reflecting equivalent resistance R_{eq}

Figure 10.15 (a) The original circuit of four resistors. (b) Step 1: The resistors R_3 and R_4 are in series and the equivalent resistance is $R_{34} = 10\,\Omega.$ (c) Step 2: The reduced circuit shows resistors R_2 and R_{34} are in parallel, with an equivalent resistance of $R_{234} = 5\,\Omega.$ (d) Step 3: The reduced circuit shows that R_1 and R_{234} are in series with an equivalent resistance of $R_{1234} = 12\,\Omega,$ which is the equivalent resistance $R_{eq}.$ (e) The reduced circuit with a voltage source of $V = 24\text{ V}$ with an equivalent resistance of $R_{eq} = 12\,\Omega.$ This results in a current of $I = 2\text{ A}$ from the voltage source.

Notice that resistors R_3 and R_4 are in series. They can be combined into a single equivalent resistance. One method of keeping track of the process is to include the resistors as subscripts. Here the equivalent resistance of R_3 and R_4 is

$$R_{34} = R_3 + R_4 = 6\,\Omega + 4\,\Omega = 10\,\Omega.$$

The circuit now reduces to three resistors, shown in Figure 10.15(c). Redrawing, we now see that resistors R_2 and R_{34} constitute a parallel circuit. Those two resistors can be reduced to an equivalent resistance:

$$R_{234} = \left(\frac{1}{R_2} + \frac{1}{R_{34}}\right)^{-1} = \left(\frac{1}{10\,\Omega} + \frac{1}{10\,\Omega}\right)^{-1} = 5\,\Omega.$$

This step of the process reduces the circuit to two resistors, shown in in Figure 10.15(d). Here, the circuit reduces to two resistors, which in this case are in series. These two resistors can be reduced to an equivalent resistance, which is the equivalent resistance of the circuit:

$$R_{eq} = R_{1234} = R_1 + R_{234} = 7\,\Omega + 5\,\Omega = 12\,\Omega.$$

The main goal of this circuit analysis is reached, and the circuit is now reduced to a single resistor and single voltage source. Now we can analyze the circuit. The current provided by the voltage source is $I = \frac{V}{R_{eq}} = \frac{24\text{ V}}{12\,\Omega} = 2\text{ A}$. This current runs through resistor R_1 and is designated as I_1. The potential drop across R_1 can be found using Ohm's law:

$$V_1 = I_1 R_1 = (2\text{ A})(7\,\Omega) = 14\text{ V}.$$

Looking at Figure 10.15(c), this leaves $24\text{ V} - 14\text{ V} = 10\text{ V}$ to be dropped across the parallel combination of R_2 and R_{34}. The current through R_2 can be found using Ohm's law:

$$I_2 = \frac{V_2}{R_2} = \frac{10\text{ V}}{10\,\Omega} = 1\text{ A}.$$

The resistors R_3 and R_4 are in series so the currents I_3 and I_4 are equal to

$$I_3 = I_4 = I - I_2 = 2\text{ A} - 1\text{ A} = 1\text{ A}.$$

Using Ohm's law, we can find the potential drop across the last two resistors. The potential drops are $V_3 = I_3 R_3 = 6\text{ V}$ and $V_4 = I_4 R_4 = 4\text{ V}$. The final analysis is to look at the power supplied by the voltage source and the power dissipated by the resistors. The power dissipated by the resistors is

$$\begin{aligned} P_1 &= I_1^2 R_1 = (2\text{ A})^2 (7\,\Omega) = 28\text{ W}, \\ P_2 &= I_2^2 R_2 = (1\text{ A})^2 (10\,\Omega) = 10\text{ W}, \\ P_3 &= I_3^2 R_3 = (1\text{ A})^2 (6\,\Omega) = 6\text{ W}, \\ P_4 &= I_4^2 R_4 = (1\text{ A})^2 (4\,\Omega) = 4\text{ W}, \\ P_{\text{dissipated}} &= P_1 + P_2 + P_3 + P_4 = 48\text{ W}. \end{aligned}$$

The total energy is constant in any process. Therefore, the power supplied by the voltage source is $P_s = IV = (2\text{ A})(24\text{ V}) = 48\text{ W}$. Analyzing the power supplied to the circuit and the power dissipated by the resistors is a good check for the validity of the analysis; they should be equal.

Example 10.4

Combining Series and Parallel Circuits

Figure 10.16 shows resistors wired in a combination of series and parallel. We can consider R_1 to be the resistance of wires leading to R_2 and R_3. (a) Find the equivalent resistance of the circuit. (b) What is the potential drop V_1 across resistor R_1? (c) Find the current I_2 through resistor R_2. (d) What power is dissipated by R_2?

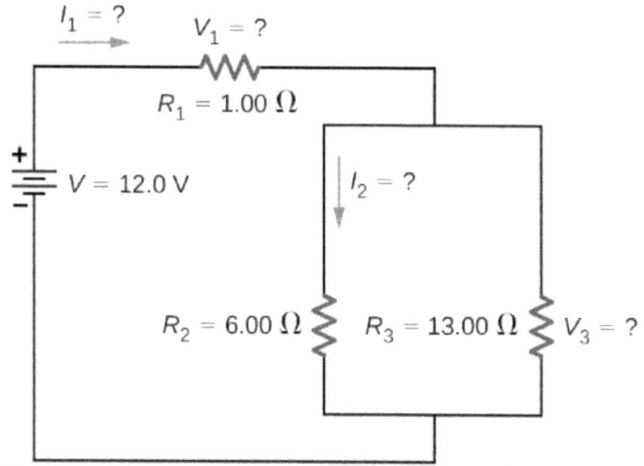

Figure 10.16 These three resistors are connected to a voltage source so that R_2 and R_3 are in parallel with one another and that combination is in series with R_1.

Strategy

(a) To find the equivalent resistance, first find the equivalent resistance of the parallel connection of R_2 and R_3. Then use this result to find the equivalent resistance of the series connection with R_1.

(b) The current through R_1 can be found using Ohm's law and the voltage applied. The current through R_1 is equal to the current from the battery. The potential drop V_1 across the resistor R_1 (which represents the resistance in the connecting wires) can be found using Ohm's law.

(c) The current through R_2 can be found using Ohm's law $I_2 = \frac{V_2}{R_2}$. The voltage across R_2 can be found using $V_2 = V - V_1$.

(d) Using Ohm's law $(V_2 = I_2 R_2)$, the power dissipated by the resistor can also be found using $P_2 = I_2^2 R_2 = \frac{V_2^2}{R_2}$.

Solution

a. To find the equivalent resistance of the circuit, notice that the parallel connection of R_2 and R_3 is in series with R_1, so the equivalent resistance is

$$R_{eq} = R_1 + \left(\frac{1}{R_2} + \frac{1}{R_3}\right)^{-1} = 1.00\,\Omega + \left(\frac{1}{6.00\,\Omega} + \frac{1}{13.00\,\Omega}\right)^{-1} = 5.10\,\Omega.$$

The total resistance of this combination is intermediate between the pure series and pure parallel values ($20.0\,\Omega$ and $0.804\,\Omega$, respectively).

b. The current through R_1 is equal to the current supplied by the battery:

$$I_1 = I = \frac{V}{R_{eq}} = \frac{12.0\text{ V}}{5.10\,\Omega} = 2.35\text{ A}.$$

The voltage across R_1 is

$$V_1 = I_1 R_1 = (2.35\text{ A})(1\,\Omega) = 2.35\text{ V}.$$

The voltage applied to R_2 and R_3 is less than the voltage supplied by the battery by an amount V_1. When wire resistance is large, it can significantly affect the operation of the devices represented by R_2 and R_3.

c. To find the current through R_2, we must first find the voltage applied to it. The voltage across the two resistors in parallel is the same:

$$V_2 = V_3 = V - V_1 = 12.0\text{ V} - 2.35\text{ V} = 9.65\text{ V}.$$

Now we can find the current I_2 through resistance R_2 using Ohm's law:

$$I_2 = \frac{V_2}{R_2} = \frac{9.65\text{ V}}{6.00\,\Omega} = 1.61\text{ A}.$$

The current is less than the 2.00 A that flowed through R_2 when it was connected in parallel to the battery in the previous parallel circuit example.

d. The power dissipated by R_2 is given by

$$P_2 = I_2^2 R_2 = (1.61\text{ A})^2(6.00\,\Omega) = 15.5\text{ W}.$$

Significance

The analysis of complex circuits can often be simplified by reducing the circuit to a voltage source and an equivalent resistance. Even if the entire circuit cannot be reduced to a single voltage source and a single equivalent resistance, portions of the circuit may be reduced, greatly simplifying the analysis.

10.5 Check Your Understanding Consider the electrical circuits in your home. Give at least two examples of circuits that must use a combination of series and parallel circuits to operate efficiently.

Practical Implications

One implication of this last example is that resistance in wires reduces the current and power delivered to a resistor. If wire resistance is relatively large, as in a worn (or a very long) extension cord, then this loss can be significant. If a large current is drawn, the *IR* drop in the wires can also be significant and may become apparent from the heat generated in the cord.

For example, when you are rummaging in the refrigerator and the motor comes on, the refrigerator light dims momentarily. Similarly, you can see the passenger compartment light dim when you start the engine of your car (although this may be due to resistance inside the battery itself).

What is happening in these high-current situations is illustrated in Figure 10.17. The device represented by R_3 has a very low resistance, so when it is switched on, a large current flows. This increased current causes a larger *IR* drop in the wires represented by R_1, reducing the voltage across the light bulb (which is R_2), which then dims noticeably.

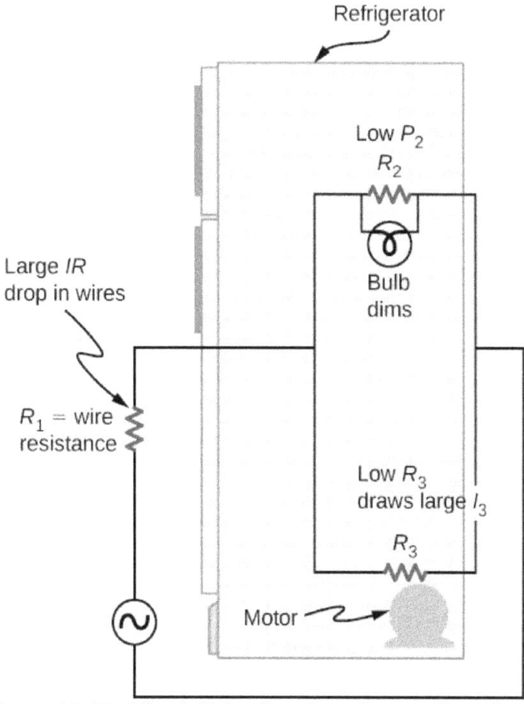

Figure 10.17 Why do lights dim when a large appliance is switched on? The answer is that the large current the appliance motor draws causes a significant *IR* drop in the wires and reduces the voltage across the light.

Problem-Solving Strategy: Series and Parallel Resistors

1. Draw a clear circuit diagram, labeling all resistors and voltage sources. This step includes a list of the known values for the problem, since they are labeled in your circuit diagram.
2. Identify exactly what needs to be determined in the problem (identify the unknowns). A written list is useful.
3. Determine whether resistors are in series, parallel, or a combination of both series and parallel. Examine the circuit diagram to make this assessment. Resistors are in series if the same current must pass sequentially through them.
4. Use the appropriate list of major features for series or parallel connections to solve for the unknowns. There is one list for series and another for parallel.
5. Check to see whether the answers are reasonable and consistent.

Example 10.5

Combining Series and Parallel Circuits

Two resistors connected in series (R_1, R_2) are connected to two resistors that are connected in parallel (R_3, R_4). The series-parallel combination is connected to a battery. Each resistor has a resistance of 10.00 Ohms. The wires connecting the resistors and battery have negligible resistance. A current of 2.00 Amps runs through resistor R_1. What is the voltage supplied by the voltage source?

Strategy

Use the steps in the preceding problem-solving strategy to find the solution for this example.

Solution

1. Draw a clear circuit diagram (Figure 10.18).

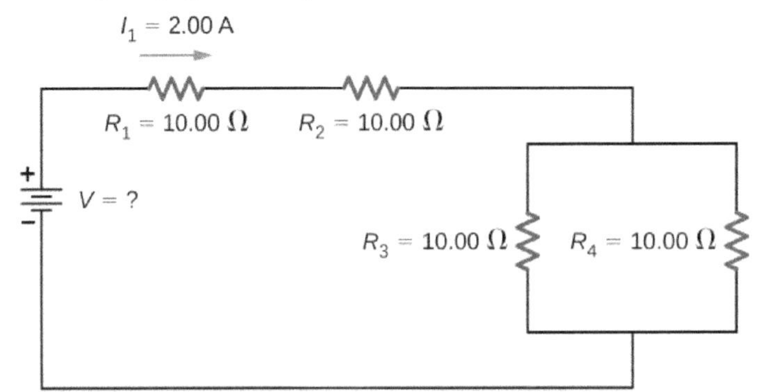

Figure 10.18 To find the unknown voltage, we must first find the equivalent resistance of the circuit.

2. The unknown is the voltage of the battery. In order to find the voltage supplied by the battery, the equivalent resistance must be found.
3. In this circuit, we already know that the resistors R_1 and R_2 are in series and the resistors R_3 and R_4 are in parallel. The equivalent resistance of the parallel configuration of the resistors R_3 and R_4 is in series with the series configuration of resistors R_1 and R_2.
4. The voltage supplied by the battery can be found by multiplying the current from the battery and the equivalent resistance of the circuit. The current from the battery is equal to the current through R_1 and is equal to 2.00 A. We need to find the equivalent resistance by reducing the circuit. To reduce the circuit, first consider the two resistors in parallel. The equivalent resistance is $R_{34} = \left(\frac{1}{10.00\,\Omega} + \frac{1}{10.00\,\Omega}\right)^{-1} = 5.00\,\Omega$. This parallel combination is in series with the other two resistors, so the equivalent resistance of the circuit is $R_{\text{eq}} = R_1 + R_2 + R_{34} = 25.00\,\Omega$. The voltage supplied by the battery is therefore $V = IR_{\text{eq}} = 2.00\text{ A}(25.00\,\Omega) = 50.00\text{ V}$.
5. One way to check the consistency of your results is to calculate the power supplied by the battery and the power dissipated by the resistors. The power supplied by the battery is $P_{\text{batt}} = IV = 100.00\text{ W}$.

   Since they are in series, the current through R_2 equals the current through R_1. Since $R_3 = R_4$, the current through each will be 1.00 Amps. The power dissipated by the resistors is equal to the sum of the power dissipated by each resistor:

$$P = I_1^2R_1 + I_2^2R_2 + I_3^2R_3 + I_4^2R_4 = 40.00\text{ W} + 40.00\text{ W} + 10.00\text{ W} + 10.00\text{ W} = 100.00\text{ W}.$$

   Since the power dissipated by the resistors equals the power supplied by the battery, our solution seems consistent.

Significance

If a problem has a combination of series and parallel, as in this example, it can be reduced in steps by using the preceding problem-solving strategy and by considering individual groups of series or parallel connections. When finding R_{eq} for a parallel connection, the reciprocal must be taken with care. In addition, units and numerical results must be reasonable. Equivalent series resistance should be greater, whereas equivalent parallel resistance should be smaller, for example. Power should be greater for the same devices in parallel compared with series, and so on.

10.3 | Kirchhoff's Rules

Learning Objectives

By the end of the section, you will be able to:

- State Kirchhoff's junction rule
- State Kirchhoff's loop rule
- Analyze complex circuits using Kirchhoff's rules

We have just seen that some circuits may be analyzed by reducing a circuit to a single voltage source and an equivalent resistance. Many complex circuits cannot be analyzed with the series-parallel techniques developed in the preceding sections. In this section, we elaborate on the use of Kirchhoff's rules to analyze more complex circuits. For example, the circuit in Figure 10.19 is known as a multi-loop circuit, which consists of junctions. A junction, also known as a node, is a connection of three or more wires. In this circuit, the previous methods cannot be used, because not all the resistors are in clear series or parallel configurations that can be reduced. Give it a try. The resistors R_1 and R_2 are in series and can be reduced to an equivalent resistance. The same is true of resistors R_4 and R_5. But what do you do then?

Even though this circuit cannot be analyzed using the methods already learned, two circuit analysis rules can be used to analyze any circuit, simple or complex. The rules are known as **Kirchhoff's rules**, after their inventor Gustav Kirchhoff (1824–1887).

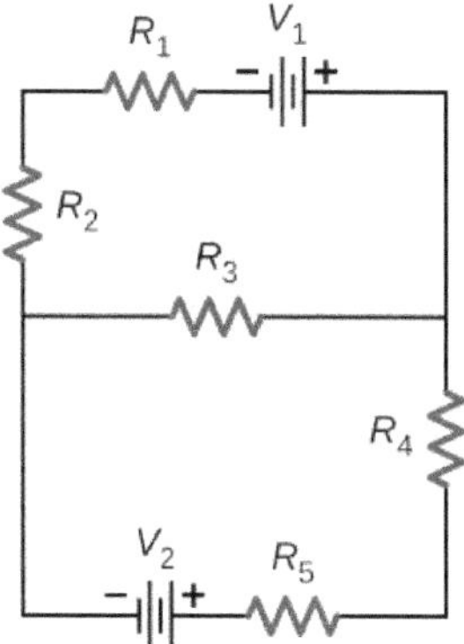

Figure 10.19 This circuit cannot be reduced to a combination of series and parallel connections. However, we can use Kirchhoff's rules to analyze it.

Kirchhoff's Rules

- Kirchhoff's first rule—the junction rule. The sum of all currents entering a junction must equal the sum of all currents leaving the junction:

$$\sum I_{\text{in}} = \sum I_{\text{out}}. \tag{10.4}$$

- Kirchhoff's second rule—the loop rule. The algebraic sum of changes in potential around any closed circuit path (loop) must be zero:

$$\sum V = 0. \tag{10.5}$$

We now provide explanations of these two rules, followed by problem-solving hints for applying them and a worked example that uses them.

Kirchhoff's First Rule

Kirchhoff's first rule (the **junction rule**) applies to the charge entering and leaving a junction (Figure 10.20). As stated earlier, a junction, or node, is a connection of three or more wires. Current is the flow of charge, and charge is conserved; thus, whatever charge flows into the junction must flow out.

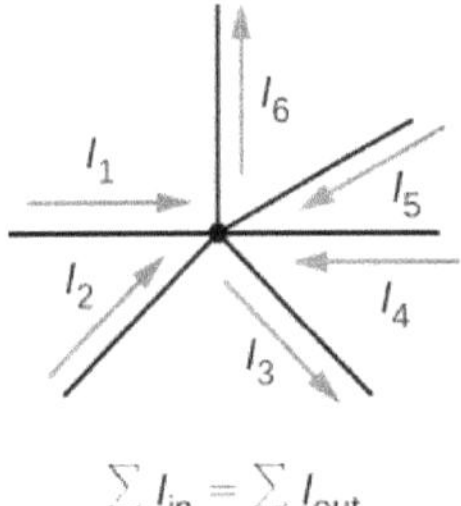

$$\sum I_{in} = \sum I_{out}$$
$$I_1 + I_2 + I_4 + I_5 = I_3 + I_6$$

Figure 10.20 Charge must be conserved, so the sum of currents into a junction must be equal to the sum of currents out of the junction.

Although it is an over-simplification, an analogy can be made with water pipes connected in a plumbing junction. If the wires in Figure 10.20 were replaced by water pipes, and the water was assumed to be incompressible, the volume of water flowing into the junction must equal the volume of water flowing out of the junction.

Kirchhoff's Second Rule

Kirchhoff's second rule (the **loop rule**) applies to potential differences. The loop rule is stated in terms of potential V rather than potential energy, but the two are related since $U = qV$. In a closed loop, whatever energy is supplied by a voltage source, the energy must be transferred into other forms by the devices in the loop, since there are no other ways in which energy can be transferred into or out of the circuit. Kirchhoff's loop rule states that the algebraic sum of potential differences, including voltage supplied by the voltage sources and resistive elements, in any loop must be equal to zero. For example, consider a simple loop with no junctions, as in Figure 10.21.

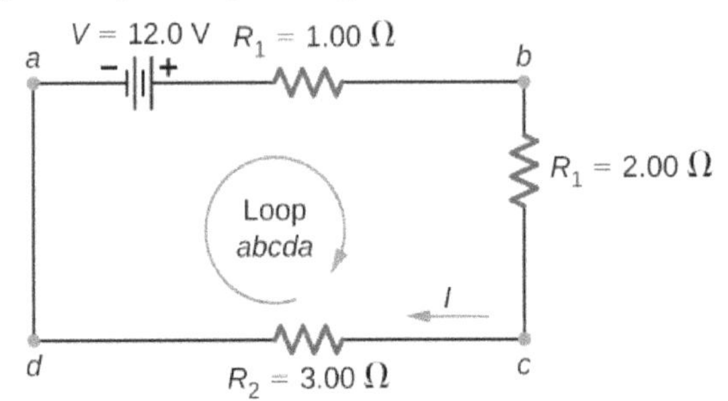

Figure 10.21 A simple loop with no junctions. Kirchhoff's loop rule states that the algebraic sum of the voltage differences is equal to zero.

The circuit consists of a voltage source and three external load resistors. The labels *a*, *b*, *c*, and *d* serve as references, and have no other significance. The usefulness of these labels will become apparent soon. The loop is designated as Loop *abcda*, and the labels help keep track of the voltage differences as we travel around the circuit. Start at point *a* and travel to point *b*. The voltage of the voltage source is added to the equation and the potential drop of the resistor R_1 is subtracted. From point *b* to *c*, the potential drop across R_2 is subtracted. From *c* to *d*, the potential drop across R_3 is subtracted. From points *d* to *a*, nothing is done because there are no components.

Figure 10.22 shows a graph of the voltage as we travel around the loop. Voltage increases as we cross the battery, whereas voltage decreases as we travel across a resistor. The potential drop, or change in the electric potential, is equal to the current through the resistor times the resistance of the resistor. Since the wires have negligible resistance, the voltage remains constant as we cross the wires connecting the components.

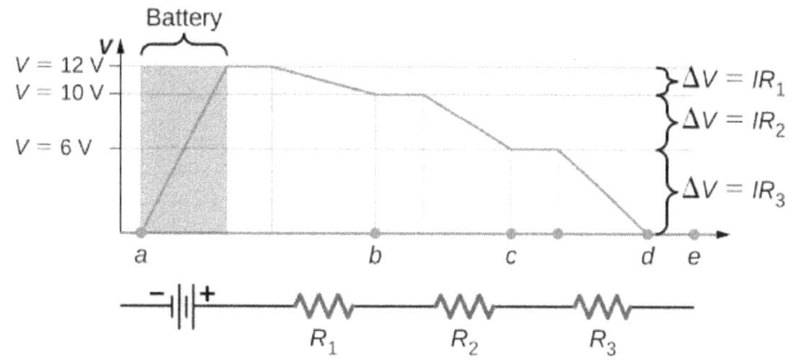

Figure 10.22 A voltage graph as we travel around the circuit. The voltage increases as we cross the battery and decreases as we cross each resistor. Since the resistance of the wire is quite small, we assume that the voltage remains constant as we cross the wires connecting the components.

Then Kirchhoff's loop rule states

$$V - IR_1 - IR_2 - IR_3 = 0.$$

The loop equation can be used to find the current through the loop:

$$I = \frac{V}{R_1 + R_2 + R_2} = \frac{12.00\ \text{V}}{1.00\ \Omega + 2.00\ \Omega + 3.00\ \Omega} = 2.00\ \text{A}.$$

This loop could have been analyzed using the previous methods, but we will demonstrate the power of Kirchhoff's method in the next section.

Applying Kirchhoff's Rules

By applying Kirchhoff's rules, we generate a set of linear equations that allow us to find the unknown values in circuits. These may be currents, voltages, or resistances. Each time a rule is applied, it produces an equation. If there are as many independent equations as unknowns, then the problem can be solved.

Using Kirchhoff's method of analysis requires several steps, as listed in the following procedure.

Problem-Solving Strategy: Kirchhoff's Rules

1. Label points in the circuit diagram using lowercase letters *a*, *b*, *c*, …. These labels simply help with orientation.
2. Locate the junctions in the circuit. The junctions are points where three or more wires connect. Label each junction with the currents and directions into and out of it. Make sure at least one current points into the junction and at least one current points out of the junction.
3. Choose the loops in the circuit. Every component must be contained in at least one loop, but a component may be contained in more than one loop.
4. Apply the junction rule. Again, some junctions should not be included in the analysis. You need only use enough nodes to include every current.
5. Apply the loop rule. Use the map in Figure 10.23.

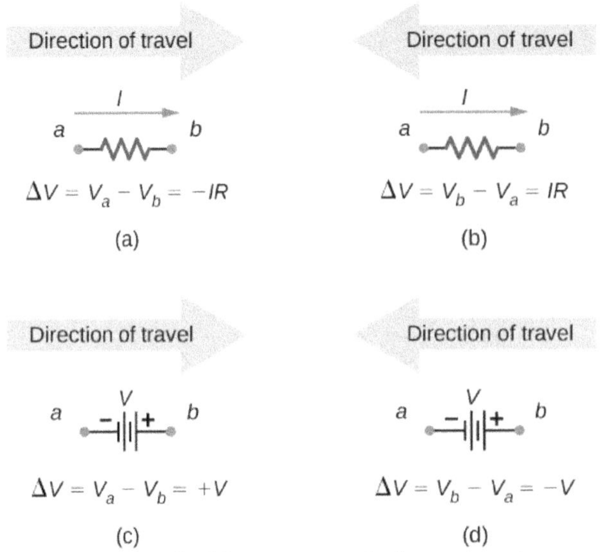

Direction of travel

$\Delta V = V_a - V_b = -IR$

(a)

Direction of travel

$\Delta V = V_b - V_a = IR$

(b)

Direction of travel

$\Delta V = V_a - V_b = +V$

(c)

Direction of travel

$\Delta V = V_b - V_a = -V$

(d)

Figure 10.23 Each of these resistors and voltage sources is traversed from a to b. (a) When moving across a resistor in the same direction as the current flow, subtract the potential drop. (b) When moving across a resistor in the opposite direction as the current flow, add the potential drop. (c) When moving across a voltage source from the negative terminal to the positive terminal, add the potential drop. (d) When moving across a voltage source from the positive terminal to the negative terminal, subtract the potential drop.

Let's examine some steps in this procedure more closely. When locating the junctions in the circuit, do not be concerned about the direction of the currents. If the direction of current flow is not obvious, choosing any direction is sufficient as long as at least one current points into the junction and at least one current points out of the junction. If the arrow is in the opposite direction of the conventional current flow, the result for the current in question will be negative but the answer will still be correct.

The number of nodes depends on the circuit. Each current should be included in a node and thus included in at least one junction equation. Do not include nodes that are not linearly independent, meaning nodes that contain the same information.

Consider Figure 10.24. There are two junctions in this circuit: Junction b and Junction e. Points a, c, d, and f are not junctions, because a junction must have three or more connections. The equation for Junction b is $I_1 = I_2 + I_3$, and the equation for Junction e is $I_2 + I_3 = I_1$. These are equivalent equations, so it is necessary to keep only one of them.

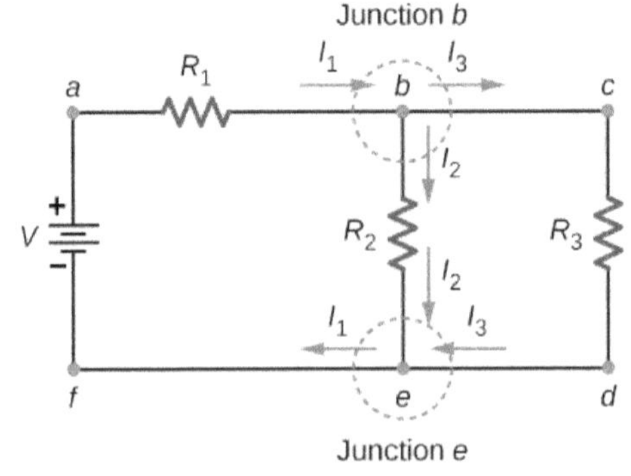

Figure 10.24 At first glance, this circuit contains two junctions, Junction b and Junction e, but only one should be considered because their junction equations are equivalent.

When choosing the loops in the circuit, you need enough loops so that each component is covered once, without repeating loops. Figure 10.25 shows four choices for loops to solve a sample circuit; choices (a), (b), and (c) have a sufficient amount of loops to solve the circuit completely. Option (d) reflects more loops than necessary to solve the circuit.

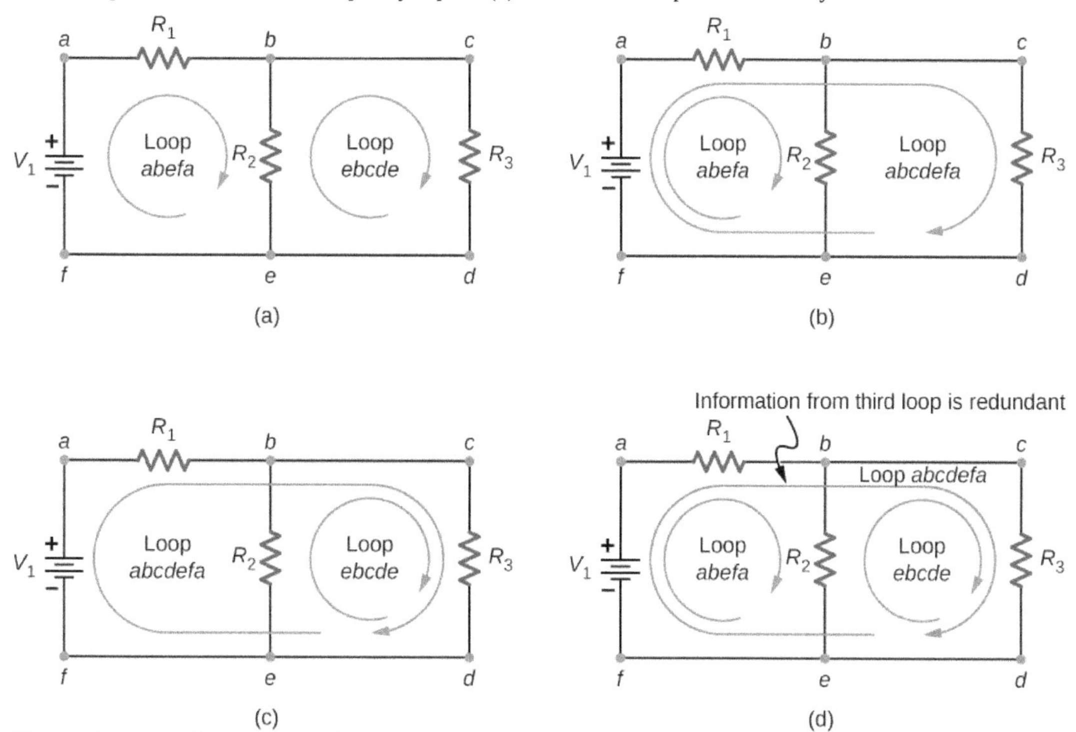

Figure 10.25 Panels (a)–(c) are sufficient for the analysis of the circuit. In each case, the two loops shown contain all the circuit elements necessary to solve the circuit completely. Panel (d) shows three loops used, which is more than necessary. Any two loops in the system will contain all information needed to solve the circuit. Adding the third loop provides redundant information.

Consider the circuit in Figure 10.26(a). Let us analyze this circuit to find the current through each resistor. First, label the circuit as shown in part (b).

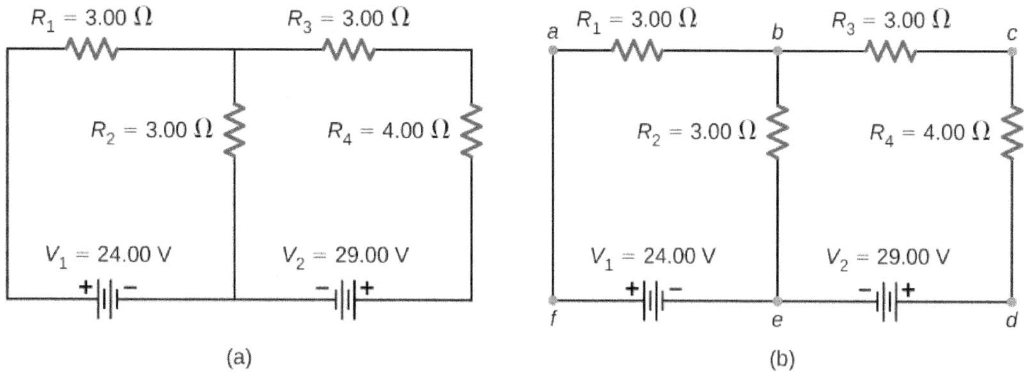

Figure 10.26 (a) A multi-loop circuit. (b) Label the circuit to help with orientation.

Next, determine the junctions. In this circuit, points *b* and *e* each have three wires connected, making them junctions. Start to apply Kirchhoff's junction rule $\left(\sum I_{\text{in}} = \sum I_{\text{out}}\right)$ by drawing arrows representing the currents and labeling each arrow, as shown in Figure 10.27(b). Junction *b* shows that $I_1 = I_2 + I_3$ and Junction *e* shows that $I_2 + I_3 = I_1$. Since Junction *e* gives the same information of Junction *b*, it can be disregarded. This circuit has three unknowns, so we need three linearly independent equations to analyze it.

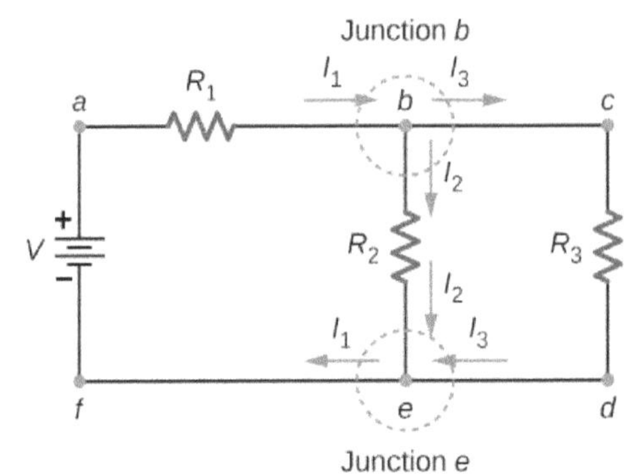

Figure 10.27 (a) This circuit has two junctions, labeled *b* and *e*, but only node *b* is used in the analysis. (b) Labeled arrows represent the currents into and out of the junctions.

Next we need to choose the loops. In Figure 10.28, Loop *abefa* includes the voltage source V_1 and resistors R_1 and R_2 . The loop starts at point *a*, then travels through points *b*, *e*, and *f*, and then back to point *a*. The second loop, Loop *ebcde*, starts at point *e* and includes resistors R_2 and R_3, and the voltage source V_2.

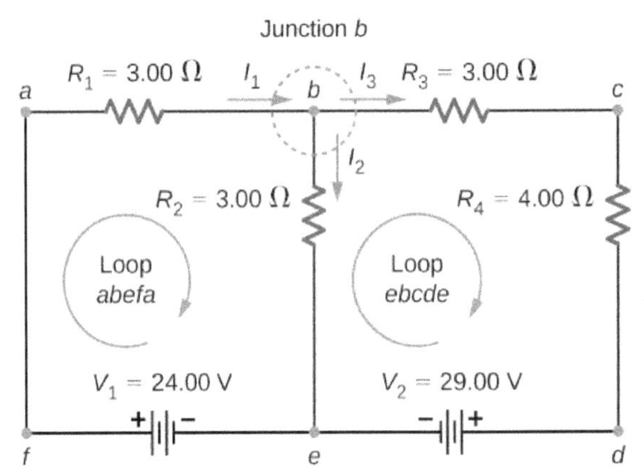

Figure 10.28 Choose the loops in the circuit.

Now we can apply Kirchhoff's loop rule, using the map in Figure 10.23. Starting at point *a* and moving to point *b*, the resistor R_1 is crossed in the same direction as the current flow I_1, so the potential drop $I_1 R_1$ is subtracted. Moving from point *b* to point *e*, the resistor R_2 is crossed in the same direction as the current flow I_2 so the potential drop $I_2 R_2$ is subtracted. Moving from point *e* to point *f*, the voltage source V_1 is crossed from the negative terminal to the positive terminal, so V_1 is added. There are no components between points *f* and *a*. The sum of the voltage differences must equal zero:

$$\text{Loop } abefa: \quad -I_1 R_1 - I_2 R_2 + V_1 = 0 \text{ or } V_1 = I_1 R_1 + I_2 R_2.$$

Finally, we check loop *ebcde*. We start at point *e* and move to point *b*, crossing R_2 in the opposite direction as the current flow I_2. The potential drop $I_2 R_2$ is added. Next, we cross R_3 and R_4 in the same direction as the current flow I_3 and subtract the potential drops $I_3 R_3$ and $I_3 R_4$. Note that the current is the same through resistors R_3 and R_4, because they are connected in series. Finally, the voltage source is crossed from the positive terminal to the negative terminal, and the voltage source V_2 is subtracted. The sum of these voltage differences equals zero and yields the loop equation

$$\text{Loop } ebcde : I_2 R_2 - I_3 (R_3 + R_4) - V_2 = 0.$$

We now have three equations, which we can solve for the three unknowns.

$$\begin{aligned}
&(1)\ \text{Junction } b : I_1 - I_2 - I_3 = 0. \\
&(2)\ \text{Loop } abefa :\ I_1 R_1 + I_2 R_2 = V_1. \\
&(3)\ \text{Loop } ebcde : I_2 R_2 - I_3 (R_3 + R_4) = V_2.
\end{aligned}$$

To solve the three equations for the three unknown currents, start by eliminating current I_2. First add Eq. (1) times R_2 to Eq. (2). The result is labeled as Eq. (4):

$$\begin{aligned}
(R_1 + R_2) I_1 - R_2 I_3 &= V_1. \\
(4)\ 6\,\Omega I_1 - 3\,\Omega I_3 &= 24\text{ V}.
\end{aligned}$$

Next, subtract Eq. (3) from Eq. (2). The result is labeled as Eq. (5):

$$\begin{aligned}
I_1 R_1 + I_3 (R_3 + R_4) &= V_1 - V_2. \\
(5)\ 3\,\Omega I_1 + 7\,\Omega I_3 &= -5\text{ V}.
\end{aligned}$$

We can solve Eqs. (4) and (5) for current I_1. Adding seven times Eq. (4) and three times Eq. (5) results in $51\,\Omega I_1 = 153\text{ V}$, or $I_1 = 3.00\text{ A}$. Using Eq. (4) results in $I_3 = -2.00\text{ A}$. Finally, Eq. (1) yields $I_2 = I_1 - I_3 = 5.00\text{ A}$. One way to check that the solutions are consistent is to check the power supplied by the voltage sources and the power dissipated by the resistors:

$$\begin{aligned}
P_{\text{in}} &= I_1 V_1 + I_3 V_2 = 130\text{ W}, \\
P_{\text{out}} &= I_1^2 R_1 + I_2^2 R_2 + I_3^2 R_3 + I_3^2 R_4 = 130\text{ W}.
\end{aligned}$$

Note that the solution for the current I_3 is negative. This is the correct answer, but suggests that the arrow originally drawn in the junction analysis is the direction opposite of conventional current flow. The power supplied by the second voltage source is 58 W and not −58 W.

Example 10.6

Calculating Current by Using Kirchhoff's Rules

Find the currents flowing in the circuit in Figure 10.29.

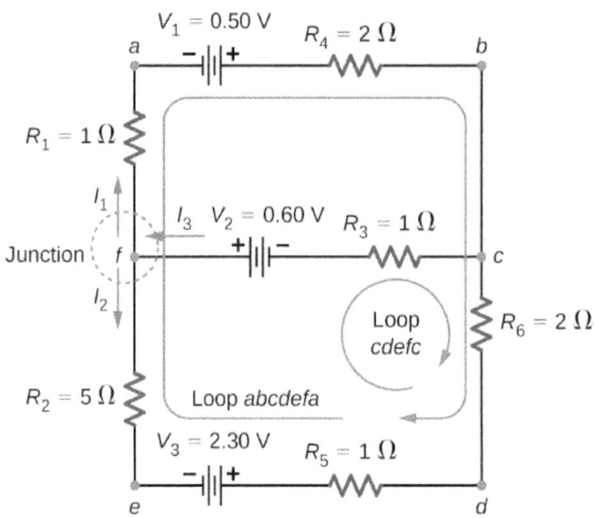

Figure 10.29 This circuit is combination of series and parallel configurations of resistors and voltage sources. This circuit cannot be analyzed using the techniques discussed in Electromotive Force but can be analyzed using Kirchhoff's rules.

Strategy

This circuit is sufficiently complex that the currents cannot be found using Ohm's law and the series-parallel techniques—it is necessary to use Kirchhoff's rules. Currents have been labeled I_1, I_2, and I_3 in the figure, and assumptions have been made about their directions. Locations on the diagram have been labeled with letters *a* through *h*. In the solution, we apply the junction and loop rules, seeking three independent equations to allow us to solve for the three unknown currents.

Solution

Applying the junction and loop rules yields the following three equations. We have three unknowns, so three equations are required.

$$\text{Junction } c: \; I_1 + I_2 = I_3.$$
$$\text{Loop } abcdefa: \; I_1(R_1 + R_4) - I_2(R_2 + R_5 + R_6) = V_1 - V_3.$$
$$\text{Loop } cdefc: \; I_2(R_2 + R_5 + R_6) + I_3 R_3 = V_2 + V_3.$$

Simplify the equations by placing the unknowns on one side of the equations.

$$\text{Junction } c: \; I_1 + I_2 - I_3 = 0.$$
$$\text{Loop } abcdefa: \; I_1(3\,\Omega) - I_2(8\,\Omega) = 0.5\text{ V} - 2.30\text{ V}.$$
$$\text{Loop } cdefc: \; I_2(8\,\Omega) + I_3(1\,\Omega) = 0.6\text{ V} + 2.30\text{ V}.$$

Simplify the equations. The first loop equation can be simplified by dividing both sides by 3.00. The second loop equation can be simplified by dividing both sides by 6.00.

$$\text{Junction } c: \; I_1 + I_2 - I_3 = 0.$$
$$\text{Loop } abcdefa: \; I_1(3\,\Omega) - I_2(8\,\Omega) = -1.8\text{ V}.$$
$$\text{Loop } cdefc: \; I_2(8\,\Omega) + I_3(1\,\Omega) = 2.9\text{ V}.$$

The results are

$$I_1 = 0.20\text{ A}, \quad I_2 = 0.30\text{ A}, \quad I_3 = 0.50\text{ A}.$$

Significance

A method to check the calculations is to compute the power dissipated by the resistors and the power supplied by the voltage sources:

$$P_{R_1} = I_1^2 R_1 = 0.04\text{ W}.$$
$$P_{R_2} = I_2^2 R_2 = 0.45\text{ W}.$$
$$P_{R_3} = I_3^2 R_3 = 0.25\text{ W}.$$
$$P_{R_4} = I_1^2 R_4 = 0.08\text{ W}.$$
$$P_{R_5} = I_2^2 R_5 = 0.09\text{ W}.$$
$$P_{R_6} = I_2^2 R_6 = 0.18\text{ W}.$$
$$P_{\text{dissipated}} = 1.09\text{ W}.$$
$$P_{\text{source}} = I_1 V_1 + I_2 V_3 + I_3 V_2 = 0.10\text{ W} + 0.69\text{ W} + 0.30\text{ W} = 1.09\text{ W}.$$

The power supplied equals the power dissipated by the resistors.

10.6 Check Your Understanding In considering the following schematic and the power supplied and consumed by a circuit, will a voltage source always provide power to the circuit, or can a voltage source consume power?

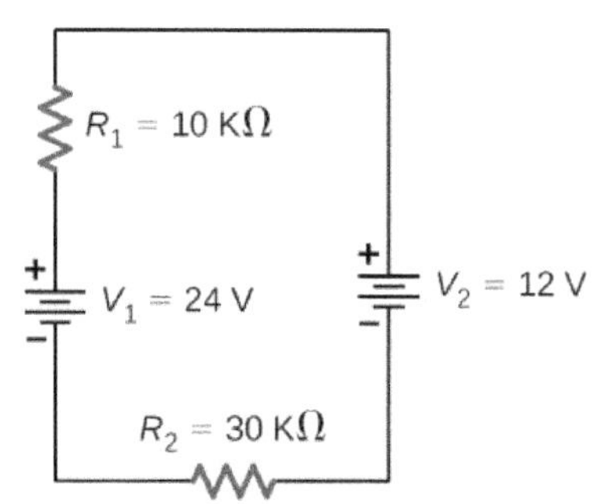

Example 10.7

Calculating Current by Using Kirchhoff's Rules

Find the current flowing in the circuit in Figure 10.30.

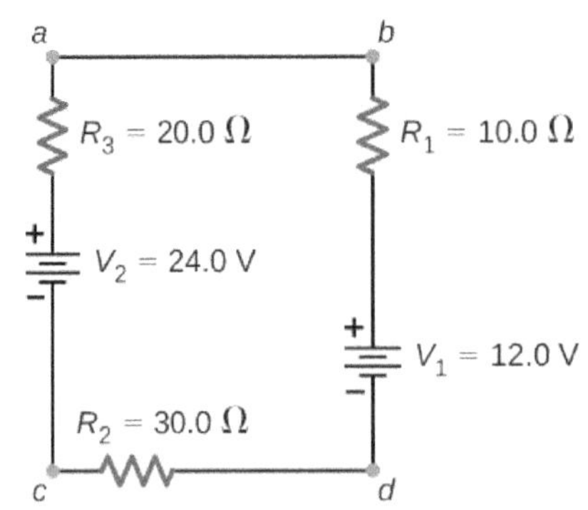

Figure 10.30 This circuit consists of three resistors and two batteries connected in series. Note that the batteries are connected with opposite polarities.

Strategy

This circuit can be analyzed using Kirchhoff's rules. There is only one loop and no nodes. Choose the direction of current flow. For this example, we will use the clockwise direction from point *a* to point *b*. Consider Loop *abcda* and use Figure 10.23 to write the loop equation. Note that according to Figure 10.23, battery V_1 will be added and battery V_2 will be subtracted.

Solution

Applying the junction rule yields the following three equations. We have one unknown, so one equation is required:

$$\text{Loop } abcda: \ -IR_1 - V_1 - IR_2 + V_2 - IR_3 = 0.$$

Simplify the equations by placing the unknowns on one side of the equations. Use the values given in the figure.

$$I(R_1 + R_2 + R_3) = V_2 - V_1.$$

$$I = \frac{V_2 - V_1}{R_1 + R_2 + R_3} = \frac{24\text{ V} - 12\text{ V}}{10.0\,\Omega + 30.0\,\Omega + 10.0\,\Omega} = 0.20\text{ A}.$$

Significance

The power dissipated or consumed by the circuit equals the power supplied to the circuit, but notice that the current in the battery V_1 is flowing through the battery from the positive terminal to the negative terminal and consumes power.

$$P_{R_1} = I^2 R_1 = 0.40\text{ W}$$
$$P_{R_2} = I^2 R_2 = 1.20\text{ W}$$
$$P_{R_3} = I^2 R_3 = 0.80\text{ W}$$
$$P_{V_1} = IV_1 = 2.40\text{ W}$$
$$P_{\text{dissipated}} = 4.80\text{ W}$$
$$P_{\text{source}} = IV_2 = 4.80\text{ W}$$

The power supplied equals the power dissipated by the resistors and consumed by the battery V_1.

10.7 Check Your Understanding When using Kirchhoff's laws, you need to decide which loops to use and the direction of current flow through each loop. In analyzing the circuit in Example 10.7, the direction of current flow was chosen to be clockwise, from point *a* to point *b*. How would the results change if the direction of the current was chosen to be counterclockwise, from point *b* to point *a*?

Multiple Voltage Sources

Many devices require more than one battery. Multiple voltage sources, such as batteries, can be connected in series configurations, parallel configurations, or a combination of the two.

In series, the positive terminal of one battery is connected to the negative terminal of another battery. Any number of voltage sources, including batteries, can be connected in series. Two batteries connected in series are shown in Figure 10.31. Using Kirchhoff's loop rule for the circuit in part (b) gives the result

$$\varepsilon_1 - Ir_1 + \varepsilon_2 - Ir_2 - IR = 0,$$
$$\left[(\varepsilon_1 + \varepsilon_2) - I(r_1 + r_2)\right] - IR = 0.$$

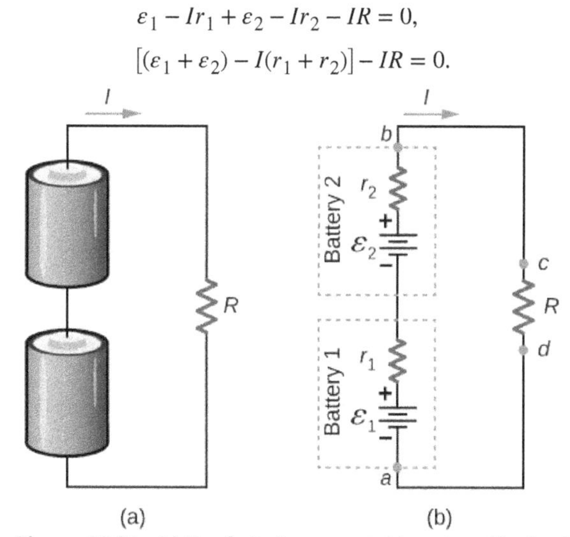

Figure 10.31 (a) Two batteries connected in series with a load resistor. (b) The circuit diagram of the two batteries and the load resistor, with each battery modeled as an idealized emf source and an internal resistance.

When voltage sources are in series, their internal resistances can be added together and their emfs can be added together to get the total values. Series connections of voltage sources are common—for example, in flashlights, toys, and other appliances. Usually, the cells are in series in order to produce a larger total emf. In Figure 10.31, the terminal voltage is

$$V_{\text{terminal}} = (\varepsilon_1 - Ir_1) + (\varepsilon_2 - Ir_2) = \left[(\varepsilon_1 + \varepsilon_2) - I(r_1 + r_2)\right] = (\varepsilon_1 + \varepsilon_2) + Ir_{\text{eq}}.$$

Note that the same current I is found in each battery because they are connected in series. The disadvantage of series connections of cells is that their internal resistances are additive.

Batteries are connected in series to increase the voltage supplied to the circuit. For instance, an LED flashlight may have two AAA cell batteries, each with a terminal voltage of 1.5 V, to provide 3.0 V to the flashlight.

Any number of batteries can be connected in series. For N batteries in series, the terminal voltage is equal to

$$V_{\text{terminal}} = (\varepsilon_1 + \varepsilon_2 + \cdots + \varepsilon_{N-1} + \varepsilon_N) - I(r_1 + r_2 + \cdots + r_{N-1} + r_N) = \sum_{i=1}^{N} \varepsilon_i - Ir_{\text{eq}} \quad (10.6)$$

where the equivalent resistance is $r_{\text{eq}} = \sum_{i=1}^{N} r_i$.

When a load is placed across voltage sources in series, as in Figure 10.32, we can find the current:

$$\begin{aligned}(\varepsilon_1 - Ir_1) + (\varepsilon_2 - Ir_2) &= IR, \\ Ir_1 + Ir_2 + IR &= \varepsilon_1 + \varepsilon_2, \\ I &= \frac{\varepsilon_1 + \varepsilon_2}{r_1 + r_2 + R}.\end{aligned}$$

As expected, the internal resistances increase the equivalent resistance.

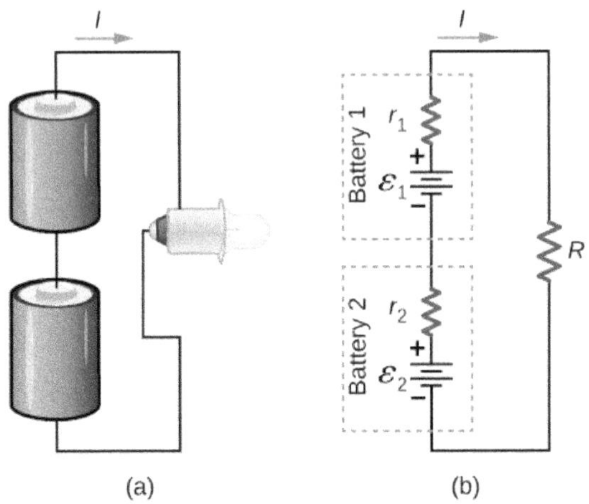

Figure 10.32 Two batteries connect in series to an LED bulb, as found in a flashlight.

Voltage sources, such as batteries, can also be connected in parallel. Figure 10.33 shows two batteries with identical emfs in parallel and connected to a load resistance. When the batteries are connect in parallel, the positive terminals are connected together and the negative terminals are connected together, and the load resistance is connected to the positive and negative terminals. Normally, voltage sources in parallel have identical emfs. In this simple case, since the voltage sources are in parallel, the total emf is the same as the individual emfs of each battery.

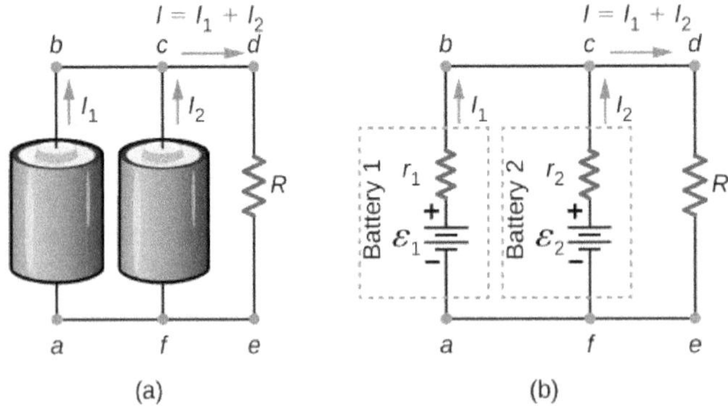

Figure 10.33 (a) Two batteries connect in parallel to a load resistor. (b) The circuit diagram shows the shows battery as an emf source and an internal resistor. The two emf sources have identical emfs (each labeled by ε) connected in parallel that produce the same emf.

Consider the Kirchhoff analysis of the circuit in Figure 10.33(b). There are two loops and a node at point b and $\varepsilon = \varepsilon_1 = \varepsilon_2$.

Node b: $I_1 + I_2 - I = 0$.

Loop $abcfa$:
$$\begin{aligned} \varepsilon - I_1 r_1 + I_2 r_2 - \varepsilon &= 0, \\ I_1 r_1 &= I_2 r_2. \end{aligned}$$

Loop $fcdef$:
$$\begin{aligned} \varepsilon_2 - I_2 r_2 - IR &= 0, \\ \varepsilon - I_2 r_2 - IR &= 0. \end{aligned}$$

Solving for the current through the load resistor results in $I = \frac{\varepsilon}{r_{\text{eq}} + R}$, where $r_{\text{eq}} = \left(\frac{1}{r_1} + \frac{1}{r_2}\right)^{-1}$. The terminal voltage is equal to the potential drop across the load resistor $IR = \left(\frac{\varepsilon}{r_{\text{eq}} + R}\right)$. The parallel connection reduces the internal resistance and thus can produce a larger current.

Any number of batteries can be connected in parallel. For N batteries in parallel, the terminal voltage is equal to

$$V_{\text{terminal}} = \varepsilon - I\left(\frac{1}{r_1} + \frac{1}{r_2} + \cdots + \frac{1}{r_{N-1}} + \frac{1}{r_N}\right)^{-1} = \varepsilon - Ir_{\text{eq}} \tag{10.7}$$

where the equivalent resistance is $r_{\text{eq}} = \left(\sum_{i=1}^{N} \frac{1}{r_i}\right)^{-1}$.

As an example, some diesel trucks use two 12-V batteries in parallel; they produce a total emf of 12 V but can deliver the larger current needed to start a diesel engine.

In summary, the terminal voltage of batteries in series is equal to the sum of the individual emfs minus the sum of the internal resistances times the current. When batteries are connected in parallel, they usually have equal emfs and the terminal voltage is equal to the emf minus the equivalent internal resistance times the current, where the equivalent internal resistance is smaller than the individual internal resistances. Batteries are connected in series to increase the terminal voltage to the load. Batteries are connected in parallel to increase the current to the load.

Solar Cell Arrays

Another example dealing with multiple voltage sources is that of combinations of solar cells—wired in both series and parallel combinations to yield a desired voltage and current. Photovoltaic generation, which is the conversion of sunlight directly into electricity, is based upon the photoelectric effect. The photoelectric effect is beyond the scope of this chapter and is covered in Photons and Matter Waves (http://cnx.org/content/m58757/latest/) , but in general, photons hitting the surface of a solar cell create an electric current in the cell.

Most solar cells are made from pure silicon. Most single cells have a voltage output of about 0.5 V, while the current output is a function of the amount of sunlight falling on the cell (the incident solar radiation known as the insolation). Under bright noon sunlight, a current per unit area of about 100 mA/cm^2 of cell surface area is produced by typical single-crystal cells.

Individual solar cells are connected electrically in modules to meet electrical energy needs. They can be wired together in series or in parallel—connected like the batteries discussed earlier. A solar-cell array or module usually consists of between 36 and 72 cells, with a power output of 50 W to 140 W.

Solar cells, like batteries, provide a direct current (dc) voltage. Current from a dc voltage source is unidirectional. Most household appliances need an alternating current (ac) voltage.

10.4 | Electrical Measuring Instruments

Learning Objectives

By the end of the section, you will be able to:

- Describe how to connect a voltmeter in a circuit to measure voltage
- Describe how to connect an ammeter in a circuit to measure current
- Describe the use of an ohmmeter

Ohm's law and Kirchhoff's method are useful to analyze and design electrical circuits, providing you with the voltages across, the current through, and the resistance of the components that compose the circuit. To measure these parameters require instruments, and these instruments are described in this section.

DC Voltmeters and Ammeters

Whereas **voltmeters** measure voltage, **ammeters** measure current. Some of the meters in automobile dashboards, digital cameras, cell phones, and tuner-amplifiers are actually voltmeters or ammeters (Figure 10.34). The internal construction of the simplest of these meters and how they are connected to the system they monitor give further insight into applications of series and parallel connections.

Figure 10.34 The fuel and temperature gauges (far right and far left, respectively) in this 1996 Volkswagen are voltmeters that register the voltage output of "sender" units. These units are proportional to the amount of gasoline in the tank and to the engine temperature. (credit: Christian Giersing)

Measuring Current with an Ammeter

To measure the current through a device or component, the ammeter is placed in series with the device or component. A series connection is used because objects in series have the same current passing through them. (See Figure 10.35, where the ammeter is represented by the symbol A.)

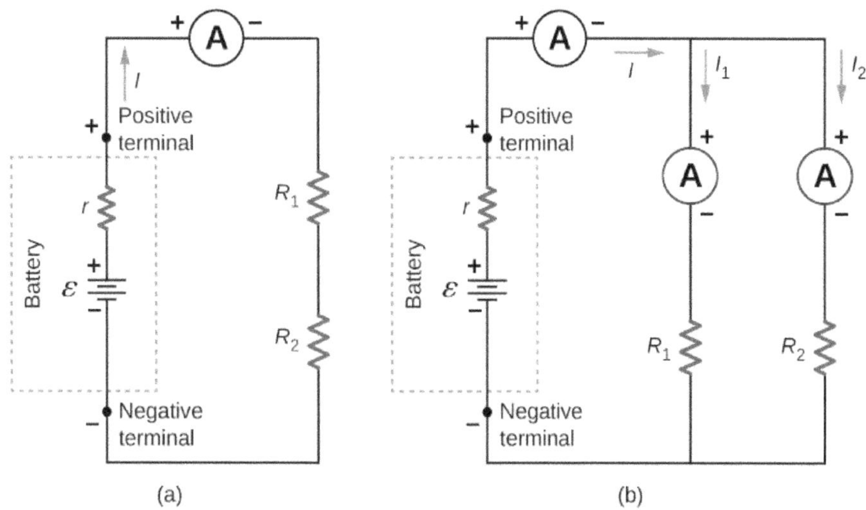

Figure 10.35 (a) When an ammeter is used to measure the current through two resistors connected in series to a battery, a single ammeter is placed in series with the two resistors because the current is the same through the two resistors in series. (b) When two resistors are connected in parallel with a battery, three meters, or three separate ammeter readings, are necessary to measure the current from the battery and through each resistor. The ammeter is connected in series with the component in question.

Ammeters need to have a very low resistance, a fraction of a milliohm. If the resistance is not negligible, placing the ammeter in the circuit would change the equivalent resistance of the circuit and modify the current that is being measured. Since the current in the circuit travels through the meter, ammeters normally contain a fuse to protect the meter from damage from currents which are too high.

Measuring Voltage with a Voltmeter

A voltmeter is connected in parallel with whatever device it is measuring. A parallel connection is used because objects in parallel experience the same potential difference. (See Figure 10.36, where the voltmeter is represented by the symbol V.)

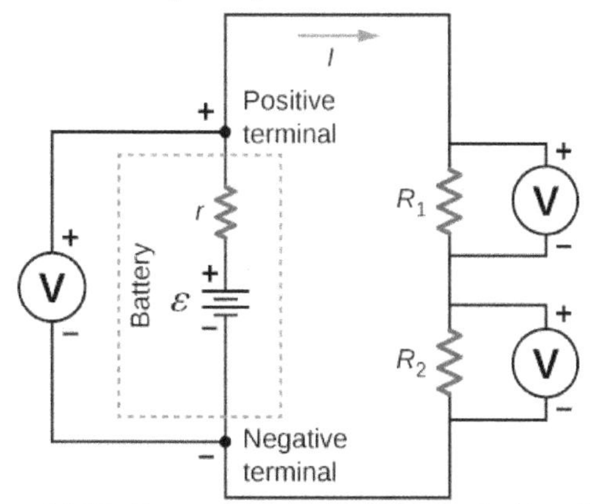

Figure 10.36 To measure potential differences in this series circuit, the voltmeter (V) is placed in parallel with the voltage source or either of the resistors. Note that terminal voltage is measured between the positive terminal and the negative terminal of the battery or voltage source. It is not possible to connect a voltmeter directly across the emf without including the internal resistance r of the battery.

Since voltmeters are connected in parallel, the voltmeter must have a very large resistance. Digital voltmeters convert the analog voltage into a digital value to display on a digital readout (Figure 10.37). Inexpensive voltmeters have resistances on the order of $R_M = 10\,M\,\Omega,$ whereas high-precision voltmeters have resistances on the order of $R_M = 10\,G\,\Omega$. The value of the resistance may vary, depending on which scale is used on the meter.

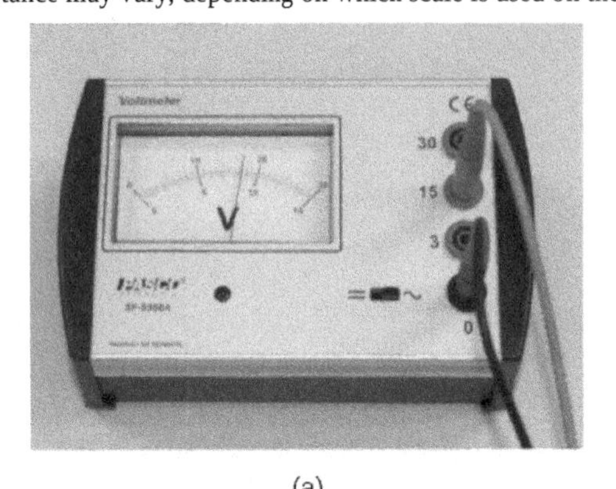

(a) (b)

Figure 10.37 (a) An analog voltmeter uses a galvanometer to measure the voltage. (b) Digital meters use an analog-to-digital converter to measure the voltage. (credit a and credit b: Joseph J. Trout)

Analog and Digital Meters

You may encounter two types of meters in the physics lab: analog and digital. The term 'analog' refers to signals or information represented by a continuously variable physical quantity, such as voltage or current. An analog meter uses a galvanometer, which is essentially a coil of wire with a small resistance, in a magnetic field, with a pointer attached that points to a scale. Current flows through the coil, causing the coil to rotate. To use the galvanometer as an ammeter, a small resistance is placed in parallel with the coil. For a voltmeter, a large resistance is placed in series with the coil. A digital meter uses a component called an analog-to-digital (A to D) converter and expresses the current or voltage as a series of the digits 0 and 1, which are used to run a digital display. Most analog meters have been replaced by digital meters.

10.8 Check Your Understanding Digital meters are able to detect smaller currents than analog meters employing galvanometers. How does this explain their ability to measure voltage and current more accurately than analog meters?

In this virtual lab (https://openstaxcollege.org/l/21cirreslabsim) simulation, you may construct circuits with resistors, voltage sources, ammeters and voltmeters to test your knowledge of circuit design.

Ohmmeters

An ohmmeter is an instrument used to measure the resistance of a component or device. The operation of the ohmmeter is based on Ohm's law. Traditional ohmmeters contained an internal voltage source (such as a battery) that would be connected across the component to be tested, producing a current through the component. A galvanometer was then used to measure the current and the resistance was deduced using Ohm's law. Modern digital meters use a constant current source to pass current through the component, and the voltage difference across the component is measured. In either case, the resistance is measured using Ohm's law $(R = V/I)$, where the voltage is known and the current is measured, or the current is known and the voltage is measured.

The component of interest should be isolated from the circuit; otherwise, you will be measuring the equivalent resistance of the circuit. An ohmmeter should never be connected to a "live" circuit, one with a voltage source connected to it and current running through it. Doing so can damage the meter.

10.5 | RC Circuits

Learning Objectives

By the end of the section, you will be able to:

- Describe the charging process of a capacitor
- Describe the discharging process of a capacitor
- List some applications of RC circuits

When you use a flash camera, it takes a few seconds to charge the capacitor that powers the flash. The light flash discharges the capacitor in a tiny fraction of a second. Why does charging take longer than discharging? This question and several other phenomena that involve charging and discharging capacitors are discussed in this module.

Circuits with Resistance and Capacitance

An ***RC* circuit** is a circuit containing resistance and capacitance. As presented in Capacitance, the capacitor is an electrical component that stores electric charge, storing energy in an electric field.

Figure 10.38(a) shows a simple *RC* circuit that employs a dc (direct current) voltage source ε, a resistor *R*, a capacitor *C*, and a two-position switch. The circuit allows the capacitor to be charged or discharged, depending on the position of the switch. When the switch is moved to position *A*, the capacitor charges, resulting in the circuit in part (b). When the switch is moved to position *B*, the capacitor discharges through the resistor.

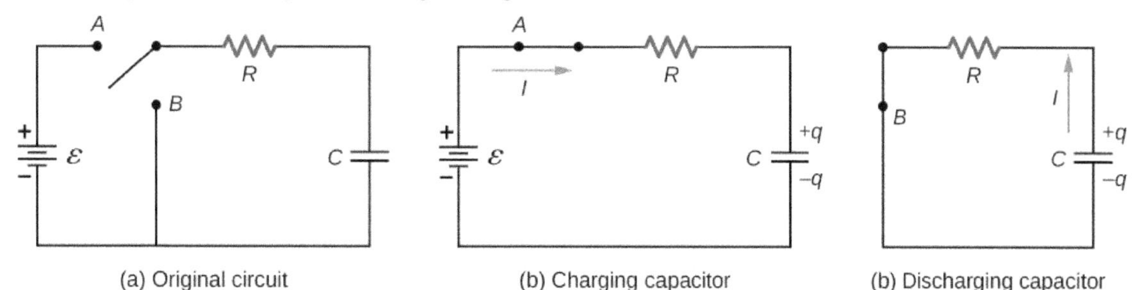

Figure 10.38 (a) An *RC* circuit with a two-pole switch that can be used to charge and discharge a capacitor. (b) When the switch is moved to position *A*, the circuit reduces to a simple series connection of the voltage source, the resistor, the capacitor, and the switch. (c) When the switch is moved to position *B*, the circuit reduces to a simple series connection of the resistor, the capacitor, and the switch. The voltage source is removed from the circuit.

Charging a Capacitor

We can use Kirchhoff's loop rule to understand the charging of the capacitor. This results in the equation $\varepsilon - V_R - V_c = 0$. This equation can be used to model the charge as a function of time as the capacitor charges. Capacitance is defined as $C = q/V$, so the voltage across the capacitor is $V_C = \frac{q}{C}$. Using Ohm's law, the potential drop across the resistor is $V_R = IR$, and the current is defined as $I = dq/dt$.

$$\varepsilon - V_R - V_c = 0,$$
$$\varepsilon - IR - \frac{q}{C} = 0,$$
$$\varepsilon - R\frac{dq}{dt} - \frac{q}{C} = 0.$$

This differential equation can be integrated to find an equation for the charge on the capacitor as a function of time.

$$\varepsilon - R\frac{dq}{dt} - \frac{q}{C} = 0,$$

$$\frac{dq}{dt} = \frac{\varepsilon C - q}{RC},$$

$$\int_0^q \frac{dq}{\varepsilon C - q} = \frac{1}{RC}\int_0^t dt.$$

Let $u = \varepsilon C - q$, then $du = -dq$. The result is

$$-\int_0^q \frac{du}{u} = \frac{1}{RC}\int_0^t dt,$$

$$\ln\left(\frac{\varepsilon C - q}{\varepsilon C}\right) = -\frac{1}{RC}t,$$

$$\frac{\varepsilon C - q}{\varepsilon C} = e^{-\frac{t}{RC}}.$$

Simplifying results in an equation for the charge on the charging capacitor as a function of time:

$$q(t) = C\varepsilon\left(1 - e^{-\frac{t}{RC}}\right) = Q\left(1 - e^{-\frac{t}{\tau}}\right). \tag{10.8}$$

A graph of the charge on the capacitor versus time is shown in Figure 10.39(a). First note that as time approaches infinity, the exponential goes to zero, so the charge approaches the maximum charge $Q = C\varepsilon$ and has units of coulombs. The units of RC are seconds, units of time. This quantity is known as the time constant:

$$\tau = RC. \tag{10.9}$$

At time $t = \tau = RC$, the charge is equal to $1 - e^{-1} = 1 - 0.368 = 0.632$ of the maximum charge $Q = C\varepsilon$. Notice that the time rate change of the charge is the slope at a point of the charge versus time plot. The slope of the graph is large at time $t = 0.0\text{ s}$ and approaches zero as time increases.

As the charge on the capacitor increases, the current through the resistor decreases, as shown in Figure 10.39(b). The current through the resistor can be found by taking the time derivative of the charge.

$$I(t) = \frac{dq}{dt} = \frac{d}{dt}\left[C\varepsilon\left(1 - e^{-\frac{t}{RC}}\right)\right],$$

$$I(t) = C\varepsilon\left(\frac{1}{RC}\right)e^{-\frac{t}{RC}} = \frac{\varepsilon}{R}e^{-\frac{t}{RC}} = I_o e^{-\frac{t}{RC}},$$

$$I(t) = I_0 e^{-t/\tau}. \tag{10.10}$$

At time $t = 0.00$ s, the current through the resistor is $I_0 = \frac{\varepsilon}{R}$. As time approaches infinity, the current approaches zero. At time $t = \tau$, the current through the resistor is $I(t = \tau) = I_0 e^{-1} = 0.368 I_0$.

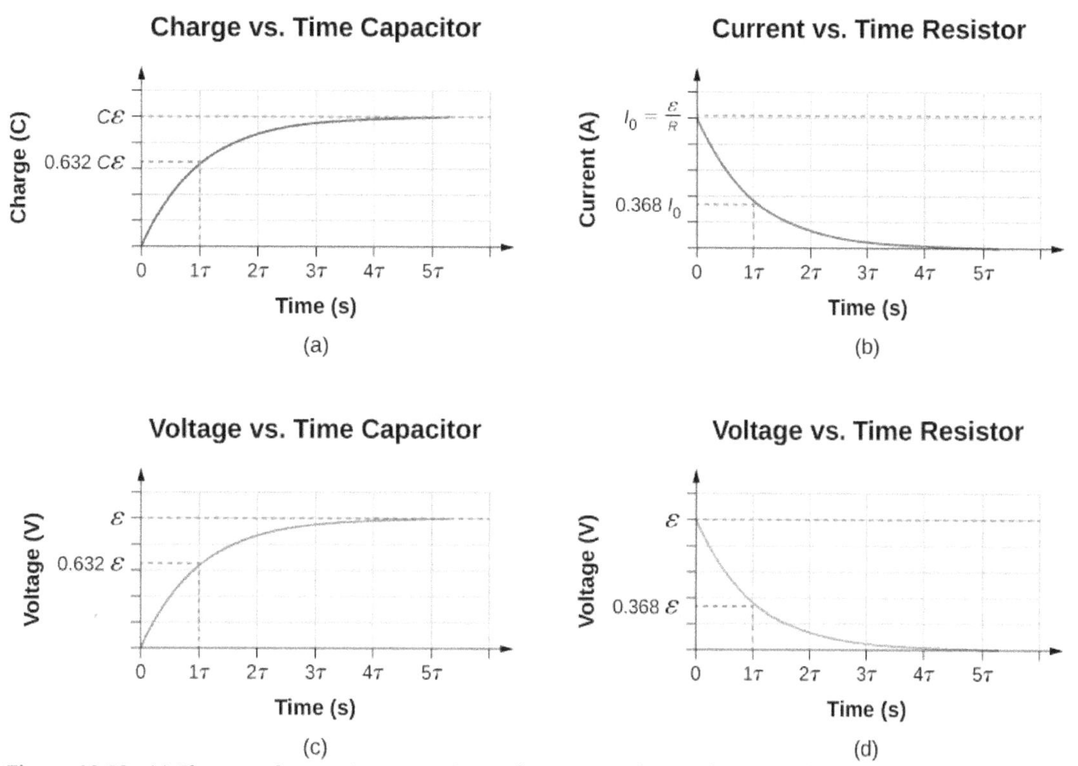

Figure 10.39 (a) Charge on the capacitor versus time as the capacitor charges. (b) Current through the resistor versus time. (c) Voltage difference across the capacitor. (d) Voltage difference across the resistor.

Figure 10.39(c) and Figure 10.39(d) show the voltage differences across the capacitor and the resistor, respectively. As the charge on the capacitor increases, the current decreases, as does the voltage difference across the resistor $V_R(t) = (I_0 R)e^{-t/\tau} = \varepsilon e^{-t/\tau}$. The voltage difference across the capacitor increases as $V_C(t) = \varepsilon\left(1 - e^{-t/\tau}\right)$.

Discharging a Capacitor

When the switch in Figure 10.38(a) is moved to position *B*, the circuit reduces to the circuit in part (c), and the charged capacitor is allowed to discharge through the resistor. A graph of the charge on the capacitor as a function of time is shown in Figure 10.40(a). Using Kirchhoff's loop rule to analyze the circuit as the capacitor discharges results in the equation $-V_R - V_c = 0$, which simplifies to $IR + \frac{q}{C} = 0$. Using the definition of current $\frac{dq}{dt}R = -\frac{q}{C}$ and integrating the loop equation yields an equation for the charge on the capacitor as a function of time:

$$q(t) = Qe^{-t/\tau}. \quad (10.11)$$

Here, *Q* is the initial charge on the capacitor and $\tau = RC$ is the time constant of the circuit. As shown in the graph, the charge decreases exponentially from the initial charge, approaching zero as time approaches infinity.

The current as a function of time can be found by taking the time derivative of the charge:

$$I(t) = -\frac{Q}{RC}e^{-t/\tau}. \tag{10.12}$$

The negative sign shows that the current flows in the opposite direction of the current found when the capacitor is charging. Figure 10.40(b) shows an example of a plot of charge versus time and current versus time. A plot of the voltage difference across the capacitor and the voltage difference across the resistor as a function of time are shown in parts (c) and (d) of the figure. Note that the magnitudes of the charge, current, and voltage all decrease exponentially, approaching zero as time increases.

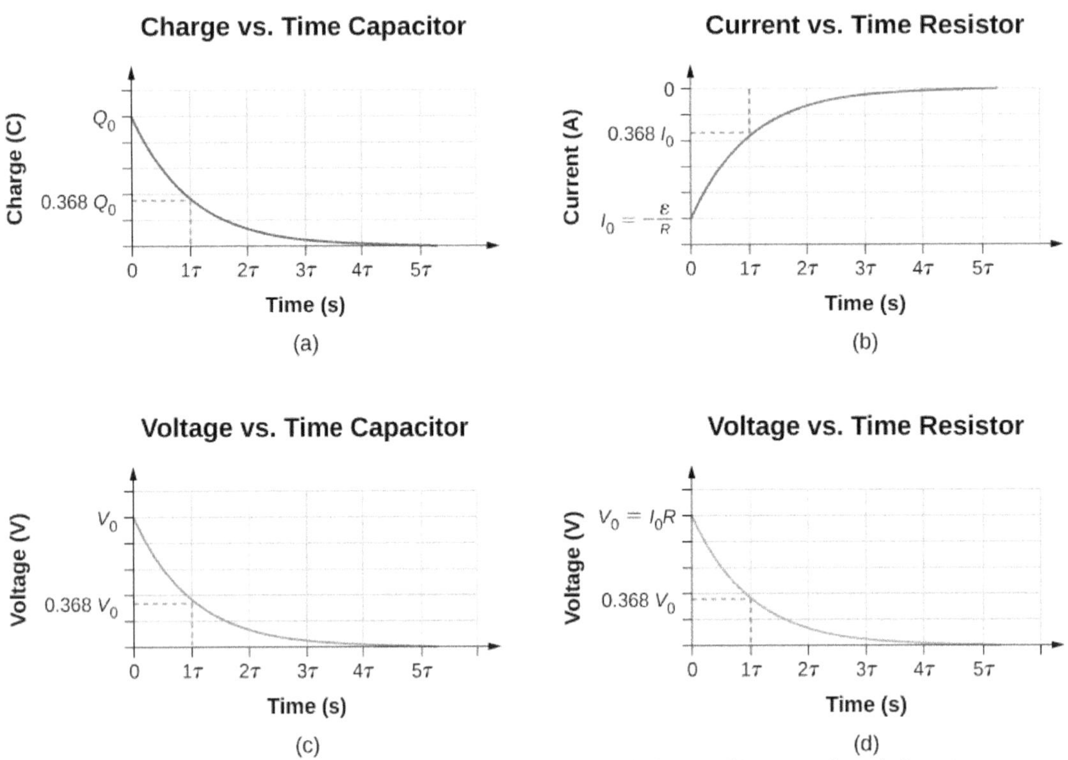

Figure 10.40 (a) Charge on the capacitor versus time as the capacitor discharges. (b) Current through the resistor versus time. (c) Voltage difference across the capacitor. (d) Voltage difference across the resistor.

Now we can explain why the flash camera mentioned at the beginning of this section takes so much longer to charge than discharge: The resistance while charging is significantly greater than while discharging. The internal resistance of the battery accounts for most of the resistance while charging. As the battery ages, the increasing internal resistance makes the charging process even slower.

Example 10.8

The Relaxation Oscillator

One application of an *RC* circuit is the relaxation oscillator, as shown below. The relaxation oscillator consists of a voltage source, a resistor, a capacitor, and a neon lamp. The neon lamp acts like an open circuit (infinite resistance) until the potential difference across the neon lamp reaches a specific voltage. At that voltage, the lamp acts like a short circuit (zero resistance), and the capacitor discharges through the neon lamp and produces light. In the relaxation oscillator shown, the voltage source charges the capacitor until the voltage across the capacitor is 80 V. When this happens, the neon in the lamp breaks down and allows the capacitor to discharge through the lamp, producing a bright flash. After the capacitor fully discharges through the neon lamp, it begins to charge again, and the process repeats. Assuming that the time it takes the capacitor to discharge is negligible, what is the time interval between flashes?

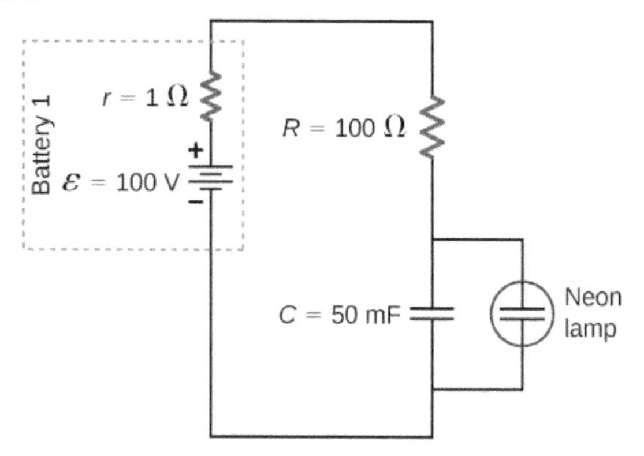

Strategy

The time period can be found from considering the equation $V_C(t) = \varepsilon\left(1 - e^{-t/\tau}\right)$, where $\tau = (R + r)C$.

Solution

The neon lamp flashes when the voltage across the capacitor reaches 80 V. The *RC* time constant is equal to $\tau = (R + r)C = (101\ \Omega)\left(50 \times 10^{-3}\,\text{F}\right) = 5.05\ \text{s}$. We can solve the voltage equation for the time it takes the capacitor to reach 80 V:

$$\begin{aligned} V_C(t) &= \varepsilon\left(1 - e^{-t/\tau}\right), \\ e^{-t/\tau} &= 1 - \frac{V_C(t)}{\varepsilon}, \\ \ln\left(e^{-t/\tau}\right) &= \ln\left(1 - \frac{V_C(t)}{\varepsilon}\right), \\ t &= -\tau\ln\left(1 - \frac{V_C(t)}{\varepsilon}\right) = -5.05\ \text{s} \cdot \ln\left(1 - \frac{80\ \text{V}}{100\ \text{V}}\right) = 8.13\ \text{s}. \end{aligned}$$

Significance

One application of the relaxation oscillator is for controlling indicator lights that flash at a frequency determined by the values for *R* and *C*. In this example, the neon lamp will flash every 8.13 seconds, a frequency of $f = \frac{1}{T} = \frac{1}{8.13\ \text{s}} = 0.55\ \text{Hz}$. The relaxation oscillator has many other practical uses. It is often used in electronic circuits, where the neon lamp is replaced by a transistor or a device known as a tunnel diode. The description of the transistor and tunnel diode is beyond the scope of this chapter, but you can think of them as voltage controlled switches. They are normally open switches, but when the right voltage is applied, the switch closes and conducts. The "switch" can be used to turn on another circuit, turn on a light, or run a small motor. A relaxation oscillator can be used to make the turn signals of your car blink or your cell phone to vibrate.

RC circuits have many applications. They can be used effectively as timers for applications such as intermittent windshield wipers, pace makers, and strobe lights. Some models of intermittent windshield wipers use a variable resistor to adjust the interval between sweeps of the wiper. Increasing the resistance increases the *RC* time constant, which increases the time between the operation of the wipers.

Another application is the pacemaker. The heart rate is normally controlled by electrical signals, which cause the muscles of the heart to contract and pump blood. When the heart rhythm is abnormal (the heartbeat is too high or too low), pace makers can be used to correct this abnormality. Pacemakers have sensors that detect body motion and breathing to increase the heart rate during physical activities, thus meeting the increased need for blood and oxygen, and an *RC* timing circuit can be used to control the time between voltage signals to the heart.

Looking ahead to the study of ac circuits (Alternating-Current Circuits), ac voltages vary as sine functions with specific frequencies. Periodic variations in voltage, or electric signals, are often recorded by scientists. These voltage signals could come from music recorded by a microphone or atmospheric data collected by radar. Occasionally, these signals can contain unwanted frequencies known as "noise." *RC* filters can be used to filter out the unwanted frequencies.

In the study of electronics, a popular device known as a 555 timer provides timed voltage pulses. The time between pulses is controlled by an *RC* circuit. These are just a few of the countless applications of *RC* circuits.

Example 10.9

Intermittent Windshield Wipers

A relaxation oscillator is used to control a pair of windshield wipers. The relaxation oscillator consists of a 10.00-mF capacitor and a $10.00\text{-k}\Omega$ variable resistor known as a rheostat. A knob connected to the variable resistor allows the resistance to be adjusted from $0.00\,\Omega$ to $10.00\,\text{k}\Omega$. The output of the capacitor is used to control a voltage-controlled switch. The switch is normally open, but when the output voltage reaches 10.00 V, the switch closes, energizing an electric motor and discharging the capacitor. The motor causes the windshield wipers to sweep once across the windshield and the capacitor begins to charge again. To what resistance should the rheostat be adjusted for the period of the wiper blades be 10.00 seconds?

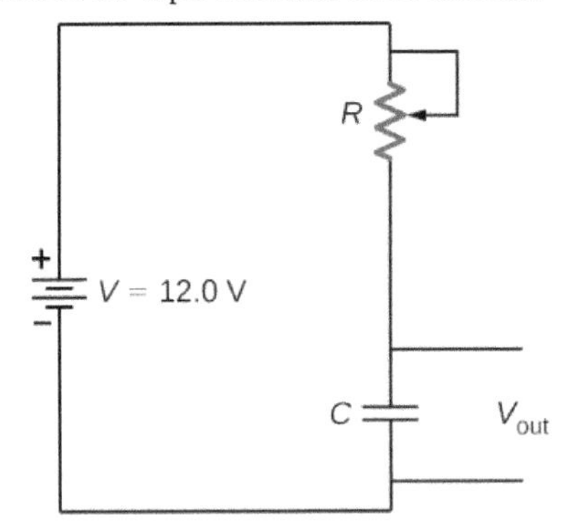

Strategy

The resistance considers the equation $V_{\text{out}}(t) = V\left(1 - e^{-t/\tau}\right)$, where $\tau = RC$. The capacitance, output voltage, and voltage of the battery are given. We need to solve this equation for the resistance.

Solution

The output voltage will be 10.00 V and the voltage of the battery is 12.00 V. The capacitance is given as 10.00 mF. Solving for the resistance yields

$$\begin{aligned}
V_{\text{out}}(t) &= V\left(1 - e^{-t/\tau}\right), \\
e^{-t/RC} &= 1 - \frac{V_{\text{out}}(t)}{V}, \\
\ln\left(e^{-t/RC}\right) &= \ln\left(1 - \frac{V_{\text{out}}(t)}{V}\right), \\
-\frac{t}{RC} &= \ln\left(1 - \frac{V_{\text{out}}(t)}{V}\right), \\
R &= \frac{-t}{C\ln\left(1 - \frac{V_C(t)}{V}\right)} = \frac{-10.00\text{ s}}{10 \times 10^{-3}\text{ F}\ln\left(1 - \frac{10\text{ V}}{12\text{ V}}\right)} = 558.11\,\Omega.
\end{aligned}$$

Significance

Increasing the resistance increases the time delay between operations of the windshield wipers. When the resistance is zero, the windshield wipers run continuously. At the maximum resistance, the period of the operation of the wipers is:

$$t = -RC\ln\left(1 - \frac{V_{\text{out}}(t)}{V}\right) = -\left(10 \times 10^{-3}\text{ F}\right)\left(10 \times 10^{3}\,\Omega\right)\ln\left(1 - \frac{10\text{ V}}{12\text{ V}}\right) = 179.18\text{ s} = 2.98\text{ min}.$$

The *RC* circuit has thousands of uses and is a very important circuit to study. Not only can it be used to time circuits, it can also be used to filter out unwanted frequencies in a circuit and used in power supplies, like the one for your computer, to help turn ac voltage to dc voltage.

10.6 | Household Wiring and Electrical Safety

Learning Objectives

By the end of the section, you will be able to:

- List the basic concepts involved in house wiring
- Define the terms thermal hazard and shock hazard
- Describe the effects of electrical shock on human physiology and their relationship to the amount of current through the body
- Explain the function of fuses and circuit breakers

Electricity presents two known hazards: thermal and shock. A **thermal hazard** is one in which an excessive electric current causes undesired thermal effects, such as starting a fire in the wall of a house. A **shock hazard** occurs when an electric current passes through a person. Shocks range in severity from painful, but otherwise harmless, to heart-stopping lethality. In this section, we consider these hazards and the various factors affecting them in a quantitative manner. We also examine systems and devices for preventing electrical hazards.

Thermal Hazards

Electric power causes undesired heating effects whenever electric energy is converted into thermal energy at a rate faster than it can be safely dissipated. A classic example of this is the short circuit, a low-resistance path between terminals of a voltage source. An example of a short circuit is shown in Figure 10.41. A toaster is plugged into a common household electrical outlet. Insulation on wires leading to an appliance has worn through, allowing the two wires to come into contact, or "short." As a result, thermal energy can quickly raise the temperature of surrounding materials, melting the insulation and perhaps causing a fire.

The circuit diagram shows a symbol that consists of a sine wave enclosed in a circle. This symbol represents an alternating current (ac) voltage source. In an ac voltage source, the voltage oscillates between a positive and negative maximum amplitude. Up to now, we have been considering direct current (dc) voltage sources, but many of the same concepts are applicable to ac circuits.

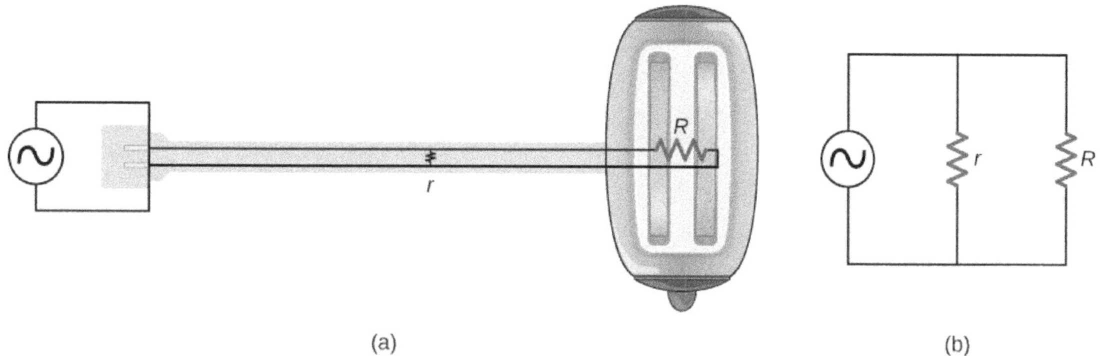

Figure 10.41 A short circuit is an undesired low-resistance path across a voltage source. (a) Worn insulation on the wires of a toaster allow them to come into contact with a low resistance r. Since $P = V^2/r$, thermal power is created so rapidly that the cord melts or burns. (b) A schematic of the short circuit.

Another serious thermal hazard occurs when wires supplying power to an appliance are overloaded. Electrical wires and appliances are often rated for the maximum current they can safely handle. The term "overloaded" refers to a condition where the current exceeds the rated maximum current. As current flows through a wire, the power dissipated in the supply wires is $P = I^2 R_W$, where R_W is the resistance of the wires and I is the current flowing through the wires. If either I or R_W is too large, the wires overheat. Fuses and circuit breakers are used to limit excessive currents.

Shock Hazards

Electric shock is the physiological reaction or injury caused by an external electric current passing through the body. The effect of an electric shock can be negative or positive. When a current with a magnitude above 300 mA passes through the heart, death may occur. Most electrical shock fatalities occur because a current causes ventricular fibrillation, a massively irregular and often fatal, beating of the heart. On the other hand, a heart attack victim, whose heart is in fibrillation, can be saved by an electric shock from a defibrillator.

The effects of an undesirable electric shock can vary in severity: a slight sensation at the point of contact, pain, loss of voluntary muscle control, difficulty breathing, heart fibrillation, and possibly death. The loss of voluntary muscle control can cause the victim to not be able to let go of the source of the current.

The major factors upon which the severity of the effects of electrical shock depend are

1. The amount of current I
2. The path taken by the current
3. The duration of the shock
4. The frequency f of the current ($f = 0$ for dc)

Our bodies are relatively good electric conductors due to the body's water content. A dangerous condition occurs when the body is in contact with a voltage source and "ground." The term "ground" refers to a large sink or source of electrons, for example, the earth (thus, the name). When there is a direct path to ground, large currents will pass through the parts of the body with the lowest resistance and a direct path to ground. A safety precaution used by many professions is the wearing of insulated shoes. Insulated shoes prohibit a pathway to ground for electrons through the feet by providing a large resistance. Whenever working with high-power tools, or any electric circuit, ensure that you do not provide a pathway for current flow (especially across the heart). A common safety precaution is to work with one hand, reducing the possibility of providing a current path through the heart.

Very small currents pass harmlessly and unfelt through the body. This happens to you regularly without your knowledge. The threshold of sensation is only 1 mA and, although unpleasant, shocks are apparently harmless for currents less than 5 mA. A great number of safety rules take the 5-mA value for the maximum allowed shock. At 5–30 mA and above, the current can stimulate sustained muscular contractions, much as regular nerve impulses do (Figure 10.42). Very large currents (above 300 mA) cause the heart and diaphragm of the lung to contract for the duration of the shock. Both the heart and respiration stop. Both often return to normal following the shock.

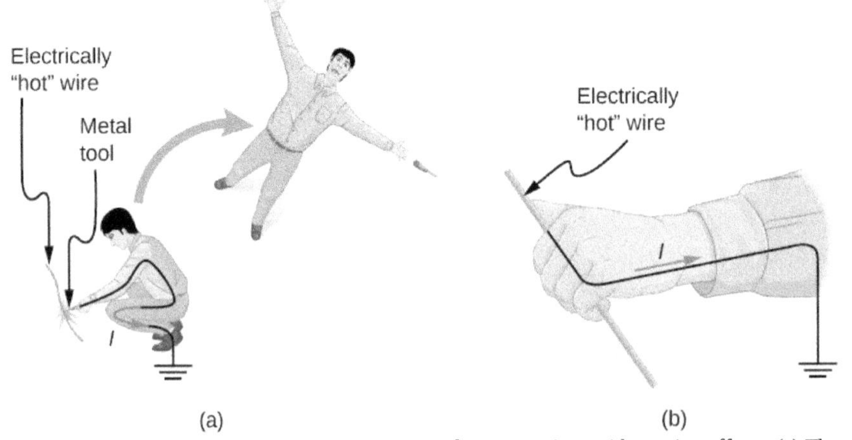

Figure 10.42 An electric current can cause muscular contractions with varying effects. (a) The victim is "thrown" backward by involuntary muscle contractions that extend the legs and torso. (b) The victim can't let go of the wire that is stimulating all the muscles in the hand. Those that close the fingers are stronger than those that open them.

Current is the major factor determining shock severity. A larger voltage is more hazardous, but since $I = V/R$, the severity of the shock depends on the combination of voltage and resistance. For example, a person with dry skin has a resistance of about $200\,k\Omega$. If he comes into contact with 120-V ac, a current

$$I = (120\,\text{V})/(200\,\text{k}\Omega) = 0.6\,\text{mA}$$

passes harmlessly through him. The same person soaking wet may have a resistance of $10.0\,k\Omega$ and the same 120 V will produce a current of 12 mA—above the "can't let go" threshold and potentially dangerous.

Electrical Safety: Systems and Devices

Figure 10.43(a) shows the schematic for a simple ac circuit with no safety features. This is not how power is distributed in practice. Modern household and industrial wiring requires the **three-wire system**, shown schematically in part (b), which has several safety features, with live, neutral, and ground wires. First is the familiar circuit breaker (or fuse) to prevent thermal overload. Second is a protective case around the appliance, such as a toaster or refrigerator. The case's safety feature is that it prevents a person from touching exposed wires and coming into electrical contact with the circuit, helping prevent shocks.

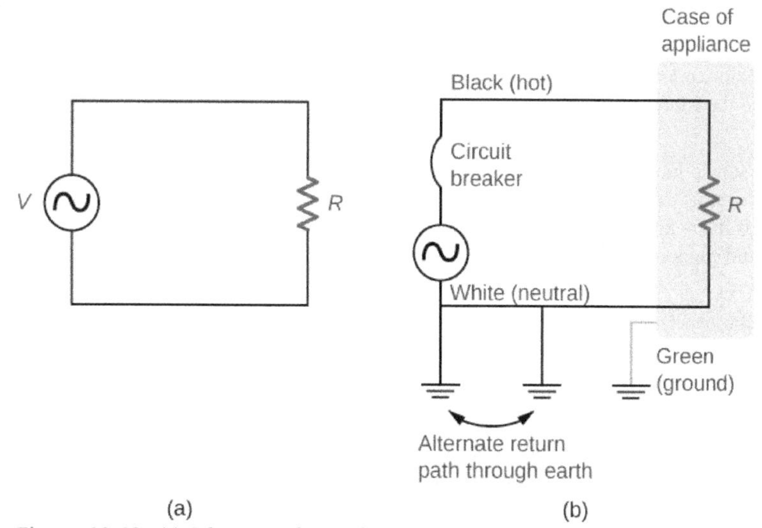

Figure 10.43 (a) Schematic of a simple ac circuit with a voltage source and a single appliance represented by the resistance *R*. There are no safety features in this circuit. (b) The three-wire system connects the neutral wire to ground at the voltage source and user location, forcing it to be at zero volts and supplying an alternative return path for the current through ground. Also grounded to zero volts is the case of the appliance. A circuit breaker or fuse protects against thermal overload and is in series on the active (live/hot) wire.

There are three connections to ground shown in Figure 10.43(b). Recall that a ground connection is a low-resistance path directly to ground. The two ground connections on the neutral wire force it to be at zero volts relative to ground, giving the wire its name. This wire is therefore safe to touch even if its insulation, usually white, is missing. The neutral wire is the return path for the current to follow to complete the circuit. Furthermore, the two ground connections supply an alternative path through ground (a good conductor) to complete the circuit. The ground connection closest to the power source could be at the generating plant, whereas the other is at the user's location. The third ground is to the case of the appliance, through the green ground wire, forcing the case, too, to be at zero volts. The live or hot wire (hereafter referred to as "live/hot") supplies voltage and current to operate the appliance. Figure 10.44 shows a more pictorial version of how the three-wire system is connected through a three-prong plug to an appliance.

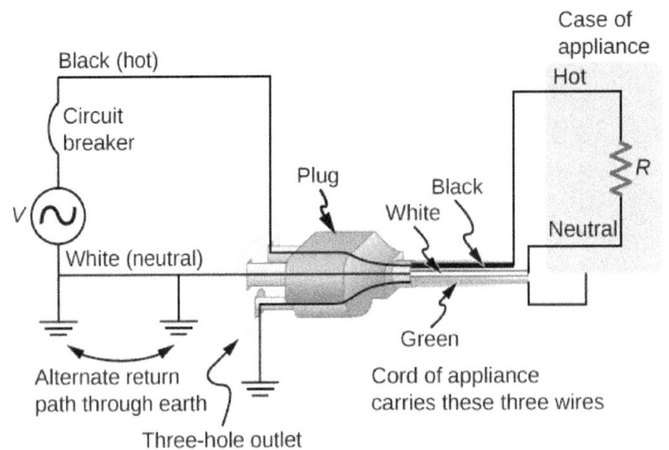

Figure 10.44 The standard three-prong plug can only be inserted in one way, to ensure proper function of the three-wire system.

Insulating plastic is color-coded to identify live/hot, neutral, and ground wires, but these codes vary around the world. It is essential to determine the color code in your region. Striped coatings are sometimes used for the benefit of those who are colorblind.

Grounding the case solves more than one problem. The simplest problem is worn insulation on the live/hot wire that allows it to contact the case, as shown in Figure 10.45. Lacking a ground connection, a severe shock is possible. This is particularly dangerous in the kitchen, where a good connection to ground is available through water on the floor or a water faucet. With the ground connection intact, the circuit breaker will trip, forcing repair of the appliance.

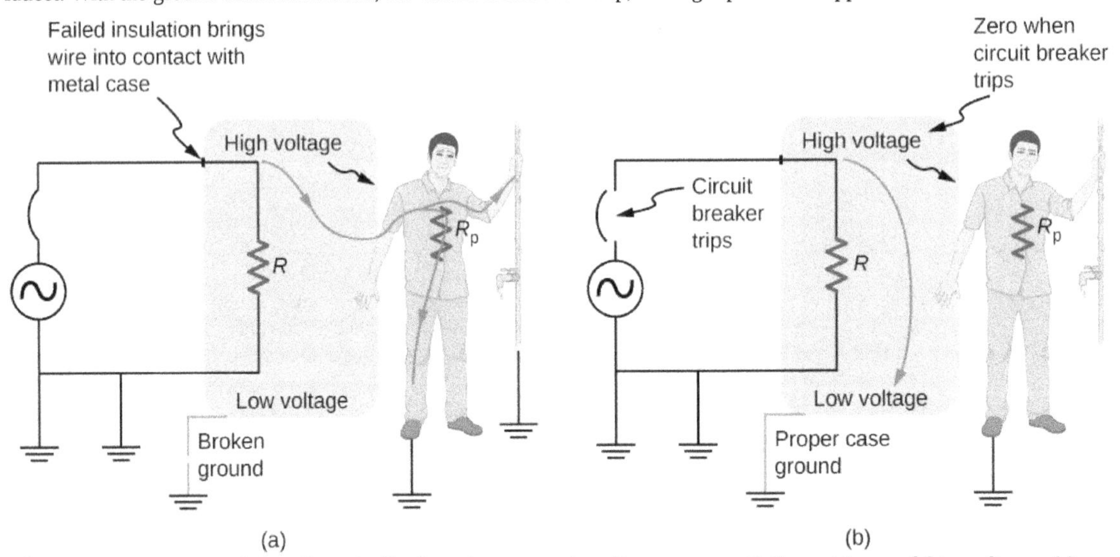

Figure 10.45 Worn insulation allows the live/hot wire to come into direct contact with the metal case of this appliance. (a) The ground connection being broken, the person is severely shocked. The appliance may operate normally in this situation. (b) With a proper ground, the circuit breaker trips, forcing repair of the appliance.

A ground fault circuit interrupter (GFCI) is a safety device found in updated kitchen and bathroom wiring that works based on electromagnetic induction. GFCIs compare the currents in the live/hot and neutral wires. When live/hot and neutral currents are not equal, it is almost always because current in the neutral is less than in the live/hot wire. Then some of the current, called a leakage current, is returning to the voltage source by a path other than through the neutral wire. It is assumed that this path presents a hazard. GFCIs are usually set to interrupt the circuit if the leakage current is greater than 5 mA, the accepted maximum harmless shock. Even if the leakage current goes safely to ground through an intact ground wire, the GFCI will trip, forcing repair of the leakage.

CHAPTER 10 REVIEW

KEY TERMS

ammeter instrument that measures current

electromotive force (emf) energy produced per unit charge, drawn from a source that produces an electrical current

equivalent resistance resistance of a combination of resistors; it can be thought of as the resistance of a single resistor that can replace a combination of resistors in a series and/or parallel circuit

internal resistance amount of resistance to the flow of current within the voltage source

junction rule sum of all currents entering a junction must equal the sum of all currents leaving the junction

Kirchhoff's rules set of two rules governing current and changes in potential in an electric circuit

loop rule algebraic sum of changes in potential around any closed circuit path (loop) must be zero

potential difference difference in electric potential between two points in an electric circuit, measured in volts

potential drop loss of electric potential energy as a current travels across a resistor, wire, or other component

***RC* circuit** circuit that contains both a resistor and a capacitor

shock hazard hazard in which an electric current passes through a person

terminal voltage potential difference measured across the terminals of a source when there is no load attached

thermal hazard hazard in which an excessive electric current causes undesired thermal effects

three-wire system wiring system used at present for safety reasons, with live, neutral, and ground wires

voltmeter instrument that measures voltage

KEY EQUATIONS

Terminal voltage of a single voltage source	$V_{\text{terminal}} = \varepsilon - Ir_{\text{eq}}$
Equivalent resistance of a series circuit	$R_{\text{eq}} = R_1 + R_2 + R_3 + \cdots + R_{N-1} + R_N = \sum_{i=1}^{N} R_i$
Equivalent resistance of a parallel circuit	$R_{\text{eq}} = \left(\frac{1}{R_1} + \frac{1}{R2} + \cdots + \frac{1}{R_N}\right)^{-1} = \left(\sum_{i=1}^{N} \frac{1}{R_i}\right)^{-1}$
Junction rule	$\sum I_{\text{in}} = \sum I_{\text{out}}$
Loop rule	$\sum V = 0$
Terminal voltage of N voltage sources in series	$V_{\text{terminal}} = \sum_{i=1}^{N} \varepsilon_i - I \sum_{i=1}^{N} r_i = \sum_{i=1}^{N} \varepsilon_i - Ir_{\text{eq}}$
Terminal voltage of N voltage sources in parallel	$V_{\text{terminal}} = \varepsilon - I \sum_{i=1}^{N} \left(\frac{1}{r_i}\right)^{-1} = \varepsilon - Ir_{\text{eq}}$
Charge on a charging capacitor	$q(t) = C\varepsilon\left(1 - e^{-\frac{t}{RC}}\right) = Q\left(1 - e^{-\frac{t}{\tau}}\right)$
Time constant	$\tau = RC$

Current during charging of a capacitor $I = \frac{\varepsilon}{R}e^{-\frac{t}{RC}} = I_o e^{-\frac{t}{RC}}$

Charge on a discharging capacitor $q(t) = Qe^{-\frac{t}{\tau}}$

Current during discharging of a capacitor $I(t) = -\frac{Q}{RC}e^{-\frac{t}{\tau}}$

SUMMARY

10.1 Electromotive Force

- All voltage sources have two fundamental parts: a source of electrical energy that has a characteristic electromotive force (emf), and an internal resistance *r*. The emf is the work done per charge to keep the potential difference of a source constant. The emf is equal to the potential difference across the terminals when no current is flowing. The internal resistance *r* of a voltage source affects the output voltage when a current flows.
- The voltage output of a device is called its terminal voltage V_{terminal} and is given by $V_{\text{terminal}} = \varepsilon - Ir$, where *I* is the electric current and is positive when flowing away from the positive terminal of the voltage source and *r* is the internal resistance.

10.2 Resistors in Series and Parallel

- The equivalent resistance of an electrical circuit with resistors wired in a series is the sum of the individual resistances: $R_s = R_1 + R_2 + R_3 + \cdots = \sum_{i=1}^{N} R_i$.
- Each resistor in a series circuit has the same amount of current flowing through it.
- The potential drop, or power dissipation, across each individual resistor in a series is different, and their combined total is the power source input.
- The equivalent resistance of an electrical circuit with resistors wired in parallel is less than the lowest resistance of any of the components and can be determined using the formula

$$R_{\text{eq}} = \left(\frac{1}{R_1} + \frac{1}{R_2} + \frac{1}{R_3} + \cdots\right)^{-1} = \left(\sum_{i=1}^{N} \frac{1}{R_i}\right)^{-1}.$$

- Each resistor in a parallel circuit has the same full voltage of the source applied to it.
- The current flowing through each resistor in a parallel circuit is different, depending on the resistance.
- If a more complex connection of resistors is a combination of series and parallel, it can be reduced to a single equivalent resistance by identifying its various parts as series or parallel, reducing each to its equivalent, and continuing until a single resistance is eventually reached.

10.3 Kirchhoff's Rules

- Kirchhoff's rules can be used to analyze any circuit, simple or complex. The simpler series and parallel connection rules are special cases of Kirchhoff's rules.
- Kirchhoff's first rule, also known as the junction rule, applies to the charge to a junction. Current is the flow of charge; thus, whatever charge flows into the junction must flow out.
- Kirchhoff's second rule, also known as the loop rule, states that the voltage drop around a loop is zero.
- When calculating potential and current using Kirchhoff's rules, a set of conventions must be followed for determining the correct signs of various terms.
- When multiple voltage sources are in series, their internal resistances add together and their emfs add together to get the total values.

- When multiple voltage sources are in parallel, their internal resistances combine to an equivalent resistance that is less than the individual resistance and provides a higher current than a single cell.
- Solar cells can be wired in series or parallel to provide increased voltage or current, respectively.

10.4 Electrical Measuring Instruments

- Voltmeters measure voltage, and ammeters measure current. Analog meters are based on the combination of a resistor and a galvanometer, a device that gives an analog reading of current or voltage. Digital meters are based on analog-to-digital converters and provide a discrete or digital measurement of the current or voltage.
- A voltmeter is placed in parallel with the voltage source to receive full voltage and must have a large resistance to limit its effect on the circuit.
- An ammeter is placed in series to get the full current flowing through a branch and must have a small resistance to limit its effect on the circuit.
- Standard voltmeters and ammeters alter the circuit they are connected to and are thus limited in accuracy.
- Ohmmeters are used to measure resistance. The component in which the resistance is to be measured should be isolated (removed) from the circuit.

10.5 RC Circuits

- An *RC* circuit is one that has both a resistor and a capacitor.
- The time constant τ for an *RC* circuit is $\tau = RC$.
- When an initially uncharged $(q = 0 \text{ at } t = 0)$ capacitor in series with a resistor is charged by a dc voltage source, the capacitor asymptotically approaches the maximum charge.
- As the charge on the capacitor increases, the current exponentially decreases from the initial current: $I_0 = \varepsilon/R$.
- If a capacitor with an initial charge Q is discharged through a resistor starting at $t = 0$, then its charge decreases exponentially. The current flows in the opposite direction, compared to when it charges, and the magnitude of the charge decreases with time.

10.6 Household Wiring and Electrical Safety

- The two types of electric hazards are thermal (excessive power) and shock (current through a person). Electrical safety systems and devices are employed to prevent thermal and shock hazards.
- Shock severity is determined by current, path, duration, and ac frequency.
- Circuit breakers and fuses interrupt excessive currents to prevent thermal hazards.
- The three-wire system guards against thermal and shock hazards, utilizing live/hot, neutral, and ground wires, and grounding the neutral wire and case of the appliance.
- A ground fault circuit interrupter (GFCI) prevents shock by detecting the loss of current to unintentional paths.

CONCEPTUAL QUESTIONS

10.1 Electromotive Force

1. What effect will the internal resistance of a rechargeable battery have on the energy being used to recharge the battery?

2. A battery with an internal resistance of r and an emf of 10.00 V is connected to a load resistor $R = r$. As the battery ages, the internal resistance triples. How much is the current through the load resistor reduced?

3. Show that the power dissipated by the load resistor is maximum when the resistance of the load resistor is equal to the internal resistance of the battery.

10.2 Resistors in Series and Parallel

4. A voltage occurs across an open switch. What is the power dissipated by the open switch?

5. The severity of a shock depends on the magnitude of the current through your body. Would you prefer to be in series or in parallel with a resistance, such as the heating element of a toaster, if you were shocked by it? Explain.

6. Suppose you are doing a physics lab that asks you to put a resistor into a circuit, but all the resistors supplied have a larger resistance than the requested value. How would you connect the available resistances to attempt to get the smaller value asked for?

7. Some light bulbs have three power settings (not including zero), obtained from multiple filaments that are individually switched and wired in parallel. What is the minimum number of filaments needed for three power settings?

10.3 Kirchhoff's Rules

8. Can all of the currents going into the junction shown below be positive? Explain.

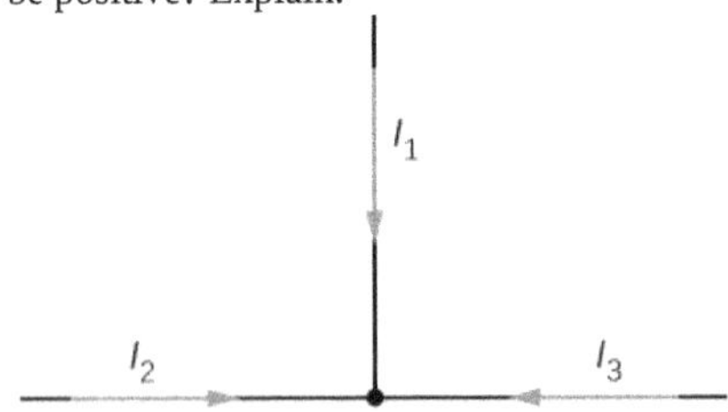

9. Consider the circuit shown below. Does the analysis of the circuit require Kirchhoff's method, or can it be redrawn to simplify the circuit? If it is a circuit of series and parallel connections, what is the equivalent resistance?

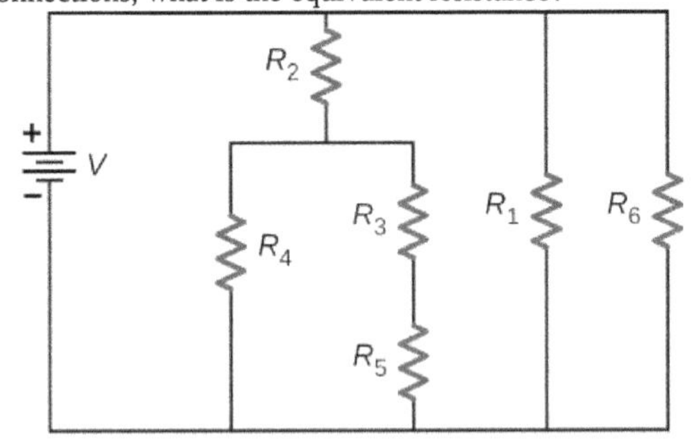

10. Do batteries in a circuit always supply power to a circuit, or can they absorb power in a circuit? Give an example.

11. What are the advantages and disadvantages of connecting batteries in series? In parallel?

12. Semi-tractor trucks use four large 12-V batteries. The starter system requires 24 V, while normal operation of the truck's other electrical components utilizes 12 V. How could the four batteries be connected to produce 24 V? To produce 12 V? Why is 24 V better than 12 V for starting the truck's engine (a very heavy load)?

10.4 Electrical Measuring Instruments

13. What would happen if you placed a voltmeter in series with a component to be tested?

14. What is the basic operation of an ohmmeter as it measures a resistor?

15. Why should you not connect an ammeter directly across a voltage source as shown below?

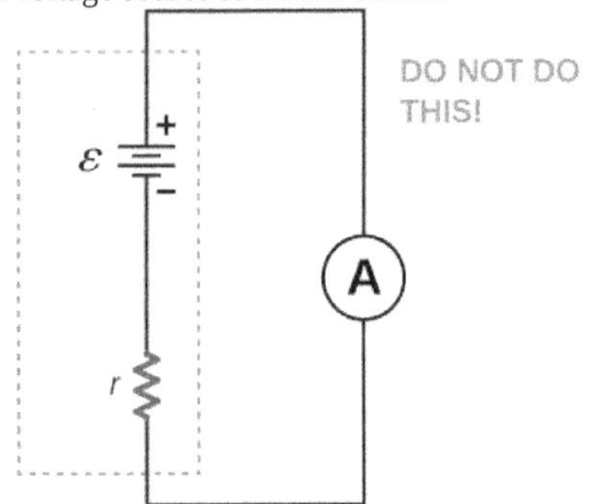

10.5 RC Circuits

16. A battery, switch, capacitor, and lamp are connected in series. Describe what happens to the lamp when the switch is closed.

17. When making an ECG measurement, it is important to measure voltage variations over small time intervals. The time is limited by the *RC* constant of the circuit—it is not possible to measure time variations shorter than *RC*. How would you manipulate *R* and *C* in the circuit to allow the necessary measurements?

10.6 Household Wiring and Electrical Safety

18. Why isn't a short circuit necessarily a shock hazard?

19. We are often advised to not flick electric switches with wet hands, dry your hand first. We are also advised to never throw water on an electric fire. Why?

PROBLEMS

10.1 Electromotive Force

20. A car battery with a 12-V emf and an internal resistance of $0.050\ \Omega$ is being charged with a current of 60 A. Note that in this process, the battery is being charged. (a) What is the potential difference across its terminals? (b) At what rate is thermal energy being dissipated in the battery? (c) At what rate is electric energy being converted into chemical energy?

21. The label on a battery-powered radio recommends the use of rechargeable nickel-cadmium cells (nicads), although they have a 1.25-V emf, whereas alkaline cells have a 1.58-V emf. The radio has a $3.20\ \Omega$ resistance. (a) Draw a circuit diagram of the radio and its batteries. Now, calculate the power delivered to the radio (b) when using nicad cells, each having an internal resistance of $0.0400\ \Omega$, and (c) when using alkaline cells, each having an internal resistance of $0.200\ \Omega$. (d) Does this difference seem significant, considering that the radio's effective resistance is lowered when its volume is turned up?

22. An automobile starter motor has an equivalent resistance of $0.0500\ \Omega$ and is supplied by a 12.0-V battery with a $0.0100\text{-}\Omega$ internal resistance. (a) What is the current to the motor? (b) What voltage is applied to it? (c) What power is supplied to the motor? (d) Repeat these calculations for when the battery connections are corroded and add $0.0900\ \Omega$ to the circuit. (Significant problems are caused by even small amounts of unwanted resistance in low-voltage, high-current applications.)

23. (a) What is the internal resistance of a voltage source if its terminal potential drops by 2.00 V when the current supplied increases by 5.00 A? (b) Can the emf of the voltage source be found with the information supplied?

24. A person with body resistance between his hands of $10.0\ \text{k}\Omega$ accidentally grasps the terminals of a 20.0-kV power supply. (Do NOT do this!) (a) Draw a circuit diagram to represent the situation. (b) If the internal resistance of the power supply is $2000\ \Omega$, what is the current through his body? (c) What is the power dissipated in his body? (d) If the power supply is to be made safe by increasing its internal resistance, what should the internal resistance be for the maximum current in this situation to be 1.00 mA or less? (e) Will this modification compromise the effectiveness of the power supply for driving low-resistance devices? Explain your reasoning.

25. A 12.0-V emf automobile battery has a terminal voltage of 16.0 V when being charged by a current of 10.0 A. (a) What is the battery's internal resistance? (b) What power is dissipated inside the battery? (c) At what rate (in $°\text{C/min}$) will its temperature increase if its mass is 20.0 kg and it has a specific heat of $0.300\ \text{kcal/kg} \cdot °\text{C}$, assuming no heat escapes?

10.2 Resistors in Series and Parallel

26. (a) What is the resistance of a $1.00 \times 10^2\text{-}\Omega$, a $2.50\text{-k}\Omega$, and a $4.00\text{-k}\Omega$ resistor connected in series? (b) In parallel?

27. What are the largest and smallest resistances you can obtain by connecting a $36.0\text{-}\Omega$, a $50.0\text{-}\Omega$, and a $700\text{-}\Omega$ resistor together?

28. An 1800-W toaster, a 1400-W speaker, and a 75-W lamp are plugged into the same outlet in a 15-A fuse and 120-V circuit. (The three devices are in parallel when plugged into the same socket.) (a) What current is drawn by each device? (b) Will this combination blow the 15-A fuse?

29. Your car's 30.0-W headlight and 2.40-kW starter are ordinarily connected in parallel in a 12.0-V system. What power would one headlight and the starter consume if connected in series to a 12.0-V battery? (Neglect any other resistance in the circuit and any change in resistance in the two devices.)

30. (a) Given a 48.0-V battery and $24.0\text{-}\Omega$ and $96.0\text{-}\Omega$ resistors, find the current and power for each when connected in series. (b) Repeat when the resistances are in parallel.

31. Referring to the example combining series and parallel circuits and Figure 10.16, calculate I_3 in the following two different ways: (a) from the known values of I and I_2; (b) using Ohm's law for R_3. In both parts, explicitly show how you follow the steps in the Problem-Solving Strategy: Series and Parallel Resistors.

32. Referring to Figure 10.16, (a) Calculate P_3 and note how it compares with P_3 found in the first two example problems in this module. (b) Find the total power supplied by the source and compare it with the sum of the powers dissipated by the resistors.

33. Refer to Figure 10.17 and the discussion of lights dimming when a heavy appliance comes on. (a) Given the voltage source is 120 V, the wire resistance is $0.800\,\Omega$, and the bulb is nominally 75.0 W, what power will the bulb dissipate if a total of 15.0 A passes through the wires when the motor comes on? Assume negligible change in bulb resistance. (b) What power is consumed by the motor?

34. Show that if two resistors R_1 and R_2 are combined and one is much greater than the other $(R_1 \gg R_2)$, (a) their series resistance is very nearly equal to the greater resistance R_1 and (b) their parallel resistance is very nearly equal to smaller resistance R_2.

35. Consider the circuit shown below. The terminal voltage of the battery is $V = 18.00\,\text{V}$. (a) Find the equivalent resistance of the circuit. (b) Find the current through each resistor. (c) Find the potential drop across each resistor. (d) Find the power dissipated by each resistor. (e) Find the power supplied by the battery.

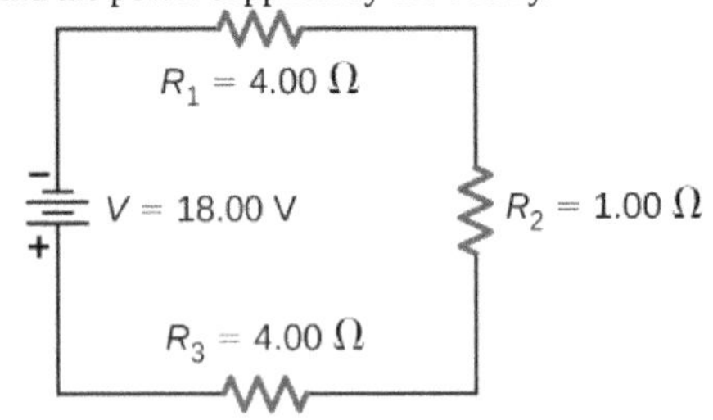

10.3 Kirchhoff's Rules

36. Consider the circuit shown below. (a) Find the voltage across each resistor. (b)What is the power supplied to the circuit and the power dissipated or consumed by the circuit?

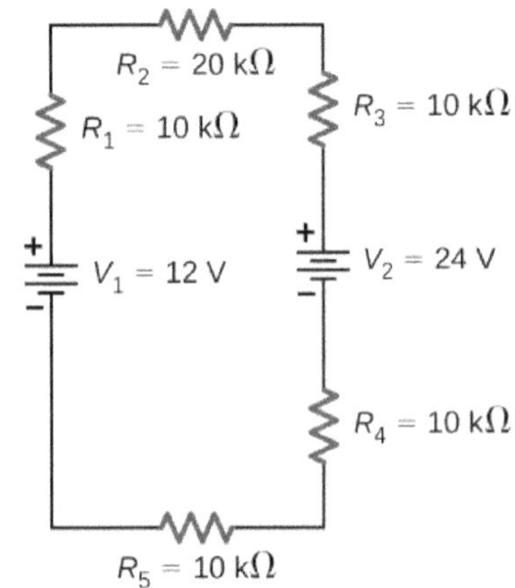

37. Consider the circuits shown below. (a) What is the current through each resistor in part (a)? (b) What is the current through each resistor in part (b)? (c) What is the power dissipated or consumed by each circuit? (d) What is the power supplied to each circuit?

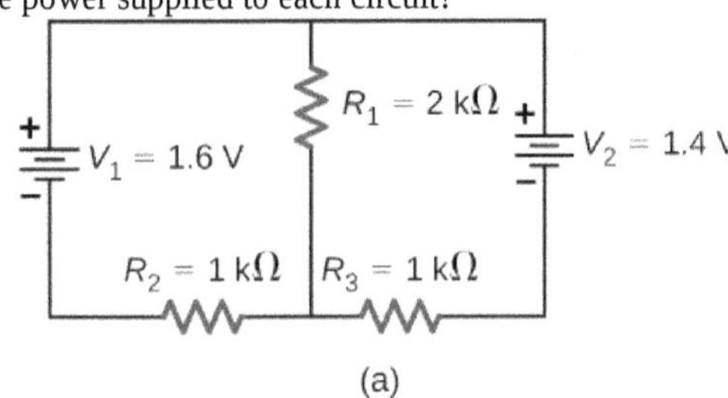

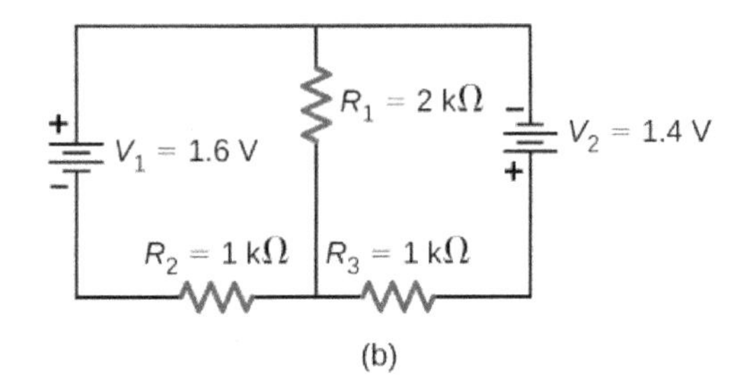

38. Consider the circuit shown below. Find V_1, I_2, and I_3.

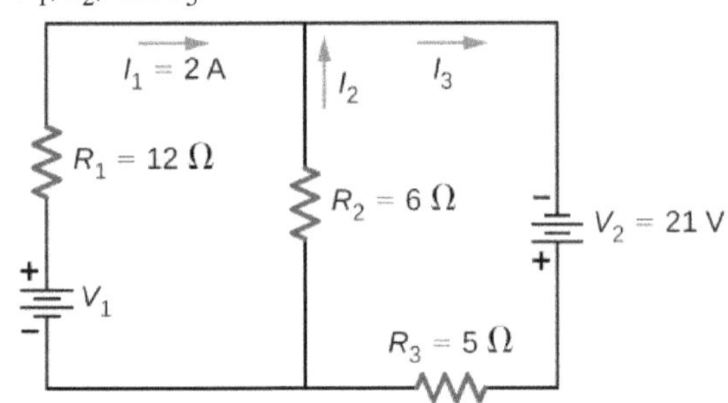

39. Consider the circuit shown below. Find V_1, V_2, and R_4.

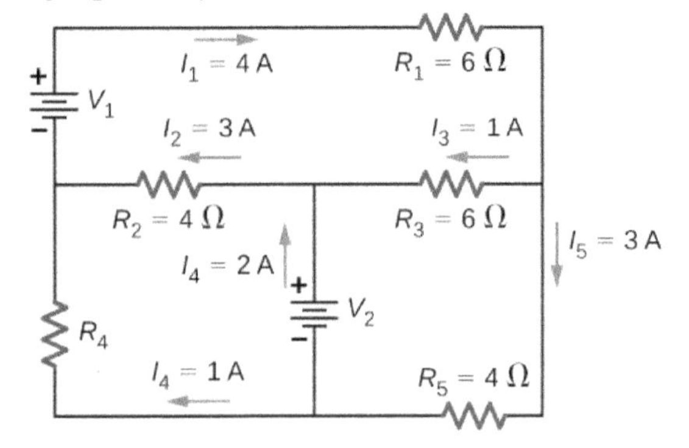

40. Consider the circuit shown below. Find $I_1, I_2,$ and I_3.

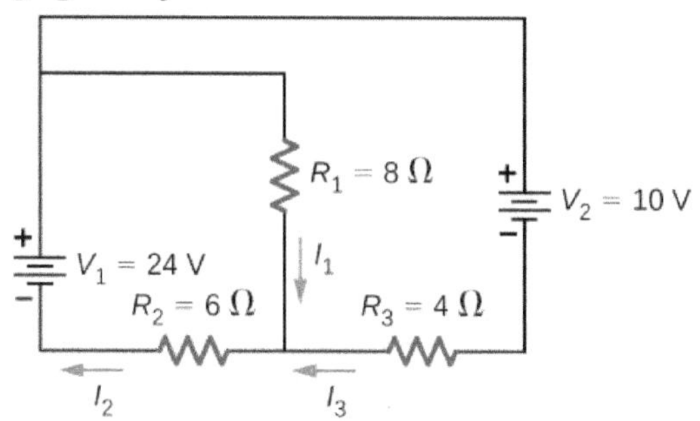

41. Consider the circuit shown below. (a) Find $I_1, I_2, I_3, I_4,$ and I_5. (b) Find the power supplied by the voltage sources. (c) Find the power dissipated by the resistors.

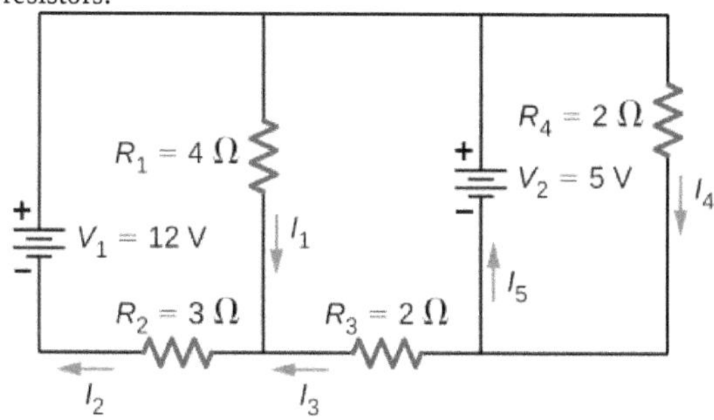

42. Consider the circuit shown below. Write the three loop equations for the loops shown.

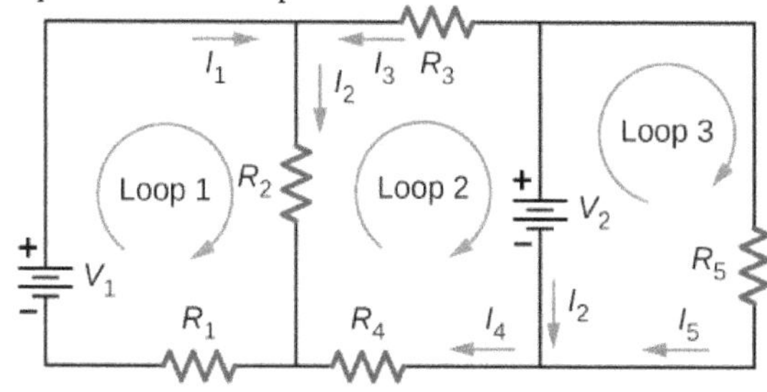

43. Consider the circuit shown below. Write equations for the three currents in terms of *R* and *V*.

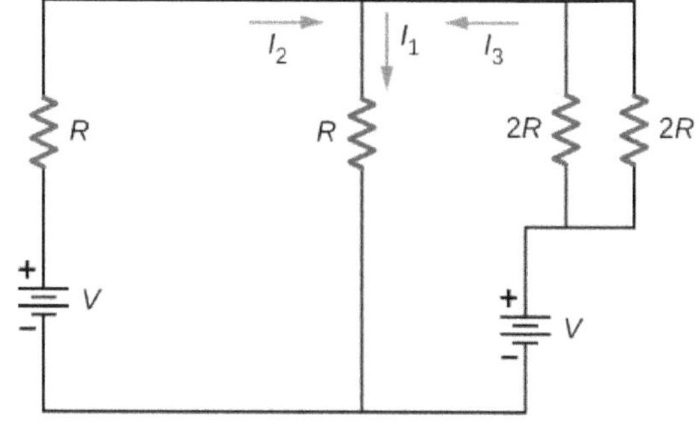

44. Consider the circuit shown in the preceding problem. Write equations for the power supplied by the voltage sources and the power dissipated by the resistors in terms of *R* and *V*.

45. A child's electronic toy is supplied by three 1.58-V alkaline cells having internal resistances of $0.0200\ \Omega$ in series with a 1.53-V carbon-zinc dry cell having a $0.100\text{-}\Omega$ internal resistance. The load resistance is $10.0\ \Omega$. (a) Draw a circuit diagram of the toy and its batteries. (b) What current flows? (c) How much power is supplied to the load? (d) What is the internal resistance of the dry cell if it goes bad, resulting in only 0.500 W being supplied to the load?

46. Apply the junction rule to Junction *b* shown below. Is any new information gained by applying the junction rule at *e*?

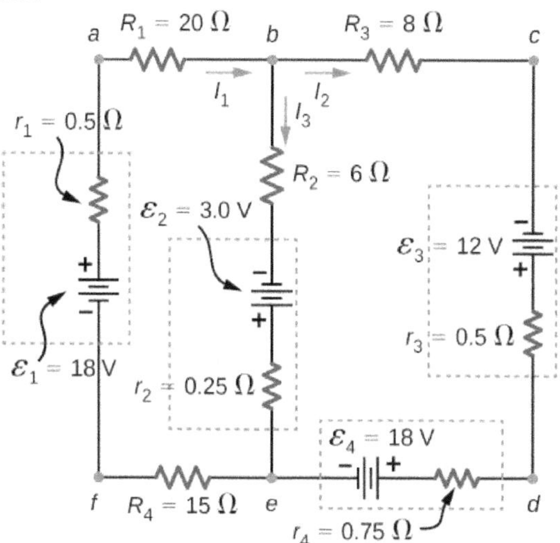

47. Apply the loop rule to Loop *afedcba* in the preceding problem.

10.4 Electrical Measuring Instruments

48. Suppose you measure the terminal voltage of a 1.585-V alkaline cell having an internal resistance of $0.100\ \Omega$ by placing a $1.00\text{-k}\Omega$ voltmeter across its terminals (see below). (a) What current flows? (b) Find the terminal voltage. (c) To see how close the measured terminal voltage is to the emf, calculate their ratio.

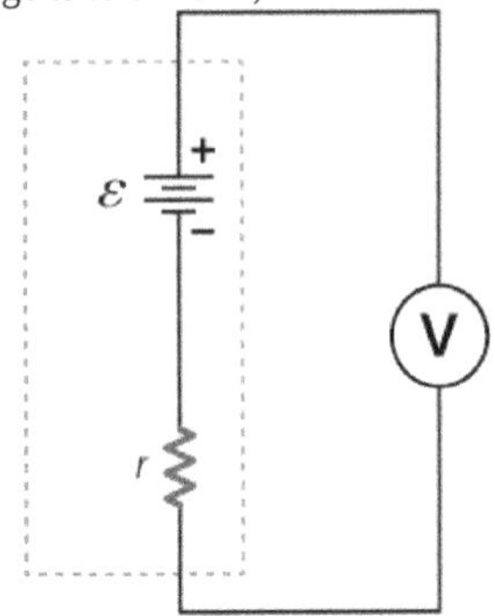

10.5 RC Circuits

49. The timing device in an automobile's intermittent wiper system is based on an *RC* time constant and utilizes a $0.500\text{-}\mu\text{F}$ capacitor and a variable resistor. Over what range must *R* be made to vary to achieve time constants from 2.00 to 15.0 s?

50. A heart pacemaker fires 72 times a minute, each time a 25.0-nF capacitor is charged (by a battery in series with a resistor) to 0.632 of its full voltage. What is the value of the resistance?

51. The duration of a photographic flash is related to an *RC* time constant, which is $0.100 \mu\text{F}$ for a certain camera. (a) If the resistance of the flash lamp is $0.0400\ \Omega$ during discharge, what is the size of the capacitor supplying its energy? (b) What is the time constant for charging the capacitor, if the charging resistance is $800\ \text{k}\Omega$?

52. A 2.00- and a $7.50\text{-}\mu\text{F}$ capacitor can be connected in series or parallel, as can a 25.0- and a $100\text{-k}\Omega$ resistor. Calculate the four *RC* time constants possible from connecting the resulting capacitance and resistance in series.

53. A $500\text{-}\Omega$ resistor, an uncharged $1.50\text{-}\mu\text{F}$ capacitor, and a 6.16-V emf are connected in series. (a) What is the initial current? (b) What is the *RC* time constant? (c) What is the current after one time constant? (d) What is the voltage on the capacitor after one time constant?

54. A heart defibrillator being used on a patient has an *RC* time constant of 10.0 ms due to the resistance of the patient and the capacitance of the defibrillator. (a) If the defibrillator has a capacitance of $8.00 \mu\text{F}$, what is the resistance of the path through the patient? (You may neglect the capacitance of the patient and the resistance of the defibrillator.) (b) If the initial voltage is 12.0 kV, how long does it take to decline to 6.00×10^2 V?

55. An ECG monitor must have an *RC* time constant less than $1.00 \times 10^2 \mu\text{s}$ to be able to measure variations in voltage over small time intervals. (a) If the resistance of the circuit (due mostly to that of the patient's chest) is $1.00\ \text{k}\Omega$, what is the maximum capacitance of the circuit? (b) Would it be difficult in practice to limit the capacitance to less than the value found in (a)?

56. Using the exact exponential treatment, determine how much time is required to charge an initially uncharged 100-pF capacitor through a $75.0\text{-M}\Omega$ resistor to 90.0% of its final voltage.

57. If you wish to take a picture of a bullet traveling at 500 m/s, then a very brief flash of light produced by an *RC* discharge through a flash tube can limit blurring. Assuming 1.00 mm of motion during one *RC* constant is acceptable, and given that the flash is driven by a $600\text{-}\mu\text{F}$ capacitor, what is the resistance in the flash tube?

10.6 Household Wiring and Electrical Safety

58. (a) How much power is dissipated in a short circuit of 240-V ac through a resistance of $0.250\ \Omega$? (b) What current flows?

59. What voltage is involved in a 1.44-kW short circuit through a $0.100\text{-}\Omega$ resistance?

60. Find the current through a person and identify the likely effect on her if she touches a 120-V ac source: (a) if she is standing on a rubber mat and offers a total resistance of $300\ \text{k}\Omega$; (b) if she is standing barefoot on wet grass and has a resistance of only $4000\ \text{k}\Omega$.

61. While taking a bath, a person touches the metal case of a radio. The path through the person to the drainpipe and ground has a resistance of $4000\ \Omega$. What is the smallest voltage on the case of the radio that could cause ventricular fibrillation?

62. A man foolishly tries to fish a burning piece of bread from a toaster with a metal butter knife and comes into contact with 120-V ac. He does not even feel it since, luckily, he is wearing rubber-soled shoes. What is the minimum resistance of the path the current follows through the person?

63. (a) During surgery, a current as small as $20.0\,\mu\text{A}$ applied directly to the heart may cause ventricular fibrillation. If the resistance of the exposed heart is $300\,\Omega$, what is the smallest voltage that poses this danger? (b) Does your answer imply that special electrical safety precautions are needed?

64. (a) What is the resistance of a 220-V ac short circuit that generates a peak power of 96.8 kW? (b) What would the average power be if the voltage were 120 V ac?

65. A heart defibrillator passes 10.0 A through a patient's torso for 5.00 ms in an attempt to restore normal beating. (a) How much charge passed? (b) What voltage was applied if 500 J of energy was dissipated? (c) What was the path's resistance? (d) Find the temperature increase caused in the 8.00 kg of affected tissue.

66. A short circuit in a 120-V appliance cord has a $0.500\text{-}\Omega$ resistance. Calculate the temperature rise of the 2.00 g of surrounding materials, assuming their specific heat capacity is $0.200\,\text{cal/g}\cdot{}^\circ\text{C}$ and that it takes 0.0500 s for a circuit breaker to interrupt the current. Is this likely to be damaging?

ADDITIONAL PROBLEMS

67. A circuit contains a D cell battery, a switch, a $20\text{-}\Omega$ resistor, and four 20-mF capacitors connected in series. (a) What is the equivalent capacitance of the circuit? (b) What is the *RC* time constant? (c) How long before the current decreases to 50% of the initial value once the switch is closed?

68. A circuit contains a D-cell battery, a switch, a $20\text{-}\Omega$ resistor, and three 20-mF capacitors. The capacitors are connected in parallel, and the parallel connection of capacitors are connected in series with the switch, the resistor and the battery. (a) What is the equivalent capacitance of the circuit? (b) What is the *RC* time constant? (c) How long before the current decreases to 50% of the initial value once the switch is closed?

69. Consider the circuit below. The battery has an emf of $\varepsilon = 30.00\text{ V}$ and an internal resistance of $r = 1.00\,\Omega$. (a) Find the equivalent resistance of the circuit and the current out of the battery. (b) Find the current through each resistor. (c) Find the potential drop across each resistor. (d) Find the power dissipated by each resistor. (e) Find the total power supplied by the batteries.

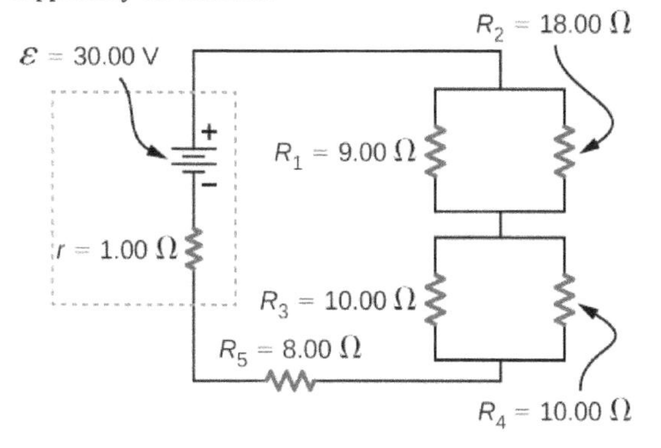

70. A homemade capacitor is constructed of 2 sheets of aluminum foil with an area of 2.00 square meters, separated by paper, 0.05 mm thick, of the same area and a dielectric constant of 3.7. The homemade capacitor is connected in series with a $100.00\text{-}\Omega$ resistor, a switch, and a 6.00-V voltage source. (a) What is the *RC* time constant of the circuit? (b) What is the initial current through the circuit, when the switch is closed? (c) How long does it take the current to reach one third of its initial value?

71. A student makes a homemade resistor from a graphite pencil 5.00 cm long, where the graphite is 0.05 mm in diameter. The resistivity of the graphite is $\rho = 1.38 \times 10^{-5}\,\Omega/\text{m}$. The homemade resistor is place in series with a switch, a 10.00-mF capacitor and a 0.50-V power source. (a) What is the *RC* time constant of the circuit? (b) What is the potential drop across the pencil 1.00 s after the switch is closed?

72. The rather simple circuit shown below is known as a voltage divider. The symbol consisting of three horizontal lines is represents "ground" and can be defined as the point where the potential is zero. The voltage divider is widely used in circuits and a single voltage source can be used to provide reduced voltage to a load resistor as shown in the second part of the figure. (a) What is the output voltage V_{out} of circuit (a) in terms of R_1, R_2, and V_{in}? (b) What is the output voltage V_{out} of circuit (b) in terms of R_1, R_2, R_L, and V_{in}?

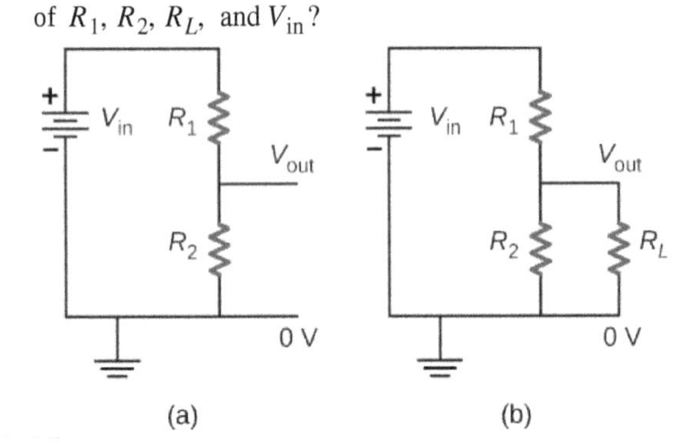

73. Three 300-Ω resistors are connect in series with an AAA battery with a rating of 3 AmpHours. (a) How long can the battery supply the resistors with power? (b) If the resistors are connected in parallel, how long can the battery last?

74. Consider a circuit that consists of a real battery with an emf ε and an internal resistance of r connected to a variable resistor R. (a) In order for the terminal voltage of the battery to be equal to the emf of the battery, what should the resistance of the variable resistor be adjusted to? (b) In order to get the maximum current from the battery, what should the resistance variable resistor be adjusted to? (c) In order for the maximum power output of the battery to be reached, what should the resistance of the variable resistor be set to?

75. Consider the circuit shown below. What is the energy stored in each capacitor after the switch has been closed for a very long time?

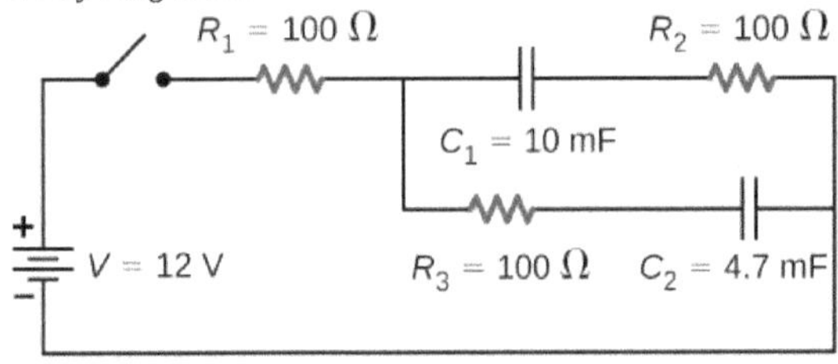

76. Consider a circuit consisting of a battery with an emf ε and an internal resistance of r connected in series with a resistor R and a capacitor C. Show that the total energy supplied by the battery while charging the battery is equal to $\varepsilon^2 C$.

77. Consider the circuit shown below. The terminal voltages of the batteries are shown. (a) Find the equivalent resistance of the circuit and the current out of the battery. (b) Find the current through each resistor. (c) Find the potential drop across each resistor. (d) Find the power dissipated by each resistor. (e) Find the total power supplied by the batteries.

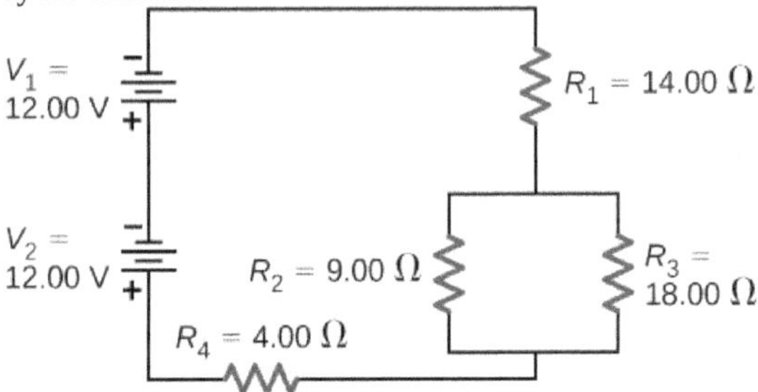

78. Consider the circuit shown below. (a) What is the terminal voltage of the battery? (b) What is the potential drop across resistor R_2?

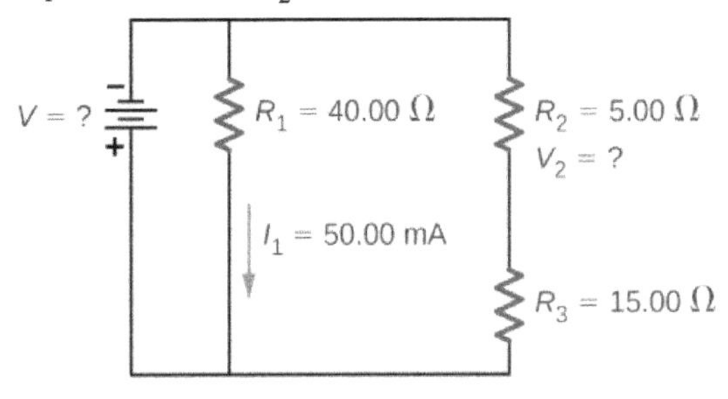

79. Consider the circuit shown below. (a)Determine the equivalent resistance and the current from the battery with switch S_1 open. (b) Determine the equivalent resistance and the current from the battery with switch S_1 closed.

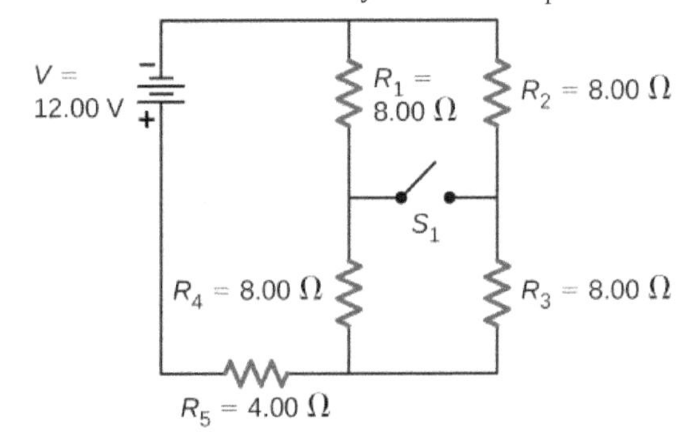

80. Two resistors, one having a resistance of $145\ \Omega$, are connected in parallel to produce a total resistance of $150\ \Omega$. (a) What is the value of the second resistance? (b) What is unreasonable about this result? (c) Which assumptions are unreasonable or inconsistent?

81. Two resistors, one having a resistance of $900\ \text{k}\Omega$, are connected in series to produce a total resistance of $0.500\ \text{M}\Omega$. (a) What is the value of the second resistance? (b) What is unreasonable about this result? (c) Which assumptions are unreasonable or inconsistent?

82. Apply the junction rule at point *a* shown below.

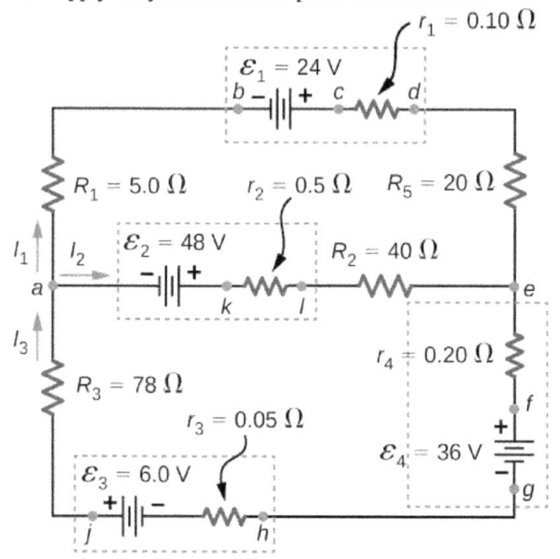

83. Apply the loop rule to Loop *akledcba* in the preceding problem.

84. Find the currents flowing in the circuit in the preceding problem. Explicitly show how you follow the steps in the Problem-Solving Strategy: Series and Parallel Resistors.

85. Consider the circuit shown below. (a) Find the current through each resistor. (b) Check the calculations by analyzing the power in the circuit.

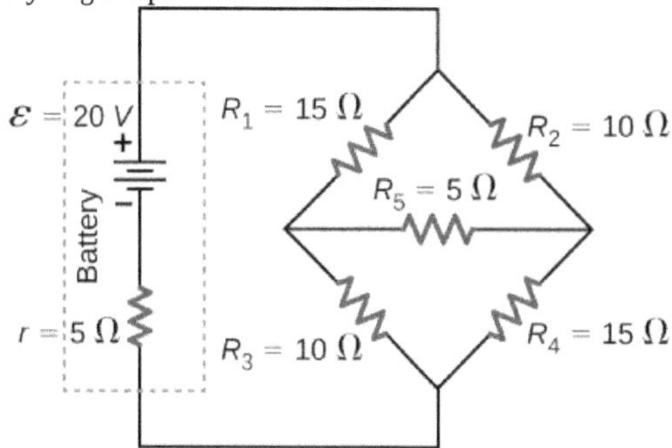

86. A flashing lamp in a Christmas earring is based on an *RC* discharge of a capacitor through its resistance. The effective duration of the flash is 0.250 s, during which it produces an average 0.500 W from an average 3.00 V. (a) What energy does it dissipate? (b) How much charge moves through the lamp? (c) Find the capacitance. (d) What is the resistance of the lamp? (Since average values are given for some quantities, the shape of the pulse profile is not needed.)

87. A 160-μF capacitor charged to 450 V is discharged through a $31.2\text{-k}\Omega$ resistor. (a) Find the time constant. (b) Calculate the temperature increase of the resistor, given that its mass is 2.50 g and its specific heat is $1.67\ \text{kJ/kg} \cdot {}^\circ\text{C}$, noting that most of the thermal energy is retained in the short time of the discharge. (c) Calculate the new resistance, assuming it is pure carbon. (d) Does this change in resistance seem significant?

CHALLENGE PROBLEMS

88. Some camera flashes use flash tubes that require a high voltage. They obtain a high voltage by charging capacitors in parallel and then internally changing the connections of the capacitors to place them in series. Consider a circuit that uses four AAA batteries connected in series to charge six 10-mF capacitors through an equivalent resistance of $100\,\Omega$. The connections are then switched internally to place the capacitors in series. The capacitors discharge through a lamp with a resistance of $100\,\Omega$. (a) What is the *RC* time constant and the initial current out of the batteries while they are connected in parallel? (b) How long does it take for the capacitors to charge to 90% of the terminal voltages of the batteries? (c) What is the *RC* time constant and the initial current of the capacitors connected in series assuming it discharges at 90% of full charge? (d) How long does it take the current to decrease to 10% of the initial value?

89. Consider the circuit shown below. Each battery has an emf of 1.50 V and an internal resistance of $1.00\,\Omega$. (a) What is the current through the external resistor, which has a resistance of 10.00 ohms? (b) What is the terminal voltage of each battery?

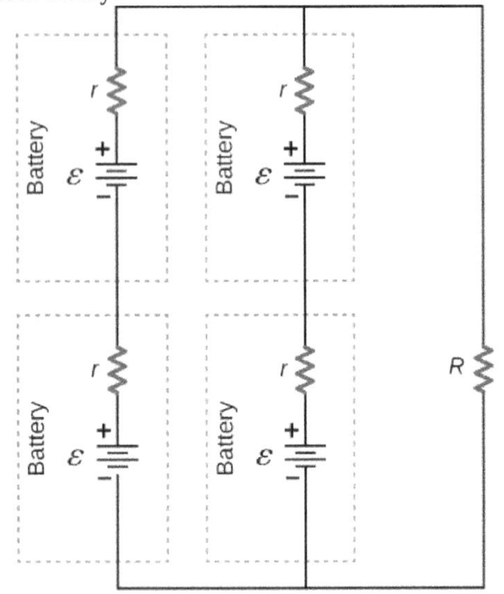

90. Analog meters use a galvanometer, which essentially consists of a coil of wire with a small resistance and a pointer with a scale attached. When current runs through the coil, the pointer turns; the amount the pointer turns is proportional to the amount of current running through the coil. Galvanometers can be used to make an ammeter if a resistor is placed in parallel with the galvanometer. Consider a galvanometer that has a resistance of $25.00\,\Omega$ and gives a full scale reading when a $50\text{-}\mu\text{A}$ current runs through it. The galvanometer is to be used to make an ammeter that has a full scale reading of 10.00 A, as shown below. Recall that an ammeter is connected in series with the circuit of interest, so all 10 A must run through the meter. (a) What is the current through the parallel resistor in the meter? (b) What is the voltage across the parallel resistor? (c) What is the resistance of the parallel resistor?

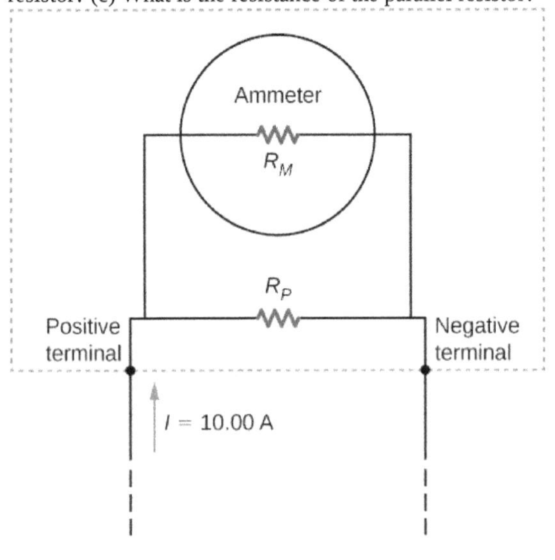

91. Analog meters use a galvanometer, which essentially consists of a coil of wire with a small resistance and a pointer with a scale attached. When current runs through the coil, the point turns; the amount the pointer turns is proportional to the amount of current running through the coil. Galvanometers can be used to make a voltmeter if a resistor is placed in series with the galvanometer. Consider a galvanometer that has a resistance of $25.00\,\Omega$ and gives a full scale reading when a $50\text{-}\mu\text{A}$ current runs through it. The galvanometer is to be used to make an voltmeter that has a full scale reading of 10.00 V, as shown below. Recall that a voltmeter is connected in parallel with the component of interest, so the meter must have a high resistance or it will change the current running through the component. (a) What is the potential drop across the series resistor in the meter? (b) What is the resistance of the parallel resistor?

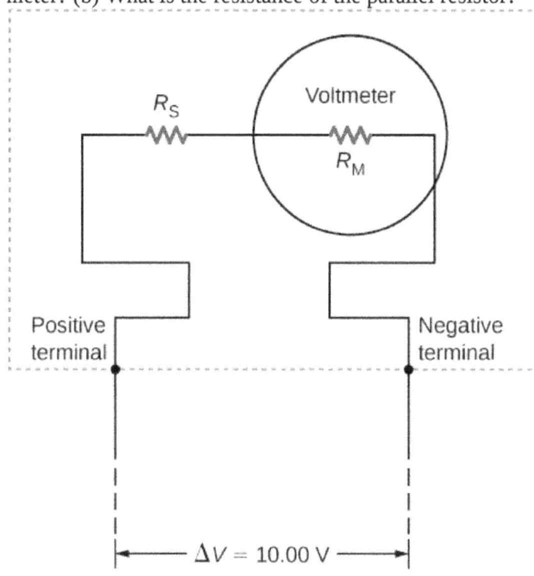

92. Consider the circuit shown below. Find I_1, V_1, I_2, and V_3.

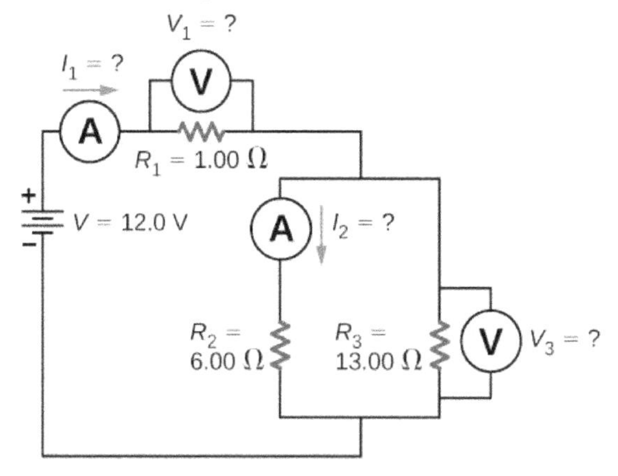

93. Consider the circuit below. (a) What is the *RC* time constant of the circuit? (b) What is the initial current in the circuit once the switch is closed? (c) How much time passes between the instant the switch is closed and the time the current has reached half of the initial current?

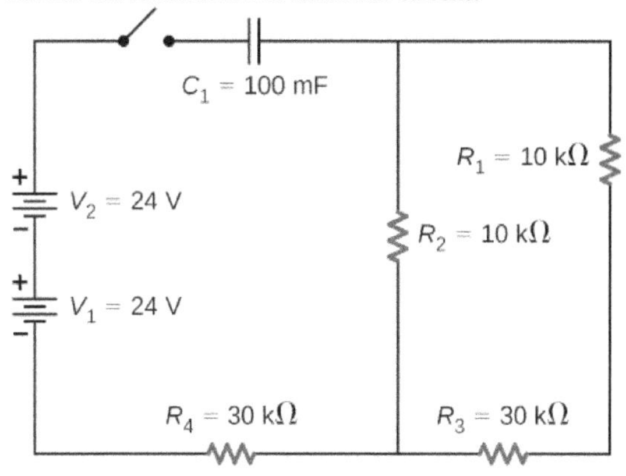

94. Consider the circuit below. (a) What is the initial current through resistor R_2 when the switch is closed? (b) What is the current through resistor R_2 when the capacitor is fully charged, long after the switch is closed? (c) What happens if the switch is opened after it has been closed for some time? (d) If the switch has been closed for a time period long enough for the capacitor to become fully charged, and then the switch is opened, how long before the current through resistor R_1 reaches half of its initial value?

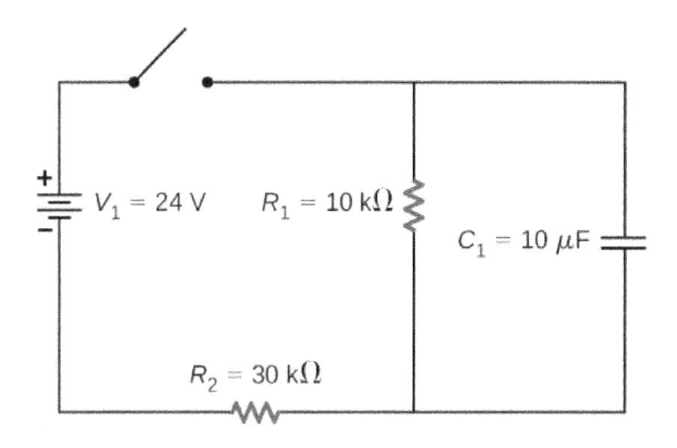

95. Consider the infinitely long chain of resistors shown below. What is the resistance between terminals *a* and *b*?

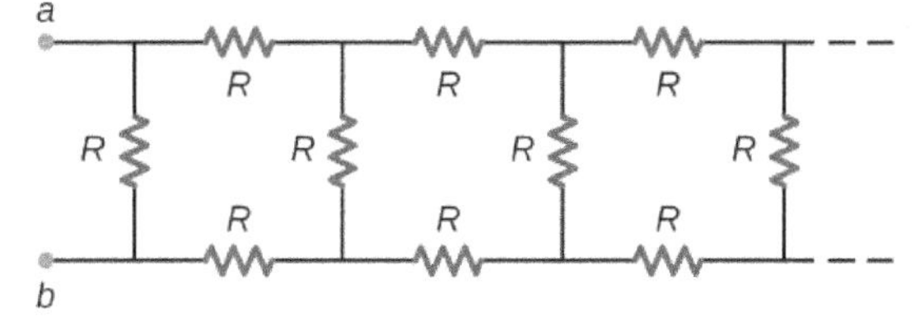

96. Consider the circuit below. The capacitor has a capacitance of 10 mF. The switch is closed and after a long time the capacitor is fully charged. (a) What is the current through each resistor a long time after the switch is closed? (b) What is the voltage across each resistor a long time after the switch is closed? (c) What is the voltage across the capacitor a long time after the switch is closed? (d) What is the charge on the capacitor a long time after the switch is closed? (e) The switch is then opened. The capacitor discharges through the resistors. How long from the time before the current drops to one fifth of the initial value?

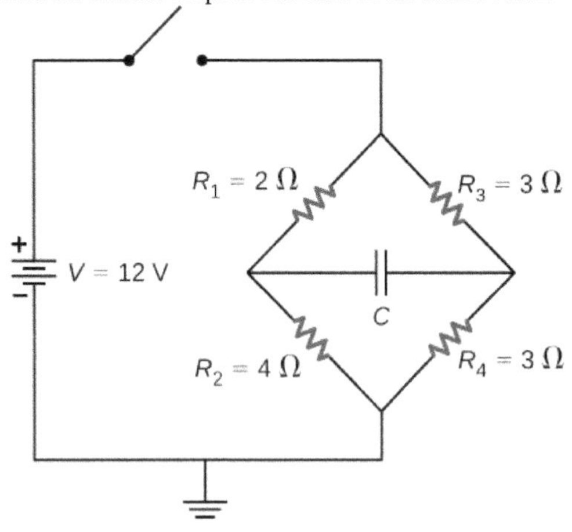

97. A 120-V immersion heater consists of a coil of wire that is placed in a cup to boil the water. The heater can boil one cup of $20.00\ °C$ water in 180.00 seconds. You buy one to use in your dorm room, but you are worried that you will overload the circuit and trip the 15.00-A, 120-V circuit breaker, which supplies your dorm room. In your dorm room, you have four 100.00-W incandescent lamps and a 1500.00-W space heater. (a) What is the power rating of the immersion heater? (b) Will it trip the breaker when everything is turned on? (c) If it you replace the incandescent bulbs with 18.00-W LED, will the breaker trip when everything is turned on?

98. Find the resistance that must be placed in series with a 25.0-Ω galvanometer having a $50.0\text{-}\mu\text{A}$ sensitivity (the same as the one discussed in the text) to allow it to be used as a voltmeter with a 3000-V full-scale reading. Include a circuit diagram with your solution.

99. Find the resistance that must be placed in parallel with a 60.0-Ω galvanometer having a 1.00-mA sensitivity (the same as the one discussed in the text) to allow it to be used as an ammeter with a 25.0-A full-scale reading. Include a circuit diagram with your solution.

11 | MAGNETIC FORCES AND FIELDS

Figure 11.1 An industrial electromagnet is capable of lifting thousands of pounds of metallic waste. (credit: modification of work by “BedfordAl”/Flickr)

Chapter Outline

Introduction

For the past few chapters, we have been studying electrostatic forces and fields, which are caused by electric charges at rest. These electric fields can move other free charges, such as producing a current in a circuit; however, the electrostatic forces and fields themselves come from other static charges. In this chapter, we see that when an electric charge moves, it generates other forces and fields. These additional forces and fields are what we commonly call magnetism.

Before we examine the origins of magnetism, we first describe what it is and how magnetic fields behave. Once we are more familiar with magnetic effects, we can explain how they arise from the behavior of atoms and molecules, and how magnetism is related to electricity. The connection between electricity and magnetism is fascinating from a theoretical point of view, but it is also immensely practical, as shown by an industrial electromagnet that can lift thousands of pounds of metal.

11.1 | Magnetism and Its Historical Discoveries

Learning Objectives

By the end of this section, you will be able to:

- Explain attraction and repulsion by magnets
- Describe the historical and contemporary applications of magnetism

Magnetism has been known since the time of the ancient Greeks, but it has always been a bit mysterious. You can see electricity in the flash of a lightning bolt, but when a compass needle points to magnetic north, you can't see any force causing it to rotate. People learned about magnetic properties gradually, over many years, before several physicists of the nineteenth century connected magnetism with electricity. In this section, we review the basic ideas of magnetism and describe how they fit into the picture of a magnetic field.

Brief History of Magnetism

Magnets are commonly found in everyday objects, such as toys, hangers, elevators, doorbells, and computer devices. Experimentation on these magnets shows that all magnets have two poles: One is labeled north (N) and the other is labeled south (S). Magnetic poles repel if they are alike (both N or both S), they attract if they are opposite (one N and the other S), and both poles of a magnet attract unmagnetized pieces of iron. An important point to note here is that you cannot isolate an individual magnetic pole. Every piece of a magnet, no matter how small, which contains a north pole must also contain a south pole.

Visit this website (https://openstaxcollege.org/l/21magnetcompass) for an interactive demonstration of magnetic north and south poles.

An example of a magnet is a compass needle. It is simply a thin bar magnet suspended at its center, so it is free to rotate in a horizontal plane. Earth itself also acts like a very large bar magnet, with its south-seeking pole near the geographic North Pole (Figure 11.2). The north pole of a compass is attracted toward Earth's geographic North Pole because the magnetic pole that is near the geographic North Pole is actually a south magnetic pole. Confusion arises because the geographic term "North Pole" has come to be used (incorrectly) for the magnetic pole that is near the North Pole. Thus, " **north magnetic pole**" is actually a misnomer—it should be called the **south magnetic pole**. [Note that the orientation of Earth's magnetic field is not permanent but changes ("flips") after long time intervals. Eventually, Earth's north magnetic pole may be located near its geographic North Pole.]

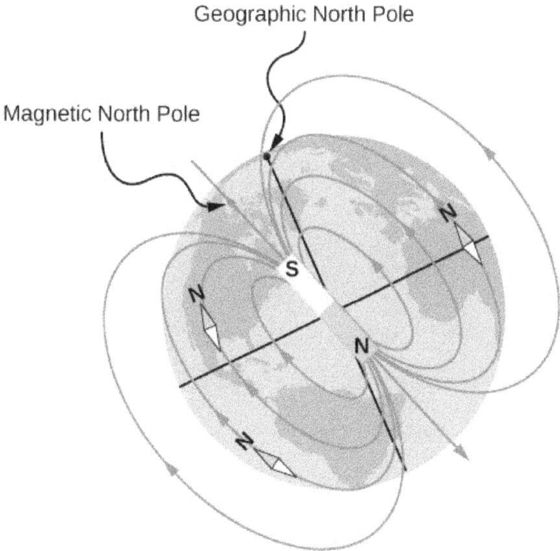

Figure 11.2 The north pole of a compass needle points toward the south pole of a magnet, which is how today's magnetic field is oriented from inside Earth. It also points toward Earth's geographic North Pole because the geographic North Pole is near the magnetic south pole.

Back in 1819, the Danish physicist Hans Oersted was performing a lecture demonstration for some students and noticed that a compass needle moved whenever current flowed in a nearby wire. Further investigation of this phenomenon convinced Oersted that an electric current could somehow cause a magnetic force. He reported this finding to an 1820 meeting of the French Academy of Science.

Soon after this report, Oersted's investigations were repeated and expanded upon by other scientists. Among those whose work was especially important were Jean-Baptiste Biot and Felix Savart, who investigated the forces exerted on magnets by currents; André Marie Ampère, who studied the forces exerted by one current on another; François Arago, who found that iron could be magnetized by a current; and Humphry Davy, who discovered that a magnet exerts a force on a wire carrying an electric current. Within 10 years of Oersted's discovery, Michael Faraday found that the relative motion of a magnet and a metallic wire induced current in the wire. This finding showed not only that a current has a magnetic effect, but that a magnet can generate electric current. You will see later that the names of Biot, Savart, Ampère, and Faraday are linked to some of the fundamental laws of electromagnetism.

The evidence from these various experiments led Ampère to propose that electric current is the source of all magnetic phenomena. To explain permanent magnets, he suggested that matter contains microscopic current loops that are somehow aligned when a material is magnetized. Today, we know that permanent magnets are actually created by the alignment of spinning electrons, a situation quite similar to that proposed by Ampère. This model of permanent magnets was developed by Ampère almost a century before the atomic nature of matter was understood. (For a full quantum mechanical treatment of magnetic spins, see Quantum Mechanics (http://cnx.org/content/m58573/latest/) and Atomic Structure (http://cnx.org/content/m58583/latest/) .)

Contemporary Applications of Magnetism

Today, magnetism plays many important roles in our lives. Physicists' understanding of magnetism has enabled the development of technologies that affect both individuals and society. The electronic tablet in your purse or backpack, for example, wouldn't have been possible without the applications of magnetism and electricity on a small scale (Figure 11.3). Weak changes in a magnetic field in a thin film of iron and chromium were discovered to bring about much larger changes in resistance, called giant magnetoresistance. Information can then be recorded magnetically based on the direction in which the iron layer is magnetized. As a result of the discovery of giant magnetoresistance and its applications to digital storage, the 2007 Nobel Prize in Physics was awarded to Albert Fert from France and Peter Grunberg from Germany.

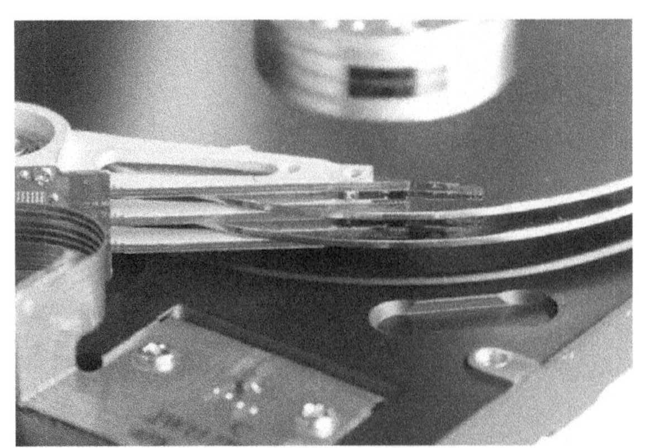

Figure 11.3 Engineering technology like computer storage would not be possible without a deep understanding of magnetism. (credit: Klaus Eifert)

All electric motors—with uses as diverse as powering refrigerators, starting cars, and moving elevators—contain magnets. Generators, whether producing hydroelectric power or running bicycle lights, use magnetic fields. Recycling facilities employ magnets to separate iron from other refuse. Research into using magnetic containment of fusion as a future energy source has been continuing for several years. Magnetic resonance imaging (MRI) has become an important diagnostic tool in the field of medicine, and the use of magnetism to explore brain activity is a subject of contemporary research and development. The list of applications also includes computer hard drives, tape recording, detection of inhaled asbestos, and levitation of high-speed trains. Magnetism is involved in the structure of atomic energy levels, as well as the motion of cosmic rays and charged particles trapped in the Van Allen belts around Earth. Once again, we see that all these disparate phenomena are linked by a small number of underlying physical principles.

11.2 | Magnetic Fields and Lines

Learning Objectives

By the end of this section, you will be able to:

- Define the magnetic field based on a moving charge experiencing a force
- Apply the right-hand rule to determine the direction of a magnetic force based on the motion of a charge in a magnetic field
- Sketch magnetic field lines to understand which way the magnetic field points and how strong it is in a region of space

We have outlined the properties of magnets, described how they behave, and listed some of the applications of magnetic properties. Even though there are no such things as isolated magnetic charges, we can still define the attraction and repulsion of magnets as based on a field. In this section, we define the magnetic field, determine its direction based on the right-hand rule, and discuss how to draw magnetic field lines.

Defining the Magnetic Field

A magnetic field is defined by the force that a charged particle experiences moving in this field, after we account for the gravitational and any additional electric forces possible on the charge. The magnitude of this force is proportional to the amount of charge q, the speed of the charged particle v, and the magnitude of the applied magnetic field. The direction of this force is perpendicular to both the direction of the moving charged particle and the direction of the applied magnetic field. Based on these observations, we define the magnetic field strength B based on the **magnetic force** $\vec{\mathbf{F}}$ on a charge q moving at velocity $\vec{\mathbf{v}}$ as the cross product of the velocity and magnetic field, that is,

$$\vec{\mathbf{F}} = q\vec{\mathbf{v}} \times \vec{\mathbf{B}}. \tag{11.1}$$

In fact, this is how we define the magnetic field $\vec{\mathbf{B}}$ —in terms of the force on a charged particle moving in a magnetic field. The magnitude of the force is determined from the definition of the cross product as it relates to the magnitudes of each of the vectors. In other words, the magnitude of the force satisfies

$$F = qvB\sin\theta \tag{11.2}$$

where θ is the angle between the velocity and the magnetic field.

The SI unit for magnetic field strength B is called the **tesla** (T) after the eccentric but brilliant inventor Nikola Tesla (1856–1943), where

$$1\,\mathrm{T} = \frac{1\,\mathrm{N}}{\mathrm{A \cdot m}}. \tag{11.3}$$

A smaller unit, called the **gauss** (G), where $1\,\mathrm{G} = 10^{-4}\,\mathrm{T}$, is sometimes used. The strongest permanent magnets have fields near 2 T; superconducting electromagnets may attain 10 T or more. Earth's magnetic field on its surface is only about $5 \times 10^{-5}\,\mathrm{T}$, or 0.5 G.

Problem-Solving Strategy: Direction of the Magnetic Field by the Right-Hand Rule

The direction of the magnetic force $\vec{\mathbf{F}}$ is perpendicular to the plane formed by $\vec{\mathbf{v}}$ and $\vec{\mathbf{B}}$, as determined by the **right-hand rule-1** (or RHR-1), which is illustrated in Figure 11.4.

1. Orient your right hand so that your fingers curl in the plane defined by the velocity and magnetic field vectors.
2. Using your right hand, sweep from the velocity toward the magnetic field with your fingers through the smallest angle possible.
3. The magnetic force is directed where your thumb is pointing.
4. If the charge was negative, reverse the direction found by these steps.

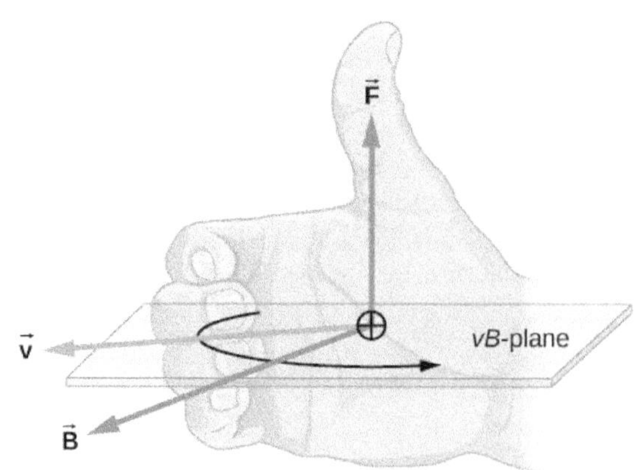

Figure 11.4 Magnetic fields exert forces on moving charges. The direction of the magnetic force on a moving charge is perpendicular to the plane formed by $\vec{\mathbf{v}}$ and $\vec{\mathbf{B}}$ and follows the right-hand rule-1 (RHR-1) as shown. The magnitude of the force is proportional to $q,\ v,\ B,$ and the sine of the angle between $\vec{\mathbf{v}}$ and $\vec{\mathbf{B}}$.

Visit this website (https://openstaxcollege.org/l/21magfields) for additional practice with the direction of magnetic fields.

There is no magnetic force on static charges. However, there is a magnetic force on charges moving at an angle to a magnetic field. When charges are stationary, their electric fields do not affect magnets. However, when charges move, they produce magnetic fields that exert forces on other magnets. When there is relative motion, a connection between electric and magnetic forces emerges—each affects the other.

Example 11.1

An Alpha-Particle Moving in a Magnetic Field

An alpha-particle $\left(q = 3.2 \times 10^{-19}\,\text{C}\right)$ moves through a uniform magnetic field whose magnitude is 1.5 T. The field is directly parallel to the positive *z*-axis of the rectangular coordinate system of Figure 11.5. What is the magnetic force on the alpha-particle when it is moving (a) in the positive *x*-direction with a speed of $5.0 \times 10^4\,\text{m/s}$? (b) in the negative *y*-direction with a speed of $5.0 \times 10^4\,\text{m/s}$? (c) in the positive *z*-direction with a speed of $5.0 \times 10^4\,\text{m/s}$? (d) with a velocity $\vec{\mathbf{v}} = \left(2.0\,\hat{\mathbf{i}} - 3.0\,\hat{\mathbf{j}} + 1.0\,\hat{\mathbf{k}}\right) \times 10^4\,\text{m/s}$?

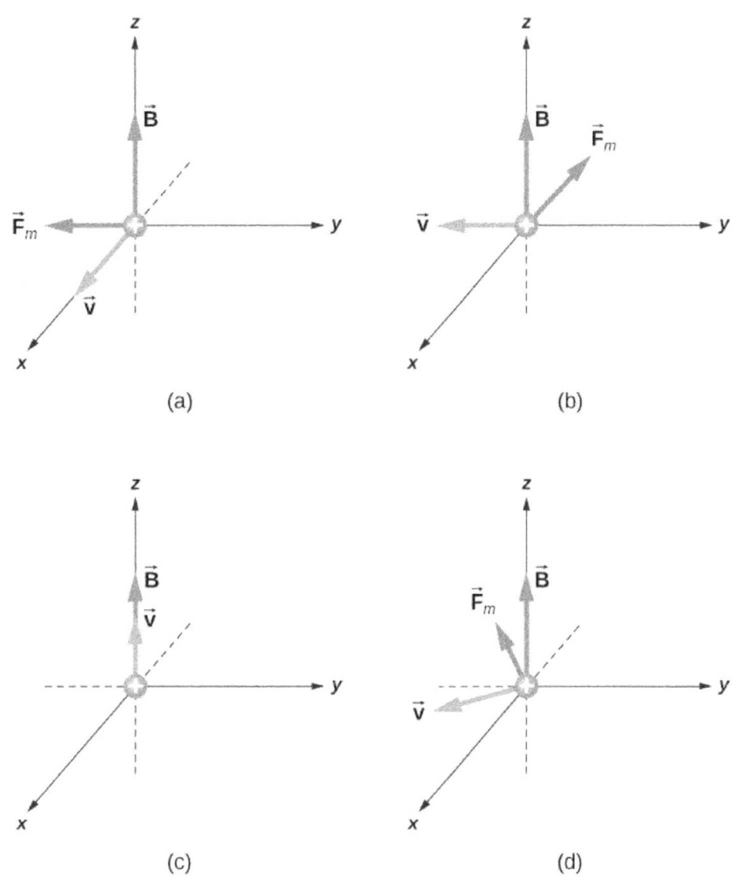

Figure 11.5 The magnetic forces on an alpha-particle moving in a uniform magnetic field. The field is the same in each drawing, but the velocity is different.

Strategy

We are given the charge, its velocity, and the magnetic field strength and direction. We can thus use the equation $\vec{\mathbf{F}} = q\vec{\mathbf{v}} \times \vec{\mathbf{B}}$ or $F = qvB\sin\theta$ to calculate the force. The direction of the force is determined by RHR-1.

Solution

a. First, to determine the direction, start with your fingers pointing in the positive x-direction. Sweep your fingers upward in the direction of magnetic field. Your thumb should point in the negative y-direction. This should match the mathematical answer. To calculate the force, we use the given charge, velocity, and magnetic field and the definition of the magnetic force in cross-product form to calculate:

$$\vec{\mathbf{F}} = q\vec{\mathbf{v}} \times \vec{\mathbf{B}} = (3.2 \times 10^{-19}\,\text{C})\left(5.0 \times 10^{4}\,\text{m/s}\,\hat{\mathbf{i}}\right) \times \left(1.5\,\text{T}\,\hat{\mathbf{k}}\right) = -2.4 \times 10^{-14}\,\text{N}\,\hat{\mathbf{j}}.$$

b. First, to determine the directionality, start with your fingers pointing in the negative y-direction. Sweep your fingers upward in the direction of magnetic field as in the previous problem. Your thumb should be open in the negative x-direction. This should match the mathematical answer. To calculate the force, we use the given charge, velocity, and magnetic field and the definition of the magnetic force in cross-product form to calculate:

$$\vec{\mathbf{F}} = q\vec{\mathbf{v}} \times \vec{\mathbf{B}} = (3.2 \times 10^{-19}\,\text{C})\left(-5.0 \times 10^{4}\,\text{m/s}\,\hat{\mathbf{j}}\right) \times \left(1.5\,\text{T}\,\hat{\mathbf{k}}\right) = -2.4 \times 10^{-14}\,\text{N}\,\hat{\mathbf{i}}.$$

An alternative approach is to use Equation 11.2 to find the magnitude of the force. This applies for both parts (a) and (b). Since the velocity is perpendicular to the magnetic field, the angle between them is 90 degrees. Therefore, the magnitude of the force is:

$$F = qvB\sin\theta = \left(3.2 \times 10^{-19}\,\text{C}\right)\left(5.0 \times 10^{4}\,\text{m/s}\right)(1.5\,\text{T})\sin(90°) = 2.4 \times 10^{-14}\,\text{N}.$$

c. Since the velocity and magnetic field are parallel to each other, there is no orientation of your hand that will result in a force direction. Therefore, the force on this moving charge is zero. This is confirmed by the cross product. When you cross two vectors pointing in the same direction, the result is equal to zero.

d. First, to determine the direction, your fingers could point in any orientation; however, you must sweep your fingers upward in the direction of the magnetic field. As you rotate your hand, notice that the thumb can point in any *x*- or *y*-direction possible, but not in the *z*-direction. This should match the mathematical answer. To calculate the force, we use the given charge, velocity, and magnetic field and the definition of the magnetic force in cross-product form to calculate:

$$\begin{aligned}\vec{\mathbf{F}} &= q\,\vec{\mathbf{v}} \times \vec{\mathbf{B}} = \left(3.2 \times 10^{-19}\,\text{C}\right)\left(\left(2.0\,\hat{\mathbf{i}} - 3.0\,\hat{\mathbf{j}} + 1.0\,\hat{\mathbf{k}}\right) \times 10^{4}\,\text{m/s}\right) \times \left(1.5\,\text{T}\,\hat{\mathbf{k}}\right) \\ &= \left(-14.4\,\hat{\mathbf{i}} - 9.6\,\hat{\mathbf{j}}\right) \times 10^{-15}\,\text{N}.\end{aligned}$$

This solution can be rewritten in terms of a magnitude and angle in the *xy*-plane:

$$\begin{aligned}\left|\vec{\mathbf{F}}\right| &= \sqrt{F_x^2 + F_y^2} = \sqrt{(-14.4)^2 + (-9.6)^2} \times 10^{-15}\,\text{N} = 1.7 \times 10^{-14}\,\text{N} \\ \theta &= \tan^{-1}\left(\frac{F_y}{F_x}\right) = \tan^{-1}\left(\frac{-9.6 \times 10^{-15}\,\text{N}}{-14.4 \times 10^{-15}\,\text{N}}\right) = 34°.\end{aligned}$$

The magnitude of the force can also be calculated using Equation 11.2. The velocity in this question, however, has three components. The *z*-component of the velocity can be neglected, because it is parallel to the magnetic field and therefore generates no force. The magnitude of the velocity is calculated from the *x*- and *y*-components. The angle between the velocity in the *xy*-plane and the magnetic field in the *z*-plane is 90 degrees. Therefore, the force is calculated to be:

$$\begin{aligned}\left|\vec{\mathbf{v}}\right| &= \sqrt{(2)^2 + (-3)^2} \times 10^{4}\,\frac{\text{m}}{\text{s}} = 3.6 \times 10^{4}\,\frac{\text{m}}{\text{s}} \\ F &= qvB\sin\theta = (3.2 \times 10^{-19}\,\text{C})(3.6 \times 10^{4}\,\text{m/s})(1.5\,\text{T})\sin(90°) = 1.7 \times 10^{-14}\,\text{N}.\end{aligned}$$

This is the same magnitude of force calculated by unit vectors.

Significance

The cross product in this formula results in a third vector that must be perpendicular to the other two. Other physical quantities, such as angular momentum, also have three vectors that are related by the cross product. Note that typical force values in magnetic force problems are much larger than the gravitational force. Therefore, for an isolated charge, the magnetic force is the dominant force governing the charge's motion.

11.1 Check Your Understanding Repeat the previous problem with the magnetic field in the *x*-direction rather than in the *z*-direction. Check your answers with RHR-1.

Representing Magnetic Fields

The representation of magnetic fields by **magnetic field lines** is very useful in visualizing the strength and direction of the magnetic field. As shown in Figure 11.6, each of these lines forms a closed loop, even if not shown by the constraints of the space available for the figure. The field lines emerge from the north pole (N), loop around to the south pole (S), and continue through the bar magnet back to the north pole.

Magnetic field lines have several hard-and-fast rules:

1. The direction of the magnetic field is tangent to the field line at any point in space. A small compass will point in the direction of the field line.

2. The strength of the field is proportional to the closeness of the lines. It is exactly proportional to the number of lines per unit area perpendicular to the lines (called the areal density).
3. Magnetic field lines can never cross, meaning that the field is unique at any point in space.
4. Magnetic field lines are continuous, forming closed loops without a beginning or end. They are directed from the north pole to the south pole.

The last property is related to the fact that the north and south poles cannot be separated. It is a distinct difference from electric field lines, which generally begin on positive charges and end on negative charges or at infinity. If isolated magnetic charges (referred to as magnetic monopoles) existed, then magnetic field lines would begin and end on them.

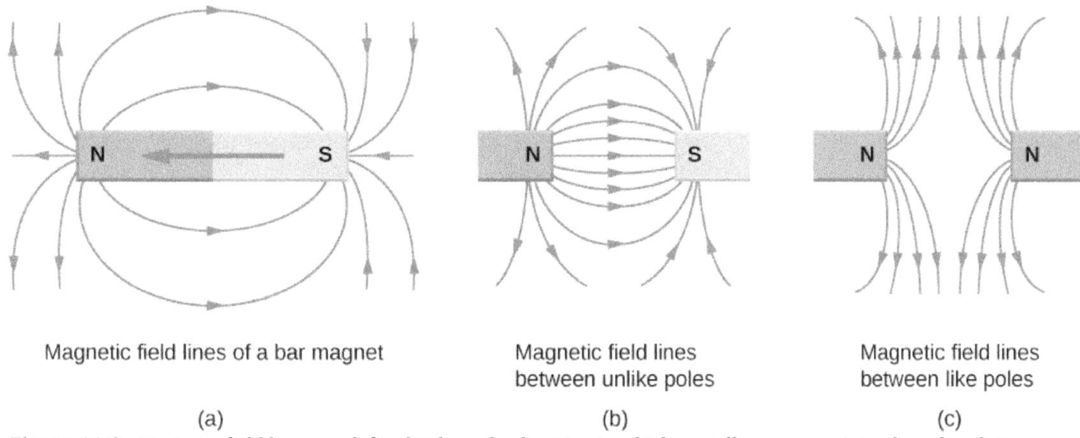

Figure 11.6 Magnetic field lines are defined to have the direction in which a small compass points when placed at a location in the field. The strength of the field is proportional to the closeness (or density) of the lines. If the interior of the magnet could be probed, the field lines would be found to form continuous, closed loops. To fit in a reasonable space, some of these drawings may not show the closing of the loops; however, if enough space were provided, the loops would be closed.

11.3 | Motion of a Charged Particle in a Magnetic Field

Learning Objectives

By the end of this section, you will be able to:

- Explain how a charged particle in an external magnetic field undergoes circular motion
- Describe how to determine the radius of the circular motion of a charged particle in a magnetic field

A charged particle experiences a force when moving through a magnetic field. What happens if this field is uniform over the motion of the charged particle? What path does the particle follow? In this section, we discuss the circular motion of the charged particle as well as other motion that results from a charged particle entering a magnetic field.

The simplest case occurs when a charged particle moves perpendicular to a uniform *B*-field (Figure 11.7). If the field is in a vacuum, the magnetic field is the dominant factor determining the motion. Since the magnetic force is perpendicular to the direction of travel, a charged particle follows a curved path in a magnetic field. The particle continues to follow this curved path until it forms a complete circle. Another way to look at this is that the magnetic force is always perpendicular to velocity, so that it does no work on the charged particle. The particle's kinetic energy and speed thus remain constant. The direction of motion is affected but not the speed.

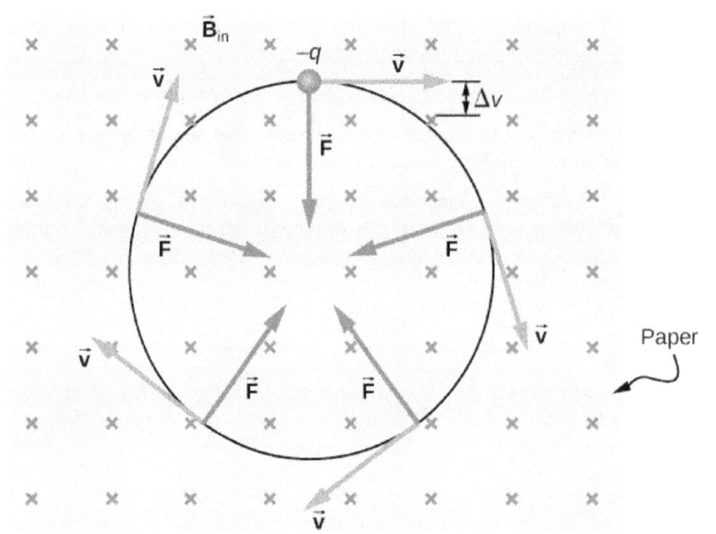

Figure 11.7 A negatively charged particle moves in the plane of the paper in a region where the magnetic field is perpendicular to the paper (represented by the small × 's—like the tails of arrows). The magnetic force is perpendicular to the velocity, so velocity changes in direction but not magnitude. The result is uniform circular motion. (Note that because the charge is negative, the force is opposite in direction to the prediction of the right-hand rule.)

In this situation, the magnetic force supplies the centripetal force $F_c = \frac{mv^2}{r}$. Noting that the velocity is perpendicular to the magnetic field, the magnitude of the magnetic force is reduced to $F = qvB$. Because the magnetic force F supplies the centripetal force F_c, we have

$$qvB = \frac{mv^2}{r}. \tag{11.4}$$

Solving for r yields

$$r = \frac{mv}{qB}. \tag{11.5}$$

Here, r is the radius of curvature of the path of a charged particle with mass m and charge q, moving at a speed v that is perpendicular to a magnetic field of strength B. The time for the charged particle to go around the circular path is defined as the period, which is the same as the distance traveled (the circumference) divided by the speed. Based on this and Equation 11.4, we can derive the period of motion as

$$T = \frac{2\pi r}{v} = \frac{2\pi}{v}\frac{mv}{qB} = \frac{2\pi m}{qB}. \tag{11.6}$$

If the velocity is not perpendicular to the magnetic field, then we can compare each component of the velocity separately with the magnetic field. The component of the velocity perpendicular to the magnetic field produces a magnetic force perpendicular to both this velocity and the field:

$$v_{\text{perp}} = v\sin\theta, \quad v_{\text{para}} = v\cos\theta. \tag{11.7}$$

where θ is the angle between v and B. The component parallel to the magnetic field creates constant motion along the same direction as the magnetic field, also shown in Equation 11.7. The parallel motion determines the *pitch p* of the helix, which is the distance between adjacent turns. This distance equals the parallel component of the velocity times the period:

$$p = v_{\text{para}} T. \tag{11.8}$$

The result is a **helical motion**, as shown in the following figure.

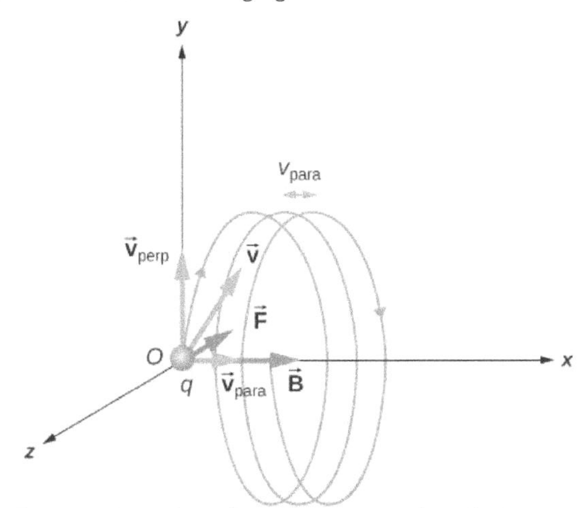

Figure 11.8 A charged particle moving with a velocity not in the same direction as the magnetic field. The velocity component perpendicular to the magnetic field creates circular motion, whereas the component of the velocity parallel to the field moves the particle along a straight line. The pitch is the horizontal distance between two consecutive circles. The resulting motion is helical.

While the charged particle travels in a helical path, it may enter a region where the magnetic field is not uniform. In particular, suppose a particle travels from a region of strong magnetic field to a region of weaker field, then back to a region of stronger field. The particle may reflect back before entering the stronger magnetic field region. This is similar to a wave on a string traveling from a very light, thin string to a hard wall and reflecting backward. If the reflection happens at both ends, the particle is trapped in a so-called magnetic bottle.

Trapped particles in magnetic fields are found in the Van Allen radiation belts around Earth, which are part of Earth's magnetic field. These belts were discovered by James Van Allen while trying to measure the flux of **cosmic rays** on Earth (high-energy particles that come from outside the solar system) to see whether this was similar to the flux measured on Earth. Van Allen found that due to the contribution of particles trapped in Earth's magnetic field, the flux was much higher on Earth than in outer space. Aurorae, like the famous aurora borealis (northern lights) in the Northern Hemisphere (Figure 11.9), are beautiful displays of light emitted as ions recombine with electrons entering the atmosphere as they spiral along magnetic field lines. (The ions are primarily oxygen and nitrogen atoms that are initially ionized by collisions with energetic particles in Earth's atmosphere.) Aurorae have also been observed on other planets, such as Jupiter and Saturn.

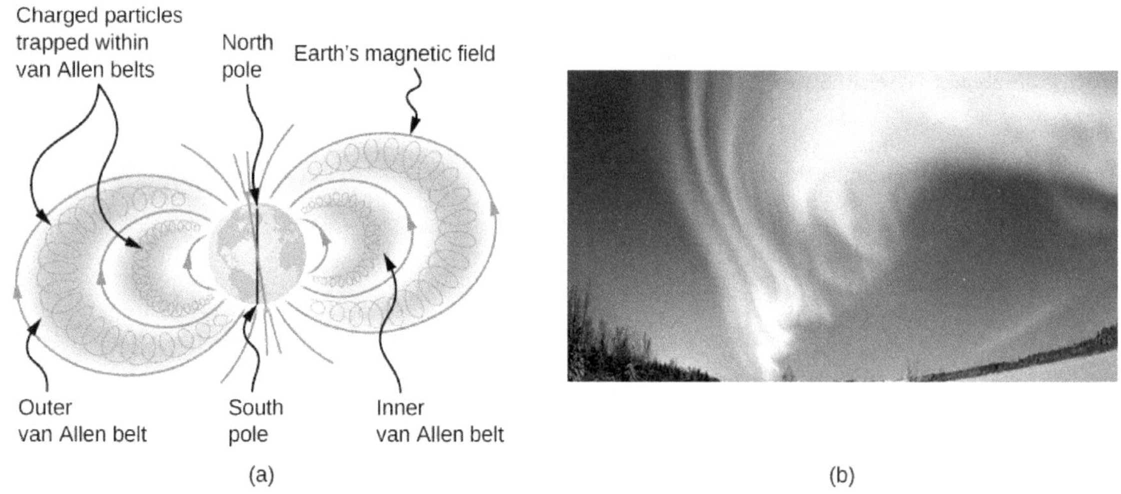

Figure 11.9 (a) The Van Allen radiation belts around Earth trap ions produced by cosmic rays striking Earth's atmosphere. (b) The magnificent spectacle of the aurora borealis, or northern lights, glows in the northern sky above Bear Lake near Eielson Air Force Base, Alaska. Shaped by Earth's magnetic field, this light is produced by glowing molecules and ions of oxygen and nitrogen. (credit b: modification of work by USAF Senior Airman Joshua Strang)

Example 11.2

Beam Deflector

A research group is investigating short-lived radioactive isotopes. They need to design a way to transport alpha-particles (helium nuclei) from where they are made to a place where they will collide with another material to form an isotope. The beam of alpha-particles $\left(m = 6.64 \times 10^{-27}\,\text{kg}, q = 3.2 \times 10^{-19}\,\text{C}\right)$ bends through a 90-degree region with a uniform magnetic field of 0.050 T (Figure 11.10). (a) In what direction should the magnetic field be applied? (b) How much time does it take the alpha-particles to traverse the uniform magnetic field region?

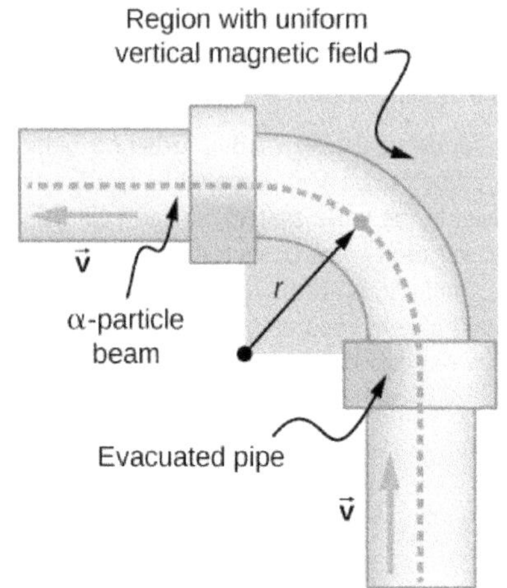

Figure 11.10 Top view of the beam deflector setup.

Strategy

a. The direction of the magnetic field is shown by the RHR-1. Your fingers point in the direction of v, and your thumb needs to point in the direction of the force, to the left. Therefore, since the alpha-particles are positively charged, the magnetic field must point down.

b. The period of the alpha-particle going around the circle is

$$T = \frac{2\pi m}{qB}. \tag{11.9}$$

Because the particle is only going around a quarter of a circle, we can take 0.25 times the period to find the time it takes to go around this path.

Solution

a. Let's start by focusing on the alpha-particle entering the field near the bottom of the picture. First, point your thumb up the page. In order for your palm to open to the left where the centripetal force (and hence the magnetic force) points, your fingers need to change orientation until they point into the page. This is the direction of the applied magnetic field.

b. The period of the charged particle going around a circle is calculated by using the given mass, charge, and magnetic field in the problem. This works out to be

$$T = \frac{2\pi m}{qB} = \frac{2\pi\left(6.64 \times 10^{-27}\,\text{kg}\right)}{\left(3.2 \times 10^{-19}\,\text{C}\right)(0.050\,\text{T})} = 2.6 \times 10^{-6}\,\text{s}.$$

However, for the given problem, the alpha-particle goes around a quarter of the circle, so the time it takes would be

$$t = 0.25 \times 2.61 \times 10^{-6}\,\text{s} = 6.5 \times 10^{-7}\,\text{s}.$$

Significance

This time may be quick enough to get to the material we would like to bombard, depending on how short-lived the radioactive isotope is and continues to emit alpha-particles. If we could increase the magnetic field applied in the region, this would shorten the time even more. The path the particles need to take could be shortened, but this may not be economical given the experimental setup.

11.2 Check Your Understanding A uniform magnetic field of magnitude 1.5 T is directed horizontally from west to east. (a) What is the magnetic force on a proton at the instant when it is moving vertically downward in the field with a speed of 4×10^7 m/s? (b) Compare this force with the weight w of a proton.

Example 11.3

Helical Motion in a Magnetic Field

A proton enters a uniform magnetic field of $1.0 \times 10^{-4}\,\text{T}$ with a speed of 5×10^5 m/s. At what angle must the magnetic field be from the velocity so that the pitch of the resulting helical motion is equal to the radius of the helix?

Strategy

The pitch of the motion relates to the parallel velocity times the period of the circular motion, whereas the radius relates to the perpendicular velocity component. After setting the radius and the pitch equal to each other, solve for the angle between the magnetic field and velocity or θ.

Solution

The pitch is given by Equation 11.8, the period is given by Equation 11.6, and the radius of circular motion is given by Equation 11.5. Note that the velocity in the radius equation is related to only the perpendicular velocity, which is where the circular motion occurs. Therefore, we substitute the sine component of the overall velocity into the radius equation to equate the pitch and radius:

$$\begin{aligned} p &= r \\ v_{\parallel}\, T &= \frac{mv_{\perp}}{qB} \\ v\cos\theta\frac{2\pi m}{qB} &= \frac{mv\sin\theta}{qB} \\ 2\pi &= \tan\theta \\ \theta &= 81.0^\circ. \end{aligned}$$

Significance

If this angle were 0°, only parallel velocity would occur and the helix would not form, because there would be no circular motion in the perpendicular plane. If this angle were 90°, only circular motion would occur and there would be no movement of the circles perpendicular to the motion. That is what creates the helical motion.

11.4 | Magnetic Force on a Current-Carrying Conductor

Learning Objectives

By the end of this section, you will be able to:

- Determine the direction in which a current-carrying wire experiences a force in an external magnetic field
- Calculate the force on a current-carrying wire in an external magnetic field

Moving charges experience a force in a magnetic field. If these moving charges are in a wire—that is, if the wire is carrying a current—the wire should also experience a force. However, before we discuss the force exerted on a current by a magnetic field, we first examine the magnetic field generated by an electric current. We are studying two separate effects here that interact closely: A current-carrying wire generates a magnetic field and the magnetic field exerts a force on the current-carrying wire.

Magnetic Fields Produced by Electrical Currents

When discussing historical discoveries in magnetism, we mentioned Oersted's finding that a wire carrying an electrical current caused a nearby compass to deflect. A connection was established that electrical currents produce magnetic fields. (This connection between electricity and magnetism is discussed in more detail in Sources of Magnetic Fields.)

The compass needle near the wire experiences a force that aligns the needle tangent to a circle around the wire. Therefore, a current-carrying wire produces circular loops of magnetic field. To determine the direction of the magnetic field generated from a wire, we use a second right-hand rule. In RHR-2, your thumb points in the direction of the current while your fingers wrap around the wire, pointing in the direction of the magnetic field produced (Figure 11.11). If the magnetic field were coming at you or out of the page, we represent this with a dot. If the magnetic field were going into the page, we represent this with an $\times$. These symbols come from considering a vector arrow: An arrow pointed toward you, from your perspective, would look like a dot or the tip of an arrow. An arrow pointed away from you, from your perspective, would look like a cross or an $\times$. A composite sketch of the magnetic circles is shown in Figure 11.11, where the field strength is shown to decrease as you get farther from the wire by loops that are farther separated.

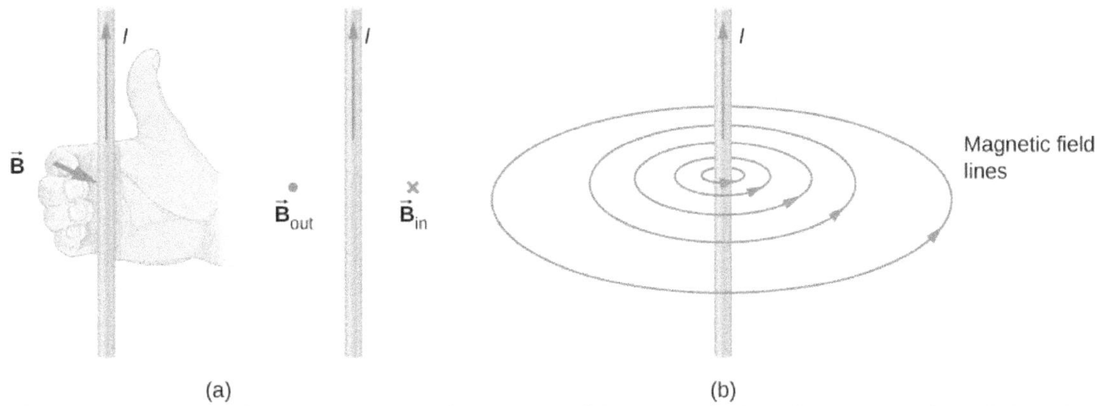

Figure 11.11 (a) When the wire is in the plane of the paper, the field is perpendicular to the paper. Note the symbols used for the field pointing inward (like the tail of an arrow) and the field pointing outward (like the tip of an arrow). (b) A long and straight wire creates a field with magnetic field lines forming circular loops.

Calculating the Magnetic Force

Electric current is an ordered movement of charge. A current-carrying wire in a magnetic field must therefore experience a force due to the field. To investigate this force, let's consider the infinitesimal section of wire as shown in Figure 11.12. The length and cross-sectional area of the section are dl and A, respectively, so its volume is $V = A \cdot dl.$ The wire is formed from material that contains n charge carriers per unit volume, so the number of charge carriers in the section is $nA \cdot dl.$ If the charge carriers move with drift velocity $\vec{\mathbf{v}}_{\text{d}},$ the current I in the wire is (from Current and Resistance)

$$I = neAv_d.$$

The magnetic force on any single charge carrier is $e\vec{\mathbf{v}}_{\text{d}} \times \vec{\mathbf{B}},$ so the total magnetic force $d\vec{\mathbf{F}}$ on the $nA \cdot dl$ charge carriers in the section of wire is

$$d\vec{\mathbf{F}} = (nA \cdot dl)e\vec{\mathbf{v}}_{\text{d}} \times \vec{\mathbf{B}}. \tag{11.10}$$

We can define dl to be a vector of length dl pointing along $\vec{\mathbf{v}}_{\text{d}},$ which allows us to rewrite this equation as

$$d\vec{\mathbf{F}} = neAv_{\text{d}}\,\vec{\mathbf{dl}} \times \vec{\mathbf{B}}, \tag{11.11}$$

or

$$d\vec{\mathbf{F}} = I\,\vec{\mathbf{dl}} \times \vec{\mathbf{B}}. \tag{11.12}$$

This is the magnetic force on the section of wire. Note that it is actually the net force exerted by the field on the charge carriers themselves. The direction of this force is given by RHR-1, where you point your fingers in the direction of the current and curl them toward the field. Your thumb then points in the direction of the force.

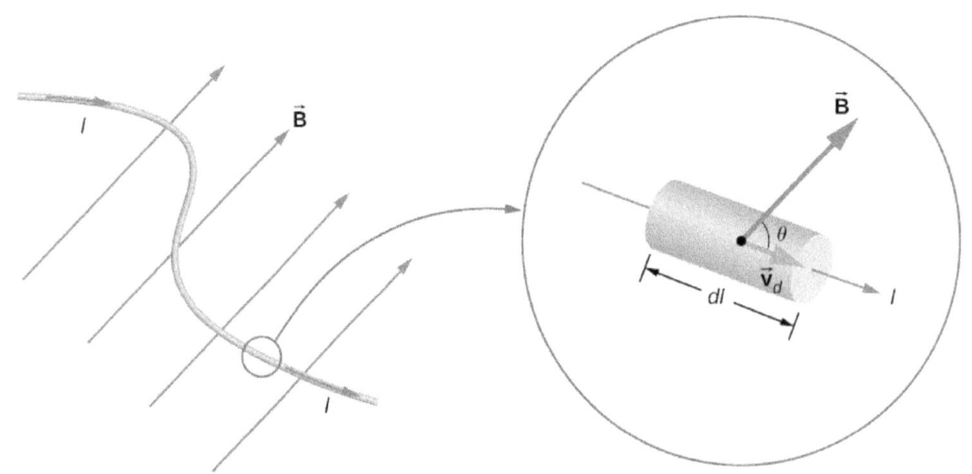

Figure 11.12 An infinitesimal section of current-carrying wire in a magnetic field.

To determine the magnetic force $\vec{F}$ on a wire of arbitrary length and shape, we must integrate Equation 11.12 over the entire wire. If the wire section happens to be straight and B is uniform, the equation differentials become absolute quantities, giving us

$$\vec{F} = I\vec{l} \times \vec{B}. \tag{11.13}$$

This is the force on a straight, current-carrying wire in a uniform magnetic field.

Example 11.4

Balancing the Gravitational and Magnetic Forces on a Current-Carrying Wire

A wire of length 50 cm and mass 10 g is suspended in a horizontal plane by a pair of flexible leads (Figure 11.13). The wire is then subjected to a constant magnetic field of magnitude 0.50 T, which is directed as shown. What are the magnitude and direction of the current in the wire needed to remove the tension in the supporting leads?

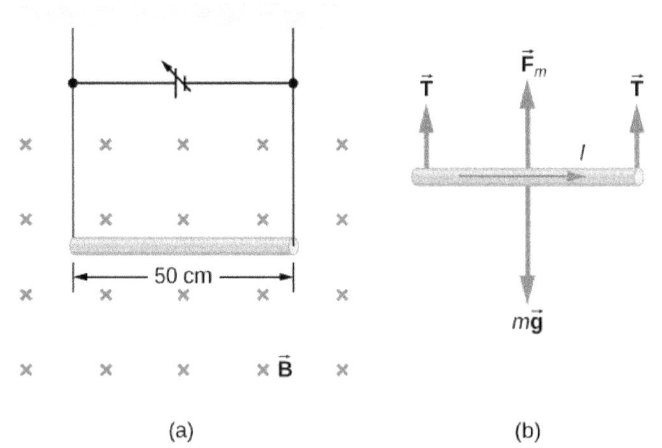

Figure 11.13 (a) A wire suspended in a magnetic field. (b) The free-body diagram for the wire.

Strategy

From the free-body diagram in the figure, the tensions in the supporting leads go to zero when the gravitational and magnetic forces balance each other. Using the RHR-1, we find that the magnetic force points up. We can then determine the current I by equating the two forces.

Solution

Equate the two forces of weight and magnetic force on the wire:

$$mg = IlB.$$

Thus,

$$I = \frac{mg}{lB} = \frac{(0.010\,\text{kg})(9.8\,\text{m/s}^2)}{(0.50\,\text{m})(0.50\,\text{T})} = 0.39\,\text{A}.$$

Significance

This large magnetic field creates a significant force on a length of wire to counteract the weight of the wire.

Example 11.5

Calculating Magnetic Force on a Current-Carrying Wire

A long, rigid wire lying along the y-axis carries a 5.0-A current flowing in the positive y-direction. (a) If a constant magnetic field of magnitude 0.30 T is directed along the positive x-axis, what is the magnetic force per unit length on the wire? (b) If a constant magnetic field of 0.30 T is directed 30 degrees from the $+x$-axis towards the $+y$-axis, what is the magnetic force per unit length on the wire?

Strategy

The magnetic force on a current-carrying wire in a magnetic field is given by $\vec{\mathbf{F}} = I\,\vec{\mathbf{l}} \times \vec{\mathbf{B}}$. For part a, since the current and magnetic field are perpendicular in this problem, we can simplify the formula to give us the magnitude and find the direction through the RHR-1. The angle θ is 90 degrees, which means $\sin\theta = 1$. Also, the length can be divided over to the left-hand side to find the force per unit length. For part b, the current times length is written in unit vector notation, as well as the magnetic field. After the cross product is taken, the directionality is evident by the resulting unit vector.

Solution

a. We start with the general formula for the magnetic force on a wire. We are looking for the force per unit length, so we divide by the length to bring it to the left-hand side. We also set $\sin\theta = 1$. The solution therefore is

$$\begin{aligned} F &= IlB\sin\theta \\ \frac{F}{l} &= (5.0\,\text{A})(0.30\,\text{T}) \\ \frac{F}{l} &= 1.5\,\text{N/m}. \end{aligned}$$

Directionality: Point your fingers in the positive y-direction and curl your fingers in the positive x-direction. Your thumb will point in the $-\vec{\mathbf{k}}$ direction. Therefore, with directionality, the solution is

$$\frac{\vec{\mathbf{F}}}{l} = -1.5\,\vec{\mathbf{k}}\,\text{N/m}.$$

b. The current times length and the magnetic field are written in unit vector notation. Then, we take the cross product to find the force:

$$\vec{\mathbf{F}} = I\,\vec{\mathbf{l}} \times \vec{\mathbf{B}} = (5.0A)l\,\hat{\mathbf{j}} \times \left(0.30T\cos(30°)\,\hat{\mathbf{i}} + 0.30T\sin(30°)\,\hat{\mathbf{j}}\right)$$
$$\vec{\mathbf{F}}/l = -1.30\hat{\mathbf{k}}\ \text{N/m}.$$

Significance

This large magnetic field creates a significant force on a small length of wire. As the angle of the magnetic field becomes more closely aligned to the current in the wire, there is less of a force on it, as seen from comparing parts a and b.

11.3 Check Your Understanding A straight, flexible length of copper wire is immersed in a magnetic field that is directed into the page. (a) If the wire's current runs in the +x-direction, which way will the wire bend? (b) Which way will the wire bend if the current runs in the –x-direction?

Example 11.6

Force on a Circular Wire

A circular current loop of radius R carrying a current I is placed in the xy-plane. A constant uniform magnetic field cuts through the loop parallel to the y-axis (Figure 11.14). Find the magnetic force on the upper half of the loop, the lower half of the loop, and the total force on the loop.

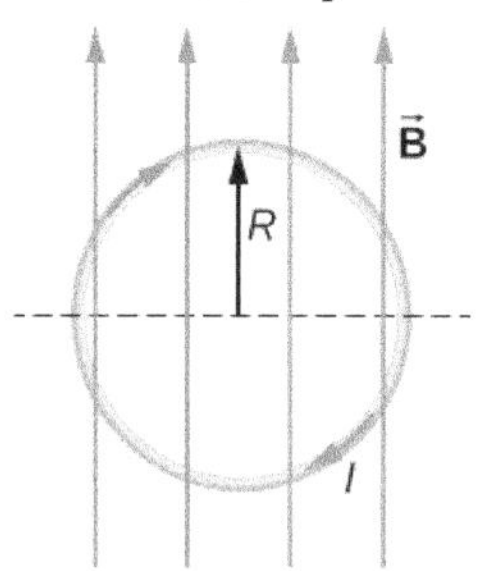

Figure 11.14 A loop of wire carrying a current in a magnetic field.

Strategy

The magnetic force on the upper loop should be written in terms of the differential force acting on each segment of the loop. If we integrate over each differential piece, we solve for the overall force on that section of the loop. The force on the lower loop is found in a similar manner, and the total force is the addition of these two forces.

Solution

A differential force on an arbitrary piece of wire located on the upper ring is:

$$dF = IB\sin\theta dl.$$

where θ is the angle between the magnetic field direction (+y) and the segment of wire. A differential segment is located at the same radius, so using an arc-length formula, we have:

$$dl = Rd\theta$$
$$dF = IBR\sin\theta d\theta.$$

In order to find the force on a segment, we integrate over the upper half of the circle, from 0 to π. This results in:

$$F = IBR\int_0^{\pi}\sin\theta d\theta = IBR(-\cos\pi + \cos 0) = 2IBR.$$

The lower half of the loop is integrated from π to zero, giving us:

$$F = IBR\int_{\pi}^{0}\sin\theta d\theta = IBR(-\cos 0 + \cos\pi) = -2IBR.$$

The net force is the sum of these forces, which is zero.

Significance

The total force on any closed loop in a uniform magnetic field is zero. Even though each piece of the loop has a force acting on it, the net force on the system is zero. (Note that there is a net torque on the loop, which we consider in the next section.)

11.5 | Force and Torque on a Current Loop

Learning Objectives

By the end of this section, you will be able to:

- Evaluate the net force on a current loop in an external magnetic field
- Evaluate the net torque on a current loop in an external magnetic field
- Define the magnetic dipole moment of a current loop

Motors are the most common application of magnetic force on current-carrying wires. Motors contain loops of wire in a magnetic field. When current is passed through the loops, the magnetic field exerts torque on the loops, which rotates a shaft. Electrical energy is converted into mechanical work in the process. Once the loop's surface area is aligned with the magnetic field, the direction of current is reversed, so there is a continual torque on the loop (Figure 11.15). This reversal of the current is done with commutators and brushes. The commutator is set to reverse the current flow at set points to keep continual motion in the motor. A basic commutator has three contact areas to avoid and dead spots where the loop would have zero instantaneous torque at that point. The brushes press against the commutator, creating electrical contact between parts of the commutator during the spinning motion.

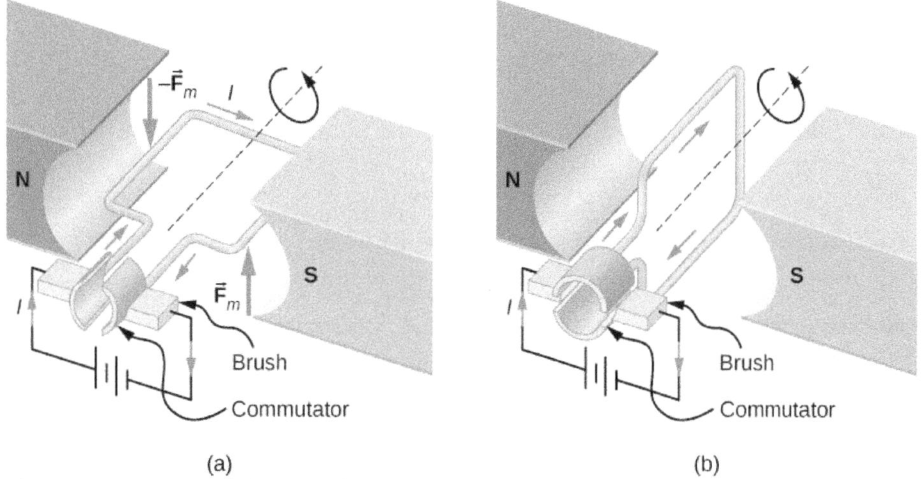

Figure 11.15 A simplified version of a dc electric motor. (a) The rectangular wire loop is placed in a magnetic field. The forces on the wires closest to the magnetic poles (N and S) are opposite in direction as determined by the right-hand rule-1. Therefore, the loop has a net torque and rotates to the position shown in (b). (b) The brushes now touch the commutator segments so that no current flows through the loop. No torque acts on the loop, but the loop continues to spin from the initial velocity given to it in part (a). By the time the loop flips over, current flows through the wires again but now in the opposite direction, and the process repeats as in part (a). This causes continual rotation of the loop.

In a uniform magnetic field, a current-carrying loop of wire, such as a loop in a motor, experiences both forces and torques on the loop. Figure 11.16 shows a rectangular loop of wire that carries a current I and has sides of lengths a and b. The loop is in a uniform magnetic field: $\vec{\mathbf{B}} = B\hat{\mathbf{j}}$. The magnetic force on a straight current-carrying wire of length l is given by $I\vec{\mathbf{l}} \times \vec{\mathbf{B}}$. To find the net force on the loop, we have to apply this equation to each of the four sides. The force on side 1 is

$$\vec{\mathbf{F}}_1 = IaB\sin(90° - \theta)\hat{\mathbf{i}} = IaB\cos\theta\hat{\mathbf{i}} \tag{11.14}$$

where the direction has been determined with the RHR-1. The current in side 3 flows in the opposite direction to that of side 1, so

$$\vec{\mathbf{F}}_3 = -IaB\sin(90° + \theta)\hat{\mathbf{i}} = -IaB\cos\theta\hat{\mathbf{i}}. \tag{11.15}$$

The currents in sides 2 and 4 are perpendicular to $\vec{\mathbf{B}}$ and the forces on these sides are

$$\vec{\mathbf{F}}_2 = IbB\hat{\mathbf{k}}, \quad \vec{\mathbf{F}}_4 = -IbB\hat{\mathbf{k}}. \tag{11.16}$$

We can now find the net force on the loop:

$$\sum \vec{\mathbf{F}}_{\text{net}} = \vec{\mathbf{F}}_1 + \vec{\mathbf{F}}_2 + \vec{\mathbf{F}}_3 + \vec{\mathbf{F}}_4 = 0. \tag{11.17}$$

Although this result $(\Sigma F = 0)$ has been obtained for a rectangular loop, it is far more general and holds for current-carrying loops of arbitrary shapes; that is, there is no net force on a current loop in a uniform magnetic field.

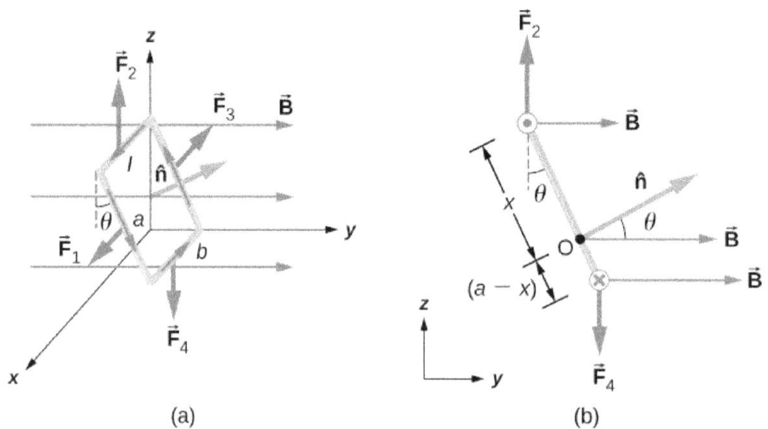

Figure 11.16 (a) A rectangular current loop in a uniform magnetic field is subjected to a net torque but not a net force. (b) A side view of the coil.

To find the net torque on the current loop shown in Figure 11.16, we first consider F_1 and F_3. Since they have the same line of action and are equal and opposite, the sum of their torques about any axis is zero (see Fixed-Axis Rotation (http://cnx.org/content/m58325/latest/)). Thus, if there is any torque on the loop, it must be furnished by F_2 and F_4. Let's calculate the torques around the axis that passes through point O of Figure 11.16 (a side view of the coil) and is perpendicular to the plane of the page. The point O is a distance x from side 2 and a distance $(a - x)$ from side 4 of the loop. The moment arms of F_2 and F_4 are $x\sin\theta$ and $(a - x)\sin\theta$, respectively, so the net torque on the loop is

$$\begin{aligned}\sum \vec{\boldsymbol{\tau}} = \vec{\boldsymbol{\tau}}_1 + \vec{\boldsymbol{\tau}}_2 + \vec{\boldsymbol{\tau}}_3 + \vec{\boldsymbol{\tau}}_4 &= F_2 x\sin\theta\hat{\mathbf{i}} - F_4(a - x)\sin(\theta)\hat{\mathbf{i}} \\ &= -IbBx\sin\theta\hat{\mathbf{i}} - IbB(a - x)\sin\theta\hat{\mathbf{i}}.\end{aligned} \tag{11.18}$$

This simplifies to

$$\vec{\tau} = -IAB\sin\theta\,\hat{\mathbf{i}} \tag{11.19}$$

where $A = ab$ is the area of the loop.

Notice that this torque is independent of x; it is therefore independent of where point O is located in the plane of the current loop. Consequently, the loop experiences the same torque from the magnetic field about any axis in the plane of the loop and parallel to the x-axis.

A closed-current loop is commonly referred to as a **magnetic dipole** and the term IA is known as its **magnetic dipole moment** μ. Actually, the magnetic dipole moment is a vector that is defined as

$$\vec{\mu} = IA\hat{\mathbf{n}} \tag{11.20}$$

where $\hat{\mathbf{n}}$ is a unit vector directed perpendicular to the plane of the loop (see Figure 11.16). The direction of $\hat{\mathbf{n}}$ is obtained with the RHR-2—if you curl the fingers of your right hand in the direction of current flow in the loop, then your thumb points along $\hat{\mathbf{n}}$. If the loop contains N turns of wire, then its magnetic dipole moment is given by

$$\vec{\mu} = NIA\hat{\mathbf{n}}. \tag{11.21}$$

In terms of the magnetic dipole moment, the torque on a current loop due to a uniform magnetic field can be written simply as

$$\vec{\tau} = \vec{\mu} \times \vec{\mathbf{B}}. \tag{11.22}$$

This equation holds for a current loop in a two-dimensional plane of arbitrary shape.

Using a calculation analogous to that found in Capacitance for an electric dipole, the potential energy of a magnetic dipole is

$$U = -\vec{\mu} \cdot \vec{\mathbf{B}}. \tag{11.23}$$

Example 11.7

Forces and Torques on Current-Carrying Loops

A circular current loop of radius 2.0 cm carries a current of 2.0 mA. (a) What is the magnitude of its magnetic dipole moment? (b) If the dipole is oriented at 30 degrees to a uniform magnetic field of magnitude 0.50 T, what is the magnitude of the torque it experiences and what is its potential energy?

Strategy

The dipole moment is defined by the current times the area of the loop. The area of the loop can be calculated from the area of the circle. The torque on the loop and potential energy are calculated from identifying the magnetic moment, magnetic field, and angle oriented in the field.

Solution

a. The magnetic moment μ is calculated by the current times the area of the loop or πr^2.

$$\mu = IA = (2.0 \times 10^{-3}\ \text{A})(\pi(0.02\ \text{m})^2) = 2.5 \times 10^{-6}\ \text{A}\cdot\text{m}^2$$

b. The torque and potential energy are calculated by identifying the magnetic moment, magnetic field, and the angle between these two vectors. The calculations of these quantities are:

$$\tau = \vec{\mu} \times \vec{B} = \mu B \sin\theta = (2.5 \times 10^{-6}\,\text{A} \cdot \text{m}^2)(0.50\text{T})\sin(30°) = 6.3 \times 10^{-7}\,\text{N} \cdot \text{m}$$

$$U = -\vec{\mu} \cdot \vec{B} = -\mu B \cos\theta = -(2.5 \times 10^{-6}\,\text{A} \cdot \text{m}^2)(0.50\text{T})\cos(30°) = -1.1 \times 10^{-6}\,\text{J}.$$

Significance

The concept of magnetic moment at the atomic level is discussed in the next chapter. The concept of aligning the magnetic moment with the magnetic field is the functionality of devices like magnetic motors, whereby switching the external magnetic field results in a constant spinning of the loop as it tries to align with the field to minimize its potential energy.

11.4 Check Your Understanding

In what orientation would a magnetic dipole have to be to produce (a) a maximum torque in a magnetic field? (b) A maximum energy of the dipole?

11.6 | The Hall Effect

Learning Objectives

By the end of this section, you will be able to:

- Explain a scenario where the magnetic and electric fields are crossed and their forces balance each other as a charged particle moves through a velocity selector
- Compare how charge carriers move in a conductive material and explain how this relates to the Hall effect

In 1879, E.H. Hall devised an experiment that can be used to identify the sign of the predominant charge carriers in a conducting material. From a historical perspective, this experiment was the first to demonstrate that the charge carriers in most metals are negative.

Visit this website (https://openstaxcollege.org/l/21halleffect) to find more information about the Hall effect.

We investigate the **Hall effect** by studying the motion of the free electrons along a metallic strip of width l in a constant magnetic field (Figure 11.17). The electrons are moving from left to right, so the magnetic force they experience pushes them to the bottom edge of the strip. This leaves an excess of positive charge at the top edge of the strip, resulting in an electric field E directed from top to bottom. The charge concentration at both edges builds up until the electric force on the electrons in one direction is balanced by the magnetic force on them in the opposite direction. Equilibrium is reached when:

$$eE = ev_d B \tag{11.24}$$

where e is the magnitude of the electron charge, v_d is the drift speed of the electrons, and E is the magnitude of the electric field created by the separated charge. Solving this for the drift speed results in

$$v_d = \frac{E}{B}. \tag{11.25}$$

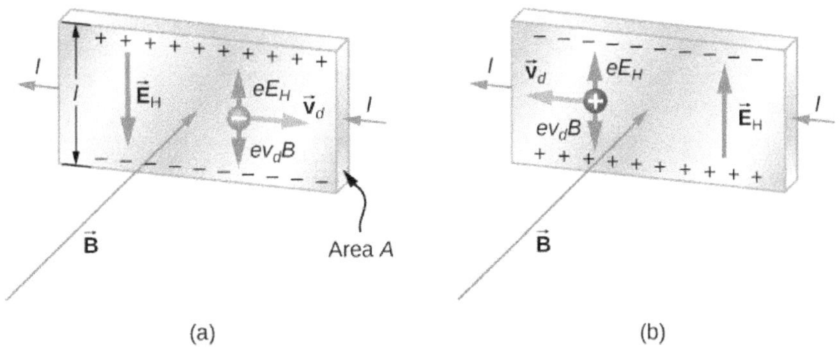

(a) (b)

Figure 11.17 In the Hall effect, a potential difference between the top and bottom edges of the metal strip is produced when moving charge carriers are deflected by the magnetic field. (a) Hall effect for negative charge carriers; (b) Hall effect for positive charge carriers.

A scenario where the electric and magnetic fields are perpendicular to one another is called a crossed-field situation. If these fields produce equal and opposite forces on a charged particle with the velocity that equates the forces, these particles are able to pass through an apparatus, called a **velocity selector**, undeflected. This velocity is represented in Equation 11.26. Any other velocity of a charged particle sent into the same fields would be deflected by the magnetic force or electric force.

Going back to the Hall effect, if the current in the strip is I, then from Current and Resistance, we know that

$$I = nev_d A \tag{11.26}$$

where n is the number of charge carriers per volume and A is the cross-sectional area of the strip. Combining the equations for v_d and I results in

$$I = ne\left(\frac{E}{B}\right)A. \tag{11.27}$$

The field E is related to the potential difference V between the edges of the strip by

$$E = \frac{V}{l}. \tag{11.28}$$

The quantity V is called the Hall potential and can be measured with a voltmeter. Finally, combining the equations for I and E gives us

$$V = \frac{IBl}{neA} \tag{11.29}$$

where the upper edge of the strip in Figure 11.17 is positive with respect to the lower edge.

We can also combine Equation 11.24 and Equation 11.28 to get an expression for the Hall voltage in terms of the magnetic field:

$$V = Blv_d. \tag{11.30}$$

What if the charge carriers are positive, as in Figure 11.17? For the same current I, the magnitude of V is still given by Equation 11.29. However, the upper edge is now negative with respect to the lower edge. Therefore, by simply measuring the sign of V, we can determine the sign of the majority charge carriers in a metal.

Hall potential measurements show that electrons are the dominant charge carriers in most metals. However, Hall potentials indicate that for a few metals, such as tungsten, beryllium, and many semiconductors, the majority of charge carriers are positive. It turns out that conduction by positive charge is caused by the migration of missing electron sites (called holes) on ions. Conduction by holes is studied later in Condensed Matter Physics (http://cnx.org/content/m58591/latest/) .

The Hall effect can be used to measure magnetic fields. If a material with a known density of charge carriers n is placed in a magnetic field and V is measured, then the field can be determined from Equation 11.29. In research laboratories where the fields of electromagnets used for precise measurements have to be extremely steady, a "Hall probe" is commonly used as part of an electronic circuit that regulates the field.

Example 11.8

Velocity Selector

An electron beam enters a crossed-field velocity selector with magnetic and electric fields of 2.0 mT and 6.0×10^3 N/C, respectively. (a) What must the velocity of the electron beam be to traverse the crossed fields undeflected? If the electric field is turned off, (b) what is the acceleration of the electron beam and (c) what is the radius of the circular motion that results?

Strategy

The electron beam is not deflected by either of the magnetic or electric fields if these forces are balanced. Based on these balanced forces, we calculate the velocity of the beam. Without the electric field, only the magnetic force is used in Newton's second law to find the acceleration. Lastly, the radius of the path is based on the resulting circular motion from the magnetic force.

Solution

a. The velocity of the unperturbed beam of electrons with crossed fields is calculated by Equation 11.25:

$$v_d = \frac{E}{B} = \frac{6 \times 10^3 \text{ N/C}}{2 \times 10^{-3} \text{ T}} = 3 \times 10^6 \text{ m/s}.$$

b. The acceleration is calculated from the net force from the magnetic field, equal to mass times acceleration. The magnitude of the acceleration is:

$$\begin{aligned} ma &= qvB \\ a &= \frac{qvB}{m} = \frac{(1.6 \times 10^{-19}\text{ C})(3 \times 10^6 \text{ m/s})(2 \times 10^{-3}\text{ T})}{9.1 \times 10^{-31}\text{ kg}} = 1.1 \times 10^{15} \text{ m/s}^2. \end{aligned}$$

c. The radius of the path comes from a balance of the circular and magnetic forces, or Equation 11.25:

$$r = \frac{mv}{qB} = \frac{(9.1 \times 10^{-31}\text{ kg})(3 \times 10^6 \text{ m/s})}{(1.6 \times 10^{-19}\text{ C})(2 \times 10^{-3}\text{ T})} = 8.5 \times 10^{-3}\text{ m}.$$

Significance

If electrons in the beam had velocities above or below the answer in part (a), those electrons would have a stronger net force exerted by either the magnetic or electric field. Therefore, only those electrons at this specific velocity would make it through.

Example 11.9

The Hall Potential in a Silver Ribbon

Figure 11.18 shows a silver ribbon whose cross section is 1.0 cm by 0.20 cm. The ribbon carries a current of 100 A from left to right, and it lies in a uniform magnetic field of magnitude 1.5 T. Using a density value of $n = 5.9 \times 10^{28}$ electrons per cubic meter for silver, find the Hall potential between the edges of the ribbon.

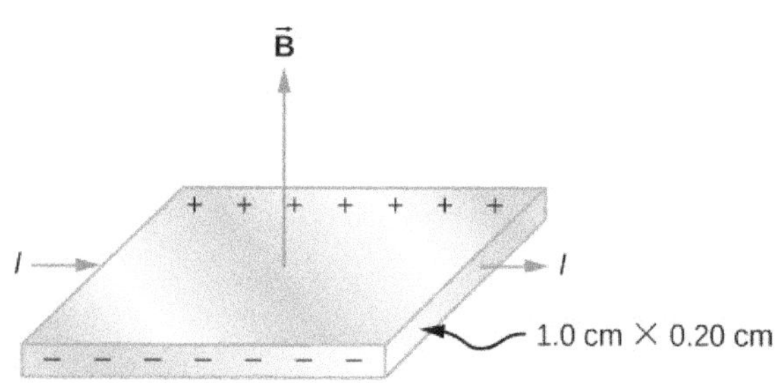

Figure 11.18 Finding the Hall potential in a silver ribbon in a magnetic field is shown.

Strategy

Since the majority of charge carriers are electrons, the polarity of the Hall voltage is that indicated in the figure. The value of the Hall voltage is calculated using Equation 11.29:

$$V = \frac{IBl}{neA}.$$

Solution

When calculating the Hall voltage, we need to know the current through the material, the magnetic field, the length, the number of charge carriers, and the area. Since all of these are given, the Hall voltage is calculated as:

$$V = \frac{IBl}{neA} = \frac{(100\text{ A})(1.5\text{ T})\left(1.0 \times 10^{-2}\text{ m}\right)}{\left(5.9 \times 10^{28}/\text{m}^3\right)\left(1.6 \times 10^{-19}\text{ C}\right)\left(2.0 \times 10^{-5}\text{ m}^2\right)} = 7.9 \times 10^{-6}\text{ V}.$$

Significance

As in this example, the Hall potential is generally very small, and careful experimentation with sensitive equipment is required for its measurement.

11.5 Check Your Understanding A Hall probe consists of a copper strip, $n = 8.5 \times 10^{28}$ electrons per cubic meter, which is 2.0 cm wide and 0.10 cm thick. What is the magnetic field when $I = 50$ A and the Hall potential is (a) $4.0\mu\text{V}$ and (b) $6.0\mu\text{V}$?

11.7 | Applications of Magnetic Forces and Fields

Learning Objectives

By the end of this section, you will be able to:

- Explain how a mass spectrometer works to separate charges
- Explain how a cyclotron works

Being able to manipulate and sort charged particles allows deeper experimentation to understand what matter is made of. We first look at a mass spectrometer to see how we can separate ions by their charge-to-mass ratio. Then we discuss cyclotrons as a method to accelerate charges to very high energies.

Mass Spectrometer

The **mass spectrometer** is a device that separates ions according to their charge-to-mass ratios. One particular version, the Bainbridge mass spectrometer, is illustrated in Figure 11.19. Ions produced at a source are first sent through a velocity selector, where the magnetic force is equally balanced with the electric force. These ions all emerge with the same speed $v = E/B$ since any ion with a different velocity is deflected preferentially by either the electric or magnetic force, and ultimately blocked from the next stage. They then enter a uniform magnetic field B_0 where they travel in a circular path whose radius R is given by Equation 11.3. The radius is measured by a particle detector located as shown in the figure.

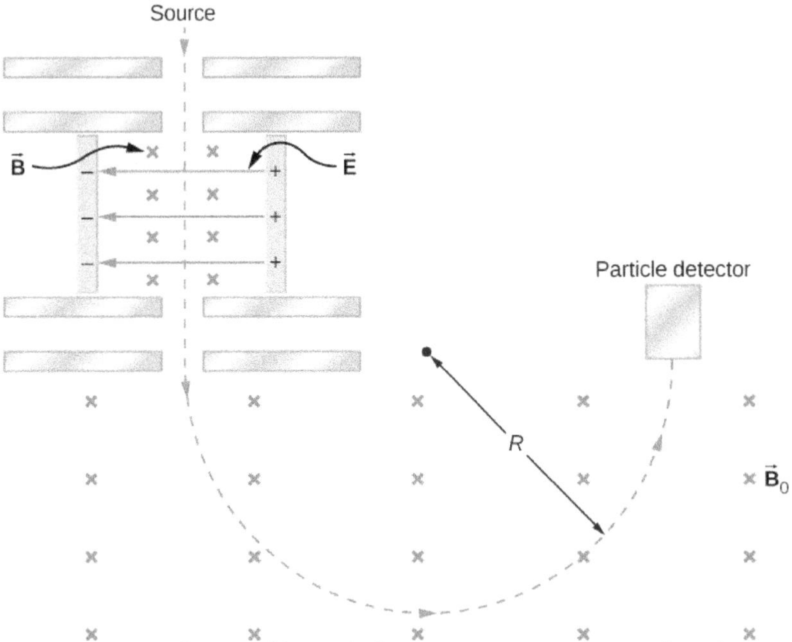

Figure 11.19 A schematic of the Bainbridge mass spectrometer, showing charged particles leaving a source, followed by a velocity selector where the electric and magnetic forces are balanced, followed by a region of uniform magnetic field where the particle is ultimately detected.

The relationship between the charge-to-mass ratio q/m and the radius R is determined by combining Equation 11.3 and Equation 11.25:

$$\frac{q}{m} = \frac{E}{BB_0 R}. \quad (11.31)$$

Since most ions are singly charged $\left(q = 1.6 \times 10^{-19}\ \text{C}\right)$, measured values of R can be used with this equation to determine the mass of ions. With modern instruments, masses can be determined to one part in 10^8.

An interesting use of a spectrometer is as part of a system for detecting very small leaks in a research apparatus. In low-temperature physics laboratories, a device known as a dilution refrigerator uses a mixture of He-3, He-4, and other cryogens to reach temperatures well below 1 K. The performance of the refrigerator is severely hampered if even a minute leak between its various components occurs. Consequently, before it is cooled down to the desired temperature, the refrigerator is subjected to a leak test. A small quantity of gaseous helium is injected into one of its compartments, while an adjacent, but supposedly isolated, compartment is connected to a high-vacuum pump to which a mass spectrometer is attached. A heated filament ionizes any helium atoms evacuated by the pump. The detection of these ions by the spectrometer then indicates a leak between the two compartments of the dilution refrigerator.

In conjunction with gas chromatography, mass spectrometers are used widely to identify unknown substances. While the gas chromatography portion breaks down the substance, the mass spectrometer separates the resulting ionized molecules. This technique is used with fire debris to ascertain the cause, in law enforcement to identify illegal drugs, in security to identify explosives, and in many medicinal applications.

Cyclotron

The **cyclotron** was developed by E.O. Lawrence to accelerate charged particles (usually protons, deuterons, or alpha-particles) to large kinetic energies. These particles are then used for nuclear-collision experiments to produce radioactive isotopes. A cyclotron is illustrated in Figure 11.20. The particles move between two flat, semi-cylindrical metallic containers D1 and D2, called **dees**. The dees are enclosed in a larger metal container, and the apparatus is placed between the poles of an electromagnet that provides a uniform magnetic field. Air is removed from the large container so that the particles neither lose energy nor are deflected because of collisions with air molecules. The dees are connected to a high-frequency voltage source that provides an alternating electric field in the small region between them. Because the dees are made of metal, their interiors are shielded from the electric field.

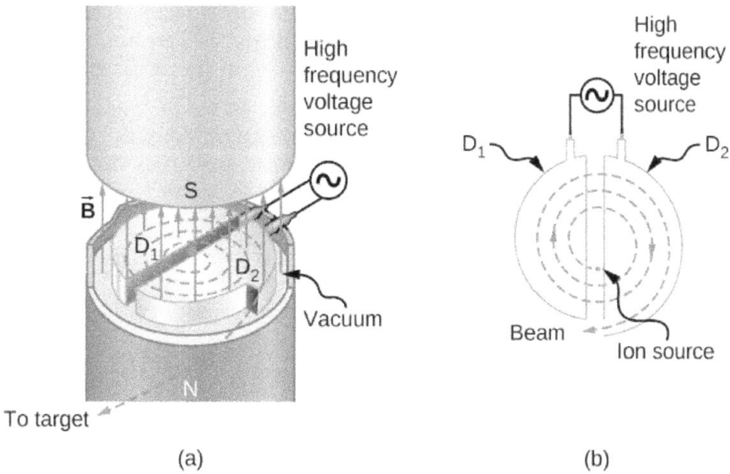

Figure 11.20 The inside of a cyclotron. A uniform magnetic field is applied as circulating protons travel through the dees, gaining energy as they traverse through the gap between the dees.

Suppose a positively charged particle is injected into the gap between the dees when D2 is at a positive potential relative to D1. The particle is then accelerated across the gap and enters D1 after gaining kinetic energy qV, where V is the average potential difference the particle experiences between the dees. When the particle is inside D1, only the uniform magnetic field $\vec{\mathbf{B}}$ of the electromagnet acts on it, so the particle moves in a circle of radius

$$r = \frac{mv}{qB} \tag{11.32}$$

with a period of

$$T = \frac{2\pi m}{qB}. \tag{11.33}$$

The period of the alternating voltage course is set at T, so while the particle is inside D1, moving along its semicircular orbit in a time $T/2$, the polarity of the dees is reversed. When the particle reenters the gap, D1 is positive with respect to D2, and the particle is again accelerated across the gap, thereby gaining a kinetic energy qV. The particle then enters D2, circulates in a slightly larger circle, and emerges from D2 after spending a time $T/2$ in this dee. This process repeats until the orbit of the particle reaches the boundary of the dees. At that point, the particle (actually, a beam of particles) is extracted from the cyclotron and used for some experimental purpose.

The operation of the cyclotron depends on the fact that, in a uniform magnetic field, a particle's orbital period is independent of its radius and its kinetic energy. Consequently, the period of the alternating voltage source need only be set at the one value given by Equation 11.33. With that setting, the electric field accelerates particles every time they are between the dees.

If the maximum orbital radius in the cyclotron is R, then from Equation 11.32, the maximum speed of a circulating particle of mass m and charge q is

$$v_{max} = \frac{qBR}{m}. \tag{11.34}$$

Thus, its kinetic energy when ejected from the cyclotron is

$$\frac{1}{2}mv_{max}{}^2 = \frac{q^2 B^2 R^2}{2m}. \tag{11.35}$$

The maximum kinetic energy attainable with this type of cyclotron is approximately 30 MeV. Above this energy, relativistic effects become important, which causes the orbital period to increase with the radius. Up to energies of several hundred MeV, the relativistic effects can be compensated for by making the magnetic field gradually increase with the radius of the orbit. However, for higher energies, much more elaborate methods must be used to accelerate particles.

Particles are accelerated to very high energies with either linear accelerators or synchrotrons. The linear accelerator accelerates particles continuously with the electric field of an electromagnetic wave that travels down a long evacuated tube. The Stanford Linear Accelerator (SLAC) is about 3.3 km long and accelerates electrons and positrons (positively charged electrons) to energies of 50 GeV. The synchrotron is constructed so that its bending magnetic field increases with particle speed in such a way that the particles stay in an orbit of fixed radius. The world's highest-energy synchrotron is located at CERN, which is on the Swiss-French border near Geneva. CERN has been of recent interest with the verified discovery of the Higgs Boson (see Particle Physics and Cosmology (http://cnx.org/content/m58767/latest/)). This synchrotron can accelerate beams of approximately 10^{13} protons to energies of about 10^3 GeV.

Example 11.10

Accelerating Alpha-Particles in a Cyclotron

A cyclotron used to accelerate alpha-particles ($m = 6.64 \times 10^{-27}\,\text{kg},\ q = 3.2 \times 10^{-19}\,\text{C}$) has a radius of 0.50 m and a magnetic field of 1.8 T. (a) What is the period of revolution of the alpha-particles? (b) What is their maximum kinetic energy?

Strategy

a. The period of revolution is approximately the distance traveled in a circle divided by the speed. Identifying that the magnetic force applied is the centripetal force, we can derive the period formula.

b. The kinetic energy can be found from the maximum speed of the beam, corresponding to the maximum radius within the cyclotron.

Solution

a. By identifying the mass, charge, and magnetic field in the problem, we can calculate the period:

$$T = \frac{2\pi m}{qB} = \frac{2\pi\left(6.64 \times 10^{-27}\,\text{kg}\right)}{\left(3.2 \times 10^{-19}\,\text{C}\right)(1.8\text{T})} = 7.3 \times 10^{-8}\,\text{s}.$$

b. By identifying the charge, magnetic field, radius of path, and the mass, we can calculate the maximum kinetic energy:

$$\frac{1}{2}mv_{max}{}^2 = \frac{q^2 B^2 R^2}{2m} = \frac{\left(3.2 \times 10^{-19}\,\text{C}\right)^2 (1.8\text{T})^2 (0.50\text{m})^2}{2(6.65 \times 10^{-27}\,\text{kg})} = 6.2 \times 10^{-12}\,\text{J} = 39\text{MeV}.$$

11.6 Check Your Understanding A cyclotron is to be designed to accelerate protons to kinetic energies of 20 MeV using a magnetic field of 2.0 T. What is the required radius of the cyclotron?

CHAPTER 11 REVIEW

KEY TERMS

cosmic rays comprised of particles that originate mainly from outside the solar system and reach Earth

cyclotron device used to accelerate charged particles to large kinetic energies

dees large metal containers used in cyclotrons that serve contain a stream of charged particles as their speed is increased

gauss G, unit of the magnetic field strength; $1\ \text{G} = 10^{-4}\ \text{T}$

Hall effect creation of voltage across a current-carrying conductor by a magnetic field

helical motion superposition of circular motion with a straight-line motion that is followed by a charged particle moving in a region of magnetic field at an angle to the field

magnetic dipole closed-current loop

magnetic dipole moment term *IA* of the magnetic dipole, also called μ

magnetic field lines continuous curves that show the direction of a magnetic field; these lines point in the same direction as a compass points, toward the magnetic south pole of a bar magnet

magnetic force force applied to a charged particle moving through a magnetic field

mass spectrometer device that separates ions according to their charge-to-mass ratios

motor (dc) loop of wire in a magnetic field; when current is passed through the loops, the magnetic field exerts torque on the loops, which rotates a shaft; electrical energy is converted into mechanical work in the process

north magnetic pole currently where a compass points to north, near the geographic North Pole; this is the effective south pole of a bar magnet but has flipped between the effective north and south poles of a bar magnet multiple times over the age of Earth

right-hand rule-1 using your right hand to determine the direction of either the magnetic force, velocity of a charged particle, or magnetic field

south magnetic pole currently where a compass points to the south, near the geographic South Pole; this is the effective north pole of a bar magnet but has flipped just like the north magnetic pole

tesla SI unit for magnetic field: 1 T = 1 N/A-m

velocity selector apparatus where the crossed electric and magnetic fields produce equal and opposite forces on a charged particle moving with a specific velocity; this particle moves through the velocity selector not affected by either field while particles moving with different velocities are deflected by the apparatus

KEY EQUATIONS

Force on a charge in a magnetic field	$\vec{\mathbf{F}} = q\vec{\mathbf{v}} \times \vec{\mathbf{B}}$
Magnitude of magnetic force	$F = qvB\sin\theta$
Radius of a particle's path in a magnetic field	$r = \frac{mv}{qB}$
Period of a particle's motion in a magnetic field	$T = \frac{2\pi m}{qB}$
Force on a current-carrying wire in a uniform magnetic field	$\vec{\mathbf{F}} = I\vec{\mathbf{l}} \times \vec{\mathbf{B}}$
Magnetic dipole moment	$\vec{\boldsymbol{\mu}} = NIA\hat{\mathbf{n}}$
Torque on a current loop	$\vec{\tau} = \vec{\boldsymbol{\mu}} \times \vec{\mathbf{B}}$

Energy of a magnetic dipole	$U = -\vec{\mu} \cdot \vec{B}$
Drift velocity in crossed electric and magnetic fields	$v_d = \frac{E}{B}$
Hall potential	$V = \frac{IBl}{neA}$
Hall potential in terms of drift velocity	$V = Blv_d$
Charge-to-mass ratio in a mass spectrometer	$\frac{q}{m} = \frac{E}{BB_0R}$
Maximum speed of a particle in a cyclotron	$v_{\text{max}} = \frac{qBR}{m}$

SUMMARY

11.1 Magnetism and Its Historical Discoveries

- Magnets have two types of magnetic poles, called the north magnetic pole and the south magnetic pole. North magnetic poles are those that are attracted toward Earth's geographic North Pole.
- Like poles repel and unlike poles attract.
- Discoveries of how magnets respond to currents by Oersted and others created a framework that led to the invention of modern electronic devices, electric motors, and magnetic imaging technology.

11.2 Magnetic Fields and Lines

- Charges moving across a magnetic field experience a force determined by $\vec{F} = q\vec{v} \times \vec{B}$. The force is perpendicular to the plane formed by $\vec{v}$ and $\vec{B}$.
- The direction of the force on a moving charge is given by the right hand rule 1 (RHR-1): Sweep your fingers in a velocity, magnetic field plane. Start by pointing them in the direction of velocity and sweep towards the magnetic field. Your thumb points in the direction of the magnetic force for positive charges.
- Magnetic fields can be pictorially represented by magnetic field lines, which have the following properties:
  1. The field is tangent to the magnetic field line.
  2. Field strength is proportional to the line density.
  3. Field lines cannot cross.
  4. Field lines form continuous, closed loops.
- Magnetic poles always occur in pairs of north and south—it is not possible to isolate north and south poles.

11.3 Motion of a Charged Particle in a Magnetic Field

- A magnetic force can supply centripetal force and cause a charged particle to move in a circular path of radius $r = \frac{mv}{qB}$.
- The period of circular motion for a charged particle moving in a magnetic field perpendicular to the plane of motion is $T = \frac{2\pi m}{qB}$.
- Helical motion results if the velocity of the charged particle has a component parallel to the magnetic field as well as a component perpendicular to the magnetic field.

11.4 Magnetic Force on a Current-Carrying Conductor

- An electrical current produces a magnetic field around the wire.

- The directionality of the magnetic field produced is determined by the right hand rule-2, where your thumb points in the direction of the current and your fingers wrap around the wire in the direction of the magnetic field.
- The magnetic force on current-carrying conductors is given by $\vec{\mathbf{F}} = I\,\vec{\mathbf{l}} \times \vec{\mathbf{B}}$ where I is the current and l is the length of a wire in a uniform magnetic field B.

11.5 Force and Torque on a Current Loop

- The net force on a current-carrying loop of any plane shape in a uniform magnetic field is zero.
- The net torque τ on a current-carrying loop of any shape in a uniform magnetic field is calculated using $\tau = \vec{\boldsymbol{\mu}} \times \vec{\mathbf{B}}$ where $\vec{\mu}$ is the magnetic dipole moment and $\vec{\mathbf{B}}$ is the magnetic field strength.
- The magnetic dipole moment μ is the product of the number of turns of wire N, the current in the loop I, and the area of the loop A or $\vec{\boldsymbol{\mu}} = NIA\hat{\mathbf{n}}$.

11.6 The Hall Effect

- Perpendicular electric and magnetic fields exert equal and opposite forces for a specific velocity of entering particles, thereby acting as a velocity selector. The velocity that passes through undeflected is calculated by $v = \frac{E}{B}$.
- The Hall effect can be used to measure the sign of the majority of charge carriers for metals. It can also be used to measure a magnetic field.

11.7 Applications of Magnetic Forces and Fields

- A mass spectrometer is a device that separates ions according to their charge-to-mass ratios by first sending them through a velocity selector, then a uniform magnetic field.
- Cyclotrons are used to accelerate charged particles to large kinetic energies through applied electric and magnetic fields.

CONCEPTUAL QUESTIONS

11.2 Magnetic Fields and Lines

1. Discuss the similarities and differences between the electrical force on a charge and the magnetic force on a charge.

2. (a) Is it possible for the magnetic force on a charge moving in a magnetic field to be zero? (b) Is it possible for the electric force on a charge moving in an electric field to be zero? (c) Is it possible for the resultant of the electric and magnetic forces on a charge moving simultaneously through both fields to be zero?

11.3 Motion of a Charged Particle in a Magnetic Field

3. At a given instant, an electron and a proton are moving with the same velocity in a constant magnetic field. Compare the magnetic forces on these particles. Compare their accelerations.

4. Does increasing the magnitude of a uniform magnetic field through which a charge is traveling necessarily mean increasing the magnetic force on the charge? Does changing the direction of the field necessarily mean a change in the force on the charge?

5. An electron passes through a magnetic field without being deflected. What do you conclude about the magnetic field?

6. If a charged particle moves in a straight line, can you conclude that there is no magnetic field present?

7. How could you determine which pole of an electromagnet is north and which pole is south?

11.4 Magnetic Force on a Current-Carrying Conductor

8. Describe the error that results from accidently using your left rather than your right hand when determining the direction of a magnetic force.

9. Considering the magnetic force law, are the velocity and magnetic field always perpendicular? Are the force and velocity always perpendicular? What about the force and magnetic field?

10. Why can a nearby magnet distort a cathode ray tube television picture?

11. A magnetic field exerts a force on the moving electrons in a current carrying wire. What exerts the force on a wire?

12. There are regions where the magnetic field of earth is almost perpendicular to the surface of Earth. What difficulty does this cause in the use of a compass?

11.6 The Hall Effect

13. Hall potentials are much larger for poor conductors than for good conductors. Why?

11.7 Applications of Magnetic Forces and Fields

14. Describe the primary function of the electric field and the magnetic field in a cyclotron.

PROBLEMS

11.2 Magnetic Fields and Lines

15. What is the direction of the magnetic force on a positive charge that moves as shown in each of the six cases?

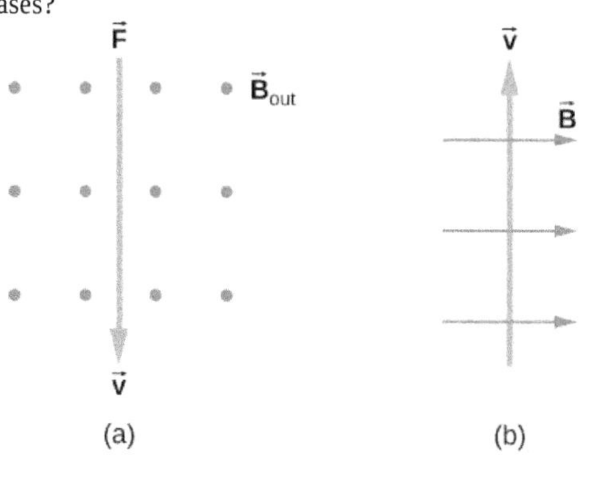

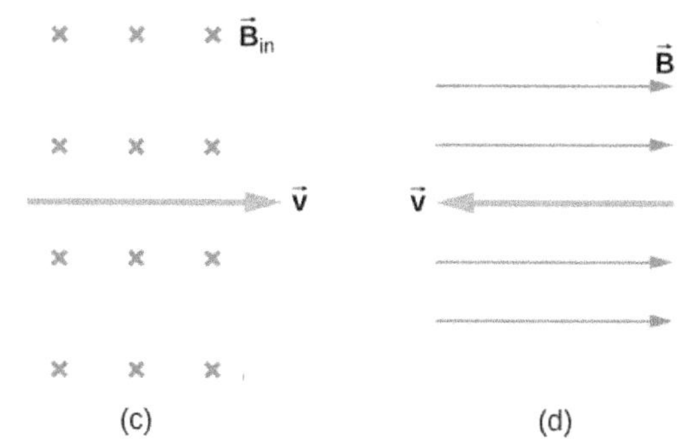

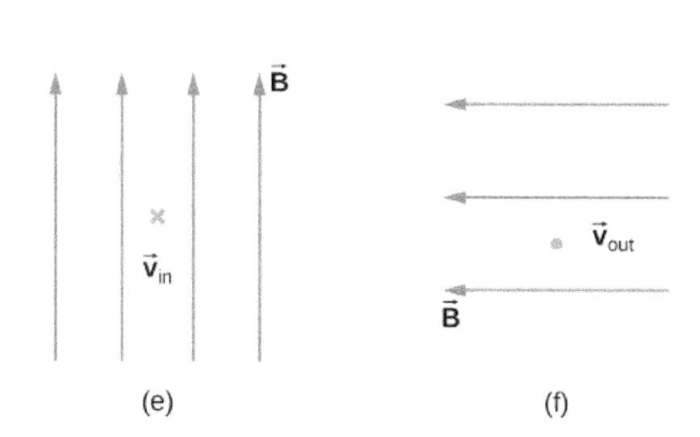

16. Repeat previous exercise for a negative charge.

17. What is the direction of the velocity of a negative charge that experiences the magnetic force shown in each of the three cases, assuming it moves perpendicular to B?

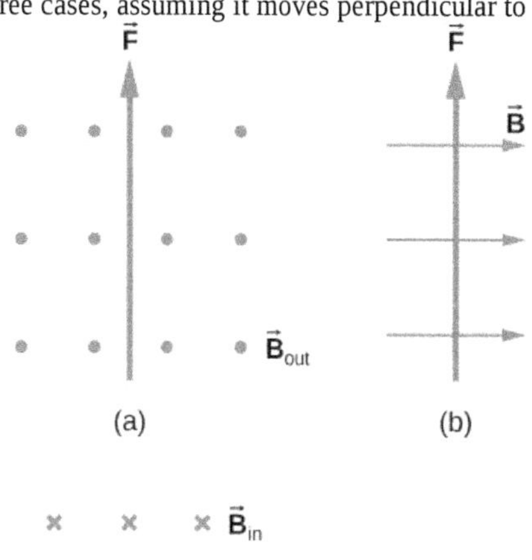

B in

F

(c)

18. Repeat previous exercise for a positive charge.

19. What is the direction of the magnetic field that produces the magnetic force on a positive charge as shown in each of the three cases, assuming $\vec{B}$ is perpendicular to $\vec{v}$?

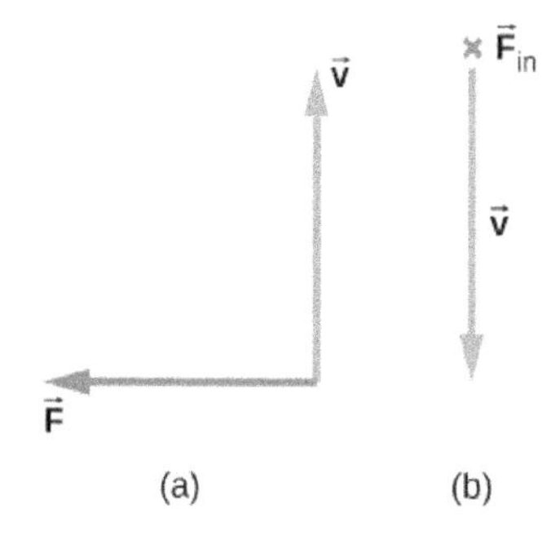

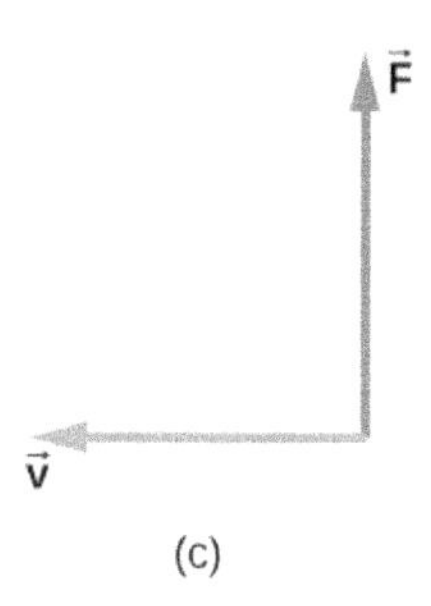

(c)

20. Repeat previous exercise for a negative charge.

21. (a) Aircraft sometimes acquire small static charges. Suppose a supersonic jet has a 0.500-μC charge and flies due west at a speed of 660. m/s over Earth's south magnetic pole, where the $8.00 \times 10^{-5} - \text{T}$ magnetic field points straight up. What are the direction and the magnitude of the magnetic force on the plane? (b) Discuss whether the value obtained in part (a) implies this is a significant or negligible effect.

22. (a) A cosmic ray proton moving toward Earth at $5.00 \times 10^{7}\,\text{m/s}$ experiences a magnetic force of $1.70 \times 10^{-16}\,\text{N}$. What is the strength of the magnetic field if there is a 45° angle between it and the proton's velocity? (b) Is the value obtained in part a. consistent with the known strength of Earth's magnetic field on its surface? Discuss.

23. An electron moving at $4.00 \times 10^{3}\,\text{m/s}$ in a 1.25-T magnetic field experiences a magnetic force of $1.40 \times 10^{-16}\,\text{N}$. What angle does the velocity of the electron make with the magnetic field? There are two answers.

24. (a) A physicist performing a sensitive measurement wants to limit the magnetic force on a moving charge in her equipment to less than $1.00 \times 10^{-12}\,\text{N}$. What is the greatest the charge can be if it moves at a maximum speed of 30.0 m/s in Earth's field? (b) Discuss whether it would be difficult to limit the charge to less than the value found in (a) by comparing it with typical static electricity and noting that static is often absent.

11.3 Motion of a Charged Particle in a Magnetic Field

25. A cosmic-ray electron moves at $7.5 \times 10^{6}\,\text{m/s}$ perpendicular to Earth's magnetic field at an altitude where the field strength is $1.0 \times 10^{-5}\,\text{T}$. What is the radius of the circular path the electron follows?

26. (a) Viewers of Star Trek have heard of an antimatter drive on the Starship *Enterprise*. One possibility for such a futuristic energy source is to store antimatter charged particles in a vacuum chamber, circulating in a magnetic field, and then extract them as needed. Antimatter annihilates normal matter, producing pure energy. What strength magnetic field is needed to hold antiprotons, moving at $5.0 \times 10^{7}\,\text{m/s}$ in a circular path 2.00 m in radius? Antiprotons have the same mass as protons but the opposite (negative) charge. (b) Is this field strength obtainable with today's technology or is it a futuristic possibility?

27. (a) An oxygen-16 ion with a mass of $2.66 \times 10^{-26}\,\text{kg}$ travels at $5.0 \times 10^{6}\,\text{m/s}$ perpendicular to a 1.20-T magnetic field, which makes it move in a circular arc with a 0.231-m radius. What positive charge is on the ion? (b) What is the ratio of this charge to the charge of an electron? (c) Discuss why the ratio found in (b) should be an integer.

28. An electron in a TV CRT moves with a speed of $6.0 \times 10^{7}\,\text{m/s}$, in a direction perpendicular to Earth's field, which has a strength of $5.0 \times 10^{-5}\,\text{T}$. (a) What strength electric field must be applied perpendicular to the Earth's field to make the electron moves in a straight line? (b) If this is done between plates separated by 1.00 cm, what is the voltage applied? (Note that TVs are usually surrounded by a ferromagnetic material to shield against external magnetic fields and avoid the need for such a correction.)

29. (a) At what speed will a proton move in a circular path of the same radius as the electron in the previous exercise? (b) What would the radius of the path be if the proton had the same speed as the electron? (c) What would the radius be if the proton had the same kinetic energy as the electron? (d) The same momentum?

30. (a) What voltage will accelerate electrons to a speed of $6.00 \times 10^{-7}\,\text{m/s}$? (b) Find the radius of curvature of the path of a proton accelerated through this potential in a 0.500-T field and compare this with the radius of curvature of an electron accelerated through the same potential.

31. An alpha-particle $(m = 6.64 \times 10^{-27}\,\text{kg},$ $q = 3.2 \times 10^{-19}\,\text{C})$ travels in a circular path of radius 25 cm in a uniform magnetic field of magnitude 1.5 T. (a) What is the speed of the particle? (b) What is the kinetic energy in electron-volts? (c) Through what potential difference must the particle be accelerated in order to give it this kinetic energy?

32. A particle of charge q and mass m is accelerated from rest through a potential difference V, after which it encounters a uniform magnetic field B. If the particle moves in a plane perpendicular to B, what is the radius of its circular orbit?

11.4 Magnetic Force on a Current-Carrying Conductor

33. What is the direction of the magnetic force on the current in each of the six cases?

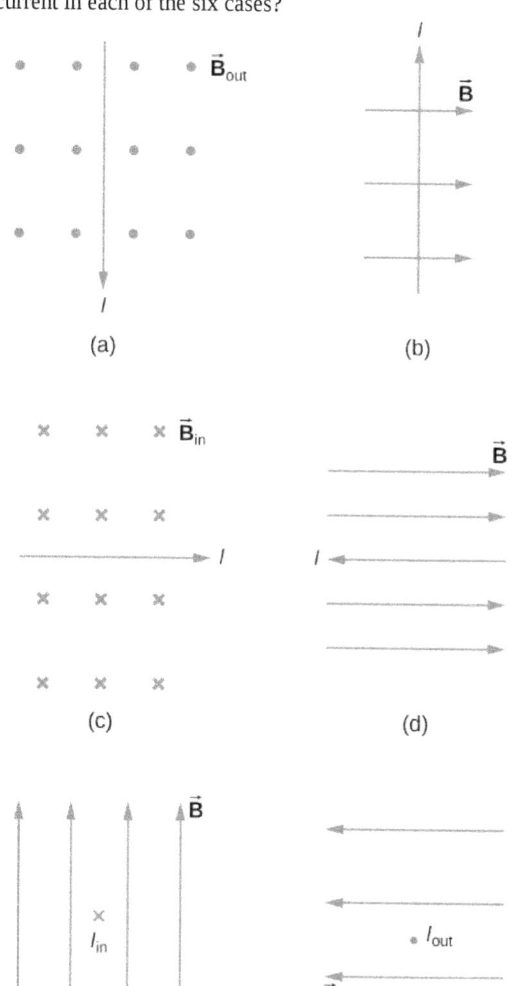

34. What is the direction of a current that experiences the magnetic force shown in each of the three cases, assuming the current runs perpendicular to $\vec{B}$?

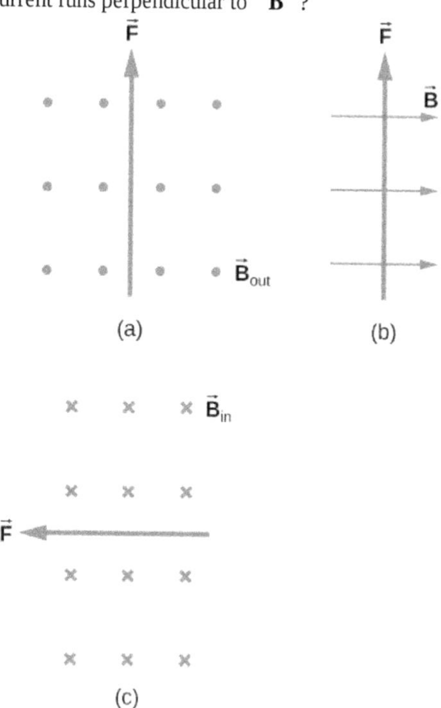

35. What is the direction of the magnetic field that produces the magnetic force shown on the currents in each of the three cases, assuming $\vec{B}$ is perpendicular to I?

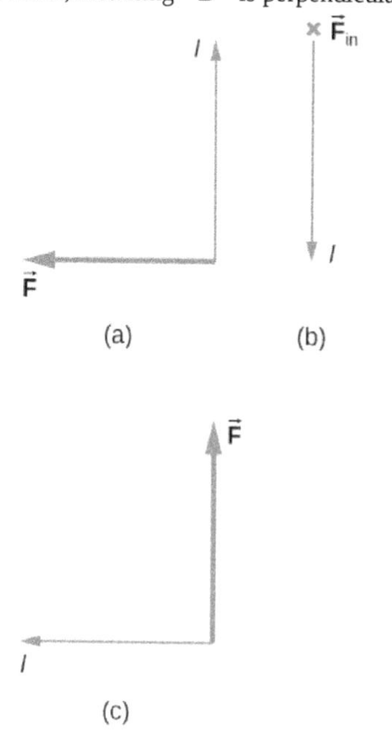

36. (a) What is the force per meter on a lightning bolt at the equator that carries 20,000 A perpendicular to Earth's $3.0 \times 10^{-5}\,\text{T}$ field? (b) What is the direction of the force if the current is straight up and Earth's field direction is due north, parallel to the ground?

37. (a) A dc power line for a light-rail system carries 1000 A at an angle of 30.0° to Earth's $5.0 \times 10^{-5}\,\text{T}$ field. What is the force on a 100-m section of this line? (b) Discuss practical concerns this presents, if any.

38. A wire carrying a 30.0-A current passes between the poles of a strong magnet that is perpendicular to its field and experiences a 2.16-N force on the 4.00 cm of wire in the field. What is the average field strength?

11.5 Force and Torque on a Current Loop

39. (a) By how many percent is the torque of a motor decreased if its permanent magnets lose 5.0% of their strength? (b) How many percent would the current need to be increased to return the torque to original values?

40. (a) What is the maximum torque on a 150-turn square loop of wire 18.0 cm on a side that carries a 50.0-A current in a 1.60-T field? (b) What is the torque when θ is 10.9°?

41. Find the current through a loop needed to create a maximum torque of $9.0\,\text{N}\cdot\text{m}$. The loop has 50 square turns that are 15.0 cm on a side and is in a uniform 0.800-T magnetic field.

42. Calculate the magnetic field strength needed on a 200-turn square loop 20.0 cm on a side to create a maximum torque of 300 N · m if the loop is carrying 25.0 A.

43. Since the equation for torque on a current-carrying loop is $\tau = NIAB \sin \theta$, the units of N · m must equal units of A · m^2 T. Verify this.

44. (a) At what angle θ is the torque on a current loop 90.0% of maximum? (b) 50.0% of maximum? (c) 10.0% of maximum?

45. A proton has a magnetic field due to its spin. The field is similar to that created by a circular current loop $0.65 \times 10^{-15}\,\text{m}$ in radius with a current of $1.05 \times 10^{4}\,\text{A}$. Find the maximum torque on a proton in a 2.50-T field. (This is a significant torque on a small particle.)

46. (a) A 200-turn circular loop of radius 50.0 cm is vertical, with its axis on an east-west line. A current of 100 A circulates clockwise in the loop when viewed from the east. Earth's field here is due north, parallel to the ground, with a strength of $3.0 \times 10^{-5}\,\text{T}$. What are the direction and magnitude of the torque on the loop? (b) Does this device have any practical applications as a motor?

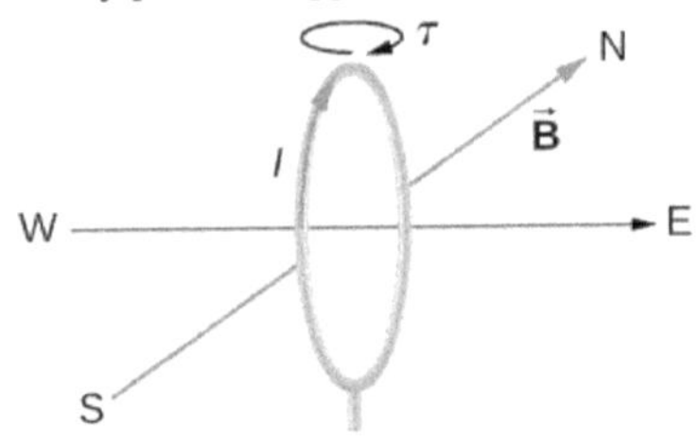

47. Repeat the previous problem, but with the loop lying flat on the ground with its current circulating counterclockwise (when viewed from above) in a location where Earth's field is north, but at an angle 45.0° below the horizontal and with a strength of $6.0 \times 10^{-5}\,\text{T}$.

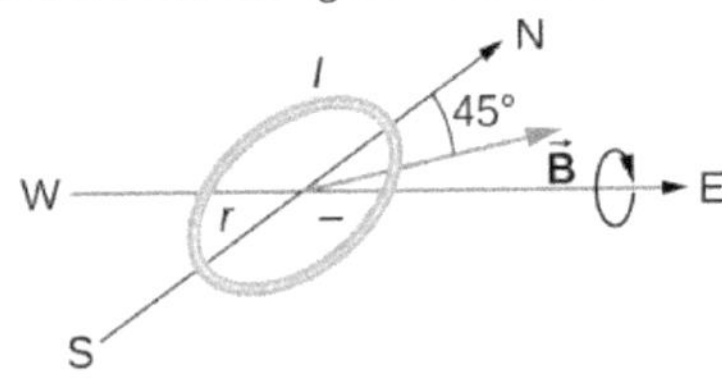

11.6 The Hall Effect

48. A strip of copper is placed in a uniform magnetic field of magnitude 2.5 T. The Hall electric field is measured to be $1.5 \times 10^{-3}\,\text{V/m}$. (a) What is the drift speed of the conduction electrons? (b) Assuming that n = 8.0×10^{28} electrons per cubic meter and that the cross-sectional area of the strip is $5.0 \times 10^{-6}\,\text{m}^2$, calculate the current in the strip. (c) What is the Hall coefficient 1/nq?

49. The cross-sectional dimensions of the copper strip shown are 2.0 cm by 2.0 mm. The strip carries a current of 100 A, and it is placed in a magnetic field of magnitude B = 1.5 T. What are the value and polarity of the Hall potential in the copper strip?

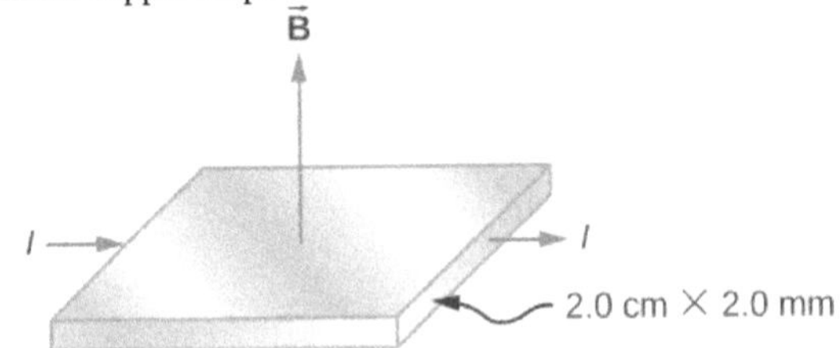

50. The magnitudes of the electric and magnetic fields in a velocity selector are 1.8×10^5 V/m and 0.080 T, respectively. (a) What speed must a proton have to pass through the selector? (b) Also calculate the speeds required for an alpha-particle and a singly ionized ${}^{s}O^{16}$ atom to pass through the selector.

51. A charged particle moves through a velocity selector at constant velocity. In the selector, $E = 1.0 \times 10^4$ N/C and $B = 0.250$ T. When the electric field is turned off, the charged particle travels in a circular path of radius 3.33 mm. Determine the charge-to-mass ratio of the particle.

52. A Hall probe gives a reading of $1.5\ \mu$ V for a current of 2 A when it is placed in a magnetic field of 1 T. What is the magnetic field in a region where the reading is $2\ \mu$ V for 1.7 A of current?

11.7 Applications of Magnetic Forces and Fields

53. A physicist is designing a cyclotron to accelerate protons to one-tenth the speed of light. The magnetic field will have a strength of 1.5 T. Determine (a) the rotational period of the circulating protons and (b) the maximum radius of the protons' orbit.

54. The strengths of the fields in the velocity selector of a Bainbridge mass spectrometer are $B = 0.500$ T and $E = 1.2 \times 10^5$ V/m, and the strength of the magnetic field that separates the ions is $B_o = 0.750$ T. A stream of singly charged Li ions is found to bend in a circular arc of radius 2.32 cm. What is the mass of the Li ions?

55. The magnetic field in a cyclotron is 1.25 T, and the maximum orbital radius of the circulating protons is 0.40 m. (a) What is the kinetic energy of the protons when they are ejected from the cyclotron? (b) What is this energy in MeV? (c) Through what potential difference would a proton have to be accelerated to acquire this kinetic energy? (d) What is the period of the voltage source used to accelerate the protons? (e) Repeat the calculations for alpha-particles.

56. A mass spectrometer is being used to separate common oxygen-16 from the much rarer oxygen-18, taken from a sample of old glacial ice. (The relative abundance of these oxygen isotopes is related to climatic temperature at the time the ice was deposited.) The ratio of the masses of these two ions is 16 to 18, the mass of oxygen-16 is 2.66×10^{-26} kg, and they are singly charged and travel at 5.00×10^6 m/s in a 1.20-T magnetic field. What is the separation between their paths when they hit a target after traversing a semicircle?

57. (a) Triply charged uranium-235 and uranium-238 ions are being separated in a mass spectrometer. (The much rarer uranium-235 is used as reactor fuel.) The masses of the ions are 3.90×10^{-25} kg and 3.95×10^{-25} kg, respectively, and they travel at 3.0×10^5 m/s in a 0.250-T field. What is the separation between their paths when they hit a target after traversing a semicircle? (b) Discuss whether this distance between their paths seems to be big enough to be practical in the separation of uranium-235 from uranium-238.

ADDITIONAL PROBLEMS

58. Calculate the magnetic force on a hypothetical particle of charge 1.0×10^{-19} C moving with a velocity of $6.0 \times 10^4\ \hat{\mathbf{i}}$ m/s in a magnetic field of $1.2\hat{\mathbf{k}}$ T.

59. Repeat the previous problem with a new magnetic field of $(0.4\ \hat{\mathbf{i}} + 1.2\hat{\mathbf{k}})$T.

60. An electron is projected into a uniform magnetic field $(0.5\ \hat{\mathbf{i}} + 0.8\hat{\mathbf{k}})$T with a velocity of $(3.0\ \hat{\mathbf{i}} + 4.0\ \hat{\mathbf{j}}) \times 10^6$ m/s. What is the magnetic force on the electron?

61. The mass and charge of a water droplet are 1.0×10^{-4} g and 2.0×10^{-8} C, respectively. If the droplet is given an initial horizontal velocity of $5.0 \times 10^5\ \hat{\mathbf{i}}$ m/s, what magnetic field will keep it moving in this direction? Why must gravity be considered here?

62. Four different proton velocities are given. For each case, determine the magnetic force on the proton in terms of e, v_0, and B_0.

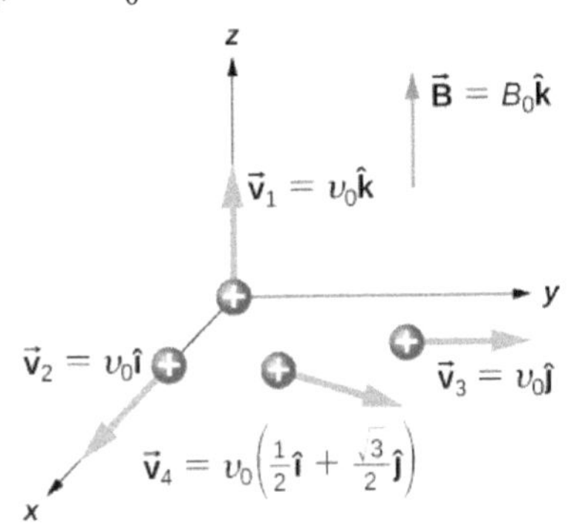

63. An electron of kinetic energy 2000 eV passes between parallel plates that are 1.0 cm apart and kept at a potential difference of 300 V. What is the strength of the uniform magnetic field B that will allow the electron to travel undeflected through the plates? Assume E and B are perpendicular.

64. An alpha-particle $\left(m = 6.64 \times 10^{-27}\,\text{kg},\ q = 3.2 \times 10^{-19}\,\text{C}\right)$ moving with a velocity $\vec{v} = (2.0\hat{i} - 4.0\hat{k}) \times 10^{6}\,\text{m/s}$ enters a region where $\vec{E} = (5.0\hat{i} - 2.0\hat{j}) \times 10^{4}\,\text{V/m}$ and $\vec{B} = (1.0\hat{i} + 4.0\hat{k}) \times 10^{-2}\,\text{T}$. What is the initial force on it?

65. An electron moving with a velocity $\vec{v} = \left(4.0\hat{i} + 3.0\hat{j} + 2.0\hat{k}\right) \times 10^{6}\,\text{m/s}$ enters a region where there is a uniform electric field and a uniform magnetic field. The magnetic field is given by $\vec{B} = \left(1.0\hat{i} - 2.0\hat{j} + 4.0\hat{k}\right) \times 10^{-2}\,\text{T}$. If the electron travels through a region without being deflected, what is the electric field?

66. At a particular instant, an electron is traveling west to east with a kinetic energy of 10 keV. Earth's magnetic field has a horizontal component of $1.8 \times 10^{-5}\,\text{T}$ north and a vertical component of $5.0 \times 10^{-5}\,\text{T}$ down. (a) What is the path of the electron? (b) What is the radius of curvature of the path?

67. Repeat the calculations of the previous problem for a proton with the same kinetic energy.

68. What magnetic field is required in order to confine a proton moving with a speed of $4.0 \times 10^{6}\,\text{m/s}$ to a circular orbit of radius 10 cm?

69. An electron and a proton move with the same speed in a plane perpendicular to a uniform magnetic field. Compare the radii and periods of their orbits.

70. A proton and an alpha-particle have the same kinetic energy and both move in a plane perpendicular to a uniform magnetic field. Compare the periods of their orbits.

71. A singly charged ion takes $2.0 \times 10^{-3}\,\text{s}$ to complete eight revolutions in a uniform magnetic field of magnitude $2.0 \times 10^{-2}\,\text{T}$. What is the mass of the ion?

72. A particle moving downward at a speed of $6.0 \times 10^{6}\,\text{m/s}$ enters a uniform magnetic field that is horizontal and directed from east to west. (a) If the particle is deflected initially to the north in a circular arc, is its charge positive or negative? (b) If B = 0.25 T and the charge-to-mass ratio (q/m) of the particle is $4.0 \times 10^{7}\,\text{C/kg}$, what is the radius of the path? (c) What is the speed of the particle after it has moved in the field for $1.0 \times 10^{-5}\,\text{s}$? for 2.0 s?

73. A proton, deuteron, and an alpha-particle are all accelerated through the same potential difference. They then enter the same magnetic field, moving perpendicular to it. Compute the ratios of the radii of their circular paths. Assume that $m_d = 2m_p$ and $m_\alpha = 4m_p$.

74. A singly charged ion is moving in a uniform magnetic field of $7.5 \times 10^{-2}\,\text{T}$ completes 10 revolutions in $3.47 \times 10^{-4}\,\text{s}$. Identify the ion.

75. Two particles have the same linear momentum, but particle A has four times the charge of particle B. If both particles move in a plane perpendicular to a uniform magnetic field, what is the ratio R_A/R_B of the radii of their circular orbits?

76. A uniform magnetic field of magnitude B is directed parallel to the z-axis. A proton enters the field with a velocity $\vec{v} = (4\hat{j} + 3\hat{k}) \times 10^{6}\,\text{m/s}$ and travels in a helical path with a radius of 5.0 cm. (a) What is the value of B? (b) What is the time required for one trip around the helix? (c) Where is the proton $5.0 \times 10^{-7}\,\text{s}$ after entering the field?

77. An electron moving at $5.0 \times 10^6\,\text{m/s}$ enters a magnetic field that makes a 75° angle with the *x*-axis of magnitude 0.20 T. Calculate the (a) pitch and (b) radius of the trajectory.

78. (a) A 0.750-m-long section of cable carrying current to a car starter motor makes an angle of 60° with Earth's $5.5 \times 10^{-5}\,\text{T}$ field. What is the current when the wire experiences a force of $7.0 \times 10^{-3}\,\text{N}$? (b) If you run the wire between the poles of a strong horseshoe magnet, subjecting 5.00 cm of it to a 1.75-T field, what force is exerted on this segment of wire?

79. (a) What is the angle between a wire carrying an 8.00-A current and the 1.20-T field it is in if 50.0 cm of the wire experiences a magnetic force of 2.40 N? (b) What is the force on the wire if it is rotated to make an angle of 90° with the field?

80. A 1.0-m-long segment of wire lies along the *x*-axis and carries a current of 2.0 A in the positive *x*-direction. Around the wire is the magnetic field of $\left(3.0\,\hat{\mathbf{i}} \times 4.0\,\hat{\mathbf{k}}\right) \times 10^{-3}\,\text{T}$. Find the magnetic force on this segment.

81. A 5.0-m section of a long, straight wire carries a current of 10 A while in a uniform magnetic field of magnitude $8.0 \times 10^{-3}\,\text{T}$. Calculate the magnitude of the force on the section if the angle between the field and the direction of the current is (a) 45°; (b) 90°; (c) 0°; or (d) 180°.

82. An electromagnet produces a magnetic field of magnitude 1.5 T throughout a cylindrical region of radius 6.0 cm. A straight wire carrying a current of 25 A passes through the field as shown in the accompanying figure. What is the magnetic force on the wire?

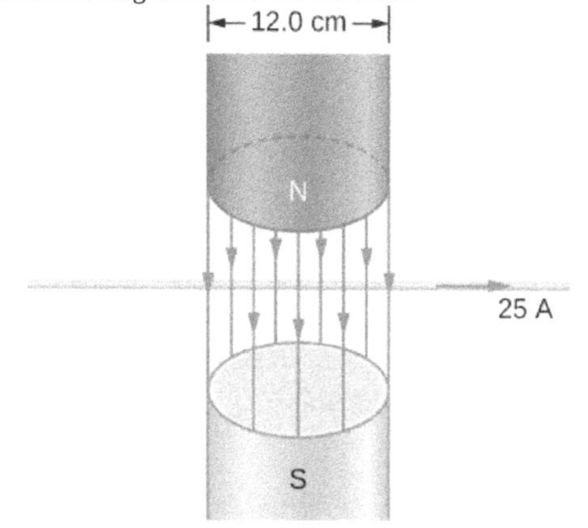

83. The current loop shown in the accompanying figure lies in the plane of the page, as does the magnetic field. Determine the net force and the net torque on the loop if *I* = 10 A and *B* = 1.5 T.

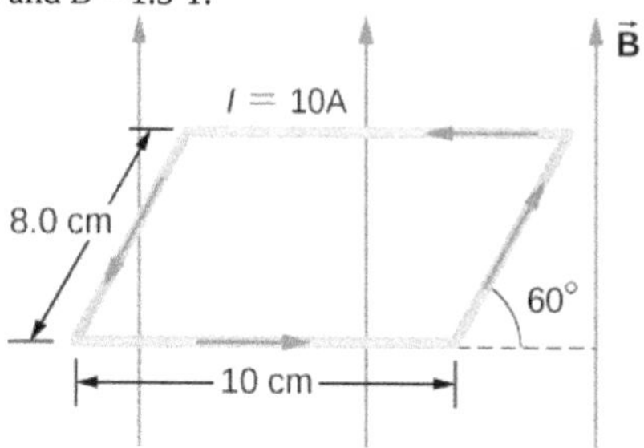

84. A circular coil of radius 5.0 cm is wound with five turns and carries a current of 5.0 A. If the coil is placed in a uniform magnetic field of strength 5.0 T, what is the maximum torque on it?

85. A circular coil of wire of radius 5.0 cm has 20 turns and carries a current of 2.0 A. The coil lies in a magnetic field of magnitude 0.50 T that is directed parallel to the plane of the coil. (a) What is the magnetic dipole moment of the coil? (b) What is the torque on the coil?

86. A current-carrying coil in a magnetic field experiences a torque that is 75% of the maximum possible torque. What is the angle between the magnetic field and the normal to the plane of the coil?

87. A 4.0-cm by 6.0-cm rectangular current loop carries a current of 10 A. What is the magnetic dipole moment of the loop?

88. A circular coil with 200 turns has a radius of 2.0 cm. (a) What current through the coil results in a magnetic dipole moment of 3.0 Am^2? (b) What is the maximum torque that the coil will experience in a uniform field of strength $5.0 \times 10^{-2}\,\text{T}$? (c) If the angle between μ and B is 45°, what is the magnitude of the torque on the coil? (d) What is the magnetic potential energy of coil for this orientation?

89. The current through a circular wire loop of radius 10 cm is 5.0 A. (a) Calculate the magnetic dipole moment of the loop. (b) What is the torque on the loop if it is in a uniform 0.20-T magnetic field such that μ and B are directed at 30° to each other? (c) For this position, what is the potential energy of the dipole?

90. A wire of length 1.0 m is wound into a single-turn planar loop. The loop carries a current of 5.0 A, and it is placed in a uniform magnetic field of strength 0.25 T. (a) What is the maximum torque that the loop will experience if it is square? (b) If it is circular? (c) At what angle relative to B would the normal to the circular coil have to be oriented so that the torque on it would be the same as the maximum torque on the square coil?

91. Consider an electron rotating in a circular orbit of radius r. Show that the magnitudes of the magnetic dipole moment μ and the angular momentum L of the electron are related by: $\frac{\mu}{L} = \frac{e}{2m}$.

92. The Hall effect is to be used to find the sign of charge carriers in a semiconductor sample. The probe is placed between the poles of a magnet so that magnetic field is pointed up. A current is passed through a rectangular sample placed horizontally. As current is passed through the sample in the east direction, the north side of the sample is found to be at a higher potential than the south side. Decide if the number density of charge carriers is positively or negatively charged.

93. The density of charge carriers for copper is 8.47×10^{28} electrons per cubic meter. What will be the Hall voltage reading from a probe made up of $3 \text{ cm} \times 2 \text{ cm} \times 1 \text{ cm} \, (\text{L} \times \text{W} \times \text{T})$ copper plate when a current of 1.5 A is passed through it in a magnetic field of 2.5 T perpendicular to the $3 \text{ cm} \times 2 \text{ cm}$.

94. The Hall effect is to be used to find the density of charge carriers in an unknown material. A Hall voltage 40 μV for 3-A current is observed in a 3-T magnetic field for a rectangular sample with length 2 cm, width 1.5 cm, and height 0.4 cm. Determine the density of the charge carriers.

95. Show that the Hall voltage across wires made of the same material, carrying identical currents, and subjected to the same magnetic field is inversely proportional to their diameters. (Hint: Consider how drift velocity depends on wire diameter.)

96. A velocity selector in a mass spectrometer uses a 0.100-T magnetic field. (a) What electric field strength is needed to select a speed of $4.0 \times 10^6 \text{ m/s}$? (b) What is the voltage between the plates if they are separated by 1.00 cm?

97. Find the radius of curvature of the path of a 25.0-MeV proton moving perpendicularly to the 1.20-T field of a cyclotron.

98. Unreasonable results To construct a non-mechanical water meter, a 0.500-T magnetic field is placed across the supply water pipe to a home and the Hall voltage is recorded. (a) Find the flow rate through a 3.00-cm-diameter pipe if the Hall voltage is 60.0 mV. (b) What would the Hall voltage be for the same flow rate through a 10.0-cm-diameter pipe with the same field applied?

99. Unreasonable results A charged particle having mass $6.64 \times 10^{-27} \text{ kg}$ (that of a helium atom) moving at $8.70 \times 10^5 \text{ m/s}$ perpendicular to a 1.50-T magnetic field travels in a circular path of radius 16.0 mm. (a) What is the charge of the particle? (b) What is unreasonable about this result? (c) Which assumptions are responsible?

100. Unreasonable results An inventor wants to generate 120-V power by moving a 1.00-m-long wire perpendicular to Earth's 5.00×10^{-5} T field. (a) Find the speed with which the wire must move. (b) What is unreasonable about this result? (c) Which assumption is responsible?

101. Unreasonable results Frustrated by the small Hall voltage obtained in blood flow measurements, a medical physicist decides to increase the applied magnetic field strength to get a 0.500-V output for blood moving at 30.0 cm/s in a 1.50-cm-diameter vessel. (a) What magnetic field strength is needed? (b) What is unreasonable about this result? (c) Which premise is responsible?

CHALLENGE PROBLEMS

102. A particle of charge $+q$ and mass m moves with velocity $\vec{\mathbf{v}}_0$ pointed in the $+y$-direction as it crosses the x-axis at $x = R$ at a particular time. There is a negative charge $-Q$ fixed at the origin, and there exists a uniform magnetic field $\vec{\mathbf{B}}_0$ pointed in the $+z$-direction. It is found that the particle describes a circle of radius R about $-Q$. Find $\vec{\mathbf{B}}_0$ in terms of the given quantities.

103. A proton of speed $v = 6 \times 10^5 \text{ m/s}$ enters a region of uniform magnetic field of $B = 0.5$ T at an angle of $q = 30°$ to the magnetic field. In the region of magnetic field proton describes a helical path with radius R and pitch p (distance between loops). Find R and p.

104. A particle's path is bent when it passes through a region of non-zero magnetic field although its speed remains unchanged. This is very useful for "beam steering" in particle accelerators. Consider a proton of speed 4×10^6 m/s entering a region of uniform magnetic field 0.2 T over a 5-cm-wide region. Magnetic field is perpendicular to the velocity of the particle. By how much angle will the path of the proton be bent? (Hint: The particle comes out tangent to a circle.)

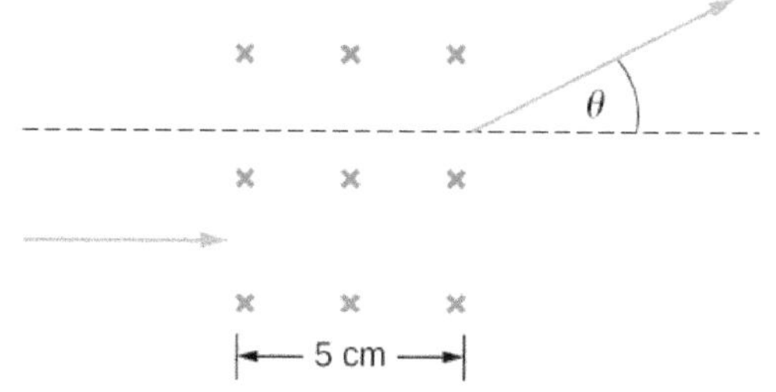

105. In a region a non-uniform magnetic field exists such that $B_x = 0, B_y = 0,\ and\ B_z = ax,$ where a is a constant. At some time t, a wire of length L is carrying a current I is located along the x-axis from origin to $x = L$. Find the magnetic force on the wire at this instant in time.

106. A copper rod of mass m and length L is hung from the ceiling using two springs of spring constant k. A uniform magnetic field of magnitude B_0 pointing perpendicular to the rod and spring (coming out of the page in the figure) exists in a region of space covering a length w of the copper rod. The ends of the rod are then connected by flexible copper wire across the terminals of a battery of voltage V. Determine the change in the length of the springs when a current I runs through the copper rod in the direction shown in figure. (Ignore any force by the flexible wire.)

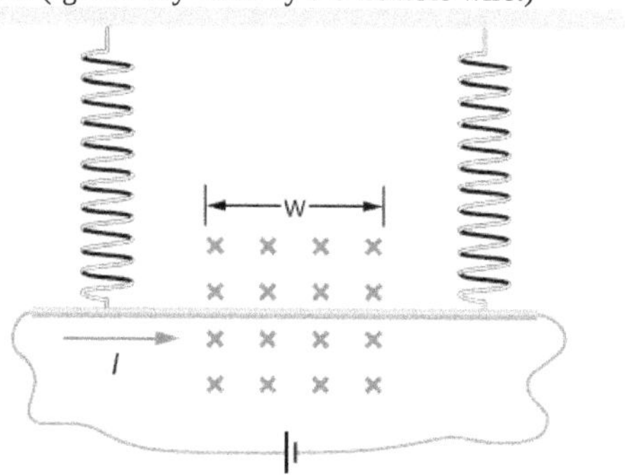

107. The accompanied figure shows an arrangement for measuring mass of ions by an instrument called the mass spectrometer. An ion of mass m and charge $+q$ is produced essentially at rest in source S, a chamber in which a gas discharge is taking place. The ion is accelerated by a potential difference V_{acc} and allowed to enter a region of constant magnetic field $\vec{\mathbf{B}}_0$. In the uniform magnetic field region, the ion moves in a semicircular path striking a photographic plate at a distance x from the entry point. Derive a formula for mass m in terms of $B_0,\ q,\ V_{acc},$ and x.

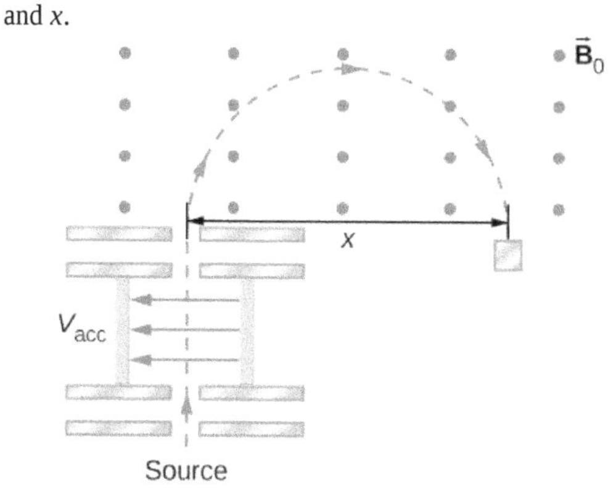

108. A wire is made into a circular shape of radius R and pivoted along a central support. The two ends of the wire are touching a brush that is connected to a dc power source. The structure is between the poles of a magnet such that we can assume there is a uniform magnetic field on the wire. In terms of a coordinate system with origin at the center of the ring, magnetic field is $B_x = B_0,\ B_y = B_z = 0,$ and the ring rotates about the z-axis. Find the torque on the ring when it is not in the xz-plane.

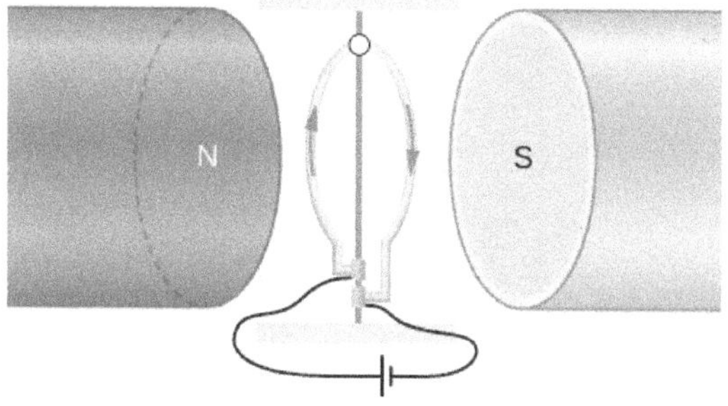

109. A long-rigid wire lies along the x-axis and carries a current of 2.5 A in the positive x-direction. Around the wire is the magnetic field $\vec{\mathbf{B}} = 2.0\,\hat{\mathbf{i}} + 5.0x^2\,\hat{\mathbf{j}},$ with x in meters and B in millitesla. Calculate the magnetic force on the segment of wire between $x = 2.0$ m and $x = 4.0$ m.

110. A circular loop of wire of area 10 cm^2 carries a current of 25 A. At a particular instant, the loop lies in the *xy*-plane and is subjected to a magnetic field $\vec{\mathbf{B}} = \left(2.0\,\hat{\mathbf{i}} + 6.0\,\hat{\mathbf{j}} + 8.0\,\hat{\mathbf{k}}\right) \times 10^{-3}$ T. As viewed from above the *xy*-plane, the current is circulating clockwise. (a) What is the magnetic dipole moment of the current loop? (b) At this instant, what is the magnetic torque on the loop?

12 | SOURCES OF MAGNETIC FIELDS

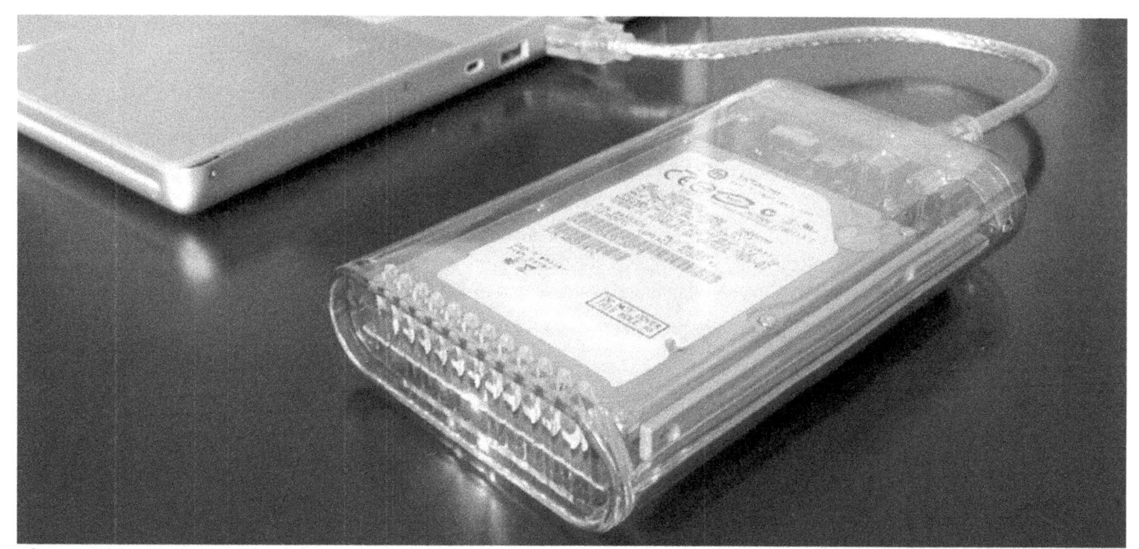

Figure 12.1 An external hard drive attached to a computer works by magnetically encoding information that can be stored or retrieved quickly. A key idea in the development of digital devices is the ability to produce and use magnetic fields in this way. (credit: modification of work by "Miss Karen"/Flickr)

Chapter Outline

Introduction

In the preceding chapter, we saw that a moving charged particle produces a magnetic field. This connection between electricity and magnetism is exploited in electromagnetic devices, such as a computer hard drive. In fact, it is the underlying principle behind most of the technology in modern society, including telephones, television, computers, and the internet.

In this chapter, we examine how magnetic fields are created by arbitrary distributions of electric current, using the Biot-Savart law. Then we look at how current-carrying wires create magnetic fields and deduce the forces that arise between two current-carrying wires due to these magnetic fields. We also study the torques produced by the magnetic fields of current loops. We then generalize these results to an important law of electromagnetism, called Ampère's law.

We examine some devices that produce magnetic fields from currents in geometries based on loops, known as solenoids and toroids. Finally, we look at how materials behave in magnetic fields and categorize materials based on their responses to magnetic fields.

12.1 | The Biot-Savart Law

Learning Objectives

By the end of this section, you will be able to:

- Explain how to derive a magnetic field from an arbitrary current in a line segment
- Calculate magnetic field from the Biot-Savart law in specific geometries, such as a current in a line and a current in a circular arc

We have seen that mass produces a gravitational field and also interacts with that field. Charge produces an electric field and also interacts with that field. Since moving charge (that is, current) interacts with a magnetic field, we might expect that it also creates that field—and it does.

The equation used to calculate the magnetic field produced by a current is known as the Biot-Savart law. It is an empirical law named in honor of two scientists who investigated the interaction between a straight, current-carrying wire and a permanent magnet. This law enables us to calculate the magnitude and direction of the magnetic field produced by a current in a wire. The **Biot-Savart law** states that at any point P (Figure 12.2), the magnetic field $d\vec{\mathbf{B}}$ due to an element $d\vec{\mathbf{l}}$ of a current-carrying wire is given by

$$d\vec{\mathbf{B}} = \frac{\mu_0}{4\pi}\frac{I d\vec{\mathbf{l}} \times \hat{\mathbf{r}}}{r^2}. \tag{12.1}$$

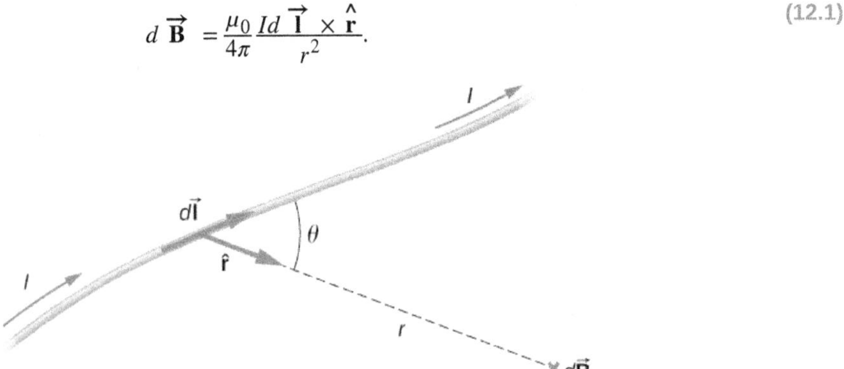

Figure 12.2 A current element $Id\vec{\mathbf{l}}$ produces a magnetic field at point P given by the Biot-Savart law.

The constant μ_0 is known as the **permeability of free space** and is exactly

$$\mu_0 = 4\pi \times 10^{-7}\,\text{T} \cdot \text{m/A} \tag{12.2}$$

in the SI system. The infinitesimal wire segment $d\vec{\mathbf{l}}$ is in the same direction as the current I (assumed positive), r is the distance from $d\vec{\mathbf{l}}$ to P and $\hat{\mathbf{r}}$ is a unit vector that points from $d\vec{\mathbf{l}}$ to P, as shown in the figure.

The direction of $d\vec{\mathbf{B}}$ is determined by applying the right-hand rule to the vector product $d\vec{\mathbf{l}} \times \hat{\mathbf{r}}$. The magnitude of $d\vec{\mathbf{B}}$ is

$$dB = \frac{\mu_0}{4\pi}\frac{I\,dl\sin\theta}{r^2} \tag{12.3}$$

where θ is the angle between $d\vec{\mathbf{l}}$ and $\hat{\mathbf{r}}$. Notice that if $\theta = 0$, then $d\vec{\mathbf{B}} = \vec{\mathbf{0}}$. The field produced by a current element $Id\vec{\mathbf{l}}$ has no component parallel to $d\vec{\mathbf{l}}$.

The magnetic field due to a finite length of current-carrying wire is found by integrating Equation 12.3 along the wire, giving us the usual form of the Biot-Savart law.

Biot-Savart law

The magnetic field $\vec{\mathbf{B}}$ due to an element $d\vec{\mathbf{l}}$ of a current-carrying wire is given by

$$\vec{\mathbf{B}} = \frac{\mu_0}{4\pi} \int_{\text{wire}} \frac{Id\vec{\mathbf{l}} \times \hat{\mathbf{r}}}{r^2}. \tag{12.4}$$

Since this is a vector integral, contributions from different current elements may not point in the same direction. Consequently, the integral is often difficult to evaluate, even for fairly simple geometries. The following strategy may be helpful.

Problem-Solving Strategy: Solving Biot-Savart Problems

To solve Biot-Savart law problems, the following steps are helpful:

1. Identify that the Biot-Savart law is the chosen method to solve the given problem. If there is symmetry in the problem comparing $\vec{\mathbf{B}}$ and $\mathbf{d}\vec{\mathbf{l}}$, Ampère's law may be the preferred method to solve the question.
2. Draw the current element length $d\vec{\mathbf{l}}$ and the unit vector $\hat{\mathbf{r}}$, noting that $d\vec{\mathbf{l}}$ points in the direction of the current and $\hat{\mathbf{r}}$ points from the current element toward the point where the field is desired.
3. Calculate the cross product $d\vec{\mathbf{l}} \times \hat{\mathbf{r}}$. The resultant vector gives the direction of the magnetic field according to the Biot-Savart law.
4. Use Equation 12.4 and substitute all given quantities into the expression to solve for the magnetic field. Note all variables that remain constant over the entire length of the wire may be factored out of the integration.
5. Use the right-hand rule to verify the direction of the magnetic field produced from the current or to write down the direction of the magnetic field if only the magnitude was solved for in the previous part.

Example 12.1

Calculating Magnetic Fields of Short Current Segments

A short wire of length 1.0 cm carries a current of 2.0 A in the vertical direction (Figure 12.3). The rest of the wire is shielded so it does not add to the magnetic field produced by the wire. Calculate the magnetic field at point *P*, which is 1 meter from the wire in the *x*-direction.

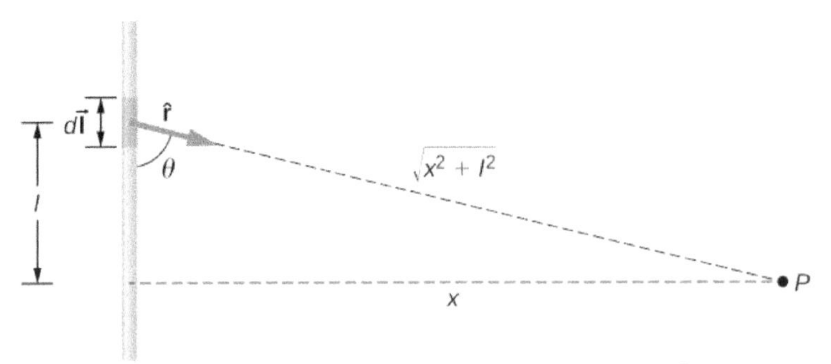

Figure 12.3 A small line segment carries a current I in the vertical direction. What is the magnetic field at a distance x from the segment?

Strategy

We can determine the magnetic field at point P using the Biot-Savart law. Since the current segment is much smaller than the distance x, we can drop the integral from the expression. The integration is converted back into a summation, but only for small dl, which we now write as Δl. Another way to think about it is that each of the radius values is nearly the same, no matter where the current element is on the line segment, if Δl is small compared to x. The angle θ is calculated using a tangent function. Using the numbers given, we can calculate the magnetic field at P.

Solution

The angle between $\Delta \vec{\mathbf{l}}$ and $\hat{\mathbf{r}}$ is calculated from trigonometry, knowing the distances l and x from the problem:

$$\theta = \tan^{-1}\left(\frac{1\text{ m}}{0.01\text{ m}}\right) = 89.4°.$$

The magnetic field at point P is calculated by the Biot-Savart law:

$$B = \frac{\mu_0}{4\pi}\frac{I\Delta l\sin\theta}{r^2} = (1 \times 10^{-7}\text{ T}\cdot\text{m/A})\left(\frac{2\text{ A}(0.01\text{ m})\sin(89.4°)}{(1\text{ m})^2}\right) = 2.0 \times 10^{-9}\text{ T}.$$

From the right-hand rule and the Biot-Savart law, the field is directed into the page.

Significance

This approximation is only good if the length of the line segment is very small compared to the distance from the current element to the point. If not, the integral form of the Biot-Savart law must be used over the entire line segment to calculate the magnetic field.

12.1 Check Your Understanding Using Example 12.1, at what distance would P have to be to measure a magnetic field half of the given answer?

Example 12.2

Calculating Magnetic Field of a Circular Arc of Wire

A wire carries a current I in a circular arc with radius R swept through an arbitrary angle θ (Figure 12.4). Calculate the magnetic field at the center of this arc at point P.

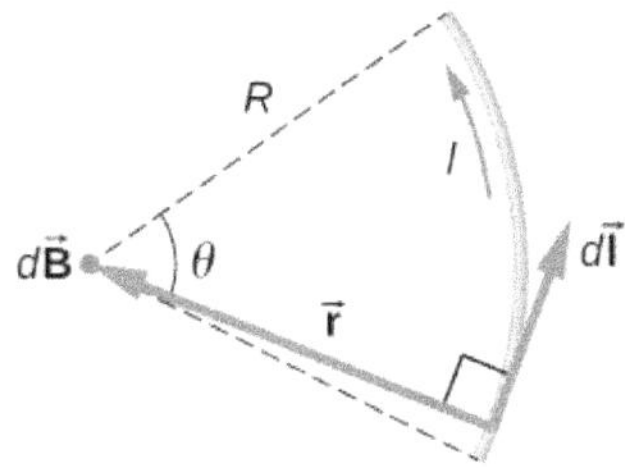

Figure 12.4 A wire segment carrying a current I. The path $d\vec{\mathbf{l}}$ and radial direction $\hat{\mathbf{r}}$ are indicated.

Strategy

We can determine the magnetic field at point P using the Biot-Savart law. The radial and path length directions are always at a right angle, so the cross product turns into multiplication. We also know that the distance along the path dl is related to the radius times the angle θ (in radians). Then we can pull all constants out of the integration and solve for the magnetic field.

Solution

The Biot-Savart law starts with the following equation:

$$\vec{\mathbf{B}} = \frac{\mu_0}{4\pi} \int_{\text{wire}} \frac{I d\vec{\mathbf{l}} \times \hat{\mathbf{r}}}{r^2}.$$

As we integrate along the arc, all the contributions to the magnetic field are in the same direction (out of the page), so we can work with the magnitude of the field. The cross product turns into multiplication because the path dl and the radial direction are perpendicular. We can also substitute the arc length formula, $dl = rd\theta$:

$$B = \frac{\mu_0}{4\pi} \int_{\text{wire}} \frac{Ir\,d\theta}{r^2}.$$

The current and radius can be pulled out of the integral because they are the same regardless of where we are on the path. This leaves only the integral over the angle,

$$B = \frac{\mu_0 I}{4\pi r} \int_{\text{wire}} d\theta.$$

The angle varies on the wire from 0 to θ; hence, the result is

$$B = \frac{\mu_0 I\theta}{4\pi r}.$$

Significance

The direction of the magnetic field at point P is determined by the right-hand rule, as shown in the previous chapter. If there are other wires in the diagram along with the arc, and you are asked to find the net magnetic field, find each contribution from a wire or arc and add the results by superposition of vectors. Make sure to pay attention to the direction of each contribution. Also note that in a symmetric situation, like a straight or circular wire, contributions from opposite sides of point P cancel each other.

12.2 Check Your Understanding The wire loop forms a full circle of radius R and current I. What is the magnitude of the magnetic field at the center?

12.2 | Magnetic Field Due to a Thin Straight Wire

Learning Objectives

By the end of this section, you will be able to:

- Explain how the Biot-Savart law is used to determine the magnetic field due to a thin, straight wire.
- Determine the dependence of the magnetic field from a thin, straight wire based on the distance from it and the current flowing in the wire.
- Sketch the magnetic field created from a thin, straight wire by using the second right-hand rule.

How much current is needed to produce a significant magnetic field, perhaps as strong as Earth's field? Surveyors will tell you that overhead electric power lines create magnetic fields that interfere with their compass readings. Indeed, when Oersted discovered in 1820 that a current in a wire affected a compass needle, he was not dealing with extremely large currents. How does the shape of wires carrying current affect the shape of the magnetic field created? We noted in Chapter 28 that a current loop created a magnetic field similar to that of a bar magnet, but what about a straight wire? We can use the Biot-Savart law to answer all of these questions, including determining the magnetic field of a long straight wire.

Figure 12.5 shows a section of an infinitely long, straight wire that carries a current I. What is the magnetic field at a point P, located a distance R from the wire?

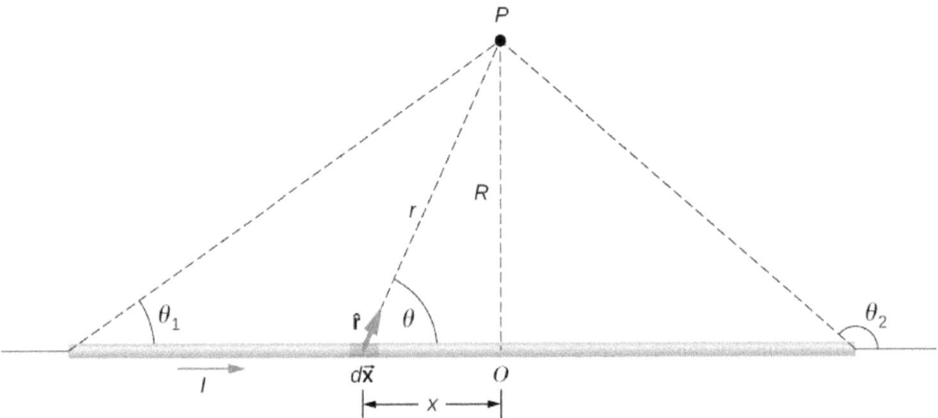

Figure 12.5 A section of a thin, straight current-carrying wire. The independent variable θ has the limits θ_1 and θ_2.

Let's begin by considering the magnetic field due to the current element $I\,d\vec{x}$ located at the position x. Using the right-hand rule 1 from the previous chapter, $d\vec{x} \times \hat{r}$ points out of the page for any element along the wire. At point P, therefore, the magnetic fields due to all current elements have the same direction. This means that we can calculate the net field there by evaluating the scalar sum of the contributions of the elements. With $\left|d\vec{x} \times \hat{r}\right| = (dx)(1)\sin\theta,$ we have from the Biot-Savart law

$$B = \frac{\mu_0}{4\pi} \int_{\text{wire}} \frac{I\sin\theta\,dx}{r^2}. \tag{12.5}$$

The wire is symmetrical about point O, so we can set the limits of the integration from zero to infinity and double the answer, rather than integrate from negative infinity to positive infinity. Based on the picture and geometry, we can write expressions for r and $\sin\theta$ in terms of x and R, namely:

$$\begin{aligned} r &= \sqrt{x^2 + R^2} \\ \sin\theta &= \frac{R}{\sqrt{x^2 + R^2}}. \end{aligned}$$

Substituting these expressions into Equation 12.5, the magnetic field integration becomes

$$B = \frac{\mu_0 I}{2\pi} \int_0^\infty \frac{R\,dx}{(x^2 + R^2)^{3/2}}. \tag{12.6}$$

Evaluating the integral yields

$$B = \frac{\mu_0 I}{2\pi R} \left[\frac{x}{(x^2 + R^2)^{1/2}} \right]_0^\infty. \tag{12.7}$$

Substituting the limits gives us the solution

$$B = \frac{\mu_0 I}{2\pi R}. \tag{12.8}$$

The magnetic field lines of the infinite wire are circular and centered at the wire (Figure 12.6), and they are identical in every plane perpendicular to the wire. Since the field decreases with distance from the wire, the spacing of the field lines must increase correspondingly with distance. The direction of this magnetic field may be found with a second form of the right-hand rule (illustrated in Figure 12.6). If you hold the wire with your right hand so that your thumb points along the current, then your fingers wrap around the wire in the same sense as $\vec{\mathbf{B}}$.

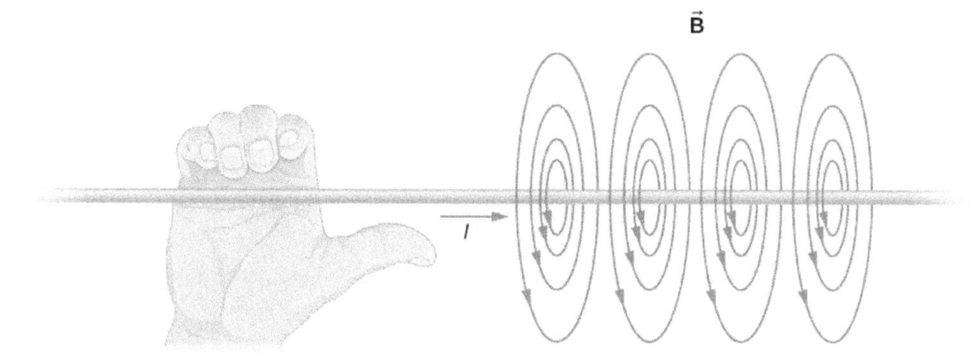

Figure 12.6 Some magnetic field lines of an infinite wire. The direction of $\vec{\mathbf{B}}$ can be found with a form of the right-hand rule.

The direction of the field lines can be observed experimentally by placing several small compass needles on a circle near the wire, as illustrated in Figure 12.7. When there is no current in the wire, the needles align with Earth's magnetic field. However, when a large current is sent through the wire, the compass needles all point tangent to the circle. Iron filings sprinkled on a horizontal surface also delineate the field lines, as shown in Figure 12.7.

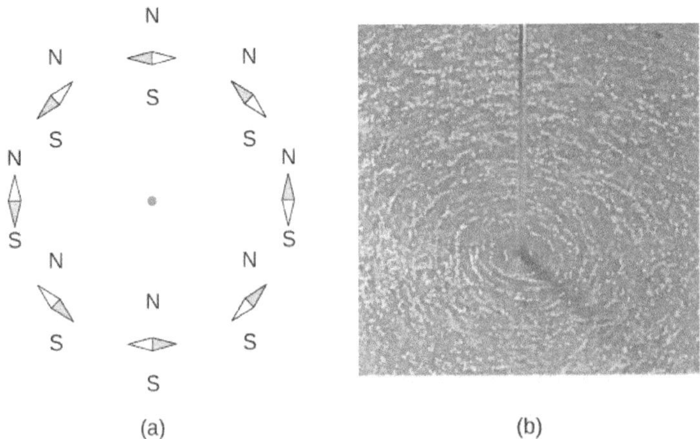

Figure 12.7 The shape of the magnetic field lines of a long wire can be seen using (a) small compass needles and (b) iron filings.

Example 12.3

Calculating Magnetic Field Due to Three Wires

Three wires sit at the corners of a square, all carrying currents of 2 amps into the page as shown in Figure 12.8. Calculate the magnitude of the magnetic field at the other corner of the square, point P, if the length of each side of the square is 1 cm.

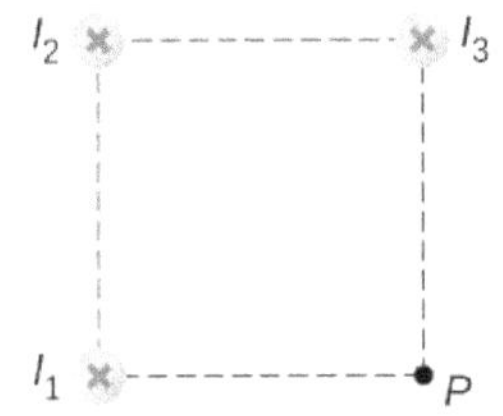

Figure 12.8 Three wires have current flowing into the page. The magnetic field is determined at the fourth corner of the square.

Strategy

The magnetic field due to each wire at the desired point is calculated. The diagonal distance is calculated using the Pythagorean theorem. Next, the direction of each magnetic field's contribution is determined by drawing a circle centered at the point of the wire and out toward the desired point. The direction of the magnetic field contribution from that wire is tangential to the curve. Lastly, working with these vectors, the resultant is calculated.

Solution

Wires 1 and 3 both have the same magnitude of magnetic field contribution at point P:

$$B_1 = B_3 = \frac{\mu_0 I}{2\pi R} = \frac{(4\pi \times 10^{-7}\,\text{T}\cdot\text{m/A})(2\,\text{A})}{2\pi(0.01\,\text{m})} = 4 \times 10^{-5}\,\text{T}.$$

Wire 2 has a longer distance and a magnetic field contribution at point P of:

$$B_2 = \frac{\mu_0 I}{2\pi R} = \frac{(4\pi \times 10^{-7}\,\text{T}\cdot\text{m/A})(2\,\text{A})}{2\pi(0.01414\,\text{m})} = 3 \times 10^{-5}\,\text{T}.$$

The vectors for each of these magnetic field contributions are shown.

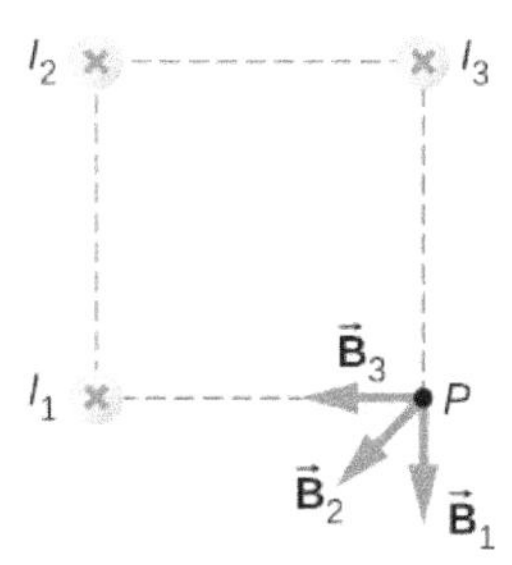

The magnetic field in the x-direction has contributions from wire 3 and the x-component of wire 2:

$$B_{\text{net } x} = -4 \times 10^{-5}\,\text{T} - 2.83 \times 10^{-5}\,\text{T}\cos(45°) = -6 \times 10^{-5}\,\text{T}.$$

The y-component is similarly the contributions from wire 1 and the y-component of wire 2:

$$B_{\text{net } y} = -4 \times 10^{-5}\,\text{T} - 2.83 \times 10^{-5}\,\text{T}\sin(45°) = -6 \times 10^{-5}\,\text{T}.$$

Therefore, the net magnetic field is the resultant of these two components:

$$B_{\text{net}} = \sqrt{B_{\text{net } x}^2 + B_{\text{net } y}^2}$$
$$B_{\text{net}} = \sqrt{(-6 \times 10^{-5}\,\text{T})^2 + (-6 \times 10^{-5}\,\text{T})^2}$$
$$B_{\text{net}} = 8.48 \times 10^{-5}\,\text{T}.$$

Significance

The geometry in this problem results in the magnetic field contributions in the x- and y-directions having the same magnitude. This is not necessarily the case if the currents were different values or if the wires were located in different positions. Regardless of the numerical results, working on the components of the vectors will yield the resulting magnetic field at the point in need.

12.3 Check Your Understanding Using Example 12.3, keeping the currents the same in wires 1 and 3, what should the current be in wire 2 to counteract the magnetic fields from wires 1 and 3 so that there is no net magnetic field at point P?

12.3 | Magnetic Force between Two Parallel Currents

Learning Objectives

By the end of this section, you will be able to:

- Explain how parallel wires carrying currents can attract or repel each other
- Define the ampere and describe how it is related to current-carrying wires
- Calculate the force of attraction or repulsion between two current-carrying wires

You might expect that two current-carrying wires generate significant forces between them, since ordinary currents produce magnetic fields and these fields exert significant forces on ordinary currents. But you might not expect that the force between wires is used to define the ampere. It might also surprise you to learn that this force has something to do with why large circuit breakers burn up when they attempt to interrupt large currents.

The force between two long, straight, and parallel conductors separated by a distance r can be found by applying what we have developed in the preceding sections. Figure 12.9 shows the wires, their currents, the field created by one wire, and the consequent force the other wire experiences from the created field. Let us consider the field produced by wire 1 and the force it exerts on wire 2 (call the force F_2). The field due to I_1 at a distance r is

$$B_1 = \frac{\mu_0 I_1}{2\pi r} \tag{12.9}$$

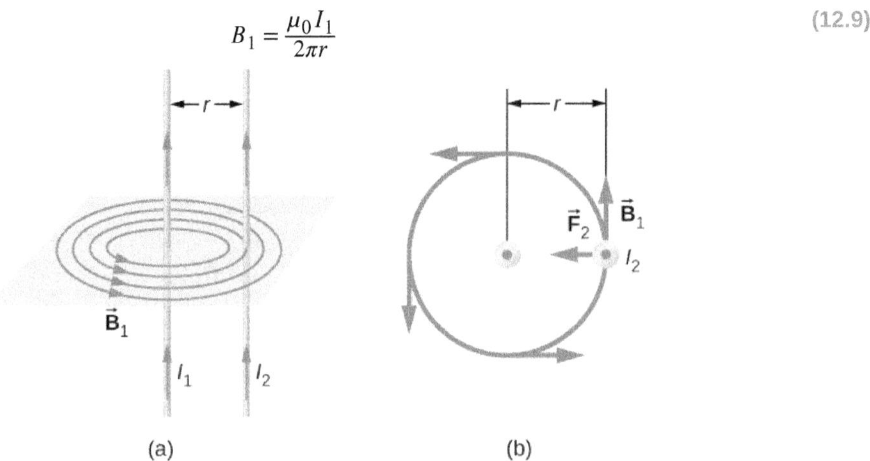

Figure 12.9 (a) The magnetic field produced by a long straight conductor is perpendicular to a parallel conductor, as indicated by right-hand rule (RHR)-2. (b) A view from above of the two wires shown in (a), with one magnetic field line shown for wire 1. RHR-1 shows that the force between the parallel conductors is attractive when the currents are in the same direction. A similar analysis shows that the force is repulsive between currents in opposite directions.

This field is uniform from the wire 1 and perpendicular to it, so the force F_2 it exerts on a length l of wire 2 is given by $F = IlB\sin\theta$ with $\sin\theta = 1$:

$$F_2 = I_2 l B_1. \tag{12.10}$$

The forces on the wires are equal in magnitude, so we just write F for the magnitude of F_2. (Note that $\vec{\mathbf{F}}_1 = -\vec{\mathbf{F}}_2$.) Since the wires are very long, it is convenient to think in terms of F/l, the force per unit length. Substituting the expression for B_1 into Equation 12.10 and rearranging terms gives

$$\frac{F}{l} = \frac{\mu_0 I_1 I_2}{2\pi r}. \tag{12.11}$$

The ratio F/l is the force per unit length between two parallel currents I_1 and I_2 separated by a distance r. The force is attractive if the currents are in the same direction and repulsive if they are in opposite directions.

This force is responsible for the *pinch effect* in electric arcs and other plasmas. The force exists whether the currents are in wires or not. It is only apparent if the overall charge density is zero; otherwise, the Coulomb repulsion overwhelms the magnetic attraction. In an electric arc, where charges are moving parallel to one another, an attractive force squeezes currents into a smaller tube. In large circuit breakers, such as those used in neighborhood power distribution systems, the pinch effect can concentrate an arc between plates of a switch trying to break a large current, burn holes, and even ignite the equipment. Another example of the pinch effect is found in the solar plasma, where jets of ionized material, such as solar flares, are shaped by magnetic forces.

The definition of the ampere is based on the force between current-carrying wires. Note that for long, parallel wires separated by 1 meter with each carrying 1 ampere, the force per meter is

$$\frac{F}{l} = \frac{\left(4\pi \times 10^{-7}\,\text{T} \cdot \text{m/A}\right)(1\text{ A})^2}{(2\pi)(1\text{ m})} = 2 \times 10^{-7}\text{ N/m}. \tag{12.12}$$

Since μ_0 is exactly $4\pi \times 10^{-7}\ \text{T} \cdot \text{m/A}$ by definition, and because $1\ \text{T} = 1\ \text{N/(A} \cdot \text{m)}$, the force per meter is exactly $2 \times 10^{-7}\ \text{N/m}$. This is the basis of the definition of the ampere.

Infinite-length wires are impractical, so in practice, a current balance is constructed with coils of wire separated by a few centimeters. Force is measured to determine current. This also provides us with a method for measuring the coulomb. We measure the charge that flows for a current of one ampere in one second. That is, $1\ \text{C} = 1\ \text{A} \cdot \text{s}$. For both the ampere and the coulomb, the method of measuring force between conductors is the most accurate in practice.

Example 12.4

Calculating Forces on Wires

Two wires, both carrying current out of the page, have a current of magnitude 5.0 mA. The first wire is located at (0.0 cm, 3.0 cm) while the other wire is located at (4.0 cm, 0.0 cm) as shown in Figure 12.10. What is the magnetic force per unit length of the first wire on the second and the second wire on the first?

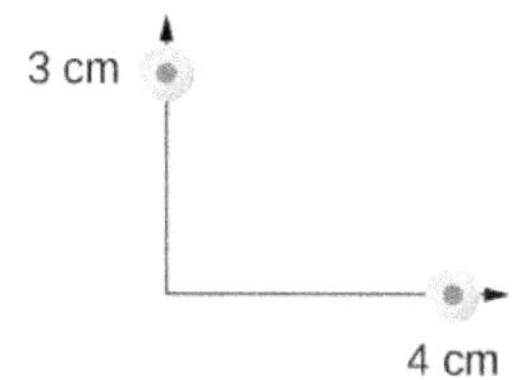

Figure 12.10 Two current-carrying wires at given locations with currents out of the page.

Strategy

Each wire produces a magnetic field felt by the other wire. The distance along the hypotenuse of the triangle between the wires is the radial distance used in the calculation to determine the force per unit length. Since both wires have currents flowing in the same direction, the direction of the force is toward each other.

Solution

The distance between the wires results from finding the hypotenuse of a triangle:

$$r = \sqrt{(3.0\ \text{cm})^2 + (4.0\ \text{cm})^2} = 5.0\ \text{cm}.$$

The force per unit length can then be calculated using the known currents in the wires:

$$\frac{F}{l} = \frac{\left(4\pi \times 10^{-7}\ \text{T} \cdot \text{m/A}\right)\left(5 \times 10^{-3}\ \text{A}\right)^2}{(2\pi)(5 \times 10^{-2}\ \text{m})} = 1 \times 10^{-10}\ \text{N/m}.$$

The force from the first wire pulls the second wire. The angle between the radius and the x-axis is

$$\theta = \tan^{-1}\left(\frac{3\ \text{cm}}{4\ \text{cm}}\right) = 36.9°.$$

The unit vector for this is calculated by

$$\cos(36.9°)\,\hat{\mathbf{i}} - \sin(36.9°)\,\hat{\mathbf{j}} = 0.8\,\hat{\mathbf{i}} - 0.6\,\hat{\mathbf{j}}.$$

Therefore, the force per unit length from wire one on wire 2 is

$$\frac{\vec{\mathbf{F}}}{l} = (1 \times 10^{-10}\ \text{N/m}) \times (0.8\,\hat{\mathbf{i}} - 0.6\,\hat{\mathbf{j}}) = (8 \times 10^{-11}\,\hat{\mathbf{i}} - 6 \times 10^{-11}\,\hat{\mathbf{j}})\ \text{N/m}.$$

The force per unit length from wire 2 on wire 1 is the negative of the previous answer:

$$\frac{\vec{\mathbf{F}}}{l} = (-8 \times 10^{-11}\,\hat{\mathbf{i}} + 6 \times 10^{-11}\,\hat{\mathbf{j}})\text{N/m}.$$

Significance

These wires produced magnetic fields of equal magnitude but opposite directions at each other's locations. Whether the fields are identical or not, the forces that the wires exert on each other are always equal in magnitude and opposite in direction (Newton's third law).

12.4 Check Your Understanding Two wires, both carrying current out of the page, have a current of magnitude 2.0 mA and 3.0 mA, respectively. The first wire is located at (0.0 cm, 5.0 cm) while the other wire is located at (12.0 cm, 0.0 cm). What is the magnitude of the magnetic force per unit length of the first wire on the second and the second wire on the first?

12.4 | Magnetic Field of a Current Loop

Learning Objectives

By the end of this section, you will be able to:

- Explain how the Biot-Savart law is used to determine the magnetic field due to a current in a loop of wire at a point along a line perpendicular to thep lane of the loop.
- Determine the magnetic field of an arc of current.

The circular loop of Figure 12.11 has a radius R, carries a current I, and lies in the xz-plane. What is the magnetic field due to the current at an arbitrary point P along the axis of the loop?

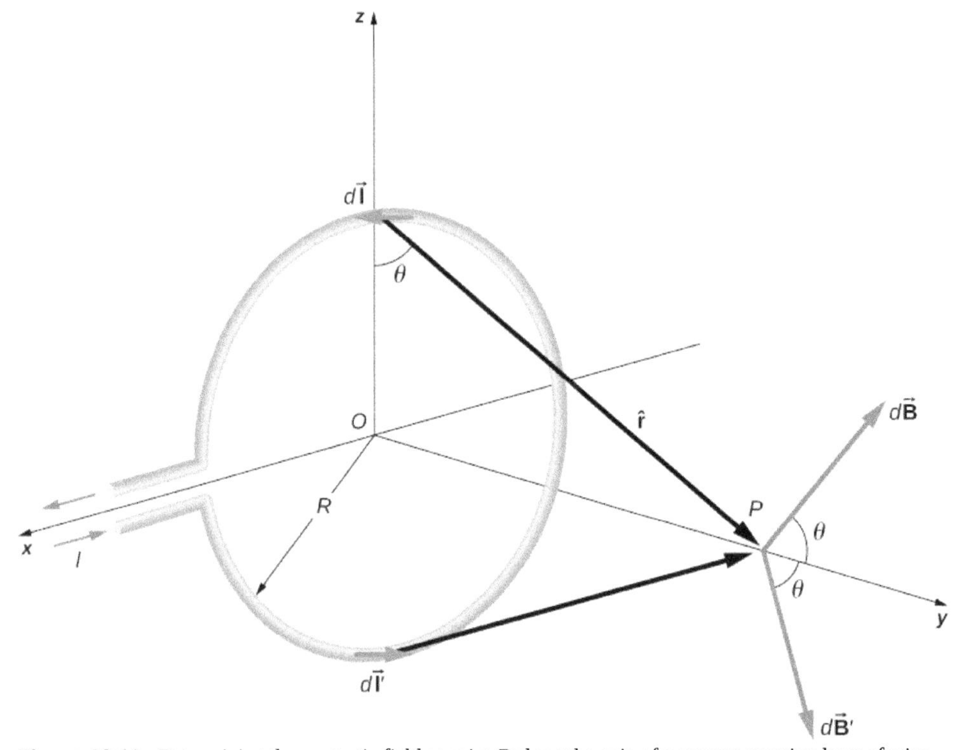

Figure 12.11 Determining the magnetic field at point P along the axis of a current-carrying loop of wire.

We can use the Biot-Savart law to find the magnetic field due to a current. We first consider arbitrary segments on opposite sides of the loop to qualitatively show by the vector results that the net magnetic field direction is along the central axis from the loop. From there, we can use the Biot-Savart law to derive the expression for magnetic field.

Let *P* be a distance *y* from the center of the loop. From the right-hand rule, the magnetic field $d\vec{\mathbf{B}}$ at *P*, produced by the current element $I\,d\vec{\mathbf{l}}$, is directed at an angle θ above the *y*-axis as shown. Since $d\vec{\mathbf{l}}$ is parallel along the *x*-axis and $\hat{\mathbf{r}}$ is in the *yz*-plane, the two vectors are perpendicular, so we have

$$dB = \frac{\mu_0}{4\pi}\frac{I\,dl\sin\theta}{r^2} = \frac{\mu_0}{4\pi}\frac{I\,dl}{y^2+R^2} \quad (12.13)$$

where we have used $r^2 = y^2 + R^2$.

Now consider the magnetic field $d\vec{\mathbf{B}}'$ due to the current element $I\,d\vec{\mathbf{l}}'$, which is directly opposite $I\,d\vec{\mathbf{l}}$ on the loop. The magnitude of $d\vec{\mathbf{B}}'$ is also given by Equation 12.13, but it is directed at an angle θ below the *y*-axis. The components of $d\vec{\mathbf{B}}$ and $d\vec{\mathbf{B}}'$ perpendicular to the *y*-axis therefore cancel, and in calculating the net magnetic field, only the components along the *y*-axis need to be considered. The components perpendicular to the axis of the loop sum to zero in pairs. Hence at point *P*:

$$\vec{\mathbf{B}} = \hat{\mathbf{j}}\int_{\text{loop}} dB\cos\theta = \hat{\mathbf{j}}\frac{\mu_0 I}{4\pi}\int_{\text{loop}}\frac{\cos\theta\,dl}{y^2+R^2}. \quad (12.14)$$

For all elements $d\vec{\mathbf{l}}$ on the wire, *y*, *R*, and $\cos\theta$ are constant and are related by

$$\cos\theta = \frac{R}{\sqrt{y^2+R^2}}.$$

Now from Equation 12.14, the magnetic field at *P* is

$$\vec{\mathbf{B}} = \hat{\mathbf{j}}\frac{\mu_0 IR}{4\pi(y^2+R^2)^{3/2}}\int_{\text{loop}} dl = \frac{\mu_0 IR^2}{2(y^2+R^2)^{3/2}}\hat{\mathbf{j}} \quad (12.15)$$

where we have used $\int_{\text{loop}} dl = 2\pi R$. As discussed in the previous chapter, the closed current loop is a magnetic dipole of moment $\vec{\mu} = IA\hat{\mathbf{n}}$. For this example, $A = \pi R^2$ and $\hat{\mathbf{n}} = \hat{\mathbf{j}}$, so the magnetic field at *P* can also be written as

$$\vec{\mathbf{B}} = \frac{\mu_0 \mu \hat{\mathbf{j}}}{2\pi(y^2+R^2)^{3/2}}. \quad (12.16)$$

By setting $y = 0$ in Equation 12.16, we obtain the magnetic field at the center of the loop:

$$\vec{\mathbf{B}} = \frac{\mu_0 I}{2R}\hat{\mathbf{j}}. \quad (12.17)$$

This equation becomes $B = \mu_0 nI/(2R)$ for a flat coil of *n* loops per length. It can also be expressed as

$$\vec{\mathbf{B}} = \frac{\mu_0 \vec{\mu}}{2\pi R^3}. \quad (12.18)$$

If we consider $y \gg R$ in Equation 12.16, the expression reduces to an expression known as the magnetic field from a dipole:

$$\vec{\mathbf{B}} = \frac{\mu_0 \vec{\mu}}{2\pi y^3}. \tag{12.19}$$

The calculation of the magnetic field due to the circular current loop at points off-axis requires rather complex mathematics, so we'll just look at the results. The magnetic field lines are shaped as shown in Figure 12.12. Notice that one field line follows the axis of the loop. This is the field line we just found. Also, very close to the wire, the field lines are almost circular, like the lines of a long straight wire.

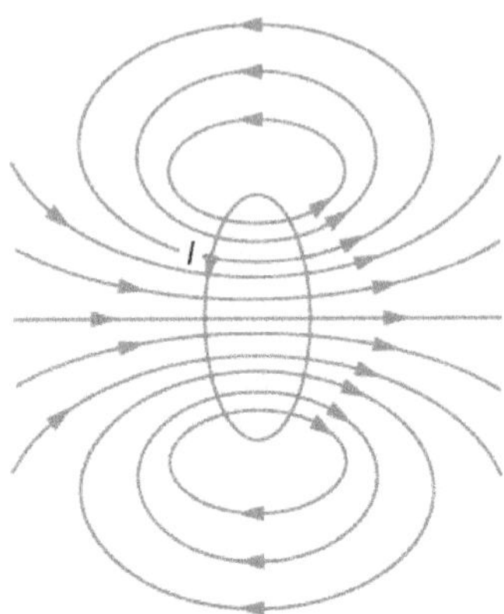

Figure 12.12 Sketch of the magnetic field lines of a circular current loop.

Example 12.5

Magnetic Field between Two Loops

Two loops of wire carry the same current of 10 mA, but flow in opposite directions as seen in Figure 12.13. One loop is measured to have a radius of $R = 50\,\text{cm}$ while the other loop has a radius of $2R = 100\,\text{cm}$. The distance from the first loop to the point where the magnetic field is measured is 0.25 m, and the distance from that point to the second loop is 0.75 m. What is the magnitude of the net magnetic field at point P?

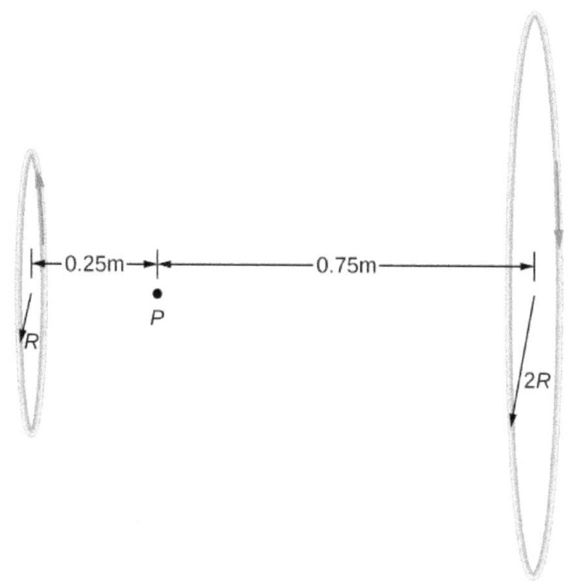

Figure 12.13 Two loops of different radii have the same current but flowing in opposite directions. The magnetic field at point P is measured to be zero.

Strategy

The magnetic field at point P has been determined in Equation 12.15. Since the currents are flowing in opposite directions, the net magnetic field is the difference between the two fields generated by the coils. Using the given quantities in the problem, the net magnetic field is then calculated.

Solution

Solving for the net magnetic field using Equation 12.15 and the given quantities in the problem yields

$$\begin{aligned} B &= \frac{\mu_0 I R_1{}^2}{2\left(y_1{}^2 + R_1{}^2\right)^{3/2}} - \frac{\mu_0 I R_2{}^2}{2\left(y_2{}^2 + R_2{}^2\right)^{3/2}} \\ B &= \frac{(4\pi \times 10^{-7}\,\text{T}\cdot\text{m/A})(0.010\,\text{A})(0.5\,\text{m})^2}{2((0.25\,\text{m})^2 + (0.5\,\text{m})^2)^{3/2}} - \frac{(4\pi \times 10^{-7}\,\text{T}\cdot\text{m/A})(0.010\,\text{A})(1.0\,\text{m})^2}{2((0.75\,\text{m})^2 + (1.0\,\text{m})^2)^{3/2}} \\ B &= 5.77 \times 10^{-9}\,\text{T to the right.} \end{aligned}$$

Significance

Helmholtz coils typically have loops with equal radii with current flowing in the same direction to have a strong uniform field at the midpoint between the loops. A similar application of the magnetic field distribution created by Helmholtz coils is found in a magnetic bottle that can temporarily trap charged particles. See Magnetic Forces and Fields for a discussion on this.

12.5 Check Your Understanding Using Example 12.5, at what distance would you have to move the first coil to have zero measurable magnetic field at point P?

12.5 | Ampère's Law

Learning Objectives

By the end of this section, you will be able to:

- Explain how Ampère's law relates the magnetic field produced by a current to the value of the current
- Calculate the magnetic field from a long straight wire, either thin or thick, by Ampère's law

A fundamental property of a static magnetic field is that, unlike an electrostatic field, it is not conservative. A conservative field is one that does the same amount of work on a particle moving between two different points regardless of the path chosen. Magnetic fields do not have such a property. Instead, there is a relationship between the magnetic field and its source, electric current. It is expressed in terms of the line integral of $\vec{\mathbf{B}}$ and is known as **Ampère's law**. This law can also be derived directly from the Biot-Savart law. We now consider that derivation for the special case of an infinite, straight wire.

Figure 12.14 shows an arbitrary plane perpendicular to an infinite, straight wire whose current I is directed out of the page. The magnetic field lines are circles directed counterclockwise and centered on the wire. To begin, let's consider $\oint \vec{\mathbf{B}} \cdot d\vec{\mathbf{l}}$ over the closed paths M and N. Notice that one path (M) encloses the wire, whereas the other (N) does not. Since the field lines are circular, $\vec{\mathbf{B}} \cdot d\vec{\mathbf{l}}$ is the product of B and the projection of dl onto the circle passing through $d\vec{\mathbf{l}}$. If the radius of this particular circle is r, the projection is $rd\theta$, and

$$\vec{\mathbf{B}} \cdot d\vec{\mathbf{l}} = Br\,d\theta.$$

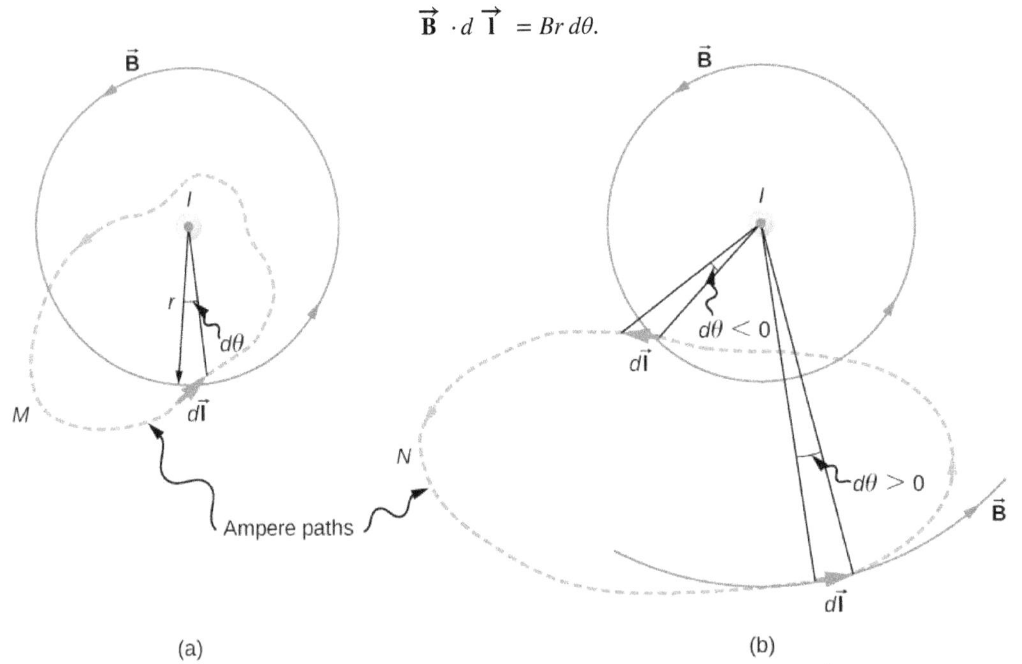

Figure 12.14 The current I of a long, straight wire is directed out of the page. The integral $\oint d\theta$ equals 2π and 0, respectively, for paths M and N.

With $\vec{\mathbf{B}}$ given by Equation 12.9,

$$\oint \vec{\mathbf{B}} \cdot d\vec{\mathbf{l}} = \oint \left(\frac{\mu_0 I}{2\pi r}\right) r\, d\theta = \frac{\mu_0 I}{2\pi} \oint d\theta. \tag{12.20}$$

For path *M*, which circulates around the wire, $\oint_M d\theta = 2\pi$ and

$$\oint_M \vec{\mathbf{B}} \cdot d\vec{\mathbf{l}} = \mu_0 I. \tag{12.21}$$

Path *N*, on the other hand, circulates through both positive (counterclockwise) and negative (clockwise) $d\theta$ (see Figure 12.14), and since it is closed, $\oint_N d\theta = 0$. Thus for path *N*,

$$\oint_N \vec{\mathbf{B}} \cdot d\vec{\mathbf{l}} = 0. \tag{12.22}$$

The extension of this result to the general case is Ampère's law.

Ampère's law

Over an arbitrary closed path,

$$\oint \vec{\mathbf{B}} \cdot d\vec{\mathbf{l}} = \mu_0 I \tag{12.23}$$

where *I* is the total current passing through any open surface *S* whose perimeter is the path of integration. Only currents inside the path of integration need be considered.

To determine whether a specific current *I* is positive or negative, curl the fingers of your right hand in the direction of the path of integration, as shown in Figure 12.14. If *I* passes through *S* in the same direction as your extended thumb, *I* is positive; if *I* passes through *S* in the direction opposite to your extended thumb, it is negative.

Problem-Solving Strategy: Ampère's Law

To calculate the magnetic field created from current in wire(s), use the following steps:

1. Identify the symmetry of the current in the wire(s). If there is no symmetry, use the Biot-Savart law to determine the magnetic field.
2. Determine the direction of the magnetic field created by the wire(s) by right-hand rule 2.
3. Chose a path loop where the magnetic field is either constant or zero.
4. Calculate the current inside the loop.
5. Calculate the line integral $\oint \vec{\mathbf{B}} \cdot d\vec{\mathbf{l}}$ around the closed loop.
6. Equate $\oint \vec{\mathbf{B}} \cdot d\vec{\mathbf{l}}$ with $\mu_0 I_{\text{enc}}$ and solve for $\vec{\mathbf{B}}$.

Example 12.6

Using Ampère's Law to Calculate the Magnetic Field Due to a Wire

Use Ampère's law to calculate the magnetic field due to a steady current *I* in an infinitely long, thin, straight wire as shown in Figure 12.15.

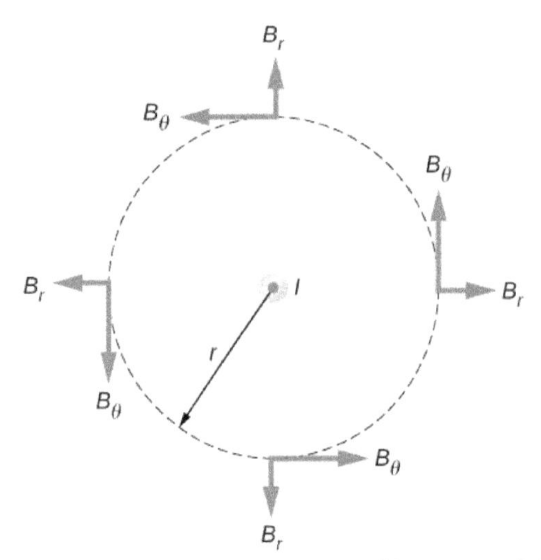

Figure 12.15 The possible components of the magnetic field B due to a current I, which is directed out of the page. The radial component is zero because the angle between the magnetic field and the path is at a right angle.

Strategy

Consider an arbitrary plane perpendicular to the wire, with the current directed out of the page. The possible magnetic field components in this plane, B_r and B_θ, are shown at arbitrary points on a circle of radius r centered on the wire. Since the field is cylindrically symmetric, neither B_r nor B_θ varies with the position on this circle. Also from symmetry, the radial lines, if they exist, must be directed either all inward or all outward from the wire. This means, however, that there must be a net magnetic flux across an arbitrary cylinder concentric with the wire. The radial component of the magnetic field must be zero because $\vec{\mathbf{B}}_r \cdot d\vec{\mathbf{l}} = 0$. Therefore, we can apply Ampère's law to the circular path as shown.

Solution

Over this path $\vec{\mathbf{B}}$ is constant and parallel to $d\vec{\mathbf{l}}$, so

$$\oint \vec{\mathbf{B}} \cdot d\vec{\mathbf{l}} = B_\theta \oint dl = B_\theta(2\pi r).$$

Thus Ampère's law reduces to

$$B_\theta(2\pi r) = \mu_0 I.$$

Finally, since B_θ is the only component of $\vec{\mathbf{B}}$, we can drop the subscript and write

$$B = \frac{\mu_0 I}{2\pi r}.$$

This agrees with the Biot-Savart calculation above.

Significance

Ampère's law works well if you have a path to integrate over which $\vec{\mathbf{B}} \cdot d\vec{\mathbf{l}}$ has results that are easy to simplify. For the infinite wire, this works easily with a path that is circular around the wire so that the magnetic field factors out of the integration. If the path dependence looks complicated, you can always go back to the Biot-Savart law and use that to find the magnetic field.

Example 12.7

Calculating the Magnetic Field of a Thick Wire with Ampère's Law

The radius of the long, straight wire of Figure 12.16 is a, and the wire carries a current I_0 that is distributed uniformly over its cross-section. Find the magnetic field both inside and outside the wire.

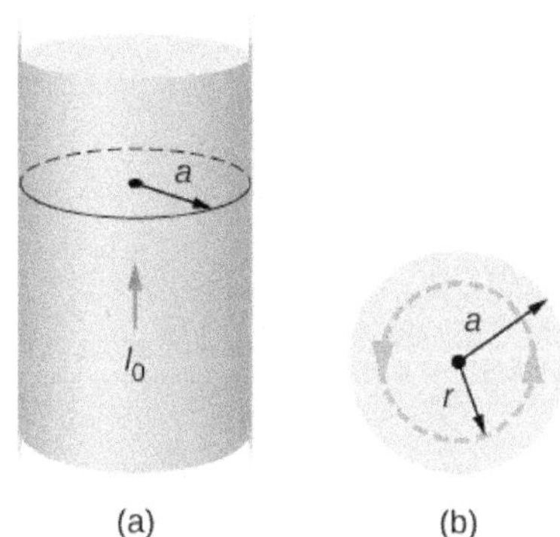

Figure 12.16 (a) A model of a current-carrying wire of radius a and current I_0. (b) A cross-section of the same wire showing the radius a and the Ampère's loop of radius r.

Strategy

This problem has the same geometry as Example 12.6, but the enclosed current changes as we move the integration path from outside the wire to inside the wire, where it doesn't capture the entire current enclosed (see Figure 12.16).

Solution

For any circular path of radius r that is centered on the wire,

$$\oint \vec{\mathbf{B}} \cdot d\vec{\mathbf{l}} = \oint B dl = B \oint dl = B(2\pi r).$$

From Ampère's law, this equals the total current passing through any surface bounded by the path of integration.

Consider first a circular path that is inside the wire $(r \leq a)$ such as that shown in part (a) of Figure 12.16. We need the current I passing through the area enclosed by the path. It's equal to the current density J times the area enclosed. Since the current is uniform, the current density inside the path equals the current density in the whole wire, which is $I_0 / \pi a^2$. Therefore the current I passing through the area enclosed by the path is

$$I = \frac{\pi r^2}{\pi a^2} I_0 = \frac{r^2}{a^2} I_0.$$

We can consider this ratio because the current density J is constant over the area of the wire. Therefore, the current density of a part of the wire is equal to the current density in the whole area. Using Ampère's law, we obtain

$$B(2\pi r) = \mu_0 \left(\frac{r^2}{a^2} \right) I_0,$$

and the magnetic field inside the wire is

$$B = \frac{\mu_0 I_0}{2\pi} \frac{r}{a^2} \, (r \leq a).$$

Outside the wire, the situation is identical to that of the infinite thin wire of the previous example; that is,

$$B = \frac{\mu_0 I_0}{2\pi r} \ (r \geq a).$$

The variation of B with r is shown in Figure 12.17.

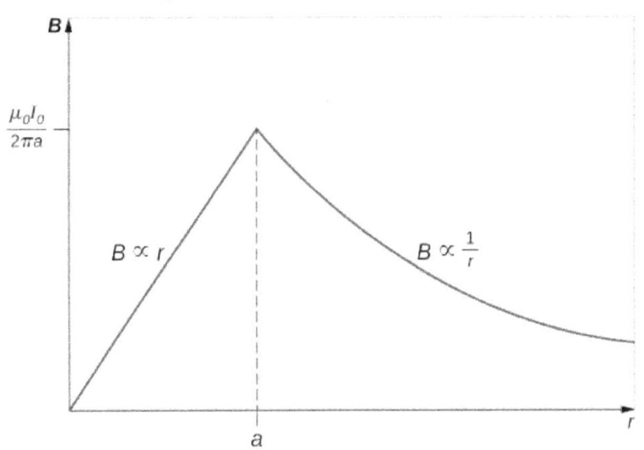

Figure 12.17 Variation of the magnetic field produced by a current I_0 in a long, straight wire of radius a.

Significance

The results show that as the radial distance increases inside the thick wire, the magnetic field increases from zero to a familiar value of the magnetic field of a thin wire. Outside the wire, the field drops off regardless of whether it was a thick or thin wire.

This result is similar to how Gauss's law for electrical charges behaves inside a uniform charge distribution, except that Gauss's law for electrical charges has a uniform volume distribution of charge, whereas Ampère's law here has a uniform area of current distribution. Also, the drop-off outside the thick wire is similar to how an electric field drops off outside of a linear charge distribution, since the two cases have the same geometry and neither case depends on the configuration of charges or currents once the loop is outside the distribution.

Example 12.8

Using Ampère's Law with Arbitrary Paths

Use Ampère's law to evaluate $\oint \vec{\mathbf{B}} \cdot d\vec{\mathbf{l}}$ for the current configurations and paths in Figure 12.18.

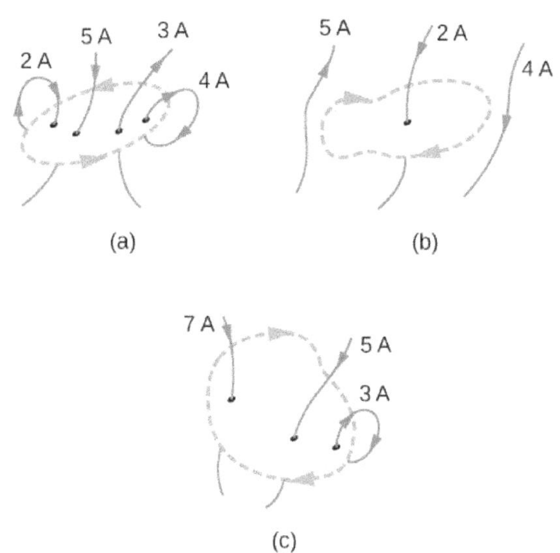

Figure 12.18 Current configurations and paths for Example 12.8.

Strategy

Ampère's law states that $\oint \vec{\mathbf{B}} \cdot d\vec{\mathbf{l}} = \mu_0 I$ where I is the total current passing through the enclosed loop. The quickest way to evaluate the integral is to calculate $\mu_0 I$ by finding the net current through the loop. Positive currents flow with your right-hand thumb if your fingers wrap around in the direction of the loop. This will tell us the sign of the answer.

Solution

(a) The current going downward through the loop equals the current going out of the loop, so the net current is zero. Thus, $\oint \vec{\mathbf{B}} \cdot d\vec{\mathbf{l}} = 0$.

(b) The only current to consider in this problem is 2A because it is the only current inside the loop. The right-hand rule shows us the current going downward through the loop is in the positive direction. Therefore, the answer is $\oint \vec{\mathbf{B}} \cdot d\vec{\mathbf{l}} = \mu_0(2\text{ A}) = 2.51 \times 10^{-6}\,\text{T} \cdot \text{m/A}$.

(c) The right-hand rule shows us the current going downward through the loop is in the positive direction. There are $7\text{A} + 5\text{A} = 12\text{A}$ of current going downward and –3 A going upward. Therefore, the total current is 9 A and $\oint \vec{\mathbf{B}} \cdot d\vec{\mathbf{l}} = \mu_0(9\text{ A}) = 5.65 \times 10^{-6}\,\text{T} \cdot \text{m/A}$.

Significance

If the currents all wrapped around so that the same current went into the loop and out of the loop, the net current would be zero and no magnetic field would be present. This is why wires are very close to each other in an electrical cord. The currents flowing toward a device and away from a device in a wire equal zero total current flow through an Ampère loop around these wires. Therefore, no stray magnetic fields can be present from cords carrying current.

12.6 Check Your Understanding Consider using Ampère's law to calculate the magnetic fields of a finite straight wire and of a circular loop of wire. Why is it not useful for these calculations?

12.6 | Solenoids and Toroids

Learning Objectives

By the end of this section, you will be able to:

- Establish a relationship for how the magnetic field of a solenoid varies with distance and current by using both the Biot-Savart law and Ampère's law
- Establish a relationship for how the magnetic field of a toroid varies with distance and current by using Ampère's law

Two of the most common and useful electromagnetic devices are called solenoids and toroids. In one form or another, they are part of numerous instruments, both large and small. In this section, we examine the magnetic field typical of these devices.

Solenoids

A long wire wound in the form of a helical coil is known as a **solenoid**. Solenoids are commonly used in experimental research requiring magnetic fields. A solenoid is generally easy to wind, and near its center, its magnetic field is quite uniform and directly proportional to the current in the wire.

Figure 12.19 shows a solenoid consisting of N turns of wire tightly wound over a length L. A current I is flowing along the wire of the solenoid. The number of turns per unit length is N/L; therefore, the number of turns in an infinitesimal length dy are $(N/L)dy$ turns. This produces a current

$$dI = \frac{NI}{L}dy. \tag{12.24}$$

We first calculate the magnetic field at the point P of Figure 12.19. This point is on the central axis of the solenoid. We are basically cutting the solenoid into thin slices that are dy thick and treating each as a current loop. Thus, dI is the current through each slice. The magnetic field $d\vec{\mathbf{B}}$ due to the current dI in dy can be found with the help of Equation 12.15 and Equation 12.24:

$$d\vec{\mathbf{B}} = \frac{\mu_0 R^2 dI}{2(y^2 + R^2)^{3/2}}\hat{\mathbf{j}} = \left(\frac{\mu_0 I R^2 N}{2L}\hat{\mathbf{j}}\right)\frac{dy}{(y^2 + R^2)^{3/2}} \tag{12.25}$$

where we used Equation 12.24 to replace dI. The resultant field at P is found by integrating $d\vec{\mathbf{B}}$ along the entire length of the solenoid. It's easiest to evaluate this integral by changing the independent variable from y to θ. From inspection of Figure 12.19, we have:

$$\sin\theta = \frac{y}{\sqrt{y^2 + R^2}}. \tag{12.26}$$

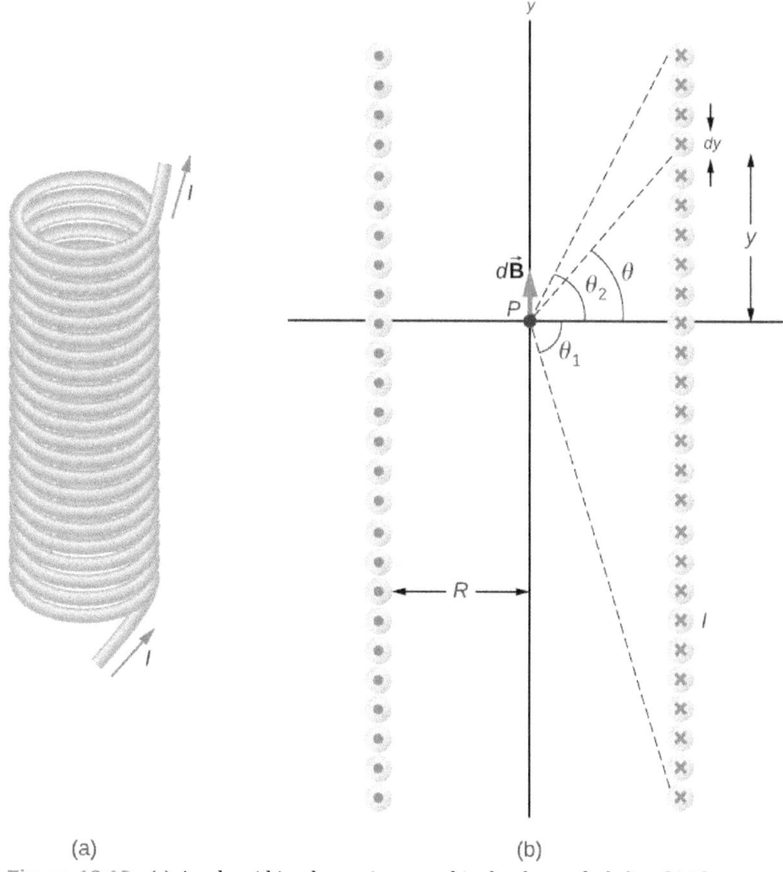

Figure 12.19 (a) A solenoid is a long wire wound in the shape of a helix. (b) The magnetic field at the point *P* on the axis of the solenoid is the net field due to all of the current loops.

Taking the differential of both sides of this equation, we obtain

$$\begin{aligned}\cos\theta\, d\theta &= \left[-\frac{y^2}{(y^2+R^2)^{3/2}}+\frac{1}{\sqrt{y^2+R^2}}\right]dy \\ &= \frac{R^2\, dy}{(y^2+R^2)^{3/2}}.\end{aligned}$$

When this is substituted into the equation for $d\vec{\mathbf{B}}$, we have

$$\vec{\mathbf{B}} = \frac{\mu I_0 N}{2L}\hat{\mathbf{j}}\int_{\theta_1}^{\theta_2}\cos\theta\, d\theta = \frac{\mu I_0 N}{2L}(\sin\theta_2 - \sin\theta_1)\hat{\mathbf{j}}, \tag{12.27}$$

which is the magnetic field along the central axis of a finite solenoid.

Of special interest is the infinitely long solenoid, for which $L \to \infty$. From a practical point of view, the infinite solenoid is one whose length is much larger than its radius ($L \gg R$). In this case, $\theta_1 = \frac{-\pi}{2}$ and $\theta_2 = \frac{\pi}{2}$. Then from Equation 12.27, the magnetic field along the central axis of an infinite solenoid is

$$\vec{\mathbf{B}} = \frac{\mu_0 IN}{2L}\hat{\mathbf{j}}[\sin(\pi/2) - \sin(-\pi/2)] = \frac{\mu_0 IN}{L}\hat{\mathbf{j}}$$

or

$$\vec{\mathbf{B}} = \mu_0 nI \hat{\mathbf{j}}, \tag{12.28}$$

where n is the number of turns per unit length. You can find the direction of $\vec{\mathbf{B}}$ with a right-hand rule: Curl your fingers in the direction of the current, and your thumb points along the magnetic field in the interior of the solenoid.

We now use these properties, along with Ampère's law, to calculate the magnitude of the magnetic field at any location inside the infinite solenoid. Consider the closed path of Figure 12.20. Along segment 1, $\vec{\mathbf{B}}$ is uniform and parallel to the path. Along segments 2 and 4, $\vec{\mathbf{B}}$ is perpendicular to part of the path and vanishes over the rest of it. Therefore, segments 2 and 4 do not contribute to the line integral in Ampère's law. Along segment 3, $\vec{\mathbf{B}} = 0$ because the magnetic field is zero outside the solenoid. If you consider an Ampère's law loop outside of the solenoid, the current flows in opposite directions on different segments of wire. Therefore, there is no enclosed current and no magnetic field according to Ampère's law. Thus, there is no contribution to the line integral from segment 3. As a result, we find

$$\oint \vec{\mathbf{B}} \cdot d\vec{\mathbf{l}} = \int_1 \vec{\mathbf{B}} \cdot d\vec{\mathbf{l}} = Bl. \tag{12.29}$$

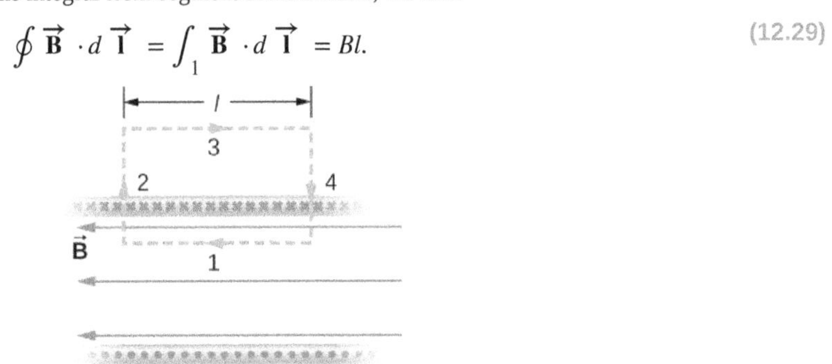

Figure 12.20 The path of integration used in Ampère's law to evaluate the magnetic field of an infinite solenoid.

The solenoid has n turns per unit length, so the current that passes through the surface enclosed by the path is nlI. Therefore, from Ampère's law,

$$Bl = \mu_0 nlI$$

and

$$B = \mu_0 nI \tag{12.30}$$

within the solenoid. This agrees with what we found earlier for B on the central axis of the solenoid. Here, however, the location of segment 1 is arbitrary, so we have found that this equation gives the magnetic field everywhere inside the infinite solenoid.

Outside the solenoid, one can draw an Ampère's law loop around the entire solenoid. This would enclose current flowing in both directions. Therefore, the net current inside the loop is zero. According to Ampère's law, if the net current is zero, the magnetic field must be zero. Therefore, for locations outside of the solenoid's radius, the magnetic field is zero.

When a patient undergoes a magnetic resonance imaging (MRI) scan, the person lies down on a table that is moved into the center of a large solenoid that can generate very large magnetic fields. The solenoid is capable of these high fields from high currents flowing through superconducting wires. The large magnetic field is used to change the spin of protons in the patient's body. The time it takes for the spins to align or relax (return to original orientation) is a signature of different tissues that can be analyzed to see if the structures of the tissues is normal (Figure 12.21).

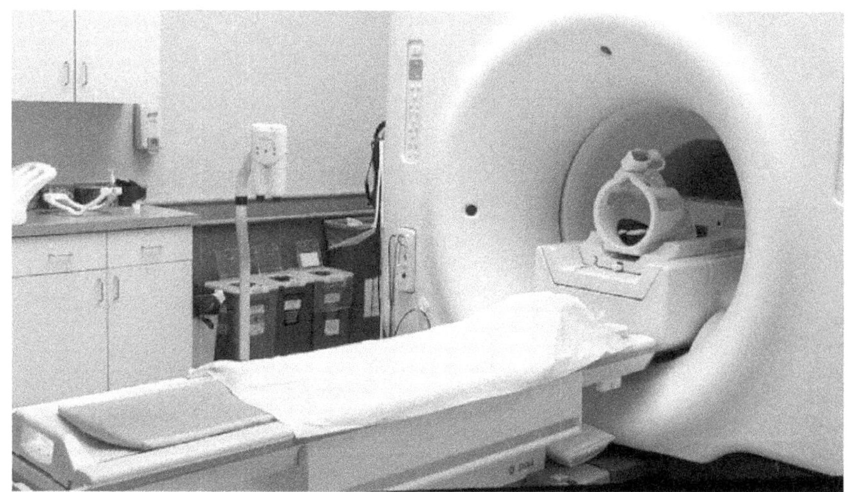

Figure 12.21 In an MRI machine, a large magnetic field is generated by the cylindrical solenoid surrounding the patient. (credit: Liz West)

Example 12.9

Magnetic Field Inside a Solenoid

A solenoid has 300 turns wound around a cylinder of diameter 1.20 cm and length 14.0 cm. If the current through the coils is 0.410 A, what is the magnitude of the magnetic field inside and near the middle of the solenoid?

Strategy

We are given the number of turns and the length of the solenoid so we can find the number of turns per unit length. Therefore, the magnetic field inside and near the middle of the solenoid is given by Equation 12.30. Outside the solenoid, the magnetic field is zero.

Solution

The number of turns per unit length is

$$n = \frac{300\text{ turns}}{0.140\text{ m}} = 2.14 \times 10^3\text{ turns/m}.$$

The magnetic field produced inside the solenoid is

$$\begin{aligned} B &= \mu_0 nI = (4\pi \times 10^{-7}\text{ T} \cdot \text{m/A})(2.14 \times 10^3\text{ turns/m})(0.410\text{ A}) \\ B &= 1.10 \times 10^{-3}\text{ T}. \end{aligned}$$

Significance

This solution is valid only if the length of the solenoid is reasonably large compared with its diameter. This example is a case where this is valid.

12.7 Check Your Understanding What is the ratio of the magnetic field produced from using a finite formula over the infinite approximation for an angle θ of (a) 85°? (b) 89°? The solenoid has 1000 turns in 50 cm with a current of 1.0 A flowing through the coils

Toroids

A toroid is a donut-shaped coil closely wound with one continuous wire, as illustrated in part (a) of Figure 12.22. If the toroid has N windings and the current in the wire is I, what is the magnetic field both inside and outside the toroid?

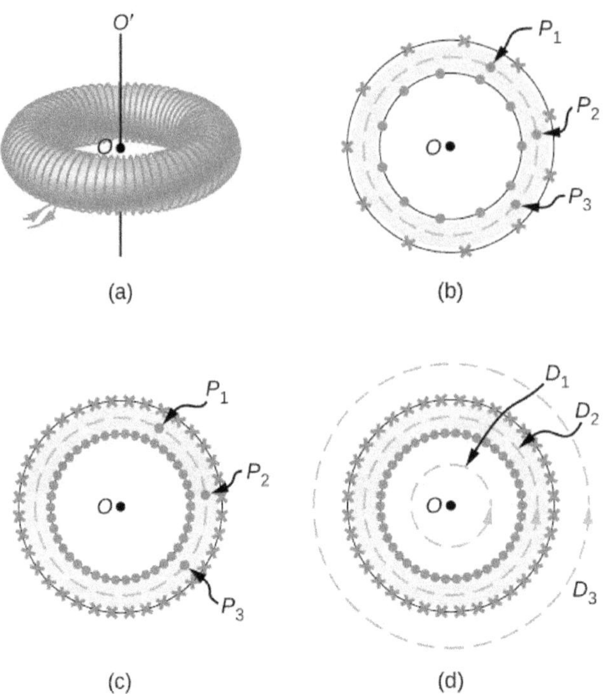

Figure 12.22 (a) A toroid is a coil wound into a donut-shaped object. (b) A loosely wound toroid does not have cylindrical symmetry. (c) In a tightly wound toroid, cylindrical symmetry is a very good approximation. (d) Several paths of integration for Ampère's law.

We begin by assuming cylindrical symmetry around the axis *OO'*. Actually, this assumption is not precisely correct, for as part (b) of Figure 12.22 shows, the view of the toroidal coil varies from point to point (for example, P_1, P_2, and P_3) on a circular path centered around *OO'*. However, if the toroid is tightly wound, all points on the circle become essentially equivalent [part (c) of Figure 12.22], and cylindrical symmetry is an accurate approximation.

With this symmetry, the magnetic field must be tangent to and constant in magnitude along any circular path centered on *OO'*. This allows us to write for each of the paths D_1, D_2, and D_3 shown in part (d) of Figure 12.22,

$$\oint \vec{\mathbf{B}} \cdot d\vec{\mathbf{l}} = B(2\pi r). \tag{12.31}$$

Ampère's law relates this integral to the net current passing through any surface bounded by the path of integration. For a path that is external to the toroid, either no current passes through the enclosing surface (path D_1), or the current passing through the surface in one direction is exactly balanced by the current passing through it in the opposite direction (path D_3). In either case, there is no net current passing through the surface, so

$$\oint B(2\pi r) = 0$$

and

$$B = 0 \quad \text{(outside the toroid)}. \tag{12.32}$$

The turns of a toroid form a helix, rather than circular loops. As a result, there is a small field external to the coil; however, the derivation above holds if the coils were circular.

For a circular path within the toroid (path D_2), the current in the wire cuts the surface *N* times, resulting in a net current *NI* through the surface. We now find with Ampère's law,

$$B(2\pi r) = \mu_0 NI$$

and

$$B = \frac{\mu_0 NI}{2\pi r} \quad \text{(within the toroid).} \tag{12.33}$$

The magnetic field is directed in the counterclockwise direction for the windings shown. When the current in the coils is reversed, the direction of the magnetic field also reverses.

The magnetic field inside a toroid is not uniform, as it varies inversely with the distance r from the axis OO'. However, if the central radius R (the radius midway between the inner and outer radii of the toroid) is much larger than the cross-sectional diameter of the coils r, the variation is fairly small, and the magnitude of the magnetic field may be calculated by Equation 12.33 where $r = R$.

12.7 | Magnetism in Matter

Learning Objectives

By the end of this section, you will be able to:

- Classify magnetic materials as paramagnetic, diamagnetic, or ferromagnetic, based on their response to a magnetic field
- Sketch how magnetic dipoles align with the magnetic field in each type of substance
- Define hysteresis and magnetic susceptibility, which determines the type of magnetic material

Why are certain materials magnetic and others not? And why do certain substances become magnetized by a field, whereas others are unaffected? To answer such questions, we need an understanding of magnetism on a microscopic level.

Within an atom, every electron travels in an orbit and spins on an internal axis. Both types of motion produce current loops and therefore magnetic dipoles. For a particular atom, the net magnetic dipole moment is the vector sum of the magnetic dipole moments. Values of μ for several types of atoms are given in Table 12.1. Notice that some atoms have a zero net dipole moment and that the magnitudes of the nonvanishing moments are typically $10^{-23}\,\text{A} \cdot \text{m}^2$.

Atom	Magnetic Moment $\left(10^{-24}\,\text{A} \cdot \text{m}^2\right)$
H	9.27
He	0
Li	9.27
O	13.9
Na	9.27
S	13.9

Table 12.1 Magnetic Moments of Some Atoms

A handful of matter has approximately 10^{26} atoms and ions, each with its magnetic dipole moment. If no external magnetic field is present, the magnetic dipoles are randomly oriented—as many are pointed up as down, as many are pointed east as west, and so on. Consequently, the net magnetic dipole moment of the sample is zero. However, if the sample is placed in a magnetic field, these dipoles tend to align with the field (see Equation 12.14), and this alignment determines how the sample responds to the field. On the basis of this response, a material is said to be either paramagnetic, ferromagnetic, or diamagnetic.

In a **paramagnetic material**, only a small fraction (roughly one-third) of the magnetic dipoles are aligned with the applied field. Since each dipole produces its own magnetic field, this alignment contributes an extra magnetic field, which enhances the applied field. When a **ferromagnetic material** is placed in a magnetic field, its magnetic dipoles also become aligned;

furthermore, they become locked together so that a permanent magnetization results, even when the field is turned off or reversed. This permanent magnetization happens in ferromagnetic materials but not paramagnetic materials. **Diamagnetic materials** are composed of atoms that have no net magnetic dipole moment. However, when a diamagnetic material is placed in a magnetic field, a magnetic dipole moment is directed opposite to the applied field and therefore produces a magnetic field that opposes the applied field. We now consider each type of material in greater detail.

Paramagnetic Materials

For simplicity, we assume our sample is a long, cylindrical piece that completely fills the interior of a long, tightly wound solenoid. When there is no current in the solenoid, the magnetic dipoles in the sample are randomly oriented and produce no net magnetic field. With a solenoid current, the magnetic field due to the solenoid exerts a torque on the dipoles that tends to align them with the field. In competition with the aligning torque are thermal collisions that tend to randomize the orientations of the dipoles. The relative importance of these two competing processes can be estimated by comparing the energies involved. From Equation 12.14, the energy difference between a magnetic dipole aligned with and against a magnetic field is $U_B = 2\mu B$. If $\mu = 9.3 \times 10^{-24}\,\text{A} \cdot \text{m}^2$ (the value of atomic hydrogen) and $B = 1.0$ T, then

$$U_B = 1.9 \times 10^{-23}\,\text{J}.$$

At a room temperature of $27\,°\text{C}$, the thermal energy per atom is

$$U_T \approx kT = (1.38 \times 10^{-23}\,\text{J/K})(300\,\text{K}) = 4.1 \times 10^{-21}\,\text{J},$$

which is about 220 times greater than U_B. Clearly, energy exchanges in thermal collisions can seriously interfere with the alignment of the magnetic dipoles. As a result, only a small fraction of the dipoles is aligned at any instant.

The four sketches of Figure 12.23 furnish a simple model of this alignment process. In part (a), before the field of the solenoid (not shown) containing the paramagnetic sample is applied, the magnetic dipoles are randomly oriented and there is no net magnetic dipole moment associated with the material. With the introduction of the field, a partial alignment of the dipoles takes place, as depicted in part (b). The component of the net magnetic dipole moment that is perpendicular to the field vanishes. We may then represent the sample by part (c), which shows a collection of magnetic dipoles completely aligned with the field. By treating these dipoles as current loops, we can picture the dipole alignment as equivalent to a current around the surface of the material, as in part (d). This fictitious surface current produces its own magnetic field, which enhances the field of the solenoid.

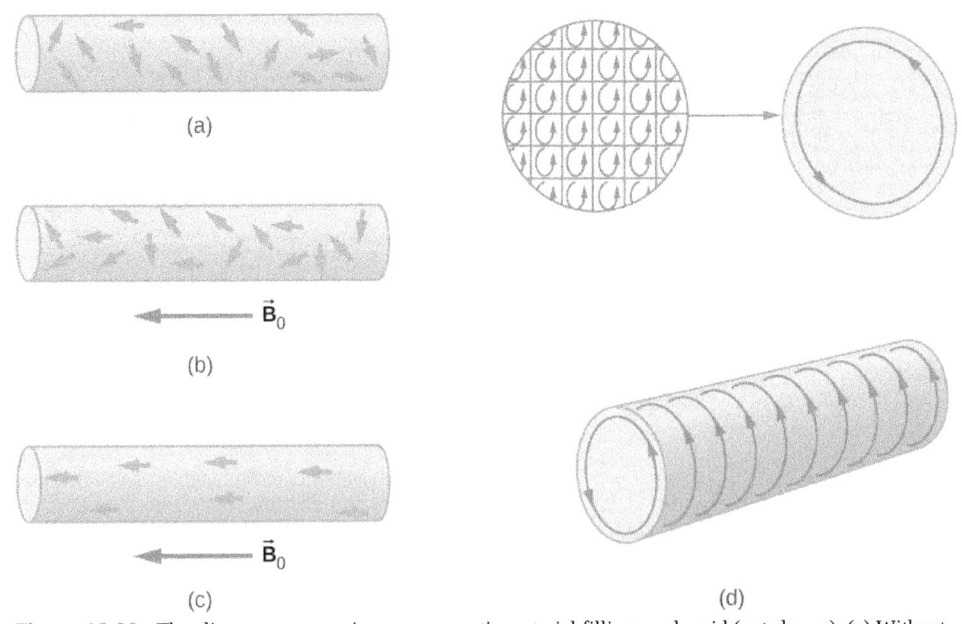

Figure 12.23 The alignment process in a paramagnetic material filling a solenoid (not shown). (a) Without an applied field, the magnetic dipoles are randomly oriented. (b) With a field, partial alignment occurs. (c) An equivalent representation of part (b). (d) The internal currents cancel, leaving an effective surface current that produces a magnetic field similar to that of a finite solenoid.

We can express the total magnetic field $\vec{\mathbf{B}}$ in the material as

$$\vec{\mathbf{B}} = \vec{\mathbf{B}}_0 + \vec{\mathbf{B}}_m, \tag{12.34}$$

where $\vec{\mathbf{B}}_0$ is the field due to the current I_0 in the solenoid and $\vec{\mathbf{B}}_m$ is the field due to the surface current I_m around the sample. Now $\vec{\mathbf{B}}_m$ is usually proportional to $\vec{\mathbf{B}}_0$, a fact we express by

$$\vec{\mathbf{B}}_m = \chi \vec{\mathbf{B}}_0, \tag{12.35}$$

where χ is a dimensionless quantity called the **magnetic susceptibility**. Values of χ for some paramagnetic materials are given in Table 12.2. Since the alignment of magnetic dipoles is so weak, χ is very small for paramagnetic materials. By combining Equation 12.34 and Equation 12.35, we obtain:

$$\vec{\mathbf{B}} = \vec{\mathbf{B}}_0 + \chi \vec{\mathbf{B}}_0 = (1+\chi)\vec{\mathbf{B}}_0. \tag{12.36}$$

For a sample within an infinite solenoid, this becomes

$$B = (1+\chi)\mu_0 nI. \tag{12.37}$$

This expression tells us that the insertion of a paramagnetic material into a solenoid increases the field by a factor of $(1+\chi)$. However, since χ is so small, the field isn't enhanced very much.

The quantity

$$\mu = (1+\chi)\mu_0. \tag{12.38}$$

is called the magnetic permeability of a material. In terms of μ, Equation 12.37 can be written as

$$B = \mu nI \tag{12.39}$$

for the filled solenoid.

Paramagnetic Materials	χ	Diamagnetic Materials	χ
Aluminum	2.2×10^{-5}	Bismuth	-1.7×10^{-5}
Calcium	1.4×10^{-5}	Carbon (diamond)	-2.2×10^{-5}
Chromium	3.1×10^{-4}	Copper	-9.7×10^{-6}
Magnesium	1.2×10^{-5}	Lead	-1.8×10^{-5}
Oxygen gas (1 atm)	1.8×10^{-6}	Mercury	-2.8×10^{-5}
Oxygen liquid (90 K)	3.5×10^{-3}	Hydrogen gas (1 atm)	-2.2×10^{-9}
Tungsten	6.8×10^{-5}	Nitrogen gas (1 atm)	-6.7×10^{-9}

Table 12.2 Magnetic Susceptibilities *Note: Unless otherwise specified, values given are for room temperature.

Paramagnetic Materials	χ	Diamagnetic Materials	χ
Air (1 atm)	3.6×10^{-7}	Water	-9.1×10^{-6}

Table 12.2 Magnetic Susceptibilities *Note: Unless otherwise specified, values given are for room temperature.

Diamagnetic Materials

A magnetic field always induces a magnetic dipole in an atom. This induced dipole points opposite to the applied field, so its magnetic field is also directed opposite to the applied field. In paramagnetic and ferromagnetic materials, the induced magnetic dipole is masked by much stronger permanent magnetic dipoles of the atoms. However, in diamagnetic materials, whose atoms have no permanent magnetic dipole moments, the effect of the induced dipole is observable.

We can now describe the magnetic effects of diamagnetic materials with the same model developed for paramagnetic materials. In this case, however, the fictitious surface current flows opposite to the solenoid current, and the magnetic susceptibility χ is negative. Values of χ for some diamagnetic materials are also given in Table 12.2.

Water is a common diamagnetic material. Animals are mostly composed of water. Experiments have been performed on frogs (https://openstaxcollege.org/l/21frogs) and mice (https://openstaxcollege.org/l/21mice) in diverging magnetic fields. The water molecules are repelled from the applied magnetic field against gravity until the animal reaches an equilibrium. The result is that the animal is levitated by the magnetic field.

Ferromagnetic Materials

Common magnets are made of a ferromagnetic material such as iron or one of its alloys. Experiments reveal that a ferromagnetic material consists of tiny regions known as **magnetic domains**. Their volumes typically range from 10^{-12} to $10^{-8}\,\text{m}^3$, and they contain about 10^{17} to 10^{21} atoms. Within a domain, the magnetic dipoles are rigidly aligned in the same direction by coupling among the atoms. This coupling, which is due to quantum mechanical effects, is so strong that even thermal agitation at room temperature cannot break it. The result is that each domain has a net dipole moment. Some materials have weaker coupling and are ferromagnetic only at lower temperatures.

If the domains in a ferromagnetic sample are randomly oriented, as shown in Figure 12.24, the sample has no net magnetic dipole moment and is said to be unmagnetized. Suppose that we fill the volume of a solenoid with an unmagnetized ferromagnetic sample. When the magnetic field $\vec{\mathbf{B}}_0$ of the solenoid is turned on, the dipole moments of the domains rotate so that they align somewhat with the field, as depicted in Figure 12.24. In addition, the aligned domains tend to increase in size at the expense of unaligned ones. The net effect of these two processes is the creation of a net magnetic dipole moment for the ferromagnet that is directed along the applied magnetic field. This net magnetic dipole moment is much larger than that of a paramagnetic sample, and the domains, with their large numbers of atoms, do not become misaligned by thermal agitation. Consequently, the field due to the alignment of the domains is quite large.

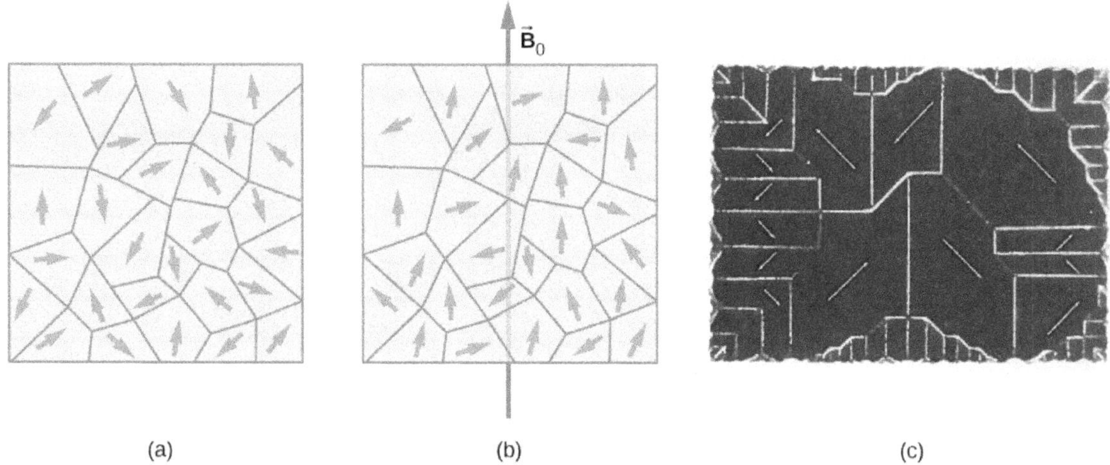

Figure 12.24 (a) Domains are randomly oriented in an unmagnetized ferromagnetic sample such as iron. The arrows represent the orientations of the magnetic dipoles within the domains. (b) In an applied magnetic field, the domains align somewhat with the field. (c) The domains of a single crystal of nickel. The white lines show the boundaries of the domains. These lines are produced by iron oxide powder sprinkled on the crystal.

Besides iron, only four elements contain the magnetic domains needed to exhibit ferromagnetic behavior: cobalt, nickel, gadolinium, and dysprosium. Many alloys of these elements are also ferromagnetic. Ferromagnetic materials can be described using Equation 12.34 through Equation 12.39, the paramagnetic equations. However, the value of χ for ferromagnetic material is usually on the order of 10^3 to 10^4, and it also depends on the history of the magnetic field to which the material has been subject. A typical plot of *B* (the total field in the material) versus B_0 (the applied field) for an initially unmagnetized piece of iron is shown in Figure 12.25. Some sample numbers are (1) for $B_0 = 1.0 \times 10^{-4}\,\text{T}, B = 0.60\,\text{T},$ and $\chi = \left(\frac{0.60}{1.0 \times 10^{-4}}\right) - 1 \approx 6.0 \times 10^3$; (2) for $B_0 = 6.0 \times 10^{-4}\,\text{T}, B = 1.5\,\text{T},$ and $\chi = \left(\frac{1.5}{6.0 \times 10^{-4}}\right) - 1 \approx 2.5 \times 10^3$.

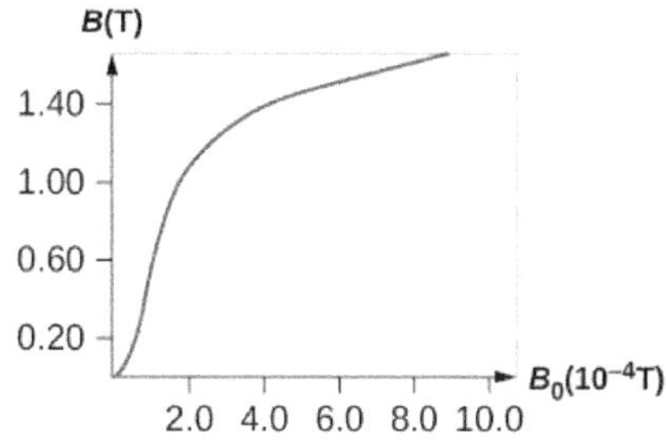

Figure 12.25 (a) The magnetic field *B* in annealed iron as a function of the applied field B_0.

When B_0 is varied over a range of positive and negative values, *B* is found to behave as shown in Figure 12.26. Note that the same B_0 (corresponding to the same current in the solenoid) can produce different values of *B* in the material. The magnetic field *B* produced in a ferromagnetic material by an applied field B_0 depends on the magnetic history of the material. This effect is called **hysteresis**, and the curve of Figure 12.26 is called a hysteresis loop. Notice that *B* does not disappear when $B_0 = 0$ (i.e., when the current in the solenoid is turned off). The iron stays magnetized, which means that it has become a permanent magnet.

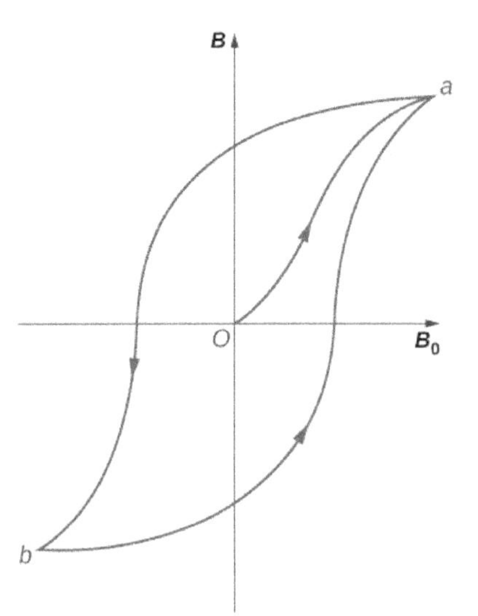

Figure 12.26 A typical hysteresis loop for a ferromagnet. When the material is first magnetized, it follows a curve from 0 to *a*. When B_0 is reversed, it takes the path shown from *a* to *b*. If B_0 is reversed again, the material follows the curve from *b* to *a*.

Like the paramagnetic sample of Figure 12.23, the partial alignment of the domains in a ferromagnet is equivalent to a current flowing around the surface. A bar magnet can therefore be pictured as a tightly wound solenoid with a large current circulating through its coils (the surface current). You can see in Figure 12.27 that this model fits quite well. The fields of the bar magnet and the finite solenoid are strikingly similar. The figure also shows how the poles of the bar magnet are identified. To form closed loops, the field lines outside the magnet leave the north (N) pole and enter the south (S) pole, whereas inside the magnet, they leave S and enter N.

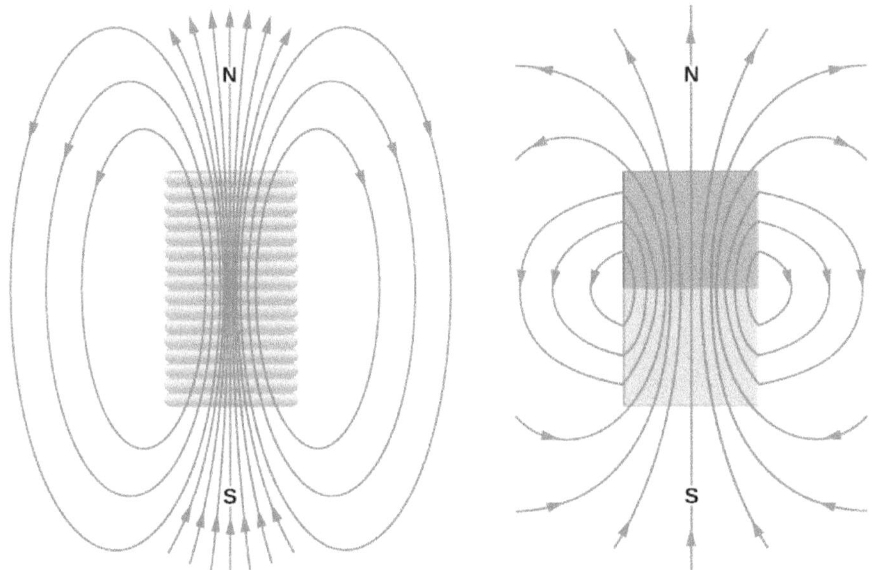

Figure 12.27 Comparison of the magnetic fields of a finite solenoid and a bar magnet.

Ferromagnetic materials are found in computer hard disk drives and permanent data storage devices (Figure 12.28). A material used in your hard disk drives is called a spin valve, which has alternating layers of ferromagnetic (aligning with the external magnetic field) and antiferromagnetic (each atom is aligned opposite to the next) metals. It was observed that a significant change in resistance was discovered based on whether an applied magnetic field was on the spin valve or not. This large change in resistance creates a quick and consistent way for recording or reading information by an applied current.

Figure 12.28 The inside of a hard disk drive. The silver disk contains the information, whereas the thin stylus on top of the disk reads and writes information to the disk.

Example 12.10

Iron Core in a Coil

A long coil is tightly wound around an iron cylinder whose magnetization curve is shown in Figure 12.25. (a) If $n = 20$ turns per centimeter, what is the applied field B_0 when $I_0 = 0.20\,\text{A}$? (b) What is the net magnetic field for this same current? (c) What is the magnetic susceptibility in this case?

Strategy

(a) The magnetic field of a solenoid is calculated using Equation 12.28. (b) The graph is read to determine the net magnetic field for this same current. (c) The magnetic susceptibility is calculated using Equation 12.37.

Solution

a. The applied field B_0 of the coil is

$$\begin{aligned} B_0 &= \mu_0 n I_0 = (4\pi \times 10^{-7}\,\text{T}\cdot\text{m/A})(2000/\text{m})(0.20\,\text{A}) \\ B_0 &= 5.0 \times 10^{-4}\,\text{T}. \end{aligned}$$

b. From inspection of the magnetization curve of Figure 12.25, we see that, for this value of B_0, $B = 1.4\,\text{T}$. Notice that the internal field of the aligned atoms is much larger than the externally applied field.

c. The magnetic susceptibility is calculated to be

$$\chi = \frac{B}{B_0} - 1 = \frac{1.4\,\text{T}}{5.0 \times 10^{-4}\,\text{T}} - 1 = 2.8 \times 10^3.$$

Significance

Ferromagnetic materials have susceptibilities in the range of 10^3 which compares well to our results here. Paramagnetic materials have fractional susceptibilities, so their applied field of the coil is much greater than the magnetic field generated by the material.

12.8 Check Your Understanding Repeat the calculations from the previous example for $I_0 = 0.040\ \text{A}$.

CHAPTER 12 REVIEW

KEY TERMS

Ampère's law physical law that states that the line integral of the magnetic field around an electric current is proportional to the current

Biot-Savart law an equation giving the magnetic field at a point produced by a current-carrying wire

diamagnetic materials their magnetic dipoles align oppositely to an applied magnetic field; when the field is removed, the material is unmagnetized

ferromagnetic materials contain groups of dipoles, called domains, that align with the applied magnetic field; when this field is removed, the material is still magnetized

hysteresis property of ferromagnets that is seen when a material's magnetic field is examined versus the applied magnetic field; a loop is created resulting from sweeping the applied field forward and reverse

magnetic domains groups of magnetic dipoles that are all aligned in the same direction and are coupled together quantum mechanically

magnetic susceptibility ratio of the magnetic field in the material over the applied field at that time; positive susceptibilities are either paramagnetic or ferromagnetic (aligned with the field) and negative susceptibilities are diamagnetic (aligned oppositely with the field)

paramagnetic materials their magnetic dipoles align partially in the same direction as the applied magnetic field; when this field is removed, the material is unmagnetized

permeability of free space μ_0, measure of the ability of a material, in this case free space, to support a magnetic field

solenoid thin wire wound into a coil that produces a magnetic field when an electric current is passed through it

toroid donut-shaped coil closely wound around that is one continuous wire

KEY EQUATIONS

Permeability of free space	$\mu_0 = 4\pi \times 10^{-7}\,\text{T} \cdot \text{m/A}$
Contribution to magnetic field from a current element	$dB = \frac{\mu_0}{4\pi}\frac{I\,dl\sin\theta}{r^2}$
Biot–Savart law	$\vec{\mathbf{B}} = \frac{\mu_0}{4\pi}\int_{\text{wire}} \frac{Id\vec{\mathbf{l}} \times \hat{\mathbf{r}}}{r^2}$
Magnetic field due to a long straight wire	$B = \frac{\mu_0 I}{2\pi R}$
Force between two parallel currents	$\frac{F}{l} = \frac{\mu_0 I_1 I_2}{2\pi r}$
Magnetic field of a current loop	$B = \frac{\mu_0 I}{2R}$ (at center of loop)
Ampère's law	$\oint \vec{\mathbf{B}} \cdot d\vec{\mathbf{l}} = \mu_0 I$
Magnetic field strength inside a solenoid	$B = \mu_0 nI$
Magnetic field strength inside a toroid	$B = \frac{\mu_o NI}{2\pi r}$

Magnetic permeability	$\mu = (1 + \chi)\mu_0$
Magnetic field of a solenoid filled with paramagnetic material	$B = \mu n I$

SUMMARY

12.1 The Biot-Savart Law

- The magnetic field created by a current-carrying wire is found by the Biot-Savart law.
- The current element $Id\vec{l}$ produces a magnetic field a distance r away.

12.2 Magnetic Field Due to a Thin Straight Wire

- The strength of the magnetic field created by current in a long straight wire is given by $B = \frac{\mu_0 I}{2\pi R}$ (long straight wire) where I is the current, R is the shortest distance to the wire, and the constant $\mu_0 = 4\pi \times 10^{-7}\ \text{T} \cdot \text{m/s}$ is the permeability of free space.
- The direction of the magnetic field created by a long straight wire is given by right-hand rule 2 (RHR-2): Point the thumb of the right hand in the direction of current, and the fingers curl in the direction of the magnetic field loops created by it.

12.3 Magnetic Force between Two Parallel Currents

- The force between two parallel currents I_1 and I_2, separated by a distance r, has a magnitude per unit length given by $\frac{F}{l} = \frac{\mu_0 I_1 I_2}{2\pi r}$.
- The force is attractive if the currents are in the same direction, repulsive if they are in opposite directions.

12.4 Magnetic Field of a Current Loop

- The magnetic field strength at the center of a circular loop is given by $B = \frac{\mu_0 I}{2R}$ (at center of loop), where R is the radius of the loop. RHR-2 gives the direction of the field about the loop.

12.5 Ampère's Law

- The magnetic field created by current following any path is the sum (or integral) of the fields due to segments along the path (magnitude and direction as for a straight wire), resulting in a general relationship between current and field known as Ampère's law.
- Ampère's law can be used to determine the magnetic field from a thin wire or thick wire by a geometrically convenient path of integration. The results are consistent with the Biot-Savart law.

12.6 Solenoids and Toroids

- The magnetic field strength inside a solenoid is

  $$B = \mu_0 n I \quad \text{(inside a solenoid)}$$

  where n is the number of loops per unit length of the solenoid. The field inside is very uniform in magnitude and direction.
- The magnetic field strength inside a toroid is

  $$B = \frac{\mu_o N I}{2\pi r} \quad \text{(within the toroid)}$$

where N is the number of windings. The field inside a toroid is not uniform and varies with the distance as $1/r$.

12.7 Magnetism in Matter

- Materials are classified as paramagnetic, diamagnetic, or ferromagnetic, depending on how they behave in an applied magnetic field.
- Paramagnetic materials have partial alignment of their magnetic dipoles with an applied magnetic field. This is a positive magnetic susceptibility. Only a surface current remains, creating a solenoid-like magnetic field.
- Diamagnetic materials exhibit induced dipoles opposite to an applied magnetic field. This is a negative magnetic susceptibility.
- Ferromagnetic materials have groups of dipoles, called domains, which align with the applied magnetic field. However, when the field is removed, the ferromagnetic material remains magnetized, unlike paramagnetic materials. This magnetization of the material versus the applied field effect is called hysteresis.

CONCEPTUAL QUESTIONS

12.1 The Biot-Savart Law

1. For calculating magnetic fields, what are the advantages and disadvantages of the Biot-Savart law?

2. Describe the magnetic field due to the current in two wires connected to the two terminals of a source of emf and twisted tightly around each other.

3. How can you decide if a wire is infinite?

4. Identical currents are carried in two circular loops; however, one loop has twice the diameter as the other loop. Compare the magnetic fields created by the loops at the center of each loop.

12.2 Magnetic Field Due to a Thin Straight Wire

5. How would you orient two long, straight, current-carrying wires so that there is no net magnetic force between them? (*Hint*: What orientation would lead to one wire not experiencing a magnetic field from the other?)

12.3 Magnetic Force between Two Parallel Currents

6. Compare and contrast the electric field of an infinite line of charge and the magnetic field of an infinite line of current.

7. Is $\vec{\mathbf{B}}$ constant in magnitude for points that lie on a magnetic field line?

12.4 Magnetic Field of a Current Loop

8. Is the magnetic field of a current loop uniform?

9. What happens to the length of a suspended spring when a current passes through it?

10. Two concentric circular wires with different diameters carry currents in the same direction. Describe the force on the inner wire.

12.5 Ampère's Law

11. Is Ampère's law valid for all closed paths? Why isn't it normally useful for calculating a magnetic field?

12.6 Solenoids and Toroids

12. Is the magnetic field inside a toroid completely uniform? Almost uniform?

13. Explain why $\vec{\mathbf{B}} = 0$ inside a long, hollow copper pipe that is carrying an electric current parallel to the axis. Is $\vec{\mathbf{B}} = 0$ outside the pipe?

12.7 Magnetism in Matter

14. A diamagnetic material is brought close to a permanent magnet. What happens to the material?

15. If you cut a bar magnet into two pieces, will you end up with one magnet with an isolated north pole and another magnet with an isolated south pole? Explain your answer.

PROBLEMS

12.1 The Biot-Savart Law

16. A 10-A current flows through the wire shown. What is the magnitude of the magnetic field due to a 0.5-mm segment of wire as measured at (a) point A and (b) point B?

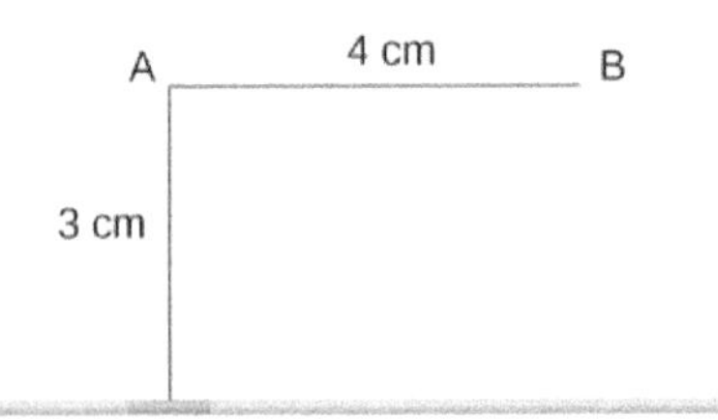

17. Ten amps flow through a square loop where each side is 20 cm in length. At each corner of the loop is a 0.01-cm segment that connects the longer wires as shown. Calculate the magnitude of the magnetic field at the center of the loop.

18. What is the magnetic field at P due to the current *I* in the wire shown?

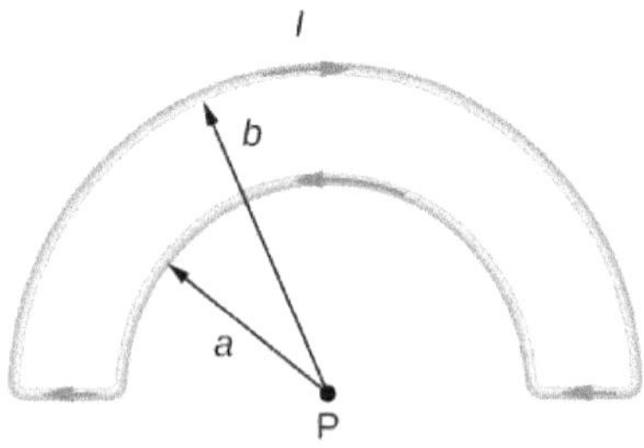

19. The accompanying figure shows a current loop consisting of two concentric circular arcs and two perpendicular radial lines. Determine the magnetic field at point P.

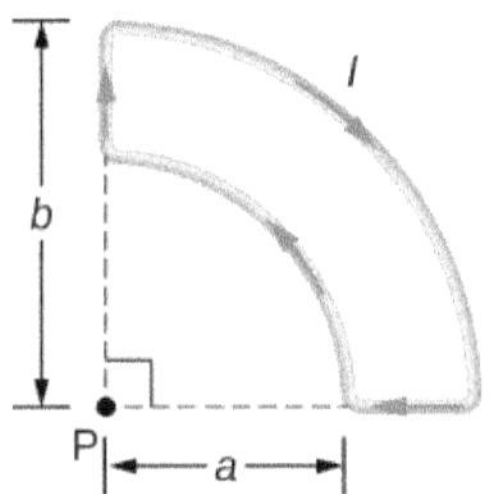

20. Find the magnetic field at the center C of the rectangular loop of wire shown in the accompanying figure.

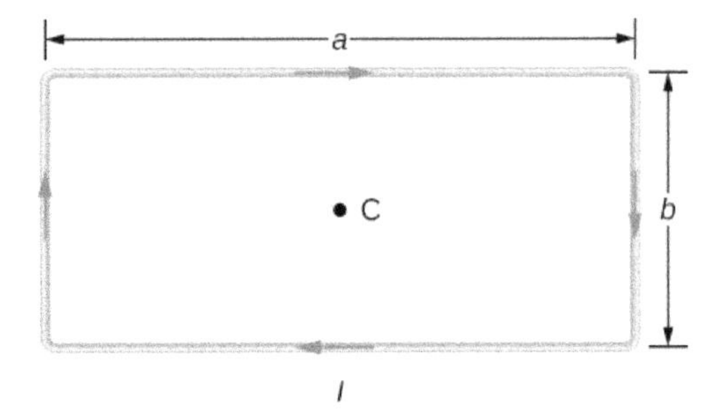

21. Two long wires, one of which has a semicircular bend of radius *R*, are positioned as shown in the accompanying figure. If both wires carry a current *I*, how far apart must their parallel sections be so that the net magnetic field at P is zero? Does the current in the straight wire flow up or down?

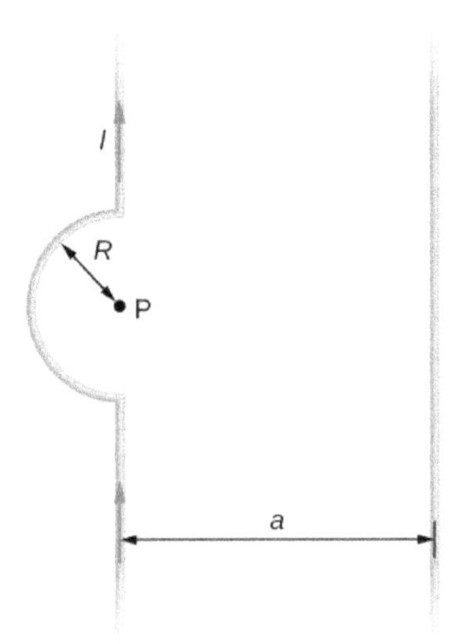

12.2 Magnetic Field Due to a Thin Straight Wire

22. A typical current in a lightning bolt is 10^4 A. Estimate the magnetic field 1 m from the bolt.

23. The magnitude of the magnetic field 50 cm from a long, thin, straight wire is $8.0\ \mu T$. What is the current through the long wire?

24. A transmission line strung 7.0 m above the ground carries a current of 500 A. What is the magnetic field on the ground directly below the wire? Compare your answer with the magnetic field of Earth.

25. A long, straight, horizontal wire carries a left-to-right current of 20 A. If the wire is placed in a uniform magnetic field of magnitude $4.0 \times 10^{-5}\,T$ that is directed vertically downward, what is the resultant magnitude of the magnetic field 20 cm above the wire? 20 cm below the wire?

26. The two long, parallel wires shown in the accompanying figure carry currents in the same direction. If $I_1 = 10\ A$ and $I_2 = 20\ A$, what is the magnetic field at point P?

27. The accompanying figure shows two long, straight, horizontal wires that are parallel and a distance $2a$ apart. If both wires carry current I in the same direction, (a) what is the magnetic field at P_1? (b) P_2?

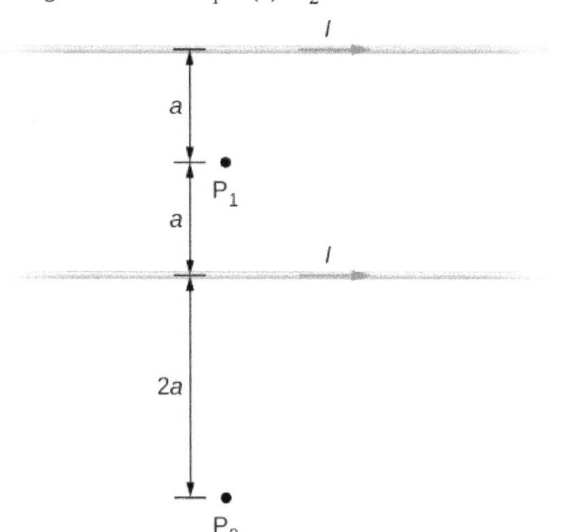

28. Repeat the calculations of the preceding problem with the direction of the current in the lower wire reversed.

29. Consider the area between the wires of the preceding problem. At what distance from the top wire is the net magnetic field a minimum? Assume that the currents are equal and flow in opposite directions.

12.3 Magnetic Force between Two Parallel Currents

30. Two long, straight wires are parallel and 25 cm apart. (a) If each wire carries a current of 50 A in the same direction, what is the magnetic force per meter exerted on each wire? (b) Does the force pull the wires together or push them apart? (c) What happens if the currents flow in opposite directions?

31. Two long, straight wires are parallel and 10 cm apart. One carries a current of 2.0 A, the other a current of 5.0 A. (a) If the two currents flow in opposite directions, what is the magnitude and direction of the force per unit length of one wire on the other? (b) What is the magnitude and direction of the force per unit length if the currents flow in the same direction?

32. Two long, parallel wires are hung by cords of length 5.0 cm, as shown in the accompanying figure. Each wire has a mass per unit length of 30 g/m, and they carry the same current in opposite directions. What is the current if the cords hang at $6.0°$ with respect to the vertical?

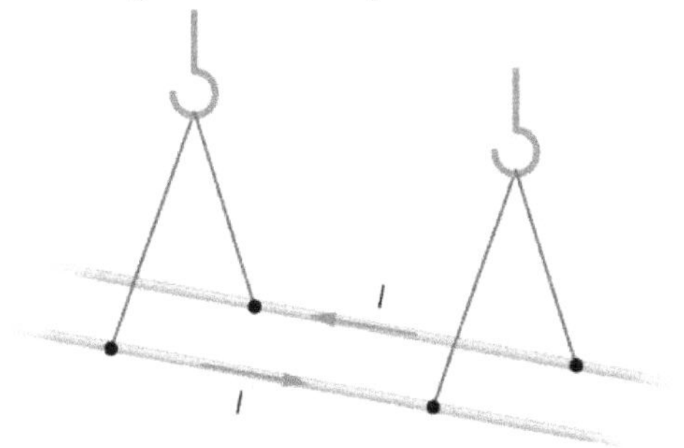

33. A circuit with current I has two long parallel wire sections that carry current in opposite directions. Find magnetic field at a point P near these wires that is a distance a from one wire and b from the other wire as shown in the figure.

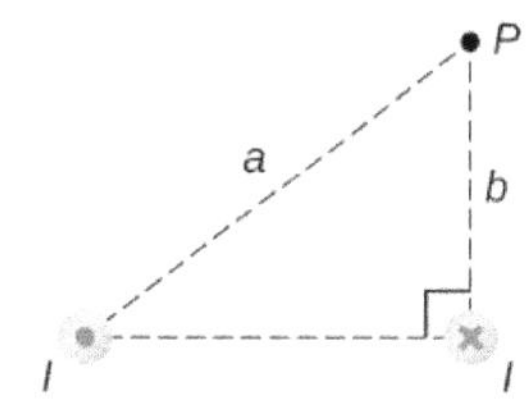

34. The infinite, straight wire shown in the accompanying figure carries a current I_1. The rectangular loop, whose long sides are parallel to the wire, carries a current I_2. What are the magnitude and direction of the force on the rectangular loop due to the magnetic field of the wire?

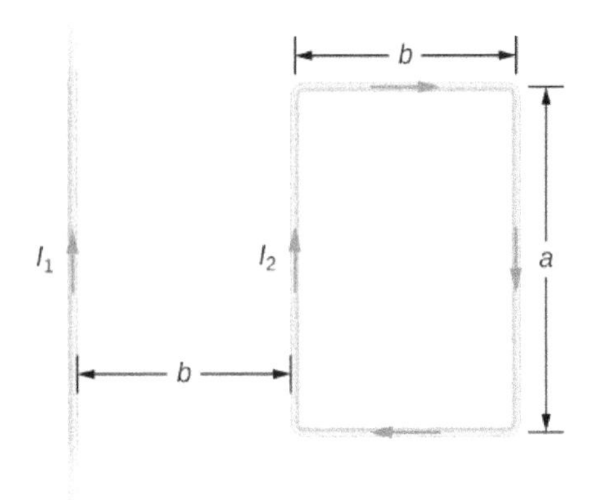

12.4 Magnetic Field of a Current Loop

35. When the current through a circular loop is 6.0 A, the magnetic field at its center is 2.0×10^{-4} T. What is the radius of the loop?

36. How many turns must be wound on a flat, circular coil of radius 20 cm in order to produce a magnetic field of magnitude 4.0×10^{-5} T at the center of the coil when the current through it is 0.85 A?

37. A flat, circular loop has 20 turns. The radius of the loop is 10.0 cm and the current through the wire is 0.50 A. Determine the magnitude of the magnetic field at the center of the loop.

38. A circular loop of radius *R* carries a current *I*. At what distance along the axis of the loop is the magnetic field one-half its value at the center of the loop?

39. Two flat, circular coils, each with a radius *R* and wound with *N* turns, are mounted along the same axis so that they are parallel a distance *d* apart. What is the magnetic field at the midpoint of the common axis if a current *I* flows in the same direction through each coil?

40. For the coils in the preceding problem, what is the magnetic field at the center of either coil?

12.5 Ampère's Law

41. A current *I* flows around the rectangular loop shown in the accompanying figure. Evaluate $\oint \vec{\mathbf{B}} \cdot d\vec{\mathbf{l}}$ for the paths *A*, *B*, *C*, and *D*.

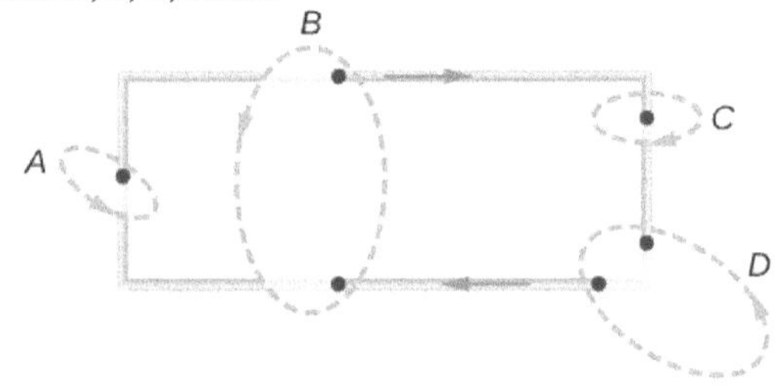

42. Evaluate $\oint \vec{\mathbf{B}} \cdot d\vec{\mathbf{l}}$ for each of the cases shown in the accompanying figure.

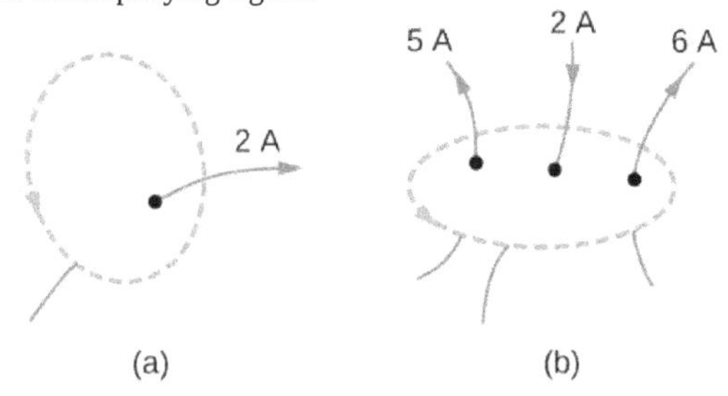

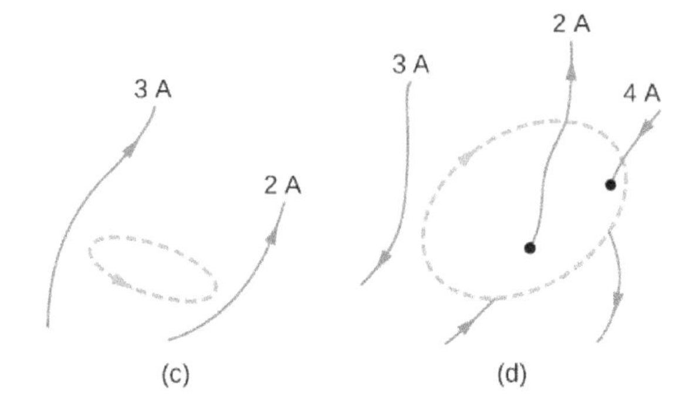

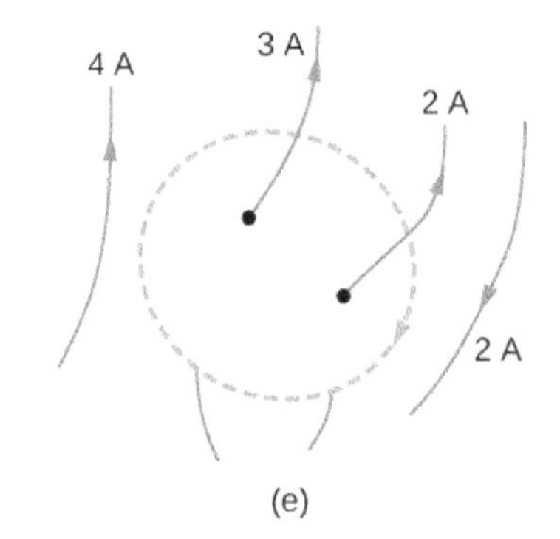

43. The coil whose lengthwise cross section is shown in the accompanying figure carries a current I and has N evenly spaced turns distributed along the length l. Evaluate $\oint \vec{\mathbf{B}} \cdot d\vec{\mathbf{l}}$ for the paths indicated.

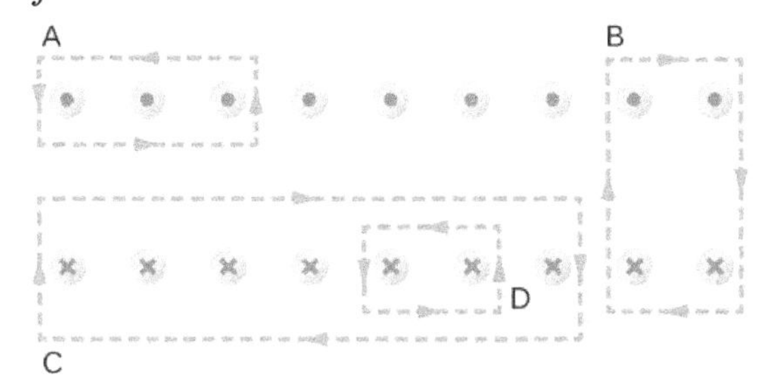

44. A superconducting wire of diameter 0.25 cm carries a current of 1000 A. What is the magnetic field just outside the wire?

45. A long, straight wire of radius R carries a current I that is distributed uniformly over the cross-section of the wire. At what distance from the axis of the wire is the magnitude of the magnetic field a maximum?

46. The accompanying figure shows a cross-section of a long, hollow, cylindrical conductor of inner radius $r_1 = 3.0\text{ cm}$ and outer radius $r_2 = 5.0\text{ cm}$. A 50-A current distributed uniformly over the cross-section flows into the page. Calculate the magnetic field at $r = 2.0\text{ cm}$, $r = 4.0\text{ cm}$, and $r = 6.0\text{ cm}$.

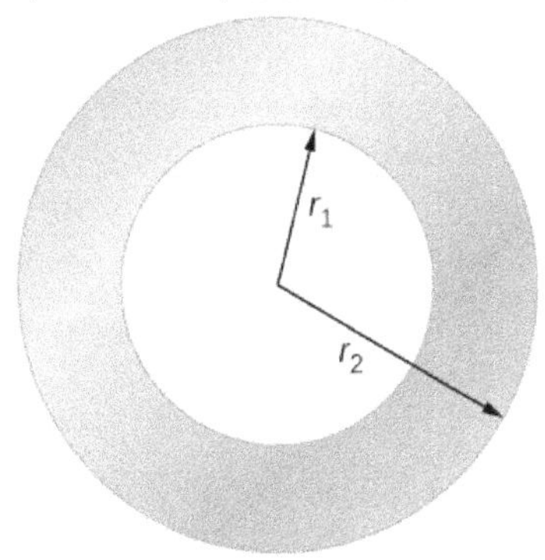

47. A long, solid, cylindrical conductor of radius 3.0 cm carries a current of 50 A distributed uniformly over its cross-section. Plot the magnetic field as a function of the radial distance r from the center of the conductor.

48. A portion of a long, cylindrical coaxial cable is shown in the accompanying figure. A current I flows down the center conductor, and this current is returned in the outer conductor. Determine the magnetic field in the regions (a) $r \le r_1$, (b) $r_2 \ge r \ge r_1$, (c) $r_3 \ge r \ge r_2$, and (d) $r \ge r_3$. Assume that the current is distributed uniformly over the cross sections of the two parts of the cable.

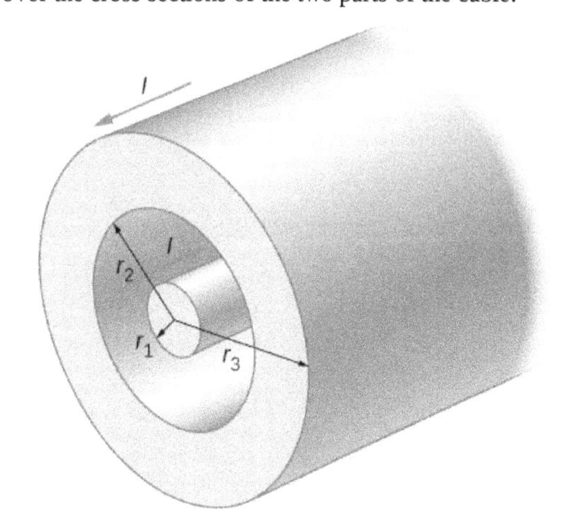

12.6 Solenoids and Toroids

49. A solenoid is wound with 2000 turns per meter. When the current is 5.2 A, what is the magnetic field within the solenoid?

50. A solenoid has 12 turns per centimeter. What current will produce a magnetic field of $2.0 \times 10^{-2}\text{ T}$ within the solenoid?

51. If a current is 2.0 A, how many turns per centimeter must be wound on a solenoid in order to produce a magnetic field of $2.0 \times 10^{-3}\text{ T}$ within it?

52. A solenoid is 40 cm long, has a diameter of 3.0 cm, and is wound with 500 turns. If the current through the windings is 4.0 A, what is the magnetic field at a point on the axis of the solenoid that is (a) at the center of the solenoid, (b) 10.0 cm from one end of the solenoid, and (c) 5.0 cm from one end of the solenoid? (d) Compare these answers with the infinite-solenoid case.

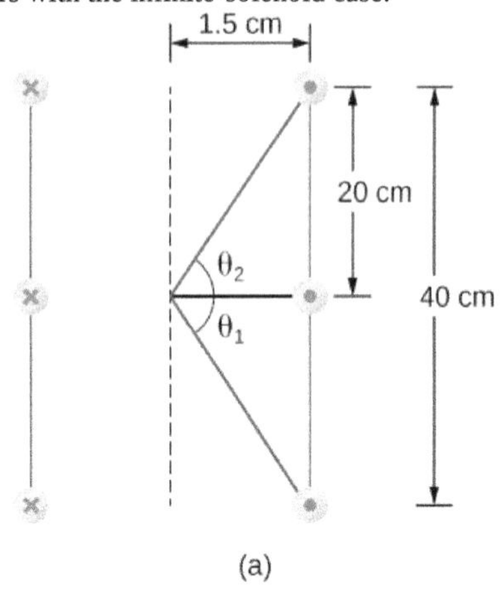

(a)

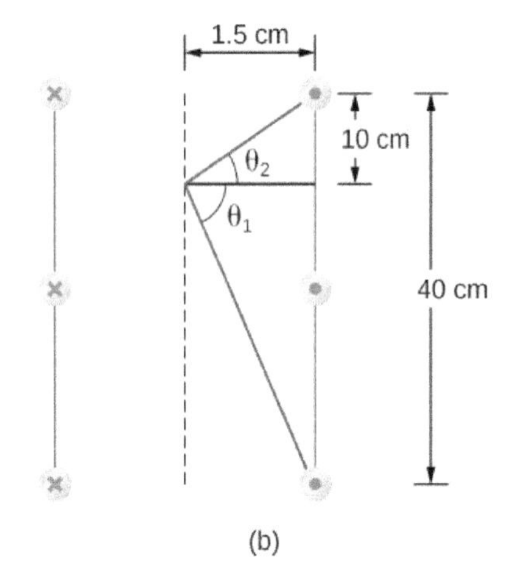

(b)

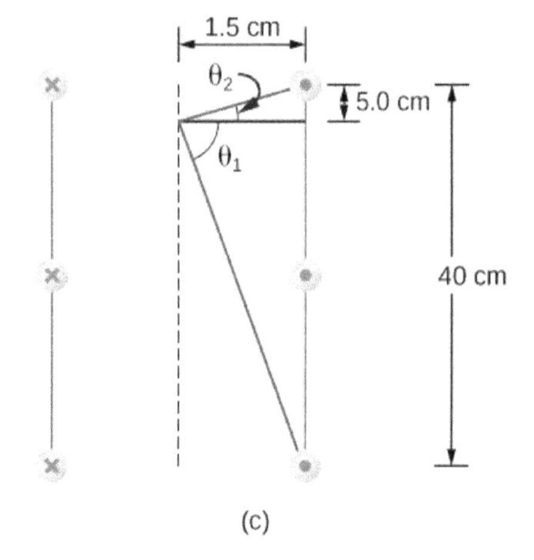

(c)

53. Determine the magnetic field on the central axis at the opening of a semi-infinite solenoid. (That is, take the opening to be at $x = 0$ and the other end to be at $x = \infty$.)

54. By how much is the approximation $B = \mu_0 nI$ in error at the center of a solenoid that is 15.0 cm long, has a diameter of 4.0 cm, is wrapped with n turns per meter, and carries a current I?

55. A solenoid with 25 turns per centimeter carries a current I. An electron moves within the solenoid in a circle that has a radius of 2.0 cm and is perpendicular to the axis of the solenoid. If the speed of the electron is $2.0 \times 10^5\ \text{m/s}$, what is I?

56. A toroid has 250 turns of wire and carries a current of 20 A. Its inner and outer radii are 8.0 and 9.0 cm. What are the values of its magnetic field at $r = 8.1,\ 8.5,$ and $8.9\ \text{cm}$?

57. A toroid with a square cross section 3.0 cm × 3.0 cm has an inner radius of 25.0 cm. It is wound with 500 turns of wire, and it carries a current of 2.0 A. What is the strength of the magnetic field at the center of the square cross section?

12.7 Magnetism in Matter

58. The magnetic field in the core of an air-filled solenoid is 1.50 T. By how much will this magnetic field decrease if the air is pumped out of the core while the current is held constant?

59. A solenoid has a ferromagnetic core, n = 1000 turns per meter, and I = 5.0 A. If B inside the solenoid is 2.0 T, what is χ for the core material?

60. A 20-A current flows through a solenoid with 2000 turns per meter. What is the magnetic field inside the solenoid if its core is (a) a vacuum and (b) filled with liquid oxygen at 90 K?

61. The magnetic dipole moment of the iron atom is about $2.1 \times 10^{-23}\ \text{A} \cdot \text{m}^2$. (a) Calculate the maximum magnetic dipole moment of a domain consisting of 10^{19} iron atoms. (b) What current would have to flow through a single circular loop of wire of diameter 1.0 cm to produce this magnetic dipole moment?

62. Suppose you wish to produce a 1.2-T magnetic field in a toroid with an iron core for which $\chi = 4.0 \times 10^3$. The toroid has a mean radius of 15 cm and is wound with 500 turns. What current is required?

63. A current of 1.5 A flows through the windings of a large, thin toroid with 200 turns per meter. If the toroid is filled with iron for which $\chi = 3.0 \times 10^3$, what is the magnetic field within it?

64. A solenoid with an iron core is 25 cm long and is wrapped with 100 turns of wire. When the current through the solenoid is 10 A, the magnetic field inside it is 2.0 T. For this current, what is the permeability of the iron? If the current is turned off and then restored to 10 A, will the magnetic field necessarily return to 2.0 T?

ADDITIONAL PROBLEMS

65. Three long, straight, parallel wires, all carrying 20 A, are positioned as shown in the accompanying figure. What is the magnitude of the magnetic field at the point *P*?

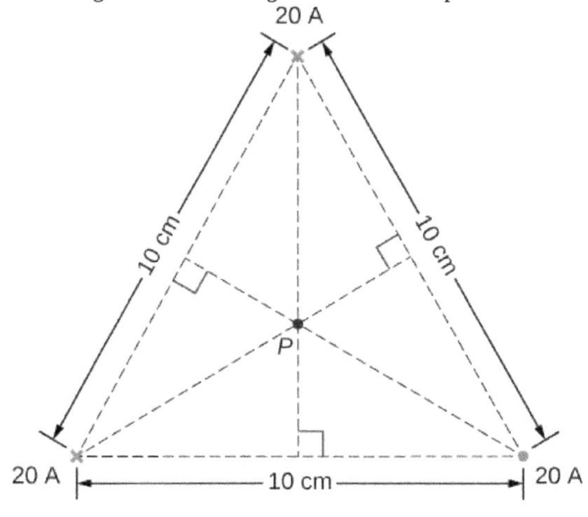

66. A current *I* flows around a wire bent into the shape of a square of side a. What is the magnetic field at the point P that is a distance z above the center of the square (see the accompanying figure)?

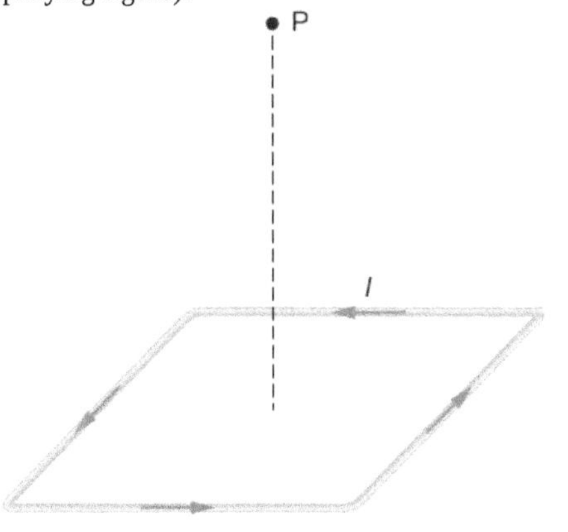

67. The accompanying figure shows a long, straight wire carrying a current of 10 A. What is the magnetic force on an electron at the instant it is 20 cm from the wire, traveling parallel to the wire with a speed of 2.0×10^5 m/s? Describe qualitatively the subsequent motion of the electron.

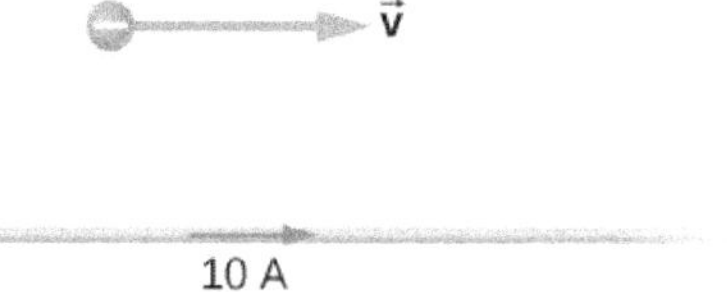

68. Current flows along a thin, infinite sheet as shown in the accompanying figure. The current per unit length along the sheet is *J* in amperes per meter. (a) Use the Biot-Savart law to show that $B = \mu_0 J / 2$ on either side of the sheet. What is the direction of $\vec{\mathbf{B}}$ on each side? (b) Now use Ampère's law to calculate the field.

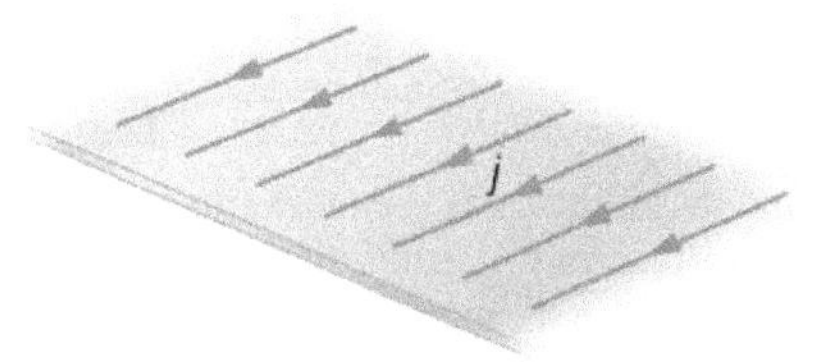

69. (a) Use the result of the previous problem to calculate the magnetic field between, above, and below the pair of infinite sheets shown in the accompanying figure. (b) Repeat your calculations if the direction of the current in the lower sheet is reversed.

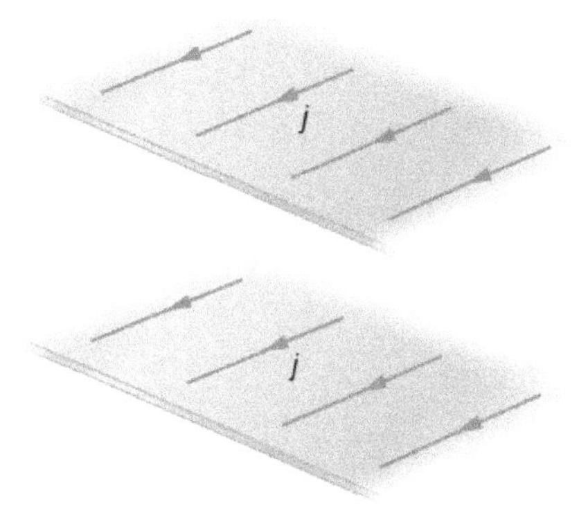

70. We often assume that the magnetic field is uniform in a region and zero everywhere else. Show that in reality it is impossible for a magnetic field to drop abruptly to zero, as illustrated in the accompanying figure. (*Hint*: Apply Ampère's law over the path shown.)

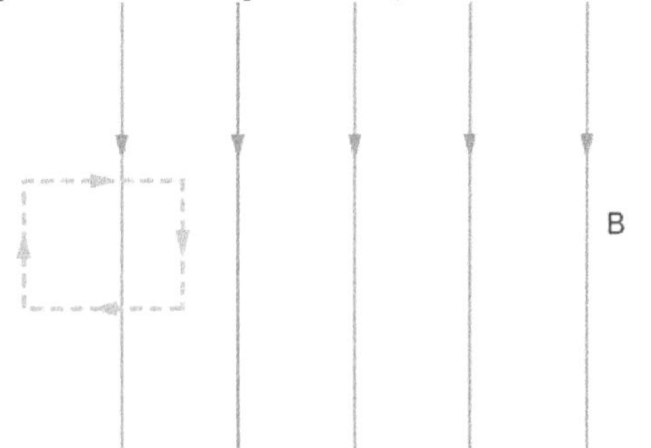

71. How is the percentage change in the strength of the magnetic field across the face of the toroid related to the percentage change in the radial distance from the axis of the toroid?

72. Show that the expression for the magnetic field of a toroid reduces to that for the field of an infinite solenoid in the limit that the central radius goes to infinity.

73. A toroid with an inner radius of 20 cm and an outer radius of 22 cm is tightly wound with one layer of wire that has a diameter of 0.25 mm. (a) How many turns are there on the toroid? (b) If the current through the toroid windings is 2.0 A, what is the strength of the magnetic field at the center of the toroid?

74. A wire element has $d\vec{\mathbf{l}}$, $Id\vec{\mathbf{l}} = \mathbf{J}Adl = \mathbf{J}dv$, where A and dv are the cross-sectional area and volume of the element, respectively. Use this, the Biot-Savart law, and $\mathbf{J} = ne\mathbf{v}$ to show that the magnetic field of a moving point charge q is given by:

$$\vec{\mathbf{B}} = \frac{\mu_0}{4\pi}\frac{q\mathbf{v} \times \hat{\mathbf{r}}}{r^2}$$

75. A reasonably uniform magnetic field over a limited region of space can be produced with the Helmholtz coil, which consists of two parallel coils centered on the same axis. The coils are connected so that they carry the same current I. Each coil has N turns and radius R, which is also the distance between the coils. (a) Find the magnetic field at any point on the z-axis shown in the accompanying figure. (b) Show that dB/dz and $\frac{d^2B}{dz^2}$ are both zero at $z = 0$. (These vanishing derivatives demonstrate that the magnetic field varies only slightly near $z = 0$.)

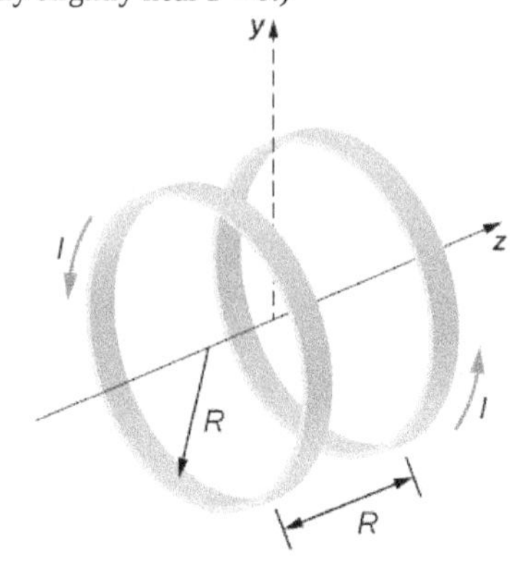

76. A charge of 4.0 μC is distributed uniformly around a thin ring of insulating material. The ring has a radius of 0.20 m and rotates at 2.0×10^4 rev/min around the axis that passes through its center and is perpendicular to the plane of the ring. What is the magnetic field at the center of the ring?

77. A thin, nonconducting disk of radius R is free to rotate around the axis that passes through its center and is perpendicular to the face of the disk. The disk is charged uniformly with a total charge q. If the disk rotates at a constant angular velocity ω, what is the magnetic field at its center?

78. Consider the disk in the previous problem. Calculate the magnetic field at a point on its central axis that is a distance y above the disk.

79. Consider the axial magnetic field $B_y = \mu_0 IR^2/2(y^2 + R^2)^{3/2}$ of the circular current loop shown below. (a) Evaluate $\int_{-a}^{a} B_y dy$. Also show that $\lim_{a \to \infty} \int_{-a}^{a} B_y dy = \mu_0 I$. (b) Can you deduce this limit without evaluating the integral? (*Hint:* See the accompanying figure.)

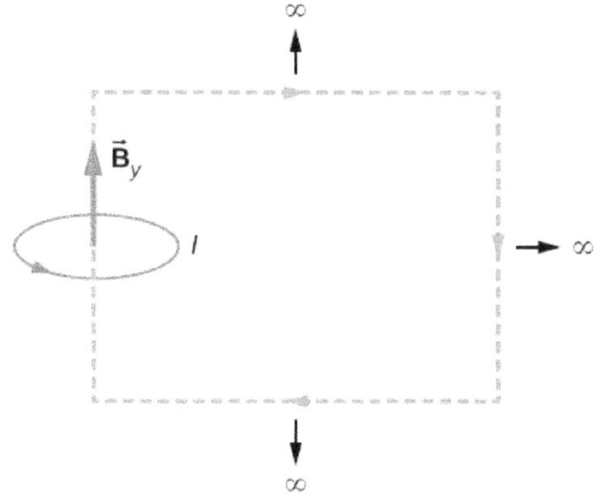

80. The current density in the long, cylindrical wire shown in the accompanying figure varies with distance r from the center of the wire according to $J = cr$, where c is a constant. (a) What is the current through the wire? (b) What is the magnetic field produced by this current for $r \leq R$? For $r \geq R$?

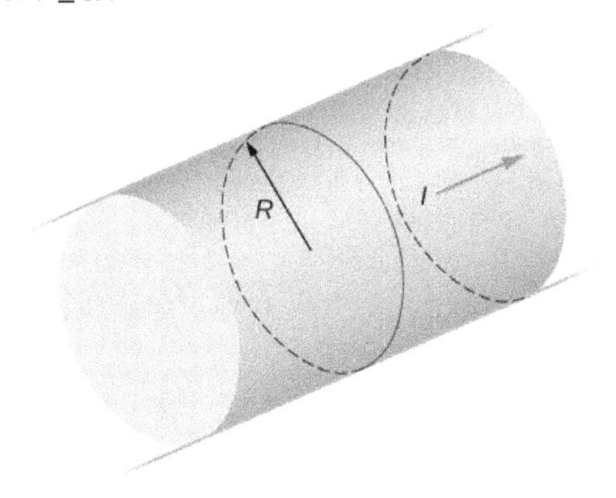

81. A long, straight, cylindrical conductor contains a cylindrical cavity whose axis is displaced by a from the axis of the conductor, as shown in the accompanying figure. The current density in the conductor is given by $\vec{\mathbf{J}} = J_0 \hat{\mathbf{k}}$, where J_0 is a constant and $\hat{\mathbf{k}}$ is along the axis of the conductor. Calculate the magnetic field at an arbitrary point P in the cavity by superimposing the field of a solid cylindrical conductor with radius R_1 and current density $\vec{\mathbf{J}}$ onto the field of a solid cylindrical conductor with radius R_2 and current density $-\vec{\mathbf{J}}$. Then use the fact that the appropriate azimuthal unit vectors can be expressed as $\hat{\theta}_1 = \hat{k} \times \hat{r}_1$ and $\hat{\theta}_2 = \hat{k} \times \hat{r}_2$ to show that everywhere inside the cavity the magnetic field is given by the constant $\vec{\mathbf{B}} = \frac{1}{2}\mu_0 J_0 \mathbf{k} \times \mathbf{a}$, where $\mathbf{a} = \mathbf{r}_1 - \mathbf{r}_2$ and $\mathbf{r}_1 = r_1 \hat{r}_1$ is the position of P relative to the center of the conductor and $\mathbf{r}_2 = r_2 \hat{r}_2$ is the position of P relative to the center of the cavity.

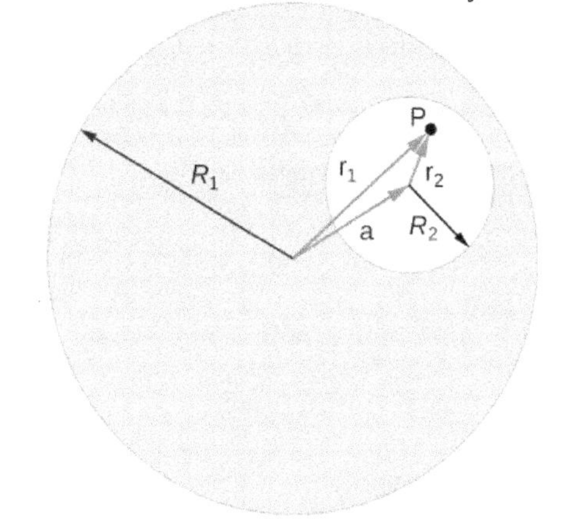

82. Between the two ends of a horseshoe magnet the field is uniform as shown in the diagram. As you move out to outside edges, the field bends. Show by Ampère's law that the field must bend and thereby the field weakens due to these bends.

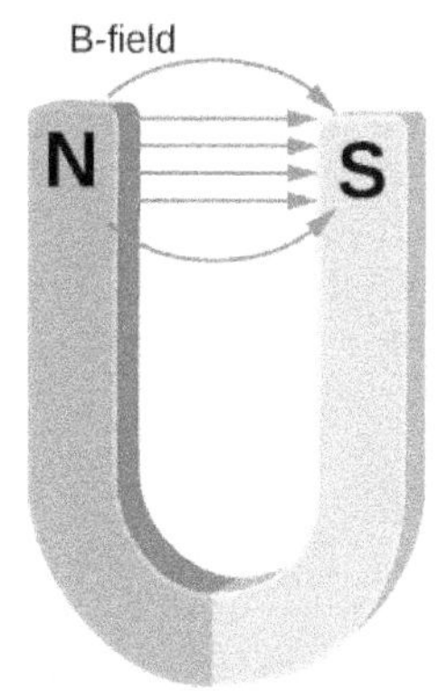

83. Show that the magnetic field of a thin wire and that of a current loop are zero if you are infinitely far away.

84. An Ampère loop is chosen as shown by dashed lines for a parallel constant magnetic field as shown by solid arrows. Calculate $\vec{\mathbf{B}} \cdot d\vec{\mathbf{l}}$ for each side of the loop then find the entire $\oint \vec{\mathbf{B}} \cdot d\vec{\mathbf{l}}$. Can you think of an Ampère loop that would make the problem easier? Do those results match these?

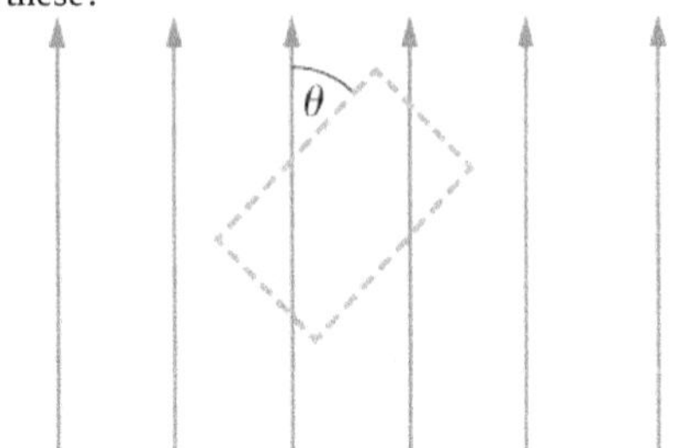

85. A very long, thick cylindrical wire of radius R carries a current density J that varies across its cross-section. The magnitude of the current density at a point a distance r from the center of the wire is given by $J = J_0 \frac{r}{R}$, where J_0 is a constant. Find the magnetic field (a) at a point outside the wire and (b) at a point inside the wire. Write your answer in terms of the net current I through the wire.

86. A very long, cylindrical wire of radius a has a circular hole of radius b in it at a distance d from the center. The wire carries a uniform current of magnitude I through it. The direction of the current in the figure is out of the paper. Find the magnetic field (a) at a point at the edge of the hole closest to the center of the thick wire, (b) at an arbitrary point inside the hole, and (c) at an arbitrary point outside the wire. (*Hint:* Think of the hole as a sum of two wires carrying current in the opposite directions.)

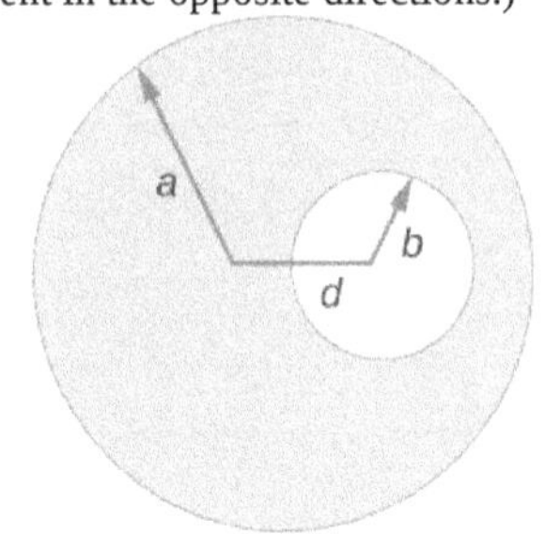

87. Magnetic field inside a torus. Consider a torus of rectangular cross-section with inner radius a and outer radius b. N turns of an insulated thin wire are wound evenly on the torus tightly all around the torus and connected to a battery producing a steady current I in the wire. Assume that the current on the top and bottom surfaces in the figure is radial, and the current on the inner and outer radii surfaces is vertical. Find the magnetic field inside the torus as a function of radial distance r from the axis.

88. Two long coaxial copper tubes, each of length L, are connected to a battery of voltage V. The inner tube has inner radius a and outer radius b, and the outer tube has inner radius c and outer radius d. The tubes are then disconnected from the battery and rotated in the same direction at angular speed of ω radians per second about their common axis. Find the magnetic field (a) at a point inside the space enclosed by the inner tube $r < a$, and (b) at a point between the tubes $b < r < c$, and (c) at a point outside the tubes $r > d$. (*Hint:* Think of copper tubes as a capacitor and find the charge density based on the voltage applied, $Q = VC$, $C = \frac{2\pi\varepsilon_0 L}{\ln(c/b)}$.)

CHALLENGE PROBLEMS

89. The accompanying figure shows a flat, infinitely long sheet of width a that carries a current I uniformly distributed across it. Find the magnetic field at the point P, which is in the plane of the sheet and at a distance x from one edge. Test your result for the limit $a \to 0$.

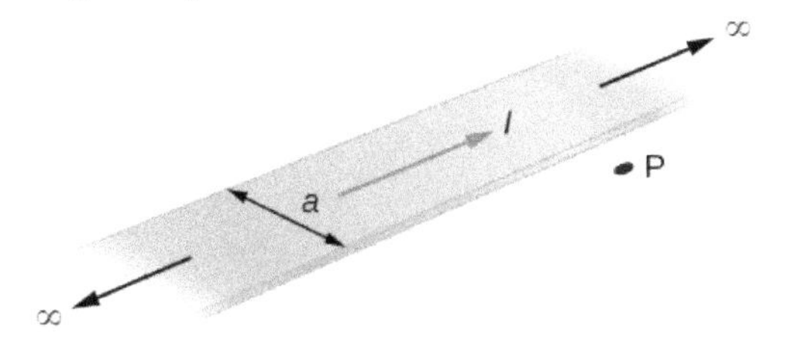

90. A hypothetical current flowing in the z-direction creates the field $\vec{\mathbf{B}} = C\left[\left(x/y^2\right)\hat{\mathbf{i}} + (1/y)\hat{\mathbf{j}}\right]$ in the rectangular region of the xy-plane shown in the accompanying figure. Use Ampère's law to find the current through the rectangle.

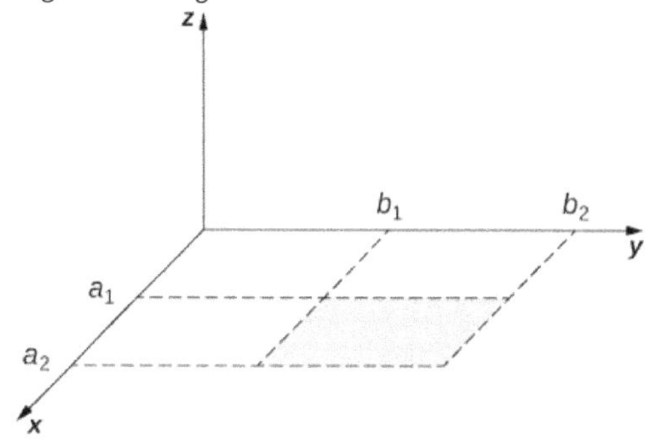

91. A nonconducting hard rubber circular disk of radius R is painted with a uniform surface charge density σ. It is rotated about its axis with angular speed ω. (a) Find the magnetic field produced at a point on the axis a distance h meters from the center of the disk. (b) Find the numerical value of magnitude of the magnetic field when $\sigma = 1\text{C/m}^2$, $R = 20\text{ cm}$, $h = 2\text{ cm}$, and $\omega = 400\text{ rad/sec}$, and compare it with the magnitude of magnetic field of Earth, which is about 1/2 Gauss.

13 | ELECTROMAGNETIC INDUCTION

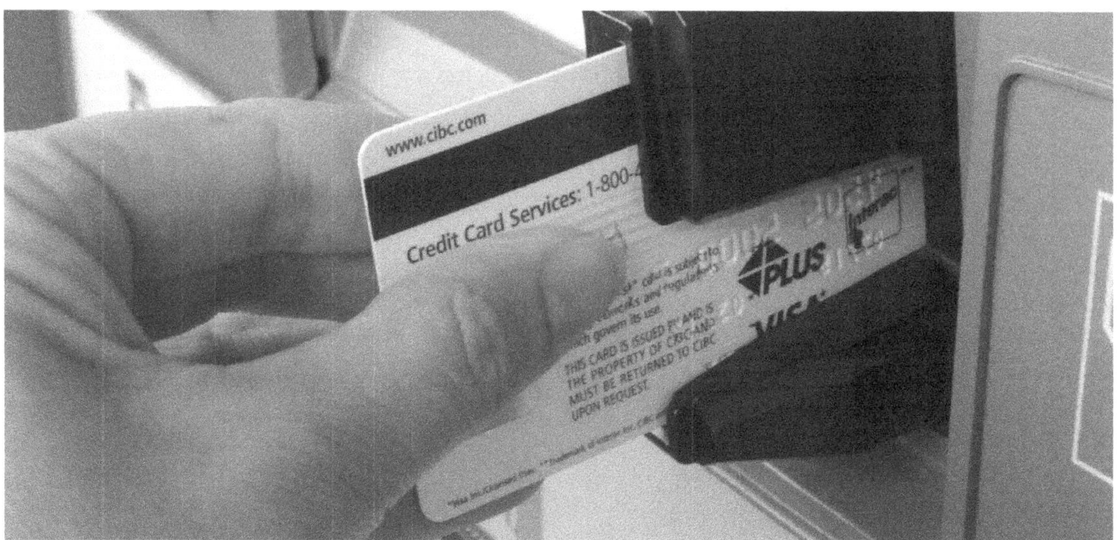

Figure 13.1 The black strip found on the back of credit cards and driver's licenses is a very thin layer of magnetic material with information stored on it. Reading and writing the information on the credit card is done with a swiping motion. The physical reason why this is necessary is called electromagnetic induction and is discussed in this chapter.

Chapter Outline

Introduction

We have been considering electric fields created by fixed charge distributions and magnetic fields produced by constant currents, but electromagnetic phenomena are not restricted to these stationary situations. Most of the interesting applications of electromagnetism are, in fact, time-dependent. To investigate some of these applications, we now remove the time-independent assumption that we have been making and allow the fields to vary with time. In this and the next several chapters, you will see a wonderful symmetry in the behavior exhibited by time-varying electric and magnetic fields. Mathematically, this symmetry is expressed by an additional term in Ampère's law and by another key equation of electromagnetism called Faraday's law. We also discuss how moving a wire through a magnetic field produces an emf or voltage. Lastly, we describe applications of these principles, such as the card reader shown above.

13.1 | Faraday's Law

Learning Objectives

By the end of this section, you will be able to:

- Determine the magnetic flux through a surface, knowing the strength of the magnetic field, the surface area, and the angle between the normal to the surface and the magnetic field
- Use Faraday's law to determine the magnitude of induced emf in a closed loop due to changing magnetic flux through the loop

The first productive experiments concerning the effects of time-varying magnetic fields were performed by Michael Faraday in 1831. One of his early experiments is represented in Figure 13.2. An emf is induced when the magnetic field in the coil is changed by pushing a bar magnet into or out of the coil. Emfs of opposite signs are produced by motion in opposite directions, and the directions of emfs are also reversed by reversing poles. The same results are produced if the coil is moved rather than the magnet—it is the relative motion that is important. The faster the motion, the greater the emf, and there is no emf when the magnet is stationary relative to the coil.

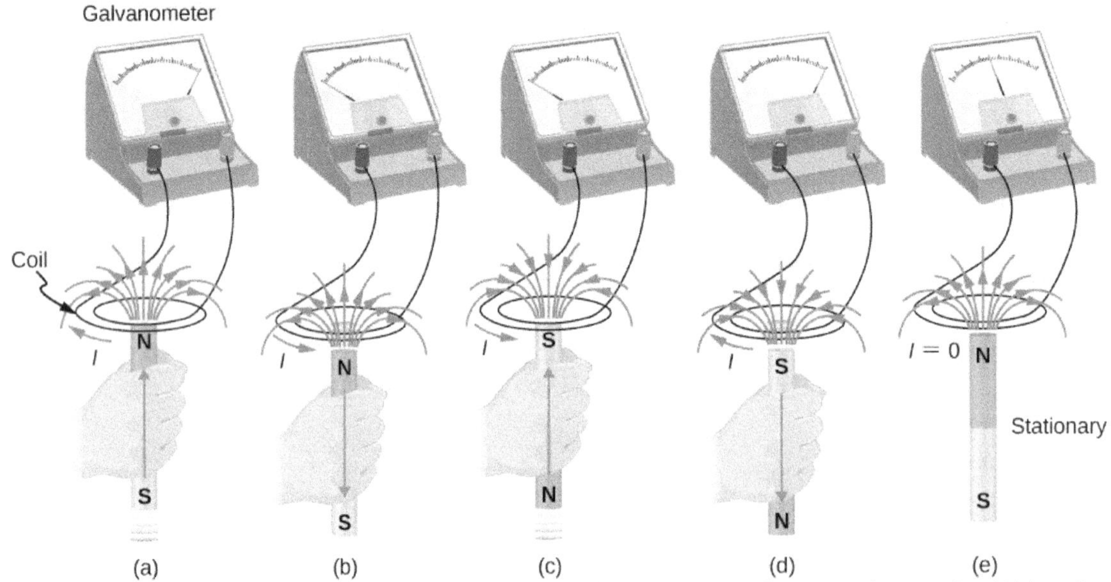

Figure 13.2 Movement of a magnet relative to a coil produces emfs as shown (a–d). The same emfs are produced if the coil is moved relative to the magnet. This short-lived emf is only present during the motion. The greater the speed, the greater the magnitude of the emf, and the emf is zero when there is no motion, as shown in (e).

Faraday also discovered that a similar effect can be produced using two circuits—a changing current in one circuit induces a current in a second, nearby circuit. For example, when the switch is closed in circuit 1 of Figure 13.3(a), the ammeter needle of circuit 2 momentarily deflects, indicating that a short-lived current surge has been induced in that circuit. The ammeter needle quickly returns to its original position, where it remains. However, if the switch of circuit 1 is now suddenly opened, another short-lived current surge in the direction opposite from before is observed in circuit 2.

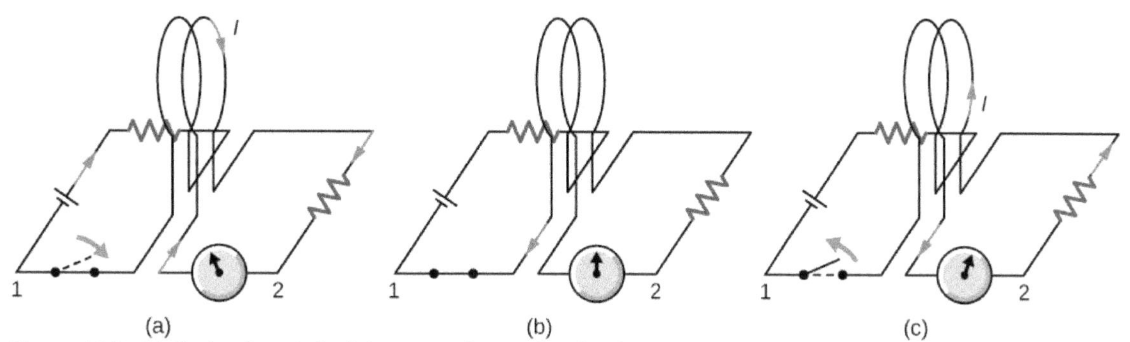

Figure 13.3 (a) Closing the switch of circuit 1 produces a short-lived current surge in circuit 2. (b) If the switch remains closed, no current is observed in circuit 2. (c) Opening the switch again produces a short-lived current in circuit 2 but in the opposite direction from before.

Faraday realized that in both experiments, a current flowed in the circuit containing the ammeter only when the magnetic field in the region occupied by that circuit was *changing*. As the magnet of the figure was moved, the strength of its magnetic field at the loop changed; and when the current in circuit 1 was turned on or off, the strength of its magnetic field at circuit 2 changed. Faraday was eventually able to interpret these and all other experiments involving magnetic fields that vary with time in terms of the following law:

Faraday's Law

The emf ε induced is the negative change in the magnetic flux Φ_{m} per unit time. Any change in the magnetic field or change in orientation of the area of the coil with respect to the magnetic field induces a voltage (emf).

The **magnetic flux** is a measurement of the amount of magnetic field lines through a given surface area, as seen in Figure 13.4. This definition is similar to the electric flux studied earlier. This means that if we have

$$\Phi_{\mathrm{m}} = \int_S \vec{\mathbf{B}} \cdot \hat{\mathbf{n}} dA, \tag{13.1}$$

then the **induced emf** or the voltage generated by a conductor or coil moving in a magnetic field is

$$\varepsilon = -\frac{d}{dt}\int_S \vec{\mathbf{B}} \cdot \hat{\mathbf{n}} dA = -\frac{d\Phi_{\mathrm{m}}}{dt}. \tag{13.2}$$

The negative sign describes the direction in which the induced emf drives current around a circuit. However, that direction is most easily determined with a rule known as Lenz's law, which we will discuss shortly.

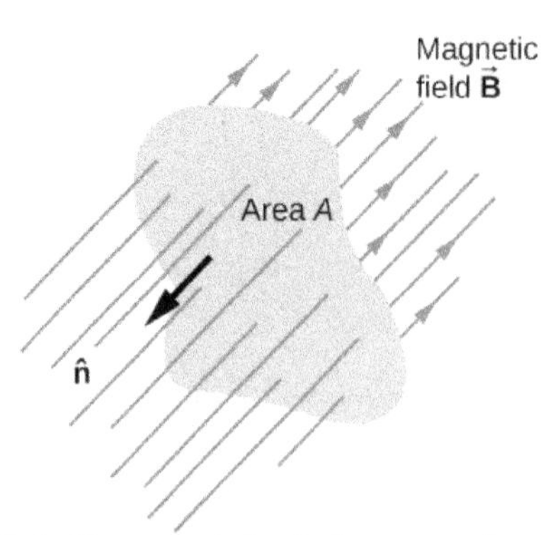

Figure 13.4 The magnetic flux is the amount of magnetic field lines cutting through a surface area *A* defined by the unit area vector $\hat{\mathbf{n}}$. If the angle between the unit area $\hat{\mathbf{n}}$ and magnetic field vector $\vec{\mathbf{B}}$ are parallel or antiparallel, as shown in the diagram, the magnetic flux is the highest possible value given the values of area and magnetic field.

Part (a) of Figure 13.5 depicts a circuit and an arbitrary surface *S* that it bounds. Notice that *S* is an *open surface*. It can be shown that *any* open surface bounded by the circuit in question can be used to evaluate Φ_m. For example, Φ_m is the same for the various surfaces $S_1, S_2, \ldots$ of part (b) of the figure.

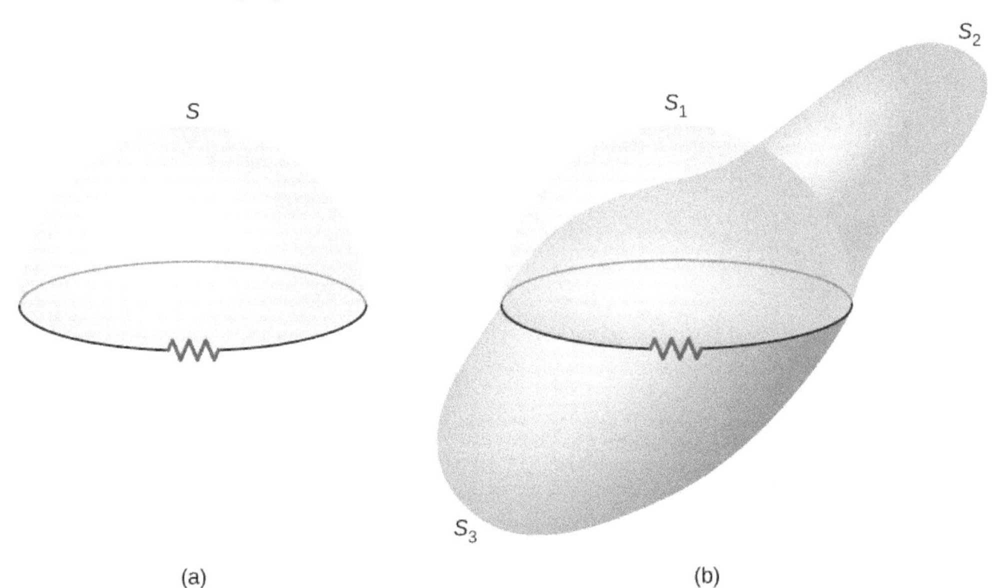

Figure 13.5 (a) A circuit bounding an arbitrary open surface *S*. The planar area bounded by the circuit is not part of *S*. (b) Three arbitrary open surfaces bounded by the same circuit. The value of Φ_m is the same for all these surfaces.

The SI unit for magnetic flux is the weber (Wb),

$$1\ \text{Wb} = 1\ \text{T} \cdot \text{m}^2.$$

Occasionally, the magnetic field unit is expressed as webers per square meter (Wb/m^2) instead of teslas, based on this definition. In many practical applications, the circuit of interest consists of a number *N* of tightly wound turns (see Figure 13.6). Each turn experiences the same magnetic flux. Therefore, the net magnetic flux through the circuits is *N* times the flux through one turn, and Faraday's law is written as

$$\varepsilon = -\frac{d}{dt}(N\Phi_m) = -N\frac{d\Phi_m}{dt}. \qquad (13.3)$$

Example 13.1

A Square Coil in a Changing Magnetic Field

The square coil of Figure 13.6 has sides $l = 0.25$ m long and is tightly wound with $N = 200$ turns of wire. The resistance of the coil is $R = 5.0\,\Omega$. The coil is placed in a spatially uniform magnetic field that is directed perpendicular to the face of the coil and whose magnitude is decreasing at a rate $dB/dt = -0.040$ T/s. (a) What is the magnitude of the emf induced in the coil? (b) What is the magnitude of the current circulating through the coil?

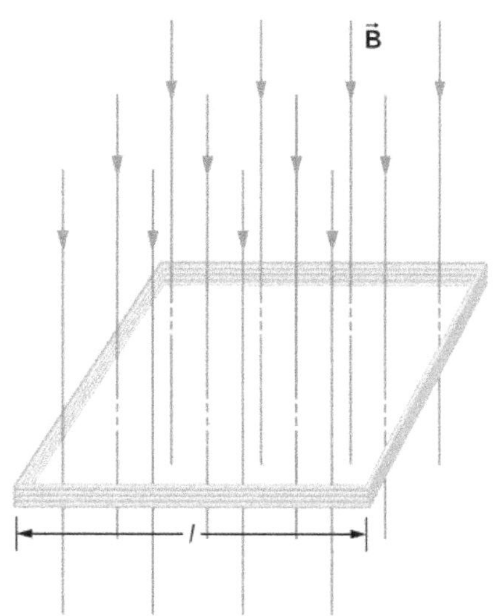

Figure 13.6 A square coil with N turns of wire with uniform magnetic field $\vec{\mathbf{B}}$ directed in the downward direction, perpendicular to the coil.

Strategy

The area vector, or $\hat{\mathbf{n}}$ direction, is perpendicular to area covering the loop. We will choose this to be pointing downward so that $\vec{\mathbf{B}}$ is parallel to $\hat{\mathbf{n}}$ and that the flux turns into multiplication of magnetic field times area. The area of the loop is not changing in time, so it can be factored out of the time derivative, leaving the magnetic field as the only quantity varying in time. Lastly, we can apply Ohm's law once we know the induced emf to find the current in the loop.

Solution

a. The flux through one turn is

$$\Phi_m = BA = Bl^2,$$

so we can calculate the magnitude of the emf from Faraday's law. The sign of the emf will be discussed in the next section, on Lenz's law:

$$\begin{aligned}|\varepsilon| &= \left|-N\frac{d\Phi_m}{dt}\right| = Nl^2\frac{dB}{dt} \\ &= (200)(0.25\text{ m})^2(0.040\text{ T/s}) = 0.50\text{ V}.\end{aligned}$$

b. The magnitude of the current induced in the coil is

$$I = \frac{\varepsilon}{R} = \frac{0.50\,\text{V}}{5.0\,\Omega} = 0.10\,\text{A}.$$

Significance

If the area of the loop were changing in time, we would not be able to pull it out of the time derivative. Since the loop is a closed path, the result of this current would be a small amount of heating of the wires until the magnetic field stops changing. This may increase the area of the loop slightly as the wires are heated.

13.1 Check Your Understanding A closely wound coil has a radius of 4.0 cm, 50 turns, and a total resistance of $40\,\Omega$. At what rate must a magnetic field perpendicular to the face of the coil change in order to produce Joule heating in the coil at a rate of 2.0 mW?

13.2 | Lenz's Law

Learning Objectives

By the end of this section, you will be able to:

- Use Lenz's law to determine the direction of induced emf whenever a magnetic flux changes
- Use Faraday's law with Lenz's law to determine the induced emf in a coil and in a solenoid

The direction in which the induced emf drives current around a wire loop can be found through the negative sign. However, it is usually easier to determine this direction with **Lenz's law**, named in honor of its discoverer, Heinrich Lenz (1804–1865). (Faraday also discovered this law, independently of Lenz.) We state Lenz's law as follows:

Lenz's Law

The direction of the induced emf drives current around a wire loop to always *oppose* the change in magnetic flux that causes the emf.

Lenz's law can also be considered in terms of conservation of energy. If pushing a magnet into a coil causes current, the energy in that current must have come from somewhere. If the induced current causes a magnetic field opposing the increase in field of the magnet we pushed in, then the situation is clear. We pushed a magnet against a field and did work on the system, and that showed up as current. If it were not the case that the induced field opposes the change in the flux, the magnet would be pulled in produce a current without anything having done work. Electric potential energy would have been created, violating the conservation of energy.

To determine an induced emf ε, you first calculate the magnetic flux Φ_m and then obtain $d\Phi_m/dt$. The magnitude of ε is given by $\varepsilon = |d\Phi_m/dt|$. Finally, you can apply Lenz's law to determine the sense of ε. This will be developed through examples that illustrate the following problem-solving strategy.

Problem-Solving Strategy: Lenz's Law

To use Lenz's law to determine the directions of induced magnetic fields, currents, and emfs:

1. Make a sketch of the situation for use in visualizing and recording directions.
2. Determine the direction of the applied magnetic field $\vec{\mathbf{B}}$.
3. Determine whether its magnetic flux is increasing or decreasing.

4. Now determine the direction of the induced magnetic field $\vec{\mathbf{B}}$. The induced magnetic field tries to reinforce a magnetic flux that is decreasing or opposes a magnetic flux that is increasing. Therefore, the induced magnetic field adds or subtracts to the applied magnetic field, depending on the change in magnetic flux.
5. Use right-hand rule 2 (RHR-2; see Magnetic Forces and Fields) to determine the direction of the induced current I that is responsible for the induced magnetic field $\vec{\mathbf{B}}$.
6. The direction (or polarity) of the induced emf can now drive a conventional current in this direction.

Let's apply Lenz's law to the system of Figure 13.7(a). We designate the "front" of the closed conducting loop as the region containing the approaching bar magnet, and the "back" of the loop as the other region. As the north pole of the magnet moves toward the loop, the flux through the loop due to the field of the magnet increases because the strength of field lines directed from the front to the back of the loop is increasing. A current is therefore induced in the loop. By Lenz's law, the direction of the induced current must be such that its own magnetic field is directed in a way to *oppose* the changing flux caused by the field of the approaching magnet. Hence, the induced current circulates so that its magnetic field lines through the loop are directed from the back to the front of the loop. By RHR-2, place your thumb pointing against the magnetic field lines, which is toward the bar magnet. Your fingers wrap in a counterclockwise direction as viewed from the bar magnet. Alternatively, we can determine the direction of the induced current by treating the current loop as an electromagnet that *opposes* the approach of the north pole of the bar magnet. This occurs when the induced current flows as shown, for then the face of the loop nearer the approaching magnet is also a north pole.

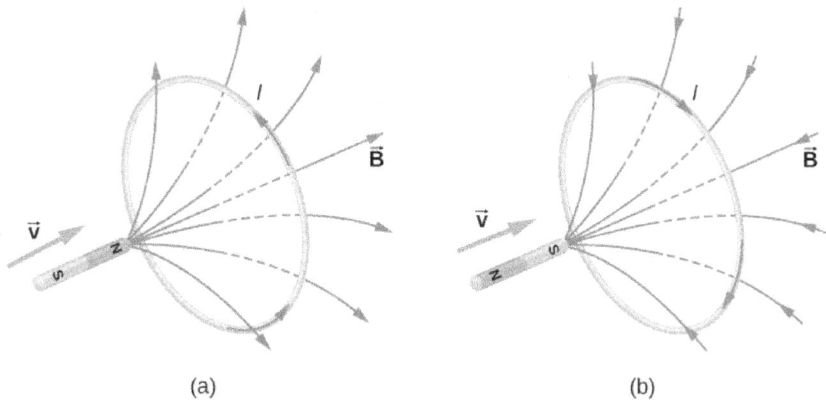

Figure 13.7 The change in magnetic flux caused by the approaching magnet induces a current in the loop. (a) An approaching north pole induces a counterclockwise current with respect to the bar magnet. (b) An approaching south pole induces a clockwise current with respect to the bar magnet.

Part (b) of the figure shows the south pole of a magnet moving toward a conducting loop. In this case, the flux through the loop due to the field of the magnet increases because the number of field lines directed from the back to the front of the loop is increasing. To oppose this change, a current is induced in the loop whose field lines through the loop are directed from the front to the back. Equivalently, we can say that the current flows in a direction so that the face of the loop nearer the approaching magnet is a south pole, which then repels the approaching south pole of the magnet. By RHR-2, your thumb points away from the bar magnet. Your fingers wrap in a clockwise fashion, which is the direction of the induced current.

Another example illustrating the use of Lenz's law is shown in Figure 13.8. When the switch is opened, the decrease in current through the solenoid causes a decrease in magnetic flux through its coils, which induces an emf in the solenoid. This emf must oppose the change (the termination of the current) causing it. Consequently, the induced emf has the polarity shown and drives in the direction of the original current. This may generate an arc across the terminals of the switch as it is opened.

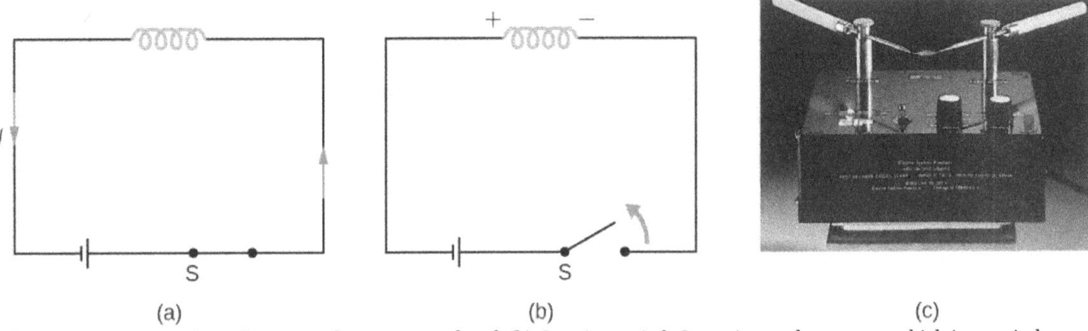

Figure 13.8 (a) A solenoid connected to a source of emf. (b) Opening switch S terminates the current, which in turn induces an emf in the solenoid. (c) A potential difference between the ends of the sharply pointed rods is produced by inducing an emf in a coil. This potential difference is large enough to produce an arc between the sharp points.

13.2 Check Your Understanding Find the direction of the induced current in the wire loop shown below as the magnet enters, passes through, and leaves the loop.

13.3 Check Your Understanding Verify the directions of the induced currents in Figure 13.3.

Example 13.2

A Circular Coil in a Changing Magnetic Field

A magnetic field $\vec{\mathbf{B}}$ is directed outward perpendicular to the plane of a circular coil of radius $r = 0.50\,\text{m}$ (Figure 13.9). The field is cylindrically symmetrical with respect to the center of the coil, and its magnitude decays exponentially according to $B = (1.5T)e^{-(5.0\text{s}^{-1})t}$, where B is in teslas and t is in seconds. (a) Calculate the emf induced in the coil at the times $t_1 = 0$, $t_2 = 5.0 \times 10^{-2}\,\text{s}$, and $t_3 = 1.0\,\text{s}$. (b) Determine the current in the coil at these three times if its resistance is $10\,\Omega$.

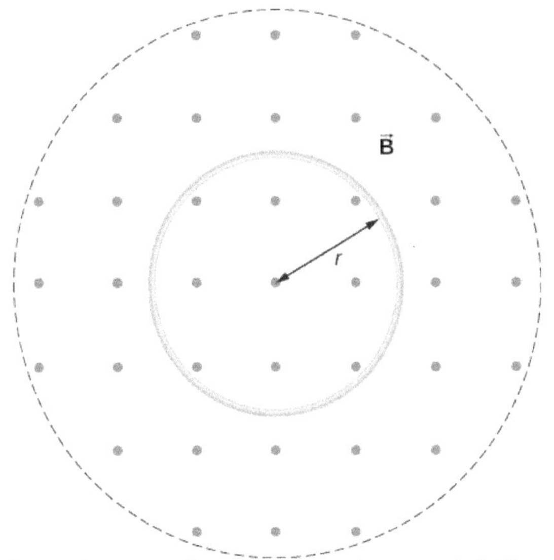

Figure 13.9 A circular coil in a decreasing magnetic field.

Strategy

Since the magnetic field is perpendicular to the plane of the coil and constant over each spot in the coil, the dot product of the magnetic field $\vec{\mathbf{B}}$ and normal to the area unit vector $\hat{\mathbf{n}}$ turns into a multiplication. The magnetic field can be pulled out of the integration, leaving the flux as the product of the magnetic field times area. We need to take the time derivative of the exponential function to calculate the emf using Faraday's law. Then we use Ohm's law to calculate the current.

Solution

a. Since $\vec{\mathbf{B}}$ is perpendicular to the plane of the coil, the magnetic flux is given by

$$\begin{aligned}\Phi_{\text{m}} &= B\pi r^2 = (1.5e^{-5.0t}\ \text{T})\pi(0.50\ \text{m})^2 \\ &= 1.2e^{-(5.0\text{s}^{-1})t}\ \text{Wb}.\end{aligned}$$

From Faraday's law, the magnitude of the induced emf is

$$\varepsilon = \left|\frac{d\Phi_m}{dt}\right| = \left|\frac{d}{dt}(1.2e^{-(5.0\text{s}^{-1})t}\ \text{Wb})\right| = 6.0\,e^{-(5.0\text{s}^{-1})t}\ \text{V}.$$

Since $\vec{\mathbf{B}}$ is directed out of the page and is decreasing, the induced current must flow counterclockwise when viewed from above so that the magnetic field it produces through the coil also points out of the page. For all three times, the sense of ε is counterclockwise; its magnitudes are

$$\varepsilon(t_1) = 6.0\ \text{V};\ \ \varepsilon(t_2) = 4.7\ \text{V};\ \ \varepsilon(t_3) = 0.040\ \text{V}.$$

b. From Ohm's law, the respective currents are

$$\begin{aligned}I(t_1) &= \frac{\varepsilon(t_1)}{R} = \frac{6.0\ \text{V}}{10\ \Omega} = 0.60\ \text{A}; \\ I(t_2) &= \frac{4.7\ \text{V}}{10\ \Omega} = 0.47\ \text{A};\end{aligned}$$

and

$$I(t_3) = \frac{0.040\ \text{V}}{10\ \Omega} = 4.0 \times 10^{-3}\ \text{A}.$$

Significance

An emf voltage is created by a changing magnetic flux over time. If we know how the magnetic field varies with time over a constant area, we can take its time derivative to calculate the induced emf.

Example 13.3

Changing Magnetic Field Inside a Solenoid

The current through the windings of a solenoid with $n = 2000$ turns per meter is changing at a rate $dI/dt = 3.0$ A/s. (See Sources of Magnetic Fields for a discussion of solenoids.) The solenoid is 50-cm long and has a cross-sectional diameter of 3.0 cm. A small coil consisting of $N = 20$ closely wound turns wrapped in a circle of diameter 1.0 cm is placed in the middle of the solenoid such that the plane of the coil is perpendicular to the central axis of the solenoid. Assuming that the infinite-solenoid approximation is valid at the location of the small coil, determine the magnitude of the emf induced in the coil.

Strategy

The magnetic field in the middle of the solenoid is a uniform value of $\mu_0 nI$. This field is producing a maximum magnetic flux through the coil as it is directed along the length of the solenoid. Therefore, the magnetic flux through the coil is the product of the solenoid's magnetic field times the area of the coil. Faraday's law involves a time derivative of the magnetic flux. The only quantity varying in time is the current, the rest can be pulled out of the time derivative. Lastly, we include the number of turns in the coil to determine the induced emf in the coil.

Solution

Since the field of the solenoid is given by $B = \mu_0 nI,$ the flux through each turn of the small coil is

$$\Phi_{\rm m} = \mu_0 nI\left(\frac{\pi d^2}{4}\right),$$

where d is the diameter of the coil. Now from Faraday's law, the magnitude of the emf induced in the coil is

$$\begin{aligned}\varepsilon &= \left|N\frac{d\Phi_{\rm m}}{dt}\right| = \left|N\mu_0 n\frac{\pi d^2}{4}\frac{dI}{dt}\right| \\ &= 20\left(4\pi \times 10^{-7}\ \text{T}\cdot\text{m/s}\right)\left(2000\ \text{m}^{-1}\right)\frac{\pi(0.010\ \text{m})^2}{4}(3.0\ \text{A/s}) \\ &= 1.2 \times 10^{-5}\ \text{V}.\end{aligned}$$

Significance

When the current is turned on in a vertical solenoid, as shown in Figure 13.10, the ring has an induced emf from the solenoid's changing magnetic flux that opposes the change. The result is that the ring is fired vertically into the air.

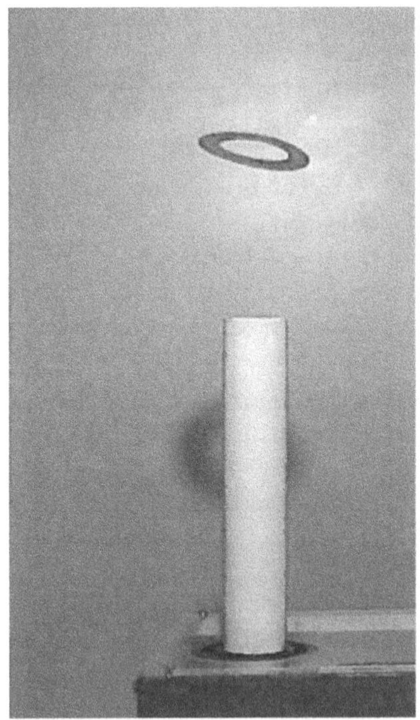

Figure 13.10 The jumping ring. When a current is turned on in the vertical solenoid, a current is induced in the metal ring. The stray field produced by the solenoid causes the ring to jump off the solenoid.

Visit this website (https://openstaxcollege.org/l/21flashmagind) for a demonstration of the jumping ring from MIT.

13.3 | Motional Emf

Learning Objectives

By the end of this section, you will be able to:

- Determine the magnitude of an induced emf in a wire moving at a constant speed through a magnetic field
- Discuss examples that use motional emf, such as a rail gun and a tethered satellite

Magnetic flux depends on three factors: the strength of the magnetic field, the area through which the field lines pass, and the orientation of the field with the surface area. If any of these quantities varies, a corresponding variation in magnetic flux occurs. So far, we've only considered flux changes due to a changing field. Now we look at another possibility: a changing area through which the field lines pass including a change in the orientation of the area.

Two examples of this type of flux change are represented in Figure 13.11. In part (a), the flux through the rectangular loop increases as it moves into the magnetic field, and in part (b), the flux through the rotating coil varies with the angle θ.

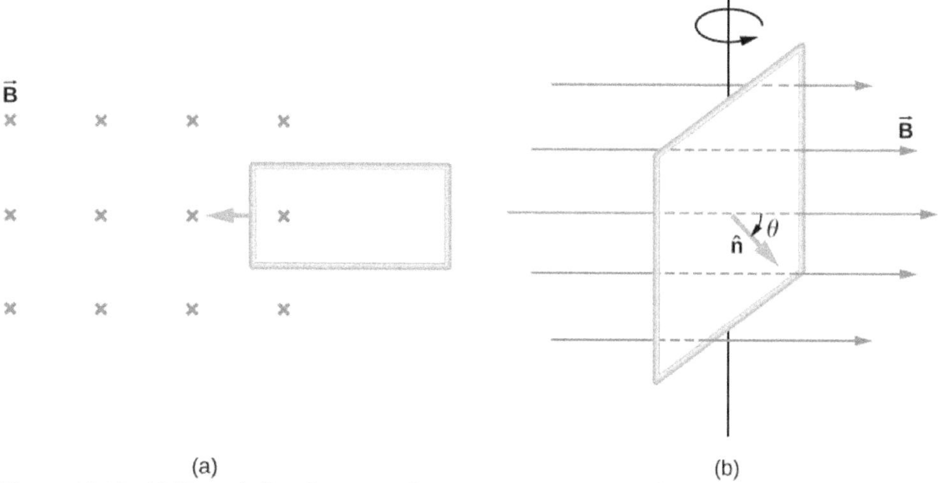

Figure 13.11 (a) Magnetic flux changes as a loop moves into a magnetic field; (b) magnetic flux changes as a loop rotates in a magnetic field.

It's interesting to note that what we perceive as the cause of a particular flux change actually depends on the frame of reference we choose. For example, if you are at rest relative to the moving coils of Figure 13.11, you would see the flux vary because of a changing magnetic field—in part (a), the field moves from left to right in your reference frame, and in part (b), the field is rotating. It is often possible to describe a flux change through a coil that is moving in one particular reference frame in terms of a changing magnetic field in a second frame, where the coil is stationary. However, reference-frame questions related to magnetic flux are beyond the level of this textbook. We'll avoid such complexities by always working in a frame at rest relative to the laboratory and explain flux variations as due to either a changing field or a changing area.

Now let's look at a conducting rod pulled in a circuit, changing magnetic flux. The area enclosed by the circuit 'MNOP' of Figure 13.12 is lx and is perpendicular to the magnetic field, so we can simplify the integration of Equation 13.1 into a multiplication of magnetic field and area. The magnetic flux through the open surface is therefore

$$\Phi_{\mathrm{m}} = Blx. \tag{13.4}$$

Since B and l are constant and the velocity of the rod is $v = dx/dt$, we can now restate Faraday's law, Equation 13.2, for the magnitude of the emf in terms of the moving conducting rod as

$$\varepsilon = \frac{d\Phi_{\text{m}}}{dt} = Bl\frac{dx}{dt} = Blv. \tag{13.5}$$

The current induced in the circuit is the emf divided by the resistance or

$$I = \frac{Blv}{R}.$$

Furthermore, the direction of the induced emf satisfies Lenz's law, as you can verify by inspection of the figure.

This calculation of motionally induced emf is not restricted to a rod moving on conducting rails. With $\vec{\mathbf{F}} = q\vec{\mathbf{v}} \times \vec{\mathbf{B}}$ as the starting point, it can be shown that $\varepsilon = -d\Phi_{\text{m}}/dt$ holds for any change in flux caused by the motion of a conductor. We saw in Faraday's Law that the emf induced by a time-varying magnetic field obeys this same relationship, which is Faraday's law. Thus Faraday's law *holds for all flux changes*, whether they are produced by a changing magnetic field, by motion, or by a combination of the two.

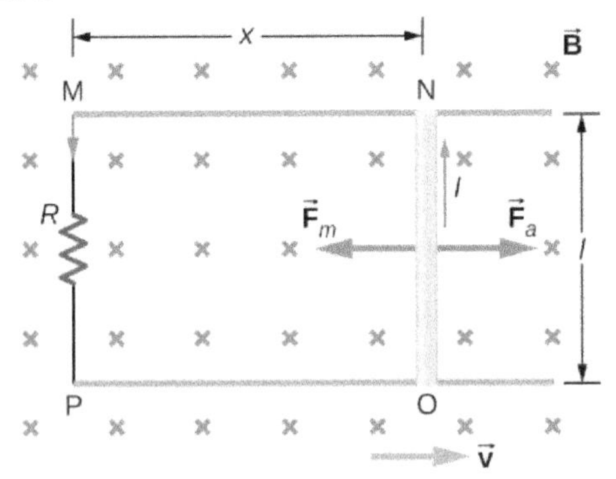

Figure 13.12 A conducting rod is pushed to the right at constant velocity. The resulting change in the magnetic flux induces a current in the circuit.

From an energy perspective, $\vec{\mathbf{F}}_a$ produces power $F_a v$, and the resistor dissipates power I^2R. Since the rod is moving at constant velocity, the applied force F_a must balance the magnetic force $F_{\text{m}} = IlB$ on the rod when it is carrying the induced current I. Thus the power produced is

$$F_a v = IlBv = \frac{Blv}{R} \cdot lBv = \frac{l^2B^2v^2}{R}. \tag{13.6}$$

The power dissipated is

$$P = I^2R = \left(\frac{Blv}{R}\right)^2 R = \frac{l^2B^2v^2}{R}. \tag{13.7}$$

In satisfying the principle of energy conservation, the produced and dissipated powers are equal.

This principle can be seen in the operation of a rail gun. A rail gun is an electromagnetic projectile launcher that uses an apparatus similar to Figure 13.12 and is shown in schematic form in Figure 13.13. The conducting rod is replaced with a projectile or weapon to be fired. So far, we've only heard about how motion causes an emf. In a rail gun, the optimal shutting off/ramping down of a magnetic field decreases the flux in between the rails, causing a current to flow in the rod (armature) that holds the projectile. This current through the armature experiences a magnetic force and is propelled forward. Rail guns, however, are not used widely in the military due to the high cost of production and high currents: Nearly one million amps is required to produce enough energy for a rail gun to be an effective weapon.

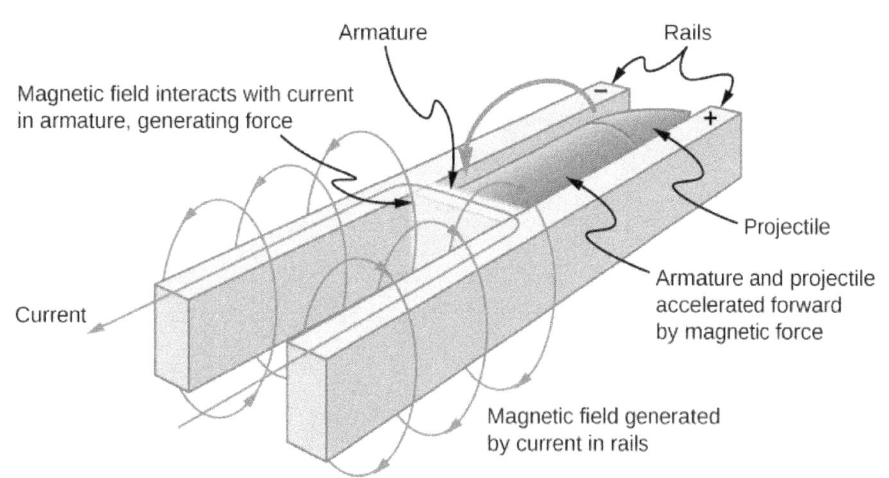

Figure 13.13 Current through two rails drives a conductive projectile forward by the magnetic force created.

We can calculate a **motionally induced emf** with Faraday's law *even when an actual closed circuit is not present.* We simply imagine an enclosed area whose boundary includes the moving conductor, calculate Φ_m, and then find the emf from Faraday's law. For example, we can let the moving rod of Figure 13.14 be one side of the imaginary rectangular area represented by the dashed lines. The area of the rectangle is lx, so the magnetic flux through it is $\Phi_m = Blx$. Differentiating this equation, we obtain

$$\frac{d\Phi_m}{dt} = Bl\frac{dx}{dt} = Blv, \tag{13.8}$$

which is identical to the potential difference between the ends of the rod that we determined earlier.

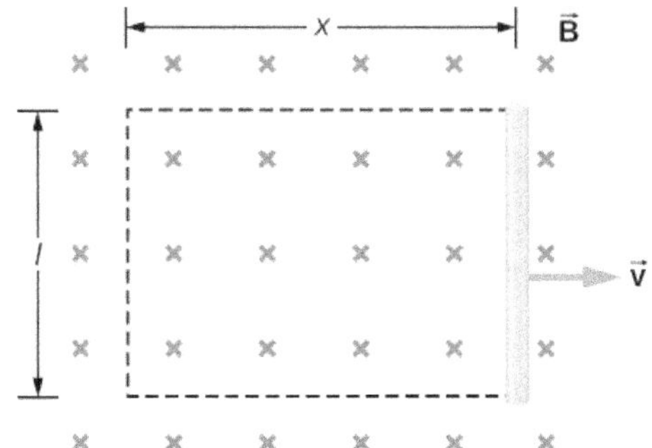

Figure 13.14 With the imaginary rectangle shown, we can use Faraday's law to calculate the induced emf in the moving rod.

Motional emfs in Earth's weak magnetic field are not ordinarily very large, or we would notice voltage along metal rods, such as a screwdriver, during ordinary motions. For example, a simple calculation of the motional emf of a 1.0-m rod moving at 3.0 m/s perpendicular to the Earth's field gives

$$\text{emf} = B\ell v = (5.0 \times 10^{-5}\ \text{T})(1.0\ \text{m})(3.0\ \text{m/s}) = 150\mu\text{V}.$$

This small value is consistent with experience. There is a spectacular exception, however. In 1992 and 1996, attempts were made with the space shuttle to create large motional emfs. The tethered satellite was to be let out on a 20-km length of wire, as shown in Figure 13.15, to create a 5-kV emf by moving at orbital speed through Earth's field. This emf could be used to convert some of the shuttle's kinetic and potential energy into electrical energy if a complete circuit could be made. To complete the circuit, the stationary ionosphere was to supply a return path through which current could flow. (The ionosphere is the rarefied and partially ionized atmosphere at orbital altitudes. It conducts because of the ionization. The ionosphere serves the same function as the stationary rails and connecting resistor in Figure 13.13, without which there would not be a complete circuit.) Drag on the current in the cable due to the magnetic force $F = I\ell B \sin\theta$ does the work that reduces the shuttle's kinetic and potential energy, and allows it to be converted into electrical energy. Both tests were unsuccessful. In the first, the cable hung up and could only be extended a couple of hundred meters; in the second, the cable broke when almost fully extended. Example 13.4 indicates feasibility in principle.

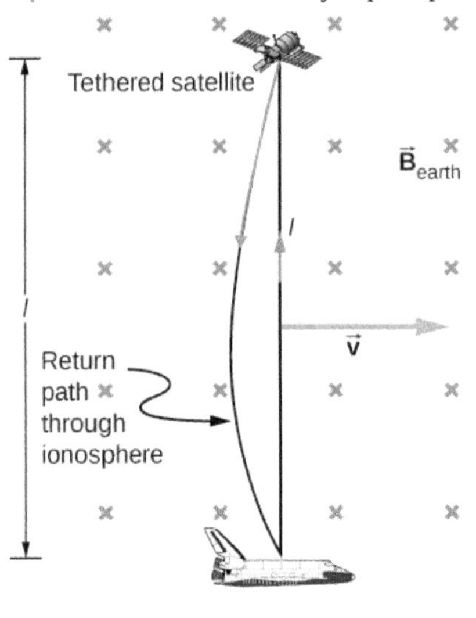

Figure 13.15 Motional emf as electrical power conversion for the space shuttle was the motivation for the tethered satellite experiment. A 5-kV emf was predicted to be induced in the 20-km tether while moving at orbital speed in Earth's magnetic field. The circuit is completed by a return path through the stationary ionosphere.

Example 13.4

Calculating the Large Motional Emf of an Object in Orbit

Calculate the motional emf induced along a 20.0-km conductor moving at an orbital speed of 7.80 km/s perpendicular to Earth's 5.00×10^{-5} T magnetic field.

Strategy

This is a great example of using the equation motional $\varepsilon = B\ell v$.

Solution

Entering the given values into $\varepsilon = B\ell v$ gives

$$\begin{aligned} \varepsilon &= B\ell v \\ &= (5.00 \times 10^{-5}\ \text{T})(2.00 \times 10^{4}\ \text{m})(7.80 \times 10^{3}\ \text{m/s}) \\ &= 7.80 \times 10^{3}\ \text{V}. \end{aligned}$$

Significance

The value obtained is greater than the 5-kV measured voltage for the shuttle experiment, since the actual orbital motion of the tether is not perpendicular to Earth's field. The 7.80-kV value is the maximum emf obtained when $\theta = 90°$ and so $\sin\theta = 1$.

Example 13.5

A Metal Rod Rotating in a Magnetic Field

Part (a) of Figure 13.16 shows a metal rod *OS* that is rotating in a horizontal plane around point *O*. The rod slides along a wire that forms a circular arc *PST* of radius *r*. The system is in a constant magnetic field $\vec{\mathbf{B}}$ that is directed out of the page. (a) If you rotate the rod at a constant angular velocity ω, what is the current *I* in the closed loop *OPSO*? Assume that the resistor *R* furnishes all of the resistance in the closed loop. (b) Calculate the work per unit time that you do while rotating the rod and show that it is equal to the power dissipated in the resistor.

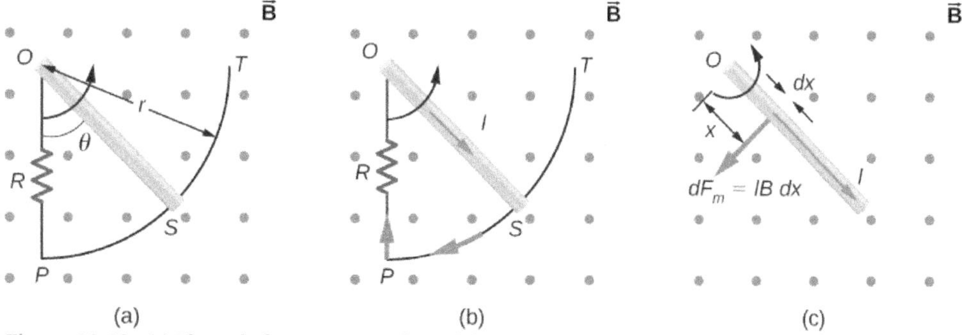

Figure 13.16 (a) The end of a rotating metal rod slides along a circular wire in a horizontal plane. (b) The induced current in the rod. (c) The magnetic force on an infinitesimal current segment.

Strategy

The magnetic flux is the magnetic field times the area of the quarter circle or $A = r^2\theta/2$. When finding the emf through Faraday's law, all variables are constant in time but θ, with $\omega = d\theta/dt$. To calculate the work per unit time, we know this is related to the torque times the angular velocity. The torque is calculated by knowing the force on a rod and integrating it over the length of the rod.

Solution

a. From geometry, the area of the loop *OPSO* is $A = \frac{r^2\theta}{2}$. Hence, the magnetic flux through the loop is

$$\Phi_{\text{m}} = BA = B\frac{r^2\theta}{2}.$$

Differentiating with respect to time and using $\omega = d\theta/dt$, we have

$$\varepsilon = \left|\frac{d\Phi_{\text{m}}}{dt}\right| = \frac{Br^2\omega}{2}.$$

When divided by the resistance *R* of the loop, this yields for the magnitude of the induced current

$$I = \frac{\varepsilon}{R} = \frac{Br^2\omega}{2R}.$$

As θ increases, so does the flux through the loop due to $\vec{\mathbf{B}}$. To counteract this increase, the magnetic field due to the induced current must be directed into the page in the region enclosed by the loop. Therefore, as part (b) of Figure 13.16 illustrates, the current circulates clockwise.

b. You rotate the rod by exerting a torque on it. Since the rod rotates at constant angular velocity, this torque is equal and opposite to the torque exerted on the current in the rod by the original magnetic field. The magnetic force on the infinitesimal segment of length dx shown in part (c) of Figure 13.16 is $dF_{\text{m}} = IBdx,$ so the magnetic torque on this segment is

$$d\tau_{\text{m}} = x \cdot dF_{\text{m}} = IBxdx.$$

The net magnetic torque on the rod is then

$$\tau_{\text{m}} = \int_0^r d\tau_{\text{m}} = IB\int_0^r x\,dx = \frac{1}{2}IBr^2.$$

The torque τ that you exert on the rod is equal and opposite to $\tau_{\text{m}},$ and the work that you do when the rod rotates through an angle $d\theta$ is $dW = \tau d\theta.$ Hence, the work per unit time that you do on the rod is

$$\frac{dW}{dt} = \tau\frac{d\theta}{dt} = \frac{1}{2}IBr^2\frac{d\theta}{dt} = \frac{1}{2}\left(\frac{Br^2\omega}{2R}\right)Br^2\omega = \frac{B^2r^4\omega^2}{4R},$$

where we have substituted for I. The power dissipated in the resister is $P = I^2R$, which can be written as

$$P = \left(\frac{Br^2\omega}{2R}\right)^2 R = \frac{B^2r^4\omega^2}{4R}.$$

Therefore, we see that

$$P = \frac{dW}{dt}.$$

Hence, the power dissipated in the resistor is equal to the work per unit time done in rotating the rod.

Significance

An alternative way of looking at the induced emf from Faraday's law is to integrate in space instead of time. The solution, however, would be the same. The motional emf is

$$|\varepsilon| = \int Bvdl.$$

The velocity can be written as the angular velocity times the radius and the differential length written as dr. Therefore,

$$|\varepsilon| = B\int vdr = B\omega\int_0^l rdr = \frac{1}{2}B\omega l^2,$$

which is the same solution as before.

Example 13.6

A Rectangular Coil Rotating in a Magnetic Field

A rectangular coil of area A and N turns is placed in a uniform magnetic field $\vec{\mathbf{B}} = B\hat{\mathbf{j}},$ as shown in Figure 13.17. The coil is rotated about the z-axis through its center at a constant angular velocity ω. Obtain an expression for the induced emf in the coil.

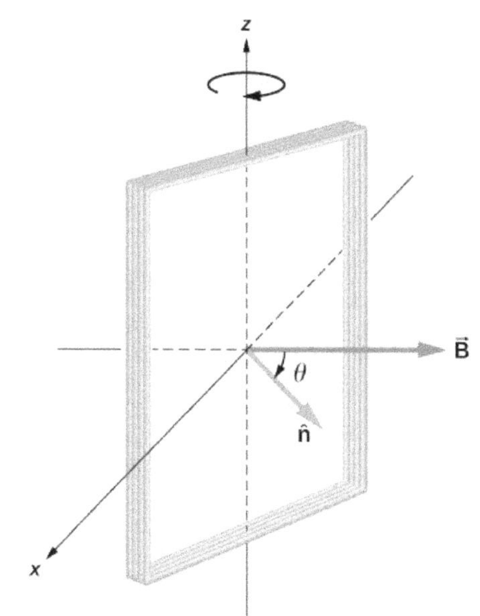

Figure 13.17 A rectangular coil rotating in a uniform magnetic field.

Strategy

According to the diagram, the angle between the perpendicular to the surface ($\hat{n}$) and the magnetic field ($\vec{\mathbf{B}}$) is θ. The dot product of $\vec{\mathbf{B}} \cdot \hat{\mathbf{n}}$ simplifies to only the $\cos\theta$ component of the magnetic field, namely where the magnetic field projects onto the unit area vector $\hat{n}$. The magnitude of the magnetic field and the area of the loop are fixed over time, which makes the integration simplify quickly. The induced emf is written out using Faraday's law.

Solution

When the coil is in a position such that its normal vector $\hat{\mathbf{n}}$ makes an angle θ with the magnetic field $\vec{\mathbf{B}}$, the magnetic flux through a single turn of the coil is

$$\Phi_{\text{m}} = \int_S \vec{\mathbf{B}} \cdot \hat{\mathbf{n}} dA = BA\cos\theta.$$

From Faraday's law, the emf induced in the coil is

$$\varepsilon = -N\frac{d\Phi_{\text{m}}}{dt} = NBA\sin\theta\frac{d\theta}{dt}.$$

The constant angular velocity is $\omega = d\theta/dt$. The angle θ represents the time evolution of the angular velocity or ωt. This is changes the function to time space rather than θ. The induced emf therefore varies sinusoidally with time according to

$$\varepsilon = \varepsilon_0 \sin\omega t,$$

where $\varepsilon_0 = NBA\omega$.

Significance

If the magnetic field strength or area of the loop were also changing over time, these variables wouldn't be able to be pulled out of the time derivative to simply the solution as shown. This example is the basis for an electric generator, as we will give a full discussion in Applications of Newton's Law (http://cnx.org/content/m58302/latest/) .

13.4 Check Your Understanding Shown below is a rod of length l that is rotated counterclockwise around the axis through O by the torque due to $m\vec{g}$. Assuming that the rod is in a uniform magnetic field $\vec{B}$, what is the emf induced between the ends of the rod when its angular velocity is ω? Which end of the rod is at a higher potential?

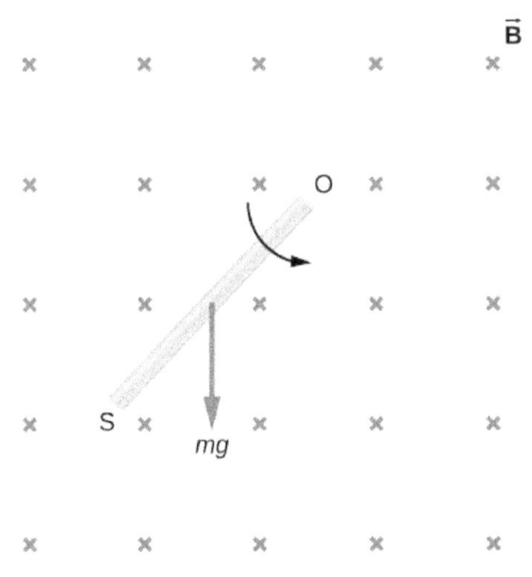

13.5 Check Your Understanding A rod of length 10 cm moves at a speed of 10 m/s perpendicularly through a 1.5-T magnetic field. What is the potential difference between the ends of the rod?

13.4 | Induced Electric Fields

Learning Objectives

By the end of this section, you will be able to:

- Connect the relationship between an induced emf from Faraday's law to an electric field, thereby showing that a changing magnetic flux creates an electric field
- Solve for the electric field based on a changing magnetic flux in time

The fact that emfs are induced in circuits implies that work is being done on the conduction electrons in the wires. What can possibly be the source of this work? We know that it's neither a battery nor a magnetic field, for a battery does not have to be present in a circuit where current is induced, and magnetic fields never do work on moving charges. The answer is that the source of the work is an electric field $\vec{E}$ that is induced in the wires. The work done by $\vec{E}$ in moving a unit charge completely around a circuit is the induced emf ε; that is,

$$\varepsilon = \oint \vec{E} \cdot d\vec{l}, \tag{13.9}$$

where $\oint$ represents the line integral around the circuit. Faraday's law can be written in terms of the **induced electric field** as

$$\oint \vec{E} \cdot d\vec{l} = -\frac{d\Phi_m}{dt}. \tag{13.10}$$

There is an important distinction between the electric field induced by a changing magnetic field and the electrostatic field produced by a fixed charge distribution. Specifically, the induced electric field is nonconservative because it does net work in moving a charge over a closed path, whereas the electrostatic field is conservative and does no net work over a closed path. Hence, electric potential can be associated with the electrostatic field, but not with the induced field. The following equations represent the distinction between the two types of electric field:

$$\oint \vec{\mathbf{E}} \cdot d\vec{\mathbf{l}} \neq 0 \text{ (induced)}; \qquad (13.11)$$
$$\oint \vec{\mathbf{E}} \cdot d\vec{\mathbf{l}} = 0 \text{ (electrostatic)}.$$

Our results can be summarized by combining these equations:

$$\varepsilon = \oint \vec{\mathbf{E}} \cdot d\vec{\mathbf{l}} = -\frac{d\Phi_m}{dt}. \qquad (13.12)$$

Example 13.7

Induced Electric Field in a Circular Coil

What is the induced electric field in the circular coil of Example 13.2 (and Figure 13.9) at the three times indicated?

Strategy

Using cylindrical symmetry, the electric field integral simplifies into the electric field times the circumference of a circle. Since we already know the induced emf, we can connect these two expressions by Faraday's law to solve for the induced electric field.

Solution

The induced electric field in the coil is constant in magnitude over the cylindrical surface, similar to how Ampere's law problems with cylinders are solved. Since $\vec{\mathbf{E}}$ is tangent to the coil,

$$\oint \vec{\mathbf{E}} \cdot d\vec{\mathbf{l}} = \oint E dl = 2\pi r E.$$

When combined with Equation 13.12, this gives

$$E = \frac{\varepsilon}{2\pi r}.$$

The direction of ε is counterclockwise, and $\vec{\mathbf{E}}$ circulates in the same direction around the coil. The values of E are

$$E(t_1) = \frac{6.0\text{ V}}{2\pi(0.50\text{ m})} = 1.9\text{ V/m};$$
$$E(t_2) = \frac{4.7\text{ V}}{2\pi(0.50\text{ m})} = 1.5\text{ V/m};$$
$$E(t_3) = \frac{0.040\text{ V}}{2\pi(0.50\text{ m})} = 0.013\text{ V/m}.$$

Significance

When the magnetic flux through a circuit changes, a nonconservative electric field is induced, which drives current through the circuit. But what happens if $dB/dt \neq 0$ in free space where there isn't a conducting path? The answer is that this case can be treated *as if a conducting path were present*; that is, nonconservative electric fields are induced wherever $dB/dt \neq 0$, whether or not there is a conducting path present.

These nonconservative electric fields always satisfy Equation 13.12. For example, if the circular coil of Figure 13.9 were removed, an electric field *in free space* at $r = 0.50\text{ m}$ would still be directed counterclockwise, and its magnitude would still be 1.9 V/m at $t = 0$, 1.5 V/m at $t = 5.0 \times 10^{-2}$ s, etc. The existence of induced electric fields is certainly *not* restricted to wires in circuits.

Example 13.8

Electric Field Induced by the Changing Magnetic Field of a Solenoid

Part (a) of Figure 13.18 shows a long solenoid with radius R and n turns per unit length; its current decreases with time according to $I = I_0 e^{-\alpha t}$. What is the magnitude of the induced electric field at a point a distance r from the central axis of the solenoid (a) when $r > R$ and (b) when $r < R$ [see part (b) of Figure 13.18]. (c) What is the direction of the induced field at both locations? Assume that the infinite-solenoid approximation is valid throughout the regions of interest.

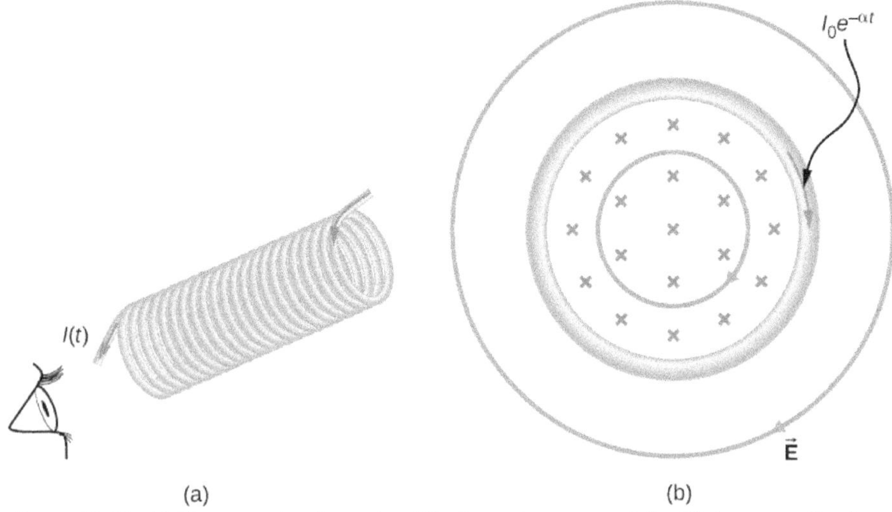

Figure 13.18 (a) The current in a long solenoid is decreasing exponentially. (b) A cross-sectional view of the solenoid from its left end. The cross-section shown is near the middle of the solenoid. An electric field is induced both inside and outside the solenoid.

Strategy

Using the formula for the magnetic field inside an infinite solenoid and Faraday's law, we calculate the induced emf. Since we have cylindrical symmetry, the electric field integral reduces to the electric field times the circumference of the integration path. Then we solve for the electric field.

Solution

a. The magnetic field is confined to the interior of the solenoid where

$$B = \mu_0 nI = \mu_0 nI_0 e^{-\alpha t}.$$

Thus, the magnetic flux through a circular path whose radius r is greater than R, the solenoid radius, is

$$\Phi_m = BA = \mu_0 nI_0 \pi R^2 e^{-\alpha t}.$$

The induced field $\vec{E}$ is tangent to this path, and because of the cylindrical symmetry of the system, its magnitude is constant on the path. Hence, we have

$$\left|\oint \vec{\mathbf{E}} \cdot d\vec{\mathbf{l}}\right| = \left|\frac{d\Phi_m}{dt}\right|,$$

$$E(2\pi r) = \left|\frac{d}{dt}(\mu_0 n I_0 \pi R^2 e^{-\alpha t})\right| = \alpha\mu_0 n I_0 \pi R^2 e^{-\alpha t},$$

$$E = \frac{\alpha\mu_0 n I_0 R^2}{2r} e^{-\alpha t} \quad (r > R).$$

b. For a path of radius r inside the solenoid, $\Phi_m = B\pi r^2$, so

$$E(2\pi r) = \left|\frac{d}{dt}(\mu_0 n I_0 \pi r^2 e^{-\alpha t})\right| = \alpha\mu_0 n I_0 \pi r^2 e^{-\alpha t},$$

and the induced field is

$$E = \frac{\alpha\mu_0 n I_0 r}{2} e^{-\alpha t} \quad (r < R).$$

c. The magnetic field points into the page as shown in part (b) and is decreasing. If either of the circular paths were occupied by conducting rings, the currents induced in them would circulate as shown, in conformity with Lenz's law. The induced electric field must be so directed as well.

Significance

In part (b), note that $\left|\vec{\mathbf{E}}\right|$ increases with r inside and decreases as $1/r$ outside the solenoid, as shown in Figure 13.19.

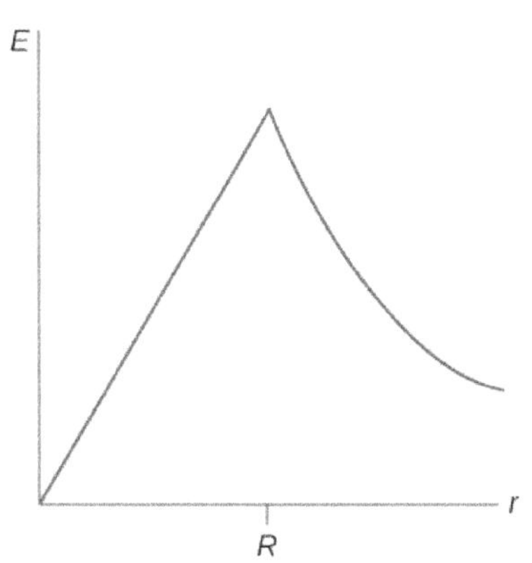

Figure 13.19 The electric field vs. distance r. When $r < R$, the electric field rises linearly, whereas when $r > R$, the electric field falls of proportional to $1/r$.

13.6 Check Your Understanding Suppose that the coil of Example 13.2 is a square rather than circular. Can Equation 13.12 be used to calculate (a) the induced emf and (b) the induced electric field?

13.7 Check Your Understanding What is the magnitude of the induced electric field in Example 13.8 at $t = 0$ if $r = 6.0\,\text{cm}$, $R = 2.0\,\text{cm}$, $n = 2000$ turns per meter, $I_0 = 2.0\,\text{A}$, and $\alpha = 200\,\text{s}^{-1}$?

13.8 Check Your Understanding The magnetic field shown below is confined to the cylindrical region shown and is changing with time. Identify those paths for which $\varepsilon = \oint \vec{\mathbf{E}} \cdot d\vec{\mathbf{l}} \neq 0$.

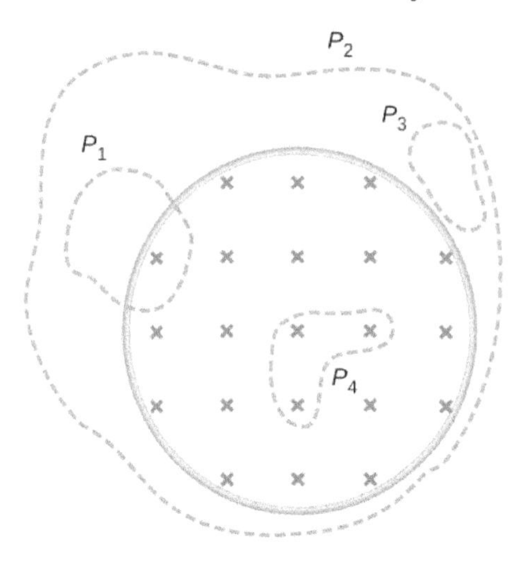

13.9 Check Your Understanding A long solenoid of cross-sectional area $5.0\,\text{cm}^2$ is wound with 25 turns of wire per centimeter. It is placed in the middle of a closely wrapped coil of 10 turns and radius 25 cm, as shown below. (a) What is the emf induced in the coil when the current through the solenoid is decreasing at a rate $dI/dt = -0.20\,\text{A/s}$? (b) What is the electric field induced in the coil?

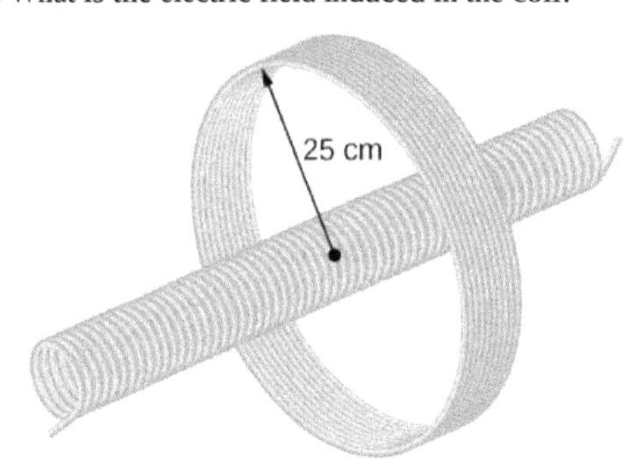

13.5 | Eddy Currents

Learning Objectives

By the end of this section, you will be able to:

- Explain how eddy currents are created in metals
- Describe situations where eddy currents are beneficial and where they are not helpful

As discussed two sections earlier, a motional emf is induced when a conductor moves in a magnetic field or when a magnetic field moves relative to a conductor. If motional emf can cause a current in the conductor, we refer to that current as an **eddy current**.

Magnetic Damping

Eddy currents can produce significant drag, called **magnetic damping**, on the motion involved. Consider the apparatus shown in Figure 13.20, which swings a pendulum bob between the poles of a strong magnet. (This is another favorite physics demonstration.) If the bob is metal, significant drag acts on the bob as it enters and leaves the field, quickly damping the motion. If, however, the bob is a slotted metal plate, as shown in part (b) of the figure, the magnet produces a much smaller effect. There is no discernible effect on a bob made of an insulator. Why does drag occur in both directions, and are there any uses for magnetic drag?

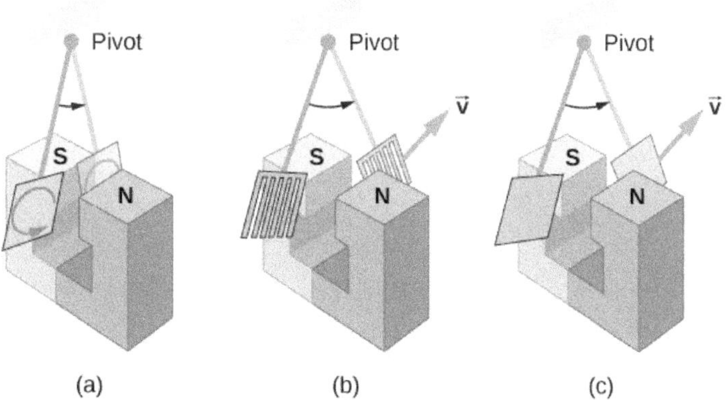

Figure 13.20 A common physics demonstration device for exploring eddy currents and magnetic damping. (a) The motion of a metal pendulum bob swinging between the poles of a magnet is quickly damped by the action of eddy currents. (b) There is little effect on the motion of a slotted metal bob, implying that eddy currents are made less effective. (c) There is also no magnetic damping on a nonconducting bob, since the eddy currents are extremely small.

Figure 13.21 shows what happens to the metal plate as it enters and leaves the magnetic field. In both cases, it experiences a force opposing its motion. As it enters from the left, flux increases, setting up an eddy current (Faraday's law) in the counterclockwise direction (Lenz's law), as shown. Only the right-hand side of the current loop is in the field, so an unopposed force acts on it to the left (RHR-1). When the metal plate is completely inside the field, there is no eddy current if the field is uniform, since the flux remains constant in this region. But when the plate leaves the field on the right, flux decreases, causing an eddy current in the clockwise direction that, again, experiences a force to the left, further slowing the motion. A similar analysis of what happens when the plate swings from the right toward the left shows that its motion is also damped when entering and leaving the field.

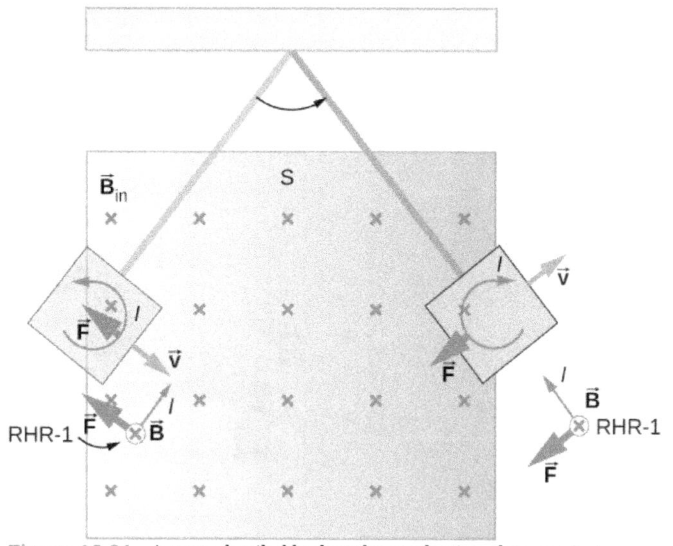

Figure 13.21 A more detailed look at the conducting plate passing between the poles of a magnet. As it enters and leaves the field, the change in flux produces an eddy current. Magnetic force on the current loop opposes the motion. There is no current and no magnetic drag when the plate is completely inside the uniform field.

When a slotted metal plate enters the field (Figure 13.22), an emf is induced by the change in flux, but it is less effective because the slots limit the size of the current loops. Moreover, adjacent loops have currents in opposite directions, and their effects cancel. When an insulating material is used, the eddy current is extremely small, so magnetic damping on insulators is negligible. If eddy currents are to be avoided in conductors, then they must be slotted or constructed of thin layers of conducting material separated by insulating sheets.

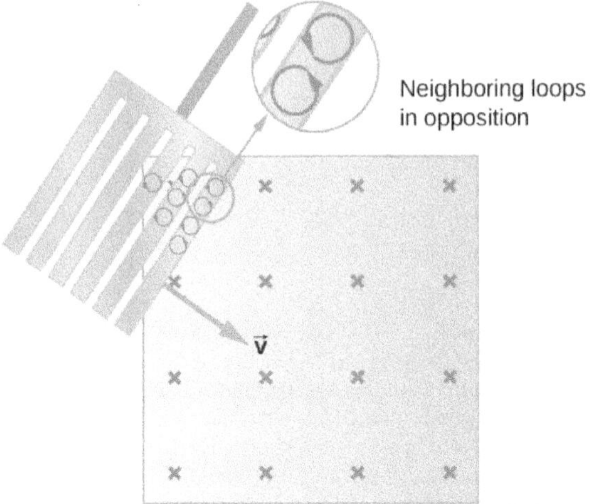

Figure 13.22 Eddy currents induced in a slotted metal plate entering a magnetic field form small loops, and the forces on them tend to cancel, thereby making magnetic drag almost zero.

Applications of Magnetic Damping

One use of magnetic damping is found in sensitive laboratory balances. To have maximum sensitivity and accuracy, the balance must be as friction-free as possible. But if it is friction-free, then it will oscillate for a very long time. Magnetic damping is a simple and ideal solution. With magnetic damping, drag is proportional to speed and becomes zero at zero velocity. Thus, the oscillations are quickly damped, after which the damping force disappears, allowing the balance to be very sensitive (Figure 13.23). In most balances, magnetic damping is accomplished with a conducting disc that rotates in a fixed field.

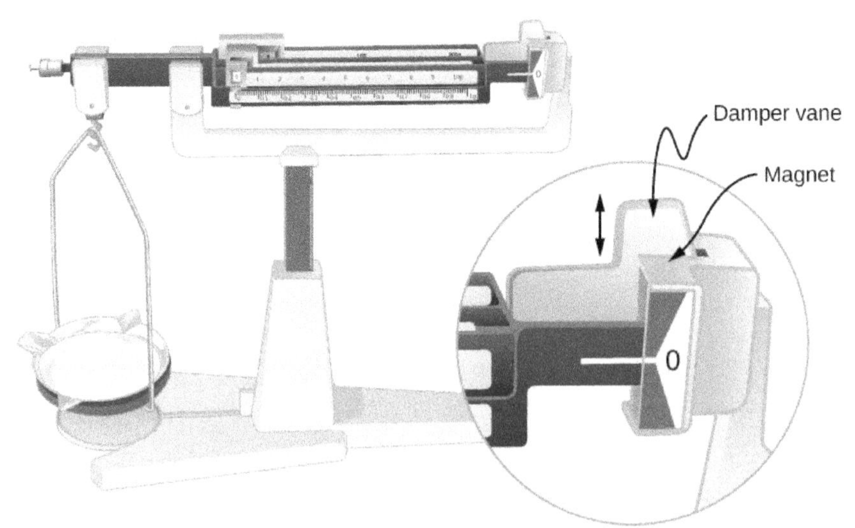

Figure 13.23 Magnetic damping of this sensitive balance slows its oscillations. Since Faraday's law of induction gives the greatest effect for the most rapid change, damping is greatest for large oscillations and goes to zero as the motion stops.

Since eddy currents and magnetic damping occur only in conductors, recycling centers can use magnets to separate metals from other materials. Trash is dumped in batches down a ramp, beneath which lies a powerful magnet. Conductors in the trash are slowed by magnetic damping while nonmetals in the trash move on, separating from the metals (Figure 13.24). This works for all metals, not just ferromagnetic ones. A magnet can separate out the ferromagnetic materials alone by acting on stationary trash.

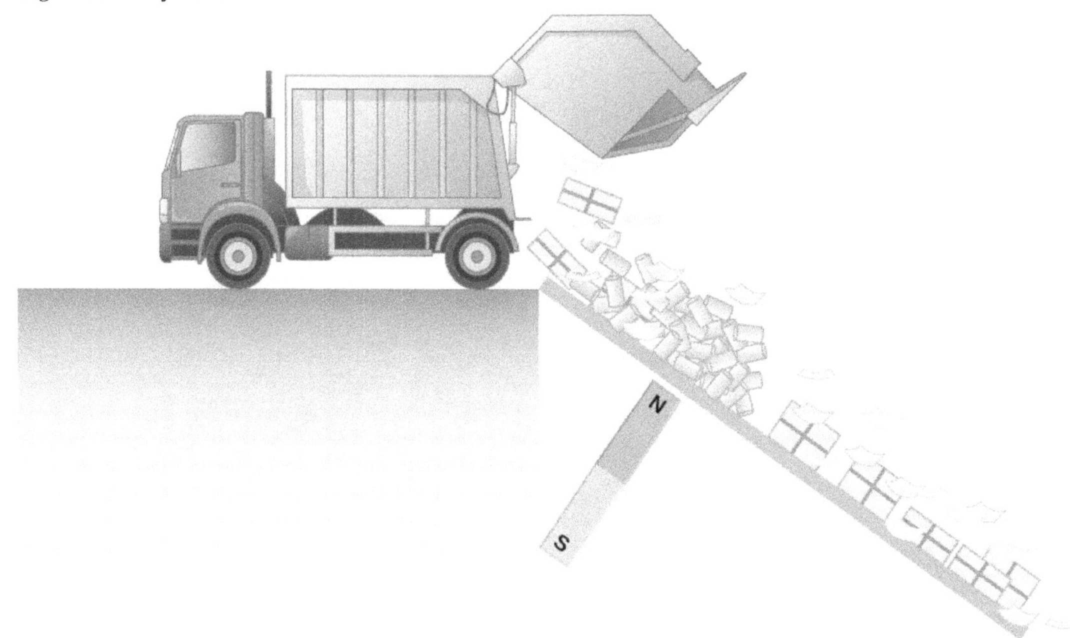

Figure 13.24 Metals can be separated from other trash by magnetic drag. Eddy currents and magnetic drag are created in the metals sent down this ramp by the powerful magnet beneath it. Nonmetals move on.

Other major applications of eddy currents appear in metal detectors and braking systems in trains and roller coasters. Portable metal detectors (Figure 13.25) consist of a primary coil carrying an alternating current and a secondary coil in which a current is induced. An eddy current is induced in a piece of metal close to the detector, causing a change in the induced current within the secondary coil. This can trigger some sort of signal, such as a shrill noise.

Figure 13.25 A soldier in Iraq uses a metal detector to search for explosives and weapons. (credit: U.S. Army)

Braking using eddy currents is safer because factors such as rain do not affect the braking and the braking is smoother. However, eddy currents cannot bring the motion to a complete stop, since the braking force produced decreases as speed is reduced. Thus, speed can be reduced from say 20 m/s to 5 m/s, but another form of braking is needed to completely stop the vehicle. Generally, powerful rare-earth magnets such as neodymium magnets are used in roller coasters. Figure 13.26 shows rows of magnets in such an application. The vehicle has metal fins (normally containing copper) that pass through the magnetic field, slowing the vehicle down in much the same way as with the pendulum bob shown in Figure 13.20.

Figure 13.26 The rows of rare-earth magnets (protruding horizontally) are used for magnetic braking in roller coasters. (credit: Stefan Scheer)

Induction cooktops have electromagnets under their surface. The magnetic field is varied rapidly, producing eddy currents in the base of the pot, causing the pot and its contents to increase in temperature. Induction cooktops have high efficiencies and good response times but the base of the pot needs to be conductors, such as iron or steel, for induction to work.

13.6 | Electric Generators and Back Emf

Learning Objectives

By the end of this section, you will be able to:

- Explain how an electric generator works
- Determine the induced emf in a loop at any time interval, rotating at a constant rate in a magnetic field
- Show that rotating coils have an induced emf; in motors this is called back emf because it opposes the emf input to the motor

A variety of important phenomena and devices can be understood with Faraday's law. In this section, we examine two of these.

Electric Generators

Electric generators induce an emf by rotating a coil in a magnetic field, as briefly discussed in Motional Emf. We now explore generators in more detail. Consider the following example.

Example 13.9

Calculating the Emf Induced in a Generator Coil

The generator coil shown in Figure 13.27 is rotated through one-fourth of a revolution (from $\theta = 0°$ to $\theta = 90°$) in 15.0 ms. The 200-turn circular coil has a 5.00-cm radius and is in a uniform 0.80-T magnetic field. What is the emf induced?

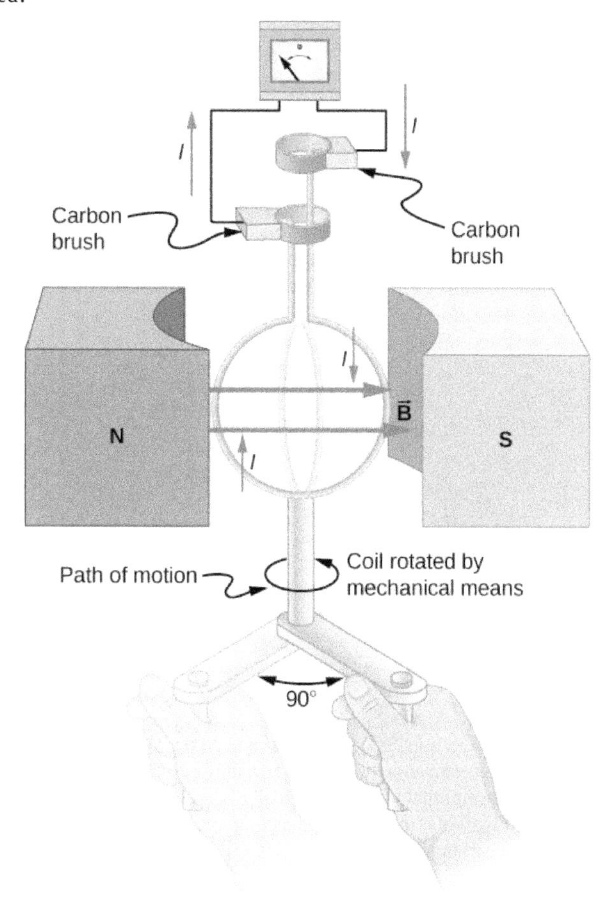

Figure 13.27 When this generator coil is rotated through one-fourth of a revolution, the magnetic flux Φ_{m} changes from its maximum to zero, inducing an emf.

Strategy

Faraday's law of induction is used to find the emf induced:

$$\varepsilon = -N\frac{d\Phi_{\text{m}}}{dt}.$$

We recognize this situation as the same one in Example 13.6. According to the diagram, the projection of the surface normal vector $\hat{\mathbf{n}}$ to the magnetic field is initially $\cos\theta$, and this is inserted by the definition of the dot product. The magnitude of the magnetic field and area of the loop are fixed over time, which makes the integration simplify quickly. The induced emf is written out using Faraday's law:

$$\varepsilon = NBA\sin\theta\frac{d\theta}{dt}.$$

Solution

We are given that $N = 200$, $B = 0.80\text{ T}$, $\theta = 90°$, $d\theta = 90° = \pi/2$, and $dt = 15.0\text{ ms}$. The area of the loop is

$$A = \pi r^2 = (3.14)(0.0500\text{ m})^2 = 7.85 \times 10^{-3}\text{ m}^2.$$

Entering this value gives

$$\varepsilon = (200)(0.80\text{ T})(7.85 \times 10^{-3}\text{ m}^2)\sin(90°)\frac{\pi/2}{15.0 \times 10^{-3}\text{ s}} = 131\text{ V}.$$

Significance

This is a practical average value, similar to the 120 V used in household power.

The emf calculated in Example 13.9 is the average over one-fourth of a revolution. What is the emf at any given instant? It varies with the angle between the magnetic field and a perpendicular to the coil. We can get an expression for emf as a function of time by considering the motional emf on a rotating rectangular coil of width *w* and height *l* in a uniform magnetic field, as illustrated in Figure 13.28.

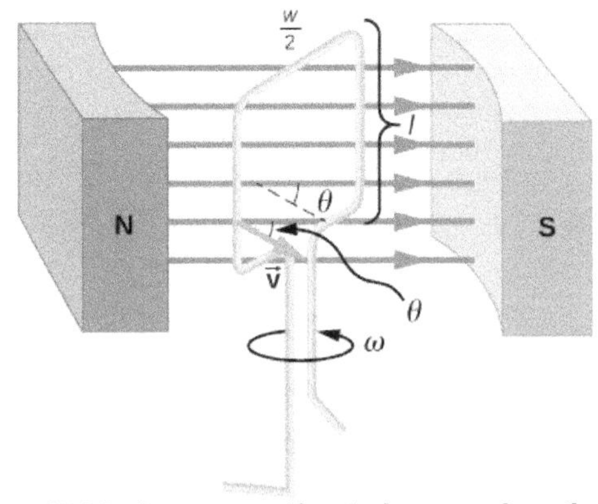

Figure 13.28 A generator with a single rectangular coil rotated at constant angular velocity in a uniform magnetic field produces an emf that varies sinusoidally in time. Note the generator is similar to a motor, except the shaft is rotated to produce a current rather than the other way around.

Charges in the wires of the loop experience the magnetic force, because they are moving in a magnetic field. Charges in the vertical wires experience forces parallel to the wire, causing currents. But those in the top and bottom segments feel a force perpendicular to the wire, which does not cause a current. We can thus find the induced emf by considering only the side wires. Motional emf is given to be $\varepsilon = Blv$, where the velocity *v* is perpendicular to the magnetic field *B*. Here the velocity is at an angle θ with *B*, so that its component perpendicular to *B* is $v \sin\theta$ (see Figure 13.28). Thus, in this case, the emf induced on each side is $\varepsilon = Blv \sin\theta$, and they are in the same direction. The total emf around the loop is then

$$\varepsilon = 2Blv \sin\theta. \tag{13.13}$$

This expression is valid, but it does not give emf as a function of time. To find the time dependence of emf, we assume the coil rotates at a constant angular velocity ω. The angle θ is related to angular velocity by $\theta = \omega t$, so that

$$\varepsilon = 2Blv \sin(\omega t). \tag{13.14}$$

Now, linear velocity *v* is related to angular velocity ω by $v = r\omega$. Here, $r = w/2$, so that $v = (w/2)\omega$, and

$$\varepsilon = 2Bl\frac{w}{2}\omega \sin\omega t = (lw)B\omega \sin\omega t. \tag{13.15}$$

Noting that the area of the loop is $A = lw$, and allowing for *N* loops, we find that

$$\varepsilon = NBA\omega \sin(\omega t). \tag{13.16}$$

This is the emf induced in a generator coil of N turns and area A rotating at a constant angular velocity ω in a uniform magnetic field B. This can also be expressed as

$$\varepsilon = \varepsilon_0 \sin \omega t, \tag{13.17}$$

where

$$\varepsilon_0 = NAB\omega \tag{13.18}$$

is the peak emf, since the maximum value of $\sin(wt) = 1$. Note that the frequency of the oscillation is $f = \omega/2\pi$ and the period is $T = 1/f = 2\pi/\omega$. Figure 13.29 shows a graph of emf as a function of time, and it now seems reasonable that ac voltage is sinusoidal.

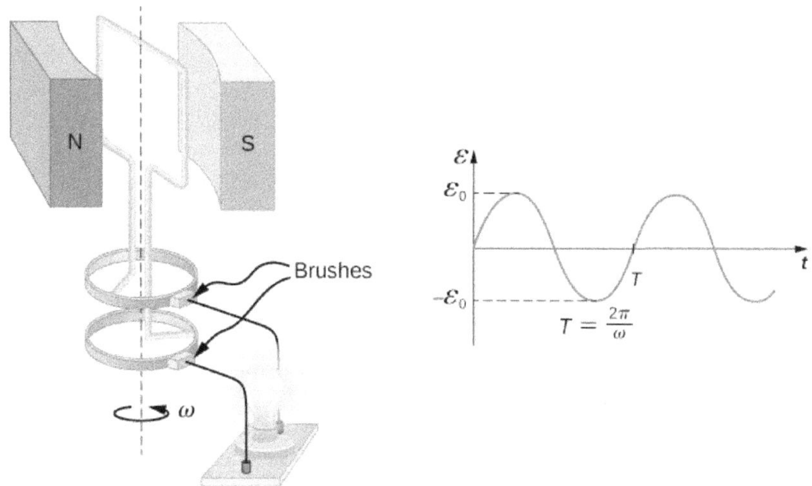

Figure 13.29 The emf of a generator is sent to a light bulb with the system of rings and brushes shown. The graph gives the emf of the generator as a function of time, where ε_0 is the peak emf. The period is $T = 1/f = 2\pi/\omega$, where f is the frequency.

The fact that the peak emf is $\varepsilon_0 = NBA\omega$ makes good sense. The greater the number of coils, the larger their area, and the stronger the field, the greater the output voltage. It is interesting that the faster the generator is spun (greater ω), the greater the emf. This is noticeable on bicycle generators—at least the cheaper varieties.

Figure 13.30 shows a scheme by which a generator can be made to produce pulsed dc. More elaborate arrangements of multiple coils and split rings can produce smoother dc, although electronic rather than mechanical means are usually used to make ripple-free dc.

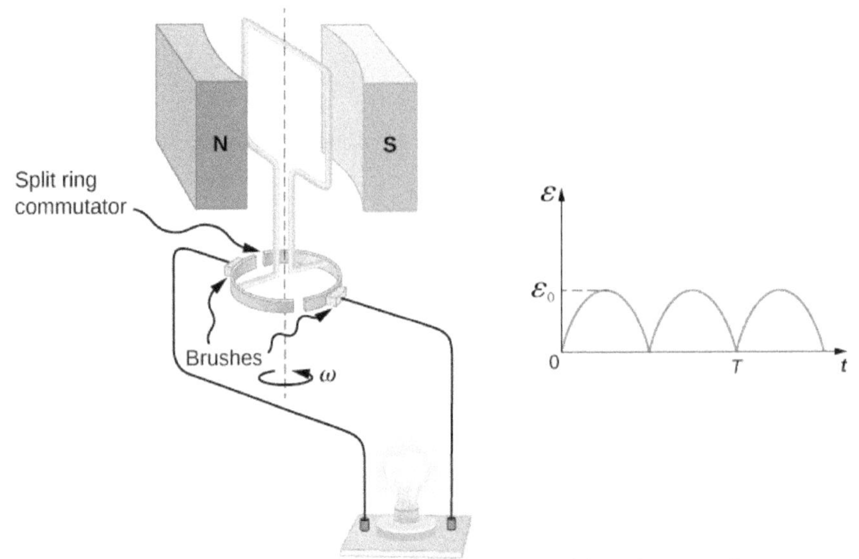

Figure 13.30 Split rings, called commutators, produce a pulsed dc emf output in this configuration.

In real life, electric generators look a lot different from the figures in this section, but the principles are the same. The source of mechanical energy that turns the coil can be falling water (hydropower), steam produced by the burning of fossil fuels, or the kinetic energy of wind. Figure 13.31 shows a cutaway view of a steam turbine; steam moves over the blades connected to the shaft, which rotates the coil within the generator. The generation of electrical energy from mechanical energy is the basic principle of all power that is sent through our electrical grids to our homes.

Figure 13.31 Steam turbine/generator. The steam produced by burning coal impacts the turbine blades, turning the shaft, which is connected to the generator.

Generators illustrated in this section look very much like the motors illustrated previously. This is not coincidental. In fact, a motor becomes a generator when its shaft rotates. Certain early automobiles used their starter motor as a generator. In the next section, we further explore the action of a motor as a generator.

Back Emf

Generators convert mechanical energy into electrical energy, whereas motors convert electrical energy into mechanical energy. Thus, it is not surprising that motors and generators have the same general construction. A motor works by sending a current through a loop of wire located in a magnetic field. As a result, the magnetic field exerts torque on the loop. This rotates a shaft, thereby extracting mechanical work out of the electrical current sent in initially. (Refer to Force and Torque on a Current Loop for a discussion on motors that will help you understand more about them before proceeding.)

When the coil of a motor is turned, magnetic flux changes through the coil, and an emf (consistent with Faraday's law) is induced. The motor thus acts as a generator whenever its coil rotates. This happens whether the shaft is turned by an external input, like a belt drive, or by the action of the motor itself. That is, when a motor is doing work and its shaft is turning, an emf is generated. Lenz's law tells us the emf opposes any change, so that the input emf that powers the motor is opposed by the motor's self-generated emf, called the **back emf** of the motor (Figure 13.32).

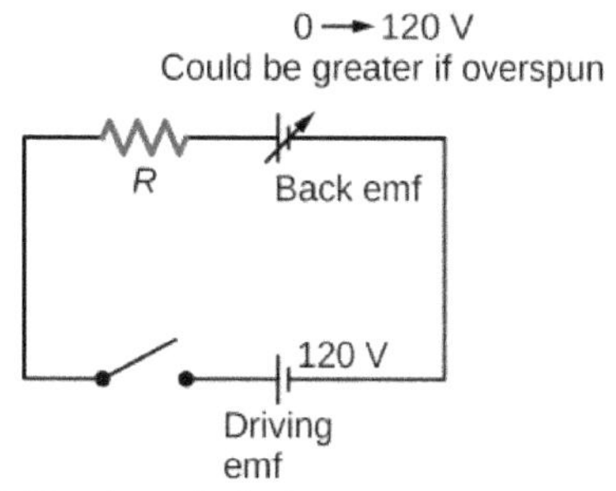

Figure 13.32 The coil of a dc motor is represented as a resistor in this schematic. The back emf is represented as a variable emf that opposes the emf driving the motor. Back emf is zero when the motor is not turning and increases proportionally to the motor's angular velocity.

The generator output of a motor is the difference between the supply voltage and the back emf. The back emf is zero when the motor is first turned on, meaning that the coil receives the full driving voltage and the motor draws maximum current when it is on but not turning. As the motor turns faster, the back emf grows, always opposing the driving emf, and reduces both the voltage across the coil and the amount of current it draws. This effect is noticeable in many common situations. When a vacuum cleaner, refrigerator, or washing machine is first turned on, lights in the same circuit dim briefly due to the *IR* drop produced in feeder lines by the large current drawn by the motor.

When a motor first comes on, it draws more current than when it runs at its normal operating speed. When a mechanical load is placed on the motor, like an electric wheelchair going up a hill, the motor slows, the back emf drops, more current flows, and more work can be done. If the motor runs at too low a speed, the larger current can overheat it (via resistive power in the coil, $P = I^2 R$), perhaps even burning it out. On the other hand, if there is no mechanical load on the motor, it increases its angular velocity ω until the back emf is nearly equal to the driving emf. Then the motor uses only enough energy to overcome friction.

Eddy currents in iron cores of motors can cause troublesome energy losses. These are usually minimized by constructing the cores out of thin, electrically insulated sheets of iron. The magnetic properties of the core are hardly affected by the lamination of the insulating sheet, while the resistive heating is reduced considerably. Consider, for example, the motor coils represented in Figure 13.32. The coils have an equivalent resistance of $0.400\ \Omega$ and are driven by an emf of 48.0 V. Shortly after being turned on, they draw a current

$$I = V/R = (48.0\ \text{V})/(0.400\ \Omega) = 120\ \text{A}$$

and thus dissipate $P = I^2 R = 5.76\ \text{kW}$ of energy as heat transfer. Under normal operating conditions for this motor, suppose the back emf is 40.0 V. Then at operating speed, the total voltage across the coils is 8.0 V (48.0 V minus the 40.0 V back emf), and the current drawn is

$$I = V/R = (8.0\ \text{V})/(0.400\ \Omega) = 20\ \text{A}.$$

Under normal load, then, the power dissipated is $P = IV = (20\text{ A})(8.0\text{ V}) = 160\text{ W}$. This does not cause a problem for this motor, whereas the former 5.76 kW would burn out the coils if sustained.

Example 13.10

A Series-Wound Motor in Operation

The total resistance $(R_f + R_a)$ of a series-wound dc motor is $2.0\,\Omega$ (Figure 13.33). When connected to a 120-V source (ε_S), the motor draws 10 A while running at constant angular velocity. (a) What is the back emf induced in the rotating coil, ε_i? (b) What is the mechanical power output of the motor? (c) How much power is dissipated in the resistance of the coils? (d) What is the power output of the 120-V source? (e) Suppose the load on the motor increases, causing it to slow down to the point where it draws 20 A. Answer parts (a) through (d) for this situation.

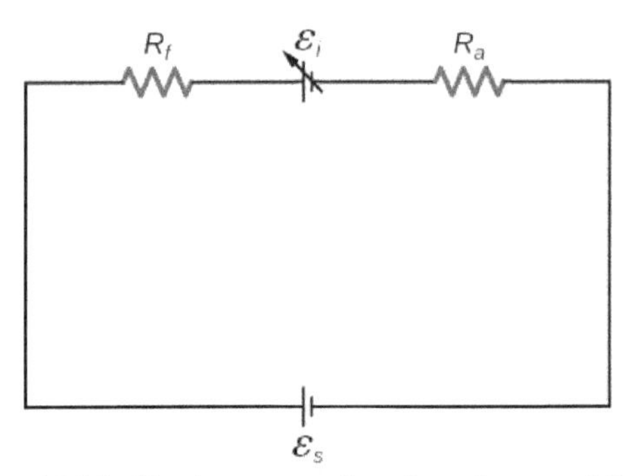

Figure 13.33 Circuit representation of a series-wound direct current motor.

Strategy

The back emf is calculated based on the difference between the supplied voltage and the loss from the current through the resistance. The power from each device is calculated from one of the power formulas based on the given information.

Solution

a. The back emf is

$$\varepsilon_i = \varepsilon_s - I(R_f + R_a) = 120\text{ V} - (10\text{ A})(2.0\,\Omega) = 100\text{ V}.$$

b. Since the potential across the armature is 100 V when the current through it is 10 A, the power output of the motor is

$$P_\text{m} = \varepsilon_i I = (100\text{ V})(10\text{ A}) = 1.0 \times 10^3\text{ W}.$$

c. A 10-A current flows through coils whose combined resistance is $2.0\,\Omega$, so the power dissipated in the coils is

$$P_R = I^2 R = (10\text{ A})^2(2.0\,\Omega) = 2.0 \times 10^2\text{ W}.$$

d. Since 10 A is drawn from the 120-V source, its power output is

$$P_s = \varepsilon_s\, I = (120\text{ V})(10\text{ A}) = 1.2 \times 10^3\text{ W}.$$

e. Repeating the same calculations with $I = 20\text{ A}$, we find

$$\varepsilon_\text{i} = 80\text{ V},\ P_\text{m} = 1.6 \times 10^3\text{ W},\ P_R = 8.0 \times 10^2\text{ W},\ \text{and}\ P_s = 2.4 \times 10^3\text{ W}.$$

The motor is turning more slowly in this case, so its power output and the power of the source are larger.

Significance

Notice that we have an energy balance in part (d): $1.2 \times 10^3 \text{ W} = 1.0 \times 10^3 \text{ W} + 2.0 \times 10^2 \text{ W}$.

13.7 | Applications of Electromagnetic Induction

Learning Objectives

By the end of this section, you will be able to:

- Explain how computer hard drives and graphic tablets operate using magnetic induction
- Explain how hybrid/electric vehicles and transcranial magnetic stimulation use magnetic induction to their advantage

Modern society has numerous applications of Faraday's law of induction, as we will explore in this chapter and others. At this juncture, let us mention several that involve recording information using magnetic fields.

Some computer hard drives apply the principle of magnetic induction. Recorded data are made on a coated, spinning disk. Historically, reading these data was made to work on the principle of induction. However, most input information today is carried in digital rather than analog form—a series of 0s or 1s are written upon the spinning hard drive. Therefore, most hard drive readout devices do not work on the principle of induction, but use a technique known as giant magnetoresistance. Giant magnetoresistance is the effect of a large change of electrical resistance induced by an applied magnetic field to thin films of alternating ferromagnetic and nonmagnetic layers. This is one of the first large successes of nanotechnology.

Graphics tablets, or tablet computers where a specially designed pen is used to draw digital images, also applies induction principles. The tablets discussed here are labeled as passive tablets, since there are other designs that use either a battery-operated pen or optical signals to write with. The passive tablets are different than the touch tablets and phones many of us use regularly, but may still be found when signing your signature at a cash register. Underneath the screen, shown in Figure 13.34, are tiny wires running across the length and width of the screen. The pen has a tiny magnetic field coming from the tip. As the tip brushes across the screen, a changing magnetic field is felt in the wires which translates into an induced emf that is converted into the line you just drew.

Figure 13.34 A tablet with a specially designed pen to write with is another application of magnetic induction.

Another application of induction is the magnetic stripe on the back of your personal credit card as used at the grocery store or the ATM machine. This works on the same principle as the audio or video tape, in which a playback head reads personal information from your card.

Check out this video (https://openstaxcollege.org/l/21flashmagind) to see how flashlights can use magnetic induction. A magnet moves by your mechanical work through a wire. The induced current charges a capacitor that stores the charge that will light the lightbulb even while you are not doing this mechanical work.

Electric and hybrid vehicles also take advantage of electromagnetic induction. One limiting factor that inhibits widespread acceptance of 100% electric vehicles is that the lifetime of the battery is not as long as the time you get to drive on a full tank of gas. To increase the amount of charge in the battery during driving, the motor can act as a generator whenever the car is braking, taking advantage of the back emf produced. This extra emf can be newly acquired stored energy in the car's battery, prolonging the life of the battery.

Another contemporary area of research in which electromagnetic induction is being successfully implemented is transcranial magnetic stimulation (TMS). A host of disorders, including depression and hallucinations, can be traced to irregular localized electrical activity in the brain. In transcranial magnetic stimulation, a rapidly varying and very localized magnetic field is placed close to certain sites identified in the brain. The usage of TMS as a diagnostic technique is well established.

Check out this Youtube video (https://openstaxcollege.org/l/21randrelectro) to see how rock-and-roll instruments like electric guitars use electromagnetic induction to get those strong beats.

CHAPTER 13 REVIEW

KEY TERMS

back emf emf generated by a running motor, because it consists of a coil turning in a magnetic field; it opposes the voltage powering the motor

eddy current current loop in a conductor caused by motional emf

electric generator device for converting mechanical work into electric energy; it induces an emf by rotating a coil in a magnetic field

Faraday's law induced emf is created in a closed loop due to a change in magnetic flux through the loop

induced electric field created based on the changing magnetic flux with time

induced emf short-lived voltage generated by a conductor or coil moving in a magnetic field

Lenz's law direction of an induced emf opposes the change in magnetic flux that produced it; this is the negative sign in Faraday's law

magnetic damping drag produced by eddy currents

magnetic flux measurement of the amount of magnetic field lines through a given area

motionally induced emf voltage produced by the movement of a conducting wire in a magnetic field

peak emf maximum emf produced by a generator

KEY EQUATIONS

Magnetic flux	$\Phi_m = \int_S \vec{\mathbf{B}} \cdot \hat{\mathbf{n}} dA$
Faraday's law	$\varepsilon = -N\frac{d\Phi_m}{dt}$
Motionally induced emf	$\varepsilon = Blv$
Motional emf around a circuit	$\varepsilon = \oint \vec{\mathbf{E}} \cdot d\vec{\mathbf{l}} = -\frac{d\Phi_m}{dt}$
Emf produced by an electric generator	$\varepsilon = NBA\omega \sin(\omega t)$

SUMMARY

13.1 Faraday's Law

- The magnetic flux through an enclosed area is defined as the amount of field lines cutting through a surface area A defined by the unit area vector.
- The units for magnetic flux are webers, where $1 \text{ Wb} = 1 \text{ T} \cdot \text{m}^2$.
- The induced emf in a closed loop due to a change in magnetic flux through the loop is known as Faraday's law. If there is no change in magnetic flux, no induced emf is created.

13.2 Lenz's Law

- We can use Lenz's law to determine the directions of induced magnetic fields, currents, and emfs.
- The direction of an induced emf always opposes the change in magnetic flux that causes the emf, a result known as Lenz's law.

13.3 Motional Emf

- The relationship between an induced emf ε in a wire moving at a constant speed v through a magnetic field B is given by $\varepsilon = Blv$.
- An induced emf from Faraday's law is created from a motional emf that opposes the change in flux.

13.4 Induced Electric Fields

- A changing magnetic flux induces an electric field.
- Both the changing magnetic flux and the induced electric field are related to the induced emf from Faraday's law.

13.5 Eddy Currents

- Current loops induced in moving conductors are called eddy currents. They can create significant drag, called magnetic damping.
- Manipulation of eddy currents has resulted in applications such as metal detectors, braking in trains or roller coasters, and induction cooktops.

13.6 Electric Generators and Back Emf

- An electric generator rotates a coil in a magnetic field, inducing an emf given as a function of time by $\varepsilon = NBA\omega \sin(\omega t)$ where A is the area of an N-turn coil rotated at a constant angular velocity ω in a uniform magnetic field $\vec{\mathbf{B}}$.
- The peak emf of a generator is $\varepsilon_0 = NBA\omega$.
- Any rotating coil produces an induced emf. In motors, this is called back emf because it opposes the emf input to the motor.

13.7 Applications of Electromagnetic Induction

- Hard drives utilize magnetic induction to read/write information.
- Other applications of magnetic induction can be found in graphics tablets, electric and hybrid vehicles, and in transcranial magnetic stimulation.

CONCEPTUAL QUESTIONS

13.1 Faraday's Law

1. A stationary coil is in a magnetic field that is changing with time. Does the emf induced in the coil depend on the actual values of the magnetic field?

2. In Faraday's experiments, what would be the advantage of using coils with many turns?

3. A copper ring and a wooden ring of the same dimensions are placed in magnetic fields so that there is the same change in magnetic flux through them. Compare the induced electric fields and currents in the rings.

4. Discuss the factors determining the induced emf in a closed loop of wire.

5. (a) Does the induced emf in a circuit depend on the resistance of the circuit? (b) Does the induced current depend on the resistance of the circuit?

6. How would changing the radius of loop D shown below affect its emf, assuming C and D are much closer together compared to their radii?

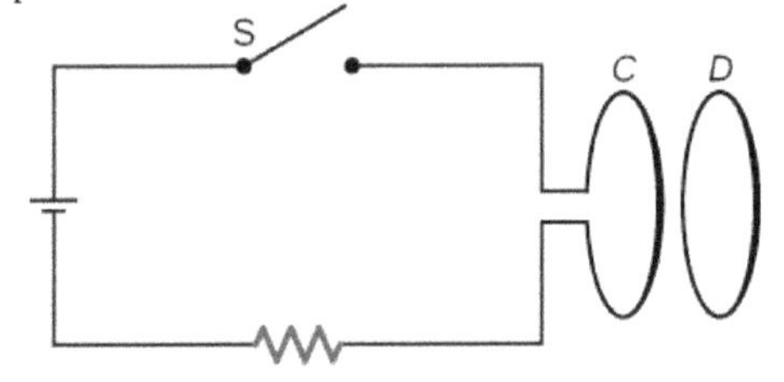

7. Can there be an induced emf in a circuit at an instant when the magnetic flux through the circuit is zero?

8. Does the induced emf always act to decrease the magnetic flux through a circuit?

9. How would you position a flat loop of wire in a changing magnetic field so that there is no induced emf in the loop?

10. The normal to the plane of a single-turn conducting loop is directed at an angle θ to a spatially uniform magnetic field $\vec{\mathbf{B}}$. It has a fixed area and orientation relative to the magnetic field. Show that the emf induced in the loop is given by $\varepsilon = (dB/dt)(A\cos\theta)$, where A is the area of the loop.

13.2 Lenz's Law

11. The circular conducting loops shown in the accompanying figure are parallel, perpendicular to the plane of the page, and coaxial. (a) When the switch S is closed, what is the direction of the current induced in *D*? (b) When the switch is opened, what is the direction of the current induced in loop *D*?

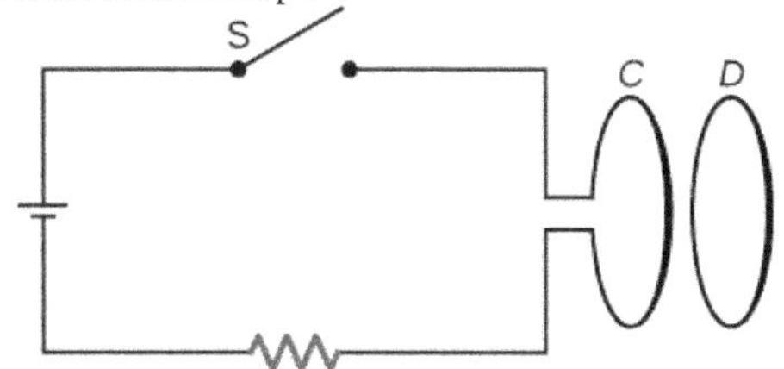

12. The north pole of a magnet is moved toward a copper loop, as shown below. If you are looking at the loop from above the magnet, will you say the induced current is circulating clockwise or counterclockwise?

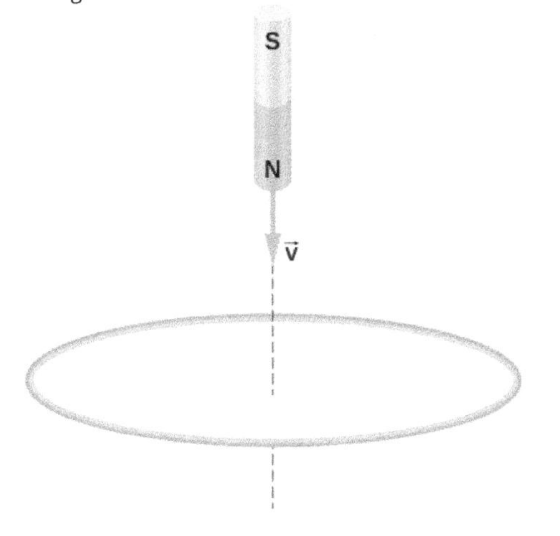

13. The accompanying figure shows a conducting ring at various positions as it moves through a magnetic field. What is the sense of the induced emf for each of those positions?

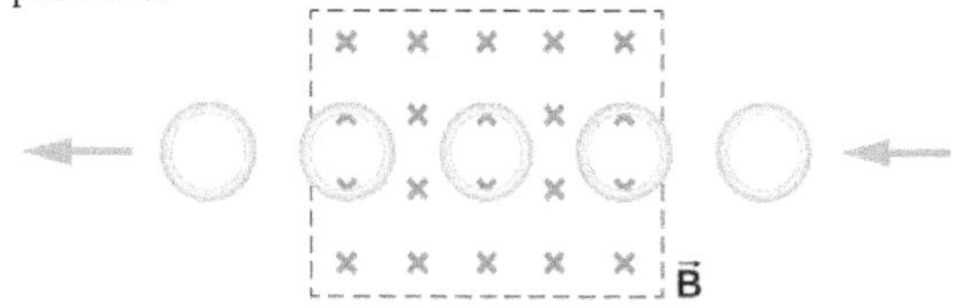

14. Show that ε and $d\Phi_m/dt$ have the same units.

15. State the direction of the induced current for each case shown below, observing from the side of the magnet.

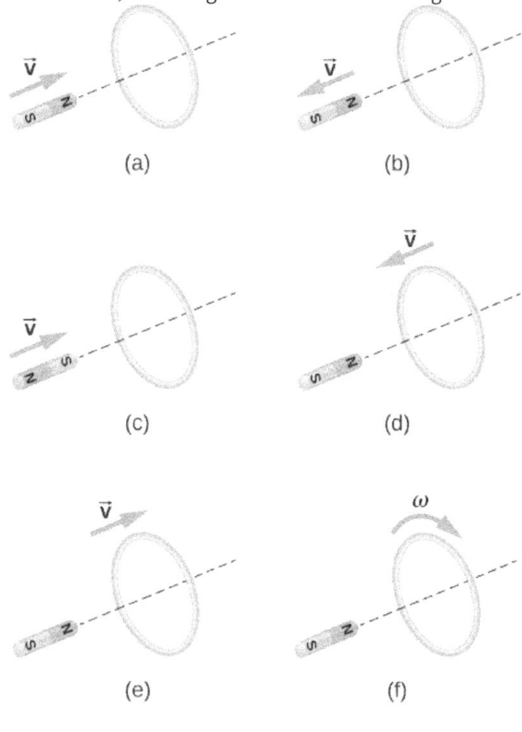

13.3 Motional Emf

16. A bar magnet falls under the influence of gravity along the axis of a long copper tube. If air resistance is negligible, will there be a force to oppose the descent of the magnet? If so, will the magnet reach a terminal velocity?

17. Around the geographic North Pole (or magnetic South Pole), Earth's magnetic field is almost vertical. If an airplane is flying northward in this region, which side of the wing is positively charged and which is negatively charged?

18. A wire loop moves translationally (no rotation) in a uniform magnetic field. Is there an emf induced in the loop?

13.4 Induced Electric Fields

19. Is the work required to accelerate a rod from rest to a speed v in a magnetic field greater than the final kinetic energy of the rod? Why?

20. The copper sheet shown below is partially in a magnetic field. When it is pulled to the right, a resisting force pulls it to the left. Explain. What happen if the sheet is pushed to the left?

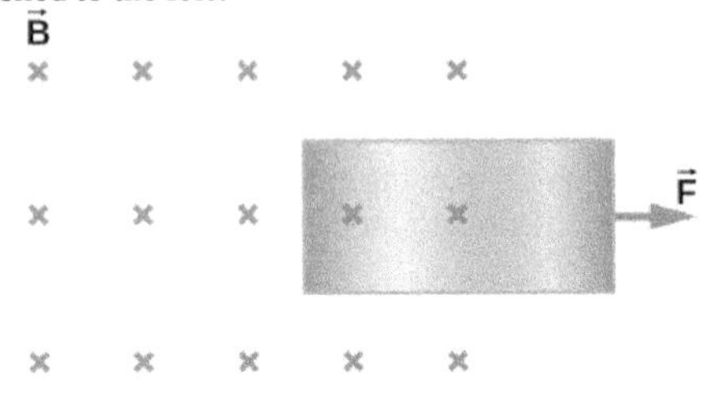

13.5 Eddy Currents

21. A conducting sheet lies in a plane perpendicular to a magnetic field $\vec{\mathbf{B}}$ that is below the sheet. If $\vec{\mathbf{B}}$ oscillates at a high frequency and the conductor is made of a material of low resistivity, the region above the sheet is effectively shielded from $\vec{\mathbf{B}}$. Explain why. Will the conductor shield this region from static magnetic fields?

22. Electromagnetic braking can be achieved by applying a strong magnetic field to a spinning metal disk attached to a shaft. (a) How can a magnetic field slow the spinning of a disk? (b) Would the brakes work if the disk was made of plastic instead of metal?

23. A coil is moved through a magnetic field as shown below. The field is uniform inside the rectangle and zero outside. What is the direction of the induced current and what is the direction of the magnetic force on the coil at each position shown?

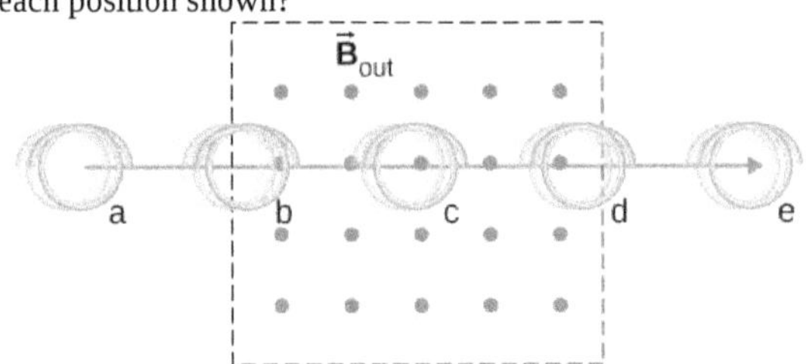

PROBLEMS

13.1 Faraday's Law

24. A 50-turn coil has a diameter of 15 cm. The coil is placed in a spatially uniform magnetic field of magnitude 0.50 T so that the face of the coil and the magnetic field are perpendicular. Find the magnitude of the emf induced in the coil if the magnetic field is reduced to zero uniformly in (a) 0.10 s, (b) 1.0 s, and (c) 60 s.

25. Repeat your calculations of the preceding problem's time of 0.1 s with the plane of the coil making an angle of (a) $30°$, (b) $60°$, and (c) $90°$ with the magnetic field.

26. A square loop whose sides are 6.0-cm long is made with copper wire of radius 1.0 mm. If a magnetic field perpendicular to the loop is changing at a rate of 5.0 mT/s, what is the current in the loop?

27. The magnetic field through a circular loop of radius 10.0 cm varies with time as shown below. The field is perpendicular to the loop. Plot the magnitude of the induced emf in the loop as a function of time.

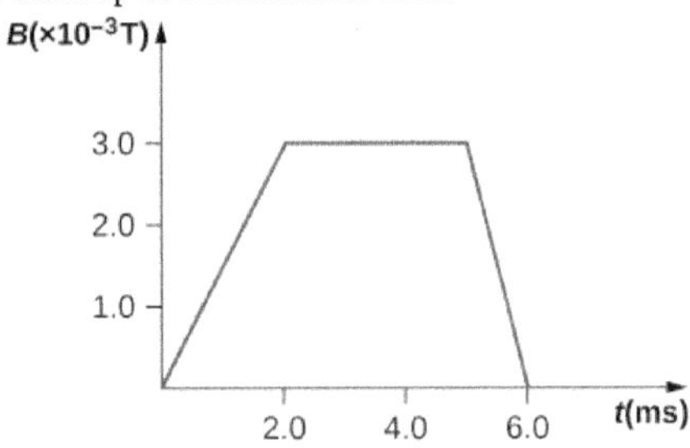

28. The accompanying figure shows a single-turn rectangular coil that has a resistance of $2.0\,\Omega$. The magnetic field at all points inside the coil varies according to $B = B_0 e^{-\alpha t}$, where $B_0 = 0.25\,\text{T}$ and $\alpha = 200\,\text{Hz}$. What is the current induced in the coil at (a) $t = 0.001\,\text{s}$, (b) 0.002 s, (c) 2.0 s?

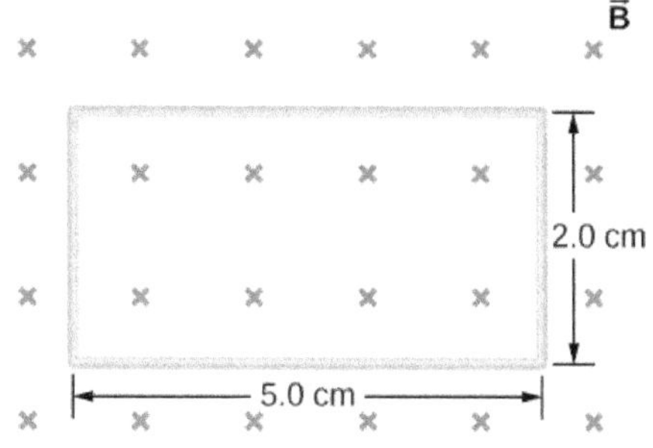

29. How would the answers to the preceding problem change if the coil consisted of 20 closely spaced turns?

30. A long solenoid with $n = 10$ turns per centimeter has a cross-sectional area of $5.0\,\text{cm}^2$ and carries a current of 0.25 A. A coil with five turns encircles the solenoid. When the current through the solenoid is turned off, it decreases to zero in 0.050 s. What is the average emf induced in the coil?

31. A rectangular wire loop with length a and width b lies in the xy-plane, as shown below. Within the loop there is a time-dependent magnetic field given by $\vec{\mathbf{B}}(t) = C\left((x\cos\omega t)\,\hat{\mathbf{i}} + (y\sin\omega t)\,\hat{\mathbf{k}}\right)$, with $\vec{\mathbf{B}}(t)$ in tesla. Determine the emf induced in the loop as a function of time.

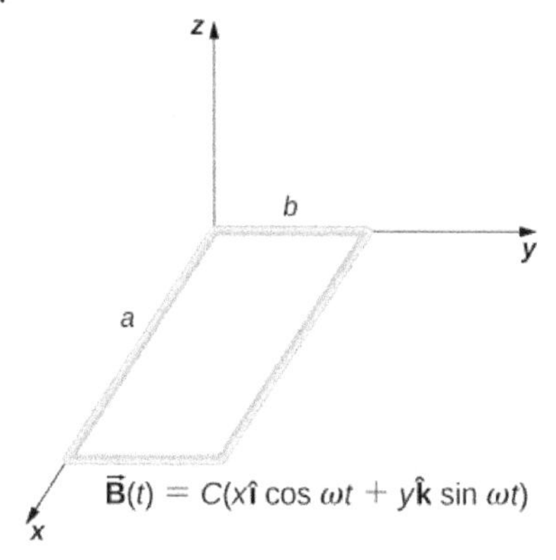

32. The magnetic field perpendicular to a single wire loop of diameter 10.0 cm decreases from 0.50 T to zero. The wire is made of copper and has a diameter of 2.0 mm and length 1.0 cm. How much charge moves through the wire while the field is changing?

13.2 Lenz's Law

33. A single-turn circular loop of wire of radius 50 mm lies in a plane perpendicular to a spatially uniform magnetic field. During a 0.10-s time interval, the magnitude of the field increases uniformly from 200 to 300 mT. (a) Determine the emf induced in the loop. (b) If the magnetic field is directed out of the page, what is the direction of the current induced in the loop?

34. When a magnetic field is first turned on, the flux through a 20-turn loop varies with time according to $\Phi_m = 5.0t^2 - 2.0t$, where Φ_m is in milliwebers, t is in seconds, and the loop is in the plane of the page with the unit normal pointing outward. (a) What is the emf induced in the loop as a function of time? What is the direction of the induced current at (b) $t = 0$, (c) 0.10, (d) 1.0, and (e) 2.0 s?

35. The magnetic flux through the loop shown in the accompanying figure varies with time according to $\Phi_m = 2.00e^{-3t}\sin(120\pi t)$, where Φ_m is in milliwebers. What are the direction and magnitude of the current through the $5.00\text{-}\Omega$ resistor at (a) $t = 0$; (b) $t = 2.17 \times 10^{-2}$ s, and (c) $t = 3.00$ s?

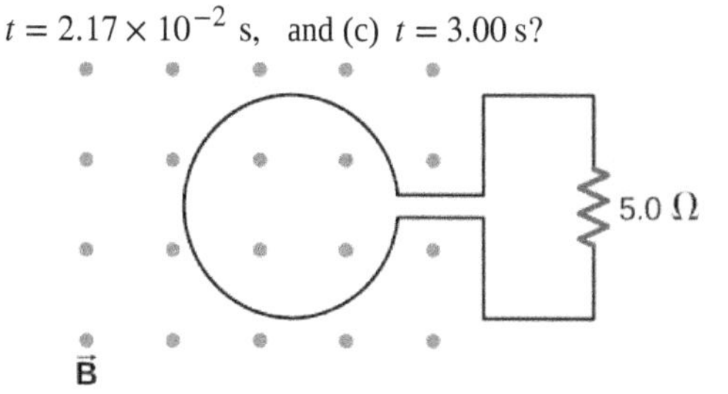

36. Use Lenz's law to determine the direction of induced current in each case.

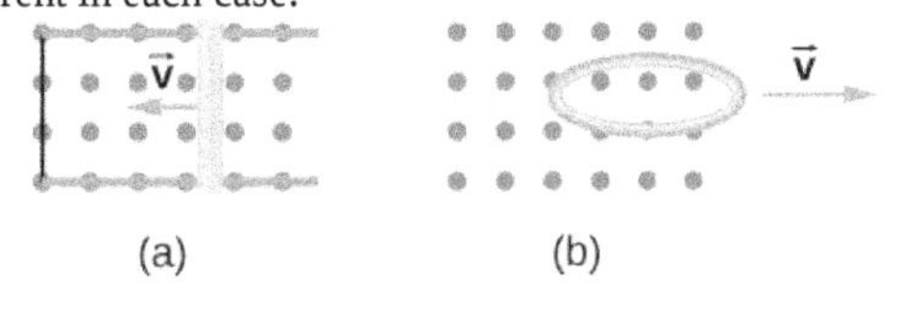

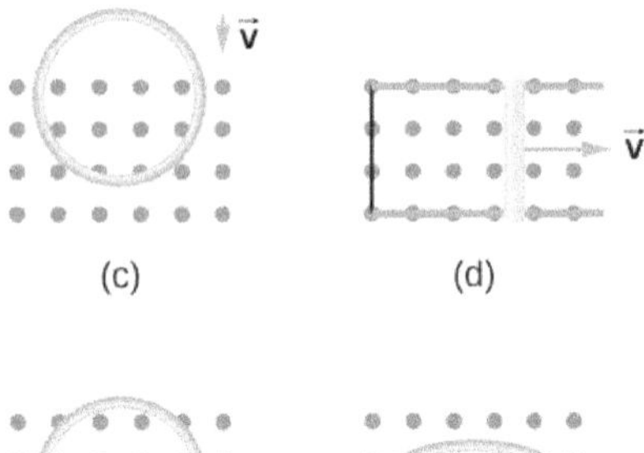

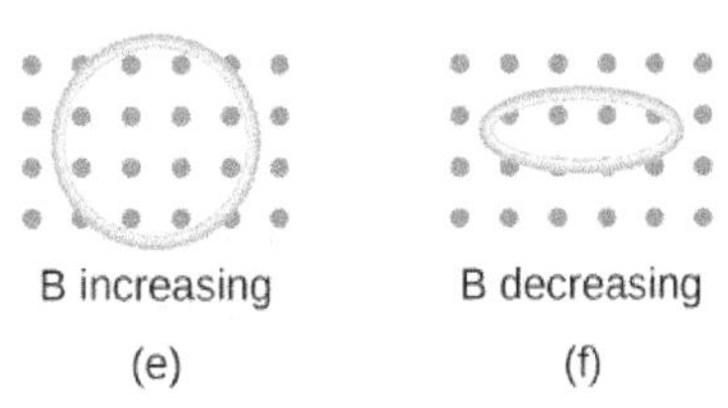

13.3 Motional Emf

37. An automobile with a radio antenna 1.0 m long travels at 100.0 km/h in a location where the Earth's horizontal magnetic field is 5.5×10^{-5} T. What is the maximum possible emf induced in the antenna due to this motion?

38. The rectangular loop of N turns shown below moves to the right with a constant velocity $\vec{v}$ while leaving the poles of a large electromagnet. (a) Assuming that the magnetic field is uniform between the pole faces and negligible elsewhere, determine the induced emf in the loop. (b) What is the source of work that produces this emf?

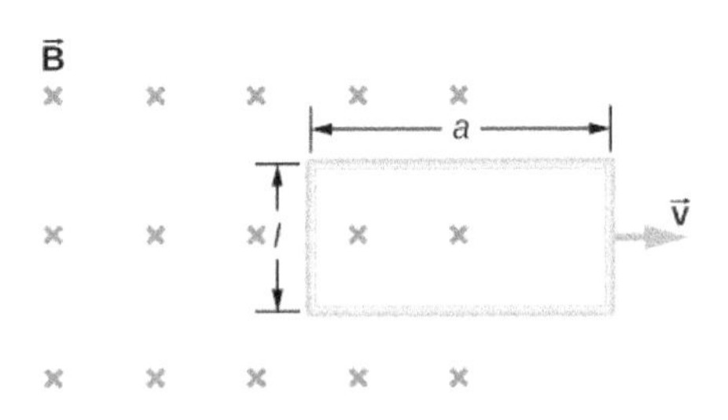

39. Suppose the magnetic field of the preceding problem oscillates with time according to $B = B_0 \sin \omega t$. What then is the emf induced in the loop when its trailing side is a distance d from the right edge of the magnetic field region?

40. A coil of 1000 turns encloses an area of $25\ \text{cm}^2$. It is rotated in 0.010 s from a position where its plane is perpendicular to Earth's magnetic field to one where its plane is parallel to the field. If the strength of the field is 6.0×10^{-5} T, what is the average emf induced in the coil?

41. In the circuit shown in the accompanying figure, the rod slides along the conducting rails at a constant velocity $\vec{v}$. The velocity is in the same plane as the rails and directed at an angle θ to them. A uniform magnetic field $\vec{B}$ is directed out of the page. What is the emf induced in the rod?

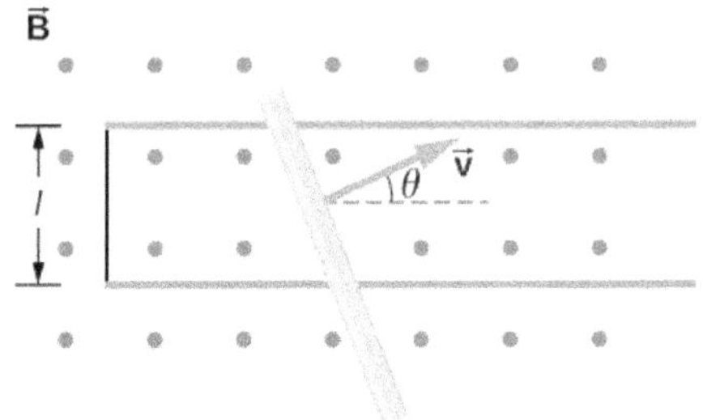

42. The rod shown in the accompanying figure is moving through a uniform magnetic field of strength $B = 0.50\ \text{T}$ with a constant velocity of magnitude $v = 8.0$ m/s. What is the potential difference between the ends of the rod? Which end of the rod is at a higher potential?

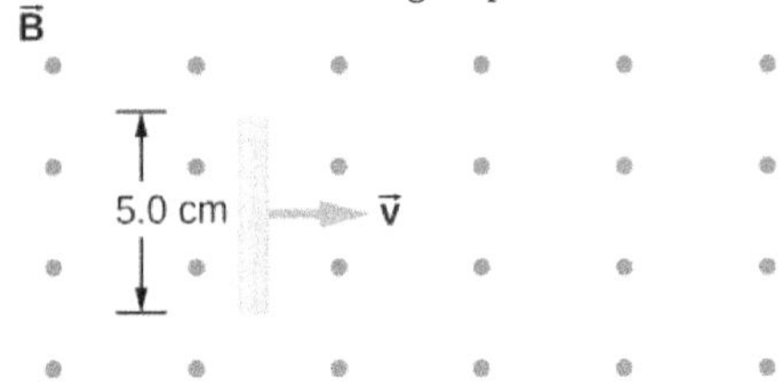

43. A 25-cm rod moves at 5.0 m/s in a plane perpendicular to a magnetic field of strength 0.25 T. The rod, velocity vector, and magnetic field vector are mutually perpendicular, as indicated in the accompanying figure. Calculate (a) the magnetic force on an electron in the rod, (b) the electric field in the rod, and (c) the potential difference between the ends of the rod. (d) What is the speed of the rod if the potential difference is 1.0 V?

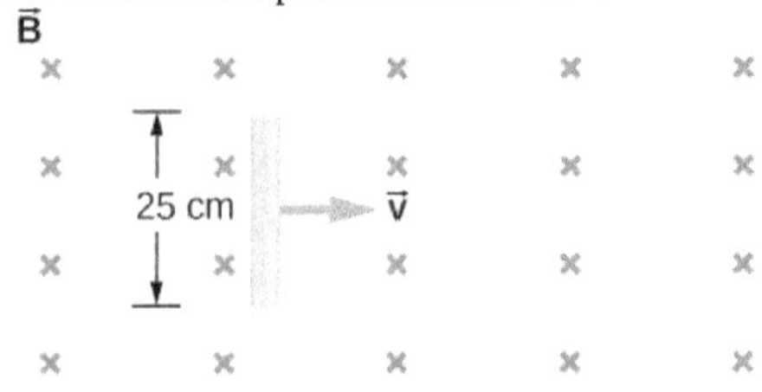

44. In the accompanying figure, the rails, connecting end piece, and rod all have a resistance per unit length of 2.0 Ω/cm. The rod moves to the left at $v = 3.0\,\text{m/s}$. If $B = 0.75\,\text{T}$ everywhere in the region, what is the current in the circuit (a) when $a = 8.0\,\text{cm}$? (b) when $a = 5.0\,\text{cm}$? Specify also the sense of the current flow.

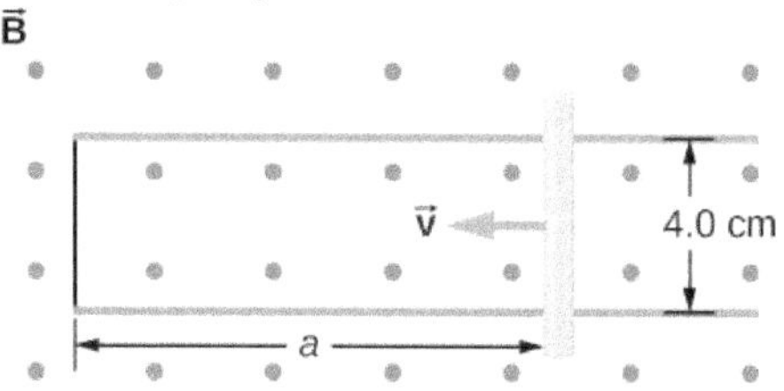

45. The rod shown below moves to the right on essentially zero-resistance rails at a speed of $v = 3.0\,\text{m/s}$. If $B = 0.75\,\text{T}$ everywhere in the region, what is the current through the 5.0-Ω resistor? Does the current circulate clockwise or counterclockwise?

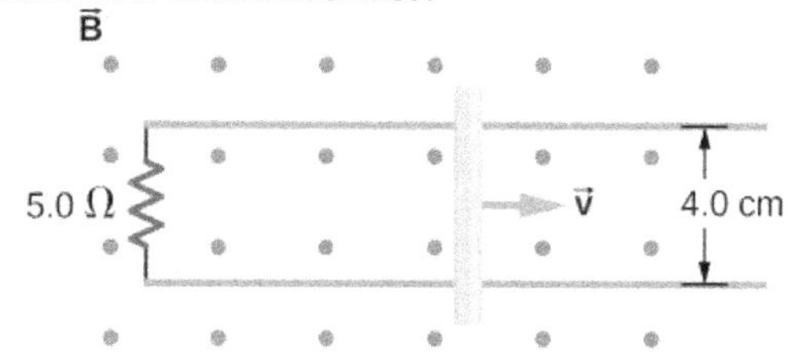

46. Shown below is a conducting rod that slides along metal rails. The apparatus is in a uniform magnetic field of strength 0.25 T, which is directly into the page. The rod is pulled to the right at a constant speed of 5.0 m/s by a force $\vec{\mathbf{F}}$. The only significant resistance in the circuit comes from the 2.0-Ω resistor shown. (a) What is the emf induced in the circuit? (b) What is the induced current? Does it circulate clockwise or counter clockwise? (c) What is the magnitude of $\vec{\mathbf{F}}$? (d) What are the power output of $\vec{\mathbf{F}}$ and the power dissipated in the resistor?

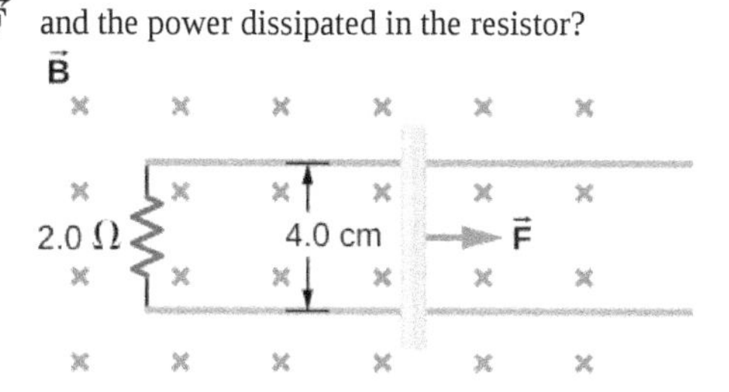

13.4 Induced Electric Fields

47. Calculate the induced electric field in a 50-turn coil with a diameter of 15 cm that is placed in a spatially uniform magnetic field of magnitude 0.50 T so that the face of the coil and the magnetic field are perpendicular. This magnetic field is reduced to zero in 0.10 seconds. Assume that the magnetic field is cylindrically symmetric with respect to the central axis of the coil.

48. The magnetic field through a circular loop of radius 10.0 cm varies with time as shown in the accompanying figure. The field is perpendicular to the loop. Assuming cylindrical symmetry with respect to the central axis of the loop, plot the induced electric field in the loop as a function of time.

49. The current *I* through a long solenoid with *n* turns per meter and radius *R* is changing with time as given by *dI*/*dt*. Calculate the induced electric field as a function of distance *r* from the central axis of the solenoid.

50. Calculate the electric field induced both inside and outside the solenoid of the preceding problem if $I = I_0 \sin \omega t$.

51. Over a region of radius *R*, there is a spatially uniform magnetic field $\vec{\mathbf{B}}$. (See below.) At $t = 0$, $B = 1.0\,\text{T}$, after which it decreases at a constant rate to zero in 30 s. (a) What is the electric field in the regions where $r \leq R$ and $r \geq R$ during that 30-s interval? (b) Assume that $R = 10.0\,\text{cm}$. How much work is done by the electric field on a proton that is carried once clock wise around a circular path of radius 5.0 cm? (c) How much work is done by the electric field on a proton that is carried once counterclockwise around a circular path of any radius $r \geq R$? (d) At the instant when $B = 0.50\,\text{T}$, a proton enters the magnetic field at *A*, moving a velocity $\vec{\mathbf{v}}$ $\left(v = 5.0 \times 10^6\ \text{m/s}\right)$ as shown. What are the electric and magnetic forces on the proton at that instant?

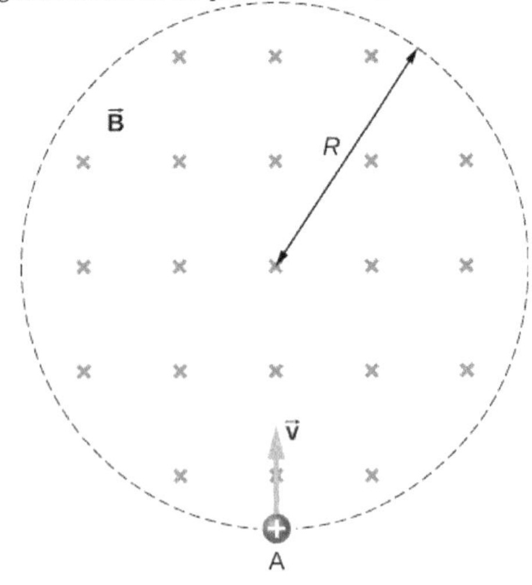

52. The magnetic field at all points within the cylindrical region whose cross-section is indicated in the accompanying figure starts at 1.0 T and decreases uniformly to zero in 20 s. What is the electric field (both magnitude and direction) as a function of *r*, the distance from the geometric center of the region?

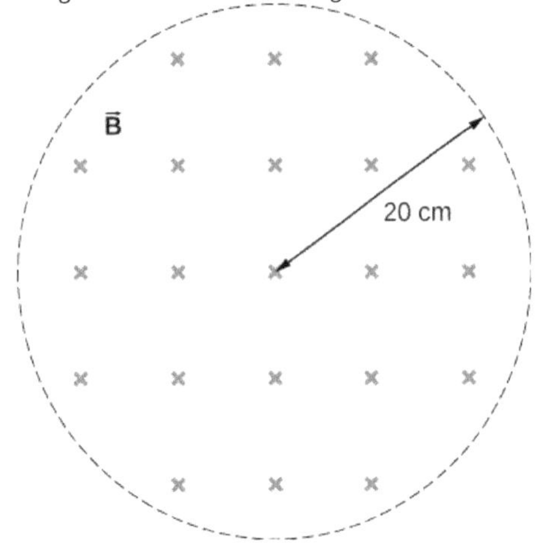

53. The current in a long solenoid of radius 3 cm is varied with time at a rate of 2 A/s. A circular loop of wire of radius 5 cm and resistance $2\,\Omega$ surrounds the solenoid. Find the electrical current induced in the loop.

54. The current in a long solenoid of radius 3 cm and 20 turns/cm is varied with time at a rate of 2 A/s. Find the electric field at a distance of 4 cm from the center of the solenoid.

13.6 Electric Generators and Back Emf

55. Design a current loop that, when rotated in a uniform magnetic field of strength 0.10 T, will produce an emf $\varepsilon = \varepsilon_0 \sin \omega t$, where $\varepsilon_0 = 110\,\text{V}$ and $\omega = 120\pi$ rad/s.

56. A flat, square coil of 20 turns that has sides of length 15.0 cm is rotating in a magnetic field of strength 0.050 T. If the maximum emf produced in the coil is 30.0 mV, what is the angular velocity of the coil?

57. A 50-turn rectangular coil with dimensions $0.15\,\text{m} \times 0.40\,\text{m}$ rotates in a uniform magnetic field of magnitude 0.75 T at 3600 rev/min. (a) Determine the emf induced in the coil as a function of time. (b) If the coil is connected to a $1000\text{-}\Omega$ resistor, what is the power as a function of time required to keep the coil turning at 3600 rpm? (c) Answer part (b) if the coil is connected to a 2000-Ω resistor.

58. The square armature coil of an alternating current generator has 200 turns and is 20.0 cm on side. When it rotates at 3600 rpm, its peak output voltage is 120 V. (a) What is the frequency of the output voltage? (b) What is the strength of the magnetic field in which the coil is turning?

59. A flip coil is a relatively simple device used to measure a magnetic field. It consists of a circular coil of *N* turns wound with fine conducting wire. The coil is attached to a ballistic galvanometer, a device that measures the total charge that passes through it. The coil is placed in a magnetic field $\vec{\mathbf{B}}$ such that its face is perpendicular to the field. It is then flipped through 180°, and the total charge *Q* that flows through the galvanometer is measured. (a) If the total resistance of the coil and galvanometer is *R*, what is the relationship between *B* and *Q*? Because the coil is very small, you can assume that $\vec{\mathbf{B}}$ is uniform over it. (b) How can you determine whether or not the magnetic field is perpendicular to the face of the coil?

60. The flip coil of the preceding problem has a radius of 3.0 cm and is wound with 40 turns of copper wire. The total resistance of the coil and ballistic galvanometer is $0.20\ \Omega$. When the coil is flipped through $180°$ in a magnetic field $\vec{\mathbf{B}}$, a change of 0.090 C flows through the ballistic galvanometer. (a) Assuming that $\vec{\mathbf{B}}$ and the face of the coil are initially perpendicular, what is the magnetic field? (b) If the coil is flipped through $90°$, what is the reading of the galvanometer?

61. A 120-V, series-wound motor has a field resistance of $80\ \Omega$ and an armature resistance of $10\ \Omega$. When it is operating at full speed, a back emf of 75 V is generated. (a) What is the initial current drawn by the motor? When the motor is operating at full speed, where are (b) the current drawn by the motor, (c) the power output of the source, (d) the power output of the motor, and (e) the power dissipated in the two resistances?

62. A small series-wound dc motor is operated from a 12-V car battery. Under a normal load, the motor draws 4.0 A, and when the armature is clamped so that it cannot turn, the motor draws 24 A. What is the back emf when the motor is operating normally?

ADDITIONAL PROBLEMS

63. Shown in the following figure is a long, straight wire and a single-turn rectangular loop, both of which lie in the plane of the page. The wire is parallel to the long sides of the loop and is 0.50 m away from the closer side. At an instant when the emf induced in the loop is 2.0 V, what is the time rate of change of the current in the wire?

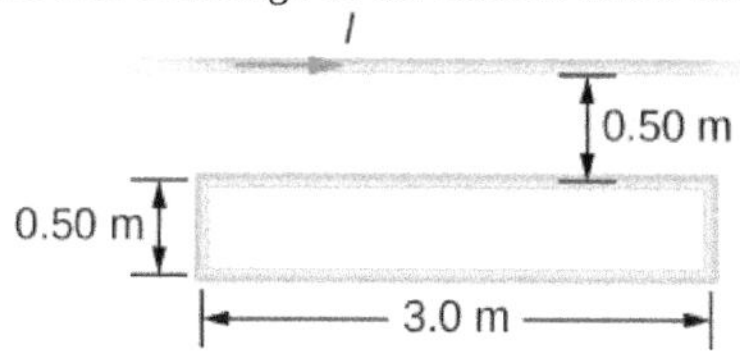

64. A metal bar of mass 500 g slides outward at a constant speed of 1.5 cm/s over two parallel rails separated by a distance of 30 cm which are part of a U-shaped conductor. There is a uniform magnetic field of magnitude 2 T pointing out of the page over the entire area. The railings and metal bar have an equivalent resistance of $150\ \Omega$. (a) Determine the induced current, both magnitude and direction. (b) Find the direction of the induced current if the magnetic field is pointing into the page. (c) Find the direction of the induced current if the magnetic field is pointed into the page and the bar moves inwards.

65. A current is induced in a circular loop of radius 1.5 cm between two poles of a horseshoe electromagnet when the current in the electromagnet is varied. The magnetic field in the area of the loop is perpendicular to the area and has a uniform magnitude. If the rate of change of magnetic field is 10 T/s, find the magnitude and direction of the induced current if resistance of the loop is $25\ \Omega$.

66. A metal bar of length 25 cm is placed perpendicular to a uniform magnetic field of strength 3 T. (a) Determine the induced emf between the ends of the rod when it is not moving. (b) Determine the emf when the rod is moving perpendicular to its length and magnetic field with a speed of 50 cm/s.

67. A coil with 50 turns and area 10 cm^2 is oriented with its plane perpendicular to a 0.75-T magnetic field. If the coil is flipped over (rotated through $180°$) in 0.20 s, what is the average emf induced in it?

68. A 2-turn planer loop of flexible wire is placed inside a long solenoid of n turns per meter that carries a constant current I_0. The area A of the loop is changed by pulling on its sides while ensuring that the plane of the loop always remains perpendicular to the axis of the solenoid. If $n = 500$ turns per meter, $I_0 = 20\text{ A}$, and $A = 20\text{ cm}^2$, what is the emf induced in the loop when $dA/dt = 100$?

69. The conducting rod shown in the accompanying figure moves along parallel metal rails that are 25-cm apart. The system is in a uniform magnetic field of strength 0.75 T, which is directed into the page. The resistances of the rod and the rails are negligible, but the section PQ has a resistance of $0.25\,\Omega$. (a) What is the emf (including its sense) induced in the rod when it is moving to the right with a speed of 5.0 m/s? (b) What force is required to keep the rod moving at this speed? (c) What is the rate at which work is done by this force? (d) What is the power dissipated in the resistor?

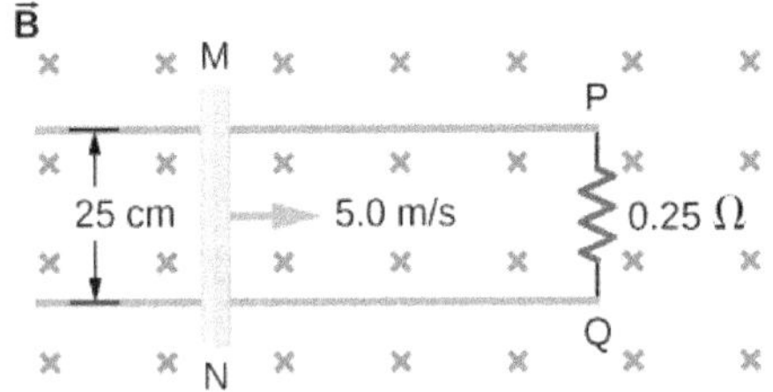

70. A circular loop of wire of radius 10 cm is mounted on a vertical shaft and rotated at a frequency of 5 cycles per second in a region of uniform magnetic field of 2 Gauss perpendicular to the axis of rotation. (a) Find an expression for the time-dependent flux through the ring. (b) Determine the time-dependent current through the ring if it has a resistance of 10 Ω.

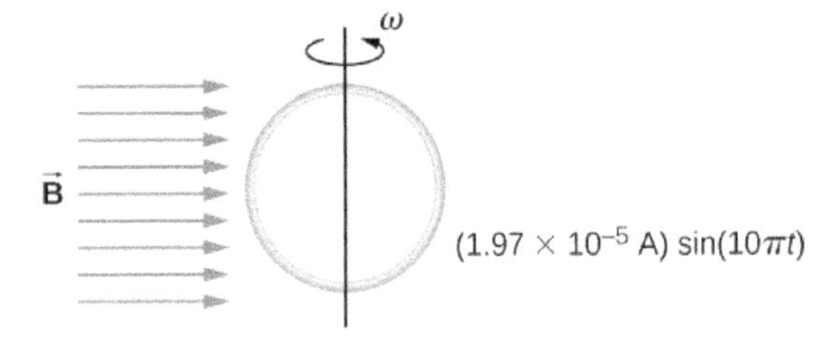

71. The magnetic field between the poles of a horseshoe electromagnet is uniform and has a cylindrical symmetry about an axis from the middle of the South Pole to the middle of the North Pole. The magnitude of the magnetic field changes as a rate of dB/dt due to the changing current through the electromagnet. Determine the electric field at a distance r from the center.

72. A long solenoid of radius a with n turns per unit length is carrying a time-dependent current $I(t) = I_0 \sin(\omega t)$, where I_0 and ω are constants. The solenoid is surrounded by a wire of resistance R that has two circular loops of radius b with $b > a$ (see the following figure). Find the magnitude and direction of current induced in the outer loops at time $t = 0$.

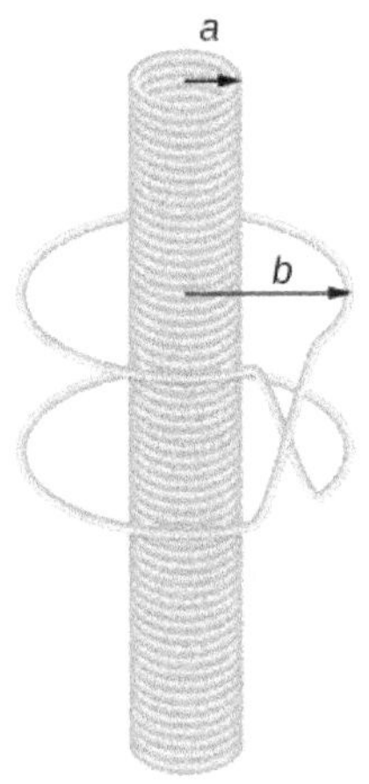

73. A 120-V, series-wound dc motor draws 0.50 A from its power source when operating at full speed, and it draws 2.0 A when it starts. The resistance of the armature coils is $10\,\Omega$. (a) What is the resistance of the field coils? (b) What is the back emf of the motor when it is running at full speed? (c) The motor operates at a different speed and draws 1.0 A from the source. What is the back emf in this case?

74. The armature and field coils of a series-wound motor have a total resistance of $3.0\,\Omega$. When connected to a 120-V source and running at normal speed, the motor draws 4.0 A. (a) How large is the back emf? (b) What current will the motor draw just after it is turned on? Can you suggest a way to avoid this large initial current?

CHALLENGE PROBLEMS

75. A copper wire of length L is fashioned into a circular coil with N turns. When the magnetic field through the coil changes with time, for what value of N is the induced emf a maximum?

76. A 0.50-kg copper sheet drops through a uniform horizontal magnetic field of 1.5 T, and it reaches a terminal velocity of 2.0 m/s. (a) What is the net magnetic force on the sheet after it reaches terminal velocity? (b) Describe the mechanism responsible for this force. (c) How much power is dissipated as Joule heating while the sheet moves at terminal velocity?

77. A circular copper disk of radius 7.5 cm rotates at 2400 rpm around the axis through its center and perpendicular to its face. The disk is in a uniform magnetic field $\vec{\mathbf{B}}$ of strength 1.2 T that is directed along the axis. What is the potential difference between the rim and the axis of the disk?

78. A short rod of length a moves with its velocity $\vec{\mathbf{v}}$ parallel to an infinite wire carrying a current I (see below). If the end of the rod nearer the wire is a distance b from the wire, what is the emf induced in the rod?

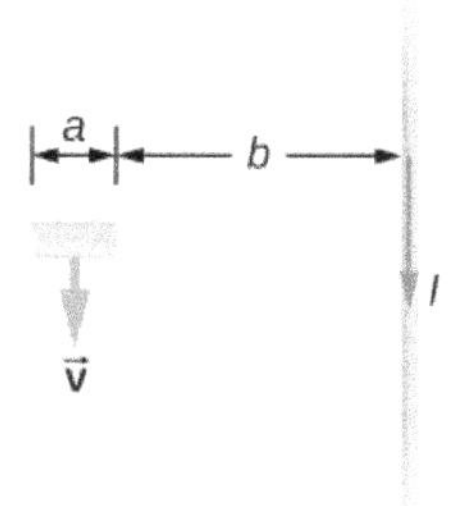

79. A rectangular circuit containing a resistance R is pulled at a constant velocity $\vec{\mathbf{v}}$ away from a long, straight wire carrying a current I_0 (see below). Derive an equation that gives the current induced in the circuit as a function of the distance x between the near side of the circuit and the wire.

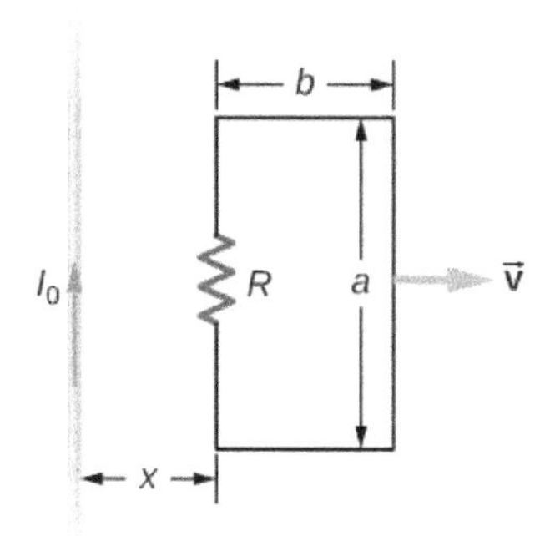

80. Two infinite solenoids cross the plane of the circuit as shown below. The radii of the solenoids are 0.10 and 0.20 m, respectively, and the current in each solenoid is changing such that $dB/dt = 50.0\ \text{T/s}$. What are the currents in the resistors of the circuit?

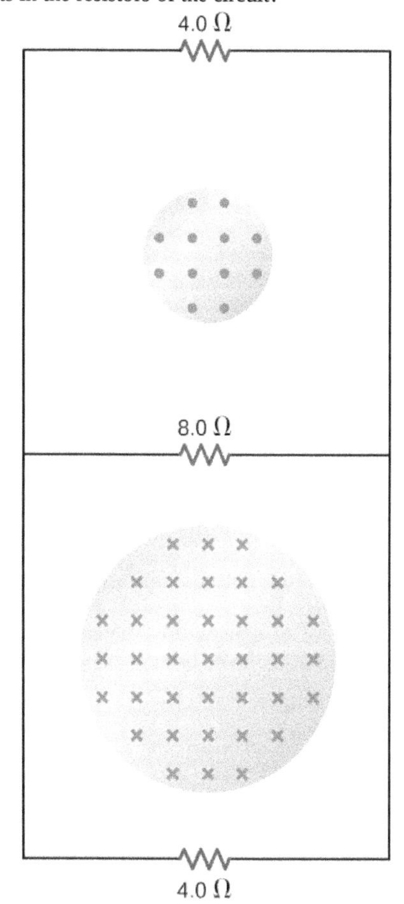

81. An eight-turn coil is *tightly wrapped* around the outside of the long solenoid as shown below. The radius of the solenoid is 2.0 cm and it has 10 turns per centimeter. The current through the solenoid increases according to $I = I_0(1 - e^{-\alpha t})$, where $I_0 = 4.0\ \text{A}$ and $\alpha = 2.0 \times 10^{-2}\ \text{s}^{-1}$. What is the emf induced in the coil when (a) $t = 0$, (b) $t = 1.0 \times 10^2\ \text{s}$, and (c) $t \to \infty$?

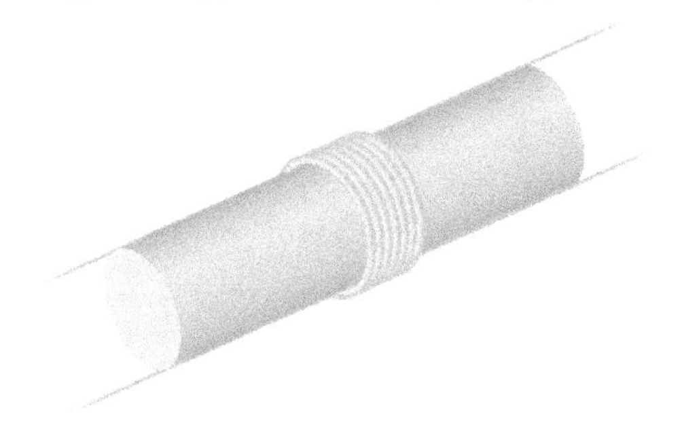

82. Shown below is a long rectangular loop of width *w*, length *l*, mass *m*, and resistance *R*. The loop starts from rest at the edge of a uniform magnetic field $\vec{\mathbf{B}}$ and is pushed into the field by a constant force $\vec{\mathbf{F}}$. Calculate the speed of the loop as a function of time.

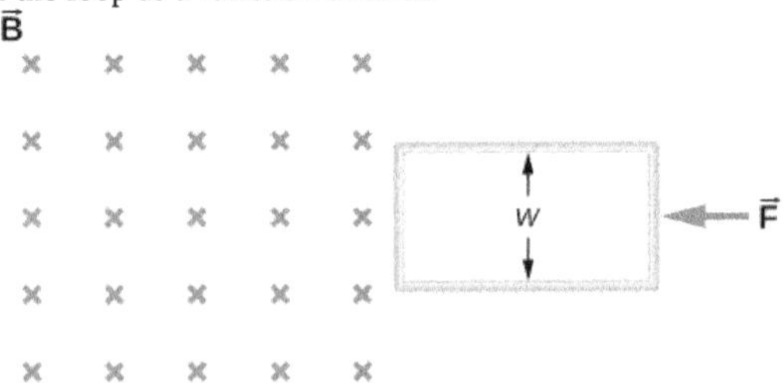

83. A square bar of mass *m* and resistance *R* is sliding without friction down very long, parallel conducting rails of negligible resistance (see below). The two rails are a distance *l* apart and are connected to each other at the bottom of the incline by a zero-resistance wire. The rails are inclined at an angle θ, and there is a uniform vertical magnetic field $\vec{\mathbf{B}}$ throughout the region. (a) Show that the bar acquires a terminal velocity given by $v = \dfrac{mgR \sin\theta}{B^2 l^2 \cos^2\theta}$. (b) Calculate the work per unit time done by the force of gravity. (c) Compare this with the power dissipated in the Joule heating of the bar. (d) What would happen if $\vec{\mathbf{B}}$ were reversed?

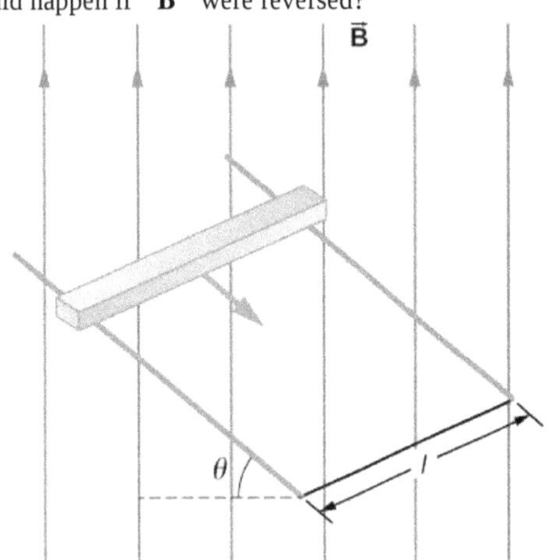

84. The accompanying figure shows a metal disk of inner radius r_1 and other radius r_2 rotating at an angular velocity $\vec{\omega}$ while in a uniform magnetic field directed parallel to the rotational axis. The brush leads of a voltmeter are connected to the dark's inner and outer surfaces as shown. What is the reading of the voltmeter?

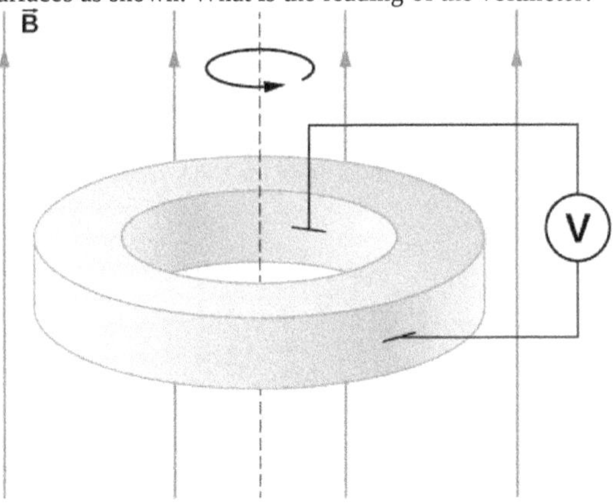

85. A long solenoid with 10 turns per centimeter is placed inside a copper ring such that both objects have the same central axis. The radius of the ring is 10.0 cm, and the radius of the solenoid is 5.0 cm. (a) What is the emf induced in the ring when the current *I* through the solenoid is 5.0 A and changing at a rate of 100 A/s? (b) What is the emf induced in the ring when $I = 2.0\,\text{A}$ and $dI/dt = 100\,\text{A/s}$? (c) What is the electric field inside the ring for these two cases? (d) Suppose the ring is moved so that its central axis and the central axis of the solenoid are still parallel but no longer coincide. (You should assume that the solenoid is still inside the ring.) Now what is the emf induced in the ring? (e) Can you calculate the electric field in the ring as you did in part (c)?

86. The current in the long, straight wire shown in the accompanying figure is given by $I = I_0 \sin\omega t$, where $I_0 = 15\,\text{A}$ and $\omega = 120\pi\,\text{rad/s}$. What is the current induced in the rectangular loop at (a) $t = 0$ and (b) $t = 2.1 \times 10^{-3}\,\text{s}$? The resistance of the loop is $2.0\,\Omega$.

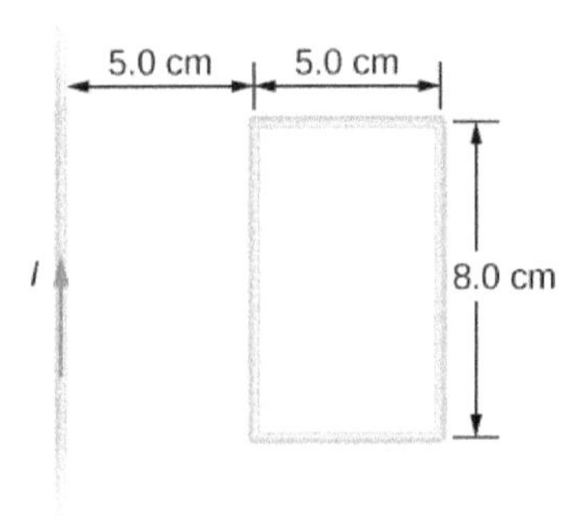

87. A 500-turn coil with a 0.250-m^2 area is spun in Earth's $5.00 \times 10^{-5}\,\text{T}$ magnetic field, producing a 12.0-kV maximum emf. (a) At what angular velocity must the coil be spun? (b) What is unreasonable about this result? (c) Which assumption or premise is responsible?

88. A circular loop of wire of radius 10 cm is mounted on a vertical shaft and rotated at a frequency of 5 cycles per second in a region of uniform magnetic field of $2 \times 10^{-4}\,T$ perpendicular to the axis of rotation. (a) Find an expression for the time-dependent flux through the ring (b) Determine the time-dependent current through the ring if it has a resistance of $10\,\Omega$.

89. A long solenoid of radius a with n turns per unit length is carrying a time-dependent current $I(t) = I_0 \sin \omega t$ where I_0 and ω are constants. The solenoid is surrounded by a wire of resistance R that has two circular loops of radius b with $b > a$. Find the magnitude and direction of current induced in the outer loops at time $t = 0$.

90. A rectangular copper loop of mass 100 g and resistance $0.2\,\Omega$ is in a region of uniform magnetic field that is perpendicular to the area enclosed by the ring and horizontal to Earth's surface (see below). The loop is let go from rest when it is at the edge of the nonzero magnetic field region. (a) Find an expression for the speed when the loop just exits the region of uniform magnetic field. (b) If it was let go at $t = 0$, what is the time when it exits the region of magnetic field for the following values: $a = 25$ cm, $b = 50$ cm, $B = 3$ T, $g = 9.8\ \text{m/s}^2$? Assume that the magnetic field of the induced current is negligible compared to 3 T.

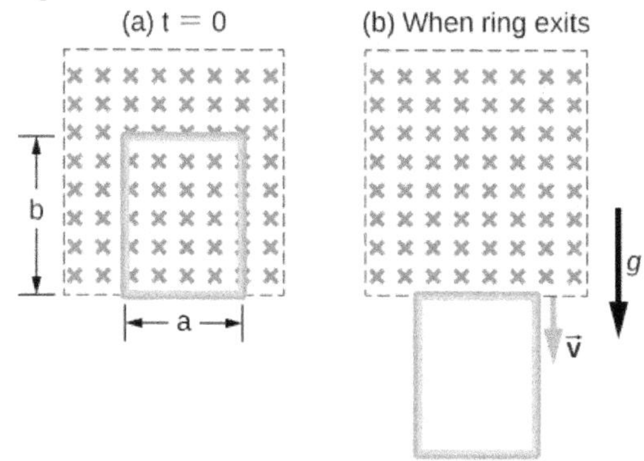

91. A metal bar of mass m slides without friction over two rails a distance D apart in the region that has a uniform magnetic field of magnitude B_0 and direction perpendicular to the rails (see below). The two rails are connected at one end to a resistor whose resistance is much larger than the resistance of the rails and the bar. The bar is given an initial speed of v_0. It is found to slow down. How far does the bar go before coming to rest? Assume that the magnetic field of the induced current is negligible compared to B_0.

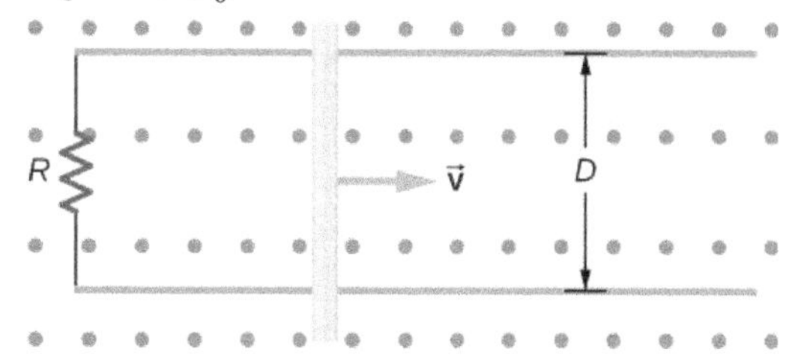

92. A time-dependent uniform magnetic field of magnitude $B(t)$ is confined in a cylindrical region of radius R. A conducting rod of length $2D$ is placed in the region, as shown below. Show that the emf between the ends of the rod is given by $\frac{dB}{dt}D\sqrt{R^2 - D^2}$. (*Hint:* To find the emf between the ends, we need to integrate the electric field from one end to the other. To find the electric field, use Faraday's law as "Ampère's law for *E*.")

14 | INDUCTANCE

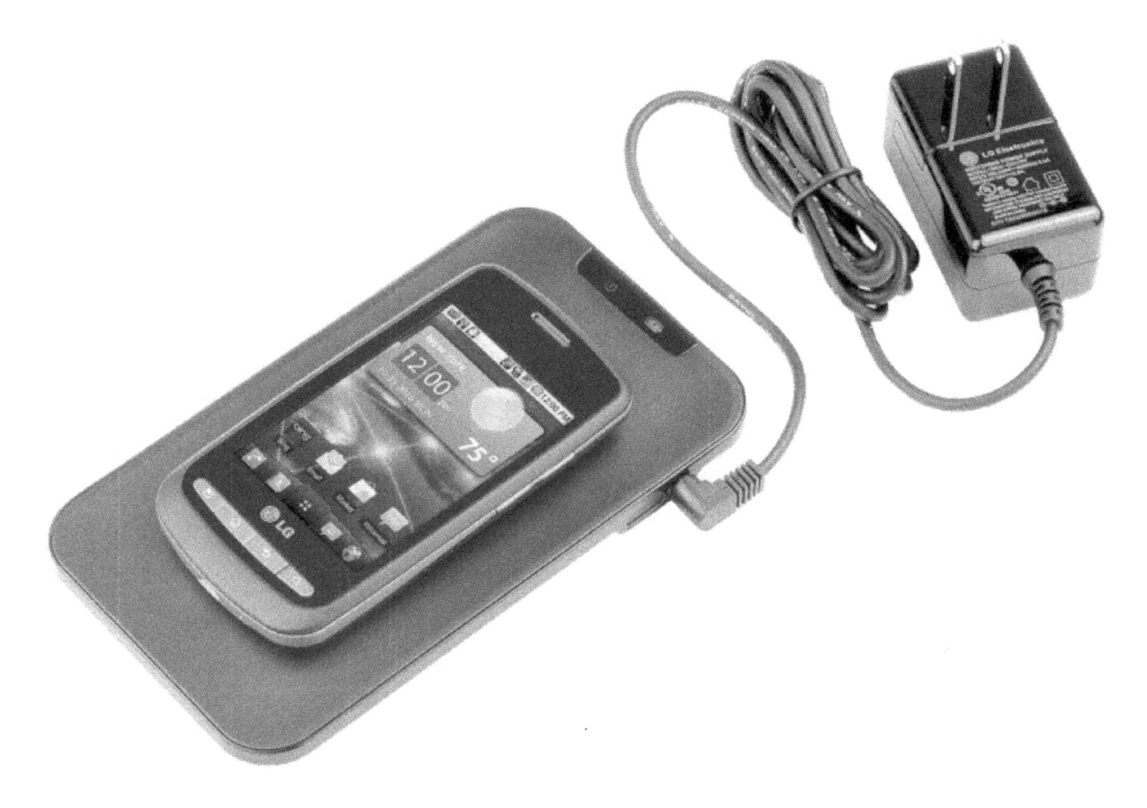

Figure 14.1 A smartphone charging mat contains a coil that receives alternating current, or current that is constantly increasing and decreasing. The varying current induces an emf in the smartphone, which charges its battery. Note that the black box containing the electrical plug also contains a transformer (discussed in Alternating-Current Circuits) that modifies the current from the outlet to suit the needs of the smartphone. (credit: modification of work by "LG"/Flickr)

Chapter Outline

Introduction

In Electromagnetic Induction, we discussed how a time-varying magnetic flux induces an emf in a circuit. In many of our calculations, this flux was due to an applied time-dependent magnetic field. The reverse of this phenomenon also occurs: The current flowing in a circuit produces its own magnetic field.

In Electric Charges and Fields, we saw that induction is the process by which an emf is induced by changing magnetic flux. So far, we have discussed some examples of induction, although some of these applications are more effective than others. The smartphone charging mat in the chapter opener photo also works by induction. Is there a useful physical quantity related to how "effective" a given device is? The answer is yes, and that physical quantity is *inductance*. In this chapter, we look at the applications of inductance in electronic devices and how inductors are used in circuits.

14.1 | Mutual Inductance

Learning Objectives

By the end of this section, you will be able to:

- Correlate two nearby circuits that carry time-varying currents with the emf induced in each circuit
- Describe examples in which mutual inductance may or may not be desirable

Inductance is the property of a device that tells us how effectively it induces an emf in another device. In other words, it is a physical quantity that expresses the effectiveness of a given device.

When two circuits carrying time-varying currents are close to one another, the magnetic flux through each circuit varies because of the changing current *I* in the other circuit. Consequently, an emf is induced in each circuit by the changing current in the other. This type of emf is therefore called a *mutually induced emf,* and the phenomenon that occurs is known as **mutual inductance (*M*)**. As an example, let's consider two tightly wound coils (Figure 14.2). Coils 1 and 2 have N_1 and N_2 turns and carry currents I_1 and $I_2,$ respectively. The flux through a single turn of coil 2 produced by the magnetic field of the current in coil 1 is $\Phi_{21},$ whereas the flux through a single turn of coil 1 due to the magnetic field of I_2 is $\Phi_{12}.$

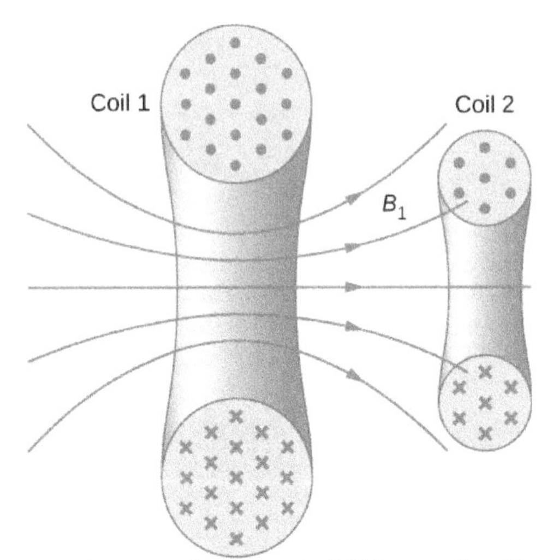

Figure 14.2 Some of the magnetic field lines produced by the current in coil 1 pass through coil 2.

The mutual inductance M_{21} of coil 2 with respect to coil 1 is the ratio of the flux through the N_2 turns of coil 2 produced by the magnetic field of the current in coil 1, divided by that current, that is,

$$M_{21} = \frac{N_2 \Phi_{21}}{I_1}. \tag{14.1}$$

Similarly, the mutual inductance of coil 1 with respect to coil 2 is

$$M_{12} = \frac{N_1 \Phi_{12}}{I_2}. \tag{14.2}$$

Like capacitance, mutual inductance is a geometric quantity. It depends on the shapes and relative positions of the two coils, and it is independent of the currents in the coils. The SI unit for mutual inductance *M* is called the **henry (H)** in honor of

Joseph Henry (1799–1878), an American scientist who discovered induced emf independently of Faraday. Thus, we have $1\,\text{H} = 1\,\text{V} \cdot \text{s/A}$. From Equation 14.1 and Equation 14.2, we can show that $M_{21} = M_{12}$, so we usually drop the subscripts associated with mutual inductance and write

$$M = \frac{N_2 \Phi_{21}}{I_1} = \frac{N_1 \Phi_{12}}{I_2}. \quad (14.3)$$

The emf developed in either coil is found by combining Faraday's law and the definition of mutual inductance. Since $N_2 \Phi_{21}$ is the total flux through coil 2 due to I_1, we obtain

$$\varepsilon_2 = -\frac{d}{dt}(N_2 \Phi_{21}) = -\frac{d}{dt}(M I_1) = -M\frac{dI_1}{dt} \quad (14.4)$$

where we have used the fact that M is a time-independent constant because the geometry is time-independent. Similarly, we have

$$\varepsilon_1 = -M\frac{dI_2}{dt}. \quad (14.5)$$

In Equation 14.5, we can see the significance of the earlier description of mutual inductance (M) as a geometric quantity. The value of M neatly encapsulates the physical properties of circuit elements and allows us to separate the physical layout of the circuit from the dynamic quantities, such as the emf and the current. Equation 14.5 defines the mutual inductance in terms of properties in the circuit, whereas the previous definition of mutual inductance in Equation 14.1 is defined in terms of the magnetic flux experienced, regardless of circuit elements. You should be careful when using Equation 14.4 and Equation 14.5 because ε_1 and ε_2 do not necessarily represent the total emfs in the respective coils. Each coil can also have an emf induced in it because of its *self-inductance* (self-inductance will be discussed in more detail in a later section).

A large mutual inductance M may or may not be desirable. We want a transformer to have a large mutual inductance. But an appliance, such as an electric clothes dryer, can induce a dangerous emf on its metal case if the mutual inductance between its coils and the case is large. One way to reduce mutual inductance is to counter-wind coils to cancel the magnetic field produced (Figure 14.3).

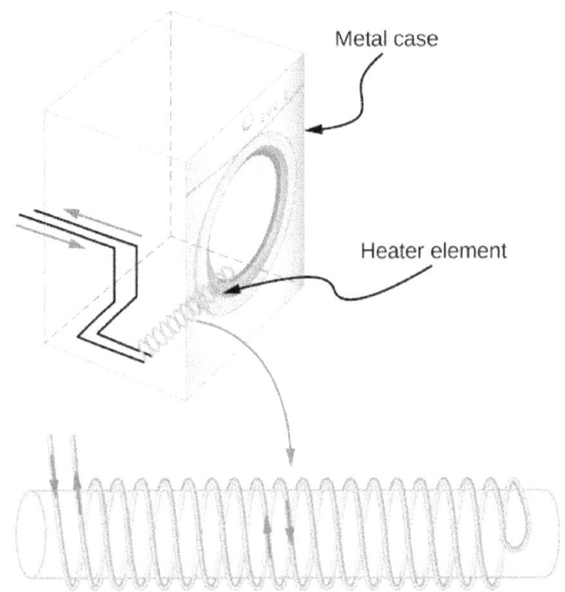

Figure 14.3 The heating coils of an electric clothes dryer can be counter-wound so that their magnetic fields cancel one another, greatly reducing the mutual inductance with the case of the dryer.

Digital signal processing is another example in which mutual inductance is reduced by counter-winding coils. The rapid on/off emf representing 1s and 0s in a digital circuit creates a complex time-dependent magnetic field. An emf can be generated in neighboring conductors. If that conductor is also carrying a digital signal, the induced emf may be large enough to switch 1s and 0s, with consequences ranging from inconvenient to disastrous.

Example 14.1

Mutual Inductance

Figure 14.4 shows a coil of N_2 turns and radius R_2 surrounding a long solenoid of length l_1, radius R_1, and N_1 turns. (a) What is the mutual inductance of the two coils? (b) If $N_1 = 500\,\text{turns}$, $N_2 = 10\,\text{turns}$, $R_1 = 3.10\,\text{cm}$, $l_1 = 75.0\,\text{cm}$, and the current in the solenoid is changing at a rate of 200 A/s, what is the emf induced in the surrounding coil?

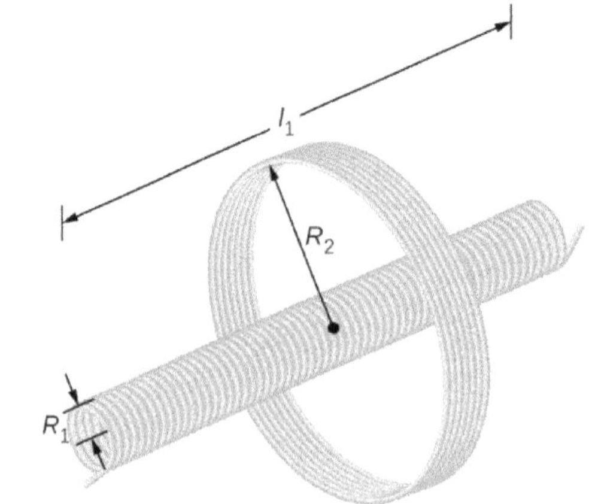

Figure 14.4 A solenoid surrounded by a coil.

Strategy

There is no magnetic field outside the solenoid, and the field inside has magnitude $B_1 = \mu_0(N_1/l_1)I_1$ and is directed parallel to the solenoid's axis. We can use this magnetic field to find the magnetic flux through the surrounding coil and then use this flux to calculate the mutual inductance for part (a), using Equation 14.3. We solve part (b) by calculating the mutual inductance from the given quantities and using Equation 14.4 to calculate the induced emf.

Solution

a. The magnetic flux Φ_{21} through the surrounding coil is

$$\Phi_{21} = B_1 \pi R_1^2 = \frac{\mu_0 N_1 I_1}{l_1} \pi R_1^2.$$

Now from Equation 14.3, the mutual inductance is

$$M = \frac{N_2 \Phi_{21}}{I_1} = \left(\frac{N_2}{I_1}\right)\left(\frac{\mu_0 N_1 I_1}{l_1}\right)\pi R_1^2 = \frac{\mu_0 N_1 N_2 \pi R_1^2}{l_1}.$$

b. Using the previous expression and the given values, the mutual inductance is

$$\begin{aligned} M &= \frac{(4\pi \times 10^{-7}\ \text{T} \cdot \text{m/A})(500)(10)\pi(0.0310\ \text{m})^2}{0.750\ \text{m}} \\ &= 2.53 \times 10^{-5}\ \text{H}. \end{aligned}$$

Thus, from Equation 14.4, the emf induced in the surrounding coil is

$$\begin{aligned} \varepsilon_2 &= -M\frac{dI_1}{dt} = -(2.53 \times 10^{-5}\ \text{H})(200\ \text{A/s}) \\ &= -5.06 \times 10^{-3}\ \text{V}. \end{aligned}$$

Significance

Notice that M in part (a) is independent of the radius R_2 of the surrounding coil because the solenoid's magnetic field is confined to its interior. In principle, we can also calculate M by finding the magnetic flux through the solenoid produced by the current in the surrounding coil. This approach is much more difficult because Φ_{12} is so complicated. However, since $M_{12} = M_{21}$, we do know the result of this calculation.

14.1 Check Your Understanding A current $I(t) = (5.0\ \text{A})\sin((120\pi\ \text{rad/s})t)$ flows through the solenoid of part (b) of Example 14.1. What is the maximum emf induced in the surrounding coil?

14.2 | Self-Inductance and Inductors

Learning Objectives

By the end of this section, you will be able to:

- Correlate the rate of change of current to the induced emf created by that current in the same circuit
- Derive the self-inductance for a cylindrical solenoid
- Derive the self-inductance for a rectangular toroid

Mutual inductance arises when a current in one circuit produces a changing magnetic field that induces an emf in another circuit. But can the magnetic field affect the current in the original circuit that produced the field? The answer is yes, and this is the phenomenon called *self-inductance*.

Inductors

Figure 14.5 shows some of the magnetic field lines due to the current in a circular loop of wire. If the current is constant, the magnetic flux through the loop is also constant. However, if the current I were to vary with time—say, immediately after switch S is closed—then the magnetic flux Φ_m would correspondingly change. Then Faraday's law tells us that an emf ε would be induced in the circuit, where

$$\varepsilon = -\frac{d\Phi_\text{m}}{dt}. \tag{14.6}$$

Since the magnetic field due to a current-carrying wire is directly proportional to the current, the flux due to this field is also proportional to the current; that is,

$$\Phi_\text{m} \propto I. \tag{14.7}$$

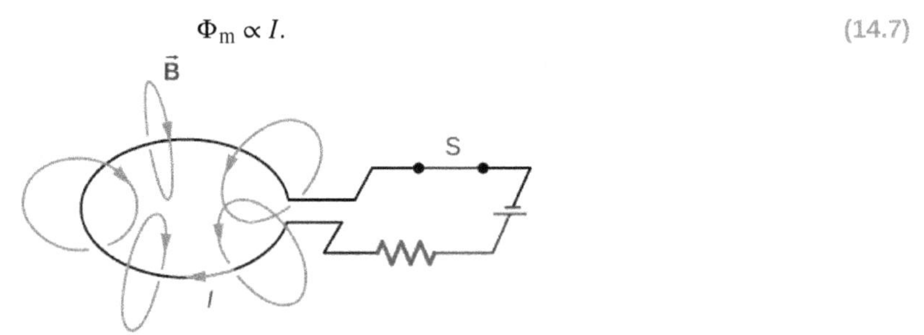

Figure 14.5 A magnetic field is produced by the current I in the loop. If I were to vary with time, the magnetic flux through the loop would also vary and an emf would be induced in the loop.

This can also be written as

$$\Phi_\text{m} = LI \tag{14.8}$$

where the constant of proportionality L is known as the **self-inductance** of the wire loop. If the loop has N turns, this equation becomes

$$N\Phi_\text{m} = LI. \tag{14.9}$$

By convention, the positive sense of the normal to the loop is related to the current by the right-hand rule, so in Figure 14.5, the normal points downward. With this convention, Φ_m is positive in Equation 14.9, so L *always has a positive value*.

For a loop with N turns, $\varepsilon = -Nd\Phi_\text{m}/dt,$ so the induced emf may be written in terms of the self-inductance as

$$\varepsilon = -L\frac{dI}{dt}. \tag{14.10}$$

When using this equation to determine L, it is easiest to ignore the signs of ε and $dI/dt,$ and calculate L as

$$L = \frac{|\varepsilon|}{|dI/dt|}.$$

Since self-inductance is associated with the magnetic field produced by a current, any configuration of conductors possesses self-inductance. For example, besides the wire loop, a long, straight wire has self-inductance, as does a coaxial cable. A coaxial cable is most commonly used by the cable television industry and may also be found connecting to your cable modem. Coaxial cables are used due to their ability to transmit electrical signals with minimal distortions. Coaxial cables have two long cylindrical conductors that possess current and a self-inductance that may have undesirable effects.

A circuit element used to provide self-inductance is known as an **inductor**. It is represented by the symbol shown in Figure 14.6, which resembles a coil of wire, the basic form of the inductor. Figure 14.7 shows several types of inductors commonly used in circuits.

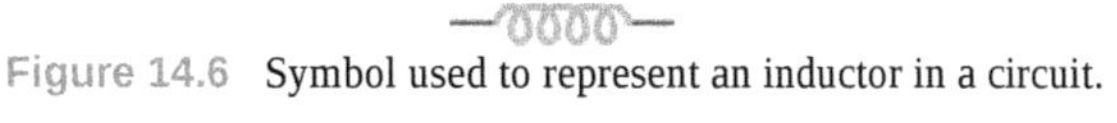

Figure 14.6 Symbol used to represent an inductor in a circuit.

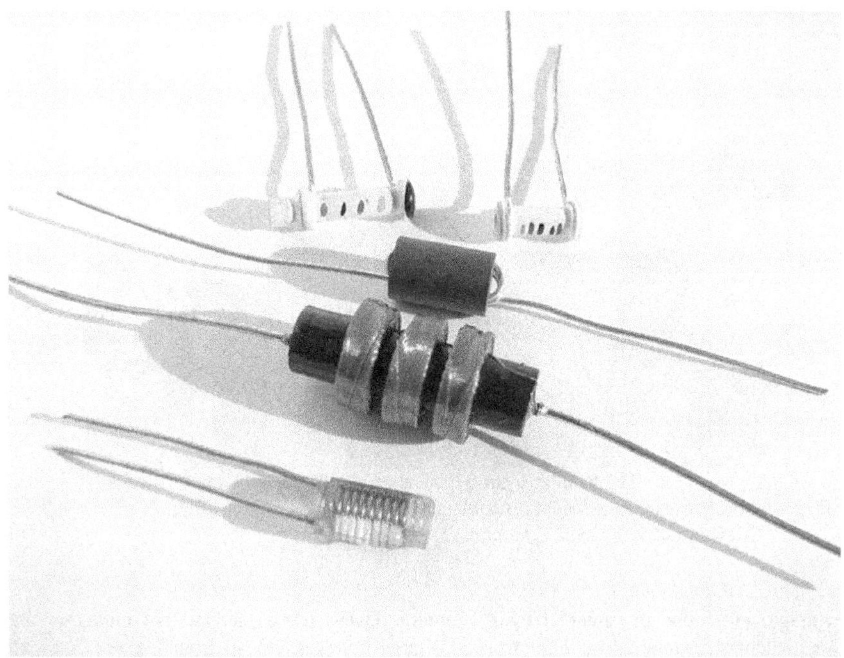

Figure 14.7 A variety of inductors. Whether they are encapsulated like the top three shown or wound around in a coil like the bottom-most one, each is simply a relatively long coil of wire. (credit: Windell Oskay)

In accordance with Lenz's law, the negative sign in Equation 14.10 indicates that the induced emf across an inductor always has a polarity that *opposes* the change in the current. For example, if the current flowing from *A* to *B* in Figure 14.8(a) were increasing, the induced emf (represented by the imaginary battery) would have the polarity shown in order to oppose the increase. If the current from *A* to *B* were decreasing, then the induced emf would have the opposite polarity, again to oppose the change in current (Figure 14.8(b)). Finally, if the current through the inductor were constant, no emf would be induced in the coil.

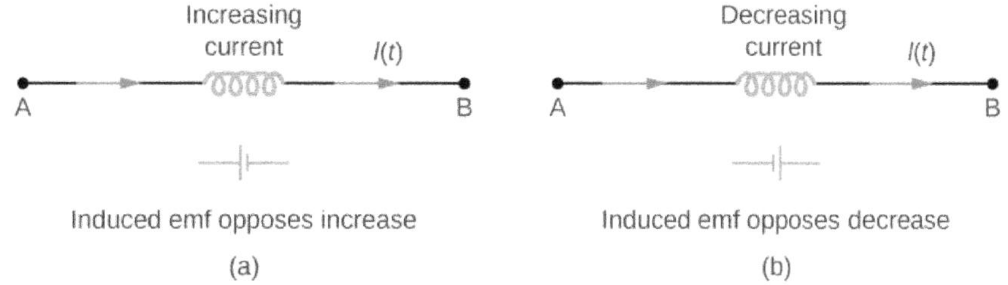

Figure 14.8 The induced emf across an inductor always acts to oppose the change in the current. This can be visualized as an imaginary battery causing current to flow to oppose the change in (a) and reinforce the change in (b).

One common application of inductance is to allow traffic signals to sense when vehicles are waiting at a street intersection. An electrical circuit with an inductor is placed in the road underneath the location where a waiting car will stop. The body of the car increases the inductance and the circuit changes, sending a signal to the traffic lights to change colors. Similarly, metal detectors used for airport security employ the same technique. A coil or inductor in the metal detector frame acts as both a transmitter and a receiver. The pulsed signal from the transmitter coil induces a signal in the receiver. The self-inductance of the circuit is affected by any metal object in the path (Figure 14.9). Metal detectors can be adjusted for sensitivity and can also sense the presence of metal on a person.

Figure 14.9 The familiar security gate at an airport not only detects metals, but can also indicate their approximate height above the floor. (credit: "Alexbuirds"/Wikimedia Commons)

Large induced voltages are found in camera flashes. Camera flashes use a battery, two inductors that function as a transformer, and a switching system or *oscillator* to induce large voltages. Recall from Oscillations (http://cnx.org/content/m58360/latest/) on oscillations that "oscillation" is defined as the fluctuation of a quantity, or repeated regular fluctuations of a quantity, between two extreme values around an average value. Also recall (from Electromagnetic Induction on electromagnetic induction) that we need a changing magnetic field, brought about by a changing current, to induce a voltage in another coil. The oscillator system does this many times as the battery voltage is boosted to over 1000 volts. (You may hear the high-pitched whine from the transformer as the capacitor is being charged.) A capacitor stores the high voltage for later use in powering the flash.

Example 14.2

Self-Inductance of a Coil

An induced emf of 2.0 V is measured across a coil of 50 closely wound turns while the current through it increases uniformly from 0.0 to 5.0 A in 0.10 s. (a) What is the self-inductance of the coil? (b) With the current at 5.0 A, what is the flux through each turn of the coil?

Strategy

Both parts of this problem give all the information needed to solve for the self-inductance in part (a) or the flux through each turn of the coil in part (b). The equations needed are Equation 14.10 for part (a) and Equation 14.9 for part (b).

Solution

a. Ignoring the negative sign and using magnitudes, we have, from Equation 14.10,

$$L = \frac{\varepsilon}{dI/dt} = \frac{2.0\text{ V}}{5.0\text{ A}/0.10\text{ s}} = 4.0 \times 10^{-2}\text{ H}.$$

b. From Equation 14.9, the flux is given in terms of the current by $\Phi_{\text{m}} = LI/N$, so

$$\Phi_m = \frac{(4.0 \times 10^{-2}\ \text{H})(5.0\ \text{A})}{50\ \text{turns}} = 4.0 \times 10^{-3}\ \text{Wb}.$$

Significance

The self-inductance and flux calculated in parts (a) and (b) are typical values for coils found in contemporary devices. If the current is not changing over time, the flux is not changing in time, so no emf is induced.

14.2 Check Your Understanding Current flows through the inductor in Figure 14.8 from *B* to *A* instead of from *A* to *B* as shown. Is the current increasing or decreasing in order to produce the emf given in diagram (a)? In diagram (b)?

14.3 Check Your Understanding A changing current induces an emf of 10 V across a 0.25-H inductor. What is the rate at which the current is changing?

A good approach for calculating the self-inductance of an inductor consists of the following steps:

Problem-Solving Strategy: Self-Inductance

1. Assume a current I is flowing through the inductor.
2. Determine the magnetic field $\vec{\mathbf{B}}$ produced by the current. If there is appropriate symmetry, you may be able to do this with Ampère's law.
3. Obtain the magnetic flux, Φ_m.
4. With the flux known, the self-inductance can be found from Equation 14.9, $L = N\Phi_m / I$.

To demonstrate this procedure, we now calculate the self-inductances of two inductors.

Cylindrical Solenoid

Consider a long, cylindrical solenoid with length l, cross-sectional area A, and N turns of wire. We assume that the length of the solenoid is so much larger than its diameter that we can take the magnetic field to be $B = \mu_0 nI$ throughout the interior of the solenoid, that is, we ignore end effects in the solenoid. With a current I flowing through the coils, the magnetic field produced within the solenoid is

$$B = \mu_0 \left(\frac{N}{l}\right) I, \tag{14.11}$$

so the magnetic flux through one turn is

$$\Phi_m = BA = \frac{\mu_0 NA}{l} I. \tag{14.12}$$

Using Equation 14.9, we find for the self-inductance of the solenoid,

$$L_{\text{solenoid}} = \frac{N\Phi_m}{I} = \frac{\mu_0 N^2 A}{l}. \tag{14.13}$$

If $n = N/l$ is the number of turns per unit length of the solenoid, we may write Equation 14.13 as

$$L = \mu_0 \left(\frac{N}{l}\right)^2 Al = \mu_0 n^2 Al = \mu_0 n^2 (V), \tag{14.14}$$

where $V = Al$ is the volume of the solenoid. Notice that *the self-inductance of a long solenoid depends only on its physical properties* (such as the number of turns of wire per unit length and the volume), and not on the magnetic field or the current. This is true for inductors in general.

Rectangular Toroid

A toroid with a rectangular cross-section is shown in Figure 14.10. The inner and outer radii of the toroid are R_1 and R_2, and h is the height of the toroid. Applying Ampère's law in the same manner as we did in Example 13.8 for a toroid with a circular cross-section, we find the magnetic field inside a rectangular toroid is also given by

$$B = \frac{\mu_0 NI}{2\pi r}, \tag{14.15}$$

where r is the distance from the central axis of the toroid. Because the field changes within the toroid, we must calculate the flux by integrating over the toroid's cross-section. Using the infinitesimal cross-sectional area element $da = h\,dr$ shown in Figure 14.10, we obtain

$$\Phi_m = \int B\,da = \int_{R_1}^{R_2} \left(\frac{\mu_0 NI}{2\pi r}\right)(hdr) = \frac{\mu_0 NhI}{2\pi} \ln \frac{R_2}{R_1}. \tag{14.16}$$

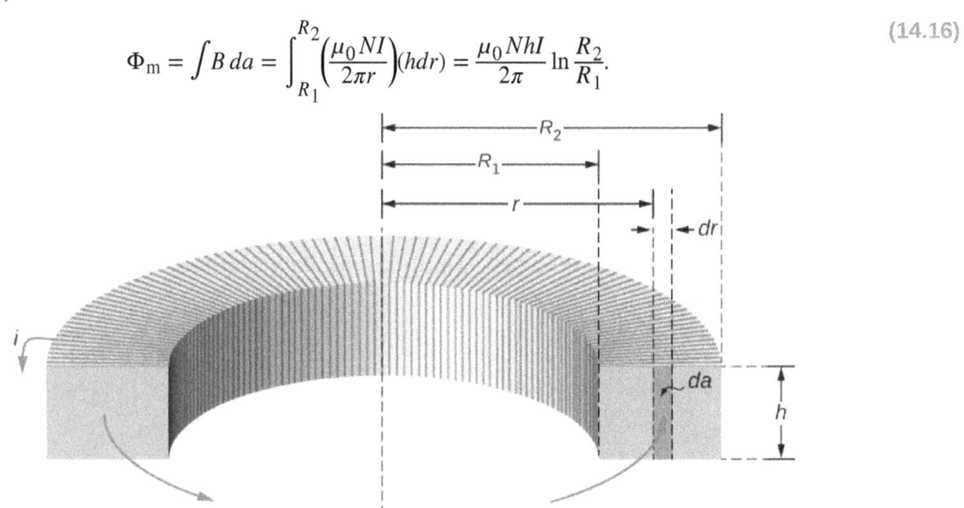

Figure 14.10 Calculating the self-inductance of a rectangular toroid.

Now from Equation 14.16, we obtain for the self-inductance of a rectangular toroid

$$L = \frac{N\Phi_m}{I} = \frac{\mu_0 N^2 h}{2\pi} \ln \frac{R_2}{R_1}. \tag{14.17}$$

As expected, the self-inductance is a constant determined by only the physical properties of the toroid.

14.4 Check Your Understanding (a) Calculate the self-inductance of a solenoid that is tightly wound with wire of diameter 0.10 cm, has a cross-sectional area of $0.90\ \text{cm}^2$, and is 40 cm long. (b) If the current through the solenoid decreases uniformly from 10 to 0 A in 0.10 s, what is the emf induced between the ends of the solenoid?

14.5 Check Your Understanding (a) What is the magnetic flux through one turn of a solenoid of self-inductance 8.0×10^{-5} H when a current of 3.0 A flows through it? Assume that the solenoid has 1000 turns and is wound from wire of diameter 1.0 mm. (b) What is the cross-sectional area of the solenoid?

14.3 | Energy in a Magnetic Field

Learning Objectives

By the end of this section, you will be able to:

- Explain how energy can be stored in a magnetic field
- Derive the equation for energy stored in a coaxial cable given the magnetic energy density

The energy of a capacitor is stored in the electric field between its plates. Similarly, an inductor has the capability to store energy, but in its magnetic field. This energy can be found by integrating the **magnetic energy density,**

$$u_m = \frac{B^2}{2\mu_0} \tag{14.18}$$

over the appropriate volume. To understand where this formula comes from, let's consider the long, cylindrical solenoid of the previous section. Again using the infinite solenoid approximation, we can assume that the magnetic field is essentially constant and given by $B = \mu_0 nI$ everywhere inside the solenoid. Thus, the energy stored in a solenoid or the magnetic energy density times volume is equivalent to

$$U = u_m(V) = \frac{(\mu_0 nI)^2}{2\mu_0}(Al) = \frac{1}{2}(\mu_0 n^2 Al)I^2. \tag{14.19}$$

With the substitution of Equation 14.14, this becomes

$$U = \frac{1}{2}LI^2. \tag{14.20}$$

Although derived for a special case, this equation gives the energy stored in the magnetic field of *any* inductor. We can see this by considering an arbitrary inductor through which a changing current is passing. At any instant, the magnitude of the induced emf is $\varepsilon = Ldi/dt,$ so the power absorbed by the inductor is

$$P = \varepsilon i = L\frac{di}{dt}i. \tag{14.21}$$

The total energy stored in the magnetic field when the current increases from 0 to I in a time interval from 0 to t can be determined by integrating this expression:

$$U = \int_0^t P dt' = \int_0^t L\frac{di}{dt'} i dt' = L\int_0^l i di = \frac{1}{2}LI^2. \tag{14.22}$$

Example 14.3

Self-Inductance of a Coaxial Cable

Figure 14.11 shows two long, concentric cylindrical shells of radii R_1 and R_2. As discussed in Capacitance on capacitance, this configuration is a simplified representation of a coaxial cable. The capacitance per unit length of the cable has already been calculated. Now (a) determine the magnetic energy stored per unit length of the coaxial cable and (b) use this result to find the self-inductance per unit length of the cable.

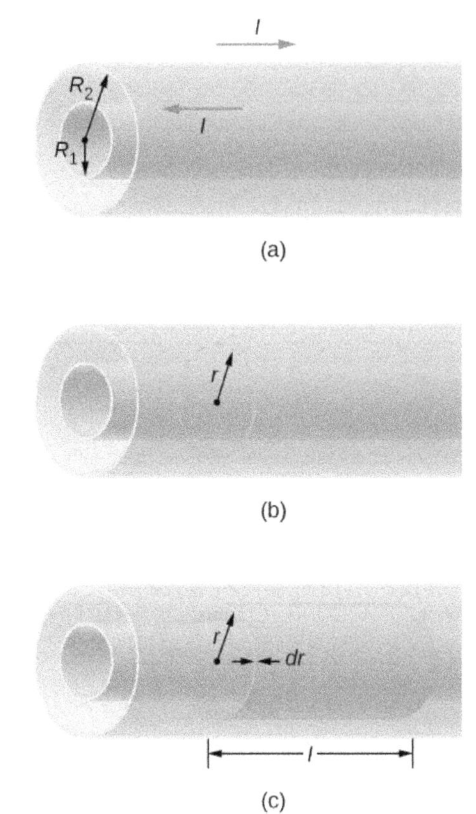

Figure 14.11 (a) A coaxial cable is represented here by two hollow, concentric cylindrical conductors along which electric current flows in opposite directions. (b) The magnetic field between the conductors can be found by applying Ampère's law to the dashed path. (c) The cylindrical shell is used to find the magnetic energy stored in a length l of the cable.

Strategy

The magnetic field both inside and outside the coaxial cable is determined by Ampère's law. Based on this magnetic field, we can use Equation 14.22 to calculate the energy density of the magnetic field. The magnetic energy is calculated by an integral of the magnetic energy density times the differential volume over the cylindrical shell. After the integration is carried out, we have a closed-form solution for part (a). The self-inductance per unit length is determined based on this result and Equation 14.22.

Solution

a. We determine the magnetic field between the conductors by applying Ampère's law to the dashed circular path shown in Figure 14.11(b). Because of the cylindrical symmetry, $\vec{\mathbf{B}}$ is constant along the path, and

$$\oint \vec{\mathbf{B}} \cdot d\vec{\mathbf{l}} = B(2\pi r) = \mu_0 I.$$

This gives us

$$B = \frac{\mu_0 I}{2\pi r}.$$

In the region outside the cable, a similar application of Ampère's law shows that $B = 0$, since no net current crosses the area bounded by a circular path where $r > R_2$. This argument also holds when $r < R_1$; that is, in the region within the inner cylinder. All the magnetic energy of the cable is therefore stored between the two conductors. Since the energy density of the magnetic field is

$$u_m = \frac{B^2}{2\mu_0} = \frac{\mu_0 I^2}{8\pi^2 r^2},$$

the energy stored in a cylindrical shell of inner radius r, outer radius $r + dr$, and length l (see part (c) of the figure) is

$$u_m = \frac{B^2}{2\mu_0} = \frac{\mu_0 I^2}{8\pi^2 r^2}.$$

Thus, the total energy of the magnetic field in a length l of the cable is

$$U = \int_{R_1}^{R_2} dU = \int_{R_1}^{R_2} \frac{\mu_0 I^2}{8\pi^2 r^2}(2\pi r l)dr = \frac{\mu_0 I^2 l}{4\pi} \ln \frac{R_2}{R_1},$$

and the energy per unit length is $(\mu_0 I^2/4\pi)\ln(R_2/R_1)$.

b. From Equation 14.22,

$$U = \frac{1}{2}LI^2,$$

where L is the self-inductance of a length l of the coaxial cable. Equating the previous two equations, we find that the self-inductance per unit length of the cable is

$$\frac{L}{l} = \frac{\mu_0}{2\pi} \ln \frac{R_2}{R_1}.$$

Significance

The inductance per unit length depends only on the inner and outer radii as seen in the result. To increase the inductance, we could either increase the outer radius (R_2) or decrease the inner radius (R_1). In the limit as the two radii become equal, the inductance goes to zero. In this limit, there is no coaxial cable. Also, the magnetic energy per unit length from part (a) is proportional to the square of the current.

14.6 Check Your Understanding How much energy is stored in the inductor of Example 14.2 after the current reaches its maximum value?

14.4 | RL Circuits

Learning Objectives

By the end of this section, you will be able to:

- Analyze circuits that have an inductor and resistor in series
- Describe how current and voltage exponentially grow or decay based on the initial conditions

A circuit with resistance and self-inductance is known as an *RL* circuit. Figure 14.12(a) shows an *RL* circuit consisting of a resistor, an inductor, a constant source of emf, and switches S_1 and S_2. When S_1 is closed, the circuit is equivalent to a single-loop circuit consisting of a resistor and an inductor connected across a source of emf (Figure 14.12(b)). When

S_1 is opened and S_2 is closed, the circuit becomes a single-loop circuit with only a resistor and an inductor (Figure 14.12(c)).

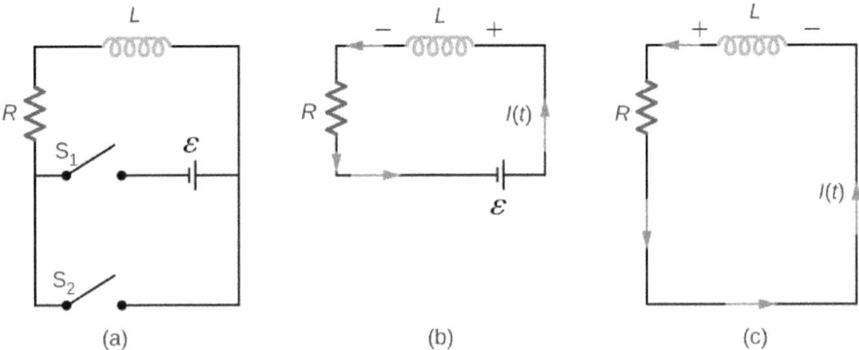

Figure 14.12 (a) An *RL* circuit with switches S_1 and S_2. (b) The equivalent circuit with S_1 closed and S_2 open. (c) The equivalent circuit after S_1 is opened and S_2 is closed.

We first consider the *RL* circuit of Figure 14.12(b). Once S_1 is closed and S_2 is open, the source of emf produces a current in the circuit. If there were no self-inductance in the circuit, the current would rise immediately to a steady value of ε/R. However, from Faraday's law, the increasing current produces an emf $V_L = -L(dI/dt)$ across the inductor. In accordance with Lenz's law, the induced emf counteracts the increase in the current and is directed as shown in the figure. As a result, *I(t)* starts at zero and increases asymptotically to its final value.

Applying Kirchhoff's loop rule to this circuit, we obtain

$$\varepsilon - L\frac{dI}{dt} - IR = 0, \quad (14.23)$$

which is a first-order differential equation for *I(t)*. Notice its similarity to the equation for a capacitor and resistor in series (See RC Circuits). Similarly, the solution to Equation 14.23 can be found by making substitutions in the equations relating the capacitor to the inductor. This gives

$$I(t) = \frac{\varepsilon}{R}\left(1 - e^{-Rt/L}\right) = \frac{\varepsilon}{R}\left(1 - e^{-t/\tau_L}\right), \quad (14.24)$$

where

$$\tau_L = L/R \quad (14.25)$$

is the **inductive time constant** of the circuit.

The current *I(t)* is plotted in Figure 14.13(a). It starts at zero, and as $t \to \infty$, *I(t)* approaches ε/R asymptotically. The induced emf $V_L(t)$ is directly proportional to *dI/dt*, or the slope of the curve. Hence, while at its greatest immediately after the switches are thrown, the induced emf decreases to zero with time as the current approaches its final value of ε/R. The circuit then becomes equivalent to a resistor connected across a source of emf.

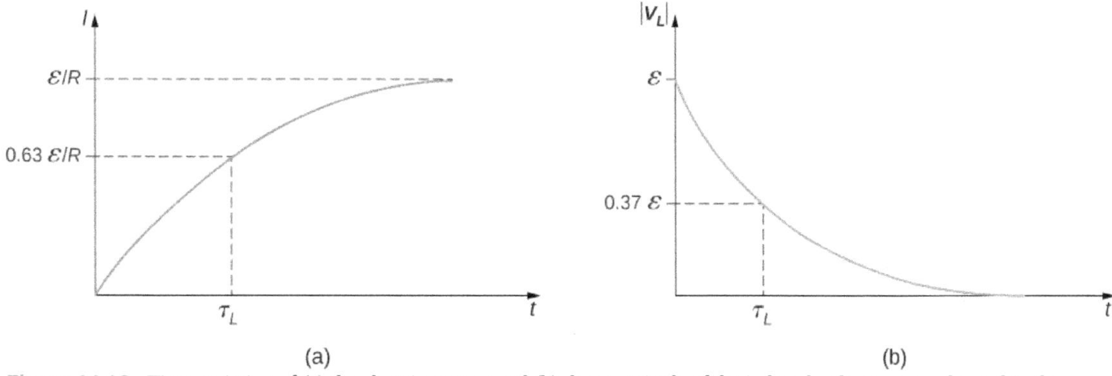

Figure 14.13 Time variation of (a) the electric current and (b) the magnitude of the induced voltage across the coil in the circuit of Figure 14.12(b).

The energy stored in the magnetic field of an inductor is

$$U_L = \frac{1}{2}LI^2. \tag{14.26}$$

Thus, as the current approaches the maximum current ε/R, the stored energy in the inductor increases from zero and asymptotically approaches a maximum of $L(\varepsilon/R)^2/2$.

The time constant τ_L tells us how rapidly the current increases to its final value. At $t = \tau_L,$ the current in the circuit is, from Equation 14.24,

$$I(\tau_L) = \frac{\varepsilon}{R}(1 - e^{-1}) = 0.63\frac{\varepsilon}{R}, \tag{14.27}$$

which is 63% of the final value ε/R. The smaller the inductive time constant $\tau_L = L/R,$ the more rapidly the current approaches ε/R.

We can find the time dependence of the induced voltage across the inductor in this circuit by using $V_L(t) = -L(dI/dt)$ and Equation 14.24:

$$V_L(t) = -L\frac{dI}{dt} = -\varepsilon e^{-t/\tau_L}. \tag{14.28}$$

The magnitude of this function is plotted in Figure 14.13(b). The greatest value of $L(dI/dt)$ is ε; it occurs when dI/dt is greatest, which is immediately after S_1 is closed and S_2 is opened. In the approach to steady state, dI/dt decreases to zero. As a result, the voltage across the inductor also vanishes as $t \to \infty$.

The time constant τ_L also tells us how quickly the induced voltage decays. At $t = \tau_L,$ the magnitude of the induced voltage is

$$|V_L(\tau_L)| = \varepsilon e^{-1} = 0.37\varepsilon = 0.37V(0). \tag{14.29}$$

The voltage across the inductor therefore drops to about 37% of its initial value after one time constant. The shorter the time constant $\tau_L,$ the more rapidly the voltage decreases.

After enough time has elapsed so that the current has essentially reached its final value, the positions of the switches in Figure 14.12(a) are reversed, giving us the circuit in part (c). At $t = 0,$ the current in the circuit is $I(0) = \varepsilon/R.$ With Kirchhoff's loop rule, we obtain

$$IR + L\frac{dI}{dt} = 0. \tag{14.30}$$

The solution to this equation is similar to the solution of the equation for a discharging capacitor, with similar substitutions. The current at time t is then

$$I(t) = \frac{\varepsilon}{R}e^{-t/\tau_L}. \quad (14.31)$$

The current starts at $I(0) = \varepsilon/R$ and decreases with time as the energy stored in the inductor is depleted (Figure 14.14).

The time dependence of the voltage across the inductor can be determined from $V_L = -L(dI/dt)$:

$$V_L(t) = \varepsilon e^{-t/\tau L}. \quad (14.32)$$

This voltage is initially $V_L(0) = \varepsilon$, and it decays to zero like the current. The energy stored in the magnetic field of the inductor, $LI^2/2$, also decreases exponentially with time, as it is dissipated by Joule heating in the resistance of the circuit.

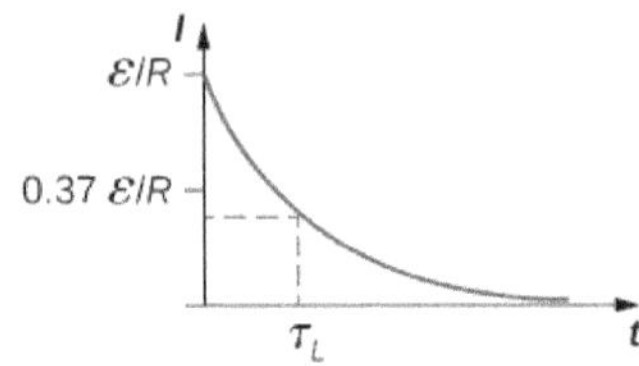

Figure 14.14 Time variation of electric current in the *RL* circuit of Figure 14.12(c). The induced voltage across the coil also decays exponentially.

Example 14.4

An *RL* Circuit with a Source of emf

In the circuit of Figure 14.12(a), let $\varepsilon = 2.0V$, $R = 4.0\,\Omega$, and $L = 4.0\,\text{H}$. With S_1 closed and S_2 open (Figure 14.12(b)), (a) what is the time constant of the circuit? (b) What are the current in the circuit and the magnitude of the induced emf across the inductor at $t = 0$, at $t = 2.0\tau_L$, and as $t \to \infty$?

Strategy

The time constant for an inductor and resistor in a series circuit is calculated using Equation 14.25. The current through and voltage across the inductor are calculated by the scenarios detailed from Equation 14.24 and Equation 14.32.

Solution

a. The inductive time constant is

$$\tau_L = \frac{L}{R} = \frac{4.0\,\text{H}}{4.0\,\Omega} = 1.0\,\text{s}.$$

b. The current in the circuit of Figure 14.12(b) increases according to Equation 14.24:

$$I(t) = \frac{\varepsilon}{R}(1 - e^{-t/\tau_L}).$$

At $t = 0$,

$$(1 - e^{-t/\tau_L}) = (1 - 1) = 0; \text{ so } I(0) = 0.$$

At $t = 2.0\tau_L$ and $t \to \infty$, we have, respectively,

$$I(2.0\tau_L) = \frac{\varepsilon}{R}(1 - e^{-2.0}) = (0.50\,\text{A})(0.86) = 0.43\,\text{A},$$

and

$$I(\infty) = \frac{\varepsilon}{R} = 0.50\,\text{A}.$$

From Equation 14.32, the magnitude of the induced emf decays as

$$|V_L(t)| = \varepsilon e^{-t/\tau_L}.$$

At $t = 0,\ t = 2.0\tau_L,$ and as $t \to \infty,$ we obtain

$$\begin{aligned} |V_L(0)| &= \varepsilon = 2.0\text{ V}, \\ |V_L(2.0\tau_L)| &= (2.0\text{ V})\,e^{-2.0} = 0.27\text{ V} \\ &\text{and} \\ |V_L(\infty)| &= 0. \end{aligned}$$

Significance

If the time of the measurement were much larger than the time constant, we would not see the decay or growth of the voltage across the inductor or resistor. The circuit would quickly reach the asymptotic values for both of these. See Figure 14.15.

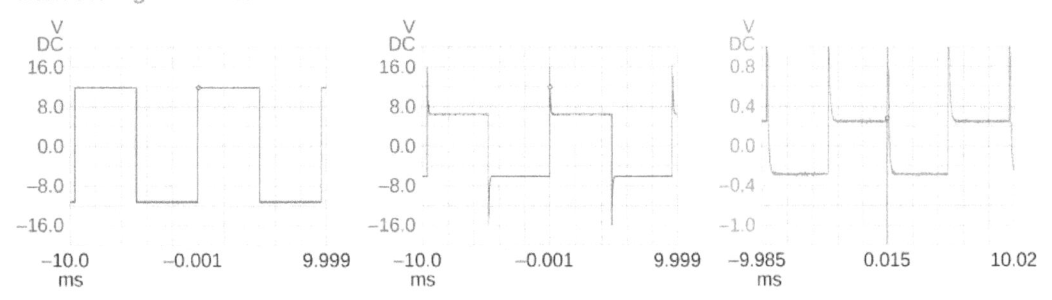

(a) Voltage across the source (b) Voltage across the inductor (c) Voltage across the resistor

Figure 14.15 A generator in an *RL* circuit produces a square-pulse output in which the voltage oscillates between zero and some set value. These oscilloscope traces show (a) the voltage across the source; (b) the voltage across the inductor; (c) the voltage across the resistor.

Example 14.5

An *RL* Circuit without a Source of emf

After the current in the *RL* circuit of Example 14.4 has reached its final value, the positions of the switches are reversed so that the circuit becomes the one shown in Figure 14.12(c). (a) How long does it take the current to drop to half its initial value? (b) How long does it take before the energy stored in the inductor is reduced to 1.0% of its maximum value?

Strategy

The current in the inductor will now decrease as the resistor dissipates this energy. Therefore, the current falls as an exponential decay. We can also use that same relationship as a substitution for the energy in an inductor formula to find how the energy decreases at different time intervals.

Solution

a. With the switches reversed, the current decreases according to

$$I(t) = \frac{\varepsilon}{R}e^{-t/\tau_L} = I(0)e^{-t/\tau_L}.$$

At a time t when the current is one-half its initial value, we have

$$I(t) = 0.50I(0) \text{ so } e^{-t/\tau_L} = 0.50,$$

and

$$t = -[\ln(0.50)]\tau_L = 0.69(1.0\text{ s}) = 0.69\text{ s},$$

where we have used the inductive time constant found in Example 14.4.

b. The energy stored in the inductor is given by

$$U_L(t) = \frac{1}{2}L[I(t)]^2 = \frac{1}{2}L\left(\frac{\varepsilon}{R}e^{-t/\tau_L}\right)^2 = \frac{L\varepsilon^2}{2R^2}e^{-2t/\tau_L}.$$

If the energy drops to 1.0% of its initial value at a time t, we have

$$U_L(t) = (0.010)U_L(0) \text{ or } \frac{L\varepsilon^2}{2R^2}e^{-2t/\tau_L} = (0.010)\frac{L\varepsilon^2}{2R^2}.$$

Upon canceling terms and taking the natural logarithm of both sides, we obtain

$$-\frac{2t}{\tau_L} = \ln(0.010),$$

so

$$t = -\frac{1}{2}\tau_L\ln(0.010).$$

Since $\tau_L = 1.0\text{ s}$, the time it takes for the energy stored in the inductor to decrease to 1.0% of its initial value is

$$t = -\frac{1}{2}(1.0\text{ s})\ln(0.010) = 2.3\text{ s}.$$

Significance

This calculation only works if the circuit is at maximum current in situation (b) prior to this new situation. Otherwise, we start with a lower initial current, which will decay by the same relationship.

14.7 Check Your Understanding Verify that *RC* and *L/R* have the dimensions of time.

14.8 Check Your Understanding (a) If the current in the circuit of in Figure 14.12(b) increases to 90% of its final value after 5.0 s, what is the inductive time constant? (b) If $R = 20\,\Omega$, what is the value of the self-inductance? (c) If the $20\text{-}\Omega$ resistor is replaced with a $100\text{-}\Omega$ resister, what is the time taken for the current to reach 90% of its final value?

14.9 Check Your Understanding For the circuit of in Figure 14.12(b), show that when steady state is reached, the difference in the total energies produced by the battery and dissipated in the resistor is equal to the energy stored in the magnetic field of the coil.

14.5 | Oscillations in an LC Circuit

Learning Objectives

By the end of this section, you will be able to:

- Explain why charge or current oscillates between a capacitor and inductor, respectively, when wired in series
- Describe the relationship between the charge and current oscillating between a capacitor and inductor wired in series

It is worth noting that both capacitors and inductors store energy, in their electric and magnetic fields, respectively. A circuit containing both an inductor (*L*) and a capacitor (*C*) can oscillate without a source of emf by shifting the energy stored in the circuit between the electric and magnetic fields. Thus, the concepts we develop in this section are directly applicable to the exchange of energy between the electric and magnetic fields in electromagnetic waves, or light. We start with an idealized circuit of zero resistance that contains an inductor and a capacitor, an ***LC* circuit.**

An *LC* circuit is shown in Figure 14.16. If the capacitor contains a charge q_0 before the switch is closed, then all the energy of the circuit is initially stored in the electric field of the capacitor (Figure 14.16(a)). This energy is

$$U_C = \frac{1}{2}\frac{q_0^2}{C}. \tag{14.33}$$

When the switch is closed, the capacitor begins to discharge, producing a current in the circuit. The current, in turn, creates a magnetic field in the inductor. The net effect of this process is a transfer of energy from the capacitor, with its diminishing electric field, to the inductor, with its increasing magnetic field.

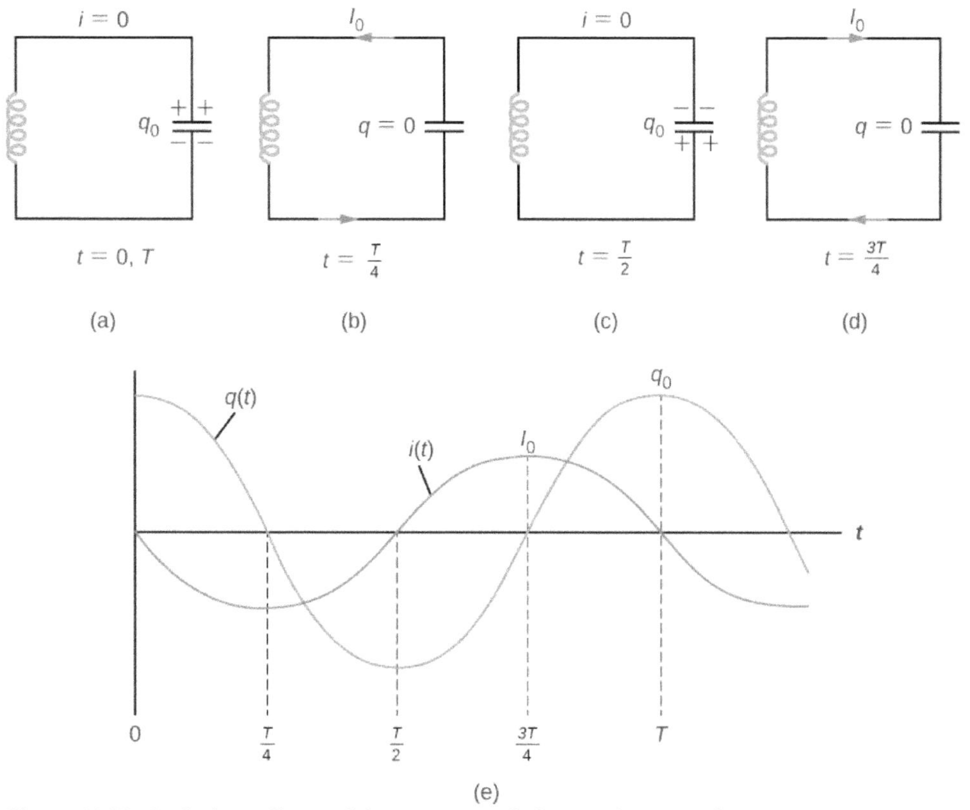

Figure 14.16 (a–d) The oscillation of charge storage with changing directions of current in an *LC* circuit. (e) The graphs show the distribution of charge and current between the capacitor and inductor.

In Figure 14.16(b), the capacitor is completely discharged and all the energy is stored in the magnetic field of the inductor. At this instant, the current is at its maximum value I_0 and the energy in the inductor is

$$U_L = \frac{1}{2}LI_0^2. \tag{14.34}$$

Since there is no resistance in the circuit, no energy is lost through Joule heating; thus, the maximum energy stored in the capacitor is equal to the maximum energy stored at a later time in the inductor:

$$\frac{1}{2}\frac{q_0^2}{C} = \frac{1}{2}LI_0^2. \tag{14.35}$$

At an arbitrary time when the capacitor charge is *q(t)* and the current is *i(t)*, the total energy *U* in the circuit is given by

$$\frac{q^2(t)}{2C} + \frac{Li^2(t)}{2}.$$

Because there is no energy dissipation,

$$U = \frac{1}{2}\frac{q^2}{C} + \frac{1}{2}Li^2 = \frac{1}{2}\frac{q_0^2}{C} = \frac{1}{2}LI_0^2. \tag{14.36}$$

After reaching its maximum I_0, the current *i(t)* continues to transport charge between the capacitor plates, thereby recharging the capacitor. Since the inductor resists a change in current, current continues to flow, even though the capacitor is discharged. This continued current causes the capacitor to charge with opposite polarity. The electric field of the capacitor increases while the magnetic field of the inductor diminishes, and the overall effect is a transfer of energy from the inductor *back* to the capacitor. From the law of energy conservation, the maximum charge that the capacitor re-acquires is q_0. However, as Figure 14.16(c) shows, the capacitor plates are charged *opposite* to what they were initially.

When fully charged, the capacitor once again transfers its energy to the inductor until it is again completely discharged, as shown in Figure 14.16(d). Then, in the last part of this cyclic process, energy flows back to the capacitor, and the initial state of the circuit is restored.

We have followed the circuit through one complete cycle. Its electromagnetic oscillations are analogous to the mechanical oscillations of a mass at the end of a spring. In this latter case, energy is transferred back and forth between the mass, which has kinetic energy $mv^2/2$, and the spring, which has potential energy $kx^2/2$. With the absence of friction in the mass-spring system, the oscillations would continue indefinitely. Similarly, the oscillations of an *LC* circuit with no resistance would continue forever if undisturbed; however, this ideal zero-resistance *LC* circuit is not practical, and any *LC* circuit will have at least a small resistance, which will radiate and lose energy over time.

The frequency of the oscillations in a resistance-free *LC* circuit may be found by analogy with the mass-spring system. For the circuit, $i(t) = dq(t)/dt$, the total electromagnetic energy *U* is

$$U = \frac{1}{2}Li^2 + \frac{1}{2}\frac{q^2}{C}. \tag{14.37}$$

For the mass-spring system, $v(t) = dx(t)/dt$, the total mechanical energy *E* is

$$E = \frac{1}{2}mv^2 + \frac{1}{2}kx^2. \tag{14.38}$$

The equivalence of the two systems is clear. To go from the mechanical to the electromagnetic system, we simply replace *m* by *L*, *v* by *i*, *k* by 1/*C*, and *x* by *q*. Now *x(t)* is given by

$$x(t) = A\cos(\omega t + \phi) \tag{14.39}$$

where $\omega = \sqrt{k/m}$. Hence, the charge on the capacitor in an *LC* circuit is given by

$$q(t) = q_0\cos(\omega t + \phi) \tag{14.40}$$

where the angular frequency of the oscillations in the circuit is

$$\omega = \sqrt{\frac{1}{LC}}. \quad (14.41)$$

Finally, the current in the LC circuit is found by taking the time derivative of $q(t)$:

$$i(t) = \frac{dq(t)}{dt} = -\omega q_0 \sin(\omega t + \phi). \quad (14.42)$$

The time variations of q and I are shown in Figure 14.16(e) for $\phi = 0$.

Example 14.6

An *LC* Circuit

In an LC circuit, the self-inductance is 2.0×10^{-2} H and the capacitance is 8.0×10^{-6} F. At $t = 0$, all of the energy is stored in the capacitor, which has charge 1.2×10^{-5} C. (a) What is the angular frequency of the oscillations in the circuit? (b) What is the maximum current flowing through circuit? (c) How long does it take the capacitor to become completely discharged? (d) Find an equation that represents $q(t)$.

Strategy

The angular frequency of the LC circuit is given by Equation 14.41. To find the maximum current, the maximum energy in the capacitor is set equal to the maximum energy in the inductor. The time for the capacitor to become discharged if it is initially charged is a quarter of the period of the cycle, so if we calculate the period of the oscillation, we can find out what a quarter of that is to find this time. Lastly, knowing the initial charge and angular frequency, we can set up a cosine equation to find $q(t)$.

Solution

a. From Equation 14.41, the angular frequency of the oscillations is

$$\omega = \sqrt{\frac{1}{LC}} = \sqrt{\frac{1}{(2.0 \times 10^{-2}\text{ H})(8.0 \times 10^{-6}\text{ F})}} = 2.5 \times 10^{3}\text{ rad/s}.$$

b. The current is at its maximum I_0 when all the energy is stored in the inductor. From the law of energy conservation,

$$\frac{1}{2}LI_0^2 = \frac{1}{2}\frac{q_0^2}{C},$$

so

$$I_0 = \sqrt{\frac{1}{LC}}q_0 = (2.5 \times 10^{3}\text{ rad/s})(1.2 \times 10^{-5}\text{ C}) = 3.0 \times 10^{-2}\text{ A}.$$

This result can also be found by an analogy to simple harmonic motion, where current and charge are the velocity and position of an oscillator.

c. The capacitor becomes completely discharged in one-fourth of a cycle, or during a time $T/4$, where T is the period of the oscillations. Since

$$T = \frac{2\pi}{\omega} = \frac{2\pi}{2.5 \times 10^{3}\text{ rad/s}} = 2.5 \times 10^{-3}\text{ s},$$

the time taken for the capacitor to become fully discharged is $(2.5 \times 10^{-3}\ \text{s})/4 = 6.3 \times 10^{-4}\ \text{s}$.

d. The capacitor is completely charged at $t = 0$, so $q(0) = q_0$. Using Equation 14.20, we obtain

$$q(0) = q_0 = q_0 \cos \phi.$$

Thus, $\phi = 0$, and

$$q(t) = (1.2 \times 10^{-5}\ \text{C})\cos(2.5 \times 10^3 t).$$

Significance

The energy relationship set up in part (b) is not the only way we can equate energies. At most times, some energy is stored in the capacitor and some energy is stored in the inductor. We can put both terms on each side of the equation. By examining the circuit only when there is no charge on the capacitor or no current in the inductor, we simplify the energy equation.

14.10 Check Your Understanding The angular frequency of the oscillations in an *LC* circuit is 2.0×10^3 rad/s. (a) If $L = 0.10\ \text{H}$, what is *C*? (b) Suppose that at $t = 0$, all the energy is stored in the inductor. What is the value of ϕ? (c) A second identical capacitor is connected in parallel with the original capacitor. What is the angular frequency of this circuit?

14.6 | RLC Series Circuits

Learning Objectives

By the end of this section, you will be able to:

- Determine the angular frequency of oscillation for a resistor, inductor, capacitor (RLC) series circuit
- Relate the RLC circuit to a damped spring oscillation

When the switch is closed in the ***RLC* circuit** of Figure 14.17(a), the capacitor begins to discharge and electromagnetic energy is dissipated by the resistor at a rate $i^2 R$. With *U* given by Equation 14.19, we have

$$\frac{dU}{dt} = \frac{q}{C}\frac{dq}{dt} + Li\frac{di}{dt} = -i^2 R \tag{14.43}$$

where *i* and *q* are time-dependent functions. This reduces to

$$L\frac{d^2 q}{dt^2} + R\frac{dq}{dt} + \frac{1}{C}q = 0. \tag{14.44}$$

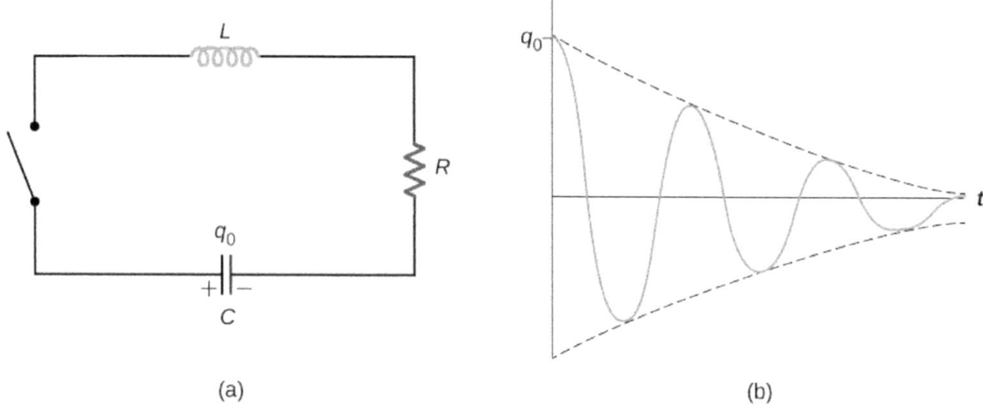

Figure 14.17 (a) An *RLC* circuit. Electromagnetic oscillations begin when the switch is closed. The capacitor is fully charged initially. (b) Damped oscillations of the capacitor charge are shown in this curve of charge versus time, or q versus t. The capacitor contains a charge q_0 before the switch is closed.

This equation is analogous to

$$m\frac{d^2x}{dt^2} + b\frac{dx}{dt} + kx = 0,$$

which is the equation of motion for a *damped mass-spring system* (you first encountered this equation in Oscillations (http://cnx.org/content/m58360/latest/)). As we saw in that chapter, it can be shown that the solution to this differential equation takes three forms, depending on whether the angular frequency of the undamped spring is greater than, equal to, or less than $b/2m$. Therefore, the result can be underdamped $(\sqrt{k/m} > b/2m)$, critically damped $(\sqrt{k/m} = b/2m)$, or overdamped $(\sqrt{k/m} < b/2m)$. By analogy, the solution $q(t)$ to the *RLC* differential equation has the same feature. Here we look only at the case of under-damping. By replacing m by L, b by R, k by $1/C$, and x by q in Equation 14.44, and assuming $\sqrt{1/LC} > R/2L$, we obtain

$$q(t) = q_0 e^{-Rt/2L} \cos(\omega' t + \phi) \tag{14.45}$$

where the angular frequency of the oscillations is given by

$$\omega' = \sqrt{\frac{1}{LC} - \left(\frac{R}{2L}\right)^2} \tag{14.46}$$

This underdamped solution is shown in Figure 14.17(b). Notice that the amplitude of the oscillations decreases as energy is dissipated in the resistor. Equation 14.45 can be confirmed experimentally by measuring the voltage across the capacitor as a function of time. This voltage, multiplied by the capacitance of the capacitor, then gives $q(t)$.

Try an interactive circuit construction kit (https://openstaxcollege.org/l/21phetcirconstr) that allows you to graph current and voltage as a function of time. You can add inductors and capacitors to work with any combination of R, L, and C circuits with both dc and ac sources.

Try out a circuit-based java applet website (https://openstaxcollege.org/l/21cirphysbascur) that has many problems with both dc and ac sources that will help you practice circuit problems.

14.11 Check Your Understanding In an *RLC* circuit, $L = 5.0\ \text{mH}$, $C = 6.0\mu\text{F}$, and $R = 200\ \Omega$. (a) Is the circuit underdamped, critically damped, or overdamped? (b) If the circuit starts oscillating with a charge of $3.0 \times 10^{-3}\ \text{C}$ on the capacitor, how much energy has been dissipated in the resistor by the time the oscillations cease?

CHAPTER 14 REVIEW

KEY TERMS

henry (H) unit of inductance, $1\ \text{H} = 1\ \Omega \cdot \text{s}$; it is also expressed as a volt second per ampere

inductance property of a device that tells how effectively it induces an emf in another device

inductive time constant denoted by τ, the characteristic time given by quantity *L/R* of a particular series *RL* circuit

inductor part of an electrical circuit to provide self-inductance, which is symbolized by a coil of wire

***LC* circuit** circuit composed of an ac source, inductor, and capacitor

magnetic energy density energy stored per volume in a magnetic field

mutual inductance geometric quantity that expresses how effective two devices are at inducing emfs in one another

***RLC* circuit** circuit with an ac source, resistor, inductor, and capacitor all in series.

self-inductance effect of the device inducing emf in itself

KEY EQUATIONS

Mutual inductance by flux	$M = \frac{N_2 \Phi_{21}}{I_1} = \frac{N_1 \Phi_{12}}{I_2}$
Mutual inductance in circuits	$\varepsilon_1 = -M\frac{dI_2}{dt}$
Self-inductance in terms of magnetic flux	$N\Phi_\text{m} = LI$
Self-inductance in terms of emf	$\varepsilon = -L\frac{dI}{dt}$
Self-inductance of a solenoid	$L_\text{solenoid} = \frac{\mu_0 N^2 A}{l}$
Self-inductance of a toroid	$L_\text{toroid} = \frac{\mu_0 N^2 h}{2\pi} \ln \frac{R_2}{R_1}.$
Energy stored in an inductor	$U = \frac{1}{2}LI^2$
Current as a function of time for a *RL* circuit	$I(t) = \frac{\varepsilon}{R}(1 - e^{-t/\tau_L})$
Time constant for a *RL* circuit	$\tau_L = L/R$
Charge oscillation in *LC* circuits	$q(t) = q_0 \cos(\omega t + \phi)$
Angular frequency in *LC* circuits	$\omega = \sqrt{\frac{1}{LC}}$
Current oscillations in *LC* circuits	$i(t) = -\omega q_0 \sin(\omega t + \phi)$
Charge as a function of time in *RLC* circuit	$q(t) = q_0 e^{-Rt/2L} \cos(\omega' t + \phi)$
Angular frequency in *RLC* circuit	$\omega' = \sqrt{\frac{1}{LC} - \left(\frac{R}{2L}\right)^2}$

SUMMARY

14.1 Mutual Inductance

- Inductance is the property of a device that expresses how effectively it induces an emf in another device.
- Mutual inductance is the effect of two devices inducing emfs in each other.
- A change in current dI_1/dt in one circuit induces an emf (ε_2) in the second:

$$\varepsilon_2 = -M\frac{dI1}{dt},$$

where M is defined to be the mutual inductance between the two circuits and the minus sign is due to Lenz's law.

- Symmetrically, a change in current dI_2/dt through the second circuit induces an emf (ε_1) in the first:

$$\varepsilon_1 = -M\frac{dI_2}{dt},$$

where M is the same mutual inductance as in the reverse process.

14.2 Self-Inductance and Inductors

- Current changes in a device induce an emf in the device itself, called self-inductance,

$$\varepsilon = -L\frac{dI}{dt},$$

where L is the self-inductance of the inductor and dI/dt is the rate of change of current through it. The minus sign indicates that emf opposes the change in current, as required by Lenz's law. The unit of self-inductance and inductance is the henry (H), where $1\ \text{H} = 1\ \Omega \cdot \text{s}$.

- The self-inductance of a solenoid is

$$L = \frac{\mu_0 N^2 A}{l},$$

where N is its number of turns in the solenoid, A is its cross-sectional area, l is its length, and $\mu_0 = 4\pi \times 10^{-7}\ \text{T} \cdot \text{m/A}$ is the permeability of free space.

- The self-inductance of a toroid is

$$L = \frac{\mu_0 N^2 h}{2\pi} \ln\frac{R_2}{R_1},$$

where N is its number of turns in the toroid, R_1 and R_2 are the inner and outer radii of the toroid, h is the height of the toroid, and $\mu_0 = 4\pi \times 10^{-7}\ \text{T} \cdot \text{m/A}$ is the permeability of free space.

14.3 Energy in a Magnetic Field

- The energy stored in an inductor U is

$$U = \frac{1}{2}LI^2.$$

- The self-inductance per unit length of coaxial cable is

$$\frac{L}{l} = \frac{\mu_0}{2\pi} \ln\frac{R_2}{R_1}.$$

14.4 RL Circuits

- When a series connection of a resistor and an inductor—an *RL* circuit—is connected to a voltage source, the time variation of the current is
  $I(t) = \frac{\varepsilon}{R}(1 - e^{-Rt/L}) = \frac{\varepsilon}{R}(1 - e^{-t/\tau_L})$ (turning on),
  where the initial current is $I_0 = \varepsilon/R$.
- The characteristic time constant τ is $\tau_L = L/R$, where *L* is the inductance and *R* is the resistance.
- In the first time constant τ, the current rises from zero to $0.632I_0$, and to 0.632 of the remainder in every subsequent time interval τ.
- When the inductor is shorted through a resistor, current decreases as
  $I(t) = \frac{\varepsilon}{R}e^{-t/\tau_L}$ (turning off).
  Current falls to $0.368I_0$ in the first time interval τ, and to 0.368 of the remainder toward zero in each subsequent time τ.

14.5 Oscillations in an LC Circuit

- The energy transferred in an oscillatory manner between the capacitor and inductor in an *LC* circuit occurs at an angular frequency $\omega = \sqrt{\frac{1}{LC}}$.
- The charge and current in the circuit are given by

$$\begin{aligned} q(t) &= q_0 \cos(\omega t + \phi), \\ i(t) &= -\omega q_0 \sin(\omega t + \phi). \end{aligned}$$

14.6 RLC Series Circuits

- The underdamped solution for the capacitor charge in an *RLC* circuit is

$$q(t) = q_0 e^{-Rt/2L} \cos(\omega' t + \phi).$$

- The angular frequency given in the underdamped solution for the *RLC* circuit is

$$\omega' = \sqrt{\frac{1}{LC} - \left(\frac{R}{2L}\right)^2}.$$

CONCEPTUAL QUESTIONS

14.1 Mutual Inductance

1. Show that $N\Phi_m/I$ and $\varepsilon/(dI/dt)$, which are both expressions for self-inductance, have the same units.

2. A 10-H inductor carries a current of 20 A. Describe how a 50-V emf can be induced across it.

3. The ignition circuit of an automobile is powered by a 12-V battery. How are we able to generate large voltages with this power source?

4. When the current through a large inductor is interrupted with a switch, an arc appears across the open terminals of the switch. Explain.

14.2 Self-Inductance and Inductors

5. Does self-inductance depend on the value of the magnetic flux? Does it depend on the current through the wire? Correlate your answers with the equation $N\Phi_m = LI$.

6. Would the self-inductance of a 1.0 m long, tightly wound solenoid differ from the self-inductance per meter of an infinite, but otherwise identical, solenoid?

7. Discuss how you might determine the self-inductance per unit length of a long, straight wire.

8. The self-inductance of a coil is zero if there is no current passing through the windings. True or false?

9. How does the self-inductance per unit length near the center of a solenoid (away from the ends) compare with its value near the end of the solenoid?

14.3 Energy in a Magnetic Field

10. Show that $LI^2/2$ has units of energy.

14.4 RL Circuits

11. Use Lenz's law to explain why the initial current in the *RL* circuit of Figure 14.12(b) is zero.

12. When the current in the *RL* circuit of Figure 14.12(b) reaches its final value $\varepsilon/R,$ what is the voltage across the inductor? Across the resistor?

13. Does the time required for the current in an *RL* circuit to reach any fraction of its steady-state value depend on the emf of the battery?

14. An inductor is connected across the terminals of a battery. Does the current that eventually flows through the inductor depend on the internal resistance of the battery? Does the time required for the current to reach its final value depend on this resistance?

15. At what time is the voltage across the inductor of the *RL* circuit of Figure 14.12(b) a maximum?

16. In the simple *RL* circuit of Figure 14.12(b), can the emf induced across the inductor ever be greater than the emf of the battery used to produce the current?

17. If the emf of the battery of Figure 14.12(b) is reduced by a factor of 2, by how much does the steady-state energy stored in the magnetic field of the inductor change?

18. A steady current flows through a circuit with a large inductive time constant. When a switch in the circuit is opened, a large spark occurs across the terminals of the switch. Explain.

19. Describe how the currents through R_1 and R_2 shown below vary with time after switch S is closed.

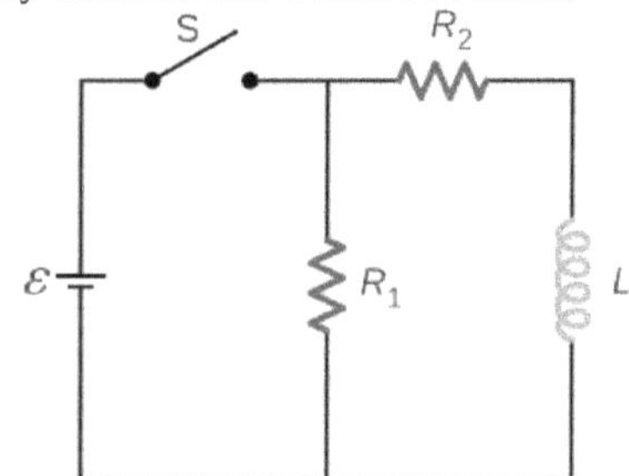

20. Discuss possible practical applications of *RL* circuits.

14.5 Oscillations in an LC Circuit

21. Do Kirchhoff's rules apply to circuits that contain inductors and capacitors?

22. Can a circuit element have both capacitance and inductance?

23. In an *LC* circuit, what determines the frequency and the amplitude of the energy oscillations in either the inductor or capacitor?

14.6 RLC Series Circuits

24. When a wire is connected between the two ends of a solenoid, the resulting circuit can oscillate like an *RLC* circuit. Describe what causes the capacitance in this circuit.

25. Describe what effect the resistance of the connecting wires has on an oscillating *LC* circuit.

26. Suppose you wanted to design an *LC* circuit with a frequency of 0.01 Hz. What problems might you encounter?

27. A radio receiver uses an *RLC* circuit to pick out particular frequencies to listen to in your house or car without hearing other unwanted frequencies. How would someone design such a circuit?

PROBLEMS

14.1 Mutual Inductance

28. When the current in one coil changes at a rate of 5.6 A/s, an emf of 6.3×10^{-3} V is induced in a second, nearby coil. What is the mutual inductance of the two coils?

29. An emf of 9.7×10^{-3} V is induced in a coil while the current in a nearby coil is decreasing at a rate of 2.7 A/s. What is the mutual inductance of the two coils?

30. Two coils close to each other have a mutual inductance of 32 mH. If the current in one coil decays according to $I = I_0 e^{-\alpha t}$, where $I_0 = 5.0\text{ A}$ and $\alpha = 2.0 \times 10^3\text{ s}^{-1}$, what is the emf induced in the second coil immediately after the current starts to decay? At $t = 1.0 \times 10^{-3}\text{ s}$?

31. A coil of 40 turns is wrapped around a long solenoid of cross-sectional area $7.5 \times 10^{-3}\text{ m}^2$. The solenoid is 0.50 m long and has 500 turns. (a) What is the mutual inductance of this system? (b) The outer coil is replaced by a coil of 40 turns whose radius is three times that of the solenoid. What is the mutual inductance of this configuration?

32. A 600-turn solenoid is 0.55 m long and 4.2 cm in diameter. Inside the solenoid, a small $(1.1\text{ cm} \times 1.4\text{ cm})$, single-turn rectangular coil is fixed in place with its face perpendicular to the long axis of the solenoid. What is the mutual inductance of this system?

33. A toroidal coil has a mean radius of 16 cm and a cross-sectional area of 0.25 cm^2; it is wound uniformly with 1000 turns. A second toroidal coil of 750 turns is wound uniformly over the first coil. Ignoring the variation of the magnetic field within a toroid, determine the mutual inductance of the two coils.

34. A solenoid of N_1 turns has length l_1 and radius R_1, and a second smaller solenoid of N_2 turns has length l_2 and radius R_2. The smaller solenoid is placed completely inside the larger solenoid so that their long axes coincide. What is the mutual inductance of the two solenoids?

14.2 Self-Inductance and Inductors

35. An emf of 0.40 V is induced across a coil when the current through it changes uniformly from 0.10 to 0.60 A in 0.30 s. What is the self-inductance of the coil?

36. The current shown in part (a) below is increasing, whereas that shown in part (b) is decreasing. In each case, determine which end of the inductor is at the higher potential.

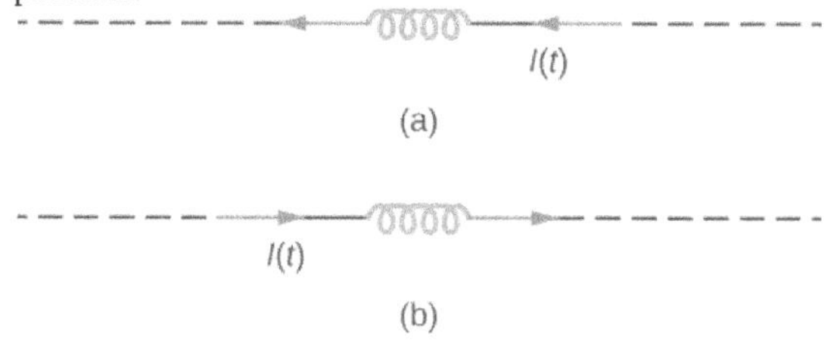

37. What is the rate at which the current though a 0.30-H coil is changing if an emf of 0.12 V is induced across the coil?

38. When a camera uses a flash, a fully charged capacitor discharges through an inductor. In what time must the 0.100-A current through a 2.00-mH inductor be switched on or off to induce a 500-V emf?

39. A coil with a self-inductance of 2.0 H carries a current that varies with time according to $I(t) = (2.0\text{ A})\sin 120\pi t$. Find an expression for the emf induced in the coil.

40. A solenoid 50 cm long is wound with 500 turns of wire. The cross-sectional area of the coil is 2.0 cm^2 What is the self-inductance of the solenoid?

41. A coil with a self-inductance of 3.0 H carries a current that decreases at a uniform rate $dI/dt = -0.050\text{ A/s}$. What is the emf induced in the coil? Describe the polarity of the induced emf.

42. The current *I(t)* through a 5.0-mH inductor varies with time, as shown below. The resistance of the inductor is $5.0\ \Omega$. Calculate the voltage across the inductor at $t = 2.0\text{ ms}$, $t = 4.0\text{ ms}$, and $t = 8.0\text{ ms}$.

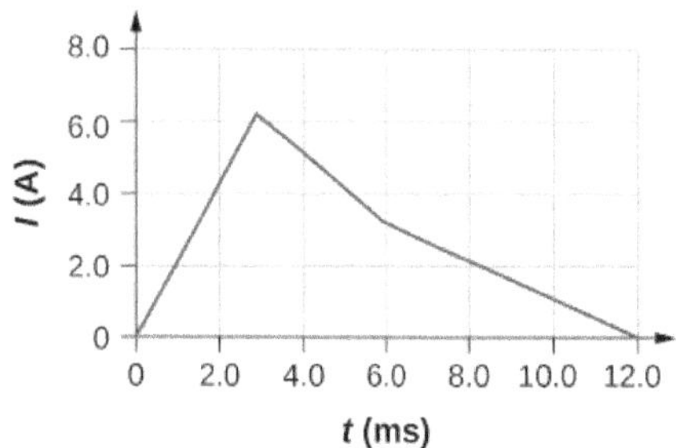

43. A long, cylindrical solenoid with 100 turns per centimeter has a radius of 1.5 cm. (a) Neglecting end effects, what is the self-inductance per unit length of the solenoid? (b) If the current through the solenoid changes at the rate 5.0 A/s, what is the emf induced per unit length?

44. Suppose that a rectangular toroid has 2000 windings and a self-inductance of 0.040 H. If $h = 0.10\,\text{m}$, what is the ratio of its outer radius to its inner radius?

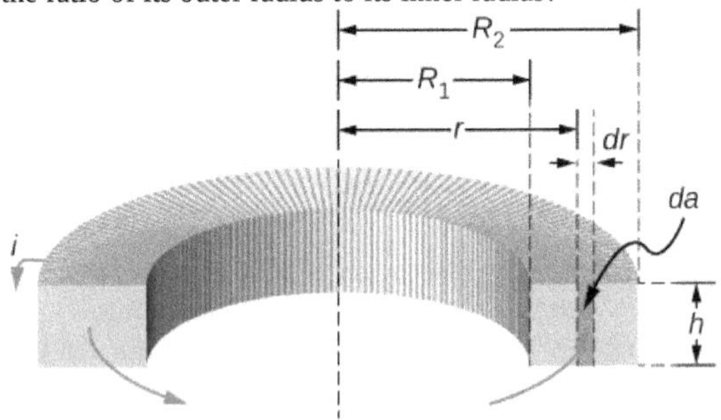

45. What is the self-inductance per meter of a coaxial cable whose inner radius is 0.50 mm and whose outer radius is 4.00 mm?

14.3 Energy in a Magnetic Field

46. At the instant a current of 0.20 A is flowing through a coil of wire, the energy stored in its magnetic field is 6.0×10^{-3} J. What is the self-inductance of the coil?

47. Suppose that a rectangular toroid has 2000 windings and a self-inductance of 0.040 H. If $h = 0.10\,\text{m}$, what is the current flowing through a rectangular toroid when the energy in its magnetic field is 2.0×10^{-6} J?

48. Solenoid *A* is tightly wound while solenoid *B* has windings that are evenly spaced with a gap equal to the diameter of the wire. The solenoids are otherwise identical. Determine the ratio of the energies stored per unit length of these solenoids when the same current flows through each.

49. A 10-H inductor carries a current of 20 A. How much ice at $0°\,\text{C}$ could be melted by the energy stored in the magnetic field of the inductor? (*Hint*: Use the value $L_\text{f} = 334\,\text{J/g}$ for ice.)

50. A coil with a self-inductance of 3.0 H and a resistance of $100\,\Omega$ carries a steady current of 2.0 A. (a) What is the energy stored in the magnetic field of the coil? (b) What is the energy per second dissipated in the resistance of the coil?

51. A current of 1.2 A is flowing in a coaxial cable whose outer radius is five times its inner radius. What is the magnetic field energy stored in a 3.0-m length of the cable?

14.4 RL Circuits

52. In Figure 14.12, $\varepsilon = 12\,\text{V}$, $L = 20\,\text{mH}$, and $R = 5.0\,\Omega$. Determine (a) the time constant of the circuit, (b) the initial current through the resistor, (c) the final current through the resistor, (d) the current through the resistor when $t = 2\tau_L$, and (e) the voltages across the inductor and the resistor when $t = 2\tau_L$.

53. For the circuit shown below, $\varepsilon = 20\,\text{V}$, $L = 4.0\,\text{mH}$, and $R = 5.0\,\Omega$. After steady state is reached with S_1 closed and S_2 open, S_2 is closed and immediately thereafter $(\text{at}\ t = 0)$ S_1 is opened. Determine (a) the current through L at $t = 0$, (b) the current through L at $t = 4.0 \times 10^{-4}$ s, and (c) the voltages across L and R at $t = 4.0 \times 10^{-4}$ s.

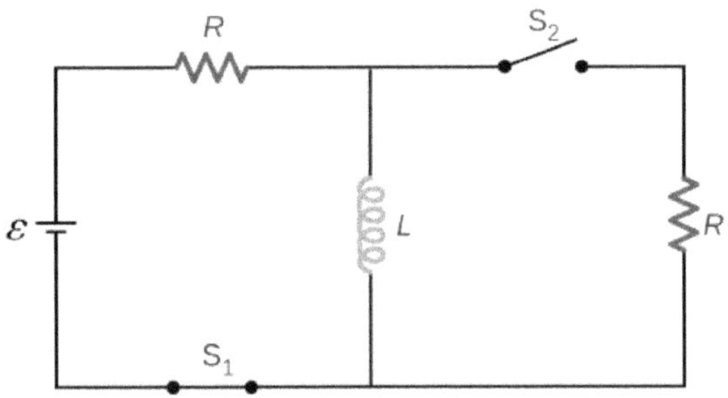

54. The current in the *RL* circuit shown here increases to 40% of its steady-state value in 2.0 s. What is the time constant of the circuit?

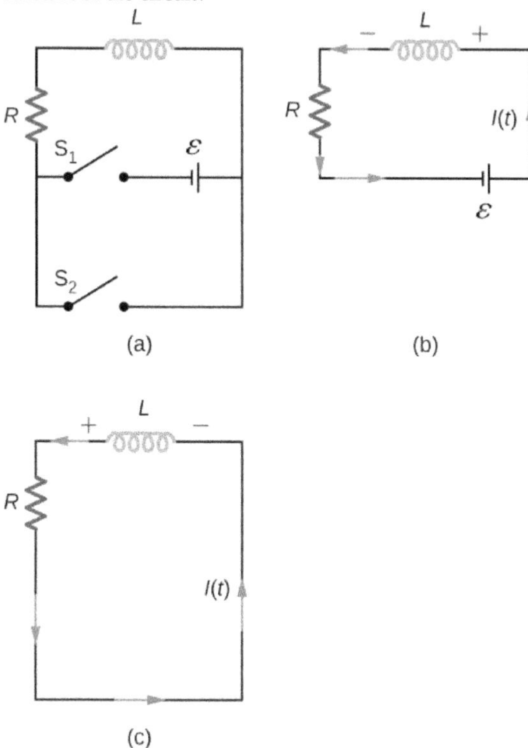

55. How long after switch S_1 is thrown does it take the current in the circuit shown to reach half its maximum value? Express your answer in terms of the time constant of the circuit.

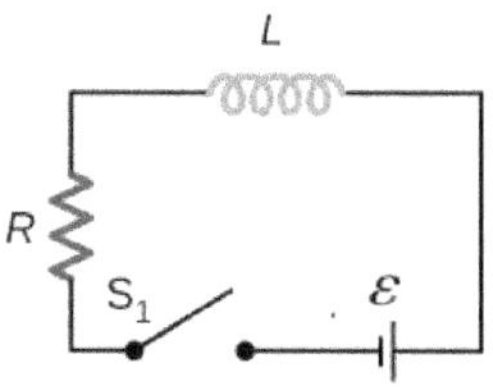

56. Examine the circuit shown below in part (a). Determine *dI/dt* at the instant after the switch is thrown in the circuit of (a), thereby producing the circuit of (b). Show that if *I* were to continue to increase at this initial rate, it would reach its maximum ε/R in one time constant.

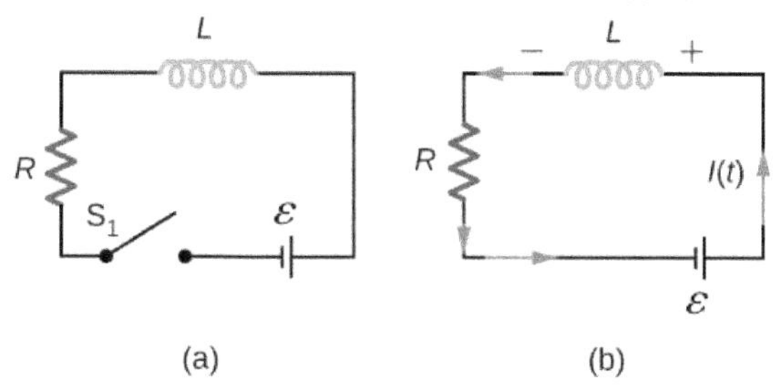

57. The current in the *RL* circuit shown below reaches half its maximum value in 1.75 ms after the switch S_1 is thrown. Determine (a) the time constant of the circuit and (b) the resistance of the circuit if $L = 250\ \text{mH}$.

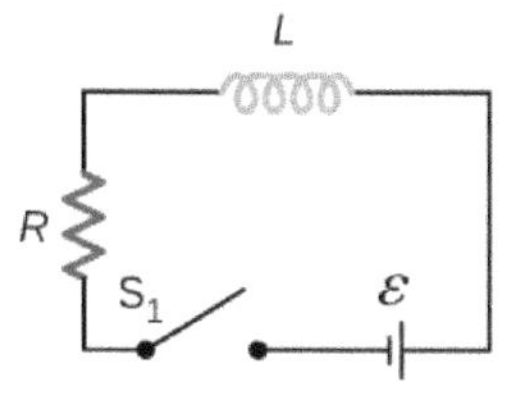

58. Consider the circuit shown below. Find I_1, I_2, and I_3 when (a) the switch S is first closed, (b) after the currents have reached steady-state values, and (c) at the instant the switch is reopened (after being closed for a long time).

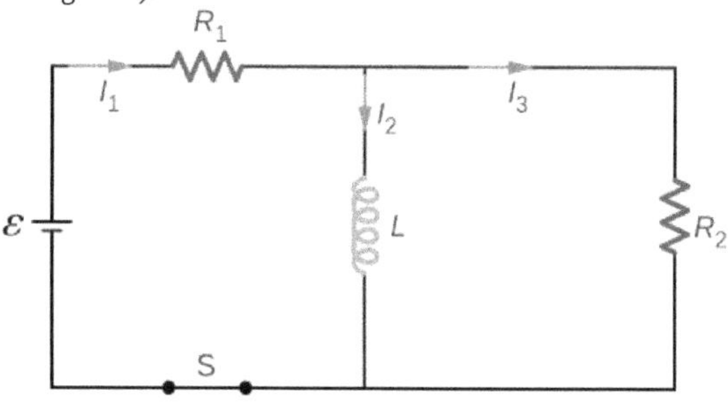

59. For the circuit shown below, $\varepsilon = 50\ \text{V}$, $R_1 = 10\ \Omega$, and $L = 2.0\ \text{mH}$. Find the values of I_1 and I_2 (a) immediately after switch S is closed, (b) a long time after S is closed, (c) immediately after S is reopened, and (d) a long time after S is reopened.

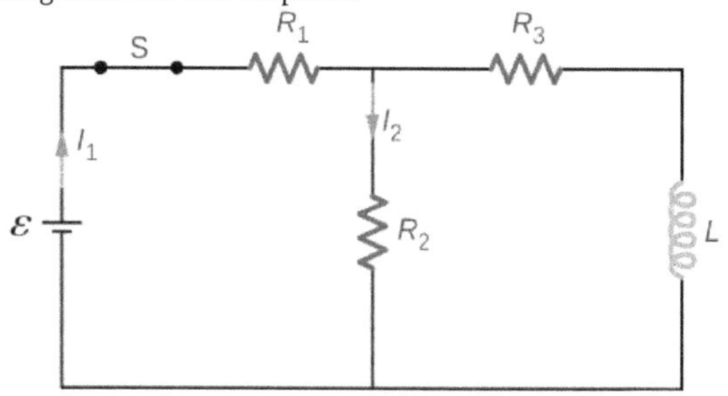

60. For the circuit shown below, find the current through the inductor 2.0×10^{-5} s after the switch is reopened.

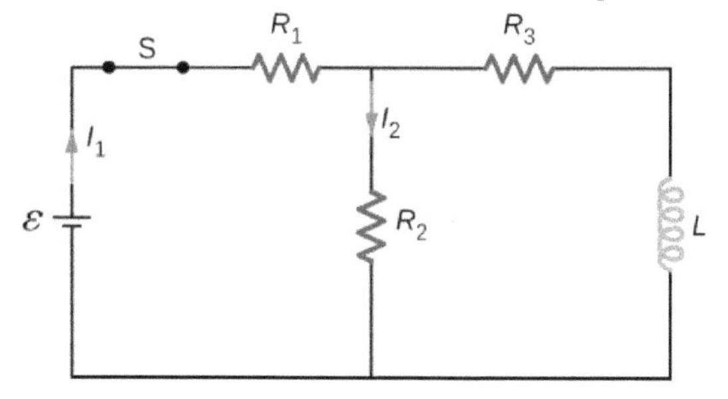

61. Show that for the circuit shown below, the initial energy stored in the inductor, $LI^2(0)/2$, is equal to the total energy eventually dissipated in the resistor, $\int_0^\infty I^2(t)Rdt$.

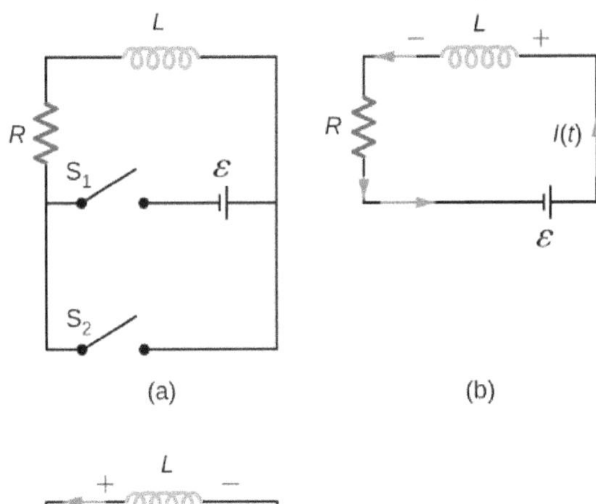

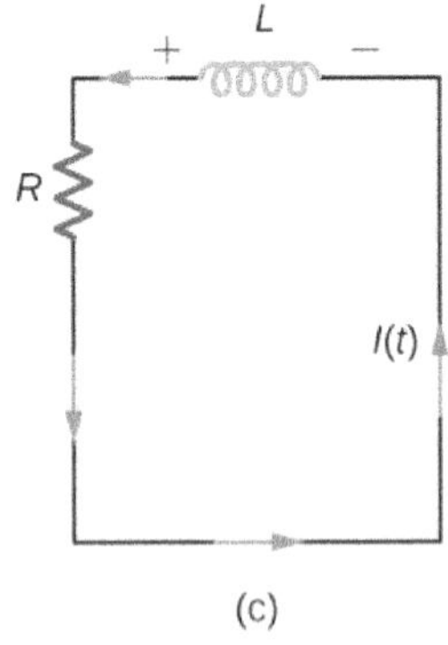

14.5 Oscillations in an LC Circuit

62. A 5000-pF capacitor is charged to 100 V and then quickly connected to an 80-mH inductor. Determine (a) the maximum energy stored in the magnetic field of the inductor, (b) the peak value of the current, and (c) the frequency of oscillation of the circuit.

63. The self-inductance and capacitance of an *LC* circuit are 0.20 mH and 5.0 pF. What is the angular frequency at which the circuit oscillates?

64. What is the self-inductance of an *LC* circuit that oscillates at 60 Hz when the capacitance is $10\ \mu\text{F}$?

65. In an oscillating *LC* circuit, the maximum charge on the capacitor is 2.0×10^{-6} C and the maximum current through the inductor is 8.0 mA. (a) What is the period of the oscillations? (b) How much time elapses between an instant when the capacitor is uncharged and the next instant when it is fully charged?

66. The self-inductance and capacitance of an oscillating *LC* circuit are $L = 20$ mH and $C = 1.0\ \mu\text{F}$, respectively. (a) What is the frequency of the oscillations? (b) If the maximum potential difference between the plates of the capacitor is 50 V, what is the maximum current in the circuit?

67. In an oscillating *LC* circuit, the maximum charge on the capacitor is q_m. Determine the charge on the capacitor and the current through the inductor when energy is shared equally between the electric and magnetic fields. Express your answer in terms of q_m, *L*, and *C*.

68. In the circuit shown below, S_1 is opened and S_2 is closed simultaneously. Determine (a) the frequency of the resulting oscillations, (b) the maximum charge on the capacitor, (c) the maximum current through the inductor, and (d) the electromagnetic energy of the oscillating circuit.

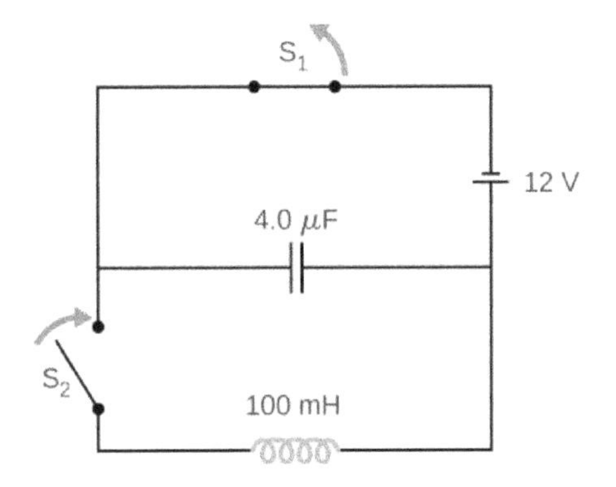

69. An *LC* circuit in an AM tuner (in a car stereo) uses a coil with an inductance of 2.5 mH and a variable capacitor. If the natural frequency of the circuit is to be adjustable over the range 540 to 1600 kHz (the AM broadcast band), what range of capacitance is required?

14.6 RLC Series Circuits

70. In an oscillating *RLC* circuit, $R = 5.0\ \Omega$, $L = 5.0$ mH, and $C = 500\ \mu\text{F}$. What is the angular frequency of the oscillations?

71. In an oscillating *RLC* circuit with $L = 10$ mH, $C = 1.5\ \mu\text{F}$, and $R = 2.0\ \Omega$, how much time elapses before the amplitude of the oscillations drops to half its initial value?

72. What resistance *R* must be connected in series with a 200-mH inductor of the resulting *RLC* oscillating circuit is to decay to 50% of its initial value of charge in 50 cycles? To 0.10% of its initial value in 50 cycles?

ADDITIONAL PROBLEMS

73. Show that the self-inductance per unit length of an infinite, straight, thin wire is infinite.

74. Two long, parallel wires carry equal currents in opposite directions. The radius of each wire is *a*, and the distance between the centers of the wires is *d*. Show that if the magnetic flux within the wires themselves can be ignored, the self-inductance of a length *l* of such a pair of wires is

$$L = \frac{\mu_0 l}{\pi} \ln \frac{d-a}{a}.$$

(*Hint*: Calculate the magnetic flux through a rectangle of length *l* between the wires and then use $L = N\Phi/I$.)

75. A small, rectangular single loop of wire with dimensions *l*, and *a* is placed, as shown below, in the plane of a much larger, rectangular single loop of wire. The two short sides of the larger loop are so far from the smaller loop that their magnetic fields over the smaller fields over the smaller loop can be ignored. What is the mutual inductance of the two loops?

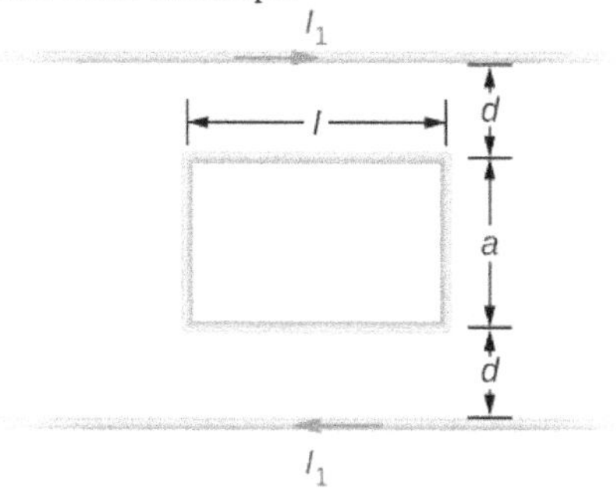

76. Suppose that a cylindrical solenoid is wrapped around a core of iron whose magnetic susceptibility is **x**. Using Equation 14.9, show that the self-inductance of the solenoid is given by

$$L = \frac{(1+x)\mu_0 N^2 A}{l},$$

where *l* is its length, *A* its cross-sectional area, and *N* its total number of turns.

77. The solenoid of the preceding problem is wrapped around an iron core whose magnetic susceptibility is 4.0×10^3. (a) If a current of 2.0 A flows through the solenoid, what is the magnetic field in the iron core? (b) What is the effective surface current formed by the aligned atomic current loops in the iron core? (c) What is the self-inductance of the filled solenoid?

78. A rectangular toroid with inner radius $R_1 = 7.0\,\text{cm}$, outer radius $R_2 = 9.0\,\text{cm}$, height $h = 3.0$, and $N = 3000$ turns is filled with an iron core of magnetic susceptibility 5.2×10^3. (a) What is the self-inductance of the toroid? (b) If the current through the toroid is 2.0 A, what is the magnetic field at the center of the core? (c) For this same 2.0-A current, what is the effective surface current formed by the aligned atomic current loops in the iron core?

79. The switch S of the circuit shown below is closed at $t = 0$. Determine (a) the initial current through the battery and (b) the steady-state current through the battery.

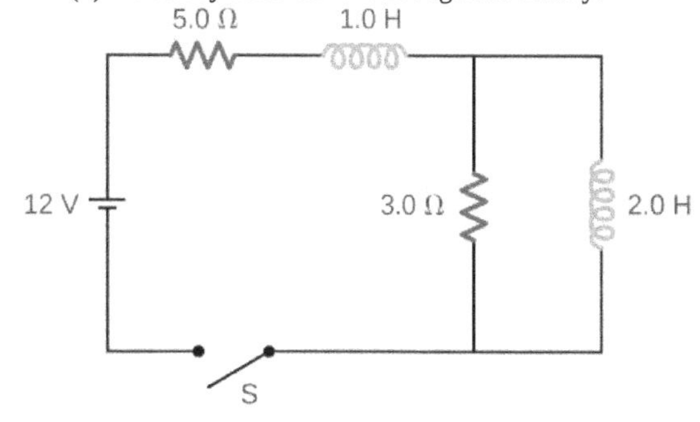

80. In an oscillating *RLC* circuit, $R = 7.0\,\Omega$, $L = 10\,\text{mH}$, and $C = 3.0\,\mu\text{F}$. Initially, the capacitor has a charge of $8.0\,\mu\text{C}$ and the current is zero. Calculate the charge on the capacitor (a) five cycles later and (b) 50 cycles later.

81. A 25.0-H inductor has 100 A of current turned off in 1.00 ms. (a) What voltage is induced to oppose this? (b) What is unreasonable about this result? (c) Which assumption or premise is responsible?

CHALLENGE PROBLEMS

82. A coaxial cable has an inner conductor of radius a, and outer thin cylindrical shell of radius b. A current *I* flows in the inner conductor and returns in the outer conductor. The self-inductance of the structure will depend on how the current in the inner cylinder tends to be distributed. Investigate the following two extreme cases. (a) Let current in the inner conductor be distributed only on the surface and find the self-inductance. (b) Let current in the inner cylinder be distributed uniformly over its cross-section and find the self-inductance. Compare with your results in (a).

83. In a damped oscillating circuit the energy is dissipated in the resistor. The *Q*-factor is a measure of the persistence of the oscillator against the dissipative loss. (a) Prove that for a lightly damped circuit the energy, *U*, in the circuit decreases according to the following equation.

$$\frac{dU}{dt} = -2\beta U, \quad \text{where } \beta = \frac{R}{2L}.$$

(b) Using the definition of the *Q*-factor as energy divided by the loss over the next cycle, prove that *Q*-factor of a lightly damped oscillator as defined in this problem is

$$Q \equiv \frac{U_{\text{begin}}}{\Delta U_{\text{one cycle}}} = \frac{1}{R}\sqrt{\frac{L}{C}}.$$

(*Hint:* For (b), to obtain *Q*, divide *E* at the beginning of one cycle by the change ΔE over the next cycle.)

84. The switch in the circuit shown below is closed at $t = 0\,\text{s}$. Find currents through (a) R_1, (b) R_2, and (c) the battery as function of time.

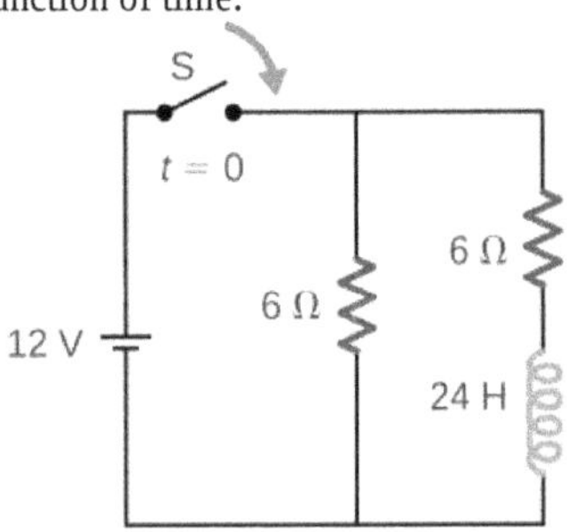

85. A square loop of side 2 cm is placed 1 cm from a long wire carrying a current that varies with time at a constant rate of 3 A/s as shown below. (a) Use Ampère's law and find the magnetic field as a function of time from the current in the wire. (b) Determine the magnetic flux through the loop. (c) If the loop has a resistance of $3\,\Omega$, how much induced current flows in the loop?

86. A rectangular copper ring, of mass 100 g and resistance $0.2\,\Omega$, is in a region of uniform magnetic field that is perpendicular to the area enclosed by the ring and horizontal to Earth's surface. The ring is let go from rest when it is at the edge of the nonzero magnetic field region (see below). (a) Find its speed when the ring just exits the region of uniform magnetic field. (b) If it was let go at $t = 0$, what is the time when it exits the region of magnetic field for the following values: $a = 25\,\text{cm}$, $b = 50\,\text{cm}$, $B = 3\,\text{T}$, and $g = 9.8\,\text{m/s}^2$? Assume the magnetic field of the induced current is negligible compared to 3 T.

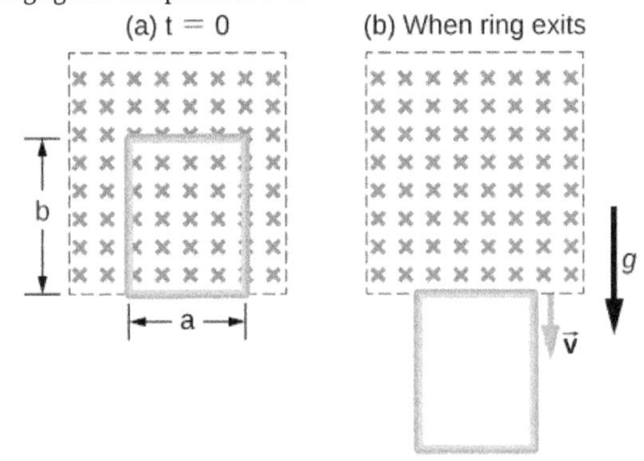

15 | ALTERNATING-CURRENT CIRCUITS

Figure 15.1 The current we draw into our houses is an alternating current (ac). Power lines transmit ac to our neighborhoods, where local power stations and transformers distribute it to our homes. In this chapter, we discuss how a transformer works and how it allows us to transmit power at very high voltages and minimal heating losses across the lines.

Chapter Outline

Introduction

Electric power is delivered to our homes by alternating current (ac) through high-voltage transmission lines. As explained in Transformers, transformers can then change the amplitude of the alternating potential difference to a more useful form. This lets us transmit power at very high voltages, minimizing resistive heating losses in the lines, and then furnish that power to homes at lower, safer voltages. Because constant potential differences are unaffected by transformers, this capability is more difficult to achieve with direct-current transmission.

In this chapter, we use Kirchhoff's laws to analyze four simple circuits in which ac flows. We have discussed the use of the resistor, capacitor, and inductor in circuits with batteries. These components are also part of ac circuits. However, because ac is required, the constant source of emf supplied by a battery is replaced by an ac voltage source, which produces an oscillating emf.

15.1 | AC Sources

Learning Objectives

By the end of the section, you will be able to:

- Explain the differences between direct current (dc) and alternating current (ac)
- Define characteristic features of alternating current and voltage, such as the amplitude or peak and the frequency

Most examples dealt with so far in this book, particularly those using batteries, have constant-voltage sources. Thus, once the current is established, it is constant. **Direct current (dc)** is the flow of electric charge in only one direction. It is the steady state of a constant-voltage circuit.

Most well-known applications, however, use a time-varying voltage source. **Alternating current (ac)** is the flow of electric charge that periodically reverses direction. An ac is produced by an alternating emf, which is generated in a power plant, as described in Induced Electric Fields. If the ac source varies periodically, particularly sinusoidally, the circuit is known as an ac circuit. Examples include the commercial and residential power that serves so many of our needs.

The ac voltages and frequencies commonly used in businesses and homes vary around the world. In a typical house, the potential difference between the two sides of an electrical outlet alternates sinusoidally with a frequency of 60 or 50 Hz and an amplitude of 156 or 311 V, depending on whether you live in the United States or Europe, respectively. Most people know the potential difference for electrical outlets is 120 V or 220 V in the US or Europe, but as explained later in the chapter, these voltages are not the peak values given here but rather are related to the common voltages we see in our electrical outlets. Figure 15.2 shows graphs of voltage and current versus time for typical dc and ac power in the United States.

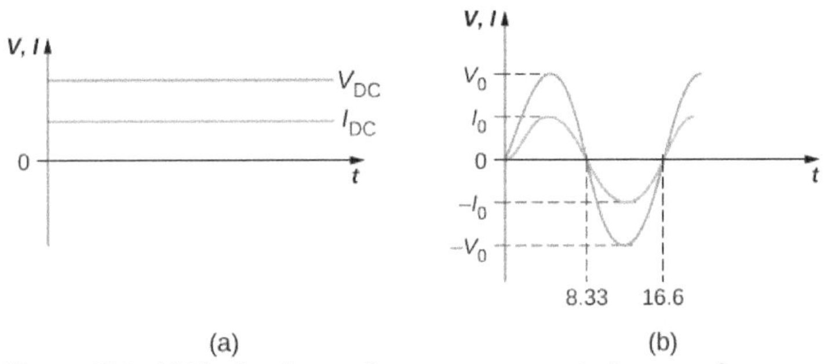

Figure 15.2 (a) The dc voltage and current are constant in time, once the current is established. (b) The voltage and current versus time are quite different for ac power. In this example, which shows 60-Hz ac power and time *t* in seconds, voltage and current are sinusoidal and are in phase for a simple resistance circuit. The frequencies and peak voltages of ac sources differ greatly.

Suppose we hook up a resistor to an ac voltage source and determine how the voltage and current vary in time across the resistor. Figure 15.3 shows a schematic of a simple circuit with an ac voltage source. The voltage fluctuates sinusoidally with time at a fixed frequency, as shown, on either the battery terminals or the resistor. Therefore, the **ac voltage**, or the "voltage at a plug," can be given by

$$v = V_0 \sin \omega t, \tag{15.1}$$

where v is the voltage at time t, V_0 is the peak voltage, and ω is the angular frequency in radians per second. For a typical house in the United States, $V_0 = 156 \text{ V}$ and $\omega = 120\pi \text{ rad/s}$, whereas in Europe, $V_0 = 311 \text{ V}$ and $\omega = 100\pi \text{ rad/s}$.

For this simple resistance circuit, $I = V/R$, so the **ac current**, meaning the current that fluctuates sinusoidally with time at a fixed frequency, is

$$i = I_0 \sin \omega t, \quad (15.2)$$

where i is the current at time t and I_0 is the peak current and is equal to V_0/R. For this example, the voltage and current are said to be in phase, meaning that their sinusoidal functional forms have peaks, troughs, and nodes in the same place. They oscillate in sync with each other, as shown in Figure 15.2(b). In these equations, and throughout this chapter, we use lowercase letters (such as i) to indicate instantaneous values and capital letters (such as I) to indicate maximum, or peak, values.

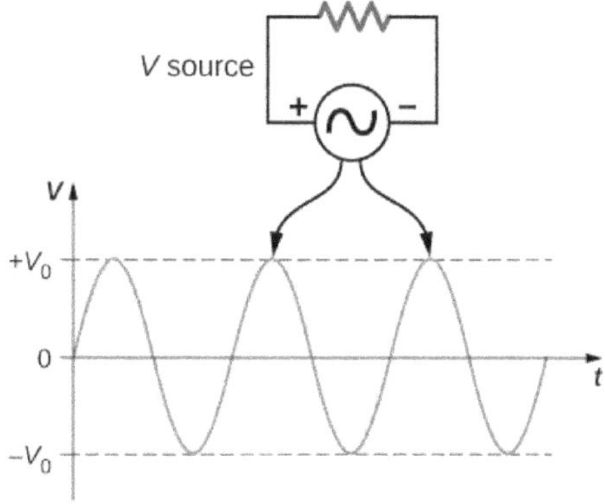

Figure 15.3 The potential difference V between the terminals of an ac voltage source fluctuates, so the source and the resistor have ac sine waves on top of each other. The mathematical expression for v is given by $v = V_0 \sin \omega t$.

Current in the resistor alternates back and forth just like the driving voltage, since $I = V/R$. If the resistor is a fluorescent light bulb, for example, it brightens and dims 120 times per second as the current repeatedly goes through zero. A 120-Hz flicker is too rapid for your eyes to detect, but if you wave your hand back and forth between your face and a fluorescent light, you will see the stroboscopic effect of ac.

15.1 Check Your Understanding If a European ac voltage source is considered, what is the time difference between the zero crossings on an ac voltage-versus-time graph?

15.2 | Simple AC Circuits

Learning Objectives

By the end of the section, you will be able to:

- Interpret phasor diagrams and apply them to ac circuits with resistors, capacitors, and inductors
- Define the reactance for a resistor, capacitor, and inductor to help understand how current in the circuit behaves compared to each of these devices

In this section, we study simple models of ac voltage sources connected to three circuit components: (1) a resistor, (2) a capacitor, and (3) an inductor. The power furnished by an ac voltage source has an emf given by

$$v(t) = V_0 \sin \omega t,$$

as shown in Figure 15.4. This sine function assumes we start recording the voltage when it is $v = 0\text{ V}$ at a time of $t = 0\text{ s}$. A phase constant may be involved that shifts the function when we start measuring voltages, similar to the phase constant in the waves we studied in Waves (http://cnx.org/content/m58367/latest/) . However, because we are free to choose when we start examining the voltage, we can ignore this phase constant for now. We can measure this voltage across the circuit components using one of two methods: (1) a quantitative approach based on our knowledge of circuits, or (2) a graphical approach that is explained in the coming sections.

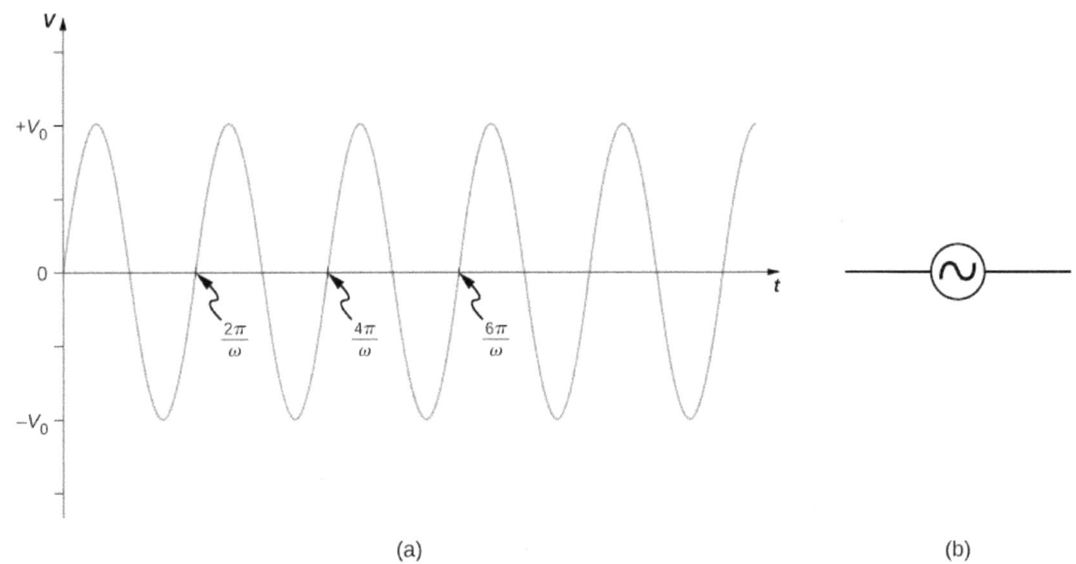

Figure 15.4 (a) The output $v(t) = V_0 \sin \omega t$ of an ac generator. (b) Symbol used to represent an ac voltage source in a circuit diagram.

Resistor

First, consider a resistor connected across an ac voltage source. From Kirchhoff's loop rule, the instantaneous voltage across the resistor of Figure 15.5(a) is

$$v_R(t) = V_0 \sin \omega t$$

and the instantaneous current through the resistor is

$$i_R(t) = \frac{v_R(t)}{R} = \frac{V_0}{R} \sin \omega t = I_0 \sin \omega t.$$

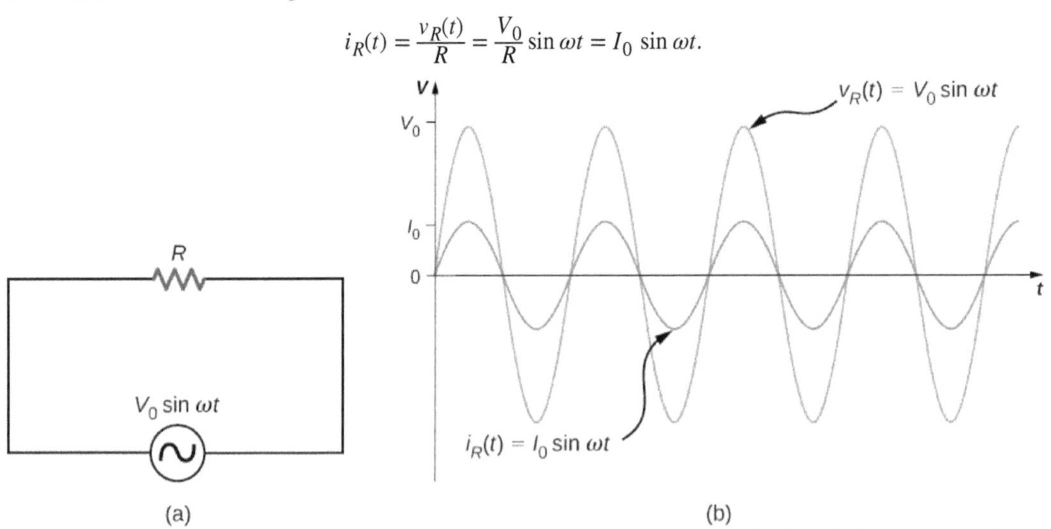

Figure 15.5 (a) A resistor connected across an ac voltage source. (b) The current $i_R(t)$ through the resistor and the voltage $v_R(t)$ across the resistor. The two quantities are in phase.

Here, $I_0 = V_0/R$ is the amplitude of the time-varying current. Plots of $i_R(t)$ and $v_R(t)$ are shown in Figure 15.5(b). Both curves reach their maxima and minima at the same times, that is, the current through and the voltage across the resistor are in phase.

Graphical representations of the phase relationships between current and voltage are often useful in the analysis of ac circuits. Such representations are called *phasor diagrams*. The phasor diagram for $i_R(t)$ is shown in Figure 15.6(a), with the current on the vertical axis. The arrow (or phasor) is rotating counterclockwise at a constant angular frequency ω, so we are viewing it at one instant in time. If the length of the arrow corresponds to the current amplitude I_0, the projection of the rotating arrow onto the vertical axis is $i_R(t) = I_0 \sin \omega t$, which is the instantaneous current.

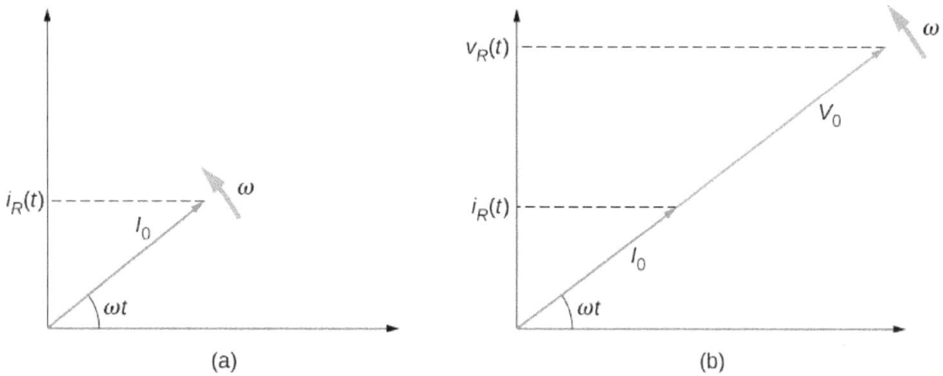

Figure 15.6 (a) The phasor diagram representing the current through the resistor of Figure 15.5. (b) The phasor diagram representing both $i_R(t)$ and $v_R(t)$.

The vertical axis on a phasor diagram could be either the voltage or the current, depending on the phasor that is being examined. In addition, several quantities can be depicted on the same phasor diagram. For example, both the current $i_R(t)$ and the voltage $v_R(t)$ are shown in the diagram of Figure 15.6(b). Since they have the same frequency and are in phase, their phasors point in the same direction and rotate together. The relative lengths of the two phasors are arbitrary because they represent different quantities; however, the ratio of the lengths of the two phasors can be represented by the resistance, since one is a voltage phasor and the other is a current phasor.

Capacitor

Now let's consider a capacitor connected across an ac voltage source. From Kirchhoff's loop rule, the instantaneous voltage across the capacitor of Figure 15.7(a) is

$$v_C(t) = V_0 \sin \omega t.$$

Recall that the charge in a capacitor is given by $Q = CV$. This is true at any time measured in the ac cycle of voltage. Consequently, the instantaneous charge on the capacitor is

$$q(t) = Cv_C(t) = CV_0 \sin \omega t.$$

Since the current in the circuit is the rate at which charge enters (or leaves) the capacitor,

$$i_C(t) = \frac{dq(t)}{dt} = \omega CV_0 \cos \omega t = I_0 \cos \omega t,$$

where $I_0 = \omega CV_0$ is the current amplitude. Using the trigonometric relationship $\cos \omega t = \sin(\omega t + \pi/2)$, we may express the instantaneous current as

$$i_C(t) = I_0 \sin\left(\omega t + \frac{\pi}{2}\right).$$

Dividing V_0 by I_0, we obtain an equation that looks similar to Ohm's law:

$$\frac{V_0}{I_0} = \frac{1}{\omega C} = X_C. \tag{15.3}$$

The quantity X_C is analogous to resistance in a dc circuit in the sense that both quantities are a ratio of a voltage to a current. As a result, they have the same unit, the ohm. Keep in mind, however, that a capacitor stores and discharges electric energy, whereas a resistor dissipates it. The quantity X_C is known as the **capacitive reactance** of the capacitor, or the opposition of a capacitor to a change in current. It depends inversely on the frequency of the ac source—high frequency leads to low capacitive reactance.

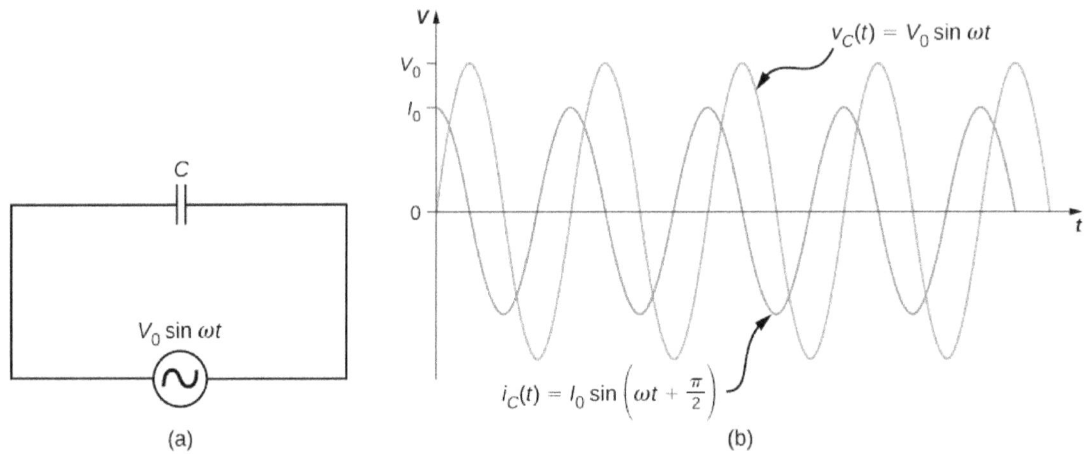

Figure 15.7 (a) A capacitor connected across an ac generator. (b) The current $i_C(t)$ through the capacitor and the voltage $v_C(t)$ across the capacitor. Notice that $i_C(t)$ leads $v_C(t)$ by $\pi/2$ rad.

A comparison of the expressions for $v_C(t)$ and $i_C(t)$ shows that there is a phase difference of $\pi/2$ rad between them. When these two quantities are plotted together, the current peaks a quarter cycle (or $\pi/2$ rad) ahead of the voltage, as illustrated in Figure 15.7(b). The current through a capacitor leads the voltage across a capacitor by $\pi/2$ rad, or a quarter of a cycle.

The corresponding phasor diagram is shown in Figure 15.8. Here, the relationship between $i_C(t)$ and $v_C(t)$ is represented by having their phasors rotate at the same angular frequency, with the current phasor leading by $\pi/2$ rad.

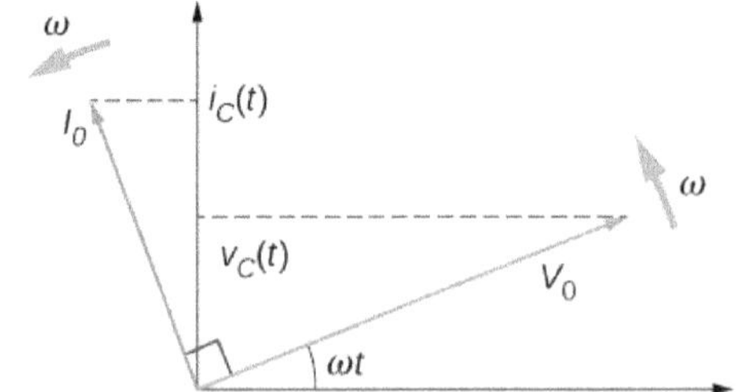

Figure 15.8 The phasor diagram for the capacitor of Figure 15.7. The current phasor leads the voltage phasor by $\pi/2$ rad as they both rotate with the same angular frequency.

To this point, we have exclusively been using peak values of the current or voltage in our discussion, namely, I_0 and V_0. However, if we average out the values of current or voltage, these values are zero. Therefore, we often use a second convention called the root mean square value, or rms value, in discussions of current and voltage. The rms operates in reverse of the terminology. First, you square the function, next, you take the mean, and then, you find the square root. As a result, the rms values of current and voltage are not zero. Appliances and devices are commonly quoted with rms values for their operations, rather than peak values. We indicate rms values with a subscript attached to a capital letter (such as I_{rms}).

Although a capacitor is basically an open circuit, an **rms current**, or the root mean square of the current, appears in a circuit with an ac voltage applied to a capacitor. Consider that

$$I_{\text{rms}} = \frac{I_0}{\sqrt{2}}, \tag{15.4}$$

where I_0 is the peak current in an ac system. The **rms voltage**, or the root mean square of the voltage, is

$$V_{\text{rms}} = \frac{V_0}{\sqrt{2}}, \tag{15.5}$$

where V_0 is the peak voltage in an ac system. The rms current appears because the voltage is continually reversing, charging, and discharging the capacitor. If the frequency goes to zero, which would be a dc voltage, X_C tends to infinity, and the current is zero once the capacitor is charged. At very high frequencies, the capacitor's reactance tends to zero—it has a negligible reactance and does not impede the current (it acts like a simple wire).

Inductor

Lastly, let's consider an inductor connected to an ac voltage source. From Kirchhoff's loop rule, the voltage across the inductor L of Figure 15.9(a) is

$$v_L(t) = V_0 \sin \omega t. \tag{15.6}$$

The emf across an inductor is equal to $\varepsilon = -L(di_L/dt)$; however, the potential difference across the inductor is $v_L(t) = Ldi_L(t)/dt$, because if we consider that the voltage around the loop must equal zero, the voltage gained from the ac source must dissipate through the inductor. Therefore, connecting this with the ac voltage source, we have

$$\frac{di_L(t)}{dt} = \frac{V_0}{L} \sin \omega t.$$

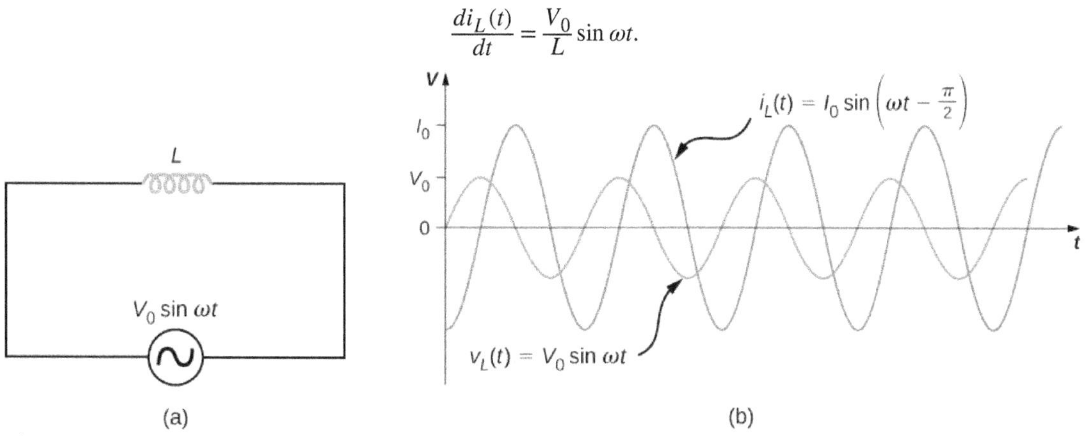

Figure 15.9 (a) An inductor connected across an ac generator. (b) The current $i_L(t)$ through the inductor and the voltage $v_L(t)$ across the inductor. Here $i_L(t)$ lags $v_L(t)$ by $\pi/2$ rad.

The current $i_L(t)$ is found by integrating this equation. Since the circuit does not contain a source of constant emf, there is no steady current in the circuit. Hence, we can set the constant of integration, which represents the steady current in the circuit, equal to zero, and we have

$$i_L(t) = -\frac{V_0}{\omega L}\cos \omega t = \frac{V_0}{\omega L}\sin\left(\omega t - \frac{\pi}{2}\right) = I_0 \sin\left(\omega t - \frac{\pi}{2}\right), \quad (15.7)$$

where $I_0 = V_0/\omega L$. The relationship between V_0 and I_0 may also be written in a form analogous to Ohm's law:

$$\frac{V_0}{I_0} = \omega L = X_L. \quad (15.8)$$

The quantity X_L is known as the **inductive reactance** of the inductor, or the opposition of an inductor to a change in current; its unit is also the ohm. Note that X_L varies directly as the frequency of the ac source—high frequency causes high inductive reactance.

A phase difference of $\pi/2$ rad occurs between the current through and the voltage across the inductor. From Equation 15.6 and Equation 15.7, the current through an inductor lags the potential difference across an inductor by $\pi/2$ rad, or a quarter of a cycle. The phasor diagram for this case is shown in Figure 15.10.

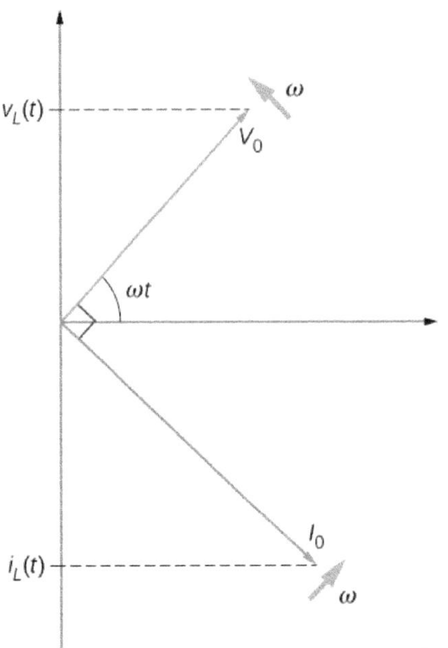

Figure 15.10 The phasor diagram for the inductor of Figure 15.9. The current phasor lags the voltage phasor by $\pi/2$ rad as they both rotate with the same angular frequency.

An animation from the University of New South Wales AC Circuits (https://openstaxcollege.org/l/21accircuits) illustrates some of the concepts we discuss in this chapter. They also include wave and phasor diagrams that evolve over time so that you can get a better picture of how each changes over time.

Example 15.1

Simple AC Circuits

An ac generator produces an emf of amplitude 10 V at a frequency $f = 60\text{ Hz}$. Determine the voltages across and the currents through the circuit elements when the generator is connected to (a) a $100\text{-}\Omega$ resistor, (b) a $10\text{-}\mu\text{F}$ capacitor, and (c) a 15-mH inductor.

Strategy

The entire AC voltage across each device is the same as the source voltage. We can find the currents by finding the reactance X of each device and solving for the peak current using $I_0 = V_0/X$.

Solution

The voltage across the terminals of the source is

$$v(t) = V_0 \sin \omega t = (10\text{ V}) \sin 120\pi t,$$

where $\omega = 2\pi f = 120\pi\text{ rad/s}$ is the angular frequency. Since $v(t)$ is also the voltage across each of the elements, we have

$$v(t) = v_R(t) = v_C(t) = v_L(t) = (10\text{ V}) \sin 120\pi t.$$

a. When $R = 100\,\Omega$, the amplitude of the current through the resistor is

$$I_0 = V_0/R = 10\text{ V}/100\,\Omega = 0.10\text{ A},$$

so

$$i_R(t) = (0.10\text{ A}) \sin 120\pi t.$$

b. From Equation 15.3, the capacitive reactance is

$$X_C = \frac{1}{\omega C} = \frac{1}{(120\pi\text{ rad/s})(10 \times 10^{-6}\text{F})} = 265\,\Omega,$$

so the maximum value of the current is

$$I_0 = \frac{V_0}{X_C} = \frac{10\text{ V}}{265\,\Omega} = 3.8 \times 10^{-2}\text{ A}$$

and the instantaneous current is given by

$$i_C(t) = (3.8 \times 10^{-2}\text{ A}) \sin\left(120\pi t + \frac{\pi}{2}\right).$$

c. From Equation 15.8, the inductive reactance is

$$X_L = \omega L = (120\pi\text{ rad/s})(15 \times 10^{-3}\text{ H}) = 5.7\,\Omega.$$

The maximum current is therefore

$$I_0 = \frac{10\text{ V}}{5.7\,\Omega} = 1.8\text{ A}$$

and the instantaneous current is

$$i_L(t) = (1.8\text{ A}) \sin\left(120\pi t - \frac{\pi}{2}\right).$$

Significance

Although the voltage across each device is the same, the peak current has different values, depending on the reactance. The reactance for each device depends on the values of resistance, capacitance, or inductance.

15.2 Check Your Understanding Repeat Example 15.1 for an ac source of amplitude 20 V and frequency 100 Hz.

15.3 | RLC Series Circuits with AC

Learning Objectives

By the end of the section, you will be able to:

- Describe how the current varies in a resistor, a capacitor, and an inductor while in series with an ac power source
- Use phasors to understand the phase angle of a resistor, capacitor, and inductor ac circuit and to understand what that phase angle means
- Calculate the impedance of a circuit

The ac circuit shown in Figure 15.11, called an *RLC* series circuit, is a series combination of a resistor, capacitor, and inductor connected across an ac source. It produces an emf of

$$v(t) = V_0 \sin \omega t.$$

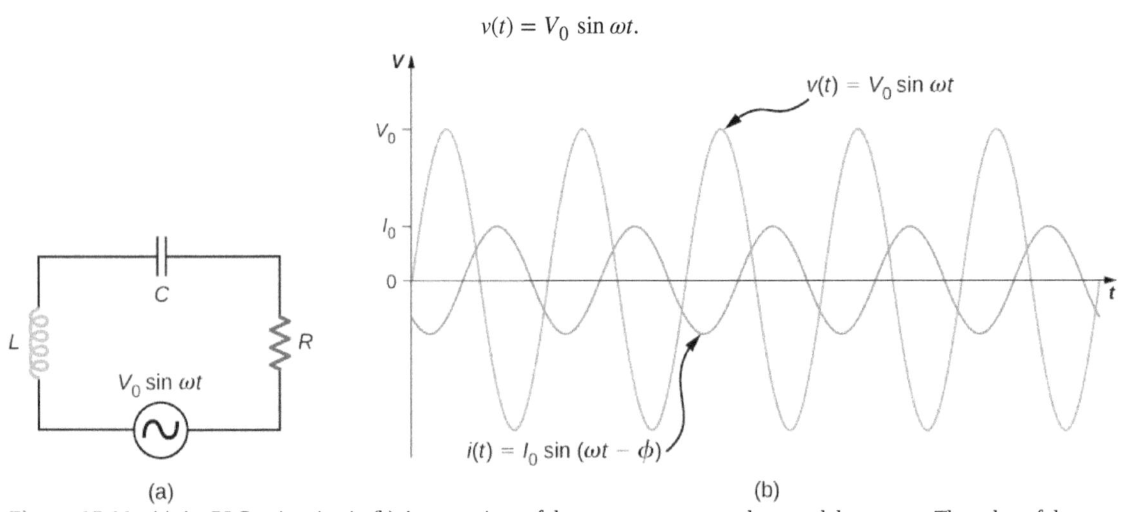

Figure 15.11 (a) An *RLC* series circuit. (b) A comparison of the generator output voltage and the current. The value of the phase difference ϕ depends on the values of *R*, *C*, and *L*.

Since the elements are in series, the same current flows through each element at all points in time. The relative phase between the current and the emf is not obvious when all three elements are present. Consequently, we represent the current by the general expression

$$i(t) = I_0 \sin (\omega t - \phi),$$

where I_0 is the current amplitude and ϕ is the **phase angle** between the current and the applied voltage. The phase angle is thus the amount by which the voltage and current are out of phase with each other in a circuit. Our task is to find I_0 and ϕ.

A phasor diagram involving $i(t)$, $v_R(t)$, $v_C(t)$, and $v_L(t)$ is helpful for analyzing the circuit. As shown in Figure 15.12, the phasor representing $v_R(t)$ points in the same direction as the phasor for $i(t)$; its amplitude is $V_R = I_0 R$. The $v_C(t)$ phasor lags the $i(t)$ phasor by $\pi/2$ rad and has the amplitude $V_C = I_0 X_C$. The phasor for $v_L(t)$ leads the $i(t)$ phasor by $\pi/2$ rad and has the amplitude $V_L = I_0 X_L$.

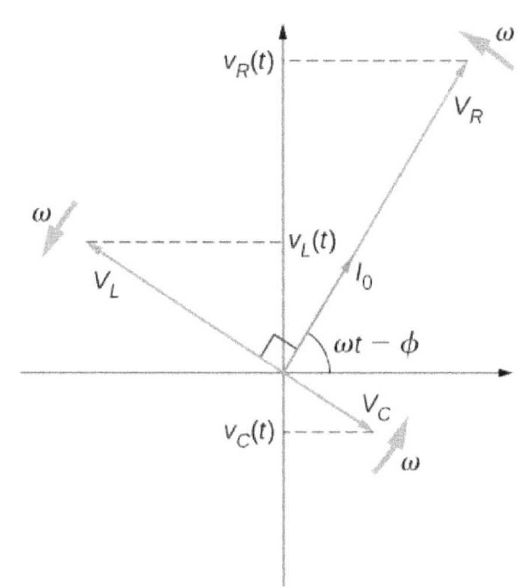

Figure 15.12 The phasor diagram for the *RLC* series circuit of Figure 15.11.

At any instant, the voltage across the *RLC* combination is $v_R(t) + v_L(t) + v_C(t) = v(t)$, the emf of the source. Since a component of a sum of vectors is the sum of the components of the individual vectors—for example, $(A + B)_y = A_y + B_y$ —the projection of the vector sum of phasors onto the vertical axis is the sum of the vertical projections of the individual phasors. Hence, if we add vectorially the phasors representing $v_R(t)$, $v_L(t)$, and $v_C(t)$ and then find the projection of the resultant onto the vertical axis, we obtain

$$v_R(t) + v_L(t) + v_C(t) = v(t) = V_0 \sin \omega t.$$

The vector sum of the phasors is shown in Figure 15.13. The resultant phasor has an amplitude V_0 and is directed at an angle ϕ with respect to the $v_R(t)$, or $i(t)$, phasor. The projection of this resultant phasor onto the vertical axis is $v(t) = V_0 \sin \omega t$. We can easily determine the unknown quantities I_0 and ϕ from the geometry of the phasor diagram. For the phase angle,

$$\phi = \tan^{-1} \frac{V_L - V_C}{V_R} = \tan^{-1} \frac{I_0 X_L - I_0 X_C}{I_0 R},$$

and after cancellation of I_0, this becomes

$$\phi = \tan^{-1} \frac{X_L - X_C}{R}. \quad (15.9)$$

Furthermore, from the Pythagorean theorem,

$$V_0 = \sqrt{V_R{}^2 + (V_L - V_C)^2} = \sqrt{(I_0 R)^2 + (I_0 X_L - I_0 X_C)^2} = I_0 \sqrt{R^2 + (X_L - X_C)^2}.$$

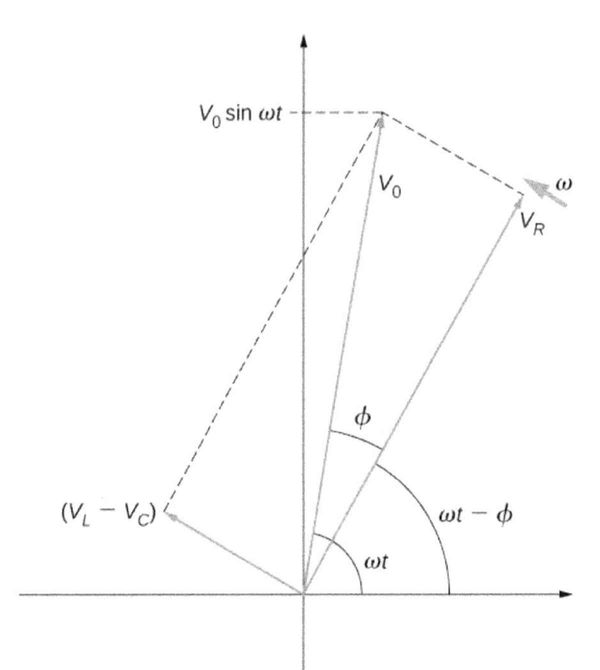

Figure 15.13 The resultant of the phasors for $v_L(t)$, $v_C(t)$, and $v_R(t)$ is equal to the phasor for $v(t) = V_0 \sin \omega t$. The $i(t)$ phasor (not shown) is aligned with the $v_R(t)$ phasor.

The current amplitude is therefore the ac version of Ohm's law:

$$I_0 = \frac{V_0}{\sqrt{R^2 + (X_L - X_C)^2}} = \frac{V_0}{Z}, \tag{15.10}$$

where

$$Z = \sqrt{R^2 + (X_L - X_C)^2} \tag{15.11}$$

is known as the **impedance** of the circuit. Its unit is the ohm, and it is the ac analog to resistance in a dc circuit, which measures the combined effect of resistance, capacitive reactance, and inductive reactance (Figure 15.14).

Figure 15.14 Power capacitors are used to balance the impedance of the effective inductance in transmission lines.

The *RLC* circuit is analogous to the wheel of a car driven over a corrugated road (Figure 15.15). The regularly spaced bumps in the road drive the wheel up and down; in the same way, a voltage source increases and decreases. The shock absorber acts like the resistance of the *RLC* circuit, damping and limiting the amplitude of the oscillation. Energy within the wheel system goes back and forth between kinetic and potential energy stored in the car spring, analogous to the shift between a maximum current, with energy stored in an inductor, and no current, with energy stored in the electric field of a capacitor. The amplitude of the wheel's motion is at a maximum if the bumps in the road are hit at the resonant frequency, which we describe in more detail in Resonance in an AC Circuit.

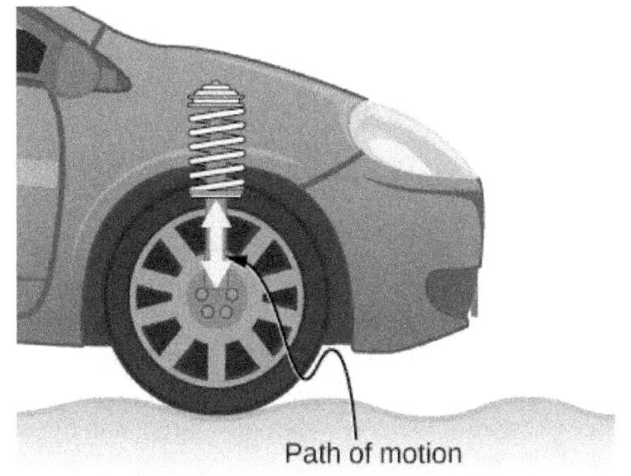

Figure 15.15 On a car, the shock absorber damps motion and dissipates energy. This is much like the resistance in an *RLC* circuit. The mass and spring determine the resonant frequency.

Problem-Solving Strategy: AC Circuits

To analyze an ac circuit containing resistors, capacitors, and inductors, it is helpful to think of each device's reactance and find the equivalent reactance using the rules we used for equivalent resistance in the past. Phasors are a great method to determine whether the emf of the circuit has positive or negative phase (namely, leads or lags other values). A mnemonic device of "ELI the ICE man" is sometimes used to remember that the emf (E) leads the current (I) in an inductor (L) and the current (I) leads the emf (E) in a capacitor (C).

Use the following steps to determine the emf of the circuit by phasors:

1. Draw the phasors for voltage across each device: resistor, capacitor, and inductor, including the phase angle in the circuit.
2. If there is both a capacitor and an inductor, find the net voltage from these two phasors, since they are antiparallel.
3. Find the equivalent phasor from the phasor in step 2 and the resistor's phasor using trigonometry or components of the phasors. The equivalent phasor found is the emf of the circuit.

Example 15.2

An *RLC* Series Circuit

The output of an ac generator connected to an *RLC* series combination has a frequency of 200 Hz and an amplitude of 0.100 V. If $R = 4.00\,\Omega$, $L = 3.00 \times 10^{-3}$ H, and $C = 8.00 \times 10^{-4}$ F, what are (a) the capacitive reactance, (b) the inductive reactance, (c) the impedance, (d) the current amplitude, and (e) the phase difference between the current and the emf of the generator?

Strategy

The reactances and impedance in (a)–(c) are found by substitutions into Equation 15.3, Equation 15.8, and Equation 15.11, respectively. The current amplitude is calculated from the peak voltage and the impedance. The phase difference between the current and the emf is calculated by the inverse tangent of the difference between the reactances divided by the resistance.

Solution

a. From Equation 15.3, the capacitive reactance is

$$X_C = \frac{1}{\omega C} = \frac{1}{2\pi(200\text{ Hz})\left(8.00 \times 10^{-4}\text{ F}\right)} = 0.995\,\Omega.$$

b. From Equation 15.8, the inductive reactance is

$$X_L = \omega L = 2\pi(200\text{ Hz})\left(3.00 \times 10^{-3}\text{ H}\right) = 3.77\,\Omega.$$

c. Substituting the values of R, X_C, and X_L into Equation 15.11, we obtain for the impedance

$$Z = \sqrt{(4.00\,\Omega)^2 + (3.77\,\Omega - 0.995\,\Omega)^2} = 4.87\,\Omega.$$

d. The current amplitude is

$$I_0 = \frac{V_0}{Z} = \frac{0.100\text{ V}}{4.87\,\Omega} = 2.05 \times 10^{-2}\text{ A}.$$

e. From Equation 15.9, the phase difference between the current and the emf is

$$\phi = \tan^{-1}\frac{X_L - X_C}{R} = \tan^{-1}\frac{2.77\,\Omega}{4.00\,\Omega} = 0.607\text{ rad}.$$

Significance

The phase angle is positive because the reactance of the inductor is larger than the reactance of the capacitor.

15.3 Check Your Understanding Find the voltages across the resistor, the capacitor, and the inductor in the circuit of Figure 15.11 using $v(t) = V_0 \sin \omega t$ as the output of the ac generator.

15.4 | Power in an AC Circuit

Learning Objectives

By the end of the section, you will be able to:

- Describe how average power from an ac circuit can be written in terms of peak current and voltage and of rms current and voltage
- Determine the relationship between the phase angle of the current and voltage and the average power, known as the power factor

A circuit element dissipates or produces power according to $P = IV$, where I is the current through the element and V is the voltage across it. Since the current and the voltage both depend on time in an ac circuit, the instantaneous power $p(t) = i(t)v(t)$ is also time dependent. A plot of $p(t)$ for various circuit elements is shown in Figure 15.16. For a resistor, $i(t)$ and $v(t)$ are in phase and therefore always have the same sign (see Figure 15.5). For a capacitor or inductor, the relative signs of $i(t)$ and $v(t)$ vary over a cycle due to their phase differences (see Figure 15.7 and Figure 15.9). Consequently, $p(t)$ is positive at some times and negative at others, indicating that capacitive and inductive elements produce power at some instants and absorb it at others.

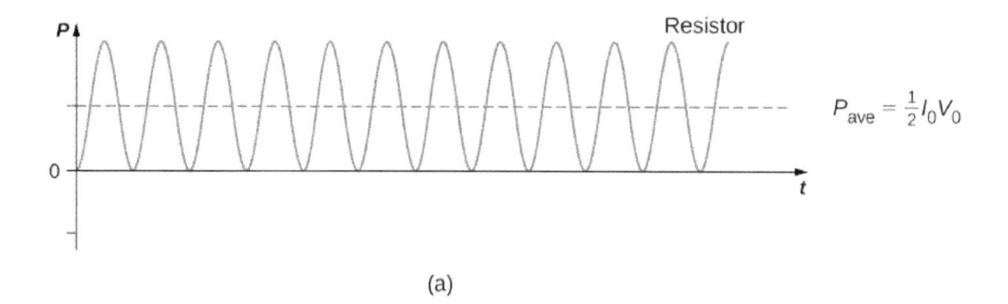

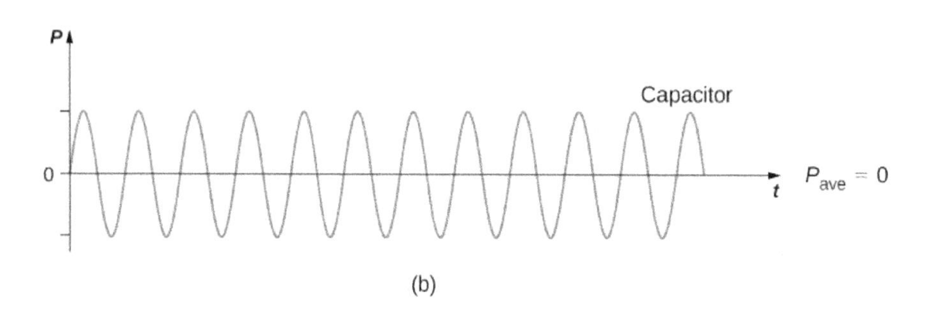

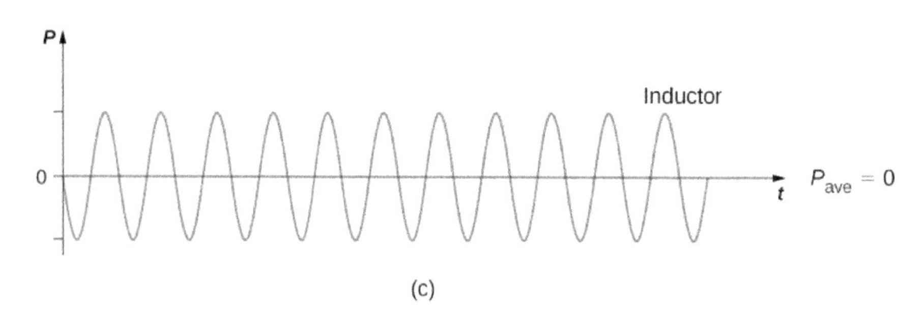

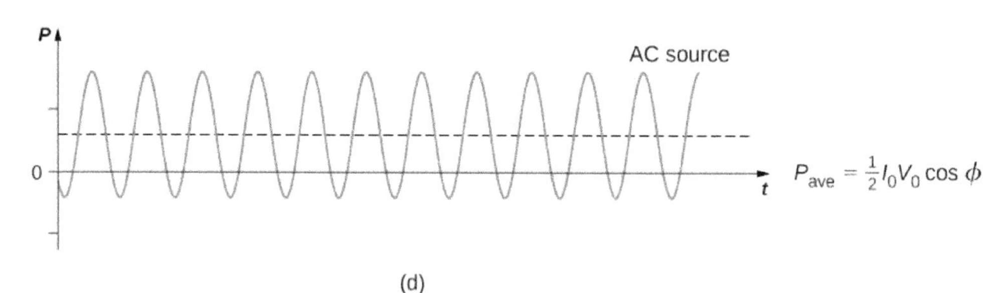

Figure 15.16 Graph of instantaneous power for various circuit elements. (a) For the resistor, $P_{ave} = I_0 V_0/2,$ whereas for (b) the capacitor and (c) the inductor, $P_{ave} = 0.$ (d) For the source, $P_{ave} = I_0 V_0(\cos \phi)/2,$ which may be positive, negative, or zero, depending on ϕ.

Because instantaneous power varies in both magnitude and sign over a cycle, it seldom has any practical importance. What we're almost always concerned with is the power averaged over time, which we refer to as the **average power**. It is defined by the time average of the instantaneous power over one cycle:

$$P_{ave} = \frac{1}{T}\int_0^T p(t)dt,$$

where $T = 2\pi/\omega$ is the period of the oscillations. With the substitutions $v(t) = V_0 \sin \omega t$ and $i(t) = I_0 \sin(\omega t - \phi)$, this integral becomes

$$P_{\text{ave}} = \frac{I_0 V_0}{T} \int_0^T \sin(\omega t - \phi) \sin \omega t \, dt.$$

Using the trigonometric relation $\sin(A - B) = \sin A \cos B - \sin B \cos A$, we obtain

$$P_{\text{ave}} = \frac{I_0 V_0 \cos \phi}{T} \int_0^T \sin^2 \omega t dt - \frac{I_0 V_0 \sin \phi}{T} \int_0^T \sin^2 \omega t \cos \omega t dt.$$

Evaluation of these two integrals yields

$$\frac{1}{T} \int_0^T \sin^2 \omega t dt = \frac{1}{2}$$

and

$$\frac{1}{T} \int_0^T \sin^2 \omega t \cos \omega t dt = 0.$$

Hence, the average power associated with a circuit element is given by

$$P_{\text{ave}} = \frac{1}{2} I_0 V_0 \cos \phi. \quad (15.12)$$

In engineering applications, $\cos \phi$ is known as the **power factor**, which is the amount by which the power delivered in the circuit is less than the theoretical maximum of the circuit due to voltage and current being out of phase. For a resistor, $\phi = 0$, so the average power dissipated is

$$P_{\text{ave}} = \frac{1}{2} I_0 V_0.$$

A comparison of $p(t)$ and P_{ave} is shown in Figure 15.16(d). To make $P_{\text{ave}} = (1/2) I_0 V_0$ look like its dc counterpart, we use the rms values I_{rms} and V_{rms} of the current and the voltage. By definition, these are

$$I_{\text{rms}} = \sqrt{i^2_{\text{ave}}} \text{ and } V_{\text{rms}} = \sqrt{v^2_{\text{ave}}},$$

where

$$i^2_{\text{ave}} = \frac{1}{T} \int_0^T i^2(t) dt \text{ and } v^2_{\text{ave}} = \frac{1}{T} \int_0^T v^2(t) dt.$$

With $i(t) = I_0 \sin(\omega t - \phi)$ and $v(t) = V_0 \sin \omega t$, we obtain

$$I_{\text{rms}} = \frac{1}{\sqrt{2}} I_0 \text{ and } V_{\text{rms}} = \frac{1}{\sqrt{2}} V_0.$$

We may then write for the average power dissipated by a resistor,

$$P_{\text{ave}} = \frac{1}{2} I_0 V_0 = I_{\text{rms}} V_{\text{rms}} = I^2_{\text{rms}} R. \quad (15.13)$$

This equation further emphasizes why the rms value is chosen in discussion rather than peak values. Both equations for average power are correct for Equation 15.13, but the rms values in the formula give a cleaner representation, so the extra factor of 1/2 is not necessary.

Alternating voltages and currents are usually described in terms of their rms values. For example, the 110 V from a household outlet is an rms value. The amplitude of this source is $110\sqrt{2}\text{ V} = 156\text{ V}$. Because most ac meters are calibrated in terms of rms values, a typical ac voltmeter placed across a household outlet will read 110 V.

For a capacitor and an inductor, $\phi = \pi/2$ and $-\pi/2$ rad, respectively. Since $\cos\pi/2 = \cos(-\pi/2) = 0$, we find from Equation 15.12 that the average power dissipated by either of these elements is $P_{\text{ave}} = 0$. Capacitors and inductors absorb energy from the circuit during one half-cycle and then discharge it back to the circuit during the other half-cycle. This behavior is illustrated in the plots of Figure 15.16, (b) and (c), which show *p*(t) oscillating sinusoidally about zero.

The phase angle for an ac generator may have any value. If $\cos\phi > 0$, the generator produces power; if $\cos\phi < 0$, it absorbs power. In terms of rms values, the average power of an ac generator is written as

$$P_{\text{ave}} = I_{\text{rms}} V_{\text{rms}} \cos\phi.$$

For the generator in an *RLC* circuit,

$$\tan\phi = \frac{X_L - X_C}{R}$$

and

$$\cos\phi = \frac{R}{\sqrt{R^2 + (X_L - X_C)^2}} = \frac{R}{Z}.$$

Hence the average power of the generator is

$$P_{\text{ave}} = I_{\text{rms}} V_{\text{rms}} \cos\phi = \frac{V_{\text{rms}}}{Z} V_{\text{rms}} \frac{R}{Z} = \frac{V_{\text{rms}}^2 R}{Z^2}. \quad (15.14)$$

This can also be written as

$$P_{\text{ave}} = I_{\text{rms}}^2 R,$$

which designates that the power produced by the generator is dissipated in the resistor. As we can see, Ohm's law for the rms ac is found by dividing the rms voltage by the impedance.

Example 15.3

Power Output of a Generator

An ac generator whose emf is given by

$$v(t) = (4.00\text{ V})\sin\left[\left(1.00 \times 10^4\text{ rad/s}\right)t\right]$$

is connected to an *RLC* circuit for which $L = 2.00 \times 10^{-3}\text{ H}$, $C = 4.00 \times 10^{-6}\text{ F}$, and $R = 5.00\,\Omega$. (a) What is the rms voltage across the generator? (b) What is the impedance of the circuit? (c) What is the average power output of the generator?

Strategy

The rms voltage is the amplitude of the voltage times $1/\sqrt{2}$. The impedance of the circuit involves the resistance and the reactances of the capacitor and the inductor. The average power is calculated by Equation 15.14, or more specifically, the last part of the equation, because we have the impedance of the circuit *Z*, the rms voltage V_{rms}, and the resistance *R*.

Solution

a. Since $V_0 = 4.00\text{ V}$, the rms voltage across the generator is

$$V_{\text{rms}} = \frac{1}{\sqrt{2}}(4.00\text{ V}) = 2.83\text{ V}.$$

b. The impedance of the circuit is

$$\begin{aligned} Z &= \sqrt{R^2 + (X_L - X_C)^2} \\ &= \left\{ (5.00\ \Omega)^2 + \left[(1.00 \times 10^4\ \text{rad/s})(2.00 \times 10^{-3}\ \text{H}) - \frac{1}{(1.00 \times 10^4\ \text{rad/s})(4.00 \times 10^{-6}\ \text{F})} \right]^2 \right\}^{1/2} \\ &= 7.07\ \Omega. \end{aligned}$$

c. From Equation 15.14, the average power transferred to the circuit is

$$P_{\text{ave}} = \frac{V_{\text{rms}}^2 R}{Z^2} = \frac{(2.83\ \text{V})^2 (5.00\ \Omega)}{(7.07\ \Omega)^2} = 0.801\ \text{W}.$$

Significance

If the resistance is much larger than the reactance of the capacitor or inductor, the average power is a dc circuit equation of $P = V^2/R$, where V replaces the rms voltage.

15.4 Check Your Understanding An ac voltmeter attached across the terminals of a 45-Hz ac generator reads 7.07 V. Write an expression for the emf of the generator.

15.5 Check Your Understanding Show that the rms voltages across a resistor, a capacitor, and an inductor in an ac circuit where the rms current is I_{rms} are given by $I_{\text{rms}}R$, $I_{\text{rms}}X_C$, and $I_{\text{rms}}X_L$, respectively. Determine these values for the components of the *RLC* circuit of Equation 15.12.

15.5 | Resonance in an AC Circuit

Learning Objectives

By the end of the section, you will be able to:

- Determine the peak ac resonant angular frequency for a RLC circuit
- Explain the width of the average power versus angular frequency curve and its significance using terms like bandwidth and quality factor

In the *RLC* series circuit of Figure 15.11, the current amplitude is, from Equation 15.10,

$$I_0 = \frac{V_0}{\sqrt{R^2 + (\omega L - 1/\omega C)^2}}. \tag{15.15}$$

If we can vary the frequency of the ac generator while keeping the amplitude of its output voltage constant, then the current changes accordingly. A plot of I_0 versus ω is shown in Figure 15.17.

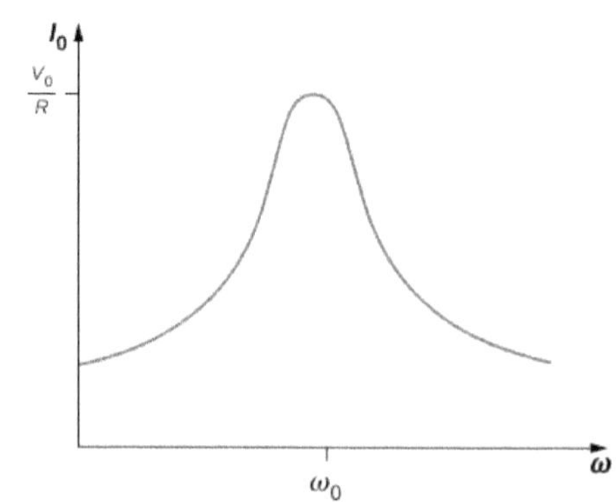

Figure 15.17 At an *RLC* circuit's resonant frequency, $\omega_0 = \sqrt{1/LC}$, the current amplitude is at its maximum value.

In Oscillations (http://cnx.org/content/m58360/latest/) , we encountered a similar graph where the amplitude of a damped harmonic oscillator was plotted against the angular frequency of a sinusoidal driving force (see m58366 (http://cnx.org/content/m58366/latest/#CNX_UPhysics_15_07_ForcDmpAmp)). This similarity is more than just a coincidence, as shown earlier by the application of Kirchhoff's loop rule to the circuit of Figure 15.11. This yields

$$L\frac{di}{dt} + iR + \frac{q}{C} = V_0 \sin \omega t, \tag{15.16}$$

or

$$L\frac{d^2q}{dt^2} + R\frac{dq}{dt} + \frac{1}{C}q = V_0 \sin \omega t,$$

where we substituted $dq(t)/dt$ for $i(t)$. A comparison of Equation 15.16 and, from Oscillations (http://cnx.org/content/m58360/latest/) , m58365 (http://cnx.org/content/m58365/latest/#fs-id1167131231570) for damped harmonic motion clearly demonstrates that the driven *RLC* series circuit is the electrical analog of the driven damped harmonic oscillator.

The **resonant frequency** f_0 of the *RLC* circuit is the frequency at which the amplitude of the current is a maximum and the circuit would oscillate if not driven by a voltage source. By inspection, this corresponds to the angular frequency $\omega_0 = 2\pi f_0$ at which the impedance *Z* in Equation 15.15 is a minimum, or when

$$\omega_0 L = \frac{1}{\omega_0 C}$$

and

$$\omega_0 = \sqrt{\frac{1}{LC}}. \tag{15.17}$$

This is the resonant angular frequency of the circuit. Substituting ω_0 into Equation 15.9, Equation 15.10, and Equation 15.11, we find that at resonance,

$$\phi = \tan^{-1}(0) = 0, \quad I_0 = V_0/R, \quad \text{and} \quad Z = R.$$

Therefore, at resonance, an *RLC* circuit is purely resistive, with the applied emf and current in phase.

What happens to the power at resonance? Equation 15.14 tells us how the average power transferred from an ac generator to the *RLC* combination varies with frequency. In addition, P_{ave} reaches a maximum when *Z*, which depends on the frequency, is a minimum, that is, when $X_L = X_C$ and $Z = R$. Thus, at resonance, the average power output of the source in an *RLC* series circuit is a maximum. From Equation 15.14, this maximum is V_{rms}^2/R.

Figure 15.18 is a typical plot of P_{ave} versus ω in the region of maximum power output. The **bandwidth** $\Delta\omega$ of the resonance peak is defined as the range of angular frequencies ω over which the average power P_{ave} is greater than one-half the maximum value of P_{ave}. The sharpness of the peak is described by a dimensionless quantity known as the **quality factor** Q of the circuit. By definition,

$$Q = \frac{\omega_0}{\Delta\omega}, \tag{15.18}$$

where ω_0 is the resonant angular frequency. A high Q indicates a sharp resonance peak. We can give Q in terms of the circuit parameters as

$$Q = \frac{\omega_0 L}{R}. \tag{15.19}$$

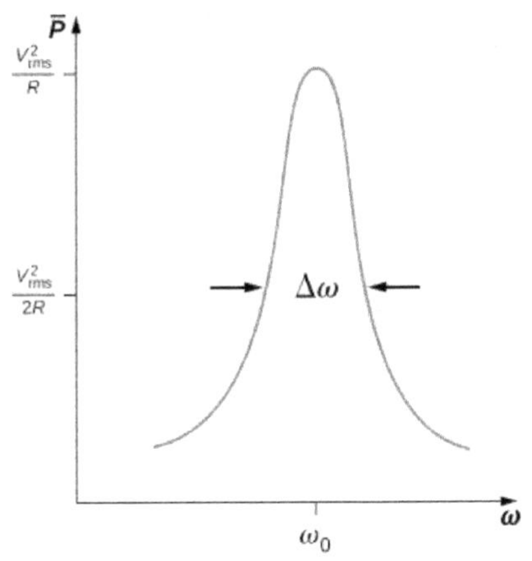

Figure 15.18 Like the current, the average power transferred from an ac generator to an *RLC* circuit peaks at the resonant frequency.

Resonant circuits are commonly used to pass or reject selected frequency ranges. This is done by adjusting the value of one of the elements and hence "tuning" the circuit to a particular resonant frequency. For example, in radios, the receiver is tuned to the desired station by adjusting the resonant frequency of its circuitry to match the frequency of the station. If the tuning circuit has a high Q, it will have a small bandwidth, so signals from other stations at frequencies even slightly different from the resonant frequency encounter a high impedance and are not passed by the circuit. Cell phones work in a similar fashion, communicating with signals of around 1 GHz that are tuned by an inductor-capacitor circuit. One of the most common applications of capacitors is their use in ac-timing circuits, based on attaining a resonant frequency. A metal detector also uses a shift in resonance frequency in detecting metals (Figure 15.19).

Figure 15.19 When a metal detector comes near a piece of metal, the self-inductance of one of its coils changes. This causes a shift in the resonant frequency of a circuit containing the coil. That shift is detected by the circuitry and transmitted to the diver by means of the headphones.

Example 15.4

Resonance in an *RLC* Series Circuit

(a) What is the resonant frequency of the circuit of Example 15.1? (b) If the ac generator is set to this frequency without changing the amplitude of the output voltage, what is the amplitude of the current?

Strategy

The resonant frequency for a *RLC* circuit is calculated from Equation 15.17, which comes from a balance between the reactances of the capacitor and the inductor. Since the circuit is at resonance, the impedance is equal to the resistor. Then, the peak current is calculated by the voltage divided by the resistance.

Solution

a. The resonant frequency is found from Equation 15.17:

$$f_0 = \frac{1}{2\pi}\sqrt{\frac{1}{LC}} = \frac{1}{2\pi}\sqrt{\frac{1}{(3.00\times10^{-3}\ \text{H})(8.00\times10^{-4}\ \text{F})}}$$
$$= 1.03\times10^{2}\ \text{Hz}.$$

b. At resonance, the impedance of the circuit is purely resistive, and the current amplitude is

$$I_0 = \frac{0.100\ \text{V}}{4.00\ \Omega} = 2.50\times10^{-2}\ \text{A}.$$

Significance

If the circuit were not set to the resonant frequency, we would need the impedance of the entire circuit to calculate the current.

Example 15.5

Power Transfer in an *RLC* Series Circuit at Resonance

(a) What is the resonant angular frequency of an *RLC* circuit with $R = 0.200\ \Omega$, $L = 4.00 \times 10^{-3}\ \text{H}$, and $C = 2.00 \times 10^{-6}\ \text{F}$? (b) If an ac source of constant amplitude 4.00 V is set to this frequency, what is the average power transferred to the circuit? (c) Determine Q and the bandwidth of this circuit.

Strategy

The resonant angular frequency is calculated from Equation 15.17. The average power is calculated from the rms voltage and the resistance in the circuit. The quality factor is calculated from Equation 15.19 and by knowing the resonant frequency. The bandwidth is calculated from Equation 15.18 and by knowing the quality factor.

Solution

a. The resonant angular frequency is

$$\omega_0 = \sqrt{\frac{1}{LC}} = \sqrt{\frac{1}{(4.00 \times 10^{-3}\ \text{H})(2.00 \times 10^{-6}\ \text{F})}} = 1.12 \times 10^4\ \text{rad/s}.$$

b. At this frequency, the average power transferred to the circuit is a maximum. It is

$$P_{\text{ave}} = \frac{V_{\text{rms}}^2}{R} = \frac{[(1/\sqrt{2})(4.00\ \text{V})]^2}{0.200\ \Omega} = 40.0\ \text{W}.$$

c. The quality factor of the circuit is

$$Q = \frac{\omega_0 L}{R} = \frac{(1.12 \times 10^4\ \text{rad/s})(4.00 \times 10^{-3}\ \text{H})}{0.200\ \Omega} = 224.$$

We then find for the bandwidth

$$\Delta\omega = \frac{\omega_0}{Q} = \frac{1.12 \times 10^4\ \text{rad/s}}{224} = 50.0\ \text{rad/s}.$$

Significance

If a narrower bandwidth is desired, a lower resistance or higher inductance would help. However, a lower resistance increases the power transferred to the circuit, which may not be desirable, depending on the maximum power that could possibly be transferred.

15.6 Check Your Understanding In the circuit of Figure 15.11, $L = 2.0 \times 10^{-3}\ \text{H}$, $C = 5.0 \times 10^{-4}\ \text{F}$, and $R = 40\ \Omega$. (a) What is the resonant frequency? (b) What is the impedance of the circuit at resonance? (c) If the voltage amplitude is 10 V, what is $i(t)$ at resonance? (d) The frequency of the AC generator is now changed to 200 Hz. Calculate the phase difference between the current and the emf of the generator.

15.7 Check Your Understanding What happens to the resonant frequency of an *RLC* series circuit when the following quantities are increased by a factor of 4: (a) the capacitance, (b) the self-inductance, and (c) the resistance?

15.8 Check Your Understanding The resonant angular frequency of an *RLC* series circuit is 4.0×10^2 rad/s. An ac source operating at this frequency transfers an average power of 2.0×10^{-2} W to the circuit. The resistance of the circuit is $0.50\ \Omega$. Write an expression for the emf of the source.

15.6 | Transformers

Learning Objectives

By the end of the section, you will be able to:

- Explain why power plants transmit electricity at high voltages and low currents and how they do this
- Develop relationships among current, voltage, and the number of windings in step-up and step-down transformers

Although ac electric power is produced at relatively low voltages, it is sent through transmission lines at very high voltages (as high as 500 kV). The same power can be transmitted at different voltages because power is the product $I_{rms} V_{rms}$. (For simplicity, we ignore the phase factor $\cos \phi$.) A particular power requirement can therefore be met with a low voltage and a high current or with a high voltage and a low current. The advantage of the high-voltage/low-current choice is that it results in lower $I_{rms}^2 R$ ohmic losses in the transmission lines, which can be significant in lines that are many kilometers long (Figure 15.20).

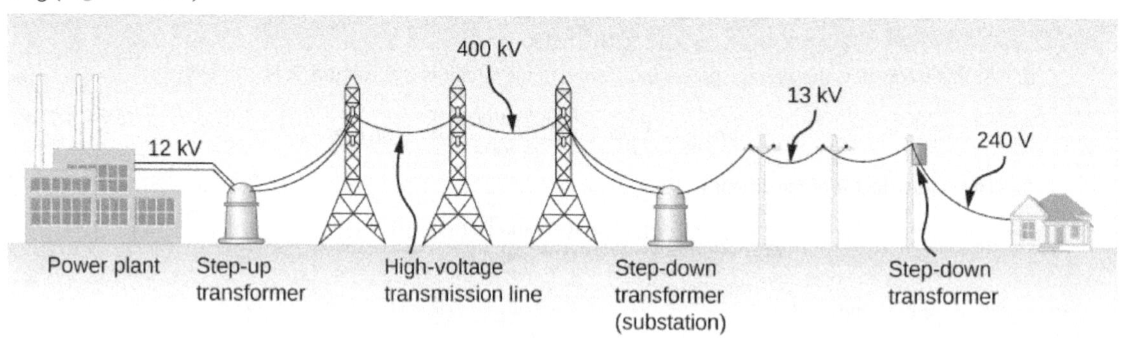

Figure 15.20 The rms voltage from a power plant eventually needs to be stepped down from 12 kV to 240 V so that it can be safely introduced into a home. A high-voltage transmission line allows a low current to be transmitted via a substation over long distances.

Typically, the alternating emfs produced at power plants are "stepped up" to very high voltages before being transmitted through power lines; then, they must be "stepped down" to relatively safe values (110 or 220 V rms) before they are introduced into homes. The device that transforms voltages from one value to another using induction is the **transformer** (Figure 15.21).

Figure 15.21 Transformers are used to step down the high voltages in transmission lines to the 110 to 220 V used in homes. (credit: modification of work by "Fortyseven"/Flickr)

As Figure 15.22 illustrates, a transformer basically consists of two separated coils, or windings, wrapped around a soft iron core. The primary winding has N_P loops, or turns, and is connected to an alternating voltage $v_P(t)$. The secondary winding has N_S turns and is connected to a load resistor R_S. We assume the ideal case for which all magnetic field lines are confined to the core so that the same magnetic flux permeates each turn of both the primary and the secondary windings. We also neglect energy losses to magnetic hysteresis, to ohmic heating in the windings, and to ohmic heating of the induced eddy currents in the core. A good transformer can have losses as low as 1% of the transmitted power, so this is not a bad assumption.

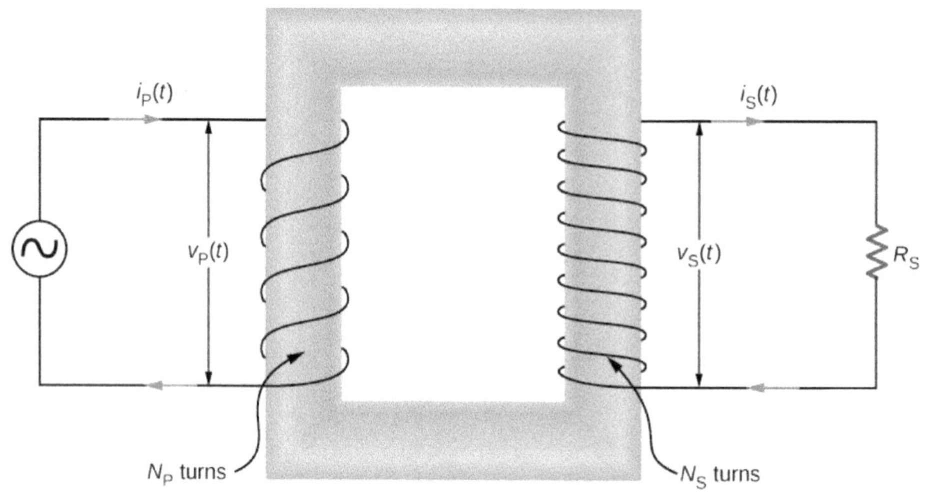

Figure 15.22 A step-up transformer (more turns in the secondary winding than in the primary winding). The two windings are wrapped around a soft iron core.

To analyze the transformer circuit, we first consider the primary winding. The input voltage $v_P(t)$ is equal to the potential difference induced across the primary winding. From Faraday's law, the induced potential difference is $-N_P(d\Phi/dt)$, where Φ is the flux through one turn of the primary winding. Thus,

$$v_P(t) = -N_P\frac{d\Phi}{dt}.$$

Similarly, the output voltage $v_S(t)$ delivered to the load resistor must equal the potential difference induced across the secondary winding. Since the transformer is ideal, the flux through every turn of the secondary winding is also Φ, and

$$v_S(t) = -N_S\frac{d\Phi}{dt}.$$

Combining the last two equations, we have

$$v_S(t) = \frac{N_S}{N_P}v_P(t). \tag{15.20}$$

Hence, with appropriate values for N_S and N_P, the input voltage $v_P(t)$ may be "stepped up" $(N_S > N_P)$ or "stepped down" $(N_S < N_P)$ to $v_S(t)$, the output voltage. This is often abbreviated as the **transformer equation**,

$$\frac{V_S}{V_P} = \frac{N_S}{N_P}, \tag{15.21}$$

which shows that the ratio of the secondary to primary voltages in a transformer equals the ratio of the number of turns in their windings. For a **step-up transformer**, which increases voltage and decreases current, this ratio is greater than one; for a **step-down transformer**, which decreases voltage and increases current, this ratio is less than one.

From the law of energy conservation, the power introduced at any instant by $v_P(t)$ to the primary winding must be equal to the power dissipated in the resistor of the secondary circuit; thus,

$$i_P(t)v_P(t) = i_S(t)v_S(t).$$

When combined with Equation 15.20, this gives

$$i_S(t) = \frac{N_P}{N_S} i_P(t). \tag{15.22}$$

If the voltage is stepped up, the current is stepped down, and vice versa.

Finally, we can use $i_S(t) = v_S(t)/R_S$, along with Equation 15.20 and Equation 15.22, to obtain

$$v_P(t) = i_P\left[\left(\frac{N_P}{N_S}\right)^2 R_S\right],$$

which tells us that the input voltage $v_P(t)$ "sees" not a resistance R_S but rather a resistance

$$R_P = \left(\frac{N_P}{N_S}\right)^2 R_S.$$

Our analysis has been based on instantaneous values of voltage and current. However, the resulting equations are not limited to instantaneous values; they hold also for maximum and rms values.

Example 15.6

A Step-Down Transformer

A transformer on a utility pole steps the rms voltage down from 12 kV to 240 V. (a) What is the ratio of the number of secondary turns to the number of primary turns? (b) If the input current to the transformer is 2.0 A, what is the output current? (c) Determine the power loss in the transmission line if the total resistance of the transmission line is $200\ \Omega$. (d) What would the power loss have been if the transmission line was at 240 V the entire length of the line, rather than providing voltage at 12 kV? What does this say about transmission lines?

Strategy

The number of turns related to the voltages is found from Equation 15.20. The output current is calculated using Equation 15.22.

Solution

a. Using Equation 15.20 with rms values V_P and V_S, we have

$$\frac{N_S}{N_P} = \frac{240\ \text{V}}{12 \times 10^3\ \text{V}} = \frac{1}{50},$$

so the primary winding has 50 times the number of turns in the secondary winding.

b. From Equation 15.22, the output rms current I_S is found using the transformer equation with current

$$I_S = \frac{N_P}{N_S} I_P \tag{15.23}$$

such that

$$I_S = \frac{N_P}{N_S} I_P = (50)(2.0\ \text{A}) = 100\ \text{A}.$$

c. The power loss in the transmission line is calculated to be

$$P_{\text{loss}} = I_P^2 R = (2.0\ \text{A})^2 (200\ \Omega) = 800\ \text{W}.$$

d. If there were no transformer, the power would have to be sent at 240 V to work for these houses, and the power loss would be

$$P_{\text{loss}} = I_S^2 R = (100 \text{ A})^2 (200 \,\Omega) = 2 \times 10^6 \text{ W}.$$

Therefore, when power needs to be transmitted, we want to avoid power loss. Thus, lines are sent with high voltages and low currents and adjusted with a transformer before power is sent into homes.

Significance

This application of a step-down transformer allows a home that uses 240-V outlets to have 100 A available to draw upon. This can power many devices in the home.

15.9 Check Your Understanding A transformer steps the line voltage down from 110 to 9.0 V so that a current of 0.50 A can be delivered to a doorbell. (a) What is the ratio of the number of turns in the primary and secondary windings? (b) What is the current in the primary winding? (c) What is the resistance seen by the 110-V source?

CHAPTER 15 REVIEW

KEY TERMS

ac current current that fluctuates sinusoidally with time at a fixed frequency

ac voltage voltage that fluctuates sinusoidally with time at a fixed frequency

alternating current (ac) flow of electric charge that periodically reverses direction

average power time average of the instantaneous power over one cycle

bandwidth range of angular frequencies over which the average power is greater than one-half the maximum value of the average power

capacitive reactance opposition of a capacitor to a change in current

direct current (dc) flow of electric charge in only one direction

impedance ac analog to resistance in a dc circuit, which measures the combined effect of resistance, capacitive reactance, and inductive reactance

inductive reactance opposition of an inductor to a change in current

phase angle amount by which the voltage and current are out of phase with each other in a circuit

power factor amount by which the power delivered in the circuit is less than the theoretical maximum of the circuit due to voltage and current being out of phase

quality factor dimensionless quantity that describes the sharpness of the peak of the bandwidth; a high quality factor is a sharp or narrow resonance peak

resonant frequency frequency at which the amplitude of the current is a maximum and the circuit would oscillate if not driven by a voltage source

rms current root mean square of the current

rms voltage root mean square of the voltage

step-down transformer transformer that decreases voltage and increases current

step-up transformer transformer that increases voltage and decreases current

transformer device that transforms voltages from one value to another using induction

transformer equation equation showing that the ratio of the secondary to primary voltages in a transformer equals the ratio of the number of turns in their windings

KEY EQUATIONS

AC voltage	$v = V_0 \sin \omega t$
AC current	$i = I_0 \sin \omega t$
capacitive reactance	$\frac{V_0}{I_0} = \frac{1}{\omega C} = X_C$
rms voltage	$V_{rms} = \frac{V_0}{\sqrt{2}}$
rms current	$I_{rms} = \frac{I_0}{\sqrt{2}}$
inductive reactance	$\frac{V_0}{I_0} = \omega L = X_L$

Phase angle of an ac circuit	$\phi = \tan^{-1}\frac{X_L - X_C}{R}$
AC version of Ohm's law	$I_0 = \frac{V_0}{Z}$
Impedance of an ac circuit	$Z = \sqrt{R^2 + (X_L - X_C)^2}$
Average power associated with a circuit element	$P_{\text{ave}} = \frac{1}{2}I_0 V_0 \cos\phi$
Average power dissipated by a resistor	$P_{\text{ave}} = \frac{1}{2}I_0 V_0 = I_{\text{rms}} V_{\text{rms}} = I_{\text{rms}}^2 R$
Resonant angular frequency of a circuit	$\omega_0 = \sqrt{\frac{1}{LC}}$
Quality factor of a circuit	$Q = \frac{\omega_0}{\Delta\omega}$
Quality factor of a circuit in terms of the circuit parameters	$Q = \frac{\omega_0 L}{R}$
Transformer equation with voltage	$\frac{V_{\text{S}}}{V_{\text{P}}} = \frac{N_{\text{S}}}{N_{\text{P}}}$
Transformer equation with current	$I_{\text{S}} = \frac{N_{\text{P}}}{N_{\text{S}}} I_{\text{P}}$

SUMMARY

15.1 AC Sources

- Direct current (dc) refers to systems in which the source voltage is constant.
- Alternating current (ac) refers to systems in which the source voltage varies periodically, particularly sinusoidally.
- The voltage source of an ac system puts out a voltage that is calculated from the time, the peak voltage, and the angular frequency.
- In a simple circuit, the current is found by dividing the voltage by the resistance. An ac current is calculated using the peak current (determined by dividing the peak voltage by the resistance), the angular frequency, and the time.

15.2 Simple AC Circuits

- For resistors, the current through and the voltage across are in phase.
- For capacitors, we find that when a sinusoidal voltage is applied to a capacitor, the voltage follows the current by one-fourth of a cycle. Since a capacitor can stop current when fully charged, it limits current and offers another form of ac resistance, called capacitive reactance, which has units of ohms.
- For inductors in ac circuits, we find that when a sinusoidal voltage is applied to an inductor, the voltage leads the current by one-fourth of a cycle.
- The opposition of an inductor to a change in current is expressed as a type of ac reactance. This inductive reactance, which has units of ohms, varies with the frequency of the ac source.

15.3 RLC Series Circuits with AC

- An *RLC* series circuit is a resistor, capacitor, and inductor series combination across an ac source.
- The same current flows through each element of an *RLC* series circuit at all points in time.
- The counterpart of resistance in a dc circuit is impedance, which measures the combined effect of resistors, capacitors, and inductors. The maximum current is defined by the ac version of Ohm's law.

- Impedance has units of ohms and is found using the resistance, the capacitive reactance, and the inductive reactance.

15.4 Power in an AC Circuit

- The average ac power is found by multiplying the rms values of current and voltage.
- Ohm's law for the rms ac is found by dividing the rms voltage by the impedance.
- In an ac circuit, there is a phase angle between the source voltage and the current, which can be found by dividing the resistance by the impedance.
- The average power delivered to an *RLC* circuit is affected by the phase angle.
- The power factor ranges from –1 to 1.

15.5 Resonance in an AC Circuit

- At the resonant frequency, inductive reactance equals capacitive reactance.
- The average power versus angular frequency plot for a *RLC* circuit has a peak located at the resonant frequency; the sharpness or width of the peak is known as the bandwidth.
- The bandwidth is related to a dimensionless quantity called the quality factor. A high quality factor value is a sharp or narrow peak.

15.6 Transformers

- Power plants transmit high voltages at low currents to achieve lower ohmic losses in their many kilometers of transmission lines.
- Transformers use induction to transform voltages from one value to another.
- For a transformer, the voltages across the primary and secondary coils, or windings, are related by the transformer equation.
- The currents in the primary and secondary windings are related by the number of primary and secondary loops, or turns, in the windings of the transformer.
- A step-up transformer increases voltage and decreases current, whereas a step-down transformer decreases voltage and increases current.

CONCEPTUAL QUESTIONS

15.1 AC Sources

1. What is the relationship between frequency and angular frequency?

15.2 Simple AC Circuits

2. Explain why at high frequencies a capacitor acts as an ac short, whereas an inductor acts as an open circuit.

15.3 RLC Series Circuits with AC

3. In an *RLC* series circuit, can the voltage measured across the capacitor be greater than the voltage of the source? Answer the same question for the voltage across the inductor.

15.4 Power in an AC Circuit

4. For what value of the phase angle ϕ between the voltage output of an ac source and the current is the average power output of the source a maximum?

5. Discuss the differences between average power and instantaneous power.

6. The average ac current delivered to a circuit is zero. Despite this, power is dissipated in the circuit. Explain.

7. Can the instantaneous power output of an ac source ever be negative? Can the average power output be negative?

8. The power rating of a resistor used in ac circuits refers to the maximum average power dissipated in the resistor. How does this compare with the maximum instantaneous power dissipated in the resistor?

15.6 Transformers

9. Why do transmission lines operate at very high voltages while household circuits operate at fairly small voltages?

10. How can you distinguish the primary winding from the secondary winding in a step-up transformer?

11. Battery packs in some electronic devices are charged using an adapter connected to a wall socket. Speculate as to the purpose of the adapter.

12. Will a transformer work if the input is a dc voltage?

13. Why are the primary and secondary coils of a transformer wrapped around the same closed loop of iron?

PROBLEMS

15.1 AC Sources

14. Write an expression for the output voltage of an ac source that has an amplitude of 12 V and a frequency of 200 Hz.

15.2 Simple AC Circuits

15. Calculate the reactance of a $5.0\text{-}\mu\text{F}$ capacitor at (a) 60 Hz, (b) 600 Hz, and (c) 6000 Hz.

16. What is the capacitance of a capacitor whose reactance is $10\,\Omega$ at 60 Hz?

17. Calculate the reactance of a 5.0-mH inductor at (a) 60 Hz, (b) 600 Hz, and (c) 6000 Hz.

18. What is the self-inductance of a coil whose reactance is $10\,\Omega$ at 60 Hz?

19. At what frequency is the reactance of a $20\text{-}\mu\text{F}$ capacitor equal to that of a 10-mH inductor?

20. At 1000 Hz, the reactance of a 5.0-mH inductor is equal to the reactance of a particular capacitor. What is the capacitance of the capacitor?

21. A $50\text{-}\Omega$ resistor is connected across the emf $v(t) = (160\text{ V})\sin(120\pi t)$. Write an expression for the current through the resistor.

22. A $25\text{-}\mu\text{F}$ capacitor is connected to an emf given by $v(t) = (160\text{ V})\sin(120\pi t)$. (a) What is the reactance of the capacitor? (b) Write an expression for the current output of the source.

23. A 100-mH inductor is connected across the emf of the preceding problem. (a) What is the reactance of the inductor? (b) Write an expression for the current through the inductor.

15.3 RLC Series Circuits with AC

24. What is the impedance of a series combination of a $50\text{-}\Omega$ resistor, a $5.0\text{-}\mu\text{F}$ capacitor, and a $10\text{-}\mu\text{F}$ capacitor at a frequency of 2.0 kHz?

25. A resistor and capacitor are connected in series across an ac generator. The emf of the generator is given by $v(t) = V_0 \cos \omega t$, where $V_0 = 120\text{ V}$, $\omega = 120\pi\text{ rad/s}$, $R = 400\,\Omega$, and $C = 4.0\mu\text{F}$. (a) What is the impedance of the circuit? (b) What is the amplitude of the current through the resistor? (c) Write an expression for the current through the resistor. (d) Write expressions representing the voltages across the resistor and across the capacitor.

26. A resistor and inductor are connected in series across an ac generator. The emf of the generator is given by $v(t) = V_0 \cos \omega t$, where $V_0 = 120\text{ V}$ and $\omega = 120\pi\text{ rad/s}$; also, $R = 400\,\Omega$ and $L = 1.5\text{ H}$. (a) What is the impedance of the circuit? (b) What is the amplitude of the current through the resistor? (c) Write an expression for the current through the resistor. (d) Write expressions representing the voltages across the resistor and across the inductor.

27. In an *RLC* series circuit, the voltage amplitude and frequency of the source are 100 V and 500 Hz, respectively, an $R = 500\,\Omega$, $L = 0.20\text{ H}$, and $C = 2.0\mu\text{F}$. (a) What is the impedance of the circuit? (b) What is the amplitude of the current from the source? (c) If the emf of the source is given by $v(t) = (100\text{ V}) \sin 1000\pi t$, how does the current vary with time? (d) Repeat the calculations with *C* changed to $0.20\mu\text{F}$.

28. An *RLC* series circuit with $R = 600\,\Omega$, $L = 30\text{ mH}$, and $C = 0.050\mu\text{F}$ is driven by an ac source whose frequency and voltage amplitude are 500 Hz and 50 V, respectively. (a) What is the impedance of the circuit? (b) What is the amplitude of the current in the circuit? (c) What is the phase angle between the emf of the source and the current?

29. For the circuit shown below, what are (a) the total impedance and (b) the phase angle between the current and the emf? (c) Write an expression for $i(t)$.

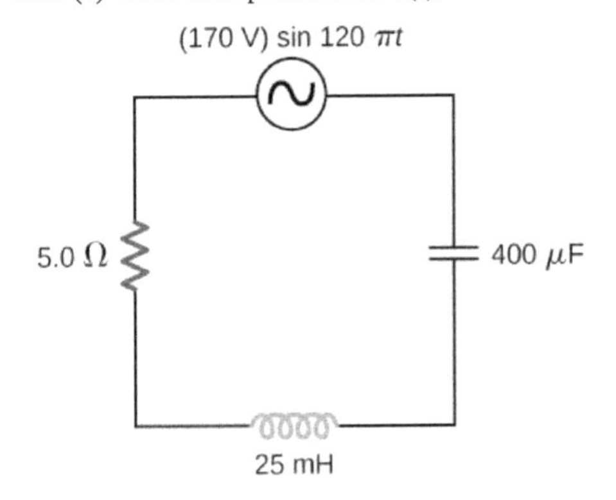

15.4 Power in an AC Circuit

30. The emf of an ac source is given by $v(t) = V_0 \sin \omega t$, where $V_0 = 100\text{ V}$ and $\omega = 200\pi\text{ rad/s}$. Calculate the average power output of the source if it is connected across (a) a $20\text{-}\mu\text{F}$ capacitor, (b) a 20-mH inductor, and (c) a $50\text{-}\Omega$ resistor.

31. Calculate the rms currents for an ac source is given by $v(t) = V_0 \sin \omega t$, where $V_0 = 100\text{ V}$ and $\omega = 200\pi\text{ rad/s}$ when connected across (a) a $20\text{-}\mu\text{F}$ capacitor, (b) a 20-mH inductor, and (c) a $50\text{-}\Omega$ resistor.

32. A 40-mH inductor is connected to a 60-Hz AC source whose voltage amplitude is 50 V. If an AC voltmeter is placed across the inductor, what does it read?

33. For an *RLC* series circuit, the voltage amplitude and frequency of the source are 100 V and 500 Hz, respectively; $R = 500\,\Omega$; and $L = 0.20\text{ H}$. Find the average power dissipated in the resistor for the following values for the capacitance: (a) $C = 2.0\mu\text{F}$ and (b) $C = 0.20\,\mu\text{F}$.

34. An ac source of voltage amplitude 10 V delivers electric energy at a rate of 0.80 W when its current output is 2.5 A. What is the phase angle ϕ between the emf and the current?

35. An *RLC* series circuit has an impedance of $60\,\Omega$ and a power factor of 0.50, with the voltage lagging the current. (a) Should a capacitor or an inductor be placed in series with the elements to raise the power factor of the circuit? (b) What is the value of the capacitance or self-inductance that will raise the power factor to unity?

15.5 Resonance in an AC Circuit

36. (a) Calculate the resonant angular frequency of an *RLC* series circuit for which $R = 20\,\Omega$, $L = 75\text{ mH}$, and $C = 4.0\mu\text{F}$. (b) If *R* is changed to $300\,\Omega$, what happens to the resonant angular frequency?

37. The resonant frequency of an *RLC* series circuit is $2.0 \times 10^3\text{ Hz}$. If the self-inductance in the circuit is 5.0 mH, what is the capacitance in the circuit?

38. (a) What is the resonant frequency of an *RLC* series circuit with $R = 20\,\Omega$, $L = 2.0\text{ mH}$, and $C = 4.0\mu\text{F}$? (b) What is the impedance of the circuit at resonance?

39. For an *RLC* series circuit, $R = 100\,\Omega$, $L = 150\,\text{mH}$, and $C = 0.25\mu\text{F}$. (a) If an ac source of variable frequency is connected to the circuit, at what frequency is maximum power dissipated in the resistor? (b) What is the quality factor of the circuit?

40. An ac source of voltage amplitude 100 V and variable frequency *f* drives an *RLC* series circuit with $R = 10\,\Omega$, $L = 2.0\,\text{mH}$, and $C = 25\mu\text{F}$. (a) Plot the current through the resistor as a function of the frequency *f*. (b) Use the plot to determine the resonant frequency of the circuit.

41. (a) What is the resonant frequency of a resistor, capacitor, and inductor connected in series if $R = 100\,\Omega$, $L = 2.0\,\text{H}$, and $C = 5.0\mu\text{F}$? (b) If this combination is connected to a 100-V source operating at the constant frequency, what is the power output of the source? (c) What is the *Q* of the circuit? (d) What is the bandwidth of the circuit?

42. Suppose a coil has a self-inductance of 20.0 H and a resistance of $200\,\Omega$. What (a) capacitance and (b) resistance must be connected in series with the coil to produce a circuit that has a resonant frequency of 100 Hz and a *Q* of 10?

43. An ac generator is connected to a device whose internal circuits are not known. We only know current and voltage outside the device, as shown below. Based on the information given, what can you infer about the electrical nature of the device and its power usage?

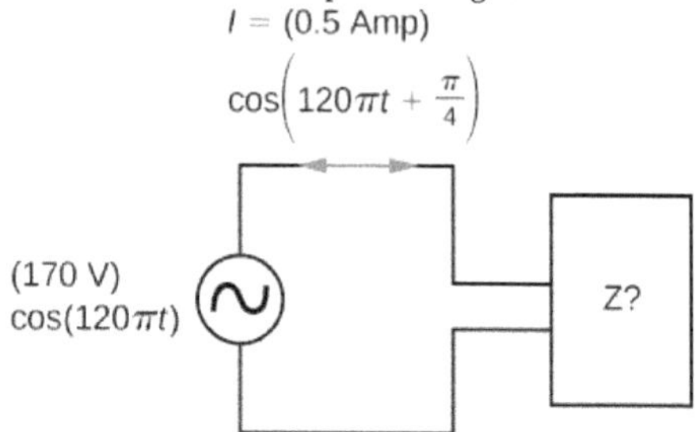

15.6 Transformers

44. A step-up transformer is designed so that the output of its secondary winding is 2000 V (rms) when the primary winding is connected to a 110-V (rms) line voltage. (a) If there are 100 turns in the primary winding, how many turns are there in the secondary winding? (b) If a resistor connected across the secondary winding draws an rms current of 0.75 A, what is the current in the primary winding?

45. A step-up transformer connected to a 110-V line is used to supply a hydrogen-gas discharge tube with 5.0 kV (rms). The tube dissipates 75 W of power. (a) What is the ratio of the number of turns in the secondary winding to the number of turns in the primary winding? (b) What are the rms currents in the primary and secondary windings? (c) What is the effective resistance seen by the 110-V source?

46. An ac source of emf delivers 5.0 mW of power at an rms current of 2.0 mA when it is connected to the primary coil of a transformer. The rms voltage across the secondary coil is 20 V. (a) What are the voltage across the primary coil and the current through the secondary coil? (b) What is the ratio of secondary to primary turns for the transformer?

47. A transformer is used to step down 110 V from a wall socket to 9.0 V for a radio. (a) If the primary winding has 500 turns, how many turns does the secondary winding have? (b) If the radio operates at a current of 500 mA, what is the current through the primary winding?

48. A transformer is used to supply a 12-V model train with power from a 110-V wall plug. The train operates at 50 W of power. (a) What is the rms current in the secondary coil of the transformer? (b) What is the rms current in the primary coil? (c) What is the ratio of the number of primary to secondary turns? (d) What is the resistance of the train? (e) What is the resistance seen by the 110-V source?

ADDITIONAL PROBLEMS

49. The emf of an dc source is given by $v(t) = V_0 \sin \omega t$, where $V_0 = 100\,\text{V}$ and $\omega = 200\pi$ rad/s. Find an expression that represents the output current of the source if it is connected across (a) a 20-μF capacitor, (b) a 20-mH inductor, and (c) a 50-Ω resistor.

50. A 700-pF capacitor is connected across an ac source with a voltage amplitude of 160 V and a frequency of 20 kHz. (a) Determine the capacitive reactance of the capacitor and the amplitude of the output current of the source. (b) If the frequency is changed to 60 Hz while keeping the voltage amplitude at 160 V, what are the capacitive reactance and the current amplitude?

51. A 20-mH inductor is connected across an AC source with a variable frequency and a constant-voltage amplitude of 9.0 V. (a) Determine the reactance of the circuit and the maximum current through the inductor when the frequency is set at 20 kHz. (b) Do the same calculations for a frequency of 60 Hz.

52. A 30-μF capacitor is connected across a 60-Hz ac source whose voltage amplitude is 50 V. (a) What is the maximum charge on the capacitor? (b) What is the maximum current into the capacitor? (c) What is the phase relationship between the capacitor charge and the current in the circuit?

53. A 7.0-mH inductor is connected across a 60-Hz ac source whose voltage amplitude is 50 V. (a) What is the maximum current through the inductor? (b) What is the phase relationship between the current through and the potential difference across the inductor?

54. What is the impedance of an *RLC* series circuit at the resonant frequency?

55. What is the resistance *R* in the circuit shown below if the amplitude of the ac through the inductor is 4.24 A?

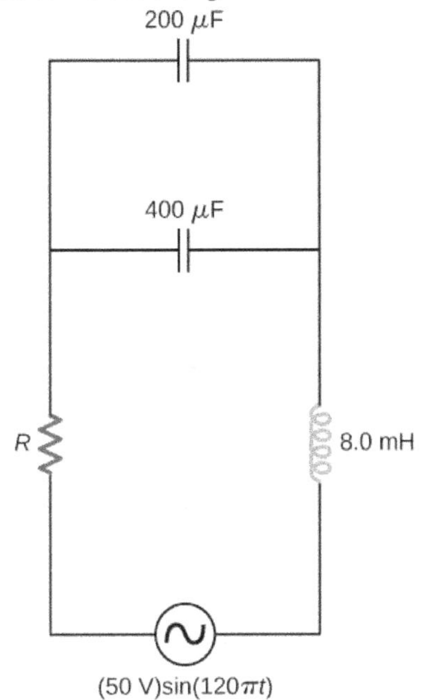

56. An ac source of voltage amplitude 100 V and frequency 1.0 kHz drives an *RLC* series circuit with $R = 20\,\Omega$, $L = 4.0\,\text{mH}$, and $C = 50\mu\text{F}$. (a) Determine the rms current through the circuit. (b) What are the rms voltages across the three elements? (c) What is the phase angle between the emf and the current? (d) What is the power output of the source? (e) What is the power dissipated in the resistor?

57. In an *RLC* series circuit, $R = 200\,\Omega$, $L = 1.0\,\text{H}$, $C = 50\mu\text{F}$, $V_0 = 120\,\text{V}$, and $f = 50\,\text{Hz}$. What is the power output of the source?

58. A power plant generator produces 100 A at 15 kV (rms). A transformer is used to step up the transmission line voltage to 150 kV (rms). (a) What is rms current in the transmission line? (b) If the resistance per unit length of the line is 8.6×10^{-8} Ω/m, what is the power loss per meter in the line? (c) What would the power loss per meter be if the line voltage were 15 kV (rms)?

59. Consider a power plant located 25 km outside a town delivering 50 MW of power to the town. The transmission lines are made of aluminum cables with a $7\,\text{cm}^2$ cross-sectional area. Find the loss of power in the transmission lines if it is transmitted at (a) 200 kV (rms) and (b) 120 V (rms).

60. Neon signs require 12-kV for their operation. A transformer is to be used to change the voltage from 220-V (rms) ac to 12-kV (rms) ac. What must the ratio be of turns in the secondary winding to the turns in the primary winding? (b) What is the maximum rms current the neon lamps can draw if the fuse in the primary winding goes off at 0.5 A? (c) How much power is used by the neon sign when it is drawing the maximum current allowed by the fuse in the primary winding?

CHALLENGE PROBLEMS

61. The 335-kV ac electricity from a power transmission line is fed into the primary winding of a transformer. The ratio of the number of turns in the secondary winding to the number in the primary winding is $N_s / N_p = 1000$.

(a) What voltage is induced in the secondary winding? (b) What is unreasonable about this result? (c) Which assumption or premise is responsible?

62. A 1.5-kΩ resistor and 30-mH inductor are connected in series, as shown below, across a 120-V (rms) ac power source oscillating at 60-Hz frequency. (a) Find the current in the circuit. (b) Find the voltage drops across the resistor and inductor. (c) Find the impedance of the circuit. (d) Find the power dissipated in the resistor. (e) Find the power dissipated in the inductor. (f) Find the power produced by the source.

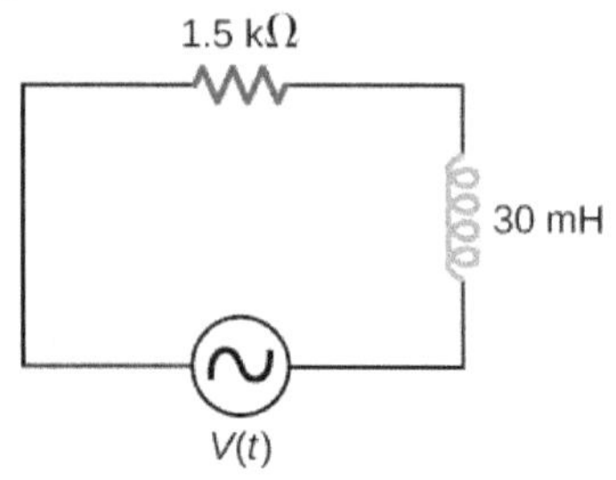

63. A 20-Ω resistor, 50-μF capacitor, and 30-mH inductor are connected in series with an ac source of amplitude 10 V and frequency 125 Hz. (a) What is the impedance of the circuit? (b) What is the amplitude of the current in the circuit? (c) What is the phase constant of the current? Is it leading or lagging the source voltage? (d) Write voltage drops across the resistor, capacitor, and inductor and the source voltage as a function of time. (e) What is the power factor of the circuit? (f) How much energy is used by the resistor in 2.5 s?

64. A 200-Ω resistor, 150-μF capacitor, and 2.5-H inductor are connected in series with an ac source of amplitude 10 V and variable angular frequency ω. (a) What is the value of the resonance frequency ω_R? (b) What is the amplitude of the current if $\omega = \omega_R$? (c) What is the phase constant of the current when $\omega = \omega_R$? Is it leading or lagging the source voltage, or is it in phase? (d) Write an equation for the voltage drop across the resistor as a function of time when $\omega = \omega_R$. (e) What is the power factor of the circuit when $\omega = \omega_R$? (f) How much energy is used up by the resistor in 2.5 s when $\omega = \omega_R$?

65. Find the reactances of the following capacitors and inductors in ac circuits with the given frequencies in each case: (a) 2-mH inductor with a frequency 60-Hz of the ac circuit; (b) 2-mH inductor with a frequency 600-Hz of the ac circuit; (c) 20-mH inductor with a frequency 6-Hz of the ac circuit; (d) 20-mH inductor with a frequency 60-Hz of the ac circuit; (e) 2-mF capacitor with a frequency 60-Hz of the ac circuit; and (f) 2-mF capacitor with a frequency 600-Hz of the AC circuit.

66. An output impedance of an audio amplifier has an impedance of 500 Ω and has a mismatch with a low-impedance 8-Ω loudspeaker. You are asked to insert an appropriate transformer to match the impedances. What turns ratio will you use, and why? Use the simplified circuit shown below.

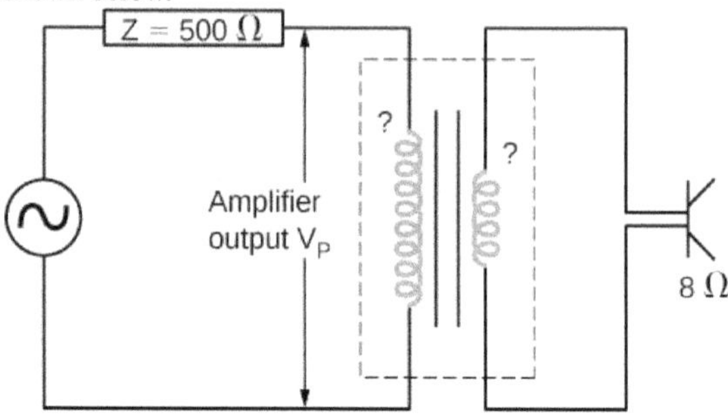

67. Show that the SI unit for capacitive reactance is the ohm. Show that the SI unit for inductive reactance is also the ohm.

68. A coil with a self-inductance of 16 mH and a resistance of 6.0 Ω is connected to an ac source whose frequency can be varied. At what frequency will the voltage across the coil lead the current through the coil by $45°$?

69. An *RLC* series circuit consists of a 50-Ω resistor, a 200-μF capacitor, and a 120-mH inductor whose coil has a resistance of 20 Ω. The source for the circuit has an rms emf of 240 V at a frequency of 60 Hz. Calculate the rms voltages across the (a) resistor, (b) capacitor, and (c) inductor.

70. An *RLC* series circuit consists of a 10-Ω resistor, an 8.0-μF capacitor, and a 50-mH inductor. A 110-V (rms) source of variable frequency is connected across the combination. What is the power output of the source when its frequency is set to one-half the resonant frequency of the circuit?

71. Shown below are two circuits that act as crude high-pass filters. The input voltage to the circuits is v_{in}, and the output voltage is v_{out}. (a) Show that for the capacitor circuit,

$$\frac{v_{out}}{v_{in}} = \frac{1}{\sqrt{1 + 1/\omega^2 R^2 C^2}},$$

and for the inductor circuit,

$$\frac{v_{out}}{v_{in}} = \frac{\omega L}{\sqrt{R^2 + \omega^2 L^2}}.$$

(b) Show that for high frequencies, $v_{out} \approx v_{in}$, but for low frequencies, $v_{out} \approx 0$.

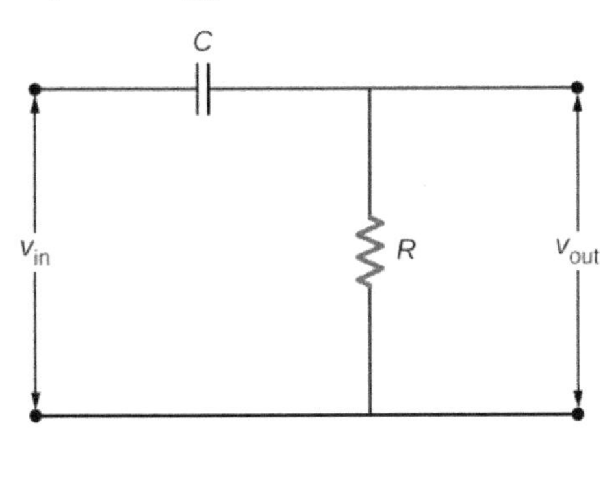

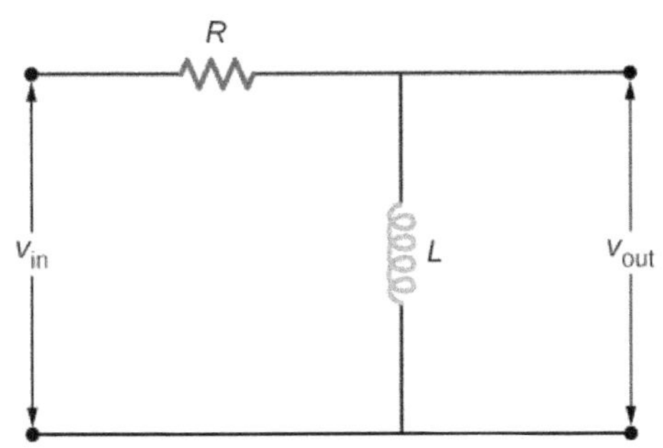

72. The two circuits shown below act as crude low-pass filters. The input voltage to the circuits is v_{in}, and the output voltage is v_{out}. (a) Show that for the capacitor circuit,

$$\frac{v_{out}}{v_{in}} = \frac{1}{\sqrt{1 + \omega^2 R^2 C^2}},$$

and for the inductor circuit,

$$\frac{v_{out}}{v_{in}} = \frac{R}{\sqrt{R^2 + \omega^2 L^2}}.$$

(b) Show that for low frequencies, $v_{out} \approx v_{in}$, but for high frequencies, $v_{out} \approx 0$.

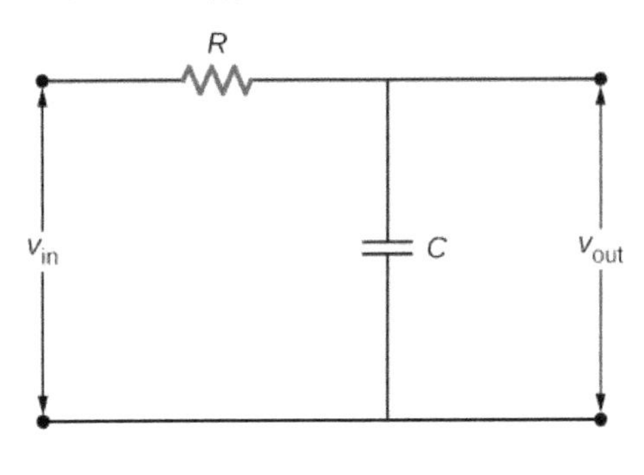

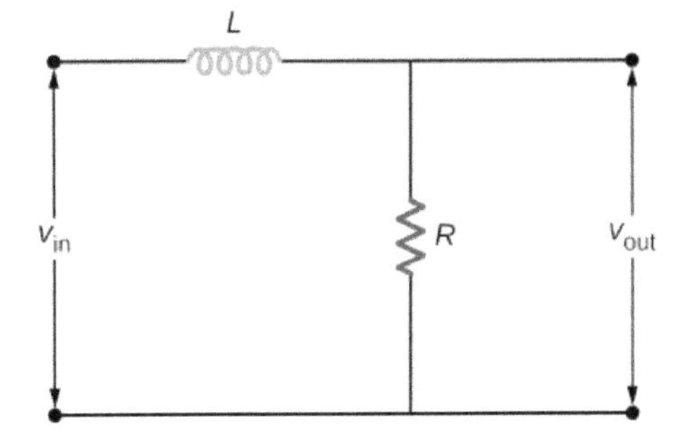

16 | ELECTROMAGNETIC WAVES

Figure 16.1 The pressure from sunlight predicted by Maxwell's equations helped produce the tail of Comet McNaught. (credit: modification of work by Sebastian Deiries—ESO)

Chapter Outline

Introduction

Our view of objects in the sky at night, the warm radiance of sunshine, the sting of sunburn, our cell phone conversations, and the X-rays revealing a broken bone—all are brought to us by electromagnetic waves. It would be hard to overstate the practical importance of electromagnetic waves, through their role in vision, through countless technological applications, and through their ability to transport the energy from the Sun through space to sustain life and almost all of its activities on Earth.

Theory predicted the general phenomenon of electromagnetic waves before anyone realized that light is a form of an electromagnetic wave. In the mid-nineteenth century, James Clerk Maxwell formulated a single theory combining all the electric and magnetic effects known at that time. Maxwell's equations, summarizing this theory, predicted the existence of electromagnetic waves that travel at the speed of light. His theory also predicted how these waves behave, and how they carry both energy and momentum. The tails of comets, such as Comet McNaught in Figure 16.1, provide a spectacular example. Energy carried by light from the Sun warms the comet to release dust and gas. The momentum carried by the light exerts a weak force that shapes the dust into a tail of the kind seen here. The flux of particles emitted by the Sun, called the solar wind, typically produces an additional, second tail, as described in detail in this chapter.

In this chapter, we explain Maxwell's theory and show how it leads to his prediction of electromagnetic waves. We use his theory to examine what electromagnetic waves are, how they are produced, and how they transport energy and momentum. We conclude by summarizing some of the many practical applications of electromagnetic waves.

16.1 | Maxwell's Equations and Electromagnetic Waves

Learning Objectives

By the end of this section, you will be able to:

- Explain Maxwell's correction of Ampère's law by including the displacement current
- State and apply Maxwell's equations in integral form
- Describe how the symmetry between changing electric and changing magnetic fields explains Maxwell's prediction of electromagnetic waves
- Describe how Hertz confirmed Maxwell's prediction of electromagnetic waves

James Clerk Maxwell (1831–1879) was one of the major contributors to physics in the nineteenth century (Figure 16.2). Although he died young, he made major contributions to the development of the kinetic theory of gases, to the understanding of color vision, and to the nature of Saturn's rings. He is probably best known for having combined existing knowledge of the laws of electricity and of magnetism with insights of his own into a complete overarching electromagnetic theory, represented by **Maxwell's equations**.

Figure 16.2 James Clerk Maxwell, a nineteenth-century physicist, developed a theory that explained the relationship between electricity and magnetism, and correctly predicted that visible light consists of electromagnetic waves.

Maxwell's Correction to the Laws of Electricity and Magnetism

The four basic laws of electricity and magnetism had been discovered experimentally through the work of physicists such as Oersted, Coulomb, Gauss, and Faraday. Maxwell discovered logical inconsistencies in these earlier results and identified the incompleteness of Ampère's law as their cause.

Recall that according to Ampère's law, the integral of the magnetic field around a closed loop C is proportional to the current I passing through any surface whose boundary is loop C itself:

$$\oint_C \vec{\mathbf{B}} \cdot d\vec{\mathbf{s}} = \mu_0 I. \tag{16.1}$$

There are infinitely many surfaces that can be attached to any loop, and Ampère's law stated in Equation 16.1 is independent of the choice of surface.

Consider the set-up in Figure 16.3. A source of emf is abruptly connected across a parallel-plate capacitor so that a time-dependent current I develops in the wire. Suppose we apply Ampère's law to loop C shown at a time before the capacitor is fully charged, so that $I \neq 0$. Surface S_1 gives a nonzero value for the enclosed current I, whereas surface S_2 gives zero for the enclosed current because no current passes through it:

$$\oint_C \vec{\mathbf{B}} \cdot d\vec{\mathbf{s}} = \begin{cases} \mu_0 I & \text{if surface } S_1 \text{ is used} \\ 0 & \text{if surface } S_2 \text{ is used} \end{cases}.$$

Clearly, Ampère's law in its usual form does not work here. This may not be surprising, because Ampère's law as applied in earlier chapters required a steady current, whereas the current in this experiment is changing with time and is not steady at all.

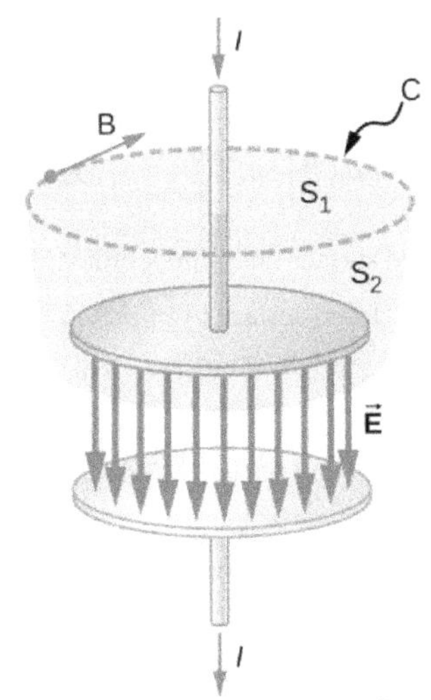

Figure 16.3 The currents through surface S_1 and surface S_2 are unequal, despite having the same boundary loop C.

How can Ampère's law be modified so that it works in all situations? Maxwell suggested including an additional contribution, called the displacement current I_d, to the real current I,

$$\oint_C \vec{\mathbf{B}} \cdot d\vec{\mathbf{s}} = \mu_0(I + I_d) \tag{16.2}$$

where the displacement current is defined to be

$$I_d = \varepsilon_0 \frac{d\Phi_E}{dt}. \tag{16.3}$$

Here ε_0 is the permittivity of free space and Φ_E is the electric flux, defined as

$$\Phi_E = \iint_{\text{Surface } S} \vec{\mathbf{E}} \cdot d\vec{\mathbf{A}}.$$

The **displacement current** is analogous to a real current in Ampère's law, entering into Ampère's law in the same way. It is produced, however, by a changing electric field. It accounts for a changing electric field producing a magnetic field, just as a real current does, but the displacement current can produce a magnetic field even where no real current is present. When this extra term is included, the modified Ampère's law equation becomes

$$\oint_C \vec{\mathbf{B}} \cdot d\vec{\mathbf{s}} = \mu_0 I + \varepsilon_0 \mu_0 \frac{d\Phi_E}{dt} \tag{16.4}$$

and is independent of the surface *S* through which the current *I* is measured.

We can now examine this modified version of Ampère's law to confirm that it holds independent of whether the surface S_1 or the surface S_2 in Figure 16.3 is chosen. The electric field $\vec{\mathbf{E}}$ corresponding to the flux Φ_E in Equation 16.3 is between the capacitor plates. Therefore, the $\vec{\mathbf{E}}$ field and the displacement current through the surface S_1 are both zero, and Equation 16.2 takes the form

$$\oint_C \vec{\mathbf{B}} \cdot d\vec{\mathbf{s}} = \mu_0 I. \tag{16.5}$$

We must now show that for surface $S_2,$ through which no actual current flows, the displacement current leads to the same value $\mu_0 I$ for the right side of the Ampère's law equation. For surface $S_2,$ the equation becomes

$$\oint_C \vec{\mathbf{B}} \cdot d\vec{\mathbf{s}} = \mu_0 \frac{d}{dt}\left[\varepsilon_0 \iint_{\text{Surface } S_2} \vec{\mathbf{E}} \cdot d\vec{\mathbf{A}}\right]. \tag{16.6}$$

Gauss's law for electric charge requires a closed surface and cannot ordinarily be applied to a surface like S_1 alone or S_2 alone. But the two surfaces S_1 and S_2 form a closed surface in Figure 16.3 and can be used in Gauss's law. Because the electric field is zero on S_1, the flux contribution through S_1 is zero. This gives us

$$\begin{aligned} \oiint_{\text{Surface } S_1 + S_2} \vec{\mathbf{E}} \cdot d\vec{\mathbf{A}} &= \iint_{\text{Surface } S_1} \vec{\mathbf{E}} \cdot d\vec{\mathbf{A}} + \iint_{\text{Surface } S_2} \vec{\mathbf{E}} \cdot d\vec{\mathbf{A}} \\ &= 0 + \iint_{\text{Surface } S_2} \vec{\mathbf{E}} \cdot d\vec{\mathbf{A}} \\ &= \iint_{\text{Surface } S_2} \vec{\mathbf{E}} \cdot d\vec{\mathbf{A}}. \end{aligned}$$

Therefore, we can replace the integral over S_2 in Equation 16.5 with the closed Gaussian surface $S_1 + S_2$ and apply Gauss's law to obtain

$$\oint_{S_1} \vec{\mathbf{B}} \cdot d\vec{\mathbf{s}} = \mu_0 \frac{dQ_{\text{in}}}{dt} = \mu_0 I. \tag{16.7}$$

Thus, the modified Ampère's law equation is the same using surface $S_2,$ where the right-hand side results from the displacement current, as it is for the surface $S_1,$ where the contribution comes from the actual flow of electric charge.

Example 16.1

Displacement current in a charging capacitor

A parallel-plate capacitor with capacitance *C* whose plates have area *A* and separation distance *d* is connected to a resistor *R* and a battery of voltage *V*. The current starts to flow at $t = 0$. (a) Find the displacement current between the capacitor plates at time *t*. (b) From the properties of the capacitor, find the corresponding real current $I = \frac{dQ}{dt}$, and compare the answer to the expected current in the wires of the corresponding *RC* circuit.

Strategy

We can use the equations from the analysis of an *RC* circuit (Alternating-Current Circuits) plus Maxwell's version of Ampère's law.

Solution

a. The voltage between the plates at time *t* is given by

$$V_C = \frac{1}{C}Q(t) = V_0\left(1 - e^{-t/RC}\right).$$

Let the z-axis point from the positive plate to the negative plate. Then the z-component of the electric field between the plates as a function of time t is

$$E_z(t) = \frac{V_0}{d}\left(1 - e^{-t/RC}\right).$$

Therefore, the z-component of the displacement current I_d between the plates is

$$I_\text{d}(t) = \varepsilon_0 A \frac{\partial E_z(t)}{\partial t} = \varepsilon_0 A \frac{V_0}{d} \times \frac{1}{RC} e^{-t/RC} = \frac{V_0}{R} e^{-t/RC},$$

where we have used $C = \varepsilon_0 \frac{A}{d}$ for the capacitance.

b. From the expression for V_C, the charge on the capacitor is

$$Q(t) = CV_C = CV_0\left(1 - e^{-t/RC}\right).$$

The current into the capacitor after the circuit is closed, is therefore

$$I = \frac{dQ}{dt} = \frac{V_0}{R} e^{-t/RC}.$$

This current is the same as I_d found in (a).

Maxwell's Equations

With the correction for the displacement current, Maxwell's equations take the form

$$\oint \vec{\mathbf{E}} \cdot d\vec{\mathbf{A}} = \frac{Q_\text{in}}{\varepsilon_0} \quad \text{(Gauss's law)} \tag{16.8}$$

$$\oint \vec{\mathbf{B}} \cdot d\vec{\mathbf{A}} = 0 \quad \text{(Gauss's law for magnetism)} \tag{16.9}$$

$$\oint \vec{\mathbf{E}} \cdot d\vec{\mathbf{s}} = -\frac{d\Phi_\text{m}}{dt} \quad \text{(Faraday's law)} \tag{16.10}$$

$$\oint \vec{\mathbf{B}} \cdot d\vec{\mathbf{s}} = \mu_0 I + \varepsilon_0 \mu_0 \frac{d\Phi_\text{E}}{dt} \quad \text{(Ampère-Maxwell law)}. \tag{16.11}$$

Once the fields have been calculated using these four equations, the Lorentz force equation

$$\vec{\mathbf{F}} = q\vec{\mathbf{E}} + q\vec{\mathbf{v}} \times \vec{\mathbf{B}} \tag{16.12}$$

gives the force that the fields exert on a particle with charge q moving with velocity $\vec{\mathbf{v}}$. The Lorentz force equation combines the force of the electric field and of the magnetic field on the moving charge. The magnetic and electric forces have been examined in earlier modules. These four Maxwell's equations are, respectively,

Maxwell's Equations

1. **Gauss's law**

The electric flux through any closed surface is equal to the electric charge Q_{in} enclosed by the surface. Gauss's law [Equation 16.7] describes the relation between an electric charge and the electric field it produces. This is often pictured in terms of electric field lines originating from positive charges and terminating on negative charges, and indicating the direction of the electric field at each point in space.

2. **Gauss's law for magnetism**

The magnetic field flux through any closed surface is zero [Equation 16.8]. This is equivalent to the statement that magnetic field lines are continuous, having no beginning or end. Any magnetic field line entering the region enclosed by the surface must also leave it. No magnetic monopoles, where magnetic field lines would terminate, are known to exist (see Magnetic Fields and Lines).

3. **Faraday's law**

A changing magnetic field induces an electromotive force (emf) and, hence, an electric field. The direction of the emf opposes the change. This third of Maxwell's equations, Equation 16.9, is Faraday's law of induction and includes Lenz's law. The electric field from a changing magnetic field has field lines that form closed loops, without any beginning or end.

4. **Ampère-Maxwell law**

Magnetic fields are generated by moving charges or by changing electric fields. This fourth of Maxwell's equations, Equation 16.10, encompasses Ampère's law and adds another source of magnetic fields, namely changing electric fields.

Maxwell's equations and the Lorentz force law together encompass all the laws of electricity and magnetism. The symmetry that Maxwell introduced into his mathematical framework may not be immediately apparent. Faraday's law describes how changing magnetic fields produce electric fields. The displacement current introduced by Maxwell results instead from a changing electric field and accounts for a changing electric field producing a magnetic field. The equations for the effects of both changing electric fields and changing magnetic fields differ in form only where the absence of magnetic monopoles leads to missing terms. This symmetry between the effects of changing magnetic and electric fields is essential in explaining the nature of electromagnetic waves.

Later application of Einstein's theory of relativity to Maxwell's complete and symmetric theory showed that electric and magnetic forces are not separate but are different manifestations of the same thing—the electromagnetic force. The electromagnetic force and weak nuclear force are similarly unified as the electroweak force. This unification of forces has been one motivation for attempts to unify all of the four basic forces in nature—the gravitational, electrical, strong, and weak nuclear forces (see Particle Physics and Cosmology (http://cnx.org/content/m58767/latest/)).

The Mechanism of Electromagnetic Wave Propagation

To see how the symmetry introduced by Maxwell accounts for the existence of combined electric and magnetic waves that propagate through space, imagine a time-varying magnetic field $\vec{\mathbf{B}}_0(t)$ produced by the high-frequency alternating current seen in Figure 16.4. We represent $\vec{\mathbf{B}}_0(t)$ in the diagram by one of its field lines. From Faraday's law, the changing magnetic field through a surface induces a time-varying electric field $\vec{\mathbf{E}}_0(t)$ at the boundary of that surface. The displacement current source for the electric field, like the Faraday's law source for the magnetic field, produces only closed loops of field lines, because of the mathematical symmetry involved in the equations for the induced electric and induced magnetic fields. A field line representation of $\vec{\mathbf{E}}_0(t)$ is shown. In turn, the changing electric field $\vec{\mathbf{E}}_0(t)$ creates a magnetic field $\vec{\mathbf{B}}_1(t)$ according to the modified Ampère's law. This changing field induces $\vec{\mathbf{E}}_1(t)$, which induces $\vec{\mathbf{B}}_2(t)$, and so on. We then have a self-continuing process that leads to the creation of time-varying electric and magnetic fields in regions farther and farther away from O. This process may be visualized as the propagation of an electromagnetic wave through space.

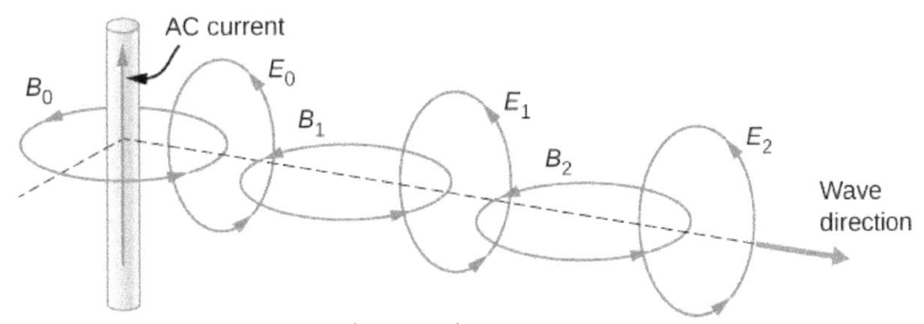

Figure 16.4 How changing $\vec{\mathbf{E}}$ and $\vec{\mathbf{B}}$ fields propagate through space.

In the next section, we show in more precise mathematical terms how Maxwell's equations lead to the prediction of electromagnetic waves that can travel through space without a material medium, implying a speed of electromagnetic waves equal to the speed of light.

Prior to Maxwell's work, experiments had already indicated that light was a wave phenomenon, although the nature of the waves was yet unknown. In 1801, Thomas Young (1773–1829) showed that when a light beam was separated by two narrow slits and then recombined, a pattern made up of bright and dark fringes was formed on a screen. Young explained this behavior by assuming that light was composed of waves that added constructively at some points and destructively at others (see Interference (http://cnx.org/content/m58536/latest/)). Subsequently, Jean Foucault (1819–1868), with measurements of the speed of light in various media, and Augustin Fresnel (1788–1827), with detailed experiments involving interference and diffraction of light, provided further conclusive evidence that light was a wave. So, light was known to be a wave, and Maxwell had predicted the existence of electromagnetic waves that traveled at the speed of light. The conclusion seemed inescapable: Light must be a form of electromagnetic radiation. But Maxwell's theory showed that other wavelengths and frequencies than those of light were possible for electromagnetic waves. He showed that electromagnetic radiation with the same fundamental properties as visible light should exist at any frequency. It remained for others to test, and confirm, this prediction.

16.1 Check Your Understanding When the emf across a capacitor is turned on and the capacitor is allowed to charge, when does the magnetic field induced by the displacement current have the greatest magnitude?

Hertz's Observations

The German physicist Heinrich Hertz (1857–1894) was the first to generate and detect certain types of electromagnetic waves in the laboratory. Starting in 1887, he performed a series of experiments that not only confirmed the existence of electromagnetic waves but also verified that they travel at the speed of light.

Hertz used an alternating-current *RLC* (resistor-inductor-capacitor) circuit that resonates at a known frequency $f_0 = \frac{1}{2\pi\sqrt{LC}}$ and connected it to a loop of wire, as shown in Figure 16.5. High voltages induced across the gap in the loop produced sparks that were visible evidence of the current in the circuit and helped generate electromagnetic waves.

Across the laboratory, Hertz placed another loop attached to another *RLC* circuit, which could be tuned (as the dial on a radio) to the same resonant frequency as the first and could thus be made to receive electromagnetic waves. This loop also had a gap across which sparks were generated, giving solid evidence that electromagnetic waves had been received.

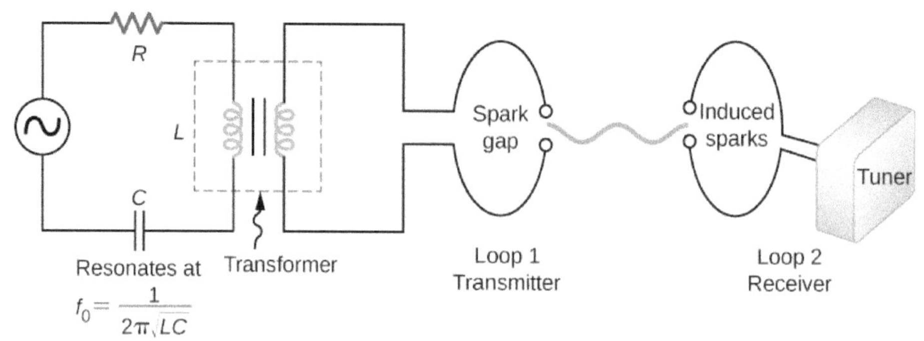

Figure 16.5 The apparatus used by Hertz in 1887 to generate and detect electromagnetic waves.

Hertz also studied the reflection, refraction, and interference patterns of the electromagnetic waves he generated, confirming their wave character. He was able to determine the wavelengths from the interference patterns, and knowing their frequencies, he could calculate the propagation speed using the equation $v = f\lambda$, where v is the speed of a wave, f is its frequency, and λ is its wavelength. Hertz was thus able to prove that electromagnetic waves travel at the speed of light. The SI unit for frequency, the hertz ($1 \text{ Hz} = 1 \text{ cycle/second}$), is named in his honor.

16.2 Check Your Understanding Could a purely electric field propagate as a wave through a vacuum without a magnetic field? Justify your answer.

16.2 | Plane Electromagnetic Waves

Learning Objectives

By the end of this section, you will be able to:

- Describe how Maxwell's equations predict the relative directions of the electric fields and magnetic fields, and the direction of propagation of plane electromagnetic waves
- Explain how Maxwell's equations predict that the speed of propagation of electromagnetic waves in free space is exactly the speed of light
- Calculate the relative magnitude of the electric and magnetic fields in an electromagnetic plane wave
- Describe how electromagnetic waves are produced and detected

Mechanical waves travel through a medium such as a string, water, or air. Perhaps the most significant prediction of Maxwell's equations is the existence of combined electric and magnetic (or electromagnetic) fields that propagate through space as electromagnetic waves. Because Maxwell's equations hold in free space, the predicted electromagnetic waves, unlike mechanical waves, do not require a medium for their propagation.

A general treatment of the physics of electromagnetic waves is beyond the scope of this textbook. We can, however, investigate the special case of an electromagnetic wave that propagates through free space along the *x*-axis of a given coordinate system.

Electromagnetic Waves in One Direction

An electromagnetic wave consists of an electric field, defined as usual in terms of the force per charge on a stationary charge, and a magnetic field, defined in terms of the force per charge on a moving charge. The electromagnetic field is assumed to be a function of only the *x*-coordinate and time. The *y*-component of the electric field is then written as $E_y(x, t)$, the *z*-component of the magnetic field as $B_z(x, t)$, etc. Because we are assuming free space, there are no free charges or currents, so we can set $Q_{\text{in}} = 0$ and $I = 0$ in Maxwell's equations.

The transverse nature of electromagnetic waves

We examine first what Gauss's law for electric fields implies about the relative directions of the electric field and the propagation direction in an electromagnetic wave. Assume the Gaussian surface to be the surface of a rectangular box whose cross-section is a square of side l and whose third side has length Δx, as shown in Figure 16.6. Because the electric field is a function only of x and t, the y-component of the electric field is the same on both the top (labeled Side 2) and bottom (labeled Side 1) of the box, so that these two contributions to the flux cancel. The corresponding argument also holds for the net flux from the z-component of the electric field through Sides 3 and 4. Any net flux through the surface therefore comes entirely from the x-component of the electric field. Because the electric field has no y- or z-dependence, $E_x(x, t)$ is constant over the face of the box with area A and has a possibly different value $E_x(x + \Delta x, t)$ that is constant over the opposite face of the box. Applying Gauss's law gives

$$\text{Net flu} \;= -E_x(x, t)A + E_x(x + \Delta x, t)A = \frac{Q_{\text{in}}}{\varepsilon_0} \tag{16.13}$$

where $A = l \times l$ is the area of the front and back faces of the rectangular surface. But the charge enclosed is $Q_{\text{in}} = 0$, so this component's net flux is also zero, and Equation 16.13 implies $E_x(x, t) = E_x(x + \Delta x, t)$ for any Δx. Therefore, if there is an x-component of the electric field, it cannot vary with x. A uniform field of that kind would merely be superposed artificially on the traveling wave, for example, by having a pair of parallel-charged plates. Such a component $E_x(x, t)$ would not be part of an electromagnetic wave propagating along the x-axis; so $E_x(x, t) = 0$ for this wave. Therefore, the only nonzero components of the electric field are $E_y(x, t)$ and $E_z(x, t)$, perpendicular to the direction of propagation of the wave.

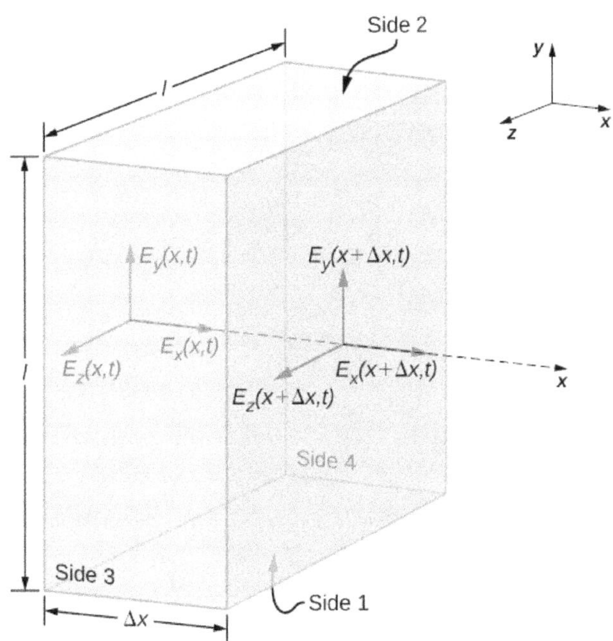

Figure 16.6 The surface of a rectangular box of dimensions $l \times l \times \Delta x$ is our Gaussian surface. The electric field shown is from an electromagnetic wave propagating along the x-axis.

A similar argument holds by substituting E for B and using Gauss's law for magnetism instead of Gauss's law for electric fields. This shows that the B field is also perpendicular to the direction of propagation of the wave. The electromagnetic wave is therefore a transverse wave, with its oscillating electric and magnetic fields perpendicular to its direction of propagation.

The speed of propagation of electromagnetic waves

We can next apply Maxwell's equations to the description given in connection with Figure 16.4 in the previous section to obtain an equation for the E field from the changing B field, and for the B field from a changing E field. We then combine the two equations to show how the changing E and B fields propagate through space at a speed precisely equal to the speed of light.

First, we apply Faraday's law over Side 3 of the Gaussian surface, using the path shown in Figure 16.7. Because $E_x(x, t) = 0$, we have

$$\oint \vec{\mathbf{E}} \cdot d\vec{\mathbf{s}} = -E_y(x, t)l + E_y(x + \Delta x, t)l.$$

Assuming Δx is small and approximating $E_y(x + \Delta x, t)$ by

$$E_y(x + \Delta x, t) = E_y(x, t) + \frac{\partial E_y(x, t)}{\partial x}\Delta x,$$

we obtain

$$\oint \vec{\mathbf{E}} \cdot d\vec{\mathbf{s}} = \frac{\partial E_y(x, t)}{\partial x}(l\Delta x).$$

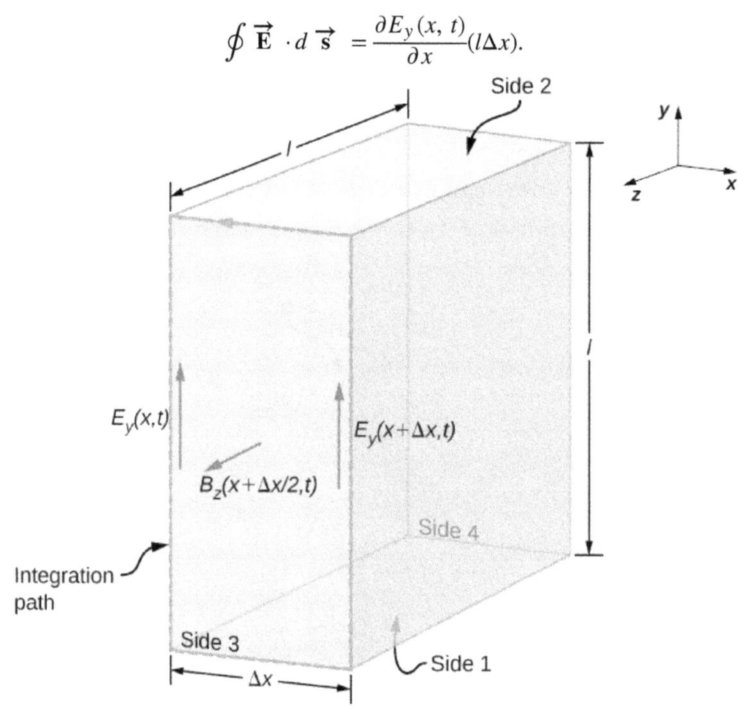

Figure 16.7 We apply Faraday's law to the front of the rectangle by evaluating $\oint \vec{\mathbf{E}} \cdot d\vec{\mathbf{s}}$ along the rectangular edge of Side 3 in the direction indicated, taking the B field crossing the face to be approximately its value in the middle of the area traversed.

Because Δx is small, the magnetic flux through the face can be approximated by its value in the center of the area traversed, namely $B_z\left(x + \frac{\Delta x}{2}, t\right)$. The flux of the B field through Face 3 is then the B field times the area,

$$\oint_S \vec{\mathbf{B}} \cdot \vec{\mathbf{n}}\, dA = B_z\left(x + \frac{\Delta x}{2}, t\right)(l\Delta x). \tag{16.14}$$

From Faraday's law,

$$\oint \vec{\mathbf{E}} \cdot d\vec{\mathbf{s}} = -\frac{d}{dt}\int_S \vec{\mathbf{B}} \cdot \vec{\mathbf{n}}\, dA. \tag{16.15}$$

Therefore, from Equation 16.13 and Equation 16.14,

$$\frac{\partial E_y(x,t)}{\partial x}(l\Delta x) = -\frac{\partial}{\partial t}\left[B_z\left(x+\frac{\Delta x}{2},t\right)\right](l\Delta x).$$

Canceling $l\Delta x$ and taking the limit as $\Delta x = 0$, we are left with

$$\frac{\partial E_y(x,t)}{\partial x} = -\frac{\partial B_z(x,t)}{\partial t}. \tag{16.16}$$

We could have applied Faraday's law instead to the top surface (numbered 2) in Figure 16.7, to obtain the resulting equation

$$\frac{\partial E_z(x,t)}{\partial x} = -\frac{\partial B_y(x,t)}{\partial t}. \tag{16.17}$$

This is the equation describing the spatially dependent *E* field produced by the time-dependent *B* field.

Next we apply the Ampère-Maxwell law (with $I = 0$) over the same two faces (Surface 3 and then Surface 2) of the rectangular box of Figure 16.7. Applying Equation 16.10,

$$\oint \vec{\mathbf{B}} \cdot d\vec{\mathbf{s}} = \mu_0\varepsilon_0(d/dt)\int_S \vec{\mathbf{E}} \cdot \mathbf{n}\,da$$

to Surface 3, and then to Surface 2, yields the two equations

$$\frac{\partial B_y(x,t)}{\partial x} = -\varepsilon_0\mu_0\frac{\partial E_z(x,t)}{\partial t}, \text{ and} \tag{16.18}$$

$$\frac{\partial B_z(x,t)}{\partial x} = -\varepsilon_0\mu_0\frac{\partial E_y(x,t)}{\partial t}. \tag{16.19}$$

These equations describe the spatially dependent *B* field produced by the time-dependent *E* field.

We next combine the equations showing the changing *B* field producing an *E* field with the equation showing the changing *E* field producing a *B* field. Taking the derivative of Equation 16.16 with respect to *x* and using Equation 16.26 gives

$$\frac{\partial^2 E_y}{\partial x^2} = \frac{\partial}{\partial x}\left(\frac{\partial E_y}{\partial x}\right) = -\frac{\partial}{\partial x}\left(\frac{\partial B_z}{\partial t}\right) = -\frac{\partial}{\partial t}\left(\frac{\partial B_z}{\partial x}\right) = \frac{\partial}{\partial t}\left(\varepsilon_0\mu_0\frac{\partial E_y}{\partial t}\right)$$

or

$$\frac{\partial^2 E_y}{\partial x^2} = \varepsilon_0\mu_0\frac{\partial^2 E_y}{\partial t^2}. \tag{16.20}$$

This is the form taken by the general wave equation for our plane wave. Because the equations describe a wave traveling at some as-yet-unspecified speed *c*, we can assume the field components are each functions of $x - ct$ for the wave traveling in the $+x$-direction, that is,

$$E_y(x,t) = f(\xi) \quad \text{where } \xi = x - ct. \tag{16.21}$$

It is left as a mathematical exercise to show, using the chain rule for differentiation, that Equation 16.17 and Equation 16.18 imply

$$1 = \varepsilon_0\mu_0 c^2.$$

The speed of the electromagnetic wave in free space is therefore given in terms of the permeability and the permittivity of free space by

$$c = \frac{1}{\sqrt{\varepsilon_0\mu_0}}. \tag{16.22}$$

We could just as easily have assumed an electromagnetic wave with field components $E_z(x, t)$ and $B_y(x, t)$. The same type of analysis with Equation 16.25 and Equation 16.24 would also show that the speed of an electromagnetic wave is $c = 1/\sqrt{\varepsilon_0 \mu_0}$.

The physics of traveling electromagnetic fields was worked out by Maxwell in 1873. He showed in a more general way than our derivation that electromagnetic waves always travel in free space with a speed given by Equation 16.18. If we evaluate the speed $c = \frac{1}{\sqrt{\varepsilon_0 \mu_0}}$, we find that

$$c = \frac{1}{\sqrt{\left(8.85 \times 10^{-12} \frac{\mathrm{C}^2}{\mathrm{N \cdot m}^2}\right)\left(4\pi \times 10^{-7} \frac{\mathrm{T \cdot m}}{\mathrm{A}}\right)}} = 3.00 \times 10^8 \text{ m/s},$$

which is the speed of light. Imagine the excitement that Maxwell must have felt when he discovered this equation! He had found a fundamental connection between two seemingly unrelated phenomena: electromagnetic fields and light.

16.3 Check Your Understanding The wave equation was obtained by (1) finding the E field produced by the changing B field, (2) finding the B field produced by the changing E field, and combining the two results. Which of Maxwell's equations was the basis of step (1) and which of step (2)?

How the *E* and *B* Fields Are Related

So far, we have seen that the rates of change of different components of the E and B fields are related, that the electromagnetic wave is transverse, and that the wave propagates at speed c. We next show what Maxwell's equations imply about the ratio of the E and B field magnitudes and the relative directions of the E and B fields.

We now consider solutions to Equation 16.16 in the form of plane waves for the electric field:

$$E_y(x, t) = E_0 \cos(kx - \omega t). \tag{16.23}$$

We have arbitrarily taken the wave to be traveling in the +x-direction and chosen its phase so that the maximum field strength occurs at the origin at time $t = 0$. We are justified in considering only sines and cosines in this way, and generalizing the results, because Fourier's theorem implies we can express any wave, including even square step functions, as a superposition of sines and cosines.

At any one specific point in space, the E field oscillates sinusoidally at angular frequency ω between $+E_0$ and $-E_0$, and similarly, the B field oscillates between $+B_0$ and $-B_0$. The amplitude of the wave is the maximum value of $E_y(x, t)$. The period of oscillation T is the time required for a complete oscillation. The frequency f is the number of complete oscillations per unit of time, and is related to the angular frequency ω by $\omega = 2\pi f$. The wavelength λ is the distance covered by one complete cycle of the wave, and the wavenumber k is the number of wavelengths that fit into a distance of 2π in the units being used. These quantities are related in the same way as for a mechanical wave:

$$\omega = 2\pi f, \quad f = \frac{1}{T}, \quad k = \frac{2\pi}{\lambda}, \quad \text{and} \quad c = f\lambda = \omega/k.$$

Given that the solution of E_y has the form shown in Equation 16.20, we need to determine the B field that accompanies it. From Equation 16.24, the magnetic field component B_z must obey

$$\frac{\partial B_Z}{\partial t} = -\frac{\partial E_y}{\partial x} \tag{16.24}$$

$$\frac{\partial B_Z}{\partial t} = -\frac{\partial}{\partial x} E_0 \cos(kx - \omega t) = kE_0 \sin(kx - \omega t).$$

Because the solution for the B-field pattern of the wave propagates in the +x-direction at the same speed c as the E-field pattern, it must be a function of $k(x - ct) = kx - \omega t$. Thus, we conclude from Equation 16.21 that B_z is

$$B_z(x, t) = \frac{k}{\omega} E_0 \cos(kx - \omega t) = \frac{1}{c} E_0 \cos(kx - \omega t).$$

These results may be written as

$$E_y(x, t) = E_0 \cos(kx - \omega t)$$
$$B_z(x, t) = B_0 \cos(kx - \omega t) \tag{16.25}$$

$$\frac{E_y}{B_z} = \frac{E_0}{B_0} = c. \tag{16.26}$$

Therefore, the peaks of the *E* and *B* fields coincide, as do the troughs of the wave, and at each point, the *E* and *B* fields are in the same ratio equal to the speed of light *c*. The plane wave has the form shown in Figure 16.8.

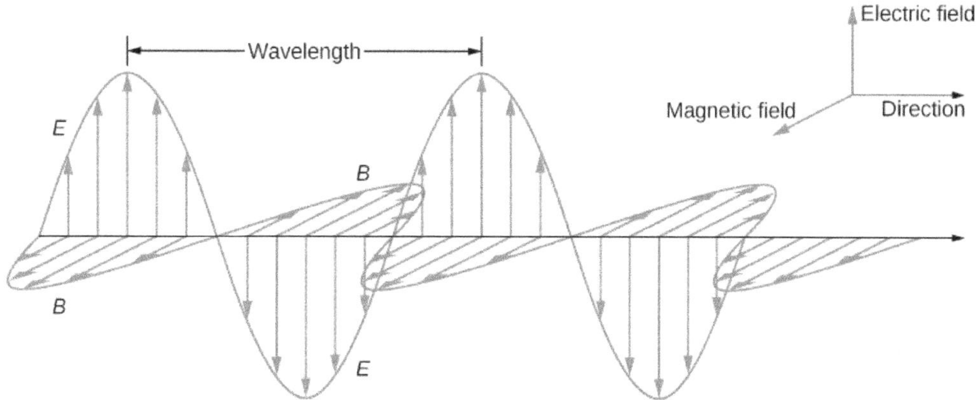

Figure 16.8 The plane wave solution of Maxwell's equations has the *B* field directly proportional to the *E* field at each point, with the relative directions shown.

Example 16.2

Calculating *B*-Field Strength in an Electromagnetic Wave

What is the maximum strength of the *B* field in an electromagnetic wave that has a maximum *E*-field strength of 1000 V/m?

Strategy

To find the *B*-field strength, we rearrange Equation 16.23 to solve for *B*, yielding

$$B = \frac{E}{c}.$$

Solution

We are given *E*, and *c* is the speed of light. Entering these into the expression for *B* yields

$$B = \frac{1000 \text{ V/m}}{3.00 \times 10^8 \text{ m/s}} = 3.33 \times 10^{-6} \text{ T}.$$

Significance

The *B*-field strength is less than a tenth of Earth's admittedly weak magnetic field. This means that a relatively strong electric field of 1000 V/m is accompanied by a relatively weak magnetic field.

Changing electric fields create relatively weak magnetic fields. The combined electric and magnetic fields can be detected in electromagnetic waves, however, by taking advantage of the phenomenon of resonance, as Hertz did. A system with the same natural frequency as the electromagnetic wave can be made to oscillate. All radio and TV receivers use this principle to pick up and then amplify weak electromagnetic waves, while rejecting all others not at their resonant frequency.

16.4 Check Your Understanding What conclusions did our analysis of Maxwell's equations lead to about these properties of a plane electromagnetic wave:
(a) the relative directions of wave propagation, of the *E* field, and of *B* field,
(b) the speed of travel of the wave and how the speed depends on frequency, and
(c) the relative magnitudes of the *E* and *B* fields.

Production and Detection of Electromagnetic Waves

A steady electric current produces a magnetic field that is constant in time and which does not propagate as a wave. Accelerating charges, however, produce electromagnetic waves. An electric charge oscillating up and down, or an alternating current or flow of charge in a conductor, emit radiation at the frequencies of their oscillations. The electromagnetic field of a *dipole antenna* is shown in Figure 16.9. The positive and negative charges on the two conductors are made to reverse at the desired frequency by the output of a transmitter as the power source. The continually changing current accelerates charge in the antenna, and this results in an oscillating electric field a distance away from the antenna. The changing electric fields produce changing magnetic fields that in turn produce changing electric fields, which thereby propagate as electromagnetic waves. The frequency of this radiation is the same as the frequency of the ac source that is accelerating the electrons in the antenna. The two conducting elements of the dipole antenna are commonly straight wires. The total length of the two wires is typically about one-half of the desired wavelength (hence, the alternative name *half-wave antenna*), because this allows standing waves to be set up and enhances the effectiveness of the radiation.

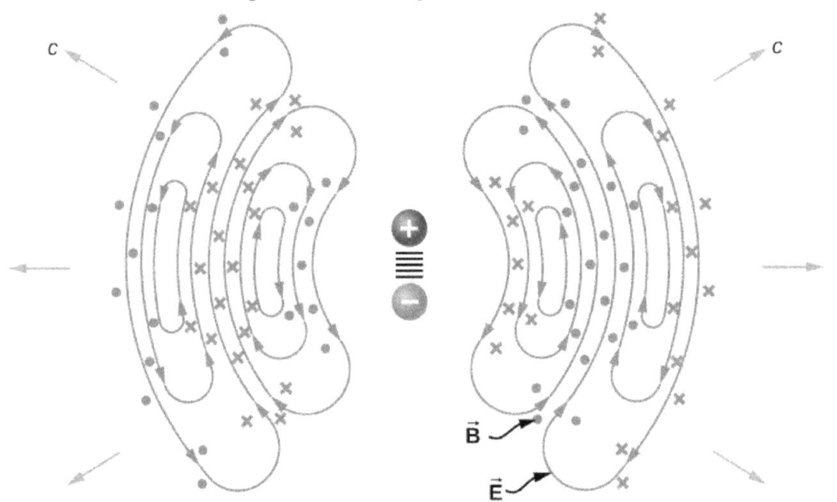

Figure 16.9 The oscillatory motion of the charges in a dipole antenna produces electromagnetic radiation.

The electric field lines in one plane are shown. The magnetic field is perpendicular to this plane. This radiation field has cylindrical symmetry around the axis of the dipole. Field lines near the dipole are not shown. The pattern is not at all uniform in all directions. The strongest signal is in directions perpendicular to the axis of the antenna, which would be horizontal if the antenna is mounted vertically. There is zero intensity along the axis of the antenna. The fields detected far from the antenna are from the changing electric and magnetic fields inducing each other and traveling as electromagnetic waves. Far from the antenna, the wave fronts, or surfaces of equal phase for the electromagnetic wave, are almost spherical. Even farther from the antenna, the radiation propagates like electromagnetic plane waves.

The electromagnetic waves carry energy away from their source, similar to a sound wave carrying energy away from a standing wave on a guitar string. An antenna for receiving electromagnetic signals works in reverse. Incoming electromagnetic waves induce oscillating currents in the antenna, each at its own frequency. The radio receiver includes a tuner circuit, whose resonant frequency can be adjusted. The tuner responds strongly to the desired frequency but not others, allowing the user to tune to the desired broadcast. Electrical components amplify the signal formed by the moving electrons. The signal is then converted into an audio and/or video format.

Use this simulation (https://openstaxcollege.org/l/21radwavsim) to broadcast radio waves. Wiggle the transmitter electron manually or have it oscillate automatically. Display the field as a curve or vectors. The strip chart shows the electron positions at the transmitter and at the receiver.

16.3 | Energy Carried by Electromagnetic Waves

Learning Objectives

By the end of this section, you will be able to:

- Express the time-averaged energy density of electromagnetic waves in terms of their electric and magnetic field amplitudes
- Calculate the Poynting vector and the energy intensity of electromagnetic waves
- Explain how the energy of an electromagnetic wave depends on its amplitude, whereas the energy of a photon is proportional to its frequency

Anyone who has used a microwave oven knows there is energy in electromagnetic waves. Sometimes this energy is obvious, such as in the warmth of the summer Sun. Other times, it is subtle, such as the unfelt energy of gamma rays, which can destroy living cells.

Electromagnetic waves bring energy into a system by virtue of their electric and magnetic fields. These fields can exert forces and move charges in the system and, thus, do work on them. However, there is energy in an electromagnetic wave itself, whether it is absorbed or not. Once created, the fields carry energy away from a source. If some energy is later absorbed, the field strengths are diminished and anything left travels on.

Clearly, the larger the strength of the electric and magnetic fields, the more work they can do and the greater the energy the electromagnetic wave carries. In electromagnetic waves, the amplitude is the maximum field strength of the electric and magnetic fields (Figure 16.10). The wave energy is determined by the wave amplitude.

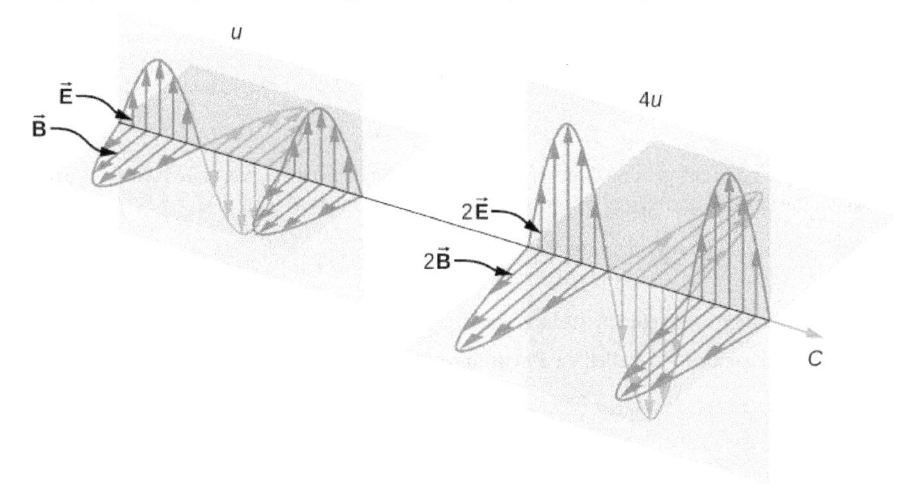

Figure 16.10 Energy carried by a wave depends on its amplitude. With electromagnetic waves, doubling the E fields and B fields quadruples the energy density u and the energy flux uc.

For a plane wave traveling in the direction of the positive x-axis with the phase of the wave chosen so that the wave maximum is at the origin at $t = 0$, the electric and magnetic fields obey the equations

$$E_y(x, t) = E_0 \cos(kx - \omega t)$$
$$B_z(x, t) = B_0 \cos(kx - \omega t).$$

The energy in any part of the electromagnetic wave is the sum of the energies of the electric and magnetic fields. This energy per unit volume, or energy density u, is the sum of the energy density from the electric field and the energy density from the magnetic field. Expressions for both field energy densities were discussed earlier (u_E in Capacitance and u_B in Inductance). Combining these the contributions, we obtain

$$u(x,\ t) = u_E + u_B = \frac{1}{2}\varepsilon_0 E^2 + \frac{1}{2\mu_0}B^2.$$

The expression $E = cB = \frac{1}{\sqrt{\varepsilon_0 \mu_0}}B$ then shows that the magnetic energy density u_B and electric energy density u_E are equal, despite the fact that changing electric fields generally produce only small magnetic fields. The equality of the electric and magnetic energy densities leads to

$$u(x,\ t) = \varepsilon_0 E^2 = \frac{B^2}{\mu_0}. \tag{16.27}$$

The energy density moves with the electric and magnetic fields in a similar manner to the waves themselves.

We can find the rate of transport of energy by considering a small time interval Δt . As shown in Figure 16.11, the energy contained in a cylinder of length $c\Delta t$ and cross-sectional area A passes through the cross-sectional plane in the interval Δt.

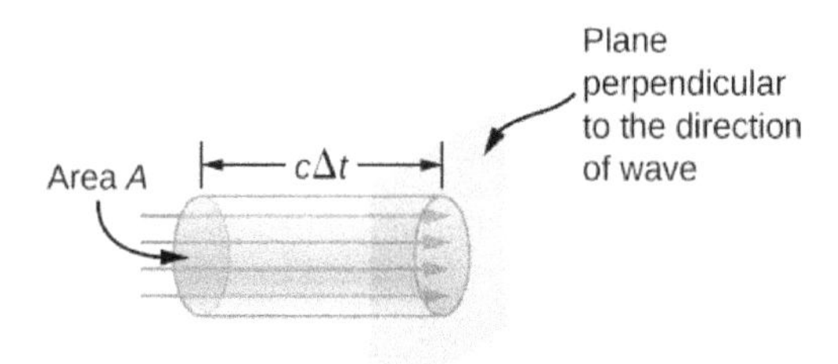

Figure 16.11 The energy $uAc\Delta t$ contained in the electric and magnetic fields of the electromagnetic wave in the volume $Ac\Delta t$ passes through the area A in time Δt .

The energy passing through area A in time Δt is

$$u \times \text{volume} = uAc\Delta t.$$

The energy per unit area per unit time passing through a plane perpendicular to the wave, called the energy flux and denoted by S, can be calculated by dividing the energy by the area A and the time interval Δt .

$$S = \frac{\text{Energy passing area } A \text{ in time } \Delta t}{A\Delta t} = uc = \varepsilon_0 cE^2 = \frac{1}{\mu_0}EB.$$

More generally, the flux of energy through any surface also depends on the orientation of the surface. To take the direction into account, we introduce a vector $\vec{\mathbf{S}}$, called the **Poynting vector**, with the following definition:

$$\vec{\mathbf{S}} = \frac{1}{\mu_0}\vec{\mathbf{E}} \times \vec{\mathbf{B}}. \tag{16.28}$$

The cross-product of $\vec{\mathbf{E}}$ and $\vec{\mathbf{B}}$ points in the direction perpendicular to both vectors. To confirm that the direction of $\vec{\mathbf{S}}$ is that of wave propagation, and not its negative, return to Figure 16.7. Note that Lenz's and Faraday's laws imply that when the magnetic field shown is increasing in time, the electric field is greater at x than at $x + \Delta x$. The electric field is decreasing with increasing x at the given time and location. The proportionality between electric and magnetic fields requires the electric field to increase in time along with the magnetic field. This is possible only if the wave is propagating to the right in the diagram, in which case, the relative orientations show that $\vec{\mathbf{S}} = \frac{1}{\mu_0}\vec{\mathbf{E}} \times \vec{\mathbf{B}}$ is specifically in the direction of propagation of the electromagnetic wave.

The energy flux at any place also varies in time, as can be seen by substituting u from Equation 16.23 into Equation 16.27.

$$S(x,\, t) = c\varepsilon_0 E_0^2 \cos^2 (kx - \omega t) \tag{16.29}$$

Because the frequency of visible light is very high, of the order of 10^{14} Hz, the energy flux for visible light through any area is an extremely rapidly varying quantity. Most measuring devices, including our eyes, detect only an average over many cycles. The time average of the energy flux is the intensity I of the electromagnetic wave and is the power per unit area. It can be expressed by averaging the cosine function in Equation 16.29 over one complete cycle, which is the same as time-averaging over many cycles (here, T is one period):

$$I = S_{\text{avg}} = c\varepsilon_0 E_0^2 \frac{1}{T}\int_0^T \cos^2\left(2\pi \frac{t}{T}\right)dt. \tag{16.30}$$

We can either evaluate the integral, or else note that because the sine and cosine differ merely in phase, the average over a complete cycle for $\cos^2(\xi)$ is the same as for $\sin^2(\xi)$, to obtain

$$\langle \cos^2 \xi \rangle = \frac{1}{2}\left[\langle \cos^2 \xi \rangle + \langle \sin^2 \xi \rangle\right] = \frac{1}{2}\langle 1 \rangle = \frac{1}{2}.$$

where the angle brackets $\langle \cdots \rangle$ stand for the time-averaging operation. The intensity of light moving at speed c in vacuum is then found to be

$$I = S_{\text{avg}} = \frac{1}{2} c\varepsilon_0 E_0^2 \tag{16.31}$$

in terms of the maximum electric field strength E_0, which is also the electric field amplitude. Algebraic manipulation produces the relationship

$$I = \frac{cB_0^2}{2\mu_0} \tag{16.32}$$

where B_0 is the magnetic field amplitude, which is the same as the maximum magnetic field strength. One more expression for I_{avg} in terms of both electric and magnetic field strengths is useful. Substituting the fact that $cB_0 = E_0$, the previous expression becomes

$$I = \frac{E_0 B_0}{2\mu_0}. \tag{16.33}$$

We can use whichever of the three preceding equations is most convenient, because the three equations are really just different versions of the same result: The energy in a wave is related to amplitude squared. Furthermore, because these equations are based on the assumption that the electromagnetic waves are sinusoidal, the peak intensity is twice the average intensity; that is, $I_0 = 2I$.

Example 16.3

A Laser Beam

The beam from a small laboratory laser typically has an intensity of about $1.0 \times 10^{-3}\ \text{W/m}^2$. Assuming that the beam is composed of plane waves, calculate the amplitudes of the electric and magnetic fields in the beam.

Strategy

Use the equation expressing intensity in terms of electric field to calculate the electric field from the intensity.

Solution

From Equation 16.31, the intensity of the laser beam is

$$I = \frac{1}{2}c\varepsilon_0 E_0^2.$$

The amplitude of the electric field is therefore

$$E_0 = \sqrt{\frac{2}{c\varepsilon_0}I} = \sqrt{\frac{2}{(3.00 \times 10^8\ \text{m/s})(8.85 \times 10^{-12}\ \text{F/m})}(1.0 \times 10^{-3}\ \text{W/m}^2)} = 0.87\ \text{V/m}.$$

The amplitude of the magnetic field can be obtained from Equation 16.20:

$$B_0 = \frac{E_0}{c} = 2.9 \times 10^{-9}\ \text{T}.$$

Example 16.4

Light Bulb Fields

A light bulb emits 5.00 W of power as visible light. What are the average electric and magnetic fields from the light at a distance of 3.0 m?

Strategy

Assume the bulb's power output P is distributed uniformly over a sphere of radius 3.0 m to calculate the intensity, and from it, the electric field.

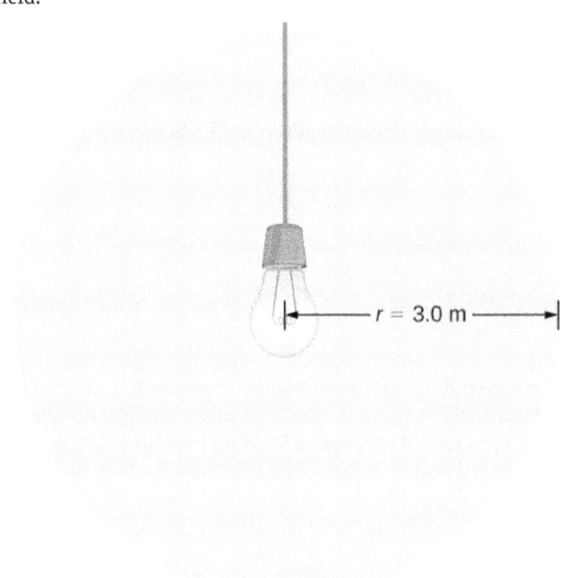

Solution

The power radiated as visible light is then

$$\begin{aligned} I &= \frac{P}{4\pi r^2} = \frac{c\varepsilon_0 E_0^2}{2}, \\ E_0 &= \sqrt{2\frac{P}{4\pi r^2 c\varepsilon_0}} = \sqrt{2\frac{5.00\text{ W}}{4\pi(3.0\text{ m})^2\left(3.00\times10^8\text{ m/s}\right)\left(8.85\times10^{-12}\text{ C}^2/\text{N}\cdot\text{m}^2\right)}} = 5.77\text{ N/C}, \\ B_0 &= E_0/c = 1.92\times10^{-8}\text{ T}. \end{aligned}$$

Significance

The intensity I falls off as the distance squared if the radiation is dispersed uniformly in all directions.

Example 16.5

Radio Range

A 60-kW radio transmitter on Earth sends its signal to a satellite 100 km away (Figure 16.12). At what distance in the same direction would the signal have the same maximum field strength if the transmitter's output power were increased to 90 kW?

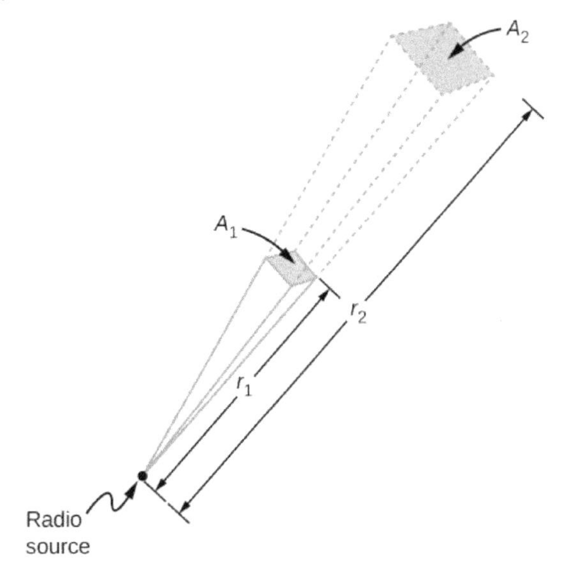

Figure 16.12 In three dimensions, a signal spreads over a solid angle as it travels outward from its source.

Strategy

The area over which the power in a particular direction is dispersed increases as distance squared, as illustrated in the figure. Change the power output P by a factor of (90 kW/60 kW) and change the area by the same factor to keep $I = \frac{P}{A} = \frac{c\varepsilon_0 E_0^2}{2}$ the same. Then use the proportion of area A in the diagram to distance squared to find the distance that produces the calculated change in area.

Solution

Using the proportionality of the areas to the squares of the distances, and solving, we obtain from the diagram

$$\frac{r_2^2}{r_1^2} = \frac{A_2}{A_1} = \frac{90\,\text{W}}{60\,\text{W}},$$

$$r_2 = \sqrt{\frac{90}{60}}(100\,\text{km}) = 122\,\text{km}.$$

Significance

The range of a radio signal is the maximum distance between the transmitter and receiver that allows for normal operation. In the absence of complications such as reflections from obstacles, the intensity follows an inverse square law, and doubling the range would require multiplying the power by four.

16.4 | Momentum and Radiation Pressure

Learning Objectives

By the end of this section, you will be able to:

- Describe the relationship of the radiation pressure and the energy density of an electromagnetic wave
- Explain how the radiation pressure of light, while small, can produce observable astronomical effects

Material objects consist of charged particles. An electromagnetic wave incident on the object exerts forces on the charged particles, in accordance with the Lorentz force, Equation 16.11. These forces do work on the particles of the object, increasing its energy, as discussed in the previous section. The energy that sunlight carries is a familiar part of every warm sunny day. A much less familiar feature of electromagnetic radiation is the extremely weak pressure that electromagnetic radiation produces by exerting a force in the direction of the wave. This force occurs because electromagnetic waves contain and transport momentum.

To understand the direction of the force for a very specific case, consider a plane electromagnetic wave incident on a metal in which electron motion, as part of a current, is damped by the resistance of the metal, so that the average electron motion is in phase with the force causing it. This is comparable to an object moving against friction and stopping as soon as the force pushing it stops (Figure 16.13). When the electric field is in the direction of the positive y-axis, electrons move in the negative y-direction, with the magnetic field in the direction of the positive z-axis. By applying the right-hand rule, and accounting for the negative charge of the electron, we can see that the force on the electron from the magnetic field is in the direction of the positive x-axis, which is the direction of wave propagation. When the E field reverses, the B field does too, and the force is again in the same direction. Maxwell's equations together with the Lorentz force equation imply the existence of radiation pressure much more generally than this specific example, however.

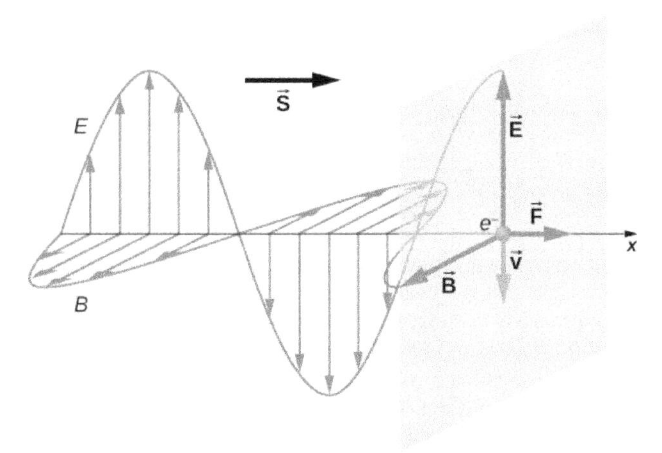

Figure 16.13 Electric and magnetic fields of an electromagnetic wave can combine to produce a force in the direction of propagation, as illustrated for the special case of electrons whose motion is highly damped by the resistance of a metal.

Maxwell predicted that an electromagnetic wave carries momentum. An object absorbing an electromagnetic wave would experience a force in the direction of propagation of the wave. The force corresponds to radiation pressure exerted on the object by the wave. The force would be twice as great if the radiation were reflected rather than absorbed.

Maxwell's prediction was confirmed in 1903 by Nichols and Hull by precisely measuring radiation pressures with a torsion balance. The schematic arrangement is shown in Figure 16.14. The mirrors suspended from a fiber were housed inside a glass container. Nichols and Hull were able to obtain a small measurable deflection of the mirrors from shining light on one of them. From the measured deflection, they could calculate the unbalanced force on the mirror, and obtained agreement with the predicted value of the force.

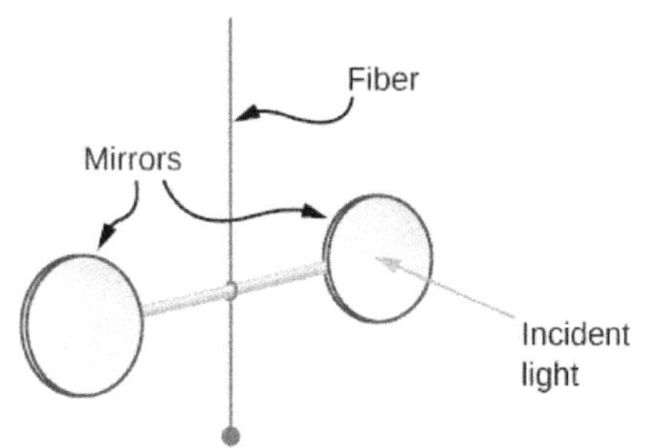

Figure 16.14 Simplified diagram of the central part of the apparatus Nichols and Hull used to precisely measure radiation pressure and confirm Maxwell's prediction.

The **radiation pressure** p_{rad} applied by an electromagnetic wave on a perfectly absorbing surface turns out to be equal to the energy density of the wave:

$$p_{\text{rad}} = u \,(\text{Perfect absorber}). \tag{16.34}$$

If the material is perfectly reflecting, such as a metal surface, and if the incidence is along the normal to the surface, then the pressure exerted is twice as much because the momentum direction reverses upon reflection:

$$p_{\text{rad}} = 2u \,(\text{Perfect reflec or}). \tag{16.35}$$

We can confirm that the units are right:

$$[u] = \frac{\text{J}}{\text{m}^3} = \frac{\text{N} \cdot \text{m}}{\text{m}^3} = \frac{\text{N}}{\text{m}^2} = \text{units of pressure.}$$

Equation 16.34 and Equation 16.35 give the instantaneous pressure, but because the energy density oscillates rapidly, we are usually interested in the time-averaged radiation pressure, which can be written in terms of intensity:

$$p = \langle p_{\text{rad}} \rangle = \begin{cases} I/c & \text{Perfect absorber} \\ 2I/c & \text{Perfect reflec or.} \end{cases} \tag{16.36}$$

Radiation pressure plays a role in explaining many observed astronomical phenomena, including the appearance of comets. Comets are basically chunks of icy material in which frozen gases and particles of rock and dust are embedded. When a comet approaches the Sun, it warms up and its surface begins to evaporate. The *coma* of the comet is the hazy area around it from the gases and dust. Some of the gases and dust form tails when they leave the comet. Notice in Figure 16.15 that a comet has *two* tails. The *ion tail* (or *gas tail* in Figure 16.15) is composed mainly of ionized gases. These ions interact electromagnetically with the solar wind, which is a continuous stream of charged particles emitted by the Sun. The force of the solar wind on the ionized gases is strong enough that the ion tail almost always points directly away from the Sun. The second tail is composed of dust particles. Because the *dust tail* is electrically neutral, it does not interact with the solar wind. However, this tail is affected by the radiation pressure produced by the light from the Sun. Although quite small, this pressure is strong enough to cause the dust tail to be displaced from the path of the comet.

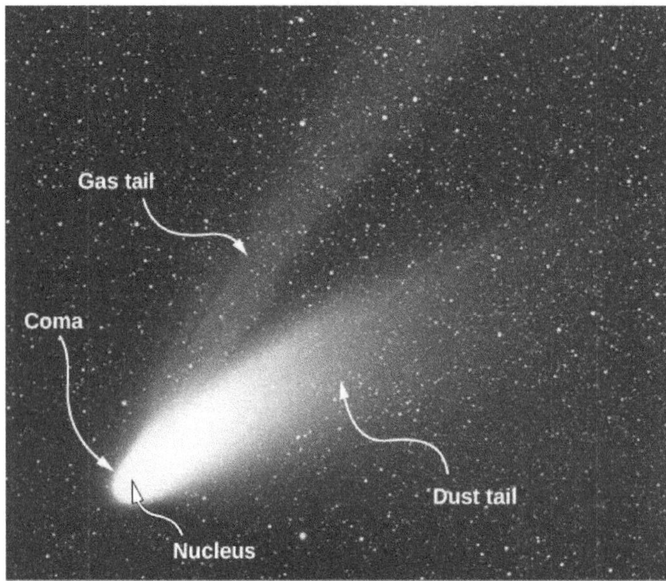

Figure 16.15 Evaporation of material being warmed by the Sun forms two tails, as shown in this photo of Comet Ison. (credit: modification of work by E. Slawik—ESO)

Example 16.6

Halley's Comet

On February 9, 1986, Comet Halley was at its closest point to the Sun, about 9.0×10^{10} m from the center of the Sun. The average power output of the Sun is 3.8×10^{26} W.

(a) Calculate the radiation pressure on the comet at this point in its orbit. Assume that the comet reflects all the incident light.

(b) Suppose that a 10-kg chunk of material of cross-sectional area 4.0×10^{-2} m^2 breaks loose from the comet. Calculate the force on this chunk due to the solar radiation. Compare this force with the gravitational force of the Sun.

Strategy

Calculate the intensity of solar radiation at the given distance from the Sun and use that to calculate the radiation pressure. From the pressure and area, calculate the force.

Solution

a. The intensity of the solar radiation is the average solar power per unit area. Hence, at 9.0×10^{10} m from the center of the Sun, we have

$$I = S_{\text{avg}} = \frac{3.8 \times 10^{26}\ \text{W}}{4\pi\left(9.0 \times 10^{10}\ \text{m}\right)^2} = 3.7 \times 10^3\ \text{W/m}^2.$$

Assuming the comet reflects all the incident radiation, we obtain from Equation 16.36

$$p = \frac{2I}{c} = \frac{2\left(3.7 \times 10^3\ \text{W/m}^2\right)}{3.00 \times 10^8\ \text{m/s}} = 2.5 \times 10^{-5}\ \text{N/m}^2.$$

b. The force on the chunk due to the radiation is

$$\begin{aligned} F &= pA = \left(2.5 \times 10^{-5}\ \text{N/m}^2\right)\left(4.0 \times 10^{-2}\ \text{m}^2\right) \\ &= 1.0 \times 10^{-6}\ \text{N}, \end{aligned}$$

whereas the gravitational force of the Sun is

$$F_{\text{g}} = \frac{GMm}{r^2} = \frac{\left(6.67 \times 10^{-11}\ \text{N} \cdot \text{m}^2/\text{kg}^2\right)\left(2.0 \times 10^{30}\ \text{kg}\right)(10\ \text{kg})}{\left(9.0 \times 10^{10}\ \text{m}\right)^2} = 0.16\ \text{N}.$$

Significance

The gravitational force of the Sun on the chunk is therefore much greater than the force of the radiation.

After Maxwell showed that light carried momentum as well as energy, a novel idea eventually emerged, initially only as science fiction. Perhaps a spacecraft with a large reflecting light sail could use radiation pressure for propulsion. Such a vehicle would not have to carry fuel. It would experience a constant but small force from solar radiation, instead of the short bursts from rocket propulsion. It would accelerate slowly, but by being accelerated continuously, it would eventually reach great speeds. A spacecraft with small total mass and a sail with a large area would be necessary to obtain a usable acceleration.

When the space program began in the 1960s, the idea started to receive serious attention from NASA. The most recent development in light propelled spacecraft has come from a citizen-funded group, the Planetary Society. It is currently testing the use of light sails to propel a small vehicle built from *CubeSats*, tiny satellites that NASA places in orbit for various research projects during space launches intended mainly for other purposes.

The *LightSail* spacecraft shown below (Figure 16.16) consists of three *CubeSats* bundled together. It has a total mass of only about 5 kg and is about the size as a loaf of bread. Its sails are made of very thin Mylar and open after launch to have a surface area of $32\ \text{m}^2$.

Figure 16.16 Two small *CubeSat* satellites deployed from the International Space Station in May, 2016. The solar sails open out when the CubeSats are far enough away from the Station.

The first *LightSail* spacecraft was launched in 2015 to test the sail deployment system. It was placed in low-earth orbit in 2015 by hitching a ride on an Atlas 5 rocket launched for an unrelated mission. The test was successful, but the low-earth orbit allowed too much drag on the spacecraft to accelerate it by sunlight. Eventually, it burned in the atmosphere, as expected. The next Planetary Society's *LightSail* solar sailing spacecraft is scheduled for 2016. An illustration (https://openstaxcollege.org/l/21lightsail) of the spacecraft, as it is expected to appear in flight, can be seen on the Planetary Society's website.

Example 16.7

LightSail Acceleration

The intensity of energy from sunlight at a distance of 1 AU from the Sun is $1370\ \text{W/m}^2$. The *LightSail* spacecraft has sails with total area of $32\ \text{m}^2$ and a total mass of 5.0 kg. Calculate the maximum acceleration LightSail spacecraft could achieve from radiation pressure when it is about 1 AU from the Sun.

Strategy

The maximum acceleration can be expected when the sail is opened directly facing the Sun. Use the light intensity to calculate the radiation pressure and from it, the force on the sails. Then use Newton's second law to calculate the acceleration.

Solution

The radiation pressure is

$$F = pA = 2uA = \frac{2I}{c}A = \frac{2\left(1370\ \text{W/m}^2\right)\left(32\ \text{m}^2\right)}{\left(3.00 \times 10^8\ \text{m/s}\right)} = 2.92 \times 10^{-4}\ \text{N}.$$

The resulting acceleration is

$$a = \frac{F}{m} = \frac{2.92 \times 10^{-4}\ \text{N}}{5.0\ \text{kg}} = 5.8 \times 10^{-5}\ \text{m/s}^2.$$

Significance

If this small acceleration continued for a year, the craft would attain a speed of 1829 m/s, or 6600 km/h.

16.5 Check Your Understanding How would the speed and acceleration of a radiation-propelled spacecraft be affected as it moved farther from the Sun on an interplanetary space flight?

16.5 | The Electromagnetic Spectrum

Learning Objectives

By the end of this section, you will be able to:

- Explain how electromagnetic waves are divided into different ranges, depending on wavelength and corresponding frequency
- Describe how electromagnetic waves in different categories are produced
- Describe some of the many practical everyday applications of electromagnetic waves

Electromagnetic waves have a vast range of practical everyday applications that includes such diverse uses as communication by cell phone and radio broadcasting, WiFi, cooking, vision, medical imaging, and treating cancer. In this module, we discuss how electromagnetic waves are classified into categories such as radio, infrared, ultraviolet, and so on. We also summarize some of the main applications for each range.

The different categories of electromagnetic waves differ in their wavelength range, or equivalently, in their corresponding frequency ranges. Their properties change smoothly from one frequency range to the next, with different applications in each range. A brief overview of the production and utilization of electromagnetic waves is found in Table 16.1.

Type of wave	Production	Applications	Issues
Radio	Accelerating charges	Communications Remote controls MRI	Requires control for band use
Microwaves	Accelerating charges and thermal agitation	Communications Ovens Radar Cell phone use	
Infrared	Thermal agitation and electronic transitions	Thermal imaging Heating	Absorbed by atmosphere Greenhouse effect
Visible light	Thermal agitation and electronic transitions	Photosynthesis Human vision	
Ultraviolet	Thermal agitation and electronic transitions	Sterilization Vitamin D production	Ozone depletion Cancer causing
X-rays	Inner electronic transitions and fast collisions	Security Medical diagnosis Cancer therapy	Cancer causing
Gamma rays	Nuclear decay	Nuclear medicine Security Medical diagnosis Cancer therapy	Cancer causing Radiation damage

Table 16.1 Electromagnetic Waves

The relationship $c = f\lambda$ between frequency f and wavelength λ applies to all waves and ensures that greater frequency means smaller wavelength. Figure 16.17 shows how the various types of electromagnetic waves are categorized according to their wavelengths and frequencies—that is, it shows the electromagnetic spectrum.

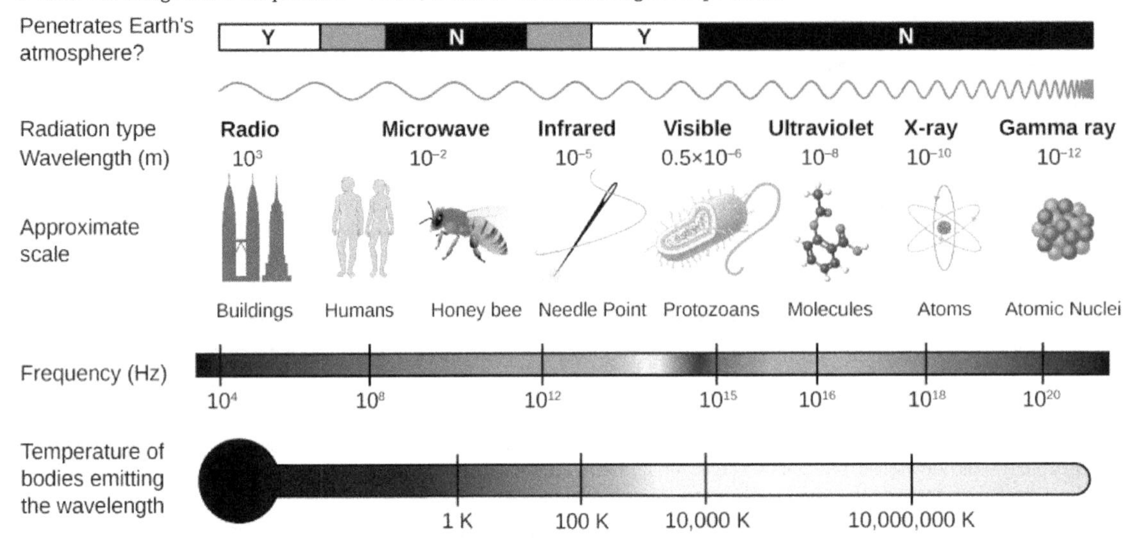

Figure 16.17 The electromagnetic spectrum, showing the major categories of electromagnetic waves.

Radio Waves

The term **radio waves** refers to electromagnetic radiation with wavelengths greater than about 0.1 m. Radio waves are commonly used for audio communications (i.e., for radios), but the term is used for electromagnetic waves in this range regardless of their application. Radio waves typically result from an alternating current in the wires of a broadcast antenna. They cover a very broad wavelength range and are divided into many subranges, including microwaves, electromagnetic waves used for AM and FM radio, cellular telephones, and TV signals.

There is no lowest frequency of radio waves, but ELF waves, or “extremely low frequency” are among the lowest frequencies commonly encountered, from 3 Hz to 3 kHz. The accelerating charge in the ac currents of electrical power lines produce electromagnetic waves in this range. ELF waves are able to penetrate sea water, which strongly absorbs electromagnetic waves of higher frequency, and therefore are useful for submarine communications.

In order to use an electromagnetic wave to transmit information, the amplitude, frequency, or phase of the wave is *modulated*, or varied in a controlled way that encodes the intended information into the wave. In AM radio transmission, the amplitude of the wave is modulated to mimic the vibrations of the sound being conveyed. Fourier’s theorem implies that the modulated AM wave amounts to a superposition of waves covering some narrow frequency range. Each AM station is assigned a specific carrier frequency that, by international agreement, is allowed to vary by $\pm 5\ \text{kHz}$. In FM radio transmission, the frequency of the wave is modulated to carry this information, as illustrated in Figure 16.18, and the frequency of each station is allowed to use 100 kHz on each side of its carrier frequency. The electromagnetic wave produces a current in a receiving antenna, and the radio or television processes the signal to produce the sound and any image. The higher the frequency of the radio wave used to carry the data, the greater the detailed variation of the wave that can be carried by modulating it over each time unit, and the more data that can be transmitted per unit of time. The assigned frequencies for AM broadcasting are 540 to 1600 kHz, and for FM are 88 MHz to108 MHz.

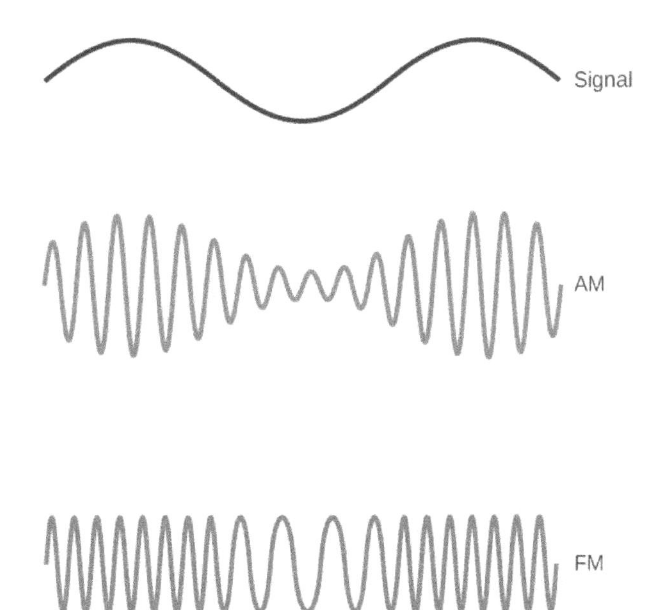

Figure 16.18 Electromagnetic waves are used to carry communications signals by varying the wave's amplitude (AM), its frequency (FM), or its phase.

Cell phone conversations, and television voice and video images are commonly transmitted as digital data, by converting the signal into a sequence of binary ones and zeros. This allows clearer data transmission when the signal is weak, and allows using computer algorithms to compress the digital data to transmit more data in each frequency range. Computer data as well is transmitted as a sequence of binary ones and zeros, each one or zero constituting one bit of data.

Microwaves

Microwaves are the highest-frequency electromagnetic waves that can be produced by currents in macroscopic circuits and devices. Microwave frequencies range from about $10^9\,\text{Hz}$ to nearly $10^{12}\,\text{Hz}$. Their high frequencies correspond to short wavelengths compared with other radio waves—hence the name "microwave." Microwaves also occur naturally as the cosmic background radiation left over from the origin of the universe. Along with other ranges of electromagnetic waves, they are part of the radiation that any object above absolute zero emits and absorbs because of **thermal agitation**, that is, from the thermal motion of its atoms and molecules.

Most satellite-transmitted information is carried on microwaves. **Radar** is a common application of microwaves. By detecting and timing microwave echoes, radar systems can determine the distance to objects as diverse as clouds, aircraft, or even the surface of Venus.

Microwaves of 2.45 GHz are commonly used in microwave ovens. The electrons in a water molecule tend to remain closer to the oxygen nucleus than the hydrogen nuclei (Figure 16.19). This creates two separated centers of equal and opposite charges, giving the molecule a dipole moment (see Electric Field). The oscillating electric field of the microwaves inside the oven exerts a torque that tends to align each molecule first in one direction and then in the other, with the motion of each molecule coupled to others around it. This pumps energy into the continual thermal motion of the water to heat the food. The plate under the food contains no water, and remains relatively unheated.

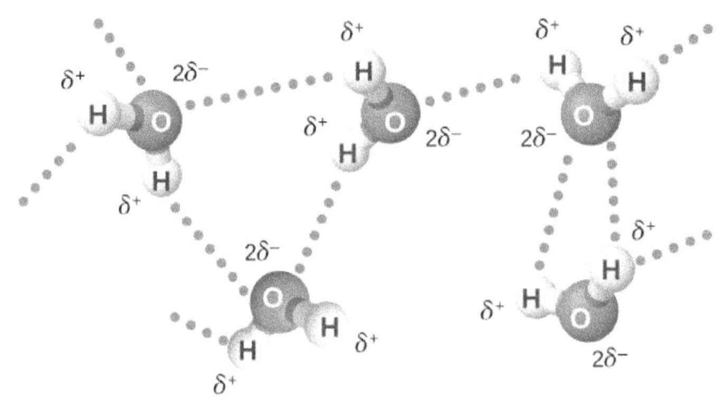

Figure 16.19 The oscillating electric field in a microwave oven exerts a torque on water molecules because of their dipole moment, and the torque reverses direction 4.90×10^9 times per second. Interactions between the molecules distributes the energy being pumped into them. The δ^+ and δ^- denote the charge distribution on the molecules.

The microwaves in a microwave oven reflect off the walls of the oven, so that the superposition of waves produces standing waves, similar to the standing waves of a vibrating guitar or violin string (see Normal Modes of a Standing Sound Wave (http://cnx.org/content/m58378/latest/)). A rotating fan acts as a stirrer by reflecting the microwaves in different directions, and food turntables, help spread out the hot spots.

Example 16.8

Why Microwave Ovens Heat Unevenly

How far apart are the hotspots in a 2.45-GHz microwave oven?

Strategy

Consider the waves along one direction in the oven, being reflected at the opposite wall from where they are generated.

Solution

The antinodes, where maximum intensity occurs, are half the wavelength apart, with separation

$$d = \frac{1}{2}\lambda = \frac{1}{2}\frac{c}{f} = \frac{3.00 \times 10^8 \text{ m/s}}{2\left(2.45 \times 10^9 \text{ Hz}\right)} = 6.02 \text{ cm}.$$

Significance

The distance between the hot spots in a microwave oven are determined by the wavelength of the microwaves.

A cell phone has a radio receiver and a weak radio transmitter, both of which can quickly tune to hundreds of specifically assigned microwave frequencies. The low intensity of the transmitted signal gives it an intentionally limited range. A ground-based system links the phone to only to the broadcast tower assigned to the specific small area, or cell, and smoothly transitions its connection to the next cell when the signal reception there is the stronger one. This enables a cell phone to be used while changing location.

Microwaves also provide the WiFi that enables owners of cell phones, laptop computers, and similar devices to connect wirelessly to the Internet at home and at coffee shops and airports. A wireless WiFi router is a device that exchanges data over the Internet through the cable or another connection, and uses microwaves to exchange the data wirelessly with devices such as cell phones and computers. The term WiFi itself refers to the standards followed in modulating and analyzing the microwaves so that wireless routers and devices from different manufacturers work compatibly with one another. The computer data in each direction consist of sequences of binary zeros and ones, each corresponding to a binary bit. The microwaves are in the range of 2.4 GHz to 5.0 GHz range.

Other wireless technologies also use microwaves to provide everyday communications between devices. Bluetooth developed alongside WiFi as a standard for radio communication in the 2.4-GHz range between nearby devices, for example, to link to headphones and audio earpieces to devices such as radios, or a driver's cell phone to a hands-free device to allow answering phone calls without fumbling directly with the cell phone.

Microwaves find use also in radio tagging, using RFID (radio frequency identification) technology. Examples are RFID tags attached to store merchandize, transponder for toll booths use attached to the windshield of a car, or even a chip embedded into a pet's skin. The device responds to a microwave signal by emitting a signal of its own with encoded information, allowing stores to quickly ring up items at their cash registers, drivers to charge tolls to their account without stopping, and lost pets to be reunited with their owners. NFC (near field communication) works similarly, except it is much shorter range. Its mechanism of interaction is the induced magnetic field at microwave frequencies between two coils. Cell phones that have NFC capability and the right software can supply information for purchases using the cell phone instead of a physical credit card. The very short range of the data transfer is a desired security feature in this case.

Infrared Radiation

The boundary between the microwave and infrared regions of the electromagnetic spectrum is not well defined (see Figure 16.17). **Infrared radiation** is generally produced by thermal motion, and the vibration and rotation of atoms and molecules. Electronic transitions in atoms and molecules can also produce infrared radiation. About half of the solar energy arriving at Earth is in the infrared region, with most of the rest in the visible part of the spectrum. About 23% of the solar energy is absorbed in the atmosphere, about 48% is absorbed at Earth's surface, and about 29% is reflected back into space.[1]

The range of infrared frequencies extends up to the lower limit of visible light, just below red. In fact, infrared means "below red." Water molecules rotate and vibrate particularly well at infrared frequencies. Reconnaissance satellites can detect buildings, vehicles, and even individual humans by their infrared emissions, whose power radiation is proportional to the fourth power of the absolute temperature. More mundanely, we use infrared lamps, including those called *quartz heaters*, to preferentially warm us because we absorb infrared better than our surroundings.

The familiar handheld "remotes" for changing channels and settings on television sets often transmit their signal by modulating an infrared beam. If you try to use a TV remote without the infrared emitter being in direct line of sight with the infrared detector, you may find the television not responding. Some remotes use Bluetooth instead and reduce this annoyance.

Visible Light

Visible light is the narrow segment of the electromagnetic spectrum between about 400 nm and about 750 nm to which the normal human eye responds. Visible light is produced by vibrations and rotations of atoms and molecules, as well as by electronic transitions within atoms and molecules. The receivers or detectors of light largely utilize electronic transitions.

Red light has the lowest frequencies and longest wavelengths, whereas violet has the highest frequencies and shortest wavelengths (Figure 16.20). Blackbody radiation from the Sun peaks in the visible part of the spectrum but is more intense in the red than in the violet, making the sun yellowish in appearance.

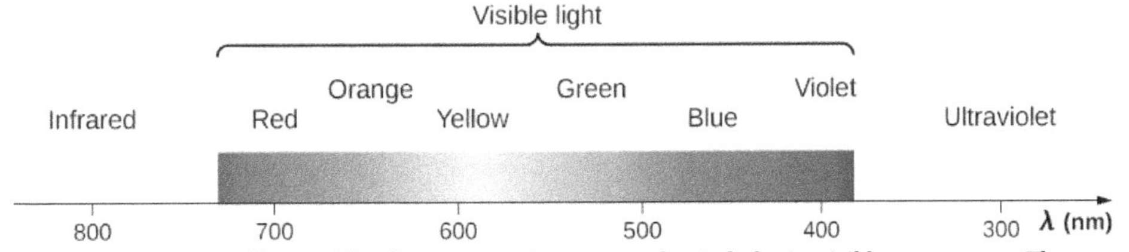

Figure 16.20 A small part of the electromagnetic spectrum that includes its visible components. The divisions between infrared, visible, and ultraviolet are not perfectly distinct, nor are those between the seven rainbow colors.

Living things—plants and animals—have evolved to utilize and respond to parts of the electromagnetic spectrum in which they are embedded. We enjoy the beauty of nature through visible light. Plants are more selective. Photosynthesis uses parts of the visible spectrum to make sugars.

1. http://earthobservatory.nasa.gov/Features/EnergyBalance/page4.php

Ultraviolet Radiation

Ultraviolet means "above violet." The electromagnetic frequencies of **ultraviolet radiation (UV)** extend upward from violet, the highest-frequency visible light. The highest-frequency ultraviolet overlaps with the lowest-frequency X-rays. The wavelengths of ultraviolet extend from 400 nm down to about 10 nm at its highest frequencies. Ultraviolet is produced by atomic and molecular motions and electronic transitions.

UV radiation from the Sun is broadly subdivided into three wavelength ranges: UV-A (320–400 nm) is the lowest frequency, then UV-B (290–320 nm) and UV-C (220–290 nm). Most UV-B and all UV-C are absorbed by ozone (O_3) molecules in the upper atmosphere. Consequently, 99% of the solar UV radiation reaching Earth's surface is UV-A.

Sunburn is caused by large exposures to UV-B and UV-C, and repeated exposure can increase the likelihood of skin cancer. The tanning response is a defense mechanism in which the body produces pigments in inert skin layers to reduce exposure of the living cells below.

As examined in a later chapter, the shorter the wavelength of light, the greater the energy change of an atom or molecule that absorbs the light in an electronic transition. This makes short-wavelength ultraviolet light damaging to living cells. It also explains why ultraviolet radiation is better able than visible light to cause some materials to glow, or *fluoresce*.

Besides the adverse effects of ultraviolet radiation, there are also benefits of exposure in nature and uses in technology. Vitamin D production in the skin results from exposure to UV-B radiation, generally from sunlight. Several studies suggest vitamin D deficiency is associated with the development of a range of cancers (prostate, breast, colon), as well as osteoporosis. Low-intensity ultraviolet has applications such as providing the energy to cause certain dyes to fluoresce and emit visible light, for example, in printed money to display hidden watermarks as counterfeit protection.

X-Rays

X-rays have wavelengths from about $10^{-8}\,\text{m}$ to $10^{-12}\,\text{m}$. They have shorter wavelengths, and higher frequencies, than ultraviolet, so that the energy they transfer at an atomic level is greater. As a result, X-rays have adverse effects on living cells similar to those of ultraviolet radiation, but they are more penetrating. Cancer and genetic defects can be induced by X-rays. Because of their effect on rapidly dividing cells, X-rays can also be used to treat and even cure cancer.

The widest use of X-rays is for imaging objects that are opaque to visible light, such as the human body or aircraft parts. In humans, the risk of cell damage is weighed carefully against the benefit of the diagnostic information obtained.

Gamma Rays

Soon after nuclear radioactivity was first detected in 1896, it was found that at least three distinct types of radiation were being emitted, and these were designated as alpha, beta, and gamma rays. The most penetrating nuclear radiation, the **gamma ray (γ ray)**, was later found to be an extremely high-frequency electromagnetic wave.

The lower end of the γ- ray frequency range overlaps the upper end of the X-ray range. Gamma rays have characteristics identical to X-rays of the same frequency—they differ only in source. The name "gamma rays" is generally used for electromagnetic radiation emitted by a nucleus, while X-rays are generally produced by bombarding a target with energetic electrons in an X-ray tube. At higher frequencies, γ rays are more penetrating and more damaging to living tissue. They have many of the same uses as X-rays, including cancer therapy. Gamma radiation from radioactive materials is used in nuclear medicine.

Use this simulation (https://openstaxcollege.org/l/21simlightmol) to explore how light interacts with molecules in our atmosphere.

Explore how light interacts with molecules in our atmosphere.

Identify that absorption of light depends on the molecule and the type of light.

Relate the energy of the light to the resulting motion.

Identify that energy increases from microwave to ultraviolet.

Predict the motion of a molecule based on the type of light it absorbs.

16.6 Check Your Understanding How do the electromagnetic waves for the different kinds of electromagnetic radiation differ?

CHAPTER 16 REVIEW

KEY TERMS

displacement current extra term in Maxwell's equations that is analogous to a real current but accounts for a changing electric field producing a magnetic field, even when the real current is present

gamma ray (γ ray) extremely high frequency electromagnetic radiation emitted by the nucleus of an atom, either from natural nuclear decay or induced nuclear processes in nuclear reactors and weapons; the lower end of the γ -ray frequency range overlaps the upper end of the X-ray range, but γ rays can have the highest frequency of any electromagnetic radiation

infrared radiation region of the electromagnetic spectrum with a frequency range that extends from just below the red region of the visible light spectrum up to the microwave region, or from $0.74\ \mu\text{m}$ to $300\ \mu\text{m}$

Maxwell's equations set of four equations that comprise a complete, overarching theory of electromagnetism

microwaves electromagnetic waves with wavelengths in the range from 1 mm to 1 m; they can be produced by currents in macroscopic circuits and devices

Poynting vector vector equal to the cross product of the electric-and magnetic fields, that describes the flow of electromagnetic energy through a surface

radar common application of microwaves; radar can determine the distance to objects as diverse as clouds and aircraft, as well as determine the speed of a car or the intensity of a rainstorm

radiation pressure force divided by area applied by an electromagnetic wave on a surface

radio waves electromagnetic waves with wavelengths in the range from 1 mm to 100 km; they are produced by currents in wires and circuits and by astronomical phenomena

thermal agitation thermal motion of atoms and molecules in any object at a temperature above absolute zero, which causes them to emit and absorb radiation

ultraviolet radiation electromagnetic radiation in the range extending upward in frequency from violet light and overlapping with the lowest X-ray frequencies, with wavelengths from 400 nm down to about 10 nm

visible light narrow segment of the electromagnetic spectrum to which the normal human eye responds, from about 400 to 750 nm

X-ray invisible, penetrating form of very high frequency electromagnetic radiation, overlapping both the ultraviolet range and the γ -ray range

KEY EQUATIONS

Displacement current	$I_{\text{d}} = \varepsilon_0 \frac{d\Phi_{\text{E}}}{dt}$
Gauss's law	$\oint \vec{\mathbf{E}} \cdot d\vec{\mathbf{A}} = \frac{Q_{\text{in}}}{\varepsilon_0}$
Gauss's law for magnetism	$\oint \vec{\mathbf{B}} \cdot d\vec{\mathbf{A}} = 0$
Faraday's law	$\oint \vec{\mathbf{E}} \cdot d\vec{\mathbf{s}} = -\frac{d\Phi_{\text{m}}}{dt}$
Ampère-Maxwell law	$\oint \vec{\mathbf{B}} \cdot d\vec{\mathbf{s}} = \mu_0 I + \varepsilon_0 \mu_0 \frac{d\Phi_{\text{E}}}{dt}$
Wave equation for plane EM wave	$\frac{\partial^2 E_y}{\partial x^2} = \varepsilon_0 \mu_0 \frac{\partial^2 E_y}{\partial t^2}$

Speed of EM waves	$c = \frac{1}{\sqrt{\varepsilon_0 \mu_0}}$
Ratio of *E* field to *B* field in electromagnetic wave	$c = \frac{E}{B}$
Energy flux (Poynting) vector	$\vec{\mathbf{S}} = \frac{1}{\mu_0} \vec{\mathbf{E}} \times \vec{\mathbf{B}}$
Average intensity of an electromagnetic wave	$I = S_{\text{avg}} = \frac{c\varepsilon_0 E_0^2}{2} = \frac{cB_0^2}{2\mu_0} = \frac{E_0 B_0}{2\mu_0}$
Radiation pressure	$p = \begin{cases} I/c & \text{Perfect absorber} \\ 2I/c & \text{Perfect reflec or} \end{cases}$

SUMMARY

16.1 Maxwell's Equations and Electromagnetic Waves

- Maxwell's prediction of electromagnetic waves resulted from his formulation of a complete and symmetric theory of electricity and magnetism, known as Maxwell's equations.
- The four Maxwell's equations together with the Lorentz force law encompass the major laws of electricity and magnetism. The first of these is Gauss's law for electricity; the second is Gauss's law for magnetism; the third is Faraday's law of induction (including Lenz's law); and the fourth is Ampère's law in a symmetric formulation that adds another source of magnetism, namely changing electric fields.
- The symmetry introduced between electric and magnetic fields through Maxwell's displacement current explains the mechanism of electromagnetic wave propagation, in which changing magnetic fields produce changing electric fields and vice versa.
- Although light was already known to be a wave, the nature of the wave was not understood before Maxwell. Maxwell's equations also predicted electromagnetic waves with wavelengths and frequencies outside the range of light. These theoretical predictions were first confirmed experimentally by Heinrich Hertz.

16.2 Plane Electromagnetic Waves

- Maxwell's equations predict that the directions of the electric and magnetic fields of the wave, and the wave's direction of propagation, are all mutually perpendicular. The electromagnetic wave is a transverse wave.
- The strengths of the electric and magnetic parts of the wave are related by $c = E/B,$ which implies that the magnetic field *B* is very weak relative to the electric field *E*.
- Accelerating charges create electromagnetic waves (for example, an oscillating current in a wire produces electromagnetic waves with the same frequency as the oscillation).

16.3 Energy Carried by Electromagnetic Waves

- The energy carried by any wave is proportional to its amplitude squared. For electromagnetic waves, this means intensity can be expressed as

$$I = \frac{c\varepsilon_0 E_0^2}{2}$$

where *I* is the average intensity in W/m^2 and E_0 is the maximum electric field strength of a continuous sinusoidal wave. This can also be expressed in terms of the maximum magnetic field strength B_0 as

$$I = \frac{cB_0^2}{2\mu_0}$$

and in terms of both electric and magnetic fields as

$$I = \frac{E_0 B_0}{2\mu_0}.$$

The three expressions for I_{avg} are all equivalent.

16.4 Momentum and Radiation Pressure

- Electromagnetic waves carry momentum and exert radiation pressure.
- The radiation pressure of an electromagnetic wave is directly proportional to its energy density.
- The pressure is equal to twice the electromagnetic energy intensity if the wave is reflected and equal to the incident energy intensity if the wave is absorbed.

16.5 The Electromagnetic Spectrum

- The relationship among the speed of propagation, wavelength, and frequency for any wave is given by $v = f\lambda,$ so that for electromagnetic waves, $c = f\lambda,$ where *f* is the frequency, λ is the wavelength, and *c* is the speed of light.
- The electromagnetic spectrum is separated into many categories and subcategories, based on the frequency and wavelength, source, and uses of the electromagnetic waves.

CONCEPTUAL QUESTIONS

16.1 Maxwell's Equations and Electromagnetic Waves

1. Explain how the displacement current maintains the continuity of current in a circuit containing a capacitor.

2. Describe the field lines of the induced magnetic field along the edge of the imaginary horizontal cylinder shown below if the cylinder is in a spatially uniform electric field that is horizontal, pointing to the right, and increasing in magnitude.

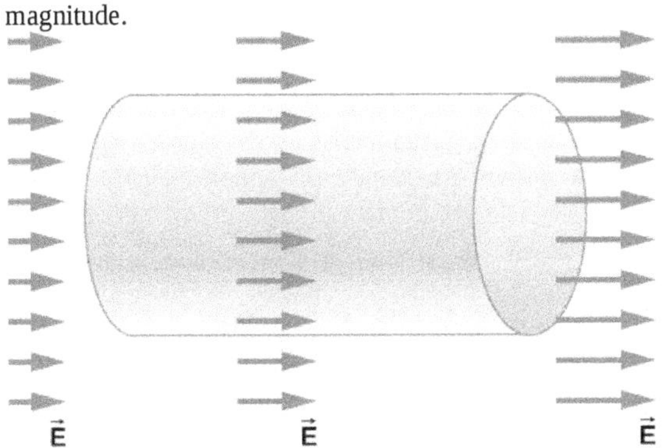

3. Why is it much easier to demonstrate in a student lab that a changing magnetic field induces an electric field than it is to demonstrate that a changing electric field produces a magnetic field?

16.2 Plane Electromagnetic Waves

4. If the electric field of an electromagnetic wave is oscillating along the *z*-axis and the magnetic field is oscillating along the *x*-axis, in what possible direction is the wave traveling?

5. In which situation shown below will the electromagnetic wave be more successful in inducing a current in the wire? Explain.

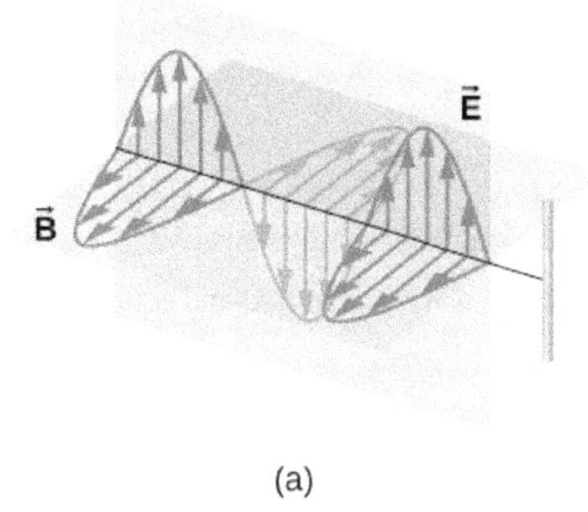

(a)

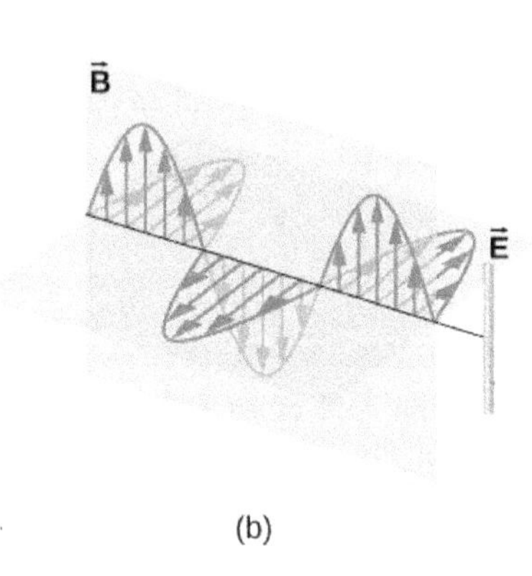

(b)

6. In which situation shown below will the electromagnetic wave be more successful in inducing a current in the loop? Explain.

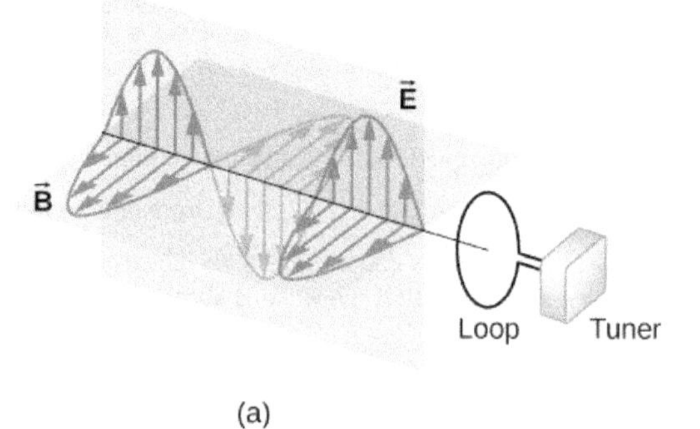

(a)

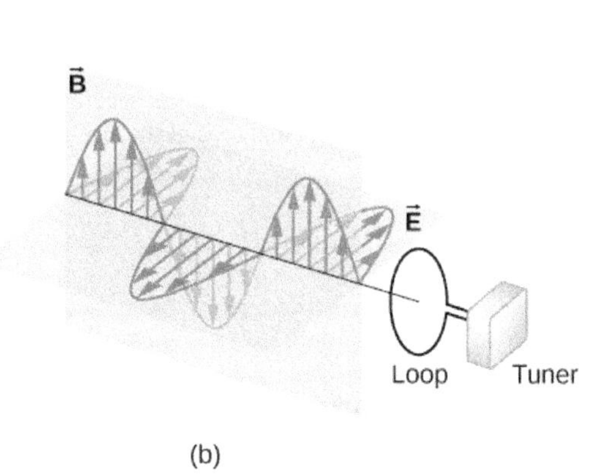

(b)

7. Under what conditions might wires in a circuit where the current flows in only one direction emit electromagnetic waves?

8. Shown below is the interference pattern of two radio antennas broadcasting the same signal. Explain how this is analogous to the interference pattern for sound produced by two speakers. Could this be used to make a directional antenna system that broadcasts preferentially in certain directions? Explain.

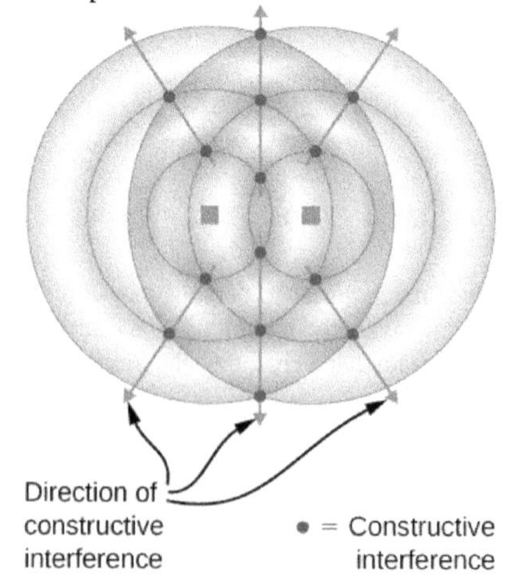

16.3 Energy Carried by Electromagnetic Waves

9. When you stand outdoors in the sunlight, why can you feel the energy that the sunlight carries, but not the momentum it carries?

10. How does the intensity of an electromagnetic wave depend on its electric field? How does it depend on its magnetic field?

11. What is the physical significance of the Poynting vector?

12. A 2.0-mW helium-neon laser transmits a continuous beam of red light of cross-sectional area $0.25\ \text{cm}^2$. If the beam does not diverge appreciably, how would its rms electric field vary with distance from the laser? Explain.

16.4 Momentum and Radiation Pressure

13. Why is the radiation pressure of an electromagnetic wave on a perfectly reflecting surface twice as large as the pressure on a perfectly absorbing surface?

14. Why did the early Hubble Telescope photos of Comet Ison approaching Earth show it to have merely a fuzzy coma around it, and not the pronounced double tail that developed later (see below)?

Figure 16.21 (credit: ESA, Hubble)

15. (a) If the electric field and magnetic field in a sinusoidal plane wave were interchanged, in which direction relative to before would the energy propagate?
(b) What if the electric and the magnetic fields were both changed to their negatives?

16.5 The Electromagnetic Spectrum

16. Compare the speed, wavelength, and frequency of radio waves and X-rays traveling in a vacuum.

17. Accelerating electric charge emits electromagnetic radiation. How does this apply in each case: (a) radio waves, (b) infrared radiation.

18. Compare and contrast the meaning of the prefix "micro" in the names of SI units in the term *microwaves*.

19. Part of the light passing through the air is scattered in all directions by the molecules comprising the atmosphere. The wavelengths of visible light are larger than molecular sizes, and the scattering is strongest for wavelengths of light closest to sizes of molecules.
(a) Which of the main colors of light is scattered the most?
(b) Explain why this would give the sky its familiar background color at midday.

20. When a bowl of soup is removed from a microwave oven, the soup is found to be steaming hot, whereas the bowl is only warm to the touch. Discuss the temperature changes that have occurred in terms of energy transfer.

21. Certain orientations of a broadcast television antenna give better reception than others for a particular station. Explain.

22. What property of light corresponds to loudness in sound?

23. Is the visible region a major portion of the electromagnetic spectrum?

24. Can the human body detect electromagnetic radiation that is outside the visible region of the spectrum?

25. Radio waves normally have their *E* and *B* fields in specific directions, whereas visible light usually has its *E* and *B* fields in random and rapidly changing directions that are perpendicular to each other and to the propagation direction. Can you explain why?

26. Give an example of resonance in the reception of electromagnetic waves.

27. Illustrate that the size of details of an object that can be detected with electromagnetic waves is related to their wavelength, by comparing details observable with two different types (for example, radar and visible light).

28. In which part of the electromagnetic spectrum are each of these waves:
(a) $f = 10.0$ kHz, (b) $f = \lambda = 750\,\text{nm}$,

(c) $f = 1.25 \times 10^8\ \text{Hz}$, (d) 0.30 nm

29. In what range of electromagnetic radiation are the electromagnetic waves emitted by power lines in a country that uses 50-Hz ac current?

30. If a microwave oven could be modified to merely tune the waves generated to be in the infrared range instead of using microwaves, how would this affect the uneven heating of the oven?

31. A leaky microwave oven in a home can sometimes cause interference with the homeowner's WiFi system. Why?

32. When a television news anchor in a studio speaks to a reporter in a distant country, there is sometimes a noticeable lag between when the anchor speaks in the studio and when the remote reporter hears it and replies. Explain what causes this delay.

PROBLEMS

16.1 Maxwell's Equations and Electromagnetic Waves

33. Show that the magnetic field at a distance r from the axis of two circular parallel plates, produced by placing charge $Q(t)$ on the plates is
$$B_{\text{ind}} = \frac{\mu_0}{2\pi r}\frac{dQ(t)}{dt}.$$

34. Express the displacement current in a capacitor in terms of the capacitance and the rate of change of the voltage across the capacitor.

35. A potential difference $V(t) = V_0 \sin \omega t$ is maintained across a parallel-plate capacitor with capacitance C consisting of two circular parallel plates. A thin wire with resistance R connects the centers of the two plates, allowing charge to leak between plates while they are charging.
(a) Obtain expressions for the leakage current $I_{\text{res}}(t)$ in the thin wire. Use these results to obtain an expression for the current $I_{\text{real}}(t)$ in the wires connected to the capacitor.
(b) Find the displacement current in the space between the plates from the changing electric field between the plates.
(c) Compare $I_{\text{real}}(t)$ with the sum of the displacement current $I_{\text{d}}(t)$ and resistor current $I_{\text{res}}(t)$ between the plates, and explain why the relationship you observe would be expected.

36. Suppose the parallel-plate capacitor shown below is accumulating charge at a rate of 0.010 C/s. What is the induced magnetic field at a distance of 10 cm from the capacitator?

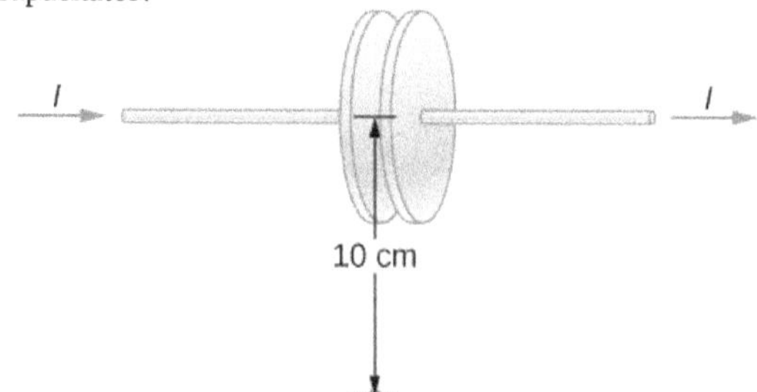

37. The potential difference $V(t)$ between parallel plates shown above is instantaneously increasing at a rate of 10^7 V/s. What is the displacement current between the plates if the separation of the plates is 1.00 cm and they have an area of $0.200\,\text{m}^2$?

38. A parallel-plate capacitor has a plate area of $A = 0.250\,\text{m}^2$ and a separation of 0.0100 m. What must be must be the angular frequency ω for a voltage $V(t) = V_0 \sin \omega t$ with $V_0 = 100\,\text{V}$ to produce a maximum displacement induced current of 1.00 A between the plates?

39. The voltage across a parallel-plate capacitor with area $A = 800\,\text{cm}^2$ and separation $d = 2\,\text{mm}$ varies sinusoidally as $V = (15\,\text{mV})\cos(150t)$, where t is in seconds. Find the displacement current between the plates.

40. The voltage across a parallel-plate capacitor with area A and separation d varies with time t as $V = at^2$, where a is a constant. Find the displacement current between the plates.

16.2 Plane Electromagnetic Waves

41. If the Sun suddenly turned off, we would not know it until its light stopped coming. How long would that be, given that the Sun is 1.496×10^{11} m away?

42. What is the maximum electric field strength in an electromagnetic wave that has a maximum magnetic field strength of 5.00×10^{-4} T (about 10 times Earth's magnetic field)?

43. An electromagnetic wave has a frequency of 12 MHz. What is its wavelength in vacuum?

44. If electric and magnetic field strengths vary sinusoidally in time at frequency 1.00 GHz, being zero at $t = 0$, then $E = E_0 \sin 2\pi ft$ and $B = B_0 \sin 2\pi ft$. (a) When are the field strengths next equal to zero? (b) When do they reach their most negative value? (c) How much time is needed for them to complete one cycle?

45. The electric field of an electromagnetic wave traveling in vacuum is described by the following wave function:

$$\vec{\mathbf{E}} = (5.00\ \text{V/m}) \cos\left[kx - \left(6.00 \times 10^9\ \text{s}^{-1}\right)t + 0.40\right]\hat{\mathbf{j}}$$

where k is the wavenumber in rad/m, x is in m, t is in s.

Find the following quantities:

(a) amplitude
(b) frequency
(c) wavelength
(d) the direction of the travel of the wave
(e) the associated magnetic field wave

46. A plane electromagnetic wave of frequency 20 GHz moves in the positive y-axis direction such that its electric field is pointed along the z-axis. The amplitude of the electric field is 10 V/m. The start of time is chosen so that at $t = 0$, the electric field has a value 10 V/m at the origin. (a) Write the wave function that will describe the electric field wave. (b) Find the wave function that will describe the associated magnetic field wave.

47. The following represents an electromagnetic wave traveling in the direction of the positive y-axis:

$$E_x = 0;\ E_y = E_0 \cos(kx - \omega t);\ E_z = 0$$
$$B_x = 0;\ B_y = 0;\ B_z = B_0 \cos(kx - \omega t).$$

The wave is passing through a wide tube of circular cross-section of radius R whose axis is along the y-axis. Find the expression for the displacement current through the tube.

16.3 Energy Carried by Electromagnetic Waves

48. While outdoors on a sunny day, a student holds a large convex lens of radius 4.0 cm above a sheet of paper to produce a bright spot on the paper that is 1.0 cm in radius, rather than a sharp focus. By what factor is the electric field in the bright spot of light related to the electric field of sunlight leaving the side of the lens facing the paper?

49. A plane electromagnetic wave travels northward. At one instant, its electric field has a magnitude of 6.0 V/m and points eastward. What are the magnitude and direction of the magnetic field at this instant?

50. The electric field of an electromagnetic wave is given by

$$E = \left(6.0 \times 10^{-3}\ \text{V/m}\right) \sin\left[2\pi\left(\frac{x}{18\ \text{m}} - \frac{t}{6.0 \times 10^{-8}\ \text{s}}\right)\right]\hat{\mathbf{j}}.$$

Write the equations for the associated magnetic field and Poynting vector.

51. A radio station broadcasts at a frequency of 760 kHz. At a receiver some distance from the antenna, the maximum magnetic field of the electromagnetic wave detected is 2.15×10^{-11} T.

(a) What is the maximum electric field? (b) What is the wavelength of the electromagnetic wave?

52. The filament in a clear incandescent light bulb radiates visible light at a power of 5.00 W. Model the glass part of the bulb as a sphere of radius $r_0 = 3.00$ cm and calculate the amount of electromagnetic energy from visible light inside the bulb.

53. At what distance does a 100-W lightbulb produce the same intensity of light as a 75-W lightbulb produces 10 m away? (Assume both have the same efficiency for converting electrical energy in the circuit into emitted electromagnetic energy.)

54. An incandescent light bulb emits only 2.6 W of its power as visible light. What is the rms electric field of the emitted light at a distance of 3.0 m from the bulb?

55. A 150-W lightbulb emits 5% of its energy as electromagnetic radiation. What is the magnitude of the average Poynting vector 10 m from the bulb?

56. A small helium-neon laser has a power output of 2.5 mW. What is the electromagnetic energy in a 1.0-m length of the beam?

57. At the top of Earth's atmosphere, the time-averaged Poynting vector associated with sunlight has a magnitude of about $1.4\ \text{kW/m}^2$.

(a) What are the maximum values of the electric and magnetic fields for a wave of this intensity? (b) What is the total power radiated by the sun? Assume that the Earth is $1.5 \times 10^{11}\ \text{m}$ from the Sun and that sunlight is composed of electromagnetic plane waves.

58. The magnetic field of a plane electromagnetic wave moving along the z axis is given by $\vec{\mathbf{B}} = B_0(\cos kz + \omega t)\,\hat{\mathbf{j}}$, where $B_0 = 5.00 \times 10^{-10}\ \text{T}$ and $k = 3.14 \times 10^{-2}\ \text{m}^{-1}$.

(a) Write an expression for the electric field associated with the wave. (b) What are the frequency and the wavelength of the wave? (c) What is its average Poynting vector?

59. What is the intensity of an electromagnetic wave with a peak electric field strength of 125 V/m?

60. Assume the helium-neon lasers commonly used in student physics laboratories have power outputs of 0.500 mW. (a) If such a laser beam is projected onto a circular spot 1.00 mm in diameter, what is its intensity? (b) Find the peak magnetic field strength. (c) Find the peak electric field strength.

61. An AM radio transmitter broadcasts 50.0 kW of power uniformly in all directions. (a) Assuming all of the radio waves that strike the ground are completely absorbed, and that there is no absorption by the atmosphere or other objects, what is the intensity 30.0 km away? (*Hint:* Half the power will be spread over the area of a hemisphere.) (b) What is the maximum electric field strength at this distance?

62. Suppose the maximum safe intensity of microwaves for human exposure is taken to be $1.00\ \text{W/m}^2$. (a) If a radar unit leaks 10.0 W of microwaves (other than those sent by its antenna) uniformly in all directions, how far away must you be to be exposed to an intensity considered to be safe? Assume that the power spreads uniformly over the area of a sphere with no complications from absorption or reflection. (b) What is the maximum electric field strength at the safe intensity? (Note that early radar units leaked more than modern ones do. This caused identifiable health problems, such as cataracts, for people who worked near them.)

63. A 2.50-m-diameter university communications satellite dish receives TV signals that have a maximum electric field strength (for one channel) of $7.50\ \mu\text{V/m}$ (see below). (a) What is the intensity of this wave? (b) What is the power received by the antenna? (c) If the orbiting satellite broadcasts uniformly over an area of $1.50 \times 10^{13}\ \text{m}^2$ (a large fraction of North America), how much power does it radiate?

64. Lasers can be constructed that produce an extremely high intensity electromagnetic wave for a brief time—called pulsed lasers. They are used to initiate nuclear fusion, for example. Such a laser may produce an electromagnetic wave with a maximum electric field strength of $1.00 \times 10^{11}\ \text{V/m}$ for a time of 1.00 ns. (a) What is the maximum magnetic field strength in the wave? (b) What is the intensity of the beam? (c) What energy does it deliver on an 1.00-mm^2 area?

16.4 Momentum and Radiation Pressure

65. A 150-W lightbulb emits 5% of its energy as electromagnetic radiation. What is the radiation pressure on an absorbing sphere of radius 10 m that surrounds the bulb?

66. What pressure does light emitted uniformly in all directions from a 100-W incandescent light bulb exert on a mirror at a distance of 3.0 m, if 2.6 W of the power is emitted as visible light?

67. A microscopic spherical dust particle of radius $2\ \mu m$ and mass $10\ \mu g$ is moving in outer space at a constant speed of 30 cm/sec. A wave of light strikes it from the opposite direction of its motion and gets absorbed. Assuming the particle decelerates uniformly to zero speed in one second, what is the average electric field amplitude in the light?

68. A Styrofoam spherical ball of radius 2 mm and mass $20\ \mu g$ is to be suspended by the radiation pressure in a vacuum tube in a lab. How much intensity will be required if the light is completely absorbed the ball?

69. Suppose that $\vec{S}_{avg}$ for sunlight at a point on the surface of Earth is $900\ W/m^2$. (a) If sunlight falls perpendicularly on a kite with a reflecting surface of area $0.75\ m^2$, what is the average force on the kite due to radiation pressure? (b) How is your answer affected if the kite material is black and absorbs all sunlight?

70. Sunlight reaches the ground with an intensity of about $1.0\ kW/m^2$. A sunbather has a body surface area of $0.8\ m^2$ facing the sun while reclining on a beach chair on a clear day. (a) how much energy from direct sunlight reaches the sunbather's skin per second? (b) What pressure does the sunlight exert if it is absorbed?

71. Suppose a spherical particle of mass *m* and radius *R* in space absorbs light of intensity *I* for time *t*. (a) How much work does the radiation pressure do to accelerate the particle from rest in the given time it absorbs the light? (b) How much energy carried by the electromagnetic waves is absorbed by the particle over this time based on the radiant energy incident on the particle?

16.5 The Electromagnetic Spectrum

72. How many helium atoms, each with a radius of about 31 pm, must be placed end to end to have a length equal to one wavelength of 470 nm blue light?

73. If you wish to detect details of the size of atoms (about 0.2 nm) with electromagnetic radiation, it must have a wavelength of about this size. (a) What is its frequency? (b) What type of electromagnetic radiation might this be?

74. Find the frequency range of visible light, given that it encompasses wavelengths from 380 to 760 nm.

75. (a) Calculate the wavelength range for AM radio given its frequency range is 540 to 1600 kHz. (b) Do the same for the FM frequency range of 88.0 to 108 MHz.

76. Radio station WWVB, operated by the National Institute of Standards and Technology (NIST) from Fort Collins, Colorado, at a low frequency of 60 kHz, broadcasts a time synchronization signal whose range covers the entire continental US. The timing of the synchronization signal is controlled by a set of atomic clocks to an accuracy of 1×10^{-12} s, and repeats every 1 minute. The signal is used for devices, such as radio-controlled watches, that automatically synchronize with it at preset local times. WWVB's long wavelength signal tends to propagate close to the ground.
(a) Calculate the wavelength of the radio waves from WWVB.
(b) Estimate the error that the travel time of the signal causes in synchronizing a radio controlled watch in Norfolk, Virginia, which is 1570 mi (2527 km) from Fort Collins, Colorado.

77. An outdoor WiFi unit for a picnic area has a 100-mW output and a range of about 30 m. What output power would reduce its range to 12 m for use with the same devices as before? Assume there are no obstacles in the way and that microwaves into the ground are simply absorbed.

78. 7. The prefix "mega" (M) and "kilo" (k), when referring to amounts of computer data, refer to factors of 1024 or 2^{10} rather than 1000 for the prefix *kilo*, and $1024^2 = 2^{20}$ rather than 1,000,000 for the prefix *Mega* (M). If a wireless (WiFi) router transfers 150 Mbps of data, how many bits per second is that in decimal arithmetic?

79. A computer user finds that his wireless router transmits data at a rate of 75 Mbps (megabits per second). Compare the average time to transmit one bit of data with the time difference between the wifi signal reaching an observer's cell phone directly and by bouncing back to the observer from a wall 8.00 m past the observer.

80. (a) The ideal size (most efficient) for a broadcast antenna with one end on the ground is one-fourth the wavelength ($\lambda/4$) of the electromagnetic radiation being sent out. If a new radio station has such an antenna that is 50.0 m high, what frequency does it broadcast most efficiently? Is this in the AM or FM band? (b) Discuss the analogy of the fundamental resonant mode of an air column closed at one end to the resonance of currents on an antenna that is one-fourth their wavelength.

81. What are the wavelengths of (a) X-rays of frequency 2.0×10^{17} Hz? (b) Yellow light of frequency 5.1×10^{14} Hz? (c) Gamma rays of frequency 1.0×10^{23} Hz?

82. For red light of $\lambda = 660\text{ nm}$, what are f, ω, and k?

83. A radio transmitter broadcasts plane electromagnetic waves whose maximum electric field at a particular location is $1.55 \times 10^{-3}\text{ V/m}$. What is the maximum magnitude of the oscillating magnetic field at that location? How does it compare with Earth's magnetic field?

84. (a) Two microwave frequencies authorized for use in microwave ovens are: 915 and 2450 MHz. Calculate the wavelength of each. (b) Which frequency would produce smaller hot spots in foods due to interference effects?

85. During normal beating, the heart creates a maximum 4.00-mV potential across 0.300 m of a person's chest, creating a 1.00-Hz electromagnetic wave. (a) What is the maximum electric field strength created? (b) What is the corresponding maximum magnetic field strength in the electromagnetic wave? (c) What is the wavelength of the electromagnetic wave?

86. Distances in space are often quoted in units of light-years, the distance light travels in 1 year. (a) How many meters is a light-year? (b) How many meters is it to Andromeda, the nearest large galaxy, given that it is 2.54×10^6 ly away? (c) The most distant galaxy yet discovered is 13.4×10^9 ly away. How far is this in meters?

87. A certain 60.0-Hz ac power line radiates an electromagnetic wave having a maximum electric field strength of 13.0 kV/m. (a) What is the wavelength of this very-low-frequency electromagnetic wave? (b) What type of electromagnetic radiation is this wave (b) What is its maximum magnetic field strength?

88. (a) What is the frequency of the 193-nm ultraviolet radiation used in laser eye surgery? (b) Assuming the accuracy with which this electromagnetic radiation can ablate (reshape) the cornea is directly proportional to wavelength, how much more accurate can this UV radiation be than the shortest visible wavelength of light?

ADDITIONAL PROBLEMS

89. In a region of space, the electric field is pointed along the x-axis, but its magnitude changes as described by

$$E_x = (10\text{ N/C})\sin(20x - 500t)$$

$$E_y = E_z = 0$$

where t is in nanoseconds and x is in cm. Find the displacement current through a circle of radius 3 cm in the $x = 0$ plane at $t = 0$.

90. A microwave oven uses electromagnetic waves of frequency $f = 2.45 \times 10^9\text{ Hz}$ to heat foods. The waves reflect from the inside walls of the oven to produce an interference pattern of standing waves whose antinodes are hot spots that can leave observable pit marks in some foods. The pit marks are measured to be 6.0 cm apart. Use the method employed by Heinrich Hertz to calculate the speed of electromagnetic waves this implies.

Use the Appendix D for the next two exercises

91. Galileo proposed measuring the speed of light by uncovering a lantern and having an assistant a known distance away uncover his lantern when he saw the light from Galileo's lantern, and timing the delay. How far away must the assistant be for the delay to equal the human reaction time of about 0.25 s?

92. Show that the wave equation in one dimension

$$\frac{\partial^2 f}{\partial x^2} = \frac{1}{v^2}\frac{\partial^2 f}{\partial t^2}$$

is satisfied by any doubly differentiable function of either the form $f(x - vt)$ or $f(x + vt)$.

93. On its highest power setting, a microwave oven increases the temperature of 0.400 kg of spaghetti by $45.0\,°\text{C}$ in 120 s. (a) What was the rate of energy absorption by the spaghetti, given that its specific heat is $3.76 \times 10^3\text{ J/kg}\cdot°\text{C}$? Assume the spaghetti is perfectly absorbing. (b) Find the average intensity of the microwaves, given that they are absorbed over a circular area 20.0 cm in diameter. (c) What is the peak electric field strength of the microwave? (d) What is its peak magnetic field strength?

94. A certain microwave oven projects 1.00 kW of microwaves onto a 30-cm-by-40-cm area. (a) What is its intensity in W/m^2? (b) Calculate the maximum electric field strength E_0 in these waves. (c) What is the maximum magnetic field strength B_0?

95. Electromagnetic radiation from a 5.00-mW laser is concentrated on a 1.00-mm^2 area. (a) What is the intensity in W/m^2? (b) Suppose a 2.00-nC electric charge is in the beam. What is the maximum electric force it experiences? (c) If the electric charge moves at 400 m/s, what maximum magnetic force can it feel?

96. A 200-turn flat coil of wire 30.0 cm in diameter acts as an antenna for FM radio at a frequency of 100 MHz. The magnetic field of the incoming electromagnetic wave is perpendicular to the coil and has a maximum strength of $1.00 \times 10^{-12}\ \text{T}$. (a) What power is incident on the coil? (b) What average emf is induced in the coil over one-fourth of a cycle? (c) If the radio receiver has an inductance of $2.50\ \mu\text{H}$, what capacitance must it have to resonate at 100 MHz?

97. Suppose a source of electromagnetic waves radiates uniformly in all directions in empty space where there are no absorption or interference effects. (a) Show that the intensity is inversely proportional to r^2, the distance from the source squared. (b) Show that the magnitudes of the electric and magnetic fields are inversely proportional to *r*.

98. A radio station broadcasts its radio waves with a power of 50,000 W. What would be the intensity of this signal if it is received on a planet orbiting Proxima Centuri, the closest star to our Sun, at 4.243 ly away?

99. The Poynting vector describes a flow of energy whenever electric and magnetic fields are present. Consider a long cylindrical wire of radius *r* with a current *I* in the wire, with resistance *R* and voltage *V*. From the expressions for the electric field along the wire and the magnetic field around the wire, obtain the magnitude and direction of the Poynting vector at the surface. Show that it accounts for an energy flow into the wire from the fields around it that accounts for the Ohmic heating of the wire.

100. The Sun's energy strikes Earth at an intensity of $1.37\ \text{kW/m}^2$. Assume as a model approximation that all of the light is absorbed. (Actually, about 30% of the light intensity is reflected out into space.)
(a) Calculate the total force that the Sun's radiation exerts on Earth.
(b) Compare this to the force of gravity between the Sun and Earth.
Earth's mass is $5.972 \times 10^{24}\ \text{kg}$.

101. If a *Lightsail* spacecraft were sent on a Mars mission, by what fraction would its propulsion force be reduced when it reached Mars?

102. Lunar astronauts placed a reflector on the Moon's surface, off which a laser beam is periodically reflected. The distance to the Moon is calculated from the round-trip time. (a) To what accuracy in meters can the distance to the Moon be determined, if this time can be measured to 0.100 ns? (b) What percent accuracy is this, given the average distance to the Moon is 384,400 km?

103. Radar is used to determine distances to various objects by measuring the round-trip time for an echo from the object. (a) How far away is the planet Venus if the echo time is 1000 s? (b) What is the echo time for a car 75.0 m from a highway police radar unit? (c) How accurately (in nanoseconds) must you be able to measure the echo time to an airplane 12.0 km away to determine its distance within 10.0 m?

104. Calculate the ratio of the highest to lowest frequencies of electromagnetic waves the eye can see, given the wavelength range of visible light is from 380 to 760 nm. (Note that the ratio of highest to lowest frequencies the ear can hear is 1000.)

105. How does the wavelength of radio waves for an AM radio station broadcasting at 1030 KHz compare with the wavelength of the lowest audible sound waves (of 20 Hz). The speed of sound in air at $20\ °\text{C}$ is about 343 m/s.

CHALLENGE PROBLEMS

106. A parallel-plate capacitor with plate separation d is connected to a source of emf that places a time-dependent voltage $V(t)$ across its circular plates of radius r_0 and area $A = \pi r_0^2$ (see below).

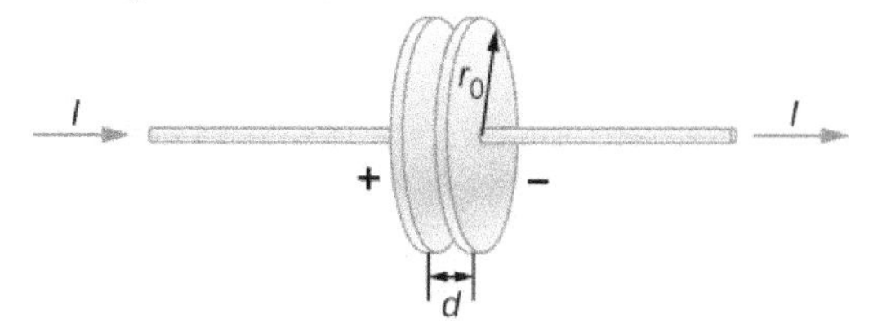

(a) Write an expression for the time rate of change of energy inside the capacitor in terms of $V(t)$ and $dV(t)/dt$.

(b) Assuming that $V(t)$ is increasing with time, identify the directions of the electric field lines inside the capacitor and of the magnetic field lines at the edge of the region between the plates, and then the direction of the Poynting vector $\vec{\mathbf{S}}$ at this location.

(c) Obtain expressions for the time dependence of $E(t)$, for $B(t)$ from the displacement current, and for the magnitude of the Poynting vector at the edge of the region between the plates.

(d) From $\vec{\mathbf{S}}$, obtain an expression in terms of $V(t)$ and $dV(t)/dt$ for the rate at which electromagnetic field energy enters the region between the plates.

(e) Compare the results of parts (a) and (d) and explain the relationship between them.

107. A particle of cosmic dust has a density $\rho = 2.0\ \text{g/cm}^3$. (a) Assuming the dust particles are spherical and light absorbing, and are at the same distance as Earth from the Sun, determine the particle size for which radiation pressure from sunlight is equal to the Sun's force of gravity on the dust particle. (b) Explain how the forces compare if the particle radius is smaller. (c) Explain what this implies about the sizes of dust particle likely to be present in the inner solar system compared with outside the Oort cloud.

APPENDIX A | UNITS

Quantity	Common Symbol	Unit	Unit in Terms of Base SI Units
Acceleration	$\vec{\mathbf{a}}$	m/s^2	m/s^2
Amount of substance	n	**mole**	mol
Angle	θ, ϕ	radian (rad)	
Angular acceleration	$\vec{\alpha}$	rad/s^2	s^{-2}
Angular frequency	ω	rad/s	s^{-1}
Angular momentum	$\vec{\mathbf{L}}$	$\text{kg} \cdot \text{m}^2/\text{s}$	$\text{kg} \cdot \text{m}^2/\text{s}$
Angular velocity	$\vec{\omega}$	rad/s	s^{-1}
Area	A	m^2	m^2
Atomic number	Z		
Capacitance	C	farad (F)	$\text{A}^2 \cdot \text{s}^4/\text{kg} \cdot \text{m}^2$
Charge	q, Q, e	coulomb (C)	$\text{A} \cdot \text{s}$
Charge density:			
Line	λ	C/m	$\text{A} \cdot \text{s/m}$
Surface	σ	C/m^2	$\text{A} \cdot \text{s/m}^2$
Volume	ρ	C/m^3	$\text{A} \cdot \text{s/m}^3$
Conductivity	σ	$1/\Omega \cdot \text{m}$	$\text{A}^2 \cdot \text{s}^3/\text{kg} \cdot \text{m}^3$
Current	I	**ampere**	A
Current density	$\vec{\mathbf{J}}$	A/m^2	A/m^2
Density	ρ	kg/m^3	kg/m^3
Dielectric constant	κ		
Electric dipole moment	$\vec{\mathbf{p}}$	$\text{C} \cdot \text{m}$	$\text{A} \cdot \text{s} \cdot \text{m}$
Electric field	$\vec{\mathbf{E}}$	N/C	$\text{kg} \cdot \text{m/A} \cdot \text{s}^3$
Electric flux	Φ	$\text{N} \cdot \text{m}^2/\text{C}$	$\text{kg} \cdot \text{m}^3/\text{A} \cdot \text{s}^3$
Electromotive force	ε	volt (V)	$\text{kg} \cdot \text{m}^2/\text{A} \cdot \text{s}^3$
Energy	E, U, K	joule (J)	$\text{kg} \cdot \text{m}^2/\text{s}^2$
Entropy	S	J/K	$\text{kg} \cdot \text{m}^2/\text{s}^2 \cdot \text{K}$

Table A1 Units Used in Physics (Fundamental units in bold)

Quantity	Common Symbol	Unit	Unit in Terms of Base SI Units
Force	$\vec{\mathbf{F}}$	newton (N)	$\text{kg} \cdot \text{m/s}^2$
Frequency	f	hertz (Hz)	s^{-1}
Heat	Q	joule (J)	$\text{kg} \cdot \text{m}^2/\text{s}^2$
Inductance	L	henry (H)	$\text{kg} \cdot \text{m}^2/\text{A}^2 \cdot \text{s}^2$
Length:	ℓ, L	**meter**	m
Displacement	$\Delta x, \Delta \vec{\mathbf{r}}$		
Distance	d, h		
Position	$x, y, z, \vec{\mathbf{r}}$		
Magnetic dipole moment	$\vec{\mu}$	$\text{N} \cdot \text{J/T}$	$\text{A} \cdot \text{m}^2$
Magnetic field	$\vec{\mathbf{B}}$	$\text{tesla(T)} = \left(\text{Wb/m}^2\right)$	$\text{kg/A} \cdot \text{s}^2$
Magnetic flux	Φ_m	weber (Wb)	$\text{kg} \cdot \text{m}^2/\text{A} \cdot \text{s}^2$
Mass	m, M	**kilogram**	kg
Molar specific heat	C	$\text{J/mol} \cdot \text{K}$	$\text{kg} \cdot \text{m}^2/\text{s}^2 \cdot \text{mol} \cdot \text{K}$
Moment of inertia	I	$\text{kg} \cdot \text{m}^2$	$\text{kg} \cdot \text{m}^2$
Momentum	$\vec{\mathbf{p}}$	$\text{kg} \cdot \text{m/s}$	$\text{kg} \cdot \text{m/s}$
Period	T	s	s
Permeability of free space	μ_0	$\text{N/A}^2 = (\text{H/m})$	$\text{kg} \cdot \text{m/A}^2 \cdot \text{s}^2$
Permittivity of free space	ε_0	$\text{C}^2/\text{N} \cdot \text{m}^2 = (\text{F/m})$	$\text{A}^2 \cdot \text{s}^4/\text{kg} \cdot \text{m}^3$
Potential	V	$\text{volt(V)} = (\text{J/C})$	$\text{kg} \cdot \text{m}^2/\text{A} \cdot \text{s}^3$
Power	P	$\text{watt(W)} = (\text{J/s})$	$\text{kg} \cdot \text{m}^2/\text{s}^3$
Pressure	p	$\text{pascal(Pa)} = \left(\text{N/m}^2\right)$	$\text{kg/m} \cdot \text{s}^2$
Resistance	R	$\text{ohm}(\Omega) = (\text{V/A})$	$\text{kg} \cdot \text{m}^2/\text{A}^2 \cdot \text{s}^3$
Specific heat	c	$\text{J/kg} \cdot \text{K}$	$\text{m}^2/\text{s}^2 \cdot \text{K}$
Speed	v	m/s	m/s
Temperature	T	**kelvin**	K
Time	t	**second**	s
Torque	$\vec{\tau}$	$\text{N} \cdot \text{m}$	$\text{kg} \cdot \text{m}^2/\text{s}^2$

Table A1 Units Used in Physics (Fundamental units in bold)

Quantity	Common Symbol	Unit	Unit in Terms of Base SI Units
Velocity	$\vec{v}$	m/s	m/s
Volume	V	m^3	m^3
Wavelength	λ	m	m
Work	W	$\text{joule(J)} = (\text{N} \cdot \text{m})$	$\text{kg} \cdot \text{m}^2/\text{s}^2$

Table A1 Units Used in Physics (Fundamental units in bold)

APPENDIX B | CONVERSION FACTORS

	m	cm	km
1 meter	1	10^2	10^{-3}
1 centimeter	10^{-2}	1	10^{-5}
1 kilometer	10^3	10^5	1
1 inch	2.540×10^{-2}	2.540	2.540×10^{-5}
1 foot	0.3048	30.48	3.048×10^{-4}
1 mile	1609	1.609×10^4	1.609
1 angstrom	10^{-10}		
1 fermi	10^{-15}		
1 light-year			9.460×10^{12}
	in.	**ft**	**mi**
1 meter	39.37	3.281	6.214×10^{-4}
1 centimeter	0.3937	3.281×10^{-2}	6.214×10^{-6}
1 kilometer	3.937×10^4	3.281×10^3	0.6214
1 inch	1	8.333×10^{-2}	1.578×10^{-5}
1 foot	12	1	1.894×10^{-4}
1 mile	6.336×10^4	5280	1

Table B1 Length

Area

$1\ \text{cm}^2 = 0.155\ \text{in.}^2$

$1\ \text{m}^2 = 10^4\ \text{cm}^2 = 10.76\ \text{ft}^2$

$1\ \text{in.}^2 = 6.452\ \text{cm}^2$

$1\ \text{ft}^2 = 144\ \text{in.}^2 = 0.0929\ \text{m}^2$

Volume

$1\ \text{liter} = 1000\ \text{cm}^3 = 10^{-3}\ \text{m}^3 = 0.03531\ \text{ft}^3 = 61.02\ \text{in.}^3$

$1\ \text{ft}^3 = 0.02832\ \text{m}^3 = 28.32\ \text{liters} = 7.477\ \text{gallons}$

$1\ \text{gallon} = 3.788\ \text{liters}$

	s	min	h	day	yr
1 second	1	1.667×10^{-2}	2.778×10^{-4}	1.157×10^{-5}	3.169×10^{-8}
1 minute	60	1	1.667×10^{-2}	6.944×10^{-4}	1.901×10^{-6}
1 hour	3600	60	1	4.167×10^{-2}	1.141×10^{-4}
1 day	8.640×10^{4}	1440	24	1	2.738×10^{-3}
1 year	3.156×10^{7}	5.259×10^{5}	8.766×10^{3}	365.25	1

Table B2 Time

	m/s	cm/s	ft/s	mi/h
1 meter/second	1	10^{2}	3.281	2.237
1 centimeter/second	10^{-2}	1	3.281×10^{-2}	2.237×10^{-2}
1 foot/second	0.3048	30.48	1	0.6818
1 mile/hour	0.4470	44.70	1.467	1

Table B3 Speed

Acceleration

$1 \text{ m/s}^2 = 100 \text{ cm/s}^2 = 3.281 \text{ ft/s}^2$

$1 \text{ cm/s}^2 = 0.01 \text{ m/s}^2 = 0.03281 \text{ ft/s}^2$

$1 \text{ ft/s}^2 = 0.3048 \text{ m/s}^2 = 30.48 \text{ cm/s}^2$

$1 \text{ mi/h} \cdot \text{s} = 1.467 \text{ ft/s}^2$

	kg	g	slug	u
1 kilogram	1	10^{3}	6.852×10^{-2}	6.024×10^{26}
1 gram	10^{-3}	1	6.852×10^{-5}	6.024×10^{23}
1 slug	14.59	1.459×10^{4}	1	8.789×10^{27}
1 atomic mass unit	1.661×10^{-27}	1.661×10^{-24}	1.138×10^{-28}	1
1 metric ton	1000			

Table B4 Mass

	N	dyne	lb
1 newton	1	10^{5}	0.2248
1 dyne	10^{-5}	1	2.248×10^{-6}
1 pound	4.448	4.448×10^{5}	1

Table B5 Force

	Pa	dyne/cm^2	atm	cmHg	lb/in.^{2}
1 pascal	1	10	9.869×10^{-6}	7.501×10^{-4}	1.450×10^{-4}
1 dyne/centimeter^2	10^{-1}	1	9.869×10^{-7}	7.501×10^{-5}	1.450×10^{-5}
1 atmosphere	1.013×10^5	1.013×10^6	1	76	14.70
1 centimeter mercury*	1.333×10^3	1.333×10^4	1.316×10^{-2}	1	0.1934
1 pound/inch^2	6.895×10^3	6.895×10^4	6.805×10^{-2}	5.171	1
1 bar	10^5				
1 torr				1 (mmHg)	

***Where the acceleration due to gravity is $9.80665\ \text{m/s}^2$ and the temperature is $0°\text{C}$**

Table B6 Pressure

	J	erg	ft.lb
1 joule	1	10^7	0.7376
1 erg	10^{-7}	1	7.376×10^{-8}
1 foot-pound	1.356	1.356×10^7	1
1 electron-volt	1.602×10^{-19}	1.602×10^{-12}	1.182×10^{-19}
1 calorie	4.186	4.186×10^7	3.088
1 British thermal unit	1.055×10^3	1.055×10^{10}	7.779×10^2
1 kilowatt-hour	3.600×10^6		
	eV	**cal**	**Btu**
1 joule	6.242×10^{18}	0.2389	9.481×10^{-4}
1 erg	6.242×10^{11}	2.389×10^{-8}	9.481×10^{-11}
1 foot-pound	8.464×10^{18}	0.3239	1.285×10^{-3}
1 electron-volt	1	3.827×10^{-20}	1.519×10^{-22}
1 calorie	2.613×10^{19}	1	3.968×10^{-3}
1 British thermal unit	6.585×10^{21}	2.520×10^2	1

Table B7 Work, Energy, Heat

Power

$1\ \text{W} = 1\ \text{J/s}$

$1\ \text{hp} = 746\ \text{W} = 550\ \text{ft} \cdot \text{lb/s}$

$1\ \text{Btu/h} = 0.293\ \text{W}$

Angle

$1 \text{ rad} = 57.30° = 180°/\pi$

$1° = 0.01745 \text{ rad} = \pi/180 \text{ rad}$

$1 \text{ revolution} = 360° = 2\pi \text{ rad}$

$1 \text{ rev/min(rpm)} = 0.1047 \text{ rad/s}$

APPENDIX C | FUNDAMENTAL CONSTANTS

Quantity	Symbol	Value
Atomic mass unit	u	$1.660\ 538\ 782\ (83) \times 10^{-27}$ kg $931.494\ 028\ (23)$ MeV/c^2
Avogadro's number	N_A	$6.022\ 141\ 79\ (30) \times 10^{23}$ particles/mol
Bohr magneton	$\mu_B = \frac{e\hbar}{2m_e}$	$9.274\ 009\ 15\ (23) \times 10^{-24}$ J/T
Bohr radius	$a_0 = \frac{\hbar^2}{m_e e^2 k_e}$	$5.291\ 772\ 085\ 9\ (36) \times 10^{-11}$ m
Boltzmann's constant	$k_B = \frac{R}{N_A}$	$1.380\ 650\ 4\ (24) \times 10^{-23}$ J/K
Compton wavelength	$\lambda_C = \frac{h}{m_e c}$	$2.426\ 310\ 217\ 5\ (33) \times 10^{-12}$ m
Coulomb constant	$k_e = \frac{1}{4\pi\varepsilon_0}$	$8.987\ 551\ 788... \times 10^{9}\ \text{N} \cdot \text{m}^2/\text{C}^2$ (exact)
Deuteron mass	m_d	$3.343\ 583\ 20\ (17) \times 10^{-27}$ kg $2.013\ 553\ 212\ 724(78)$ u $1875.612\ 859$ MeV/c^2
Electron mass	m_e	$9.109\ 382\ 15\ (45) \times 10^{-31}$ kg $5.485\ 799\ 094\ 3(23) \times 10^{-4}$ u $0.510\ 998\ 910\ (13)$ MeV/c^2
Electron volt	eV	$1.602\ 176\ 487\ (40) \times 10^{-19}$ J
Elementary charge	e	$1.602\ 176\ 487\ (40) \times 10^{-19}$ C
Gas constant	R	$8.314\ 472\ (15)$ J/mol · K
Gravitational constant	G	$6.674\ 28\ (67) \times 10^{-11}\ \text{N} \cdot \text{m}^2/\text{kg}^2$

Table C1 Fundamental Constants *Note:* These constants are the values recommended in 2006 by CODATA, based on a least-squares adjustment of data from different measurements. The numbers in parentheses for the values represent the uncertainties of the last two digits.

Quantity	Symbol	Value
Neutron mass	m_n	$1.674\ 927\ 211\ (84) \times 10^{-27}$ kg $1.008\ 664\ 915\ 97\ (43)$ u $939.565\ 346\ (23)\ \text{MeV}/c^2$
Nuclear magneton	$\mu_n = \frac{e\hbar}{2m_p}$	$5.050\ 783\ 24\ (13) \times 10^{-27}$ J/T
Permeability of free space	μ_0	$4\pi \times 10^{-7}\ \text{T} \cdot \text{m/A}$(exact)
Permittivity of free space	$\varepsilon_0 = \frac{1}{\mu_0 c^2}$	$8.854\ 187\ 817... \times 10^{-12}\ \text{C}^2/\text{N} \cdot \text{m}^2$(exact)
Planck's constant	h $\hbar = \frac{h}{2\pi}$	$6.626\ 068\ 96\ (33) \times 10^{-34}\ \text{J} \cdot \text{s}$ $1.054\ 571\ 628\ (53) \times 10^{-34}\ \text{J} \cdot \text{s}$
Proton mass	m_p	$1.672\ 621\ 637\ (83) \times 10^{-27}$ kg $1.007\ 276\ 466\ 77\ (10)$ u $938.272\ 013\ (23)\ \text{MeV}/c^2$
Rydberg constant	R_H	$1.097\ 373\ 156\ 852\ 7\ (73) \times 10^7\ \text{m}^{-1}$
Speed of light in vacuum	c	$2.997\ 924\ 58 \times 10^8$ m/s (exact)

Table C1 Fundamental Constants *Note:* These constants are the values recommended in 2006 by CODATA, based on a least-squares adjustment of data from different measurements. The numbers in parentheses for the values represent the uncertainties of the last two digits.

Useful combinations of constants for calculations:

$hc = 12{,}400\ \text{eV} \cdot \text{Å} = 1240\ \text{eV} \cdot \text{nm} = 1240\ \text{MeV} \cdot \text{fm}$

$\hbar c = 1973\ \text{eV} \cdot \text{Å} = 197.3\ \text{eV} \cdot \text{nm} = 197.3\ \text{MeV} \cdot \text{fm}$

$k_e e^2 = 14.40\ \text{eV} \cdot \text{Å} = 1.440\ \text{eV} \cdot \text{nm} = 1.440\ \text{MeV} \cdot \text{fm}$

$k_\text{B} T = 0.02585\ \text{eV}$ at $T = 300$ K

APPENDIX D | ASTRONOMICAL DATA

Celestial Object	Mean Distance from Sun (million km)	Period of Revolution (d = days) (y = years)	Period of Rotation at Equator	Eccentricity of Orbit
Sun	–	–	27 d	–
Mercury	57.9	88 d	59 d	0.206
Venus	108.2	224.7 d	243 d	0.007
Earth	149.6	365.26 d	23 h 56 min 4 s	0.017
Mars	227.9	687 d	24 h 37 min 23 s	0.093
Jupiter	778.4	11.9 y	9 h 50 min 30 s	0.048
Saturn	1426.7	29.5 6	10 h 14 min	0.054
Uranus	2871.0	84.0 y	17 h 14 min	0.047
Neptune	4498.3	164.8 y	16 h	0.009
Earth's Moon	149.6 (0.386 from Earth)	27.3 d	27.3 d	0.055
Celestial Object	**Equatorial Diameter (km)**	**Mass (Earth = 1)**	**Density (g/cm^3)**	
Sun	1,392,000	333,000.00	1.4	
Mercury	4879	0.06	5.4	
Venus	12,104	0.82	5.2	
Earth	12,756	1.00	5.5	
Mars	6794	0.11	3.9	
Jupiter	142,984	317.83	1.3	
Saturn	120,536	95.16	0.7	
Uranus	51,118	14.54	1.3	
Neptune	49,528	17.15	1.6	
Earth's Moon	3476	0.01	3.3	

Table D1 Astronomical Data

Other Data:

Mass of Earth: 5.97×10^{24} kg

Mass of the Moon: 7.36×10^{22} kg

Mass of the Sun: 1.99×10^{30} kg

APPENDIX E | MATHEMATICAL FORMULAS

Quadratic formula

If $ax^2 + bx + c = 0$, then $x = \frac{-b \pm \sqrt{b^2 - 4ac}}{2a}$

Triangle of base b and height h	Area $= \frac{1}{2}bh$	
Circle of radius r	Circumference $= 2\pi r$	Area $= \pi r^2$
Sphere of radius r	Surface area $= 4\pi r^2$	Volume $= \frac{4}{3}\pi r^3$
Cylinder of radius r and height h	Area of curved surface $= 2\pi rh$	Volume $= \pi r^2 h$

Table E1 Geometry

Trigonometry

Trigonometric Identities

1. $\sin\theta = 1/\csc\theta$
2. $\cos\theta = 1/\sec\theta$
3. $\tan\theta = 1/\cot\theta$
4. $\sin(90^0 - \theta) = \cos\theta$
5. $\cos(90^0 - \theta) = \sin\theta$
6. $\tan(90^0 - \theta) = \cot\theta$
7. $\sin^2\theta + \cos^2\theta = 1$
8. $\sec^2\theta - \tan^2\theta = 1$
9. $\tan\theta = \sin\theta/\cos\theta$
10. $\sin(\alpha \pm \beta) = \sin\alpha\cos\beta \pm \cos\alpha\sin\beta$
11. $\cos(\alpha \pm \beta) = \cos\alpha\cos\beta \mp \sin\alpha\sin\beta$
12. $\tan(\alpha \pm \beta) = \frac{\tan\alpha \pm \tan\beta}{1 \mp \tan\alpha\tan\beta}$
13. $\sin 2\theta = 2\sin\theta\cos\theta$
14. $\cos 2\theta = \cos^2\theta - \sin^2\theta = 2\cos^2\theta - 1 = 1 - 2\sin^2\theta$

15. $\sin\alpha + \sin\beta = 2\sin\frac{1}{2}(\alpha+\beta)\cos\frac{1}{2}(\alpha-\beta)$

16. $\cos\alpha + \cos\beta = 2\cos\frac{1}{2}(\alpha+\beta)\cos\frac{1}{2}(\alpha-\beta)$

Triangles

1. Law of sines: $\frac{a}{\sin\alpha} = \frac{b}{\sin\beta} = \frac{c}{\sin\gamma}$
2. Law of cosines: $c^2 = a^2 + b^2 - 2ab\cos\gamma$

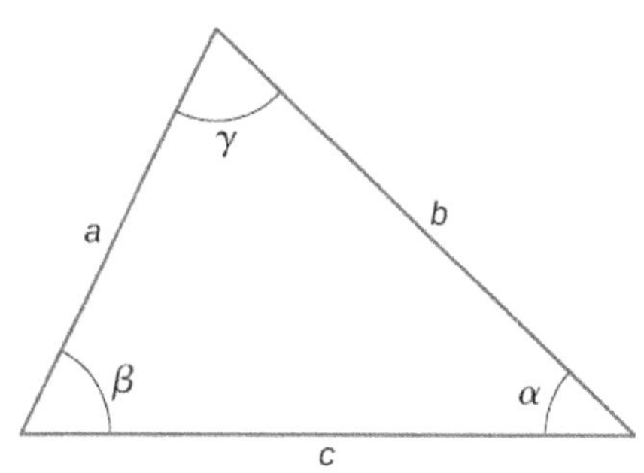

3. Pythagorean theorem: $a^2 + b^2 = c^2$

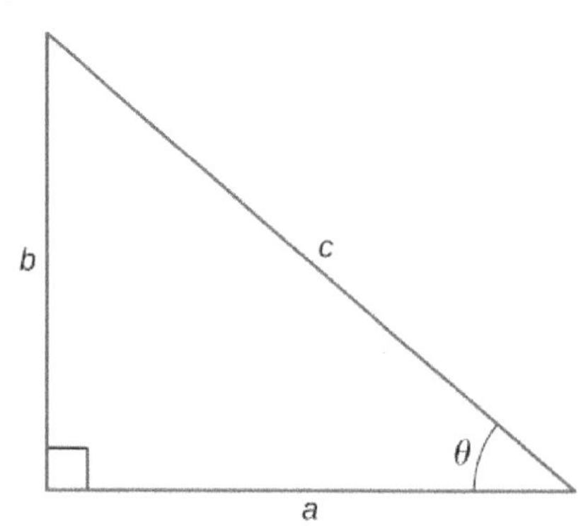

Series expansions

1. Binomial theorem: $(a+b)^n = a^n + na^{n-1}b + \frac{n(n-1)a^{n-2}b^2}{2!} + \frac{n(n-1)(n-2)a^{n-3}b^3}{3!} + \cdots$
2. $(1 \pm x)^n = 1 \pm \frac{nx}{1!} + \frac{n(n-1)x^2}{2!} \pm \cdots \left(x^2 < 1\right)$
3. $(1 \pm x)^{-n} = 1 \mp \frac{nx}{1!} + \frac{n(n+1)x^2}{2!} \mp \cdots \left(x^2 < 1\right)$
4. $\sin x = x - \frac{x^3}{3!} + \frac{x^5}{5!} - \cdots$
5. $\cos x = 1 - \frac{x^2}{2!} + \frac{x^4}{4!} - \cdots$
6. $\tan x = x + \frac{x^3}{3} + \frac{2x^5}{15} + \cdots$
7. $e^x = 1 + x + \frac{x^2}{2!} + \cdots$
8. $\ln(1+x) = x - \frac{1}{2}x^2 + \frac{1}{3}x^3 - \cdots (|x| < 1)$

Derivatives

1. $\frac{d}{dx}[af(x)] = a\frac{d}{dx}f(x)$
2. $\frac{d}{dx}[f(x) + g(x)] = \frac{d}{dx}f(x) + \frac{d}{dx}g(x)$
3. $\frac{d}{dx}[f(x)g(x)] = f(x)\frac{d}{dx}g(x) + g(x)\frac{d}{dx}f(x)$
4. $\frac{d}{dx}f(u) = \left[\frac{d}{du}f(u)\right]\frac{du}{dx}$
5. $\frac{d}{dx}x^m = mx^{m-1}$
6. $\frac{d}{dx}\sin x = \cos x$
7. $\frac{d}{dx}\cos x = -\sin x$
8. $\frac{d}{dx}\tan x = \sec^2 x$
9. $\frac{d}{dx}\cot x = -\csc^2 x$
10. $\frac{d}{dx}\sec x = \tan x \sec x$
11. $\frac{d}{dx}\csc x = -\cot x \csc x$
12. $\frac{d}{dx}e^x = e^x$
13. $\frac{d}{dx}\ln x = \frac{1}{x}$
14. $\frac{d}{dx}\sin^{-1} x = \frac{1}{\sqrt{1-x^2}}$
15. $\frac{d}{dx}\cos^{-1} x = -\frac{1}{\sqrt{1-x^2}}$
16. $\frac{d}{dx}\tan^{-1} x = -\frac{1}{1+x^2}$

Integrals

1. $\int af(x)dx = a\int f(x)dx$
2. $\int[f(x) + g(x)]dx = \int f(x)dx + \int g(x)dx$
3. $$\begin{aligned}\int x^m dx &= \frac{x^{m+1}}{m+1} (m \neq -1)\\ &= \ln x (m = -1)\end{aligned}$$
4. $\int \sin x\, dx = -\cos x$
5. $\int \cos x\, dx = \sin x$
6. $\int \tan x\, dx = \ln|\sec x|$

7. $\int \sin^2 ax\,dx = \frac{x}{2} - \frac{\sin 2ax}{4a}$

8. $\int \cos^2 ax\,dx = \frac{x}{2} + \frac{\sin 2ax}{4a}$

9. $\int \sin ax \cos ax\,dx = -\frac{\cos 2ax}{4a}$

10. $\int e^{ax}\,dx = \frac{1}{a}e^{ax}$

11. $\int xe^{ax}\,dx = \frac{e^{ax}}{a^2}(ax - 1)$

12. $\int \ln ax\,dx = x \ln ax - x$

13. $\int \frac{dx}{a^2 + x^2} = \frac{1}{a}\tan^{-1}\frac{x}{a}$

14. $\int \frac{dx}{a^2 - x^2} = \frac{1}{2a}\ln\left|\frac{x + a}{x - a}\right|$

15. $\int \frac{dx}{\sqrt{a^2 + x^2}} = \sinh^{-1}\frac{x}{a}$

16. $\int \frac{dx}{\sqrt{a^2 - x^2}} = \sin^{-1}\frac{x}{a}$

17. $\int \sqrt{a^2 + x^2}\,dx = \frac{x}{2}\sqrt{a^2 + x^2} + \frac{a^2}{2}\sinh^{-1}\frac{x}{a}$

18. $\int \sqrt{a^2 - x^2}\,dx = \frac{x}{2}\sqrt{a^2 - x^2} + \frac{a^2}{2}\sin^{-1}\frac{x}{a}$

APPENDIX F | CHEMISTRY

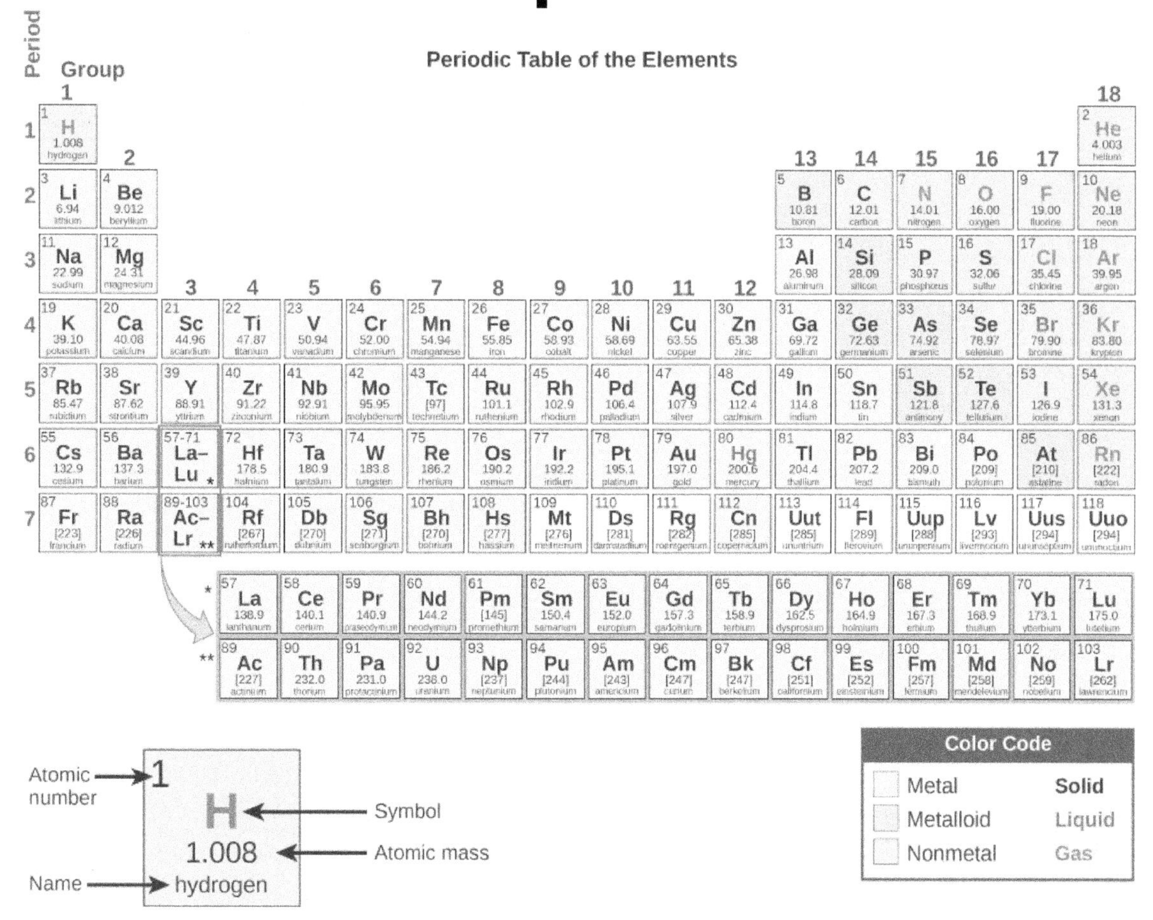

APPENDIX G | THE GREEK ALPHABET

Name	Capital	Lowercase	Name	Capital	Lowercase
Alpha	A	α	Nu	N	ν
Beta	B	β	Xi	Ξ	ξ
Gamma	Γ	γ	Omicron	O	o
Delta	Δ	δ	Pi	Π	π
Epsilon	E	ε	Rho	P	ρ
Zeta	Z	ζ	Sigma	Σ	σ
Eta	H	η	Tau	T	τ
Theta	Θ	θ	Upsilon	Υ	υ
Iota	I	ι	Phi	Φ	ϕ
Kappa	K	κ	Chi	X	χ
Lambda	Λ	λ	Psi	Ψ	ψ
Mu	M	μ	Omega	Ω	ω

Table G1 The Greek Alphabet

ANSWER KEY

CHAPTER 1

CHECK YOUR UNDERSTANDING

1.1. The actual amount (mass) of gasoline left in the tank when the gauge hits "empty" is less in the summer than in the winter. The gasoline has the same volume as it does in the winter when the "add fuel" light goes on, but because the gasoline has expanded, there is less mass.
1.2. Not necessarily, as the thermal stress is also proportional to Young's modulus.
1.3. To a good approximation, the heat transfer depends only on the temperature difference. Since the temperature differences are the same in both cases, the same 25 kJ is necessary in the second case. (As we will see in the next section, the answer would have been different if the object had been made of some substance that changes phase anywhere between $30\,°C$ and $50\,°C$.)
1.4. The ice and liquid water are in thermal equilibrium, so that the temperature stays at the freezing temperature as long as ice remains in the liquid. (Once all of the ice melts, the water temperature will start to rise.)
1.5. Snow is formed from ice crystals and thus is the solid phase of water. Because enormous heat is necessary for phase changes, it takes a certain amount of time for this heat to be transferred from the air, even if the air is above $0\,°C$.
1.6. Conduction: Heat transfers into your hands as you hold a hot cup of coffee. Convection: Heat transfers as the barista "steams" cold milk to make hot cocoa. Radiation: Heat transfers from the Sun to a jar of water with tea leaves in it to make "Sun tea." A great many other answers are possible.
1.7. Because area is the product of two spatial dimensions, it increases by a factor of four when each dimension is doubled $\left(A_{\text{fina}} = (2d)^2 = 4d^2 = 4A_{\text{initial}}\right)$. The distance, however, simply doubles. Because the temperature difference and the coefficient of thermal conductivity are independent of the spatial dimensions, the rate of heat transfer by conduction increases by a factor of four divided by two, or two:

$$P_{\text{fina}} = \frac{kA_{\text{fina}}(T_{\text{h}} - T_{\text{c}})}{d_{\text{fina}}} = \frac{k(4A_{\text{fina}}(T_{\text{h}} - T_{\text{c}}))}{2d_{\text{initial}}} = 2\frac{kA_{\text{fina}}(T_{\text{h}} - T_{\text{c}})}{d_{\text{initial}}} = 2P_{\text{initial}}.$$

1.8. Using a fan increases the flow of air: Warm air near your body is replaced by cooler air from elsewhere. Convection increases the rate of heat transfer so that moving air "feels" cooler than still air.
1.9. The radiated heat is proportional to the fourth power of the *absolute temperature*. Because $T_1 = 293\text{ K}$ and $T_2 = 313\text{ K}$, the rate of heat transfer increases by about 30% of the original rate.

CONCEPTUAL QUESTIONS

1. They are at the same temperature, and if they are placed in contact, no net heat flows between them.
3. The reading will change.
5. The cold water cools part of the inner surface, making it contract, while the rest remains expanded. The strain is too great for the strength of the material. Pyrex contracts less, so it experiences less strain.
7. In principle, the lid expands more than the jar because metals have higher coefficients of expansion than glass. That should make unscrewing the lid easier. (In practice, getting the lid and jar wet may make gripping them more difficult.)
9. After being heated, the length is ($1 + 300\alpha$) (1 m). After being cooled, the length is $(1 - 300\,\alpha)(1 + 300\,\alpha)(1\text{ m})$. That answer is not 1 m, but it should be. The explanation is that even if α is exactly constant, the relation $\Delta L = \alpha L \Delta T$ is strictly true only in the limit of small ΔT. Since α values are small, the discrepancy is unimportant in practice.
11. Temperature differences cause heat transfer.
13. No, it is stored as thermal energy. A thermodynamic system does not have a well-defined quantity of heat.
15. It raises the boiling point, so the water, which the food gains heat from, is at a higher temperature.
17. Yes, by raising the pressure above 56 atm.
19. work
21. $0\,°C$ (at or near atmospheric pressure)
23. Condensation releases heat, so it speeds up the melting.
25. Because of water's high specific heat, it changes temperature less than land. Also, evaporation reduces temperature rises. The air tends to stay close to equilibrium with the water, so its temperature does not change much where there's a lot of water around, as in San Francisco but not Sacramento.
27. The liquid is oxygen, whose boiling point is above that of nitrogen but whose melting point is below the boiling point of liquid nitrogen. The crystals that sublime are carbon dioxide, which has no liquid phase at atmospheric pressure. The crystals that melt are water, whose melting point is above carbon dioxide's sublimation point. The water came from the instructor's breath.
29. Increasing circulation to the surface will warm the person, as the temperature of the water is warmer than human body temperature. Sweating will cause no evaporative cooling under water or in the humid air immediately above the tub.
31. It spread the heat over the area above the heating elements, evening the temperature there, but does not spread the heat much beyond the heating elements.

33. Heat is conducted from the fire through the fire box to the circulating air and then convected by the air into the room (forced convection).
35. The tent is heated by the Sun and transfers heat to you by all three processes, especially radiation.
37. If shielded, it measures the air temperature. If not, it measures the combined effect of air temperature and net radiative heat gain from the Sun.
39. Turn the thermostat down. To have the house at the normal temperature, the heating system must replace all the heat that was lost. For all three mechanisms of heat transfer, the greater the temperature difference between inside and outside, the more heat is lost and must be replaced. So the house should be at the lowest temperature that does not allow freezing damage.
41. Air is a good insulator, so there is little conduction, and the heated air rises, so there is little convection downward.

PROBLEMS

43. That must be Celsius. Your Fahrenheit temperature is $102\ °\text{F}$. Yes, it is time to get treatment.

45. a. $\Delta T_\text{C} = 22.2\ °\text{C}$; b. We know that $\Delta T_\text{F} = T_\text{F2} - T_\text{F1}$. We also know that $T_\text{F2} = \frac{9}{5}T_\text{C2} + 32$ and $T_\text{F1} = \frac{9}{5}T_\text{C1} + 32$. So, substituting, we have $\Delta T_\text{F} = \left(\frac{9}{5}T_\text{C2} + 32\right) - \left(\frac{9}{5}T_\text{C1} + 32\right)$. Partially solving and rearranging the equation, we have $\Delta T_\text{F} = \frac{9}{5}(T_\text{C2} - T_\text{C1})$. Therefore, $\Delta T_\text{F} = \frac{9}{5}\Delta T_\text{C}$.

47. a. $-40°$; b. 575 K

49. Using Table 1.2 to find the coefficient of thermal expansion of marble:
$$L = L_0 + \Delta L = L_0(1 + \alpha\Delta T) = 170\ \text{m}\left[1 + \left(2.5 \times 10^{-6}/°\text{C}\right)(-45.0\ °\text{C})\right] = 169.98\ \text{m}.$$
(Answer rounded to five significant figures to show the slight difference in height.)

51. Using Table 1.2 to find the coefficient of thermal expansion of mercury:
$$\Delta L = \alpha L\Delta T = \left(6.0 \times 10^{-5}/°\text{C}\right)(0.0300\ \text{m})(3.00\ °\text{C}) = 5.4 \times 10^{-6}\ \text{m}.$$

53. On the warmer day, our tape measure will expand linearly. Therefore, each measured dimension will be smaller than the actual dimension of the land. Calling these measured dimensions l' and w', we will find a new area, *A*. Let's calculate these measured dimensions:
$$l' = l_0 - \Delta l = (20\ \text{m}) - (20\ °\text{C})(20\ \text{m})\left(\frac{1.2 \times 10^{-5}}{°\text{C}}\right) = 19.9952\ \text{m};$$
$$A' = l \times w' = (29.9928\ \text{m})(19.9952\ \text{m}) = 599.71\ \text{m}^2;$$
$$\text{Cost change} = (A - A')\left(\frac{\$60{,}000}{\text{m}^2}\right) = \left((600 - 599.71)\text{m}^2\right)\left(\frac{\$60{,}000}{\text{m}^2}\right) = \$17{,}000.$$
Because the area gets smaller, the price of the land *decreases* by about \$17,000.

55. a. Use Table 1.2 to find the coefficients of thermal expansion of steel and aluminum. Then
$$\Delta L_\text{Al} - \Delta L_\text{steel} = (\alpha_\text{Al} - \alpha_\text{steel})L_0\Delta T = \left(\frac{2.5 \times 10^{-5}}{°\text{C}} - \frac{1.2 \times 10^{-5}}{°\text{C}}\right)(1.00\ \text{m})(22\ °\text{C}) = 2.9 \times 10^{-4}\ \text{m}.$$
b. By the same method with $L_0 = 30.0\ \text{m}$, we have $\Delta L = 8.6 \times 10^{-3}\ \text{m}$.

57. $\Delta V = 0.475\ \text{L}$

59. If we start with the freezing of water, then it would expand to $\left(1\ \text{m}^3\right)\left(\frac{1000\ \text{kg/m}^3}{917\ \text{kg/m}^3}\right) = 1.09\ \text{m}^3 = 1.98 \times 10^8\ \text{N/m}^2$ of ice.

61. $m = 5.20 \times 10^8\ \text{J}$

63. $Q = mc\Delta T \Rightarrow \Delta T = \frac{Q}{mc}$; a. $21.0\ °\text{C}$; b. $25.0\ °\text{C}$; c. $29.3\ °\text{C}$; d. $50.0\ °\text{C}$

65. $Q = mc\Delta T \Rightarrow c = \frac{Q}{m\Delta T} = \frac{1.04\ \text{kcal}}{(0.250\ \text{kg})(45.0\ °\text{C})} = 0.0924\ \text{kcal/kg} \cdot °\text{C}$. It is copper.

67. a. $Q = m_\text{w}c_\text{w}\Delta T + m_\text{Al}c_\text{Al}\Delta T = (m_\text{w}c_\text{w} + m_\text{Al}c_\text{Al})\Delta T$;
$$Q = \begin{bmatrix}(0.500\ \text{kg})(1.00\ \text{kcal/kg} \cdot °\text{C}) + \\ (0.100\ \text{kg})(0.215\ \text{kcal/kg} \cdot °\text{C})\end{bmatrix}(54.9\ °\text{C}) = 28.63\ \text{kcal};$$
$\frac{Q}{m_p} = \frac{28.63\ \text{kcal}}{5.00\ \text{g}} = 5.73\ \text{kcal/g}$; b. $\frac{Q}{m_p} = \frac{200\ \text{kcal}}{33\ \text{g}} = 6\ \text{kcal/g}$, which is consistent with our results to part (a), to one significant figure.

69. $0.139\ °C$

71. It should be lower. The beaker will not make much difference: $16.3\ °C$

73. a. 1.00×10^5 J ; b. 3.68×10^5 J ; c. The ice is much more effective in absorbing heat because it first must be melted, which requires a lot of energy, and then it gains the same amount of heat as the bag that started with water. The first 2.67×10^5 J of heat is used to melt the ice, then it absorbs the 1.00×10^5 J of heat as water.

75. 58.1 g

77. Let *M* be the mass of pool water and *m* be the mass of pool water that evaporates.

$$Mc\Delta T = mL_{V(37\ °C)} \Rightarrow \frac{m}{M} = \frac{c\Delta T}{L_{V(37\ °C)}} = \frac{(1.00\ \text{kcal/kg}\cdot °C)(1.50\ °C)}{580\ \text{kcal/kg}} = 2.59 \times 10^{-3};$$

(Note that L_V for water at $37\ °C$ is used here as a better approximation than L_V for $100\ °C$ water.)

79. a. 1.47×10^{15} kg ; b. 4.90×10^{20} J ; c. 48.5 y

81. a. 9.67 L; b. Crude oil is less dense than water, so it floats on top of the water, thereby exposing it to the oxygen in the air, which it uses to burn. Also, if the water is under the oil, it is less able to absorb the heat generated by the oil.

83. a. 319 kcal; b. $2.00\ °C$

85. First bring the ice up to $0\ °C$ and melt it with heat Q_1 : 4.74 kcal. This lowers the temperature of water by ΔT_2 : $23.15\ °C$. Now, the heat lost by the hot water equals that gained by the cold water (T_f is the final temperature): $20.6\ °C$

87. Let the subscripts r, e, v, and w represent rock, equilibrium, vapor, and water, respectively.

$$m_r c_r (T_1 - T_e) = m_V L_V + m_W c_W (T_e - T_2);$$

$$\begin{aligned} m_r &= \frac{m_V L_V + m_W c_W (T_e - T_2)}{c_r (T_1 - T_e)} \\ &= \frac{(0.0250\ \text{kg})(2256 \times 10^3\ \text{J/kg}) + (3.975\ \text{kg})(4186 \times 10^3\ \text{J/kg}\cdot °C)(100\ °C - 15\ °C)}{(840\ \text{J/kg}\cdot °C)(500\ °C - 100\ °C)} \\ &= 4.38\ \text{kg} \end{aligned}$$

89. a. 1.01×10^3 W ; b. One 1-kilowatt room heater is needed.

91. 84.0 W

93. 2.59 kg

95. a. 39.7 W; b. 820 kcal

97. $\frac{Q}{t} = \frac{kA(T_2 - T_1)}{d}$, so that

$$\frac{(Q/t)_{wall}}{(Q/t)_{window}} = \frac{k_{wall} A_{wall} d_{window}}{k_{window} A_{window} d_{wall}} = \frac{(2 \times 0.042\ \text{J/s}\cdot \text{m}\cdot °C)(10.0\ \text{m}^2)(0.750 \times 10^{-2}\ \text{m})}{(0.84\ \text{J/s}\cdot \text{m}\cdot °C)(2.00\ \text{m}^2)(13.0 \times 10^{-2}\ \text{m})}$$

This gives 0.0288 wall: window, or 35:1 window: wall

99. $\frac{Q}{t} = \frac{kA(T_2 - T_1)}{d} = \frac{kA\Delta T}{d} \Rightarrow$

$$\Delta T = \frac{d(Q/t)}{kA} = \frac{(6.00 \times 10^{-3}\ \text{m})(2256\ \text{W})}{(0.84\ \text{J/s}\cdot \text{m}\cdot °C)(1.54 \times 10^{-2}\ \text{m}^2)} = 1046\ °C = 1.05 \times 10^3\ \text{K}$$

101. We found in the preceding problem that $P = 126\Delta T\ \text{W}\cdot °C$ as baseline energy use. So the total heat loss during this period is $Q = (126\ \text{J/s}\cdot °C)(15.0\ °C)(120\ \text{days})(86.4 \times 10^3\ \text{s/day}) = 1960 \times 10^6\ \text{J}$. At the cost of $1/MJ, the cost is $1960. From an earlier problem, the savings is 12% or $235/y. We need $150\ \text{m}^2$ of insulation in the attic. At $\$4/\text{m}^2$, this is a $500 cost. So the payback period is $\$600/(\$235/\text{y}) = 2.6$ years (excluding labor costs).

ADDITIONAL PROBLEMS

103. 7.39%

105. $\frac{F}{A} = (210 \times 10^9\ \text{Pa})(12 \times 10^{-6}/°C)(40\ °C - (-15\ °C)) = 1.4 \times 10^8\ \text{N/m}^2$.

107. a. 1.06 cm; b. 1.11 cm

109. $1.7\,\text{kJ/(kg}\cdot{}^\circ\text{C)}$

111. a. $1.57\times10^4\,\text{kcal}$; b. $18.3\,\text{kW}\cdot\text{h}$; c. $1.29\times10^4\,\text{kcal}$

113. $6.3\,{}^\circ\text{C}$. All of the ice melted.

115. $63.9\,{}^\circ\text{C}$, all the ice melted

117. a. 83 W; b. $1.97\times10^3\,\text{W}$; The single-pane window has a rate of heat conduction equal to 1969/83, or 24 times that of a double-pane window.

119. The rate of heat transfer by conduction is 20.0 W. On a daily basis, this is 1,728 kJ/day. Daily food intake is $2400\,\text{kcal/d}\times4186\,\text{J/kcal}=10{,}050\,\text{kJ/day}$. So only 17.2% of energy intake goes as heat transfer by conduction to the environment at this ΔT .

121. 620 K

CHALLENGE PROBLEMS

123. Denoting the period by *P*, we know $P=2\pi\sqrt{L/g}$. When the temperature increases by *dT*, the length increases by αLdT . Then the new length is a. $P=2\pi\sqrt{\frac{L+\alpha LdT}{g}}=2\pi\sqrt{\frac{L}{g}(1+\alpha dT)}=2\pi\sqrt{\frac{L}{g}}\left(1+\frac{1}{2}\alpha dT\right)=P\left(1+\frac{1}{2}\alpha dT\right)$ by the binomial expansion. b. The clock runs slower, as its new period is 1.00019 s. It loses 16.4 s per day.

125. The amount of heat to melt the ice and raise it to $100\,{}^\circ\text{C}$ is not enough to condense the steam, but it is more than enough to lower the steam's temperature by $50\,{}^\circ\text{C}$, so the final state will consist of steam and liquid water in equilibrium, and the final temperature is $100\,{}^\circ\text{C}$; 9.5 g of steam condenses, so the final state contains 49.5 g of steam and 40.5 g of liquid water.

127. a. $dL/dT=kT/\rho L$; b. $L=\sqrt{2kTt/\rho L_f}$; c. yes

129. a. $\sigma(\pi R^2)T_s{}^4$; b. $e\sigma\pi R^2T_s{}^4$; c. $2e\sigma\pi R^2T_e{}^4$; d. $T_s^4=2T_e^4$; e. $e\sigma T_s^4+\frac{1}{4}(1-A)S=\sigma T_s^4$; f. 288 K

CHAPTER 2

CHECK YOUR UNDERSTANDING

2.1. We first need to calculate the molar mass (the mass of one mole) of niacin. To do this, we must multiply the number of atoms of each element in the molecule by the element's molar mass.

$$(6\text{ mol of carbon})(12.0\text{ g/mol})+(5\text{ mol hydrogen})(1.0\text{ g/mol})$$
$$+(1\text{ mol nitrogen})(14\text{ g/mol})+(2\text{ mol oxygen})(16.0\text{ g/mol})=123\text{ g/mol}$$

Then we need to calculate the number of moles in 14 mg.

$$\left(\frac{14\text{ mg}}{123\text{ g/mol}}\right)\left(\frac{1\text{ g}}{1000\text{ mg}}\right)=1.14\times10^{-4}\text{ mol.}$$

Then, we use Avogadro's number to calculate the number of molecules:

$$N=nN_A=\left(1.14\times10^{-4}\text{ mol}\right)\left(6.02\times10^{23}\text{ molecules/mol}\right)=6.85\times10^{19}\text{ molecules.}$$

2.2. The density of a gas is equal to a constant, the average molecular mass, times the number density *N*/*V*. From the ideal gas law, $pV=Nk_BT,$ we see that $N/V=p/k_BT.$ Therefore, at constant temperature, if the density and, consequently, the number density are reduced by half, the pressure must also be reduced by half, and $p_f=0.500\,\text{atm}$.

2.3. Density is mass per unit volume, and volume is proportional to the size of a body (such as the radius of a sphere) cubed. So if the distance between molecules increases by a factor of 10, then the volume occupied increases by a factor of 1000, and the density decreases by a factor of 1000. Since we assume molecules are in contact in liquids and solids, the distance between their centers is on the order of their typical size, so the distance in gases is on the order of 10 times as great.

2.4. Yes. Such fluctuations actually occur for a body of any size in a gas, but since the numbers of molecules are immense for macroscopic bodies, the fluctuations are a tiny percentage of the number of collisions, and the averages spoken of in this section vary imperceptibly. Roughly speaking, the fluctuations are inversely proportional to the square root of the number of collisions, so for small bodies, they can become significant. This was actually observed in the nineteenth century for pollen grains in water and is known as Brownian motion.

2.5. In a liquid, the molecules are very close together, constantly colliding with one another. For a gas to be nearly ideal, as air is under ordinary conditions, the molecules must be very far apart. Therefore the mean free path is much longer in the air.

2.6. As the number of moles is equal and we know the molar heat capacities of the two gases are equal, the temperature is halfway between the initial temperatures, 300 K.

CONCEPTUAL QUESTIONS

1. 2 moles, as that will contain twice as many molecules as the 1 mole of oxygen
3. pressure
5. The flame contains hot gas (heated by combustion). The pressure is still atmospheric pressure, in mechanical equilibrium with the air around it (or roughly so). The density of the hot gas is proportional to its number density *N*/*V* (neglecting the difference in composition between the gas in the flame and the surrounding air). At higher temperature than the surrounding air, the ideal gas law says that $N/V = p/k_B T$ is less than that of the surrounding air. Therefore the hot air has lower density than the surrounding air and is lifted by the buoyant force.
7. The mean free path is inversely proportional to the square of the radius, so it decreases by a factor of 4. The mean free time is proportional to the mean free path and inversely proportional to the rms speed, which in turn is inversely proportional to the square root of the mass. That gives a factor of $\sqrt{8}$ in the numerator, so the mean free time decreases by a factor of $\sqrt{2}$.
9. Since they're more massive, their gravity is stronger, so the escape velocity from them is higher. Since they're farther from the Sun, they're colder, so the speeds of atmospheric molecules including hydrogen and helium are lower. The combination of those facts means that relatively few hydrogen and helium molecules have escaped from the outer planets.
11. One where nitrogen is stored, as excess CO_2 will cause a feeling of suffocating, but excess nitrogen and insufficient oxygen will not.
13. Less, because at lower temperatures their heat capacity was only 3*RT*/2.
15. a. false; b. true; c. true; d. true
17. 1200 K

PROBLEMS

19. a. 0.137 atm; b. $p_g = (1\text{ atm})\frac{T_2 V_1}{T_1 V_2} - 1\text{ atm}$. Because of the expansion of the glass, $V_2 = 0.99973$. Multiplying by that factor does not make any significant difference.
21. a. 1.79×10^{-3} mol; b. 0.227 mol; c. 1.08×10^{21} molecules for the nitrogen, 1.37×10^{23} molecules for the carbon dioxide
23. 7.84×10^{-2} mol
25. 1.87×10^3
27. 2.47×10^7 molecules
29. 6.95×10^5 Pa; 6.86 atm
31. a. 9.14×10^6 Pa; b. 8.22×10^6 Pa; c. 2.15 K; d. no
33. 40.7 km
35. a. 0.61 N; b. 0.20 Pa
37. a. 5.88 m/s; b. 5.89 m/s
39. 177 m/s
41. 4.54×10^3
43. a. 0.0352 mol; b. 5.65×10^{-21} J; c. 139 J
45. 21.1 kPa
47. 458 K
49. 3.22×10^3 K
51. a. 1.004; b. 764 K; c. This temperature is equivalent to 915 °F, which is high but not impossible to achieve. Thus, this process is feasible. At this temperature, however, there may be other considerations that make the process difficult. (In general, uranium enrichment by gaseous diffusion is indeed difficult and requires many passes.)
53. 65 mol
55. a. 0.76 atm; b. 0.29 atm; c. The pressure there is barely above the quickly fatal level.
57. 4.92×10^5 K; Yes, that's an impractically high temperature.
59. polyatomic
61. 3.08×10^3 J
63. 29.2 °C
65. −1.6 °C
67. 0.00157
69. About 0.072. Answers may vary slightly. A more accurate answer is 0.074.
71. a. 419 m/s; b. 472 m/s; c. 513 m/s
73. 541 K

75. 2400 K for all three parts

ADDITIONAL PROBLEMS

77. a. $1.20\ \text{kg/m}^3$; b. $65.9\ \text{kg/m}^3$

79. 7.9 m

81. a. supercritical fluid; b. 3.00×10^7 Pa

83. 40.18%

85. a. 2.21×10^{27} molecules/m^3; b. 3.67×10^3 mol/m^3

87. 8.2 mm

89. a. $1080\ \text{J/kg}\ ^\circ\text{C}$; b. 12%

91. $2\sqrt{e}/3$ or about 1.10

93. a. 411 m/s; b. According to Table 2.3, the C_V of H_2S is significantly different from the theoretical value, so the ideal gas model does not describe it very well at room temperature and pressure, and the Maxwell-Boltzmann speed distribution for ideal gases may not hold very well, even less well at a lower temperature.

CHALLENGE PROBLEMS

95. 29.5 N/m

97. Substituting $v = \sqrt{\frac{2k_B T}{m}}u$ and $dv = \sqrt{\frac{2k_B T}{m}}du$ gives

$$\int_0^\infty \frac{4}{\sqrt{\pi}}\left(\frac{m}{2k_B T}\right)^{3/2} v^2 e^{-mv^2/2k_B T} dv = \int_0^\infty \frac{4}{\sqrt{\pi}}\left(\frac{m}{2k_B T}\right)^{3/2}\left(\frac{2k_B T}{m}\right)u^2 e^{-u^2}\sqrt{\frac{2k_B T}{m}}du$$

$$= \int_0^\infty \frac{4}{\sqrt{\pi}}u^2 e^{-u^2} du = \frac{4}{\sqrt{\pi}}\frac{\sqrt{\pi}}{4} = 1$$

99. Making the scaling transformation as in the previous problems, we find that

$$\bar{v^2} = \int_0^\infty \frac{4}{\sqrt{\pi}}\left(\frac{m}{2k_B T}\right)^{3/2} v^2 v^2 e^{-mv^2/2k_B T} dv = \int_0^\infty \frac{4}{\sqrt{\pi}}\frac{2k_B T}{m}u^4 e^{-u^2} du.$$

As in the previous problem, we integrate by parts:

$$\int_0^\infty u^4 e^{-u^2} du = \left[-\frac{1}{2}u^3 e^{-u^2}\right]_0^\infty + \frac{3}{2}\int_0^\infty u^2 e^{-u^2} du.$$

Again, the first term is 0, and we were given in an earlier problem that the integral in the second term equals $\frac{\sqrt{\pi}}{4}$. We now have

$$\bar{v^2} = \frac{4}{\sqrt{\pi}}\frac{2k_B T}{m}\frac{3}{2}\frac{\sqrt{\pi}}{4} = \frac{3k_B T}{m}.$$

Taking the square root of both sides gives the desired result: $v_{\text{rms}} = \sqrt{\frac{3k_B T}{m}}$.

CHAPTER 3

CHECK YOUR UNDERSTANDING

3.1. $p_2(V_2 - V_1)$

3.2. Line 1, $\Delta E_{\text{int}} = 40\ \text{J}$; line 2, $W = 50\ \text{J}$ and $\Delta E_{\text{int}} = 40\ \text{J}$; line 3, $Q = 80\ \text{J}$ and $\Delta E_{\text{int}} = 40\ \text{J}$; and line 4, $Q = 0$ and $\Delta E_{\text{int}} = 40\ \text{J}$

3.3. So that the process is represented by the curve $p = nRT/V$ on the pV plot for the evaluation of work.

3.4. 1.26×10^3 J.

CONCEPTUAL QUESTIONS

1. a. SE; b. ES; c. ES

3. Some of the energy goes into changing the phase of the liquid to gas.

5. Yes, as long as the work done equals the heat added there will be no change in internal energy and thereby no change in temperature. When water freezes or when ice melts while removing or adding heat, respectively, the temperature remains constant.
7. If more work is done on the system than heat added, the internal energy of the system will actually decrease.
9. The system must be in contact with a heat source that allows heat to flow into the system.
11. Isothermal processes must be slow to make sure that as heat is transferred, the temperature does not change. Even for isobaric and isochoric processes, the system must be in thermal equilibrium with slow changes of thermodynamic variables.
13. Typically C_p is greater than C_V because when expansion occurs under constant pressure, it does work on the surroundings. Therefore, heat can go into internal energy and work. Under constant volume, all heat goes into internal energy. In this example, water contracts upon heating, so if we add heat at constant pressure, work is done on the water by surroundings and therefore, C_p is less than C_V.
15. No, it is always greater than 1.
17. An adiabatic process has a change in temperature but no heat flow. The isothermal process has no change in temperature but has heat flow.

PROBLEMS

19. $p(V-b) = -c_T$ is the temperature scale desired and mirrors the ideal gas if under constant volume.
21. $V - bpT + cT^2 = 0$
23. 74 K
25. 1.4 times
27. pVln(4)
29. a. 160 J; b. –160 J
31.

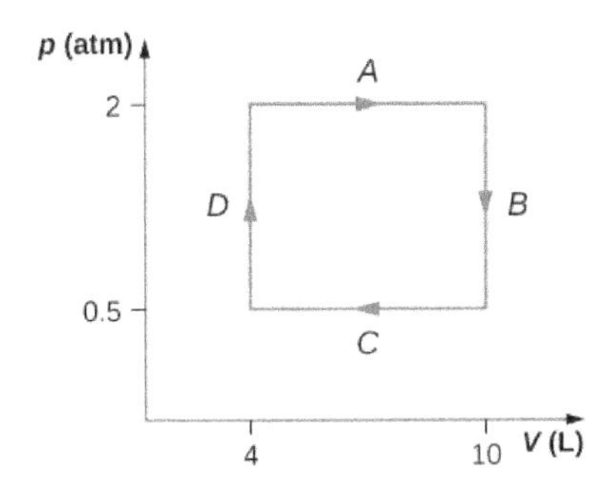

$W = 900\text{ J}$

33. 3.53×10^4 J
35. a. 1:1; b. 10:1
37. a. 600 J; b. 0; c. 500 J; d. 200 J; e. 800 J; f. 500 J
39. 580 J
41. a. 600 J; b. 600 J; c. 800 J
43. a. 0; b. 160 J; c. –160 J
45. a. –150 J; b. –400 J
47. No work is done and they reach the same common temperature.
49. 54,500 J
51. a. $(p_1 + 3V_1^2)(V_2 - V_1) - 3V_1(V_2^2 - V_1^2) + (V_2^3 - V_1^3)$; b. $\frac{3}{2}(p_2V_2 - p_1V_1)$; c. the sum of parts (a) and (b); d. $T_1 = \frac{p_1V_1}{nR}$ and $T_2 = \frac{p_2V_2}{nR}$

53. a.

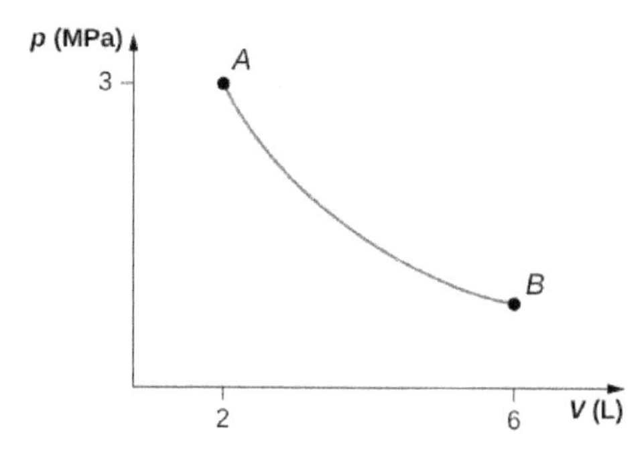

;

b. $W = 4.39\,\text{kJ}, \Delta E_{\text{int}} = -4.39\,\text{kJ}$

55. a. 1660 J; b. −2730 J; c. It does not depend on the process.
57. a. 700 J; b. 500 J
59. a. −3 400 J; b. 3400 J enters the gas
61. 100 J
63. a. 370 J; b. 100 J; c. 500 J
65. 850 J
67. pressure decreased by 0.31 times the original pressure
69.

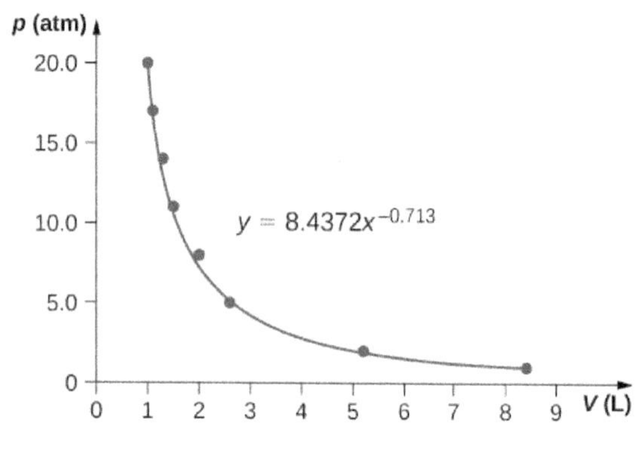

;

$\gamma = 0.713$

71. 84 K
73. An adiabatic expansion has less work done and no heat flow, thereby a lower internal energy comparing to an isothermal expansion which has both heat flow and work done. Temperature decreases during adiabatic expansion.
75. Isothermal has a greater final pressure and does not depend on the type of gas.
77.

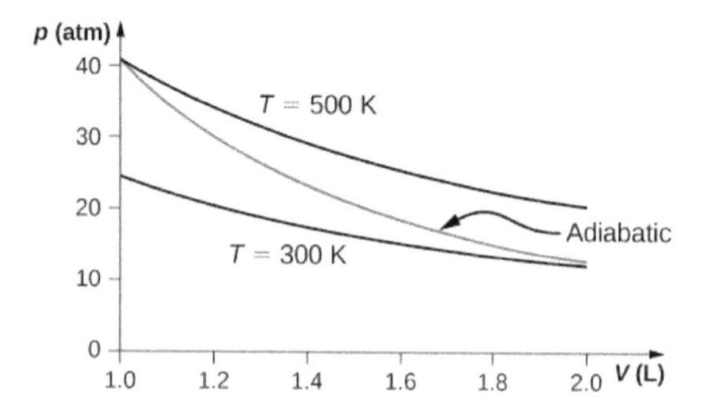

ADDITIONAL PROBLEMS

79. a. $W_{AB} = 0,\ W_{BC} = 2026\,\text{J},\ W_{AD} = 810.4\,\text{J},\ W_{DC} = 0;$ b. $\Delta E_{AB} = 3600\,\text{J},\ \Delta E_{BC} = 374\,\text{J};$ c. $\Delta E_{AC} = 3974\,\text{J};$ d. $Q_{ADC} = 4784\,\text{J};$ e. No, because heat was added for both parts *AD* and *DC*. There is not enough information to figure out how much is from each segment of the path.
81. 300 J
83. a. 59.5 J; b. 170 N
85. 2.4×10^3 J
87. a. 15,000 J; b. 10,000 J; c. 25,000 J
89. 78 J
91. A cylinder containing three moles of nitrogen gas is heated at a constant pressure of 2 atm. a. −1220 J; b. +1220 J
93. a. 7.6 L, 61.6 K; b. 81.3 K; c. $3.63\,\text{L}\cdot\text{atm} = 367\,\text{J}$; d. −367 J

CHALLENGE PROBLEMS

95. a. 1700 J; b. 1200 J; c. 2400 J
97. a. 2.2 mol; b. $V_A = 6.7 \times 10^{-2}\text{m}^3,\ V_B = 3.3 \times 10^{-2}\text{m}^3$; c. $T_A = 2400\,\text{K},\ T_B = 397\,\text{K}$; d. 26,000 J

CHAPTER 4

CHECK YOUR UNDERSTANDING

4.1. A perfect heat engine would have $Q_c = 0$, which would lead to $e = 1 - Q_c/Q_h = 1$. A perfect refrigerator would need zero work, that is, $W = 0$, which leads to $K_R = Q_c/W \to \infty$.
4.2. From the engine on the right, we have $W = Q'_h - Q'_c$. From the refrigerator on the right, we have $Q_h = Q_c + W$. Thus, $W = Q'_h - Q'_c = Q_h - Q_c$.
4.3. a. $e = 1 - T_c/T_h = 0.55$; b. $Q_h = eW = 9.1\,\text{J}$; c. $Q_c = Q_h - W = 4.1\,\text{J}$; d. $-273\,°\text{C}$ and $400\,°\text{C}$
4.4. a. $K_R = T_c/(T_h - T_c) = 10.9$; b. $Q_c = K_R W = 2.18\,\text{kJ}$; c. $Q_h = Q_c + W = 2.38\,\text{kJ}$
4.5. When heat flows from the reservoir to the ice, the internal (mainly kinetic) energy of the ice goes up, resulting in a higher average speed and thus an average greater position variance of the molecules in the ice. The reservoir does become more ordered, but due to its much larger amount of molecules, it does not offset the change in entropy in the system.
4.6. $-Q/T_h$; Q/T_c; and $Q(T_h - T_c)/(T_h T_c)$
4.7. a. 4.71 J/K; b. −4.18 J/K; c. 0.53 J/K

CONCEPTUAL QUESTIONS

1. Some possible solutions are frictionless movement; restrained compression or expansion; energy transfer as heat due to infinitesimal temperature nonuniformity; electric current flow through a zero resistance; restrained chemical reaction; and mixing of two samples of the same substance at the same state.
3. The temperature increases since the heat output behind the refrigerator is greater than the cooling from the inside of the refrigerator.
5. If we combine a perfect engine and a real refrigerator with the engine converting heat *Q* from the hot reservoir into work $W = Q$ to drive the refrigerator, then the heat dumped to the hot reservoir by the refrigerator will be $W + \Delta Q$, resulting in a perfect refrigerator transferring heat ΔQ from the cold reservoir to hot reservoir without any other effect.
7. Heat pumps can efficiently extract heat from the ground to heat on cooler days or pull heat out of the house on warmer days. The disadvantage of heat pumps are that they are more costly than alternatives, require maintenance, and will not work efficiently when temperature differences between the inside and outside are very large. Electric heating is much cheaper to purchase than a heat pump; however, it may be more costly to run depending on the electric rates and amount of usage.
9. A nuclear reactor needs to have a lower temperature to operate, so its efficiency will not be as great as a fossil-fuel plant. This argument does not take into consideration the amount of energy per reaction: Nuclear power has a far greater energy output than fossil fuels.
11. In order to increase the efficiency, the temperature of the hot reservoir should be raised, and the cold reservoir should be lowered as much as possible. This can be seen in Equation 4.8.
13. adiabatic and isothermal processes
15. Entropy will not change if it is a reversible transition but will change if the process is irreversible.
17. Entropy is a function of disorder, so all the answers apply here as well.

PROBLEMS

19. 4.53×10^3 J

21. $4.5\, pV_0$
23. 0.667
25. a. 0.556; b. 125.0 J
27. a. 0.50; b. 100 J; c. 50 J
29. a. 600 J; b. 800 J
31. a. 69 J; b. 11 J
33. 2.0
35. 50 J
37.

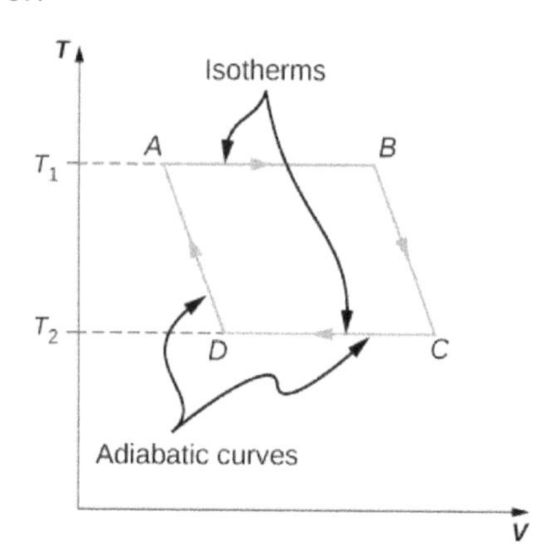

39. a. 900 J; b. 100 J
41. a. 546 K; b. 137 K
43. –1 J/K
45. –13 J(K mole)
47. $-\frac{Q}{T_h}, \frac{Q}{T_c}, Q\left(\frac{1}{T_c} - \frac{1}{T_h}\right)$
49. a. –540 J/K; b. 1600 J/K; c. 1100 J/K
51. a. $Q = nR\Delta T$; b. $S = nR \ln(T_2/T_1)$
53. 3.78×10^{-3} W/K
55. 430 J/K
57. $80\,°\text{C}$, $80\,°\text{C}$, 6.70×10^4 J, 215 J/K, –190 J/K, 25 J/K
59. $\Delta S_{H_2O} = 215$ J/K, $\Delta S_R = -208$ J/K, $\Delta S_U = 7$ J/K
61. a. 1200 J; b. 600 J; c. 600 J; d. 0.50
63. $\Delta S = nC_V \ln\left(\frac{T_2}{T_1}\right) + nC_p \ln\left(\frac{T_3}{T_2}\right)$
65. a. 0.33, 0.39; b. 0.91

ADDITIONAL PROBLEMS

67. 1.45×10^7 J
69. a. $V_B = 0.042\ \text{m}^3$, $V_D = 0.018\ \text{m}^3$; b. 13,000 J; c. 13,000 J; d. –8,000 J; e. –8,000 J; f. 6200 J; g. –6200 J; h. 39% ; with temperatures efficiency is 40% , which is off likely by rounding errors.
71. –670 J/K
73. a. –570 J/K; b. 570 J/K
75. 82 J/K
77. a. 2000 J; b. 40%
79. 60%
81. 64.4%

CHALLENGE PROBLEMS

83. derive
85. derive
87. 18 J/K
89. proof
91. $K_R = \frac{3(p_1 - p_2)V_1}{5p_2V_3 - 3p_1V_1 - p_2V_1}$
93. $W = 110{,}000\text{ J}$

CHAPTER 5

CHECK YOUR UNDERSTANDING

5.1. The force would point outward.
5.2. The net force would point $58°$ below the $-x$-axis.
5.3. $\vec{\mathbf{E}} = \frac{1}{4\pi\varepsilon_0}\frac{q}{r^2}\hat{r}$
5.4. We will no longer be able to take advantage of symmetry. Instead, we will need to calculate each of the two components of the electric field with their own integral.
5.5. The point charge would be $Q = \sigma ab$ where *a* and *b* are the sides of the rectangle but otherwise identical.
5.6. The electric field would be zero in between, and have magnitude $\frac{\sigma}{\varepsilon_0}$ everywhere else.

CONCEPTUAL QUESTIONS

1. There are mostly equal numbers of positive and negative charges present, making the object electrically neutral.
3. a. yes; b. yes
5. Take an object with a known charge, either positive or negative, and bring it close to the rod. If the known charged object is positive and it is repelled from the rod, the rod is charged positive. If the positively charged object is attracted to the rod, the rod is negatively charged.
7. No, the dust is attracted to both because the dust particle molecules become polarized in the direction of the silk.
9. Yes, polarization charge is induced on the conductor so that the positive charge is nearest the charged rod, causing an attractive force.
11. Charging by conduction is charging by contact where charge is transferred to the object. Charging by induction first involves producing a polarization charge in the object and then connecting a wire to ground to allow some of the charge to leave the object, leaving the object charged.
13. This is so that any excess charge is transferred to the ground, keeping the gasoline receptacles neutral. If there is excess charge on the gasoline receptacle, a spark could ignite it.
15. The dryer charges the clothes. If they are damp, the presence of water molecules suppresses the charge.
17. There are only two types of charge, attractive and repulsive. If you bring a charged object near the quartz, only one of these two effects will happen, proving there is not a third kind of charge.
19. a. No, since a polarization charge is induced. b. Yes, since the polarization charge would produce only an attractive force.
21. The force holding the nucleus together must be greater than the electrostatic repulsive force on the protons.
23. Either sign of the test charge could be used, but the convention is to use a positive test charge.
25. The charges are of the same sign.
27. At infinity, we would expect the field to go to zero, but because the sheet is infinite in extent, this is not the case. Everywhere you are, you see an infinite plane in all directions.
29. The infinite charged plate would have $E = \frac{\sigma}{2\varepsilon_0}$ everywhere. The field would point toward the plate if it were negatively charged and point away from the plate if it were positively charged. The electric field of the parallel plates would be zero between them if they had the same charge, and *E* would be $E = \frac{\sigma}{\varepsilon_0}$ everywhere else. If the charges were opposite, the situation is reversed, zero outside the plates and $E = \frac{\sigma}{\varepsilon_0}$ between them.
31. yes; no
33. At the surface of Earth, the gravitational field is always directed in toward Earth's center. An electric field could move a charged particle in a different direction than toward the center of Earth. This would indicate an electric field is present.
35. 10

PROBLEMS

37. a. $2.00 \times 10^{-9}\ \text{C}\left(\frac{1}{1.602 \times 10^{-19}}\ \text{e/C}\right) = 1.248 \times 10^{10}$ electrons ;

b. $0.500 \times 10^{-6}\ \text{C}\left(\frac{1}{1.602 \times 10^{-19}}\ \text{e/C}\right) = 3.121 \times 10^{12}$ electrons

39. $\frac{3.750 \times 10^{21}\ \text{e}}{6.242 \times 10^{18}\ \text{e/C}} = 600.8\ \text{C}$

41. a. $2.0 \times 10^{-9}\ \text{C}\ (6.242 \times 10^{18}\ \text{e/C}) = 1.248 \times 10^{10}\ \text{e}$;

b. $9.109 \times 10^{-31}\ \text{kg}\ (1.248 \times 10^{10}\ \text{e}) = 1.137 \times 10^{-20}\ \text{kg}$,

$\frac{1.137 \times 10^{-20}\ \text{kg}}{2.5 \times 10^{-3}\ \text{kg}} = 4.548 \times 10^{-18}$ or $4.545 \times 10^{-16}\,\%$

43. $5.00 \times 10^{-9}\ \text{C}\ (6.242 \times 10^{18}\ \text{e/C}) = 3.121 \times 10^{19}\ \text{e}$;

$3.121 \times 10^{19}\ \text{e} + 1.0000 \times 10^{12}\ \text{e} = 3.1210001 \times 10^{19}\ \text{e}$

45. atomic mass of copper atom times $1\ \text{u} = 1.055 \times 10^{-25}\ \text{kg}$;

number of copper atoms $= 4.739 \times 10^{23}$ atoms ;

number of electrons equals 29 times number of atoms or 1.374×10^{25} electrons ;

$\frac{2.00 \times 10^{-6}\ \text{C}(6.242 \times 10^{18}\ \text{e/C})}{1.374 \times 10^{25}\ \text{e}} = 9.083 \times 10^{-13}$ or $9.083 \times 10^{-11}\,\%$

47. $244.00\ \text{u}(1.66 \times 10^{-27}\ \text{kg/u}) = 4.050 \times 10^{-25}\ \text{kg}$;

$\frac{4.00\ \text{kg}}{4.050 \times 10^{-25}\ \text{kg}} = 9.877 \times 10^{24}$ atoms $\quad 9.877 \times 10^{24}(94) = 9.284 \times 10^{26}$ protons ;

$9.284 \times 10^{26}(1.602 \times 10^{-19}\ \text{C/p}) = 1.487 \times 10^{8}\ \text{C}$

49. a. charge 1 is $3\ \mu\text{C}$; charge 2 is $12\ \mu\text{C}$, $F_{31} = 2.16 \times 10^{-4}\ \text{N}$ to the left,

$F_{32} = 8.63 \times 10^{-4}\ \text{N}$ to the right,

$F_{net} = 6.47 \times 10^{-4}\ \text{N}$ to the right;

b. $F_{31} = 2.16 \times 10^{-4}\ \text{N}$ to the right,

$F_{32} = 9.59 \times 10^{-5}\ \text{N}$ to the right,

$F_{net} = 3.12 \times 10^{-4}\ \text{N}$ to the right,

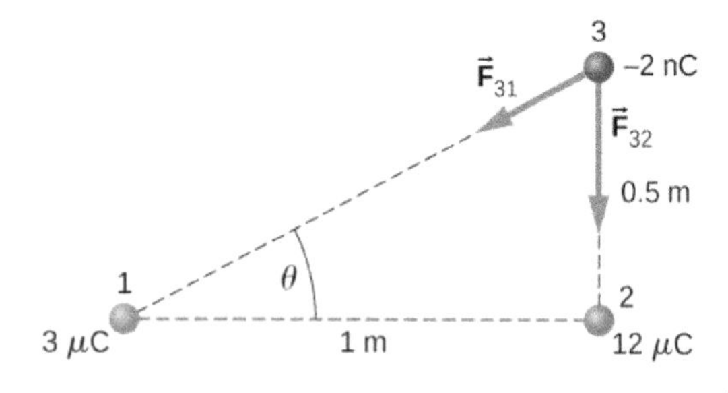

;

c. $\vec{F}_{31x} = -2.76 \times 10^{-5}\ \text{N}\ \hat{i}$,

$\vec{F}_{31y} = -1.38 \times 10^{-5}\ \text{N}\ \hat{j}$,

$\vec{F}_{32y} = -8.63 \times 10^{-4}\ \text{N}\ \hat{j}$

$\vec{F}_{net} = -2.76 \times 10^{-5}\ \text{N}\ \hat{i} - 8.77 \times 10^{-4}\ \text{N}\ \hat{j}$

51. $F = 230.7\,\text{N}$

53. $F = 53.94\,\text{N}$

55. The tension is $T = 0.049\,\text{N}$. The horizontal component of the tension is $0.0043\,\text{N}$

$d = 0.088\,\text{m}, \quad q = 6.1 \times 10^{-8}\,\text{C}$.

The charges can be positive or negative, but both have to be the same sign.

57. Let the charge on one of the spheres be rQ, where r is a fraction between 0 and 1. In the numerator of Coulomb's law, the term involving the charges is $rQ(1-r)Q$. This is equal to $(r-r^2)Q^2$. Finding the maximum of this term gives

$1 - 2r = 0 \Rightarrow r = \frac{1}{2}$

59. Define right to be the positive direction and hence left is the negative direction, then $F = -0.05\,\text{N}$

61. The particles form triangle of sides 13, 13, and 24 cm. The x-components cancel, whereas there is a contribution to the y-component from both charges 24 cm apart. The y-axis passing through the third charge bisects the 24-cm line, creating two right triangles of sides 5, 12, and 13 cm.

$F_y = 2.56\,\text{N}$ in the negative y-direction since the force is attractive. The net force from both charges is $\vec{\mathbf{F}}_{\text{net}} = -5.12\,\text{N}\,\hat{\mathbf{j}}$

.

63. The diagonal is $\sqrt{2}a$ and the components of the force due to the diagonal charge has a factor $\cos\theta = \frac{1}{\sqrt{2}}$;

$$\vec{\mathbf{F}}_{\text{net}} = \left[k\frac{q^2}{a^2} + k\frac{q^2}{2a^2}\frac{1}{\sqrt{2}}\right]\hat{\mathbf{i}} - \left[k\frac{q^2}{a^2} + k\frac{q^2}{2a^2}\frac{1}{\sqrt{2}}\right]\hat{\mathbf{j}}$$

65. a. $E = 2.0 \times 10^{-2}\,\frac{\text{N}}{\text{C}}$;

b. $F = 2.0 \times 10^{-19}\,\text{N}$

67. a. $E = 2.88 \times 10^{11}\,\text{N/C}$;

b. $E = 1.44 \times 10^{11}\,\text{N/C}$;

c. $F = 4.61 \times 10^{-8}\,\text{N}$ on alpha particle;

$F = 4.61 \times 10^{-8}\,\text{N}$ on electron

69. $E = \left(-2.0\,\hat{\mathbf{i}} + 3.0\,\hat{\mathbf{j}}\right)\text{N}$

71. $F = 3.204 \times 10^{-14}\,\text{N}$,

$a = 3.517 \times 10^{16}\,\text{m/s}^2$

73. $q = 2.78 \times 10^{-9}\,\text{C}$

75. a. $E = 1.15 \times 10^{12}\,\text{N/C}$;

b. $F = 1.47 \times 10^{-6}\,\text{N}$

77. If the q_2 is to the right of q_1, the electric field vector from both charges point to the right. a. $E = 2.70 \times 10^6\,\text{N/C}$;

b. $F = 54.0\,\text{N}$

79. There is $45°$ right triangle geometry. The x-components of the electric field at $y = 3\,\text{m}$ cancel. The y-components give

$E(y = 3\,\text{m}) = 2.83 \times 10^3\,\text{N/C}$.

At the origin we have a a negative charge of magnitide

$q = -2.83 \times 10^{-6}\,\text{C}$.

81. $\vec{\mathbf{E}}(z) = 3.6 \times 10^4\,\text{N/C}\,\hat{\mathbf{k}}$

83. $dE = \frac{1}{4\pi\varepsilon_0}\frac{\lambda dx}{(x+a)^2}, \quad E = \frac{\lambda}{4\pi\varepsilon_0}\left[\frac{1}{l+a} - \frac{1}{a}\right]$

85. $\sigma = 0.02\,\text{C/m}^2 \quad E = 2.26 \times 10^9\,\text{N/C}$

87. At P_1: $\vec{\mathbf{E}}(y) = \frac{1}{4\pi\varepsilon_0}\frac{\lambda L}{y\sqrt{y^2 + \frac{L^2}{4}}}\hat{\mathbf{j}} \Rightarrow \frac{1}{4\pi\varepsilon_0}\frac{q}{\frac{a}{2}\sqrt{(\frac{a}{2})^2 + \frac{L^2}{4}}}\hat{\mathbf{j}} = \frac{1}{\pi\varepsilon_0}\frac{q}{a\sqrt{a^2 + L^2}}\hat{\mathbf{j}}$

At P_2: Put the origin at the end of L.

$dE = \frac{1}{4\pi\varepsilon_0}\frac{\lambda dx}{(x+a)^2}$, $\quad \vec{\mathbf{E}} = -\frac{q}{4\pi\varepsilon_0 l}\left[\frac{1}{l+a} - \frac{1}{a}\right]\hat{\mathbf{i}}$

89. a. $\vec{\mathbf{E}}(\vec{\mathbf{r}}) = \frac{1}{4\pi\varepsilon_0}\frac{2\lambda_x}{a}\hat{\mathbf{i}} + \frac{1}{4\pi\varepsilon_0}\frac{2\lambda_y}{b}\hat{\mathbf{j}}$; b. $\frac{1}{4\pi\varepsilon_0}\frac{2(\lambda_x + \lambda_y)}{c}\hat{\mathbf{k}}$

91. a. $\vec{\mathbf{F}} = 3.2 \times 10^{-17}\ \text{N}\,\hat{\mathbf{i}}$,

$\vec{\mathbf{a}} = 1.92 \times 10^{10}\ \text{m/s}^2\,\hat{\mathbf{i}}$;

b. $\vec{\mathbf{F}} = -3.2 \times 10^{-17}\ \text{N}\,\hat{\mathbf{i}}$,

$\vec{\mathbf{a}} = -3.51 \times 10^{13}\ \text{m/s}^2\,\hat{\mathbf{i}}$

93. $m = 6.5 \times 10^{-11}$ kg,

$E = 1.6 \times 107$ N/C

95. $E = 1.70 \times 10^6$ N/C,

$F = 1.53 \times 10^{-3}$ N $T\cos\theta = mg$ $T\sin\theta = qE$,

$\tan\theta = 0.62 \Rightarrow \theta = 32.0°$,

This is independent of the length of the string.

97. circular arc $dE_x(-\hat{\mathbf{i}}) = \frac{1}{4\pi\varepsilon_0}\frac{\lambda ds}{r^2}\cos\theta(-\hat{\mathbf{i}})$,

$\vec{\mathbf{E}}_x = \frac{\lambda}{4\pi\varepsilon_0 r}(-\hat{\mathbf{i}})$,

$dE_y(-\hat{\mathbf{i}}) = \frac{1}{4\pi\varepsilon_0}\frac{\lambda ds}{r^2}\sin\theta(-\hat{\mathbf{j}})$,

$\vec{\mathbf{E}}_y = \frac{\lambda}{4\pi\varepsilon_0 r}(-\hat{\mathbf{j}})$;

y-axis: $\vec{\mathbf{E}}_x = \frac{\lambda}{4\pi\varepsilon_0 r}(-\hat{\mathbf{i}})$;

x-axis: $\vec{\mathbf{E}}_y = \frac{\lambda}{4\pi\varepsilon_0 r}(-\hat{\mathbf{j}})$,

$\vec{\mathbf{E}} = \frac{\lambda}{2\pi\varepsilon_0 r}(-\hat{\mathbf{i}}) + \frac{\lambda}{2\pi\varepsilon_0 r}(-\hat{\mathbf{j}})$

99. a. $W = \frac{1}{2}m(v^2 - v_0^2)$, $\frac{Qq}{4\pi\varepsilon_0}\left(\frac{1}{r} - \frac{1}{r_0}\right) = \frac{1}{2}m(v^2 - v_0^2) \Rightarrow r_0 - r = \frac{4\pi\varepsilon_0}{Qq}\frac{1}{2}rr_0 m(v^2 - v_0^2)$; b. $r_0 - r$ is negative; therefore, $v_0 > v$, $r \to \infty$, and $v \to 0$: $\frac{Qq}{4\pi\varepsilon_0}\left(-\frac{1}{r_0}\right) = -\frac{1}{2}mv_0^2 \Rightarrow v_0 = \sqrt{\frac{Qq}{2\pi\varepsilon_0 m r_0}}$

101.

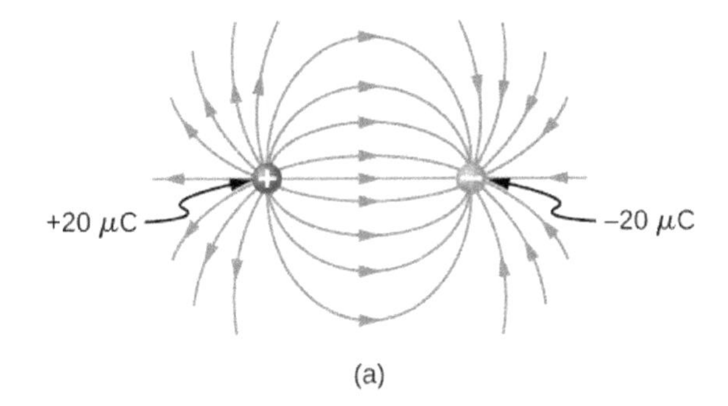

(a)

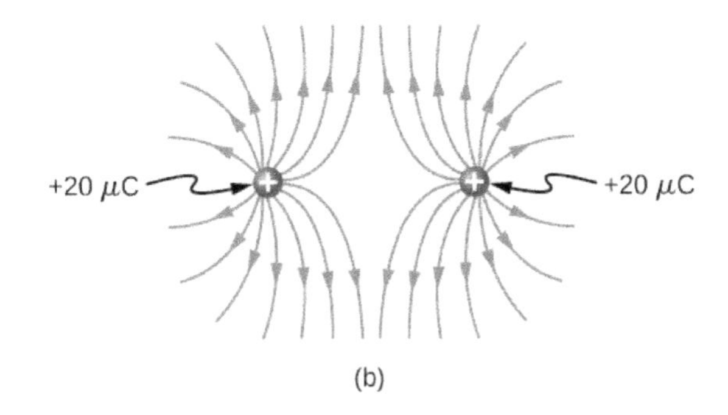

(b)

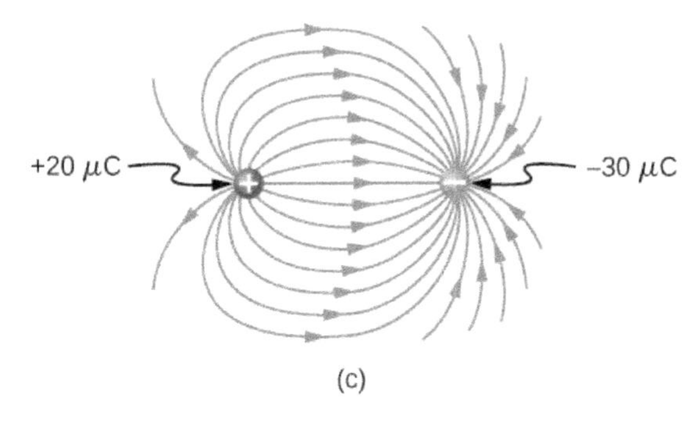

(c)

103.

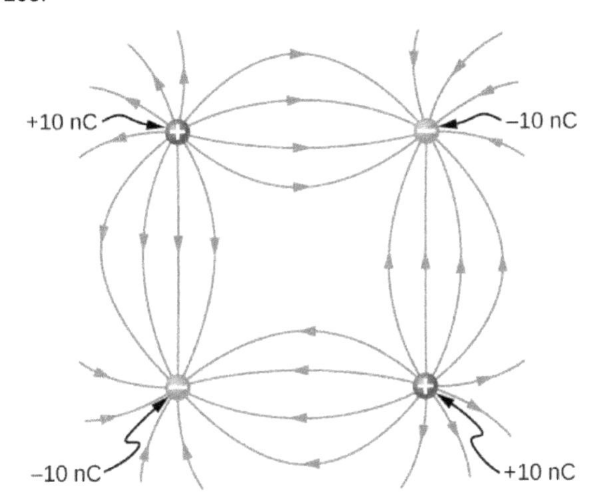

105. $E_x = 0$,

$$E_y = \frac{1}{4\pi\varepsilon_0}\left[\frac{2q}{(x^2+a^2)}\frac{a}{\sqrt{(x^2+a^2)}}\right]$$

$$\Rightarrow x \gg a \Rightarrow \frac{1}{2\pi\varepsilon_0}\frac{qa}{x^3},$$

$$E_y = \frac{q}{4\pi\varepsilon_0}\left[\frac{2ya+2ya}{(y-a)^2(y+a)^2}\right]$$

$$\Rightarrow y \gg a \Rightarrow \frac{1}{\pi\varepsilon_0}\frac{qa}{y^3}$$

107. The net dipole moment of the molecule is the vector sum of the individual dipole moments between the two O-H. The separation O-H is 0.9578 angstroms:

$$\vec{\mathbf{p}} = 1.889 \times 10^{-29}\ \text{Cm}\ \hat{\mathbf{i}}$$

ADDITIONAL PROBLEMS

109. $\vec{\mathbf{F}}_{\text{net}} = [-8.99 \times 10^9 \frac{3.0 \times 10^{-6}(5.0 \times 10^{-6})}{(3.0\ \text{m})^2} - 8.99 \times 10^9 \frac{9.0 \times 10^{-6}(5.0 \times 10^{-6})}{(3.0\ \text{m})^2}]\hat{\mathbf{i}}$,

$$-8.99 \times 10^9 \frac{6.0 \times 10^{-6}(5.0 \times 10^{-6})}{(3.0\ \text{m})^2}\hat{\mathbf{j}} = -0.06\ \text{N}\ \hat{\mathbf{i}} - 0.03\ \text{N}\ \hat{\mathbf{j}}$$

111. Charges Q and q form a right triangle of sides 1 m and $3+\sqrt{3}$ m. Charges $2Q$ and q form a right triangle of sides 1 m and $\sqrt{3}$ m.

$F_x = 0.036\ \text{N}$,

$F_y = 0.09\ \text{N}$,

$$\vec{\mathbf{F}}_{\text{net}} = 0.036\ \text{N}\ \hat{\mathbf{i}} + 0.09\ \text{N}\ \hat{\mathbf{j}}$$

113. $W = 0.054\ \text{J}$

115. a. $\vec{\mathbf{E}} = \frac{1}{4\pi\varepsilon_0}(\frac{q}{(2a)^2} - \frac{q}{a^2})\hat{\mathbf{i}}$; b. $\vec{\mathbf{E}} = \frac{\sqrt{3}}{4\pi\varepsilon_0}\frac{q}{a^2}(-\hat{\mathbf{j}})$; c. $\vec{\mathbf{E}} = \frac{2}{\pi\varepsilon_0}\frac{q}{a^2}\frac{1}{\sqrt{2}}(-\hat{\mathbf{j}})$

117. $\vec{\mathbf{E}} = 6.4 \times 10^6(\hat{\mathbf{i}}) + 1.5 \times 10^7(\hat{\mathbf{j}})\ \text{N/C}$

119. $F = qE_0(1 + x/a)$ $W = \frac{1}{2}m(v^2 - v_0^2)$,

$$\frac{1}{2}mv^2 = qE_0(\frac{15a}{2})\ \text{J}$$

121. Electric field of wire at x: $\vec{\mathbf{E}}(x) = \frac{1}{4\pi\varepsilon_0}\frac{2\lambda_y}{x}\hat{\mathbf{i}}$,

$$dF = \frac{\lambda_y\lambda_x}{2\pi\varepsilon_0}(\ln b - \ln a)$$

123.

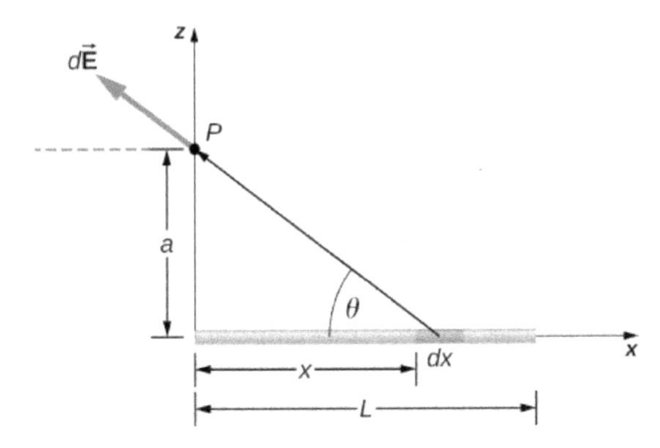

$$dE_x = \frac{1}{4\pi\varepsilon_0} \frac{\lambda dx}{(x^2 + a^2)} \frac{x}{\sqrt{x^2 + a^2}},$$

$$\vec{E}_x = \frac{\lambda}{4\pi\varepsilon_0}\left[\frac{1}{\sqrt{L^2 + a^2}} - \frac{1}{a}\right]\hat{i},$$

$$dE_z = \frac{1}{4\pi\varepsilon_0} \frac{\lambda dx}{(x^2 + a^2)} \frac{a}{\sqrt{x^2 + a^2}},$$

$$\vec{E}_z = \frac{\lambda}{4\pi\varepsilon_0 a} \frac{L}{\sqrt{L^2 + a^2}}\hat{k},$$

Substituting z for a, we have:

$$\vec{E}(z) = \frac{\lambda}{4\pi\varepsilon_0}\left[\frac{1}{\sqrt{L^2 + z^2}} - \frac{1}{z}\right]\hat{i} + \frac{\lambda}{4\pi\varepsilon_0 z} \frac{L}{\sqrt{L^2 + z^2}}\hat{k}$$

125. There is a net force only in the y-direction. Let θ be the angle the vector from dx to q makes with the x-axis. The components along the x-axis cancel due to symmetry, leaving the y-component of the force.

$$dF_y = \frac{1}{4\pi\varepsilon_0} \frac{aq\lambda dx}{(x^2 + a^2)^{3/2}},$$

$$F_y = \frac{1}{2\pi\varepsilon_0} \frac{q\lambda}{a}\left[\frac{l/2}{((l/2)^2 + a^2)^{1/2}}\right]$$

CHAPTER 6

CHECK YOUR UNDERSTANDING

6.1. Place it so that its unit normal is perpendicular to $\vec{E}$.

6.2. $mab^2/2$

6.3. a. $3.4 \times 10^5\ \text{N} \cdot \text{m}^2/\text{C}$; b. $-3.4 \times 10^5\ \text{N} \cdot \text{m}^2/\text{C}$; c. $3.4 \times 10^5\ \text{N} \cdot \text{m}^2/\text{C}$; d. 0

6.4. In this case, there is only $\vec{E}_{\text{out}}$. So, yes.

6.5. $\vec{E} = \frac{\lambda_0}{2\pi\varepsilon_0} \frac{1}{d}\hat{r}$; This agrees with the calculation of Example 5.5 where we found the electric field by integrating over the charged wire. Notice how much simpler the calculation of this electric field is with Gauss's law.

6.6. If there are other charged objects around, then the charges on the surface of the sphere will not necessarily be spherically symmetrical; there will be more in certain direction than in other directions.

CONCEPTUAL QUESTIONS

1. a. If the planar surface is perpendicular to the electric field vector, the maximum flux would be obtained. b. If the planar surface were parallel to the electric field vector, the minimum flux would be obtained.

3. true

5. Since the electric field vector has a $\frac{1}{r^2}$ dependence, the fluxes are the same since $A = 4\pi r^2$.

7. a. no; b. zero

9. Both fields vary as $\frac{1}{r^2}$. Because the gravitational constant is so much smaller than $\frac{1}{4\pi\varepsilon_0}$, the gravitational field is orders of magnitude weaker than the electric field.

11. No, it is produced by all charges both inside and outside the Gaussian surface.

13. yes, using superposition

15. Any shape of the Gaussian surface can be used. The only restriction is that the Gaussian integral must be calculable; therefore, a box or a cylinder are the most convenient geometrical shapes for the Gaussian surface.

17. yes

19. Since the electric field is zero inside a conductor, a charge of $-2.0\,\mu\text{C}$ is induced on the inside surface of the cavity. This will put a charge of $+2.0\,\mu C$ on the outside surface leaving a net charge of $-3.0\,\mu\text{C}$ on the surface.

PROBLEMS

21. $\Phi = \vec{\mathbf{E}} \cdot \vec{\mathbf{A}} \rightarrow EA\cos\theta = 2.2 \times 10^4\,\text{N}\cdot\text{m}^2/\text{C}$ electric field in direction of unit normal; $\Phi = \vec{\mathbf{E}} \cdot \vec{\mathbf{A}} \rightarrow EA\cos\theta = -2.2 \times 10^4\,\text{N}\cdot\text{m}^2/\text{C}$ electric field opposite to unit normal

23. $\frac{3 \times 10^{-5}\ \text{N}\cdot\text{m}^2/\text{C}}{(0.05\ \text{m})^2} = E \Rightarrow \sigma = 2.12 \times 10^{-13}\ \text{C/m}^2$

25. a. $\Phi = 0.17\,\text{N}\cdot\text{m}^2/\text{C}$;

b. $\Phi = 0$; c. $\Phi = EA\cos 0° = 1.0 \times 10^3\ \text{N/C}(2.0 \times 10^{-4}\ \text{m})^2\ \cos 0° = 0.20\,\text{N}\cdot\text{m}^2/\text{C}$

27. $\Phi = 3.8 \times 10^4\ \text{N}\cdot\text{m}^2/\text{C}$

29. $\vec{\mathbf{E}}(z) = \frac{1}{4\pi\varepsilon_0}\frac{2\lambda}{z}\hat{\mathbf{k}},\ \int \vec{\mathbf{E}} \cdot \hat{\mathbf{n}}\,dA = \frac{\lambda}{\varepsilon_0}l$

31. a. $\Phi = 3.39 \times 10^3\ \text{N}\cdot\text{m}^2/\text{C}$; b. $\Phi = 0$;

c. $\Phi = -2.25 \times 10^5\ \text{N}\cdot\text{m}^2/\text{C}$;

d. $\Phi = 90.4\,\text{N}\cdot\text{m}^2/\text{C}$

33. $\Phi = 1.13 \times 10^6\ \text{N}\cdot\text{m}^2/\text{C}$

35. Make a cube with q at the center, using the cube of side a. This would take four cubes of side a to make one side of the large cube. The shaded side of the small cube would be 1/24th of the total area of the large cube; therefore, the flux through the shaded area would be

$\Phi = \frac{1}{24}\frac{q}{\varepsilon_0}$.

37. $q = 3.54 \times 10^{-7}\ \text{C}$

39. zero, also because flux in equals flux out

41. $r > R,\ E = \frac{Q}{4\pi\varepsilon_0 r^2};\ r < R,\ E = \frac{qr}{4\pi\varepsilon_0 R^3}$

43. $EA = \frac{\lambda l}{\varepsilon_0} \Rightarrow E = 4.50 \times 10^7\ \text{N/C}$

45. a. 0; b. 0; c. $\vec{\mathbf{E}} = 6.74 \times 10^6\ \text{N/C}(-\hat{\mathbf{r}})$

47. a. 0; b. $E = 2.70 \times 10^6\ \text{N/C}$

49. a. Yes, the length of the rod is much greater than the distance to the point in question. b. No, The length of the rod is of the same order of magnitude as the distance to the point in question. c. Yes, the length of the rod is much greater than the distance to the point in question. d. No. The length of the rod is of the same order of magnitude as the distance to the point in question.

51. a. $\vec{\mathbf{E}} = \frac{R\sigma_0}{\varepsilon_0}\frac{1}{r}\hat{\mathbf{r}} \Rightarrow \sigma_0 = 5.31 \times 10^{-11}\ \text{C/m}^2$,

$\lambda = 3.33 \times 10^{-12}\ \text{C/m}$;

b. $\Phi = \frac{q_{\text{enc}}}{\varepsilon_0} = \frac{3.33 \times 10^{-12}\ \text{C/m}(0.05\ \text{m})}{\varepsilon_0} = 0.019\,\text{N}\cdot\text{m}^2/\text{C}$

53. $E2\pi rl = \frac{\rho\pi r^2 l}{\varepsilon_0} \Rightarrow E = \frac{\rho r}{2\varepsilon_0}(r \le R)$;

$E2\pi rl = \frac{\rho\pi R^2 l}{\varepsilon_0} \Rightarrow E = \frac{\rho R^2}{2\varepsilon_0 r}(r \ge R)$

55. $\Phi = \frac{q_{\text{enc}}}{\varepsilon_0} \Rightarrow q_{\text{enc}} = -4.45 \times 10^{-10}\ \text{C}$

57. $q_{\text{enc}} = \frac{4}{5}\pi\alpha r^5$,

$E4\pi r^2 = \frac{4\pi\alpha r^5}{5\varepsilon_0} \Rightarrow E = \frac{\alpha r^3}{5\varepsilon_0}(r \le R)$,

$q_{\text{enc}} = \frac{4}{5}\pi\alpha R^5,\ E4\pi r^2 = \frac{4\pi\alpha R^5}{5\varepsilon_0} \Rightarrow E = \frac{\alpha R^5}{5\varepsilon_0 r^2}(r \ge R)$

59. integrate by parts: $q_{\text{enc}} = 4\pi\rho_0\left[-e^{-\alpha r}(\frac{(r)^2}{\alpha} + \frac{2r}{\alpha^2} + \frac{2}{\alpha^3}) + \frac{2}{\alpha^3}\right] \Rightarrow E = \frac{\rho_0}{r^2\varepsilon_0}\left[-e^{-\alpha r}(\frac{(r)^2}{\alpha} + \frac{2r}{\alpha^2} + \frac{2}{\alpha^3}) + \frac{2}{\alpha^3}\right]$

61.

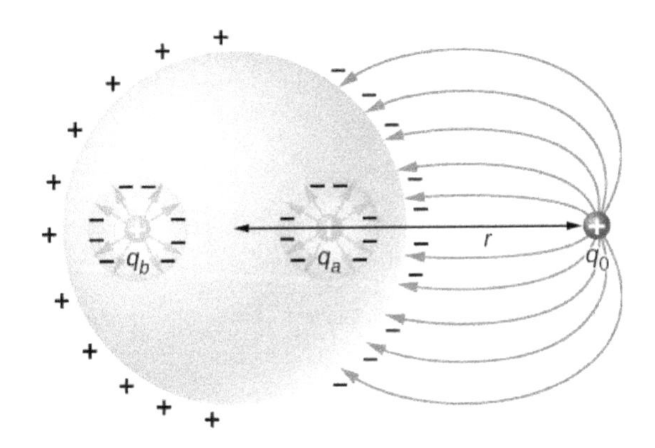

63. a. Outside: $E2\pi rl = \frac{\lambda l}{\varepsilon_0} \Rightarrow E = \frac{3.0\ \text{C/m}}{2\pi\varepsilon_0 r}$; Inside $E_{\text{in}} = 0$; b.

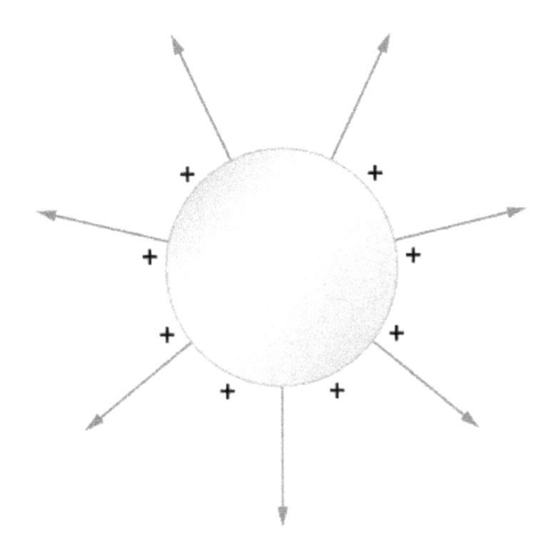

65. a. $E2\pi rl = \frac{\lambda l}{\varepsilon_0} \Rightarrow E = \frac{\lambda}{2\pi\varepsilon_0 r} r \geq R$ E inside equals 0; b.

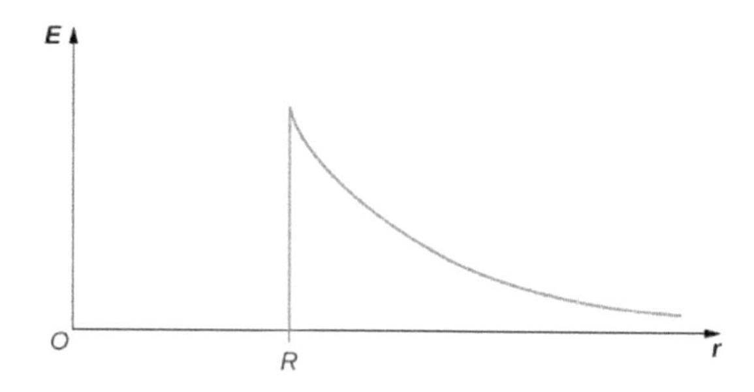

67. $E = 5.65 \times 10^4$ N/C

69. $\lambda = \frac{\lambda l}{\varepsilon_0} \Rightarrow E = \frac{a\sigma}{\varepsilon_0 r} r \geq a$, $E = 0$ inside since q enclosed $= 0$

71. a. $E = 0$; b. $E2\pi rL = \frac{Q}{\varepsilon_0} \Rightarrow E = \frac{Q}{2\pi\varepsilon_0 rL}$; c. $E = 0$ since r would be either inside the second shell or if outside then q enclosed equals 0.

ADDITIONAL PROBLEMS

73. $\int \vec{\mathbf{E}} \cdot \hat{\mathbf{n}}\, dA = a^4$

75. a. $\int \vec{\mathbf{E}} \cdot \hat{\mathbf{n}}\, dA = E_0 r^2 \pi$; b. zero, since the flux through the upper half cancels the flux through the lower half of the sphere

77. $\Phi = \frac{q_{enc}}{\varepsilon_0}$; There are two contributions to the surface integral: one at the side of the rectangle at $x = 0$ and the other at the side at $x = 2.0$ m;

$-E(0)[1.5\,\text{m}^2] + E(2.0\,\text{m})[1.5\,\text{m}^2] = \frac{q_{enc}}{\varepsilon_0} = -100\,\text{Nm}^2/\text{C}$

where the minus sign indicates that at $x = 0$, the electric field is along positive x and the unit normal is along negative x. At $x = 2$, the unit normal and the electric field vector are in the same direction: $q_{enc} = \varepsilon_0\,\Phi = -8.85 \times 10^{-10}$ C.

79. didn't keep consistent directions for the area vectors, or the electric fields

81. a. $\sigma = 3.0 \times 10^{-3}\ \text{C/m}^2$, $+3 \times 10^{-3}\ \text{C/m}^2$ on one and $-3 \times 10^{-3}\ \text{C/m}^2$ on the other; b. $E = 3.39 \times 10^8$ N/C

83. Construct a Gaussian cylinder along the z-axis with cross-sectional area A.

$|z| \geq \frac{a}{2}\ q_{enc} = \rho A a,\ \Phi = \frac{\rho A a}{\varepsilon_0} \Rightarrow E = \frac{\rho a}{2\varepsilon_0},$

$|z| \leq \frac{a}{2}\ q_{enc} = \rho A 2z,\ E(2A) = \frac{\rho A 2z}{\varepsilon_0} \Rightarrow E = \frac{\rho z}{\varepsilon_0}$

85. a. $r > b_2\ E4\pi r^2 = \frac{\frac{4}{3}\pi[\rho_1(b_1^3 - a_1^3) + \rho_2(b_2^3 - a_2^3)}{\varepsilon_0} \Rightarrow E = \frac{\rho_1(b_1^3 - a_1^3) + \rho_2(b_2^3 - a_2^3)}{3\varepsilon_0 r^2}$;

b. $a_2 < r < b_2\quad E4\pi r^2 = \frac{\frac{4}{3}\pi[\rho_1(b_1^3 - a_1^3) + \rho_2(r^3 - a_2^3)]}{\varepsilon_0} \Rightarrow E = \frac{\rho_1(b_1^3 - a_1^3) + \rho_2(r^3 - a_2^3)}{3\varepsilon_0 r^2}$;

c. $b_1 < r < a_2\quad E4\pi r^2 = \frac{\frac{4}{3}\pi\rho_1(b_1^3 - a_1^3)}{\varepsilon_0} \Rightarrow E = \frac{\rho_1(b_1^3 - a_1^3)}{3\varepsilon_0 r^2}$;

d. $a_1 < r < b_1\quad E4\pi r^2 = \frac{\frac{4}{3}\pi\rho_1(r^3 - a_1^3)}{\varepsilon_0} \Rightarrow E = \frac{\rho_1(r^3 - a_1^3)}{3\varepsilon_0 r^2}$; e. 0

87. Electric field due to plate without hole: $E = \frac{\sigma}{2\varepsilon_0}$.

Electric field of just hole filled with $-\sigma$ $E = \frac{-\sigma}{2\varepsilon_0}\left(1 - \frac{z}{\sqrt{R^2 + z^2}}\right)$.

Thus, $E_{\textbf{net}} = \frac{\sigma}{2\varepsilon_0} \frac{h}{\sqrt{R^2 + h^2}}$.

89. a. $E = 0$; b. $E = \frac{q_1}{4\pi\varepsilon_0 r^2}$; c. $E = \frac{q_1 + q_2}{4\pi\varepsilon_0 r^2}$; d. $0\, q_1 - q_1,\ q_1 + q_2$

CHALLENGE PROBLEMS

91. Given the referenced link, using a distance to Vega of 237×10^{15} m[1] and a diameter of 2.4 m for the primary mirror,[2] we find that at a wavelength of 555.6 nm, Vega is emitting 1.1×10^{25} J/s at that wavelength. Note that the flux through the mirror is essentially constant.

93. The symmetry of the system forces $\vec{\mathbf{E}}$ to be perpendicular to the sheet and constant over any plane parallel to the sheet. To calculate the electric field, we choose the cylindrical Gaussian surface shown. The cross-section area and the height of the cylinder are *A* and 2*x*, respectively, and the cylinder is positioned so that it is bisected by the plane sheet. Since *E* is perpendicular to each end and parallel to the side of the cylinder, we have *EA* as the flux through each end and there is no flux through the side. The charge enclosed by the cylinder is σA, so from Gauss's law, $2EA = \frac{\sigma A}{\varepsilon_0}$, and the electric field of an infinite sheet of charge is $E = \frac{\sigma}{2\varepsilon_0}$, in agreement with the calculation of in the text.

95. There is *Q*/2 on each side of the plate since the net charge is *Q*: $\sigma = \frac{Q}{2A}$,

$$\oint_S \vec{\mathbf{E}} \cdot \hat{\mathbf{n}}\, dA = \frac{2\sigma\Delta A}{\varepsilon_0} \Rightarrow E_P = \frac{\sigma}{\varepsilon_0} = \frac{Q}{\varepsilon_0 2A}$$

CHAPTER 7

CHECK YOUR UNDERSTANDING

7.1. $K = \frac{1}{2}mv^2,\ v = \sqrt{2\frac{K}{m}} = \sqrt{2\frac{4.5 \times 10^{-7}\ \text{J}}{4.00 \times 10^{-9}\ \text{kg}}} = 15\ \text{m/s}$

7.2. It has kinetic energy of 4.5×10^{-7} J at point r_2 and potential energy of 9.0×10^{-7} J, which means that as *Q* approaches infinity, its kinetic energy totals three times the kinetic energy at r_2, since all of the potential energy gets converted to kinetic.

7.3. positive, negative, and these quantities are the same as the work you would need to do to bring the charges in from infinity

7.4. $\Delta U = q\Delta V = (100\ \text{C})(1.5\ \text{V}) = 150\ \text{J}$

7.5. –2.00 C, $n_e = 1.25 \times 10^{19}$ electrons

7.6. It would be going in the opposite direction, with no effect on the calculations as presented.

7.7. Given a fixed maximum electric field strength, the potential at which a strike occurs increases with increasing height above the ground. Hence, each electron will carry more energy. Determining if there is an effect on the total number of electrons lies in the future.

7.8. $V = k\frac{q}{r} = \left(8.99 \times 10^9\ \text{N} \cdot \text{m}^2/\text{C}^2\right)\left(\frac{-3.00 \times 10^{-9}\ \text{C}}{5.00 \times 10^{-3}\ \text{m}}\right) = -5390\ \text{V}$; recall that the electric field inside a conductor is zero. Hence, any path from a point on the surface to any point in the interior will have an integrand of zero when calculating the change in potential, and thus the potential in the interior of the sphere is identical to that on the surface.

7.9. The *x*-axis the potential is zero, due to the equal and opposite charges the same distance from it. On the *z*-axis, we may superimpose the two potentials; we will find that for $z > > d$, again the potential goes to zero due to cancellation.

7.10. It will be zero, as at all points on the axis, there are equal and opposite charges equidistant from the point of interest. Note that this distribution will, in fact, have a dipole moment.

---

1. http://webviz.u-strasbg.fr/viz-bin/VizieR-5?-source=I/311&HIP=91262
2. http://ntrs.nasa.gov/archive/nasa/casi.ntrs.nasa.gov/19910003124.pdf

7.11. Any, but cylindrical is closest to the symmetry of a dipole.
7.12. infinite cylinders of constant radius, with the line charge as the axis

CONCEPTUAL QUESTIONS

1. No. We can only define potential energies for conservative fields.
3. No, though certain orderings may be simpler to compute.
5. The electric field strength is zero because electric potential differences are directly related to the field strength. If the potential difference is zero, then the field strength must also be zero.
7. Potential difference is more descriptive because it indicates that it is the difference between the electric potential of two points.
9. They are very similar, but potential difference is a feature of the system; when a charge is introduced to the system, it will have a potential energy which may be calculated by multiplying the magnitude of the charge by the potential difference.
11. An electron-volt is a volt multiplied by the charge of an electron. Volts measure potential difference, electron-volts are a unit of energy.
13. The second has 1/4 the dipole moment of the first.
15. The region outside of the sphere will have a potential indistinguishable from a point charge; the interior of the sphere will have a different potential.
17. No. It will be constant, but not necessarily zero.
19. no
21. No; it might not be at electrostatic equilibrium.
23. Yes. It depends on where the zero reference for potential is. (Though this might be unusual.)
25. So that lightning striking them goes into the ground instead of the television equipment.
27. They both make use of static electricity to stick small particles to another surface. However, the precipitator has to charge a wide variety of particles, and is not designed to make sure they land in a particular place.

PROBLEMS

29. a. $U = 3.4\text{ J}$;

b. $\frac{1}{2}mv^2 = kQ_1Q_2\left(\frac{1}{r_i} - \frac{1}{r_f}\right) \rightarrow v = 750\text{ m/s}$

31. $U = 4.36 \times 10^{-18}\text{ J}$

33.
$$\begin{aligned} \tfrac{1}{2}m_e v_e^2 &= qV,\ \tfrac{1}{2}m_\text{H} v_\text{H}^2 = qV,\ \text{so that} \\ \frac{m_e v_e^2}{m_\text{H} v_\text{H}^2} &= 1 \text{ or } \frac{v_e}{v_\text{H}} = 42.8 \end{aligned}$$

35. $1\text{ V} = 1\text{ J/C};\ 1\text{ J} = 1\text{ N}\cdot\text{m} \rightarrow 1\text{ V/m} = 1\text{ N/C}$

37. a. $V_{AB} = 3.00\text{ kV}$; b. $V_{AB} = 7.50\text{ kV}$

39. a. $V_{AB} = Ed \rightarrow E = 5.63\text{ kV/m}$;

b. $V_{AB} = 563\text{ V}$

41. a.
$$\begin{aligned} \Delta K &= q\Delta V \text{ and } V_{AB} = Ed,\ \text{so that} \\ \Delta K &= 800\text{ keV}; \end{aligned}$$

b. $d = 25.0\text{ km}$

43. One possibility is to stay at constant radius and go along the arc from P_1 to P_2, which will have zero potential due to the path being perpendicular to the electric field. Then integrate from *a* to *b*: $V_{ab} = \alpha \ln\left(\frac{b}{a}\right)$

45. $V = 144\text{ V}$

47. $V = \frac{kQ}{r} \rightarrow Q = 8.33 \times 10^{-7}\text{ C}$;

The charge is positive because the potential is positive.
49. a. $V = 45.0\text{ MV}$;

b. $V = \frac{kQ}{r} \rightarrow r = 45.0\text{ m}$;

c. $\Delta U = 132\text{ MeV}$

51. $V = kQ/r$; a. Relative to origin, find the potential at each point and then calculate the difference.

$\Delta V = 135 \times 10^3\text{ V}$;

b. To double the potential difference, move the point from 20 cm to infinity; the potential at 20 cm is halfway between zero and that at 10 cm.

53. a. $V_{P1} = 7.4 \times 10^5$ V

and $V_{P2} = 6.9 \times 10^3$ V ;

b. $V_{P1} = 6.9 \times 10^5$ V and $V_{P2} = 6.9 \times 10^3$ V

55. The problem is describing a uniform field, so $E = 200$ V/m in the –z-direction.

57. Apply $\vec{\mathbf{E}} = -\vec{\nabla} V$ with $\vec{\nabla} = \hat{\mathbf{r}}\frac{\partial}{\partial r} + \hat{\varphi}\frac{1}{r}\frac{\partial}{\partial \varphi} + \hat{\mathbf{z}}\frac{\partial}{\partial z}$ to the potential calculated earlier,

$V = -2k\lambda \ln s$: $\vec{\mathbf{E}} = 2k\lambda\frac{1}{r}\hat{\mathbf{r}}$ as expected.

59. a. decreases; the constant (negative) electric field has this effect, the reference point only matters for magnitude; b. they are planes parallel to the sheet; c. 0.59 m

61. a. from the previous chapter, the electric field has magnitude $\frac{\sigma}{\varepsilon_0}$ in the region between the plates and zero outside; defining the negatively charged plate to be at the origin and zero potential, with the positively charged plate located at $+5$ mm in the z-direction, $V = 1.7 \times 10^4$ V so the potential is 0 for $z < 0$, $1.7 \times 10^4 \text{ V}\left(\frac{z}{5 \text{ mm}}\right)$ for $0 \le z \le 5$ mm, 1.7×10^4 V for $z > 5$ mm;

b. $qV = \frac{1}{2}mv^2 \rightarrow v = 7.7 \times 10^7$ m/s

63. $V = 85$ V

65. In the region $a \le r \le b$, $\vec{\mathbf{E}} = \frac{kQ}{r^2}\hat{\mathbf{r}}$, and E is zero elsewhere; hence, the potential difference is $V = kQ\left(\frac{1}{a} - \frac{1}{b}\right)$.

67. From previous results $V_P - V_R = -2k\lambda \ln\frac{s_P}{s_R}$. , note that b is a very convenient location to define the zero level of potential:

$\Delta V = -2k\frac{Q}{L}\ln\frac{a}{b}$.

69. a. $F = 5.58 \times 10^{-11}$ N/C ; The electric field is towards the surface of Earth. b. The coulomb force is much stronger than gravity.

71. We know from the Gauss's law chapter that the electric field for an infinite line charge is $\vec{\mathbf{E}}_P = 2k\lambda\frac{1}{s}\hat{\mathbf{s}}$, and from earlier in this chapter that the potential of a wire-cylinder system of this sort is $V_P = -2k\lambda \ln\frac{s_P}{R}$ by integration. We are not given λ, but we are given a fixed V_0; thus, we know that $V_0 = -2k\lambda \ln\frac{a}{R}$ and hence $\lambda = -\frac{V_0}{2k \ln\left(\frac{a}{R}\right)}$. We may substitute this back in to find a. $\vec{\mathbf{E}}_P = -\frac{V_0}{\ln\left(\frac{a}{R}\right)}\frac{1}{s}\hat{\mathbf{s}}$; b. $V_P = V_0\frac{\ln\left(\frac{s_P}{R}\right)}{\ln\left(\frac{a}{R}\right)}$; c. 4.74×10^4 N/C

73. a. $U_1 = -7.68 \times 10^{-18}$ J, $U_2 = -5.76 \times 10^{-18}$ J ;

b. $U_1 + U_2 = -1.34 \times 10^{-17}$ J

75. a. $U = 2.30 \times 10^{-16}$ J ;

b. $\bar{K} = \frac{3}{2}kT \rightarrow T = 1.11 \times 10^7$ K

77. a. 1.9×10^6 m/s ; b. 4.2×10^6 m/s ; c. 5.9×10^6 m/s ; d. 7.3×10^6 m/s ; e. 8.4×10^6 m/s

79. a. $E = 2.5 \times 10^6 \text{ V/m} < 3 \times 10^6 \text{ V/m}$
No, the field trength is smaller than the breakdown strength for air. ;

b. $d = 1.7$ mm

81.
$$K_f = qV_{AB} = qEd \rightarrow$$
$$E = 8.00 \times 10^5 \text{ V/m}$$

83. a. Energy $= 2.00 \times 10^9$ J ;

b. $$\begin{aligned} Q &= m(c\Delta T + L_\nabla) \\ m &= 766 \text{ kg} \end{aligned};$$

c. The expansion of the steam upon boiling can literally blow the tree apart.

85. a. $V = \frac{kQ}{r} \rightarrow r = 1.80$ km ; b. A 1-C charge is a very large amount of charge; a sphere of 1.80 km is impractical.

87. The alpha particle approaches the gold nucleus until its original energy is converted to potential energy. $5.00\text{ MeV} = 8.00 \times 10^{-13}$ J , so

$$\begin{aligned} E_0 &= \frac{qkQ}{r} \rightarrow \\ r &= 4.54 \times 10^{-14} \text{ m} \end{aligned}$$

(Size of gold nucleus is about 7×10^{-15} m).

ADDITIONAL PROBLEMS

89. $$\begin{aligned} E_{\text{tot}} &= 4.67 \times 10^7 \text{ J} \\ E_{\text{tot}} &= qV \rightarrow q = \frac{E_{\text{tot}}}{V} = 3.89 \times 10^6 \text{ C} \end{aligned}$$

91. $V_P = k\frac{q_{\text{tot}}}{\sqrt{z^2 + R^2}} \rightarrow q_{\text{tot}} = -3.5 \times 10^{-11}$ C

93. $V_P = -2.2$ GV

95. Recall from the previous chapter that the electric field $E_P = \frac{\sigma_0}{2\varepsilon_0}$ is uniform throughout space, and that for uniform fields we have $E = -\frac{\Delta V}{\Delta z}$ for the relation. Thus, we get $\frac{\sigma}{2\varepsilon_0} = \frac{\Delta V}{\Delta z} \rightarrow \Delta z = 0.22$ m for the distance between 25-V equipotentials.

97. a. Take the result from Example 7.13, divide both the numerator and the denominator by x, take the limit of that, and then apply a Taylor expansion to the resulting log to get: $V_P \approx k\lambda\frac{L}{x}$; b. which is the result we expect, because at great distances, this should look like a point charge of $q = \lambda L$

99. a. $V = 9.0 \times 10^3$ V ; b. $-9.0 \times 10^3 \text{ V}\left(\frac{1.25 \text{ cm}}{2.0 \text{ cm}}\right) = -5.7 \times 10^3$ V

101. a. $E = \frac{KQ}{r^2} \rightarrow Q = -6.76 \times 10^5$ C ;

b. $$\begin{aligned} F &= ma = qE \rightarrow \\ a &= \frac{qE}{m} = 2.63 \times 10^{13} \text{ m/s}^2 \text{ (upwards)} \end{aligned};$$

c. $F = -mg = qE \rightarrow m = \frac{-qE}{g} = 2.45 \times 10^{-18}$ kg

103. If the electric field is zero ¼ from the way of q_1 and q_2 , then we know from $E = k\frac{Q}{r^2}$ that $|E_1| = |E_2| \rightarrow \frac{Kq_1}{x^2} = \frac{Kq_2}{(3x)^2}$ so that $\frac{q_2}{q_1} = \frac{(3x)^2}{x^2} = 9$; the charge q_2 is 9 times larger than q_1 .

105. a. The field is in the direction of the electron's initial velocity.

b. $$\begin{aligned} v^2 &= v_0^2 + 2ax \rightarrow x = -\frac{v_0^2}{2a}(v = 0). \text{ Also, } F = ma = qE \rightarrow a = \frac{qE}{m}, \\ x &= 3.56 \times 10^{-4} \text{ m;} \end{aligned}$$

c. $$\begin{aligned} v_2 &= v_0 + at \rightarrow t = -\frac{v_0 m}{qE}(v = 0), \\ \therefore t &= 1.42 \times 10^{-10} \text{ s;} \end{aligned}$$

d. $v = -\left(\frac{2qEx}{m}\right)^{1/2}$ -5.00×10^6 m/s (opposite its initial velocity)

CHALLENGE PROBLEMS

107. Answers will vary. This appears to be proprietary information, and ridiculously difficult to find. Speeds will be 20 m/s or less, and there are claims of $\sim 10^{-7}$ grams for the mass of a drop.

109. Apply $\vec{\mathbf{E}} = -\vec{\nabla} V$ with $\vec{\nabla} = \hat{\mathbf{r}}\frac{\partial}{\partial r} + \hat{\theta}\frac{1}{r}\frac{\partial}{\partial \theta} + \hat{\varphi}\frac{1}{r\sin\theta}\frac{\partial}{\partial \varphi}$ to the potential calculated earlier, $V_P = k\frac{\vec{\mathbf{p}} \cdot \hat{\mathbf{r}}}{r^2}$ with $\vec{\mathbf{p}} = q\vec{\mathbf{d}}$, and assume that the axis of the dipole is aligned with the *z*-axis of the coordinate system. Thus, the potential is $V_P = k\frac{q\vec{\mathbf{d}} \cdot \hat{\mathbf{r}}}{r^2} = k\frac{qd\cos\theta}{r^2}$.

$$\vec{\mathbf{E}} = 2kqd\left(\frac{\cos\theta}{r^3}\right)\hat{\mathbf{r}} + kqd\left(\frac{\sin\theta}{r^3}\right)\hat{\theta}$$

CHAPTER 8

CHECK YOUR UNDERSTANDING

8.1. 1.1×10^{-3} m
8.3. 3.59 cm, 17.98 cm
8.4. a. 25.0 pF; b. 9.2
8.5. a. $C = 0.86\,\text{pF}, Q_1 = 10\,\text{pC}, Q_2 = 3.4\,\text{pC}, Q_3 = 6.8\,\text{pC}$;
b. $C = 2.3\,\text{pF}, Q_1 = 12\,\text{pC}, Q_2 = Q_3 = 16\,\text{pC}$;
c. $C = 2.3\,\text{pF}, Q_1 = 9.0\,\text{pC}, Q_2 = 18\,\text{pC}, Q_3 = 12\,\text{pC}, Q_4 = 15\,\text{pC}$
8.6. a. $4.0 \times 10^{-13}\,\text{J}$; b. 9 times
8.7. a. 3.0; b. $C = 3.0\,C_0$
8.9. a. $C_0 = 20\,\text{pF}$, $C = 42\,\text{pF}$; b. $Q_0 = 0.8\,\text{nC}$, $Q = 1.7\,\text{nC}$; c. $V_0 = V = 40\,\text{V}$; d. $U_0 = 16\,\text{nJ}$, $U = 34\,\text{nJ}$

CONCEPTUAL QUESTIONS

1. no; yes
3. false
5. no
7. $3.0\,\mu\text{F}, 0.33\,\mu\text{F}$
9. answers may vary
11. Dielectric strength is a critical value of an electrical field above which an insulator starts to conduct; a dielectric constant is the ratio of the electrical field in vacuum to the net electrical field in a material.
13. Water is a good solvent.
15. When energy of thermal motion is large (high temperature), an electrical field must be large too in order to keep electric dipoles aligned with it.
17. answers may vary

PROBLEMS

19. 21.6 mC
21. 1.55 V
23. 25.0 nF
25. $1.1 \times 10^{-3}\,\text{m}^2$
27. 500 μC
29. 1:16
31. a. 1.07 nC; b. 267 V, 133 V
33. $0.29\,\mu\text{F}$
34. 500 capacitors; connected in parallel
35. $3.08\,\mu\text{F}$ (series) and $13.0\,\mu\text{F}$ (parallel)
37. $11.4\,\mu\text{F}$
39. 0.89 mC; 1.78 mC; 444 V
41. $7.5\,\mu\text{J}$
43. a. 405 J; b. 90.0 mC

45. 1.15 J

47. a. 4.43×10^{-12} F; b. 452 V; c. 4.52×10^{-7} J; d. no

49. 0.7 mJ

51. a. 7.1 pF; b. 42 pF

53. a. before 3.00 V; after 0.600 V; b. before 1500 V/m; after 300 V/m

55. a. 3.91; b. 22.8 V

57. a. 37 nC; b. 0.4 MV/m; c. 19 nC

59. a. $4.4\ \mu F$; b. 4.0×10^{-5} C

61. $0.0135\ m^2$

63. $0.185\ \mu J$

ADDITIONAL PROBLEMS

65. a. 0.277 nF; b. 27.7 nC; c. 50 kV/m

67. a. 0.065 F; b. 23,000 C; c. 4.0 GJ

69. a. $75.6\ \mu C$; b. 10.8 V

71. a. 0.13 J; b. no, because of resistive heating in connecting wires that is always present, but the circuit schematic does not indicate resistors

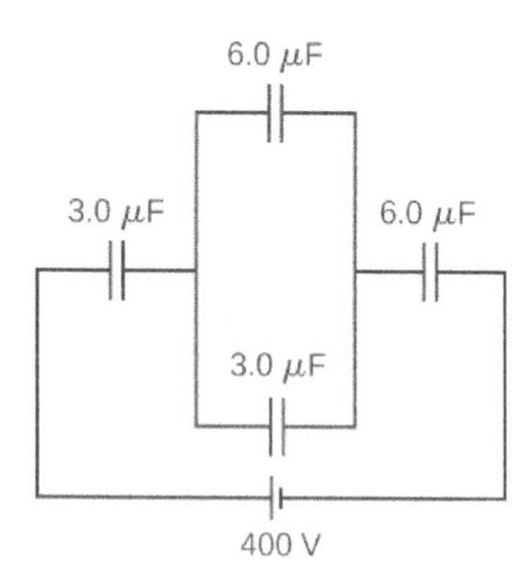

73. a. $-3.00\ \mu F$; b. You cannot have a negative C_2 capacitance. c. The assumption that they were hooked up in parallel, rather than in series, is incorrect. A parallel connection always produces a greater capacitance, while here a smaller capacitance was assumed. This could only happen if the capacitors are connected in series.

75. a. 14.2 kV; b. The voltage is unreasonably large, more than 100 times the breakdown voltage of nylon. c. The assumed charge is unreasonably large and cannot be stored in a capacitor of these dimensions.

CHALLENGE PROBLEMS

77. a. 89.6 pF; b. 6.09 kV/m; c. 4.47 kV/m; d. no

79. a. 421 J; b. 53.9 mF

81. $C = \varepsilon_0 A/(d_1 + d_2)$

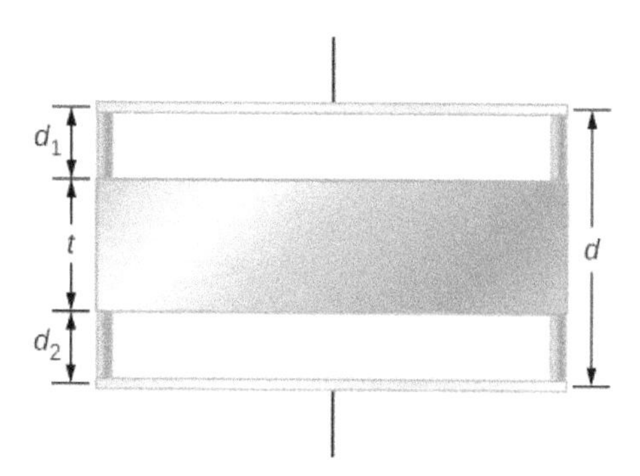

83. proof

CHAPTER 9

CHECK YOUR UNDERSTANDING

9.1. The time for 1.00 C of charge to flow would be $\Delta t = \frac{\Delta Q}{I} = \frac{1.00\text{ C}}{0.300 \times 10^{-3}\text{ C/s}} = 3.33 \times 10^{3}\text{ s}$, slightly less than an hour. This is quite different from the 5.55 ms for the truck battery. The calculator takes a very small amount of energy to operate, unlike the truck's starter motor. There are several reasons that vehicles use batteries and not solar cells. Aside from the obvious fact that a light source to run the solar cells for a car or truck is not always available, the large amount of current needed to start the engine cannot easily be supplied by present-day solar cells. Solar cells can possibly be used to charge the batteries. Charging the battery requires a small amount of energy when compared to the energy required to run the engine and the other accessories such as the heater and air conditioner. Present day solar-powered cars are powered by solar panels, which may power an electric motor, instead of an internal combustion engine.

9.2. The total current needed by all the appliances in the living room (a few lamps, a television, and your laptop) draw less current and require less power than the refrigerator.

9.3. The diameter of the 14-gauge wire is smaller than the diameter of the 12-gauge wire. Since the drift velocity is inversely proportional to the cross-sectional area, the drift velocity in the 14-gauge wire is larger than the drift velocity in the 12-gauge wire carrying the same current. The number of electrons per cubic meter will remain constant.

9.4. The current density in a conducting wire increases due to an increase in current. The drift velocity is inversely proportional to the current $\left(v_d = \frac{nqA}{I}\right)$, so the drift velocity would decrease.

9.5. Silver, gold, and aluminum are all used for making wires. All four materials have a high conductivity, silver having the highest. All four can easily be drawn into wires and have a high tensile strength, though not as high as copper. The obvious disadvantage of gold and silver is the cost, but silver and gold wires are used for special applications, such as speaker wires. Gold does not oxidize, making better connections between components. Aluminum wires do have their drawbacks. Aluminum has a higher resistivity than copper, so a larger diameter is needed to match the resistance per length of copper wires, but aluminum is cheaper than copper, so this is not a major drawback. Aluminum wires do not have as high of a ductility and tensile strength as copper, but the ductility and tensile strength is within acceptable levels. There are a few concerns that must be addressed in using aluminum and care must be used when making connections. Aluminum has a higher rate of thermal expansion than copper, which can lead to loose connections and a possible fire hazard. The oxidation of aluminum does not conduct and can cause problems. Special techniques must be used when using aluminum wires and components, such as electrical outlets, must be designed to accept aluminum wires.

9.6. The foil pattern stretches as the backing stretches, and the foil tracks become longer and thinner. Since the resistance is calculated as $R = \rho\frac{L}{A}$, the resistance increases as the foil tracks are stretched. When the temperature changes, so does the resistivity of the foil tracks, changing the resistance. One way to combat this is to use two strain gauges, one used as a reference and the other used to measure the strain. The two strain gauges are kept at a constant temperature

9.7. The longer the length, the smaller the resistance. The greater the resistivity, the higher the resistance. The larger the difference between the outer radius and the inner radius, that is, the greater the ratio between the two, the greater the resistance. If you are attempting to maximize the resistance, the choice of the values for these variables will depend on the application. For example, if the cable must be flexible, the choice of materials may be limited.

9.8. Yes, Ohm's law is still valid. At every point in time the current is equal to $I(t) = V(t)/R$, so the current is also a function of time, $I(t) = \frac{V_{\text{max}}}{R}\sin(2\pi f t)$.

9.9. Even though electric motors are highly efficient 10–20% of the power consumed is wasted, not being used for doing useful work. Most of the 10–20% of the power lost is transferred into heat dissipated by the copper wires used to make the coils of the motor. This heat adds to the heat of the environment and adds to the demand on power plants providing the power. The demand on the power plant can lead to increased greenhouse gases, particularly if the power plant uses coal or gas as fuel.

9.10. No, the efficiency is a very important consideration of the light bulbs, but there are many other considerations. As mentioned above, the cost of the bulbs and the life span of the bulbs are important considerations. For example, CFL bulbs contain mercury, a neurotoxin, and must be disposed of as hazardous waste. When replacing incandescent bulbs that are being controlled by a dimmer switch with LED, the dimmer switch may need to be replaced. The dimmer switches for LED lights are comparably priced to the incandescent light switches, but this is an initial cost which should be considered. The spectrum of light should also be considered, but there is a broad range of color temperatures available, so you should be able to find one that fits your needs. None of these considerations mentioned are meant to discourage the use of LED or CFL light bulbs, but they are considerations.

CONCEPTUAL QUESTIONS

1. If a wire is carrying a current, charges enter the wire from the voltage source's positive terminal and leave at the negative terminal, so the total charge remains zero while the current flows through it.

3. Using one hand will reduce the possibility of "completing the circuit" and having current run through your body, especially current running through your heart.

5. Even though the electrons collide with atoms and other electrons in the wire, they travel from the negative terminal to the positive terminal, so they drift in one direction. Gas molecules travel in completely random directions.
7. In the early years of light bulbs, the bulbs are partially evacuated to reduce the amount of heat conducted through the air to the glass envelope. Dissipating the heat would cool the filament, increasing the amount of energy needed to produce light from the filament. It also protects the glass from the heat produced from the hot filament. If the glass heats, it expands, and as it cools, it contacts. This expansion and contraction could cause the glass to become brittle and crack, reducing the life of the bulbs. Many bulbs are now partially filled with an inert gas. It is also useful to remove the oxygen to reduce the possibility of the filament actually burning. When the original filaments were replaced with more efficient tungsten filaments, atoms from the tungsten would evaporate off the filament at such high temperatures. The atoms collide with the atoms of the inert gas and land back on the filament.
9. In carbon, resistivity increases with the amount of impurities, meaning fewer free charges. In silicon and germanium, impurities decrease resistivity, meaning more free electrons.
11. Copper has a lower resistivity than aluminum, so if length is the same, copper must have the smaller diameter.
13. Device *B* shows a linear relationship and the device is ohmic.
15. Although the conductors have a low resistance, the lines from the power company can be kilometers long. Using a high voltage reduces the current that is required to supply the power demand and that reduces line losses.
17. The resistor would overheat, possibly to the point of causing the resistor to burn. Fuses are commonly added to circuits to prevent such accidents.
19. Very low temperatures necessitate refrigeration. Some materials require liquid nitrogen to cool them below their critical temperatures. Other materials may need liquid helium, which is even more costly.

PROBLEMS

21. a. $v = 4.38 \times 10^5 \frac{\text{m}}{\text{s}}$;

b. $\Delta q = 5.00 \times 10^{-3}\,\text{C}, \quad \text{no. of protons} = 3.13 \times 10^{16}$

23. $I = \frac{\Delta Q}{\Delta t}, \quad \Delta Q = 12.00\,\text{C}$

$\text{no. of electrons} = 7.46 \times 10^{15}$

25.
$$I(t) = 0.016 \frac{\text{C}}{\text{s}^4} t^3 - 0.001 \frac{\text{C}}{\text{s}}$$
$$I(3.00\,\text{s}) = 0.431\,\text{A}$$

27. $I(t) = -I_{\text{max}} \sin(\omega t + \phi)$

29. $|J| = 15.92\,\text{A/m}^2$

31. $I = 40\,\text{mA}$

33. a. $|J| = 7.60 \times 10^5 \frac{\text{A}}{\text{m}^2}$; b. $v_\text{d} = 5.60 \times 10^{-5} \frac{\text{m}}{\text{s}}$

35. $R = 6.750\,\text{k}\,\Omega$

37. $R = 0.10\,\Omega$

39.
$$R = \rho \frac{L}{A}$$
$$L = 3\,\text{mm}$$

41.
$$\frac{\frac{R_\text{Al}}{L_\text{Al}}}{\frac{R_\text{Cu}}{L_\text{Cu}}} = \frac{\rho_\text{Al} \frac{1}{\pi \left(\frac{D_\text{Al}}{2}\right)^2}}{\rho_\text{Cu} \frac{1}{\pi \left(\frac{D_\text{Cu}}{2}\right)^2}} = \frac{\rho_\text{Al}}{\rho_\text{Cu}} \left(\frac{D_\text{Cu}}{D_\text{Al}}\right)^2 = 1, \quad \frac{D_\text{Al}}{D_\text{Cu}} = \sqrt{\frac{\rho_\text{Al}}{\rho_{Cu}}}$$

43. a. $R = R_0(1 + \alpha \Delta T), \quad 2 = 1 + \alpha \Delta T, \quad \Delta T = 256.4\,°\text{C}, \quad T = 276.4\,°\text{C}$;

b. Under normal conditions, no it should not occur.

45.
$$R = R_0(1 + \alpha \Delta T)$$
$$\alpha = 0.006\,°\text{C}^{-1}$$, iron

47. a. $R = \rho\frac{L}{A}, \quad \rho = 2.44 \times 10^{-8}\ \Omega \cdot \text{m}$, gold;
$R = \rho\frac{L}{A}(1 + \alpha\Delta T)$

b. $R = 2.44 \times 10^{-8}\ \Omega \cdot \text{m}\left(\frac{25\ \text{m}}{\pi\left(\frac{0.100 \times 10^{-3}\ \text{m}}{2}\right)^2}\right)\left(1 + 0.0034\ °\text{C}^{-1}(150\ °\text{C} - 20\ °\text{C})\right)$

$R = 112\ \Omega$

49. $R_{\text{Fe}} = 0.525\ \Omega, \quad R_{\text{Cu}} = 0.500\ \Omega, \quad \alpha_{\text{Fe}} = 0.0065\ °\text{C}^{-1} \quad \alpha_{\text{Cu}} = 0.0039\ °\text{C}^{-1}$
$R_{\text{Fe}} = R_{\text{Cu}}$
$R_{0\,\text{Fe}}(1 + \alpha_{\text{Fe}}(T - T_0)) = R_{0\,\text{Cu}}(1 + \alpha_{\text{Cu}}(T - T_0))$
$\frac{R_{0\,\text{Fe}}}{R_{0\,\text{Cu}}}(1 + \alpha_{\text{Fe}}(T - T_0)) = 1 + \alpha_{\text{Cu}}(T - T_0)$
$T = 2.91\ °\text{C}$

51. $R_{\min} = 2.375 \times 10^5\ \Omega, \quad I_{\min} = 12.63\ \mu\text{A}$
$R_{\max} = 2.625 \times 10^5\ \Omega, \quad I_{\max} = 11.43\ \mu\text{A}$

53. $R = 100\ \Omega$

55. a. $I = 0.30\ \text{mA}$; b. $P = 0.90\ \text{mW}$; c. $P = 0.90\ \text{mW}$; d. It is converted into heat.

57. $P = \frac{V^2}{R}$, $R = 40\ \Omega$; $A = 2.08\text{mm}^2$, $\rho = 100 \times 10^{-8}\ \Omega \cdot \text{m}$, $R = \rho\frac{L}{A}$, $L = 83\ \text{m}$

59. $I = 0.1\ \text{A}, \quad V = 14\ \text{V}$

61. a. $I \approx 3.00\ \text{A} + \frac{100\ \text{W}}{110\ \text{V}} + \frac{60\ \text{W}}{110\ \text{V}} + \frac{3.00\ \text{W}}{110\ \text{V}} = 4.48\ \text{A}$
$P = 493\ \text{W}$
$R = 9.91\ \Omega,$
$P_{\text{loss}} = 200.\ \text{W}$
$\%\text{loss} = 40\%$

b. $P = 493\ \text{W}$
$I = 0.0045\ \text{A}$
$R = 9.91\ \Omega$
$P_{\text{loss}} = 201\mu\ \text{W}$
$\%\text{loss} = 0.00004\%$

63. $R_{\text{copper}} = 0.24\ \Omega$
$P = 2.377 \times 10^3\ \text{W}$

65. $R = R_0(1 + \alpha(T - T_0))$
$0.82R_0 = R_0(1 + \alpha(T - T_0)), \quad 0.82 = 1 - 0.06(T - 37\ °\text{C}), \quad T = 40\ °\text{C}$

67. a. $R_{\text{Au}} = R_{\text{Ag}}, \quad \rho_{\text{Au}}\frac{L_{\text{Au}}}{A_{\text{Au}}} = \rho_{\text{Ag}}\frac{L_{\text{Ag}}}{A_{\text{Ag}}}, \quad L_{\text{Ag}} = 1.53\ \text{m}$;

b. $R_{\text{Au},20\ °\text{C}} = 0.0074\ \Omega, \quad R_{\text{Au},100\ °\text{C}} = 0.0094\ \Omega, \quad R_{\text{Ag},\ 100\ °\text{C}} = 0.0096\ \Omega$

ADDITIONAL PROBLEMS

$$dR = \frac{\rho}{2\pi r L}dr$$

69. $R = \frac{\rho}{2\pi L}\ln\frac{r_o}{r_i}$

$R = 2.21 \times 10^{11}\ \Omega$

71. a.

$R_0 = 3.00 \times 10^6\ \Omega$; b.

$T_c = 37.0\ °C$

$R = 3.02 \times 10^{-6}\ \Omega$

73. $\rho = 5.00 \times 10^{-8}\ \Omega \cdot m$

75. $\rho = 1.71 \times 10^{-8}\ \Omega \cdot m$

77. a. $V = 6000\ V$; b. $V = 60\ V$

79. $P = \frac{W}{t}, \quad W = 8.64\ J$

CHALLENGE PROBLEMS

81.

$V = 7.09\ cm^3$

$n = 8.49 \times 10^{28}\ \frac{electrons}{m^2}$

$v_d = 7.00 \times 10^{-5}\ \frac{m}{s}$

83. a. $v = 5.83 \times 10^{13}\ \frac{protons}{m^3}$

85. $E = 75\ kJ$

87. a. $P = 52\ W$; b. $V = 43.54\ V$

$R = 36\ \Omega$

89. a. $R = \frac{\rho}{2\pi L}\ln\left(\frac{R_0}{R_i}\right)$; b. $R = 2.5\ m\Omega$

91. a. $I = 8.69\ A$; b. # electrons $= 2.61 \times 10^{25}$; c. $R = 13.23\ \Omega$; d. $q = 4.68 \times 10^6\ J$

93. $P = 1045\ W, \quad P = \frac{V^2}{R}, \quad R = 12.27\ \Omega$

CHAPTER 10

CHECK YOUR UNDERSTANDING

10.1. If a wire is connected across the terminals, the load resistance is close to zero, or at least considerably less than the internal resistance of the battery. Since the internal resistance is small, the current through the circuit will be large, $I = \frac{\varepsilon}{R + r} = \frac{\varepsilon}{0 + r} = \frac{\varepsilon}{r}$. The large current causes a high power to be dissipated by the internal resistance $\left(P = I^2 r\right)$. The power is dissipated as heat.

10.2. The equivalent resistance of nine bulbs connected in series is 9*R*. The current is $I = V/9\,R$. If one bulb burns out, the equivalent resistance is 8*R*, and the voltage does not change, but the current increases $(I = V/8\,R)$. As more bulbs burn out, the current becomes even higher. Eventually, the current becomes too high, burning out the shunt.

10.3. The equivalent of the series circuit would be $R_{eq} = 1.00\ \Omega + 2.00\ \Omega + 2.00\ \Omega = 5.00\ \Omega$, which is higher than the equivalent resistance of the parallel circuit $R_{eq} = 0.50\ \Omega$. The equivalent resistor of any number of resistors is always higher than the equivalent resistance of the same resistors connected in parallel. The current through for the series circuit would be $I = \frac{3.00\ V}{5.00\ \Omega} = 0.60\ A$, which is lower than the sum of the currents through each resistor in the parallel circuit, $I = 6.00\ A$. This is not surprising since the equivalent resistance of the series circuit is higher. The current through a series connection of any number of resistors will always be lower than the current into a parallel connection of the same resistors, since the equivalent resistance of the series circuit will be higher than the parallel circuit. The power dissipated by the resistors in series would be $P = 1.80\text{ W},$ which is lower than the power dissipated in the parallel circuit $P = 18.00\text{ W}.$

10.4. A river, flowing horizontally at a constant rate, splits in two and flows over two waterfalls. The water molecules are analogous to the electrons in the parallel circuits. The number of water molecules that flow in the river and falls must be equal to the number of molecules that flow over each waterfall, just like sum of the current through each resistor must be equal to the current flowing into the parallel circuit. The water molecules in the river have energy due to their motion and height. The potential energy of the water molecules in the river is constant due to their equal heights. This is analogous to the constant change in voltage across a parallel circuit. Voltage is the potential energy across each resistor.
The analogy quickly breaks down when considering the energy. In the waterfall, the potential energy is converted into kinetic energy of the water molecules. In the case of electrons flowing through a resistor, the potential drop is converted into heat and light, not into the kinetic energy of the electrons.

10.5. 1. All the overhead lighting circuits are in parallel and connected to the main supply line, so when one bulb burns out, all the overhead lighting does not go dark. Each overhead light will have at least one switch in series with the light, so you can turn it on and off. 2. A refrigerator has a compressor and a light that goes on when the door opens. There is usually only one cord for the refrigerator to plug into the wall. The circuit containing the compressor and the circuit containing the lighting circuit are in parallel, but there is a switch in series with the light. A thermostat controls a switch that is in series with the compressor to control the temperature of the refrigerator.

10.6. The circuit can be analyzed using Kirchhoff's loop rule. The first voltage source supplies power: $P_{\text{in}} = IV_1 = 7.20\text{ mW}.$

The second voltage source consumes power: $P_{\text{out}} = IV_2 + I^2 R_1 + I^2 R_2 = 7.2\text{ mW}.$

10.7. The current calculated would be equal to $I = -0.20\text{ A}$ instead of $I = 0.20\text{ A}.$ The sum of the power dissipated and the power consumed would still equal the power supplied.

10.8. Since digital meters require less current than analog meters, they alter the circuit less than analog meters. Their resistance as a voltmeter can be far greater than an analog meter, and their resistance as an ammeter can be far less than an analog meter. Consult Figure 10.36 and Figure 10.35 and their discussion in the text.

CONCEPTUAL QUESTIONS

1. Some of the energy being used to recharge the battery will be dissipated as heat by the internal resistance.

3.

$$P = I^2 R = \left(\frac{\varepsilon}{r+R}\right)^2 R = \varepsilon^2 R(r+R)^{-2}, \quad \frac{dP}{dR} = \varepsilon^2\left[(r+R)^{-2} - 2R(r+R)^{-3}\right] = 0,$$

$$\left[\frac{(r+R) - 2R}{(r+R)^3}\right] = 0, \quad r = R$$

5. It would probably be better to be in series because the current will be less than if it were in parallel.

7. two filaments, a low resistance and a high resistance, connected in parallel

9. It can be redrawn.

$$R_{\text{eq}} = \left[\frac{1}{R_6} + \frac{1}{R_1} + \frac{1}{R_2 + \left(\frac{1}{R_4} + \frac{1}{R_3 + R_5}\right)^{-1}}\right]^{-1}$$

11. In series the voltages add, but so do the internal resistances, because the internal resistances are in series. In parallel, the terminal voltage is the same, but the equivalent internal resistance is smaller than the smallest individual internal resistance and a higher current can be provided.

13. The voltmeter would put a large resistance in series with the circuit, significantly changing the circuit. It would probably give a reading, but it would be meaningless.

15. The ammeter has a small resistance; therefore, a large current will be produced and could damage the meter and/or overheat the battery.

17. The time constant can be shortened by using a smaller resistor and/or a smaller capacitor. Care should be taken when reducing the resistance because the initial current will increase as the resistance decreases.

19. Not only might water drip into the switch and cause a shock, but also the resistance of your body is lower when you are wet.

PROBLEMS

21. a.

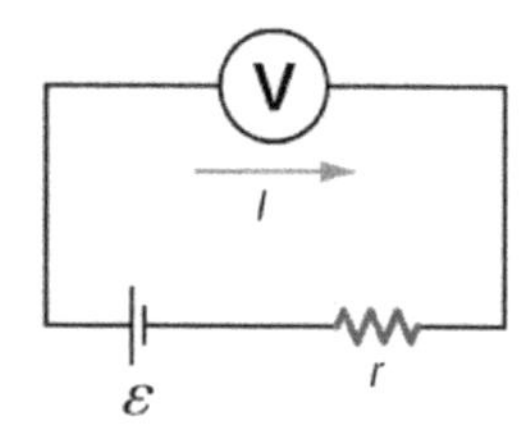

b. 0.476W; c. 0.691 W; d. As R_L is lowered, the power difference decreases; therefore, at higher volumes, there is no significant difference.

23. a. $0.400\ \Omega$; b. No, there is only one independent equation, so only *r* can be found.

25. a. $0.400\ \Omega$; b. 40.0 W; c. $0.0956\ °\text{C/min}$

27. largest, $786\ \Omega$, smallest, $20.32\ \Omega$

29. 29.6 W

31. a. 0.74 A; b. 0.742 A

33. a. 60.8 W; b. 3.18 kW

35. a. $R_s = 9.00\ \Omega$; b. $I_1 = I_2 = I_3 = 2.00\ \text{A}$;

c. $V_1 = 8.00\ \text{V},\ V_2 = 2.00\ \text{V},\ V_3 = 8.00\ \text{V}$; d. $P_1 = 16.00\ \text{W},\ P_2 = 4.00\ \text{W},\ P_3 = 16.00\ \text{W}$; e. $P = 36.00\ \text{W}$

37. a. $I_1 = 0.6\ \text{mA},\quad I_2 = 0.4\ \text{mA},\quad I_3 = 0.2\ \text{mA}$;

b. $I_1 = 0.04\ \text{mA},\quad I_2 = 1.52\ \text{mA},\quad I_3 = -1.48\ \text{mA}$; c. $P_{\text{out}} = 0.92\ \text{mW},\quad P_{\text{out}} = 4.50\ \text{mW}$;

d. $P_{\text{in}} = 0.92\ \text{mW},\quad P_{\text{in}} = 4.50\ \text{mW}$

39. $V_1 = 42\ \text{V},\ V_2 = 6\ \text{V},\ R_4 = 6\ \Omega$

41. a. $I_1 = 1.5\ \text{A},\ I_2 = 2\ \text{A},\ I_3 = 0.5\ \text{A},\ I_4 = 2.5\ \text{A},\ I_5 = 2\ \text{A}$; b. $P_{\text{in}} = I_2 V_1 + I_5 V_5 = 34\ \text{W}$;

c. $P_{\text{out}} = I_1^2 R_1 + I_2^2 R_2 + I_3^2 R_3 + I_4^2 R_4 = 34\ \text{W}$

43. $I_1 = \frac{3V}{5R},\ I_2 = \frac{2V}{5R},\ I_3 = \frac{1V}{5R}$

45. a.

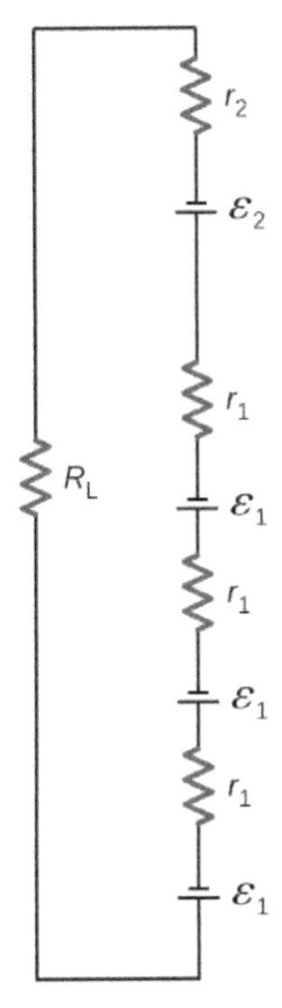

;
b. 0.617 A; c. 3.81 W; d. $18.0\,\Omega$

47. $I_1 r_1 - \varepsilon_1 + I_1 R_4 + \varepsilon_4 + I_2 r_4 + I_4 r_3 - \varepsilon_3 + I_2 R_3 + I_1 R_1 = 0$

49. 4.00 to $30.0\,\text{M}\Omega$

51. a. $2.50\,\mu\text{F}$; b. 2.00 s

53. a. 12.3 mA; b. $7.50 \times 10^{-4}\,\text{s}$; c. 4.53 mA; d. 3.89 V

55. a. 1.00×10^{-7} F; b. No, in practice it would not be difficult to limit the capacitance to less than 100 nF, since typical capacitors range from fractions of a picofarad (pF) to milifarad (mF).

57. $3.33 \times 10^{-3}\,\Omega$

59. 12.0 V

61. 400 V

63. a. 6.00 mV; b. It would not be necessary to take extra precautions regarding the power coming from the wall. However, it is possible to generate voltages of approximately this value from static charge built up on gloves, for instance, so some precautions are necessary.

65. a. 5.00×10^{-2} C; b. 10.0 kV; c. $1.00\,\text{k}\Omega$; d. $1.79 \times 10^{-2}\,°\text{C}$

ADDITIONAL PROBLEMS

67. a. $C_{\text{eq}} = 4.00\,\text{mF}$; b. $\tau = 80\,\text{ms}$; c. 55.45 ms

69. a. $R_{\text{eq}} = 20.00\,\Omega$;

b. $I_r = 1.50\,\text{A},\ I_1 = 1.00\,\text{A},\ I_2 = 0.50\,\text{A},\ I_3 = 0.75\,\text{A},\ I_4 = 0.75\,\text{A},\ I_5 = 1.50\,\text{A}$;

c. $V_r = 1.50\,\text{V},\ V_1 = 9.00\,\text{V},\ V_2 = 9.00\,\text{V},\ V_3 = 7.50\,\text{V},\ V_4 = 7.50\,\text{V},\ V_5 = 12.00\,\text{V}$;

d. $P_r = 2.25\,\text{W},\ P_1 = 9.00\,\text{W},\ P_2 = 4.50\,\text{W},\ P_3 = 5.625\,\text{W},\ P_4 = 5.625\,\text{W},\ P_5 = 18.00\,\text{W}$;

e. $P = 45.00\,\text{W}$

71. a. $\tau = \left(\left|1.38 \times 10^{-5}\,\frac{\Omega}{\text{m}}\left(\frac{5.00 \times 10^{-2}\,\text{m}}{3.14\left(\frac{0.05 \times 10^{-3}}{2}\right)^2}\right)\right|\right) 10 \times 10^{-3}\,\text{F} = 3.52\,\text{s}$; b. $V = 0.017\,\text{A}\left(e^{-\frac{1.00\,\text{s}}{3.52\,\text{s}}}\right) 351.59\,\Omega = 4.55\,\text{V}$

73. a. $t = \frac{3\,\text{A}\cdot\text{h}}{\frac{1.5\,\text{V}}{900\,\Omega}} = 1800\,\text{h}$; b. $t = \frac{3\,\text{A}\cdot\text{h}}{\frac{1.5\,\text{V}}{100\,\Omega}} = 200\,\text{h}$

75. $U_1 = C_1 V_1^2 = 0.16\text{J},\quad U_2 = C_2 V_2^2 = 0.075\,\text{J}$

77. a. $R_{\text{eq}} = 24.00\,\Omega$; b. $I_1 = 1.00\,\text{A},\ I_2 = 0.67\,\text{A},\ I_3 = 0.33\,\text{A},\ I_4 = 1.00\,\text{A}$;
c. $V_1 = 14.00\,\text{V},\ V_2 = 6.00\,\text{V},\ V_3 = 6.00\,\text{V},\ V_4 = 4.00\,\text{V}$;
d. $P_1 = 14.00\,\text{W},\ P_2 = 4.04\,\text{W},\ P_3 = 1.96\,\text{W},\ P_4 = 4.00\,\text{W}$; e. $P = 24.00\,\text{W}$

79. a. $R_{\text{eq}} = 12.00\,\Omega,\ I = 1.00\,\text{A}$; b. $R_{\text{eq}} = 12.00\,\Omega,\ I = 1.00\,\text{A}$

81. a. $-400\,\text{k}\Omega$; b. You cannot have negative resistance. c. The assumption that $R_{\text{eq}} < R_1$ is unreasonable. Series resistance is always greater than any of the individual resistances.

83. $E_2 - I_2 r_2 - I_2 R_2 + I_1 R_5 + I_1 r_1 - E_1 + I_1 R_1 = 0$

85. a. $I = 1.17\,\text{A},\ I_1 = 0.50\,\text{A},\ I_2 = 0.67\,\text{A},\ I_3 = 0.67\,\text{A},\ I_4 = 0.50\,\text{A},\ I_5 = 0.17\,\text{A}$;
b. $P_{\text{output}} = 23.4\,\text{W},\ P_{\text{input}} = 23.4\,\text{W}$

87. a. 4.99 s; b. $3.87\,°\text{C}$; c. $3.11 \times 10^4\ \Omega$; d. No, this change does not seem significant. It probably would not be noticed.

CHALLENGE PROBLEMS

89. a. 0.273 A; b. $V_T = 1.36\,\text{V}$

91. a. $V_s = V - I_M R_M = 9.99875\,\text{V}$; b. $R_S = \frac{V_P}{I_M} = 199.975\,\text{k}\Omega$

93. a. $\tau = 3800\,\text{s}$; b. 1.26 A; c. $t = 2633.96\,\text{s}$

95. $R_{\text{eq}} = (1 + \sqrt{3})R$

97. a. $P_{\text{imheater}} = \frac{1\text{cup}\left(\frac{0.000237\,\text{m}^3}{\text{cup}}\right)\left(\frac{1000\,\text{kg}}{\text{m}^3}\right)\left(4186\frac{\text{J}}{\text{kg}\,°\text{C}}\right)(100\,°\text{C} - 20\,°\text{C})}{180.00\,\text{s}} \approx 441\,\text{W}$;

b. $I = \frac{441\,\text{W}}{120\,\text{V}} + 4\left(\frac{100\,\text{W}}{120\,\text{V}}\right) + \frac{1500\,\text{W}}{120\,\text{V}} = 19.51\,\text{A}$; Yes, the breaker will trip.

c. $I = \frac{441\,\text{W}}{120\,\text{V}} + 4\left(\frac{18\,\text{W}}{120\,\text{V}}\right) + \frac{1500\,\text{W}}{120\,\text{V}} = 13.47\,\text{A}$; No, the breaker will not trip.

99.

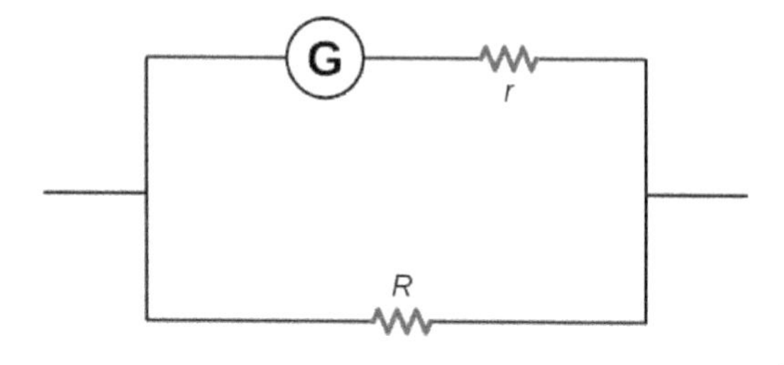

,

$2.40 \times 10^{-3}\ \Omega$

CHAPTER 11

CHECK YOUR UNDERSTANDING

11.1. a. 0 N; b. $2.4 \times 10^{-14}\,\hat{\mathbf{k}}\text{N}$; c. $2.4 \times 10^{-14}\,\hat{\mathbf{j}}\ \text{N}$; d. $(7.2\hat{\mathbf{j}} + 2.2\hat{\mathbf{k}}) \times 10^{-15}\text{N}$

11.2. a. $9.6 \times 10^{-12}\text{N}$ toward the south; b. $\frac{w}{F_m} = 1.7 \times 10^{-15}$

11.3. a. bends upward; b. bends downward
11.4. a. aligned or anti-aligned; b. perpendicular
11.5. a. 1.1 T; b. 1.6 T
11.6. 0.32 m

CONCEPTUAL QUESTIONS

1. Both are field dependent. Electrical force is dependent on charge, whereas magnetic force is dependent on current or rate of charge flow.
3. The magnitude of the proton and electron magnetic forces are the same since they have the same amount of charge. The direction of these forces however are opposite of each other. The accelerations are opposite in direction and the electron has a larger acceleration than the proton due to its smaller mass.
5. The magnetic field must point parallel or anti-parallel to the velocity.
7. A compass points toward the north pole of an electromagnet.
9. Velocity and magnetic field can be set together in any direction. If there is a force, the velocity is perpendicular to it. The magnetic field is also perpendicular to the force if it exists.
11. A force on a wire is exerted by an external magnetic field created by a wire or another magnet.
13. Poor conductors have a lower charge carrier density, *n*, which, based on the Hall effect formula, relates to a higher Hall potential. Good conductors have a higher charge carrier density, thereby a lower Hall potential.

PROBLEMS

15. a. left; b. into the page; c. up the page; d. no force; e. right; f. down
17. a. right; b. into the page; c. down
19. a. into the page; b. left; c. out of the page
21. a. 2.64×10^{-8} N; b. The force is very small, so this implies that the effect of static charges on airplanes is negligible.
23. $10.1°$; $169.9°$
25. 4.27 m
27. a. 4.80×10^{-19} C; b. 3; c. This ratio must be an integer because charges must be integer numbers of the basic charge of an electron. There are no free charges with values less than this basic charge, and all charges are integer multiples of this basic charge.
29. a. 4.09×10^3 m/s; b. 7.83×10^3 m; c. 1.75×10^5 m/s, then, 1.83×10^2 m; d. 4.27 m
31. a. 1.8×10^7 m/s; b. 6.8×10^6 eV; c. 6.8×10^6 V
33. a. left; b. into the page; c. up; d. no force; e. right; f. down
35. a. into the page; b. left; c. out of the page
37. a. 2.50 N; b. This means that the light-rail power lines must be attached in order not to be moved by the force caused by Earth's magnetic field.
39. a. $\tau = NIAB$, so τ decreases by 5.00% if *B* decreases by 5.00%; b. 5.26% increase
41. 10.0 A
43. $\mathrm{A \cdot m^2 \cdot T = A \cdot m^2 \cdot \frac{N}{A \cdot m} = N \cdot m}$
45. 3.48×10^{-26} N · m
47. 0.666 N · m
49. 5.8×10^{-7} V
51. 4.8×10^7 C/kg
53. a. 4.4×10^{-8} s; b. 0.21 m
55. a. 1.8×10^{-12} J; b. 11.5 MeV; c. 11.5 MV; d. 5.2×10^{-8} s; e. 0.45×10^{-12} J, 2.88 MeV, 2.88 V, 10.4×10^{-8} s
57. a. 2.50×10^{-2} m; b. Yes, this distance between their paths is clearly big enough to separate the U-235 from the U-238, since it is a distance of 2.5 cm.

ADDITIONAL PROBLEMS

59. $-7.2 \times 10^{-15}\,\mathrm{N}\,\hat{\mathbf{j}}$
61. $9.8 \times 10^{-5}\,\hat{\mathbf{j}}\,\mathrm{T}$; the magnetic and gravitational forces must balance to maintain dynamic equilibrium
63. 1.13×10^{-3} T
65. $(1.6\,\hat{\mathbf{i}} - 1.4\,\hat{\mathbf{j}} - 1.1\,\hat{\mathbf{k}}) \times 10^5$ V/m
67. a. circular motion in a north, down plane; b. $(1.61\,\hat{\mathbf{j}} - 0.58\,\hat{\mathbf{k}}) \times 10^{-14}$ N

69. The proton has more mass than the electron; therefore, its radius and period will be larger.
71. $1.3 \times 10^{-25}\,\text{kg}$
73. 1:0.707:1
75. 1/4
77. a. $2.3 \times 10^{-4}\,\text{m}$; b. $1.37 \times 10^{-4}\,\text{T}$
79. a. $30.0°$; b. 4.80 N
81. a. 0.283 N; b. 0.4 N; c. 0 N; d. 0 N
83. 0 N and 0.010 Nm
85. a. $0.31\,\text{Am}^2$; b. 0.16 Nm
87. $0.024\,\text{Am}^2$
89. a. $0.16\,\text{Am}^2$; b. 0.016 Nm; c. 0.028 J
91. (Proof)
93. $4.65 \times 10^{-7}\,\text{V}$
95. Since $E = Blv$, where the width is twice the radius, $I = 2r$, $I = nqAv_d$,
$v_{\text{d}} = \frac{I}{nqA} = \frac{I}{nq\pi r^2}$ so $E = B \times 2r \times \frac{\text{I}}{nq\pi r^2} = \frac{2IB}{nq\pi r} \propto \frac{1}{r} \propto \frac{1}{d}$.
The Hall voltage is inversely proportional to the diameter of the wire.
97. 6.92×10^7 m/s; 0.602 m
99. a. $2.4 \times 10^{-19}\,\text{C}$; b. not an integer multiple of e; c. need to assume all charges have multiples of e, could be other forces not accounted for
101. a. B = 5 T; b. very large magnet; c. applying such a large voltage

CHALLENGE PROBLEMS

103. $R = (mv\sin\theta)/qB$; $p = \left(\frac{2\pi m}{eB}\right)v\cos\theta$
105. $IaL^2/2$
107. $m = \frac{qB_0{}^2}{8V_{\text{acc}}}x^2$
109. 0.01 N

CHAPTER 12

CHECK YOUR UNDERSTANDING

12.1. 1.41 meters
12.2. $\frac{\mu_0 I}{2R}$
12.3. 4 amps flowing out of the page
12.4. Both have a force per unit length of 9.23×10^{-12} N/m
12.5. 0.608 meters
12.6. In these cases the integrals around the Ampèrian loop are very difficult because there is no symmetry, so this method would not be useful.
12.7. a. 1.00382; b. 1.00015
12.8. a. $1.0 \times 10^{-4}\,\text{T}$; b. 0.60 T; c. 6.0×10^3

CONCEPTUAL QUESTIONS

1. Biot-Savart law's advantage is that it works with any magnetic field produced by a current loop. The disadvantage is that it can take a long time.
3. If you were to go to the start of a line segment and calculate the angle θ to be approximately $0°$, the wire can be considered infinite. This judgment is based also on the precision you need in the result.
5. You would make sure the currents flow perpendicular to one another.
7. A magnetic field line gives the direction of the magnetic field at any point in space. The density of magnetic field lines indicates the strength of the magnetic field.

9. The spring reduces in length since each coil with have a north pole-produced magnetic field next to a south pole of the next coil.
11. Ampère's law is valid for all closed paths, but it is not useful for calculating fields when the magnetic field produced lacks symmetry that can be exploited by a suitable choice of path.
13. If there is no current inside the loop, there is no magnetic field (see Ampère's law). Outside the pipe, there may be an enclosed current through the copper pipe, so the magnetic field may not be zero outside the pipe.
15. The bar magnet will then become two magnets, each with their own north and south poles. There are no magnetic monopoles or single pole magnets.

PROBLEMS

17. $1 \times 10^{-8}\,\text{T}$

19. $B = \frac{\mu_o I}{8}\left(\frac{1}{a} - \frac{1}{b}\right)$ out of the page

21. $a = \frac{2R}{\pi}$; the current in the wire to the right must flow up the page.

23. 20 A
25. Both answers have the magnitude of magnetic field of $4.5 \times 10^{-5}\,\text{T}$.

27. At P1, the net magnetic field is zero. At P2, $B = \dfrac{3\mu_o I}{8\pi a}$ into the page.

29. The magnetic field is at a minimum at distance a from the top wire, or half-way between the wires.
31. a. $F/l = 2 \times 10^{-5}\,\text{N/m}$ away from the other wire; b. $F/l = 2 \times 10^{-5}\,\text{N/m}$ toward the other wire

33. $B = \dfrac{\mu_o I a}{2\pi b^2}$ into the page

35. 0.019 m
37. $6.28 \times 10^{-5}\,\text{T}$

39. $$B = \frac{\mu_o I R^2}{\left(\left(\frac{d}{2}\right)^2 + R^2\right)^{3/2}}$$

41. a. $\mu_0 I$; b. 0; c. $\mu_0 I$; d. 0

43. a. $3\mu_0 I$; b. 0; c. $7\mu_0 I$; d. $-2\mu_0 I$

45. at the radius R

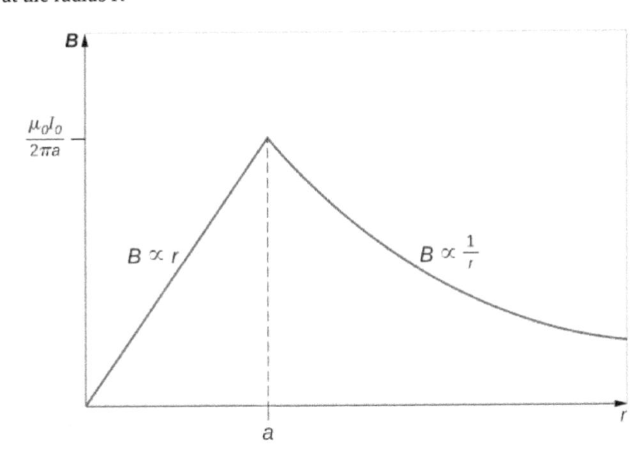

47.
49. $B = 1.3 \times 10^{-2}\,\text{T}$
51. roughly eight turns per cm
53. $B = \frac{1}{2}\mu_0 nI$

55. 0.0181 A
57. 0.0008 T
59. 317.31

61. $2.1 \times 10^{-4} \text{A} \cdot \text{m}^2$
2.7 A

63. 0.18 T

ADDITIONAL PROBLEMS

65. $B = 6.93 \times 10^{-5}\text{T}$

67. $3.2 \times 10^{-19} N$ in an arc away from the wire

69. a. above and below $B = \mu_0 j$, in the middle $B = 0$; b. above and below $B = 0$, in the middle $B = \mu_0 j$

71. $\frac{dB}{B} = -\frac{dr}{r}$

73. a. 52778 turns; b. 0.10 T

75. $B_1(x) = \dfrac{\mu_0 I R^2}{2\left(R^2 + z^2\right)^{3/2}}$

77. $B = \frac{\mu_0 \sigma \omega}{2} R$

79. derivation

81. derivation

83. As the radial distance goes to infinity, the magnetic fields of each of these formulae go to zero.

85. a. $B = \frac{\mu_0 I}{2\pi r}$; b. $B = \frac{\mu_0 J_0 r^2}{3R}$

87. $B(r) = \mu_0 NI/2\pi r$

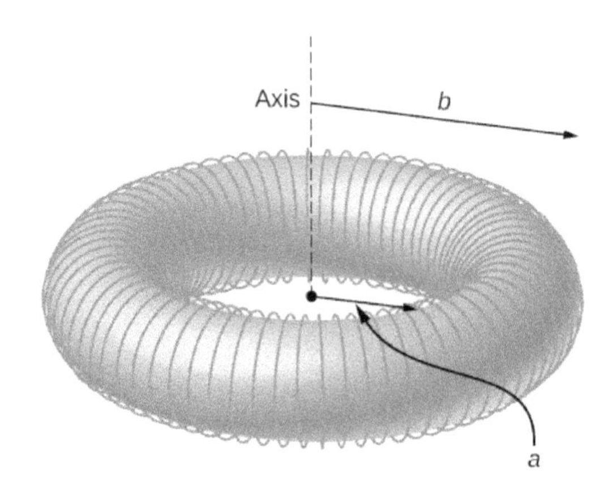

CHALLENGE PROBLEMS

89. $B = \frac{\mu_0 I}{2\pi x}$.

91. a. $B = \frac{\mu_0 \sigma \omega}{2}\left[\frac{2h^2 + R^2}{\sqrt{R^2 + h^2}} - 2h\right]$; b. $B = 4.09 \times 10^{-5}\text{T}$, 82% of Earth's magnetic field

CHAPTER 13

CHECK YOUR UNDERSTANDING

13.1. 1.1 T/s

13.2. To the observer shown, the current flows clockwise as the magnet approaches, decreases to zero when the magnet is centered in the plane of the coil, and then flows counterclockwise as the magnet leaves the coil.

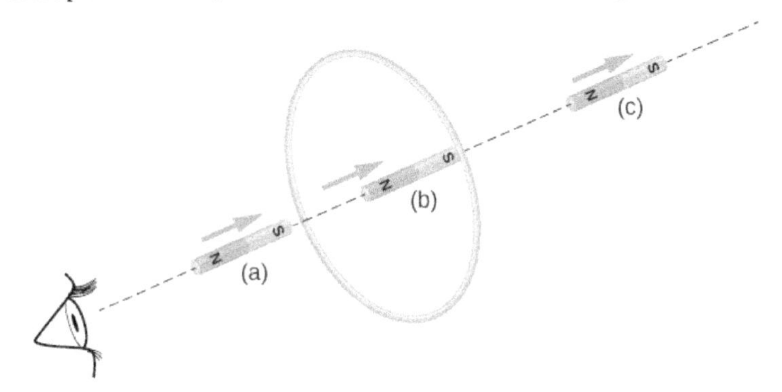

13.4. $\varepsilon = Bl^2 \omega/2$, with O at a higher potential than S

13.5. 1.5 V

13.6. a. yes; b. Yes; however there is a lack of symmetry between the electric field and coil, making $\oint \vec{\mathbf{E}} \cdot d\vec{\mathbf{l}}$ a more complicated relationship that can't be simplified as shown in the example.

13.7. 3.4×10^{-3} V/m

13.8. P_1, P_2, P_4

13.9. a. 3.1×10^{-6} V; b. 2.0×10^{-7} V/m

CONCEPTUAL QUESTIONS

1. The emf depends on the rate of change of the magnetic field.

3. Both have the same induced electric fields; however, the copper ring has a much higher induced emf because it conducts electricity better than the wooden ring.

5. a. no; b. yes

7. As long as the magnetic flux is changing from positive to negative or negative to positive, there could be an induced emf.

9. Position the loop so that the field lines run perpendicular to the area vector or parallel to the surface.

11. a. CW as viewed from the circuit; b. CCW as viewed from the circuit

13. As the loop enters, the induced emf creates a CCW current while as the loop leaves the induced emf creates a CW current. While the loop is fully inside the magnetic field, there is no flux change and therefore no induced current.

15. a. CCW viewed from the magnet; b. CW viewed from the magnet; c. CW viewed from the magnet; d. CCW viewed from the magnet; e. CW viewed from the magnet; f. no current

17. Positive charges on the wings would be to the west, or to the left of the pilot while negative charges would be pulled east or to the right of the pilot. Thus, the left hand tips of the wings would be positive and the right hand tips would be negative.

19. The work is greater than the kinetic energy because it takes energy to counteract the induced emf.

21. The conducting sheet is shielded from the changing magnetic fields by creating an induced emf. This induced emf creates an induced magnetic field that opposes any changes in magnetic fields from the field underneath. Therefore, there is no net magnetic field in the region above this sheet. If the field were due to a static magnetic field, no induced emf will be created since you need a changing magnetic flux to induce an emf. Therefore, this static magnetic field will not be shielded.

23. a. zero induced current, zero force; b. clockwise induced current, force is to the left; c. zero induced current, zero force; d. counterclockwise induced current, force is to the left; e. zero induced current, zero force.

PROBLEMS

25. a. 3.8 V; b. 2.2 V; c. 0 V

27.

$$\begin{aligned} B &= 1.5t,\ 0 \le t < 2.0\ \text{ms},\ B = 3.0\ \text{mT},\ 2.0\ \text{ms} \le t \le 5.0\ \text{ms}, \\ B &= -3.0t + 18\ \text{mT},\ 5.0\ \text{ms} < t \le 6.0\ \text{ms}, \\ \varepsilon &= -\frac{d\Phi_\text{m}}{dt} = -\frac{d(BA)}{dt} = -A\frac{dB}{dt}, \\ \varepsilon &= -\pi(0.100\ \text{m})^2(1.5\ \text{T/s}) \\ &= -47\ \text{mV}(0 \le t < 2.0\ \text{ms}), \\ \varepsilon &= \pi(0.100\ \text{m})^2(0) = 0(2.0\ \text{ms} \le t \le 5.0\ \text{ms}), \\ \varepsilon &= -\pi(0.100\ \text{m})^2(-3.0\ \text{T/s}) = 94\ \text{mV}\ (5.0\ \text{ms} < t < 6.0\ \text{ms}). \end{aligned}$$

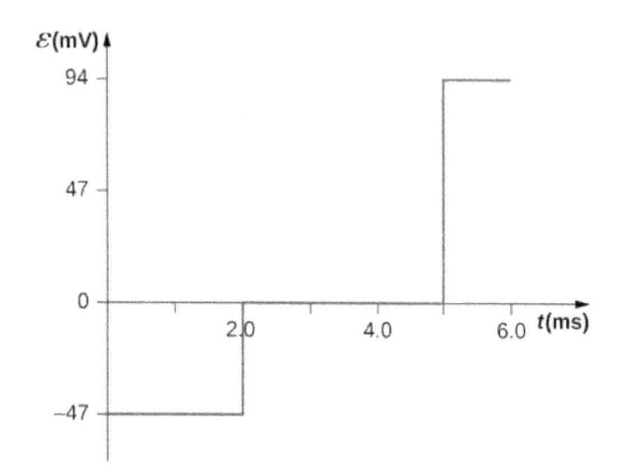

29. Each answer is 20 times the previously given answers.

31.

$$\begin{aligned} \hat{\mathbf{n}} &= \hat{k},\ d\Phi_\text{m} = Cy\sin(\omega t)dxdy, \\ \Phi_\text{m} &= \frac{Cab^2\sin(\omega t)}{2}, \\ \varepsilon &= -\frac{Cab^2\omega\cos(\omega t)}{2}. \end{aligned}$$

33. a. $7.8 \times 10^{-3}\ \text{V}$; b. CCW from the same view as the magnetic field

35. a. 150 A downward through the resistor; b. 232 A upward through the resistor; c. 0.093 A downward through the resistor

37. 0.0015 V

39. $\varepsilon = -B_0 ld\omega\cos\omega t$

41. $\varepsilon = Blv\cos\theta$

43. a. $2 \times 10^{-19}\,T$; b. 1.25 V/m; c. 0.3125 V; d. 16 m/s

45. 0.018 A, CW as seen in the diagram

47. 4.67 V/m

49. Inside, $B = \mu_0 nI$, $\oint \vec{\mathbf{E}} \cdot d\vec{\mathbf{l}} = (\pi r^2)\mu_0 n\frac{dI}{dt}$, so, $E = \frac{\mu_0 nr}{2} \cdot \frac{dI}{dt}$ (inside). Outside, $E(2\pi r) = \pi R^2 \mu_0 n\frac{dI}{dt}$, so, $E = \frac{\mu_0 nR^2}{2r} \cdot \frac{dI}{dt}$ (outside)

51. a. $E_\text{inside} = \frac{r}{2}\frac{dB}{dt}$, $E_\text{outside} = \frac{r^2}{2R}\frac{dB}{dt}$; b. $W = 4.19 \times 10^{-23}\ \text{J}$; c. 0 J; d. $F_\text{mag} = 4 \times 10^{-13}\ \text{N}$, $F_\text{elec} = 2.7 \times 10^{-22}\ \text{N}$

53. $7.1\ \mu\text{A}$

55. three turns with an area of 1 m^2

57. a. $\omega = 120\pi$ rad/s,
$\varepsilon = 850 \sin 120\,\pi t$ V;

b. $P = 720 \sin^2 120\,\pi t$ W;

c. $P = 360 \sin^2 120\,\pi t$ W

59. a. B is proportional to Q; b. If the coin turns easily, the magnetic field is perpendicular. If the coin is at an equilibrium position, it is parallel.
61. a. 1.33 A; b. 0.50 A; c. 60 W; d. 22.5 W; e. 2.5W

ADDITIONAL PROBLEMS

63. 3.0 A/s
65. 2.83×10^{-4} A , the direction as follows for increasing magnetic field:

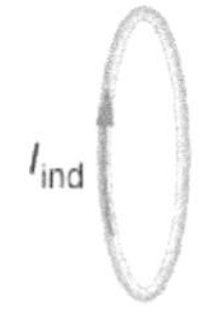

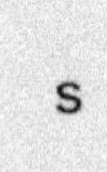

67. 0.375 V
69. a. 0.94 V; b. 0.70 N; c. 3.52 J/s; d. 3.52 W
71. $\left(\frac{dB}{dt}\right)\frac{A}{2\pi r}$

73. a. $R_f + R_a = \frac{120\text{ V}}{2.0\text{ A}} = 60\,\Omega$, so $R_f = 50\,\Omega$;

b. $I = \frac{\varepsilon_s - \varepsilon_i}{R_f + R_a}$, $\Rightarrow \varepsilon_i = 90$ V ;

c. $\varepsilon_i = 60$ V

CHALLENGE PROBLEMS

75. N is a maximum number of turns allowed.
77. 5.3 V

79.
$$\Phi = \frac{\mu_0 I_0 a}{2\pi}\ln\left(1 + \frac{b}{x}\right), \quad \varepsilon = \frac{\mu_0 I_0 abv}{2\pi x(x+b)},$$
$$\text{so } I = \frac{\mu_0 I_0 abv}{2\pi Rx(x+b)}$$

81. a. 1.26×10^{-7} V ; b. 1.71×10^{-8} V ; c. 0 V

83. a. $v = \frac{mgR \sin\theta}{B^2 l^2 \cos^2\theta}$; b. $mgv \sin\,\theta$; c. $mc\Delta T$; d. current would reverse direction but bar would still slide at the same speed

85. a.

$$B = \mu_0 nI, \ \Phi_m = BA = \mu_0 nIA,$$
$$\varepsilon = 9.9 \times 10^{-4}\text{ V};$$

b. 9.9×10^{-4} V ;

c. $\oint \vec{\mathbf{E}} \cdot d\vec{\mathbf{l}} = \varepsilon, \quad \Rightarrow \quad E = 1.6 \times 10^{-3}$ V/m ; d. 9.9×10^{-4} V ;

e. no, because there is no cylindrical symmetry
87. a. 1.92×10^6 rad/s $= 1.83 \times 10^7$ rpm ; b. This angular velocity is unreasonably high, higher than can be obtained for any mechanical system. c. The assumption that a voltage as great as 12.0 kV could be obtained is unreasonable.
89. $\frac{2\mu_0 \pi a^2 I_0 n\omega}{R}$

91. $\frac{mRv_o}{B^2D^2}$

CHAPTER 14

CHECK YOUR UNDERSTANDING

14.1. 4.77×10^{-2} V

14.2. a. decreasing; b. increasing; Since the current flows in the opposite direction of the diagram, in order to get a positive emf on the left-hand side of diagram (a), we need to decrease the current to the left, which creates a reinforced emf where the positive end is on the left-hand side. To get a positive emf on the right-hand side of diagram (b), we need to increase the current to the left, which creates a reinforced emf where the positive end is on the right-hand side.
14.3. 40 A/s
14.4. a. 4.5×10^{-5} H; b. 4.5×10^{-3} V
14.5. a. 2.4×10^{-7} Wb; b. 6.4×10^{-5} m^2
14.6. 0.50 J
14.8. a. 2.2 s; b. 43 H; c. 1.0 s
14.10. a. $2.5\mu F$; b. $\pi/2$ rad or $3\pi/2$ rad; c. 1.4×10^3 rad/s
14.11. a. overdamped; b. 0.75 J

CONCEPTUAL QUESTIONS

1. $\frac{Wb}{A} = \frac{T \cdot m^2}{A} = \frac{V \cdot s}{A} = \frac{V}{A/s}$
3. The induced current from the 12-V battery goes through an inductor, generating a large voltage.
5. Self-inductance is proportional to the magnetic flux and inversely proportional to the current. However, since the magnetic flux depends on the current *I*, these effects cancel out. This means that the self-inductance does not depend on the current. If the emf is induced across an element, it does depend on how the current changes with time.
7. Consider the ends of a wire a part of an *RL* circuit and determine the self-inductance from this circuit.
9. The magnetic field will flare out at the end of the solenoid so there is less flux through the last turn than through the middle of the solenoid.
11. As current flows through the inductor, there is a back current by Lenz's law that is created to keep the net current at zero amps, the initial current.
13. no
15. At $t = 0$, or when the switch is first thrown.
17. 1/4
19. Initially, $I_{R1} = \frac{\varepsilon}{R_1}$ and $I_{R2} = 0$, and after a long time has passed, $I_{R1} = \frac{\varepsilon}{R_1}$ and $I_{R2} = \frac{\varepsilon}{R_2}$.
21. yes
23. The amplitude of energy oscillations depend on the initial energy of the system. The frequency in a *LC* circuit depends on the values of inductance and capacitance.
25. This creates an *RLC* circuit that dissipates energy, causing oscillations to decrease in amplitude slowly or quickly depending on the value of resistance.
27. You would have to pick out a resistance that is small enough so that only one station at a time is picked up, but big enough so that the tuner doesn't have to be set at exactly the correct frequency. The inductance or capacitance would have to be varied to tune into the station however practically speaking, variable capacitors are a lot easier to build in a circuit.

PROBLEMS

29. $M = 3.6 \times 10^{-3}$ H
31. a. 3.8×10^{-4} H; b. 3.8×10^{-4} H
33. $M_{21} = 2.3 \times 10^{-5}$ H
35. 0.24 H
37. 0.4 A/s
39. $\varepsilon = 480\pi \sin(120\pi t - \pi/2)$ V
41. 0.15 V. This is the same polarity as the emf driving the current.
43. a. 0.089 H/m; b. 0.44 V/m
45. $\frac{L}{l} = 4.16 \times 10^{-7}$ H/m

47. 0.01 A
49. 6.0 g
51. $U_m = 7.0 \times 10^{-7}$ J
53. a. 4.0 A; b. 2.4 A; c. on R: $V = 12$ V ; on L: $V = 7.9$ V
55. 0.69τ
57. a. 2.52 ms; b. 99.2 Ω
59. a. $I_1 = I_2 = 1.7\,A$; b. $I_1 = 2.73\,A,\ I_2 = 1.36\,A$; c. $I_1 = 0,\ I_2 = 0.54\,A$; d. $I_1 = I_2 = 0$
61. proof
63. $\omega = 3.2 \times 10^{-7}$ rad/s
65. a. 7.9×10^{-4} s ; b. 4.0×10^{-4} s
67. $q = \frac{q_m}{\sqrt{2}},\ I = \frac{q_m}{\sqrt{2LC}}$
69.
$$\begin{aligned} C &= \frac{1}{4\pi^2 f^2 L} \\ f_1 &= 540\,\text{Hz}; \quad C_1 = 3.5 \times 10^{-11}\,\text{F} \\ f_2 &= 1600\,\text{Hz}; \quad C_2 = 4.0 \times 10^{-12}\,\text{F} \end{aligned}$$
71. 6.9 ms

ADDITIONAL PROBLEMS

73. proof
$$\begin{aligned} \text{Outside,} \quad B &= \frac{\mu_0 I}{2\pi r} \quad \text{Inside,} \quad B = \frac{\mu_0 I r}{2\pi a^2} \\ U &= \frac{\mu_0 I^2 l}{4\pi}\left(\frac{1}{4} + \ln\frac{R}{a}\right) \\ \text{So,} \quad \frac{2U}{I^2} &= \frac{\mu_0 l}{2\pi}\left(\frac{1}{4} + \ln\frac{R}{a}\right) \text{ and } L = \infty \end{aligned}$$
75. $M = \frac{\mu_0 l}{\pi} \ln \frac{d+a}{d}$
77. a. 100 T; b. 2 A; c. 0.50 H
79. a. 0 A; b. 2.4 A
81. a. 2.50×10^6 V ; (b) The voltage is so extremely high that arcing would occur and the current would not be reduced so rapidly. (c) It is not reasonable to shut off such a large current in such a large inductor in such an extremely short time.

CHALLENGE PROBLEMS

83. proof
85. a. $\frac{dB}{dt} = 6 \times 10^{-6}$ T/s; b. $\Phi = \frac{\mu_0 a I}{2\pi} \ln\left(\frac{a+b}{b}\right)$; c. 4.0 nA

CHAPTER 15

CHECK YOUR UNDERSTANDING

15.1. 10 ms
15.2. a. $(20\text{ V}) \sin 200\pi t,\ (0.20\text{ A}) \sin 200\pi t$; b. $(20\text{ V}) \sin 200\pi t,\ (0.13\text{ A}) \sin(200\pi t + \pi/2)$; c. $(20\text{ V}) \sin 200\pi t,\ (2.1\text{ A}) \sin(200\pi t - \pi/2)$
15.3. $v_R = (V_0 R/Z) \sin(\omega t - \phi)$; $v_C = (V_0 X_C/Z) \sin(\omega t - \phi + \pi/2) = -(V_0 X_C/Z)\cos(\omega t - \phi)$; $v_L = (V_0 X_L/Z) \sin(\omega t - \phi + \pi/2) = (V_0 X_L/Z)\cos(\omega t - \phi)$
15.4. $v(t) = (10.0\text{ V}) \sin 90\pi t$
15.5. 2.00 V; 10.01 V; 8.01 V
15.6. a. 160 Hz; b. 40 Ω ; c. $(0.25\text{ A}) \sin 10^3 t$; d. 0.023 rad
15.7. a. halved; b. halved; c. same
15.8. $v(t) = (0.14\text{ V}) \sin(4.0 \times 10^2 t)$

15.9. a. 12:1; b. 0.042 A; c. $2.6 \times 10^3\ \Omega$

CONCEPTUAL QUESTIONS

1. Angular frequency is 2π times frequency.

3. yes for both

5. The instantaneous power is the power at a given instant. The average power is the power averaged over a cycle or number of cycles.

7. The instantaneous power can be negative, but the power output can't be negative.

9. There is less thermal loss if the transmission lines operate at low currents and high voltages.

11. The adapter has a step-down transformer to have a lower voltage and possibly higher current at which the device can operate.

13. so each loop can experience the same changing magnetic flux

PROBLEMS

15. a. $530\ \Omega$; b. $53\ \Omega$; c. $5.3\ \Omega$

17. a. $1.9\ \Omega$; b. $19\ \Omega$; c. $190\ \Omega$

19. 360 Hz

21. $i(t) = (3.2\text{ A})\sin(120\pi t)$

23. a. $38\ \Omega$; b. $i(t) = (4.24\text{A})\sin(120\pi t - \pi/2)$

25. a. $770\ \Omega$; b. 0.16 A; c. $I = (0.16\text{ A})\cos(120\pi t)$; d. $v_R = 120\cos(120\pi t)$; $v_C = 120\cos(120\pi t - \pi/2)$

27. a. $690\ \Omega$; b. 0.15 A; c. $I = (0.15\text{A})\sin(1000\pi t - 0.753)$; d. $1100\ \Omega$, 0.092 A, $I = (0.092\text{A})\sin(1000\pi t + 1.09)$

29. a. $5.7\ \Omega$; b. $29°$; c. $I = (30.\text{ A})\cos(120\pi t)$

31. a. 0.89 A; b. 5.6A; c. 1.4 A

33. a. 7.3 W; b. 6.3 W

35. a. inductor; b. $X_L = 52\ \Omega$

37. $1.3 \times 10^{-7}\text{ F}$

39. a. 820 Hz; b. 7.8

41. a. 50 Hz; b. 50 W; c. 13; d. 25 rad/s

43. The reactance of the capacitor is larger than the reactance of the inductor because the current leads the voltage. The power usage is 30 W.

45. a. 45:1; b. 0.68 A, 0.015 A; c. $160\ \Omega$

47. a. 41 turns; b. 40.9 mA

ADDITIONAL PROBLEMS

49. a. $i(t) = (1.26\text{A})\sin(200\pi t + \pi/2)$; b. $i(t) = (12.6\text{A})\sin(200\pi t - \pi/2)$; c. $i(t) = (2\text{A})\sin(200\pi t)$

51. a. $2.5 \times 10^3\ \Omega,\ 3.6 \times 10^{-3}\text{ A}$; b. $7.5\ \Omega,\ 1.2\text{A}$

53. a. 19 A; b. inductor leads by $90°$

55. $11.7\ \Omega$

57. 36 W

59. a. $5.9 \times 10^4\text{ W}$; b. $1.64 \times 10^{11}\text{ W}$

CHALLENGE PROBLEMS

61. a. 335 MV; b. the result is way too high, well beyond the breakdown voltage of air over reasonable distances; c. the input voltage is too high

63. a. $20\ \Omega$; b. 0.5 A; c. $5.4°$, lagging;
d. $V_R = (9.96\text{ V})\cos(250\pi t + 5.4°),\ V_C = (12.7\text{ V})\cos(250\pi t + 5.4° - 90°),$ $V_L = (11.8\text{ V})\cos(250\pi t + 5.4° + 90°),\ V_{\text{source}} = (10.0\text{ V})\cos(250\pi t);$ e. 0.995; f. 6.25 J

65. a. $0.75\ \Omega$; b. $7.5\ \Omega$; c. $0.75\ \Omega$; d. $7.5\ \Omega$; e. $1.3\ \Omega$; f. $0.13\ \Omega$

67. The units as written for inductive reactance Equation 15.16 are $\frac{\text{rad}}{\text{s}}\text{H}$. Radians can be ignored in unit analysis. The Henry can be defined as $\text{H} = \frac{\text{V}\cdot\text{s}}{\text{A}} = \Omega\cdot\text{s}$. Combining these together results in a unit of Ω for reactance.

69. a. 156 V; b. 42 V; c. 154 V

71. a. $\frac{v_{out}}{v_{in}} = \frac{1}{\sqrt{1 + 1/\omega^2 R^2 C^2}}$ and $\frac{v_{out}}{v_{in}} = \frac{\omega L}{\sqrt{R^2 + \omega^2 L^2}}$; b. $v_{out} \approx v_{in}$ and $v_{out} \approx 0$

CHAPTER 16

CHECK YOUR UNDERSTANDING

16.1. It is greatest immediately after the current is switched on. The displacement current and the magnetic field from it are proportional to the rate of change of electric field between the plates, which is greatest when the plates first begin to charge.
16.2. No. The changing electric field according to the modified version of Ampère's law would necessarily induce a changing magnetic field.
16.3. (1) Faraday's law, (2) the Ampère-Maxwell law
16.4. a. The directions of wave propagation, of the E field, and of B field are all mutually perpendicular. b. The speed of the electromagnetic wave is the speed of light $c = 1/\sqrt{\varepsilon_0 \mu_0}$ independent of frequency. c. The ratio of electric and magnetic field amplitudes is $E/B = c$.
16.5. Its acceleration would decrease because the radiation force is proportional to the intensity of light from the Sun, which decreases with distance. Its speed, however, would not change except for the effects of gravity from the Sun and planets.
16.6. They fall into different ranges of wavelength, and therefore also different corresponding ranges of frequency.

CONCEPTUAL QUESTIONS

1. The current into the capacitor to change the electric field between the plates is equal to the displacement current between the plates.
3. The first demonstration requires simply observing the current produced in a wire that experiences a changing magnetic field. The second demonstration requires moving electric charge from one location to another, and therefore involves electric currents that generate a changing electric field. The magnetic fields from these currents are not easily separated from the magnetic field that the displacement current produces.
5. in (a), because the electric field is parallel to the wire, accelerating the electrons
7. A steady current in a dc circuit will not produce electromagnetic waves. If the magnitude of the current varies while remaining in the same direction, the wires will emit electromagnetic waves, for example, if the current is turned on or off.
9. The amount of energy (about $100\ \text{W/m}^2$) is can quickly produce a considerable change in temperature, but the light pressure (about $3.00 \times 10^{-7}\ \text{N/m}^2$) is much too small to notice.
11. It has the magnitude of the energy flux and points in the direction of wave propagation. It gives the direction of energy flow and the amount of energy per area transported per second.
13. The force on a surface acting over time Δt is the momentum that the force would impart to the object. The momentum change of the light is doubled if the light is reflected back compared with when it is absorbed, so the force acting on the object is twice as great.
15. a. According to the right hand rule, the direction of energy propagation would reverse. b. This would leave the vector $\vec{\mathbf{S}}$, and therefore the propagation direction, the same.
17. a. Radio waves are generally produced by alternating current in a wire or an oscillating electric field between two plates; b. Infrared radiation is commonly produced by heated bodies whose atoms and the charges in them vibrate at about the right frequency.
19. a. blue; b. Light of longer wavelengths than blue passes through the air with less scattering, whereas more of the blue light is scattered in different directions in the sky to give it is blue color.
21. A typical antenna has a stronger response when the wires forming it are orientated parallel to the electric field of the radio wave.
23. No, it is very narrow and just a small portion of the overall electromagnetic spectrum.
25. Visible light is typically produced by changes of energies of electrons in randomly oriented atoms and molecules. Radio waves are typically emitted by an ac current flowing along a wire, that has fixed orientation and produces electric fields pointed in particular directions.
27. Radar can observe objects the size of an airplane and uses radio waves of about 0.5 cm in wavelength. Visible light can be used to view single biological cells and has wavelengths of about 10^{-7} m.
29. ELF radio waves
31. The frequency of 2.45 GHz of a microwave oven is close to the specific frequencies in the 2.4 GHz band used for WiFi.

PROBLEMS

33. $$B_{\text{ind}} = \frac{\mu_0}{2\pi r}I_{\text{ind}} = \frac{\mu_0}{2\pi r}\varepsilon_0\frac{\partial \Phi_E}{\partial t} = \frac{\mu_0}{2\pi r}\varepsilon_0\left(A\frac{\partial E}{\partial t}\right) = \frac{\mu_0}{2\pi r}\varepsilon_0 A\left(\frac{1}{d}\frac{dV(t)}{dt}\right)$$
$$= \frac{\mu_0}{2\pi r}\left[\frac{\varepsilon_0 A}{d}\right]\left[\frac{1}{C}\frac{dQ(t)}{dt}\right] = \frac{\mu_0}{2\pi r}\frac{dQ(t)}{dt} \quad \text{because } C = \frac{\varepsilon_0 A}{d}$$

35. a. $I_{\text{res}} = \frac{V_0 \sin \omega t}{R}$; b. $I_d = CV_0\omega\cos\omega t$;

c. $I_{\text{real}} = I_{\text{res}} + \frac{dQ}{dt} = \frac{V_0 \sin \omega t}{R} + CV_0\frac{d}{dt}\sin\omega t = \frac{V_0\sin\omega t}{R} + CV_0\omega\cos\omega t$; which is the sum of I_{res} and I_{real}, consistent with how the displacement current maintaining the continuity of current.

37. 1.77×10^{-3} A

39. $I_d = (7.97 \times 10^{-10}\text{ A})\sin(150\,t)$

41. 499 s

43. 25 m

45. a. 5.00 V/m; b. 9.55×10^8 Hz; c. 31.4 cm; d. toward the +*x*-axis;

e. $B = (1.67 \times 10^{-8}\text{ T})\cos\left[kx - (6 \times 10^9\text{ s}^{-1})t + 0.40\right]\hat{\mathbf{k}}$

47. $I_d = \pi\varepsilon_0\omega R^2 E_0 \sin(kx - \omega t)$

49. The magnetic field is downward, and it has magnitude 2.00×10^{-8} T.

51. a. 6.45×10^{-3} V/m; b. 394 m

53. 11.5 m

55. $5.97 \times 10^{-3}\text{ W/m}^2$

57. a. $E_0 = 1027$ V/m, $B_0 = 3.42 \times 10^{-6}$ T; b. 3.96×10^{26} W

59. 20.8 W/m^2

61. a. $4.42 \times 10^{-6}\text{ W/m}^2$; b. 5.77×10^{-2} V/m

63. a. $7.47 \times 10^{-14}\text{ W/m}^2$; b. 3.66×10^{-13} W; c. 1.12 W

65. $1.99 \times 10^{-11}\text{ N/m}^2$

67. $$F = ma = (p)(\pi r^2),\ p = \frac{ma}{\pi r^2} = \frac{\varepsilon_0}{2}E_0^2$$
$$E_0 = \sqrt{\frac{2ma}{\varepsilon_0\pi r^2}} = \sqrt{\frac{2(10^{-8}\text{ kg})(0.30\text{ m/s}^2)}{(8.854 \times 10^{-12}\text{ C}^2/\text{N}\cdot\text{m}^2)(\pi)(2 \times 10^{-6}\text{ m})^2}}$$
$$E_0 = 7.34 \times 10^6\text{ V/m}$$

69. a. 4.50×10^{-6} N; b. it is reduced to half the pressure, 2.25×10^{-6} N

71. a. $W = \frac{1}{2}\frac{\pi^2 r^4}{mc^2}I^2 t^2$; b. $E = \pi r^2 It$

73. a. 1.5×10^{18} Hz; b. X-rays

75. a. The wavelength range is 187 m to 556 m. b. The wavelength range is 2.78 m to 3.41 m.

77. $P' = \left(\frac{12\text{ m}}{30\text{ m}}\right)^2(100\text{ mW}) = 16\text{ mW}$

79. time for 1 bit $= 1.27 \times 10^{-8}$ s, difference in travel time is 2.67×10^{-8} s

81. a. 1.5×10^{-9} m; b. 5.9×10^{-7} m; c. 3.0×10^{-15} m

83. 5.17×10^{-12} T, the non-oscillating geomagnetic field of 25–65 μT is much larger

85. a. 1.33×10^{-2} V/m; b. 4.34×10^{-11} T; c. 3.00×10^8 m

87. a. 5.00×10^6 m; b. radio wave; c. 4.33×10^{-5} T

ADDITIONAL PROBLEMS

89. $I_{\text{d}} = (10\ \text{N/C})\left(8.845 \times 10^{-12}\ \text{C}^2/\text{N}\cdot\text{m}^2\right)\pi(0.03\ \text{m})^2(5000) = 1.25 \times 10^{-5}\ \text{mA}$

91. $6.0 \times 10^5\ \text{km}$, which is much greater than Earth's circumference

93. a. 564 W; b. $1.80 \times 10^4\ \text{W/m}^2$; c. $3.68 \times 10^3\ \text{V/m}$; d. $1.23 \times 10^{-5}\ \text{T}$

95. a. $5.00 \times 10^3\ \text{W/m}^2$; b. $3.88 \times 10^{-6}\ \text{N}$; c. $5.18 \times 10^{-12}\ \text{N}$

97. a. $I = \frac{P}{A} = \frac{P}{4\pi r^2} \propto \frac{1}{r^2}$; b. $I \propto E_0^2$, $B_0^2 \Rightarrow E_0^2$, $B_0^2 \propto \frac{1}{r^2} \Rightarrow E_0, B_0 \propto \frac{1}{r}$

99.

$$\begin{aligned}\text{Power into the wire} &= \int \vec{\mathbf{S}} \cdot d\vec{\mathbf{A}} = \left(\frac{1}{\mu_0}EB\right)(2\pi rL)\\ &= \frac{1}{\mu_0}\left(\frac{V}{L}\right)\left(\frac{\mu_0 i}{2\pi r}\right)(2\pi rL) = iV = i^2R\end{aligned}$$

101. 0.431

103. a. $1.5 \times 10^{11}\ \text{m}$; b. $5.0 \times 10^{-7}\ \text{s}$; c. 33 ns

105.

$$\begin{aligned}\text{sound: } \lambda_{\text{sound}} &= \frac{v_s}{f} = \frac{343\ \text{m/s}}{20.0\ \text{Hz}} = 17.2\ \text{m}\\ \text{radio: } \lambda_{\text{radio}} &= \frac{c}{f} = \frac{3.00 \times 10^8\ \text{m/s}}{1030 \times 10^3\ \text{Hz}} = 291\ \text{m; or } 17.1\,\lambda_{\text{sound}}\end{aligned}$$

CHALLENGE PROBLEMS

107. a. $0.29\ \mu\text{m}$; b. The radiation pressure is greater than the Sun's gravity if the particle size is smaller, because the gravitational force varies as the radius cubed while the radiation pressure varies as the radius squared. c. The radiation force outward implies that particles smaller than this are less likely to be near the Sun than outside the range of the Sun's radiation pressure.

INDEX

M

N

O

P

Q

R

S

CPSIA information can be obtained
at www.ICGtesting.com
Printed in the USA
LVHW101952111218
600076LV00014B/433/P

9 781680 92042